Elementary STATISTICS

Fourth Edition

William Navidi
Colorado School of Mines

Barry Monk
Middle Georgia State University

ELEMENTARY STATISTICS, FOURTH EDITION

Published by McGraw Hill LLC, 1325 Avenue of the Americas, New York, NY 10121.

Some ancillaries, including electronic and print components, may not be available to customers outside the United States.

This book is printed on acid-free paper.

1 2 3 4 5 6 7 8 9 LWI 25 24 23 22 21

ISBN 978-1-260-72787-6 (bound edition)
MHID 1-260-72787-4 (bound edition)

ISBN 978-1-264-13640-7 (loose-leaf edition)
MHID 1-264-13640-4 (loose-leaf edition)

ISBN 978-1-264-13635-3 (instructor's edition)
MHID 1-264-13635-8 (instructor's edition)

Portfolio Manager: *Chrissy Skogen*
Product Developer: *Megan Platt*
Marketing Manager: *Emily Digiovanna*
Content Project Managers: *Jeni McAtee, Rachael Hillebrand*
Buyer: *Laura Fuller*
Designer: *Beth Blech*
Content Licensing Specialist: *Jacob Sullivan*
Cover Image: *©RTimages/Shutterstock*
Compositor: *Aptara®, Inc.*

All credits appearing on page or at the end of the book are considered to be an extension of the copyright page.

Library of Congress Cataloging-in-Publication Data

Names: Navidi, William Cyrus, author. | Monk, Barry (Barry J.) author.
Title: Elementary statistics / William Navidi, Colorado School of Mines, Barry Monk, Middle Georgia State University
Description: Fourth edition. | Dubuque : McGraw-Hill Education, [2022] | Includes index.
Identifiers: LCCN 2020024274 (print) | LCCN 2020024275 (ebook) | ISBN 9781260727876 (hardcover) | ISBN 9781264136407 (spiral bound) | ISBN 9781264136353 (hardcover) | ISBN 9781264136384 (ebook) | ISBN 9781264136414 (ebook other)
Subjects: LCSH: Mathematical statistics—Textbooks.
Classification: LCC QA276.12 .N385 2022 (print) | LCC QA276.12 (ebook) | DDC 519.5—dc23
LC record available at https://lccn.loc.gov/2020024274
LC ebook record available at https://lccn.loc.gov/2020024275

mheducation.com/highered

To Catherine, Sarah, and Thomas

—William Navidi

To Shaun

—Barry Monk

About the Authors

©William Navidi

William Navidi is a professor of Applied Mathematics and Statistics at the Colorado School of Mines in Golden, Colorado. He received a Bachelor's degree in Mathematics from New College, a Master's degree in Mathematics from Michigan State University, and a Ph.D. in Statistics from the University of California at Berkeley. Bill began his teaching career at the County College of Morris, a two-year college in Dover, New Jersey. He has taught mathematics and statistics at all levels, from developmental through the graduate level. Bill has written two Engineering Statistics textbooks for McGraw Hill. In his spare time, he likes to play racquetball.

©Dawn Sherry

Barry Monk is a Professor of Mathematics at Middle Georgia State University in Macon, Georgia. Barry received a Bachelor of Science in Mathematical Statistics, a Master of Arts in Mathematics specializing in Optimization and Statistics, and a Ph.D. in Applied Mathematics, all from the University of Alabama. Barry has been teaching Introductory Statistics since 1992 in the classroom and online environments. Barry has a minor in Creative Writing and is a skilled jazz pianist.

Brief Contents

Contents

CHAPTER 4 Summarizing Bivariate Data 165

Stockbyte/PunchStock

CHAPTER 5 Probability 213

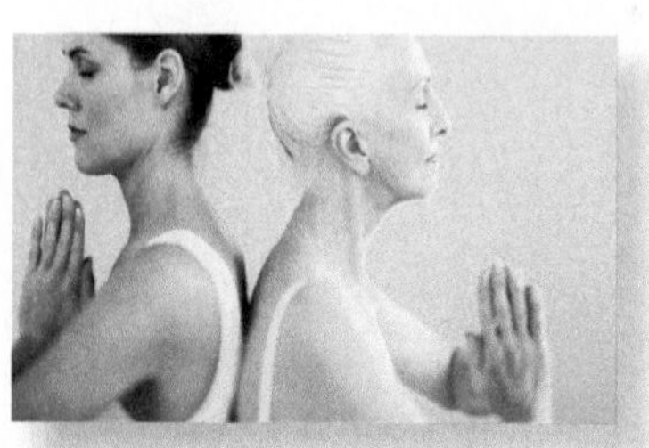
Ryan McVay/Getty Images

CHAPTER 6 Discrete Probability Distributions 263

Andriy Popov/123RF

CHAPTER 7 The Normal Distribution 303

Steve Allen/Getty Images

CHAPTER 8 Confidence Intervals 371

Schroptschop/E+/Getty Images

CHAPTER 9 Hypothesis Testing 431

Murat Taner/Getty Images

CHAPTER 10 Two-Sample Confidence Intervals 505

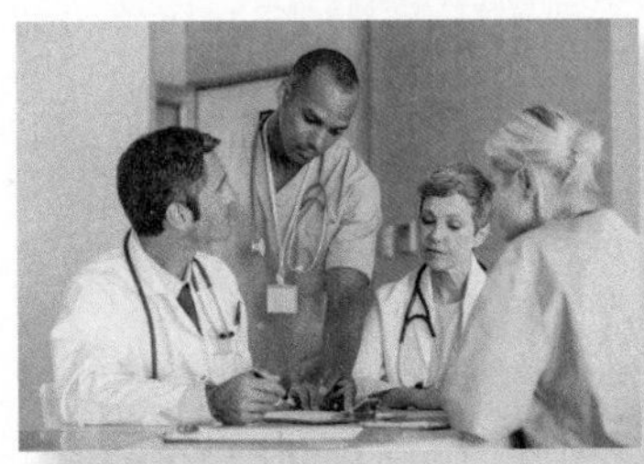
Ridofranz/Getty Images

CHAPTER 11 Two-Sample Hypothesis Tests 539

Malcolm Fife/age fotostock

CHAPTER 12 Tests with Qualitative Data 587

Mark Scott/Getty Images

CHAPTER 13 Inference in Linear Models 613

Vincent Ting/Getty Images

CHAPTER 14 Analysis of Variance 661

crstrbrt/123RF

CHAPTER 15 Nonparametric Statistics 697

CUHRIG/Getty Images

Preface

This book is designed for an introductory course in statistics. In addition to presenting the mechanics of the subject, we have endeavored to explain the concepts behind them, in a writing style as straightforward, clear, and engaging as we could make it. As practicing statisticians, we have done everything possible to ensure that the material presented is accurate and correct. We believe that this book will enable instructors to explore statistical concepts in depth yet remain easy for students to read and understand.

To achieve this goal, we have incorporated a number of useful pedagogical features:

Features

- **Check Your Understanding Exercises:** After each concept is explained, one or more exercises are immediately provided for students to be sure they are following the material. These exercises provide students with confidence that they are ready to go on, or alert them to the need to review the material just covered.
- **Explain It Again:** Many important concepts are reinforced with additional explanation in these marginal notes.
- **Real Data:** Statistics instructors universally agree that the use of real data engages students and convinces them of the usefulness of the subject. A great many of the examples and exercises use real data. Some data sets explore topics in health or social sciences, while others are based in popular culture such as movies, contemporary music, or video games.
- **Integration of Technology:** Many examples contain screenshots from the TI-84 Plus calculator, MINITAB, and Excel. Each section contains detailed, step-by-step instructions, where applicable, explaining how to use these forms of technology to carry out the procedures explained in the text.
- **Interpreting Technology:** Many exercises present output from technology and require the student to interpret the results.
- **Write About It:** These exercises, found at the end of each chapter, require students to explain statistical concepts in their own words.
- **Case Studies:** Each chapter begins with a discussion of a real problem. At the end of the chapter, a case study demonstrates applications of chapter concepts to the problem.
- **In-Class Activities:** At the end of each chapter, activities are suggested that reinforce some concepts presented in the chapter.

Flexibility

We have endeavored to make our book flexible enough to work effectively with a wide variety of instructor styles and preferences. We cover both the *P*-value and critical value approaches to hypothesis testing, so instructors can choose to cover either or both of these methods. The material on two-sample inference is divided into two chapters—Chapter 10 on two-sample confidence intervals, and Chapter 11 on two-sample hypothesis tests. This gives instructors the option of covering all the material on confidence intervals before starting hypothesis testing, by covering Chapter 10 immediately after Chapter 8.

We have placed the material on descriptive statistics for bivariate data immediately following descriptive statistics for univariate data. Those who wish to cover bivariate description and inference together may postpone Chapter 4 until sometime before covering Chapter 13.

Instructors differ widely in their preferences regarding the depth of coverage of probability. A light treatment of the subject may be obtained by covering Section 5.1 and skipping the rest of the chapter. More depth can be obtained by covering Sections 5.2 and 5.3. Section 5.4 on counting can be included for an even more comprehensive treatment.

Supplements

Supplements, including a Corequisite Workbook, online homework, videos, guided student notes, and PowerPoint presentations, play an increasingly important role in the educational process. As authors, we have adopted a hands-on approach to the development of our supplements, to make sure that they are consistent with the style of the text and that they work effectively with a variety of instructor preferences. In particular, our online homework package offers instructors the flexibility to choose whether the solutions that students view are based on tables or technology, where applicable.

New in This Edition

The fourth edition of the book is intended to extend the strengths of the third. Some of the changes are:

- Discussions of the investigative process of statistics have been added, in accordance with recommendations of the GAISE report.
- In-class activities have been added to each chapter.
- Material on the ratio and interval levels of measurement have been added.
- Material on bell-shaped histograms has been added.
- A discussion of the use of sample means to estimate population means has been added.
- Material on the uniform distribution has been added.
- A new objective on the reasoning used in hypothesis testing has been added.
- New conceptual exercises regarding assumptions in constructing confidence intervals and performing hypothesis tests have been added.
- Additional material on Type I and Type II errors has been added.
- Objectives on the relationship between confidence intervals and the margin of error, calculating the sample size needed for a confidence interval of a given width, and the difference between confidence and probability are now presented in a context where the population standard deviation is unknown.
- Objectives on the relationship between confidence intervals and hypothesis tests, the relationship between the level of a test and the probability of error, the importance of reporting *P*-values, and the difference between statistical and practical significance are now presented in a context where the population standard deviation is unknown.
- Material on confidence intervals and hypothesis tests for paired samples now immediately follows the corresponding material for independent samples.
- Material on hypothesis tests for the population correlation has been added.
- A large number of new exercises have been included, many of which involve real data from recent sources.
- A large number of new exercises have been added to the online homework system. These include new conceptual questions and stepped-out solutions for the TI-84 Plus calculator and Excel.
- Several of the case studies have been updated.
- The exposition has been improved in a number of places.

William Navidi

Barry Monk

Acknowledgments

We are indebted to many people for contributions at every stage of development. Colleagues and students who reviewed the evolving manuscript provided many valuable suggestions. In particular, Charla Baker, John Trimboli, Don Brown, and Duane Day contributed to the supplements, and Mary Wolfe helped create the video presentations. Ashlyn Munson contributed a number of exercises, Abby Noble, Lindsay Lewis, and Ryan Melendez contributed many valuable ideas for revisions, and Tim Chappell played an important role in the development of our digital content.

The staff at McGraw Hill has been extremely capable and supportive. Project Manager Jeni McAtee was always patient and helpful. We owe a debt of thanks to Emily DiGiovanna, Debbie McFarland, and Mary Ellen Rahn for their creative marketing and diligence in spreading the word about our book. We appreciate the guidance of our editors, Chrissy Skogen and Megan Platt, whose input has considerably improved the final product.

William Navidi
Barry Monk

Feedback from Statistics Instructors

Paramount to the development of ***Elementary Statistics*** was the invaluable feedback provided by the instructors from around the country who reviewed the manuscript while it was in development.

A Special Thanks to All of the Symposia and Focus Group Attendees

James Adair, *Dyersburg State Community College*
Andrea Adlman, *Ventura College*
Leandro Alvarez, *Miami Dade College*
Simon Aman, *City Colleges of Chicago*
Diane Benner, *Harrisburg Area Community College*
Karen Brady, *Columbus State Community College*
Liliana Brand, *Northern Essex Community College*
Denise Brown, *Collin College-Spring Creek*
Don Brown, *Middle Georgia State University*
Mary Brown, *Harrisburg Area Community College*
Gerald Busald, *San Antonio College*
Anna Butler, *Polk State College*
Robert Cappetta, *College of DuPage*
Joe Castillo, *Broward College*
Michele Catterton, *Harford Community College*
Tim Chappell, *Metropolitan Community College-Penn Valley*
Ivette Chuca, *El Paso Community College*
James Coker, *University of Maryland Global Campus*
James Condor, *State College of Florida*
Milena Cuellar, *LaGuardia Community College*
Phyllis Curtiss, *Grand Valley State University*
Hema Deshmukh, *Mercyhurst University*
Mitra Devkota, *Shawnee State University*
Sue Jones Dobbyn, *Pellissippi State Community College*
Nataliya Doroshenko, *University of Memphis*
Rob Eby, *Blinn College-Bryan Campus*
Charles Wayne Ehler, *Anne Arundel Community College*
Franco Fedele, *University of West Florida*
Robert Fusco, *Broward College*
Wojciech Golik, *Lindenwood University*
Tim Grant, *Southwestern Illinois College*
Nina Greenberg, *University of New Mexico*
Todd Hendricks, *Georgia State University, Perimeter College*
Mary Hill, *College of DuPage*
Steward Huang, *University of Arkansas-Fort Smith*
Vera Hu-Hyneman, *Suffolk County Community College*
Laura Iossi, *Broward College*
Vera Ioudina, *Texas State University*
Brittany Juraszek, *Santa Fe College*
Maryann Justinger, *Erie Community College-South Campus*
Joseph Karnowski, *Norwalk Community College*
Esmarie Kennedy, *San Antonio College*
Lynette Kenyon, *Collin College-Plano*

Raja Khoury, *Collin College-Plano*
Alexander Kolesnik, *Ventura College*
Holly Kresch, *Diablo Valley College*
JoAnn Kump, *West Chester University*
Dan Kumpf, *Ventura College*
Erica Kwiatkowski-Egizio, *Joliet Junior College*
Hui Liu, *Georgia State University*
Pam Lowry, *Bellevue College*
Corey Manchester, *Grossmont College*
Scott McDaniel, *Middle Tennessee State University*
Mikal McDowell, *Cedar Valley College*
Ryan Melendez, *Arizona State University*
Lynette Meslinsky, *Erie Community College*
Penny Morris, *Polk State College*
Brittany Mosby, *Pellissippi State Community College*
Cindy Moss, *Skyline College*
Kris Mudunuri, *Long Beach City College*
Linda Myers, *Harrisburg Area Community College*
Sean Nguyen, *San Francisco State University*
Kim Page, *Middle Tennessee State University*
Ronald Palcic, *Johnson County Community College*
Matthew Pragel, *Harrisburg Area Community College*
Blanche Presley, *Middle Georgia State University*
Ahmed Rashed, *Richland College*
Cyndi Roemer, *Union County College*
Ginger Rowell, *Middle Tennessee State University*
Sudipta Roy, *Kanakee Community College*
Douglas Ruvolo, *University of Akron*
Ligo Samuel, *Austin Peay State University*
Jamal Salahat, *Owens State Community College*
Kathy Shay, *Middlesex County College*
Laura Shick, *Clemson University*
Larry Shrewsbury, *Southern Oregon University*
Shannon Solis, *San Jacinto College-North*
Tommy Thompson, *Cedar Valley College*
John Trimboli, *Middle Georgia State University*
Rita Sowell, *Volunteer State Community College*
Chris Turner, *Pensacola State College*
Jo Tucker, *Tarrant County College*
Dave Vinson, *Pellissippi State Community College*
Laurie Waites, *University of South Alabama*
Henry Wakhungu, *Indiana University*
Bin Wang, *University of South Alabama*
Daniel Wang, *Central Michigan University*
Rachel Webb, *Portland State University*
Jennifer Zeigenfuse, *Anne Arundel Community College*

Manuscript Review Panels

Alisher Abdullayev, *American River College*
Andrea Adlman, *Ventura College*
Olcay Akman, *Illinois State University*
Raid Amin, *University of West Florida*
Wesley L. Anderson, *Northwest Vista College*
Peter Arvanites, *Rockland Community College*
Diana Asmus, *Greenville Technical College*
John Avioli, *Christopher Newport University*
Heather A. Barker, *Elon University*
Robert Bass, *Gardner-Webb University*
Robbin Bates-Yelverton, *Park University*
Lynn Beckett-Lemus, *El Camino College*
Diane Benner, *Harrisburg Area Community College*
Abraham Biggs, *Broward College*
Wes Black, *Illinois Valley Community College*
Gregory Bloxom, *Pensacola State College*
Gabi Booth, *Daytona State College*
Dale Bowman, *University of Memphis*
Brian Bradie, *Christopher Newport University*
Tonia Broome, *Gaston College*
Donna Brouilette, *Georgia State University, Perimeter College*
Allen Brown, *Wabash Valley College*
Denise Brown, *Collin Community College*
Don Brown, *Middle Georgia State University*
Mary Brown, *Harrisburg Area Community College*
Jennifer Bryan, *Oklahoma Christian University*
William Burgin, *Gaston College*
Gerald Busald, *San Antonio College*
David Busekist, *Southeastern Louisiana University*
Lynn Cade, *Pensacola State College*
Elizabeth Carrico, *Illinois Central College*
Connie Carroll, *Guilford Technical Community College*
Joseph Castillo, *Broward College*
Linda Chan, *Mount San Antonio College & Pasadena City College*
Ayona Chatterjee, *University of West Georgia*
Chand Chauhan, *Indiana University Purdue University Fort Wayne*
Pinyuen Chen, *Syracuse University*
Askar Choudhury, *Illinois State University*
Lee Clendenning, *University of North Georgia*
James Condor, *State College of Florida-Manatee*
Natalie Creed, *Gaston College*
John Curran, *Eastern Michigan University*
John Daniels, *Central Michigan University*
Shibasish Dasgupta, *University of South Alabama*
Nataliya Doroshenko, *University of Memphis*
Brandon Doughery, *Montgomery County Community College*
Larry Dumais, *American River College*
Christina Dwyer, *State College of Florida-Manatee*
Wayne Ehler, *Anne Arundel Community College*
Mark Ellis, *Central Piedmont Community College*
Dr. Angela Everett, *Chattanooga State Community College*
Franco Fedele, *University of West Florida*
Harshini Fernando, *Purdue University—North Central*

Art Fortgang, *Southern Oregon University*
Thomas Fox, *Cleveland State Community College*
Marnie Francisco, *Foothill College*
Robert Fusco, *Broward College*
Linda Galloway, *Kennesaw State University*
David Garth, *Truman State University*
Sharon Giles, *Grossmont Community College*
Mary Elizabeth Gore, *Community College of Baltimore County*
Carrie Grant, *Flagler College*
Delbert Greear, *University of North Georgia*
Jason Greshman, *Nova Southeastern University*
David Gurney, *Southeastern Louisiana University*
Chris Hail, *Union University-Jackson*
Ryan Harper, *Spartanburg Community College*
Phillip Harris, *Illinois Central College*
James Harrington, *Adirondack Community College*
Matthew He, *Nova Southeastern University*
Mary Beth Headlee, *State College of Florida-Manatee*
James Helmreich, *Marist College*
Todd A. Hendricks, *Perimeter Georgia State University, College*
Jada Hill, *Richland College*
Mary Hill, *College of DuPage*
William Huepenbecker, *Bowling Green State University-Firelands*
Jenny Hughes, *Columbia Basin College*
Patricia Humphrey, *Georgia Southern University*
Nancy Johnson, *State College of Florida-Manatee*
Maryann Justinger, *Erie Community College-South Campus*
Joseph Karnowski, *Norwalk Community College*
Susitha Karunaratne, *Purdue University—North Central*
Dr. Ryan H. Kasha, Ph.D., *Valencia College*
Joseph Kazimir, *East Los Angeles College*
Esmarie Kennedy, *San Antonio College*
Lynette Kenyon, *Collin College*
Gary Kersting, *North Central Michigan College*
Raja Khoury, *Collin College*
Heidi Kiley, *Suffolk County Community College-Selden*
Daniel Kim, *Southern Oregon University*
Ann Kirkpatrick, *Southeastern Louisiana University*
John Klages, *County College of Morris*
Karon Klipple, *San Diego City College*
Matthew Knowlen, *Horry Georgetown Tech College*
Alex Kolesnik, *Ventura College*
JoAnn Kump, *West Chester University*
Bohdan Kunciw, *Salisbury University*
Erica Kwiatkowski-Egizio, *Joliet Junior College*
William Langston, *Finger Lakes Community College*
Tracy Leshan, *Baltimore City Community College*
Nicole Lewis, *East Tennessee State University*
Jiawei Liu, *Georgia State University*
Fujia Lu, *Endicott College*
Habid Maagoul, *Northern Essex Community College*
Timothy Maharry, *Northwestern Oklahoma State University*
Aldo Maldonado, *Park University*
Kenneth Mann, *Catawba Valley Community College*
James Martin, *Christopher Newport University*
Erin Martin-Wilding, *Parkland College*
Amina Mathias, *Cecil College*
Angie Matthews, *Broward College*
Catherine Matos, *Clayton State University*
Mark McFadden, *Montgomery County Community College*
Karen McKarnin, *Allen Community College*
Penny Morris, *Polk Community College*
B. K. Mudunuri, *Long Beach City College-CalPoly Pomona*
Linda Myers, *Harrisburg Area Community College*
Miroslaw Mystkowski, *Gardner-Webb University*
Shai Neumann, *Brevard Community College*
Francis Kyei Nkansah, *Bunker Hill Community College*
Karen Orr, *Roane State Community College*
Richard Owens, *Park University*
Irene Palacios, *Grossmont College*
Luca Petrelli, *Mount Saint Mary's University*
Blanche Presley, *Middle Georgia State University*
Robert Prince, *Berry College*
Richard Puscas, *Georgia State University, Perimeter College*
Ramaswamy Radhakrishnan, *Illinois State University*
Leela Rakesh, *Central Michigan University*
Gina Reed, *University of North Georgia*
Andrea Reese, *Daytona State College-Daytona Beach*
Jim Robison-Cox, *Montana State University*
Alex Rolon, *Northampton Community College*
Jason Rosenberry, *Harrisburg Area Community College*
Yolanda Rush, *Illinois Central College*
Loula Rytikova, *George Mason University*
Fary Sami, *Harford Community College*
O. Dale Saylor, *University of South Carolina Upstate*
Vicki Schell, *Pensacola State College*
Angela Schirck, *Broward College*
Carol Schoen, *University of Wisconsin-Eau Claire*
Pali Sen, *University of North Florida*
Rosa Seyfried, *Harrisburg Area Community College*
Larry Shrewsbury, *Southern Oregon University*
Abdallah Shuaibi, *Truman College*
Rick Silvey, *University of Saint Mary*
Russell Simmons, *Brookhaven College*
Peggy Slavik, *University of Saint Mary*
Karen Smith, *University of West Georgia*
Aileen Solomon, *Trident Technical College*
Pam Stogsdill, *Bossier Parish Community College*
Susan Surina, *George Mason University*

Victor Swaim, *Southeastern Louisiana University*
Scott Sykes, *University of West Georgia*
Van Tran, *San Francisco State University*
John Trimboli, *Middle Georgia State University*
Barbara Tucker, *Tarrant County College South East*
Steven Forbes Tuckey, *Jackson Community College*
Christopher Turner, *Pensacola State College*
Anke Van Zuylen, *College of William and Mary*
Dave Vinson, *Pellissippi State Community College*
Erwin Walker, *Clemson University*
Joseph Walker, *Georgia State University*
James Wan, *Long Beach City College*
Xiaohong Wang, *Central Michigan University*
Jason Willis, *Gardner-Webb University*
Fuzhen Zhang, *Nova Southeastern University*
Yichuan Zhao, *Georgia State University*
Deborah Ziegler, *Hannibal LaGrange University*
Bashar Zogheib, *Nova Southeastern University*
Stephanie Zwyghuizen, *Jamestown Community College*

Personalized for your students, flexible for your approach.

Every student has different needs and enters your course with varied levels of preparation. ALEKS® pinpoints what students already know, what they don't and, most importantly, what they're ready to learn next. ALEKS® provides flexibility and control when shaping your course so you can reach every student.

ALEKS® Keeps Students Focused with a Personalized Learning Path

ALEKS® creates an optimized path with an ongoing cycle of learning and assessment, celebrating students' small wins along the way with positive real-time feedback. Rooted in research and analytics, ALEKS® improves student outcomes by fostering better preparation, increased motivation and knowledge retention.

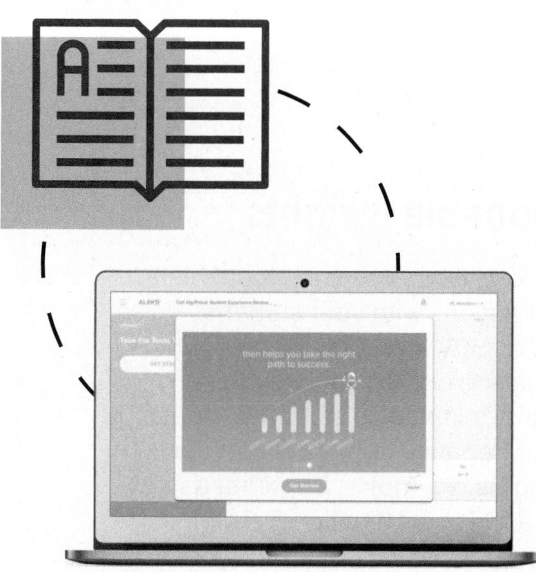

Your Course, Your Way

In a single platform, ALEKS® provides a balance of personalized, adaptive practice and non-adaptive assignments for application and assessment. This supports students individually while allowing you to create assignments that align to your course goals.

By differentiating instruction for students, ALEKS® unlocks the power of instructional time, group work, and applications and prepares students to make real-world connections.

2019 SIIA CODiE WINNER

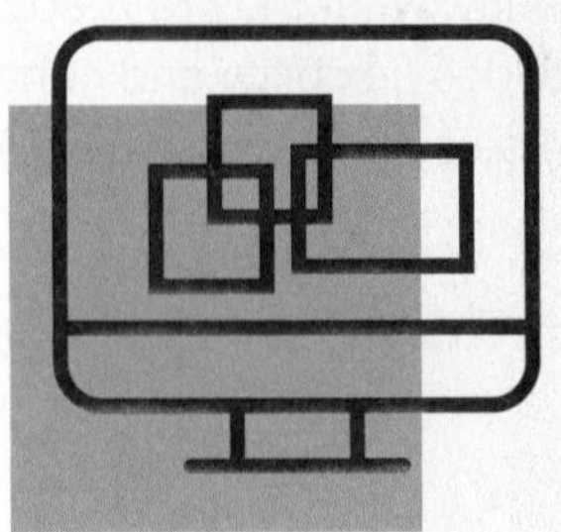

Flexible Implementation

ALEKS® enables you to structure your course regardless of your instruction style and format. From a traditional classroom, to various corequisite models, to an online prep course before the start of the term, ALEKS® can supplement your instruction or play a lead role in delivering the content.

Teaching online? Respondus integration with ALEKS® supports secure testing with LockDown Browser and Respondus Monitor for Tests, Quizzes, and Scheduled Knowledge Checks.

Outcomes & Efficacy

Our commitment to improve student outcomes services a wide variety of implementation models and best practices, from lecture-based to labs and corequisites to summer prep courses. Our case studies illustrate our commitment to help you reach your course goals and our research demonstrates our drive to support all students, regardless of their math background and preparation level.

*visit **bit.ly/outcomesandefficacy** to review empirical data from ALEKS® users around the country

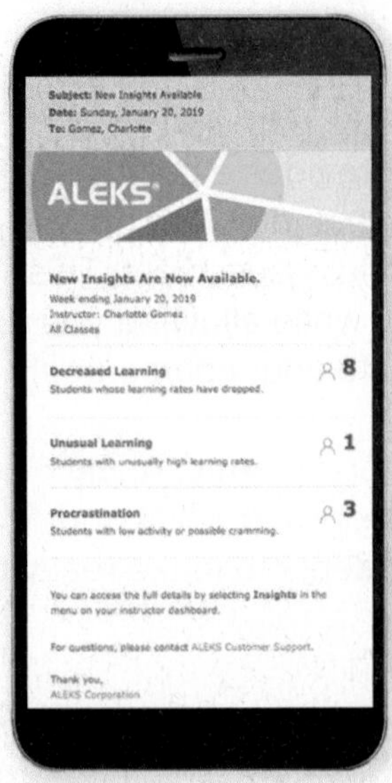

Turn Data Into Actionable Insights

ALEKS® Reports are designed to inform your instruction and create more meaningful interactions with your students when they need it the most. ALEKS® Insights alert you when students might be at risk of falling behind so that you can take immediate action. Insights summarize students exhibiting at least one of four negative behaviors that may require intervention including Failed Topics, Decreased Learning, Unusual Learning and Procrastination & Cramming.

Winner of 2019 Digital Edge 50 Award for Data Analytics!

bit.ly/ALEKS_MHE

Supplements

This text is available with the adaptive online learning platform ALEKS. Rooted in 20 years of research and analytics, ALEKS identifies what students know, what they don't know, and what they are ready to learn, then personalizes a learning path tailored to each student. This personalization keeps students focused on what they need to learn and prepares them to make real-world connections. ALEKS also delivers textbook-aligned exercises, videos, and the many author-created digital supplements that accompany this text.

With **McGraw Hill Create**™, you can easily rearrange chapters, combine material from other content sources, and quickly upload content you have written like your course syllabus or teaching notes. Find the content you need in Create by searching through thousands of leading McGraw Hill textbooks. Arrange your book to fit your teaching style. Create even allows you to personalize your book's appearance by selecting the cover and adding your name, school, and course information. Assemble a Create book, and you'll receive a complimentary print review copy in 3–5 business days or a complimentary electronic review copy (eComp) via email in minutes. Go to www.mcgrawhillcreate.com today and experience how McGraw Hill Create™ empowers you to teach *your* students *your* way.

Videos

Author-produced lecture videos introduce concepts, definitions, formulas, and problem-solving procedures to help students better comprehend the topic at hand. Exercise videos illustrate the authors working through selected exercises, following the solution methodology employed in the text. These videos are closed-captioned for the hearing-impaired and meet the Americans with Disabilities Act Standards for Accessible Design.

Computerized Test Bank Online (instructors only)

This computerized test bank, available online to adopting instructors, utilizes TestGen® cross-platform test generation software to quickly and easily create customized exams. Using hundreds of test items taken directly from the text, TestGen allows rapid test creation and flexibility for instructors to create their own questions from scratch with the ability to randomize number values. Powerful search and sort functions help quickly locate questions and arrange them in any order, and built-in mathematical templates let instructors insert stylized text, symbols, graphics, and equations directly into questions without need for a separate equation editor.

TI-84 Plus Graphing Calculator Manual

This friendly, author-influenced manual teaches students to learn about statistics and solve problems by using this calculator while following each text chapter.

Excel Manual

This workbook, specially designed to accompany the text by the authors, provides additional practice in applying the chapter concepts while using Excel.

MINITAB 17 Manual

With guidance from the authors, this manual includes material from the book to provide seamless use from one to the other, providing additional practice in applying the chapter concepts while using the MINITAB program.

Guided Student Notes

Guided notes provide instructors with the framework of day-by-day class activities for each section in the book. Each lecture guide can help instructors make more efficient use of class time and can help keep students focused on active learning. Students who use the lecture guides have the framework of well-organized notes that can be completed with the instructor in class.

Data Sets

Data sets from selected exercises have been pre-populated into MINITAB, TI-Graph Link, Excel, SPSS, and comma-delimited ASCII formats for student and instructor use. These files are available on the text's website.

Print Supplements

Annotated Instructor's Edition (instructors only)

The Annotated Instructor's Edition contains answers to all exercises. The answers to most questions are printed in blue next to each problem. Answers not appearing on the page can be found in the Answer Appendix at the end of the book.

Statistics Corequisite Workbook

This workbook, co-written by author Barry Monk, is designed to provide corequisite remediation of the necessary skills for an introductory statistics course. The included topics are largely independent of one another and may be used in any order that works best for the instructor. The workbook is available online or can be ordered in print format through Create.

Index of Applications

Miscellaneous applications

Motor vehicle applications. *See* Automotive/motor vehicle applications

Nuclear applications

Political applications

Psychological applications. *See also* Behavioral study applications

Safety applications

School applications. *See* Education/school applications

Sociological applications

Sports/exercise/fitness applications

Travel applications

Weather applications

chapter

1

Basic Ideas

Tatiana Grozetskaya/Shutterstock

Introduction

How does air pollution affect your health? Over the past several decades, scientists have become increasingly convinced that air pollution is a serious health hazard. The World Health Organization has estimated that air pollution causes 2.4 million deaths each year. The health effects of air pollution have been investigated by measuring air pollution levels and rates of disease, then using statistical methods to determine whether higher levels of pollution lead to higher rates of disease.

Many air pollution studies have been conducted in the United States. For example, the town of Libby, Montana, was the focus of a recent study of the effect of particulate matter—air pollution that consists of microscopic particles—on the respiratory health of children. As part of this study, parents were asked to fill out a questionnaire about their children's respiratory symptoms. It turned out that children exposed to higher levels of particulate pollution were more likely to exhibit symptoms of wheezing, as shown in the following table.

Level of Exposure	Percentage with Symptoms
High	8.89%
Low	4.56%

The rate of symptoms was almost twice as high among those exposed to higher levels of pollution. At first, it might seem easy to conclude that higher levels of pollution cause symptoms of wheezing. However, drawing accurate conclusions from information like this is rarely that simple. The case study at the end of this chapter will present more complete information and will show that additional factors must be considered.

Section 1.1 Sampling

Objectives

1. Describe the investigative process of statistics
2. Construct a simple random sample
3. Determine when samples of convenience are acceptable
4. Describe stratified sampling, cluster sampling, systematic sampling, and voluntary response sampling
5. Distinguish between statistics and parameters

Objective 1 Describe the investigative process of statistics

In the months leading up to an election, polls often tell us the percentages of voters that prefer each of the candidates. How do pollsters obtain this information? The ideal poll would be one in which every registered voter were asked his or her opinion. Of course, it is impossible to conduct such an ideal poll, because it is impossible to contact every voter. Instead, pollsters contact a relatively small number of voters, usually no more than a couple of thousand, and use the information from these voters to predict the preferences of the entire group of voters.

The process of polling requires two major steps. First, the voters to be polled must be selected and interviewed. In this way the pollsters collect information. In the second step, the pollsters analyze the information to make predictions about the upcoming election. Both the collection and the analysis of the information must be done properly for the results to be reliable. The field of statistics provides appropriate methods for the collection, description, and analysis of information.

NOTE TO INSTRUCTOR

We describe statistics as a process to emphasize that there is more to it than learning formulas. Our description of statistics as a four-step investigative process is in line with recommendations in the Guidelines for Assessment and Instruction in Statistics Education (GAISE) report.

DEFINITION

Statistics is the study of procedures for collecting, describing, and drawing conclusions from information.

Polling follows a process that is typical in statistics. We formulate questions (Which candidate is most likely to win?), then collect and analyze data to address the questions. In general, statistics is an investigative process that involves the following four steps:

- Formulate questions.
- Collect data needed to answer the questions.
- Describe the data.
- Draw conclusions, using appropriate methods.

In this section, we will focus on the second step in the process—the collection of data. The polling problem is typical of a data collection problem. We want some information about a large group of individuals, but we are able to collect information on only a small part of that group. In statistical terminology, the large group is called a *population*, and the part of the group on which we collect information is called a *sample*.

EXPLAIN IT AGAIN

Why do we draw samples? It's usually impossible to examine every member of a large population. So we select a group of a manageable size to examine. This group is called a sample.

DEFINITION

- A **population** is the entire collection of individuals about which information is sought.
- A **sample** is a subset of a population, containing the individuals that are actually observed.

NOTE TO INSTRUCTOR

It may be helpful to refer students to the **Statistics Corequisite Workbook** that accompanies this text. The following topics may be reviewed early in the course.

1.1 - Decimals, Fractions, and Percentages

2.1 - Introduction to Statistical Notation

Ideally, we would like our sample to represent the population as closely as possible. For example, in a political poll, we would like the proportions of voters preferring each of the candidates to be the same in the sample as in the population. Unfortunately, there are no methods that can guarantee that a sample will represent the population well. The best we can do is to use a method that makes it very likely that the sample will be similar to the population. The best sampling methods all involve some kind of random selection. The most basic, and in many cases the best, sampling method is the method of **simple random sampling**.

Objective 2 Construct a simple random sample

Simple Random Sampling

To understand the nature of a simple random sample, think of a lottery. Imagine that 10,000 lottery tickets have been sold, and that 5 winners are to be chosen. What is the fairest way to choose the winners? The fairest way is to put the 10,000 tickets in a drum, mix them thoroughly, then reach in and draw 5 tickets out one by one. These 5 winning tickets are a simple random sample from the population of 10,000 lottery tickets. Each ticket is equally likely to be one of the 5 tickets drawn. More importantly, each collection of 5 tickets that can be formed from the 10,000 is equally likely to comprise the group of 5 that is drawn.

DEFINITION

A **simple random sample** of size n is a sample chosen by a method in which each collection of n population items is equally likely to make up the sample, just as in a lottery.

Since a simple random sample is analogous to a lottery, it can often be drawn by the same method now used in many lotteries: with a computer random number generator. Suppose there are N items in the population. We number the items 1 through N. Then we generate a list of random integers between 1 and N and choose the corresponding population items to comprise the simple random sample.

Example 1.1

Choosing a simple random sample

There are 300 employees in a certain company. The Human Resources department wants to draw a simple random sample of 20 employees to fill out a questionnaire about their attitudes toward their jobs. Describe how technology can be used to draw this sample.

Solution

Step 1: Make a list of all 300 employees, and number them from 1 to 300.

Step 2: Use a random number generator on a computer or a calculator to generate 20 random numbers between 1 and 300. The employees who correspond to these numbers comprise the sample.

Example 1.2

Determining whether a sample is a simple random sample

A physical education professor wants to study the physical fitness levels of students at her university. There are 20,000 students enrolled at the university, and she wants to draw a sample of size 100 to take a physical fitness test. She obtains a list of all 20,000 students, numbered from 1 to 20,000. She uses a computer random number generator to generate 100 random integers between 1 and 20,000, then invites the 100 students corresponding to those numbers to participate in the study. Is this a simple random sample?

Solution

Yes, this is a simple random sample because any group of 100 students would have been equally likely to have been chosen.

Example 1.3

Determining whether a sample is a simple random sample

The professor in Example 1.2 now wants to draw a sample of 50 students to fill out a questionnaire about which sports they play. The professor's 10:00 A.M. class has 50 students. She uses the first 20 minutes of class to have the students fill out the questionnaire. Is this a simple random sample?

Solution

No. A simple random sample is like a lottery, in which each student in the population has an equal chance to be part of the sample. In this case, only the students in a particular class had a chance to be in the sample.

Example 1.4

In a simple random sample, all samples are equally likely

To play the Colorado Lottery Lotto game, you must select six numbers from 1 to 42. Then lottery officials draw a simple random sample of six numbers from 1 to 42. If your six numbers match the ones in the simple random sample, you win the jackpot. Sally plays the lottery and chooses the numbers 1, 2, 3, 4, 5, 6. Her friend George says that this isn't a good choice, since it is very unlikely that a random sample will turn up the first six numbers. Is he right?

Solution

No. It is true that the combination 1, 2, 3, 4, 5, 6 is unlikely, but every other combination is equally unlikely. In a simple random sample of size 6, every collection of six numbers is equally likely (or equally unlikely) to come up. So Sally has the same chance as anyone to win the jackpot.

Example 1.5

Using technology to draw a simple random sample

Use technology to draw a simple random sample of five employees from the following list.

1. Dan Aaron	11. Johnny Gaines	21. Jorge Ibarra	31. Edward Shingleton
2. Annie Bienh	12. Carlos Garcia	22. Maurice Jones	32. Michael Speciale
3. Oscar Bolivar	13. Julio Gonzalez	23. Jared Kerns	33. Andrew Steele
4. Dominique Bonnaud	14. Jacqueline Gordon	24. Kevin King	34. Neil Swain
5. Paul Campbell	15. James Graves	25. Frank Lipka	35. Sherry Thomas
6. Jeffrey Carnahan	16. Ronald Harrison	26. Carl Luther	36. Shequiea Thompson
7. Joel Chae	17. Andrew Huang	27. Laverne Mitchell	37. Barbara Tilford
8. Dustin Chen	18. Anthony Hunter	28. Zachary Quesada	38. Jermaine Tryon
9. Steven Coleman	19. Jonathan Jackson	29. Donnell Romaine	39. Lizbet Valdez
10. Richard Davis	20. Bruce Johnson	30. Gary Sanders	40. Katelyn Yu

NOTE TO INSTRUCTOR

This is a good place for a discussion of the difference between sampling with replacement and sampling without replacement. In real-life situations, sampling is usually done without replacement. One point to make is that when the sample is only a small fraction of the population, there is little practical difference between the two methods.

Solution

We will use the TI-84 Plus graphing calculator. The step-by-step procedure is presented in the Using Technology section on page 9. We begin by choosing a **seed**, which is a number that the calculator uses to get the random number generator started. Display (a) shows the seed being set to 21. (The seed can be chosen in almost any way; this number was chosen by looking at the seconds display on a digital watch.) Display (b) presents the five numbers in the sample.

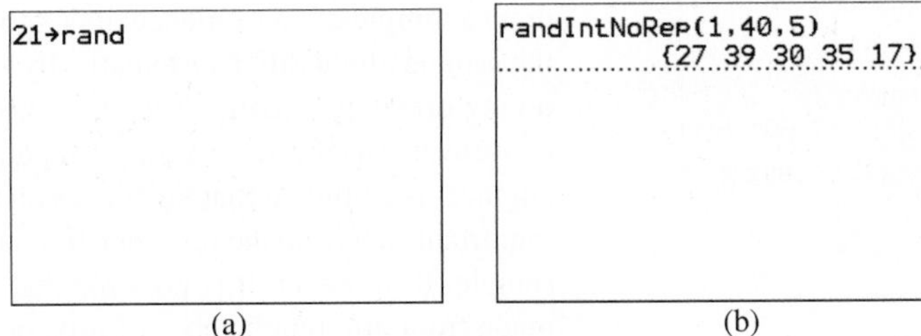

(a) (b)

The simple random sample consists of the employees with numbers 27, 39, 30, 35, and 17. These are Laverne Mitchell, Lizbet Valdez, Gary Sanders, Sherry Thomas, and Andrew Huang.

CAUTION

If you use a different type of calculator, a different statistical package, or a different seed, you will get a different random sample. This is perfectly all right. So long as the sample is drawn by using a correct procedure, it is a valid random sample.

Check Your Understanding

1. A pollster wants to estimate the proportion of voters in a certain town who are Democrats. He goes to a large shopping mall and approaches people to ask whether they are Democrats. Is this a simple random sample? Explain. *No*
2. A telephone company wants to estimate the proportion of customers who are satisfied with their service. They use a computer to generate a list of random phone numbers and call those people to ask them whether they are satisfied. Is this a simple random sample? Explain. *Yes*

Answers are on page 12.

Objective 3 Determine when samples of convenience are acceptable

Samples of Convenience

In some cases, it is difficult or impossible to draw a sample in a truly random way. In these cases, the best one can do is to sample items by some convenient method. A sample obtained in such a way is called a *sample of convenience*.

DEFINITION

A **sample of convenience** is a sample that is not drawn by a well-defined random method.

Example 1.6

Drawing a sample of convenience

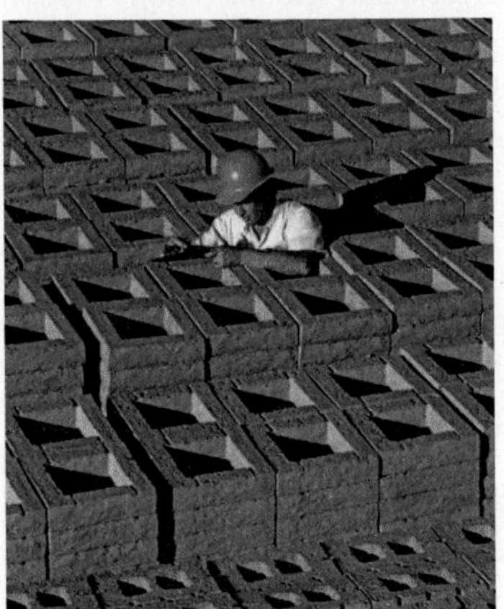
Creatas Images/Jupiterimages

A construction engineer has just received a shipment of 1000 concrete blocks, each weighing approximately 50 pounds. The blocks have been delivered in a large pile. The engineer wishes to investigate the crushing strength of the blocks by measuring the strengths in a sample of 10 blocks. Explain why it might be difficult to draw a simple random sample of blocks. Describe how the engineer might draw a sample of convenience.

Solution

To draw a simple random sample would require removing blocks from the center and bottom of the pile, which might be quite difficult. One way to draw a sample of convenience would be to simply take 10 blocks off the top of the pile.

Problems with samples of convenience

The big problem with samples of convenience is that they may differ systematically in some way from the population. For this reason, samples of convenience should not be used, except in some situations where it is not feasible to draw a random sample. When it is necessary to

CAUTION

Don't use a sample of convenience when it is possible to draw a simple random sample.

draw a sample of convenience, it is important to think carefully about all the ways in which the sample might differ systematically from the population. If it is reasonable to believe that no important systematic difference exists, then it may be acceptable to treat the sample of convenience as if it were a simple random sample. With regard to the concrete blocks, if the engineer is confident that the blocks on the top of the pile do not differ systematically in any important way from the rest, then he can treat the sample of convenience as a simple random sample. If, however, it is possible that blocks in different parts of the pile may have been made from different batches of mix, or may have different curing times or temperatures, a sample of convenience could give misleading results.

SUMMARY

- A sample of convenience may be acceptable when it is reasonable to believe that there is no systematic difference between the sample and the population.
- A sample of convenience is not acceptable when it is possible that there is a systematic difference between the sample and the population.

Objective 4 Describe stratified sampling, cluster sampling, systematic sampling, and voluntary response sampling

Some Other Sampling Methods

Stratified sampling

In **stratified sampling**, the population is divided into groups, called **strata**, where the members of each stratum are similar in some way. Then a simple random sample is drawn from each stratum. Stratified sampling is useful when the strata differ from one another, but the individuals within a stratum tend to be alike.

Example 1.7

Drawing a stratified sample

A company has 1000 employees, of whom 800 are full-time and 200 are part-time. The company wants to survey 50 employees about their opinions regarding benefits. Attitudes toward benefits may differ considerably between full-time and part-time employees. Why might it be a good idea to draw a stratified sample? Describe how one might be drawn.

Solution

If a simple random sample is drawn from the entire population of 1000 employees, it is possible that the sample will contain only a few part-time employees, and their attitudes will not be well represented. For this reason, it might be advantageous to draw a stratified sample. To draw a stratified sample, one would use two strata. One stratum would consist of the full-time employees, and the other would consist of the part-time employees. A simple random sample would be drawn from the full-time employees, and another simple random sample would be drawn from the part-time employees. This method guarantees that both full-time and part-time employees will be well represented.

EXPLAIN IT AGAIN

Example of a cluster sample: Imagine drawing a simple random sample of households and interviewing every member of each household. This would be a cluster sample, with the households as the clusters.

Cluster sampling

In **cluster sampling**, items are drawn from the population in groups, or clusters. Cluster sampling is useful when the population is too large and spread out for simple random sampling to be feasible. Cluster sampling is used extensively by U.S. government agencies in sampling the U.S. population to measure sociological factors such as income and unemployment.

Example 1.8

Drawing a cluster sample

To estimate the unemployment rate in a county, a government agency draws a simple random sample of households in the county. Someone visits each household and asks how many adults live in the household and how many of them are unemployed. What are the clusters? Why is this a cluster sample?

Solution

The clusters are the groups of adults in each of the households in the county. This is a cluster sample because a simple random sample of clusters is selected, and every individual in each selected cluster is part of the sample.

EXPLAIN IT AGAIN

The difference between cluster sampling and stratified sampling: In both cluster sampling and stratified sampling, the population is divided into groups. In stratified sampling, a simple random sample is chosen from each group. In cluster sampling, a random sample of groups is chosen, and every member of the chosen groups is sampled.

Systematic sampling

Imagine walking alongside a line of people and choosing every third one. That would produce a **systematic sample**. In a systematic sample, the population items are ordered. It is decided how frequently to sample items; for example, one could sample every third item, or every fifth item, or every hundredth item. Let k represent the sampling frequency. To begin the sampling, choose a starting place at random. Select the item in the starting place, along with every kth item after that.

Systematic sampling is sometimes used to sample products as they come off an assembly line, in order to check that they meet quality standards.

Example 1.9

Describe a systematic sample

Automobiles are coming off an assembly line. It is decided to draw a systematic sample for a detailed check of the steering system. The starting point will be the third car, then every fifth car after that will be sampled. Which cars will be sampled?

Solution

We start with the third car, then count by fives to determine which cars will be sampled. The sample will consist of cars numbered 3, 8, 13, 18, and so on.

Digital Vision/Punchstock

Voluntary response sampling

Voluntary response samples are often used by the media to try to engage the audience. For example, a news commentator will invite people to tweet an opinion, or a radio announcer will invite people to call the station to say what they think. How reliable are voluntary response samples? To put it simply, *voluntary response samples are never reliable*. People who go to the trouble to volunteer an opinion tend to have stronger opinions than is typical of the population. In addition, people with negative opinions are often more likely to volunteer their responses than those with positive opinions.

Figures 1.1–1.4 illustrate several valid methods of sampling.

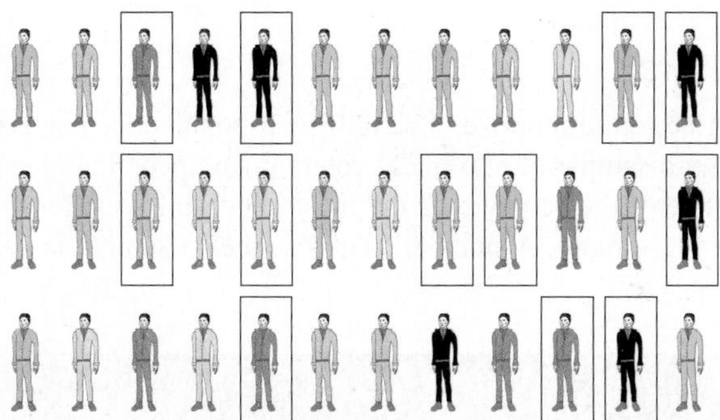

Figure 1.1 Simple random sampling

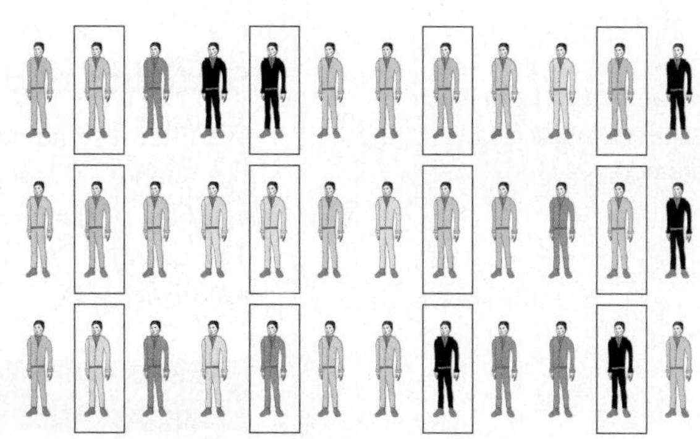

Figure 1.2 Systematic sampling

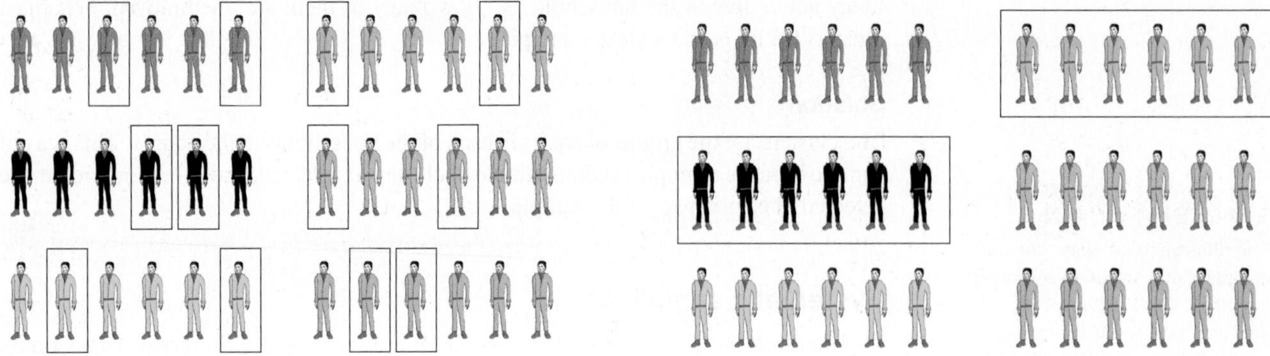

Figure 1.3 Stratified sampling

Figure 1.4 Cluster sampling

Check Your Understanding

3. A radio talk show host invites listeners to send an email to express their opinions on an upcoming election. More than 10,000 emails are received. What kind of sample is this? *Voluntary response*

4. Every 10 years, the U.S. Census Bureau attempts to count every person living in the United States. To check the accuracy of their count in a certain city, they draw a sample of census districts (roughly equivalent to a city block) and recount everyone in the sampled districts. What kind of sample is formed by the people who are recounted? *Cluster*

5. A public health researcher is designing a study of the effect of diet on heart disease. The researcher knows that the diets of men and women tend to differ and that men are more susceptible to heart disease. To be sure that both men and women are well represented, the study comprises a simple random sample of 100 men and another simple random sample of 100 women. What kind of sample do these 200 people represent? *Stratified*

6. A college basketball team held a promotion at one of its games in which every twentieth person who entered the arena won a free basketball. What kind of sample do the winners represent? *Systematic*

Answers are on page 12.

Simple random sampling is the most basic method

Simple random sampling is not the only valid method of random sampling. But it is the most basic, and we will focus most of our attention on this method. From now on, unless otherwise stated, the terms *sample* and *random sample* will be taken to mean *simple random sample*.

Objective 5 Distinguish between statistics and parameters

Statistics and Parameters

We often use numbers to describe, or summarize, a sample or a population. For example, suppose that a pollster draws a sample of 500 likely voters in an upcoming election, and 68% of them say that the state of the economy is the most important issue for them. The quantity "68%" describes the sample. A number that describes a sample is called a *statistic*.

DEFINITION

A **statistic** is a number that describes a sample.

EXPLAIN IT AGAIN

Statistic and parameter: An easy way to remember these terms is that "statistic" and "sample" both begin with "s," and "parameter" and "population" both begin with "p."

Now imagine that the election takes place and that one of the items on the ballot is a proposition to raise the sales tax to pay for the development of a new park downtown. Let's say that 53% of the voters vote in favor of the proposition. The quantity "53%" describes the population of voters who voted in the election. A number that describes a population is called a *parameter*.

DEFINITION

A **parameter** is a number that describes a population.

Example 1.10

Distinguishing between a statistic and a parameter

Which of the following is a statistic and which is a parameter?

a. 57% of the teachers at Central High School are female.

b. In a sample of 100 surgery patients who were given a new pain reliever, 78% of them reported significant pain relief.

NOTE TO INSTRUCTOR

The terminology in this section is used extensively throughout the text. It may be helpful to have students describe the investigative process of statistics using the terms *sample*, *population*, *statistic*, and *parameter*.

Solution

a. The number 57% is a parameter, because it describes the entire population of teachers in the school.

b. The number 78% is a statistic, because it describes a sample.

Using Technology

We use Example 1.5 to illustrate the technology steps.

TI-84 PLUS

Drawing a simple random sample

Step 1. Enter any nonzero number on the HOME screen as the seed.

Step 2. Press **STO >**.

Step 3. Press **MATH**, select **PRB**, then **1: rand**, and then press **ENTER**. This enters the seed into the calculator memory. See Figure A, which uses the number 21 as the seed.

Step 4. Press **MATH**, select **PRB**, then **5: randIntNoRep**. Then enter **1, N, n**, where **N** is the population size and **n** is the desired sample size. In Example 1.5, we use **N** = 40 and **n** = 5 (Figure B).

Step 5. Press **ENTER**. The five values in the random sample for Example 1.5 are **27, 39, 30, 35, 17** (Figure C).

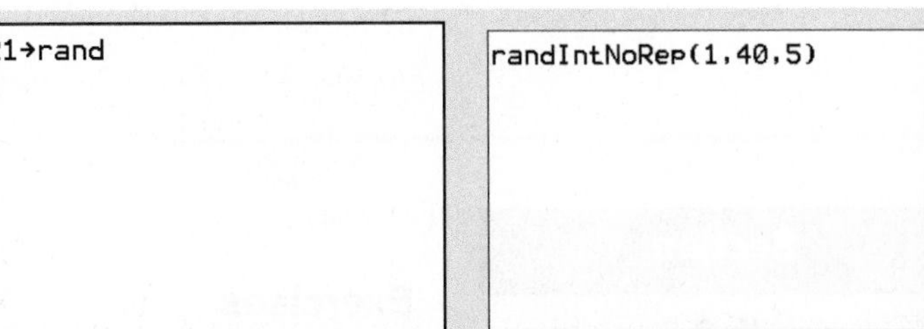

Figure A **Figure B**

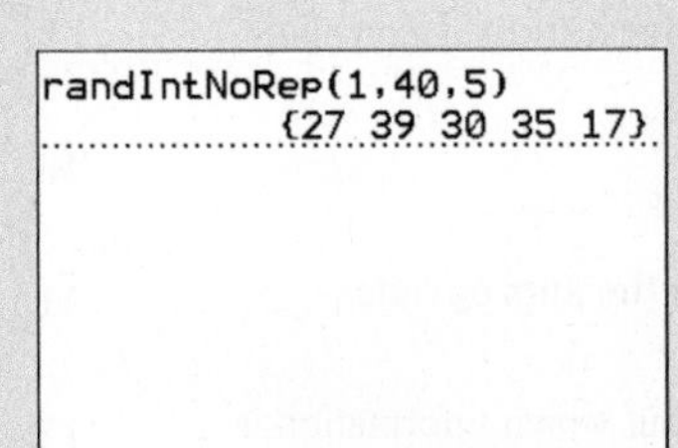

Figure C

EXCEL

Drawing a simple random sample

Step 1. In **Column A**, enter the values **1** through the population size **N**. For Example 1.5, **N** = 40.

Step 2. In **Column B**, next to each value in **Column A**, enter the command **=RAND( )**. This results (Figure D) in a randomly generated number between 0 and 1 in each cell in **Column B**.

Step 3. Select all values in **Columns A** and **B** and then click on the **Data** menu and select **Sort**.

Step 4. In the **Sort by** field, enter **Column B** and select **Smallest to Largest** in the **Order** field. Press **OK**. **Column A** now contains the random sample. Our random sample begins with **17, 12, 28, 20, 6, ...** (Figure E).

	A	B	C
1	1	0.09259	
2	2	0.93582	=rand()
3	3	0.01524	
4	4	0.74766	

Figure D

	A	B
1	17	0.9198
2	12	0.83333
3	28	0.1054
4	20	0.8635
5	6	0.82038
6	11	0.51474
7	36	0.05115
8	19	0.75697
9	39	0.82656
10	26	0.31556
11	8	0.11958
12	30	0.32979
13	35	0.59734

Figure E

MINITAB

Drawing a simple random sample

Step 1. Click **Calc**, then **Random Data**, then **Integer...**

Step 2. In the **Number of rows of data to generate** field, enter twice the desired sample size. For example, if the desired sample size is 10, enter 20. The reason for this is that some sample items may be repeated, and these will need to be deleted.

Step 3. In the **Store in column(s)** field, enter **C1**.

Step 4. Enter **1** for the **Minimum value** and the population size **N** for the **Maximum value**. We use **Maximum value** = 40 for Example 1.5. Click **OK**.

Step 5. **Column C1** of the worksheet will contain a list of randomly selected numbers between **1** and **N**. If any number appears more than once in **Column C1**, delete the replicates so that the number appears only once. For Example 1.5, our random sample begins with **16, 14, 30, 28, 17, ...** (Figure F).

↓	C1
1	16
2	14
3	30
4	28
5	17
6	13
7	4
8	8
9	6
10	15
11	35
12	5

Figure F

Section 1.1 Exercises

Exercises 1–6 are the Check Your Understanding exercises located within the section.

Understanding the Concepts

In Exercises 7–12, fill in each blank with the appropriate word or phrase.

7. The entire collection of individuals about which information is sought is called a ________. *population*

8. A ________ is a subset of a population. *sample*

9. A ________ is a type of sample that is analogous to a lottery. *simple random sample*

10. A sample that is not drawn by a well-defined random method is called a ________. *sample of convenience*

11. A ________ sample is one in which the population is divided into groups and a random sample of groups is drawn. *cluster*

12. A ________ sample is one in which the population is divided into groups and a random sample is drawn from each group. *stratified*

In Exercises 13–16, determine whether the statement is true or false. If the statement is false, rewrite it as a true statement.

13. A sample of convenience is never acceptable. *False*

14. In a cluster sample, the population is divided into groups, and a random sample from each group is drawn. *False*

15. Both stratified sampling and cluster sampling divide the population into groups. *True*

16. One reason that voluntary response sampling is unreliable is that people with stronger views tend to express them more readily. *True*

Practicing the Skills

In Exercises 17–22, determine whether the number described is a statistic or a parameter.

17. In a recent poll, 57% of the respondents supported a school bond issue. *Statistic*

18. The average age of the employees in a certain company is 35 years. *Parameter*

19. Of the students enrolled in a certain college, 80% are full-time. *Parameter*

20. In a survey of 500 high school students, 60% of them said that they intended to go to college. *Statistic*

21. The U.S. Census reports that 95% of California residents live in urban areas. *Parameter*

22. In a survey of parents with children under the age of six, 90% said that their child had been seen by a pediatrician within the past year. *Statistic*

Exercises 23–26 refer to the population of animals in the following table. The population is divided into four groups: mammals, birds, reptiles, and fish.

Mammals		Birds	
1. Aardvark	6. Lion	11. Flamingo	16. Hawk
2. Buffalo	7. Zebra	12. Swan	17. Owl
3. Elephant	8. Pig	13. Sparrow	18. Chicken
4. Squirrel	9. Dog	14. Parrot	19. Duck
5. Rabbit	10. Horse	15. Pelican	20. Turkey
Reptiles		**Fish**	
21. Gecko	26. Python	31. Catfish	36. Shark
22. Iguana	27. Turtle	32. Tuna	37. Trout
23. Chameleon	28. Tortoise	33. Cod	38. Perch
24. Rattlesnake	29. Alligator	34. Salmon	39. Guppy
25. Boa constrictor	30. Crocodile	35. Goldfish	40. Minnow

23. Simple random sample: Draw a simple random sample of eight animals from the list of 40 animals in the table.

24. Another sample: Draw a sample of eight animals by drawing a simple random sample of two animals from each group. What kind of sample is this? *Stratified*

25. Another sample: Draw a simple random sample of two groups of animals from the four groups, and construct a sample of 20 animals by including all the animals in the sampled groups. What kind of sample is this? *Cluster*

26. Another sample: Choose a random number between 1 and 5. Include the animal with that number in your sample, along with every fifth animal thereafter, to construct a sample of eight animals. What kind of sample is this? *Systematic*

In Exercises 27–42, identify the kind of sample that is described.

27. Parking on campus: A college faculty consists of 400 men and 250 women. The college administration wants to draw a sample of 65 faculty members to ask their opinion about a new parking fee. They draw a simple random sample of 40 men and another simple random sample of 25 women. *Stratified*

28. Cruising the mall: A pollster walks around a busy shopping mall and approaches people passing by to ask them how often they shop at the mall. *Sample of convenience*

29. What's on TV? A pollster obtains a list of all the residential addresses in a certain town and uses a computer random number generator to choose 150 of them. The pollster visits each of the 150 households and interviews all the adults in each household about their television viewing habits. *Cluster*

30. Don't drink and drive: Police at a sobriety checkpoint pull over every fifth car to determine whether the driver is sober. *Systematic*

31. Tell us your opinion: A television newscaster invites viewers to tweet their opinions on a proposed bill on immigration policy. More than 50,000 people express their opinions in this way. *Voluntary response*

32. Reading program: The superintendent of a large school district wants to test the effectiveness of a new program designed to improve reading skills among elementary school children. There are 30 elementary schools in the district. The superintendent chooses a simple random sample of five schools and institutes the new reading program in those schools. A total of 4700 children attend these five schools. *Cluster*

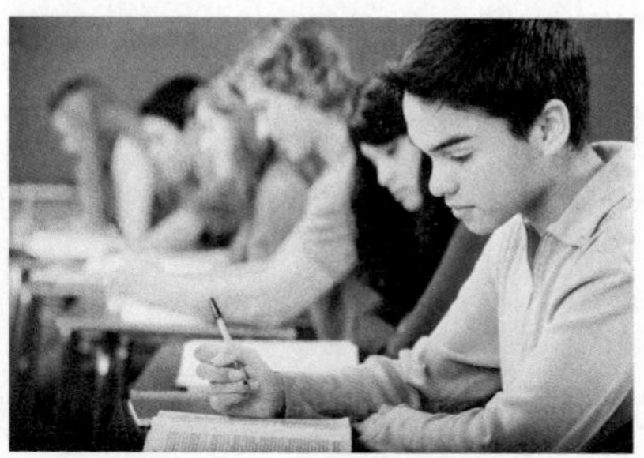

Somos Images/Alamy

33. Customer survey: All the customers who entered a store on a particular day were given a survey to fill out concerning their opinions of the service at the store. *Sample of convenience*

34. Raffle: Five hundred people attend a charity event, and each buys a raffle ticket. The 500 ticket stubs are put in a drum and thoroughly mixed, and 10 of them are drawn. The 10 people whose tickets are drawn win a prize. *Simple random sample*

35. Hospital survey: The director of a hospital pharmacy chooses at random 100 people age 60 or older from each of three surrounding counties to ask their opinions of a new prescription drug program. *Stratified*

36. Bus schedule: Officials at a metropolitan transit authority want to get input from people who use a certain bus route about a possible change in the schedule. They randomly select 5 buses during a certain week and poll all riders on those buses about the change. *Cluster*

37. How much did you spend? A retailer samples 25 receipts from the past week by numbering all the receipts, generating

25 random numbers, and sampling the receipts that correspond to these numbers. *Simple random sample*

38. **Phone features:** A phone company wants to draw a sample of 600 customers to gather opinions about potential new features on upcoming phone models. The company draws a random sample of 200 from customers with iPhones, a random sample of 100 from customers with LG phones, a random sample of 100 from customers with Samsung phones, and a random sample of 200 from customers with other phones. *Stratified*
39. **Computer network:** Every third day, a computer network administrator analyzes the company's network logs to check for signs of computer viruses. *Systematic*
40. **Apps:** An app produces a message requesting customers to click on a link to rate the app. *Voluntary response*
41. **Survey:** A nutritionist randomly chooses 100 people who have been using a certain weight loss program and asks them how much weight they have lost. *Simple random sample*
42. **Good service:** Receipts at a department store indicate a website where customers can take a survey indicating their level of satisfaction with the service they received. *Voluntary response sample*

Working with the Concepts

43. **You're giving me a headache:** A pharmaceutical company wants to test a new drug that is designed to provide superior relief from headaches. They want to select a sample of headache sufferers to try the drug. Do you think that it is feasible to draw a simple random sample of headache sufferers, or will it be necessary to use a sample of convenience? Explain your reasoning. *Sample of convenience*
44. **Pay more for recreation?** The director of the recreation center at a large university wants to sample 100 students to ask them whether they would support an increase in their recreation fees in order to expand the hours that the center is open. Do you think that it is feasible to draw a simple random sample of students, or will it be necessary to use a sample of convenience? Explain your reasoning. *Simple random sample*
45. **Voter preferences:** A pollster wants to sample 500 voters in a town to ask them who they plan to vote for in an upcoming election. Describe a sampling method that would be appropriate in this situation. Explain your reasoning.
46. **Quality control:** Products come off an assembly line at the rate of several hundred per hour. It is desired to sample 10% of them to check whether they meet quality standards. Describe a sampling method that would be appropriate in this situation. Explain your reasoning.
47. **On-site day care:** A large company wants to sample 200 employees to ask their opinions about providing a day care center for the employees' children. They want to be sure to sample equal numbers of men and women. Describe a sampling method that would be appropriate in this situation. Explain your reasoning.
48. **The tax man cometh:** The Internal Revenue Service wants to sample 1000 tax returns that were submitted last year to determine the percentage of returns that had a refund. Describe a sampling method that would be appropriate in this situation. Explain your reasoning.

Extending the Concepts

49. **Draw a sample:** Imagine that you are asked to determine students' opinions at your school about a potential change in library hours. Describe how you could go about getting a sample of each of the following types: simple random sample, sample of convenience, voluntary response sample, stratified sample, cluster sample, systematic sample.
50. **A systematic sample is a cluster sample:** Explain how a systematic sample is actually a type of cluster sample.

Answers to Check Your Understanding Exercises for Section 1.1

1. No; this sample consists only of people in the town who visit the mall.
2. Yes; every group of n customers, where n is the sample size, is equally likely to be chosen.
3. Voluntary response sample
4. Cluster sample
5. Stratified sample
6. Systematic sample

Section 1.2 Types of Data

Objectives

1. Understand the structure of a typical data set
2. Distinguish between qualitative and quantitative variables
3. Distinguish between ordinal and nominal variables
4. Distinguish between discrete and continuous variables
5. Distinguish between ratio and interval levels of measurement

Objective 1 Understand the structure of a typical data set

Data Sets

In Section 1.1, we described various methods of collecting information by sampling. Once the information has been collected, the collection is called a **data set**. A simple example of a data set is presented in Table 1.1, which shows the major, final exam score, and grade for several students in a certain statistics class.

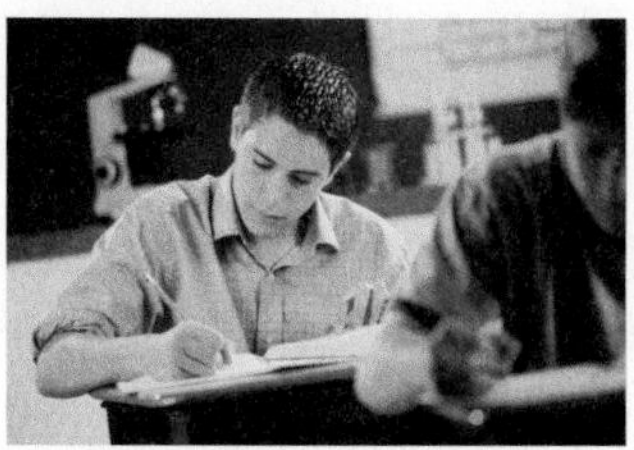
Comstock/Stockbyte/Getty Images

Table 1.1 Major, Final Exam Score, and Grade for Several Students

Student	Major	Exam Score	Grade
1	Psychology	92	A
2	Business	75	B
3	Communications	82	B
4	Psychology	72	C
5	Art	85	B

NOTE TO INSTRUCTOR

The following topic from the **Statistics Corequisite Workbook** is aligned with the material in this section.

1.2 - Qualitative and Quantitative Data, Sets of Numbers, and the Number Line.

Table 1.1 illustrates some basic features that are found in most data sets. Information is collected on **individuals**. In this example, the individuals are students. In many cases, individuals are people; in other cases, they can be animals, plants, or things. The characteristics of the individuals about which we collect information are called **variables**. In this example, the variables are major, exam score, and grade. Finally, the values of the variables that we obtain are the **data**. So, for example, the data for individual #1 are Major = Psychology, Exam score = 92, and Grade = A.

Check Your Understanding

1. A pollster asks a group of six voters about their political affiliation (Republican, Democrat, or Independent), their age, and whether they voted in the last election. The results are shown in the following table.

Voter	Political Affiliation	Age	Voted in Last Election?
1	Republican	34	Yes
2	Democrat	56	Yes
3	Democrat	21	No
4	Independent	28	Yes
5	Republican	61	No
6	Independent	46	Yes

a. How many individuals are there? *6*

b. Identify the variables. *Political affiliation, age, voted in last election*

c. What are the data for individual #3? *Democrat, 21, no*

Answers are on page 19.

Objective 2 Distinguish between qualitative and quantitative variables

Qualitative and Quantitative Variables

Variables can be divided into two types: qualitative and quantitative. **Qualitative variables**, also called **categorical variables**, classify individuals into categories. For example, college major and gender are qualitative variables. **Quantitative variables** are numerical and tell how much or how many of something there is. Height and score on an exam are examples of quantitative variables.

EXPLAIN IT AGAIN

Another way to distinguish qualitative from quantitative variables: Quantitative variables are counts or measurements, whereas qualitative variables are descriptions.

SUMMARY

- Qualitative variables classify individuals into categories.
- Quantitative variables tell how much or how many of something there is.

Example 1.11

Distinguishing between qualitative and quantitative variables

Which of the following variables are qualitative and which are quantitative?

a. A person's age

b. A person's place of birth

c. The mileage (in miles per gallon) of a car

d. The color of a car

Solution

a. Age is quantitative. It tells how much time has elapsed since the person was born.

b. Place is qualitative. It includes categories such as "New York," "Atlanta," "Denver," and "Los Angeles."

c. Mileage is quantitative. It tells how many miles a car will go on a gallon of gasoline.

d. Color is qualitative. It includes categories such as "blue," "red," and "yellow."

Objective 3 Distinguish between ordinal and nominal variables

Ordinal and Nominal Variables

Qualitative variables come in two types: **ordinal variables** and **nominal variables.** An ordinal variable is one whose categories have a natural ordering. The letter grade received in a class, such as A, B, C, D, or F, is an ordinal variable. A nominal variable is one whose categories have no natural ordering. Gender is an example of a nominal variable. Figure 1.5 illustrates how qualitative variables are divided into nominal and ordinal variables.

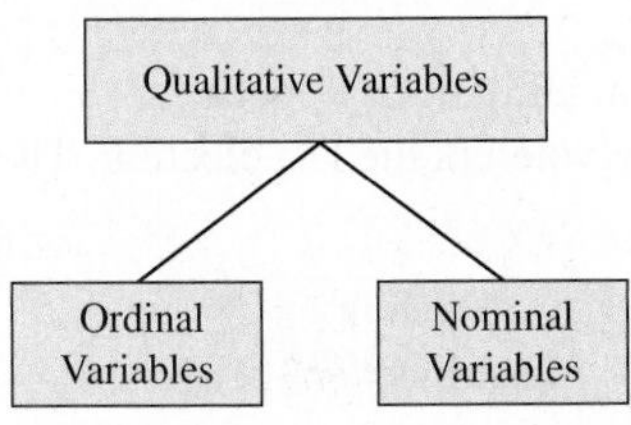

Figure 1.5 Qualitative variables come in two types: ordinal variables and nominal variables.

SUMMARY

- Ordinal variables are qualitative variables whose categories have a natural ordering.
- Nominal variables are qualitative variables whose categories have no natural ordering.

Example 1.12

Distinguishing between ordinal and nominal variables

Which of the following variables are ordinal and which are nominal?

a. State of residence

b. Gender

c. Letter grade in a statistics class (A, B, C, D, or F)

d. Size of soft drink ordered at a fast-food restaurant (small, medium, or large)

Solution

a. State of residence is nominal. There is no natural ordering to the states.

b. Gender is nominal.

c. Letter grade in a statistics class is ordinal. The order, from high to low, is A, B, C, D, F.

d. Size of soft drink is ordinal.

Objective 4 Distinguish between discrete and continuous variables

Discrete and Continuous Variables

Quantitative variables can be either *discrete* or *continuous.* **Discrete variables** are those whose possible values can be listed. Often, discrete variables result from counting something, so the possible values of the variable are 0, 1, 2, and so forth. **Continuous variables** can, in principle, take on any value within some interval. For example, height is a continuous variable because someone's height can be 68, or 68.1, or 68.1452389 inches. The possible values for height are not restricted to a list. Figure 1.6 illustrates how quantitative variables are divided into discrete and continuous variables.

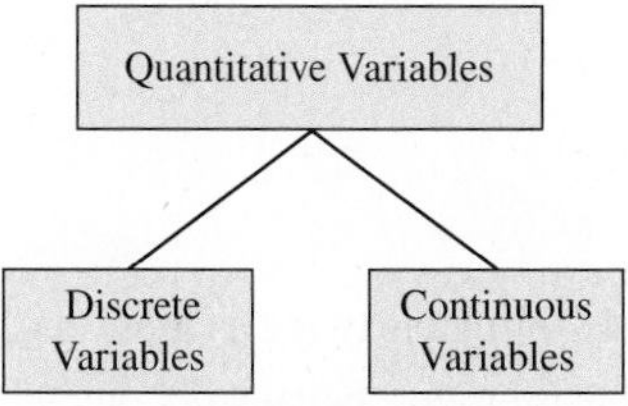

Figure 1.6 Quantitative variables come in two types: discrete variables and continuous variables.

SUMMARY

- Discrete variables are quantitative variables whose possible values can be listed. The list may be infinite—for example, the list of all whole numbers.
- Continuous variables are quantitative variables that can take on any value in some interval. The possible values of a continuous variable are not restricted to any list.

Example 1.13

Distinguishing between discrete and continuous variables

Which of the following variables are discrete, and which are continuous?

a. The age of a person at his or her last birthday
b. The height of a person
c. The number of siblings a person has
d. The distance a person commutes to work

Solution

a. Age at a person's last birthday is discrete. The possible values are 0, 1, 2, and so forth.
b. Height is continuous. A person's height is not restricted to any list of values.
c. Number of siblings is discrete. The possible values are 0, 1, 2, and so forth.
d. Distance commuted to work is continuous. It is not restricted to any list of values.

Objective 5 Distinguish between ratio and interval levels of measurement

Ratio and Interval Levels of Measurement

Quantitative variables can be categorized as having a ratio or an interval level of measurement. A variable has a **ratio level of measurement** if a value of zero indicates that none of the quantity is present, and if ratios of values of the variable are meaningful. For example, money is measured at the ratio level. An amount of $0 represents no money. Ratios are meaningful as well: for example, $20 is twice as much as $10.

NOTE TO INSTRUCTOR

The material on ratio and interval levels of measurement can be skipped without loss of continuity. In general, methods of analysis do not differ for interval as opposed to ratio variables.

A variable has the **interval level of measurement** if a value of zero does not indicate that none of the quantity is present. Two commonly encountered interval variables are dates and temperature. There is no value of 0 for dates; we do not denote the beginning of time as the year 0. Similarly, when measured in °F or °C, a temperature of 0° does not indicate an absence of heat. Ratios are not meaningful for interval variables; for example, a temperature of 100° is not twice as hot as a temperature of 50°. Differences are meaningful, however; for example, the year 2030 is 10 years later than the year 2020.

SUMMARY

- A variable has the ratio level of measurement if zero represents the absence of the quantity, and ratios are meaningful.
- A variable has the interval level of measurement if zero does not represent the absence of the quantity, and ratios are not meaningful. Differences are meaningful, however.

Example 1.14

Distinguishing between interval and ratio levels of measurement

Which of the following variables are at the interval and which are at the ratio level of measurement?

1. The number of siblings you have
2. The outdoor temperature in °C

3. The year of the next presidential election
4. The price of a pair of shoes

Solution

a. Ratio. Zero siblings means the absence of siblings, and 4 siblings is twice as many as 2.

b. Interval. Temperature in °C has the interval level of measurement, because 0° does not represent the absence of heat.

c. Interval. Dates have the interval level of measurement.

d. Ratio. A price of $0 indicates that the shoes cost no money, and a $100 pair of shoes is twice as expensive as a $50 pair of shoes.

Check Your Understanding

2. Which are qualitative and which are quantitative?
 a. The number of patients admitted to a hospital on a given day *Quantitative*
 b. The model of car last sold by a particular car dealer *Qualitative*
 c. The name of your favorite song *Qualitative*
 d. The seating capacity of an auditorium *Quantitative*

3. Which are nominal and which are ordinal?
 a. The names of the streets in a town *Nominal*
 b. The movie ratings G, PG, PG-13, R, and NC-17 *Ordinal*
 c. The winners of the gold, silver, and bronze medals in an Olympic swimming competition *Ordinal*

4. Which are discrete and which are continuous?
 a. The number of female members of the U.S. House of Representatives *Discrete*
 b. The amount of water used by a household during a given month *Continuous*
 c. The number of stories in an apartment building *Discrete*
 d. A person's body temperature *Continuous*

5. Which are at the interval and which are at the ratio level of measurement?
 a. The year you started school *Interval*
 b. Your age in years *Ratio*
 c. The time that your first class starts *Interval*
 d. The price of a loaf of bread *Ratio*

Answers are on page 19.

Section 1.2 Exercises

Exercises 1–5 are the Check Your Understanding exercises located within the section.

Understanding the Concepts

In Exercises 6–12, fill in each blank with the appropriate word or phrase.

6. The characteristics of individuals about which we collect information are called ________. *variables*

7. The values of variables are called ________. *data*

8. Variables that classify individuals into categories are called ________. *qualitative*

9. ________ variables are always numerical. *Quantitative*

10. Qualitative variables can be divided into two types: ________ and ________. *nominal; ordinal*

11. A ________ variable is a quantitative variable whose possible values can be listed. *discrete*

12. ________ variables can take on any value in some interval. *Continuous*

In Exercises 13–16, determine whether the statement is true or false. If the statement is false, rewrite it as a true statement.

13. Qualitative variables describe how much or how many of something there is. *False*

14. A nominal variable is a qualitative variable with no natural ordering. *True*

15. A discrete variable is one whose possible values can be listed. *True*

16. A person's height is an example of a continuous variable. *True*

Practicing the Skills

In Exercises 17–26, determine whether the data described are qualitative or quantitative.

17. Your best friend's name *Qualitative*

18. Your best friend's age *Quantitative*

19. The number of touchdowns in a football game *Quantitative*

20. The title of your statistics book *Qualitative*

21. The number of files on a computer *Quantitative*

22. The waist size of a pair of jeans *Quantitative*

23. The ingredients in a recipe *Qualitative*

24. Your school colors *Qualitative*

25. The makes of cars sold by a particular car dealer *Qualitative*

26. The number of cars sold by a car dealer last month *Quantitative*

In Exercises 27–34, determine whether the data described are nominal or ordinal.

27. The categories Strongly disagree, Disagree, Neutral, Agree, and Strongly agree on a survey *Ordinal*

28. The names of the counties in a state *Nominal*

29. The shirt sizes of Small, Medium, Large, and X-Large *Ordinal*

30. I got an A in statistics, a B in biology, and C's in history and English. *Ordinal*

31. This semester, I am taking statistics, biology, history, and English. *Nominal*

32. I ordered a pizza with pepperoni, mushrooms, olives, and onions. *Nominal*

33. In the track meet, I competed in the high jump and the pole vault. *Nominal*

34. I finished first in the high jump and third in the pole vault. *Ordinal*

In Exercises 35–42, determine whether the data described are discrete or continuous.

35. The amount of caffeine in a cup of Starbucks coffee *Continuous*

36. The distance from a student's home to his school *Continuous*

37. The number of steps in a stairway *Discrete*

38. The number of students enrolled at a college *Discrete*

39. The amount of charge left in a phone battery *Continuous*

40. The number of patients who reported that a new drug had relieved their pain *Discrete*

41. The number of electrical outlets in a coffee shop *Discrete*

42. The time it takes for a text message to be delivered *Continuous*

In Exercises 43–50, determine whether the data described are at the interval level or the ratio level of measurement.

43. The year of your birth *Interval*

44. The sales price of a car *Ratio*

45. The weight in pounds of a sack of potatoes *Ratio*

46. The score on an SAT exam (range is 200 to 800 points) *Interval*

47. The water temperature, in °C, in a swimming pool *Interval*

48. The number of Snapchat friends someone has *Ratio*

49. A student's GPA *Interval*

50. The amount of data, in gigabytes, remaining on your phone *Ratio*

Working with the Concepts

51. Popular Videos: Following are the ten most-viewed YouTube videos, as of January 2020:

1. Despacito—Luis Fonsi featuring Daddy Yankee
2. Shape of You—Ed Sheeran
3. Baby Shark Dance—Pinkfong Kids' Songs & Stories
4. See You Again—Wiz Khalifa featuring Charlie Puth
5. Masha and the Bear: "Recipe for Disaster"—Get Movies
6. Uptown Funk—Mark Ronson featuring Bruno Mars
7. Gangnam Style—Psy
8. Sorry—Justin Bieber
9. Sugar—Maroon 5
10. Roar—Katy Perry

Source: Wikipedia

a. Are these data nominal or ordinal? *Ordinal*

b. Are these data qualitative or quantitative? *Qualitative*

52. More Videos: The following table presents the number of views (in millions) for the videos in Exercise 51.

Video	Number of views (millions)
1. Despacito—Luis Fonsi featuring Daddy Yankee	6610
2. Shape of You—Ed Sheeran	4600
3. Baby Shark Dance—Pinkfong Kids' Songs & Stories	4470
4. See You Again—Wiz Khalifa featuring Charlie Puth	4390
5. Masha and the Bear: "Recipe for Disaster"—Get Movies	4220
6. Uptown Funk—Mark Ronson featuring Bruno Mars	3770
7. Gangnam Style—Psy	3500
8. Sorry—Justin Bieber	3240
9. Sugar—Maroon 5	3120
10. Roar—Katy Perry	2990

Source: Wikipedia

a. Are these data discrete or continuous? *Discrete*

b. Are these data qualitative or quantitative? *Quantitative*

53. How's the economy? A poll conducted by the American Research Group asked individuals their views on how the economy will be a year from now. Respondents were given four choices: Better than today, Same as today, Worse than today, and Undecided. Are these choices nominal or ordinal? *Ordinal*

54. Global warming: A recent Pew poll asked people between the ages of 18 and 29 how serious a problem global warming is. Of those who responded, 43% thought it was very serious, 24% thought it was somewhat serious, 15% thought it was not too serious, and 17% thought it was not a problem. Are these percentages qualitative or quantitative? *Quantitative*

55. Graphic Novels: According to *Time* magazine, some of the best graphic novels of all time are:

Watchmen by Alan Moore and Dave Gibbons
Sandman by Neil Gaiman
Jimmy Corrigan, the Smartest Kid on Earth by Chris Ware
Maus by Art Spiegelman
The Adventures of Tintin: The Black Island by Hergé
Miracleman: The Golden Age by Neil Gaiman and Mark Buckingham
Fun Home: A Family Tragicomic by Allison Bechdel
The Greatest of Marlys by Lynda Berry

Are these data nominal or ordinal? *Nominal*

56. Watch your language: According to MerriamWebster Online, the top ten Funny Sounding and Interesting words are:

1. Bumfuzzle
2. Cattywampus
3. Gardyloo
4. Taradiddle
5. Billingsgate
6. Snickersnee
7. Widdershins
8. Collywobbles
9. Gubbins
10. Diphthong

Are these data nominal or ordinal? *Ordinal*

57. Top ten video games: According to *Wikipedia*, the following are the top ten selling video games of all time.

Game Title	Initial Release Year	Developer	Copies Sold (millions)
1. Tetris	1984	Elektronorgtechnica	170.0
2. Minecraft	2011	Mojang	154.0
3. Grand Theft Auto V	2013	Rockstar North	100.0
4. Wii Sports	2008	Nintendo	82.9
5. PlayerUnknown's Battlegrounds	2017	PUBG Corporation	50.0
6. Pokemon Red/Green/Blue/Yellow	1996	Game Freak	47.5
7. Wii Fit and Wii Fit Plus	2007	Nintendo	43.8
8. Super Mario Bros.	1985	Nintendo	43.2
9. Mario Kart Wii	2008	Wii	37.1
10. Wii Sports Resort	2009	Nintendo	33.1

a. Which of the columns represent qualitative variables? *Game Title, Developer*
b. Which of the columns represent quantitative variables? *Initial Release Year, Copies Sold*
c. Which of the columns represent nominal variables? *Developer*
d. Which of the columns represent ordinal variables? *Game Title*

58. At the movies: The following table provides information about the top ten grossing movies of all time.

Movie Title	Release Year	Studio	Ticket Sales (millions of $)	Running Time (minutes)
1. Avengers: Endgame	2019	Disney	2797.8	182
2. Avatar	2009	Fox	2790.4	162
3. Titanic	1997	Paramount	2194.4	194
4. Star Wars: The Force Awakens	2015	Disney	2068.2	136
5. Avengers: Infinity War	2018	Disney	2048.4	160
6. Jurassic World	2015	Univision	1670.4	124
7. The Lion King	2019	Disney	1656.9	118
8. Marvel's The Avengers	2012	Disney	1518.8	143
9. Furious 7	2015	Univision	1515.0	140
10. Frozen II	2019	Disney	1420.8	103

Source: Box Office Mojo

a. Which of the columns represent qualitative variables? *Movie Title, Studio*
b. Which of the columns represent quantitative variables? *Release Year, Ticket Sales, Running Time*
c. Which of the columns represent nominal variables? *Studio*
d. Which of the columns represent ordinal variables? *Movie Title*

Extending the Concepts

59. What do the numbers mean? A survey is administered by a marketing firm. Two of the people surveyed are Brenda and Jason. Three of the questions are as follows:

i. Do you favor the construction of a new shopping mall?
(1) Strongly oppose (2) Somewhat oppose (3) Neutral (4) Somewhat favor (5) Strongly favor

ii. How many cars do you own?

iii. What is your marital status?
(1) Married (2) Single (3) Divorced (4) Domestically partnered (5) Other

a. Are the responses for question (i) nominal or ordinal? *Ordinal*

b. On question (i), Brenda answers (2) and Jason answers (4). Jason's answer (4) is greater than Brenda's answer (2). Does Jason's answer reflect more of something? *Yes*

c. Jason's answer to question (i) is twice as large as Brenda's answer. Does Jason's answer reflect twice as much of something? Explain. *No*

d. Are the responses for question (ii) qualitative or quantitative? *Quantitative*

e. On question (ii), Brenda answers 2 and Jason answers 1. Does Brenda's answer reflect more of something? Does Brenda's answer reflect twice as much of something? Explain. *Yes; yes*

f. Are the responses for question (iii) nominal or ordinal? *Nominal*

g. On question (iii), Brenda answers (4) and Jason answers (2). Does Brenda's answer reflect more of something? Does Brenda's answer reflect twice as much of something? Explain. *No; no*

Answers to Check Your Understanding Exercises for Section 1.2

1. a. 6 **b.** Political affiliation, Age, and Voted in last election
c. Political affiliation = Democrat, Age = 21, Voted in last election = no

2. a. Quantitative **b.** Qualitative **c.** Qualitative **d.** Quantitative

3. a. Nominal **b.** Ordinal **c.** Ordinal

4. a. Discrete **b.** Continuous **c.** Discrete **d.** Continuous

5. a. Interval **b.** Ratio **c.** Interval **d.** Ratio

Section 1.3 Design of Experiments

Objectives

1. Distinguish between a randomized experiment and an observational study
2. Understand the advantages of randomized experiments
3. Understand how confounding can affect the results of an observational study
4. Describe various types of observational studies

Objective 1 Distinguish between a randomized experiment and an observational study

Experiments and Observational Studies

Will a new drug help prevent heart attacks? Does one type of seed produce a larger wheat crop than another? Does exercise lower blood pressure? To illustrate how scientists address questions like these, we describe how a study might be conducted to determine which of three types of seed will result in the largest wheat yield.

- Prepare three identically sized plots of land, with similar soil types.
- Plant each type of seed on a different plot, choosing the plots at random.
- Water and fertilize the plots in the same way.
- Harvest the wheat, and measure the amount grown on each plot.
- If one type of seed produces substantially more (or less) wheat than the others, then scientists will conclude that it is better (or worse) than the others.

The following terminology is used for studies like this.

DEFINITION

The **experimental units** are the individuals that are studied. These can be people, animals, plants, or things. When the experimental units are people, they are sometimes called **subjects**.

In the wheat study just described, the experimental units are the three plots of land.

DEFINITION

The **outcome**, or **response**, is what is measured on each experimental unit.

In the wheat study, the outcome is the amount of wheat produced.

DEFINITION

The **treatments** are the procedures applied to each experimental unit. There are always two or more treatments. The purpose is to determine whether the choice of treatment affects the outcome.

In the wheat study, the treatments are the three types of seed.

In general, studies fall into two categories: *randomized experiments* and *observational studies*.

DEFINITION

A **randomized experiment** is a study in which the investigator assigns the treatments to the experimental units at random.

The wheat study described above is a randomized experiment. In some situations, randomized experiments cannot be performed, because it isn't possible to randomly assign the treatments. For example, in studies to determine how smoking affects health, people cannot be assigned to smoke. Instead, people choose for themselves whether to smoke, and scientists observe differences in health outcomes between groups of smokers and nonsmokers. Studies like this are called *observational studies*.

DEFINITION

An **observational study** is one in which the assignment to treatment groups is not made by the investigator.

When possible, it is better to assign treatments at random and perform a randomized experiment. As we will see, the results of randomized experiments are generally easier to interpret than the results of observational studies.

Objective 2 Understand the advantages of randomized experiments

Randomized Experiments

An article in *The New England Journal of Medicine* (359:339–354) reported the results of a study to determine whether a drug called raltegravir is effective in reducing levels of virus in patients with human immunodeficiency virus (HIV). A total of 699 patients participated in the experiment. These patients were divided into two groups. One group was given raltegravir. The other group was given a placebo. (A placebo is a harmless tablet, such as a sugar tablet, that looks like the drug but has no medical effect.) Thus there were two treatments in this experiment, raltegravir and placebo.

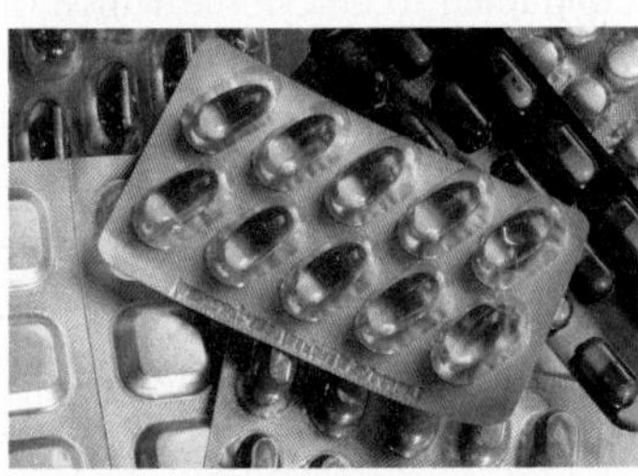

Ingram Publishing/SuperStock

The experimenters had decided to give raltegravir to about two-thirds of the subjects and the placebo to the others. To determine which patients would be assigned to which group, a simple random sample consisting of 462 of the 699 patients was drawn; this sample constituted the raltegravir group. The remaining 237 patients were assigned to the placebo group.

It was decided to examine subjects after 16 weeks and measure the levels of virus in their blood. Thus the outcome for this experiment was the number of copies of virus per milliliter of blood. Patients were considered to have a successful outcome if they had fewer than 50 copies of the virus per milliliter of blood. In the raltegravir group, 62% of the subjects had a successful outcome, but only 35% of the placebo group did. The conclusion was that raltegravir was effective in lowering the concentration of virus in HIV patients. We will examine this study and determine why it was reasonable to reach this conclusion.

The raltegravir study was a randomized experiment, because the treatments were assigned to the patients at random. What are the advantages of randomized experiments? In a perfect study, the treatment groups would not differ from each other in any important way except that they receive different treatments. Then, if the outcomes differ among the groups, we may be confident that the differences in outcome must have been caused by differences in treatment. In practice, it is impossible to construct treatment groups that are exactly alike. But randomization does the next best thing. In a randomized experiment, any differences between the groups are likely to be small. In addition, the differences are due only to chance.

Because the raltegravir study was a randomized experiment, it is reasonable to conclude that the higher success rate in the raltegravir group was actually due to raltegravir.

SUMMARY

In a randomized experiment, if there are large differences in outcomes among the treatment groups, we can conclude that the differences are due to the treatments.

Example 1.15

Identifying a randomized experiment

To assess the effectiveness of a new method for teaching arithmetic to elementary school children, a simple random sample of 30 first graders was taught with the new method, and another simple random sample of 30 first graders was taught with the currently used method. At the end of eight weeks, the children were given a test to assess their knowledge. What are the treatments in this study? Explain why this is a randomized experiment.

Solution

The treatments are the two methods of teaching. This is a randomized experiment because children were assigned to the treatment groups at random.

Double-blind experiments

We have described the advantages of assigning treatments at random. It is a further advantage if the assignment can be done in such a way that neither the experimenters nor the subjects know which treatment has been assigned to which subject. Experiments like this are called *double-blind* experiments. The raltegravir experiment was a double-blind experiment, because neither the patients nor the doctors treating them knew which patients were receiving the drug and which were receiving the placebo.

DEFINITION

An experiment is **double-blind** if neither the investigators nor the subjects know who has been assigned to which treatment.

Experiments should be run double-blind whenever possible, because when investigators or subjects know which treatment is being given, they may tend to report the results differently. For example, in an experiment to test the effectiveness of a new pain reliever, patients who know they are getting the drug may report their pain levels differently than those who know they are taking a placebo. Doctors can be affected as well; a doctor's diagnosis may be influenced by a knowledge of which treatment a patient received.

In some situations, it is not possible to run a double-blind experiment. For example, in an experiment that compares a treatment that involves taking medication to a treatment that involves surgery, both patients and doctors will know who got which treatment.

Example 1.16

Determining whether an experiment is double-blind

Is the experiment described in Example 1.15 a double-blind experiment? Explain.

Solution

This experiment is not double-blind, because the teachers know whether they are using the new method or the old method.

Randomized block experiments

The type of randomized experiment we have discussed is sometimes called a **completely randomized experiment**, because there is no restriction on which subjects may be assigned which treatment. In some situations, it is desirable to restrict the randomization a bit. For example, imagine that two reading programs are to be tested in an elementary school that has children in grades 1 through 4. If children are assigned at random to the programs, it is possible that one of the programs will end up with more fourth graders while the other one will end up with more first graders. Since fourth graders tend to be better readers, this will give an advantage to the program that happens to end up with more of them. This possibility can be avoided by randomizing the students within each grade separately. In other words, we randomly assign exactly half of the students within each grade to each reading program.

This type of experiment is called a **randomized block experiment**. In the example just discussed, each grade constitutes a block. In a randomized block experiment, the subjects are divided into blocks in such a way that the subjects in each block are the same or similar with regard to a variable that is related to the outcome. Age and gender are commonly used blocking variables. Then the subjects within each block are randomly assigned a treatment.

Observational Studies

NOTE TO INSTRUCTOR

It is useful to emphasize that randomized experiments provide more reliable information than do observational studies. We conduct observational studies to address questions such as the effects of smoking, where randomized experiments would be impossible or unethical.

Recall that an observational study is one in which the investigators do not assign the treatments. In most observational studies, the subjects choose their own treatments. Observational studies are less reliable than randomized experiments. To see why, imagine a study that is intended to determine whether smoking increases the risk of heart attack. Imagine that a group of smokers and a group of nonsmokers are observed for several years, and during that time a higher percentage of the smoking group experiences a heart attack. Does this prove that smoking increases the risk of heart attack? No. The problem is that the smoking group will differ from the nonsmoking group in many ways other than smoking, and these other differences may be responsible for differences in the rate of heart attacks. For example, smoking is more prevalent among men than among women. Therefore, the smoking group will contain a higher percentage of men than the nonsmoking group. It is known that men have a higher risk of heart attack than women. So the higher rate of heart attacks in the smoking group could be due to the fact that there are more men in the smoking group, and not to the smoking itself.

Confounding

Objective 3 Understand how confounding can affect the results of an observational study

The preceding example illustrates the major problem with observational studies. It is difficult to tell whether a difference in the outcome is due to the treatment or to some other difference between the treatment and control groups. This is known as **confounding**. In the preceding example, gender was a *confounder*. Gender is related to smoking (men are more likely to smoke) and to heart attacks (men are more likely to have heart attacks). For this reason, it is difficult to determine whether the difference in heart attack rates is due to differences in smoking (the treatment) or differences in gender (the confounder).

EXPLAIN IT AGAIN

Another way to describe a confounder: A confounder is something other than the treatment that can cause the treatment groups to have different outcomes.

SUMMARY

A **confounder** is a variable that is related to both the treatment and the outcome. When a confounder is present, it is difficult to determine whether differences in the outcome are due to the treatment or to the confounder.

How can we prevent confounding? One way is to design a study so that the confounder isn't a factor. For example, to determine whether smoking increases the risk of heart attack, we could compare a group of male smokers to a group of male nonsmokers, and a group of female smokers to a group of female nonsmokers. Gender wouldn't be a confounder here, because there would be no differences in gender between the smoking and nonsmoking groups. Of course, there are other possible confounders. Smoking rates vary among ethnic groups, and rates of heart attacks do, too. If people in ethnic groups that are more susceptible to heart attacks are also more likely to smoke, then ethnicity becomes a confounder. This can be dealt with by comparing smokers of the same gender and ethnic group to nonsmokers of that gender and ethnic group.

Designing observational studies that are relatively free of confounding is difficult. In practice, many studies must be conducted over a long period of time. In the case of smoking, this has been done, and we can be confident that smoking does indeed increase the risk of heart attack, along with other diseases. If you don't smoke, you have a much better chance to live a long and healthy life.

SUMMARY

In an observational study, when there are differences in the outcomes among the treatment groups, it is often difficult to determine whether the differences are due to the treatments or to confounding.

Example 1.17

Determining the effect of confounding

In a study of the effects of blood pressure on health, a large group of people of all ages were given regular blood pressure checkups for a period of one year. It was found that people with high blood pressure were more likely to develop cancer than people with lower blood pressure. Explain how this result might be due to confounding.

Solution

Age is a likely confounder. Older people tend to have higher blood pressure than younger people, and older people are more likely to get cancer than younger people. Therefore people with high blood pressure may have higher cancer rates than younger people, even though high blood pressure does not cause cancer.

Check Your Understanding

1. To study the effect of air pollution on respiratory health, a group of people in a city with high levels of air pollution and another group in a rural area with low levels of pollution are examined to determine their lung capacity. Is this a randomized experiment or an observational study? *Observational study*
2. It is known that drinking alcohol increases the risk of contracting liver cancer. Assume that in an observational study, a group of smokers has a higher rate of liver cancer than a group of nonsmokers. Explain how this result might be due to confounding. *Smokers may drink more than nonsmokers.*

Answers are on page 27.

Objective 4 Describe various types of observational studies

Types of Observational Studies

There are two main types of observational studies: cohort studies and case-control studies. Cohort studies can be further divided into prospective, cross-sectional, and retrospective studies.

Cohort studies

In a **cohort study**, a group of subjects (the cohort) is studied to determine whether various factors of interest are associated with an outcome.

In a **prospective** cohort study, the subjects are followed over time. One of the most famous prospective cohort studies is the Framingham Heart Study. This study began in 1948 with 5209 men and women from the town of Framingham, Massachusetts. Every two years, these subjects are given physical exams and lifestyle interviews, which are studied to discover factors that increase the risk of heart disease. Much of what is known about the effects of diet and exercise on heart disease is based on this study.

Prospective studies are among the best observational studies. Because subjects are repeatedly examined, the quality of the data is often quite good. Information on potential confounders can be collected as well. Results from prospective studies are generally more reliable than those from other observational studies. The disadvantages of prospective studies are that they are expensive to run and that it takes a long time to develop results.

In a **cross-sectional** study, measurements are taken at one point in time. For example, in a study published in the *Journal of the American Medical Association* (300:1303–1310), I. Lang and colleagues studied the health effects of bisphenol A, a chemical found in the linings of food and beverage containers. They measured the levels of bisphenol A in urine samples from 1455 adults. They found that people with higher levels of bisphenol A were more likely to have heart disease and diabetes.

Cross-sectional studies are relatively inexpensive, and results can be obtained quickly. The main disadvantage is that the exposure is measured at only one point in time, so there is little information about how past exposures may have contributed to the outcome. Another disadvantage is that because measurements are made at only one time, it is impossible to determine a time sequence of events. For example, in the bisphenol A study just described, it is possible that higher levels of bisphenol A cause heart disease and diabetes. But it is also possible that the onset of heart disease or diabetes causes levels of bisphenol A to increase. There is no way to determine which happened first.

In a **retrospective** cohort study, subjects are sampled after the outcome has occurred. The investigators then look back over time to determine whether certain factors are related to the outcome. For example, in a study published in *The New England Journal of Medicine* (357:753–761), T. Adams and colleagues sampled 9949 people who had undergone gastric bypass surgery between 5 and 15 years previously, along with 9668 obese patients who had not had bypass surgery. They looked back in time to see which patients were still alive. They found that the survival rates for the surgery patients were greater than for those who had not undergone surgery.

Retrospective cohort studies are less expensive than prospective cohort studies, and results can be obtained quickly. A disadvantage is that it is often impossible to obtain data on potential confounders.

One serious limitation of all cohort studies is that they cannot be used to study rare diseases. Even in a large cohort, very few people will contract a particular rare disease. To study rare diseases, case-control studies must be used.

Case-control studies

In a **case-control** study, two samples are drawn. One sample consists of people who have the disease of interest (the cases), and the other consists of people who do not have the disease (the controls). The investigators look back in time to determine whether a particular factor of interest differs between the two groups. For example, S. S. Nielsen and colleagues conducted a case-control study to determine whether exposure to pesticides is related to brain cancer in children (*Environmental Health Perspectives*, 118:144–149). They sampled 201 children under the age of 10 who had been diagnosed with brain cancer, and 285 children who did not have brain cancer. They interviewed the parents of the children to estimate the extent to which the children had been exposed to pesticides. They did not find a clear relationship between pesticide exposure and brain cancer. This study could not have been conducted as a cohort study, because even in a large cohort of children, very few will get brain cancer.

Case-control studies are always retrospective, because the outcome (case or control) has occurred before the sampling is done. Case-control studies have the same advantages and disadvantages as retrospective cohort studies. In addition, case-control studies have the advantage that they can be used to study rare diseases.

Check Your Understanding

3. In a study conducted at the University of Southern California, J. Peters and colleagues studied elementary school students in 12 California communities. Each year for 10 years, they measured the respiratory function of the children and the levels of air pollution in the communities.
   a. Was this a cohort study or a case-control study? *Cohort*
   b. Was the study prospective, cross-sectional, or retrospective? *Prospective*

4. In a study conducted at the University of Colorado, J. Ruttenber and colleagues studied people who had worked at the Rocky Flats nuclear weapons production facility near Denver, Colorado. They studied a group of workers who had contracted lung cancer and another group who had not contracted lung cancer. They looked back at plant records to determine the amount of radiation exposure for each worker. The purpose of the study was to determine whether the people with lung cancer had been exposed to higher levels of radiation than those who had not gotten lung cancer.
   a. Was this a cohort study or a case-control study? *Case-control*
   b. Was the study prospective, cross-sectional, or retrospective? *Retrospective*

Answers are on page 27.

Section 1.3 Exercises

Exercises 1–4 are the Check Your Understanding exercises located within the section.

Understanding the Concepts

In Exercises 5–10, fill in each blank with the appropriate word or phrase.

5. In a ________ experiment, subjects do not decide for themselves which treatment they will get. *randomized*

6. In a ________ study, neither the investigators nor the subjects know who is getting which treatment. *double-blind*

7. A study in which the assignment to treatment groups is not made by the investigator is called ________. *observational*

8. A ________ is a variable related to both the treatment and the outcome. *confounder*

9. In a ________ study, the subjects are followed over time. *prospective*

10. In a ________ study, a group of subjects is studied to determine whether various factors of interest are associated with an outcome. *cohort*

In Exercises 11–16, determine whether the statement is true or false. If the statement is false, rewrite it as a true statement.

11. In a randomized experiment, the treatment groups do not differ in any systematic way except that they receive different treatments. *True*

12. A confounder makes it easier to draw conclusions from a study. *False*

13. In an observational study, subjects are assigned to treatment groups at random. *False*

14. Observational studies are generally more reliable than randomized experiments. *False*

15. In a case-control study, the outcome has occurred before the subjects are sampled. *True*

16. In a cross-sectional study, measurements are made at only one point in time. *True*

Practicing the Skills

17. To determine the effectiveness of a new pain reliever, a randomly chosen group of pain sufferers is assigned to take the new drug, and another randomly chosen group is assigned to take a placebo.
   a. Is this a randomized experiment or an observational study? *Randomized experiment*
   b. The subjects taking the new drug experienced substantially more pain relief than those taking the placebo. The research team concluded that the new drug is effective in relieving pain. Is this conclusion well justified? Explain. *Yes*

18. A medical researcher wants to determine whether exercising can lower blood pressure. At a health fair, he measures the blood pressure of 100 individuals and interviews them about their exercise habits. He divides the individuals into two categories: those whose typical level of exercise is low, and those whose level of exercise is high.
   a. Is this a randomized experiment or an observational study? *Observational study*

b. The subjects in the low-exercise group had considerably higher blood pressure, on the average, than subjects in the high-exercise group. The researcher concluded that exercise decreases blood pressure. Is this conclusion well justified? Explain. *No*

19. A medical researcher wants to determine whether exercising can lower blood pressure. She recruits 100 people with high blood pressure to participate in the study. She assigns a random sample of 50 of them to pursue an exercise program that includes daily swimming and jogging. She assigns the other 50 to refrain from vigorous activity. She measures the blood pressure of each of the 100 individuals both before and after the study.

a. Is this a randomized experiment or an observational study? *Randomized experiment*

b. On the average, the subjects in the exercise group substantially reduced their blood pressure, while the subjects in the no-exercise group did not experience a reduction. The researcher concluded that exercise decreases blood pressure. Is this conclusion well justified? Explain. *Yes*

20. An agricultural scientist wants to determine the effect of fertilizer type on the yield of tomatoes. There are four types of fertilizer under consideration. She plants tomatoes on four plots of land. Each plot is treated identically except for receiving a different type of fertilizer.

a. What are the treatments? *Four types of fertilizer*

b. Is this a randomized experiment or an observational study? *Randomized experiment*

c. The yields differ substantially among the four plots. Can you conclude that the differences in yield are due to the differences in fertilizer? Explain. *Yes*

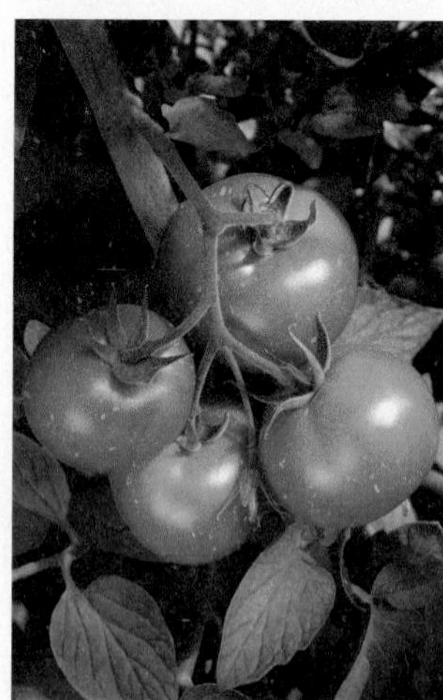
hüseyin harmandaźlž/Getty Images

Working with the Concepts

21. Air pollution and colds: A scientist wants to determine whether people who live in places with high levels of air pollution get more colds than people in areas with little air pollution. Do you think it is possible to design a randomized experiment to study this question, or will an observational study be necessary? Explain. *Observational study necessary*

22. Cold medications: A scientist wants to determine whether a new cold medicine relieves symptoms more effectively than a currently used medicine. Do you think it is possible to design a randomized experiment to study this question, or will an observational study be necessary? Explain. *Randomized experiment possible*

23. Taxicabs and crime: A sociologist discovered that regions that have more taxicabs tend to have higher crime rates. Does increasing the number of taxicabs cause the crime rate to increase, or could the result be due to confounding? Explain. *Could be due to confounding*

24. Recovering from heart attacks: In a study of people who had suffered heart attacks, it was found that those who lived in smaller houses were more likely to recover than those who lived in larger houses. Does living in a smaller house increase the likelihood of recovery from a heart attack, or could the result be due to confounding? Explain. *Could be due to confounding*

25. Eat your vegetables: In an observational study, people who ate four or more servings of fresh fruits and vegetables each day were less likely to develop colon cancer than people who ate little fruit or vegetables. True or false:

a. The results of the study show that eating more fruits and vegetables reduces your risk of contracting colon cancer. *False*

b. The results of the study may be due to confounding, since the lifestyles of people who eat large amounts of fruits and vegetables may differ in many ways from those of people who do not. *True*

26. Vocabulary and height: A vocabulary test was given to students at an elementary school. The students' ages ranged from 5 to 11 years old. It was found that the students with larger vocabularies tended to be taller than the students with smaller vocabularies. Explain how this result might be due to confounding.

27. Secondhand smoke: A recent study compared the heart rates of 19 infants born to nonsmoking mothers with those of 17 infants born to mothers who smoked an average of 15 cigarettes a day while pregnant and after giving birth. The heart rates of the infants at one year of age were 20% slower on the average for the smoking mothers.

a. What is the outcome variable? *Heart rate*

b. What is the treatment variable? *Maternal smoking*

c. Was this a cohort study or a case-control study? *Cohort*

d. Was the study prospective, cross-sectional, or retrospective? *Prospective*

e. Could the results be due to confounding? Explain. *Yes*

Source: *Environmental Health Perspectives* 118:a158–a159

28. Pollution in China: In a recent study, Z. Zhao and colleagues measured the levels of formaldehyde in the air in 34 classrooms in the schools in the city of Taiyuan, China. On the same day, they gave questionnaires to 1993 students aged 11–15 in those schools, asking them whether they had experienced respiratory problems (such as asthma attacks, wheezing, or shortness of breath). They found that the students in the classrooms with higher levels of formaldehyde reported more respiratory problems.

a. What is the outcome variable?

b. What is the treatment variable? *Formaldehyde level*

c. Was this a cohort study or a case-control study? *Cohort*

d. Was the study prospective, cross-sectional, or retrospective? *Cross-sectional*

e. Could the results be due to confounding? Explain. *Unlikely*

Source: *Environmental Health Perspectives* 116:90–97

Extending the Concepts

29. The Salk Vaccine Trial: In 1954, the first vaccine against polio, known as the Salk vaccine, was tested in a large randomized double-blind study. Approximately 750,000 children were asked to enroll in the study. Of these, approximately 350,000 did not participate, because their parents refused permission. The children who did participate were randomly divided into two groups of about 200,000 each. One group, the treatment group, got the vaccine, while the other group, the control group, got a placebo. The rate of polio in the treatment group was less than half of that in the control group.

a. Is it reasonable to conclude that the Salk vaccine was effective in reducing the rate of polio? *Yes*

b. Polio is sometimes difficult to diagnose, as its early symptoms are similar to those of the flu. Explain why it was important for the doctors in the study not to know which children were getting the vaccine.

c. Perhaps surprisingly, polio was more common among upper-income and middle-income children than among lower-income children. The reason is that lower-income children tended to live in less hygienic surroundings. They would contract mild cases of polio in infancy while still protected by their mother's antibodies, and thereby develop a resistance to the disease. The children who did not participate in the study were more likely to come from lower-income families. The rate of polio in this group was substantially lower than the rate in the placebo group. Does this prove that the placebo caused polio, or could this be due to confounding? Explain. *Could be due to confounding*

30. Another Salk Vaccine Trial: Another study of the Salk vaccine, conducted at the same time as the trial described in Exercise 29, used a different design. In this study, approximately 350,000 second graders were invited to participate. About 225,000 did so, and the other 125,000 refused. All of the participating second graders received the vaccine. The control group consisted of approximately 725,000 first and third graders. They were not given any placebo, so no consent was necessary.

a. Was this a randomized experiment? *No*

b. Was it double-blind? *No*

c. The treatment group consisted of children who had consent to participate. The control group consisted of all first and third graders. It turned out that the results of this study seriously underestimated the effectiveness of the vaccine. Use the information provided in Exercise 29(c) to explain why.

31. Smoking and health: A study was performed by the Public Health Service on the health of smokers and former smokers. The results showed that current smokers were healthier than people of the same age who had recently quit smoking. Is it reasonable to conclude that smokers become less healthy after they quit? If not, what other explanation can you give?

32. Alcohol and liver disease: It is known that alcohol consumption increases the risk of liver disease. A study finds that the rate of liver disease is higher among smokers than nonsmokers. Is it reasonable to conclude that smoking increases the risk of liver disease? What other explanation can you give?

Answers to Check Your Understanding Exercises for Section 1.3

1. Observational study

2. People who smoke may be more likely to drink alcohol than people who do not smoke. Therefore, it might be possible for smokers to have higher rates of liver cancer without it being caused by smoking.

3. a. Cohort study **b.** Prospective

4. a. Case-control study **b.** Retrospective

Section 1.4 Bias in Studies

Objectives

1. Define bias
2. Identify sources of bias

Objective 1 Define bias

Defining Bias

No study is perfect, and even a properly conducted study will generally not give results that are exactly correct. For example, imagine that you were to draw a simple random sample of students at a certain college to estimate the percentage of students who are Democrats. Your sample would probably contain a somewhat larger or smaller percentage of Democrats than the entire population of students, just by chance. However, imagine drawing many simple random samples. Some would have a greater percentage of Democrats than in the population, and some would have a smaller percentage of Democrats than in the population. But on the average, the percentage of Democrats in a simple random sample will be the same as the percentage in the population. A study conducted by a procedure that produces the correct result on the average is said to be *unbiased*.

Now imagine that you tried to estimate the percentage of Democrats in the population by selecting students who attended a speech made by a Democratic politician. On the average, studies conducted in this way would overestimate the percentage of Democrats in the population. Studies conducted with methods that tend to overestimate or underestimate a population value are said to be *biased*.

DEFINITION

Bias is the degree to which a procedure systematically overestimates or underestimates a population value.

- A study conducted by a procedure that tends to overestimate or underestimate a population value is said to be **biased**.
- A study conducted by a procedure that produces the correct result on the average is said to be **unbiased**.

Objective 2 Identify sources of bias

Identifying Sources of Bias

In practice, it is important to design studies to have as little bias as possible. Unfortunately, some studies are highly biased, and the conclusions drawn from them are not reliable. Here are some common types of bias.

Voluntary response bias

Recall that a voluntary response survey is one in which people are invited to log on to a website, send a text message, or call a phone number, in order to express their opinions on an issue. In many cases, the opinions of the people who choose to participate in such surveys do not reflect those of the population as a whole. In particular, people with strong opinions are more likely to participate. In general, voluntary response surveys are highly biased.

Self-interest bias

Many advertisements contain data that claim to show that the product being advertised is superior to its competitors. Of course, the advertiser will not report any data that tend to show that the product is inferior. Even more seriously, many people are concerned about a trend for companies to pay scientists to conduct studies involving their products. In particular, physicians are sometimes paid by drug companies to test their drugs and to publish the results of these tests in medical journals. People who have an interest in the outcome of an experiment have an incentive to use biased methods.

Social acceptability bias

People are reluctant to admit to behavior that may reflect negatively on them. This characteristic of human nature affects many surveys. For example, in political polls it is important for the pollster to determine whether the person being interviewed is likely to vote. A good way to determine whether someone is likely to vote in the next election is to find out whether they voted in the last election. It might seem reasonable to ask the following question:

"Did you vote in the last presidential election?"

The problem with this direct approach is that people are reluctant to answer "No," because they are concerned that not voting is socially less acceptable than voting. Here is how the question might be asked more indirectly:

"In the 2016 presidential election between Hillary Clinton and Donald Trump, did things come up that kept you from voting, or did you happen to vote?"

People are more likely to answer this version of the question truthfully.

Leading question bias

Sometimes questions are worded in a way that suggests a particular response. For example, a political group that supports lowering taxes sent out a survey that included the following question:

> "Do you favor decreasing the heavy tax burden on middle-class families?"

The words "heavy" and "burden" suggest that taxes are too high, and encourage a "Yes" response. A better way to ask this question is to present it as a multiple choice:

> "What is your opinion on decreasing taxes for middle-class families?
> Choices: Strongly disagree, Somewhat disagree, Neither agree nor disagree, Somewhat agree, Strongly agree."

Nonresponse bias

People cannot be forced to answer questions or to participate in a study. In any study, a certain proportion of people who are asked to participate refuse to do so. These people are called **nonresponders**. In many cases, the opinions of nonresponders tend to differ from the opinions of those who do respond. As a result, surveys with many nonresponders are often biased.

Sampling bias

Sampling bias occurs when some members of the population are more likely to be included in the sample than others. For example, samples of convenience almost always have sampling bias, because people who are easy to sample are more likely to be included. It is almost impossible to avoid sampling bias completely, but modern survey organizations work hard to keep it at a minimum.

It is important to note that a question may exhibit more than one type of bias. Here is a hypothetical example: Parents of school children were asked, "Do you support improving public schools by issuing school bonds?" Only 30% responded. This example exhibits sampling bias, because parents of school children may be more likely to support increased spending on schools. There is also leading question bias, because the question is worded to suggest a positive response. Finally, there is nonresponse bias, because only 30% of those sampled responded to the question.

NOTE TO INSTRUCTOR

The point that a big data set can be misleading is important to emphasize. Now that the term "big data" has become popular, it is easy to forget that even very large data sets can be misleading if not assembled correctly.

A Big Sample Size Doesn't Make Up for Bias

A sample is useful only if it is drawn by a method that is likely to represent the population well. If you use a biased method to draw a sample, then drawing a big sample doesn't help; a big nonrepresentative sample does not describe a population any better than a small nonrepresentative sample. In particular, voluntary response surveys often draw several hundred thousand people to participate. Although the sample is large, it is unlikely to represent the population well, so the results are meaningless.

Check Your Understanding

1. Eighty thousand people attending a professional football game filled out surveys asking their opinions on using tax money to upgrade the football stadium. Seventy percent said that they supported the use of tax money. Then a pollster surveyed a simple random sample of 500 voters, and only 30% of the voters in this sample supported the use of tax money. The owner of the football team claims that the survey done at the football stadium is more reliable, because the sample size was much larger. Is he right? Explain. *No*
2. A polling organization placed telephone calls to 1000 people in a certain city to ask them whether they favor a tax increase to build a new school. Two hundred people answered the phone, and 150 of them opposed the tax. Can you conclude that a majority of people in the city oppose the tax, or is it likely that this result is due to bias? Explain. *Likely due to bias*

Answers are on page 31.

Section 1.4 Exercises

Exercises 1 and 2 are the Check Your Understanding exercises located within the section.

Understanding the Concepts

In Exercises 3–5, fill in each blank with the appropriate word or phrase.

3. __________ are highly unreliable in part because people who have strong opinions are more likely to participate. *Voluntary response surveys*
4. People who are asked to participate in a study but refuse to do so are called __________. *nonresponders*
5. A large sample is useful only if it is drawn by a method that is likely to represent the __________ well. *population*

In Exercises 6–8, determine whether the statement is true or false. If the statement is false, rewrite it as a true statement.

6. The way that a question in a survey is worded rarely has an effect on the responses. *False*
7. Surveys with many nonresponders often provide misleading results. *True*
8. Large samples usually give reasonably accurate results, no matter how they are drawn. *False*

Practicing the Skills

In Exercises 9–16, specify the type of bias involved.

9. A bank sent out questionnaires to a simple random sample of 500 customers asking whether they would like the bank to extend its hours. Eighty percent of those returning the questionnaire said they would like the bank to extend its hours. Of the 500 questionnaires, 20 were returned. *Nonresponse*
10. To determine his constituents' feelings about election reform, a politician sends a survey to people who have subscribed to his newsletter. More than 1000 responses are received. *Voluntary response*
11. An e-store that sells phone accessories reports that 98% of its customers are satisfied with the speed of delivery. *Self-interest*
12. A sign in a restaurant claims that 95% of their customers believe them to have the best food in the world. *Self-interest*
13. A television newscaster invites viewers to tweet their opinions about whether the U.S. Congress is doing a good job in handling the economy. More than 100,000 people send in an opinion. *Voluntary response*
14. A police department conducted a survey in which police officers interviewed members of their community to ask their opinions on the effectiveness of the police department. The police chief reported that 90% of the people interviewed said that they were satisfied with the performance of the police department. *Social acceptability*
15. In a study of the effectiveness of wearing seat belts, a group of people who had survived car accidents in which they had not worn seat belts reported that seat belts would not have helped them. *Nonresponse*
16. To estimate the prevalence of illegal drug use in a certain high school, the principal interviewed a simple random sample of 100 students and asked them about their drug use. Five percent of the students acknowledged using illegal drugs. *Social acceptability*

Working with the Concepts

17. **Nuclear power, anyone?** In a survey conducted by representatives of the nuclear power industry, people were asked the question: "Do you favor the construction of nuclear power plants in order to reduce our dependence on foreign oil?" A group opposed to the use of nuclear power conducted a survey with the question: "Do you favor the construction of nuclear power plants that can kill thousands of people in an accident?"
    - **a.** Do you think that the percentage of people favoring the construction of nuclear power plants would be about the same in both surveys? *No*
    - **b.** Would either of the two surveys produce reliable results? Explain. *No*
18. **Who's calling, please?** Random-digit dialing is a sampling method in which a computer generates phone numbers at random to call. Do you think that caller ID increases the bias in random digit dialing? Explain. *Yes*
19. **Who's calling, please?** Many polls are conducted over the telephone. Suppose that a pollster chooses a sample of phone numbers to call from lists that include landline phone numbers only, and do not include phones. Do you think this increases the bias in phone polls? Explain. *Yes*
20. **Order of choices:** When multiple-choice questions are asked, the order of the choices is usually changed each time the question is asked. For example, in the 2016 presidential election, a pollster would ask one person, "Who do you prefer for president, Hillary Clinton or Donald Trump?" For the next person, the order of the names would be reversed: "Donald Trump or Hillary Clinton?" If the choices were given in the same order each time, do you think that might introduce bias? Explain. *Yes*

Extending the Concepts

21. ***Literary Digest* poll:** In the 1936 presidential election, Republican candidate Alf Landon challenged President Franklin Roosevelt. The *Literary Digest* magazine conducted a poll in which they mailed questionnaires to more than 10 million voters. The people who received the questionnaires were drawn from lists of automobile owners and people with telephones. The magazine received 2.3 million responses and predicted that Landon would win the election in a landslide with 57% of the vote. In fact, Roosevelt won in a landslide with 62% of the vote. Soon afterward, the *Literary Digest* folded.

a. In 1936 most people did not own automobiles, and many did not have telephones. Explain how this could have caused the results of the poll to be mistaken.
b. What can be said about the response rate? Explain how this could have caused the results of the poll to be mistaken.
c. The *Literary Digest* believed that its poll would be accurate, because it received 2.3 million responses, which is a very large number. Explain how the poll could be wrong, even with such a large sample.

Answers to Check Your Understanding Exercises for Section 1.4

1. No. The sample taken at the football stadium is biased, because football fans are more likely to be sampled than others. The fact that the sample is big doesn't make it any better.
2. No. There is a high degree of nonresponse bias in this sample.

Chapter 1 Summary

Section 1.1: Most populations are too large to allow us to study each member, so we draw samples and study those. Samples must be drawn by an appropriate method. Simple random sampling, stratified sampling, cluster sampling, and systematic sampling are all valid methods. When none of these methods are feasible, a sample of convenience may be used, so long as it is reasonable to believe that there is no systematic difference between the sample and the population.

Section 1.2: Data sets contain values of variables. Qualitative variables place items in categories, whereas quantitative variables are counts or measurements. Qualitative variables can be either ordinal or nominal. An ordinal variable is one for which the categories have a natural ordering. For nominal variables, the categories have no natural ordering. Quantitative variables can be discrete or continuous. Discrete variables are ones whose possible values can be listed, whereas continuous variables can take on values anywhere within an interval.

Section 1.3: Scientists conduct studies to determine whether different treatments produce different outcomes. The most reliable studies are randomized experiments, in which subjects are assigned to treatments at random. When randomized experiments are not feasible, observational studies may be performed. Results of observational studies may be hard to interpret, because of the potential for confounding.

Section 1.4: Some studies produce more reliable results than others. A study that is conducted by a method that tends to produce an incorrect result is said to be biased. Some of the most common forms of bias are voluntary response bias, self-interest bias, social acceptability bias, leading question bias, nonresponse bias, and sampling bias.

Vocabulary and Notation

Chapter Quiz

1. Provide an example of a qualitative variable and an example of a quantitative variable.
2. Is the name of your favorite author a qualitative variable or a quantitative variable? *Qualitative*
3. True or false: Nominal variables do not have a natural ordering. *True*
4. ________ variables are quantitative variables that can take on any value in some interval. *Continuous*
5. True or false: Ideally, a sample should represent the population as little as possible. *False*
6. A utility company sends surveys to 200 of its customers in such a way that 100 surveys are sent to customers who pay their bills on time, 50 surveys are sent to customers whose bills are less than 30 days late, and 50 surveys are sent to customers whose bills are more than 30 days late. Which type of sample does this represent? *Stratified sample*
7. A sample of convenience is ________ when it is reasonable to believe that there is no systematic difference between the sample and the population. (*Choices:* acceptable, not acceptable) *acceptable*
8. The manager of a restaurant walks around and asks selected customers about the service they have received. Which type of sample does this represent? *Sample of convenience*
9. True or false: An experiment where neither the investigators nor the subjects know who has been assigned to which treatment is called a double-blind experiment. *True*
10. A poll is conducted of 3500 households close to major national airports and another 2000 that are not close to an airport, in order to study whether living in a noisier environment results in health effects. Is this a randomized experiment or an observational study? *Observational study*
11. In a study, 200 patients with skin cancer are randomly divided into two groups. One group receives an experimental drug and the other group receives a placebo. Is this a randomized experiment or an observational study? *Randomized experiment*
12. In a randomized experiment, if there are large differences in outcomes among treatment groups, we can conclude that the differences are due to the ________. *differences in treatment*
13. In analyzing the course grades of students in an elementary statistics course, a professor notices that students who are seniors performed better than students who are sophomores. The professor is tempted to conclude that older students perform better than younger ones. Describe a possible confounder in this situation. *Seniors may be better prepared.*
14. True or false: The way that questions are worded on a survey may have an effect on the responses. *True*
15. A podcaster invites listeners to call the show to express their opinions about a political issue. How reliable is this survey? Explain. *Not reliable*

Review Exercises

1. **Qualitative or quantitative?** Is the number of points scored in a football game qualitative or quantitative? *Quantitative*
2. **Nominal or ordinal?** Is the color of a phone nominal or ordinal? *Nominal*
3. **Discrete or continuous?** Is the area of a college campus discrete or continuous? *Continuous*
4. **Which type of variable is it?** A theater concession stand sells soft drink and popcorn combos that come in sizes small, medium, large, and jumbo. True or false:
   a. Size is a qualitative variable. *True*
   b. Size is an ordinal variable. *True*
   c. Size is a continuous variable. *False*

In Exercises 5–8, identify the kind of sample that is described.

5. **Website ratings:** A popular website is interested in conducting a survey of 400 visitors to the site in such a way that 200 of them will be under age 30, 150 will be aged 30–55, and 50 will be over 55. *Stratified sample*
6. **Favorite performer:** Viewers of a television show are asked to vote for their favorite performer by sending a text message to the show. *Voluntary response sample*
7. **School days:** A researcher selects 4 of 12 high schools in a certain region and surveys all of the administrative staff members in each school about a potential change in the ordering of supplies. Which type of sample does this represent? *Cluster sample*
8. **Political polling:** A pollster obtains a list of registered voters and uses a computer random number generator to choose 100 of them to ask which candidate they prefer in an upcoming election. *Simple random sample*

9. Fluoride and tooth decay: Researchers examine the association between the fluoridation of water and the prevention of tooth decay by comparing the prevalence of tooth decay in countries that have fluoridated water with the prevalence in countries that do not.

a. Is this a randomized experiment or an observational study? *Observational study*

b. Assume that tooth decay was seen to be less common in countries with fluoridated water. Could this result be due to confounding? Explain. *Yes*

10. Better gas mileage: A rental company in a large city put a new type of tire with a special tread on a random sample of 50 cars, and the regular type of tire on another random sample of 50 cars. After a month, the gas mileage of each car was measured.

a. Is this a randomized experiment or an observational study? *Randomized experiment*

b. Assume that one of the samples had noticeably better gas mileage than the other. Could this result be due to confounding? Explain. *Unlikely*

11. Phones and driving: Almost all states in the United States prohibit texting while driving, and several states prohibit all phone use while driving. To determine the extent to which using a phone increases the risk of a traffic accident, a researcher examines accident reports to obtain data about the number of accidents in which a driver was using a phone.

a. Is this a randomized experiment or an observational study? *Observational study*

b. Assume that the accident reports show that people were more likely to have an accident while using a phone. Could this result be due to confounding? Explain. *Yes*

12. Turn in your homework: The English department at a local college is considering using electronic-based assignment submission in its English composition classes. To study its effects, each section of the class is divided into two groups at random. In one group, assignments are submitted by turning them in to the professor on paper. In the other group, assignments are submitted electronically.

a. Is this a randomized experiment or an observational study? *Randomized experiment*

b. Assume that the electronically submitted assignments had many fewer typographical errors, on average, than the ones submitted on paper. Could this result be due to confounding? Explain. *Unlikely*

In Exercises 13–15, explain why the results of the studies described are unreliable.

13. Which TV station do you watch? The TV columnist for a local newspaper invites readers to log on to a website to vote for their favorite TV newscaster. *Voluntary response sample*

14. Longevity: A life insurance company wants to study the life expectancy of people born in 1950. The company's actuaries examine death certificates of people born in that year to determine how long they lived. *Nonresponse bias*

15. Political opinion: A congressman sent out questionnaires to 10,000 constituents to ask their opinions on a new health-care proposal. A total of 200 questionnaires were returned, and 70% of those responding supported the proposal. *Nonresponse bias*

Write About It

1. Describe the difference between a stratified sample and a cluster sample.

2. Explain why it is better, when possible, to draw a simple random sample rather than a sample of convenience.

3. Describe circumstances under which each of the following samples could be used: simple random sample, a sample of convenience, a stratified sample, a cluster sample, a systematic sample.

4. Suppose that you were asked to collect some information about students in your class for a statistics project. Give some examples of variables you might collect that are ordinal, nominal, discrete, and continuous.

5. Quantitative variables are numerical. Are some qualitative variables numerical as well? If not, explain why not. If so, provide an example.

6. What are the primary differences between a randomized experiment and an observational study?

7. What are the advantages of a double-blind study? Are there any disadvantages?

8. Provide an example of a study, either real or hypothetical, that is conducted by people who have an interest in the outcome. Explain how the results might possibly be misleading.

9. Explain why each of the following questions is leading. Provide a more appropriate wording.

a. Should Americans save more money or continue their wasteful spending?

b. Do you support more funding for reputable organizations like the Red Cross?

In-Class Activities

NOTE TO INSTRUCTOR

The biased sampling activity illustrates how stratified sampling can ensure representation of hard-to-sample groups. The attempts to draw simple random samples are likely to overrepresent the larger items and thus be biased upward. Stratified sampling also decreases the variation in the results. This can be mentioned as an aside, although students may not be ready for a detailed treatment of this concept.

1. **Biased survey questions:** Construct survey questions that exhibit the following types of bias: self-interest bias, social acceptability bias, and leading question bias. Then construct versions of each question in which the bias is reduced or eliminated.
2. **Biased sampling:** This activity requires a bag of 50 objects of two sizes, and a small scale. Assorted miniature candy bars can be used for the larger objects, while M&Ms or Skittles can be used for the smaller ones. The number of objects of each size is known; say there are 20 large ones and 30 small ones. Students are divided into groups. Some groups are instructed to attempt to draw a simple random sample of five items, while others draw a stratified sample containing two large items and three small ones. Each group weighs its sample and multiplies the result by 10 to estimate the total weight of the population of objects. Which sampling method is more accurate in this situation? Why?

Case Study: Air Pollution And Respiratory Symptoms

Air pollution is a serious problem in many places. One form of air pollution that is suspected to cause respiratory illness is particulate matter (PM), which consists of tiny particles in the air. Particulate matter can come from many sources, most commonly ash from burning, but also from other sources such as tiny particles of rubber that wear off of automobile and truck tires.

The town of Libby, Montana, was recently the focus of a study on the effect of PM on the respiratory health of children. Many houses in Libby are heated by wood stoves, which produce a lot of particulate pollution. The level of PM is greatest in the winter when more stoves are being used and declines as the weather becomes warmer. The study attempted to determine whether higher levels of PM affect the respiratory health of children. In one part of the study, schoolchildren were given a questionnaire to bring home to their parents. Among other things, the questionnaire asked whether the child had experienced symptoms of wheezing during the past 60 days. Most parents returned the questionnaire within a couple of weeks. Parents who did not respond promptly were sent another copy of the questionnaire through the mail. Many of these parents responded to this mailed version.

Table 1.2 presents, for each day, the number of questionnaires that were returned by parents of children who wheezed, the number returned by those who did not wheeze, the average concentration of particulate matter in the atmosphere during the past 60 days (in units of micrograms per cubic meter), and whether the questionnaires were delivered in school or through the mail.

Table 1.2

Date	PM Level	Number of People Returning Questionnaires	Number Who Wheezed	School/Mail
March 5	19.815	3	0	School
March 6	19.885	72	9	School
March 7	20.006	69	5	School
March 8	19.758	30	1	School
March 9	19.827	44	7	School
March 10	19.686	31	1	School
March 11	19.823	38	3	School
March 12	19.697	66	5	School
March 13	19.505	42	4	School
March 14	19.359	31	1	School
March 15	19.348	19	4	School
March 16	19.318	3	1	School
March 17	19.124	2	0	School
April 12	14.422	10	1	Mail
April 13	14.418	9	1	Mail
April 14	14.405	8	0	Mail
April 15	14.141	3	0	Mail
April 16	13.910	4	0	Mail
April 17	13.951	2	0	Mail
April 18	13.545	2	0	Mail
April 20	13.326	3	0	Mail
April 22	13.154	2	0	Mail

We will consider a PM level of 17 or more to be high exposure and a PM level of less than 17 to be low exposure.

1. How many people had high exposure to PM? *450*
2. How many of the high-exposure people had wheeze symptoms? *41*
3. What percentage of the high-exposure people had wheeze symptoms? *9.1%*
4. How many people had low exposure to PM? *43*
5. How many of the low-exposure people had wheeze symptoms? *2*
6. What percentage of the low-exposure people had wheeze symptoms? *4.7%*
7. Is there a large difference between the percentage of high-exposure people with wheeze symptoms and the percentage of low-exposure people with wheeze symptoms? *Yes*
8. Explain why the percentage of high-exposure people with wheeze symptoms is the same as the percentage of school-return people with wheeze symptoms.
9. Explain why the percentage of low-exposure people with wheeze symptoms is the same as the percentage of mail-return people with wheeze symptoms.
10. As the weather gets warmer, PM goes down because wood stoves are used less. Explain how this causes the mode of response (school or mail) to be related to PM.
11. It is generally the case in epidemiologic studies that people who have symptoms are often eager to participate, while those who are unaffected are less interested. Explain how this may cause the mode of response (school or mail) to be related to the outcome.
12. Rather than send out questionnaires, the investigators could have telephoned a random sample of people over a period of days. Explain how this might have reduced the confounding.

chapter

2

Graphical Summaries of Data

Mike Flippo/123RF

Introduction

Are hybrid cars more fuel efficient than non-hybrid cars? Increasing prices of gasoline, along with concerns about the environment, have made fuel efficiency an important concern. To determine whether hybrid cars get better mileage than non-hybrid cars, we will compare the U.S. EPA mileage ratings for model year 2019 hybrid cars with the mileages for model year 2019 small non-hybrid cars. The following tables present the results, in miles per gallon.

Mileage Ratings for 2019 Hybrid Cars							
40	28	21	43	23	56	22	28
28	48	31	41	19	26	39	48
26	46	19	30	55	23	29	41
33	21	42	50	22	42	50	29
26	29	34	46	52	42	30	41
43	58	25	25	44	44	22	19
27	49	21	23	52	41	24	52
24	23	46					

Source: www.fueleconomy.gov

Mileage Ratings for 2019 Small Non-hybrid Cars							
33	32	28	32	34	33	27	25
31	26	30	33	32	32	29	27
33	30	29	31	30	37	34	30
35	34	36	33	35	33	39	36
32	36	22	35	36	33	28	31
27	28	35	27	31	32	29	34
30	33	29	32	32	30	32	29
26	30	32	29	31	31	32	33

Source: www.fueleconomy.gov

NOTE TO INSTRUCTOR

The following topic from the **Statistics Corequisite Workbook** is aligned with the material in this chapter.
1.3 - Frequency, Relative Frequency, Bar Graphs, and Pie Charts

It is hard to tell from the lists of numbers whether the mileages differ substantially between hybrid and non-hybrid cars. What is needed are methods to summarize the data, so that their most important features stand out. One way to do this is by constructing graphs that allow us to visualize the important features of the data. In this chapter, we will learn how to construct many of the most commonly used graphical summaries. In the case study at the end of the chapter, you will be asked to use graphical methods to compare the mileages between hybrid and non-hybrid cars.

Section 2.1 Graphical Summaries for Qualitative Data

Objectives

1. Construct frequency distributions for qualitative data
2. Construct bar graphs
3. Construct pie charts

Frequency Distributions for Qualitative Data

Objective 1 Construct frequency distributions for qualitative data

How do retailers analyze their sales data to determine which methods of payment are most popular? Table 2.1 presents a list compiled by a retailer. The retailer accepts four types of credit cards: Visa, MasterCard, American Express, and Discover. The list contains the types of credit cards used by the last 50 customers.

NOTE TO INSTRUCTOR

It is important for students to understand how to extract information from graphical summaries, as well as how to construct them.

Table 2.1 Types of Credit Cards Used

Discover	Visa	Visa	Am. Express	Visa
Visa	Visa	Am. Express	MasterCard	Visa
Am. Express	MasterCard	Visa	Visa	Visa
Visa	Am. Express	Am. Express	MasterCard	Visa
MasterCard	Visa	Discover	Am. Express	Discover
Visa	Am. Express	Discover	Visa	MasterCard
Visa	Visa	Visa	Visa	MasterCard
MasterCard	Am. Express	Visa	MasterCard	Visa
MasterCard	Discover	MasterCard	Visa	Visa
MasterCard	Discover	Am. Express	Discover	Visa

Table 2.1 is typical of data in raw form. It is a big list, and it's hard to gather much information simply by looking at it. To make the important features of the data stand out, we construct summaries. The starting point for many summaries is a frequency distribution.

DEFINITION

- The **frequency** of a category is the number of times it occurs in the data set.
- A **frequency distribution** is a table that presents the frequency for each category.

Example 2.1

Construct a frequency distribution

Construct a frequency distribution for the data in Table 2.1.

Solution

To construct a frequency distribution, we begin by tallying the number of observations in each category and recording the totals in a table. Table 2.2 (page 39) presents a frequency distribution for the credit card data. We have included the tally marks in this table, but in practice it is permissible to omit them.

Table 2.2 Frequency Distribution for Credit Cards

Credit Card	Tally	Frequency
MasterCard	𝍸 𝍸 |	11
Visa	𝍸 𝍸 𝍸 𝍸 |||	23
Am. Express	𝍸 ||||	9
Discover	𝍸 ||	7

CAUTION

When constructing a frequency distribution, be sure to check that the sum of all the frequencies is equal to the total number of observations.

It's a good idea to perform a check by adding the frequencies, to be sure that they add up to the total number of observations. In Table 2.2, the frequencies add up to 50, as they should.

Relative frequency distributions

A frequency distribution tells us exactly how many observations are in each category. Sometimes we are interested in the proportion of observations in each category. The proportion of observations in a category is called the *relative frequency* of the category.

EXPLAIN IT AGAIN

Difference between frequency and relative frequency: The frequency of a category is the number of items in the category. The relative frequency is the proportion of items in the category.

DEFINITION

The **relative frequency** of a category is the frequency of the category divided by the sum of all the frequencies.

$$\text{Relative frequency} = \frac{\text{Frequency}}{\text{Sum of all frequencies}}$$

We can add a column of relative frequencies to the frequency distribution. The resulting table is called a *relative frequency distribution*.

DEFINITION

A **relative frequency distribution** is a table that presents the relative frequency of each category. Often the frequency is presented as well.

Example 2.2

Constructing a relative frequency distribution

Construct a relative frequency distribution for the data in Table 2.2.

Solution

We compute the relative frequencies for each type of credit card in Table 2.2 by using the following steps.

Step 1: Find the total number of observations by summing the frequencies.

$$\text{Sum of frequencies} = 11 + 23 + 9 + 7 = 50$$

Step 2: Find the relative frequency for the first category, MasterCard.

$$\text{Relative frequency for MasterCard} = \frac{11}{50} = 0.22$$

Step 3: Find the relative frequencies for the remaining categories.

$$\text{Relative frequency for Visa} = \frac{23}{50} = 0.46$$

$$\text{Relative frequency for Am. Express} = \frac{9}{50} = 0.18$$

$$\text{Relative frequency for Discover} = \frac{7}{50} = 0.14$$

Table 2.3 on page 40 presents a relative frequency distribution for the data in Table 2.2.

Table 2.3 Relative Frequency Distribution for Credit Cards

Credit Card	Frequency	Relative Frequency
MasterCard	11	0.22
Visa	23	0.46
Am. Express	9	0.18
Discover	7	0.14

Check Your Understanding

1. The following table lists the types of aircraft for the landings that occurred during a day at a small airport. ("Single" refers to single-engine and "Twin" refers to twin-engine.)

Types of Aircraft Landing at an Airport

Twin	Single	Helicopter	Turboprop	Twin	Single
Turboprop	Jet	Jet	Turboprop	Turboprop	Single
Jet	Single	Single	Twin	Twin	Turboprop
Helicopter	Single	Single	Single	Twin	Single
Jet	Twin	Twin	Single	Twin	Twin

a. Construct a frequency distribution for these data.
b. Construct a relative frequency distribution for these data.

Answers are on page 50.

Bar Graphs

Objective 2 Construct bar graphs

A **bar graph** is a graphical representation of a frequency distribution. A bar graph consists of rectangles of equal width, with one rectangle for each category. The heights of the rectangles represent the frequencies or relative frequencies of the categories. Example 2.3 shows how to construct a bar graph for the credit card data in Table 2.3.

Example 2.3

Constructing bar graphs

Construct a frequency bar graph and a relative frequency bar graph for the credit card data in Table 2.3.

NOTE TO INSTRUCTOR

It is best for students to learn to construct graphs with technology in addition to drawing them by hand.

Solution

Step 1: Construct a horizontal axis. Place the category names along this axis, evenly spaced.
Step 2: Construct a vertical axis to represent the frequency or the relative frequency.
Step 3: Construct a bar for each category, with the heights of the bars equal to the frequencies or relative frequencies of their categories. The bars should not touch and should all be of the same width.

Figure 2.1(a) presents a frequency bar graph, and Figure 2.1(b) presents a relative frequency bar graph. The graphs are identical except for the scale on the vertical axis.

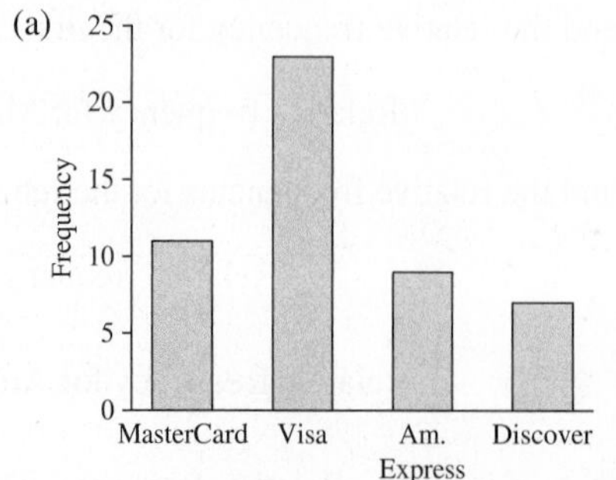

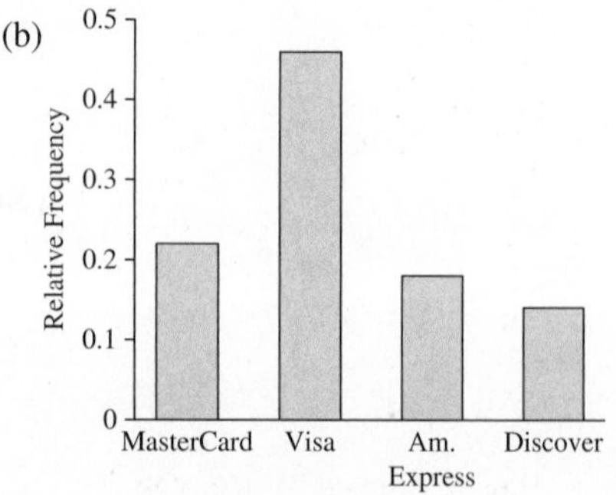

Figure 2.1 (a) Frequency bar graph. (b) Relative frequency bar graph.

Pareto charts

Sometimes it is desirable to construct a bar graph in which the categories are presented in order of frequency or relative frequency, with the largest frequency or relative frequency on the left and the smallest one on the right. Such a graph is called a **Pareto chart**. Pareto charts are useful when it is important to see clearly which are the most frequently occurring categories.

Example 2.4

Constructing a Pareto chart

Construct a relative frequency Pareto chart for the data in Table 2.3.

Solution

Figure 2.2 presents the result. It is just like Figure 2.1(b) except that the bars are ordered from tallest to shortest.

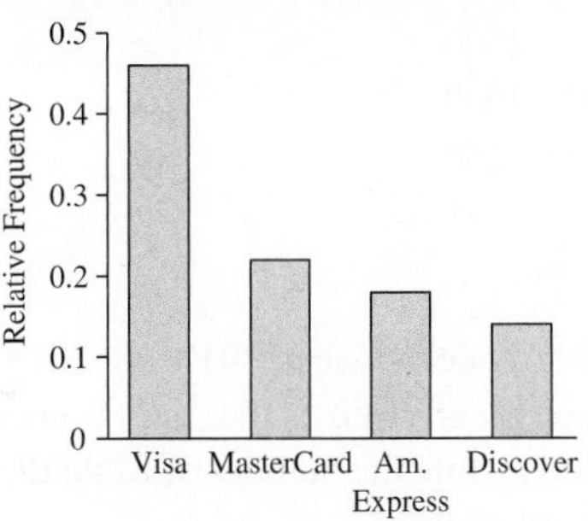

Figure 2.2 Pareto chart for the credit card data

Horizontal bars

The bars in a bar graph can be either horizontal or vertical. Horizontal bars are sometimes more convenient when the categories have long names.

Example 2.5

Constructing bar graphs with horizontal bars

The following relative frequency distribution categorizes employed U.S. residents by type of employment in a recent year. Construct a relative frequency bar graph.

Type of Employment	Relative Frequency
Farming, forestry, fishing	0.007
Manufacturing, extraction, transportation, and crafts	0.203
Managerial, professional, and technical	0.373
Sales and office	0.242
Other services	0.176

Source: CIA—*The World Factbook*

Solution

The bar graph follows. We use horizontal bars, because the category names are long.

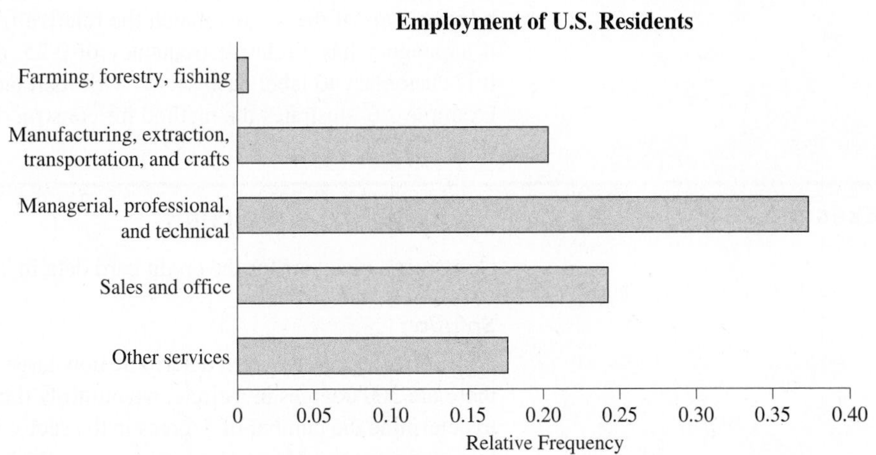

Side-by-side bar graphs

Sometimes we want to compare two bar graphs that have the same categories. The best way to do this is to construct both bar graphs on the same axes, putting bars that correspond to the same category next to each other. The result is called a **side-by-side bar graph**. As an illustration, Table 2.4 presents the number of active users, in millions, of several popular websites in 2017 and 2018.

Table 2.4 Monthly Active Users, in Millions

Website	2017	2018
Facebook	2129	2320
Gmail	1200	1500
Instagram	800	1000
LinkedIn	260	260
Twitter	330	321

Sources: Statista, Wikipedia, omnireagency,com

We would like to visualize the changes in the number of users between 2017 and 2018. Figure 2.3 presents a side-by-side bar graph. The bar graph clearly shows that the number of users of Facebook, Gmail, and Instagram grew noticeably, while the number of LinkedIn users held steady and the number of Twitter users decreased slightly.

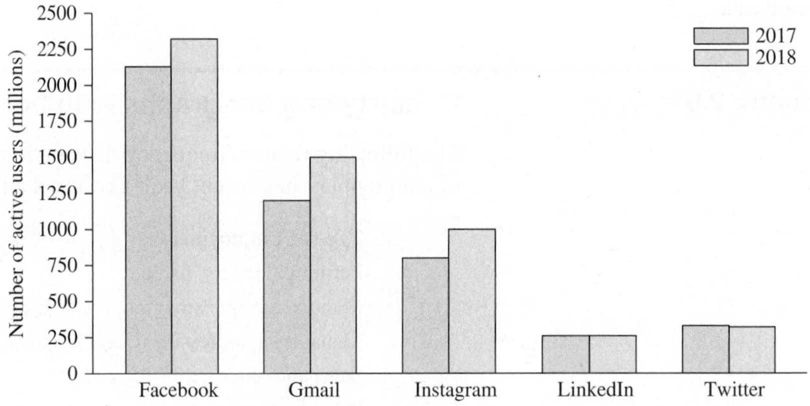

Figure 2.3 Side-by-side bar graph for the number of monthly active users of various websites

Pie Charts

Objective 3 Construct pie charts

A **pie chart** is an alternative to the bar graph for displaying relative frequency information. A pie chart is a circle. The circle is divided into sectors, one for each category. The relative sizes of the sectors match the relative frequencies of the categories. For example, if a category has a relative frequency of 0.25, then its sector takes up 25% of the circle. It is customary to label each sector with its relative frequency, expressed as a percentage. Example 2.6 illustrates the method for constructing a pie chart.

Example 2.6

Constructing a pie chart

Construct a pie chart for the credit card data in Table 2.3.

Solution

For each category, we must determine how large the sector for that category must be. Since there are 360 degrees in a circle, we multiply the relative frequency of the category by 360 to determine the number of degrees in the sector for that category. For example, the relative frequency for the MasterCard category is 0.22. Therefore, the size of the sector for this category is $0.22 \cdot 360° = 79°$. Table 2.5 (page 43) presents the results for all the categories.

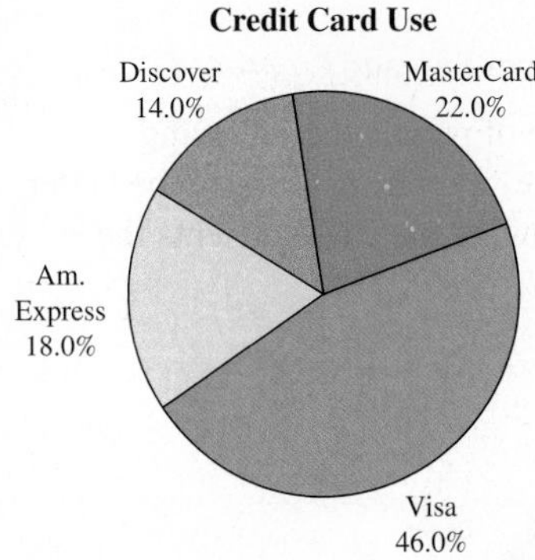

Figure 2.4 Pie chart for the credit card data in Table 2.5

Table 2.5 Sizes of Sectors for Pie Chart of Credit Card Data

Credit Card	Frequency	Relative Frequency	Size of Sector
MasterCard	11	0.22	79°
Visa	23	0.46	166°
Am. Express	9	0.18	65°
Discover	7	0.14	50°

Figure 2.4 presents the pie chart.

Constructing pie charts by hand is tedious. However, many software packages, such as MINITAB and Excel, can draw them. Step-by-step instructions for constructing a pie chart in MINITAB and Excel are presented in the Using Technology section on page 45.

Check Your Understanding

2. The following table presents a frequency distribution for the number of cars and light trucks sold in a recent month.

Type of Vehicle	Frequency
Small car	271,716
Midsize car	268,127
Luxury car	82,824
Minivan	56,772
SUV	151,703
Pickup truck	225,110
Crossover truck	416,104

Source: *Wall Street Journal*

a. Construct a bar graph.
b. Construct a relative frequency distribution.
c. Construct a relative frequency bar graph.
d. Construct a pie chart.

3. The following bar graph presents the areas (in thousands of square kilometers) of the six largest islands in the world.

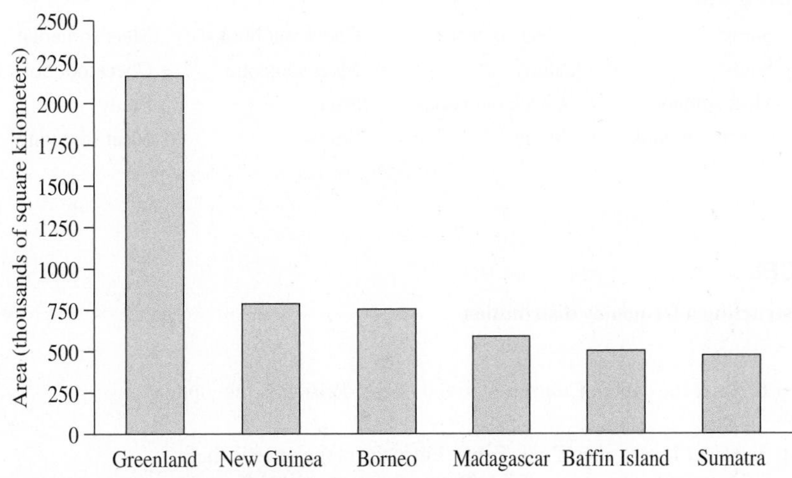

Source: CIA—*The World Factbook*

a. Which island is the largest in the world? *Greenland*
b. Someone says that Madagascar and Baffin Island together are larger than New Guinea. Is this correct? Explain how you can tell. *Yes*

c. Approximately how large is Borneo? *750,000 km²*

d. Approximately how much larger is New Guinea than Sumatra? *300,000 km²*

4. The CBS News/New York Times poll asked a sample of people the following question: Do you think things in the United States five years from now will be better, worse, or about the same as they are today? The following pie chart presents the percentages of people who gave each response.

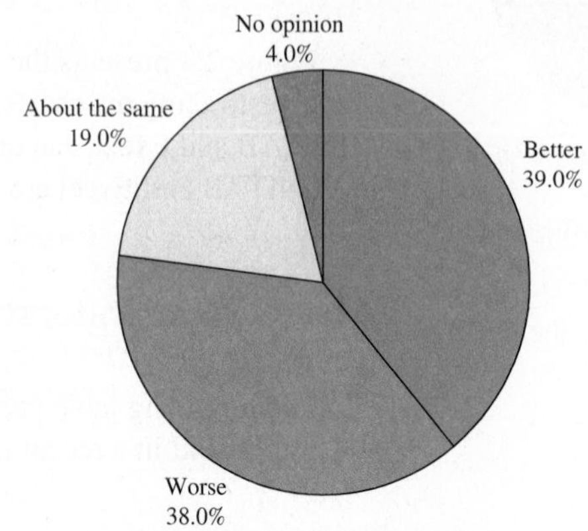

a. Which was the most common response? *Better*

b. What percentage of people said that things would be the same or worse in five years? *57%*

c. True or false: More than half of the people surveyed said that things would be the same or better in five years. *True*

Answers are on page 50.

Using Technology

We use the data in Table 2.6 to illustrate the technology steps. Table 2.6 lists 20 responses to a survey question about the reason for visiting a local library.

Table 2.6

Study	Meet someone	Check out books	Meet someone	Check out books
Study	Study	Meet someone	Check out books	Check out books
Meet someone	Check out books	Study	Study	Check out books
Check out books	Study	Study	Meet someone	Study

EXCEL

Constructing a frequency distribution

Step 1. Enter the data in **Column A** with the label *Reason* in the topmost cell.

Step 2. Select **Insert**, then **Pivot Table**. Enter the range of cells that contain the data in the **Table/Range** field and click **OK**.

Step 3. In **Choose fields to add to report**, check Reason.

Step 4. Click on *Reason* and drag to the **Values** box. The result is shown in Figure A.

Row Labels	Count of Reason
Check out books	7
Meet someone	5
Study	8
Grand Total	**20**

Figure A

Constructing bar graphs and pie charts

Step 1. Enter the categories in **Column A** and the frequencies or relative frequencies in **Column B**.

Step 2. Highlight the values in **Column A** and **Column B**, and select **Insert**. For a bar graph, select **Column**. For a pie chart, select **Pie**.

MINITAB

Constructing a frequency distribution

Step 1. Name your variable *Reason*, and enter the data into **Column C1**.

Step 2. Click on **Stat**, then **Tables**, then **Tally Individual Variables...**

Step 3. Double-click on the *Reason* variable, and check the **Counts** and **Percents** boxes.

Step 4. Press **OK** (Figure B).

Tally

Reason	Count	Percent
Check Out Books	7	35.00
Meet Someone	5	25.00
Study	8	40.00
N=	20	

Figure B

Constructing a bar graph

Step 1. Name your variable *Reason*, and enter the data into **Column C1**.

Step 2. Click on **Graph**, then **Bar Chart**. If given raw data as in Table 2.6, select **Bars Represent: Counts of Unique Values**. For the **Bar Chart type**, select **Simple**. Click **OK**. (If given a frequency distribution, select **Bars Represent: Values from a Table**.)

Step 3. Double-click on the *Reason* variable, and click on any of the options desired.

Step 4. Press **OK** (Figure C).

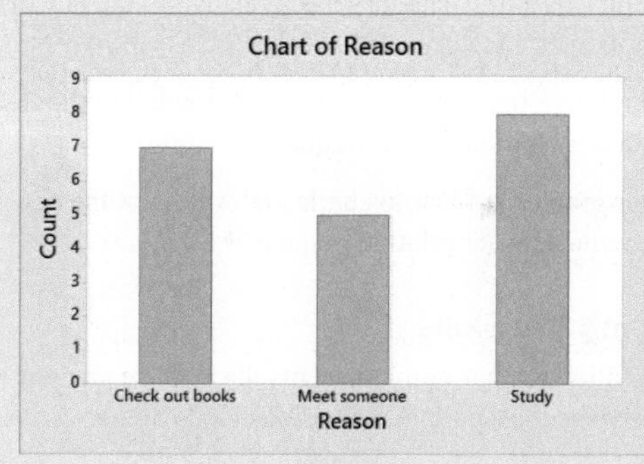

Figure C

Constructing a pie chart

Step 1. Name your variable *Reason*, and enter the data into **Column C1**.

Step 2. Click on **Graph**, then **Pie Chart**. If given raw data as in Table 2.6, select **Chart counts of unique values**, and click **OK**. (If given a frequency distribution, select **Chart Values from a Table**.)

Step 3. Double-click on the *Reason* variable, and click on any of the options desired.

Step 4. Press **OK** (Figure D).

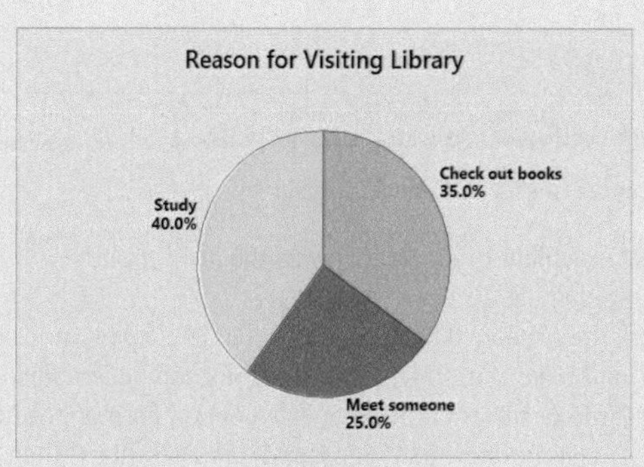

Figure D

Section 2.1 Exercises

Exercises 1–4 are the Check Your Understanding exercises located within the section.

Understanding the Concepts

In Exercises 5–8, fill in each blank with the appropriate word or phrase.

5. In a data set, the number of items that are in a particular category is called the ________. *frequency*

6. In a data set, the proportion of items that are in a particular category is called the ________. *relative frequency*

7. A ________ is a bar graph in which the bars are ordered by size. *Pareto chart*

8. A ________ is represented by a circle in which the sizes of the sectors match the relative frequencies of the categories. *pie chart*

In Exercises 9–12, determine whether the statement is true or false. If the statement is false, rewrite it as a true statement.

9. In a frequency distribution, the sum of all frequencies is less than the total number of observations. *False*

10. In a pie chart, if a category has a relative frequency of 30%, then its sector takes up 30% of the circle. *True*

11. The relative frequency of a category is equal to the frequency divided by the sum of all frequencies. *True*

12. In bar graphs and Pareto charts, the widths of the bars represent the frequencies or relative frequencies. *False*

Practicing the Skills

13. The following bar graph presents the average amount a U.S. family spent, in dollars, on various food categories in a recent year.

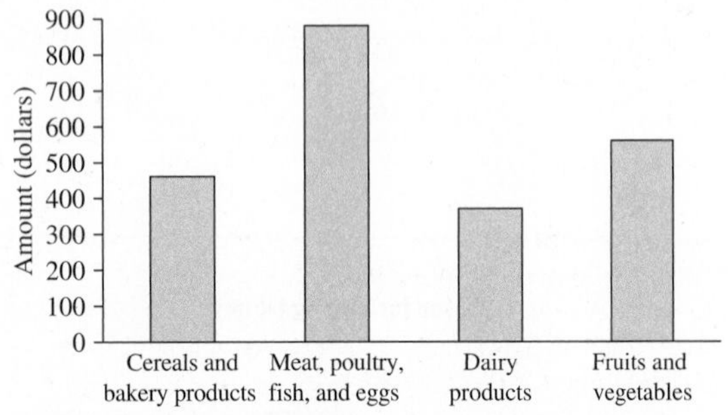

Source: Consumer Expenditure Survey

a. On which food category was the most money spent? *Meat, poultry, fish, and eggs*

b. True or false: On the average, families spent more on cereals and bakery products than on fruits and vegetables. *False*

c. True or false: Families spent more on animal products (meat, poultry, fish, eggs, and dairy products) than on plant products (cereals, bakery products, fruits, and vegetables). *True*

14. The most common blood typing system divides human blood into four groups: A, B, O, and AB. The following bar graph presents the frequencies of these types in a sample of 150 blood donors.

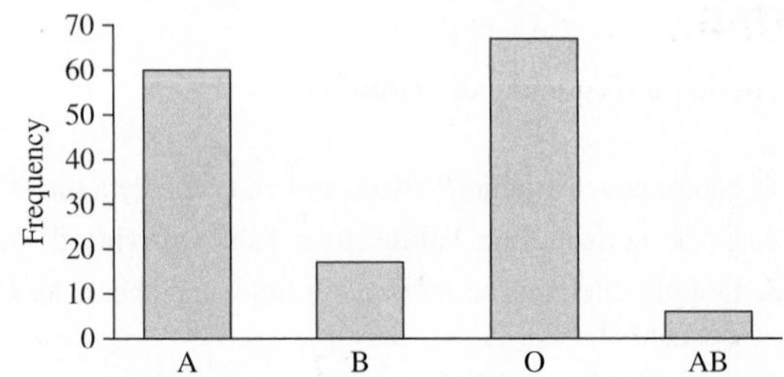

a. Which is the most frequent type? *Type O*

b. True or false: More than half of the individuals in the sample had type O blood. *False*

c. True or false: More than twice as many people had type A blood as had type B blood. *True*

15. Following is a pie chart that presents the percentages of video games sold in each of four rating categories.

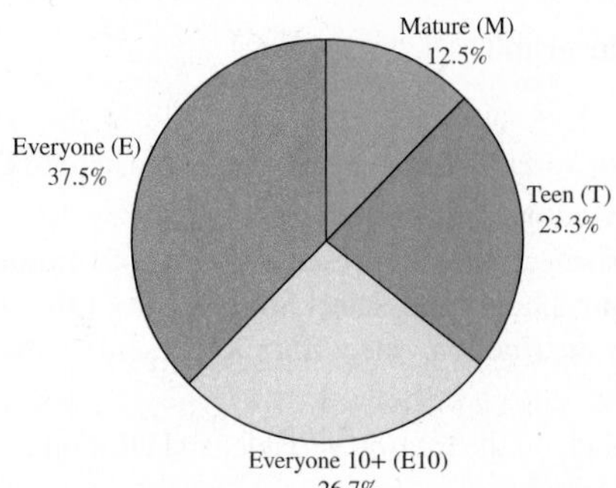

Source: Entertainment Software Association

a. Construct a relative frequency bar graph for these data.

b. Construct a relative frequency Pareto chart for these data.

c. In which rating category are the most games sold? *Everyone*

d. True or false: More than twice as many T-rated games are sold as M-rated games. *False*

e. True or false: Fewer than one in five games sold is an M-rated game. *True*

16. Government spending: The following pie chart presents the percentages of the U.S. federal budget spent in various categories during a recent year.

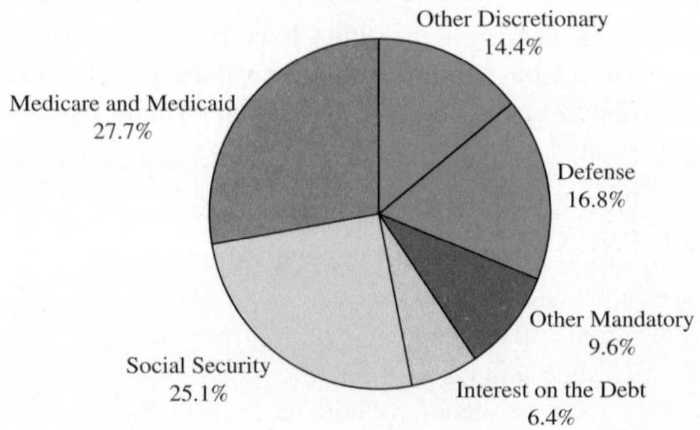

Source: nationalpriorities.org

a. Construct a relative frequency bar graph for these data.
b. Construct a relative frequency Pareto chart for these data.
c. In which category was the largest amount spent?
d. Social Security, Medicare and Medicaid, and interest on the debt are considered to be mandatory spending because they fulfill promises made by the government ahead of time. Including other mandatory spending with these categories, what percentage of the spending was mandatory? *68.8%*

17. **U.S. population:** The following side-by-side bar graph presents the proportions of people residing in various geographic regions of the United States in 1990 and 2018.

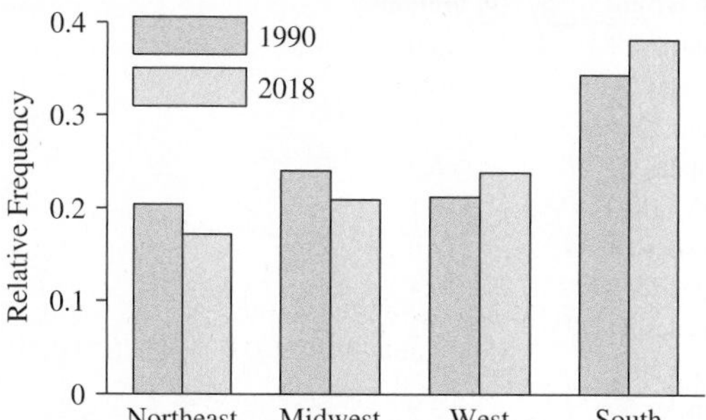

Source: U.S. Census Bureau

a. Which regions increased as a proportion of the total from 1990 to 2018? *West, South*
b. Which regions decreased as a proportion of the total from 1990 to 2018? *Northeast, Midwest*
c. True or false: The South had the largest population in both 1990 and 2018. *True*
d. True or false: All four regions had 20% or more of the population in 2018. *False*

18. **Super Bowl:** The following side-by-side bar graph presents the results of a survey in which men and women were asked to name their favorite thing about watching the Super Bowl.

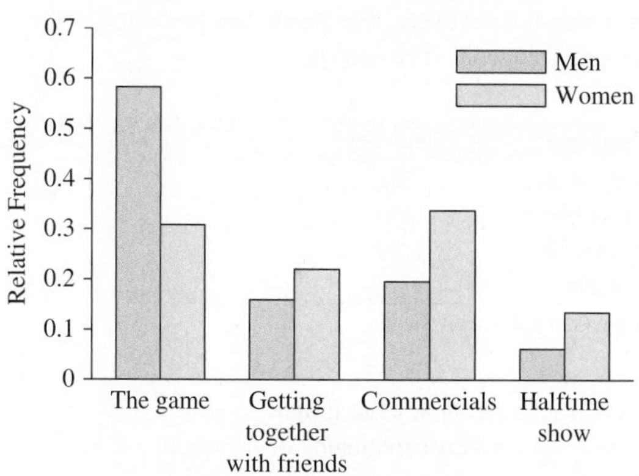

Source: BIGInsight

a. Which part of the Super Bowl do a greater proportion of men than women have as their favorite? *The game*
b. True or false: For both men and women, the smallest proportion have the halftime show as their favorite. *True*
c. True or false: About twice as many men as women have the commercials as their favorite. *False*
d. True or false: The proportion of men for whom the game or the half time show is their favorite is about the same as the proportion of women for whom the game or the commercials is their favorite. *True*

Working with the Concepts

19. **Phone sales:** The following frequency distribution presents the number of phones (in millions) shipped in each quarter of each year from 2015 through 2018.

Quarter	Number Sold (in millions)
Jan.–Mar. 2015	334.4
Apr.–Jun. 2015	336.0
Jul.–Sep. 2015	359.7
Oct.–Dec. 2015	400.7
Jan.–Mar. 2016	332.2
Apr.–Jun. 2016	346.1
Jul.–Sep. 2016	363.4
Oct.–Dec. 2016	430.7
Jan.–Mar. 2017	344.4
Apr.–Jun. 2017	348.2
Jul.–Sep. 2017	377.8
Oct.–Dec. 2017	394.6
Jan.–Mar. 2018	344.3
Apr.–Jun. 2018	342.0
Jul.–Sep. 2018	355.2
Oct.–Dec. 2018	375.4

Source: Statista

a. Construct a frequency bar graph.
b. Construct a relative frequency distribution.
c. Construct a relative frequency bar graph.
d. True or false: In each year, the quarter with the largest sales was October to December. *True*
e. True or false: Phone shipments increased each year from 2015 through 2018. *False*

20. **Popular video games:** The following frequency distribution presents the number of copies sold at retail in the United States in 2018 for each of the ten best-selling video games.

Game	Sales (millions)
Spider-Man (PS4)	5.2
God of War (PS4)	5.0
FIFA 19 (PS4)	5.0
Monster Hunter: World (PS4)	4.4
Far Cry 5 (PS4)	3.7
Mario Kart 8 Deluxe (Switch)	3.3
Super Mario Odyssey (Switch)	2.6
Call of Duty: Black Ops III (PS4)	2.5
Legend of Zelda: Breath of the Wild (Switch)	2.2
Splatoon 2	2.1

Source: http://www.msn.com

a. Construct a frequency bar graph.
b. Construct a relative frequency distribution.
c. Construct a relative frequency bar graph.
d. True or false: More than 10% of the games sold were Spider-Man. *True*

21. **More Phones:** Using the data in Exercise 19:
a. Construct a frequency distribution for the total number of phones sold in each of the four quarters Jan.–Mar., Apr.–Jun., Jul.–Sep., and Oct.–Dec.
b. Construct a frequency bar graph.
c. Construct a relative frequency distribution.
d. Construct a relative frequency bar graph.
e. Construct a pie chart.
f. True or false: More than half of phones were sold between October and December. *False*

22. More video games: Using the data in Exercise 20:

a. Construct a frequency distribution that presents the total sales for each of the platforms among the top ten games.
b. Construct a frequency bar graph.
c. Construct a relative frequency distribution.
d. Construct a relative frequency bar graph.
e. Construct a pie chart.
f. True or false: More than half of the games sold among the top ten were for the PS4. *True*

23. Hospital admissions: The following frequency distribution presents the five most frequent reasons for hospital admissions in U.S. community hospitals in a recent year.

Reason	Frequency (in thousands)
Congestive heart failure	990
Coronary atherosclerosis	1400
Heart attack	744
Infant birth	3800
Pneumonia	1200

Source: Agency for Health Care Policy and Research

a. Construct a frequency bar graph.
b. Construct a relative frequency distribution.
c. Construct a relative frequency bar graph.
d. Construct a relative frequency Pareto chart.
e. Construct a pie chart.
f. The categories coronary atherosclerosis, congestive heart failure, and heart attack refer to diseases of the circulatory system. True or false: There were more hospital admissions for infant birth than for diseases of the circulatory system. *True*

January Smith/iStockphoto/Getty Images

24. World population: Following are the populations of the continents of the world (not including Antarctica) in the year 2019.

Continent	Population (in millions)
Africa	1320
Asia	4585
Europe	743
North America	366
Oceania	42
South America	432

Source: worldpopulationreview.com

a. Construct a frequency bar graph.
b. Construct a relative frequency distribution.
c. Construct a relative frequency bar graph.
d. Construct a relative frequency Pareto chart.
e. Construct a pie chart.
f. True or false: In the year 2019, more than half of the people in the world lived in Asia. *True*
g. True or false: In the year 2019, there were more people in Europe than in North and South America combined. *False*

25. Ages of video gamers: The Nielsen Company estimated the numbers of people in various gender and age categories who used a video game console. The results are presented in the following frequency distribution.

Gender and Age Group	Frequency (in millions)
Males 2–11	13.0
Females 2–11	10.1
Males 12–17	9.6
Females 12–17	6.2
Males 18–34	16.1
Females 18–34	11.6
Males 35–49	10.4
Females 35–49	9.3
Males 50+	3.5
Females 50+	3.9

Source: The Nielsen Company

a. Construct a frequency bar graph.
b. Construct a relative frequency distribution.
c. Construct a relative frequency bar graph.
d. Construct a pie chart.
e. True or false: More than half of video gamers are male. *True*
f. True or false: More than 40% of video gamers are female. *True*
g. What proportion of video gamers are 35 or over? *0.289*

26. How secure is your job? In a survey, employed adults were asked how likely they thought it was that they would lose their jobs within the next year. The results are presented in the following frequency distribution.

Response	Frequency
Very likely	741
Fairly likely	859
Not too likely	3789
Not likely	9773

Source: General Social Survey

a. Construct a frequency bar graph.
b. Construct a relative frequency distribution.
c. Construct a relative frequency bar graph.
d. Construct a pie chart.
e. True or false: More than half of the people surveyed said that it was not likely that they would lose their job. *True*
f. What proportion of the people in the survey said that it was very likely or fairly likely that they would lose their job? *0.106*

27. Back up your data: In a survey commissioned by the Maxtor Corporation, U.S. computer users were asked how often they backed up their computer's hard drive. The following frequency distribution presents the results.

Response	Frequency
More than once per month	338
Once every 1–3 months	424
Once every 4–6 months	212
Once every 7–11 months	127
Once per year or less	311
Never	620

Source: The Maxtor Corporation

a. Construct a frequency bar graph.
b. Construct a relative frequency distribution.
c. Construct a relative frequency bar graph.
d. Construct a pie chart.
e. True or false: More than 30% of the survey respondents never back up their data. *True*
f. True or false: Less than 50% of the survey respondents back up their data more than once per year. *False*

28. Education levels: The following frequency distribution categorizes U.S. adults aged 18 and over by educational attainment in a recent year.

Educational Attainment	Frequency (in thousands)
None	834
1–4 years	1,764
5–6 years	3,618
7–8 years	4,575
9 years	4,068
10 years	4,814
11 years	11,429
High school graduate	70,441
Some college but no degree	45,685
Associate's degree (occupational)	9,380
Associate's degree (academic)	12,100
Bachelor's degree	43,277
Master's degree	16,625
Professional degree	3,099
Doctoral degree	3,191

Source: U.S. Census Bureau

a. Construct a frequency bar graph.
b. Construct a relative frequency distribution.
c. Construct a relative frequency bar graph.
d. Construct a frequency distribution with the following categories: 8 years or less, 9–11 years, High school graduate, Some college but no degree, College degree (Associate's or Bachelor's), Graduate degree (Master's, Professional, or Doctoral).
e. Construct a pie chart for the frequency distribution in part (d).
f. What proportion of people did not graduate from high school? *0.132*

29. Twitter followers: The following frequency distribution presents the number of Twitter followers in 2019 for each of five well-known singers.

Singer	Followers (millions)
Katy Perry	107.3
Justin Bieber	105.4
Rihanna	90.7
Taylor Swift	83.2
Lady Gaga	78.5

Source: www.twittercounter.com

a. Construct a frequency bar graph.
b. Construct a relative frequency distribution.
c. Construct a relative frequency bar graph.
d. Construct a pie chart.
e. What proportion are following Taylor Swift? *0.179*

30. Music sales: The following frequency distribution presents the sales, in millions of dollars, for several categories of music in the years 2017 and 2108.

Type of Music	Sales 2017	Sales 2018
Digital subscription	4092.1	5403.1
Digital streaming	1572.4	1963.7
Digital downloads	1330.7	1039.1
Physical	1495.5	1154.8

Source: Recording Industry Association of America

a. Construct a relative frequency distribution for the 2017 sales.
b. Construct a relative frequency distribution for the 2018 sales.
c. Construct a side-by-side relative frequency bar graph to compare the sales in 2017 and 2018.
d. True or false: Sales in every category increased from 2017 to 2018. *False*

31. Keeping up with the Kardashians: The following frequency distribution presents the number of Twitter and Instagram followers in a recent year for five members of the Kardashian family.

Kardashian	Followers (in millions) Twitter	Instagram
Kim	60.5	127
Kendall	27.7	104
Kylie	27.1	127
Khloé	27.0	86.9
Kourtney	24.3	73.5

Sources: twitter.com, www.cheatsheet.com

a. Construct a relative frequency distribution for the number of Twitter followers.
b. Construct a relative frequency distribution for the number of Instagram followers.
c. Construct a side-by-side relative frequency bar graph to compare the number of Twitter followers to the number of Instagram followers.
d. True or false: Each member of the family has more Instagram followers than Twitter followers. *True*
e. True or false: Kim has more Twitter followers than her sisters Khloé and Kourtney combined. *True*

32. Bought a new car lately? The following table presents the number of cars sold by several manufacturers in a recent month.

Manufacturer	Sales (in thousands)
General Motors	82.2
Toyota	96.6
Ford	68.4
Chrysler	26.6
Honda	69.9
Nissan	78.0
Hyundai	42.0
Others	159.0

Source: *Wall Street Journal*

a. Construct a frequency bar graph.
b. Construct a relative frequency distribution.
c. Construct a relative frequency bar graph.
d. Construct a pie chart.

e. What proportion of sales were for General Motors cars? *0.132*

33. Bought a new truck lately? The following table presents the number of light trucks sold by several manufacturers in a recent month.

Manufacturer	Sales (in thousands)
General Motors	173.0
Toyota	101.6
Ford	170.6
Chrysler	167.9
Honda	68.8
Nissan	62.5
Hyundai	25.5
Others	121.0

Source: *Wall Street Journal*

a. Construct a frequency bar graph.
b. Construct a relative frequency distribution.
c. Construct a relative frequency bar graph.
d. Construct a pie chart.
e. True or false: More light trucks were sold by Chrysler than by Honda, Nissan, and Hyundai combined. *True*

34. Happy Halloween: The following table presents proportions of people who get ideas for Halloween costumes from various sources. Is this a valid relative frequency distribution? Why or why not? *No*

Source	Proportion
Twitter	0.048
Facebook	0.152
Friends and Family	0.237
Retail Stores	0.357
Magazines	0.193
Pinterest	0.071

Source: National Retail Federation

Extending the Concepts

35. Native languages: The following frequency distribution presents the number of households (in thousands) categorized by the language spoken at home, for the cities of New York and Los Angeles in a recent year. The Total column presents the numbers of households in both cities combined.

Language	New York	Los Angeles	Total
English	4098	1339	5437
Spanish	1870	1555	3425
Other Indo-European	1037	237	1274
Asian and Pacific Island	618	301	919

Source: U.S. Census Bureau

a. Construct a frequency bar graph for each city.
b. Construct a frequency bar graph for the total.
c. Construct a relative frequency bar graph for each city.
d. Construct a relative frequency bar graph for the total.
e. Explain why the heights of the bars for the frequency bar graph for the total are equal to the sums of the heights for the individual cities.
f. Explain why the heights of the bars for the relative frequency bar graph for the total are not equal to the sums of the heights for the individual cities.

36. Proportion of females: Following are the proportions of the United States population that is female for five age groups.

Age Group	Proportion
0–19	0.488
20–39	0.497
40–59	0.508
60–79	0.534
over 79	0.628

Source: U.S. Census Bureau

a. Is this a relative frequency table? Explain why or why not? *No*
b. Would it be appropriate to construct a pie chart for these data? Why or why not? *No*

Answers to Check Your Understanding Exercises for Section 2.1

1. a.

Aircraft	Frequency
Twin	9
Single	10
Helicopter	2
Turboprop	5
Jet	4

b.

Aircraft	Frequency	Relative Frequency
Twin	9	0.300
Single	10	0.333
Helicopter	2	0.067
Turboprop	5	0.167
Jet	4	0.133

2. a.

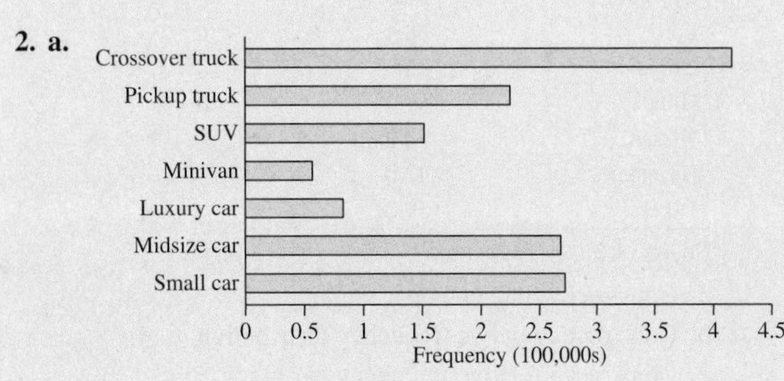

b.

Type of Vehicle	Frequency	Relative Frequency
Small car	271,716	0.185
Midsize car	268,127	0.182
Luxury car	82,824	0.056
Minivan	56,772	0.039
SUV	151,703	0.103
Pickup truck	225,110	0.153
Crossover truck	416,104	0.283

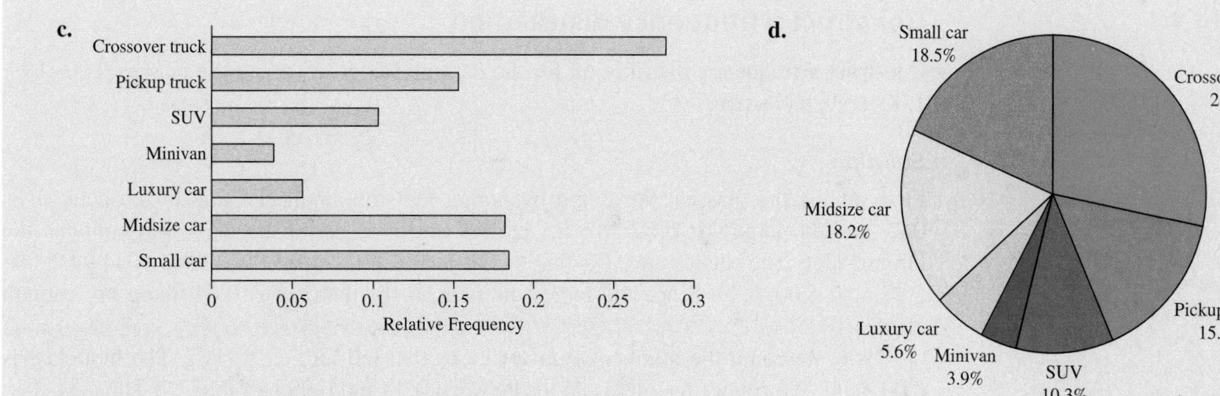

3. a. Greenland

b. Yes, the height of the bar for New Guinea is less than the heights of the bars for Madagascar and Baffin Island put together.

c. 750,000 **d.** 300,000

4. a. Better **b.** 57% **c.** True

Section 2.2 Frequency Distributions and Their Graphs

Objectives

1. Construct frequency distributions for quantitative data
2. Construct histograms
3. Determine the shape of a distribution from a histogram
4. Construct frequency polygons and ogives

Objective 1 Construct frequency distributions for quantitative data

Frequency Distributions for Quantitative Data

How much air pollution is caused by motor vehicles? This question was addressed in a study by Dr. Janet Yanowitz at the Colorado School of Mines. She studied the emissions of particulate matter, a form of pollution consisting of tiny particles, that has been associated with respiratory disease. The emissions for 65 vehicles, in units of grams of particles per gallon of fuel, are presented in Table 2.7.

Table 2.7 Particulate Emissions for 65 Vehicles

1.50	0.87	1.12	1.25	3.46	1.11	1.12	0.88	1.29	0.94	0.64	1.31	2.49
1.48	1.06	1.11	2.15	0.86	1.81	1.47	1.24	1.63	2.14	6.64	4.04	2.48
1.40	1.37	1.81	1.14	1.63	3.67	0.55	2.67	2.63	3.03	1.23	1.04	1.63
3.12	2.37	2.12	2.68	1.17	3.34	3.79	1.28	2.10	6.55	1.18	3.06	0.48
0.25	0.53	3.36	3.47	2.74	1.88	5.94	4.24	3.52	3.59	3.10	3.33	4.58

To summarize these data, we will construct a frequency distribution. Since these data are quantitative, there are no natural categories. We therefore divide the data into **classes**. The classes are intervals of equal width that cover all the values that are observed. For example, for the data in Table 2.7, we could choose the classes to be 0.00–0.99, 1.00–1.99, and so forth. We then count the number of observations that fall into each class, to obtain the class frequencies.

Example 2.7

Construct a frequency distribution

Construct a frequency distribution for the data in Table 2.7, using the classes 0.00–0.99, 1.00–1.99, and so on.

Solution

First we list the classes. We begin by noting that the smallest value in the data set is 0.25 and the largest is 6.64. We list classes until we get to the class that contains the largest value. The classes are 0.00–0.99, 1.00–1.99, 2.00–2.99, 3.00–3.99, 4.00–4.99, 5.00–5.99, and 6.00–6.99. Since the largest number in the data set is 6.64, these are enough classes.

Now we count the number of observations that fall into each class. The first class is 0.00–0.99. We count nine observations between 0.00 and 0.99 in Table 2.7. The next class is 1.00–1.99. We count 26 observations in this class. We repeat this procedure with classes 2.00–2.99, 3.00–3.99, 4.00–4.99, 5.00–5.99, and 6.00–6.99. The results are presented in Table 2.8. This is a frequency distribution for the data in Table 2.7.

EXPLAIN IT AGAIN

Frequency distributions for quantitative and qualitative data: Frequency distributions for quantitative data are just like those for qualitative data, except that the data are divided into classes rather than categories.

Table 2.8 Frequency Distribution for Particulate Data

Class	Frequency
0.00–0.99	9
1.00–1.99	26
2.00–2.99	11
3.00–3.99	13
4.00–4.99	3
5.00–5.99	1
6.00–6.99	2

We can also construct a relative frequency distribution. As with qualitative data, the relative frequency of a class is the frequency of that class, divided by the sum of all the frequencies.

DEFINITION

The **relative frequency** of a class is given by

$$\text{Relative frequency} = \frac{\text{Frequency}}{\text{Sum of all frequencies}}$$

Example 2.8

Construct a relative frequency distribution

Construct a relative frequency distribution for the data in Table 2.7, using the classes 0.00–0.99, 1.00–1.99, and so on.

Solution

The frequency distribution is presented in Table 2.8. We compute the sum of all the frequencies:

$$\text{Sum of all frequencies} = 9 + 26 + 11 + 13 + 3 + 1 + 2 = 65$$

We can now compute the relative frequency for each class. For the class 0.00–0.99, the frequency is 9. The relative frequency is therefore

$$\text{Relative frequency} = \frac{\text{Frequency}}{\text{Sum of all frequencies}} = \frac{9}{65} = 0.138$$

Table 2.9 is a relative frequency distribution for the data in Table 2.7. The frequencies are shown as well.

Table 2.9 Relative Frequency Distribution for Particulate Data

Class	Frequency	Relative Frequency
0.00–0.99	9	0.138
1.00–1.99	26	0.400
2.00–2.99	11	0.169
3.00–3.99	13	0.200
4.00–4.99	3	0.046
5.00–5.99	1	0.015
6.00–6.99	2	0.031

Choosing the classes

In Examples 2.7 and 2.8, we chose the classes to be 0.00–0.99, 1.00–1.99, and so on. There are many other choices we could have made. For example, we could have chosen the classes to be 0.00–1.99, 2.00–3.99, 4.00–5.99, and 6.00–7.99. As another example, we could have chosen them to be 0.00–0.49, 0.50–0.99, and so on, up to 6.50–6.99. We now define some of the terminology that we will use when discussing classes.

DEFINITION

- The **lower class limit** of a class is the smallest value that can appear in that class.
- The **upper class limit** of a class is the largest value that can appear in that class.
- The **class width** is the difference between consecutive lower class limits.

CAUTION

The class width is the difference between the lower limit and the lower limit of the next class, not the difference between the lower limit and the upper limit.

Class limits should be expressed with the same number of decimal places as the data. The data in Table 2.7 are rounded to two decimal places, so the class limits for these data are expressed with two decimal places as well.

Example 2.9

Find the class limits and widths

Find the lower class limits, the upper class limits, and the class widths for the relative frequency distribution in Table 2.9.

Solution

The classes are 0.00–0.99, 1.00–1.99, and so on, up to 6.00–6.99. The lower class limits are therefore 0.00, 1.00, 2.00, 3.00, 4.00, 5.00, and 6.00. The upper class limits are 0.99, 1.99, 2.99, 3.99, 4.99, 5.99, and 6.99.

We find the class width for the first class by subtracting consecutive lower limits:

$$\begin{aligned}\text{Class width} &= \text{Lower limit for second class} - \text{Lower limit for first class}\\ &= 1.00 - 0.00\\ &= 1.00\end{aligned}$$

Similarly, we find that all the classes have a width of 1.

When constructing a frequency distribution, there is no one right way to choose the classes. However, there are some requirements that must be satisfied:

Requirements for Choosing Classes

- Every observation must fall into one of the classes.
- The classes must not overlap.
- The classes must be of equal width.
- There must be no gaps between classes. Even if there are no observations in a class, it must be included in the frequency distribution.

The following procedure will produce a frequency distribution whose classes meet these requirements.

Procedure for Constructing a Frequency Distribution for Quantitative Data

Step 1: Choose a class width.

Step 2: Choose a lower class limit for the first class. This should be a convenient number that is slightly less than the minimum data value.

Step 3: Compute the lower limit for the second class by adding the class width to the lower limit for the first class:

Lower limit for second class = Lower limit for first class + Class width

Step 4: Compute the lower limits for each of the remaining classes by adding the class width to the lower limit of the preceding class. Stop when the largest data value is included in a class.

Step 5: Count the number of observations in each class, and construct the frequency distribution.

Example 2.10

Constructing a frequency distribution

Construct a frequency distribution for the data in Table 2.7, using a class width of 1.50.

Solution

Step 1: The class width is given to be 1.50.

Step 2: The smallest value in the data is 0.25. A convenient number that is smaller than 0.25 is 0.00. We will choose 0.00 to be the lower limit for the first class.

Step 3: The lower class limit for the second class is $0.00 + 1.50 = 1.50$.

Step 4: Continuing, the lower limits for the remaining classes are

$$1.50 + 1.50 = 3.00$$
$$3.00 + 1.50 = 4.50$$
$$4.50 + 1.50 = 6.00$$
$$6.00 + 1.50 = 7.50$$

Since the largest data value is 6.64, every data value is now contained in a class.

Step 5: We count the number of observations in each class to obtain the following frequency distribution.

Frequency Distribution for Particulate Data Using a Class Width of 1.5

Class	Frequency
0.00–1.49	28
1.50–2.99	18
3.00–4.49	15
4.50–5.99	2
6.00–7.49	2

Check Your Understanding

1. Using the data in Table 2.7, construct a frequency distribution with classes of width 0.5.

Answer is on page 71.

Computing the class width for a given number of classes

In Example 2.10, the first step in computing the frequency distribution was to choose a class width. Sometimes we begin by choosing an approximate number of classes instead. In these cases, we compute the class width as follows:

Step 1: Decide approximately how many classes to have.

Step 2: Compute the class width as follows:

$$\text{Class width} = \frac{\text{Largest data value} - \text{Smallest data value}}{\text{Number of classes}}$$

Step 3: Round the class width to a convenient value. It is usually better to round up.

Once the class width is determined, we proceed just as in the case where the class width is given. We choose a lower limit for the first class by choosing a convenient number that is slightly less than the minimum data value. We then compute the lower limits for the remaining classes, count the number of observations in each class, and construct the frequency distribution. Note that the actual number of classes may differ somewhat from the chosen number, because the class width is rounded and because the lower limit of the first class will generally be less than the smallest data value.

Example 2.11

Computing the class width

Find the class width for a frequency distribution for the data in Table 2.7, if it is desired to have approximately seven classes.

Solution

Step 1: We will have approximately seven classes.

Step 2: The smallest data value is 0.25 and the largest is 6.64. We compute the class width:

$$\text{Class width} = \frac{6.64 - 0.25}{7} = 0.91$$

Step 3: We round 0.91 up to 1, since this is the nearest convenient number. We will use a class width of 1.

A reasonable choice for the lower limit of the first class is 0. This choice will give us the frequency distribution in Table 2.8.

Objective 2 Construct histograms

Histograms

Once we have a frequency distribution or a relative frequency distribution, we can put the information in graphical form by constructing a **histogram**. Histograms based on frequency distributions are called **frequency histograms**, and histograms based on relative frequency distributions are called **relative frequency histograms**. Histograms are related to bar graphs and are appropriate for quantitative data. A histogram is constructed by drawing a rectangle for each class. The heights of the rectangles are equal to the frequencies or the relative frequencies, and the widths are equal to the class width. The left edge of each rectangle corresponds to the lower class limit, and the right edge touches the left edge of the next rectangle.

Example 2.12

Construct a histogram

Table 2.10 presents a frequency distribution and the relative frequency distribution for the particulate emissions data.

Construct a frequency histogram based on the frequency distribution in Table 2.10. Construct a relative frequency histogram based on the relative frequency distribution in Table 2.10.

NOTE TO INSTRUCTOR

It is helpful to emphasize that histograms are used for quantitative variables, while bar charts are used for qualitative variables.

Table 2.10 Frequency and Relative Frequency Distributions for Particulate Data

Class	Frequency	Relative Frequency
0.00–0.99	9	0.138
1.00–1.99	26	0.400
2.00–2.99	11	0.169
3.00–3.99	13	0.200
4.00–4.99	3	0.046
5.00–5.99	1	0.015
6.00–6.99	2	0.031

Solution

We construct a rectangle for each class. The first rectangle has its left edge at the lower limit of the first class, which is 0.00, and its right edge at the lower limit of the next class, which is 1.00. The second rectangle has its left edge at 1.00 and its right edge at the lower limit of the next class, which is 2.00, and so on.

For the frequency histogram, the heights of the rectangles are equal to the frequencies. For the relative frequency histogram, the heights of the rectangles are equal to the relative frequencies.

Figure 2.5 presents a frequency histogram, and Figure 2.6 presents a relative frequency histogram, for the data in Table 2.10. Note that the two histograms have the same shape. The only difference is the scale on the vertical axis.

NOTE TO INSTRUCTOR

You may point out that relative frequency histograms can be more useful for comparing two distributions, because they are not affected by the size of the data set.

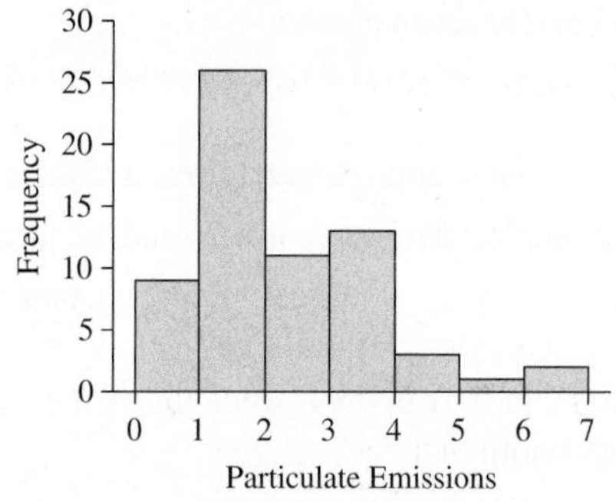

Figure 2.5 Frequency histogram for the frequency distribution in Table 2.10

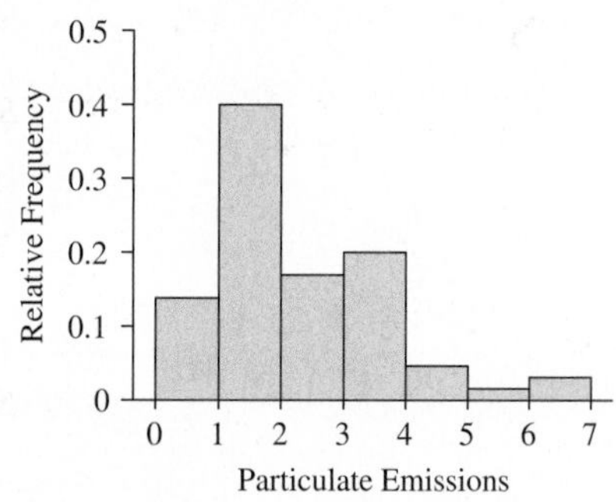

Figure 2.6 Relative frequency histogram for the relative frequency distribution in Table 2.10

How should I choose the number of classes for a histogram?

EXPLAIN IT AGAIN

Choosing the number of classes: There is no single right way to choose classes for a histogram. Use your best judgment to construct a histogram with an appropriate amount of detail.

There are no hard-and-fast rules for choosing the number of classes. In general, it is good to have more classes rather than fewer, but it is also good to have reasonably large frequencies in some of the classes. The following two principles can guide the choice:

- Too many classes produce a histogram with too much detail, so that the main features of the data are obscured.
- Too few classes produce a histogram lacking in detail.

Figures 2.7 and 2.8 (page 57) illustrate these principles. Figure 2.7 presents a histogram for the particulate data where the class width is 0.1. This narrow class width results in a

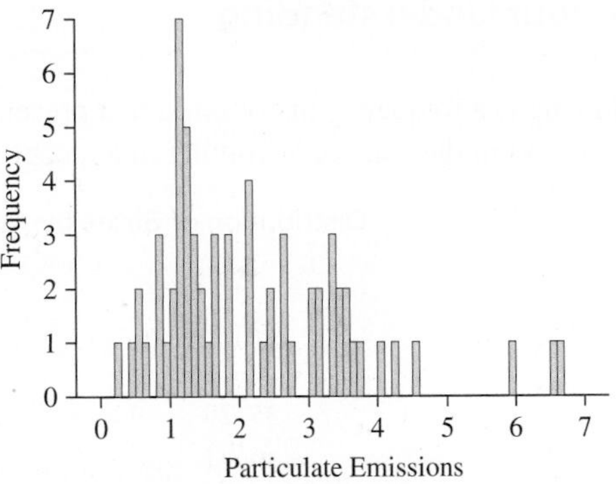

Figure 2.7 The class width is too narrow. The jagged appearance distracts from the overall shape of the data.

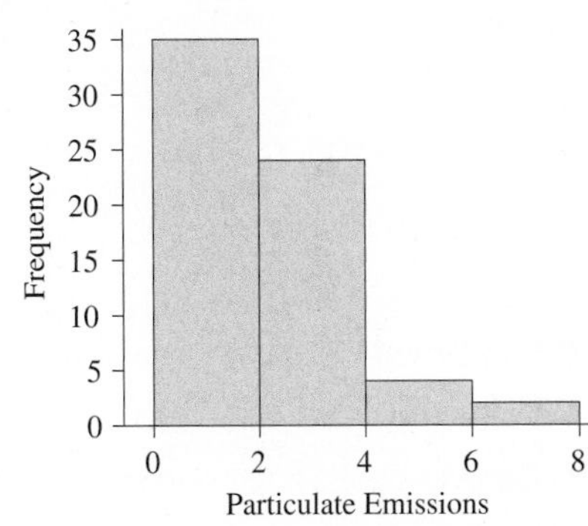

Figure 2.8 The class width is too wide. Only the most basic features of the data are visible.

large number of classes. The histogram has a jagged appearance, which distracts from the overall shape of the data. On the other extreme, Figure 2.8 presents a histogram for these data with a class width of 2.0. The number of classes is too small, so only the most basic features of the data are visible in this overly simple histogram.

Choosing a large number of classes will produce a narrow class width, and choosing a smaller number will produce a wider class width. It is appropriate to experiment with various choices for the number of classes, in order to find a good balance. The following guidelines are helpful.

Guidelines for Selecting the Number of Classes

- For many data sets, the number of classes should be at least 5 but no more than 20.
- For very large data sets, a larger number of classes may be appropriate.

Example 2.13

Constructing a histogram with technology

Use technology to construct a frequency histogram for the emissions data in Table 2.7 on page 51.

Solution

The following figure shows the histogram constructed in MINITAB. Note that MINITAB has chosen a class width of 0.5. With this class width, there are two empty classes. These show up as a gap that separates the last two rectangles on the right from the rest of the histogram.

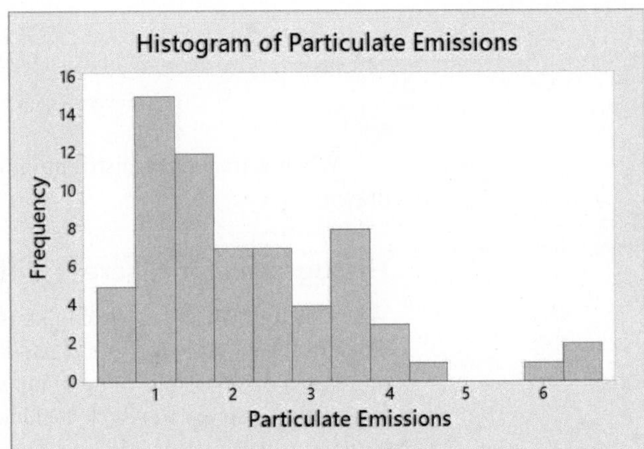

Step-by-step instructions for constructing histograms with the TI-84 Plus and with MINITAB are given in the Using Technology section on pages 63 and 64.

Check Your Understanding

2. Following is a frequency distribution that presents the number of live births to women aged 15–44 in the state of Wyoming in a recent year.

Distribution of Births by Age of Mother

Age	Frequency
15–19	795
20–24	2410
25–29	2190
30–34	1208
35–39	499
40–44	109

Source: Wyoming Department of Health

a. List the lower class limits. *15, 20, 25, 30, 35, 40*
b. What is the class width? *5*
c. Construct a frequency histogram.
d. Construct a relative frequency distribution.
e. Construct a relative frequency histogram.

Answers are on page 71.

Open-ended classes

It is sometimes necessary for the first class to have no lower limit or for the last class to have no upper limit. Such a class is called **open-ended**. Table 2.11 presents a frequency distribution for the number of deaths in the United States due to pneumonia in a recent year for various age groups. Note that the last age group is "85 and older," an open-ended class.

Table 2.11 Deaths Due to Pneumonia

Age	Number of Deaths
5–14	69
15–24	178
25–34	299
35–44	875
45–54	1872
55–64	3099
65–74	6283
75–84	17,775
85 and older	27,758

Source: U.S. Census Bureau

When a frequency distribution contains an open-ended class, a histogram cannot be drawn.

Histograms for discrete data

When data are discrete, we can construct a frequency distribution in which each possible value of the variable forms a class. Then we can draw a histogram in which each rectangle represents one possible value of the variable. Table 2.12 (page 59) presents the results of a hypothetical survey in which 1000 adult women were asked how many children they had. Number of children is a discrete variable, and in this data set, the values of this variable are 0 through 8. To construct a histogram, we draw rectangles of equal width, centered at the values of the variables. The rectangles should be just wide enough to touch. Figure 2.9 (page 59) presents a histogram.

Table 2.12 Women with a Given Number of Children

Number of Children	Frequency
0	435
1	175
2	222
3	112
4	38
5	9
6	7
7	0
8	2

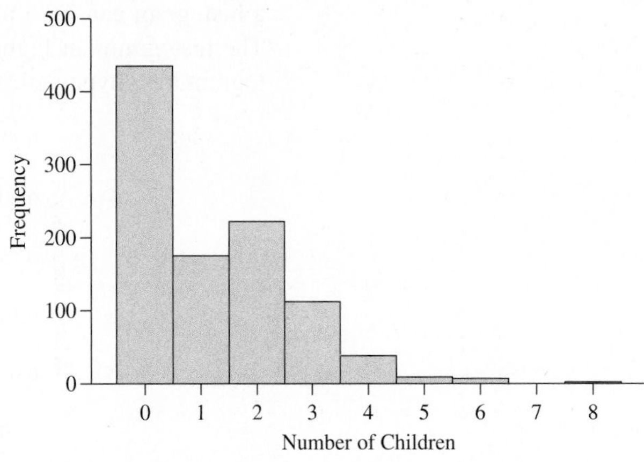

Figure 2.9 Histogram for data in Table 2.12

Objective 3 Determine the shape of a distribution from a histogram

Shapes of Histograms

The purpose of a histogram is to give a visual impression of the "shape" of a data set. Statisticians have developed terminology to describe some of the commonly observed shapes. A histogram is **skewed** if one side, or tail, is longer than the other. A histogram with a long right-hand tail is said to be **skewed to the right**, or **positively skewed**. A histogram with a long left-hand tail is said to be **skewed to the left**, or **negatively skewed**. A histogram is **symmetric** if its right half is a mirror image of its left half. Very few histograms are perfectly symmetric, but many are approximately symmetric. There are two special cases of symmetric histograms. A symmetric histogram with a peak in the middle is referred to as a **bell-shaped** histogram. A histogram in which all the classes have equal frequencies is said to be **uniformly distributed**. These terms apply to both frequency histograms and relative frequency histograms. Figure 2.10 presents some histograms for hypothetical samples. As another example, the histogram for particulate concentration, shown in Figure 2.5, is skewed to the right.

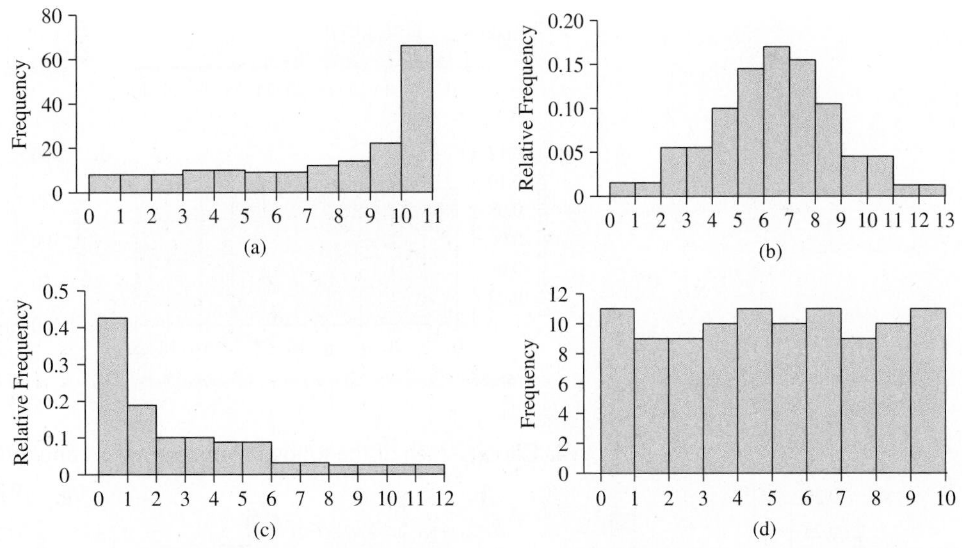

Figure 2.10 (a) A histogram skewed to the left. (b) A bell-shaped histogram. (c) A histogram skewed to the right. (d) An approximately uniformly distributed histogram.

NOTE TO INSTRUCTOR

It is useful to discuss how the shape of a data set can help make inferences. Point out that when a data set is bell-shaped, typical observations are close to the middle, with occasional observations on either side. For a data set that is skewed right, typical observations are smaller, with occasional large ones.

The examples in Figure 2.10 are straightforward to categorize. In real life, the classification is not always clear-cut, and people may sometimes disagree on how to describe the shape of a particular histogram.

Modes

A peak, or high point, of a histogram is referred to as a **mode**. A histogram is **unimodal** if it has only one mode, and **bimodal** if it has two clearly distinct modes. In principle, a histogram can have more than two modes, but this does not happen often in practice. The histograms in Figure 2.10(a)–(c) are unimodal. Figure 2.11 presents a bimodal histogram for a hypothetical sample.

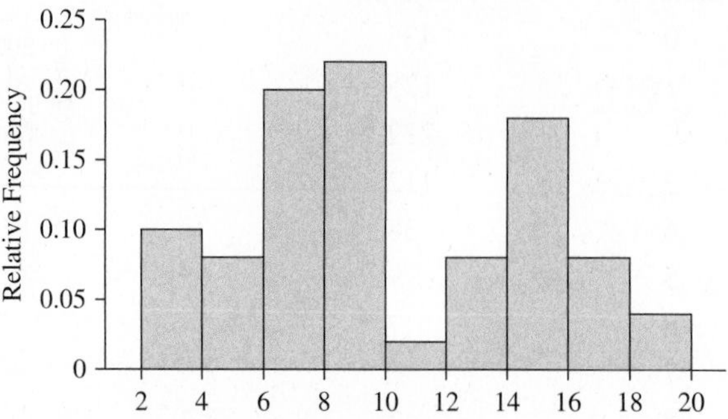

Figure 2.11 A bimodal histogram

As another example, it is reasonable to classify the histogram for particulate emissions, shown in Figure 2.5, as unimodal, with the rectangle above the class 1–2 as the only mode. While some might say that the rectangle above the class 3–4 is another mode, most would agree that it is too small a peak to count as a second mode.

Check Your Understanding

3. Classify each of the following histograms as skewed to the left, skewed to the right, approximately bell-shaped, or approximately uniformly distributed.

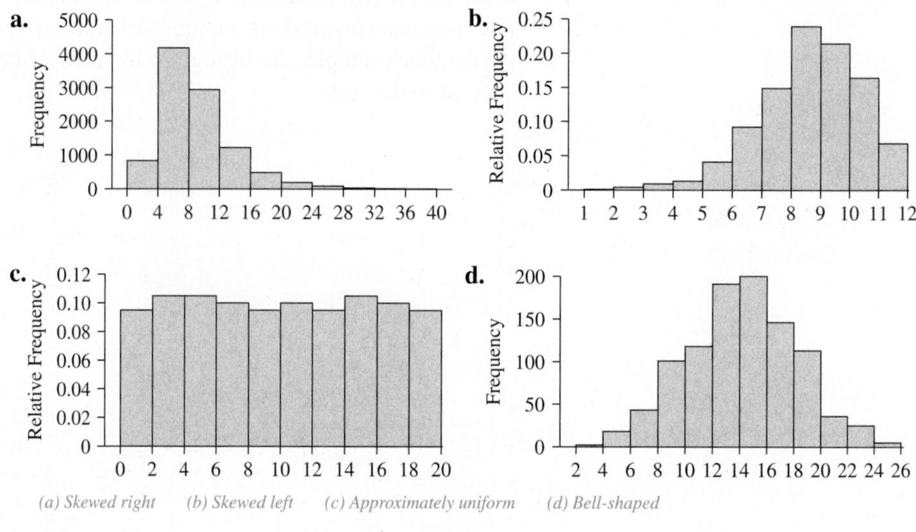

4. Classify each of the following histograms as unimodal or bimodal.

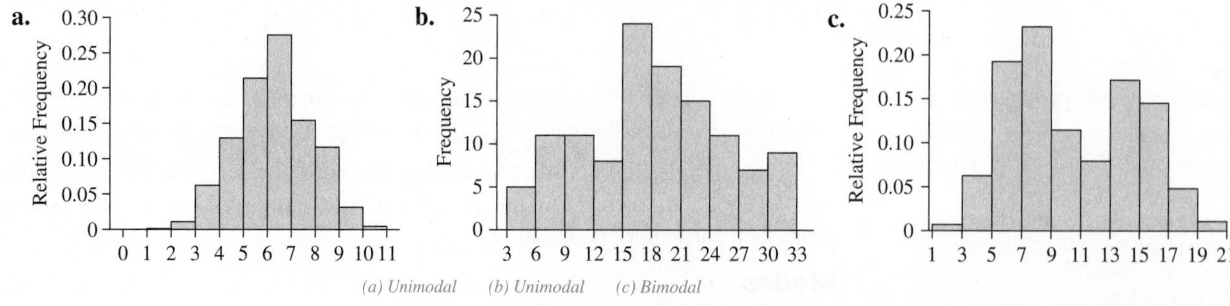

Answers are on page 71.

Objective 4 Construct frequency polygons and ogives

Frequency Polygons and Ogives

Histograms are the most commonly used graphs for representing a frequency or relative frequency distribution. Here we discuss two others. First is the **frequency polygon**. To construct a frequency polygon, we must compute the **class midpoints**, which we now define.

NOTE TO INSTRUCTOR

The material in the Frequency Polygons and Ogives section can be skipped without loss of continuity.

DEFINITION

The **midpoint** of a class is the average of its lower class limit and the lower class limit of the next class.

$$\text{Class midpoint} = \frac{\text{Lower limit} + \text{Lower limit of next class}}{2}$$

We construct a frequency polygon by plotting a point for each class. The x-coordinate of the point is the class midpoint, and the y-coordinate is the frequency of the class. We then connect all the points with straight lines.

Example 2.14

Construct a frequency polygon

Table 2.13 presents frequencies and relative frequencies for the particulate emissions data. These were first given in Table 2.9. Construct a frequency polygon for the frequencies in Table 2.13.

Table 2.13 Relative Frequency Distribution for Particulate Data

Class	Frequency	Relative Frequency
0.00–0.99	9	0.138
1.00–1.99	26	0.400
2.00–2.99	11	0.169
3.00–3.99	13	0.200
4.00–4.99	3	0.046
5.00–5.99	1	0.015
6.00–6.99	2	0.031

Solution

We first compute the midpoints of the classes. For example, to compute the midpoint of the first class, we find its lower class limit, 0.00, and the lower class limit of the next class, which is 1.00.

$$\text{Class midpoint of first class} = \frac{0.00 + 1.00}{2} = 0.50$$

Table 2.14 presents the class midpoints for the data in Table 2.13 and shows how the first two and the last of them are computed. Note that to compute the midpoint for the last class, we pretend that there is a class following it whose lower limit is 7.00.

Table 2.14 Relative Frequency Distribution for Particulate Data

Class	Midpoint	Frequency	Relative Frequency
0.00–0.99	$\frac{0+1}{2} = 0.5$	9	0.138
1.00–1.99	$\frac{1+2}{2} = 1.5$	26	0.400
2.00–2.99	2.5	11	0.169
3.00–3.99	3.5	13	0.200
4.00–4.99	4.5	3	0.046
5.00–5.99	5.5	1	0.015
6.00–6.99	$\frac{6+7}{2} = 6.5$	2	0.031

We now plot the points whose x-coordinates are the class midpoints and whose y-coordinates are the frequencies. We then connect the points with straight lines to obtain the frequency polygon. The result is shown in Figure 2.12.

We can also construct a **relative frequency polygon**. This is the same as a frequency polygon, except that the frequencies are replaced by relative frequencies. Figure 2.13 presents a relative frequency polygon for the data in Table 2.13.

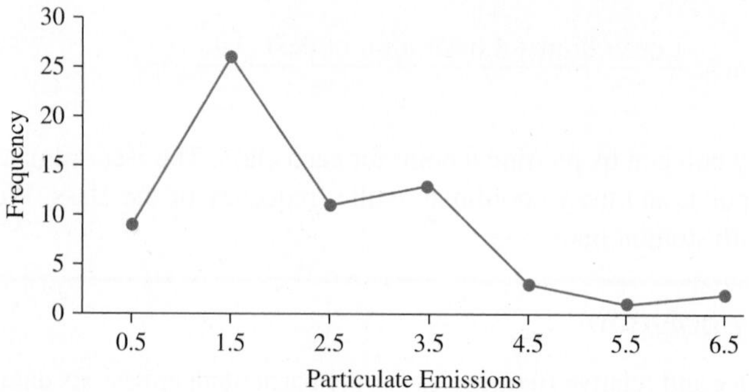

Figure 2.12 Frequency polygon for the particulate emissions data in Table 2.13

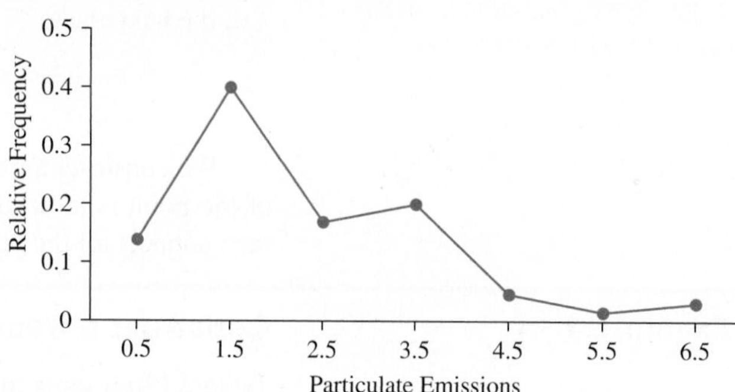

Figure 2.13 Relative frequency polygon for the particulate emissions data in Table 2.13

EXPLAIN IT AGAIN

Cumulative Frequencies: The cumulative frequency of a class is the number of data items whose values are less than or equal to the upper class limit. The cumulative relative frequency of a class is the proportion of data items whose values are less than or equal to the upper class limit.

Ogives

An **ogive** (pronounced "oh jive") is another type of polygon. A **frequency ogive** plots *cumulative frequencies*, and a **relative frequency ogive** plots *cumulative relative frequencies*. We now define these terms.

DEFINITION

The **cumulative frequency** of a class is the sum of the frequencies of that class and all previous classes.

The **cumulative relative frequency** of a class is the cumulative frequency divided by the sum of all the frequencies.

$$\text{Cumulative relative frequency} = \frac{\text{Cumulative frequency}}{\text{Sum of all frequencies}}$$

EXPLAIN IT AGAIN

Largest Cumulative Frequency: The largest cumulative frequency is equal to the sum of all the frequencies.

Table 2.15 presents the cumulative frequencies and cumulative relative frequencies for the data in Table 2.13 and shows how they are calculated. Note that the sum of all the frequencies is 65, which is the largest cumulative frequency.

NOTE TO INSTRUCTOR

Emphasize that the dots go at the class midpoints for frequency polygons and at the upper class limits for ogives.

Table 2.15

Class	Frequency	Relative Frequency	Cumulative Frequency	Cumulative Relative Frequency
0.00–0.99	9	0.138	9	$9/65 = 0.138$
1.00–1.99	26	0.400	$9 + 26 = 35$	$35/65 = 0.538$
2.00–2.99	11	0.169	$9 + 26 + 11 = 46$	$46/65 = 0.708$
3.00–3.99	13	0.200	59	$59/65 = 0.908$
4.00–4.99	3	0.046	62	$62/65 = 0.954$
5.00–5.99	1	0.015	63	$63/65 = 0.969$
6.00–6.99	2	0.031	65	$65/65 = 1.000$

CAUTION

Use upper class limits, not class midpoints, for ogives.

To construct an ogive, plot the cumulative frequencies or cumulative relative frequencies on the y-axis and the upper class limits on the x-axis. Then connect the points with straight lines.

Example 2.15

Construct an ogive

Construct a frequency ogive and a relative frequency ogive for the data in Table 2.15.

Solution

The cumulative frequencies, cumulative relative frequencies, and upper class limits are shown in Table 2.15. Figures 2.14 and 2.15 present the ogives.

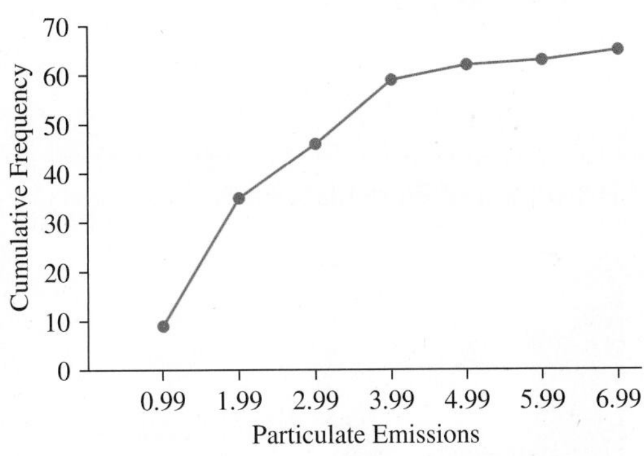

Figure 2.14 Frequency ogive for the data in Table 2.15

Figure 2.15 Relative frequency ogive for the data in Table 2.15

Using Technology

We use the data in Table 2.7 to illustrate the technology steps.

TI-84 PLUS

Entering Data

Step 1. We enter the data into **L1** in the data editor. To clear out any data that may be in the list, press **STAT**, then **4: ClrList**, then enter **L1** by pressing **2nd, L1** (Figure A). Then press **ENTER**.

Step 2. Enter the data into **L1** in the data editor by pressing **STAT** then **1: Edit...** For the data in Table 2.7, we begin with **1.5, .87, 1.12, 1.25, 3.46, ...** (Figure B).

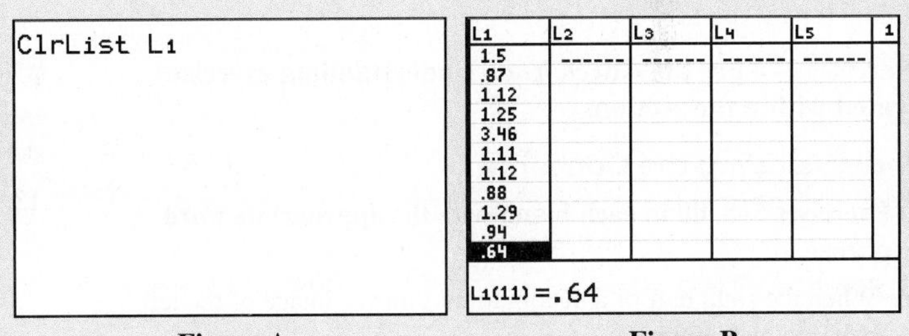

Figure A **Figure B**

Constructing a Histogram

Step 1. Press **2nd, Y=** to access the STAT PLOTS menu and select **Plot1** by pressing **1**.

Step 2. Select **On** and the histogram icon (Figure C).

Step 3. Press **WINDOW** and:

- Set **Xmin** to the lower class limit of the first class. We use 0 for our example.
- Set **Xmax** to the lower class limit of the class following the one containing the largest data value. We use 7.
- Set **Xscl** to the class width. We use 1.
- Set **Ymin** to 0.
- Set **Ymax** to a value greater than the largest frequency of all classes. We use 30.

Step 4. Press **GRAPH** to view the histogram (Figure D).

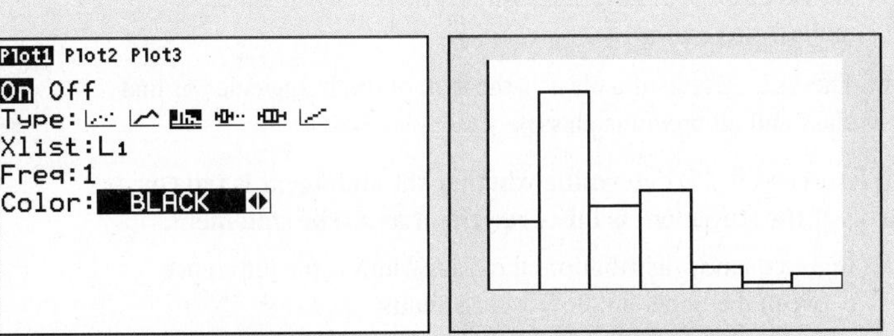

Figure C **Figure D**

EXCEL

Constructing a Histogram

Step 1. Enter the *Particulate Emissions* data from Table 2.7 in **Column A**.
Step 2. Press **Data**, then **Data Analysis**. Select **Histogram** and click **OK**.
Step 3. Enter the range of cells that contain the data in the **Input Range** field, and check the **Chart Output** box.
Step 4. Click **OK**.

MINITAB

Constructing a Histogram

Step 1. Name your variable *Particulate Emissions* and enter the data from Table 2.7 into Column **C1**.
Step 2. Click on **Graph**. Select **Histogram**. Choose the **Simple** option. Press **OK**.
Step 3. Double-click on the *Particulate Emissions* variable and press **OK** (Figure E).

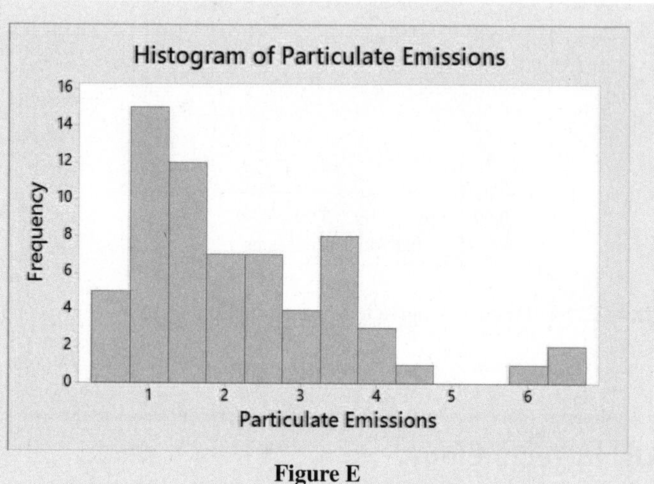

Figure E

Section 2.2 Exercises

Exercises 1–4 are the Check Your Understanding exercises located within the section.

Understanding the Concepts

In Exercises 5–8, fill in each blank with the appropriate word or phrase.

5. When the right half of a histogram is a mirror image of the left half, the histogram is ________ . *symmetric*

6. A histogram is skewed to the left if its ________ tail is longer than its ________ tail. *left, right*

7. A histogram is ________ if it has two clearly distinct modes. *bimodal*

8. The ________ of a class is the sum of the frequencies of that class and all previous classes. *cumulative frequency*

In Exercises 9–12, determine whether the statement is true or false. If the statement is false, rewrite it as a true statement.

9. In a frequency distribution, the class width is the difference between the upper and lower class limits. *False*

10. The number of classes used has little effect on the shape of the histogram. *False*

11. There is no one right way to choose the classes for a frequency distribution. *True*

12. A mode occurs at the peak of a histogram. *True*

Practicing the Skills

In Exercises 13–16, classify the histogram as skewed to the left, skewed to the right, bell-shaped, or approximately uniform.

13.

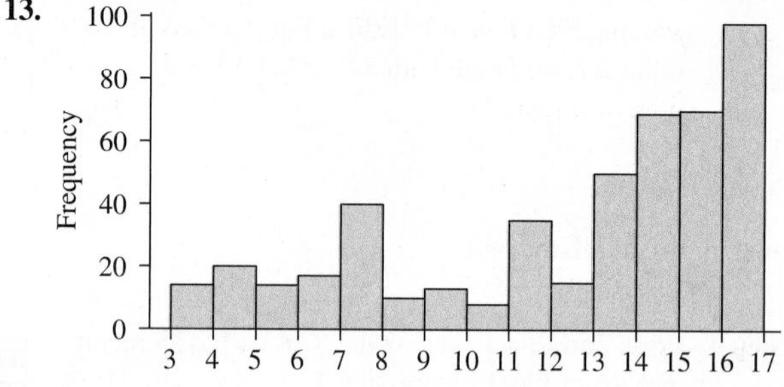

Skewed left

14.

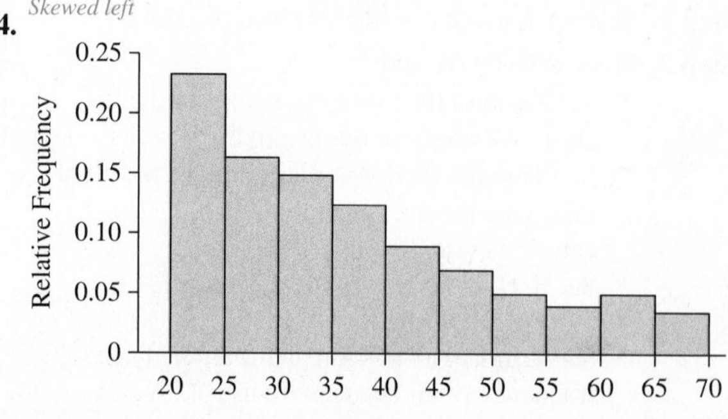

Skewed right

15.

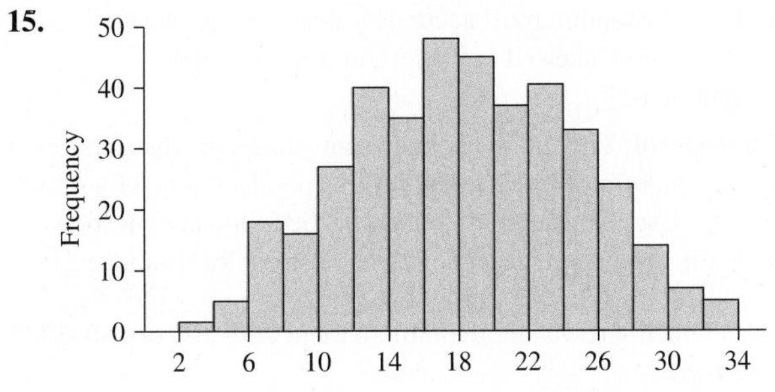

Bell-shaped

16.

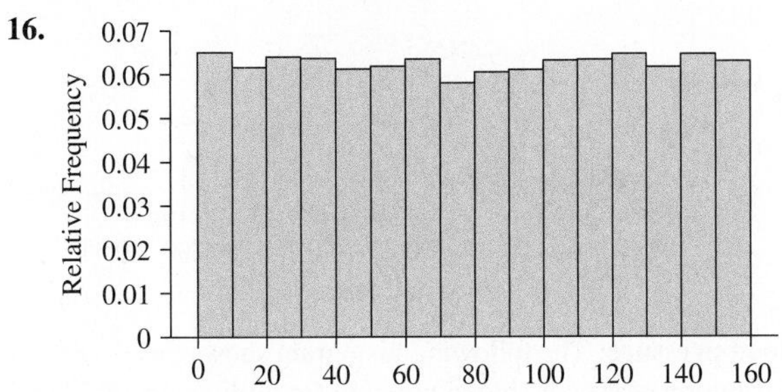

Approximately uniform

17.

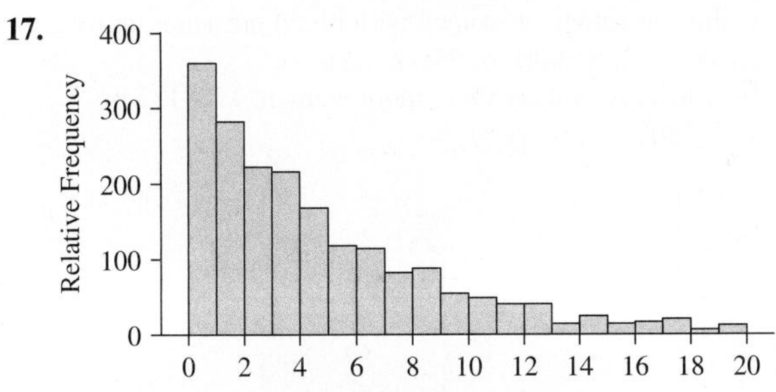

Skewed right

18.

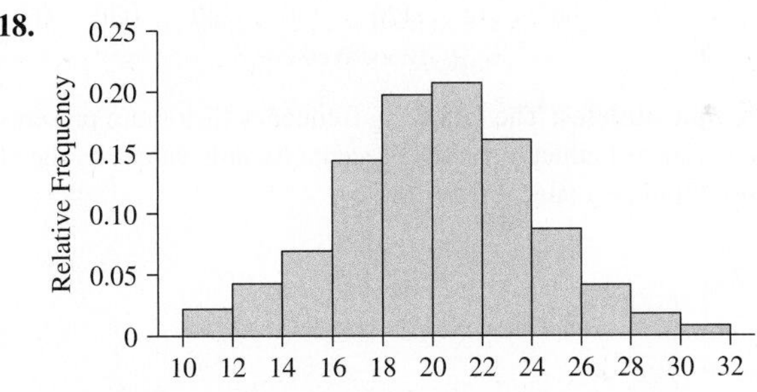

Bell-shaped

In Exercises 19 and 20, classify the histogram as unimodal or bimodal.

19.

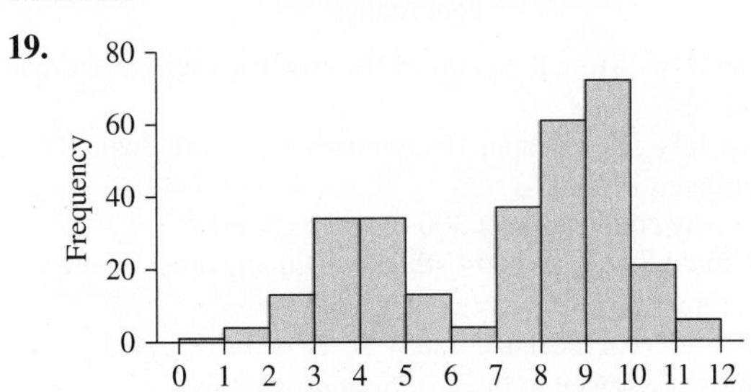

Bimodal

20.

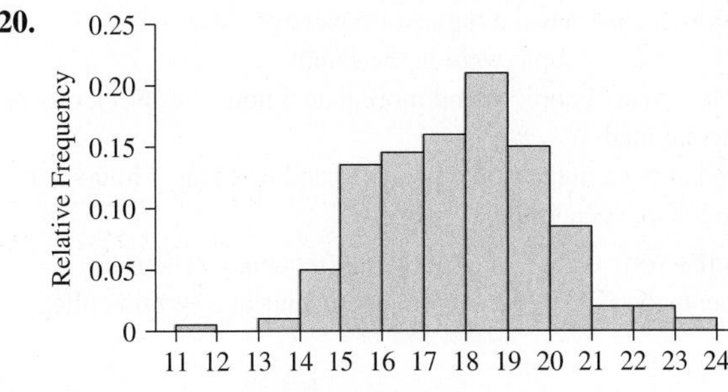

Unimodal

Working with the Concepts

21. Movies: A sample of people were asked how many movies they had seen in a movie theater during the past month. Following is a frequency histogram of their responses.

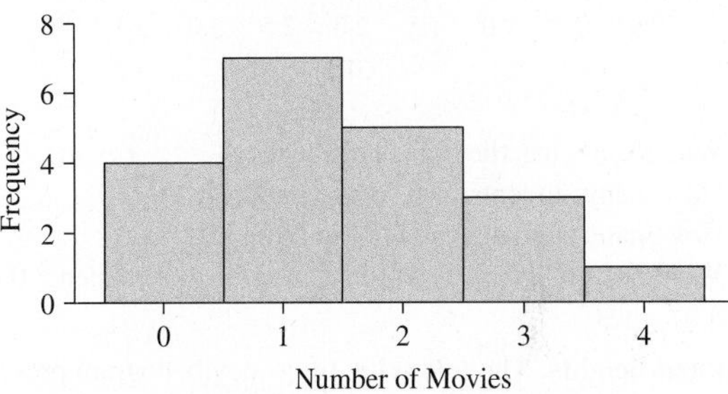

a. What is the most frequent number of movies seen? *1*

b. How many people were in the sample? *20*

c. How many people saw more than two movies? *4*

d. What percentage of the people saw fewer than two movies? *55%*

22. College courses: The following frequency histogram presents the number of courses taken by a sample of students in a certain semester.

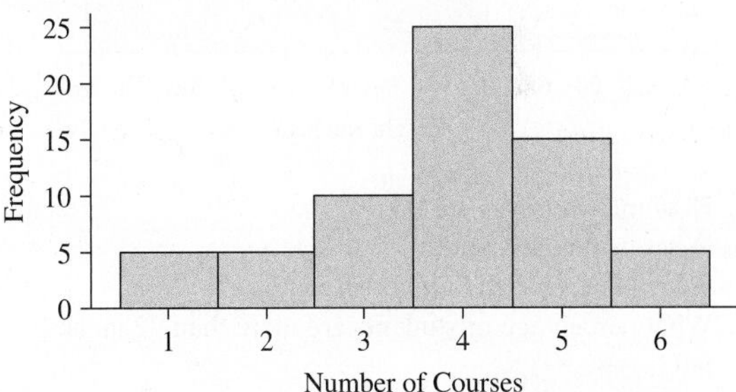

a. What is the most frequent number of courses taken? *4*

b. How many students were in the sample? *65*

c. How many students were taking more than four courses? *20*

d. What percentage of the students were taking three or fewer courses? *30.8%*

23. Time on social media: A survey was conducted to determine the average number of hours per week people spend on social media. The following frequency histogram presents the results.

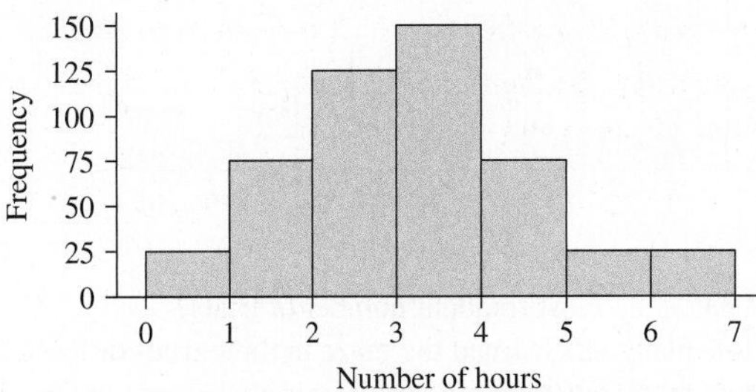

a. Which class has the highest frequency? *3–4*
b. How many people were in the sample? *500*
c. How many people spend more than 5 hours per week on social media? *50*
d. What percentage of the people spend less than 2 hours per week on social media? *20%*

24. What's your GPA? The following frequency histogram presents the GPAs of a sample of students at a certain college.

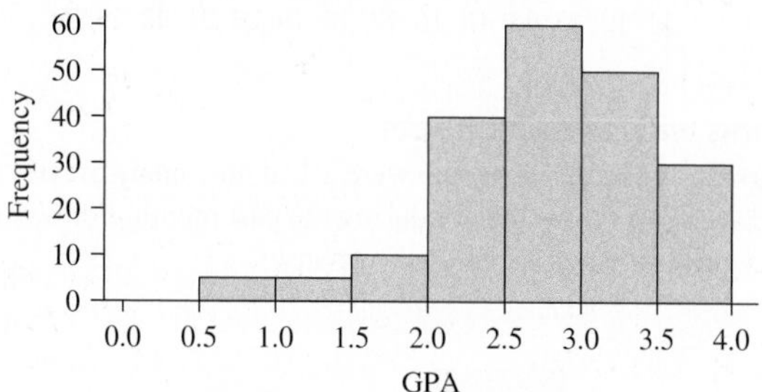

a. Which class has the highest frequency? *2.5–3.0*
b. How many students were in the sample? *200*
c. How many students had GPAs above 3.0? *80*
d. What percentage of the students had GPAs between 2.0 and 3.0? *50%*

25. Student heights: The following frequency histogram presents the heights, in inches, of a random sample of 100 male college students.

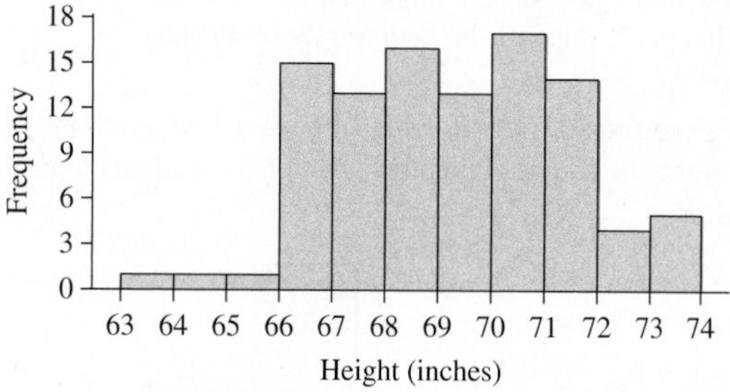

a. How many classes are there? *11*
b. What is the class width? *1*
c. Which class has the highest frequency? *70–71*
d. What percentage of students are more than 72 inches tall? *9%*
e. Is the histogram most accurately described as skewed to the right, skewed to the left, or approximately symmetric? *Approximately symmetric*

26. Trained rats: Forty rats were trained to run a maze. The following frequency histogram presents the numbers of trials it took each rat to learn the maze.

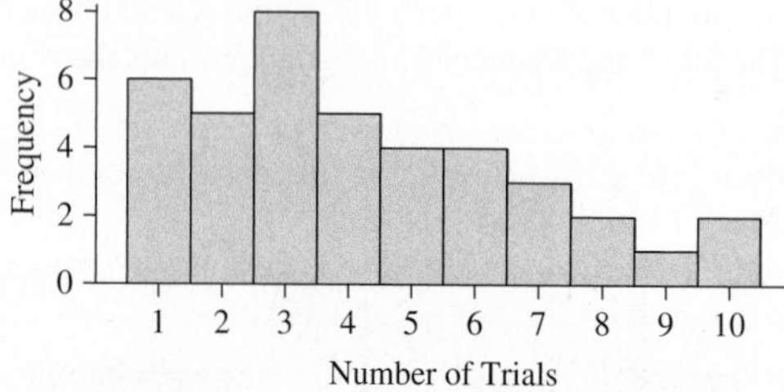

a. What is the most frequent number of trials? *3*
b. How many rats learned the maze in three trials or less? *19*
c. How many rats took nine trials or more to learn the maze? *3*
d. Is the histogram most accurately described as skewed to the right, skewed to the left, or approximately symmetric? *Skewed right*

27. Cholesterol: The following histogram shows the distribution of serum cholesterol level (in milligrams per deciliter) for a sample of men. Use the histogram to answer the following questions:
a. Is the percentage of men with cholesterol levels above 240 closest to 30%, 50%, or 70%? *30%*
b. In which interval are there more men: 240–260 or 280–340? *240–260*

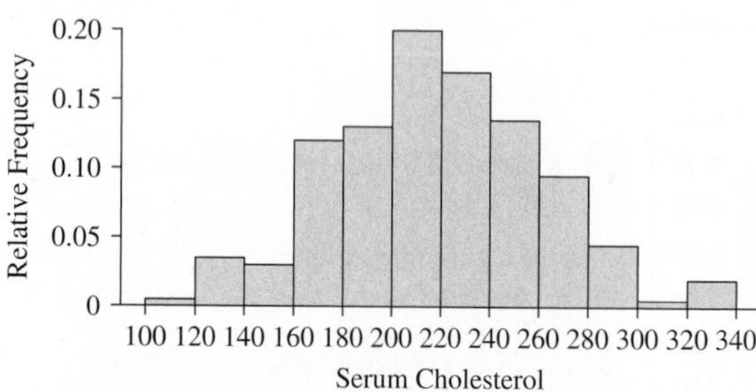

28. Blood pressure: The following histogram shows the distribution of systolic blood pressure (in millimeters of mercury) for a sample of women. Use the histogram to answer the following questions:
a. Is the percentage of women with blood pressures above 120 closest to 25%, 50%, or 75%? *50%*
b. In which interval are there more women: 130–135 or 140–150? *130–135*

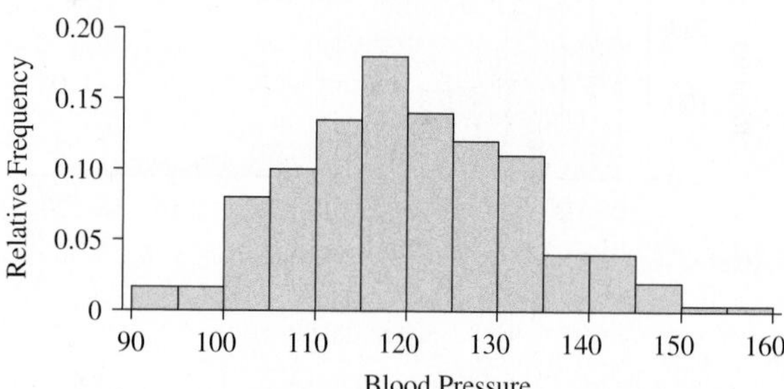

29. Olympic athletes: The following frequency histogram presents the number of athletes sent to a recent Olympic games by the 50 most populous countries.

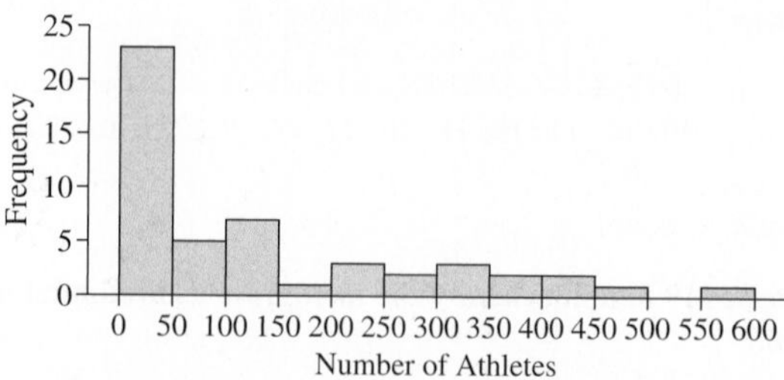

a. True or false: More than half of the countries sent fewer than 50 athletes. *False*
b. True or false: Fewer than 15 countries sent more than 300 athletes. *True*
c. How many countries sent 300 or more athletes? *9*
d. The United States sent 554 athletes. Did any other country send more than 500? *No*
e. Is the histogram most accurately described as skewed to the right, skewed to the left, or approximately symmetric? *Skewed to the right*

30. How's the weather? The following relative frequency histogram presents the average temperatures, in °F, for each of the 50 states of the United States, plus the District of Columbia. Use the histogram to answer the following questions.

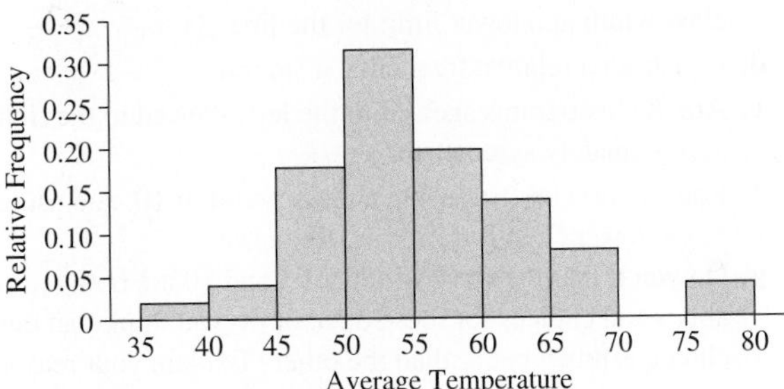

a. In which interval are the largest number of states? *50–55*
b. Is the percentage of states with average temperatures above 60° closest to 30%, 40%, or 50%? *30%*
c. In which interval are there more states, 55–60 or 60–70? *60–70*

31. Skewed which way? For which of the following data sets would you expect a histogram to be skewed to the right? For which would it be skewed to the left?
a. The lengths of the words in a book *right*
b. Dates of coins in circulation *left*
c. Scores of students on an easy exam *left*

32. Skewed which way? For which of the following data sets would you expect a histogram to be skewed to the right? For which would it be skewed to the left?
a. Annual incomes for residents of a town *right*
b. Amounts of time taken by students on a one-hour exam *left*
c. Ages of residents of a town *right*

33. Batting average: The following frequency distribution presents the batting averages of Major League Baseball players who had 300 or more plate appearances during a recent season.

Batting Average	Frequency
0.180–0.199	4
0.200–0.219	14
0.220–0.239	41
0.240–0.259	59
0.260–0.279	58
0.280–0.299	51
0.300–0.319	29
0.320–0.339	10
0.340–0.359	1

Source: sports.espn.go.com

a. How many classes are there? *9*
b. What is the class width? *0.020*
c. What are the class limits?
d. Construct a frequency histogram.
e. Construct a relative frequency distribution.
f. Construct a relative frequency histogram.
g. What percentage of players had batting averages of 0.300 or more? *15.0%*
h. What percentage of players had batting averages less than 0.220? *6.7%*

34. Batting average: The following frequency distribution presents the batting averages of Major League Baseball players in both the American League and the National League who had 300 or more plate appearances during a recent season.

Batting Average	American League Frequency	National League Frequency
0.180–0.199	2	2
0.200–0.219	7	7
0.220–0.239	21	20
0.240–0.259	30	29
0.260–0.279	26	32
0.280–0.299	21	30
0.300–0.319	12	17
0.320–0.339	5	5
0.340–0.359	0	1

Source: sports.espn.go.com

a. Construct a frequency histogram for the American League.
b. Construct a frequency histogram for the National League.
c. Construct a relative frequency distribution for the American League.
d. Construct a relative frequency distribution for the National League.
e. Construct a relative frequency histogram for the American League.
f. Construct a relative frequency histogram for the National League.
g. What percentage of American League players had batting averages of 0.300 or more? *13.7%*
h. What percentage of National League players had batting averages of 0.300 or more? *16.1%*
i. Compare the relative frequency histograms. What is the main difference between the distributions of batting averages in the two leagues?

35. Time spent playing video games: A sample of 200 college freshmen was asked how many hours per week they spent playing video games. The following frequency distribution presents the results.

Number of Hours	Frequency
1.0–3.9	25
4.0–6.9	34
7.0–9.9	48
10.0–12.9	29
13.0–15.9	23
16.0–18.9	17
19.0–21.9	13
22.0–24.9	7
25.0–27.9	3
28.0–30.9	1

a. How many classes are there? *10*
b. What is the class width? *3.0*
c. What are the class limits?
d. Construct a frequency histogram.
e. Construct a relative frequency distribution.
f. Construct a relative frequency histogram.
g. What percentage of students play video games less than 10 hours per week? *53.5%*
h. What percentage of students play video games 19 or more hours per week? *12.0%*

36. Murder, she wrote: The following frequency distribution presents the number of murders (including negligent manslaughter) per 100,000 population for each U.S. city with population over 250,000 in a recent year.

Murder Rate	Frequency
0.0–4.9	21
5.0–9.9	23
10.0–14.9	12
15.0–19.9	6
20.0–24.9	5
25.0–29.9	0
30.0–34.9	2
35.0–39.9	2
40.0–44.9	0
45.0–49.9	0
50.0–54.9	2

Source: Federal Bureau of Investigation

a. How many classes are there? *11*
b. What is the class width? *5*
c. What are the class limits?
d. Construct a frequency histogram.
e. Construct a relative frequency distribution.
f. Construct a relative frequency histogram.
g. What percentage of cities had murder rates less than 10 per 100,000 population? *60.3%*
h. What percentage of cities had murder rates of 30 or more per 100,000 population? *8.2%*

37. BMW prices: The following table presents the manufacturer's suggested retail price (in $1000s) for base models and styles of BMW automobiles.

35.3	37.3	41.1	43.1	45.0	47.0
53.1	55.1	44.8	46.8	51.4	53.4
57.7	60.4	51.2	53.2	86.5	89.5
95.6	102.7	58.9	45.8	47.8	50.4
52.4	54.0	56.0	69.2	77.7	103.1
157.7	69.9	76.9	73.4	80.4	63.7

Source: Car Finder

a. Construct a frequency distribution using a class width of 10, and using 30 as the lower class limit for the first class.
b. Construct a frequency histogram from the frequency distribution in part (a).
c. Construct a relative frequency distribution using the same class width and lower limit for the first class.
d. Construct a relative frequency histogram.
e. Are the histograms unimodal or bimodal? *Unimodal*
f. Repeat parts (a)–(d), using a class width of 20, and using 30 as the lower class limit for the first class.
g. Do you think that class widths of 10 and 20 are both reasonably good choices for these data, or do you think that one choice is much better than the other? Explain your reasoning.

38. Geysers: The geyser Old Faithful in Yellowstone National Park alternates periods of eruption, which typically last from 1.5 to 4 minutes, with periods of dormancy, which are considerably longer. The following table presents the durations, in minutes, of 60 dormancy periods that occurred during a recent year.

91	99	99	83	99	85	90	96	88	93
88	88	92	116	59	101	90	71	103	97
82	91	89	89	94	94	61	96	66	105
90	93	88	92	86	93	95	83	90	99
89	94	90	95	93	105	96	92	101	91
94	92	94	86	88	99	90	99	84	92

a. Construct a frequency distribution using a class width of 5, and using 55 as the lower class limit for the first class.
b. Construct a frequency histogram from the frequency distribution in part (a).
c. Construct a relative frequency distribution using the same class width and lower limit for the first class.
d. Construct a relative frequency histogram.
e. Are the histograms skewed to the left, skewed to the right, or approximately symmetric? *Skewed left*
f. Repeat parts (a)–(d), using a class width of 10, and using 50 as the lower class limit for the first class.
g. Do you think that class widths of 5 and 10 are both reasonably good choices for these data, or do you think that one choice is much better than the other? Explain your reasoning.

39. Hail to the chief: There have been 58 presidential inaugurations in U.S. history. At each one, the president has made an inaugural address. Following are the number of words spoken in each of these addresses.

1431	135	2321	1730	2166	1177	1211	3375
4472	2915	1128	1176	3843	8460	4809	1090
3336	2831	3637	700	1127	1339	2486	2979
1686	4392	2015	3968	2218	984	5434	1704
1526	3329	4055	3672	1880	1808	1359	559
2273	2459	1658	1366	1507	2128	1803	1229
2427	2561	2320	1598	2155	1592	2071	2395
2096	1433						

Source: The American Presidency Project

a. Construct a frequency distribution with approximately five classes.
b. Construct a frequency histogram from the frequency distribution in part (a).
c. Construct a relative frequency distribution using the same classes as in part (a).
d. Construct a relative frequency histogram from this relative frequency distribution.
e. Are the histograms skewed to the left, skewed to the right, or approximately symmetric? *Skewed right*
f. Construct a frequency distribution with approximately nine classes.
g. Repeat parts (b)–(d), using the frequency distribution constructed in part (f).
h. Do you think that five and nine classes are both reasonably good choices for these data, or do you think that one choice is much better than the other? Explain your reasoning.

40. Internet radio: The following table presents the number of hours a sample of 40 subscribers listened to Pandora Radio in a given week.

52	18	2	20	9	9	11	6	18	6
4	12	9	16	10	37	15	18	8	23
4	3	17	19	12	20	11	14	10	37
21	36	17	3	23	28	19	20	29	12

a. Construct a frequency distribution with approximately eleven classes.
b. Construct a frequency histogram from this frequency distribution.
c. Construct a relative frequency distribution using the same classes.
d. Construct a relative frequency histogram from this relative frequency distribution.

e. Are the histograms skewed to the left, skewed to the right, or approximately symmetric? *Skewed right*

f. Construct a frequency distribution with approximately four classes.

g. Repeat parts (b)–(d), using the frequency distribution constructed in part (f).

h. Do you think that four and eleven classes are both reasonably good choices for these data, or do you think that one choice is much better than the other? Explain your reasoning.

41. Brothers and sisters: Thirty students in a first-grade class were asked how many siblings they have. Following are the results.

1	1	2	1	2	3	7	1	1	5
1	1	3	0	1	1	1	2	5	0
0	1	2	2	4	2	2	3	3	4

a. Construct a frequency histogram.

b. Construct a relative frequency histogram.

c. Are the histograms skewed to the left, skewed to the right, or approximately symmetric? *Skewed right*

42. Cough, cough: The following table presents the number of days a sample of patients reported a cough lasting from an acute cough illness.

16	20	21	20	19	17	18	12	22
17	16	24	15	13	21	16	21	20
20	19	18	16	19	20	21	21	14
16	16	14	17	18	14	15	20	20
13	15	18	20	19	21	19	18	20

Based on data from *Annals of Family Medicine*

a. Construct a frequency histogram.

b. Construct a relative frequency histogram.

c. Are the histograms skewed to the left, skewed to the right, or approximately symmetric? *Skewed left*

43. Which histogram is which? One of the following histograms represents the age at death from natural causes (heart attack, cancer, etc.), and the other represents the age at death from accidents. Which represents the age at death from accidents? How can you tell? *B*

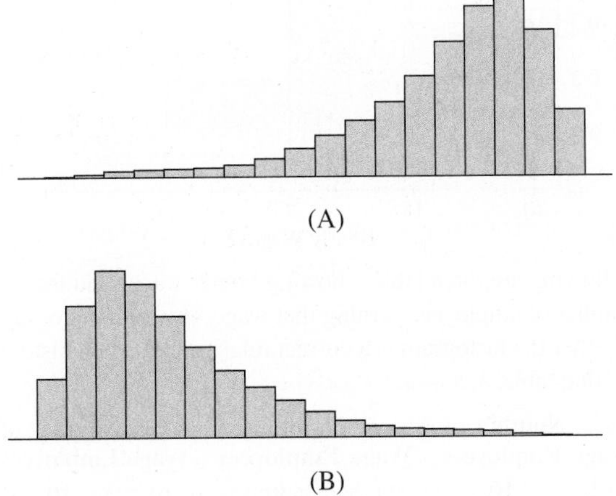

44. Test scores: In a certain city, applicants for engineering jobs are given an exam, and the ones with the top 15 scores are hired. The following frequency histogram presents the scores for the applicants on a recent exam. The examiners were charged with rigging the exam. Describe the evidence that supports this charge.

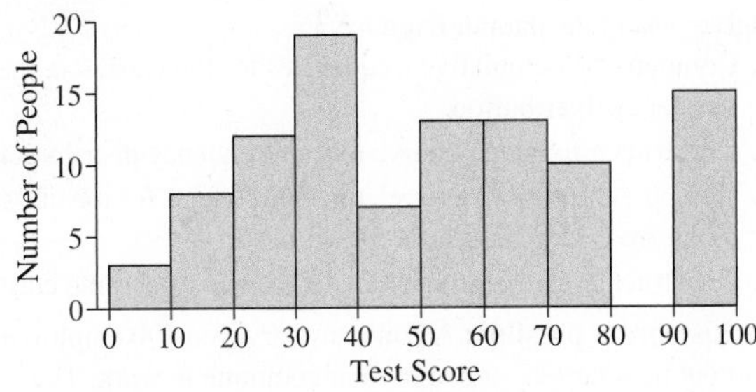

45. Frequency polygon: Using the data in Exercise 33:

a. Construct a frequency polygon for the frequency distribution.

b. Construct a relative frequency polygon, using the same classes.

46. Frequency polygon: Using the data in Exercise 34:

a. Construct a frequency polygon for the American League frequency distribution.

b. Construct a relative frequency polygon for the American League, using the same classes.

c. Construct a frequency polygon for the National League frequency distribution.

d. Construct a relative frequency polygon for the National League, using the same classes.

47. Frequency polygon: Using the data in Exercise 35:

a. Construct a frequency polygon for the frequency distribution.

b. Construct a relative frequency polygon, using the same classes.

48. Frequency polygon: Using the data in Exercise 36:

a. Construct a frequency polygon for the frequency distribution.

b. Construct a relative frequency polygon, using the same classes.

49. Ogive: Using the data in Exercise 33:

a. Compute the cumulative frequencies for the classes in the frequency distribution.

b. Construct a frequency ogive for the frequency distribution.

c. Compute the cumulative relative frequencies for the classes in the frequency distribution.

d. Construct a relative frequency ogive, using the same classes.

50. Ogive: Using the data in Exercise 34:

a. Compute the cumulative frequencies for the classes in the American League frequency distribution.

b. Construct a frequency ogive for the American League frequency distribution.

c. Compute the cumulative relative frequencies for the classes in the American League frequency distribution.

d. Construct a relative frequency ogive for the American League, using the same classes.

e. Compute the cumulative frequencies for the classes in the National League frequency distribution.

f. Construct a frequency ogive for the National League frequency distribution.

g. Compute the cumulative relative frequencies for the classes in the National League frequency distribution.

h. Construct a relative frequency ogive for the National League, using the same classes.

51. Ogive: Using the data in Exercise 35:

a. Compute the cumulative frequencies for the classes in the frequency distribution.

b. Construct a frequency ogive for the frequency distribution.

c. Compute the cumulative relative frequencies for the classes in the frequency distribution.

d. Construct a relative frequency ogive, using the same classes.

52. Ogive: Using the data in Exercise 36:

a. Compute the cumulative frequencies for the classes in the frequency distribution.

b. Construct a frequency ogive for the frequency distribution.

c. Compute the cumulative relative frequencies for the classes in the frequency distribution.

d. Construct a relative frequency ogive, using the same classes.

53. No histogram possible: A company surveyed 100 employees to find out how far they travel in their commute to work. The results are presented in the following frequency distribution.

Distance in Miles	Frequency
0.0–4.9	18
5.0–9.9	26
10.0–14.9	15
15.0–19.9	13
20.0–24.9	12
25.0–29.9	9
30 or more	7

Explain why it is not possible to construct a histogram for this data set. *Open-ended class*

54. Histogram possible? Refer to Exercise 53: Suppose you found out that none of the employees traveled more than 34 miles. Would it be possible to construct a histogram? If so, construct a histogram. If not, explain why not. *Yes*

Extending the Concepts

55. Silver ore: The following histogram presents the amounts of silver (in parts per million) found in a sample of rocks. One rectangle from the histogram is missing. What is its height? *0.15*

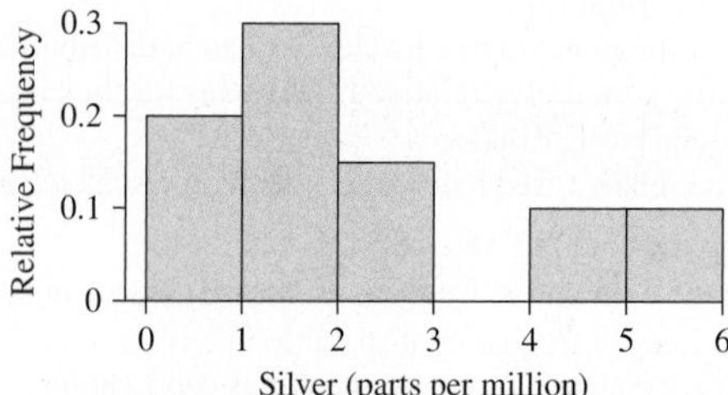

56. Classes of differing widths: Consider the following relative frequency distribution for the data in Table 2.7, in which the classes have differing widths.

Class	Frequency	Relative Frequency
0.00–0.99	9	0.138
1.00–1.49	19	0.292
1.50–1.99	7	0.108
2.00–2.99	11	0.169
3.00–3.99	13	0.200
4.00–6.99	6	0.092

a. Compute the class width for each of the classes.

b. Construct a relative frequency histogram. Compare it to the relative frequency histogram in Figure 2.6, in which the classes all have the same width. Explain why using differing widths gives a distorted picture of the data.

c. The *density* of a class is the relative frequency divided by the class width. For each class, divide the relative frequency by the class width to obtain the density.

d. Construct a histogram in which the height of each rectangle is equal to the density of the class. This is called a *density histogram.*

e. Compare the density histogram to the relative frequency histogram in Figure 2.6, in which the classes all have the same width. Explain why differing class widths in a density histogram do not distort the data.

57. Detecting skewness from ogives: Which of the following two ogives represents a skewed distribution and which represents a distribution that is approximately symmetric? *(i) Skewed; (ii) Approximately symmetric*

(i)

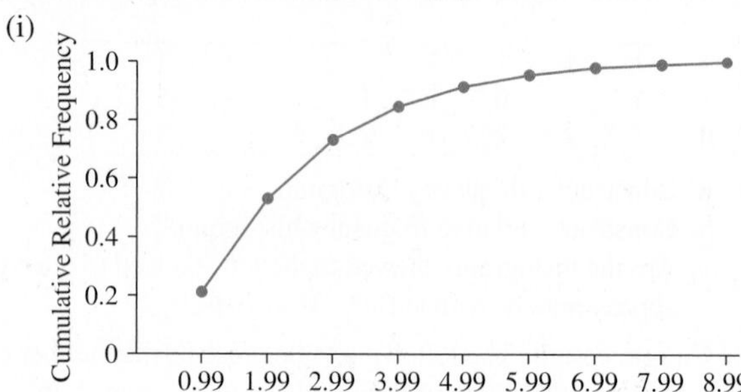

(ii)

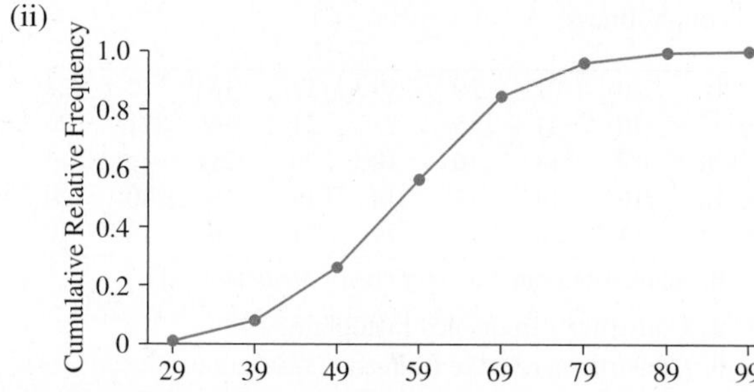

58. Skewed which way? Refer to Exercise 57. For the ogive that represents the skewed distribution, is it skewed to the right or to the left? *Skewed right*

59. Frequencies and relative frequencies: The following relative frequency histogram presents the hourly wages for employees at a certain company.

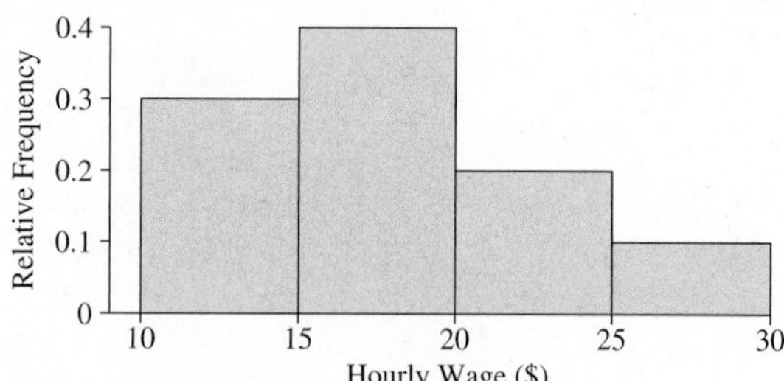

Following are three tables showing hourly wages and the number of employees earning that wage. For each table, say whether the histogram is a correct relative frequency histogram for that table. *A, B, correct; C not correct*

Wage (A)	Number of Employees (A)	Wage (B)	Number of Employees (B)	Wage (C)	Number of Employees (C)
12	10	11	9	14	10
14	5	16	3	16	5
17	10	18	7	17	10
18	10	19	2	18	10
22	10	23	6	22	10
26	3	26	2	26	3
28	2	29	1	28	2

(A) (B) (C)

Answers to Check Your Understanding Exercises for Section 2.2

1.

Class	Frequency
0.00–0.49	2
0.50–0.99	7
1.00–1.49	19
1.50–1.99	7
2.00–2.49	7
2.50–2.99	4
3.00–3.49	9
3.50–3.99	4
4.00–4.49	2
4.50–4.99	1
5.00–5.49	0
5.50–5.99	1
6.00–6.49	0
6.50–6.99	2

2. a. 15, 20, 25, 30, 35, 40 **b.** 5 **c.**

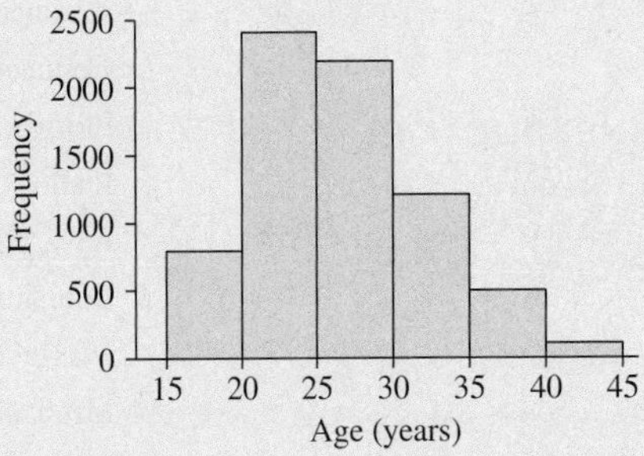

Frequency
2500
2000
1500
1000
500
0
15 20 25 30 35 40 45
Age (years)

d.

Age	Frequency	Relative Frequency
15–19	795	0.110
20–24	2410	0.334
25–29	2190	0.304
30–34	1208	0.168
35–39	499	0.069
40–44	109	0.015

e.

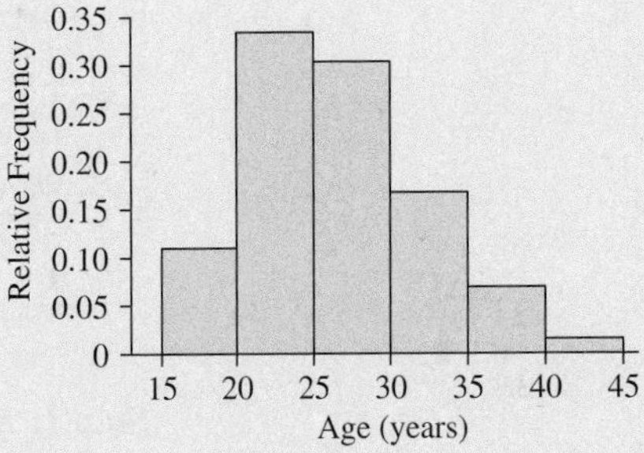

Relative Frequency
0.35
0.30
0.25
0.20
0.15
0.10
0.05
0
15 20 25 30 35 40 45
Age (years)

3. a. Skewed right **b.** Skewed left **c.** Approximately uniform **d.** Bell-shaped

4. a. Unimodal **b.** Unimodal **c.** Bimodal

Section 2.3 More Graphs for Quantitative Data

Objectives

1. Construct stem-and-leaf plots
2. Construct dotplots
3. Construct time-series plots

Histograms and other graphs that are based on frequency distributions can be used to summarize both small and large data sets. For small data sets, it is sometimes useful to have a summary that is more detailed than a histogram. In this section, we describe some commonly used graphs that provide more detailed summaries of smaller data sets. These graphs illustrate the shape of the data set, while allowing every value in the data set to be seen.

Objective 1 Construct stem-and-leaf plots

Stem-and-Leaf Plots

Stem-and-leaf plots are a simple way to display small data sets. For example, Table 2.16 (page 72) presents the U.S. Census Bureau data for the percentage of the population aged 65 and over for each state and the District of Columbia.

In a stem-and-leaf plot, the rightmost digit is the leaf, and the remaining digits form the stem. For example, the stem for Alabama is 14, and the leaf is 1. We construct a stem-and-leaf plot for the data in Table 2.16 by using the following three-step process:

NOTE TO INSTRUCTOR

It is helpful to point out that stem-and-leaf plots are similar to histograms. They provide more detail, because each data value is shown. They are inconvenient for large data sets, however.

Table 2.16 Percentage of Population Aged 65 and Over, by State

Alabama	14.1	Alaska	8.1	Arizona	13.9
Arkansas	14.3	California	11.5	Colorado	10.7
Connecticut	14.4	Delaware	14.1	District of Columbia	11.5
Florida	17.8	Georgia	10.2	Hawaii	14.3
Idaho	12.0	Illinois	12.4	Indiana	12.7
Iowa	14.9	Kansas	13.4	Kentucky	13.1
Louisiana	12.6	Maine	15.6	Maryland	12.2
Massachusetts	13.7	Michigan	12.8	Minnesota	12.4
Mississippi	12.8	Missouri	13.9	Montana	15.0
Nebraska	13.8	Nevada	12.3	New Hampshire	12.6
New Jersey	13.7	New Mexico	14.1	New York	13.6
North Carolina	12.4	North Dakota	15.3	Ohio	13.7
Oklahoma	13.8	Oregon	13.0	Pennsylvania	15.5
Rhode Island	14.1	South Carolina	13.6	South Dakota	14.6
Tennessee	13.3	Texas	10.5	Utah	9.0
Vermont	14.3	Virginia	12.4	Washington	12.2
West Virginia	16.0	Wisconsin	13.5	Wyoming	14.0

Source: U.S. Census Bureau

Step 1: Make a vertical list of all the stems in increasing order, and draw a vertical line to the right of this list. The smallest stem in Table 2.16 is 8, belonging to Alaska, and the largest is 17, belonging to Florida. The list of stems is shown in Figure 2.16(a).

Step 2: Go through the data set, and for each value, write its leaf next to its stem. For example, the first value is 14.1, for Alabama. We write a "1" next to the stem 14. The next value is 8.1 for Alaska, so we write a "1" next to the stem 8. When we are finished, we have the result shown in Figure 2.16(b).

Step 3: For each stem, arrange its leaves in increasing order. The result is the stem-and-leaf plot, shown in Figure 2.16(c).

(a)

```
 8 |
 9 |
10 |
11 |
12 |
13 |
14 |
15 |
16 |
17 |
```

(b)

```
 8 | 1
 9 | 0
10 | 725
11 | 55
12 | 0476284836442
13 | 94179876780635
14 | 13413911630
15 | 6035
16 | 0
17 | 8
```

(c)

```
 8 | 1
 9 | 0
10 | 257
11 | 55
12 | 0223444466788
13 | 01345667778899
14 | 01111333469
15 | 0356
16 | 0
17 | 8
```

Figure 2.16 Steps in the construction of a stem-and-leaf plot

Rounding data for a stem-and-leaf plot

Table 2.17 (page 73) presents the particulate emissions for 65 vehicles. The first digits range from 0 to 6, and we would like to construct a stem-and-leaf plot with these digits as the stems. The problem is that this leaves two digits for the leaf, but the leaf must consist of only one digit. The solution to this problem is to round the data so that there will be only one digit for the leaf. Table 2.18 (page 73) presents the particulate data rounded to two digits.

Table 2.17 Particulate Emissions for 65 Vehicles

1.50	0.87	1.12	1.25	3.46	1.11	1.12	0.88	1.29	0.94	0.64	1.31	2.49
1.48	1.06	1.11	2.15	0.86	1.81	1.47	1.24	1.63	2.14	6.64	4.04	2.48
1.40	1.37	1.81	1.14	1.63	3.67	0.55	2.67	2.63	3.03	1.23	1.04	1.63
3.12	2.37	2.12	2.68	1.17	3.34	3.79	1.28	2.10	6.55	1.18	3.06	0.48
0.25	0.53	3.36	3.47	2.74	1.88	5.94	4.24	3.52	3.59	3.10	3.33	4.58

Table 2.18 Particulate Emissions for 65 Vehicles, Rounded to Two Digits

1.5	0.9	1.1	1.3	3.5	1.1	1.1	0.9	1.3	0.9	0.6	1.3	2.5
1.5	1.1	1.1	2.2	0.9	1.8	1.5	1.2	1.6	2.1	6.6	4.0	2.5
1.4	1.4	1.8	1.1	1.6	3.7	0.6	2.7	2.6	3.0	1.2	1.0	1.6
3.1	2.4	2.1	2.7	1.2	3.3	3.8	1.3	2.1	6.6	1.2	3.1	0.5
0.3	0.5	3.4	3.5	2.7	1.9	5.9	4.2	3.5	3.6	3.1	3.3	4.6

We now follow the three-step process to obtain the stem-and-leaf plot. The result is shown in Figure 2.17.

0	355669999
1	011111122223333445556666889
2	11124556777
3	0111334555678
4	026
5	9
6	66

Figure 2.17 Stem-and-leaf plot for the data in Table 2.18

Split stems

Sometimes one or two stems contain most of the leaves. When this happens, we often use two or more lines for each stem. The plot is then called a **split stem-and-leaf plot**. We will use the data in Table 2.19 to illustrate the method. These data consist of scores on a final examination in a statistics class, arranged in order.

Table 2.19 Scores on a Final Examination

58	66	68	70	70	71	71	72	73	73
75	76	78	78	79	80	80	80	81	82
82	82	82	83	84	86	86	86	87	88
89	89	89	90	92	93	95	97		

Figure 2.18 presents a stem-and-leaf plot for these data, using the stems 5, 6, 7, 8, and 9.

5	8
6	68
7	001123356889
8	00012222346667899 9
9	02357

Figure 2.18 Stem-and-leaf plot for the data in Table 2.19

Most of the leaves are on two stems, 7 and 8. For this reason, the stem-and-leaf plot does not reveal much detail about the data. To remedy this situation, we will assign each stem two lines on the plot instead of one. Leaves with values 0–4 will go on the first line, and leaves with values 5–9 will go on the second line. So, for example, we will do the following with the stem 7:

7 | 001123356889 will become

7	0011233
7	56889

The split stem-and-leaf plot is shown in Figure 2.19. Note that every stem is given two lines, even those that have only a few leaves. Each stem in a split stem-and-leaf plot must receive the same number of lines.

CAUTION

In a split stem-and-leaf plot, each stem must be given the same number of lines.

```
5 |
5 | 8
6 |
6 | 68
7 | 0011233
7 | 56889
8 | 0001222234
8 | 66678999
9 | 023
9 | 57
```

Figure 2.19 Split stem-and-leaf plot for the data in Table 2.19

Check Your Understanding

1. **Weights of college students:** The following table presents weights in pounds for a group of male college freshmen.

136	163	157	195	150	149	151	155	163	145
124	124	156	148	195	192	133	129	160	158
166	155	171	157	182	124	160	172	161	143

a. List the stems for a stem-and-leaf plot. *12, 13, 14, 15, 16, 17, 18, 19*
b. For each item in the data set, write its leaf next to its stem.
c. Rearrange the leaves in numerical order to create a stem-and-leaf plot.

Answers are on page 83.

Back-to-back stem-and-leaf plots

When two data sets have values similar enough so that the same stems can be used, we can compare their shapes with a **back-to-back stem-and-leaf plot**. In a back-to-back stem-and-leaf plot, the stems go down the middle. The leaves for one of the data sets go off to the right, and the leaves for the other go off to the left.

Example 2.16

Constructing a back-to-back stem-and-leaf plot

In Table 2.18, we presented particulate emissions for 65 vehicles. In a related experiment carried out at the Colorado School of Mines, particulate emissions were measured for 35 vehicles driven at high altitude. Table 2.20 presents the results. Construct a back-to-back stem-and-leaf plot to compare the emission levels of vehicles driven at high altitude with those of vehicles driven at sea level.

Table 2.20 Particulate Emissions for 35 Vehicles Driven at High Altitude

8.9	4.4	3.6	4.4	3.8	2.4	3.8	5.3	5.8	2.9	4.7	1.9	9.1
8.7	9.5	2.7	9.2	7.3	2.1	6.3	6.5	6.3	2.0	5.9	5.6	5.6
1.5	6.5	5.3	5.6	2.1	1.1	3.3	1.8	7.6				

Solution

Figure 2.20 (page 75) presents the results. It is clear that vehicles driven at high altitude tend to produce higher emissions.

High Altitude		Sea Level
	0	355669999
9851	1	011111122223333344555666889
974110	2	11124556777
8863	3	0111334555678
744	4	026
9866633	5	9
5533	6	66
63	7	
97	8	
521	9	

Figure 2.20 Back-to-back stem-and-leaf plots comparing the emissions in vehicles driven at high altitude with emissions from vehicles driven at sea level

Objective 2 Construct dotplots

Dotplots

A **dotplot** is a graph that can be used to give a rough impression of the shape of a data set. It is useful when the data set is not too large, and when there are some repeated values. As an example, Table 2.21 presents the number of children of each of the presidents of the United States and their wives.

NOTE TO INSTRUCTOR

Dotplots and stem-and-leaf plots are useful for detecting outliers.

Table 2.21 Numbers of Children of U.S. Presidents and Their Wives

0	2	10	2	5	3	6	2	2	4	1
5	4	15	3	4	5	3	2	3	4	2
6	0	0	0	8	3	3	6	2	4	2
0	4	6	4	7	2	0	1	2	6	5

Figure 2.21 presents a dotplot for the data in Table 2.21. For each value in the data set, a vertical column of dots is drawn, with the number of dots in the column equal to the number of times the value appears in the data set. The dotplot gives a good indication of where the values are concentrated and where the gaps are. For example, it is immediately apparent from Figure 2.21 that the most frequent number of children is 2, and only four presidents had more than 6. (John Tyler holds the record with 15.)

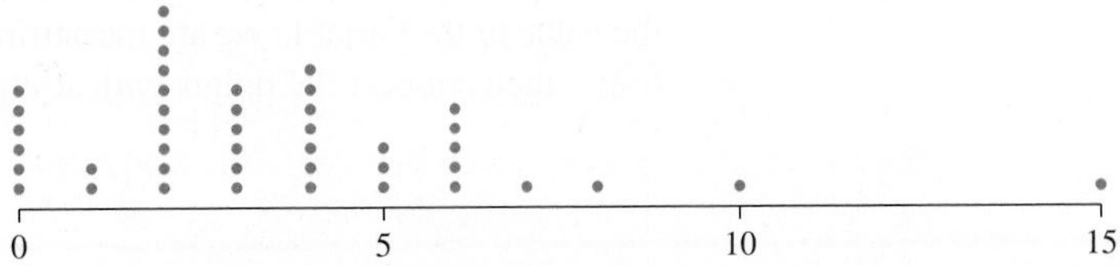

Figure 2.21 Dotplot for the data in Table 2.21

Example 2.17

Constructing a dotplot with technology

Use technology to construct a dotplot for the exam score data in Table 2.19 on page 73.

Solution

The following figure shows the dotplot constructed in MINITAB. Step-by-step instructions for constructing dotplots with MINITAB are given in the Using Technology section on page 78.

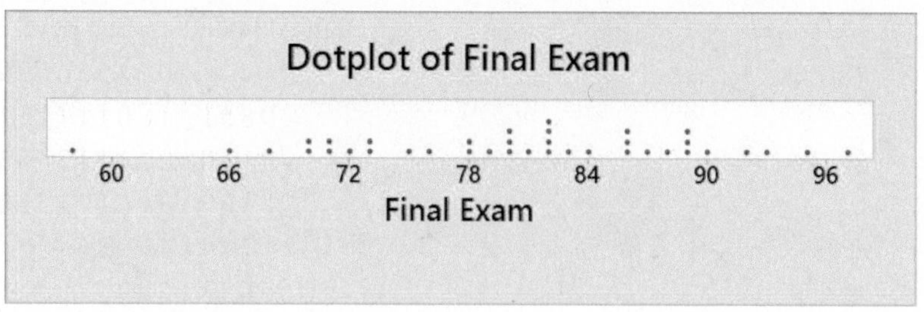

Objective 3 Construct time-series plots

Time-Series Plots

A **time-series plot** may be used when the data consist of values of a variable measured at different points in time. As an example, we consider the Dow Jones Industrial Average, which reflects the prices of 30 large stocks. Table 2.22 presents the closing value of the Dow Jones Industrial Average at the end of each year from 2005 to 2018.

Table 2.22 Dow Jones Industrial Average

Year	Average
2005	10,717.50
2006	12,463.15
2007	13,264.82
2008	8,776.39
2009	10,428.05
2010	11,557.51
2011	12,217.56
2012	13,104.14
2013	16,576.66
2014	17,823.07
2015	17,425.03
2016	19,762.60
2017	24,719.22
2018	23,327.46

In a time-series plot, the horizontal axis represents time, and the vertical axis represents the value of the variable we are measuring. We plot the values of the variable at each of the times, then connect the points with straight lines. Example 2.18 shows how.

Example 2.18

Constructing a time-series plot

Construct a time-series plot for the data in Table 2.22.

Solution

Step 1: Label the horizontal axis with the times at which measurements were made.
Step 2: Plot the value of the Dow Jones Industrial Average for each year.
Step 3: Connect the points with straight lines.

The result is shown in Figure 2.22 (page 77). It is clear that the average generally increased from 2005 to 2007, dropped sharply in 2008, generally increased from 2008 to 2017, and dropped in 2018.

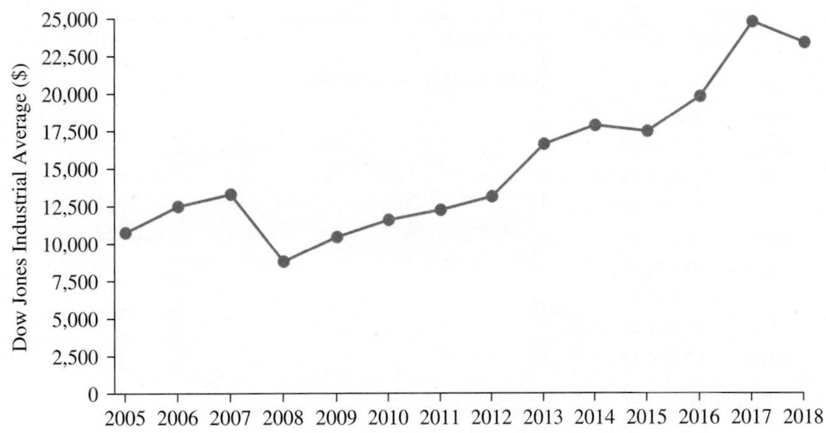

Figure 2.22

Check Your Understanding

2. The following time-series plot presents the percentage of high school seniors who smoked cigarettes every two years from 1994 through 2018.

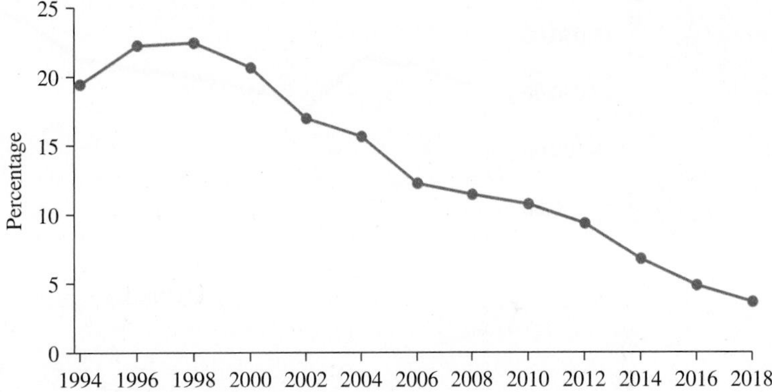

Source: Department of Health and Human Services

a. In what year was the percentage the lowest? *2018*

b. What was the first year that the percentage was below 15%? *2006*

c. True or false: Since 1994, the percentage has never been lower than 5%. *False*

d. During what period of time was the percentage decreasing? *1998–2018*

e. During what period of time was the percentage increasing? *1994–1998*

Answers are on page 83.

Using Technology

We use the data in Tables 2.19 and 2.22 to illustrate the technology steps.

EXCEL

Constructing a time-series plot

Step 1. Enter the data from Table 2.22 into **Columns A** and **B** (Figure A).

Step 2. Select the **Insert** tab and click on the **Line Graph** icon.

Step 3. Click on the **Select Data** icon. For the **Horizontal (Category) Axis Labels**, click on **Edit** and select the year values in **Column A** (Figure B).

Step 4. Click **OK** (Figure C).

	A	B
1	**Year**	**Average**
2	2005	10,717.50
3	2006	12,463.15
4	2007	13,264.82
5	2008	8,776.39
6	2009	10,428.05
7	2010	11,557.51
8	2011	12,217.56
9	2012	13,104.14

Figure A

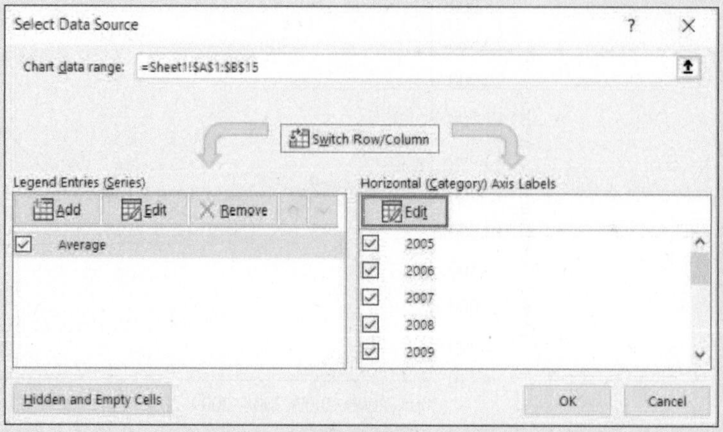

Figure B

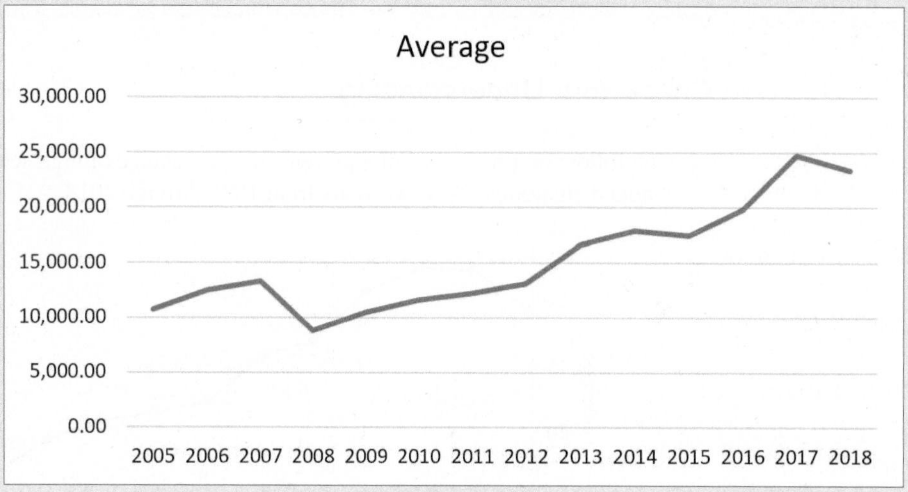

Figure C

MINITAB

Constructing a stem-and-leaf plot and dotplot

Step 1. Name your variable *Final Exam*, and enter the data from Table 2.19 into **Column C1**.

Step 2. Click on **Graph**. Select **Stem-and-Leaf** or **Dotplot**. For **Dotplot**, choose the **Simple** option. Press **OK**.

Step 3. Double-click on the *Final Exam* variable and press **OK** (See Figures D and E).

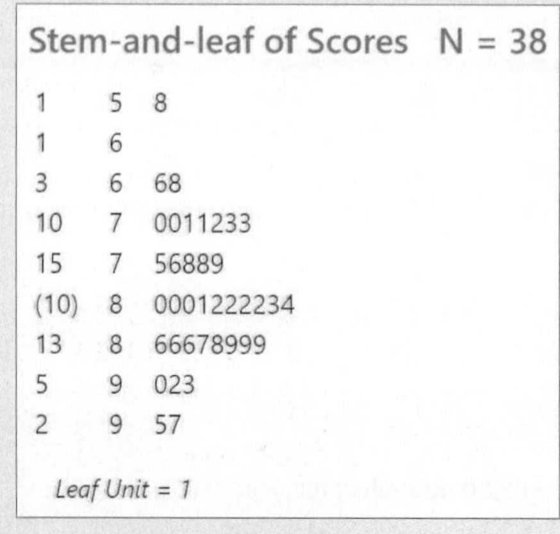

Stem-and-leaf of Scores N = 38

1	5	8
1	6	
3	6	68
10	7	0011233
15	7	56889
(10)	8	0001222234
13	8	66678999
5	9	023
2	9	57

Leaf Unit = 1

Figure D

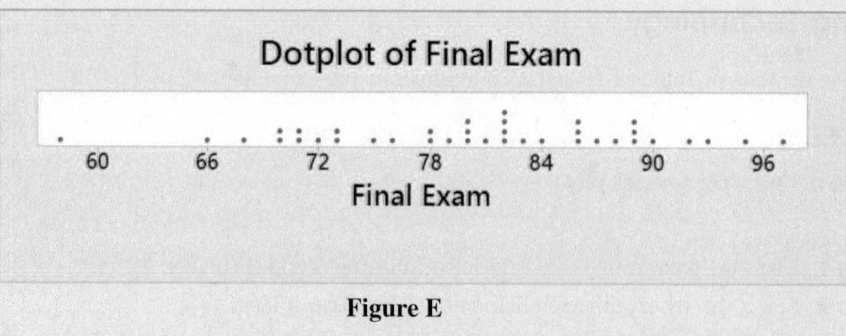

Figure E

Section 2.3 Exercises

Exercises 1 and 2 are the Check Your Understanding exercises located within the section.

Understanding the Concepts

In Exercises 3–6, fill in each blank with the appropriate word or phrase.

3. In a stem-and-leaf plot, the rightmost digit of each data value is the ________. *leaf*

4. In a back-to-back stem-and-leaf plot, each of the two data sets plotted must have the same ________. *stems*

5. A ________ is useful when the data consist of values measured at different points in time. *time-series plot*

6. In a time-series plot, the horizontal axis represents ________. *time*

In Exercises 7–10, determine whether the statement is true or false. If the statement is false, rewrite it as a true statement.

7. Stem-and-leaf plots and dotplots provide a simple way to display data for small data sets. *True*

8. In a stem-and-leaf plot, each stem must be a single digit. *False*

9. In a dotplot, the number of dots in a vertical column represents the number of times a certain value appears in a data set. *True*

10. In a time-series plot, the vertical axis represents time. *False*

Practicing the Skills

11. Construct a stem-and-leaf plot for the following data.

57	20	27	16	11	12	29	39	45	52	58	15
46	27	22	21	15	50	16	45	20	55	12	31

12. Construct a stem-and-leaf plot for the following data, in which the leaf represents the hundredths digit.

5.03	4.99	4.95	5.01	4.99	5.03	4.91	5.25	4.80
5.24	4.94	5.04	5.17	4.81	5.22	4.92	5.05	4.89
5.19	5.17	5.25	5.14	5.10	4.94	5.19	4.99	

13. List the data in the following stem-and-leaf plot. The leaf represents the ones digit.

3	0 0 1 2
3	5 6 7 7 9
4	2 3 4
4	5 6 7 7 7 7 8 8 9
5	0 1 1 1 2 2 2 2 4
5	6 7 8 8 9
6	1 3
6	

14. List the data in the following stem-and-leaf plot. The leaf represents the tenths digit.

14	4 6 8 9
15	1 2 2 4 5 7 7 8
16	0 1 1 1 2 3 7 7 9
17	
18	2 3 8

15. Construct a dotplot for the data in Exercise 11.

16. Construct a dotplot for the data in Exercise 12.

Working with the Concepts

17. BMW prices: The following table presents the manufacturers suggested retail price (in $1000s) for 2020 base models and styles of BMW automobiles.

35.3	37.3	41.1	43.1	45.0	47.0
53.1	55.1	44.8	46.8	51.4	53.4
57.7	60.4	51.2	53.2	86.5	89.5
95.6	102.7	58.9	45.8	47.8	50.4
52.4	54.0	56.0	69.2	77.7	103.1
157.7	69.9	76.9	73.4	80.4	63.7

Source: Car Finder

a. Round the data to the nearest whole number (round .5 up) and construct a stem-and-leaf plot, using the numbers 3 through 15 as the stems.

b. Repeat part (a), but split the stems, using two lines for each stem.

c. Which stem-and-leaf plot do you think is more appropriate for these data, the one in part (a) or the one in part (b)? Why?

18. How's the weather? The following table presents the daily high temperatures for the city of Macon, Georgia, in degrees Fahrenheit, for the winter months of January and February, 2019.

73	66	64	66	63	72	70	72	57	54	55	50
52	50	61	61	61	73	55	46	55	46	73	63
52	57	54	61	50	48	54	64	64	68	68	72
72	81	64	51	48	55	75	61	64	66	75	66
66	54	52	70	82	59	73	66	72	70	64	

a. Construct a stem-and-leaf plot, using the digits 4, 5, 6, 7, and 8 as the stems.

b. Repeat part (a), but split the stems, using two lines for each stem.

c. Which stem-and-leaf plot do you think is more appropriate for these data, the one in part (a) or the one in part (b)? Why?

19. Air pollution: The following table presents amounts of particulate emissions for 65 vehicles. These data also appear in Table 2.18.

1.5	0.9	1.1	1.3	3.5	1.1	1.1	0.9	1.3	0.9	0.6	1.3	2.5
1.5	1.1	1.1	2.2	0.9	1.8	1.5	1.2	1.6	2.1	6.6	4.0	2.5
1.4	1.4	1.8	1.1	1.6	3.7	0.6	2.7	2.6	3.0	1.2	1.0	1.6
3.1	2.4	2.1	2.7	1.2	3.3	3.8	1.3	2.1	6.6	1.2	3.1	0.5
0.3	0.5	3.4	3.5	2.7	1.9	5.9	4.2	3.5	3.6	3.1	3.3	4.6

a. Construct a split stem-and-leaf plot in which each stem appears twice, once for leaves 0–4 and again for leaves 5–9.

b. Compare the split stem-and-leaf plot to the plot in Figure 2.17. Comment on the advantages and disadvantages of the split stem-and-leaf plot for these data.

20. Blood pressure: The following table presents systolic blood pressures for 50 adults.

98	122	114	123	142	174	130	136	109	139
103	119	114	119	117	92	124	107	115	167
116	120	128	122	117	113	120	118	109	113
156	130	119	128	114	127	104	118	111	121
94	118	110	125	121	101	102	110	99	106

a. Construct a stem-and-leaf plot, using the numbers 9 through 17 as the stems.

b. Repeat part (a), but split the stems, using two lines for each stem.

c. Which stem-and-leaf plot do you think is more appropriate for these data, the one in part (a) or the one in part (b)? Why?

21. Tennis and golf: Following are the ages of the winners of the men's Wimbledon tennis championship and the Masters golf championship for the years 1972 through 2019.

Ages of Wimbledon Winners

25	27	21	31	20	21	22	23	24	22	29	24
25	17	18	22	22	21	24	22	22	21	22	23
26	25	26	27	28	31	21	21	22	23	24	25
22	27	24	24	30	26	27	28	29	35	31	32

Ages of Masters Winners

32	36	38	35	33	27	42	27	23	31	28	26
32	27	46	28	30	31	32	33	32	35	28	43
38	23	41	33	37	25	26	32	31	29	33	31
28	39	39	26	33	32	35	21	28	37	27	43

a. Construct back-to-back split stem-and-leaf plots for these data sets.

b. How do the ages of Wimbledon champions differ from the ages of Masters champions? *Wimbledon champions are younger.*

22. Pass the popcorn: Following are the running times (in minutes) for the top 15 domestic grossing movies of all time rated PG-13 and the top 15 domestic grossing movies of all time rated R.

Movies Rated PG-13	
Star Wars: The Force Awakens	136
Avengers: Endgame	182
Avatar	162
Black Panther	135
Avengers: Infinity War	160
Titanic	194
Jurassic World	124
Marvel's The Avengers	143
Star Wars: The Last Jedi	153
The Dark Knight	152
Rogue One: A Star Wars Story	133
Avengers: Age of Ultron	141
The Dark Knight Rises	164
Captain Marvel	125
The Hunger Games: Catching Fire	146

Source: Box Office Mojo

Movies Rated R	
The Passion of the Christ	126
Deadpool	106
American Sniper	132
It	135
Deadpool 2	134
The Matrix Reloaded	138
The Hangover	96
The Hangover Part II	102
Beverly Hills Cop	105
The Exorcist	122
Logan	141
Ted	106
Saving Private Ryan	170
A Star is Born	134
300	117

Source: Box Office Mojo

a. Construct back-to-back stem-and-leaf plots for these data sets.

b. Do the running times of R-rated movies differ greatly from the running times of movies rated PG or PG-13, or are they roughly similar? *They differ greatly.*

23. More weather: Construct a dotplot for the data in Exercise 18. Are there any gaps in the data? *Yes*

24. Safety first: Following are the numbers of hospitals in each of the 50 U.S. states plus the District of Columbia that won Patient Safety Excellence Awards. Construct a dotplot for these data and describe its shape. *Skewed right*

2	0	9	3	24	6	1	0	1	14	3
0	2	10	10	11	3	1	4	0	5	12
5	12	0	3	4	0	0	0	5	1	7
11	2	15	3	5	20	1	2	1	5	16
2	0	8	6	0	8	0				

25. Looking for a job: The following table presents the U.S. unemployment rate for each of the years 1996 through 2019.

Year	Unemployment	Year	Unemployment
1996	5.4	2008	5.8
1997	4.9	2009	9.3
1998	4.5	2010	9.6
1999	4.2	2011	8.9
2000	4.0	2012	8.1
2001	4.7	2013	7.4
2002	5.8	2014	6.1
2003	6.0	2015	5.4
2004	5.5	2016	4.9
2005	5.1	2017	4.4
2006	4.6	2018	3.9
2007	4.6	2019	3.8

Source: National Bureau of Labor Statistics

a. Construct a time-series plot of the unemployment rate.

b. For which periods of time was the unemployment rate increasing? For which periods was it decreasing? *Increasing: 2000–2003, 2007–2010. Decreasing: 1996–2000, 2003–2007, 2010–2019*

26. Vacant apartments: The following table presents the percentage of U.S. residential rental units that were vacant during June and December of each year from 2011 through 2018.

Quarter	Vacancy Rate	Quarter	Vacancy Rate
Jun. 2011	9.2	Jun. 2015	6.8
Dec. 2011	9.4	Dec. 2015	7.0
Jun. 2012	8.6	Jun. 2016	6.7
Dec. 2012	8.7	Dec. 2016	6.9
Jun. 2013	8.2	Jun. 2017	7.3
Dec. 2013	8.2	Dec. 2017	6.9
Jun. 2014	7.5	Jun. 2018	6.8
Dec. 2014	7.0	Dec. 2018	6.8

Source: Current Population Survey

a. Construct a time-series plot for these data.
b. From 2011 through 2014, the proportion of Americans who owned a home declined. What was the trend in the vacancy rate during this time period? *Decreasing*

27. Military spending: The following table presents the amount spent, in billions of dollars, on national defense by the U.S. government every other year for the years 1951 through 2019. The amounts are adjusted for inflation, and represent 2019 dollars.

Year	Spending	Year	Spending
1951	524.4	1987	617.2
1953	554.5	1989	593.0
1955	429.4	1991	583.8
1957	456.8	1993	494.7
1959	450.7	1995	448.0
1961	456.2	1997	427.2
1963	490.9	1999	434.6
1965	466.1	2001	461.4
1967	593.4	2003	614.0
1969	596.9	2005	660.3
1971	497.0	2007	748.8
1973	455.4	2009	790.6
1975	421.1	2011	756.8
1977	446.0	2013	634.3
1979	442.4	2015	647.5
1981	495.2	2017	677.6
1983	592.4	2019	686.1
1985	634.1		

Source: Department of Defense

a. Construct a time-series plot for these data.
b. The plot covers seven decades, from the 1950s through the 2010s. During which of these decades did national defense spending increase, and during which decades did it decrease? *Increased: 60s, 80s, 00s; decreased: 50s, 70s, 90s, 10s*
c. The United States fought in the Korean War, which ended in 1953. What effect did the end of the war have on military spending after 1953? *Decreased*
d. During the period 1965–1968, the United States steadily increased the number of troops in Vietnam from 23,000 at the beginning of 1965 to 537,000 at the end of 1968. Beginning in 1969, the number of Americans in Vietnam was steadily reduced, with the last of them leaving in 1975. How is this reflected in the national defense spending from 1965 to 1975? *Increased 65–69, then decreased*

28. College students: The following table presents the numbers of male and female students (in thousands) enrolled in college in the United States as undergraduates for each of the years 2000 through 2019.

Year	Male	Female	Year	Male	Female
2000	5778	7377	2010	7836	10246
2001	6004	7711	2011	7823	10254
2002	6192	8065	2012	7715	10021
2003	6227	8253	2013	7660	9815
2004	6340	8441	2014	7586	9707
2005	6409	8555	2015	7502	9544
2006	6514	8671	2016	7417	9458
2007	6728	8876	2017	7347	9413
2008	7067	9299	2018	7372	9441
2009	7563	9901	2019	7399	9478

Source: National Center for Educational Statistics

a. Construct a time-series plot for the male enrollment; then on the same axes, construct a time-series plot for the female enrollment.
b. Which grew faster from 2000–2010, male enrollment or female enrollment? *Female*

29. House prices: The following time-series plot presents the average price of houses sold in the United States during the first quarter of each of the years 2004–2019.

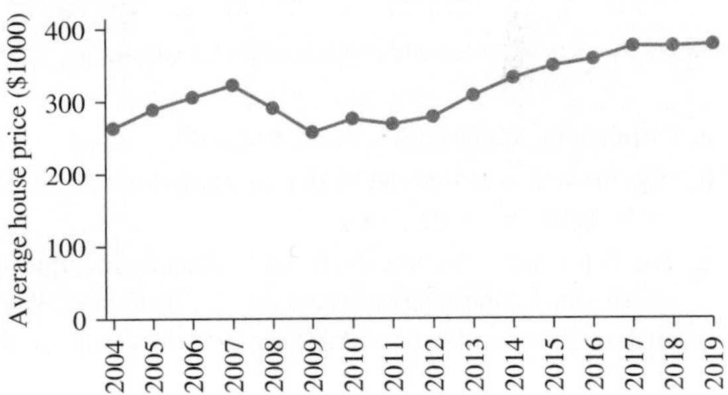

Source: Federal Reserve Bank of St. Louis

a. Estimate the average house price in 2005. *$300,000*
b. Was the average price in 2006 greater than, less than, or about the same as the average price in 2013? *About the same*
c. True or false: The average price in 2019 exceeded the average price in 2004 by more than $100,000. *True*
d. In 2008, an economic downturn known as the Great Recession occurred. What was the effect on the average house price? *It decreased*

30. Going for gold: The following time-series plot presents the number of Summer Olympic events in which the United States won a gold medal in each Olympic year from 1952 through 2016.

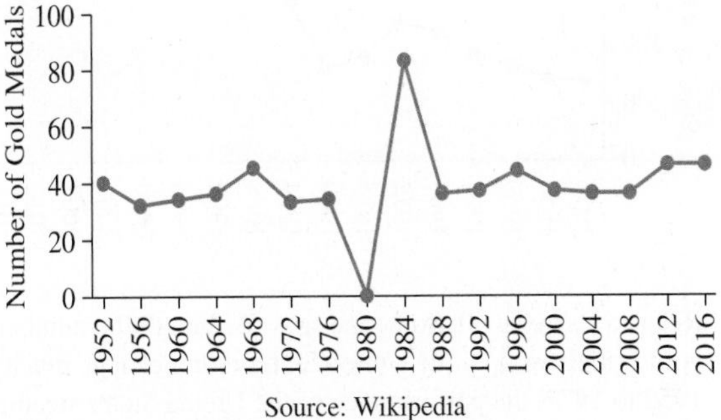

Source: Wikipedia

a. In one year, the United States did not participate in Summer games that were held in Moscow, in protest of the invasion of Afghanistan by the Soviet Union. Which year was this? *1980*

b. In 1984, the Soviet Union did not participate in the Summer games held in Los Angeles, citing "undisguised threats" against their athletes. Estimate the number of gold medals won by the United States in that year. *85*

c. Other than 1980 and 1984, has the number of gold medals won by the United States been generally increasing, generally decreasing, or staying about the same? *About the same*

31. Birth rates: The following time-series plot presents the birth rate (number of births per 1000 population) for the United States for the years 1909–2018.

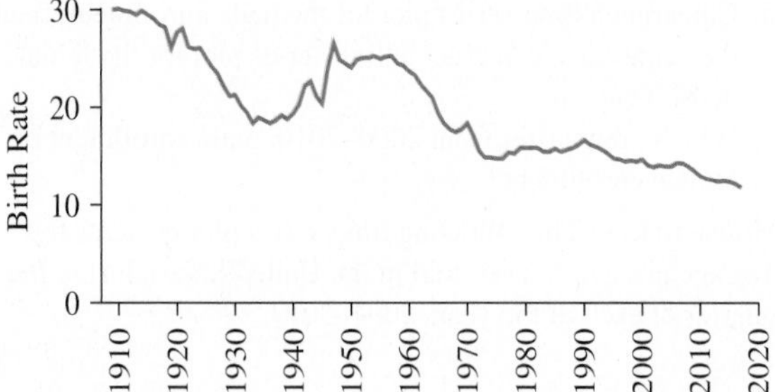

Sources: Centers for Disease Control, Statista, Infoplease

a. Estimate the highest birth rate since 1909. *30 or 31*

b. Was the first year that the birth rate was below 20 closer to 1930, 1950, or 1970? *1930*

c. The 1930s were the time of the Great Depression, during which many Americans were financially distressed. Was the birth rate during this period greater than, less than, or about the same as in the 1920s and 1940s? *Less than*

d. In 1945 World War II ended, and many soldiers returned home to their families. What was the effect on the birth rate? (*Hint:* This period is known as the "baby boom.") *It increased*

e. True or false: The birth rate in 2018 was lower than at any point in the previous 100 years. *True*

32. More gold: The following time-series plot presents the number of countries participating in the Summer Olympic games in each Olympic year from 1952 through 2016.

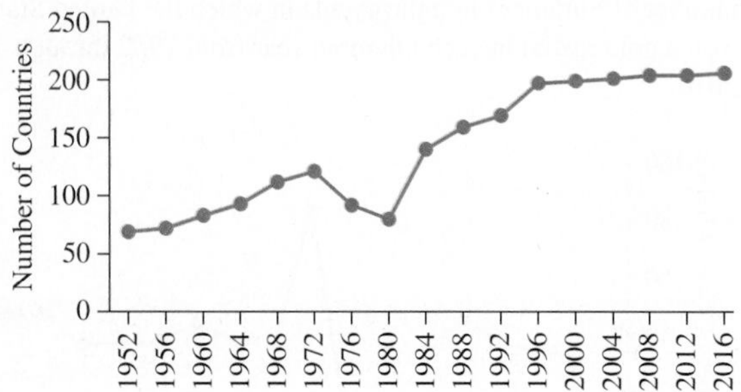

Refer to Exercise 30. Someone says "Although the number of gold medals won by the United States didn't change much from 1952 to 1972, the performance of the United States steadily improved during that period." Which feature of the plot of the number of participating countries justifies that statement?

33. Let's go skiing: The following time-series plot presents the number of inches of snow falling in Denver each year from 1882 through 2018.

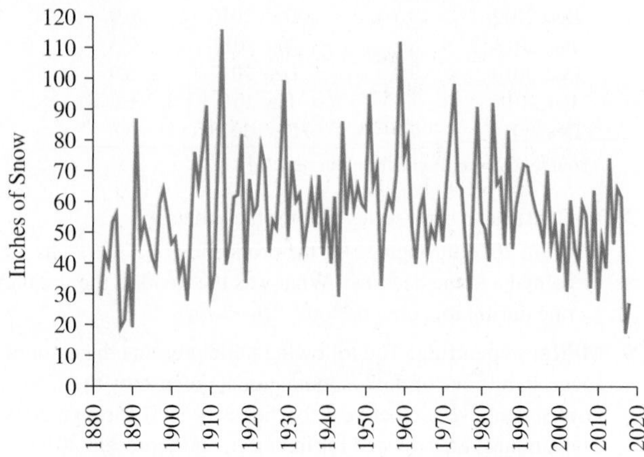

Source: National Weather Service

a. Estimate the greatest annual snowfall ever recorded in Denver. *115*

b. Was the year of the greatest annual snowfall closest to 1900, 1910, or 1920? *1910*

c. Was the amount of snowfall in the years 2000–2010 greater than, less than, or about equal to the snowfall in most other years? *Less than*

d. True or false: The year with the least snowfall ever recorded in Denver was in the 2010s. *True*

e. True or false: It usually snows more than 80 inches per year in Denver. *False*

samot/ 123RF

34. Three-point shot: The following time-series plot presents the average number of three-point shots made in a National Basketball Association game for seasons ending in 1980 through 2019.

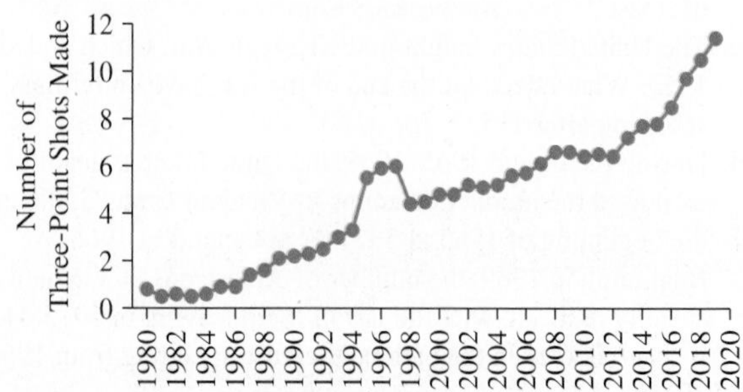

Source: Basketball-Reference.com

a. In 1997 the average number of three-point shots per game was greater than six for the first time. What was the next year that it was greater than six? *2007*

b. True or false: Since the year 2000, the average number of three-point shots made per game has increased every year. *False*

c. In 1995 the distance from the three-point line to the basket was reduced from 23 feet 9 inches to 22 feet. In 1998 the distance was restored to 23 feet 9 inches. What was the effect of these rule changes? *The number of shots increased, then decreased.*

35. Vote: The following time-series plot presents the percentage of the total U.S. population that voted in each presidential election from 1824 through 2016.

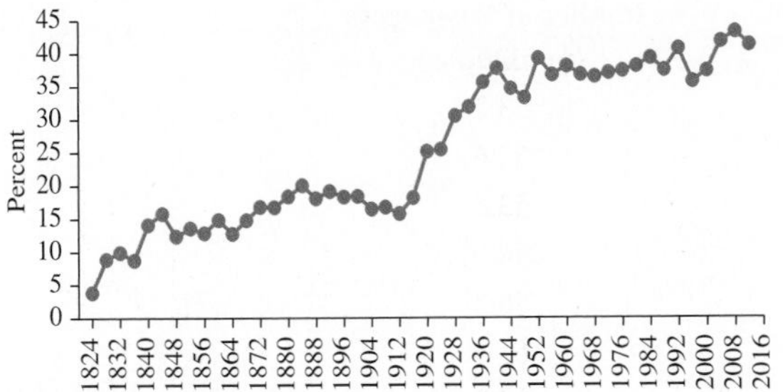

a. In 1824, the only people eligible to vote were white men who owned property. Approximately what percentage of the total population voted in 1824? *4%*

b. In 1828, voting privileges were extended to all white men, whether or not they owned property. What was the effect on the percentage of the population that voted? *It increased*

c. During the 1890s, many Southern states passed laws that effectively prevented most African Americans and many poor whites from voting. What was the effect on the percentage of the population that voted? *It decreased*

d. In 1920, women obtained the right to vote. What was the effect on the percentage of the population that voted? *It increased*

36. Arctic ice sheet: The following table presents the extent of ice coverage (in millions of square kilometers) in the Arctic region in September of each year from 2002 through 2018.

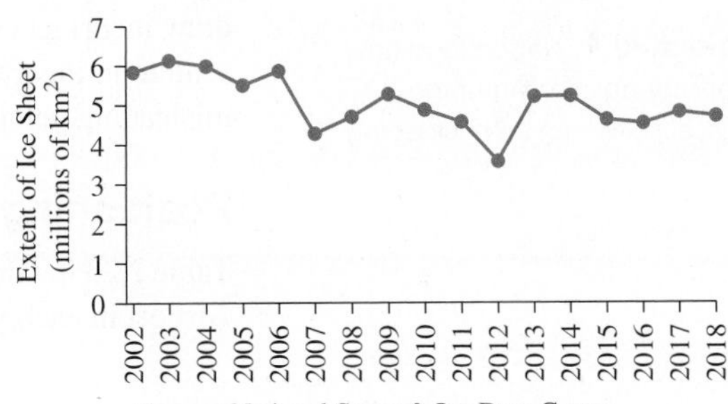

Source: National Snow & Ice Data Center

a. What was the first year that the coverage dropped below 5 million square kilometers? *2007*

b. What was the first year that the coverage dropped below 4 million square kilometers? *2012*

c. True or false: The coverage has been less than 5 million square kilometers in every year since 2011. *False*

d. True or false: The coverage has decreased in every year since 2011. *False*

Extending the Concepts

37. Elections: In U.S. presidential elections, each of the 50 states casts a number of electoral votes equal to its number of senators (2) plus its number of members of the House of Representatives. In addition, the District of Columbia casts three electoral votes. Following are the numbers of electoral votes cast for president for each of the 50 states and the District of Columbia in the election of 2020.

9	3	11	6	55	9	7	3	29	16	4
4	20	11	6	6	8	8	4	10	11	16
10	6	10	3	5	6	4	14	5	29	15
3	18	7	7	20	4	9	3	11	38	6
3	13	12	5	10	3	3				

a. Construct a split stem-and-leaf plot for these data, using two lines for each stem.

b. Construct a frequency histogram, with the classes chosen so that there are two classes for each stem.

c. Explain why the stem-and-leaf plot and the histogram have the same shape.

Answers to Check Your Understanding Exercises for Section 2.3

1. a.

12	
13	
14	
15	
16	
17	
18	
19	

b.

12	4494
13	63
14	9583
15	70156857
16	330601
17	12
18	2
19	552

c.

12	4449
13	36
14	3589
15	01556778
16	001336
17	12
18	2
19	255

2. a. 2018 **b.** 2006 **c.** False **d.** 1998–2018 **e.** 1994–1998

Section 2.4 Graphs Can Be Misleading

Objectives

1. Understand how improper positioning of the vertical scale can be misleading
2. Understand the area principle for constructing statistical graphs
3. Understand how three-dimensional graphs can be misleading

Statistical graphs, when properly used, are powerful forms of communication. Unfortunately, when graphs are improperly used, they can misrepresent the data and lead people to

Objective 1 Understand how improper positioning of the vertical scale can be misleading

draw incorrect conclusions. We discuss here three of the most common forms of misrepresentation: incorrect position of the vertical scale, incorrect sizing of graphical images, and misleading perspective for three-dimensional graphs.

Positioning the Vertical Scale

Table 2.23 is a distribution of the number of passengers, in millions, at Denver International Airport in each year from 2012 through 2018.

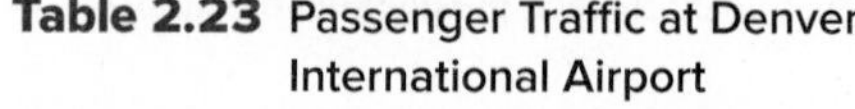
Table 2.23 Passenger Traffic at Denver International Airport

Year	Number of Passengers (in millions)
2012	53.2
2013	52.6
2014	53.5
2015	54.0
2016	58.3
2017	61.4
2018	64.5

Source: Denver International Airport

Antonio Saba

In order to get a better picture of the data, we can make a bar graph. Figures 2.23 and 2.24 present two different bar graphs of the same data. Figure 2.23 presents a clear picture of the data. We can see that the number of passengers has been fairly steady, with just a slight increase from 2012 through 2018. Now imagine that someone was eager to persuade us that passenger traffic had increased greatly since 2012. If they were to show us Figure 2.23, we wouldn't be convinced. So they might show us a misleading picture like Figure 2.24 instead. Figure 2.24 gives the impression of a truly dramatic increase.

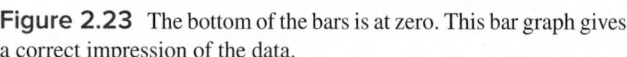

Figure 2.23 The bottom of the bars is at zero. This bar graph gives a correct impression of the data.

Figure 2.24 The bottom of the bars is not at zero. This bar graph exaggerates the differences between the bars.

Figures 2.23 and 2.24 are based on the same data. Why do they give such different impressions? The reason is that the baseline (the value corresponding to the bottom of the bars) is at zero in Figure 2.23, but not at zero in Figure 2.24. This exaggerates the differences between the bars. For example, in Figure 2.24, the bar for the year 2018 is more than three times as long as the bar for the year 2012, but the actual increase in passenger traffic is much less than that.

This sort of misleading information can be created with time-series plots as well. Figures 2.25 and 2.26 (page 85) present two different time-series plots of the data. In Figure 2.25, the baseline is at zero, so an accurate impression is given. In Figure 2.26, the baseline is larger than zero, so the rate of increase is exaggerated.

When a graph or plot represents how much or how many of something, check the baseline. If it isn't at zero, the graph may be misleading.

Figure 2.25 The baseline is at zero. This plot gives an accurate picture of the data.

Figure 2.26 The baseline is not at zero. This plot exaggerates the rate of increase.

Objective 2 Understand the area principle for constructing statistical graphs

The Area Principle

We often use images to compare amounts. Larger images correspond to greater amounts. To use images properly in this way, we must follow a rule known as the **area principle**.

The Area Principle

When amounts are compared by constructing an image for each amount, the *areas* of the images must be proportional to the amounts. For example, if one amount is twice as much as another, its image should have twice as much area as the other image.

When the area principle is violated, the images give a misleading impression of the data.

Bar graphs, when constructed properly, follow the area principle. The reason is that all the bars have the same width; only their height varies. Therefore, the areas of the bars are proportional to the amounts. For example, Figure 2.27 presents a bar graph that illustrates a comparison of the cost of jet fuel in 2016 and 2019. In 2016, the cost of jet fuel was \$0.93 per gallon, and in 2019 it had increased to \$1.78 per gallon.

The bars in the bar graph differ in only one dimension—their height. The widths are the same. For this reason, the bar graph presents an accurate comparison of the two prices. The price in 2016 is about half the price in 2019, and the area of the bar for 2016 is about half the area of the bar for 2019.

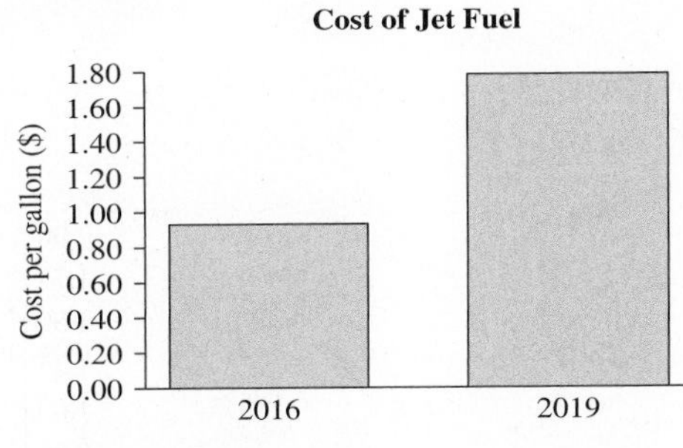

Figure 2.27 Price per gallon of jet fuel in 2016 and 2019. The bar graph accurately represents the difference.

Unfortunately, people often mistakenly vary both dimensions of an image when making a comparison. This exaggerates the difference. Following is a comparison of the cost of jet fuel in the years 2016 and 2019 that uses a picture of an airplane to illustrate the difference.

The pictures of the planes make the difference appear much larger than the correctly drawn bar graph does. The reason is that both the height and the width of the airplane have approximately doubled. Thus the area of the larger plane is about four times the area of the smaller plane. This graph violates the area principle and gives a misleading impression of the comparison.

Check Your Understanding

1. The population of country A is twice as large as the population of country B. True or false: If images are used to represent the populations, both the height and width of the image for country A should be twice as large as the height and width of the image for country B. *False*

2. If the baseline of a bar graph or time-series plot is not at zero, then the differences may appear to be ________ than they actually are. (*Choices: larger, smaller*) *larger*

Answers are on page 90.

Objective 3 Understand how three-dimensional graphs can be misleading

Three-Dimensional Graphs and Perspective

The bar graph in Figure 2.27 presents an accurate picture of the prices of jet fuel in the years 2016 and 2019. Newspapers and magazines often prefer to present three-dimensional bar graphs, because they are visually more impressive. Unfortunately, in order to make the tops of the bars visible, these graphs are often drawn as though the reader is looking down on them. This can make the bars look shorter than they really are.

Figure 2.28 presents a three-dimensional bar graph of the sort often seen in publications. The data are the same as in Figure 2.27: The price in 2016 is \$0.93, and the price in 2019 is \$1.78. However, because you are looking down on the bars, they appear shorter than they really are.

Beware of three-dimensional bar graphs. If you can see the tops of the bars, they may look shorter than they really are.

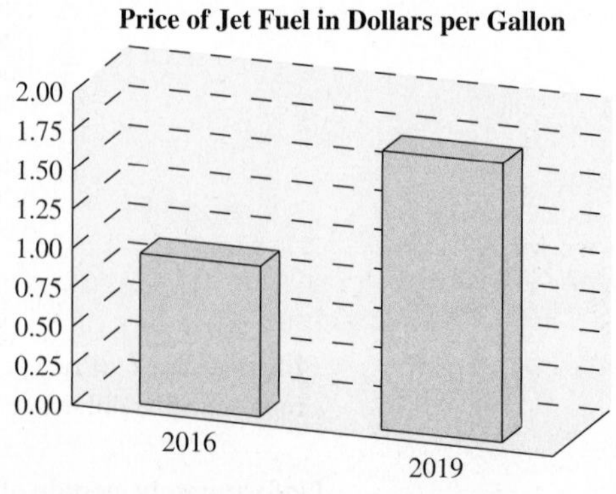

Figure 2.28 Price per gallon of jet fuel in 2016 and 2019. The bars appear shorter than they really are, because you are looking down at them.

Section 2.4 Exercises

Exercises 1 and 2 are the Check Your Understanding exercises located within the section.

Understanding the Concepts

In Exercises 3 and 4, fill in each blank with the appropriate word or phrase.

3. A plot that represents how much of something there is may be misleading if the baseline is not at ________. *0*

4. The area principle says that when images are used to compare amounts, the areas of the images should be ________ to the amounts. *proportional*

Working with the Concepts

5. **Cable subscriptions decline:** The number of cable television subscribers has been declining in recent years. Following are two bar graphs that illustrate the decline. (Source: Business Insider)

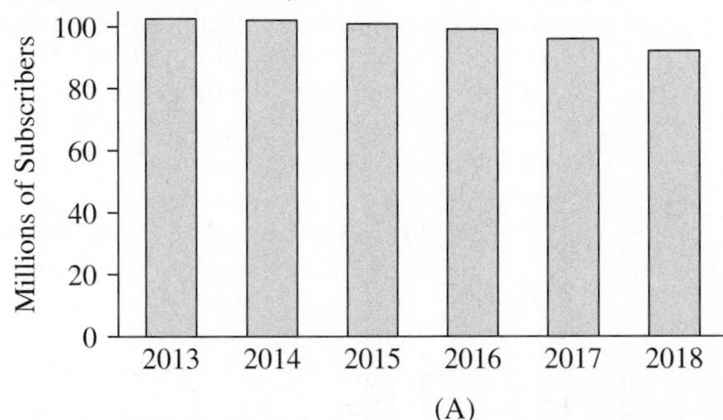

(A)

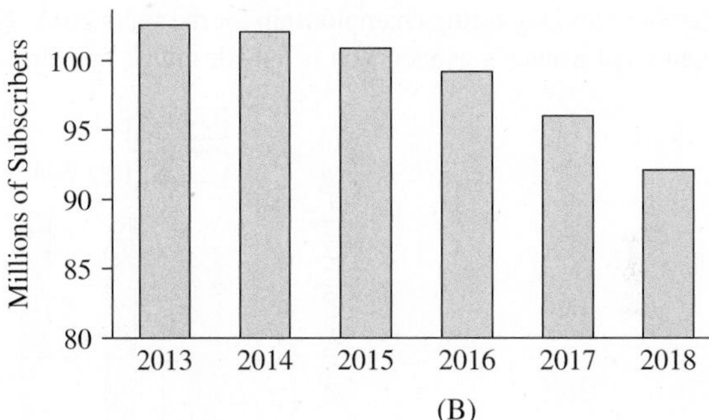

(B)

Choose one of the following options, and explain why it is correct: *(i)*

(i) Graph A presents an accurate picture, and graph B exaggerates the decline.

(ii) Graph B presents an accurate picture, and graph A understates the decline.

6. **Music sales:** The following time-series plot and bar graph both present the sales of digital music for the years 2013–2018. Which of the graphs presents the more accurate picture? Why? *Bar graph*

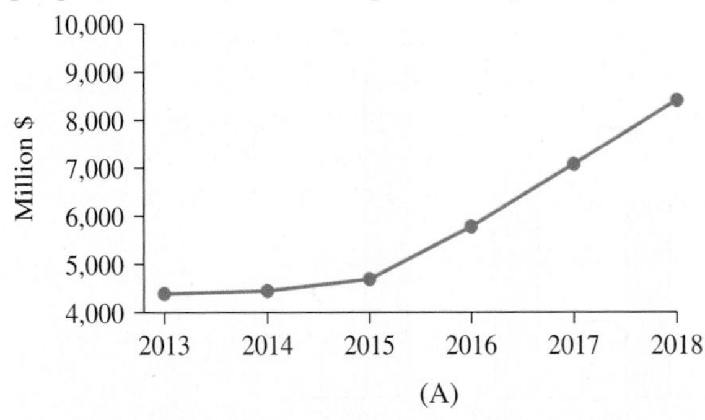

(A)

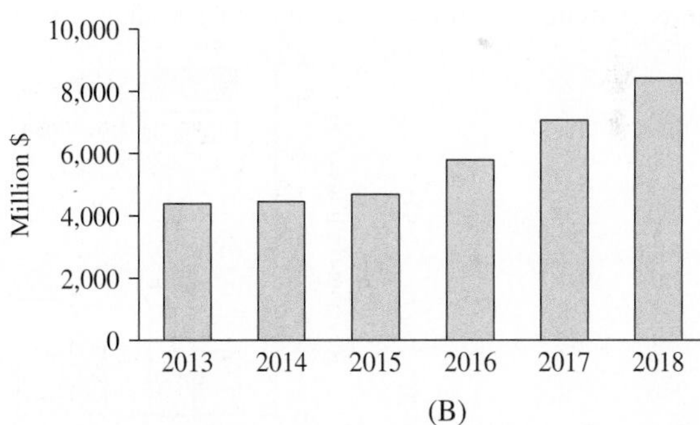

(B)

7. **Stock market prices:** The Dow Jones Industrial Average reached its lowest point in recent history on October 9, 2008, when it closed at \$8,579. Ten years later, on October 9, 2018, the average had risen to \$26,486.78. Which of the following graphs accurately represents the magnitude of the increase? Which one exaggerates it? *Bar graph is more accurate*

October 9, 2008

October 9, 2018

McGraw Hill Education/Ken Cavanagh

(A)

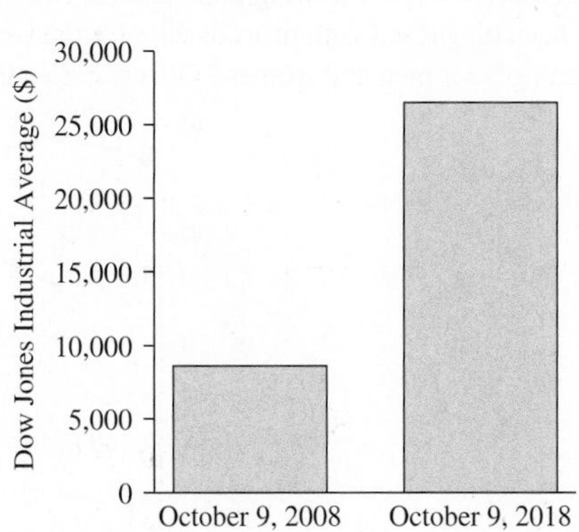

(B)

8. Gross domestic product: The gross domestic product (GDP) of the United States is the total value of all goods and services produced in the country. In 2000, the GDP was $10.0 trillion. In 2019, the GDP was $21.0 trillion, slightly more than twice as much. Which of the following graphs compares these totals more accurately, and why? (Source: St. Louis Federal Reserve) *B*

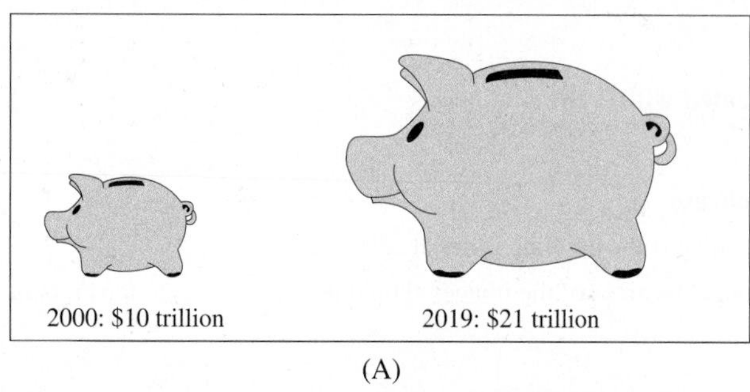

(A)

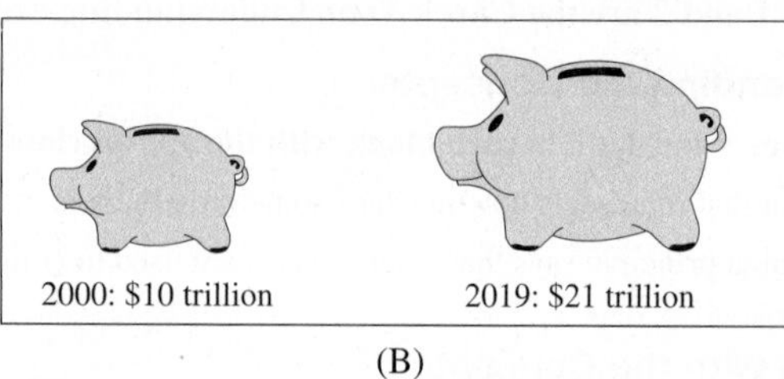

(B)

9. I'll take mine with mustard: The following bar graph presents the number of hot dogs eaten by the men's and women's winner of Nathans Famous Hot Dog eating championship for the years 2011–2019. Does the graph present an accurate picture of the difference between the men's and women's winners? Or is it misleading? Explain *Misleading*

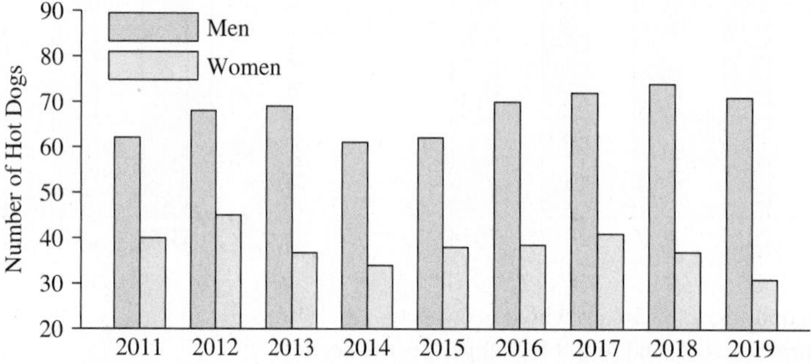

10. Stream or download? The following bar graph presents the revenue (in billions of $) for the music industry from music streaming (including subscriptions) and music downloading for the years 2013–2018. Does the graph present an accurate picture of the differences in revenue from these two sources? Or is it misleading? Explain *Accurate*

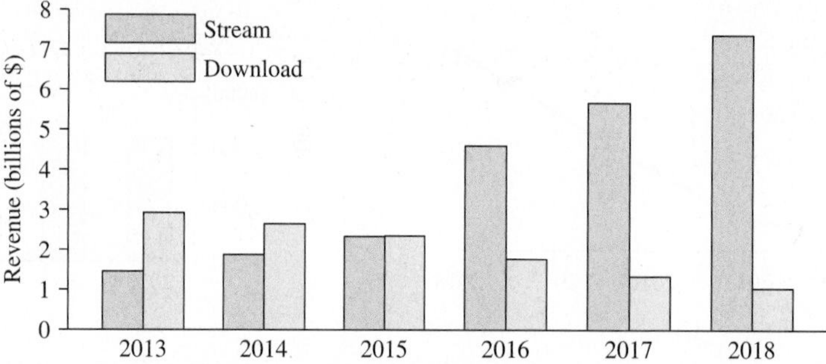

11. Heart disease: The following plot contains two time-series graphs presenting the percentages of men and women over the age of 65 who have been diagnosed with heart disease for the years 2006–2016. Does the plot present an accurate picture of the differences between the percentages for men and women? Or is it misleading? Explain? *Accurate*

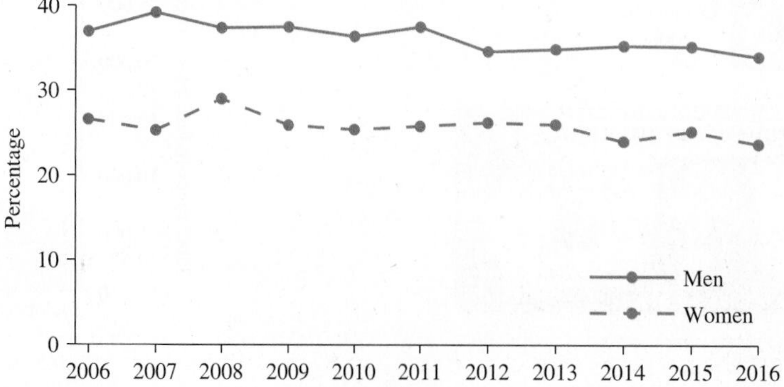

12. Strike three: The plot contains two time-series graphs presenting the number of strikeouts in both the American and National League for the years 2001–2018. There have been more strikeouts in the National League in each of those years. Does the plot present an accurate picture of the differences in the numbers of strikeouts? Or is it misleading? Explain. *Misleading*

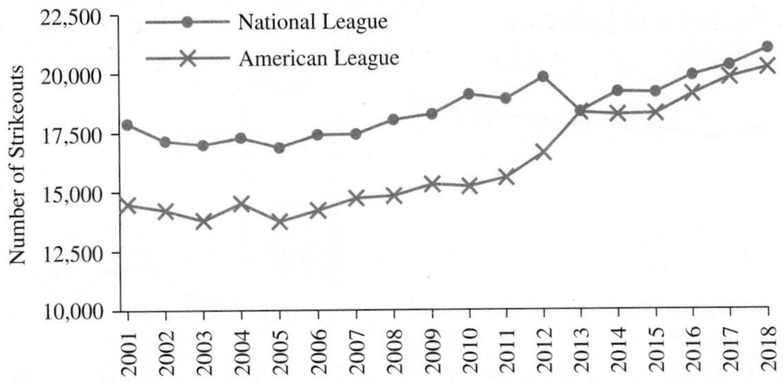

13. Female senators: Of the 100 members of the United States Senate recently, 75 were men and 25 were women. The following three-dimensional bar graph attempts to present this information.

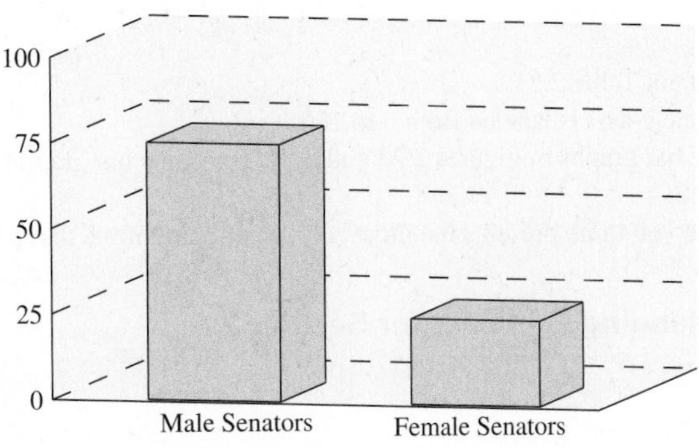

a. Explain how this graph is misleading.
b. Construct a graph (not necessarily three-dimensional) that presents this information accurately.

14. Age at marriage: Data compiled by the U.S. Census Bureau suggests that the age at which women first marry has increased over time. The following time-series plot presents the average age at which women first marry for the years 2005–2018. Does the plot present an accurate picture of the increase, or is it misleading? Explain. *Misleading*

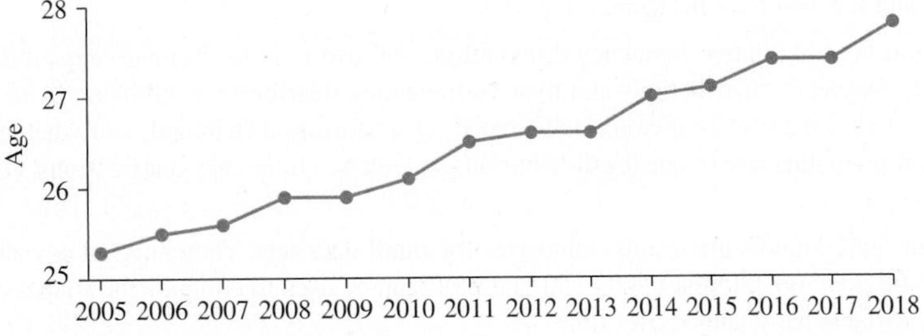

15. College degrees: Both of the following time-series plots present the percentage of U.S. adults who have earned college degrees for the years 2009–2017.

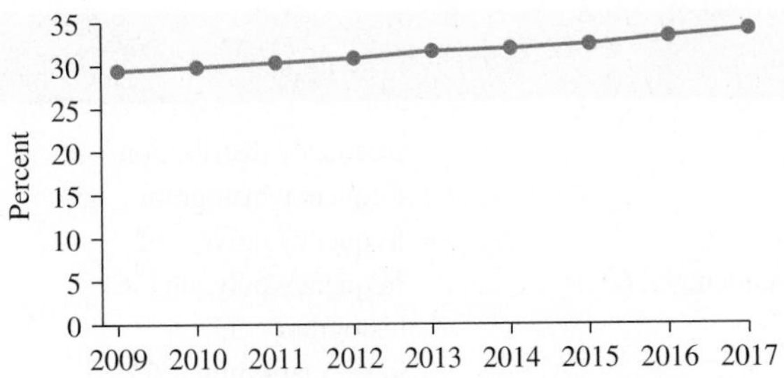

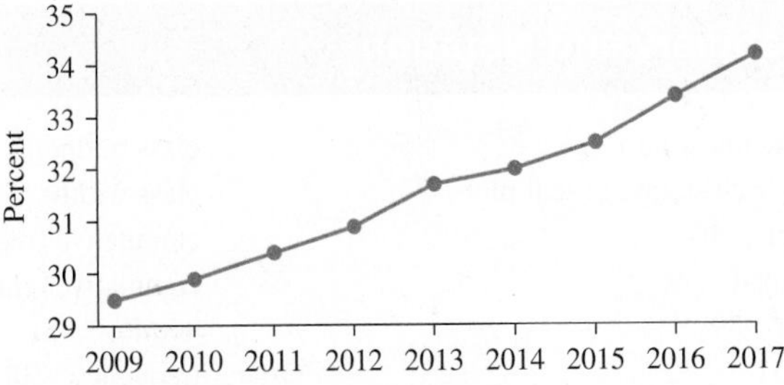

Which of the following statements is true and why? *i*

i. The percentage of U.S. adults with college degrees increased slightly between 2009 and 2017.

ii. The percentage of U.S. adults with college degrees increased considerably between 2009 and 2017.

16. Food expenditures: Both of the following time-series plots present the average amount spent on food by U.S. residents for the years 2006 through 2017. (Source: U.S. Department of Agriculture)

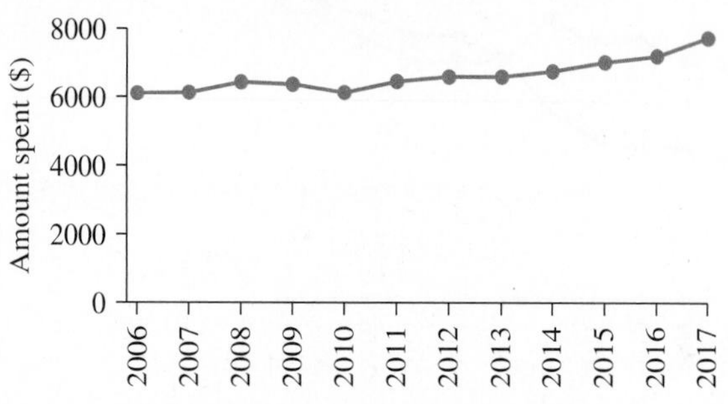

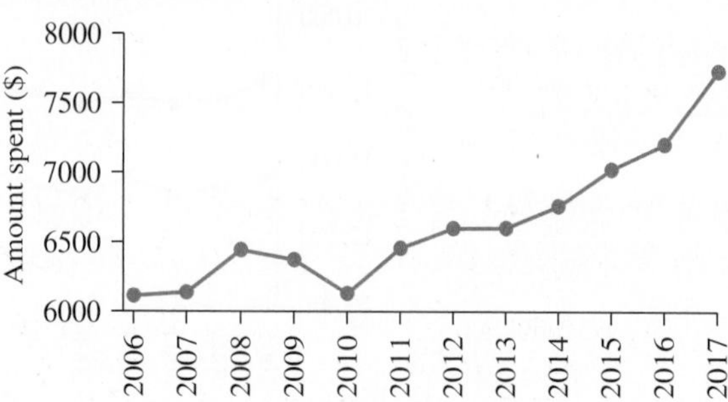

Which of the following statements is true and why? *ii*

(i) The amount spent on food increased considerably between 2006 and 2017.

(ii) The amount spent on food increased slightly between 2006 and 2017.

Extending the Concepts

17. Manipulating the *y*-axis: For the data in Table 2.23:

a. Construct a bar graph in which the *y*-axis is labeled from 0 to 100.

b. Compare this bar graph with the bar graphs in Figures 2.23 and 2.24. Does this bar graph tend to make the difference seem smaller than the other bar graphs do? *Yes*

c. Which of the three bar graphs do you think presents the most accurate picture of the data? Why?

Answers to Check Your Understanding Exercises for Section 2.4

1. False **2.** Larger

Chapter 2 Summary

Section 2.1: The first step in summarizing qualitative data is to construct a frequency distribution or relative frequency distribution. Then a bar graph or pie chart can be constructed. Bar graphs can illustrate either frequencies or relative frequencies. Side-by-side bar graphs can be used to compare two qualitative data sets that have the same categories.

Section 2.2: Frequency distributions and relative frequency distributions are also used to summarize quantitative data. Histograms are graphical summaries that illustrate frequency distributions and relative frequency distributions, allowing us to visualize the shape of a data set. Histograms can show us whether a data set is skewed, bell-shaped, or uniformly distributed, and whether it is unimodal or bimodal. Frequency polygons can also be used to illustrate frequency distributions and relative frequency distributions. Ogives illustrate cumulative relative frequency distributions.

Section 2.3: Stem-and-leaf plots and dotplots are useful summaries for small data sets. They have an advantage over histograms: They allow every point in the data set to be seen. Back-to-back stem-and-leaf plots can be used to compare the shapes of two data sets. Time-series plots illustrate how the value of a variable has changed over time.

Section 2.4: To avoid constructing a misleading graph, be sure to start the vertical scale at zero. When images are used to compare amounts, the area principle should be followed. This principle states that the areas of the images should be proportional to the amounts. Three-dimensional bar graphs are often misleading, because the bars look shorter than they really are.

Vocabulary and Notation

area principle 85
back-to-back stem-and-leaf plot 74
bar graph 40
bell-shaped 59
bimodal 59
class 51
class midpoint 61
class width 53
cumulative frequency 62
cumulative relative frequency 62
dotplot 75
frequency 38
frequency distribution 38
frequency histogram 55
frequency ogive 62
frequency polygon 61
histogram 55
lower class limit 53

mode 59
negatively skewed 59
ogive 62
open-ended class 58
Pareto chart 41
pie chart 42
positively skewed 59
relative frequency 39
relative frequency distribution 39
relative frequency histogram 55
relative frequency ogive 62
relative frequency polygon 62
side-by-side bar graph 42
skewed 59
skewed to the left 59
skewed to the right 59
split stem-and-leaf plot 73
stem-and-leaf plot 71
symmetric 59
time-series plot 76
uniformly distributed 59
unimodal 59
upper class limit 53

Chapter Quiz

1. Following is the list of letter grades for students in an algebra class: A, B, F, A, C, C, A, B, D, F, D, A, A, B, C, F, B, D, C, A, A, A, F, B, C, A, C. Construct a frequency distribution for these data.

2. Construct a relative frequency distribution for the data in Exercise 1.

3. Construct a frequency bar graph for the data in Exercise 1.

4. Construct a pie chart for the data in Exercise 1.

5. The first class in a relative frequency distribution is 2.0–4.9, and there are six classes. Find the remaining five classes. What is the class width? *3*

6. True or false: A histogram can have more than one mode. *True*

7. A sample of 100 students was asked how many hours per week they spent studying. The following frequency distribution shows the results.

Number of Hours	Frequency
1.0–4.9	14
5.0–8.9	34
9.0–12.9	29
13.0–16.9	15
17.0–20.9	8

a. Construct a frequency histogram for these data.

b. Construct a relative frequency histogram for these data.

8. Construct a frequency polygon for the data in Exercise 7.

9. Construct a relative frequency ogive for the data in Exercise 7.

10. List the data in the following stem-and-leaf plot. The leaf represents the ones digit.

1	1 1 5 5 9 9 9
2	2 2 3 5 7 8
3	0 0 8
4	4 5 7 8
5	0 1 3 3 5 6 8

11. Following are the prices (in dollars) for a sample of coffee makers.

19 22 29 68 35 37 28 22 41 39 28

Construct a stem-and-leaf plot for these data.

12. Following are the prices (in dollars) for a sample of espresso makers.

99 50 31 65 50 99 70 40 25 56 30 77

Construct a back-to-back stem-and-leaf plot for these data and the data in Exercise 11.

13. Construct a dotplot for the data in Exercise 11.

14. The following table presents the percentage of Americans who use a phone exclusively, with no landline phone, for the years 2014–2017. Construct a time-series plot for these data.

Time Period	Percent
January–June 2014	43.1
July–December 2014	44.1
January–June 2015	46.7
July–December 2015	47.7
January–June 2016	49.0
July–December 2016	50.5
January–June 2017	52.0
July–December 2017	53.3

Source: National Health Interview Survey

15. According to the area principle, if one amount is twice as much as another, its image should have ________ as much area as the other image. *twice*

Review Exercises

1. **Trust your doctor:** The General Social Survey recently surveyed people to ask, "How much would you trust your doctor to put your health above costs?" The following relative frequency bar graph presents the results.

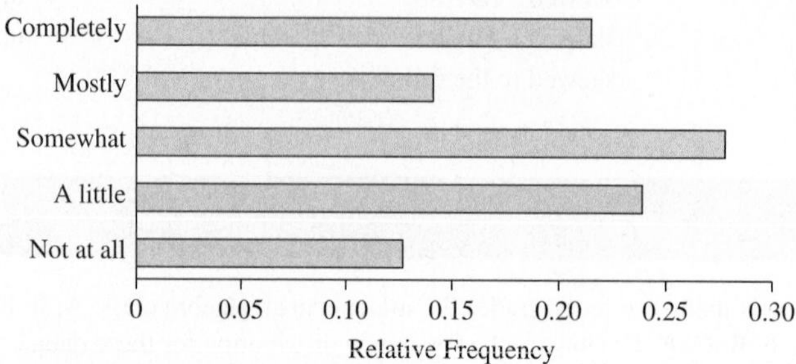

Source: General Social Survey

   a. Which was the most frequently given answer? *Somewhat*

   b. True or false: Less than one-fourth of the respondents said that they trusted their doctor completely. *True*

   c. True or false: More than half of the respondents said that they trusted their doctor either a little or not at all. *False*

   d. A total of 2719 people responded to this question. True or false: More than 500 of them said that they completely trusted or mostly trusted their doctor. *True*

2. **Phones:** The following relative frequency distribution presents the U.S. market share, in percent, of various phone companies in a recent year.

Company	Market Share
Apple	47
Samsung	22
LG	12
Motorola	6
Others	13

Source: Counterpoint Research

   a. Construct a relative frequency bar graph.

   b. Construct a pie chart.

   c. True or false: There are more than twice as many Apple users than Samsung users. *True*

3. **Poverty rates:** The following table presents the percentage of people who lived in poverty in the various regions of the United States in the years 2014 and 2017.

Region	Percent in 2014	Percent in 2017
Northeast	12.6	12.0
Midwest	13.0	12.7
South	16.5	14.8
West	15.2	12.9

Source: United States Census Bureau

   a. Construct a side-by-side bar graph for these data.

   b. True or false: The poverty rate was lower in 2017 than in 2014 in each region. *True*

   c. Which region had the greatest decrease? *West*

4. **Do your homework:** The National Survey of Student Engagement asked a sample of college freshmen how often they came to class without completing their assignments. Following are the results:

Response	Percent
Never	35
Sometimes	48
Often	12
Very often	5

   a. Construct a relative frequency bar graph.

   b. Construct a pie chart.

   c. True or false: More than half of the students reported that they sometimes come to class without completing their assignments. *False*

5. Quiz scores: The following frequency histogram presents the scores on a recent statistics quiz in a class of 50 students.

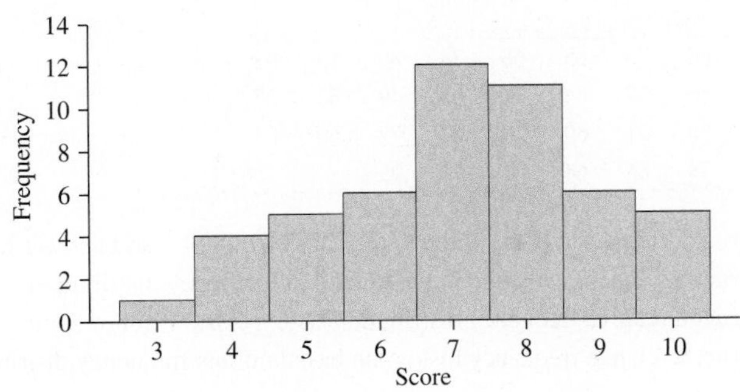

a. What is the most frequent score? *7*
b. How many students scored less than 6? *10*
c. What percentage of students scored 10? *10%*
d. Is the histogram more accurately described as unimodal or as bimodal? *Unimodal*

6. House freshmen: Newly elected members of the U.S. House of Representatives are referred to as "freshmen." The following frequency distribution presents the number of freshmen elected in each election from 1912 to 2016.

Number of Freshmen	Frequency
20–39	2
40–59	15
60–79	10
80–99	14
100–119	7
120–139	3
140–159	1
160–179	1

Source: Library of Congress

a. How many classes are there? *8*
b. What is the class width? *20*
c. What are the class limits?
d. Construct a frequency histogram.
e. Construct a relative frequency distribution.
f. Construct a relative frequency histogram.
g. In what percentage of elections were 100 or more freshmen elected? *22.6%*
h. In what percentage of elections were fewer than 60 freshmen elected? *32.1%*

7. More freshmen: For the data in Exercise 6:
a. Construct a frequency polygon.
b. Construct a relative frequency polygon.
c. Construct a frequency ogive.
d. Construct a relative frequency ogive.

8. Royalty: Following are the ages at death for all English and British monarchs since 1066.

59	40	67	58	56	28	41	49	65	68
43	64	33	46	35	49	40	12	32	52
55	15	42	69	58	48	54	67	51	49
67	76	81	67	71	81	68	70	77	56

a. Construct a frequency distribution with approximately eight classes.
b. Construct a frequency histogram based on this frequency distribution.
c. Construct a relative frequency distribution with approximately eight classes.
d. Construct a relative frequency histogram based on this frequency distribution.

9. More royalty: Construct a stem-and-leaf plot for the data in Exercise 8.

10. Presidents: Following are the ages at deaths for all U.S. presidents.

67	83	90	73	85	68	78	80	53	65
71	79	56	77	64	74	66	49	63	57
70	67	58	71	60	57	67	72	60	63
46	90	78	88	64	81	93	93		

a. Construct a frequency distribution with a class width of 5 and a lower limit of 45 for the first class.
b. Construct a frequency histogram based on this frequency distribution.
c. Construct a relative frequency distribution with a class width of 5 and a lower limit of 45 for the first class.
d. Construct a relative frequency histogram based on this frequency distribution.

11. Royalty and presidents: For the data in Exercises 8 and 10:
a. Construct a back-to-back stem-and-leaf plot.
b. Construct a back-to-back stem-and-leaf plot with split stems.
c. Which plot do you think is more appropriate for these data?

12. Dotplot: Construct a dotplot for the data in Exercise 10.

13. Pandora vs. Spotify: Following are the numbers of subscribers (in millions) to the music streaming services Pandora and Spotify for the years 2013 through 2019:

Year	Pandora	Spotify
2013	65.6	6.0
2014	76.2	10.0
2015	81.5	20.0
2016	81.1	30.0
2017	76.7	52.0
2018	72.3	75.0
2019	66.0	100.0

a. Construct a time-series plot for the number of Pandora subscribers.
b. Construct a time-series plot for the number of Spotify subscribers.
c. Describe the trends in the number of subscribers for both services.

14. Cancer rates: Cancer has long been the second most common cause of death (behind heart disease) in the United States. Cancer risk increases with age, so that older people are much more likely to be diagnosed with cancer than younger people. The following time-series plots present the number of deaths from cancer per 100,000 people, and the life expectancy, for the years 1930–2010.

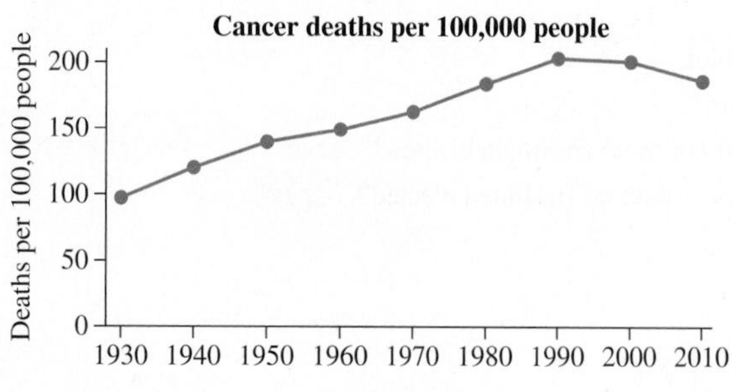

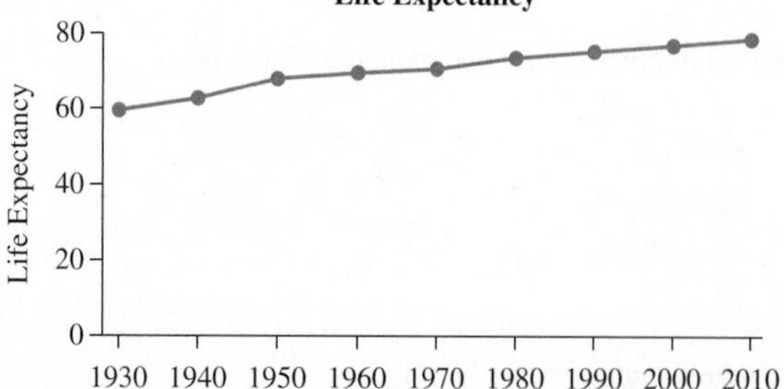

a. In which year was the death rate from cancer the highest? In which year was it the lowest? *1990, 1930*
b. Has life expectancy been increasing or decreasing during the years 1930–2010? *Increasing*

c. Treatments for cancer have been improving since 1930. Yet the death rate from cancer increased during the period 1930–1990. How can this be explained? *Life expectancy increased*

15. Falling birth rate: The following time-series plots both present estimates for the number of births per 1000 people worldwide for the years 2001–2017. (Source: The World Bank)

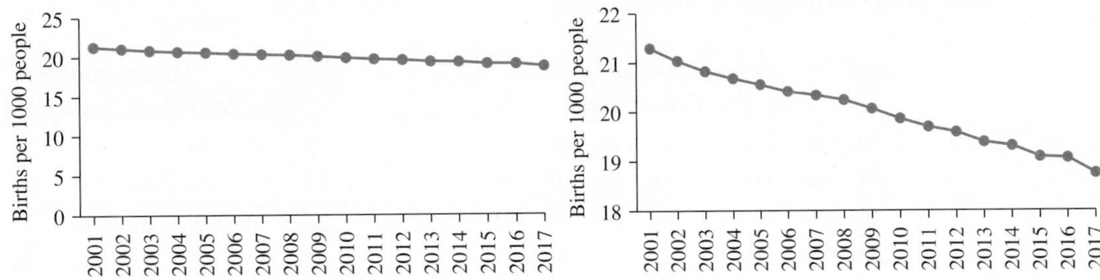

Which of the following statements is more accurate? Explain your reasoning. *i*

(i) The birth rate decreased somewhat between 2001–2017.

(ii) The birth rate decreased dramatically between 2001–2017.

Write About It

1. Explain why the frequency bar graph and the relative frequency bar graph for a data set have a similar appearance.
2. In what ways do frequency distributions for qualitative data differ from those for quantitative data?
3. Provide an example of a data set whose histogram you would expect to be skewed to the right. Explain why you would expect the histogram to be skewed to the right.
4. Time-series data are discrete when observations are made at regularly spaced time intervals. The time-series data sets in this chapter are all discrete. Time-series data are continuous when there are observations at extremely closely spaced intervals that are connected to provide values at every instant of time. An example of continuous time-series data is an electrocardiogram. Provide some examples of time-series data that are discrete and some that are continuous.

In-Class Activities

NOTE TO INSTRUCTOR

The activity involving drawing a histogram of dice sums can be used to foreshadow the Central Limit Theorem. Point out that the data consist of sums and such data sets generally have bell-shaped histograms when a sufficiently large number of values are summed. The notion of independence can be mentioned, although students may not be ready for a rigorous treatment of this concept.

1. **Shapes of histograms:** Each student tosses five dice and adds the numbers. Then each student tosses two dice and multiplies the numbers. Construct histograms for the results of each experiment. What are their shapes?
2. **Different graphs:** Collect data on variables for each student. Potential variables include height, number of siblings, number of pets, and number of classes currently taken. For each variable, construct a dotplot, a stem-and-leaf plot, and a histogram. Discuss the advantages and disadvantages of each graph.
3. **Misleading graphs:** Look through the internet, newspapers, or magazines for misleading graphs. Explain how they are misleading. Then find some that present accurate comparisons, and explain why you believe they are accurate.

Case Study: Do Hybrid Cars Get Better Gas Mileage?

In the chapter introduction, we presented gas mileage data for 2019 model year hybrid and small non-hybrid cars. We will use histograms and back-to-back stem-and-leaf plots to compare the mileages between these two groups of cars. The following tables present the mileages, in miles per gallon.

Mileage Ratings for 2019 Hybrid Cars							
40	28	21	43	23	56	22	28
28	48	31	41	19	26	39	48
26	46	19	30	55	23	29	41
33	21	42	50	22	42	50	29
26	29	34	46	52	42	30	41
43	58	25	25	44	44	22	19
27	49	21	23	52	41	24	52
24	23	46					

Source: www.fueleconomy.gov

Mileage Ratings for 2019 Small Non-hybrid Cars							
33	32	28	32	34	33	27	25
31	26	30	33	32	32	29	27
33	30	29	31	30	37	34	30
35	34	36	33	35	33	39	36
32	36	22	35	36	33	28	31
27	28	35	27	31	32	29	34
30	33	29	32	32	30	32	29
26	30	32	29	31	31	32	33

Source: www.fueleconomy.gov

1. Construct a frequency distribution for the hybrid cars with a class width of 2.
2. Explain why a class width of 2 is too narrow for these data.
3. Construct a relative frequency distribution for the hybrid cars with a class width of 3, where the first class has a lower limit of 18.
4. Construct a histogram based on this relative frequency distribution. Is the histogram unimodal or bimodal? Describe the skewness, if any, in these data. *Bimodal, little skewness*
5. Construct a frequency distribution for the non-hybrid cars with an appropriate class width.
6. Using this class width, construct a relative frequency distribution for the non-hybrid cars.
7. Construct a histogram based on this relative frequency distribution. Is the histogram unimodal or bimodal? Describe the skewness, if any, in these data. *Unimodal, skewed right*
8. Compare the histogram for the hybrid cars with the histogram for the non-hybrid cars. For which cars do the mileages vary more? *Hybrid*
9. Construct a back-to-back stem-and-leaf plot for these data, using two lines for each stem. Which do you think illustrates the comparison better, the histograms or the back-to-back stem-and-leaf plot? Why?

chapter

3

Numerical Summaries of Data

adistock/123RF

Introduction

How do manufacturers increase quality and reduce costs? Companies continually consider new ideas to produce higher-quality, lower-cost products. To determine whether a new idea can lead to higher quality or lower cost, data must be collected and analyzed.

The following tables present data produced by a manufacturer of computer chips, as described in the book *Statistical Case Studies for Industrial Process Improvement* by V. Czitrom and P. Spagon. Computer chips contain electronic circuits and are sealed with a thin layer of silicon dioxide. For the manufacturing process to work, the thickness of the layer must be carefully controlled. The manufacturer considered using recycled silicon wafers rather than new ones. Recycled wafers are much cheaper, so if the idea were feasible, it would lead to a reduction in cost. It must be determined whether the thicknesses of the oxide layers for recycled wafers are similar to those for the new wafers. The following tables present thickness measurements (in tenths of a nanometer) from some test runs.

New								
90.0	92.2	94.9	92.7	91.6	88.2	92.0	98.2	96.0
91.1	89.8	91.5	91.5	90.6	93.1	88.9	92.5	92.4
96.7	93.7	93.9	87.9	90.4	92.0	90.5	95.2	94.3
92.0	94.6	93.7	94.0	89.3	90.1	91.3	92.7	94.5

Recycled								
91.8	94.5	93.9	77.3*	92.0	89.9	87.9	92.8	93.3
92.6	90.3	92.8	91.6	92.7	91.7	89.3	95.5	93.6
92.4	91.7	91.6	91.1	88.0	92.4	88.7	92.9	92.6
91.7	97.4	95.1	96.7	77.5*	91.4	90.5	95.2	93.1

*Measurement is in error due to a defective gauge.

NOTE TO INSTRUCTOR

The following topic from the **Statistics Corequisite Workbook** is aligned with the material in this section.

2.2 - Order of Operations in Measures of Center

It is difficult to determine by looking at the tables whether the thicknesses tend to differ between new and recycled wafers. To interpret these data sets, we need to summarize them in ways that will reveal the important features. Histograms, stem-and-leaf plots, and dotplots are graphical summaries of data sets. While graphs are excellent tools for visualizing the important features of a data set, they have limitations. In particular, graphs often cannot measure a feature precisely; for precise descriptions, we need to use numbers.

In this chapter, we will learn about several of the most commonly used numerical summaries of data. Some of these describe the center of the data; these are called **measures of center**. Others describe how spread out the data values are; these are called **measures of spread**. Still others, called **measures of position**, specify the proportion of the data that is less than a given value.

In the case study at the end of the chapter, you will be asked to use some of the summaries introduced in the chapter to help determine which type of wafer will produce better results.

Section 3.1 Measures of Center

Objectives

1. Compute the mean of a data set
2. Compute the median of a data set
3. Compare the properties of the mean and median
4. Find the mode of a data set
5. Approximate the mean with grouped data
6. Compute a weighted mean

Objective 1 Compute the mean of a data set

The Mean

How do instructors determine your final grade? It's the end of the semester, and you have just finished your statistics class. During the semester, you took five exams, and your scores were 78, 83, 92, 68, and 85. Your instructor must find a single number to give a summary of your performance. The quantity he or she is most likely to use is the **arithmetic mean**, which is often simply called the **mean**. To find the mean of a list of numbers, add the numbers, then divide by how many numbers there are.

Example 3.1

Computing the mean

Find the mean of the exam scores 78, 83, 92, 68, and 85.

Solution

EXPLAIN IT AGAIN

The mean and the average: Some people refer to the mean as the "average." In fact, there are many kinds of averages; the mean is just one of them.

Step 1: Add the numbers.

$$78 + 83 + 92 + 68 + 85 = 406$$

Step 2: Divide the sum by the number of observations. There were five observations. Therefore, the mean is

$$\text{Mean} = \frac{406}{5} = 81.2$$

In Example 3.1, we rounded the mean to one more decimal place than the data. We will follow this practice in general.

SUMMARY

We will round the mean to one more decimal place than the data.

NOTATION ROUNDUP

Sample mean: $\bar{x}$
Population mean: μ

Notation for the mean

Computing a mean involves adding a list of numbers. It is useful to have some notation that will allow us to discuss lists of numbers in general.

When we wish to write down a list of n numbers without specifying what the numbers are, we often write $x_1, x_2, ..., x_n$. To indicate that we are adding these numbers, we write $\sum x$. (The symbol Σ is the uppercase Greek letter sigma.)

NOTE TO INSTRUCTOR

It is worth mentioning that the word "average" is often used to describe any number that appears typical, whereas the word "mean" has a precise definition.

NOTATION

- A list of n numbers is denoted $x_1, x_2, ..., x_n$.
- $\sum x$ represents the sum of these numbers: $\sum x = x_1 + x_2 + \cdots + x_n$

Sample means and population means

Recall that a population consists of an entire collection of individuals about which information is sought, and a sample consists of a smaller group drawn from the population. The method for calculating the mean is the same for both samples and populations, except for the notation. If $x_1, x_2, ..., x_n$ is a sample, then the mean is called the *sample mean* and is denoted with the symbol $\bar{x}$. The mean of a population is called the *population mean* and is denoted by μ (the Greek letter mu).

EXPLAIN IT AGAIN

Sample size and population size: In general, we will use a lowercase n to denote a sample size and an uppercase N to denote a population size.

DEFINITION

If $x_1, ..., x_n$ is a sample, the **sample mean** is

$$\bar{x} = \frac{x_1 + x_2 + \cdots + x_n}{n} = \frac{\sum x}{n}$$

If $x_1, ..., x_N$ is a population, the **population mean** is

$$\mu = \frac{x_1 + x_2 + \cdots + x_N}{N} = \frac{\sum x}{N}$$

Sample means are used to estimate population means

It is usually impossible to compute a population mean, because we don't know all the values in the population. Instead, we draw a sample and use the sample mean to estimate the population mean. Sample means are generally not exactly equal to their population means. Usually, they either overestimate or underestimate the population mean.

We illustrate the relationship between sample means and population means with an example in which the population mean can be computed. There were 440 players in the National Basketball Association (NBA) at the beginning of the 2018–2019 season. Their heights ranged from 69 to 87 inches. The population mean height was $\mu = 79.2$ inches. Here are two samples of five heights.

Sample 1: 81, 73, 82, 84, 81. Sample mean is $\bar{x} = \frac{81 + 73 + 82 + 84 + 81}{5} = 80.2.$

Sample 2: 77, 81, 81, 76, 75. Sample mean is $\bar{x} = \frac{77 + 81 + 81 + 76 + 75}{5} = 78.0.$

The first sample mean (80.2) overestimates the population mean of 79.2, while the second sample mean (78.0) underestimates it.

These two samples were small ($n = 5$). Means of larger samples tend to be closer to the population mean than those of smaller samples. To illustrate this, we drew ten samples of 50 NBA players and computed the mean of each. The means are:

79.10 78.92 79.00 78.82 79.30 79.38 78.98 79.10 79.36 79.38

All of these sample means are fairly close to the population mean of 79.2.

SUMMARY

- Sample means are often used to estimate population means.
- The larger the sample, the more likely it is that the sample mean is close to the population mean.

How the mean measures the center of the data

The mean is a measure of center. Figure 3.1 presents the exam scores in Example 3.1 on a number line and shows the position of the mean. If we imagine each data value to be a weight, then the mean is the point at which the data set would balance.

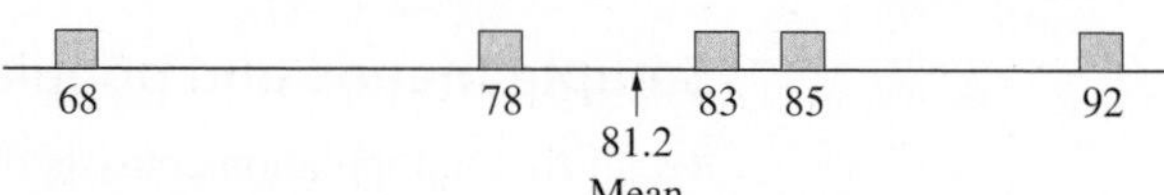

Figure 3.1 The mean is the point where the data set would balance, if each data value were represented by an equal weight.

EXPLAIN IT AGAIN

The mean may not be a value in the data set: The mean is not necessarily a typical value for the data. In fact, the mean may be a value that could not possibly appear in the data set.

A misconception about the mean

Some people believe that the mean represents a "typical" data value. In fact, this is not necessarily so. This is shown in Example 3.1, where we computed the mean of five exam scores and obtained a result of 81.2. If, like most exams, the scores are always whole numbers, then 81.2 is certainly not a "typical" data value; in fact, it could not possibly be a data value.

Objective 2 Compute the median of a data set

The Median

The basic idea behind the **median** is simple: We try to find a number that splits the data set in half, so that half of the data values are less than the median and half of the data values are greater than the median. The procedure for computing the median differs, depending on whether the number of observations in the data set is even or odd.

EXPLAIN IT AGAIN

Finding the middle number: For large data sets, it may not be easy to find the middle number just by looking. If n is odd, the middle number is in position $\frac{n+1}{2}$, and if n is even, the two middle numbers are in positions $\frac{n}{2}$ and $\frac{n}{2}+1$.

Procedure for Computing the Median

Step 1: Arrange the data values in increasing order.

Step 2: Determine the number of data values, n.

Step 3: *If n is odd:* The median is the middle number.

If n is even: The median is the average of the middle two numbers.

In Example 3.1, we found the mean of five exam scores. In Example 3.2, we will find the median.

Example 3.2

Computing the median

Find the median of the exam scores 78, 83, 92, 68, and 85.

Solution

Step 1: We arrange the data values in increasing order to obtain

68 78 83 85 92

Step 2: There are $n = 5$ values in the data set, so n is odd.

Step 3: The middle number is 83, so the median is 83.

Example 3.3

Computing the median

One of the goals of medical research is to develop treatments that reduce the time spent in recovery. Eight patients undergo a new surgical procedure, and the number of days spent in recovery for each is as follows.

20 15 12 27 13 19 13 21

Find the median time spent in recovery.

Solution

Step 1: We arrange the numbers in increasing order to obtain

12 13 13 15 19 20 21 27

Step 2: There are $n = 8$ numbers in the data set, so n is even.

Step 3: The middle two numbers are 15 and 19. The median is the average of these two numbers.

$$\text{Median} = \frac{15 + 19}{2} = 17$$

The median time spent in recovery is 17 days.

Using technology to compute the mean and median

In practice, technology is often used to compute means and medians, as Example 3.4 shows.

Example 3.4

Using technology to compute the mean and median

Use technology to compute the mean and median of the recovery times in Example 3.3.

Descriptive Statistics	
Mean	17.5
Standard Error	1.832251
Median	17
Mode	13
Standard Deviation	5.182388
Sample Variance	26.857143
Kurtosis	-0.092768
Skewness	0.759528
Range	15
Minimum	12
Maximum	27
Sum	140
Count	8

Figure 3.3

Solution

Figure 3.2 presents the TI-84 Plus display. The mean is $\bar{x} = 17.5$ and the median (denoted "Med") is 17. Step-by-step instructions for computing the mean and median with the TI-84 Plus are presented in the Using Technology section on page 107.

```
1-Var Stats
x̄=17.5
Σx=140
Σx²=2638
Sx=5.182387756
σx=4.847679857
n=8
minX=12
↓Q1=13
```

```
1-Var Stats
↑Sx=5.182387756
σx=4.847679857
n=8
minX=12
Q1=13
Med=17
Q3=20.5
maxX=27
```

Figure 3.2 TI-84 Plus display showing the mean and median for the data in Example 3.3

Figure 3.3 presents Excel output and Figure 3.4 (page 102) presents MINITAB output. The mean and median are highlighted in bold. Step-by-step instructions for computing the mean and median in Excel and MINITAB are presented in the Using Technology section on page 108.

Variable	N	**Mean**	SE Mean	StDev	Minimum	Q1	**Median**	Q3	Maximum
Time	8	**17.5**	1.832	5.182	12.00	13.00	**17.00**	20.25	27.00

Figure 3.4

Objective 3 Compare the properties of the mean and median

Comparing the Properties of the Mean and Median

Both the mean and the median are frequently used as measures of center. It is important to know how their properties differ.

NOTE TO INSTRUCTOR

Students can often relate to how the mean of exam grades can be influenced by a single perfect score.

The mean is more influenced by extreme values than the median is

One important difference between the mean and the median is that the formula for the mean uses every value in the data set, but the formula for the median depends only on the middle number or the middle two numbers. This is particularly important for data sets in which one or more numbers are unusually large or unusually small. In most cases, these extreme values will have a large influence on the mean, but little or no influence on the median. Example 3.5 illustrates this principle.

Example 3.5

Determining that the mean is more influenced by extreme values than the median is

Five families, named Smith, Jones, Gonzales, Brown, and Jackson, live in an apartment building. Their annual incomes, in dollars, are 25,000, 31,000, 34,000, 44,000, and 56,000. The Smith family, whose income is 25,000, wins a million-dollar lottery, so their income increases to 1,025,000. Find the mean and median income both before and after the Smiths win the lottery. Which measure of center is more influenced by the large number, the mean or the median?

Solution

We compute the mean and median before the lottery win. The mean income is

$$\text{Mean} = \frac{25{,}000 + 31{,}000 + 34{,}000 + 44{,}000 + 56{,}000}{5} = 38{,}000$$

The median is the middle number:

$$\text{Median} = 34{,}000$$

After the lottery win, the mean is

$$\text{Mean} = \frac{1{,}025{,}000 + 31{,}000 + 34{,}000 + 44{,}000 + 56{,}000}{5} = 238{,}000$$

To find the median, we arrange the numbers in order, obtaining

31,000 34,000 44,000 56,000 1,025,000

The median is the middle number:

$$\text{Median} = 44{,}000$$

The extreme value of 1,025,000 has influenced the mean quite a lot, increasing it from 38,000 to 238,000. In comparison, the median has been influenced much less, increasing only from 34,000 to 44,000.

Because the median is not much influenced by extreme values, we say that the median is *resistant*.

NOTE TO INSTRUCTOR

A number line can be used to visualize how the shape of a data set affects the mean and median. Draw a number line, and mark five equally spaced values, say 1,2,3,4,5. This is a symmetric data set. The middle number (median) is 3, and the mean is easily calculated to be 3 as well. Now replace 5 with a larger number. This is a skewed data set. Clearly the middle number is still 3, but the mean has increased.

DEFINITION

A statistic is **resistant** if its value is not affected much by extreme values (large or small) in the data set.

We can summarize the results of Example 3.5 as follows.

SUMMARY

The median is resistant, but the mean is not.

The mean and median can help describe the shape of a data set

The mean and median measure the center of a data set in different ways. The mean is the point at which a data set balances (see Figure 3.1). The median is the middle number, so that half of the data values are less than the median and half are greater. It turns out that when a data set is symmetric, the mean and median are equal.

When a data set is skewed, however, the mean and median are often quite different. When a data set is skewed to the right, there are some large values in the right tail. Because the median is resistant while the mean is not, the mean is generally more affected by these large values than the median is. Therefore, for a data set that is skewed to the right, the mean is often greater than the median.

CAUTION

The relationship between the mean and median and the shape of the data set holds for most data sets, but not all.

Figure 3.5 illustrates the idea. For most data sets that are skewed to the left, the mean will be to the left of, or less than, the median. For most data sets that are skewed to the right, the mean will be to the right of, or greater than, the median. When a data set is approximately symmetric, the balancing point is near the middle of the data, so the mean and the median will be approximately equal.

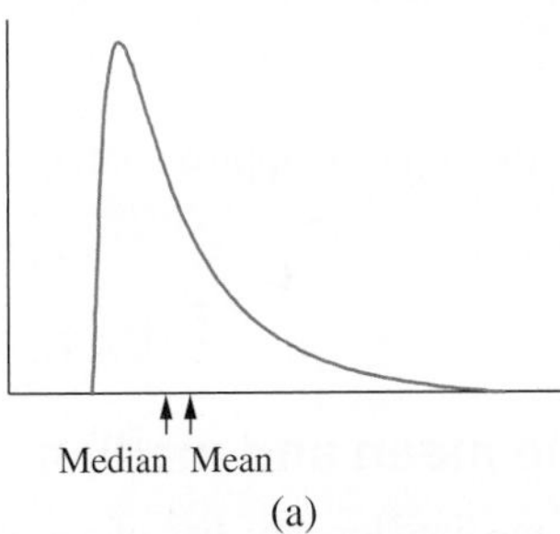

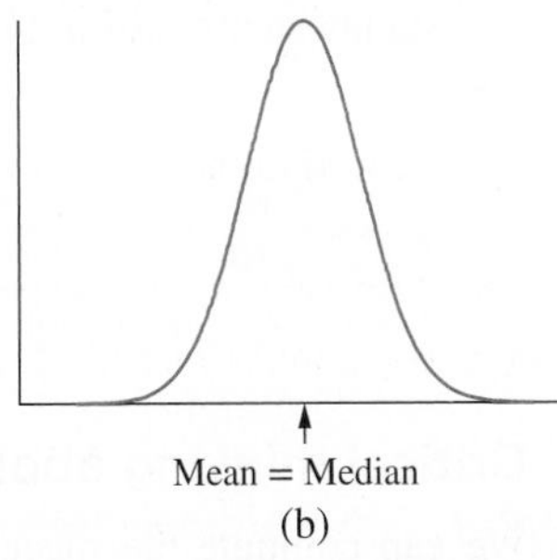

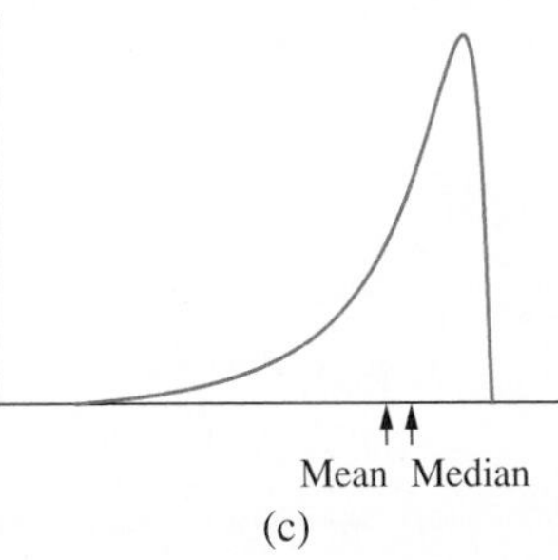

Figure 3.5 (a) When a data set is skewed to the right, the mean is generally greater than the median. (b) When a data set is approximately symmetric, the mean and median will be approximately equal. (c) When a data set is skewed to the left, the mean is generally less than the median.

SUMMARY

In most cases, the shape of a histogram reflects the relationship between the mean and median as follows:

Shape	Relationship Between Mean and Median
Skewed to the right	Mean is noticeably greater than median
Approximately symmetric	Mean is approximately equal to median
Skewed to the left	Mean is noticeably less than median

For an exception to this rule, see Exercise 88.

Which is a better measure of center, the mean or the median?

The short answer is that neither one is better than the other. They both measure the center in different, but appropriate, ways. When the data are highly skewed or contain extreme values, some people prefer to use the median, because the median is more representative of a typical value. However, the mean is sometimes an appropriate measure of center even when the data are highly skewed (see Exercises 77 and 78).

The following table summarizes the features of the mean and median.

	Advantages	Disadvantages
Mean	Takes every value into account	Highly influenced by extreme values: not resistant
Median	Not much influenced by extreme values: resistant	Depends only on middle value or middle two values

Check Your Understanding

1. Compute the mean and median of the following sample:
   74 87 36 97 60 58 46 *Mean: 65.4; median: 60*
2. Compute the mean and median of the following sample:
   69 17 75 96 74 80 *Mean: 68.5; median: 74.5*
3. Someone surveys the families in a certain town and reports that the mean number of children in a family is 2.1. Someone else says that this must be wrong, because it is impossible for a family to have 2.1 children. Comment. *Not necessarily wrong*
4. A data set has a mean of 6 and a median of 4. Would you expect this data set to be skewed to the right, skewed to the left, or approximately symmetric? *Skewed right*
5. A data set has a mean of 5 and a median of 7. Would you expect this data set to be skewed to the right, skewed to the left, or approximately symmetric? *Skewed left*
6. A data set has a mean of 8 and a median of 8.1. Would you expect this data set to be skewed to the left, skewed to the right, or approximately symmetric? *Approximately symmetric*
7. A data set has a mean of 4 and a median of 3.8. Would you expect this data set to be skewed to the left, skewed to the right, or approximately symmetric? *Approximately symmetric*

Answers are on page 117.

Critical thinking about the mean and median

We can compute the mean and median for any list of numbers. However, they do not always produce meaningful results. The mean and median are useful for numbers that measure or count something. They are not useful for numbers that are used simply as labels. Example 3.6 illustrates the idea.

Example 3.6

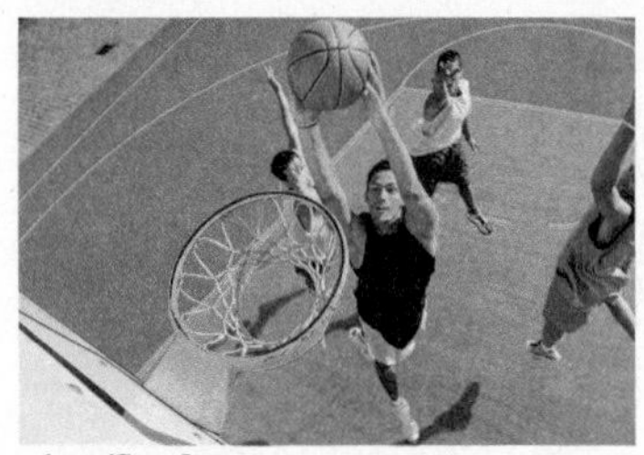

technotr/Getty Images

Determining whether the mean and median make sense

Following is information about the five starting players on a certain college basketball team:

Their heights, in inches, are 74, 76, 79, 80, and 82.

Their uniform numbers are 15, 32, 4, 43, and 26.

Will we obtain meaningful information by computing the mean and median height? How about the mean and median uniform number? Explain.

Solution

The mean and median height are meaningful, because heights are measurements.

The mean and median uniform numbers are not meaningful, because these numbers are just labels. They don't measure or count anything.

Objective 4 Find the mode of a data set

The Mode

In Section 2.2, we defined a mode to be the highest point of a histogram. There is another definition of this term. The **mode** of a data set is the value that appears most frequently.

Example 3.7

Finding the mode

Ten students were asked how many siblings they had. The results, arranged in order, were

0 1 1 1 1 2 2 3 3 6

Find the mode of this data set.

Solution

The value that appears most frequently is 1. Therefore, the mode of this data set is 1.

EXPLAIN IT AGAIN

The mode isn't really a measure of center: The mode is sometimes classified as a measure of center. However, this isn't really accurate. The mode can be the largest value in a data set, or the smallest, or anywhere in between.

When two or more values are tied for the most frequent, they are all considered to be modes. If the values all have the same frequency, we say that the data set has no mode.

SUMMARY

- The mode of a data set is the value that appears most frequently.
- If two or more values are tied for the most frequent, they are all considered to be modes.
- If the values all have the same frequency, we say that the data set has no mode.

Computing the mode for qualitative data

The mean and median can be computed only for quantitative data. The mode, on the other hand, can be computed for qualitative data as well. For qualitative data, the mode is the most frequently appearing category.

Example 3.8

Finding the mode for qualitative data

Following is a list of the makes of all the cars rented by an automobile rental company on a particular day. Which make of car is the mode?

Honda	Toyota	Toyota	Honda	Ford
Chevrolet	Nissan	Ford	Chevrolet	Chevrolet
Honda	Dodge	Ford	Ford	Toyota
Chevrolet	Toyota	Toyota	Toyota	Nissan

Solution

The most frequent category is "Toyota," which appears six times. Therefore, the mode is "Toyota."

Check Your Understanding

8. Find the mode or modes, if they exist:
  a. The sample is 3, 6, 0, 1, 1, 8, 0, 1, 1. *1*
  b. The sample is 4, 7, 4, 1, 6, 5, 6. *4 and 6*
  c. The sample is 4, 8, 5, 9, 6, 3. *No mode*

Answers are on page 117.

Objective 5 Approximate the mean with grouped data

Approximating the Mean with Grouped Data

Sometimes we don't have access to the raw data in a data set, but we are given a frequency distribution. In these cases we can approximate the mean. We use Table 3.1 (page 106) to

Thomas Northcut/Getty Images

Table 3.1 Number of Text Messages Sent by High School Students

Number of Text Messages Sent	Frequency
0–49	10
50–99	5
100–149	13
150–199	11
200–249	7
250–299	4

illustrate the method. This table presents the number of text messages sent by a sample of 50 high school students. We will approximate the mean number of messages sent.

We present the method for approximating the mean with grouped data.

Procedure for Approximating the Mean with Grouped Data

Step 1: Compute the midpoint of each class. The midpoint of a class is found by taking the average of the lower class limit and the lower limit of the next larger class. For the last class, there is no next larger class, but we use the lower limit that the next larger class would have.

Step 2: For each class, multiply the class midpoint by the class frequency.

Step 3: Add the products Midpoint × Frequency over all classes.

Step 4: Divide the sum obtained in Step 3 by the sum of the frequencies.

Example 3.9

Approximating the mean with grouped data

Compute the approximate mean number of messages sent, using Table 3.1.

Solution

The calculations are summarized in Table 3.2.

Step 1: Compute the midpoints: For the first class, the lower class limit is 0. The lower limit of the next class is 50. The midpoint is therefore

$$\frac{0 + 50}{2} = 25$$

We continue in this manner to compute the midpoint of each class. Note that for the last class, we average the lower limit of 250 with 300, which is the lower limit that the next class would have.

Step 2: Multiply the midpoints by the frequencies as shown in the column in Table 3.2 labeled "Midpoint × Frequency."

Step 3: Add the products Midpoint × Frequency, to obtain 6850.

Step 4: The sum of the frequencies is 50. The mean is approximated by $6850/50 = 137$.

Table 3.2 Calculating the Mean Number of Text Messages Sent by High School Students

Class	Midpoint	Frequency	Midpoint × Frequency
0–49	25	10	$25 \times 10 = 250$
50–99	75	5	$75 \times 5 = 375$
100–149	125	13	$125 \times 13 = 1625$
150–199	175	11	$175 \times 11 = 1925$
200–249	225	7	$225 \times 7 = 1575$
250–299	275	4	$275 \times 4 = 1100$
		Sum = 50	Sum = 6850

$$\text{Mean} \approx \frac{6850}{50} = 137$$

The Weighted Mean

Objective 6 Compute a weighted mean

A weighted mean is a mean in which some numbers count more than others. To compute a weighted mean we assign a positive number, called a weight, to each number, with the numbers that count more getting the larger weights.

DEFINITION

If $x_1, ..., x_n$ are data values, and $w_1, ..., w_n$ are weights, the **weighted mean** is

$$\bar{x}_w = \frac{w_1x_1 + w_2x_2 + \cdots + w_nx_n}{w_1 + w_2 + \cdots + w_n} = \frac{\sum w_ix_i}{\sum w_i}$$

A commonly used weighted mean is a student's grade point average (GPA). A GPA is a weighted mean of grades in courses, with each grade weighted by the number of credits in the course.

Example 3.10

Calculating a grade point average

Last semester Joe took four courses. He got an A in statistics (5 credits), a B in psychology (4 credits), a C in English (3 credits), and an A in music (2 credits). An A is worth 4 quality points, a B is worth 3 points and a C is worth 2 points. Compute Joe's GPA.

Solution

The data values are the points for each grade: 4, 3, 2, and 4. The weights are the numbers of credits: 5, 4, 3, 2. The GPA is the weighted mean

$$\frac{5(4) + 4(3) + 3(2) + 2(4)}{5 + 4 + 3 + 2} = 3.29$$

Using Technology

We use Example 3.1 to illustrate the technology steps.

TI-84 PLUS

Computing the mean and median

Step 1. Enter the data into **L1** in the data editor by pressing **STAT**, then **1: Edit...** For Example 3.1, we use **78, 83, 92, 68, 85** (Figure A).

Step 2. Press **STAT** and highlight the **CALC** menu.

Step 3. Select **1–Var Stats** and press **ENTER**. The **1–Var Stats** command is now shown on the home screen.

Step 4. Enter the list name **L1** next to the **1–Var Stats** command by pressing **2nd**, then **L1** (Figure B).

Step 5. Press **ENTER**. The descriptive statistics are displayed on the screen (Figures C and D).

L1	L2	L3	L4	L5	1
78	------	------	------	------	
83					
92					
68					
85					

L1(6)=

Figure A

```
1-Var Stats L1
```

Figure B

```
1-Var Stats
x̄=81.2
Σx=406
Σx²=33286
Sx=8.927485648
σx=7.984985911
n=5
minX=68
↓Q1=73
```

Figure C

```
1-Var Stats
↑Sx=8.927485648
σx=7.984985911
n=5
minX=68
Q1=73
Med=83
Q3=88.5
maxX=92
```

Figure D

Using the TI-84 PLUS Stat Wizards (see Appendix B for more information)

Step 1. Enter the data into **L1** in the data editor by pressing **STAT**, then **1: Edit...** For Example 3,1, we use **78, 83, 92, 68, 85** (Figure A).

Step 2. Press **STAT** and highlight the **CALC** menu.

Step 3. Select **1–Var Stats** and press **ENTER**. Enter **L1** next to the **List** field. Keep the **FreqList** field blank.

Step 4. Select **Calculate** and press **ENTER**. The descriptive statistics are displayed on the screen (Figures C and D).

EXCEL

Computing the mean and median

Step 1. Enter the data in **Column A**. For Example 3.1, we use **78, 83, 92, 68, 85**.

Step 2. Select **Data**, then **Data Analysis.** Highlight **Descriptive Statistics** and press **OK**.

Step 3. Enter the range of cells that contain the data in the **Input Range** field, and check the **Summary Statistics** box.

Step 4. Press **OK** (Figure E).

Alternatively, the mean can be computed using the **=AVERAGE**(*data_array*) command and the median can be computed using the **=MEDIAN**(*data_array*) command (Figure F).

Descriptive Statistics	
Mean	81.2
Standard Error	3.992493
Median	83
Mode	#N/A
Standard Deviation	8.927486
Sample Variance	79.7
Kurtosis	0.715308
Skewness	-0.592815
Range	24
Minimum	68
Maximum	92
Sum	406
Count	5

Figure E

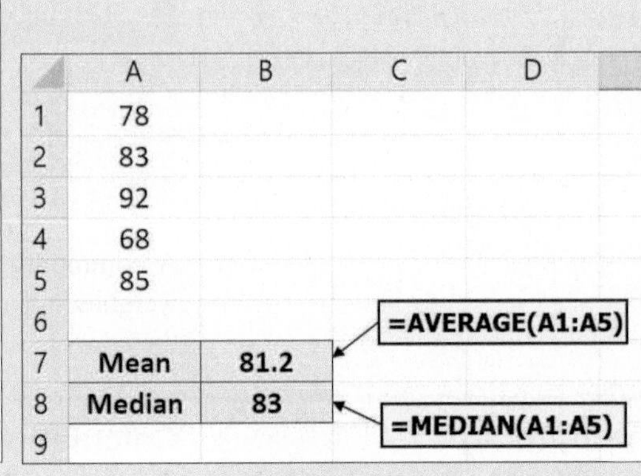

	A	B	C	D
1	78			
2	83			
3	92			
4	68			
5	85			
6				
7	Mean	81.2		
8	Median	83		
9				

Figure F

MINITAB

Computing the mean and median

Step 1. Enter the data in **Column C1**. For Example 3.1, we use **78, 83, 92, 68, 85**.

Step 2. Click on **Stat**, then **Basic Statistics**, then **Display Descriptive Statistics...**.

Step 3. Enter **C1** in the **Variables** field.

Step 4. Click **Statistics** and select the desired statistics. Press **OK**.

Step 5. Press **OK** (Figure G).

Statistics

Variable	Mean	StDev	Minimum	Q1	Median	Q3	Maximum
C1	81.20	8.93	68.00	73.00	83.00	88.50	92.00

Figure G

Section 3.1 Exercises

Exercises 1–8 are the Check Your Understanding exercises located within the section.

Understanding the Concepts

In Exercises 9–12, fill in each blank with the appropriate word or phrase.

9. The ________ is calculated by summing all data values and dividing by how many there are. *mean*

10. The ________ is a number that splits the data set in half. *median*

11. The median is resistant because it is not affected much by ________. *extreme values*

12. The ________ is the value in the data set that appears most frequently. *mode*

In Exercises 13–16, determine whether the statement is true or false. If the statement is false, rewrite it as a true statement.

13. Every data set contains at least one mode. *False*

14. The mean is resistant. *False*

15. For most data sets that are skewed to the right, the mean is less than the median. *False*

16. A mode is always a value that is in the data set. *True*

Practicing the Skills

17. Find the mean, median, and mode for the following data set: *Mean: 23.4; median: 26; mode: 27*

12 27 26 27 25

18. Find the mean, median, and mode for the following data set: *Mean: 3; median: 15; mode: −20*

−20 15 21 −20 19

19. Find the mean, median, and mode for the following data set: *Mean: 5.5; median: 14; mode: 28*

28 −31 28 0 31 −23

20. Find the mean, median, and mode for the following data set: *Mean: 81.5; median: 93.5; mode: 98*

83 98 22 89 99 98

21. Consider the following population:

3 18 22 15 11 42 22 45 21 39

The population mean is 23.8.

a. A sample from this population is 11, 42, 15. Compute the sample mean. Does the sample mean underestimate or overestimate the population mean? *22.7, underestimates*

b. Another sample from this population is 15, 45, 22. Compute the sample mean. Does the sample mean underestimate or overestimate the population mean? *27.3, overestimates*

c. Draw a sample of size 3 from this population, and compute the sample mean. Does it underestimate or overestimate the population mean?

22. Consider the following population:

264 399 430 416 227 314
136 145 155 371 298 176

The population mean is 277.58.

a. A sample from this population is 136, 430, 399. Compute the sample mean. Does the sample mean underestimate or overestimate the population mean? *321.67, overestimates*

b. Another sample from this population is 298, 145, 176. Compute the sample mean. Does the sample mean underestimate or overestimate the population mean? *206.33, underestimates*

c. Draw a sample of size 4 from this population, and compute the sample mean. Does it underestimate or overestimate the population mean?

In Exercises 23–26, use the given frequency distribution to approximate the mean.

23.

Class	Frequency
0–9	13
10–19	7
20–29	10
30–39	9
40–49	11

24.6

24.

Class	Frequency
0–15	2
16–31	14
32–47	6
48–63	13
64–79	15

48.0

25.

Class	Frequency
0–49	17
50–99	26
100–149	14
150–199	34
200–249	26
250–299	8

145.0

26.

Class	Frequency
0–19	18
20–39	11
40–59	6
60–79	6
80–99	10
100–119	5

47.9

27. Use the properties of the mean and median to determine which are the correct mean and median for the following histogram. *(ii)*

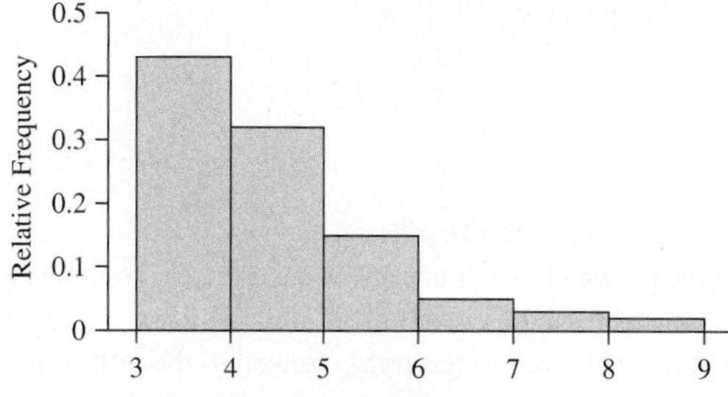

(i) Mean is 4.6; median is 5.0
(ii) Mean is 4.5; median is 4.2
(iii) Mean is 3.5; median is 4.3
(iv) Mean is 6.1; median is 5.6

28. Use the properties of the mean and median to determine which are the correct mean and median for the following histogram. *(iv)*

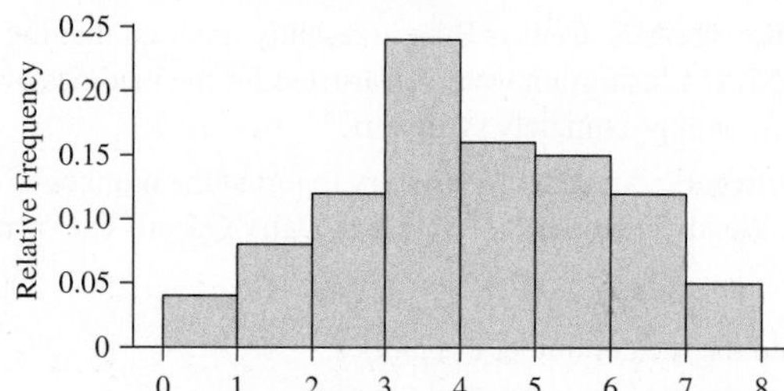

(i) Mean is 3.6; median is 4.8
(ii) Mean is 4.8; median is 3.6
(iii) Mean is 3.5; median is 3.1
(iv) Mean is 4.2; median is 4.1

29. Use the properties of the mean and median to determine which are the correct mean and median for the following histogram. *(i)*

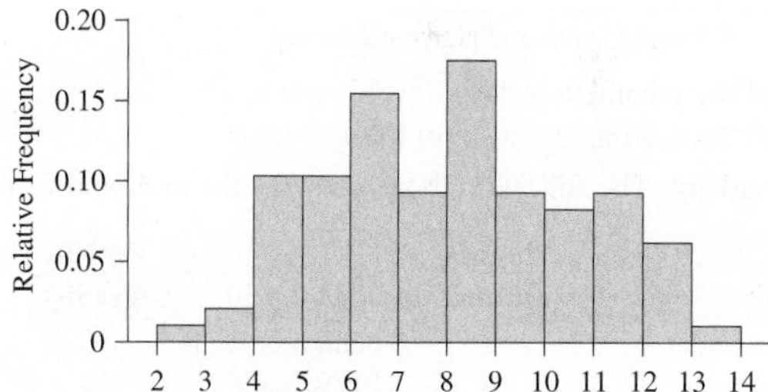

(i) Mean is 8.0; median is 8.1
(ii) Mean is 8.5; median is 6.5
(iii) Mean is 7.0; median is 8.3
(iv) Mean is 6.2; median is 6.1

30. Use the properties of the mean and median to determine which are the correct mean and median for the following histogram. *(iii)*

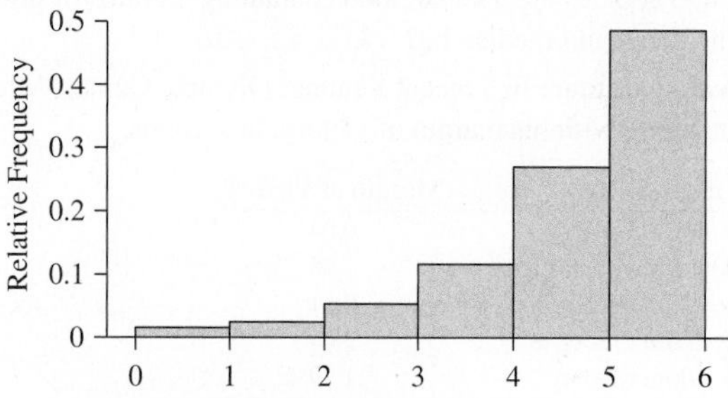

(i) Mean is 3.0; median is 4.1
(ii) Mean is 4.2; median is 3.2
(iii) Mean is 4.4; median is 5.0
(iv) Mean is 4.3; median is 4.1

31. Find the mean, median, and mode of the data in the following stem-and-leaf plot. The leaf represents the ones digit. *Mean: 30.4; median: 29; mode: 27*

1	2
2	0779
3	78
4	13

32. Find the mean, median, and mode of the data in the following stem-and-leaf plot. The leaf represents the ones digit. *Mean: 20.9; median: 24; mode: 25*

0	8
1	16
2	3559
3	0

Working with the Concepts

33. Facebook friends: In a study of Facebook users conducted by the Pew Research Center, the mean number of Facebook friends per user was 245 and the median was 111. If a histogram were constructed for the numbers of Facebook friends for all Facebook users, would you expect it to be skewed to the right, skewed to the left, or approximately symmetric? Explain. *Skewed right*

34. Mean and median height: The National Center for Health Statistics reported in a recent year that the mean height for U.S. women aged 20–29 was 64.3 inches and the median was 64.2 inches. If a histogram were constructed for the heights of all U.S. women aged 20–29, would you expect it to be skewed to the right, skewed to the left, or approximately symmetric? Explain. *Approximately symmetric*

35. Life expectancy: According to the National Vital Statistics Reports, the mean life expectancy for the United States was 78.7 years and the median was 83.5. If a histogram were constructed for the lifespans of all people in the United States, would you expect it to be skewed to the right, skewed to the left, or approximately symmetric? *Skewed left*

36. Income: The U.S. Census Bureau recently reported that the mean household income in the United States was $72,641 and the median was $53,657. If a histogram were constructed for the incomes of all U.S. households, would you expect it to be skewed to the right, skewed to the left, or approximately symmetric? *Skewed right*

37. Hamburgers: An ABC News story reported the number of calories in hamburgers from six fast-food restaurants: McDonald's, Burger King, Wendy's, Hardee's, Sonic, and Dairy Queen. The results are

250 290 230 310 310 350

a. Find the mean number of calories. *290*
b. Find the median number of calories. *300*
c. The Impossible Burger, a plant-based burger sold at some restaurants, has 240 calories. Find the mean and median if the Impossible Burger is included. *Median: 290; mean: 282.9*

38. Cholesterol levels: Following are measurements of total serum cholesterol (in milligrams per deciliter) for 15 adults.

192	279	216	160	219	153	215	140
214	156	199	174	255	186	245	

Source: National Health and Nutrition Survey

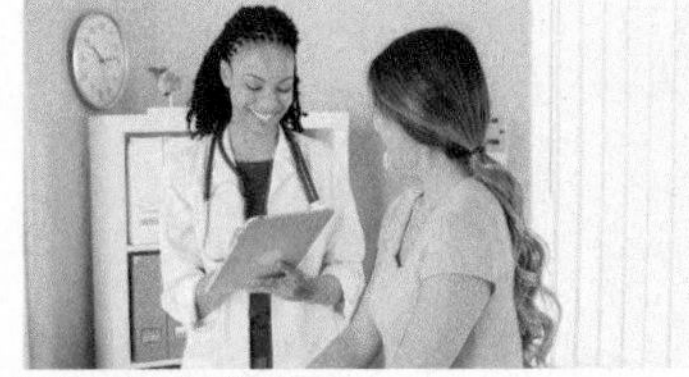
Rocketclips, Inc./Shutterstock

a. Find the mean cholesterol level. *200.2*
b. Find the median cholesterol level. *199.0*

39. Eat healthy: The following table presents the number of grams of protein in a single serving for several vegan foods.

Food	Protein (grams)	Food	Protein (grams)
Peanut butter	8	Sunflower seeds	6
Tofu	10	Black beans	15
Cooked peas	9	Pinto beans	15
Veggie baked beans	12	Spaghetti	7
Cooked bulgur	6	Bagel	11

Source: USDA Nutrient Database

a. Find the mean number of grams of protein. *9.9*
b. Find the median number of grams of protein. *9.5*
c. Oreo cookies are a vegan food containing 2 grams of protein per serving (3 cookies). If Oreo cookies are added to this list, what would the mean and median be? *Mean: 9.2; median: 9*

40. Great swimmer: In a recent Summer Olympic Games, Michael Phelps won five gold medals in swimming. Following are the events he won, along with his margin of victory, in seconds.

Event	Margin of Victory
200m Butterfly	0.04
200m Individual medley	1.95
4 × 100m Freestyle	0.61
4 × 200m Freestyle	2.47
4 × 100m Medley	1.29

a. What was the mean margin of victory? *1.272*
b. What was the median margin of victory? *1.29*
c. Assume the margin of victory in the 200m individual medley was incorrectly entered as 19.5. Which would increase more, the mean or the median? Explain why. *Mean*

41. Mobile apps: The following table presents the number of monthly users for the 15 most popular mobile apps as of March 2019.

Application	Monthly Users (millions)	Application	Monthly Users (millions)
Facebook	168.6	Tumblr	18.4
Instagram	121.2	Messenger by Google	15.7
Facebook Messenger	108.7	Google Hangouts	15.0
Twitter	76.4	Discord	12.5
Pinterest	64.9	GroupMe	11.0
Snapchat	49.4	Skype	7.0
Reddit	42.0	Google+	6.8
WhatsApp	24.9		

Source: Verto Analytics

a. Find the mean number of monthly users. *49.5*
b. Find the median number of monthly users. *24.9*
c. Would you expect these data to be skewed to the right, skewed to the left, or approximately symmetric? *Skewed to the right*

42. It's raining: The following table presents the average yearly precipitation, in inches, for the ten largest cities in the United States.

City	Precipitation (inches)	City	Precipitation (inches)
New York	49.9	Philadelphia	41.5
Los Angeles	12.8	San Antonio	23.3
Chicago	36.9	San Diego	10.3
Houston	49.8	Dallas	37.6
Phoenix	8.2	San Jose	15.8

Source: NOAA National Climatic Data Center

a. Find the mean annual precipitation. *28.6*
b. Find the median annual precipitation. *30.1*
c. Would you expect these data to be skewed to the right, skewed to the left, or approximately symmetric? *Approximately symmetric*

43. Outer space: The following table presents the distance from the sun, in millions of miles, of the eight planets in the solar system.

Planet	Distance (millions of miles)	Planet	Distance (millions of miles)
Mercury	36.0	Jupiter	483.6
Venus	67.2	Saturn	890.8
Earth	93.0	Uranus	1783.9
Mars	141.6	Neptune	2795.1

a. Find the mean distance from the sun. *786.4*
b. Find the median distance from the sun. *312.6*
c. Until 2006, Pluto was considered to be a planet. The distance from Pluto to the sun is 3670.1 million miles. If Pluto were still a planet, what would be the mean and median distance from the sun? *Mean: 1106.8; median: 483.6*

44. Big companies: Following are revenues for the ten U.S. companies with the highest earnings in 2018.

Company	Revenue (billion $)	Company	Revenue (billion $)
Walmart	514.4	UnitedHealth Group	226.2
ExxonMobil	290.2	McKesson	208.4
Apple	265.6	CVS Health	194.6
Berkshire Hathaway	247.8	AT&T	170.8
Amazon.com	232.9	AmerisourceBergen	167.9

Source: *Fortune*

a. Find the mean revenue. *251.88*
b. Find the median revenue. *229.55*
c. ExxonMobil, Berkshire Hathaway, McKesson, and AT&T have their roots in companies that were founded in the 19th century. Compute the mean and median for these older companies, and separately for the newer companies. Are the mean and median revenue for the older companies greater than, less than, or about the same for the newer companies?

45. What's your favorite TV show? The following tables present the numbers of viewers, in millions, for the top 15 prime-time shows for the 2013–2014 and 2018–2019 seasons. The numbers of viewers include those who watched the program on any platform, including time-shifting up to seven days after the original telecast.

Top-Rated TV Programs: 2013–2014

Program	Rating
NBC Sunday Night Football	21.5
The Big Bang Theory	20.0
NCIS	19.8
NCIS: Los Angeles	16.0
Dancing with the Stars	15.2
The Blacklist	15.0
The Voice	14.6
Person of Interest	14.0
The Voice—Tuesday	13.9
Blue Bloods	13.6
Resurrection	13.0
Criminal Minds	12.7
Castle	12.6
60 Minutes	12.2
Grey's Anatomy	12.1

Source: Nielsen Media Research

Top-Rated TV Programs: 2018–2019

Program	Rating
NFL Sunday Night Football	19.3
The Big Bang Theory	17.4
NCIS	15.9
Game of Thrones	15.3
Young Sheldon	14.6
NFL Thursday Night Football	14.4
This Is Us	13.8
Blue Bloods	12.8
FBI	12.7
The Good Doctor	12.6
Manifest	12.6
America's Got Talent: Champions	12.4
NFL Monday Night Football	11.9
Chicago Fire	11.7
The Masked Singer	11.6

Source: Nielsen Media Research

a. Find the mean and median ratings for 2013–2014. *Mean: 15.08; median: 14.0*

b. Find the mean and median ratings for 2018–2019. *Mean: 13.93; median: 12.8*

c. Is there a substantial difference in the numbers of people watching the top 15 shows between 2013–2014 and 2018–2019, or are they about the same? *More people watched in 2013–2014.*

46. Beer: The following table presents the number of active breweries for samples of states located east and west of the Mississippi River.

East		West	
State	**Number of Breweries**	**State**	**Number of Breweries**
Connecticut	18	Alaska	17
Delaware	10	Arizona	31
Florida	47	California	305
Georgia	22	Colorado	111
Illinois	52	Iowa	21
Kentucky	13	Louisiana	6
Maine	38	Minnesota	41
Maryland	23	Montana	30
Massachusetts	40	South Dakota	5
New Hampshire	16	Texas	37
New Jersey	20	Utah	15
New York	76		
North Carolina	46		
South Carolina	14		
Tennessee	19		
Vermont	20		

Source: http://www.beerinstitute.org/

a. Find the mean number of breweries for states east of the Mississippi. *29.6*

b. Find the mean number of breweries for states west of the Mississippi. *56.3*

c. Find the median number of breweries for states east of the Mississippi. *21*

d. Find the median number of breweries for states west of the Mississippi. *30*

e. Does one region have a lot more breweries per state than the other, or are they about the same? *West has a lot more*

f. The sample of western states happens to include California. Remove California from the sample of western states, and compute the mean and median for the remaining western states. *Mean: 31.4; median: 25.5*

g. How does the removal of California affect the comparison of the number of breweries between eastern and western states? *Means and medians become more similar.*

47. Gas prices: The following table presents the average price, in U.S. dollars per gallon, of gasoline in the years 2016 and 2019.

Country	2016	2019
Belgium	5.49	5.91
France	5.53	6.44
Germany	5.53	6.32
Italy	6.09	6.81
Netherlands	6.28	7.12
United Kingdom	5.49	6.09
United States	2.46	2.50

Source: GlobalPetrolPrices.com

a. Find the mean and median gas price for 2016. *Mean: 5.267; median: 5.53*

b. Find the mean and median gas price for 2019. *Mean: 5.884; median: 6.32*

c. Which increased more between 2016 and 2019, the mean or the median? *Median*

48. House prices: The following table presents prices, in thousands of dollars, of single-family homes for some of the largest metropolitan areas in the United States for March of 2018 and March of 2019.

Metro Area	2018	2019	Metro Area	2018	2019
Atlanta, GA	204.8	214.2	New York, NY	384.0	384.4
Baltimore, MD	250.1	247.9	Philadelphia, PA	198.9	213.0
Boston, MA	412.2	438.6	Phoenix, AZ	229.6	251.7
Chicago, IL	220.2	242.9	Pittsburgh, PA	142.8	149.3
Cincinnati, OH	165.7	172.2	Portland, OR	396.6	386.4
Cleveland, OH	136.5	140.9	Riverside, CA	353.0	357.1
Dallas, TX	247.1	244.5	St. Louis, MO	158.0	167.5
Denver, CO	388.1	402.1	San Diego, CA	546.3	553.9
Houston, TX	221.1	219.0	San Francisco, CA	763.8	795.0
Los Angeles, CA	608.2	629.2	Seattle, WA	450.2	467.4
Miami, FL	259.1	310.0	Tampa, FL	203.4	212.4
Minneapolis, MN	274.4	264.9	Washington, DC	381.0	393.7

Source: Zillow

a. Find the mean and median price for 2018. *Mean: 316.46; median: 254.6*
b. Find the mean and median price for 2019. *Mean: 327.42; median: 258.30*
c. Did the mean price increase or decrease from 2018 to 2019? *Increase*
d. Did the median price increase or decrease from 2018 to 2019? *Increase*

49. Heavy football players: Following are the weights, in pounds, for offensive and defensive linemen on the New York Giants National Football League team at the beginning of a recent year.

Offense:	315	253	255	319	252	300	327	303	299	310	317	302
Defense:	304	309	306	305	265	270	264	317	255	274	261	296

a. Find the mean and median weight for the offensive linemen. *Mean: 296; median: 302.5*
b. Find the mean and median weight for the defensive linemen. *Mean: 285.5; median: 285*
c. Do offensive or defensive linemen tend to be heavier, or are they about the same? *Offensive linemen are heavier.*

50. Stock prices: Following are the closing prices of Google stock for each trading day in May and June of a recent year.

May					
880.37	877.07	873.65	866.20	869.79	829.61
880.93	884.74	900.68	900.62	886.25	820.43
875.04	877.00	871.98	879.81	890.22	
879.73	864.64	859.70	859.10	867.63	

June				
871.22	870.76	868.31	881.27	873.32
882.79	889.42	906.97	908.53	909.18
903.87	915.89	887.10	877.53	880.23
871.48	873.63	857.23	861.55	845.72

a. Find the mean and median price in May. *Mean: 872.51; median: 876.02*
b. Find the mean and median price in June. *Mean: 881.80; median: 878.88*
c. Does there appear to be a substantial difference in price between May and June, or are the prices about the same? *About the same*

51. Flu season: The following tables present the number of specimens that tested positive for Type A and Type B influenza in the United States during the first 15 weeks of a recent flu season.

Type A				
36	99	177	200	258
384	584	999	1539	2748
3764	4841	5346	5405	5795

Source: Centers for Disease Control

Type B				
56	98	121	139	151
230	263	359	495	674
821	1096	1162	1106	1156

Source: Centers for Disease Control

a. Find the mean and median number of Type A cases in the first 15 weeks of the flu season. *Mean: 2145; median: 999*
b. Find the mean and median number of Type B cases in the first 15 weeks of the flu season. *Mean: 528.5; median: 359*
c. A public health official says that there are more than twice as many cases of Type A influenza as Type B. Do these data support this claim? *Yes*

52. News flash: The following table presents the circulation (in thousands) for the top 25 U.S. daily newspapers in both print and digital editions.

Print				
1480.7	731.4	1424.4	432.9	360.5
300.0	431.1	184.8	213.8	368.1
190.6	265.8	231.2	159.4	180.3
241.0	216.1	184.8	227.7	285.9
125.7	126.3	192.8	172.0	149.5

Source: Alliance for Audited Media

Digital				
898.1	1133.9	249.9	177.7	155.7
200.6	42.3	77.7	192.8	46.8
65.9	112.0	102.3	15.5	160.5
17.1	95.5	68.0	73.7	7.0
69.0	16.0	21.6	73.5	6.7

Source: Alliance for Audited Media

a. Find the mean and median circulation for print editions. *Mean: 355.07; median: 227.7*
b. Find the mean and median circulation for digital editions. *Mean: 163.19; median: 73.7*
c. The editor of an internet news source says that digital circulation is more than half of print circulation. Do the data support this claim? *No*

53. Commercial break: Following are the amounts spent (in millions of dollars) on media advertising in the United States by a sample of 10 companies.

295.9	722.5	280.3	286.9	362.9	331.2	394.6	257.0	463.5	340.1

a. Find the mean amount spent on advertising. *373.49*
b. Find the median amount spent on advertising. *335.65*
c. If the amount of 722.5 was incorrectly listed as 7225, how would this affect the mean? How would it affect the median? *Mean becomes 1023.7; median is unchanged.*

54. Magazines: The following data represent the annual costs in dollars of a sample of 22 popular magazine subscriptions.

14.98	9.97	20.99	7.97	9.99	11.97	116.99	14.95	11.97	16.00	9.97
17.98	38.95	14.98	14.98	16.00	24.99	14.98	10.99	18.00	12.99	11.99

Source: http://magazines.com

a. Find the mean annual subscription cost. *20.117*
b. Find the median annual subscription cost. *14.98*
c. What is the mode? *14.98*
d. Which value in this data set is most accurately described as an extreme value? *116.99*
e. How would the mean, median, and mode be affected if the extreme value were removed from the list? *Mean becomes 15.504; median and mode are unchanged.*

55. Don't drink and drive: The Insurance Institute for Highway Safety reported that there were 5037 fatalities among drivers in auto accidents in a recent year. Following is a frequency distribution of their ages.

Age	Number of Fatalities
11–20	485
21–30	1736
31–40	988
41–50	868
51–60	585
61–70	249
71–80	126

Source: Insurance Institute for Highway Safety

a. Approximate the mean age. *37.2*
b. The first class consists of ages 11 through 20, but most drivers are at least 16 years old. Does this tend to make the approximate mean too large or too small? Explain. *Too small*

56. Age distribution: The ages of residents of Banks City, Oregon, are given in the following frequency distribution.

Age	Frequency
0–9	283
10–19	203
20–29	217
30–39	256
40–49	176
50–59	92
60–69	21
70–79	23
80–89	12
90–99	3

Source: U.S. Census Bureau

a. Approximate the mean age. *28.2*
b. Assume all three people aged 90–99 were in fact 90 years old. Would this tend to make the approximate mean too large or too small? Explain. *Too large*

57. Be my Valentine: The following frequency distribution presents the amounts, in dollars, spent for Valentine's Day gifts in a survey of 123 U.S. adults.

Amount	Frequency
0.00–19.99	19
20.00–39.99	13
40.00–59.99	21
60.00–79.99	19
80.00–99.99	12
100.00–119.99	10
120.00–139.99	7
140.00–159.99	8
160.00–179.99	7
180.00–199.99	1
200.00–219.99	3
220.00–239.99	2
240.00–259.99	1

Source: Based on data from the National Retail Federation

a. Approximate the mean amount spent on Valentine's Day gifts. *81.1*

b. If the majority of the people in the category 0.00–19.99 actually didn't spend any money, would this tend to make the approximate mean too large or too small? *Too large*

58. Get your degree: The following frequency distribution presents the number of U.S. adults (in thousands) ages 25–74 who have earned a Bachelor's degree in a recent year.

Age	Frequency
25–29	5501
30–34	4643
35–39	4322
40–44	4564
45–49	4464
50–54	4262
55–59	3838
60–64	3261
65–69	2195
70–74	1335

Source: U.S. Census Bureau

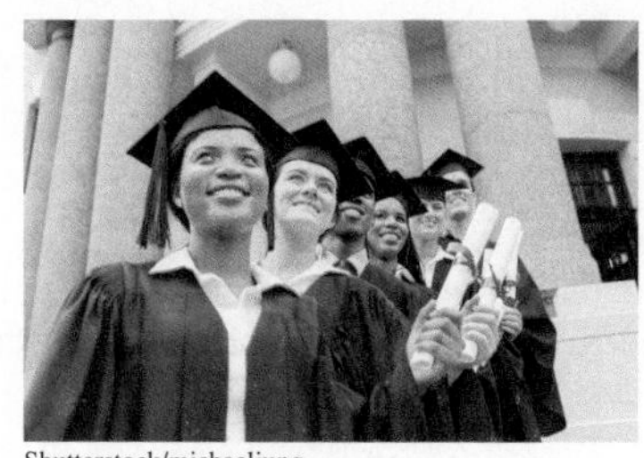
Shutterstock/michaeljung

a. Approximate the mean age. *45.9*

b. If the majority of the people ages 70–74 were in fact 74, would this tend to make the approximate mean too large or too small? *Too small*

59. Income: The personal income per capita of a state is the total income of all adults in the state, divided by the number of adults. The following table presents the personal income per capita (in thousands of dollars) for each of the 50 states and the District of Columbia.

32	40	33	30	42	41	54	41	61	38	33	39	31	40	34	35	37
31	35	34	46	49	35	41	29	34	32	36	40	42	49	31	47	34
35	35	34	35	39	39	31	34	33	37	31	37	41	38	30	36	43

Source: U.S. Bureau of Economic Analysis

a. What is the mean state income? *37.5*

b. What is the median state income? *36*

c. Based on the mean and median, would you expect the data to be skewed to the left, skewed to the right, or approximately symmetric? Explain. *Skewed right*

d. Construct a frequency histogram. Do the results agree with your expectation?

60. Take in a show: The following table presents the weekly attendance, in thousands, at Broadway shows during a recent season.

240.9	241.0	240.2	245.9	251.0	260.2	252.1	256.0	269.4	222.4	213.6	196.6	192.2
191.3	185.5	162.5	180.9	183.0	196.4	187.7	235.2	292.4	240.9	231.6	232.1	220.6
246.5	236.8	219.7	149.4	214.9	228.3	214.0	205.3	184.2	183.1	189.9	181.6	190.4
197.0	208.5	218.2	238.5	235.8	239.3	240.4	234.9	244.1	254.9	269.2	265.6	257.5

Source: Broadway League

a. Find the mean and median weekly attendance. *Mean: 222.49; median: 229.95*

b. Based on the mean and median, would you expect that a histogram would be skewed to the left, skewed to the right, or approximately symmetric? Explain. *Approximately symmetric*

c. Construct a frequency histogram. Do the results agree with your expectation?

61. Read any good books lately? The following data represent the responses of 24 library patrons when asked about their favorite type of book. Which type of book is the mode? *Fiction*

Biography	Fiction	Biography	Historical	Fiction	Biography
Fiction	Nonfiction	Fiction	Fiction	Historical	Biography
How-to guide	Nonfiction	How-to guide	Nonfiction	Historical	Nonfiction
Fiction	Fiction	How-to guide	How-to guide	Biography	How-to guide

62. Sources of news: A sample of 32 U.S. adults was surveyed and asked, "Do you get most of your information about current events from newspapers, magazines, the internet, television, radio, or some other source?" The results are shown below. What is the mode? *Television*

Television	Internet	Television	Other	Newspapers	Internet	Television	Radio
Magazines	Newspapers	Other	Radio	Radio	Internet	Other	Television
Internet	Magazines	Other	Television	Internet	Newspapers	Newspapers	Internet
Newspapers	Other	Television	Newspapers	Television	Magazines	Television	Television

Source: General Social Survey

63. Find the mean: The National Center for Health Statistics sampled 5844 American women over the age of 20 and found that their median weight was 157.2 pounds. A histogram of the data set was skewed to the right. Which of the following values for the mean weight is most plausible: 150 pounds, 155 pounds, or 160 pounds? Explain. *160*

64. Find the median: According to a recent Current Population Survey, the mean personal income for American adults was $38,337. A histogram for incomes is skewed to the right. Which of the following is a possible value for the median income, $26,000, $40,000, or $50,000? *$26,000*

65. Find the median: In a recent year, approximately 1.7 million people took the mathematics section of the SAT. Of these, 42% scored less than 490, 45% scored less than 500, 49% scored less than 510, 52% scored less than 520, and 55% scored less than 530. In what interval is the median score? *iii.*

i. Between 490 and 500
ii. Between 500 and 510
iii. Between 510 and 520
iv. Between 520 and 530

66. Find the median: The National Health and Nutrition Examination Survey measured the heights, in inches, of a large number of American adult men. Of these heights, 33% were less than 68, 42% were less than 69, 59% were less than 70, and 71% were less than 71. In what interval is the median height? *ii.*

i. Between 68 and 69
ii. Between 69 and 70
iii. Between 70 and 71
iv. Between 71 and 72

67. Heights: There are 2500 women and 2000 men enrolled in a certain college. The mean height for the women is 64.4 inches, and the mean height for the men is 69.8 inches.

a. Find the sum of the heights of the women. *161,000*
b. Find the sum of the heights of the men. *139,600*
c. Find the mean height for all 4500 people. *66.8*
d. The average of the means for men and women is $(64.4 + 69.8)/2 = 67.1$. Explain why this is not equal to the mean height for all the people.

68. Exam scores: There are two sections of an introductory statistics course. Section A has 25 students and section B has 30 students. All students took the same final exam. The mean score in section A was 82, and the mean score in section B was 78.

a. Find the sum of the scores in section A. *2050*
b. Find the sum of the scores in section B. *2340*
c. Find the mean score for all 55 students. *79.8*
d. The average of the means in the two sections is $(82 + 78)/2 = 80$. Explain why this is not equal to the mean score for all the students.

69. Heights: There are 35 students in a class. Fifteen of them are men and 20 of them are women. The mean height of the men is 69.65 inches, and the mean height of all 35 students is 66.85 inches. What is the mean height of the women? *64.75 inches*

70. Exam scores: There are two sections of an elementary statistics course. Section A has 25 students and section B has 35 students. All students took the same final exam. The mean score in section A was 72, and the mean score for all 60 students was 79. What was the mean score in section B? *84*

71. How many numbers? A data set has a median of 17, and six of the numbers in the data set are less than 17. The data set contains a total of n numbers.

a. If n is odd, and exactly one number in the data set is equal to 17, what is the value of n? *13*
b. If n is even, and none of the numbers in the data set are equal to 17, what is the value of n? *12*

72. How many numbers? A data set has a median of 10, and eight of the numbers in the data set are less than 10. The data set contains a total of n numbers.

a. If n is odd, and exactly one of the numbers in the data set is equal to 10, what is the value of n? *17*
b. If n is even, and two of the numbers in the data set are equal to 10, what is the value of n? *18*

73. What's the score? Jermaine has entered a bowling tournament. To prepare, he bowls five games each day and writes down the score of each game, along with the mean of the five scores. He is looking at the scores from one day last week and finds that one of the numbers has accidentally been erased. The four remaining scores are 201, 193, 221, and 187. The mean score is 202. What is the missing score? *208*

74. What's your grade? Addison has been told that her average on six homework assignments in her history class is 85. She can find only five of the six assignments, which have scores of 91, 72, 96, 88, and 75. What is the score on the lost homework assignment? *88*

75. Weighted mean: Rachel worked at three part-time jobs last week. At one job, she worked 20 hours at a salary of $12 per hour, at another she worked 10 hours at $15 per hour, and at the third she worked 12 hours at $10 per hour. What was her mean hourly wage? *$12.14*

76. Weighted mean: In Jacob's statistics class, the final grade is a weighted mean of a homework grade, three midterm exam grades, and a final exam grade. The homework counts for 10% of the final grade, each midterm counts for 20%, and the final exam counts for 30%. Jacob got a score of 70 on the homework, 60, 75, and 85 on the three midterms, and 80 on the final. What is Jacob's final grade? *75*

77. Mean or median? The Smith family in Example 3.5 had the good fortune to win a million-dollar prize in a lottery. Their annual income for each of the five years leading up to their lottery win are as follows:

15,000 18,000 20,000 25,000 1,025,000

a. Compute the mean annual income. *220,600*
b. Compute the median annual income. *20,000*
c. Which provides a more appropriate description of the Smiths' financial position, the mean or the median? Explain. *Mean*

78. Mean or median? An insurance company examines the driving records of 100 policy holders. They find that 80 of them had no accidents in the past year, 15 had one accident, 4 had two, and 1 had three.
a. Find the mean number of accidents. *0.26*
b. Find the median number of accidents. *0*
c. Which quantity is more useful to the insurance company, the mean or the median? Explain. *Mean*

79. Properties of the mean: Make up a data set in which the mean is equal to one of the numbers in the data set.

80. Properties of the median: Make up a data set in which the median is equal to one of the numbers in the data set.

81. Properties of the mean: Make up a data set in which the mean is *not* equal to one of the numbers in the data set.

82. Properties of the median: Make up a data set in which the median is *not* equal to one of the numbers in the data set.

83. The midrange: The **midrange** is a measure of center that is computed by averaging the largest and smallest values in a data set. In other words,

$$\text{Midrange} = \frac{\text{Largest value} + \text{Smallest value}}{2}$$

Is the midrange resistant? Explain. *No*

84. Mean, median, and midrange: A data set contains only two values. Are the mean, median, and midrange all equal? Explain. *Yes*

Extending the Concepts

85. Changing units: A sample of five college students have heights, in inches, of 65, 72, 68, 67, and 70.
a. Compute the sample mean. *68.4*
b. Compute the sample median. *68*
c. Convert each of the heights to units of feet, by dividing by 12. *5.417, 6.000, 5.667, 5.583, 5.833*
d. Compute the sample mean of the heights in feet. Is this equal to the sample mean in inches divided by 12? *5.7; yes*
e. Compute the sample median of the heights in feet. Is this equal to the sample median in inches divided by 12? *5.667; yes*

86. Effect on the mean and median: Four employees in an office have annual salaries of \$30,000, \$35,000, \$45,000, and \$70,000.
a. Compute the mean salary. *45,000*
b. Compute the median salary. *40,000*
c. Each employee gets a \$1000 raise. Compute the new mean. Does the mean increase by \$1000? *46,000; yes*
d. Each employee gets a 5% raise. Compute the new mean. Does the mean increase by 5%? *47,250; yes*

87. Nonresistant median: Consider the following data set:

0 0 1 1 1 1 1 1 2 2 8 8 9 9 9 9 9 9 10 10

a. Show that the mean and median are both equal to 5.
b. Suppose that a value of 26 is added to this data set. Which is affected more, the mean or the median? *Median*
c. Suppose that a value of 100 is added to this data set. Which is affected more, the mean or the median? *Mean*
d. It is possible for an extreme value to affect the median more than the mean, but if the value is extreme enough, the mean will be affected more than the median. Explain.

88. Exception to the skewness rule: Consider the following data set:

0 0 0 0 0 0 0 0 0 0 1 1 1 1 1 1 1 2 2 2 3

a. Compute the mean and median. *Mean: 0.76; median: 1*
b. Based on the mean and median, would you expect the data set to be skewed to the left, skewed to the right, or approximately symmetric? Explain. *Skewed left*
c. Construct a frequency histogram. Does the histogram have the shape you expected? *No*

Answers to Check Your Understanding Exercises for Section 3.1

1. Mean is 65.4; median is 60.
2. Mean is 68.5; median is 74.5.
3. The mean does not have to be a value that could possibly appear in the data set.
4. Skewed to the right
5. Skewed to the left
6. Approximately symmetric
7. Approximately symmetric
8. a. 1 **b.** 4 and 6 **c.** No mode

Section 3.2 Measures of Spread

Objectives

1. Compute the range of a data set
2. Compute the variance of a population and a sample
3. Compute the standard deviation of a population and a sample
4. Approximate the standard deviation with grouped data
5. Use the Empirical Rule to summarize data that are unimodal and approximately symmetric
6. Use Chebyshev's Inequality to describe a data set
7. Compute the coefficient of variation

NOTE TO INSTRUCTOR

The following topic from the **Statistics Corequisite Workbook** is aligned with the material in this section.

2.3 - Exponents and Square Roots in Measures of Spread

Would you rather live in San Francisco or St. Louis? If you had to choose between these two cities, one factor you might consider is the weather. Table 3.3 presents the average monthly temperatures, in degrees Fahrenheit, for both cities.

Table 3.3 Temperatures in San Francisco and St. Louis

	Jan	Feb	Mar	Apr	May	Jun	Jul	Aug	Sep	Oct	Nov	Dec
San Francisco	51	54	55	56	58	60	60	61	63	62	58	52
St. Louis	30	35	44	57	66	75	79	78	70	59	45	35

Source: National Weather Service

To compare the temperatures, we will compute their means.

$$\text{Mean for San Francisco} = \frac{51+54+55+56+58+60+60+61+63+62+58+52}{12} = 57.5$$

$$\text{Mean for St. Louis} = \frac{30+35+44+57+66+75+79+78+70+59+45+35}{12} = 56.1$$

The means are similar: 57.5° for San Francisco and 56.1° for St. Louis. Does this mean that the temperatures are similar in both cities? Definitely not. St. Louis has a cold winter and a hot summer, while the temperature in San Francisco is much the same all year round. Another way to say this is that the temperatures in St. Louis are more spread out than the temperatures in San Francisco. The dotplots in Figure 3.6 illustrate the difference in spread.

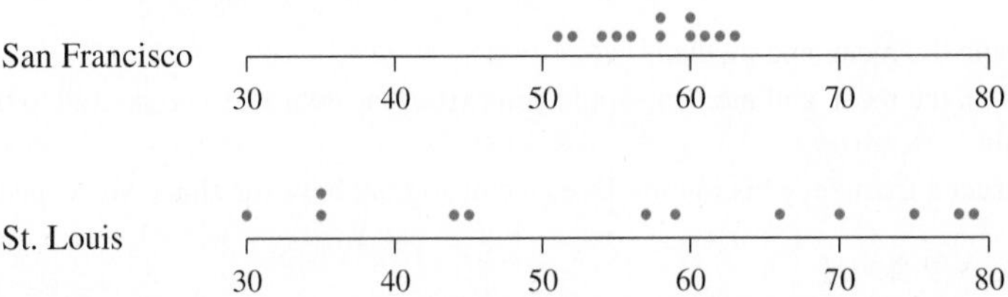

Figure 3.6 The monthly temperatures for St. Louis are more spread out than those for San Francisco.

The mean does not tell us anything about how spread out the data are; it only gives us a measure of the center. It is clear that the mean by itself is not adequate to describe a data set. We must also have a way to describe the amount of spread. Dotplots allow us to visualize the spread, but we need a numerical summary to measure it precisely.

Objective 1 Compute the range of a data set

The Range

The simplest measure of the spread of a data set is the *range*.

DEFINITION

The **range** of a data set is the difference between the largest value and the smallest value.

$$\text{Range} = \text{Largest value} - \text{Smallest value}$$

Example 3.11

Compute the range of a data set

Compute the range of the temperature data for San Francisco and for St. Louis, and interpret the results.

Solution

The largest value for San Francisco is 63 and the smallest is 51. The range for San Francisco is $63 - 51 = 12$.

The largest value for St. Louis is 79 and the smallest is 30. The range for St. Louis is $79 - 30 = 49$.

The range is much larger for St. Louis, which indicates that the spread in the temperatures is much greater there.

Although the range is easy to compute, it is not often used in practice. The reason is that the range involves only two values from the data set—the largest and the smallest. The measures of spread that are most often used are the variance and the standard deviation, which use every value in the data set.

Objective 2 Compute the variance of a population and a sample

The Variance

When a data set has a small amount of spread, like the San Francisco temperatures, most of the values will be close to the mean. When a data set has a larger amount of spread, more of the data values will be far from the mean. The **variance** is a measure of how far the values in a data set are from the mean, on the average. We will describe how to compute the variance of a population.

The difference between a population value, x, and the population mean, μ, is $x - \mu$. This difference is called a **deviation**. Values less than the mean will have negative deviations, and values greater than the mean will have positive deviations. If we were simply to add the deviations, the positive and the negative ones would cancel out. So we square the deviations to make them all positive. Data sets with a lot of spread will have many large squared deviations, while those with less spread will have smaller squared deviations. The average of the squared deviations is the *population variance*.

EXPLAIN IT AGAIN

Another formula for the population variance: An alternate formula for the population variance is $\sigma^2 = \dfrac{\sum x^2 - N\mu^2}{N}$

This formula always gives the same result as the one in the definition.

DEFINITION

Let $x_1, \ldots, x_N$ denote the values in a population of size N. Let μ denote the population mean. The **population variance**, denoted by σ^2, is

$$\sigma^2 = \frac{\sum(x - \mu)^2}{N}$$

We present the procedure for computing the population variance.

CAUTION

The population variance will *never* be negative. It will be equal to zero if all the values in a population are the same. Otherwise, the population variance will be positive.

Procedure for Computing the Population Variance

Step 1: Compute the population mean μ.
Step 2: For each population value x, compute the deviation $x - \mu$.
Step 3: Square the deviations, to obtain quantities $(x - \mu)^2$.
Step 4: Sum the squared deviations, obtaining $\sum(x - \mu)^2$.
Step 5: Divide the sum obtained in Step 4 by the population size N to obtain the population variance σ^2.

In practice, variances are usually calculated with technology. It is a good idea to compute a few by hand, however, to get a feel for the procedure.

Example 3.12

Computing the population variance

Compute the population variance for the San Francisco temperatures.

Solution

The calculations are shown in Table 3.4.

Step 1: Compute the population mean:

$$\mu = \frac{51 + 54 + 55 + 56 + 58 + 60 + 60 + 61 + 63 + 62 + 58 + 52}{12} = 57.5$$

Step 2: Subtract μ from each value to obtain the deviations $x - \mu$. These calculations are shown in the second column of Table 3.4.

NOTE TO INSTRUCTOR

It may be helpful to point out that the variance is unusual in that it is calculated differently for samples than for populations.

Table 3.4 Calculations for the Population Variance in Example 3.12

x	$x - \mu$	$(x - \mu)^2$
51	−6.5	$(-6.5)^2 = 42.25$
54	−3.5	$(-3.5)^2 = 12.25$
55	−2.5	$(-2.5)^2 = 6.25$
56	−1.5	$(-1.5)^2 = 2.25$
58	0.5	$0.5^2 = 0.25$
60	2.5	$2.5^2 = 6.25$
60	2.5	$2.5^2 = 6.25$
61	3.5	$3.5^2 = 12.25$
63	5.5	$5.5^2 = 30.25$
62	4.5	$4.5^2 = 20.25$
58	0.5	$0.5^2 = 0.25$
52	−5.5	$(-5.5)^2 = 30.25$

$\mu = 57.5$ $\qquad$ $\sum(x - \mu)^2 = 169$

$$\sigma^2 = \frac{169}{12} = 14.083$$

Step 3: Square the deviations. These calculations are shown in the third column of Table 3.4.

Step 4: Sum the squared deviations to obtain

$$\sum(x - \mu)^2 = 169$$

Step 5: The population size is $N = 12$. Divide the sum obtained in Step 4 by N to obtain the population variance σ^2.

$$\sigma^2 = \frac{\sum(x - \mu)^2}{N} = \frac{169}{12} = 14.083$$

In Example 3.12, note how important it is to make all the deviations positive, which we do by squaring them. If we simply add the deviations without squaring them, the positive and negative ones will cancel each other out, leaving 0.

Check Your Understanding

1. Compute the population variance for the St. Louis temperatures. Compare the result with the variance for the San Francisco temperatures, and interpret the result. *291.9; greater than San Francisco variance*

Answer is on page 139.

The sample variance

When the data values come from a sample rather than a population, the variance is called the *sample variance*. The procedure for computing the sample variance is a bit different from the one used to compute a population variance.

CAUTION

The sample variance will *never* be negative. It will be equal to zero if all the values in a sample are the same. Otherwise, the sample variance will be positive.

DEFINITION

Let $x_1, ..., x_n$ denote the values in a sample of size n. The **sample variance**, denoted by s^2, is

$$s^2 = \frac{\sum(x - \bar{x})^2}{n - 1}$$

The formula is the same as for the population variance, except that we replace the population mean μ with the sample mean $\bar{x}$, and we divide by $n - 1$ rather than N.

We present the procedure for computing the sample variance.

EXPLAIN IT AGAIN

Another formula for the sample variance: An alternate formula for the sample variance is

$$s^2 = \frac{\sum x^2 - n\bar{x}^2}{n - 1}$$

This formula will always give the same result as the one in the definition.

Procedure for Computing the Sample Variance

Step 1: Compute the sample mean $\bar{x}$.

Step 2: For each sample value x, compute the difference $x - \bar{x}$. This quantity is called a deviation.

Step 3: Square the deviations, to obtain quantities $(x - \bar{x})^2$.

Step 4: Sum the squared deviations, obtaining $\sum(x - \bar{x})^2$.

Step 5: Divide the sum obtained in Step 4 by $n - 1$ to obtain the sample variance s^2.

Example 3.13

Computing the sample variance

A company that manufactures batteries is testing a new type of battery designed for laptop computers. They measure the lifetimes, in hours, of six batteries, and the results are 3, 4, 6, 5, 4, and 2. Find the sample variance of the lifetimes.

CAUTION

Don't round off the value of $\bar{x}$ when computing the sample variance.

Solution

The calculations are shown in Table 3.5 (page 122).

Step 1: Compute the sample mean:

$$\bar{x} = \frac{3 + 4 + 6 + 5 + 4 + 2}{6} = 4$$

Step 2: Subtract $\bar{x}$ from each value to obtain the deviations $x - \bar{x}$. These calculations are shown in the second column of Table 3.5.

Step 3: Square the deviations. These calculations are shown in the third column of Table 3.5.

Step 4: Sum the squared deviations to obtain

$$\sum(x - \bar{x})^2 = 10$$

Table 3.5 Calculations for the Sample Variance in Example 3.13

x	$x-\bar{x}$	$(x-\bar{x})^2$
3	−1	$(-1)^2 = 1$
4	0	$0^2 = 0$
6	2	$2^2 = 4$
5	1	$1^2 = 1$
4	0	$0^2 = 0$
2	−2	$(-2)^2 = 4$
$\bar{x} = 4$		$\sum(x-\bar{x})^2 = 10$
		$s^2 = \frac{10}{6-1} = 2$

Step 5: The sample size is $n = 6$. Divide the sum obtained in Step 4 by $n - 1$ to obtain the sample variance s^2.

$$s^2 = \frac{\sum(x-\bar{x})^2}{n-1} = \frac{10}{6-1} = 2$$

Why do we divide by $n - 1$ rather than n?

It is natural to wonder why we divide by $n - 1$ rather than n when computing the sample variance. When computing the sample variance, we use the sample mean to compute the deviations $x-\bar{x}$. For the population variance, we use the population mean for the deviations $x - \mu$. Now it turns out that the deviations using the sample mean tend to be a bit smaller than the deviations using the population mean. If we were to divide by n when computing a sample variance, the value would tend to be a bit smaller than the population variance. It can be shown mathematically that the appropriate correction is to divide the sum of the squared deviations by $n - 1$ rather than n.

EXPLAIN IT AGAIN

Degrees of freedom and sample size: The number of degrees of freedom for the sample variance is one less than the sample size.

The quantity $n-1$ is sometimes called the **degrees of freedom** for the sample standard deviation. The reason is that the deviations $x - \bar{x}$ will always sum to 0. Thus, if we know the first $n - 1$ deviations, we can compute the nth one. For example, if our sample consists of the four numbers 2, 4, 9, and 13, the sample mean is

$$\bar{x} = \frac{2+4+9+13}{4} = 7$$

The first three deviations are

$$2-7=-5, \quad 4-7=-3, \quad 9-7=2$$

The sum of the first three deviations is

$$-5+(-3)+2=-6$$

We can now determine that the last deviation must be 6, in order to make the sum of all four deviations equal to 0. When we know the first three deviations, we can determine the fourth one. Thus, for a sample of size 4, there are 3 degrees of freedom. In general, for a sample of size n, there will be $n - 1$ degrees of freedom.

Check Your Understanding

2. A credit card company is studying the levels of credit card debt held by college students. They select eight students at random. Their balances, in dollars, are 342, 541, 1652, 43, 87, 137, 0, 418. Find the sample variance of the balances. *292,230*

3. The average temperatures in Denver, Colorado, for a sample of six days one summer were 77, 80, 78, 80, 83, and 85. Find the sample variance of the temperatures. *9.1*

Answers are on page 139.

Objective 3 Compute the standard deviation of a population and a sample

The Standard Deviation

There is a problem with using the variance as a measure of spread. Because the variance is computed using squared deviations, the units of the variance are the squared units of the data. For example, in Example 3.13, the units of the data are hours, and the units of variance are squared hours. In most situations, it is better to use a measure of spread that has the same units as the data. We do this simply by taking the square root of the variance. The quantity thus obtained is called the **standard deviation**. The standard deviation of a sample is denoted s, and the standard deviation of a population is denoted σ.

NOTATION ROUNDUP

Sample standard deviation: s
Population standard deviation: σ

DEFINITION

- The **sample standard deviation** s is the square root of the sample variance s^2.

$$s = \sqrt{s^2}$$

- The **population standard deviation** σ is the square root of the population variance σ^2.

$$\sigma = \sqrt{\sigma^2}$$

Example 3.14

Computing the standard deviation

The lifetimes, in hours, of six batteries (first presented in Example 3.13) were 3, 4, 6, 5, 4, and 2. Find the standard deviation of the battery lifetimes.

CAUTION

Don't round off the variance when computing the standard deviation.

Solution

In Example 3.13, we computed the sample variance to be $s^2 = 2$. The sample standard deviation is therefore

$$s = \sqrt{s^2} = \sqrt{2} = 1.414$$

Example 3.15

Computing standard deviations with technology

Compute the standard deviation for the data in Example 3.14.

Solution

Following is the display from the TI-84 Plus.

```
1-Var Stats
x̄=4
Σx=24
Σx²=106
Sx=1.414213562
σx=1.290994449
n=6
minX=2
↓Q1=3
```

NOTE TO INSTRUCTOR

It is worth emphasizing that the standard deviation is a more useful measure of spread than the variance. The reason is that the standard deviation has the same units as the data.

The TI-84 Plus calculator denotes the sample standard deviation by Sx. The display shows that the sample standard deviation is equal to 1.414213562. The calculator does not know whether the data set represents a sample or an entire population. Therefore, the display also presents the population standard deviation, which is denoted σx.

Check Your Understanding

4. Find the sample variance and standard deviation for the following samples.

a. 2 2 1 5 4 5 0 1 *Variance: 3.7143; standard deviation: 1.9272*

b. 22 27 13 53 47 50 *Variance: 281.8667; standard deviation: 16.7889*

5. Find the population variance and standard deviation for the population values

a. 7 3 2 5 4 9 6 3 *Variance: 4.8594; standard deviation: 2.2044*

b. 79 20 52 19 56 12 *Variance = 590.8889, standard deviation = 24.30821*

Answers are on page 139.

RECALL

A statistic is resistant if its value is not affected much by extreme data values.

The standard deviation is not resistant

Example 3.16 shows that the standard deviation is *not* resistant.

Example 3.16

Show that the standard deviation is not resistant

In Example 3.15, we found that the sample standard deviation for six battery lifetimes was $s = 1.414$. The six lifetimes were 3, 4, 6, 5, 4, and 2. Assume that the battery with a lifetime of 6 actually had a lifetime of 20, so that the lifetimes are 3, 4, 20, 5, 4, and 2. Compute the standard deviation.

Solution

Following is the display from the TI-84 Plus.

```
1-Var Stats
x̄=6.333333333
Σx=38
Σx²=470
Sx=6.772493386
σx=6.18241233
n=6
minX=2
↓Q1=3
```

Including the extreme value of 20 in the data set has increased the sample standard deviation from 1.414 to 6.772. Clearly, the standard deviation is not resistant.

Objective 4 Approximate the standard deviation with grouped data

Approximating the Standard Deviation with Grouped Data

Sometimes we don't have access to the raw data in a data set, but we are given a frequency distribution. In Section 3.1, we learned how to approximate the sample mean from a frequency distribution. We now show how to approximate the standard deviation.

We use Table 3.6 to illustrate the method. Table 3.6 presents the number of text messages sent by a sample of 50 high school students. In Section 3.1, we computed the approximate sample mean to be $\bar{x} = 137$. We will now approximate the standard deviation.

Table 3.6 Number of Text Messages Sent by High School Students

Number of Text Messages Sent	Frequency
0–49	10
50–99	5
100–149	13
150–199	11
200–249	7
250–299	4

We present the procedure for approximating the standard deviation from grouped data.

Procedure for Approximating the Standard Deviation with Grouped Data

Step 1: Compute the midpoint of each class. The midpoint of a class is found by taking the average of the lower class limit and the lower limit of the next larger class. Then compute the mean as described in Section 3.1.

Step 2: For each class, subtract the mean from the class midpoint to obtain Midpoint − Mean.

Step 3: For each class, square the difference obtained in Step 2 to obtain (Midpoint − Mean)^2 and multiply by the frequency to obtain (Midpoint − Mean)^2 × Frequency.

Step 4: Add the products (Midpoint − Mean)^2 × Frequency over all classes.

Step 5: Compute the sum of the frequencies n. To compute the *population* variance, divide the sum obtained in Step 4 by n. To compute the *sample* variance, divide the sum obtained in Step 4 by $n - 1$.

Step 6: Take the square root of the variance obtained in Step 5. The result is the standard deviation.

Example 3.17

Computing the standard deviation for grouped data

Compute the approximate sample standard deviation of the number of messages sent, using the data given in Table 3.6.

Solution

The calculations are summarized in Table 3.7.

Table 3.7 Calculating the Variance and Standard Deviation of the Number of Text Messages

Class	Midpoint	Frequency	Mean	Midpoint − Mean	(Midpoint − Mean)^2 × Frequency
0–49	25	10	137	−112	12,544 × 10 = 125,440
50–99	75	5	137	−62	3,844 × 5 = 19,220
100–149	125	13	137	−12	144 × 13 = 1,872
150–199	175	11	137	38	1,444 × 11 = 15,884
200–249	225	7	137	88	7,744 × 7 = 54,208
250–299	275	4	137	138	19,044 × 4 = 76,176
		Sum = 50			Sum = 292,800

$$\text{Variance} = \frac{292{,}800}{50 - 1} = 5975.51020 \qquad \text{Standard deviation} = \sqrt{5975.51020} = 77.3014$$

Step 1: Compute the midpoints: For the first class, the lower class limit is 0. The lower limit of the next class is 50. The midpoint is therefore

$$\frac{0 + 50}{2} = 25$$

We continue in this manner to compute the midpoints of each of the classes. Note that for the last class, we average the lower limit of 250 with 300, which is the lower limit that the next class would have. We computed the sample mean in Example 3.9 in Section 3.1. The sample mean is $\bar{x} = 137$.

Step 2: For each class, subtract the mean from the class midpoint as shown in the column labeled "Midpoint − Mean."

Step 3: For each class, square the difference obtained in Step 2 and multiply by the frequency as shown in the column labeled "(Midpoint − Mean)^2 × Frequency."

Step 4: Add the products $(\text{Midpoint} - \text{Mean})^2 \times \text{Frequency}$ over all classes, to obtain the sum 292,800.

Step 5: The sum of the frequencies is 50. Since we are considering the data to be a sample, we subtract 1 from this sum to obtain 49. The sample variance is $292{,}800/49 = 5975.51020$.

Step 6: The sample standard deviation is $\sqrt{5975.51020} = 77.3014$.

Objective 5 Use the Empirical Rule to summarize data that are unimodal and approximately symmetric

The Empirical Rule

Many histograms have a single mode near the center of the data and are approximately symmetric. Such histograms are often referred to as *bell-shaped*. Other histograms are strongly skewed; these are not bell-shaped. When a data set has a bell-shaped histogram, it is often possible to use the standard deviation to provide an approximate description of the data using a rule known as the **Empirical Rule**.

The Empirical Rule

When a population has a histogram that is approximately bell-shaped, then

- Approximately 68% of the data will be within one standard deviation of the mean. In other words, approximately 68% of the data will be between $\mu - \sigma$ and $\mu + \sigma$.
- Approximately 95% of the data will be within two standard deviations of the mean. In other words, approximately 95% of the data will be between $\mu - 2\sigma$ and $\mu + 2\sigma$.
- All, or almost all, of the data will be within three standard deviations of the mean. In other words, all, or almost all, of the data will be between $\mu - 3\sigma$ and $\mu + 3\sigma$.

CAUTION

The Empirical Rule should not be used for data sets that are not approximately bell-shaped.

The Empirical Rule holds for many bell-shaped data sets. Figure 3.7 illustrates the Empirical Rule.

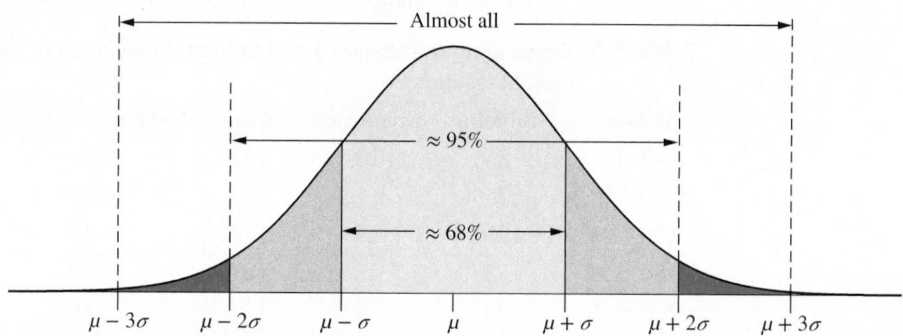

Figure 3.7 The Empirical Rule. Approximately 68% of the data values are between $\mu - \sigma$ and $\mu + \sigma$, approximately 95% are between $\mu - 2\sigma$ and $\mu + 2\sigma$, and almost all are between $\mu - 3\sigma$ and $\mu + 3\sigma$.

Example 3.18

Using the Empirical Rule to describe a data set

Table 3.8 presents the percentage of the population aged 65 and over in each state and the District of Columbia. Figure 3.8 (page 127) presents a histogram of these data. Compute the mean and standard deviation, and use the Empirical Rule to describe the data.

Table 3.8 Percentage of People Aged 65 and Over in Each of the 50 States and District of Columbia

14.1	8.1	13.9	14.3	11.5	10.7	14.4	14.1	11.5	17.8	14.0
10.2	14.3	12.0	12.4	12.7	14.9	13.4	13.1	12.6	15.6	
12.2	13.7	12.8	12.4	12.8	13.9	15.0	13.8	12.3	12.6	
13.7	14.1	13.6	12.4	15.3	13.7	13.8	13.0	15.5	14.1	
13.6	14.6	13.3	10.5	9.0	14.3	12.4	12.2	16.0	13.5	

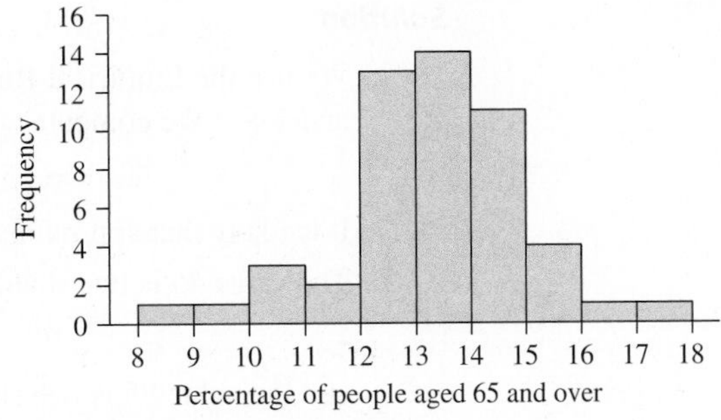

Figure 3.8 Histogram for the data in Table 3.8

Solution

Step 1: Figure 3.8 shows that the histogram is approximately bell-shaped, so we may use the Empirical Rule.

Step 2: We use the TI-84 Plus to compute the mean and standard deviation. The display is shown here:

```
1-Var Stats
x̄=13.24901961
Σx=675.7
Σx²=9096.77
Sx=1.699455507
σx=1.682711694
n=51
minX=8.1
↓Q1=12.4
```

Note that the 51 entries (corresponding to 50 states plus the District of Columbia) are an entire population. Therefore, we will interpret the quantity $\bar{x} = 13.24901961$ (rounded to 13.249) produced by the TI-84 Plus as the population mean μ, and we will use $\sigma = 1.682711694$ (rounded to 1.683) for the standard deviation.

Step 3: We compute the quantities $\mu - \sigma$, $\mu + \sigma$, $\mu - 2\sigma$, $\mu + 2\sigma$, $\mu - 3\sigma$, and $\mu + 3\sigma$.

$$\mu - \sigma = 13.249 - 1.683 = 11.57$$

$$\mu + \sigma = 13.249 + 1.683 = 14.93$$

$$\mu - 2\sigma = 13.249 - 2(1.683) = 9.88$$

$$\mu + 2\sigma = 13.249 + 2(1.683) = 16.62$$

$$\mu - 3\sigma = 13.249 - 3(1.683) = 8.20$$

$$\mu + 3\sigma = 13.249 + 3(1.683) = 18.30$$

We conclude that the percentage of the population aged 65 and over is between 11.57 and 14.93 in approximately 68% of the states, between 9.88 and 16.62 in approximately 95% of the states, and between 8.20 and 18.30 in almost all the states.

The Empirical Rule can be used for samples as well as populations. When we work with a sample, we use $\bar{x}$ in place of μ and s in place of σ.

Example 3.19

Using the Empirical Rule to describe a data set

A large class of 200 students took an exam. The scores had sample mean $\bar{x} = 65$ and sample standard deviation $s = 10$. The histogram is approximately bell-shaped.

a. Find an interval that is likely to contain approximately 68% of the scores.

b. Approximately what percentage of the scores were between 45 and 85?

c. Approximately how many students had scores between 45 and 85?

Solution

a. We use the Empirical Rule. Approximately 68% of the data will be between $\bar{x} - s$ and $\bar{x} + s$. We compute

$$\bar{x} - s = 65 - 10 = 55 \qquad \bar{x} + s = 65 + 10 = 75$$

It is likely that approximately 68% of the scores were between 55 and 75.

b. The value 45 is two standard deviations below the mean, since

$$\bar{x} - 2s = 65 - 20 = 45$$

and the value 85 is two standard deviations above the mean, since

$$\bar{x} + 2s = 65 + 20 = 85$$

Therefore, it is likely that approximately 95% of the scores were between 45 and 85.

c. From part (b), we know that approximately 95% of the students had scores between 45 and 85. There were 200 students in the class. Therefore, the number that had scores between 45 and 85 is approximately (0.95)(200) = 190.

Example 3.20

Determining whether the Empirical Rule is appropriate

Following is a histogram for a data set. Should the Empirical Rule be used?

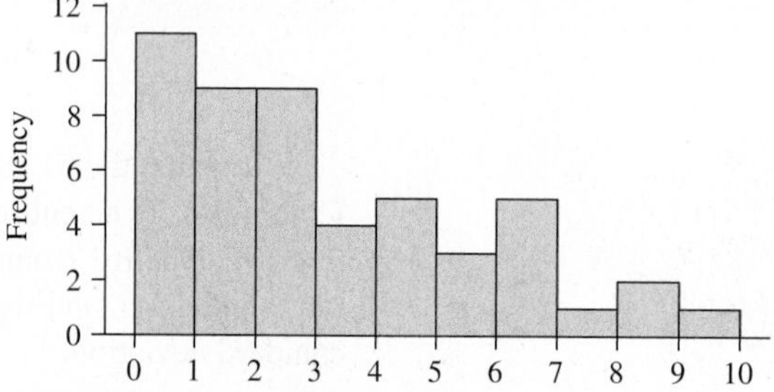

Solution

No. The distribution is skewed, rather than bell-shaped. Therefore, the Empirical Rule should *not* be used.

Check Your Understanding

6. A data set has a mean of 20 and a standard deviation of 3. A histogram is shown here. Is it appropriate to use the Empirical Rule to approximate the proportion of the data between 14 and 26? If so, find the approximation. If not, explain why not. *≈95%*

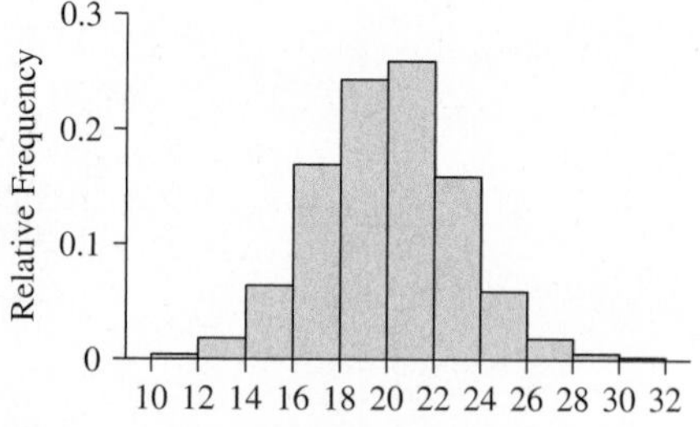

7. A data set has a mean of 50 and a standard deviation of 8. A histogram is shown here. Is it appropriate to use the Empirical Rule to approximate the proportion of the data between 42 and 58? If so, find the approximation. If not, explain why not. *Not appropriate*

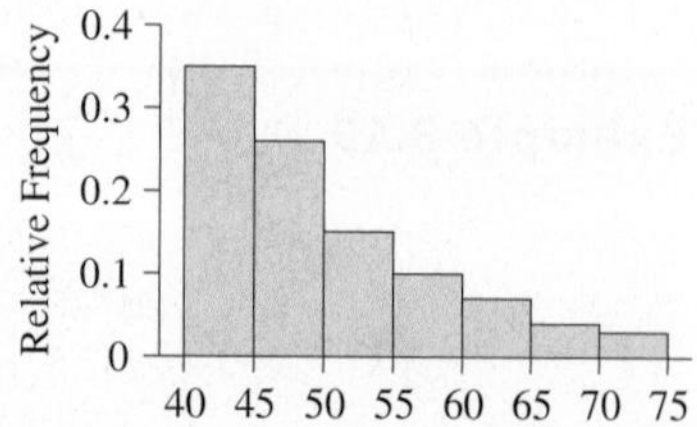

Answers are on page 139.

Objective 6 Use Chebyshev's Inequality to describe a data set

Chebyshev's Inequality

When a distribution is bell-shaped, the Empirical Rule gives us an approximation to the proportion of data that will be within one or two standard deviations of the mean. **Chebyshev's Inequality** is a rule that holds for any data set.

NOTE TO INSTRUCTOR

The material on Chebyshev's inequality and the Coefficient of Variation can be skipped without loss of continuity.

Chebyshev's Inequality

In any data set, the proportion of the data that will be within K standard deviations of the mean is at least $1 - 1/K^2$. Specifically, by setting $K = 2$ or $K = 3$, we obtain the following results:

- At least $3/4$ (75%) of the data will be within two standard deviations of the mean.
- At least $8/9$ (88.9%) of the data will be within three standard deviations of the mean.

EXPLAIN IT AGAIN

Chebyshev's Inequality is always valid: Chebyshev's Inequality can be used with any data set, whether or not it is bell-shaped.

Example 3.21

Using Chebyshev's Inequality

As part of a public health study, systolic blood pressure was measured for a large group of people. The mean was $\bar{x} = 120$ and the standard deviation was $s = 10$. What information does Chebyshev's Inequality provide about these data?

Solution

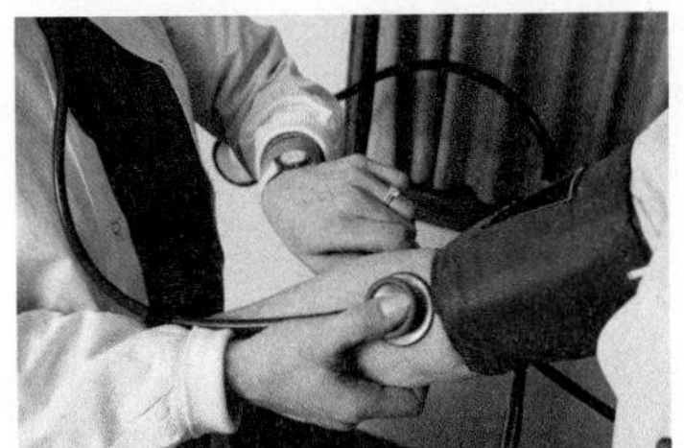

Jack Star/PhotoLink/Getty Images

We compute:

$$\bar{x} - 2s = 120 - 2(10) = 100 \qquad \bar{x} + 2s = 120 + 2(10) = 140$$

$$\bar{x} - 3s = 120 - 3(10) = 90 \qquad \bar{x} + 3s = 120 + 3(10) = 150$$

We conclude:

- At least 75% of the people had systolic blood pressures between 100 and 140.
- At least 88.9% of the people had systolic blood pressures between 90 and 150.

Comparing Chebyshev's Inequality to the Empirical Rule

Both Chebyshev's Inequality and the Empirical Rule provide information about the proportion of a data set that is within a given number of standard deviations of the mean. An advantage of Chebyshev's Inequality is that it applies to any data set, whereas the Empirical Rule applies only to data sets that are approximately bell-shaped. A disadvantage of Chebyshev's Inequality is that for most data sets, it provides only a very rough approximation. Chebyshev's Inequality produces a minimum value for the proportion of the data that will be within a given number of standard deviations of the mean. For most data sets, the actual proportions are much larger than the values given by Chebyshev's Inequality.

NOTE TO INSTRUCTOR

Mention that the value of Chebyshev's inequality is in its generality—it is valid for any data set. It is not precise, however; the bounds are extremely conservative for most data sets.

Check Your Understanding

8. A group of elementary school students took a standardized reading test. The mean score was 70 and the standard deviation was 10. Someone says that only 50% of the students scored between 50 and 90. Is this possible? Explain. *No*

9. A certain type of bolt used in an aircraft must have a length between 122 and 128 millimeters in order to be acceptable. The manufacturing process produces bolts whose mean length is 125 millimeters with a standard deviation of 1 millimeter. Can you be sure that more than 85% of the bolts are acceptable? Explain. *Yes*

Answers are on page 139.

Objective 7 Compute the coefficient of variation

The Coefficient of Variation

The coefficient of variation (CV for short) tells how large the standard deviation is relative to the mean. It can be used to compare the spreads of data sets whose values have different units.

DEFINITION

The **coefficient of variation** is found by dividing the standard deviation by the mean.

$$\text{CV} = \frac{\sigma}{\mu}$$

Example 3.22

Computing the coefficient of variation

National Weather Service records show that over a 30-year period, the annual precipitation in Atlanta, Georgia, had a mean of 49.8 inches with a standard deviation of 7.6 inches, and the annual temperature had a mean of 62.2°F with a standard deviation of 1.3 degrees. Compute the coefficient of variation for precipitation and for temperature. Which has greater spread relative to its mean?

Solution

The coefficient of variation for precipitation is

$$\text{CV for precipitation} = \frac{\text{Standard deviation of precipitation}}{\text{Mean precipitation}} = \frac{7.6}{49.8} = 0.153$$

The coefficient of variation for temperature is

$$\text{CV for temperature} = \frac{\text{Standard deviation of temperature}}{\text{Mean temperature}} = \frac{1.3}{62.2} = 0.021$$

The CV for precipitation is larger than the CV for temperature. Therefore, precipitation has a greater spread relative to its mean.

Note that we cannot compare the standard deviations of precipitation and temperature because they have different units. It does not make sense to ask whether 7.6 inches is greater than 1.3 degrees. The CV is unitless, however, so we can compare the CVs.

Check Your Understanding

10. Lengths of newborn babies have a mean of 20.1 inches with a standard deviation of 1.9 inches. Find the coefficient of variation of newborn lengths. *0.095*

Answer is on page 139.

Using Technology

TI-84 Plus

Computing the sample standard deviation

The TI-84 PLUS procedure to compute the mean and median, described on page 107, will also compute the standard deviation.

EXCEL

Computing measures of spread

The Excel procedure to compute the mean and median, described on page 108, will also compute measures of spread. Alternatively, the following commands may be used.

- The =**VAR.P**(*data_array*) and =**STDEV.P**(*data_array*) commands compute the population variance and population standard deviation, respectively.
- The =**VAR.S**(*data_array*) and =**STDEV.S**(*data_array*) commands compute the sample variance and sample standard deviation, respectively.

MINITAB

Computing the sample standard deviation

The MINITAB procedure to compute the mean and median, described on page 108, will also compute the standard deviation.

Section 3.2 Exercises

Exercises 1–10 are the Check Your Understanding exercises located within the section.

Understanding the Concepts

In Exercises 11–14, fill in each blank with the appropriate word or phrase.

11. If all values in a data set are the same, then the sample variance is equal to ________. *0*

12. The standard deviation is the square root of the ________.

13. For a bell-shaped data set, approximately ________ of the data will be in the interval $\mu - \sigma$ to $\mu + \sigma$. *68%*

14. Chebyshev's Inequality states that for any data set, the proportion of data within K standard deviations of the mean is at least ________. $1 - 1/K^2$

In Exercises 15–18, determine whether the statement is true or false. If the statement is false, rewrite it as a true statement.

15. The variance and standard deviation are measures of center. *False*

16. The range of a data set is the difference between the largest value and the smallest value. *True*

17. In a bell-shaped data set with $\mu = 15$ and $\sigma = 5$, approximately 95% of the data will be between 10 and 20. *False*

18. For some data sets, Chebyshev's Inequality may be used but the Empirical Rule should not be. *True*

Practicing the Skills

19. Find the sample variance and standard deviation for the following sample: *Variance: 100; standard deviation: 10*

17 40 24 18 16

20. Find the sample variance and standard deviation for the following sample: *Variance: 225; standard deviation: 15*

59 25 12 29 16 8 26 30 17

21. Find the sample variance and standard deviation for the following sample: *Variance: 49; standard deviation: 7*

15 9 5 12 9 21 4 24 18

22. Find the population variance and standard deviation for the following population: *Variance: 64; standard deviation: 8*

16 6 18 3 25 22

23. Find the population variance and standard deviation for the following population: *Variance: 289; standard deviation: 17*

20 8 11 23 27 29 62 4

24. Find the population variance and standard deviation for the following population: *Variance: 36; standard deviation: 6*

26 25 29 23 14 20 12 18 24 31 22 32

25. A population has variance 16. What is the population standard deviation? *4*

26. A population has variance 2.25. What is the population standard deviation? *1.5*

27. A population has standard deviation 3. What is the population variance? *9*

28. A population has standard deviation 3.5. What is the population variance? *12.25*

29. A population has mean 25 and standard deviation 6. What value is 1.5 standard deviations above the mean? *34*

30. A population has mean 10 and standard deviation 2. What value is 0.5 standard deviations below the mean? *9*

31. A population has mean 36 and standard deviation 4. What values are 2 standard deviations from the mean? *28, 44*

32. A population has mean 57 and standard deviation 6. What values are 3 standard deviations from the mean? *39, 75*

33. Approximate the sample variance and standard deviation given the following frequency distribution:

Class	Frequency
0–9	13
10–19	7
20–29	10
30–39	9
40–49	11

Variance: 228.41; standard deviation: 15.11

34. Approximate the sample variance and standard deviation given the following frequency distribution:

Class	Frequency
0–15	2
16–31	14
32–47	6
48–63	13
64–79	15

Variance: 431.02; standard deviation: 20.76

35. Approximate the population variance and standard deviation given the following frequency distribution:

Class	Frequency
0–49	17
50–99	26
100–149	14
150–199	34
200–249	26
250–299	8

Variance: 5680; standard deviation: 75.37

36. Approximate the population variance and standard deviation given the following frequency distribution:

Class	Frequency
0–19	18
20–39	11
40–59	6
60–79	6
80–99	10
100–119	5

Variance: 1238.27; standard deviation: 35.19

37. Distances of the eight planets from the sun, in millions of miles, were measured. The following TI-84 Plus display presents some results.

```
1-Var Stats
x̄=785.8125
Σx=6286.5
Σx²=12045944.6
Sx=1007.538325
σx=942.4658037
n=8
minX=36
↓Q1=80.1
```

a. Is this a population or a sample? *Population*

b. Find the appropriate standard deviation. *942.4658037*

c. Find the appropriate variance. *888,241.79*

38. A survey was taken in which people were asked how many siblings they had. The following TI-84 Plus display presents some results.

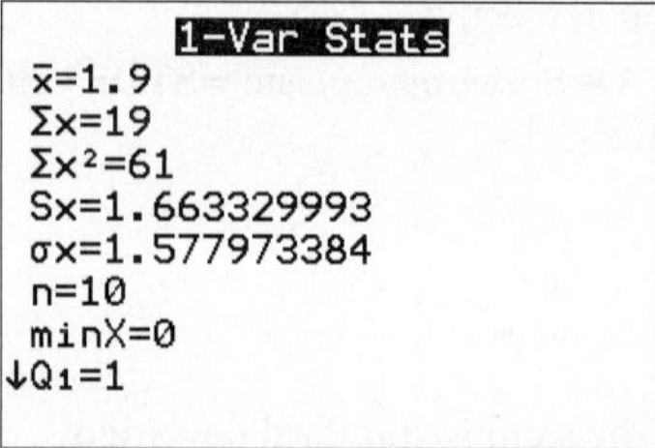

a. Is this a population or a sample? *Sample*

b. Find the appropriate standard deviation. *1.663329993*

c. Find the appropriate variance. *2.7666667*

39. The following TI-84 Plus display presents some population parameters.

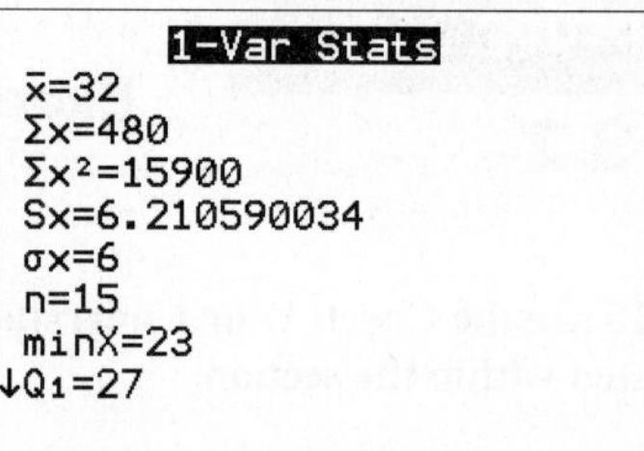

a. Assume the population is bell-shaped. Approximately what percentage of the population values are between 26 and 38? *68%*

b. Assume the population is bell-shaped. Between what two values will approximately 95% of the population be? *20 and 44*

c. If we do not assume that the population is bell-shaped, at least what percentage of the population will be between 20 and 44? *75%*

40. The following TI-84 Plus display presents some population parameters.

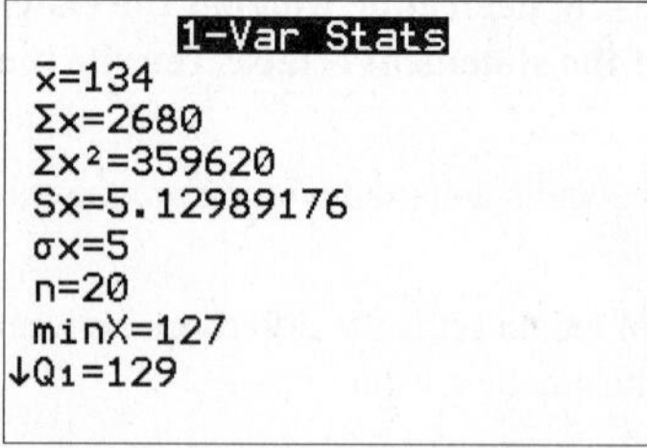

a. Assume the population is bell-shaped. Approximately what percentage of the population values are between 124 and 144? *95%*

b. Assume the population is bell-shaped. Between what two values will approximately 68% of the population be? *129 and 139*

c. If we do not assume that the population is bell-shaped, at least what percentage of the population will be between 119 and 149? *88.9%*

41. The following TI-84 Plus display presents some sample statistics.

```
1-Var Stats
x̄=176
Σx=6336
Σx²=1160496
Sx=36
σx=35.4964787
n=36
minX=20
↓Q1=176
```

a. Assume that a histogram of the sample is bell-shaped. Approximately what percentage of the sample values are between 104 and 248? 95%

b. Assume that a histogram for the sample is bell-shaped. Between what two values will approximately 68% of the sample be? 140 and 212

c. If we do not assume that the histogram is bell-shaped, at least what percentage of the sample values will be between 68 and 284? 88.9%

42. The following TI-84 Plus display presents some sample statistics.

```
1-Var Stats
x̄=79
Σx=2291
Σx²=182361
Sx=7
σx=6.878251582
n=29
minX=69
↓Q1=72
```

a. Assume that a histogram of the sample is bell-shaped. Approximately what percentage of the sample values are between 72 and 86? 68%

b. Assume that a histogram for the sample is bell-shaped. Between what two values will approximately 95% of the sample be? 65 and 93

c. If we do not assume that the histogram is bell-shaped, at least what percentage of the sample values will be between 65 and 93? 75%

Working with the Concepts

43. Touchdown! The following table presents the number of points scored by teams in the American Football Conference (AFC) during the regular season in 2017 and 2018.

Team	2017	2018
New England	458	436
Pittsburgh	406	428
LA Chargers	355	428
Kansas City	415	565
Jacksonville	417	245
Denver	289	329
Oakland	301	290
Houston	338	402
Tennessee	334	310
Cleveland	234	359
Miami	281	319
Baltimore	395	389
NY Jets	298	333
Buffalo	302	269
Indianapolis	263	433
Cincinnati	290	368

Source: ESPN

a. Find the sample standard deviation of the number of points scored in 2017. 64.98

b. Find the sample standard deviation of the number of points scored in 2018. 80.02

c. Which sample has greater spread? 2018

44. Sports car or convertible? The following table presents the fuel efficiency, in miles per gallon, for a sample of convertibles and a sample of sports cars.

Convertible Model	MPG	Sports Model	MPG
Volkswagen Eos	25	BMW 135i	23
Mini Cooper	25	Mazda3 Mazdaspeed	24
Saab 9-3	24	Subaru Impreza WRX STi	21
BMW 328i	21	Mazda RX-8	18
Toyota Camry Solara	21	Mitsubishi Lancer Evolution	21
Volvo C70	21	Volkswagen GTI	25
Ford Mustang V6	20	Honda Civic Si	27

Source: *Consumer Reports*

a. Find the sample standard deviation of the mileage for the sample of convertibles. 2.1

b. Find the sample standard deviation of the mileage for the sample of sports cars. 3.0

c. Which sample has greater spread? *Sports cars*

45. Blood pressure: The following table presents systolic blood pressure readings (in millimeters of mercury) for samples of 10 young adults (age 18–29) and 10 older adults (age 60+).

Younger:	114	103	119	117	92	116	122	117	120	109
Older:	98	130	136	139	124	167	128	113	113	156

Source: National Health and Nutrition Survey

a. Find the sample standard deviation of the blood pressures for the younger adults. *9.22*

b. Find the sample standard deviation of the blood pressures for the older adults. *20.55*

c. Which sample has greater spread? *Older*

46. Good cholesterol: High-density lipoprotein cholesterol (HDL) is sometimes called "good cholesterol," because it carries excess cholesterol to the liver where it is removed from the body. High HDL levels are desirable. Following are HDL measurements, in milligrams per deciliter, for samples of 10 men and 10 women between the ages of 18 and 29.

Men:	30	36	39	42	45	47	50	54	58	68
Women:	36	41	45	49	53	57	61	66	74	89

Source: National Health and Nutrition Survey

a. Find the sample standard deviation of HDL levels for men. *11.17*

b. Find the sample standard deviation of HDL levels for women. *16.09*

c. Which sample has greater spread? *Women*

47. Heavy football players: Following are the weights, in pounds, for samples of offensive and defensive linemen in the National Football League.

Offense:	335	301	307	252	260	307	325	310	305	305	264	325
Defense:	284	290	286	355	305	295	297	325	310	297	314	348

a. Find the sample standard deviation for the weights for the offensive linemen. *26.8*

b. Find the sample standard deviation for the weights for the defensive linemen. *23.2*

c. Is there greater spread in the weights of the offensive or the defensive linemen? *Offensive*

48. Beer: The following table presents the number of active breweries for samples of states located east and west of the Mississippi River.

East		West	
State	**Number of Breweries**	**State**	**Number of Breweries**
Connecticut	18	Alaska	17
Delaware	10	Arizona	31
Florida	47	California	305
Georgia	22	Colorado	111
Illinois	52	Iowa	21
Kentucky	13	Louisiana	6
Maine	38	Minnesota	41
Maryland	23	Montana	30
Massachusetts	40	South Dakota	5
New Hampshire	16	Texas	37
New Jersey	20	Utah	15
New York	76		
North Carolina	46		
South Carolina	14		
Tennessee	19		
Vermont	20		

Source: http://www.beerinstitute.org/

a. Compute the sample standard deviation for the number of breweries east of the Mississippi River. *18.3*

b. Compute the sample standard deviation for the number of breweries west of the Mississippi River. *87.4*

c. Compute the range for each data set. *East: 66; West: 300*

d. Based on the standard deviations, which region has the greater spread in the number of breweries? *West*

e. Based on the ranges, which region has the greater spread in the number of breweries? *West*

f. The sample of western states happens to include California. Remove California from the sample of western states, and compute the sample standard deviation for the remaining western states. Does the result show that the standard deviation is not resistant? Explain. *30.5; yes*

g. Compute the range for the western states with California removed. Is the range resistant? Explain. *106; no*

49. What's your favorite TV show? The following tables present the numbers of viewers, in millions for the top 15 prime-time shows for the 2013–2014 and 2018–2019 seasons. The numbers of viewers include those who watched the program on any platform, including time-shifting up to seven days after the original telecast.

Top Rated TV Programs: 2013–2014

Program	Rating
NBC Sunday Night Football	21.5
The Big Bang Theory	20.0
NCIS	19.8
NCIS: Los Angeles	16.0
Dancing with the Stars	15.2
The Blacklist	15.0
The Voice	14.6
Person of Interest	14.0
The Voice – Tuesday	13.9
Blue Bloods	13.6
Resurrection	13.0
Criminal Minds	12.7
Castle	12.6
60 Minutes	12.2
Grey's Anatomy	12.1

Source: Nielsen Media Research

Top Rated TV Programs: 2018–2019

Program	Rating
NFL Sunday Night Football	19.3
The Big Bang Theory	17.4
NCIS	15.9
Game of Thrones	15.3
Young Sheldon	14.6
NFL Thursday Night Football	14.4
This Is Us	13.8
Blue Bloods	12.8
FBI	12.7
The Good Doctor	12.6
Manifest	12.6
America's Got Talent: Champions	12.4
NFL Monday Night Football	11.9
Chicago Fire	11.7
The Masked Singer	11.6

Source: Nielsen Media Research

a. Find the population standard deviation of the ratings for 2013–2014. *2.91*

b. Find the population standard deviation of the ratings for 2018–2019. *2.17*

c. Compute the range for the ratings for both seasons. *2013–2014: 9.4; 2018–2019: 7.7*

d. Based on the standard deviations, did the spread in ratings increase or decrease over the two seasons? *Decrease*

e. Based on the ranges, did the spread in ratings increase or decrease over the two seasons? *Decrease*

50. House prices: The following table presents prices, in thousands of dollars, of single-family homes for some of the largest metropolitan areas in the United States for March of 2018 and March of 2019.

Metro Area	2018	2019	Metro Area	2018	2019
Atlanta, GA	204.8	214.2	New York, NY	384.0	384.4
Baltimore, MD	250.1	247.9	Philadelphia, PA	198.9	213.0
Boston, MA	412.2	438.6	Phoenix, AZ	229.6	251.7
Chicago, IL	220.2	242.9	Pittsburgh, PA	142.8	149.3
Cincinnati, OH	165.7	172.2	Portland, OR	396.6	386.4
Cleveland, OH	136.5	140.9	Riverside, CA	353.0	357.1
Dallas, TX	247.1	244.5	St. Louis, MO	158.0	167.5
Denver, CO	388.1	402.1	San Diego, CA	546.3	553.9
Houston, TX	221.1	219.0	San Francisco, CA	763.8	795.0
Los Angeles, CA	608.2	629.2	Seattle, WA	450.2	467.4
Miami, FL	259.1	310.0	Tampa, FL	203.4	212.4
Minneapolis, MN	274.4	264.9	Washington, DC	381.0	393.7

Source: Zillow

a. Find the population standard deviation for 2018. *155.0*

b. Find the population standard deviation for 2019. *159.2*

c. House prices increased in most areas from 2018 to 2019. Did the spread in house prices increase as well, or did it decrease? *Increased*

51. Stock prices: Following are the closing prices of Google stock for each trading day in May and June of a recent year.

May					
880.37	877.07	873.65	866.20	869.79	829.61
880.93	884.74	900.68	900.62	886.25	820.43
875.04	877.00	871.98	879.81	890.22	
879.73	864.64	859.70	859.10	867.63	

June				
871.22	870.76	868.31	881.27	873.32
882.79	889.42	906.97	908.53	909.18
903.87	915.89	887.10	877.53	880.23
871.48	873.63	857.23	861.55	845.72

a. Find the population standard deviation for the prices in May. *18.528*

b. Find the population standard deviation for the prices in June. *18.482*

c. Financial analysts use the word *volatility* to refer to the variation in prices of assets such as stocks. In which month was the price of Google stock more volatile? *May*

52. Stocks or bonds? Following are the annual percentage returns for the years 1999–2018 for three categories of investment: stocks, Treasury bills, and Treasury bonds. Stocks are represented by the Dow Jones Industrial Average.

Year	Stocks	Bills	Bonds	Year	Stocks	Bills	Bonds
1999	25.22	4.51	−8.25	2009	18.82	0.14	−11.12
2000	−6.18	5.76	16.66	2010	11.02	0.13	8.46
2001	−7.10	3.67	5.57	2011	5.53	0.03	16.04
2002	−16.76	1.66	15.12	2012	7.26	0.05	2.97
2003	25.32	1.03	0.38	2013	26.50	0.07	−9.10
2004	3.15	1.23	4.49	2014	7.52	0.05	10.75
2005	−0.61	3.01	2.87	2015	−2.23	0.21	1.28
2006	16.29	4.68	1.96	2016	13.42	0.51	0.69
2007	6.43	4.64	10.21	2017	25.08	1.39	2.80
2008	−33.84	1.59	20.10	2018	−5.63	2.37	−0.02

Source: Federal Reserve

a. The standard deviation of the return is a measure of the risk of an investment. Compute the population standard deviation for each type of investment. Which is the riskiest? Which is least risky? *Stocks: 15.10; bills: 1.84; bonds: 8.35; stocks riskiest, bills least risky*

b. Treasury bills are short-term (1 year or less) loans to the U.S. government. Treasury bonds are long-term (30-year) loans to the government. Finance theory states that long-term loans are riskier than short-term loans. Do the results agree with the theory? *Yes*

c. Finance theory states that the more risk an investment has, the higher its mean return must be. Compute the mean return for each class of investment. Do the results follow the theory? *Stocks: 5.96; bills: 1.84; bonds: 4.59; yes*

53. Time to review: The following table presents the time taken to review articles that were submitted for publication to the journal *Technometrics* during a recent year. A few articles took longer than 9 months to review; these are omitted from the table. Consider the data to be a population.

Time (Months)	Number of Articles
0.0–0.9	45
1.0–1.9	17
2.0–2.9	18
3.0–3.9	19
4.0–4.9	12
5.0–5.9	14
6.0–6.9	13
7.0–7.9	22
8.0–8.9	11

a. Approximate the variance of the times. *7.58*

b. Approximate the standard deviation of the times. *2.75*

54. Age distribution: The ages of residents of Banks City, Oregon, are given in the following frequency distribution. Consider these data to be a population.

Age	Frequency
0–9	283
10–19	203
20–29	217
30–39	256
40–49	176
50–59	92
60–69	21
70–79	23
80–89	12
90–99	3

Source: U.S. Census Bureau

a. Approximate the variance of the ages. *348.7*

b. Approximate the standard deviation of the ages. *18.7*

55. Lunch break: In a recent survey of 655 working Americans ages 25–34, the average weekly amount spent on lunch was \$44.60 with standard deviation \$2.81. The weekly amounts are approximately bell-shaped.

a. Estimate the percentage of amounts that are between \$36.17 and \$53.03. *Almost all*

b. Estimate the percentage of amounts that are between \$41.79 and \$47.41. *68%*

c. Between what two values will approximately 95% of the amounts be? *\$38.98 and \$50.22*

56. Pay your bills: In a large sample of customer accounts, a utility company determined that the average number of days between when a bill was sent out and when the payment was made is 32 with a standard deviation of 7 days. Assume the data to be approximately bell-shaped.

a. Between what two values will approximately 68% of the numbers of days be? *25 and 39*

b. Estimate the percentage of customer accounts for which the number of days is between 18 and 46. *95%*

c. Estimate the percentage of customer accounts for which the number of days is between 11 and 53. *Almost all*

57. Newborn babies: A study conducted by the Center for Population Economics at the University of Chicago studied the birth weights of 621 babies born in New York. The mean weight was 3234 grams with a standard deviation of 871 grams. Assume that birth weight data are approximately bell-shaped. Estimate the number of newborns who weighed between

a. 2363 grams and 4105 grams *422*

b. 1492 grams and 4976 grams *590*

58. Internet providers: In a survey of 600 homeowners with high-speed internet, the average monthly cost of a high-speed internet plan was $64.20 with standard deviation $11.77. Assume the plan costs to be approximately bell-shaped. Estimate the number of plans that cost between

a. $40.66 and $87.74 *570*

b. $52.43 and $75.97 *408*

59. Lunch break: For the data in Exercise 55, estimate the percentage of amounts that were

a. less than $41.79 *16%*

b. greater than $50.22 *2.5%*

c. between $44.60 and $47.41 *34%*

60. Pay your bills: For the data in Exercise 56, estimate the percentage of bills for which the number of days between when a bill was sent and when payment was made was

a. greater than 39 *16%*

b. less than 18 *2.5%*

c. between 18 and 32 *47.5%*

61. Newborn babies: For the data in Exercise 57, estimate the number of newborns whose weight was

a. less than 4105 grams *521 or 522*

b. greater than 1492 grams *605 or 606*

c. between 3234 and 4976 grams *295*

62. Internet providers: For the data in Exercise 58, estimate the number of internet plans whose cost is

a. greater than $52.43 *504*

b. less than $87.74 *585*

c. between $52.43 and $64.20 *204*

63. Empirical Rule OK? The following histogram presents a data set with a mean of 4.5 and a standard deviation of 2. Is it appropriate to use the Empirical Rule to approximate the proportion of the data between 0.5 and 8.5? If so, find the approximation. If not, explain why not. *Not appropriate*

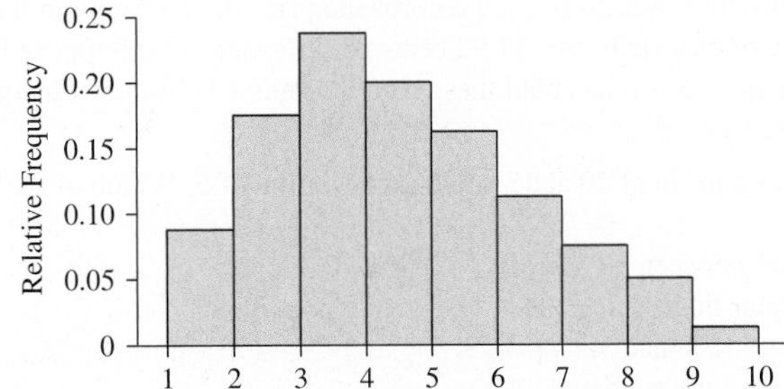

64. Empirical Rule OK? The following histogram presents a data set with a mean of 62 and a standard deviation of 17. Is it appropriate to use the Empirical Rule to approximate the proportion of the data between 45 and 79? If so, find the approximation. If not, explain why not. *Not appropriate*

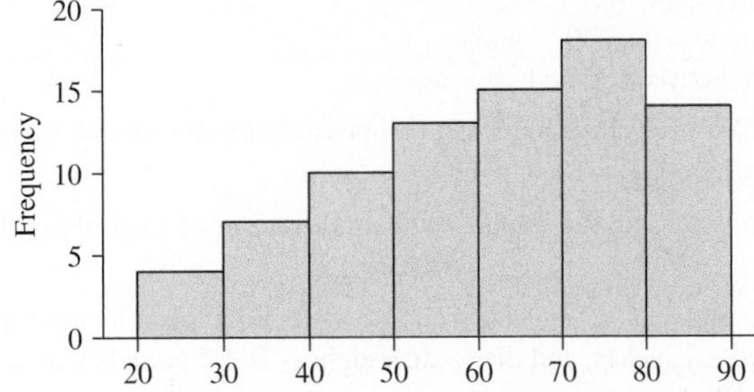

65. Empirical Rule OK? The following histogram presents a data set with a mean of 35 and a standard deviation of 9. Is it appropriate to use the Empirical Rule to approximate the proportion of the data between 26 and 44? If so, find the approximation. If not, explain why not. *68%*

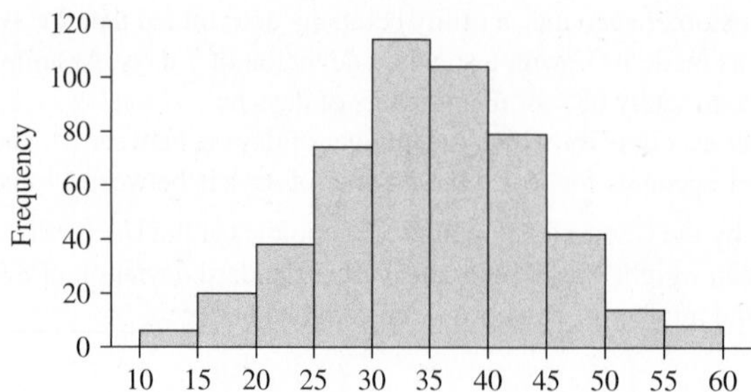

66. **Empirical Rule OK?** The following histogram presents a data set with a mean of 16 and a standard deviation of 2. Is it appropriate to use the Empirical Rule to approximate the proportion of the data between 12 and 20? If so, find the approximation. If not, explain why not. *95%*

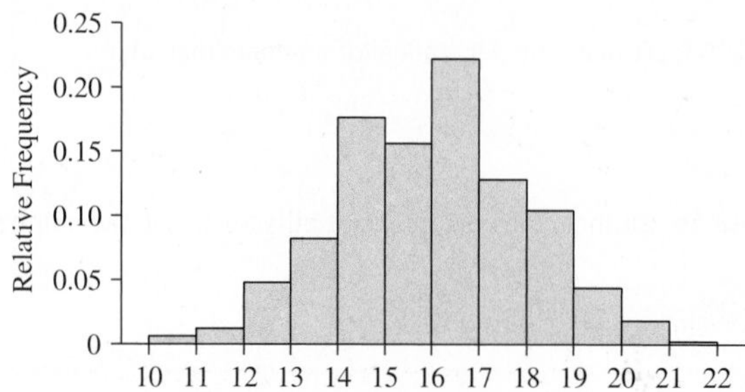

67. **What's the temperature?** The temperature in a certain location was recorded each day for two months. The mean temperature was 62.4°F with a standard deviation of 3.1°F. What can you determine about these data by using Chebyshev's Inequality with $K = 2$? *At least 75% between 56.2°F and 68.6°F*

68. **Find the standard deviation:** The National Center for Health Statistics sampled a number of women aged 20–29 and found that the mean height was 64.2 inches and the histogram for the data set was approximately bell-shaped. Assume the heights in the data set ranged from 55 to 74 inches. Is the standard deviation of the data closest to 2, 3, or 4? Explain. *3*

69. **Find the standard deviation:** The National Center for Health Statistics sampled a number of men aged 20–29 and found that the mean height was 69.4 inches and the histogram for the data set was approximately bell-shaped. Assume the heights in the data set ranged from 62 to 78 inches. Is the standard deviation of the data closest to 2.5, 3.5, or 4.5? Explain. *2.5*

70. **Price of electricity:** The Energy Information Administration records the price of electricity in the United States each month. In one recent month, the average price of electricity was 11.92 cents per kilowatt-hour. Suppose that the standard deviation is 2.1 cents per kilowatt-hour. What can you determine about these data by using Chebyshev's Inequality with $K = 3$? *At least 88.9% between 5.62 and 18.22 cents*

71. **Possible or impossible?** A data set has a mean of 20 and a standard deviation of 5. Which of the following might possibly be true, and which are impossible?
    a. Less than 50% of the data values are between 10 and 30. *Impossible*
    b. Only 1% of the data values are greater than 35. *Possible*
    c. More than 15% of the data values are less than 5. *Impossible*
    d. More than 90% of the data values are between 5 and 35. *Possible*

72. **Possible or impossible?** A data set has a mean of 50 and a standard deviation of 10. Which of the following might possibly be true, and which are impossible?
    a. More than 10% of the data values are negative. *Impossible*
    b. Only 5% of the data values are greater than 70. *Possible*
    c. More than 20% of the data values are less than 30. *Possible*
    d. Less than 75% of the data values are between 30 and 70. *Impossible*

73. **Standard deviation and mean:** For a list of positive numbers, is it possible for the sample standard deviation to be greater than the mean? If so, give an example. If not, explain why not. *Yes*

74. **Standard deviation equal to 0?** Is it possible for the sample standard deviation of a list of numbers to equal 0? If so, give an example. If not, explain why not. *Yes*

75. **Height and weight:** A National Center for Health Statistics study states that the mean height for adult men in the United States is 69.4 inches with a standard deviation of 3.1 inches, and the mean weight is 194.7 pounds with a standard deviation of 68.3 pounds.
    a. Compute the coefficient of variation for height. *0.045*
    b. Compute the coefficient of variation for weight. *0.351*
    c. Which has greater spread relative to its mean, height or weight? *Weight*

76. Test scores: Scores on a statistics exam had a mean of 75 with a standard deviation of 10. Scores on a calculus exam had a mean of 60 with a standard deviation of 9.

a. Compute the coefficient of variation for statistics exam scores. *0.13*

b. Compute the coefficient of variation for calculus exam scores. *0.15*

c. Which has greater spread relative to their mean, statistics scores or calculus scores? *Calculus scores*

Extending the Concepts

77. Mean absolute deviation: A measure of spread that is an alternative to the standard deviation (SD) is the **mean absolute deviation** (MAD). For a data set containing values $x_1, ..., x_n$, the mean absolute deviation is given by

$$\text{Mean absolute deviation} = \frac{\sum |x - \bar{x}|}{n}$$

a. Compute the mean $\bar{x}$ for the data set 1, 3, 4, 7, 9. *4.8*

b. Construct a table like Table 3.5 that contains an additional column for the values $|x - \bar{x}|$.

c. Use the table to compute the SD and the MAD. *SD: 3.1937; MAD: 2.56*

d. Now consider the data set 1, 3, 4, 7, 9, 30. Compute the SD and the MAD for this data set. *SD: 10.677; MAD: 7*

e. Which measure of spread is more resistant, the SD or the MAD? Explain. *MAD*

78. Effect on standard deviation: Four employees in an office have annual salaries of \$30,000, \$35,000, \$45,000, and \$70,000.

a. Compute the sample standard deviation of the salaries. *17,795.13*

b. Each employee gets a \$1000 raise. Compute the new standard deviation. Does the standard deviation increase by \$1000? *17,795.13; no*

c. Each employee gets a 5% raise. Compute the new standard deviation. Does the standard deviation increase by 5%? *18,684.89; yes*

Answers to Check Your Understanding Exercises for Section 3.2

1. The variance of the St. Louis temperatures is 291.9. This is greater than the variance of the San Francisco temperatures, which indicates that there is greater spread in the St. Louis temperatures.

2. 292,230

3. 9.1

4. a. Variance is 3.7143; standard deviation is 1.9272.
b. Variance is 281.8667; standard deviation is 16.7889.

5. a. Variance is 4.8594; standard deviation is 2.2044.
b. Variance is 590.8889; standard deviation is 24.30821.

6. Approximately 95% of the data values are between 14 and 26.

7. The Empirical Rule should not be used because the data are skewed.

8. No. The interval between 50 and 90 is the interval within two standard deviations of the mean. At least 75% of the data must be between 50 and 90.

9. Yes. The interval between 122 and 128 is the interval within three standard deviations of the mean. At least 8/9 (88.9%) of the data must be between 122 and 128.

10. 0.0945

Section 3.3 Measures of Position

Objectives

1. Compute and interpret *z*-scores
2. Compute the quartiles of a data set
3. Compute the percentiles of a data set
4. Compute the five-number summary for a data set
5. Understand the effects of outliers
6. Construct boxplots to visualize the five-number summary and outliers

Objective 1 Compute and interpret *z*-scores

The *z*-Score

Who is taller, a man 73 inches tall or a woman 68 inches tall? The obvious answer is that the man is taller. However, men are taller than women on the average. Let's ask the question this way: Who is taller relative to their gender, a man 73 inches tall or a woman 68 inches tall? One way to answer this question is with a *z-score*.

The *z*-score of an individual data value tells how many standard deviations that value is from its population mean. So, for example, a value one standard deviation above the mean has a *z*-score of 1. A value two standard deviations below the mean has a *z*-score of −2.

NOTE TO INSTRUCTOR

The following topics from the **Statistics Corequisite Workbook** are aligned with the material in this section.

2.4 - Variables and Formulas in Measures of Position

2.5 - Solving for an Unknown in Numerical Summaries

DEFINITION

Let x be a value from a population with mean μ and standard deviation σ. The **z-score** for x is

$$z = \frac{x - \mu}{\sigma}$$

Example 3.23

Computing and interpreting z-scores

A National Center for Health Statistics study states that the mean height for adult men in the United States is $\mu = 69.4$ inches, with a standard deviation of $\sigma = 3.1$ inches. The mean height for adult women is $\mu = 63.8$ inches, with a standard deviation of $\sigma = 2.8$ inches. Who is taller relative to their gender, a man 73 inches tall or a woman 68 inches tall?

NOTE TO INSTRUCTOR

To help students understand comparing values from different populations, ask them to consider the difference between a grade of 95 on a challenging exam with a mean of 60 and the same grade on an easy exam with a mean of 90.

Solution

We compute the z-scores for the two heights:

$$z\text{-score for man's height} = \frac{x - \mu}{\sigma} = \frac{73 - 69.4}{3.1} = 1.16$$

$$z\text{-score for woman's height} = \frac{x - \mu}{\sigma} = \frac{68 - 63.8}{2.8} = 1.50$$

The height of the 73-inch man is 1.16 standard deviations above the mean height for men. The height of the 68-inch woman is 1.50 standard deviations above the mean height for women. Therefore, the woman is taller, relative to the population of women, than the man is, relative to the population of men.

Sometimes we are given a z-score and need to find the value that has that z-score. To see how to do this, recall that the z-score tells us how many standard deviations a value is from the mean. Therefore, to find the value that has a given z-score, multiply the standard deviation by z and add that quantity to the mean.

In a population with mean μ and standard deviation σ, the value x with a given z-score is

$$x = \mu + z\sigma$$

Example 3.24

Computing a value with a given z-score

Data from the National Health and Examination survey estimates the mean total cholesterol level, in milligrams per deciliter, in the adult population to be $\mu = 200$, with a standard deviation of $\sigma = 44$. What is the cholesterol level of a person with a z-score of 1.5? What is the cholesterol level of a person with a z-score of -1?

Solution

The cholesterol level for a person with a z-score of 1.5 is 1.5 standard deviations above the mean. This person's cholesterol level is

$$200 + 1.5(44) = 266$$

The cholesterol level for a person with a z-score of -1 is 1 standard deviation below the mean. This person's cholesterol level is

$$200 - 1(44) = 156$$

z-scores and the Empirical Rule

z-scores work best for populations whose histograms are approximately bell-shaped—that is, for populations for which we can use the Empirical Rule. Recall that the Empirical Rule says that for a bell-shaped population, approximately 68% of the data will be within one standard deviation of the mean, approximately 95% will be within two standard deviations, and almost all will be within three standard deviations. Since the z-score is the number of standard deviations from the mean, we can easily interpret the z-score for bell-shaped populations.

z-Scores and the Empirical Rule

When a population has a histogram that is approximately bell-shaped, then

- Approximately 68% of the data will have z-scores between -1 and 1.
- Approximately 95% of the data will have z-scores between -2 and 2.
- All, or almost all, of the data will have z-scores between -3 and 3.

The z-score is less useful for populations that are not bell-shaped. For example, in some skewed populations there will be no values with z-scores greater than 1, while in others, values with z-scores greater than 1 occur frequently. We can't be sure how to interpret z-scores when the population is skewed. It is best, therefore, to use z-scores only for populations that are approximately bell-shaped. See Exercise 57 for an illustration.

Check Your Understanding

1. A population has mean $\mu = 10$ and standard deviation $\sigma = 4$.
  a. Find the z-score for a population value of 14. *1*
  b. Find the z-score for a population value of 8. *−0.5*
  c. What number has a z-score of 1.5? *16*

2. A population has mean $\mu = 30$ and standard deviation $\sigma = 6$.
  a. Find the z-score for a population value of 21. *−1.5*
  b. Find the z-score for a population value of 42. *2*
  c. What number has a z-score of 0.5? *33*

3. According to the *Wall Street Journal*, the maximum distance from check-in to the gate at top U.S. airports has a mean of 0.79 miles with a standard deviation of 0.42 miles.
  a. The maximum distance at the Seattle-Tacoma International Airport is 0.57 miles. What is the z-score for this distance? *−0.52*
  b. The maximum distance at the Hartsfield-Jackson International Airport is 1.67 miles. What is the z-score for this distance? *2.10*
  c. What distance would have a z-score of 1? *1.21*

Answers are on page 158.

Objective 2 Compute the quartiles of a data set

Quartiles

The weather in Los Angeles is dry most of the time, but it can be quite rainy in the winter. The rainiest month of the year is February. Table 3.9 (page 142) presents the annual rainfall in Los Angeles, in inches, for each February from 1975–2019.

There is a lot of spread in the amount of rainfall in Los Angeles in February. For example, in 1984 there was no measurable rain at all, while in 1998 it rained more than 13 inches.

In Section 3.1 we learned how to compute the mean and median of a data set, which describe the center of a distribution. For data sets like the Los Angeles rainfall data, which exhibit a lot of spread, it is useful to compute measures of positions other than the center, to get a more detailed description of the distribution. **Quartiles** provide a way to do this. Quartiles divide a data set into four approximately equal pieces.

Table 3.9 Annual Rainfall in Los Angeles During February

Year	Rainfall	Year	Rainfall	Year	Rainfall	Year	Rainfall	Year	Rainfall
1975	3.54	1984	0.00	1993	6.61	2002	0.29	2011	3.29
1976	3.71	1985	2.84	1994	3.21	2003	4.64	2012	0.16
1977	0.17	1986	6.10	1995	1.30	2004	4.89	2013	0.20
1978	8.91	1987	1.22	1996	4.94	2005	11.02	2014	3.58
1979	3.06	1988	1.72	1997	0.08	2006	2.37	2015	0.83
1980	12.75	1989	1.90	1998	13.68	2007	0.92	2016	0.79
1981	1.48	1990	3.12	1999	0.56	2008	1.64	2017	4.17
1982	0.70	1991	4.13	2000	5.54	2009	3.57	2018	0.03
1983	4.37	1992	7.96	2001	8.87	2010	4.27	2019	5.59

DEFINITION

Every data set has three quartiles:

- The **first quartile**, denoted Q_1, separates the lowest 25% of the data from the highest 75%.
- The **second quartile**, denoted Q_2, separates the lower 50% of the data from the upper 50%. Q_2 is the same as the median.
- The **third quartile**, denoted Q_3, separates the lowest 75% of the data from the highest 25%.

EXPLAIN IT AGAIN

The second quartile is the same as the median: The second quartile, Q_2, divides the data in half. Therefore Q_2 is the same as the median.

There are several methods for computing quartiles, all of which give similar results. We present a fairly straightforward method here.

NOTE TO INSTRUCTOR

A common mistake is to forget to arrange the data in increasing order.

Procedure for Computing the Quartiles of a Data Set

Step 1: Arrange the data in increasing order.

Step 2: Let n be the number of values in the data set. To compute the second quartile, simply compute the median. For the first or third quartiles, proceed as follows.

For the first quartile, compute $L = 0.25n$.
For the third quartile, compute $L = 0.75n$.

Step 3: *If L is a whole number*, the quartile is the average of the number in position L and the number in position $L + 1$.

If L is not a whole number, round it *up* to the next higher whole number. The quartile is the number in the position corresponding to the rounded-up value.

Example 3.25

Computing quartiles

Compute the first and third quartiles of the Los Angeles rainfall data.

Solution

Step 1: Table 3.10 (page 143) presents the data in increasing order.

Step 2: There are $n = 45$ values in the data set. For the first quartile we compute

$$L = 0.25(45) = 11.25$$

Step 3: Since $L = 11.25$ is not a whole number, we round it *up* to 12. The first quartile, Q_1, is the number in the 12th position. From Table 3.10 we can see that the first quartile is $Q_1 = 0.92$.

CAUTION

Always round L *up*. Do not round down.

Step 4: For the third quartile we compute

$$L = 0.75(45) = 33.75$$

Step 5: Since $L = 33.75$ is not a whole number, we round it *up* to 34. The third quartile, Q_3, is the number in the 34th position. From Table 3.10 we can see that the third quartile is $Q_3 = 4.89$.

Table 3.10 Annual Rainfall in Los Angeles During February, in Increasing Order

Year	Rainfall	Year	Rainfall	Year	Rainfall	Year	Rainfall	Year	Rainfall
1984	(1) 0.00	2016	(10) 0.79	2006	(19) 2.37	1976	(28) 3.71	2019	(37) 5.59
2018	(2) 0.03	2015	(11) 0.83	1985	(20) 2.84	1991	(29) 4.13	1986	(38) 6.10
1997	(3) 0.08	2007	(12) 0.92	1979	(21) 3.06	2017	(30) 4.17	1993	(39) 6.61
2012	(4) 0.16	1987	(13) 1.22	1990	(22) 3.12	2010	(31) 4.27	1992	(40) 7.96
1977	(5) 0.17	1995	(14) 1.30	1994	(23) 3.21	1983	(32) 4.37	2001	(41) 8.87
2013	(6) 0.20	1981	(15) 1.48	2011	(24) 3.29	2003	(33) 4.64	1978	(42) 8.91
2002	(7) 0.29	2008	(16) 1.64	1975	(25) 3.54	2004	(34) 4.89	2005	(43) 11.02
1999	(8) 0.56	1988	(17) 1.72	2009	(26) 3.57	1996	(35) 4.94	1980	(44) 12.75
1982	(9) 0.70	1989	(18) 1.90	2014	(27) 3.58	2000	(36) 5.54	1998	(45) 13.68

Figure 3.9 presents a dotplot of the Los Angeles rainfall data set with the quartiles indicated. The quartiles divide the data set into four parts, with approximately 25% of the data in each part. Recall that the median is the same as the second quartile. Since there are 45 values in this data set, the median is the 23rd value when the data are arranged in order. From Table 3.10, we can see that the median is 3.21.

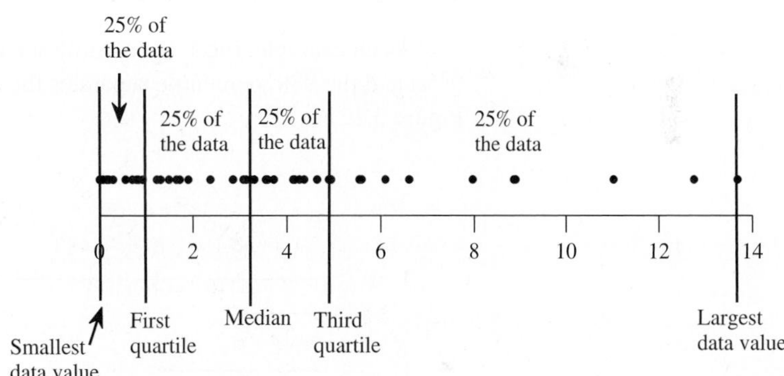

Figure 3.9 The quartiles of the Los Angeles rainfall data set

Example 3.26

Using technology to compute quartiles

Use technology to compute the first and third quartiles for the Los Angeles rainfall data presented in Table 3.9.

Solution

Figure 3.10 presents MINITAB output. The quartiles are highlighted in bold.

Variable	N	Mean	SE Mean	StDev	Minimum	Q1	Median	Q3	Maximum
Rainfall	45	3.660	0.501	3.363	0.000	**0.875**	3.210	**4.915**	13.680

Figure 3.10

NOTE TO INSTRUCTOR

The computation of percentiles may be omitted without loss of continuity. However, the *definition* of the *p*th percentile as a value that separates the lowest *p*% of the data appears in later sections.

The **1–Var Stats** command for the TI-84 Plus calculator described in Section 3.1 also computes the quartiles of a data set. Figure 3.11 presents the results for the Los Angeles rainfall data.

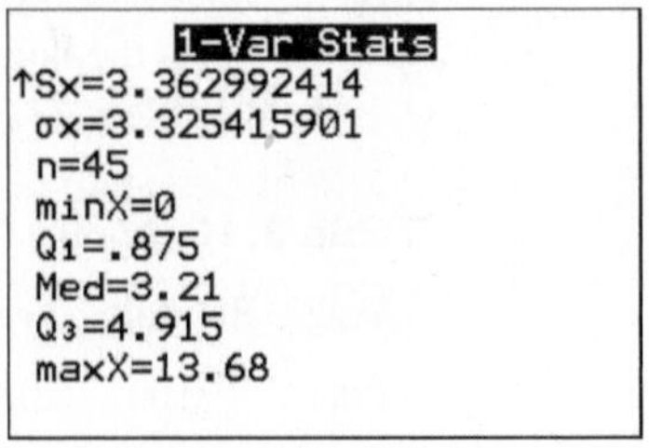

```
1-Var Stats
↑Sx=3.362992414
 σx=3.325415901
 n=45
 minX=0
 Q1=.875
 Med=3.21
 Q3=4.915
 maxX=13.68
```

Figure 3.11

Note that the values produced by both MINITAB and the TI-84 Plus differ slightly from the results obtained in Example 3.25, because they use a slightly different procedure than the one we describe here. Step-by-step instructions are presented in the Using Technology section on page 152.

Objective 3 Compute the percentiles of a data set

Percentiles

Quartiles describe the shape of a distribution by dividing it into fourths. Sometimes it is useful to divide a data set into a greater number of pieces to get a more detailed description of the distribution. *Percentiles* provide a way to do this. Percentiles divide a data set into hundredths.

NOTE TO INSTRUCTOR

Mention that percentiles other than quartiles are useful for large data sets, but not so useful for small ones.

DEFINITION

For a number p between 1 and 99, the pth **percentile** separates the lowest $p\%$ of the data from the highest $(100 - p)\%$.

As an example, the 1st percentile separates the lowest 1% of the data from the highest 99%, and the 98th percentile separates the lowest 98% of the data from the highest 2% (see Figure 3.12).

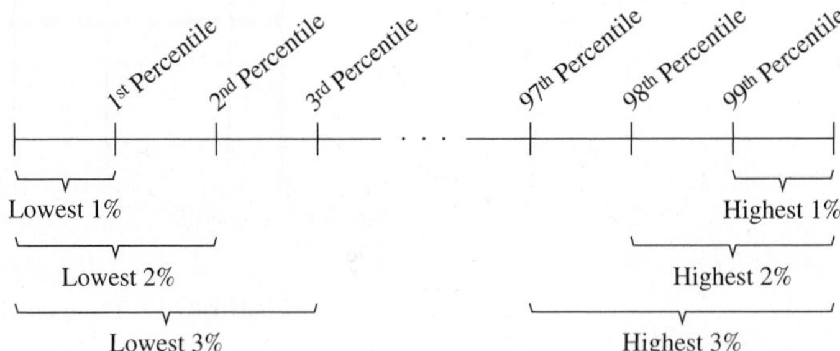

Figure 3.12

There are several methods for computing percentiles, all of which give similar results. We present a fairly straightforward method here.

Procedure for Computing the Data Value Corresponding to a Given Percentile

Step 1: Arrange the data in increasing order.

Step 2: Let n be the number of values in the data set. For the pth percentile, compute the value

$$L = \frac{p}{100} \cdot n$$

Step 3: *If L is a whole number*, then the pth percentile is the average of the number in position L and the number in position $L+1$.

If L is not a whole number, round it *up* to the next higher whole number. The pth percentile is the number in the position corresponding to the rounded-up value.

Example 3.27

Computing a percentile

Compute the 60th percentile of the Los Angeles rainfall data.

Solution

Step 1: Table 3.10 presents the data in increasing order.

Step 2: There are $n = 45$ values in the data set. For the 60th percentile, we take $p = 60$ and compute

$$L = \frac{60}{100} \cdot 45 = 27$$

CAUTION

Always round *L up*. Do not round down.

Step 3: Since $L = 27$ is a whole number, the 60th percentile is the average of the numbers in the 27th and 28th positions. From Table 3.10 we can see that the 60th percentile is $\frac{3.58 + 3.71}{2} = 3.645$.

Computing the percentile corresponding to a given data value

Sometimes we are given a value from a data set and wish to compute the percentile corresponding to that value. Following is a simple procedure for doing this.

Procedure for Computing the Percentile Corresponding to a Given Data Value

Step 1: Arrange the data in increasing order.

Step 2: Let x be the data value whose percentile is to be computed. Use the following formula to compute the percentile:

$$\text{Percentile} = 100 \cdot \frac{(\text{Number of values less than } x) + 0.5}{\text{Number of values in the data set}}$$

Round the result to the nearest whole number.

Example 3.28

Computing the percentile corresponding to a given data value

In 1989, the rainfall in Los Angeles during the month of February was 1.90. What percentile does this correspond to?

Solution

Step 1: Table 3.10 presents the data in increasing order.

Step 2: There are 45 values in the data set. There are 18 values less than 1.90. Therefore,

$$\text{Percentile} = 100 \cdot \frac{18 + 0.5}{45} = 41.1$$

We round the result to 41. The value 1.90 corresponds to the 41st percentile.

Check Your Understanding

4. Following are final exam scores, arranged in increasing order, for 28 students in an introductory statistics course.

58	59	62	64	67	68	69	71	73	74	74	75	76	76
76	77	78	78	78	82	82	84	86	87	87	88	91	97

a. Find the first quartile of the scores. *70*
b. Find the third quartile of the scores. *83*
c. Fred got a 73 on the exam. On what percentile is this score? *30th*
d. Students whose scores are on the 80th percentile or above will get a grade of A. Louisa got an 88 on the exam. Will she get an A? *Yes*

5. For the years 1882–2019, the 90th percentile of annual snowfall in Denver was 80.3 inches. Approximately what percentage of years had snowfall less than 80.3 inches? *90%*

6. In a recent year, the 65th percentile of daily mean temperatures in the city of Macon, Georgia was 78 degrees. For approximately what percentage of days that year was the mean temperature greater than 78 degrees? *35%*

7. In a recent year, 17% of players in the National Football League weighed less than 200 pounds. On approximately what percentile is a player who weighs 200 pounds? *17th*

8. In a sample of adult women participating in the National Health and Nutrition survey, 75% of them were taller than 65.2 inches. On approximately what percentile is a woman who is 65.2 inches tall? *75th*

Answers are on page 158.

Objective 4 Compute the five-number summary for a data set

The Five-Number Summary

The five-number summary of a data set consists of the median, the first quartile, the third quartile, the minimum value, and the maximum value. These values are generally arranged in order.

DEFINITION

The **five-number summary** of a data set consists of the following quantities:

Minimum First quartile Median Third quartile Maximum

Example 3.29

Constructing a five-number summary

Table 3.11 presents the number of students absent in a middle school in northwestern Montana for each school day in January of a recent year. Construct the five-number summary.

NOTE TO INSTRUCTOR

In the TI-84 Plus calculator, the last five entries in the output of the 1-VarStats command comprise the five-number summary.

Table 3.11 Number of Absences

Date	Number Absent	Date	Number Absent	Date	Number Absent
Jan. 2	65	Jan. 14	59	Jan. 23	42
Jan. 3	67	Jan. 15	49	Jan. 24	45
Jan. 4	71	Jan. 16	42	Jan. 25	46
Jan. 7	57	Jan. 17	56	Jan. 28	100
Jan. 8	51	Jan. 18	45	Jan. 29	59
Jan. 9	49	Jan. 21	77	Jan. 30	53
Jan. 10	44	Jan. 22	44	Jan. 31	51
Jan. 11	41				

Solution

Step 1: We arrange the numbers in increasing order. The ordered numbers are:

41 42 42 44 44 45 45 46 49 49 51 51 53 56 57 59 59 65 67 71 77 100

Step 2: The minimum is 41 and the maximum is 100.

Step 3: We use the methods described in Example 3.25 to compute the first and third quartiles. The first quartile is $Q_1 = 45$, and the third quartile is $Q_3 = 59$.

Step 4: We use the method described in Section 3.1 to compute the median. The median is 51.

Step 5: The five-number summary is

41 45 51 59 100

Objective 5 Understand the effects of outliers

Outliers

An **outlier** is a value that is considerably larger or considerably smaller than most of the values in a data set. Some outliers result from errors; for example, a misplaced decimal point may cause a number to be much larger or smaller than the other values in a data set. Some outliers are correct values, and simply reflect the fact that the population contains some extreme values.

CAUTION

Do not delete an outlier unless it is certain that it is an error.

When it is certain that an outlier resulted from an error, the value should be corrected or deleted. However, if it is possible that the value of an outlier is correct, it should remain in the data set. Deleting an outlier that is not an error will produce misleading results.

Example 3.30

Determining whether an outlier should be deleted

The temperature in a downtown location in a certain city is measured for eight consecutive days during the summer. The readings, in degrees Fahrenheit, are 81.2, 85.6, 89.3, 91.0, 83.2, 8.45, 79.5, and 87.8. Which reading is an outlier? Is it certain that the outlier is an error, or is it possible that it is correct? Should the outlier be deleted?

Solution

The outlier is 8.45, which is much smaller than the rest of the data. This outlier is certainly an error; it is likely that a decimal point was misplaced. The outlier should be corrected if possible, or deleted.

Example 3.31

Determining whether an outlier should be deleted

The following table presents the populations, as of June 2019, of the five largest cities in the United States.

City	Population in millions
New York	8.6
Los Angeles	4.1
Chicago	2.7
Houston	2.4
Phoenix	1.7

Source: U.S. Census Bureau

Which value is an outlier? Is it certain that the outlier is an error, or is it possible that it is correct? Should the outlier be deleted?

Solution

The population of New York, 8.6 million, is an outlier because it is much larger than the other values. This outlier is not an error. It should not be deleted. If it were deleted, the data would indicate that the largest city in the United States is Los Angeles, which would be incorrect.

The interquartile range

The *interquartile range* (IQR for short) is a measure of spread that is often used to detect outliers. The IQR is the difference between the first and third quartiles.

DEFINITION

The **interquartile range** is found by subtracting the first quartile from the third quartile.

$$\text{IQR} = Q_3 - Q_1$$

The IQR method for finding outliers

In Examples 3.30 and 3.31, we determined the outlier just by looking at the data and finding an extreme value. In many cases, this is a good way to find outliers. There are some formal methods for finding outliers as well. The most frequently used method is the **IQR method**.

The IQR Method for Finding Outliers

Step 1: Find the first quartile, Q_1, and the third quartile, Q_3, of the data set.

Step 2: Compute the interquartile range.

$$\text{IQR} = Q_3 - Q_1$$

Step 3: Compute the **outlier boundaries**. These boundaries are the cutoff points for determining outliers.

$$\text{Lower outlier boundary} = Q_1 - 1.5\,\text{IQR}$$
$$\text{Upper outlier boundary} = Q_3 + 1.5\,\text{IQR}$$

Step 4: Any data value that is less than the lower outlier boundary or greater than the upper outlier boundary is considered to be an outlier.

Example 3.32

Detecting outliers

Use the IQR method to determine which values, if any, in the absence data in Table 3.11 are outliers.

Solution

Step 1: In Example 3.29, we computed the first and third quartiles: $Q_1 = 45$ and $Q_3 = 59$.

Step 2: $\text{IQR} = Q_3 - Q_1 = 59 - 45 = 14$

Step 3: The outlier boundaries are:

$$\text{Lower outlier boundary} = 45 - 1.5(14) = 24$$
$$\text{Upper outlier boundary} = 59 + 1.5(14) = 80$$

Step 4: There are no values in the data set less than the lower boundary of 24. There is one value, 100, that is greater than the upper boundary of 80. Thus there is one outlier, 100.

Check Your Understanding

9. Table 3.12 presents the number of Executive Orders issued by U.S. presidents from March 1861 through June 2019.

Table 3.12 Number of Executive Orders by U.S. Presidents

President	Orders	President	Orders	President	Orders
Abraham Lincoln	48	Theodore Roosevelt	1081	Lyndon Johnson	325
Andrew Johnson	79	William Howard Taft	724	Richard Nixon	346
Ulysses S. Grant	217	Woodrow Wilson	1803	Gerald Ford	169
Rutherford B. Hayes	92	Warren G. Harding	522	Jimmy Carter	320
James Garfield	6	Calvin Coolidge	1203	Ronald Reagan	381
Chester Arthur	96	Herbert Hoover	968	George H. W. Bush	166
Grover Cleveland–I	113	Franklin D. Roosevelt	3721	William Clinton	364
Benjamin Harrison	143	Harry S. Truman	907	George W. Bush	291
Grover Cleveland–II	140	Dwight D. Eisenhower	484	Barack Obama	277
William McKinley	185	John F. Kennedy	214	Donald Trump	114

Source: The America Presidency Project

a. Construct the five-number summary. *6, 140, 284, 522, 3721*
b. Find the IQR. *382*
c. Find the upper and lower outlier boundaries. *Lower: −433; upper: 1095*
d. Which values, if any, are outliers? *1203, 1803, 3721*

Answers are on page 158.

Objective 6 Construct boxplots to visualize the five-number summary and outliers

EXPLAIN IT AGAIN

Another name for boxplots: Boxplots are sometimes called *box-and-whisker diagrams.*

NOTE TO INSTRUCTOR

Emphasize that the median, not the mean, is used as the measure of center in a boxplot.

Boxplots

A **boxplot** is a graph that presents the five-number summary along with some additional information about a data set. There are several kinds of boxplots. The one we describe here is sometimes called a **modified boxplot**.

Procedure for Constructing a Boxplot

Step 1: Compute the first quartile, the median, and the third quartile.

Step 2: Draw vertical lines at the first quartile, the median, and the third quartile. Draw horizontal lines between the first and third quartiles to complete the box.

Step 3: Compute the lower and upper outlier boundaries.

Step 4: Find the largest data value that is less than the upper outlier boundary. Draw a horizontal line from the third quartile to this value. This horizontal line is called a **whisker**.

Step 5: Find the smallest data value that is greater than the lower outlier boundary. Draw a horizontal line (whisker) from the first quartile to this value.

Step 6: Determine which values, if any, are outliers. Plot each outlier separately.

Example 3.33

Constructing a boxplot

Construct a boxplot for the absence data in Table 3.11.

Solution

Step 1: In Example 3.29, we computed the median to be 51 and the first and third quartiles to be $Q_1 = 45$ and $Q_3 = 59$.

Step 2: We draw vertical lines at 45, 51, and 59, then horizontal lines to complete the box, as follows:

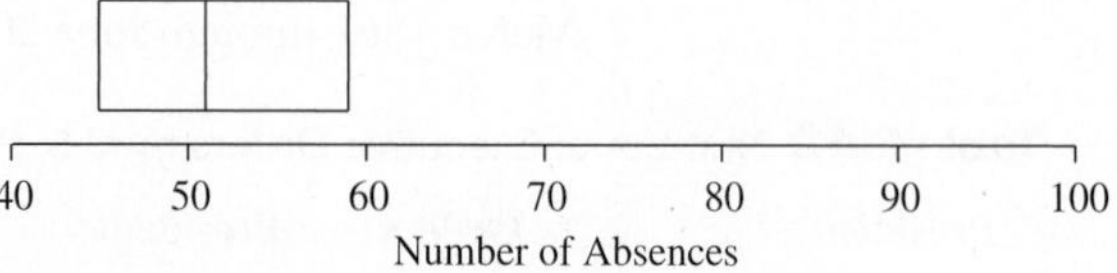

Step 3: We compute the outlier boundaries as shown in Example 3.32:

$$\text{Lower outlier boundary} = 45 - 1.5(14) = 24$$

$$\text{Upper outlier boundary} = 59 + 1.5(14) = 80$$

Step 4: The largest data value that is less than the upper boundary is 77. We draw a horizontal line from 59 up to 77, as follows:

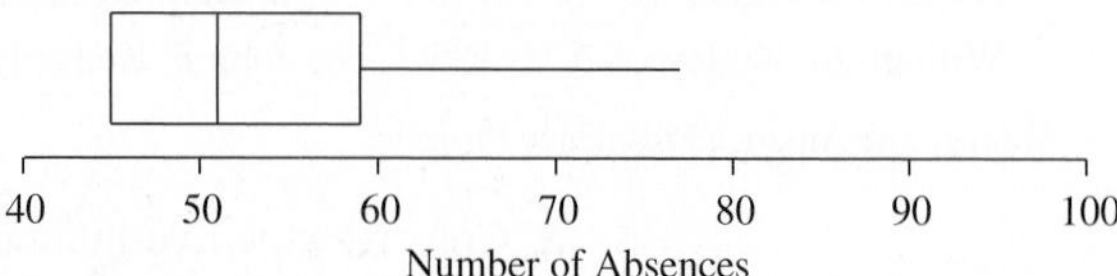

Step 5: The smallest data value that is greater than the lower boundary is 41. We draw a horizontal line from 45 down to 41, as follows:

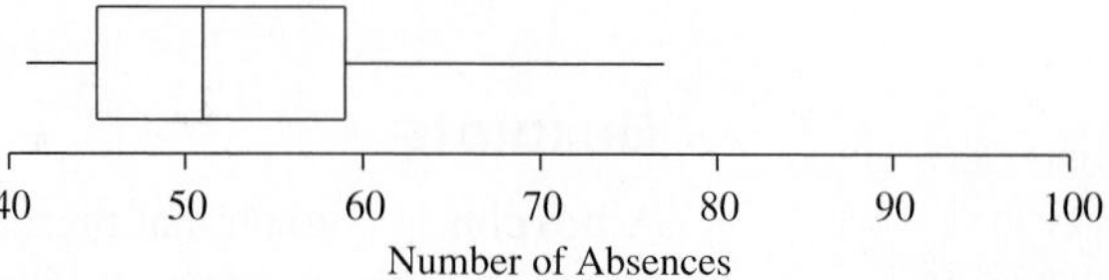

Step 6: We determine, as shown in Example 3.32, that the value 100 is the only outlier. We plot this point separately, to produce the boxplot shown in Figure 3.13.

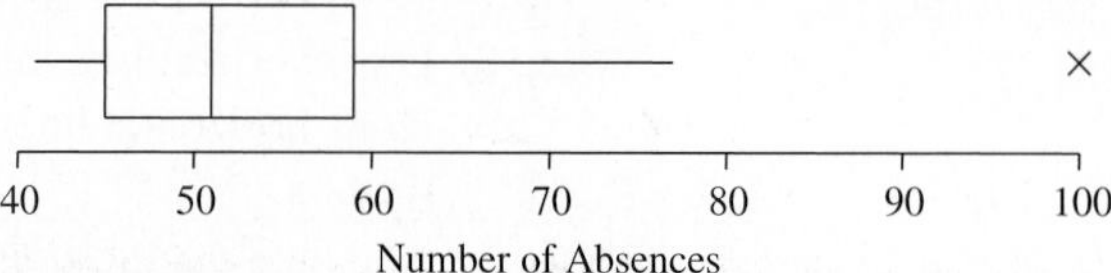

Figure 3.13 Boxplot for the absence data in Table 3.11

Check Your Understanding

10. Construct a boxplot for the data in Table 3.12.

Answer is on page 158.

Determining the shape of a data set from a boxplot

In Section 2.2, we learned how to determine from a histogram whether a data set is symmetric or skewed. In many cases, a boxplot can give us the same information. For example, in the boxplot for the absence data (Figure 3.13), the median is closer to the first quartile than to the third quartile, and the upper whisker is longer than the lower one. This indicates that the data are skewed to the right.

Figure 3.14 presents a histogram of the absence data. The skewness is clearly apparent.

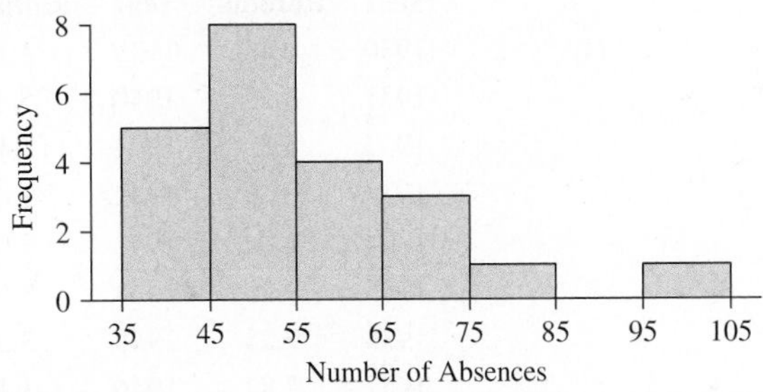

Figure 3.14 Histogram for the absence data in Table 3.11

Determining Skewness from a Boxplot

- If the median is closer to the first quartile than to the third quartile, or the upper whisker is longer than the lower whisker, the data are skewed to the right.
- If the median is closer to the third quartile than to the first quartile, or the lower whisker is longer than the upper whisker, the data are skewed to the left.
- If the median is approximately halfway between the first and third quartiles, and the two whiskers are approximately equal in length, the data are approximately symmetric.

Figures 3.15–3.17 illustrate the way in which boxplots reflect skewness and symmetry.

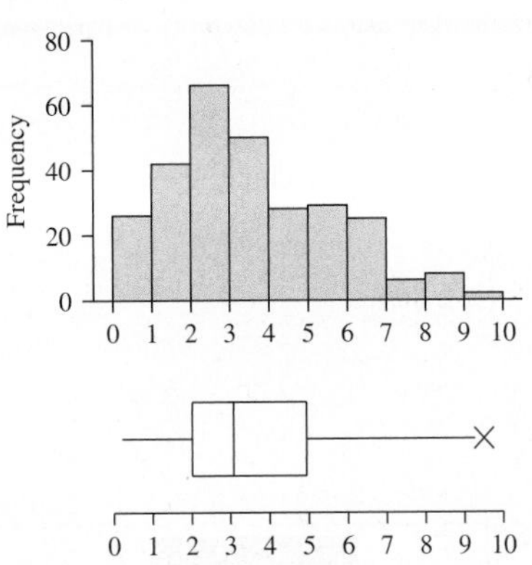

Figure 3.15 Skewed to the right

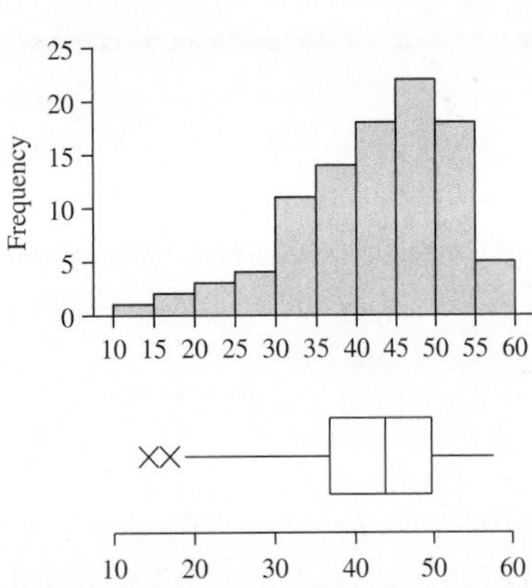

Figure 3.16 Skewed to the left

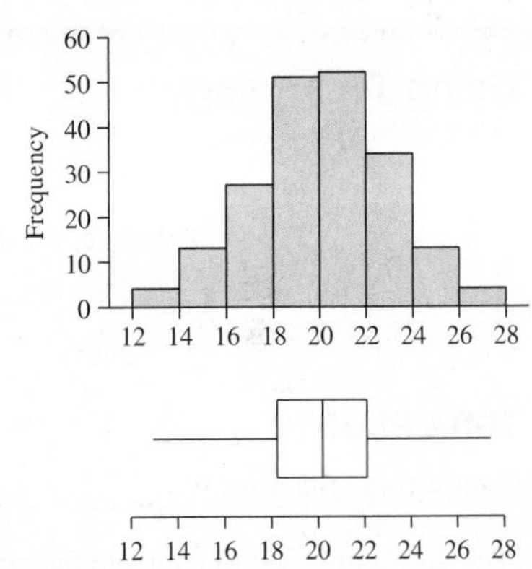

Figure 3.17 Approximately symmetric

Comparative boxplots

Boxplots do not provide as much detail as histograms do regarding the shape of a data set. However, they provide an excellent method for comparing data sets. We can plot two or more boxplots, one above another, to provide an easy visual comparison.

As an example, Table 3.13 (page 152) presents the rainfall in Los Angeles each February for the years 1930–1974. We would like to compare these data with the February rainfall for the years 1975–2019, which was presented in Table 3.9.

Figure 3.18 (page 152) presents **comparative boxplots** for February rainfall during the years 1930–1974 and 1975–2019.

We can see that the boxplot for 1975–2019 extends farther to the right than the boxplot for 1930–1974. This tells us that 1975–2019 was, on the whole, rainier than 1930–1974. We can also see the boxplot for 1975–2019 is longer than the one for 1930–1974. This tells us that the rainfall was more variable during 1975–2019. Finally, the upper whisker is longer than the lower whisker in both boxplots. This tells us that both rainfall data sets are skewed to the right. Finally, we note that there were more outliers in 1930–1974 than in 1975–2019.

Table 3.13 Annual Rainfall in Los Angeles During February: 1930–1974

Year	Rainfall	Year	Rainfall	Year	Rainfall	Year	Rainfall	Year	Rainfall
1930	0.45	1939	1.13	1948	1.29	1957	1.47	1966	1.51
1931	3.25	1940	5.43	1949	1.41	1958	6.46	1967	0.11
1932	5.33	1941	12.42	1950	1.67	1959	3.32	1968	0.49
1933	0.00	1942	1.05	1951	1.48	1960	2.26	1969	8.03
1934	2.04	1943	3.07	1952	0.63	1961	0.15	1970	2.58
1935	2.23	1944	8.65	1953	0.33	1962	11.57	1971	0.67
1936	7.25	1945	3.34	1954	2.98	1963	2.88	1972	0.13
1937	7.87	1946	1.52	1955	0.68	1964	0.00	1973	7.89
1938	9.81	1947	0.86	1956	0.59	1965	0.23	1974	0.14

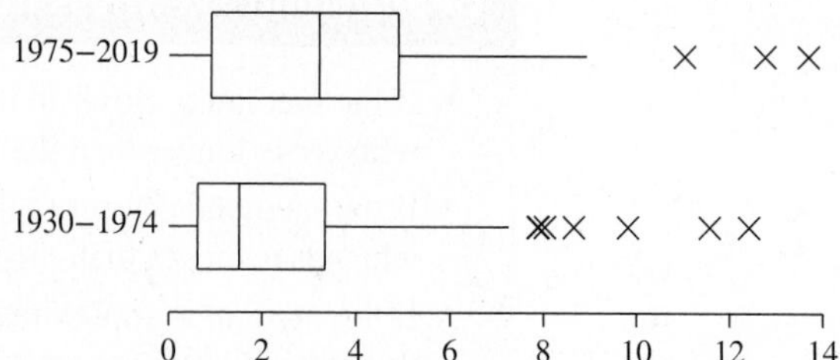

Figure 3.18 Comparative boxplots for February rainfall in Los Angeles, 1930–1974 and 1975–2019

Using Technology

Table 3.14 lists the number of calories in 11 fast-food restaurant menu items. We use these data to illustrate the technology steps.

Table 3.14 Number of Calories in Fast-Food Menu Items

840	1090	680	950	1070	860
940	1285	900	1120	720	

TI-84 PLUS

Computing Quartiles

The procedure used to compute the mean and median, presented on page 107, will also compute the quartiles. The quartiles for the data in Table 3.14 are shown in Figure A.

```
1-Var Stats
↑Sx=179.9229633
 σx=171.5498145
 n=11
 minX=680
 Q1=840
 Med=940
 Q3=1090
 maxX=1285
```

Figure A

Drawing Boxplots

Step 1. Enter the data from Table 3.14 into **L1** in the data editor.

Step 2. Press **2nd, Y=** to access the STAT PLOTS menu, and select **Plot1** by pressing **1**.

Step 3. Select **On** and the boxplot icon in the lower left. Press **ENTER** (Figure B).

Step 4. Press **ZOOM** and then **9: ZoomStat** (Figure C).

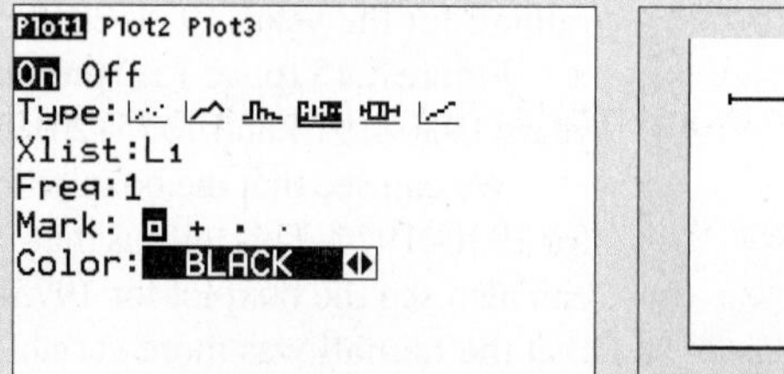

Figure B

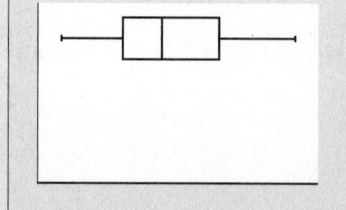

Figure C

EXCEL

Computing Quartiles

Step 1. Enter the data from Table 3.14 in Column **A**.

Step 2. Select the **Insert Function** icon f_x and highlight **Statistical** in the category field.

Step 3. Highlight **QUARTILE.EXC** and press **OK**. Enter the range of cells that contain the data from Table 3.14 in the **Array** field. In the **Quart** field, enter **1** for Q_1, **2** for Q_2, or **3** for Q_3.

Step 4. Click **OK**.

MINITAB

Computing Quartiles

The MINITAB procedure used to compute the mean and median, described on page 108, will also compute the quartiles. The quartiles for the data in Table 3.14 are shown in Figure D.

Statistics

Variable	Mean	SE Mean	StDev	Minimum	Q1	Median	Q3	Maximum
Menu Items	950.5	54.2	179.9	680.0	840.0	940.0	1090.0	1285.0

Figure D

Drawing Boxplots

Step 1. Enter the data from Table 3.14 into **Column C1**.

Step 2. Click on **Graph** and select **Boxplot**. Choose the **One Y, Simple option** and press **OK**.

Step 3. Double-click on **C1** and press **OK**.

Section 3.3 Exercises

Exercises 1–10 are the Check Your Understanding exercises located within the section.

Understanding the Concepts

In Exercises 11–14, fill in each blank with the appropriate word or phrase.

11. ________ divide the data set approximately into quarters. *Quartiles*

12. The median is the same as the ________ quartile. *second*

13. The quantity $Q_3 - Q_1$ is known as the ________.

14. A value that is considerably larger or smaller than most of the values in a data set is called an ________. *outlier*

In Exercises 15–18, determine whether the statement is true or false. If the statement is false, rewrite it as a true statement.

15. The third quartile, Q_3, separates the lowest 25% of the data from the highest 75%. *False*

16. The 25th percentile is the same as the first quartile. *True*

17. The five-number summary consists of the minimum, the first quartile, the mode, the third quartile, and the maximum. *False*

18. In a boxplot, if the lower whisker is much longer than the upper whisker, then the data are skewed to the left. *True*

Practicing the Skills

19. A population has mean $\mu = 7$ and standard deviation $\sigma = 2$.

a. Find the z-score for a population value of 5. *−1*

b. Find the z-score for a population value of 10. *1.5*

c. What number has a z-score of 2? *11*

20. A population has mean $\mu = 25$ and standard deviation $\sigma = 4$.

a. Find the z-score for a population value of 16. *−2.25*

b. Find the z-score for a population value of 31. *1.5*

c. What number has a z-score of 2.5? *35*

In Exercises 21 and 22, identify the outlier. Then tell whether the outlier seems certain to be due to an error, or whether it could conceivably be correct.

21. A rock is weighed five times. The readings in grams are 48.5, 47.2, 4.91, 49.5, and 46.3. *4.91; error*

22. A sociologist samples five families in a certain town and records their annual income. The incomes are \$34,000, \$57,000, \$13,000, \$1,200,000, and \$62,000. *\$1,2000,000; correct*

23. For the data set

37 82 20 25 31 10 41 44 4 36 68

a. Find the first and third quartiles. *$Q_1 = 20$; $Q_3 = 44$*

b. Find the IQR. *24*

c. Find the upper and lower outlier boundaries. *Lower: −16; upper: 80*

d. List all the values, if any, that are classified as outliers. *82*

24. For the data set

15	7	2	4	4	3	4	3	4	25	4	9	3
12	2	8	3	2	2	6	7	3	10	4	5	4

a. Find the first and third quartiles. *$Q_1 = 3, Q_3 = 7$*

b. Find the IQR. *4*

c. Find the upper and lower outlier boundaries. *Lower: −3, upper: 13*

d. List all the values, if any, that are classified as outliers. *15, 25*

25. For the data set

2	2	2	2	5	7	8	8	9	9	14	14
14	16	19	20	21	22	22	24	24	27	32	33
33	33	34	34	35	35	35	37	38	38	38	40
40	40	41	42	46	47	48	48	48	48	48	49

a. Find the 58th percentile. *34*

b. Find the 22nd percentile. *14*

c. Find the 78th percentile. *40*

d. Find the 15th percentile. *8*

26. For the data set

1	5	8	8	8	11	13	14	15	15	16	17
20	23	24	25	25	26	26	29	31	34	35	35
38	44	45	47	47	51	53	53	54	55	55	57
57	59	60	62	65	69	70	75	75	76	78	79
81	83	83	84	89	91	92	93	93	96	96	99

a. Find the 80th percentile. *80*

b. Find the 43rd percentile. *44*

c. Find the 18th percentile. *16*

d. Find the 65th percentile. *61*

27. The following TI-84 Plus display presents the five-number summary for a data set. Are there any outliers in this data set? *Yes*

```
1-Var Stats
↑Sx=31.72871115
 σx=30.10049833
 n=10
 minX=1
 Q1=9
 Med=16.5
 Q3=38
 maxX=110
```

28. The following TI-84 Plus display presents the five-number summary for a data set. Are there any outliers in this data set? *Yes*

```
1-Var Stats
↑Sx=17.77305767
 σx=17.3230338
 n=20
 minX=25
 Q1=59.5
 Med=64
 Q3=78.5
 maxX=85
```

Working with the Concepts

29. Standardized tests: In a recent year, the mean score on the ACT test was 21.1 and the standard deviation was 5.2. The mean score on the SAT mathematics test was 514 and the standard deviation was 117. The distributions of both scores were approximately bell-shaped.

a. Find the z-score for an ACT score of 27. *1.13*

b. Find the z-score for an SAT math score of 650. *1.16*

c. Which score is higher, relative to its population of scores? *SAT*

d. Jose's ACT score had a z-score of 0.75. What was his ACT score? *25*

e. Emma's SAT score had a z-score of −2.0. What was her SAT score? *280*

30. A fish story: The mean length of one-year-old spotted flounder, in millimeters, is 126 with standard deviation of 18, and the mean length of two-year-old spotted flounder is 162 with a standard deviation of 28. The distribution of flounder lengths is approximately bell-shaped.

a. Anna caught a one-year-old flounder that was 150 millimeters in length. What is the z-score for this length? *1.33*

b. Luis caught a two-year-old flounder that was 190 millimeters in length. What is the z-score for this length? *1*

c. Whose fish is longer, relative to fish the same age? *Anna's*

d. Joe caught a one-year-old flounder whose length had a z-score of 1.2. How long was this fish? *147.6*

e. Terry caught a two-year-old flounder whose length had a z-score of −0.5. How long was this fish? *148*

Source: *Turkish Journal of Veterinary and Animal Science*, 29:1013–1018

31. Blood pressure in men: The three quartiles for systolic blood pressure in a sample of 3179 men were $Q_1 = 108$, $Q_2 = 116$, and $Q_3 = 127$.

a. Find the IQR. *19*

b. Find the upper and lower outlier boundaries. *79.5; 155.5*

c. A systolic blood pressure greater than 140 is considered high. Would a blood pressure of 140 be an outlier? *No*

Source: *Journal of Human Hypertension*, 16:305–312

32. Blood pressure in women: The article referred to in Exercise 31 reported that the three quartiles for systolic blood pressure in a sample of 1213 women between the ages of 20 and 29 were $Q_1 = 100$, $Q_2 = 108$, and $Q_3 = 115$.

a. Find the IQR. *15*

b. Find the upper and lower outlier boundaries. *77.5; 137.5*

c. A systolic blood pressure greater than 140 is considered high. Would a blood pressure of 140 be an outlier? *Yes*

33. Hazardous waste: Following is a list of the number of hazardous waste sites in each of the 50 states of the United States in a recent year. The list has been sorted into numerical order.

0	1	2	2	3	5	6	9	9	9
9	9	11	12	12	12	12	12	13	13
14	14	14	14	15	15	15	16	19	19
20	21	25	26	29	30	32	32	32	38
40	48	49	49	52	67	86	97	97	116

Source: U.S. Environmental Protection Agency

a. Find the first and third quartiles of these data. *11; 32*

b. Find the median of these data. *15*

c. Find the upper and lower outlier boundaries. *20.5; 63.5*

d. Are there any outliers? If so, list them. *67, 86, 97, 97, 116*
e. Construct a boxplot for these data.
f. Describe the shape of this distribution. *Skewed right*
g. What is the 30th percentile? *12*
h. What is the 85th percentile? *49*
i. The state of Georgia has 16 hazardous waste sites. What percentile is this? *55th*

34. Cholesterol levels: The National Health and Nutrition Examination Survey (NHANES) measured the serum HDL cholesterol levels in a large number of women. Following is a sample of 40 HDL levels (in milligrams per deciliter) that are based on the results of that survey. They have been sorted into numerical order.

27	28	30	32	34	36	37	37	37	37
37	40	45	47	48	49	53	53	54	56
57	58	61	62	63	63	64	64	64	65
66	70	72	73	73	74	80	80	81	84

Source: NHANES

a. Find the first and third quartiles of these data. *37; 65.5*
b. Find the median of these data. *56.5*
c. Find the upper and lower outlier boundaries. *5.75; 108.5*
d. Are there any outliers? If so, list them. *No outliers*
e. Construct a boxplot for these data.
f. Describe the shape of this distribution. *Approximately symmetric*
g. What is the 20th percentile? *37*
h. What is the 67th percentile? *64*
i. One woman had a cholesterol level of 58. What percentile is this? *54th*

35. Commuting to work: Jamie drives to work every weekday morning. She keeps track of the time it takes, in minutes, for 35 days. The results follow.

15	17	17	17	17	18	19
19	19	19	19	19	20	20
20	20	20	21	21	21	21
21	21	21	21	21	22	23
23	24	26	31	36	38	39

a. Find the first and third quartiles of these data. *19; 22*
b. Find the median of these data. *21*
c. Find the upper and lower outlier boundaries. *14.5; 26.5*
d. Are there any outliers? If so, list them. *31, 36, 38, 39*
e. Construct a boxplot for these data.
f. Describe the shape of this distribution. *Skewed right*
g. What is the 14th percentile? *17*
h. What is the 87th percentile? *26*
i. One day, the commute time was 31 minutes. What percentile is this? *90th*

36. Windy city by the bay: Following are wind speeds (in mph) for 29 randomly selected days in San Francisco.

13.4	23.1	27.2	31.8	36.3	40.3	14.1	24.6
27.7	32.6	38.1	40.9	18.3	25.2	29.7	33.6
38.7	44.2	20.8	26.8	30.1	34.5	40.2	46.8
22.9	26.9	30.8	35.8	40.3			

a. Find the first and third quartiles of these data. *25.2; 38.1*
b. Find the median of these data. *30.8*
c. Find the upper and lower outlier boundaries. *25.2; 38.1*
d. Are there any outliers? If so, list them. *No outliers*
e. Construct a boxplot for these data.
f. Describe the shape of this distribution. *Approximately symmetric*
g. What is the 40th percentile? *27.7*
h. What is the 10th percentile? *18.3*
i. One day, the wind speed was 30.1 mph. What percentile is this? *47th*

37. Caffeine: Following are the number of grams of carbohydrates in 12-ounce espresso beverages offered at Starbucks.

14	43	38	44	31	27	39	59	9	10	54
14	25	26	9	46	30	24	41	26	27	14

Source: www.starbucks.com

a. Find the first and third quartiles of these data. *14; 41*
b. Find the median of these data. *27*
c. Find the upper and lower outlier boundaries. *26.5; 81.5*
d. The beverage with the most carbohydrates is a Peppermint White Chocolate Mocha, with 59 grams. Is this an outlier? *No*
e. The beverages with the least carbohydrates are an Iced Skinny Flavored Latte, and a Cappuccino, each with 9 grams. Are these outliers? *No*
f. Construct a boxplot for these data.
g. Describe the shape of this distribution. *Skewed to the right*
h. What is the 31st percentile? *24*
i. What is the 71st percentile? *39*
j. There are 38 grams of carbohydrates in an Iced Dark Cherry Mocha. What percentile is this? *66th*

38. Nuclear power: The following table presents the number of nuclear reactors in a recent year, in each country that had one or more reactors.

Country	Number of Reactors	Country	Number of Reactors
Argentina	2	South Korea	23
Armenia	1	Mexico	2
Belgium	7	Netherlands	1
Brazil	2	Pakistan	3
Bulgaria	2	Romania	2
Canada	19	Russia	33
China	18	Slovakia	4
Czech Republic	6	Slovenia	1
Finland	4	South Africa	2
France	58	Spain	8
Germany	9	Sweden	10
Hungary	4	Switzerland	5
India	20	Ukraine	15
Iran	1	United Kingdom	16
Japan	50	United States	100

Source: International Atomic Energy Agency

a. Find the first and third quartiles of these data. *2; 18*
b. Find the median of these data. *5.5*
c. Find the upper and lower outlier boundaries. *−22; 42*
d. Which countries are outliers? *France, Japan, United States*
e. Construct a boxplot for these data.
f. Describe the shape of this distribution. *Skewed right*
g. What is the 45th percentile? *4*
h. What is the 88th percentile? *33*
i. India has 20 nuclear reactors. What percentile is this? *82nd*

39. Place your bets: In a recent year, 28 states in the United States had one or more tribal gambling casinos. The following table presents the number of tribal casinos in each of those states.

3	2	26	70	2	2	8
7	3	4	3	22	39	3
14	7	3	21	8	2	11
114	8	14	1	34	31	4

Source: American Gaming Association

a. Find the first and third quartiles of these data. $Q_1 = 3; Q_3 = 21.5$
b. Find the median of these data. 7.5
c. Find the upper and lower outlier boundaries. −24.75; 49.25
d. Which values, if any, are outliers? 70, 114
e. Construct a boxplot for these data.
f. Describe the shape of this distribution. Skewed right
g. What is the 40th percentile? 4
h. What is the 83rd percentile? 31
i. North Dakota has 11 tribal casinos. What percentile is this? 63rd

40. Hail to the chief: There have been 58 presidential inaugurations in U.S. history. At each one, the president has made an inaugural address. Following are the number of words spoken in each of these addresses.

1431	135	2321	1730	2166	1177	1211	3375
4472	2915	1128	1176	3843	8460	4809	1090
3336	2831	3637	700	1127	1339	2486	2979
1686	4392	2015	3968	2218	984	5434	1704
1526	3329	4055	3672	1880	1808	1359	559
2273	2459	1658	1366	1507	2128	1803	1229
2427	2561	2320	1598	2155	1592	2071	2395
2096	1433						

Source: The American Presidency Project

a. Find the first and third quartiles of these data. $Q_1 = 1431; Q_3 = 2915$
b. Find the median of these data. 2083.5
c. Find the upper and lower outlier boundaries. Lower: −795; upper: 5141
d. The two shortest speeches were 135 words, by George Washington in 1793, and 559 words, by Franklin Roosevelt in 1945. Are either of these outliers? Neither
e. The two longest speeches were 8460 words, by William Henry Harrison in 1841, and 5434 words, by William Howard Taft in 1909. Are either of these outliers? Both are outliers.
f. Construct a boxplot for these data.
g. Describe the shape of this distribution. Skewed right
h. What is the 15th percentile? 1177
i. What is the 65th percentile? 2395
j. Donald Trump used 1433 words in his inaugural address in 2017. What percentile is this? 27th

41. Gotta catch 'em all: There are 151 Generation I Pokémon characters. Each Pokémon has a base number of hit points, which determines the health of the Pokémon.

a. Charmander is a Pokémon with 39 hit points. There are 19 Pokémon with fewer hit points and 131 Pokémon with more. On what percentile is Charmander? 13th
b. Charmander evolves into a Pokémon called Charmeleon. Charmeleon has 58 hit points. There are 62 Pokémon with fewer hit points and 88 Pokémon with more. On what percentile is Charmeleon? 41st
c. Charmeleon evolves into a Pokémon called Charizard. Charizard has 78 hit points. There are 110 Pokémon with fewer hit points and 40 Pokémon with more. On what percentile is Charizard? 73rd

42. Pigskin: There are 77 football stadiums (both college and professional) in the United States with seating capacities of 60,000 or more.

a. The largest stadium to host an NFL team is MetLife Stadium in East Rutherford, New Jersey, which is the home of the New York Giants and New York Jets. The capacity of MetLife Stadium is 82,566. There are 60 stadiums with smaller capacities and 16 with greater capacities. On what percentile is MetLife stadium? 79th
b. The smallest stadium to host an NFL team is Soldier Field in Chicago, Illinois, which is the home of the Chicago Bears. The capacity of Soldier Field is 61,500. There are 8 stadiums with smaller capacities and 68 with greater capacities. On what percentile is Soldier Field? 11th
c. The largest stadium with no permanent home team is the Cotton Bowl in Dallas, Texas. The capacity of the Cotton Bowl is 92,100. There are 65 stadiums with smaller capacities and 11 with greater capacities. On what percentile is the Cotton Bowl? 85th

43. Bragging rights: After learning his score on a recent statistics exam, Ed bragged to his friends: "My score is the first quartile of the class." Did Ed have a good reason to brag? Explain. No

44. Who scored the highest? On a final exam in a large statistics class, Tom's score was the tenth percentile, Dick's was the median, and Harry's was the third quartile. Which of the three scores was the highest? Which was the lowest? Harry's; Tom's

45. Baseball salaries: In 2018, the Boston Red Sox defeated the Los Angeles Dodgers to become the champions of Major League Baseball. Following are the salaries, in millions of dollars, of the players on each of these teams.

Red Sox						
31.00	23.75	21.00	20.00	15.00	12.00	8.55
5.00	4.30	3.58	2.85	2.48	1.60	1.15
0.72	0.62	0.58	0.56	0.56	0.56	0.56
0.56	0.56	0.56	0.56			

Dodgers						
31.00	20.00	18.00	18.00	17.90	5.00	4.50
3.73	3.50	3.00	3.00	2.10	0.71	0.61
0.60	0.58	0.58	0.57	0.57	0.57	0.56
0.56	0.56	0.56	0.56	0.56		

a. Find the median, the first quartile, and the third quartile of the Red Sox' salaries. Median: 1.60; $Q_1 = 0.56; Q_3 = 8.55$
b. Find the median, the first quartile, and the third quartile of the Dodgers' salaries. Median: 0.71; $Q_1 = 0.57; Q_3 = 4.50$
c. Find the upper and lower outlier bounds for the Red Sox' salaries. Lower: −11.425; upper: 20.535
d. Find the upper and lower outlier bounds for the Dodgers' salaries. Lower: −5.325; upper: 10.395
e. Construct comparative boxplots for the two data sets. What conclusions can you draw?

46. Automotive emissions: Following are levels of particulate emissions for 65 vehicles driven at sea level and for 35 vehicles driven at high altitude.

Sea Level									
1.5	0.9	1.1	1.3	3.5	1.1	1.1	0.9	1.3	0.9
0.6	1.3	2.5	1.5	1.1	1.1	2.2	0.9	1.8	1.5
1.2	1.6	2.1	6.6	4.0	2.5	1.4	1.4	1.8	1.1
1.6	3.7	0.6	2.7	2.6	3.0	1.2	1.0	1.6	3.1
2.4	2.1	2.7	1.2	3.3	3.8	1.3	2.1	6.6	1.2
3.1	0.5	0.3	0.5	3.4	3.5	2.7	1.9	5.9	4.2
3.5	3.6	3.1	3.3	4.6					

High Altitude						
8.9	4.4	3.6	4.4	3.8	2.4	3.8
5.3	5.8	2.9	4.7	1.9	9.1	8.7
9.5	2.7	9.2	7.3	2.1	6.3	6.5
6.3	2.0	5.9	5.6	5.6	1.5	6.5
5.3	5.6	2.1	1.1	3.3	1.8	7.6

a. Find the median, the first quartile, and the third quartile of the sea-level emissions. *Median: 1.8; $Q_1 = 1.2$; $Q_3 = 3.1$*

b. Find the median, the first quartile, and the third quartile of the high-altitude emissions. *Median: 5.3; $Q_1 = 2.7$; $Q_3 = 6.5$*

c. Find the upper and lower outlier bounds for the sea-level emissions. *Lower: −1.65; upper: 5.95*

d. Find the upper and lower outlier bounds for the high-altitude emissions. *Lower: −3; upper: 12.2*

e. Construct comparative boxplots for the two data sets. What conclusions can you draw?

47. **Comparative boxplots:** Following are boxplots of the level of fine particle air pollution, in micrograms per cubic meter, in the cities of Denver and Greeley, Colorado, during a recent winter.

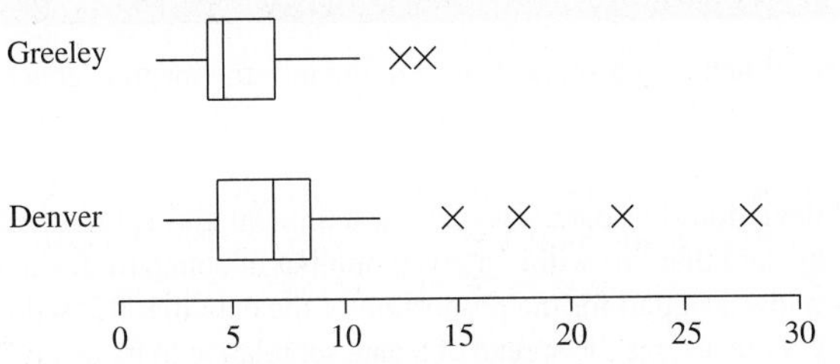

a. In which city is the pollution level generally higher? *Denver*

b. Approximately what percentage of the values for Greeley are greater than the median value for Denver? Is it closest to 25%, 50%, 75%, or 90%? *25%*

c. In which city is there more spread in the pollution levels? *Denver*

d. Are the pollution levels for Greeley skewed right, skewed left, or approximately symmetric? *Skewed right*

e. Are the pollution levels for Denver skewed right, skewed left, or approximately symmetric? *Skewed right*

48. **Comparative boxplots:** Following are boxplots of diastolic blood pressure, in millimeters, for a sample of people aged 18–39 years, and a sample over 60 years old. The data are consistent with results reported by the National Health Statistics Reports.

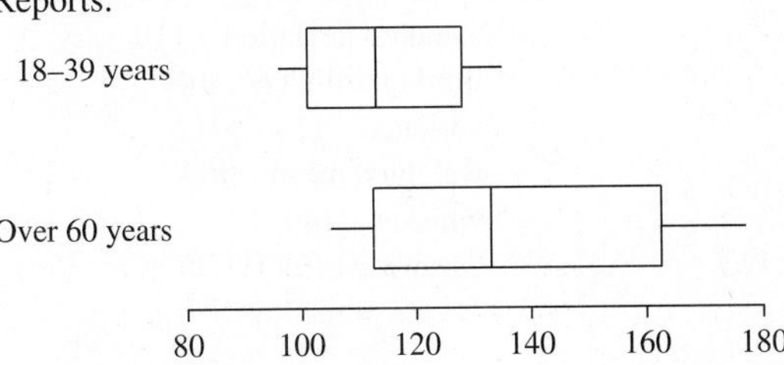

a. Which group generally has higher blood pressure? *Over 60*

b. Approximately what percentage of people over 60 have blood pressure higher than the median for people aged 18–39? Is it closest to 25%, 50%, 75%, or 90%? *75%*

c. For which group are the blood pressures more spread out? *Over 60*

d. Are the blood pressures for people 18–39 skewed right, skewed left, or approximately symmetric? *Approximately symmetric*

e. Are the blood pressures for people over 60 skewed right, skewed left, or approximately symmetric? *Approximately symmetric*

49. **Boxplot possible?** The most recent U.S. Census reported the per capita incomes for each of the 50 states. The five-number summary, in thousands of dollars, is

Minimum: 20.0, Q_1: 24.1, Median: 25.7, Q_3: 29.0, Maximum: 34.8

a. Does the five-number summary provide enough information to construct a boxplot? If so, construct the boxplot. If not, explain why not. *Yes*

b. Are these data skewed to the right, skewed to the left, or approximately symmetric? *Skewed right*

50. **Boxplot possible?** Following is the five-number summary for the populations, in millions, for the 50 states of the United States.

Minimum: 0.58, Q_1: 1.86, Median: 4.49, Q_3: 6.90, Maximum: 38.04

a. Does the five-number summary provide enough information to construct a boxplot? If so, construct the boxplot. If not, explain why not. *No*

b. Are these data skewed to the right, skewed to the left, or approximately symmetric? *Skewed right*

51. **Unusual boxplot:** Ten residents of a town were asked how many children they had. The responses were as follows.

0 0 3 1 0 0 4 0 7 0

a. Explain why the median and first quartile are the same.

b. Construct a boxplot for these data.

c. Explain why the boxplot has no left whisker.

52. **Clean-shaven boxplot:** Make up a list of numbers whose boxplot has no whiskers.

53. **Highly skewed data:** Make up a data set in which the mean is greater than the third quartile.

54. **Highly skewed data:** Make up a data set in which the mean is less than the first quartile.

Extending the Concepts

55. **The vanishing outlier:** Seven families live on a small street in a certain town. Their annual incomes (in $1000s) are 15, 20, 30, 35, 50, 60, and 150.

a. Find the first and third quartiles, and the IQR. *$Q_1 = 20$; $Q_3 = 60$; $IQR = 40$*

b. Show that 150 is an outlier.

A big new house is built on the street, and the income (in $1000s) of the family that moves in is 200.

c. Find the first and third quartiles, and the IQR of the eight incomes. *$Q_1 = 25$; $Q_3 = 105$; $IQR = 80$*

d. Are there any outliers now? *No*

e. Explain how adding the value 200 to the data set eliminated the outliers.

56. Beyond quartiles and percentiles: If we divide a data set into four approximately equal parts, the three dividing points are called quartiles. If we divide a data set into 100 approximately equal parts, the 99 dividing points are called percentiles. In general, if we divide a data set into k approximately equal parts, we can call the dividing points k-tiles. How would you find the ith k-tile of a data set of size n?

57. z-scores and skewed data: Table 3.9 presents the February rainfalls in Los Angeles for the period 1975–2019.

a. Show that the mean of these data is $\mu = 3.660$ and the population standard deviation is $\sigma = 3.325$.

b. Show that the z-score for a rainfall of 0 (rounded to two decimal places) is $z = -1.10$.

c. Show that the z-score for a rainfall of 7.32 (rounded to two decimal places) is $z = 1.10$.

d. What percentage of the years had rainfalls of 0? *2.2%*

e. What percentage of the years had rainfalls of 7.32 or more? *13.3%*

f. The z-scores indicate that a rainfall of 0 and a rainfall of 7.32 are about equally extreme. Is a rainfall of 7.32 really as extreme as a rainfall of 0, or is it less extreme? *Less extreme*

g. These data are skewed to the right. Explain how skewness causes the z-score to give misleading results.

Answers to Check Your Understanding Exercises for Section 3.3

1. a. 1 **b.** −0.5 **c.** 16

2. a. −1.5 **b.** 2 **c.** 33

3. a. −0.52 **b.** 2.10 **c.** 1.21

4. a. 70 **b.** 83 **c.** 30th **d.** Yes

5. 90%

6. 35%

7. 17th

8. 75th

9. a. 6 143 291 522 3721 **b.** 379
c. Lower outlier bound is −425.5; upper bound is 1090.5.
d. 1203, 1803, and 3721 are outliers.

10.

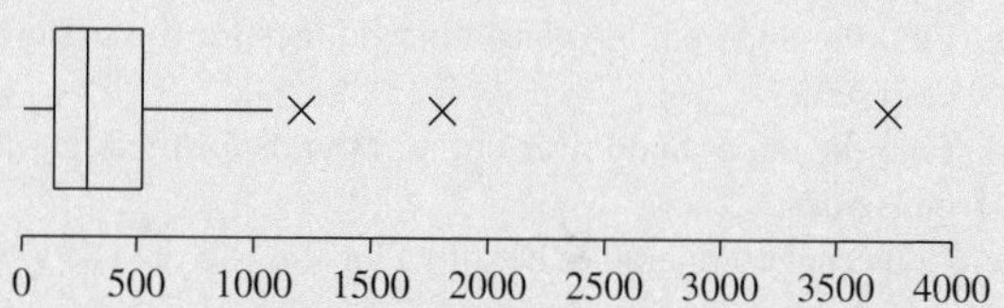

Chapter 3 Summary

Section 3.1: We can describe the center of a data set with the mean or the median. When a data set is skewed to the left, the mean is generally less than the median, and when a data set is skewed to the right, the mean is generally greater than the median. The mode of a data set is the most frequently occurring value.

Section 3.2: The spread of a data set is most often measured with the standard deviation. For data sets that are unimodal and approximately symmetric, the Empirical Rule can be used to approximate the proportion of the data that lies within a given number of standard deviations of the mean. Chebyshev's Inequality, which is valid for all data sets, provides a lower bound for the proportion of the data that lies within a given number of standard deviations of the mean. The coefficient of variation (CV) measures the spread of a data set relative to its mean. The CV provides a way to compare spreads of data sets whose values are in different units.

Section 3.3: For bell-shaped data sets, the z-score gives a good description of the position of a value in a data set. Quartiles and percentiles can be used to describe the positions for any data set. Quartiles are used to compute the five-number summary, which consists of the minimum value, the first quartile, the median, the third quartile, and the maximum value. Outliers are values that are considerably larger or smaller than most of the values in a data set. Boxplots are graphs that allow us to visualize the five-number summary, along with any outliers. Comparative boxplots allow us to visually compare the shapes of two or more data sets.

Vocabulary and Notation

$\sum x = x_1 + \cdots + x_n$ 99
arithmetic mean 98
boxplot 149
Chebyshev's Inequality 129
coefficient of variation (CV) 130
comparative boxplots 151
degrees of freedom 122
deviation 119
Empirical Rule 126
first quartile Q_1 142
five-number summary 146
interquartile range (IQR) 148
IQR method 148
mean 98
mean absolute deviation (MAD) 139
measure of center 98
measure of position 98
measure of spread 98
median 100
mode 105
modified boxplot 149
outlier 147
outlier boundaries 148
percentile 144
population mean μ 99
population standard deviation σ 123
population variance σ^2 119
quartile 141
range 119
resistant 103
sample mean $\bar{x}$ 99
sample standard deviation s 123
sample variance s^2 121
second quartile Q_2 135
standard deviation 118
third quartile Q_3 142
variance 119
weighted mean 102
whisker 149
z-score 140

Important Formulas

Sample mean:

$$\bar{x} = \frac{\sum x}{n}$$

Population mean:

$$\mu = \frac{\sum x}{N}$$

Weighted mean:

$$\bar{x}_w = \frac{w_1x_1 + w_2x_2 + \cdots + w_nx_n}{w_1 + w_2 + \cdots + w_n}$$

Range:

Range = largest value − smallest value

Population variance:

$$\sigma^2 = \frac{\sum(x - \mu)^2}{N}$$

Sample variance:

$$s^2 = \frac{\sum(x - \bar{x})^2}{n - 1}$$

Coefficient of variation:

$$CV = \frac{\sigma}{\mu}$$

***z*-score:**

$$z = \frac{x - \mu}{\sigma}$$

Interquartile range:

$IQR = Q_3 - Q_1$ = third quartile − first quartile

Lower outlier boundary:

$Q_1 - 1.5\,IQR$

Upper outlier boundary:

$Q_3 + 1.5\,IQR$

Chapter Quiz

1. Of the mean, median, and mode, which must be a value that actually appears in the data set? *Mode*

2. The prices (in dollars) for a sample of personal computers are: 550, 700, 420, 580, 550, 450, 690, 390, 350. Calculate the mean, median, and mode for this sample. *Mean: 520; median: 550; mode: 550*

3. If a computer with a price of $2000 were added to the list in Exercise 2, which would be affected more, the mean or the median? *Mean*

4. In general, a histogram is skewed to the left if the ________ is noticeably less than the ________. *mean, median*

5. A sample of 100 students was asked how many hours per week they spent studying. The following frequency table shows the results:

Number of Hours	Frequency
1.0–4.9	14
5.0–8.9	34
9.0–12.9	29
13.0–16.9	15
17.0–20.9	8

a. Approximate the mean time this sample of students spent studying. *9.76*

b. Approximate the standard deviation of the time this sample of students spent studying. *4.51*

6. A sample has a variance of 16. What is the standard deviation? *4*

7. Each of the following histograms represents a data set with mean 20. One has a standard deviation of 3.96, and the other has a standard deviation of 2.28. Which is which? Fill in the blanks: Histogram I has a standard deviation of ________ and histogram II has a standard deviation of ________. *3.96, 2.28*

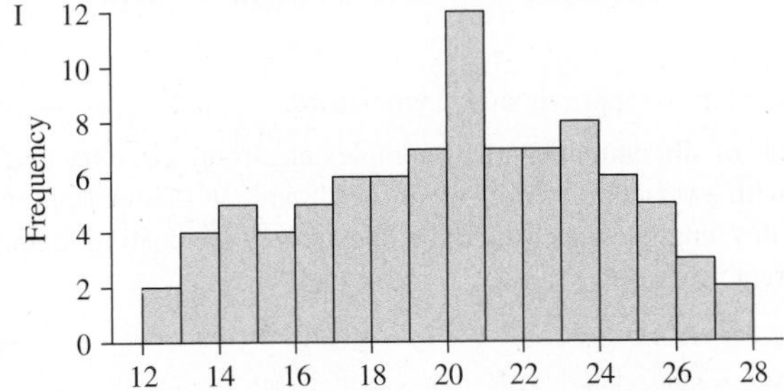

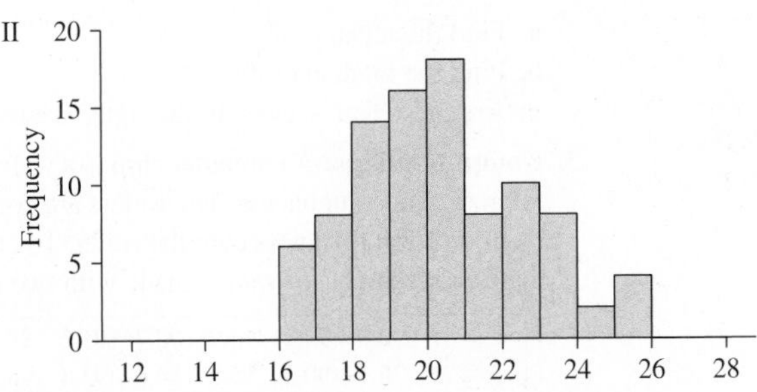

In Exercises 8–11, suppose that the mean starting salary of social workers in a specific region is $37,480 with a standard deviation of $1,400.

8. Assume that the histogram of starting salaries is approximately bell-shaped. Approximately what percentage of the salaries will be between $34,680 and $40,280? *95%*

9. Assume it is not known whether the histogram of starting salaries is bell-shaped. Fill in the blank: At least __________ percent of the salaries will be between $34,680 and $40,280. *75*

10. John's starting salary is $38,180. What is the z-score of his salary? *0.5*

11. Find the coefficient of variation of the salaries. *0.037*

12. True or false: If a student's exam grade is on the 55th percentile, then approximately 45% of the scores are below his or her grade. *False*

13. The five-number summary for a sample is 7, 18, 35, 62, 85. What is the IQR? *44*

14. The prices (in dollars) for a sample of coffee makers are:

19 22 29 68 35 37 28 22 41 39 28

a. Find the first and third quartiles. $Q_1 = 22$; $Q_3 = 39$

b. Find the upper and lower outlier boundaries. *Lower: −3.5; upper: 64.5*

c. Are there any outliers? If so, list them. *68*

15. Construct a boxplot for the data in Exercise 14.

Review Exercises

1. Support your local artist: Following are the annual amounts of federal support (in millions of dollars) for National Endowment for the Arts programs for the years 2010–2019.

Year	Amount
2010	167.5
2011	154.7
2012	146.0
2013	138.4
2014	146.0
2015	146.0
2016	147.9
2017	149.9
2018	152.9
2019	155.0

Source: National Endowment for the Arts

a. Find the mean annual amount of federal aid from 2010–2019. *150.43*

b. Find the median annual amount of federal aid from 2010–2019. *148.9*

2. Corporate profits: The following table presents the profit, in a recent year, in billions of dollars, for each of the 15 largest U.S. corporations in terms of revenue.

Corporation	Profit	Corporation	Profit
Exxon Mobil	41.1	Ford Motor	20.2
Walmart	15.7	Hewlett-Packard	7.1
Chevron	26.9	AT&T	3.9
ConocoPhillips	12.4	Valero Energy	2.1
General Motors	9.2	Bank of America Corp.	1.5
General Electric	14.2	McKesson	1.2
Berkshire Hathaway	10.2	Verizon Communications	2.4
Fannie Mae	−16.9		

Source: CNNMoney

a. Find the mean profit. *10.08*

b. Find the median profit. *9.2*

c. Are these data skewed to the right, skewed to the left, or approximately symmetric? Explain. *Approximately symmetric*

3. Computer chips: A computer chip is a wafer made of silicon that contains complex electronic circuitry made up of microscopic components. The wafers are coated with a very thin coating of silicon dioxide. It is important that the coating be of uniform thickness over the wafer. To check this, engineers measured the thickness of the coating, in millionths of a meter, for samples of wafers made with two different processes.

Process 1:	90.0	92.2	94.9	92.7	91.6	88.2	92.0	98.2	96.0
Process 2:	76.1	90.2	96.8	84.6	93.3	95.7	90.9	100.3	95.2

a. Find the mean of the thicknesses for each process. *Process 1: 92.87; process 2: 91.46*
b. Find the median of the thicknesses for each process. *Process 1: 92.2; process 2: 93.3*
c. If it is desired to obtain as thin a coating as possible, is one process much better than the other? Or are they about the same? *About the same*

4. **More computer chips:** Using the data in Exercise 3:
a. Find the sample variance of the thicknesses for each process. *Process 1: 9.40; process 2: 53.36*
b. Find the sample standard deviation of the thicknesses for each process. *Process 1: 3.07; process 2: 7.30*
c. Which process appears to be better in producing a uniform thickness? Explain. *Process 1*

5. **Stock prices:** Following are the closing prices of Microsoft stock for each trading day in May and June of a recent year.

May				
34.54	34.62	34.35	33.67	33.72
33.27	33.49	34.59	34.98	35.00
34.40	34.72	35.00	34.84	35.47
35.67	34.96	34.78	34.99	35.59
33.16	32.72			

June				
34.90	35.03	34.88	35.02	34.27
34.15	34.61	34.85	35.08	34.87
34.08	33.85	33.53	33.03	32.69
32.66	32.99	33.31	33.75	33.49

a. Find the mean and median price in May. *Mean: 34.479; median: 34.67*
b. Find the mean and median price in June. *Mean: 34.052; median: 34.115*
c. Does there appear to be a substantial difference in price between May and June? Or are the prices about the same? *About the same*

6. **More stock prices:** Using the data in Exercise 5:
a. Find the population standard deviation of the prices in May. *0.791*
b. Find the population standard deviation of the prices in June. *0.814*
c. Financial analysts use the word *volatility* to refer to the variation in stock prices. Was the volatility for the price of Microsoft stock greater in May or June? *June*

7. **Measure that ball:** Each of 16 students measured the circumference of a tennis ball by two different methods:
A: Estimate the circumference by eye.
B: Measure the circumference by rolling the ball along a ruler.
The results (in centimeters) are given below, in increasing order for each method:

A:	18.0	18.0	18.0	20.0	22.0	22.0	22.5	23.0	24.0	24.0	25.0	25.0	25.0	25.0	26.0	26.4
B:	20.0	20.0	20.0	20.0	20.2	20.5	20.5	20.7	20.7	20.7	21.0	21.1	21.5	21.6	22.1	22.3

a. Compute the sample standard deviation of the measurements for each method. *A: 2.87; B: 0.75*
b. For which method is the sample standard deviation larger? Why should one expect this method to have the larger standard deviation? *A*
c. Other things being equal, is it better for a measurement method to have a smaller standard deviation or a larger standard deviation? Or doesn't it matter? Explain. *Smaller*

8. **Time in surgery:** Records at a hospital show that a certain surgical procedure takes an average of 162.8 minutes with a standard deviation of 4.9 minutes. If the data are approximately bell-shaped, between what two values will about 95% of the data fall? *153 and 172.6*

9. **Rivets:** A machine makes rivets that are used in the manufacture of airplanes. To be acceptable, the length of a rivet must be between 0.9 centimeter and 1.1 centimeters. The mean length of a rivet is 1.0 centimeter, with a standard deviation of 0.05 centimeter. What is the maximum possible percentage of rivets that are unacceptable? *25%*

10. **How long can you talk?** A manufacturer of phone batteries determines that the average length of talk time for one of its batteries is 470 minutes. Suppose that the standard deviation is known to be 32 minutes and that the data are approximately bell-shaped. Estimate the percentage of batteries that have z-scores between -1 and 1. *68%*

11. **Paying rent:** The monthly rents for apartments in a certain town have a mean of $800 with a standard deviation of $150. What can you determine about these data by using Chebyshev's Inequality with $K = 3$? *At least 88.9% between 350 and 1250*

12. **Advertising costs:** The amounts spent (in billions) on media advertising in the United States for a sample of categories are presented in the following table.

Advertising Category	Amount Spent
Retail	16.35
Automotive	14.84
Local Services	8.98
Telecom	8.66
Financial Services	7.89
Personal Care Products	6.84
Food & Candy	6.57
Direct Response	6.34
Restaurants	6.19
Insurance	4.86

Source: Kantar Media

a. Find the mean amount spent on advertising. 8.752
b. Find the median amount spent on advertising. 7.365
c. Find the sample variance of the advertising amounts. 14.616
d. Find the sample standard deviation of the advertising amounts. 3.82
e. Find the first quartile of the advertising amounts. 6.34
f. Find the third quartile of the advertising amounts. 8.98
g. Find the 40th percentile of the advertising amounts. 6.705
h. Find the 65th percentile of the advertising amounts. 8.66

13. Matching: Match each histogram to the boxplot that represents the same data set.

a.

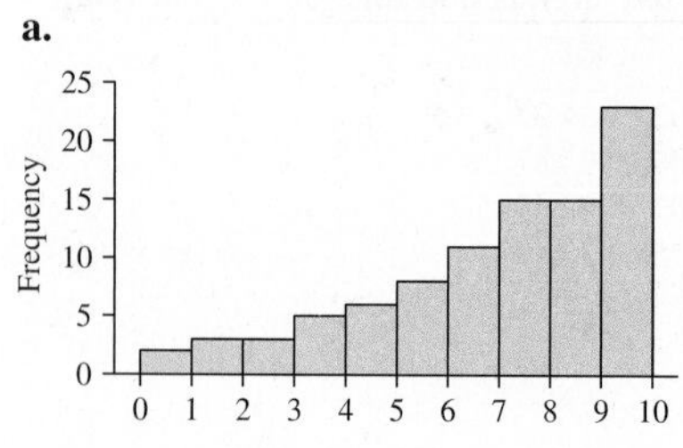

(3)

b.

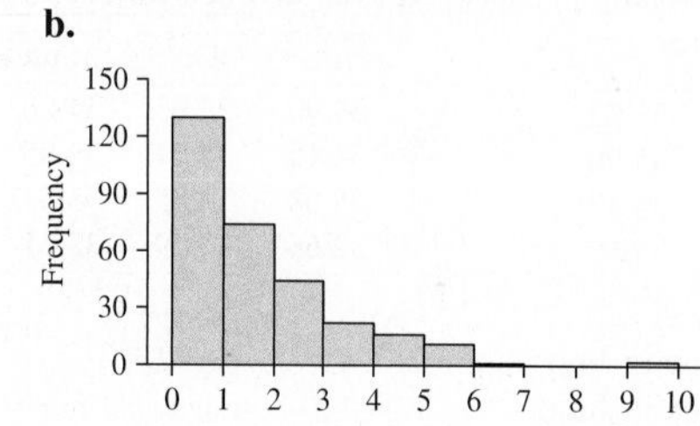

(1)

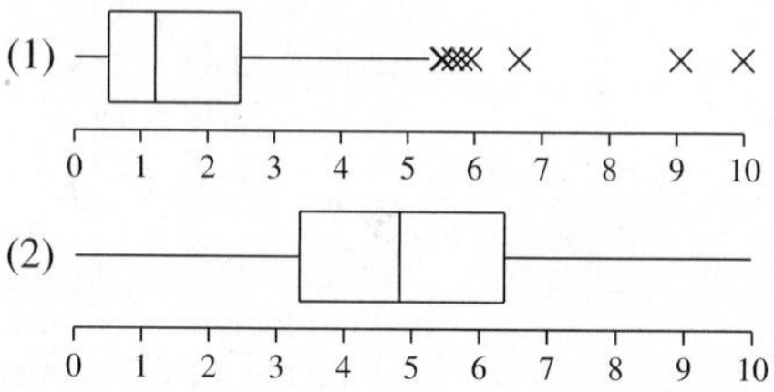

c.

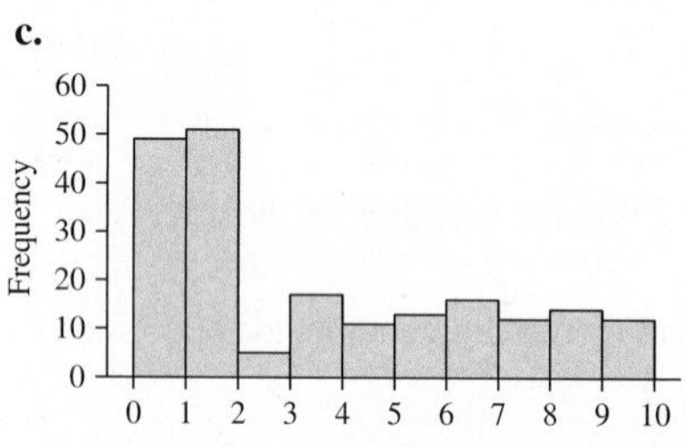

(4)

d.

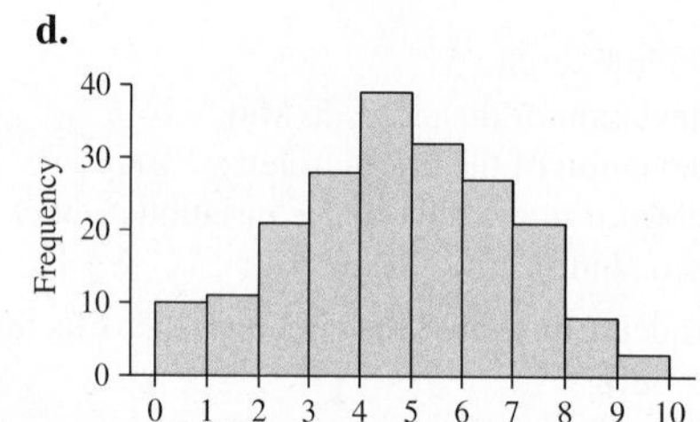

(2)

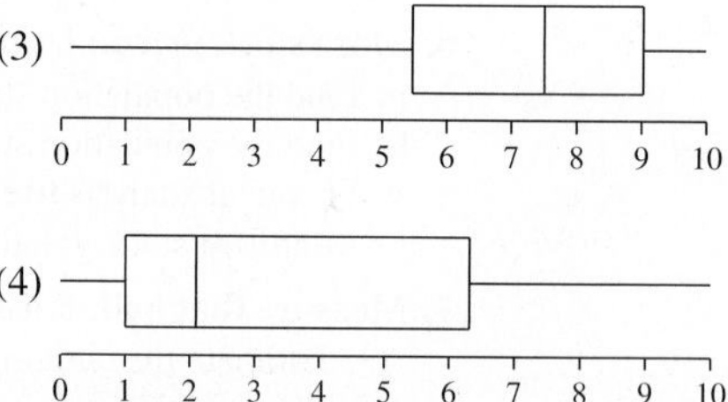

14. Weights of soap: As part of a quality control study aimed at improving a production line, the weights (in ounces) of 50 bars of soap are measured. The results are shown below, sorted from smallest to largest.

11.6	12.6	12.7	12.8	13.1	13.3	13.6	13.7
13.8	14.1	14.3	14.3	14.6	14.8	15.1	15.2
15.6	15.6	15.7	15.8	15.8	15.9	15.9	16.1
16.2	16.2	16.3	16.4	16.5	16.5	16.5	16.6
17.0	17.1	17.3	17.3	17.4	17.4	17.4	17.6
17.7	18.1	18.3	18.3	18.3	18.5	18.5	18.8
19.2	20.3						

a. Find the first and third quartiles of these data. $Q_1 = 14.6; Q_3 = 17.4$
b. Find the median of these data. 16.2
c. Find the upper and lower outlier boundaries. Lower: 10.4; upper: 21.6
d. Are there any outliers? If so, list them. No outliers
e. Construct a boxplot for these data.

15. More corporate profits: Using the data in Exercise 2:
a. Find the first and third quartiles of the profit. $Q_1 = 2.1; Q_3 = 15.7$
b. Find the median profit. 9.2
c. Find the upper and lower outlier boundaries. Lower: −18.3; upper: 36.1
d. Are there any outliers? If so, list them. 41.1
e. Construct a boxplot for these data.

Write About It

1. The U.S. Department of Labor annually publishes an Occupational Outlook Handbook, which reports the job outlook, working conditions, and earnings for thousands of different occupations. The handbook reports both the mean and median annual earnings. For most occupations, which is larger, the mean or the median? Why do you think so?
2. Explain why the Empirical Rule is more useful than Chebyshev's Inequality for bell-shaped distributions. Explain why Chebyshev's Inequality is more useful for distributions that are not bell-shaped.
3. Does Chebyshev's Inequality provide useful information when $K = 1$? Explain why or why not.
4. Is Chebyshev's Inequality true when $K < 1$? Is it useful? Explain why or why not.
5. Percentiles are values that divide a data set into hundredths. The values that divide a data set into tenths are called deciles, denoted $D_1, D_2, ..., D_9$. Describe the relationship between percentiles and deciles.

In-Class Activities

NOTE TO INSTRUCTOR

Anscombe's quartet illustrates the important point that there is much information about the shape of a data set that is not contained in numerical summaries.

1. **Estimating the population mean:** Two exams were given to a class. The scores on the first exam ranged from 30 to 90, and the scores on the second exam ranged from 70 to 75. You are asked to guess the mean score for each exam. Which of your guesses do you think is more likely to be closer to the true mean? Why?
2. **Sample mean and population mean:** Following are two populations, both of which have mean $\mu = 50$.

Population 1:	45	46	47	48	49	50	51	52	53	54	55
Population 2:	40	42	44	46	48	50	52	54	56	58	60

   Each student should draw a sample of size 3 from each population and compute the sample mean. For which population does the sample mean tend to be closer to the population mean? Compute the standard deviation of each population. What is the relationship between the population standard deviation and the closeness of the sample mean to the population mean?
3. **Anscombe's quartet:** Statistician Francis Anscombe developed four data sets that demonstrate the importance of graphing data in addition to computing summary statistics such as the mean and variance. Following are the data sets. Verify that the means and variance are the same for each, to two decimal places. Construct boxplots for each, and discuss the differences among the data sets, which are not detected by the mean and variance.

Data set 1:	8.04	6.95	7.58	8.81	8.33	9.96	7.24	4.26	10.84	4.82	5.68
Data set 2:	9.14	8.14	8.74	8.77	9.26	8.10	6.13	3.10	9.13	7.26	4.74
Data set 3:	7.46	6.77	12.74	7.11	7.81	8.84	6.08	5.39	8.15	6.42	5.73
Data set 4:	6.58	5.76	7.71	8.84	8.47	7.04	5.25	12.50	5.56	7.91	6.89

Case Study: Can Recycled Materials Be Used In Electronic Devices?

Electronic devices contain electric circuits etched into wafers made of silicon. These silicon wafers are sealed with an ultrathin layer of silicon dioxide, in a process known as oxidation. This can be done with either new or recycled wafers.

In a study described in the book *Statistical Case Studies for Industrial Process Improvement* by V. Czitrom and P. Spagon, both new and recycled wafers were oxidized, and the thicknesses of the layers were measured to determine whether they tended to differ between the two types of wafers. Recycled wafers are cheaper than new wafers, so the hope was that they would perform at least as well as the new wafers. Following are 36 thickness measurements (in tenths of a nanometer) for both new and recycled wafers.

New								
90.0	92.2	94.9	92.7	91.6	88.2	92.0	98.2	96.0
91.1	89.8	91.5	91.5	90.6	93.1	88.9	92.5	92.4
96.7	93.7	93.9	87.9	90.4	92.0	90.5	95.2	94.3
92.0	94.6	93.7	94.0	89.3	90.1	91.3	92.7	94.5

Recycled								
91.8	94.5	93.9	77.3*	92.0	89.9	87.9	92.8	93.3
92.6	90.3	92.8	91.6	92.7	91.7	89.3	95.5	93.6
92.4	91.7	91.6	91.1	88.0	92.4	88.7	92.9	92.6
91.7	97.4	95.1	96.7	77.5*	91.4	90.5	95.2	93.1

*Measurement is in error due to a defective gauge.

1. Construct comparative boxplots for the thicknesses of new wafers and recycled wafers.
2. Identify all outliers. *Recycled: 77.3, 77.5, 97.4; new: none*
3. Should any of the outliers be deleted? If so, delete them and redraw the boxplots. *Yes, 77.3 and 77.5*
4. Identify any outliers in the redrawn boxplots. Should any of these be deleted? Explain. *Recycled: 87.9, 88.0, 96.7, 97.4; new: none; no*
5. Are the distributions of thicknesses skewed, or approximately symmetric? *Approximately symmetric*
6. Delete outliers as appropriate, and compute the mean thickness for new and for recycled wafers. *New: 92.33; recycled: 92.31*
7. Delete outliers as appropriate, and compute the median thickness for new and for recycled wafers. *New: 92.1; recycled: 92.4*
8. Delete outliers as appropriate, and compute the standard deviation of the thicknesses for new and for recycled wafers. *New: 2.36; recycled: 2.21*
9. Suppose that it is desired to use the type of wafer whose distribution has less spread. Write a brief paragraph that explains which type of wafer to use and why. Which measure is more useful for spread in this case, the standard deviation or the interquartile range? Explain. *Interquartile range*

chapter

4

Summarizing Bivariate Data

Stockbyte/Punchstock

Introduction

Inflation and unemployment are measures of the health of the economy. Inflation is the percentage increase in prices over the course of a year, and unemployment is the percentage of the labor force that is out of work. The following table presents levels of inflation and unemployment, as reported by the Bureau of Labor Statistics, for the years 1991 through 2018.

Year	Inflation	Unemployment	Year	Inflation	Unemployment
1991	3.1	6.8	2005	3.4	5.1
1992	2.9	7.5	2006	2.5	4.6
1993	2.7	6.9	2007	4.1	4.6
1994	2.7	6.1	2008	0.1	5.8
1995	2.5	5.6	2009	2.7	9.3
1996	3.3	5.4	2010	1.5	9.6
1997	1.7	4.9	2011	3.0	8.9
1998	1.6	4.5	2012	1.7	8.1
1999	2.7	4.2	2013	1.5	7.4
2000	3.4	4.0	2014	0.8	6.2
2001	1.6	4.7	2015	0.7	5.3
2002	2.4	5.8	2016	2.1	4.9
2003	1.9	6.0	2017	2.1	4.4
2004	3.3	5.5	2018	1.9	3.9

Source: Bureau of Labor Statistics

Economists have long studied the relationship between inflation and unemployment. One theory states that inflation and unemployment follow a pattern called "the Phillips curve," in which higher inflation leads to lower unemployment, while lower inflation leads to higher unemployment. This theory is now widely regarded as too simple, and economists continue to study data, looking for more complex relationships. In the case study at the end of this chapter, we will examine some methods for predicting unemployment.

NOTE TO INSTRUCTOR

The Guidelines for Assessment and Instruction in Statistics Education (GAISE) College Report recommends that students be given experience with multivariable thinking. As stated in the report, students must be prepared to investigate and explore relationships among many variables. This chapter introduces students to multivariable thinking by describing relationships between two variables. Chapters 12 and 13 describe relationships among more than two variables.

Questions about relationships between variables arise frequently in science, business, public policy, and other areas where informed decisions need to be made. For example, how does a person's level of education affect his or her income? How does the amount of time spent studying for an exam affect an exam score? Data used to study questions like these consist of **ordered pairs**. An ordered pair consists of values of two variables for each individual in the data set. In the preceding table, the ordered pairs are (inflation rate, unemployment rate). To study the relationship between education and income, the ordered pair might be (number of years of education, annual income).

Data that consist of ordered pairs are called **bivariate data**. The basic graphical tool used to study bivariate data is the *scatterplot*, in which each ordered pair is plotted as a point. In many cases, the points on a scatterplot tend to cluster around a straight line. In these cases, the summary statistic most often used to measure the closeness of the relationship between the two variables is the *correlation coefficient*, which we will study in Section 4.1. When two variables are closely related to each other, it is often of interest to try to predict the value of one of them when given the value of the other. This is done with the equation of the *least-squares regression line*, which we will study in Sections 4.2 and 4.3.

Section 4.1 Correlation

Objectives

1. Construct scatterplots for bivariate data
2. Compute the correlation coefficient
3. Interpret the correlation coefficient
4. Understand that correlation is not the same as causation

Objective 1 Construct scatterplots for bivariate data

Scatterplots

A real estate agent wants to study the relationship between the size of a house and its selling price. Table 4.1 presents the size in square feet and the selling price in thousands of dollars for a sample of houses in a suburban Denver neighborhood.

NOTE TO INSTRUCTOR

The following topics from the **Statistics Corequisite Workbook** are aligned with the material in this section.

7.1 - Plotting Points, Scatterplots, and Types of Association

A.1 - Independent and Dependent Variables and Creating a Table of Values

Table 4.1 Size and Selling Price for a Sample of Houses

Size (square feet)	Selling Price ($1000s)
2521	400
2555	426
2735	428
2846	435
3028	469
3049	475
3198	488
3198	455

Source: Sue Bays Realty

It is reasonable to suspect that the selling price is related to the size of the house. Specifically, we expect that houses with larger sizes are more likely to have higher selling prices. A good way to visualize a relationship like this is with a **scatterplot**. In a scatterplot, each individual in the data set contributes an ordered pair of numbers, and each ordered pair is plotted on a set of axes.

Example 4.1

Construct a scatterplot

Construct a scatterplot for the data in Table 4.1.

Solution

Each of the eight houses in Table 4.1 contributes an ordered pair of numbers. We will take the size as the first number and selling price as the second number. So the ordered pairs to be plotted are (2521, 400), (2555, 426), and so on. Figure 4.1 presents the scatterplot. We can see that houses with larger sizes tend to have larger selling prices, and houses with smaller sizes tend to have smaller (lower) selling prices.

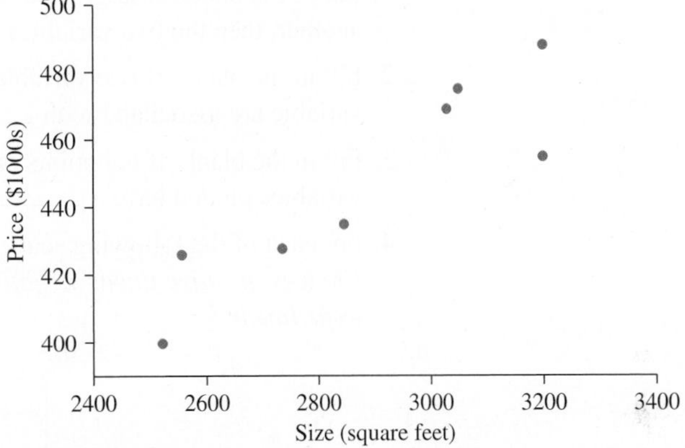

Figure 4.1 Scatterplot of selling price versus size for a sample of houses

Let's take a closer look at Figure 4.1. We have observed that larger sizes tend to be associated with larger prices, and smaller sizes tend to be associated with smaller prices. We refer to this as a *positive association* between size and selling price. In addition, the points tend to cluster around a straight line from lower left to upper right. We describe this by saying that the relationship between the two variables is **linear**. Therefore, we can say that the scatterplot in Figure 4.1 exhibits a positive linear association between size and selling price.

In some cases, large values of one variable are associated with small values of another. An example is the weight of a car and its gas mileage. Large weights are associated with small gas mileages, and small weights are associated with large gas mileages. Therefore, we say that weight and gas mileage have a *negative association*.

Figure 4.2 presents examples of scatterplots that exhibit various kinds of association.

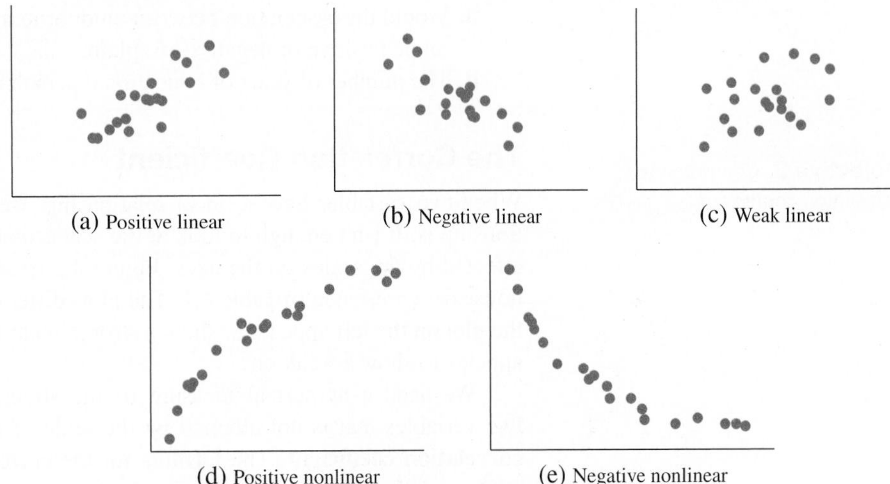

Figure 4.2 Scatterplots exhibiting various types of association

SUMMARY

- Two variables have a **positive association** if large values of one variable are associated with large values of the other.
- Two variables have a **negative association** if large values of one variable are associated with small values of the other.
- Two variables have a **linear relationship** if the data tend to cluster around a straight line when plotted on a scatterplot.

Check Your Understanding

1. Fill in the blank: If large values of one variable are associated with small values of another, then the two variables have a ________ association. *negative*
2. Fill in the blank: If two variables have a positive association, then large values of one variable are associated with ________ values of the other. *large*
3. Fill in the blank: If the points on a scatterplot tend to cluster around a straight line, the variables plotted have a ________ association. *linear*
4. For each of the following scatterplots, state the type of association that is exhibited: *Choices: positive linear, negative linear, positive nonlinear, negative nonlinear, weak linear.*

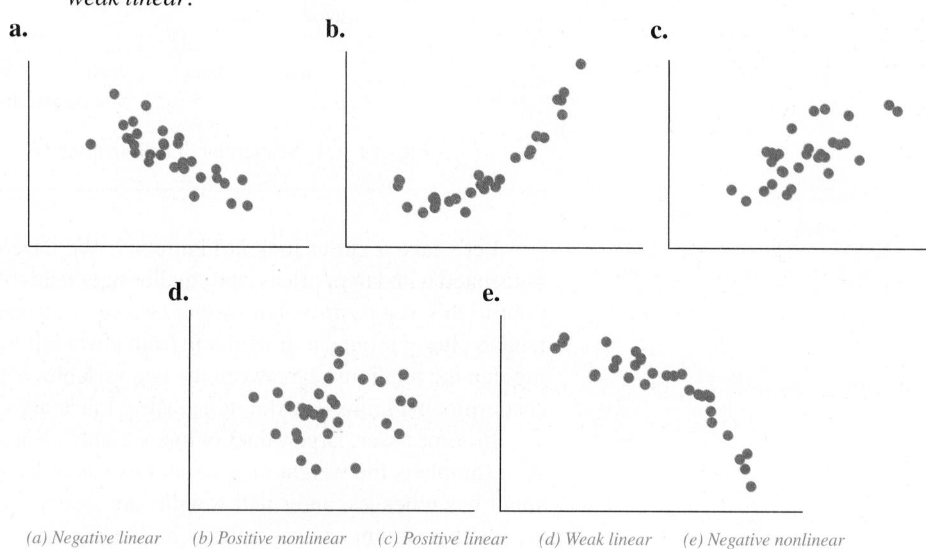

(a) Negative linear (b) Positive nonlinear (c) Positive linear (d) Weak linear (e) Negative nonlinear

5. For each of the following pairs of variables, determine whether the association is positive or negative, and explain why.
   a. Would the association between outdoor temperature and consumption of heating oil be positive or negative? Explain. *Negative*
   b. The number of years of education a person has and the person's income *Positive*

Answers are on page 178.

Objective 2 Compute the correlation coefficient

The Correlation Coefficient

When two variables have a linear relationship, we want to measure how strong the relationship is. It isn't enough to look at the scatterplot, because the visual impression can be affected by the scales on the axes. Figure 4.3 (page 169) presents two scatterplots of the house data presented in Table 4.1. The plots differ only in the scale used on the *y*-axis, yet the plot on the left appears to show a strong linear relationship, while the plot on the right appears to show a weak one.

We need a numerical measure of the strength of the linear relationship between two variables that is not affected by the scale of a plot. The appropriate quantity is the **correlation coefficient**. The formula for the correlation coefficient is a bit complicated, although calculating it does not involve much more than calculating sample means and standard deviations as was done in Chapter 3.

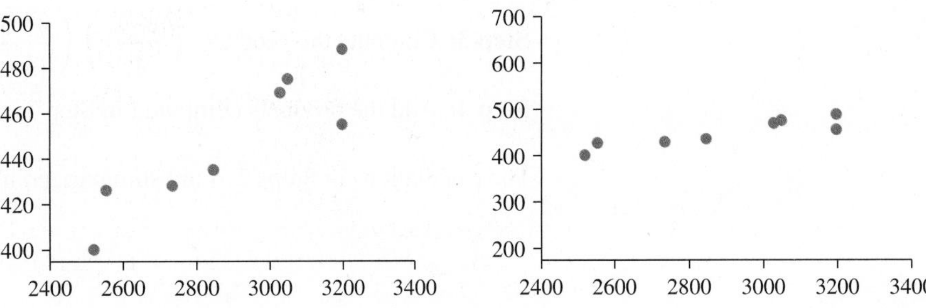

Figure 4.3 Both plots present the same data, yet the plot on the right appears to show a stronger linear relationship than the plot on the left. The reason is that the scales on the y-axis are different.

EXPLAIN IT AGAIN

Appearances can be misleading: The appearance of a scatterplot depends on the scales chosen for the axes. To get a reliable measure of the strength of a linear relationship, we must compute the correlation coefficient.

The Correlation Coefficient

Given ordered pairs (x, y), with sample means $\bar{x}$ and $\bar{y}$, sample standard deviations s_x and s_y, and sample size n, the correlation coefficient r is given by

$$r = \frac{1}{n-1} \sum \left(\frac{x - \bar{x}}{s_x}\right)\left(\frac{y - \bar{y}}{s_y}\right)$$

We often refer to r as the correlation between x and y.

Not surprisingly, most people nowadays use technology to compute the correlation coefficient. Procedures for the TI-84 Plus calculator, MINITAB, and Excel are presented in the Using Technology section on pages 173–174.

The correlation coefficient has several important properties, which we list.

NOTE TO INSTRUCTOR

There is an exception to the rule that the correlation coefficient measures the strength of the linear relationship. When $r = 0$, and the points lie on a horizontal line, it can be said that there is a perfect linear relationship. This situation rarely occurs in practice.

Properties of the Correlation Coefficient

- The correlation coefficient is always between −1 and 1, inclusive. In other words, $-1 \le r \le 1$.
- The value of the correlation coefficient does not depend on the units of the variables. If we measure x and y in different units, the correlation will still be the same.
- It does not matter which variable is x and which is y.
- The correlation coefficient measures only the strength of the *linear* relationship between variables and can be misleading when the relationship is nonlinear.
- The correlation coefficient is sensitive to outliers and can be misleading when outliers are present.

Note that if points lie exactly on a horizontal line, the correlation coefficient will be equal to 0, even though the relationship is linear. This rare case is an exception to the rule that the correlation coefficient measures the strength of the linear relationship.

Example 4.2

Compute the correlation coefficient

Use the data in Table 4.1 to compute the correlation between size and selling price.

Solution

We will denote size by x and selling price by y. We compute the correlation coefficient using the following steps:

Step 1: Compute the sample means and standard deviations. We obtain $\bar{x} = 2891.25$, $\bar{y} = 447.0$, $s_x = 269.49357$, $s_y = 29.68405$.

Step 2: Compute the quantities $\dfrac{x - \bar{x}}{s_x}$ and $\dfrac{y - \bar{y}}{s_y}$.

Step 3: Compute the products $\left(\frac{x-\bar{x}}{s_x}\right)\left(\frac{y-\bar{y}}{s_y}\right)$.

Step 4: Add the products computed in Step 3, and divide the sum by $n - 1$.

The calculations in Steps 2–4 are summarized in the following table.

x	y	$\frac{x-\bar{x}}{s_x}$	$\frac{y-\bar{y}}{s_y}$	$\left(\frac{x-\bar{x}}{s_x}\right)\left(\frac{y-\bar{y}}{s_y}\right)$
2521	400	−1.3738732	−1.5833419	2.1753110
2555	426	−1.2477106	−0.7074506	0.8826936
2735	428	−0.5797912	−0.6400744	0.3711095
2846	435	−0.1679075	−0.4042575	0.0678779
3028	469	0.5074333	0.7411387	0.3760785
3049	475	0.5853572	0.9432675	0.5521484
3198	488	1.1382461	1.3812131	1.5721604
3198	455	1.1382461	0.2695050	0.3067630

$$\frac{\sum\left(\frac{x-\bar{x}}{s_x}\right)\left(\frac{y-\bar{y}}{s_y}\right)}{n-1} = \frac{6.3041423}{7}$$
$$= 0.9005918$$

We round our final answer to three decimal places. The correlation coefficient is $r = 0.901$.

CAUTION

Do not round the intermediate values used to calculate the correlation coefficient. Round only the final result.

In general, we will round the correlation coefficient to three decimal places when it is the final result.

Objective 3 Interpret the correlation coefficient

Interpreting the correlation coefficient

The correlation coefficient measures the strength of the *linear* relationship between two variables. For this reason, it is meaningful only when the variables are linearly related. It can be misleading in other situations.

Interpreting the Correlation Coefficient

When two variables have a linear relationship, the correlation coefficient can be interpreted as follows:

- If r is positive, the two variables have a positive linear association.
- If r is negative, the two variables have a negative linear association.
- If r is close to 0, the linear association is weak.
- The closer r is to 1, the more strongly positive the linear association is.
- The closer r is to -1, the more strongly negative the linear association is.
- If $r = 1$, then the points lie exactly on a straight line with positive slope; in other words, the variables have a perfect positive linear association.
- If $r = -1$, then the points lie exactly on a straight line with negative slope; in other words, the variables have a perfect negative linear association.

When two variables are not linearly related, the correlation coefficient does not provide a reliable description of the relationship between the variables.

CAUTION

Be sure that the relationship is linear before interpreting the correlation coefficient.

In Example 4.2, the two variables do have a linear relationship, as verified by the scatterplot in Figure 4.1. The value of the correlation coefficient is $r = 0.901$, which indicates a strong positive linear association.

Check Your Understanding

6. The National Assessment for Educational Progress (NAEP) is a U.S. government organization that assesses the performance of students and schools at all levels across the United States. The following table presents the percentage of eighth-grade students who were found to be proficient in mathematics, and the percentage who were found to be proficient in reading in each of the 10 most populous states.

State	Percentage Proficient in Reading	Percentage Proficient in Mathematics
California	60	59
Texas	73	78
New York	75	70
Florida	66	68
Illinois	75	70
Pennsylvania	79	77
Ohio	79	76
Michigan	73	66
Georgia	67	64
North Carolina	71	73

Source: National Assessment for Educational Progress

a. Construct a scatterplot with reading proficiency on the horizontal axis and math proficiency on the vertical axis. Is there a linear relationship?

b. Compute the correlation between reading proficiency and math proficiency. Is the linear association positive or negative? Weak or strong? *$r = 0.809$, positive, strong*

Answers are on page 178.

RECALL
A statistic is resistant if its value is not affected much by extreme data values.

The correlation coefficient is not resistant

Figure 4.4 presents a scatterplot of the amount of farmland (including ranches) plotted against the total land area, for a selection of U.S. states. It is reasonable to suspect that states with larger land area would tend to have more farmland, and the scatterplot shows that, in general, this is true.

RECALL
An outlier is a point that is detached from the main bulk of the data.

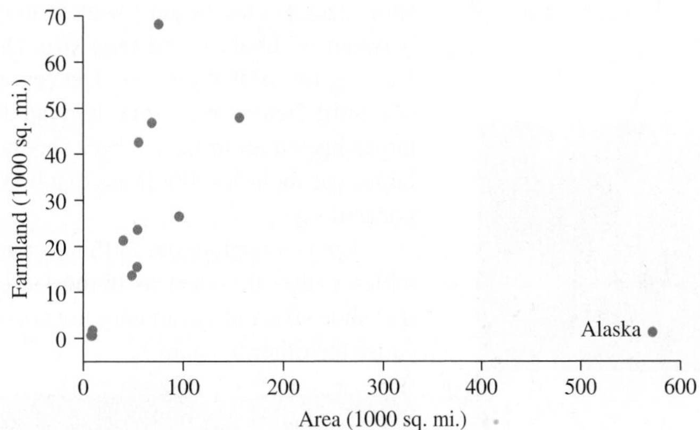

Figure 4.4 Area of farmland versus total land area for a selection of U.S. states. Alaska is an outlier. Because of the outlier, the correlation coefficient for this plot is −0.119, which is misleading. If the outlier is removed, the correlation coefficient for the remaining points is $r = 0.710$.

There is an outlier in the lower right corner of the plot, corresponding to the state of Alaska. Alaska is an outlier because it has a huge land area but very little farming. The correlation coefficient for the scatterplot in Figure 4.4 is $r = -0.119$. This suggests that there is actually a weak *negative* association between the total land area and the area of farmland. If we ignore the Alaska point and compute the correlation coefficient for the remaining points, we get $r = 0.710$, a big difference.

With the Alaska outlier in the plot, the correlation is misleading. For the states other than Alaska, there is a strong positive association between total land area and farmland area. Because Alaska is such a big exception to this rule, it throws the correlation coefficient way off.

SUMMARY

The correlation coefficient is not resistant. It may be misleading when outliers are present.

Check Your Understanding

7. For which of the following scatterplots is the correlation coefficient an appropriate summary? *(a)*

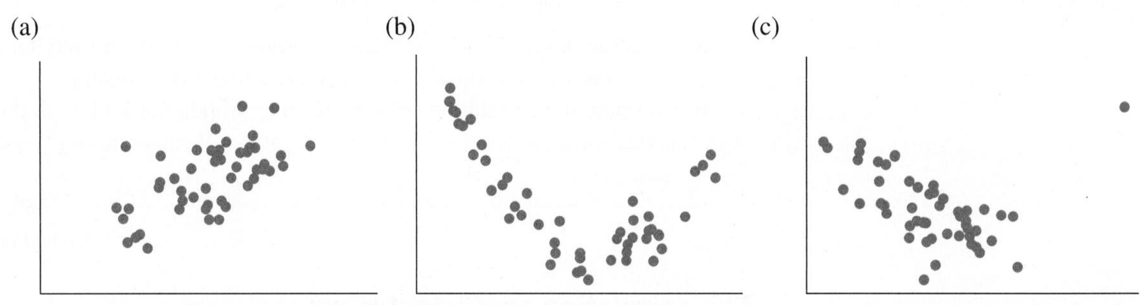

Answer is on page 178.

Objective 4 Understand that correlation is not the same as causation

Correlation Is Not Causation

A group of elementary school children took a vocabulary test. It turned out that children with larger shoe sizes tended to get higher scores on the test, and those with smaller shoe sizes tended to get lower scores. As a result, there was a large positive correlation between vocabulary and shoe size. Does this mean that learning new words causes one's feet to grow, or that growing feet cause one's vocabulary to increase? Obviously not. There is a third factor—age—that is related to both shoe size and vocabulary. Individuals with larger ages tend to have larger shoe sizes. Individuals with larger ages also tend to have larger vocabularies. It follows that individuals with larger shoe sizes will tend to have larger vocabularies.

Age is a confounder in this example. Age is related to both shoe size and vocabulary, which makes it appear as if shoe size and vocabulary are related to each other. The fact that shoe size and vocabulary are correlated does not mean that changing one variable will cause the other to change.

Getty Images

NOTE TO INSTRUCTOR

It is useful to point out that a strong correlation is no more likely to reflect causation than a weak one.

SUMMARY

Correlation is not the same as causation. In general, when two variables are correlated, we cannot conclude that changing the value of one variable will cause a change in the value of the other.

Check Your Understanding

8. An economist discovers that over the past several years, both the salaries of U.S. college professors and the amount of beer consumed in the United States have gone up. Thus, there is a positive correlation between the average salary of college professors and the amount of beer consumed. The economist concludes that the increase in beer consumption must be caused by professors spending their additional money on beer. Explain why this conclusion is not necessarily true.

Answer is on page 178.

Using Technology

We use the data in Table 4.1 to illustrate the technology steps.

TI-84 PLUS

Constructing a scatterplot

Step 1. Enter the x-values from Table 4.1 into **L1** and the y-values into **L2**.

Step 2. Press **2nd, Y=** to access the STAT PLOTS menu, and select Plot1 by pressing **1**.

Step 3. Select **On** and the scatterplot icon (Figure A).

Step 4. Press **ZOOM** and then **9: ZoomStat** (Figure B).

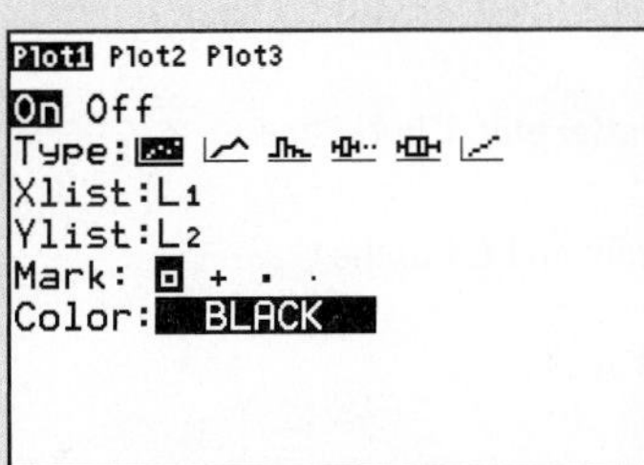

Figure A

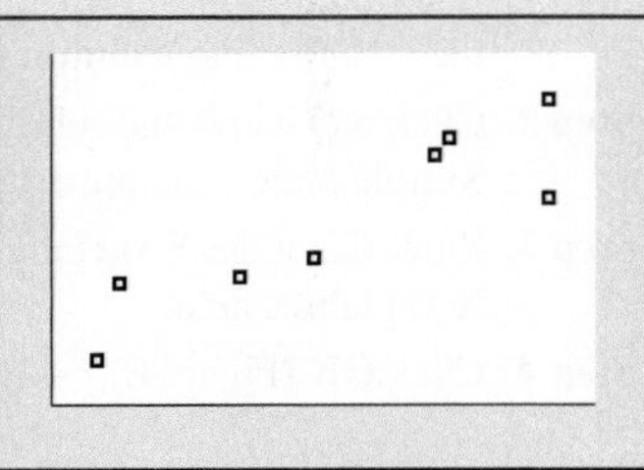

Figure B

Computing the correlation coefficient

The correlation coefficient is calculated as part of the procedure for computing the least-squares regression line and is presented at the end of Section 4.2 on page 185.

EXCEL

Constructing a scatterplot

Step 1. Enter the x-values from Table 4.1 into **Column A** and the y-values into **Column B**.

Step 2. Highlight all values in **Columns A** and **B**.

Step 3. Click on **Insert** and then **Scatter**.

Step 4. Click **OK** (Figure C).

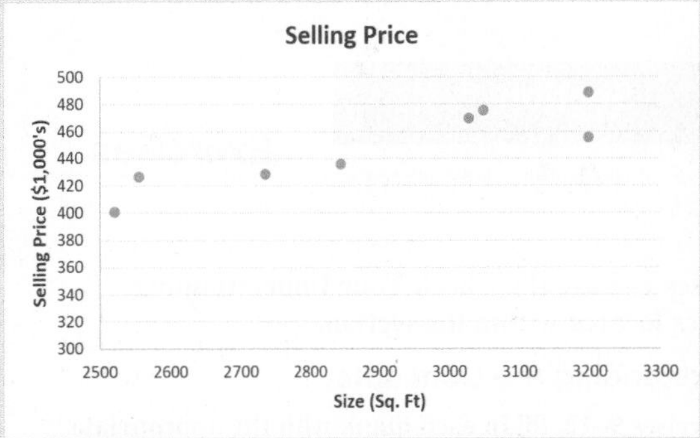

Figure C

Computing the correlation coefficient

Step 1. Enter the data from Table 4.1 into the worksheet.

Step 2. In an empty cell, select the **Insert Function** icon and highlight **Statistical** in the category field.

Step 3. Click on the **CORREL** function and press **OK**.

Step 4. Enter the range of cells that contain the x-values from Table 4.1 in the **Array1** field, and enter the range of cells that contain the y-values in the **Array2** field.

Step 5. Click **OK** (Figure D).

	A	B	C
1	Size	Selling Price	
2	2521	400	
3	2555	426	
4	2735	428	
5	2846	435	
6	3028	469	
7	3049	475	
8	3198	488	
9	3198	455	
10			
11		=CORREL(A2:A9,B2:B9)	
12	Correlation	0.900591753	

Figure D

MINITAB

Constructing a scatterplot

Step 1. Enter the x-values from Table 4.1 into **Column C1** and the y-values into **Column C2**.

Step 2. Click on **Graph** and select **Scatterplot**. Choose the **Simple** option and press **OK**.

Step 3. Enter **C2** in the **Y variables** field and **C1** in the **X variables** field.

Step 4. Click **OK** (Figure E).

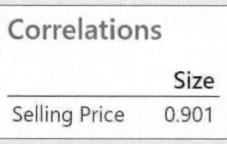

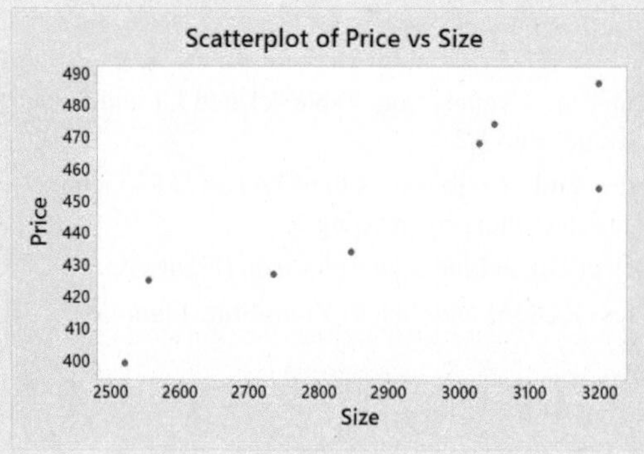

Figure E

Computing the correlation coefficient

Step 1. Enter the x-values from Table 4.1 into **Column C1** and the y-values into **Column C2**.

Step 2. Click on **Stat**, then **Basic Statistics**, then **Correlation**.

Step 3. Double-click **C1** and **C2** to enter these variables in the **Variables** field.

Step 4. Click **OK** (Figure F).

Correlations

	Size
Selling Price	0.901

Figure F

Section 4.1 Exercises

Exercises 1–8 are the Check Your Understanding exercises located within the section.

Understanding the Concepts

In Exercises 9–12, fill in each blank with the appropriate word or phrase.

9. Bivariate data consist of ordered __________. *pairs*

10. In a __________, ordered pairs are plotted on a set of axes. *scatterplot*

11. Two variables have a __________ relationship if the data tend to cluster around a straight line. *linear*

12. The correlation coefficient measures only the strength of the __________ relationship between variables. *linear*

In Exercises 13–16, determine whether the statement is true or false. If the statement is false, rewrite it as a true statement.

13. Two variables are negatively associated if large values of one variable are associated with large values of the other. *False*

14. If the correlation coefficient r equals 1, then the points on a scatterplot lie exactly on a straight line. *True*

15. The correlation coefficient is not resistant. *True*

16. When two variables are correlated, changing the value of one variable will cause a change in value of the other variable. *False*

Practicing the Skills

In Exercises 17–20, compute the correlation coefficient.

17. *0.824*

x	1	2	3	4	5
y	2	1	4	3	7

18. *−0.647*

x	24	13	8	81	63	36	5
y	44	52	42	5	1	48	15

19. *0.515*

x	5.5	4.2	4.7	5.6	6.0	3.9	6.3	5.7
y	4.9	4.8	4.8	4.7	5.5	5.1	5.8	6.5

20. *0.339*

x	5	−8	−2	6	9	−10	13	7
y	−1	−3	−6	−7	−1	5	13	22

In Exercises 21–24, determine whether the correlation coefficient is an appropriate summary for the scatterplot and explain your reasoning.

21.

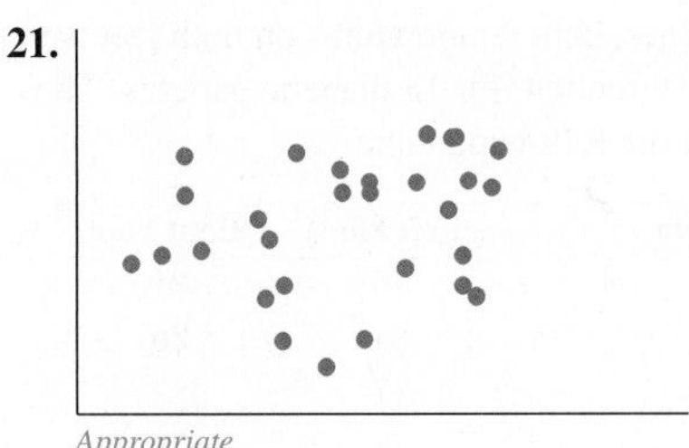

Appropriate

22.

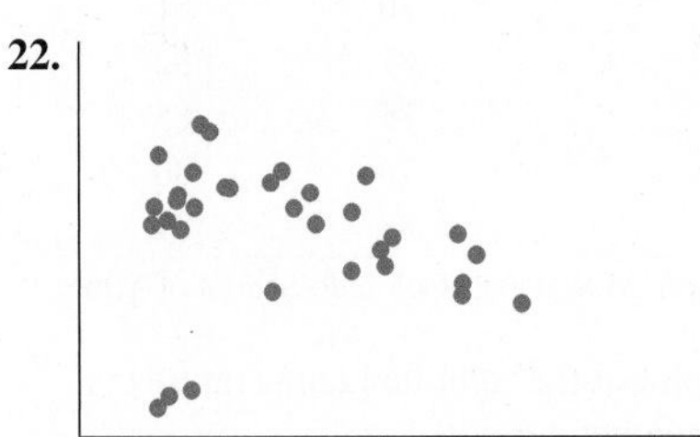

Not appropriate

23.

Not appropriate

24.

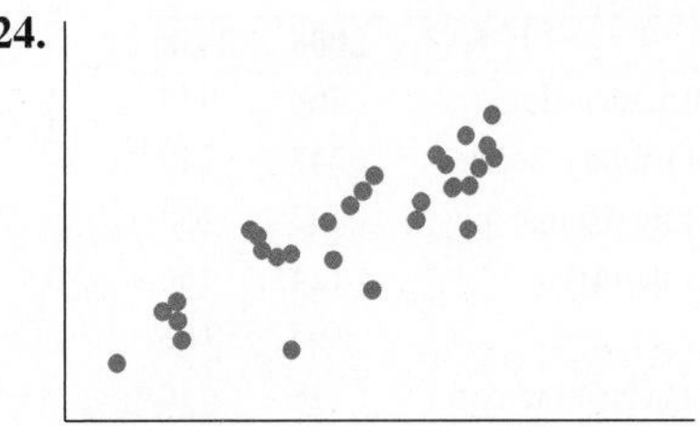

Appropriate

In Exercises 25–30, determine whether the association between the two variables is positive or negative.

25. A person's age and the person's income *Positive*

26. The age of a car and its resale value *Negative*

27. The age of a car and the number of miles on its odometer *Positive*

28. The number of times a pencil is sharpened and its length *Negative*

29. The diameter of an apple and its weight *Positive*

30. Weekly ice cream sales and weekly average temperature *Positive*

Working with the Concepts

31. Price of eggs and milk: The following table presents the average price in dollars for a dozen eggs and a gallon of milk for each month in a recent year.

Dozen Eggs	Gallon of Milk
1.94	3.58
1.80	3.52
1.77	3.50
1.83	3.47
1.69	3.43
1.67	3.40
1.65	3.43
1.88	3.47
1.89	3.47
1.96	3.52
1.96	3.54
2.01	3.58

Source: Bureau of Labor Statistics

a. Construct a scatterplot of the price of milk (y) versus the price of eggs (x).

b. Compute the correlation coefficient between the price of eggs and the price of milk. *0.845*

c. In a month where the price of eggs is above average, would you expect the price of milk to be above average or below average? Explain. *Above*

d. Which of the following is the best interpretation of the correlation coefficient? *iii*

i. When the price of eggs rises, it causes the price of milk to rise.

ii. When the price of milk rises, it causes the price of eggs to rise.

iii. Changes in the price of eggs or milk do not cause changes in the price of the other; the correlation indicates that the prices of milk and eggs tend to go up and down together.

32. Government funding: The following table presents the budget (in millions of dollars) for selected organizations that received U.S. government funding for arts and culture in both 2006 and 2018.

Organization	2006	2018
Corporation for Public Broadcasting	460	445
Institute of Museum and Library Services	247	240
National Endowment for the Humanities	142	153
National Endowment for the Arts	124	166
National Gallery of Art	95	166
Kennedy Center for the Performing Arts	18	40
Commission of Fine Arts	2	3

Source: National Endowment for the Arts

a. Construct a scatterplot of the funding in 2018 (y) versus the funding in 2006 (x).

b. Compute the correlation coefficient between the funding in 2006 and the funding in 2018. *0.984*

c. For an organization whose funding in 2006 was above the average, would you expect their funding in 2018 to be above or below average? Explain. *Above*

d. Which of the following is the best interpretation of the correlation coefficient? *iii*

i. If we increase the funding for an organization in 2006, this will cause the funding in 2018 to increase.

ii. If we increase the funding for an organization in 2018, this will cause the funding in 2006 to increase.

iii. Some organizations get more funding than others, and those that were more highly funded in 2006 were generally more highly funded in 2018 as well.

33. Pass the ball: The following table lists the heights (inches) and weights (pounds) of 15 National Football League quarterbacks in the 2018 season.

Name	Height	Weight
Patrick Mahomes	74	225
Andrew Luck	76	234
Deshaun Watson	74	221
Aaron Rogers	77	230
Matt Ryan	76	220
Tom Brady	76	225
Jameis Winston	76	231
Kirk Cousins	75	214
Russell Wilson	71	206
Cam Newton	77	244
Baker Mayfield	73	215
Dak Prescott	74	226
Mitchell Trubisky	74	222
Lamar Jackson	75	200
Carson Wentz	77	237

Source: *FFToday*

a. Construct a scatterplot of the weight (y) versus the height (x).

b. Compute the correlation coefficient between the height and weight of the quarterbacks. *0.674*

c. If a quarterback is below average in height, would you expect him to be above average or below average in weight? Explain. *Below*

d. Which of the following is the best interpretation of the correlation coefficient? *ii*

i. If a quarterback gains weight, he will grow taller.

ii. Given two quarterbacks, the taller one is likely to be heavier than the shorter one.

iii. Given two quarterbacks, the heavier one is likely to be shorter than the lighter one.

34. Noisy streets: How much noisier are streets where cars travel faster? An article in the *Journal of Transportation Engineering* reported average speeds and noise levels for eight streets in urban areas. The following table presents the results.

Speed (mph)	Noise (decibels)
17.7	78.1
22.6	79.6
24.2	81.0
18.2	78.7
18.9	78.6
18.9	78.5
18.1	78.4
20.7	79.6

a. Construct a scatterplot of speed (y) versus noise (x).

b. Compute the correlation coefficient between speed and noise. *0.958*

c. On a street where the noise level is above average, would you expect the speed to be above average or below average? *Above*

d. Which of the following is the best interpretation of the correlation coefficient? *iii*

i. If a street becomes noisier, the cars will move faster.

ii. Given two streets, the noisier one is likely to have slower moving cars.

iii. Given two streets, the one with faster moving cars is likely to be noisier.

35. Foot temperatures: Foot ulcers are a common problem for people with diabetes. Higher skin temperatures on the foot indicate an increased risk of ulcers. In a study performed at the Colorado School of Mines, skin temperatures on both feet were measured, in degrees Fahrenheit, for 18 diabetic patients. The results are presented in the following table.

Left Foot	Right Foot	Left Foot	Right Foot
80	80	76	81
85	85	89	86
75	80	87	82
88	86	78	78
89	87	80	81
87	82	87	82
78	78	86	85
88	89	76	80
89	90	88	89

Source: Kimberly Anderson, M.S. thesis, Colorado School of Mines

a. Construct a scatterplot of the right foot temperature (y) versus the left foot temperature (x).

b. Compute the correlation coefficient between the temperatures of the left and right feet. *0.812*

c. If a patient's left foot is cooler than the average, would you expect the patient's right foot to be warmer or cooler than average? Explain. *Cooler*

d. Which of the following is the best interpretation of the correlation coefficient? *i*

i. Some patients have warmer feet than others. Those who have warmer left feet generally have warmer right feet as well.

ii. If we warm a patient's left foot, the patient's right foot will become warmer.

iii. If we cool a patient's left foot, the patient's right foot will become warmer.

36. Mortgage payments: The following table presents monthly interest rates, in percent, for 30-year and 15-year fixed-rate mortgages, for a recent year.

30-Year	15-Year	30-Year	15-Year
3.92	3.20	3.55	2.85
3.89	3.16	3.60	2.86
3.95	3.20	3.47	2.78
3.91	3.14	3.38	2.69
3.80	3.03	3.35	2.66
3.68	2.95	3.35	2.66

Source: Freddie Mac

a. Construct a scatterplot of the 15-year rate (y) versus the 30-year rate (x).

b. Compute the correlation coefficient between 30-year and 15-year rates. 0.996

c. When the 30-year rate is below average, would you expect the 15-year rate to be above or below average? Explain. Below

d. Which of the following is the best interpretation of the correlation coefficient? ii

i. When a bank increases the 30-year rate, that causes the 15-year rate to rise as well.

ii. Interest rates are determined by economic conditions. When economic conditions cause 30-year rates to increase, these same conditions cause 15-year rates to increase as well.

iii. When a bank increases the 15-year rate, that causes the 30-year rate to rise as well.

37. Blood pressure: A blood pressure measurement consists of two numbers: the systolic pressure, which is the maximum pressure taken when the heart is contracting, and the diastolic pressure, which is the minimum pressure taken at the beginning of the heartbeat. Blood pressures were measured, in millimeters of mercury, for a sample of 16 adults. The following table presents the results.

Systolic	Diastolic	Systolic	Diastolic
134	87	133	91
115	83	112	75
113	77	107	71
123	77	110	74
119	69	108	69
118	88	105	66
130	76	157	103
116	70	154	94

Based on results published in the *Journal of Human Hypertension*

a. Construct a scatterplot of the diastolic blood pressure (y) versus the systolic blood pressure (x).

b. Compute the correlation coefficient between systolic and diastolic blood pressure. 0.857

c. If someone's diastolic pressure is above average, would you expect that person's systolic pressure to be above or below average? Explain. Above

38. Butterfly wings: Do larger butterflies live longer? The wingspan (in millimeters) and the lifespan in the adult state (in days) were measured for 22 species of butterfly. Following are the results.

Wingspan	Lifespan	Wingspan	Lifespan
35.5	19.8	25.9	32.5
30.6	17.3	31.3	27.5
30.0	27.5	23.0	31.0
32.3	22.4	26.3	37.4
23.9	40.7	23.7	22.6
27.7	18.3	27.1	23.1
28.8	25.9	28.1	18.5
35.9	23.1	25.9	32.3
25.4	24.0	28.8	29.1
24.6	38.8	31.4	37.0
28.1	36.5	28.5	33.7

Source: *Oikos Journal of Ecology* 105:41–54

a. Construct a scatterplot of the lifespan (y) versus the wingspan (x).

b. Compute the correlation coefficient between wingspan and lifespan. −0.412

c. Do larger butterflies tend to live for a longer or shorter time than smaller butterflies? Explain. Shorter

Shutterstock

39. Police and crime: In a survey of cities in the United States, it is discovered that there is a positive correlation between the number of police officers hired by the city and the number of crimes committed. Do you believe that increasing the number of police officers causes the crime rate to increase? Why or why not? No

40. Age and education: A survey of U.S. adults showed that there is a negative correlation between age and education level. Does this mean that people become less educated as they become older? Why or why not? No

41. What's the correlation? In a sample of adults, would the correlation between age and year graduated from high school be closest to −1, −0.5, 0, 0.5, or 1? Explain. −1

42. What's the correlation? In a sample of adults, would the correlation between year of birth and year graduated from high school be closest to −1, −0.5, 0, 0.5, or 1? Explain. 1

Extending the Concepts

43. Changing means and standard deviations: A small company has five employees. The following table presents the number of years each has been employed (x) and the hourly wage in dollars (y).

x (years)	0.5	1.0	1.75	2.5	3.0
y (dollars)	9.51	8.23	10.95	12.70	12.75

a. Compute $\bar{x}$, $\bar{y}$, s_x, and s_y.

b. Compute the correlation coefficient between years of service and hourly wage. 0.906

c. Each employee is given a raise of $1.00 per hour, so each y-value is increased by 1. Using these new y-values, compute the sample mean $\bar{y}$ and the sample standard deviation s_y. $\bar{y} = 11.828$; $s_y = 1.981$

d. In part (c), each y-value was increased by 1. What was the effect on $\bar{y}$? What was the effect on s_y? *Increased by 1; no change*

e. Compute the correlation coefficient between years of service and the increased hourly wage. Explain why the correlation coefficient is unchanged even though the y-values have changed. *0.906*

f. Convert x to months by multiplying each x-value by 12. So the new x-values are 6, 12, 21, 30, and 36. Compute the sample mean $\bar{x}$ and the sample standard deviation s_x. $\bar{x} = 21$; $s_x = 12.369$

g. In part (f), each x-value was multiplied by 12. What was the effect on $\bar{x}$ and s_x? *Each multiplied by 12*

h. Compute the correlation coefficient between months of service and hourly wage in dollars. Explain why the correlation coefficient is unchanged even though the x-values, $\bar{x}$, and s_x have changed. *0.906*

i. Use the results of parts (a)–(h) to fill in the blank: If a constant is added to each x-value or to each y-value, the correlation coefficient is __________. *unchanged*

j. Use the results of parts (a)–(h) to fill in the blank: If each x-value or each y-value is multiplied by a positive constant, the correlation coefficient is __________. *unchanged*

Answers to Check Your Understanding Exercises for Section 4.1

1. negative

2. large

3. linear

4. a. Negative linear
b. Positive nonlinear **c.** Positive linear
d. Weak linear
e. Negative nonlinear

5. a. negative. When the temperature is higher, the amount of heating oil consumed tends to be lower.
b. positive. People with more years of education tend to have larger incomes.

6. a.

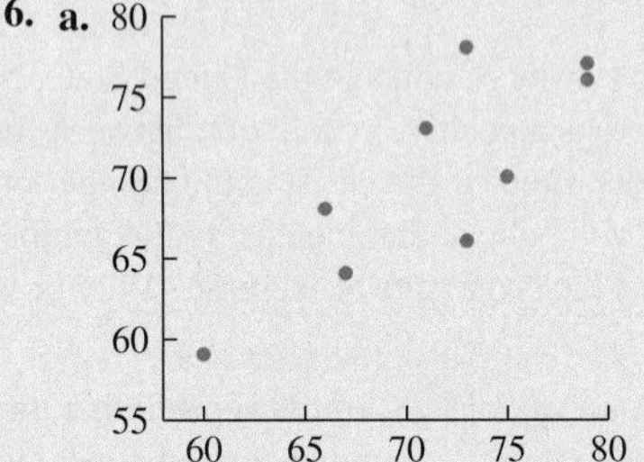

The scatterplot shows a linear relationship.

b. $r = 0.809$. The linear relationship is positive and fairly strong.

7. Scatterplot (a) is the only one for which the correlation coefficient is an appropriate summary.

8. Correlation is not causation. It is possible that the college professors are drinking the same amount of beer as before, while people other than college professors are drinking more beer.

Section 4.2 The Least-Squares Regression Line

Objectives

1. Compute the least-squares regression line
2. Use the least-squares regression line to make predictions
3. Interpret predicted values, the slope, and the y-intercept of the least-squares regression line

Objective 1 Compute the least-squares regression line

The Least-Squares Regression Line

Table 4.2 (page 179) presents the size in square feet and selling price in thousands of dollars for a sample of houses. These data were first presented in Section 4.1. A scatterplot of these data showed that they tend to cluster around a line, and we computed the correlation to be 0.901. We concluded that there is a strong positive linear association between size and price.

We can use these data to predict the selling price of a house based on its size. The key is to summarize the data with a straight line. We want to find the line that fits the data best.

NOTE TO INSTRUCTOR

The following topics from the **Statistics Corequisite Workbook** are aligned with the material in this section.

7.2 - Linear Equations and the Graph of a Line
7.3 - Finding the Slope and y-Intercept and Interpretation of the Slope
A.6 - Determining the Equation of a Line

Table 4.2 Size and Selling Price for a Sample of Houses

Size (square feet)	Selling Price ($1000s)
2521	400
2555	426
2735	428
2846	435
3028	469
3049	475
3198	488
3198	455

Source: Sue Bays Realty

We now explain what we mean by "best." Figure 4.5 presents scatterplots of the data in Table 4.2, each with a different line superimposed. For each line, we have drawn the vertical distances from the points to the line. It is clear that the line in Figure 4.5(a) fits better than the line in Figure 4.5(b). The reason is that the vertical distances are, on the whole, smaller for the line in Figure 4.5(a). We determine exactly how well a line fits the data by squaring the vertical distances and adding them up. The line that fits best is the line for which this sum of squared distances is as small as possible. This line is called the **least-squares regression line**.

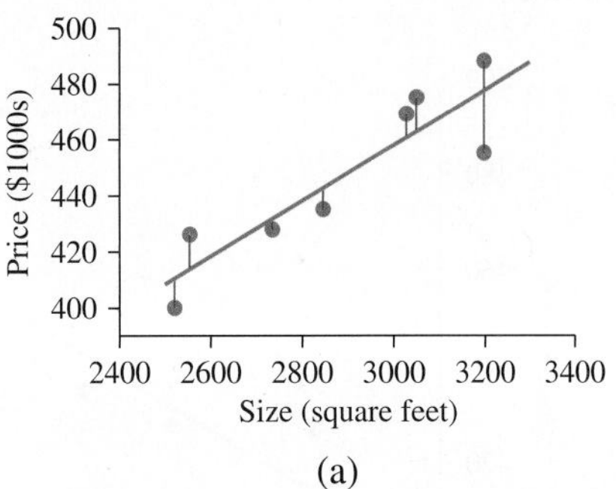

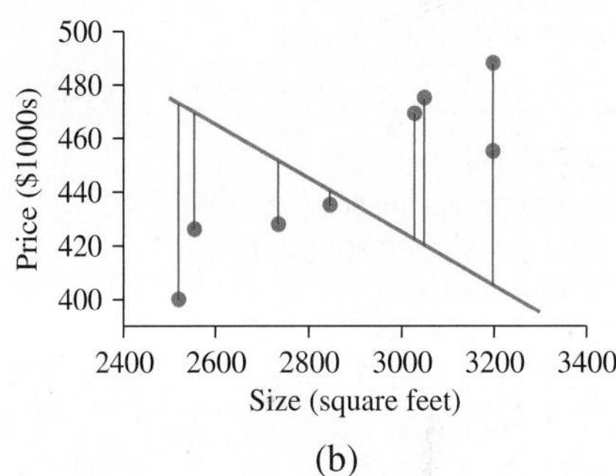

Figure 4.5 The line in (a) fits better than the line in (b), because the vertical distances are generally smaller.

Fortunately, we don't have to worry about vertical distances when we calculate the least-squares regression line. We can use the following formula.

NOTE TO INSTRUCTOR

It is important to stress that the correlation is not the same as the slope of the regression line. The equation $b = rs_y/s_x$ makes this clear.

Equation of the Least-Squares Regression Line

Given ordered pairs (x, y), with sample means $\bar{x}$ and $\bar{y}$, sample standard deviations s_x and s_y, and correlation coefficient r, the equation of the least-squares regression line for predicting y from x is

$$\hat{y} = b_0 + b_1 x$$

where $b_1 = r\dfrac{s_y}{s_x}$ is the **slope** and $b_0 = \bar{y} - b_1\bar{x}$ is the **y-intercept**.

EXPLAIN IT AGAIN

Explanatory variable and outcome variable: The explanatory variable is used to explain or predict the value of the outcome variable.

In general, the variable we want to predict (in this case, selling price) is called the **outcome variable**, or **response variable**, and the variable we are given is called the **explanatory variable**, or **predictor variable**. In the equation of the least-squares regression line, x represents the explanatory variable and y represents the outcome variable.

Example 4.3

Compute the least-squares regression line

Use the data in Table 4.2 to compute the least-squares regression line for predicting price from size.

CAUTION

When computing the least-squares regression line, be sure that x represents the variable you are given (the explanatory variable), and y represents the variable you want to predict (the outcome variable).

Solution

We first find $\bar{x}$, $\bar{y}$, s_x, s_y, and r. We computed these quantities in Section 4.1:

$$\bar{x} = 2891.25, \quad \bar{y} = 447.0, \quad s_x = 269.49357, \quad s_y = 29.68405, \quad r = 0.9005918.$$

The slope of the least-squares regression line is

$$b_1 = r\frac{s_y}{s_x} = (0.9005918)\frac{29.68405}{269.49357} = 0.09919796$$

We use the value of b_1 just found to compute b_0, the y-intercept of the least-squares regression line:

$$b_0 = \bar{y} - b_1\bar{x} = 447.0 - (0.09919796)(2891.25) = 160.1939$$

The equation of the least-squares regression line is $\hat{y} = 160.1939 + 0.0992x$.

CAUTION

We don't round the value of r in this calculation because it is an intermediate value.

In general, we will round the slope and y-intercept values to four decimal places.

Figure 4.6 presents a scatterplot of the data in Table 4.2 with the least-squares regression line superimposed. It can be seen that the points tend to cluster around the least-squares regression line.

CAUTION

Do not confuse the slope b_1 of the least-squares regression line with the correlation coefficient r. In most cases, they are not equal.

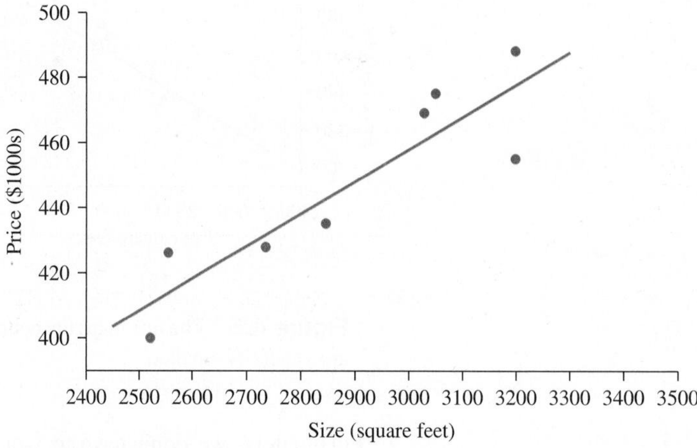

Figure 4.6 Scatterplot of the data in Table 4.2 with the least-squares regression line superimposed. The points tend to cluster around the least-squares regression line.

Computing the least-squares regression line with technology

Nowadays, least-squares regression lines are usually computed with technology rather than by hand. Figure 4.7 (page 181) shows the least-squares regression line for predicting house price from house size, as presented by the TI-84 Plus calculator. There are four numbers in the output, labeled "a", "b", "r^2", and "r". Of these, a is the y-intercept of the least-squares regression line and b is the slope. The calculator provides many more digits than are generally necessary for these quantities. If you round a and b, you will get a = 160.1939 and b = 0.0992, which agree with the results calculated by hand in Example 4.3. The last quantity, r, is the correlation coefficient. When rounded, this value is 0.901, which agrees with the value computed by hand in Example 4.2. Finally, the value r^2 is computed by squaring the correlation coefficient r. This is an important quantity that we will discuss in Section 4.3.

Step-by-step instructions for using the TI-84 Plus calculator to compute the least-squares line and correlation coefficient are presented in the Using Technology section on page 185.

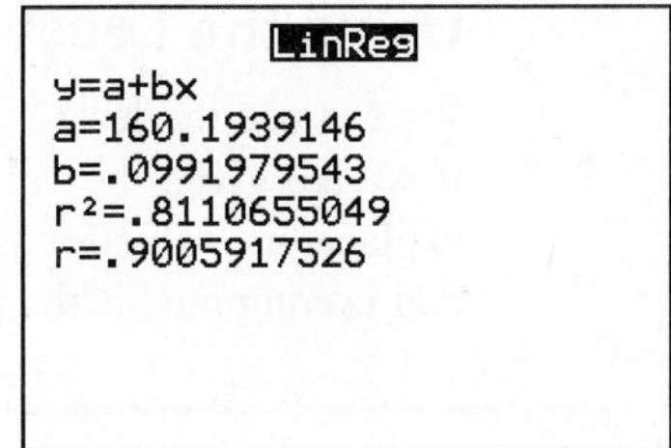

Figure 4.7 The least-squares regression line as presented by the TI-84 Plus calculator. The y-intercept is denoted by a, and the slope of the line is denoted by b. The value r, in the last line, is the correlation coefficient.

Following is MINITAB output for the least-squares regression line for predicting house price from house size. The equation is given near the top of the output as Selling Price = 160.19 + 0.09920 Size. In the table below that, the values of the slope and intercept are presented again in the column labeled Coef.

Step-by-step instructions for using MINITAB to compute the least-squares line are presented in the Using Technology section on page 186.

```
The regression equation is
Selling Price = 160.19 + 0.09920 Size

Predictor        Coef    SE Coef      T      P
Constant       160.19      56.73   2.82  0.030
Size          0.09920    0.01955   5.08  0.002
```

Finally, we present the output from Excel for this least-squares regression line. The slope and intercept are found in the column labeled "Coefficients." The slope is labeled X Variable 1. Step-by-step instructions for using Excel to compare the least-squares line are presented in the Using Technology section on page 185.

	Coefficients	Standard Error	t Stat	P-value
Intercept	160.1939146	56.72636198	2.823976525	0.030195859
X Variable 1	0.099197954	0.019545859	5.075139241	0.00227642

Check Your Understanding

1. The following table presents the percentage of students who tested proficient in reading and the percentage who tested proficient in math for each of the 10 most populous states in the United States. Compute the least-squares regression line for predicting math proficiency from reading proficiency. *$\hat{y} = 11.0358 + 0.8226x$*

State	Percent Proficient in Reading	Percent Proficient in Mathematics
California	60	59
Texas	73	78
New York	75	70
Florida	66	68
Illinois	75	70
Pennsylvania	79	77
Ohio	79	76
Michigan	73	66
Georgia	67	64
North Carolina	71	73

Source: National Assessment for Educational Progress

Answer is on page 190.

Objective 2 Use the least-squares regression line to make predictions

Using the Least-Squares Regression Line for Prediction

We can use the least-squares regression line to predict the value of the outcome variable if we are given the value of the explanatory variable. Simply substitute the value of the explanatory value for x in the equation of the least-squares regression line. The value of $\hat{y}$ that is computed is the **predicted value**.

Example 4.4

Use the least-squares regression line for prediction

Use the least-squares regression line computed in Example 4.3 to predict the selling price of a house of size 2800 square feet.

Solution

The equation of the least-squares regression line is $\hat{y} = 160.1939 + 0.0992x$. Substituting 2800 for x yields

$$\hat{y} = 160.1939 + 0.0992(2800) = 438.0$$

We predict that the selling price of the house will be 438.0 thousand dollars, or \$438,000.

We will round predicted values to one more decimal place than the outcome variable.

Figure 4.8 presents a scatterplot of the data with the point for the prediction added. The given value for the explanatory variable was $x = 2800$, and the predicted price is the y-value on the least-squares regression line corresponding to $x = 2800$.

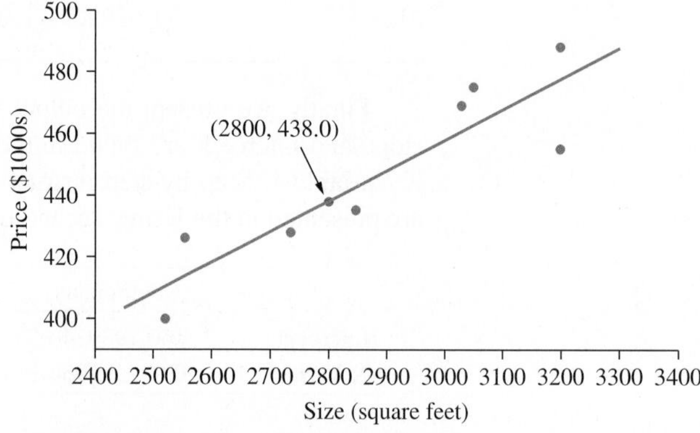

Figure 4.8 The predicted price for a house with size $x = 2800$ is 438.0, which is the y-value on the least-squares regression line.

EXPLAIN IT AGAIN

Locating the predicted value on the least-squares regression line: For a given value of the explanatory variable x, the predicted value $\hat{y}$ is the y-value on the least-squares regression line corresponding to the given value of x.

The least-squares regression line goes through the point of averages

In the house data, the average size is $\bar{x} = 2891.25$ and the average selling price is $\bar{y} = 447.0$. What selling price do we predict for a house of average size? We substitute 2891.25 for x in the equation of the least-squares regression line to obtain

$$\hat{y} = 160.1939 + 0.0992(2891.25) = 447.0$$

For a house of average size, we predict that the selling price will be the average selling price. In general, when the explanatory variable x is equal to $\bar{x}$, the predicted value $\hat{y}$ is equal to $\bar{y}$.

SUMMARY

The least-squares regression line goes through the **point of averages** $(\bar{x}, \bar{y})$.

It is reasonable that the least-squares regression line goes through $(\bar{x}, \bar{y})$. To see this, note that if the correlation coefficient is positive, then above-average values of x are associated with above-average values of y, and below-average values of x are associated with below-average values of y. So it's not surprising that the average value of x would be associated with the average value of y.

Objective 3 Interpret predicted values, the slope, and the y-intercept of the least-squares regression line

Interpreting the Least-Squares Regression Line

The least-squares regression line has slope b_1 and y-intercept b_0. We use the least-squares regression line to compute a predicted value $\hat{y}$. We explain how to interpret these quantities.

Interpreting the predicted value $\hat{y}$

The predicted value $\hat{y}$ can be used to estimate the average outcome for a given value of the explanatory variable x. For any given value of x, the value $\hat{y}$ is an estimate of the average y-value for all points with that x-value.

Example 4.5

Use the least-squares regression line to estimate the average outcome

Use the least-squares regression line computed in Example 4.3 to estimate the average price of all houses whose size is 3000 square feet.

Solution

The equation of the least-squares regression line is $\hat{y} = 160.1939 + 0.0992x$. Given $x = 3000$, we predict

$$\hat{y} = 160.1939 + 0.0992(3000) = 457.8$$

We estimate the average price for a house of 3000 square feet to be 457.8 thousand dollars, or \$457,800.

Interpreting the y-intercept b_0

The y-intercept b_0 is the point where the line crosses the y-axis. This has a practical interpretation only when the data contain both positive and negative values of x—in other words, only when the scatterplot contains points on both sides of the y-axis.

- If the data contain both positive and negative x-values, then the y-intercept is the estimated outcome when the value of the explanatory variable x is 0.
- If the x-values are all positive or all negative, then the y-intercept b_0 does not have a useful interpretation. The reason is that the intercept corresponds to an x-value of 0, which is outside the range of the data. In general, we do not interpret values outside the range of the data. See page 190 for more discussion.

Interpreting the slope b_1

If the x-values of two points on a line differ by 1, their y-values will differ by an amount equal to the slope of the line. For example, if a line has a slope of 4, then two points whose x-values differ by 1 will have y-values that differ by 4. This fact enables us to interpret the slope b_1 of the least-squares regression line. If the values of the explanatory variable for two individuals differ by 1, their predicted values will differ by b_1. If the values of the explanatory variable differ by an amount d, then their predicted values will differ by $b_1 \cdot d$.

Example 4.6

Compute the predicted difference in outcomes

Two houses are up for sale. One house is 1900 square feet in size and the other is 1750 square feet. By how much should we predict their prices to differ?

Solution

The difference in size is $1900 - 1750 = 150$. The slope of the least-squares regression line is $b_1 = 0.0992$. We predict the prices to differ by $(0.0992)(150) = 14.9$ thousand dollars, or \$14,900.

Check Your Understanding

2. At the final exam in a statistics class, the professor asks each student to indicate how many hours he or she studied for the exam. After grading the exam, the professor computes the least-squares regression line for predicting the final exam score from the number of hours studied. The equation of the line is $\hat{y} = 50 + 5x$.

a. Antoine studied for 6 hours. What do you predict his exam score to be? *80*

b. Emma studied for 3 hours longer than Jeremy did. How much higher do you predict Emma's score to be? *15*

3. For each of the following plots, interpret the y-intercept of the least-squares regression line if possible. If not possible, explain why not.

a. The least-squares regression line is $\hat{y} = 1.98 + 0.039x$, where x is the temperature in a freezer in degrees Fahrenheit, and y is the time it takes to freeze a certain amount of water into ice. *Possible*

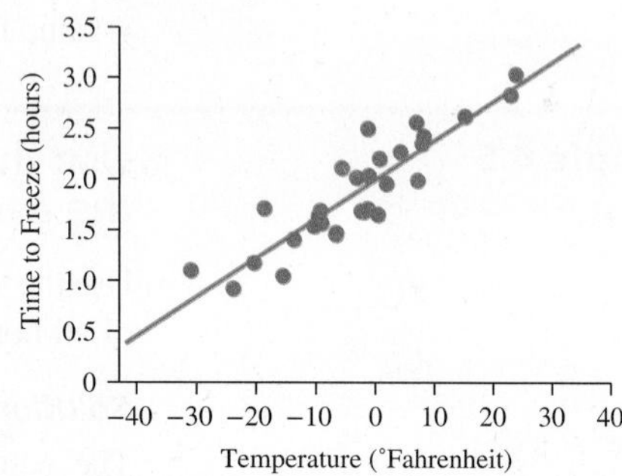

b. The least-squares regression line is $\hat{y} = -13.586 + 4.340x$, where x represents the age of an elementary school student and y represents the score on a standardized test. *Not possible*

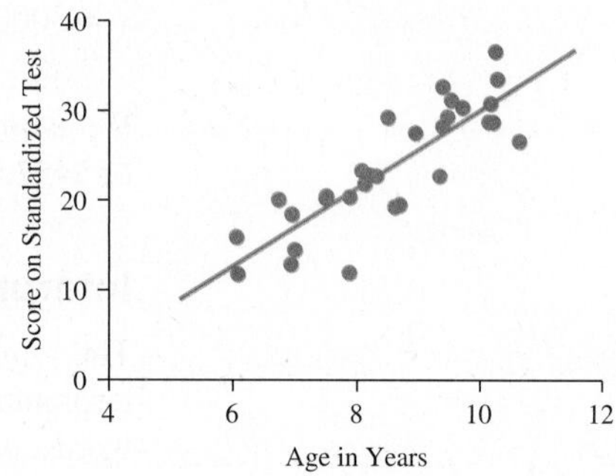

Answers are on page 190.

The least-squares regression line does not predict the effect of changing the explanatory variable for a single individual

NOTE TO INSTRUCTOR

The reason that the regression line does not predict the effect of changing the explanatory variable is that correlation does not imply causation.

We have worded the interpretation of the slope of the least-squares regression line very carefully. The slope is the estimated difference in y-values for *two different* individuals whose x-values differ by 1. This does *not* mean that changing the value of x for a particular individual will cause that individual's y-value to change. The following example will help make this clear.

Example 4.7

The least-squares regression line doesn't predict the result of changing the explanatory variable for a single individual

A study is done in which a sample of men were weighed, and then each man was tested to see how much weight he could lift. The explanatory variable (x) was the man's weight, and the outcome (y) was the amount he could lift. The least-squares regression line was computed to be $\hat{y} = 50 + 0.6x$.

Joe is a weightlifter. After looking at the equation of the least-squares regression line, he reasons as follows: "The slope of the least-squares regression line is 0.6. Therefore, if I gain 10 pounds, I'll be able to lift 6 pounds more, because (0.6)(10) = 6." Is he right?

Solution

No, he is not right. You can't improve your weightlifting ability simply by putting on weight. What the slope of 0.6 tells us is that if two men have weights that differ by 10 pounds, then, on the average, the heavier man will be able to lift 6 pounds more than the lighter man. This does not mean that an individual man can increase his weightlifting ability by increasing his weight.

Check Your Understanding

4. The Sanchez family wants to sell their house, which is 2800 square feet in size. Mr. Sanchez notices that the slope of the least-squares regression line for predicting price from size is 0.0992. He says that if they put a 250-square-foot addition on their house, the selling price will increase by (0.0992)(250) = 24.8 thousand dollars, or $24,800. Is this necessarily a good prediction? Explain. *Not necessarily good*

Answer is on page 190.

Using Technology

We use the data in Table 4.2 to illustrate the technology steps.

TI-84 PLUS

Computing the least-squares regression line

Note: Before computing the least-squares regression line, a one-time calculator setting should be modified to correctly configure the calculator to display the correlation coefficient. The following steps describe how to do this.

Step 1. Press **MODE** to access the mode menu.

Step 2. Scroll down to the **STAT DIAGNOSTICS** option, and select **On**.

The following steps describe how to compute the least-squares regression line.

Step 1. Enter the x-values from Table 4.2 into **L1** and the y-values into **L2**.

Step 2. Press **STAT** and then the right arrow key to access the **CALC** menu.

Step 3. Select **8: LinReg(a+bx)** (Figure A) and press **ENTER** (Figure B).

Using the TI-84 PLUS Stat Wizards (see Appendix B for more information)

Step 1. Enter the x-values from Table 4.2 into **L1** and the y-values into **L2**.

Step 2. Press **STAT** and then the right arrow key to access the **CALC** menu.

Step 3. Select **8:LinReg(a+bx)** (Figure A). Enter **L1** in the **Xlist** field and **L2** in the **Ylist** field. Select **Calculate** and press **ENTER** (Figure B).

```
EDIT CALC TESTS
1:1-Var Stats
2:2-Var Stats
3:Med-Med
4:LinReg(ax+b)
5:QuadReg
6:CubicReg
7:QuartReg
8:LinReg(a+bx)
9↓LnReg
```

Figure A

```
LinReg
y=a+bx
a=160.1939146
b=.0991979543
r²=.8110655049
r=.9005917526
```

Figure B

EXCEL

Computing the least-squares regression line

Step 1. Enter the x-values from Table 4.2 into **Column A** and the y-values into **Column B**.

Step 2. Select **Data**, then **Data Analysis**. Highlight **Regression** and press **OK**.

Step 3. Enter the range of cells that contain the x-values in the **Input X Range** field and the range of cells that contain the y-values in the **Input Y Range** field.

Step 4. Click **OK** (Figure C).

The least-squares regression line is given by
$\hat{y}$ = **(X Variable 1)x + Intercept**

	Coefficients
Intercept	160.1939146
X Variable 1	0.099197954

Figure C

MINITAB

Computing the least-squares regression line

Step 1. Enter the x-values from Table 4.2 into **Column C1**, and label the column **Size**. Enter the y-values into **Column C2**, and label the column **Price**.

Step 2. Click **STAT**, then **Regression**, then **Regression** again. Then select **Fit Regression Model**.

Step 3. Select the y-variable **(Price)** as the **Response** and the x-variable **(Size)** as the **Continuous predictor** and click **OK** (Figure D).

Regression Equation

Selling Price = 160.2 + 0.0992 Size

Figure D

Section 4.2 Exercises

Exercises 1–4 are the Check Your Understanding exercises located within the section.

Understanding the Concepts

In Exercises 5–7, fill in each blank with the appropriate word or phrase.

5. When we are given the value of the _________ variable, we can use the least-squares regression line to predict the value of the _________ variable. *explanatory, outcome*

6. If the correlation coefficient is equal to 0, the slope of the least-squares regression line will be equal to _________. *0*

7. If the least-squares regression line has slope $b_1 = 5$, and two x-values differ by 3, the predicted difference in the y-values is _________. *15*

In Exercises 8–12, determine whether the statement is true or false. If the statement is false, rewrite it as a true statement.

8. Substituting the value of the explanatory variable for x in the equation of the least-squares regression line results in a prediction for y. *True*

9. The least-squares regression line passes through the point of averages $(\bar{x}, \bar{y})$. *True*

10. In general, the slope of the least-squares regression line is equal to the correlation coefficient. *False*

11. The least-squares regression line predicts the result of changing the value of the explanatory variable. *False*

12. The y-intercept b_0 of a least-squares regression line has a useful interpretation only if the x-values are either all positive or all negative. *False*

Practicing the Skills

In Exercises 13–16, compute the least-squares regression line for the given data set.

13.

x	1	2	3	4	5
y	5	6	9	8	7

$\hat{y} = 5.2 + 0.6x$

14.

x	9	5	7	13	−8	−2	6	−10
y	3	3	31	36	0	3	−2	−14

$\hat{y} = 3.8441 + 1.4623x$

15.

x	42	36	14	18	23	36	17
y	72	68	25	31	42	65	32

$\hat{y} = 0.8951 + 1.7674x$

16.

x	5.7	4.1	6.2	4.4	6.5	5.8	4.9
y	1.9	4.8	0.8	3.9	1.2	1.7	3.0

$\hat{y} = 10.9386 - 1.5763x$

17. Compute the least-squares regression line for predicting y from x given the following summary statistics: *$\hat{y} = 1175 + 35x$*

$$\bar{x} = 5 \quad s_x = 2 \quad \bar{y} = 1350$$
$$s_y = 100 \quad r = 0.70$$

18. Compute the least-squares regression line for predicting y from x given the following summary statistics: *$\hat{y} = 41.301 - 1.3458x$*

$$\bar{x} = 8.1 \quad s_x = 1.2 \quad \bar{y} = 30.4$$
$$s_y = 1.9 \quad r = -0.85$$

19. In a hypothetical study of the relationship between the income of parents (x) and the IQs of their children (y), the following summary statistics were obtained:

$$\bar{x} = 45{,}000 \quad s_x = 20{,}000 \quad \bar{y} = 100$$
$$s_y = 15 \quad r = 0.40$$

Find the equation of the least-squares regression line for predicting IQ from income. $\hat{y} = 86.5 + 0.0003x$

20. Assume that in a study of educational level in years (x) and income (y), the following summary statistics were obtained:

$$\bar{x} = 12.8 \qquad s_x = 2.3 \qquad \bar{y} = 41{,}000$$
$$s_y = 15{,}000 \qquad r = 0.60$$

Find the equation of the least-squares regression line for predicting income from educational level.

Working with the Concepts

21. Price of eggs and milk: The following table presents the average price in dollars for a dozen eggs and a gallon of milk for each month in a recent year.

Dozen Eggs	Gallon of Milk
1.94	3.58
1.80	3.52
1.77	3.50
1.83	3.47
1.69	3.43
1.67	3.40
1.65	3.43
1.88	3.47
1.89	3.47
1.96	3.52
1.96	3.54
2.01	3.58

Source: Bureau of Labor Statistics

a. Compute the least-squares regression line for predicting the price of milk from the price of eggs. $\hat{y} = 2.7592 + 0.3991x$

b. If the price of eggs differs by \$0.25 from one month to the next, by how much would you expect the price of milk to differ? 0.10

c. Predict the price of milk in a month when the price of eggs is \$1.95. 3.54

22. Government funding: The following table presents the budget (in millions of dollars) for selected organizations that received U.S. government funding for arts and culture in both 2006 and 2018.

Organization	2006	2018
Corporation for Public Broadcasting	460	445
Institute of Museum and Library Services	247	240
National Endowment for the Humanities	142	153
National Endowment for the Arts	124	166
National Gallery of Art	95	166
Kennedy Center for the Performing Arts	18	40
Commission of Fine Arts	2	3

Source: National Endowment for the Arts

a. Compute the least-squares regression line for predicting the 2018 budget from the 2006 budget. $\hat{y} = 32.672 + 0.90468x$

b. If two institutions have budgets that differ by 10 million dollars in 2006, by how much would you predict their budgets to differ in 2018? 9.05

c. Predict the 2018 budget for an organization whose 2006 budget was 100 million dollars. 123.14

23. Pass the ball: The following table lists the heights (inches) and weights (pounds) of 15 National Football League quarterbacks in the 2018 season.

Name	Height	Weight
Patrick Mahomes	74	225
Andrew Luck	76	234
Deshaun Watson	74	221
Aaron Rogers	77	230
Matt Ryan	76	220
Tom Brady	76	225
Jameis Winston	76	231
Kirk Cousins	75	214
Russell Wilson	71	206
Cam Newton	77	244
Baker Mayfield	73	215
Dak Prescott	74	226
Mitchell Trubisky	74	222
Lamar Jackson	75	200
Carson Wentz	77	237

Source: *FFToday*

a. Compute the least-squares regression line for predicting weight from height. $\hat{y} = -121.67 + 4.6000x$

b. Is it possible to interpret the y-intercept? Explain. No

c. If two quarterbacks differ in height by 2 inches, by how much would you predict their weights to differ? 9.2

d. Predict the weight of a quarterback who is 74.5 inches tall. 221.03

e. Kirk Cousins is 75 inches tall and weighs 214 pounds. Does he weigh more or less than the weight predicted by the least-squares regression line? Less

24. Great Pumpkins: When growing giant pumpkins for competitions, growers need to keep track of the weights of the pumpkins while they are growing. It is difficult to weigh a large pumpkin before it is harvested, so a method has been developed for estimating the weight. The grower measures around the pumpkin both horizontally and vertically, then adds the results. This is called the OTT (over the top) measurement and is used to predict the weight of the pumpkin. Following are the OTT measurements and actual weights of the 20 largest pumpkins entered into official competitions in the year 2018.

OTT (inches)	Weight (pounds)	OTT (inches)	Weight (pounds)
490.0	2528.0	465.0	2138.0
469.0	2469.0	452.0	2136.0
490.0	2433.9	451.0	2114.0
480.0	2416.5	456.0	2091.0
455.0	2283.0	457.0	2079.0
463.0	2170.0	456.0	2077.0
473.0	2166.0	462.0	2070.1
477.0	2157.5	436.0	2027.0
465.0	2152.0	454.0	2020.5
454.0	2138.9	450.0	2017.5

Source: The Great Pumpkin Commonwealth

a. Compute the least-squares line for predicting weight (y) from OTT (x). $\hat{y} = -1917.1 + 8.8629x$

b. Is it possible to interpret the y-intercept? No

c. If two pumpkins differ in OTT by 10 inches, by how much would you predict their weights to differ? 88.629

d. Predict the weight of a pumpkin whose OTT is 460 inches. 2160

25. Foot temperatures: Foot ulcers are a common problem for people with diabetes. Higher skin temperatures on the foot indicate an increased risk of ulcers. In a study carried out at the Colorado School of Mines, skin temperatures on both feet were measured, in degrees Fahrenheit, for 18 diabetic patients. The results are presented in the following table.

Left Foot	Right Foot	Left Foot	Right Foot
80	80	76	81
85	85	89	86
75	80	87	82
88	86	78	78
89	87	80	81
87	82	87	82
78	78	86	85
88	89	76	80
89	90	88	89

Source: Kimberly Anderson, M.S. thesis, Colorado School of Mines

a. Compute the least-squares regression line for predicting the right foot temperature from the left foot temperature.

b. Construct a scatterplot of the right foot temperature (y) versus the left foot temperature (x). Graph the least-squares regression line on the same axes.

c. If the left foot temperatures of two patients differ by 2 degrees, by how much would you predict their right foot temperatures to differ? 1.186

d. Predict the right foot temperature for a patient whose left foot temperature is 81 degrees. 81.8

26. Mortgage payments: The following table presents interest rates, in percent, for 30-year and 15-year fixed-rate mortgages, for January through December, 2018.

30-Year	15-Year	30-Year	15-Year
4.03	3.48	4.53	4.01
4.33	3.79	4.55	4.02
4.44	3.91	4.63	4.08
4.48	3.93	4.83	4.25
4.59	4.07	4.87	4.28
4.57	4.04	4.69	4.09

Source: Freddie Mac

a. Compute the least-squares regression line for predicting the 15-year rate from the 30-year rate. $\hat{y} = -0.2859 + 0.9421x$

b. Construct a scatterplot of the 15-year rate (y) versus the 30-year rate (x). Graph the least-squares regression line on the same axes.

c. Is it possible to interpret the y-intercept? Explain. No

d. If the 30-year rate differs by 0.3 percent from one month to the next, by how much would you predict the 15-year rate to differ? 0.2826

e. Predict the 15-year rate for a month when the 30-year rate is 3.5 percent. 3.011

27. Blood pressure: A blood pressure measurement consists of two numbers: the systolic pressure, which is the maximum pressure taken when the heart is contracting, and the diastolic pressure, which is the minimum pressure taken at the beginning of the heartbeat. Blood pressures were measured, in millimeters of mercury (mmHg), for a sample of 16 adults. The following table presents the results.

Systolic	Diastolic	Systolic	Diastolic
134	87	133	91
115	83	112	75
113	77	107	71
123	77	110	74
119	69	108	69
118	88	105	66
130	76	157	103
116	70	154	94

Based on results published in the *Journal of Human Hypertension*

a. Compute the least-squares regression line for predicting the diastolic pressure from the systolic pressure.

b. Is it possible to interpret the y-intercept? Explain. No

c. If the systolic pressures of two patients differ by 10 mmHg, by how much would you predict their diastolic pressures to differ? 5.748

d. Predict the diastolic pressure for a patient whose systolic pressure is 125 mmHg. 81.0

28. Butterfly wings: Do larger butterflies live longer? The wingspan (in millimeters) and the lifespan in the adult state (in days) were measured for 22 species of butterfly. Following are the results.

Wingspan	Lifespan	Wingspan	Lifespan
35.5	19.8	25.9	32.5
30.6	17.3	31.3	27.5
30.0	27.5	23.0	31.0
32.3	22.4	26.3	37.4
23.9	40.7	23.7	22.6
27.7	18.3	27.1	23.1
28.8	25.9	28.1	18.5
35.9	23.1	25.9	32.3
25.4	24.0	28.8	29.1
24.6	38.8	31.4	37.0
28.1	36.5	28.5	33.7

Source: *Oikos Journal of Ecology* 105:41–54

a. Compute the least-squares regression line for predicting the lifespan from the wingspan. $\hat{y} = 52.0434 - 0.8445x$

b. Is it possible to interpret the y-intercept? Explain. No

c. If the wingspans of two butterflies differ by 2 millimeters, by how much would you predict their lifespans to differ? 1.689

d. Predict the lifespan for a butterfly whose wingspan is 28.5 millimeters. 27.98

29. Interpreting technology: The following display from the TI-84 Plus calculator presents the least-squares regression line for predicting a student's score on a statistics exam (y) from the number of hours spent studying (x).

```
LinReg
y=a+bx
a=49.7124
b=4.288759685
r²=.8439112806
r=.9186464394
```

a. Write the equation of the least-squares regression line.
b. What is the correlation between the score and the time spent studying? *0.9186464394*
c. Predict the score for a student who studies for 10 hours. *92.6*

30. Interpreting technology: The following display from the TI-84 Plus calculator presents the least-squares regression line for predicting the price of a certain stock (y) from the prime interest rate in percent (x).

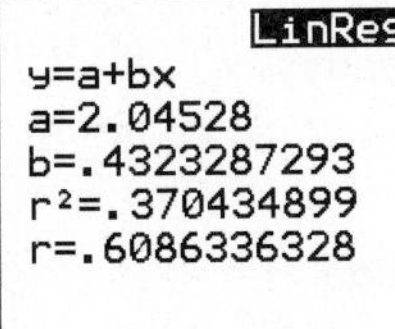

a. Write the equation of the least-squares regression line. *$\hat{y} = 2.04528 + 0.4323287293x$*
b. What is the correlation between the interest rate and the yield of the stock? *0.6086336328*
c. Predict the price when the prime interest rate is 5%. *4.2069*

31. Interpreting technology: The following MINITAB output presents the least-squares regression line for predicting the concentration of ozone in the atmosphere from the concentration of oxides of nitrogen (NOx).

```
The regression equation is
Ozone = 33.8127 + 1.21015 NOx

Predictor       Coef    SE Coef      T      P
Constant     33.8127    1.06035  31.89  0.000
NOx          1.21015    0.09047  12.38  0.000
```

a. Write the equation of the least-squares regression line.
b. Predict the ozone concentration when the NOx concentration is 21.4. *59.7*

32. Interpreting technology: The following MINITAB output presents the least-squares regression line for predicting the score on a final exam from the score on a midterm exam.

```
The regression equation is
Final = 23.7789 + 0.71384 Midterm

Predictor       Coef    SE Coef      T      P
Constant     23.7789    4.46723  5.323  0.000
Midterm      0.71384    0.13507  5.285  0.000
```

a. Write the equation of the least-squares regression line.
b. Predict the final exam score for a student who scored 75 on the midterm. *77.3*

33. Interpreting technology: A business school professor computed a least-squares regression line for predicting the salary in \$1000s for a graduate from the number of years of experience. The results are presented in the following Excel output.

	Coefficients
Intercept	55.91275257
Experience	2.58289361

a. Write the equation of the least-squares regression line.
b. Predict the salary for a graduate with 5 years of experience. *68.827*

34. Interpreting technology: A biologist computed a least-squares regression line for predicting the brain weight in grams of a bird from its body weight in grams. The results are presented in the following Excel output.

	Coefficients
Intercept	3.79229595
Body weight	0.08063922

a. Write the equation of the least-squares regression line.
b. Predict the brain weight for a bird whose body weight is 300 grams. *27.984*

Extending the Concepts

35. Least-squares regression line for z-scores: The following table presents math and verbal SAT scores for six freshmen.

Verbal (x)	428	386	653	316	438	323
Math (y)	373	571	686	319	607	440

a. Compute the correlation coefficient between math and verbal SAT score. *0.750*
b. Compute the mean $\bar{x}$ and the standard deviation s_x for the verbal scores. *$\bar{x} = 424$; $s_x = 123.26$*
c. Compute the mean $\bar{y}$ and the standard deviation s_y for the math scores. *$\bar{y} = 499.33$; $s_y = 143.93$*
d. Compute the least-squares regression line for predicting math score from verbal score. *$\hat{y} = 128.1812 + 0.8754x$*
e. Compute the z-score for each x-value:

$$z_x = \frac{x - \bar{x}}{s_x}$$

f. Compute the z-score for each y-value:

$$z_y = \frac{y - \bar{y}}{s_y}$$

g. Compute the correlation coefficient r between z_x and z_y. Is it the same as the correlation between math and verbal SAT scores? *0.750*
h. Compute the least-squares regression line for predicting z_y from z_x. Explain why the equation of the line is $\hat{z}_y = r z_x$. *$\hat{z}_y = 0.750z_x$*

Answers to Check Your Understanding Exercises for Section 4.2

1. $\hat{y} = 11.0358 + 0.8226x$
2. **a.** 80 **b.** 15 points higher
3. **a.** The length of time it takes to freeze water in a freezer set to 0°F
   b. No interpretation
4. No. The least-squares line does not predict the effect of changing the explanatory variable.

Section 4.3 Features and Limitations of the Least-Squares Regression Line

Objectives

1. Understand the importance of avoiding extrapolation
2. Compute residuals and state the least-squares property
3. Construct and interpret residual plots
4. Determine whether outliers are influential
5. Compute and interpret the coefficient of determination

Objective 1 Understand the importance of avoiding extrapolation

Don't Make Predictions Outside the Range of the Data

Table 4.2 in Section 4.2 presents the sizes and selling prices of houses in a certain neighborhood. The sizes of the houses range from 2521 to 3198 square feet. The least-squares regression line was computed to be $\hat{y} = 160.1939 + 0.0992x$. A new house of size 4000 square feet is built. Should we use the least-squares regression line to predict its selling price? The answer is no. We can compute the least-squares estimate, which is $\hat{y} = 160.1939 + 0.0992(4000) = 557.0$ thousand dollars, or \$557,000. But this prediction is unreliable because the value 4000 is outside the range of the data. While we know that the relationship between size and selling price is linear within the range 2521 to 3198 square feet, it may not be linear outside that range (see Figure 4.9).

Making predictions for values of the explanatory variable that are outside the range of the data is called **extrapolation**. In general, extrapolation can lead to unreliable predictions.

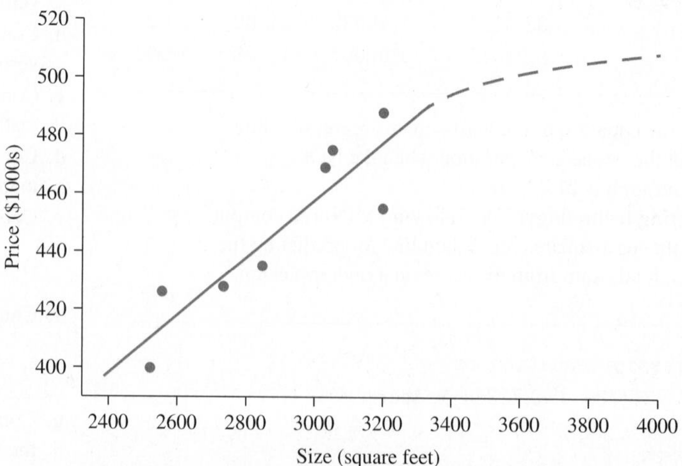

Figure 4.9 The size and selling price have a linear relationship within the range of observed values, but the relationship may become nonlinear outside that range. Therefore, predictions based on values outside the range of the data are not reliable.

SUMMARY

Do not use the least-squares regression line to make predictions for x-values that are outside the range of the data. The linear relationship may not hold there.

Check Your Understanding

1. A sample of adults was studied to determine the relationship between education level and annual income. The least-squares regression line for predicting income from education level was computed to be $\hat{y} = 2812 + 3375x$, where x is the number of years of education and $\hat{y}$ is the predicted annual income. The number of years of education among the people in the sample ranged from 8 to 18.
   a. If possible, use the least-squares regression line to predict the annual income of a person with 12 years of education. If not possible, explain why not. *$43,312*
   b. If possible, use the least-squares regression line to predict the annual income of a person with 6 years of education. If not possible, explain why not. *Not possible*

Answers are on page 204.

Objective 2 Compute residuals and state the least-squares property

Computing residuals

Figure 4.10 presents a scatterplot of selling price versus size for houses, with the least-squares regression line superimposed. For one point, (2555, 426), we have drawn a vertical line from the point to the least-squares regression line. The difference between the y-value of the point and the y-value on the line is called the *residual*. Note that residuals are positive for points above the line and negative for points below the line.

DEFINITION

Given a point (x, y) on a scatterplot, and the least-squares regression line $\hat{y} = b_0 + b_1x$, the **residual** for the point (x, y) is the difference between the observed value y and the predicted value $\hat{y}$.

$$\text{Residual} = y - \hat{y}$$

Example 4.8

Compute a residual

Compute the residual for the point (2555, 426) shown in Figure 4.10.

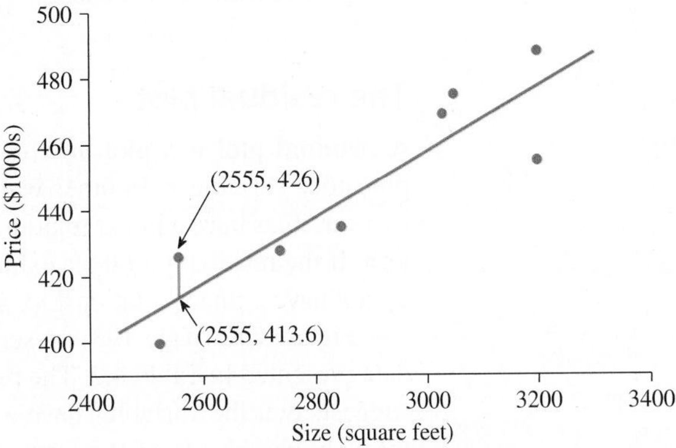

Figure 4.10 The residual is the difference between the observed value $y = 426$ and the predicted value $\hat{y} = 413.6$, which is $426 - 413.6 = 12.4$.

Solution

The y-value of the point is 426. The equation of the least-squares regression line was computed in Section 4.2 to be $\hat{y} = 160.1939 + 0.0992x$. The predicted value $\hat{y}$ is therefore $\hat{y} = 160.1939 + (0.0992)(2555) = 413.6$. The residual is

$$y - \hat{y} = 426 - 413.6 = 12.4$$

EXPLAIN IT AGAIN

What the sum of squared residuals measures: A residual for a point measures how close the predicted value is to the observed value for that point. The sum of squared residuals measures how close the predicted values are to the observed values overall. The smaller the sum of squared residuals, the better the predictions, on the whole.

Note that the magnitude of the residual is just the vertical distance from the point to the least-squares line. In Section 4.2, we described the least-squares line as the line for which the sum of the squared vertical distances is as small as possible. It follows that if we square each residual and add up the squares, the sum is less for the least-squares regression line than for any other line. This is the **least-squares property**.

The Least-Squares Property

The least-squares regression line satisfies the least-squares property. This means that the sum of the squared residuals is less for the least-squares regression line than for any other line.

Objective 3 Construct and interpret residual plots

Construct a Residual Plot to Determine Whether a Relationship Is Linear

In Section 4.1, we learned that the correlation coefficient should be used only when the relationship between x and y is linear. The same holds true for the least-squares regression line. When the scatterplot follows a curved pattern, it does not make sense to summarize it with a straight line. To illustrate this, Figure 4.11 presents a plot of a relationship that is nonlinear. The least-squares regression line does not fit the data well.

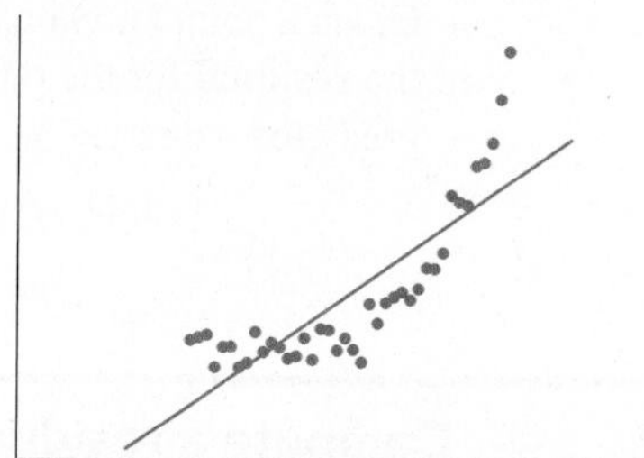

Figure 4.11 The relationship between the explanatory and outcome variables in this plot is not linear. The least-squares regression line does not fit the data well and should not be used for prediction.

The residual plot

A **residual plot** is a plot in which the residuals are plotted against the values of the explanatory variable x. In other words, the points on the residual plot are $(x, y - \hat{y})$. When two variables have a linear relationship, the residual plot will not exhibit any noticeable pattern. If the residual plot does exhibit a pattern, such as a curved pattern, then the variables do not have a linear relationship, and the least-squares regression line should not be used.

Figure 4.12 (page 193) presents a residual plot, produced by MINITAB, for the house data presented in Table 4.2. The residual plot does not exhibit any particular pattern, which suggests that the variables have a linear relationship. Note that a horizontal line is drawn through the value 0 on the y-axis. This can help make a pattern stand out, if there is one. Step-by-step instructions for constructing residual plots with the TI-84 Plus, with Excel, and with MINITAB are presented in the Using Technology section on page 198.

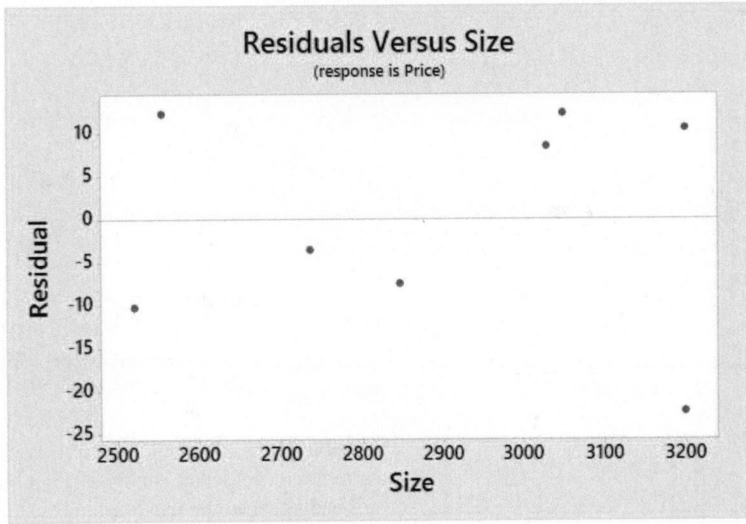

Figure 4.12 Residual plot for the data in Table 4.2. The plot exhibits no noticeable pattern, so use of the least-squares regression line is appropriate.

A nonlinear relationship can often be detected from a scatterplot, without constructing a residual plot. In some cases, however, curvature is easier to detect on a residual plot. As an example, Figure 4.13 presents a scatterplot with the least-squares regression line superimposed. The line appears to fit almost perfectly. Figure 4.14 presents the residual plot. The curvature is clearly evident.

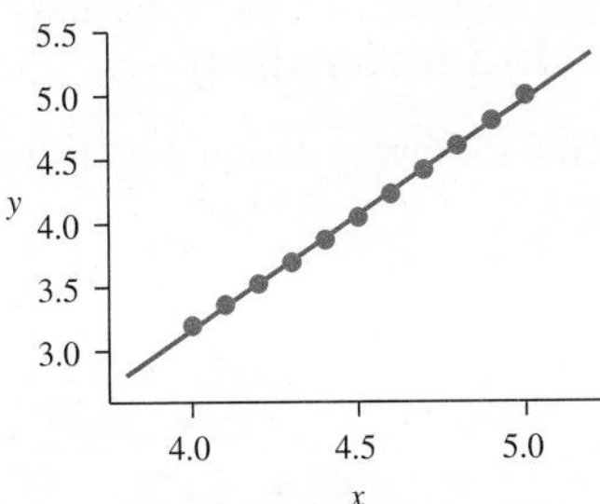

Figure 4.13 Scatterplot showing what appears to be a nearly perfect linear relationship.

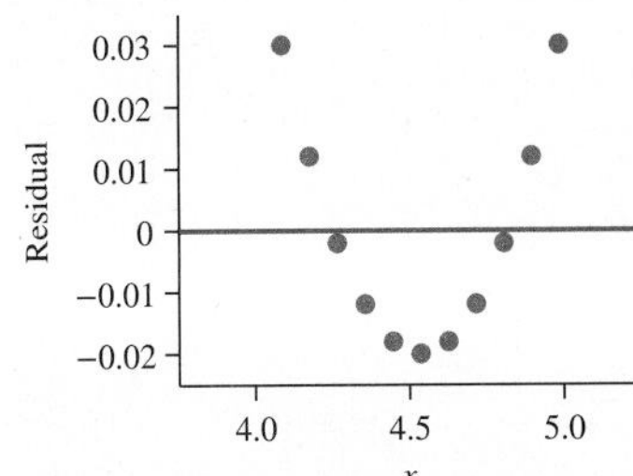

Figure 4.14 The residual plot for the scatterplot in Figure 4.13. The nonlinearity of the relationship is now clear.

SUMMARY

- When a residual plot exhibits a noticeable pattern, the variables do not have a linear relationship, and the least-squares regression line should not be used.
- When a residual plot exhibits no noticeable pattern, the least-squares line may be used to describe the relationship between the variables.

The relationship can be nonlinear even when the correlation is large

If the correlation coefficient r is close to 1 or -1, it is tempting to assume that x and y have a linear relationship. However, this is not necessarily the case. Sometimes r can be close to 1 or -1 when the relationship is far from linear.

For example, consider the scatterplot in Figure 4.15 (page 194). The least-squares regression line is superimposed and clearly does not describe the data well. However, the correlation coefficient is $r = 0.928$, which is fairly close to 1. The correlation coefficient is intended to be used only when variables have a linear relationship and can be misleading otherwise. The only reliable way to determine whether two variables have a linear relationship is to look at a scatterplot or a residual plot.

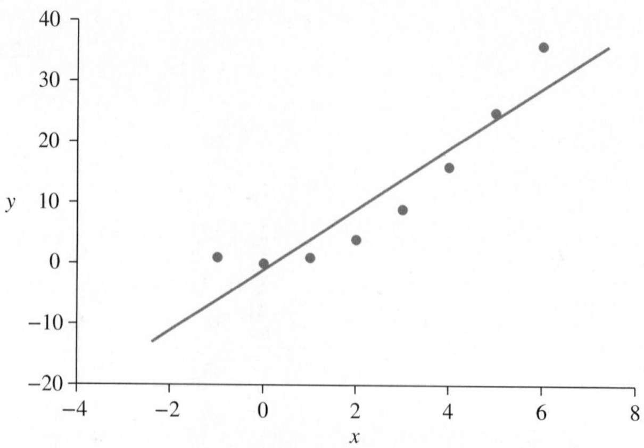

Figure 4.15 The relationship between x and y is clearly nonlinear. However, the correlation coefficient is $r = 0.928$, which is large. The correlation coefficient can be misleading when the relationship is not linear.

SUMMARY

Do not rely on the correlation coefficient to determine whether two variables have a linear relationship. Even when the correlation is close to 1 or to −1, the relationship may not be linear. To determine whether two variables have a linear relationship, construct a scatterplot or a residual plot.

Check Your Understanding

2. For each of the following residual plots, determine whether a linear model is appropriate.

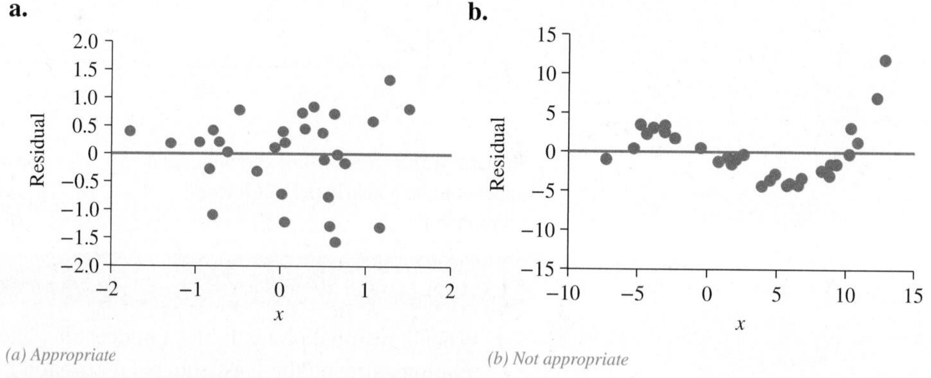

(a) Appropriate (b) Not appropriate

Answers are on page 204.

Objective 4 Determine whether outliers are influential

Digital Vision/Punchstock

Outliers and Influential Points

Recall that an outlier is a point in a data set that is detached from the bulk of the data. Figure 4.16 (page 195) presents a scatterplot of the total area of farmland (including ranches) plotted against the total land area, for each of the 50 states of the United States. Both Texas and Alaska are outliers. The blue line on the plot is the least-squares regression line computed for the 48 states not including Texas or Alaska. The red line is the least-squares regression line obtained when Texas is added to the 48 states. Including Texas moves the line somewhat. The green line is the least-squares regression line obtained when Alaska is added to the 48 states. Including Alaska causes a big shift in the position of the line.

An outlier that causes a big shift in the position of the line is said to be an *influential point*. In Figure 4.16, Alaska is an influential point, and Texas is somewhat influential as well. Influential points are troublesome, because the least-squares regression line is supposed to summarize all the data, rather than reflect the position of a single point.

NOTE TO INSTRUCTOR

There are rules for detecting outliers; for example, the rule used in boxplots defines an outlier as a point whose distance from the nearest quartile is more than 1.5 times the interquartile range. Another rule sometimes used is that any point that is more than two standard deviations from the mean is an outlier. These rules are not always reliable, so it's best to have students focus on visual methods of detecting outliers.

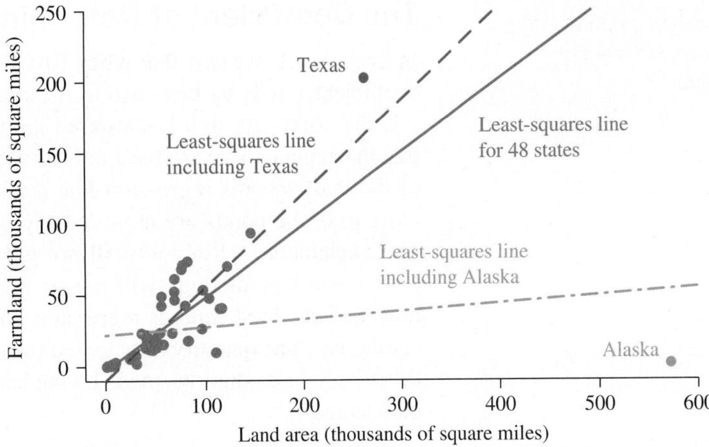

Figure 4.16 Scatterplot of farm area versus total land area for U.S. states. The blue solid line on the plot is the least-squares regression line computed for the 48 states not including Texas or Alaska. The red dashed line is the least-squares regression line for 49 states including Texas. Including Texas moves the line somewhat. The green dash-dot line is the least-squares regression line for 49 states including Alaska. Including Alaska causes a big shift in the position of the line.

DEFINITION

An **influential point** is a point that, when included in a scatterplot, strongly affects the position of the least-squares regression line.

When a scatterplot contains outliers, the least-squares regression line should be computed both with and without each outlier, to determine which outliers are influential. If there is an influential point, the least-squares regression line should be computed both with and without the point, and the equations of both lines should be reported.

NOTE TO INSTRUCTOR

It is important to distinguish between outliers that are influential and those that are not.

SUMMARY

When a scatterplot contains outliers:

- Compute the least-squares regression line both with and without each outlier to determine which outliers are influential.
- Report the equations of the least-squares regression line both with and without each influential point.

Check Your Understanding

3. In each of the following plots, one point is an outlier. The blue solid line is the least-squares regression line computed without using the outlier, and the red dashed line is the least-squares regression line computed by including the outlier. State whether the outlier is influential.

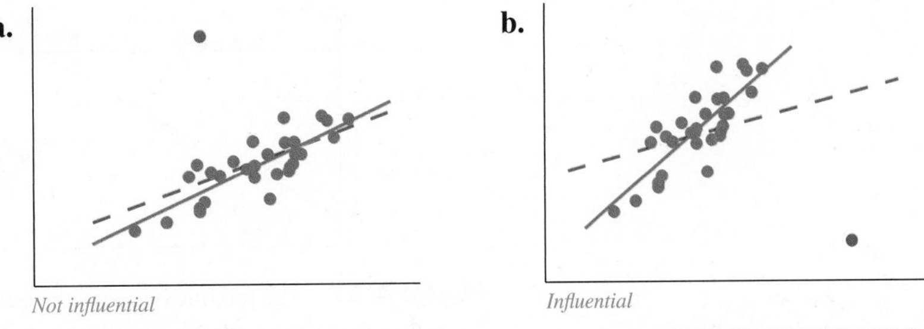

Answers are on page 204.

Objective 5 Compute and interpret the coefficient of determination

The Coefficient of Determination

In Section 4.1, we saw that when two variables have a linear relationship, the correlation coefficient r tells us how strong the relationship is. Recall that when r is close to 1 or -1, the points are tightly clustered around the least-squares regression line. This means that the explanatory variable x tells us a lot about the outcome variable y, so the predictions of the least-squares regression line are likely to be close to the actual values. When r is close to 0, the points are more loosely clustered around the least-squares regression line. The explanatory variable x will not tell us much about the outcome variable y, and the least-squares predictions will not be as good. The quantity most often used to measure how well the least-squares regression line fits the data is r^2, the square of the correlation coefficient. The quantity r^2 is called the *coefficient of determination*. The closer r^2 is to 1, the closer the predictions made by the least-squares regression line are to the actual values, on average.

NOTATION ROUNDUP

Correlation Coefficient: r

Coefficient of Determination: r^2

DEFINITION

The **coefficient of determination** is r^2, the square of the correlation coefficient r.

Explained variation and unexplained variation

Why is the coefficient of determination the best way to measure how close the predictions made by the least-squares regression line are to the actual values? As we will see, it turns out that some of the variation in the outcome variable is explained by the least-squares regression line and some is not. The coefficient of determination can be thought of as the proportion of the variation in the outcome variable that is explained by the least-squares regression line.

To illustrate this idea, Figure 4.17 presents a scatterplot on which two lines have been superimposed. One is the least-squares regression line and the other is the horizontal line $y = \bar{y}$, which goes through the sample mean $\bar{y}$. The sample mean $\bar{y}$ represents the center of the y-values. We have highlighted one particular point (x, y). Its y-value differs from the central value $\bar{y}$; the difference is $y - \bar{y}$. This difference can be divided into two parts. The first part, between the points $(x, \hat{y})$ and $(x, \bar{y})$ in Figure 4.17, is the difference between the central value $\bar{y}$ and the value $\hat{y}$ predicted for this particular x-value. This difference, $\hat{y} - \bar{y}$, is called the difference explained by the least-squares regression line, or the **explained difference**. The second part, between the points (x, y) and $(x, \hat{y})$ in Figure 4.17, is the difference between the observed data value y and the predicted value $\hat{y}$. This difference, which is the residual $y - \hat{y}$, is caused by factors unrelated to the least-squares regression line and is called the difference unexplained by the least-squares regression line, or the **unexplained difference**.

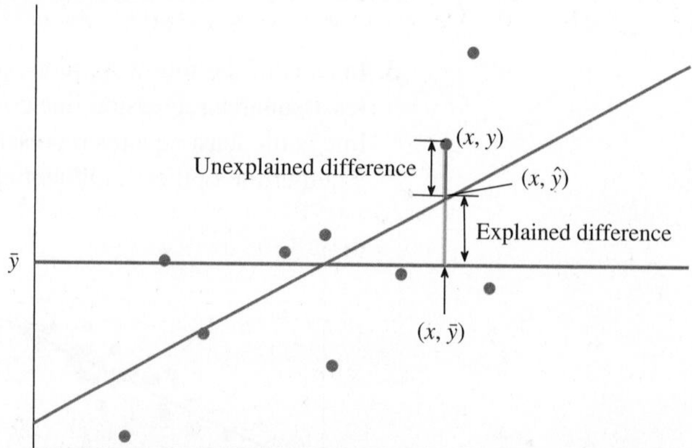

Figure 4.17 The red line between the points (x, y) and $(x, \hat{y})$ represents the difference $y - \hat{y}$, which is the unexplained difference. The green line between the points $(x, \hat{y})$ and $(x, \bar{y})$ represents the difference $\hat{y} - \bar{y}$, which is the explained difference.

Now the better the least-squares predictions are, the smaller the unexplained differences will be, on the whole. We measure the size of the unexplained differences by squaring them and adding them together. The quantity thus obtained is called the **unexplained variation**. Then we do the same for the explained differences to obtain the **explained variation**. The sum of the explained variation and the unexplained variation is called the **total variation**.

The explained variation is

$$\text{Explained variation} = \sum(\hat{y} - \bar{y})^2$$

The unexplained variation is

$$\text{Unexplained variation} = \sum(y - \hat{y})^2$$

The total variation is

$$\text{Total variation} = \text{Unexplained variation} + \text{Explained variation}$$

It can be shown by advanced methods that the coefficient of determination is

$$r^2 = \frac{\text{Explained variation}}{\text{Total variation}}$$

Thus, r^2 measures the proportion of the total variation that is explained by the least-squares regression line.

Fortunately, we don't have to compute the explained and unexplained variation in order to compute r^2. All we need to do is to compute the correlation coefficient r and square it.

SUMMARY

- The coefficient of determination r^2 measures the proportion of the variation in the outcome variable that is explained by the least-squares regression line.
- The larger the value of r^2, the closer the predictions made by the least-squares regression line are to the actual values, on average.
- To compute the coefficient of determination, first compute the correlation coefficient r, then square it to obtain r^2.

Example 4.9

Computing and interpreting the coefficient of determination

CAUTION

Don't round the correlation coefficient when computing the coefficient of determination.

The correlation between size and selling price for the data in Table 4.1 was computed in Example 4.2 to be $r = 0.9005918$.

a. What is the coefficient of determination?

b. How much of the variation in selling price is explained by the least-squares regression line?

Solution

a. To find the coefficient of determination, we square the correlation coefficient.

$$\text{Coefficient of determination} = r^2 = (0.9005918)^2 = 0.811$$

b. The coefficient of determination is 0.811. Therefore, 81.1% of the variation in selling price is explained by the least-squares regression line.

Check Your Understanding

4. For each of the following values of the correlation coefficient, determine how much of the variation in the outcome variable is explained by the least-squares regression line.
   a. $r = 0.6$ 36%
   b. $r = -0.9$ 81%
   c. $r = 0$ 0%
   d. $r = 1$ 100%

Answers are on page 204.

Using Technology

We use the data in Table 4.2 to illustrate the technology steps.

TI-84 PLUS

Constructing residual plots

Step 1. Enter the x-values from Table 4.2 into **L1** and the y-values into **L2**. Run the **LinReg(a+bx)** command, using the steps on page 185.

Step 2. Press **2nd, Y=** to access the STAT PLOTS menu, and select Plot1 by pressing **1**.

Step 3. Select **On** and the scatterplot icon. Enter the residuals for the **Ylist** option by pressing **2nd, STAT** and then **7: RESID** (Figure A).

Step 4. Press **ZOOM** and then **9: ZoomStat** (Figure B).

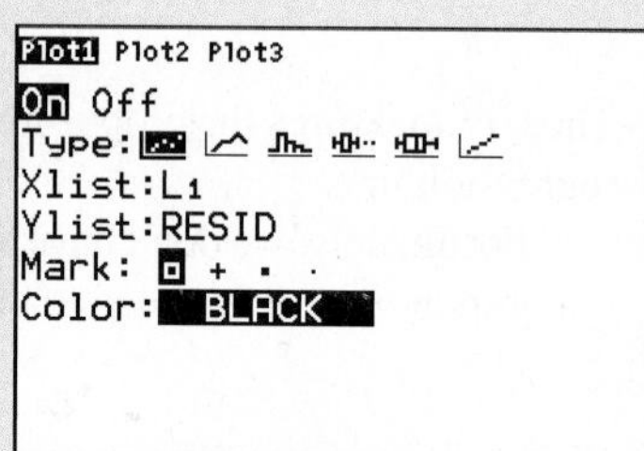

Figure A

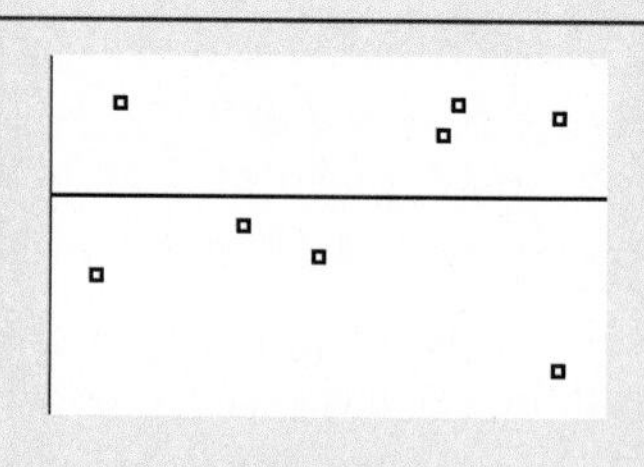

Figure B

EXCEL

Constructing residual plots

Step 1. Enter the x-values from Table 4.2 into **Column A** and the y-values into **Column B**.

Step 2. Select **Data**, then **Data Analysis**. Highlight **Regression** and press **OK**.

Step 3. Enter the range of cells that contain the x-values in the **Input X Range** field and the range of cells that contain the y-values in the **Input Y Range** field.

Step 4. Select the **Residual Plots** option.

Step 5. Click **OK** (Figure C).

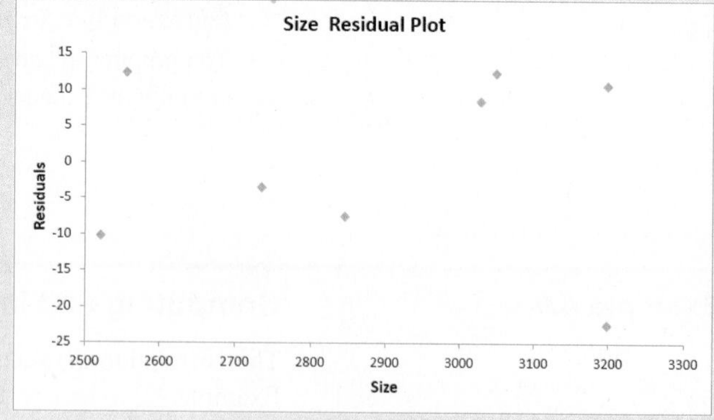

Figure C

MINITAB

Constructing residual plots

Step 1. Enter the x-values from Table 4.2 into **Column C1**, and label the column **Size**. Enter the y-values into **Column C2**, and label the column **Price**.

Step 2. Click **STAT**, then **Regression**, then **Regression** again. Then select **Fit Regression Model**.

Step 3. Select the y-variable **(Price)** as the **Response** and the x-variable **(Size)** as the **Continuous predictor**, and click **OK**.

Step 4. Click on **Graphs**, select *Regular* for **Residuals for plots**, and enter the x-variable **(Size)** in the **Residual versus the variables** field.

Step 5. Click **OK** and then **OK**.

Section 4.3 Exercises

Exercises 1–4 are the Check Your Understanding exercises located within the section.

Understanding the Concepts

In Exercises 5–10, fill in each blank with the appropriate word or phrase.

5. Making predictions for values of the explanatory variable that are outside of the range of the data is called ________. *extrapolation*
6. A ________ is the difference between an observed value and a predicted value of the outcome variable. *residual*
7. The least-squares property says that the ________ is smaller for the least-squares regression line than for any other line. *sum of squared residuals*
8. An outlier that strongly affects the position of a least-squares regression line is said to be ________. *an influential point*
9. The coefficient of determination is the square of the ________. *correlation coefficient*
10. The total variation is the sum of the ________ and the ________. *explained variation, unexplained variation*

In Exercises 11–14, determine whether the statement is true or false. If the statement is false, rewrite it as a true statement.

11. When two values have a nonlinear relationship, the residual plot will exhibit a noticeable pattern. *True*
12. When the correlation coefficient is close to 1 or −1, there is not necessarily a linear relationship between the variables. *True*
13. The closer r^2 is to 0, the closer the predictions made by the least-squares regression line are to the actual values, on average. *False*
14. The coefficient of determination may be interpreted as the proportion of variation in the outcome variable explained by the least-squares regression line. *True*

Practicing the Skills

15. For the following data set:

x	1	2	3	4	5
y	5	6	9	8	7

a. Compute the coefficient of determination. *0.36*
b. How much of the variation in the outcome variable is explained by the least-squares regression line? *36%*

16. For the following data set:

x	9	5	7	13	−8	−2	6	−10
y	3	3	31	36	0	3	−2	−14

a. Compute the coefficient of determination. *0.503*
b. How much of the variation in the outcome variable is explained by the least-squares regression line? *50.3%*

17. For the following data set:

x	73	86	25	3	52	45	52
y	62	66	41	21	33	37	47

a. Compute the least-squares regression line. $\hat{y} = 19.3734 + 0.5101x$
b. Which point is an outlier? *(3, 21)*
c. Remove the outlier and compute the least-squares regression line. $\hat{y} = 19.2354 + 0.5123x$
d. Is the outlier influential? Explain. *No*

18. For the following data set:

x	3.9	5.8	4.8	3.3	1.8	3.4	3.2
y	4.6	4.1	5.2	4.5	9.2	5.6	4.3

a. Compute the least-squares regression line. $\hat{y} = 8.9888 - 0.9703x$
b. Which point is an outlier? *(1.8, 9.2)*
c. Remove the outlier and compute the least-squares regression line. $\hat{y} = 5.2915 - 0.1413x$
d. Is the outlier influential? Explain. *Yes*

19. **Least-squares OK?** Following is a residual plot produced by MINITAB. Was it appropriate to compute the least-squares regression line? Explain. *No*

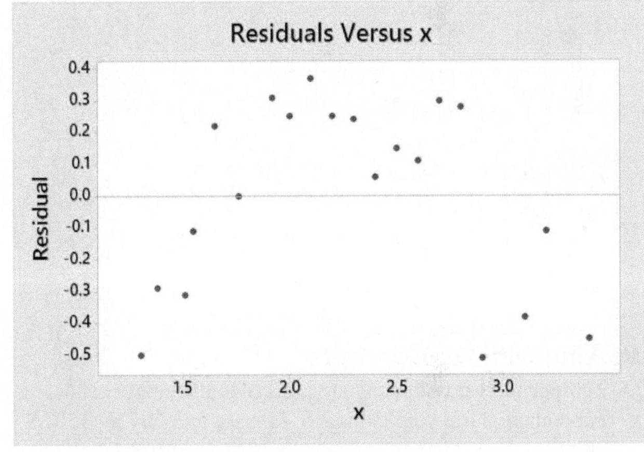

20. **Least-squares OK?** Following is a residual plot produced by MINITAB. Was it appropriate to compute the least-squares regression line? Explain. *Yes*

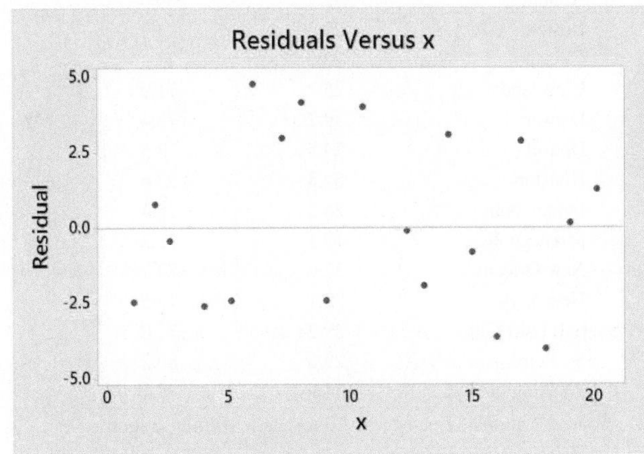

21. **Least-squares OK?** Following is a residual plot produced by MINITAB. Was it appropriate to compute the least-squares regression line? Explain. *Yes*

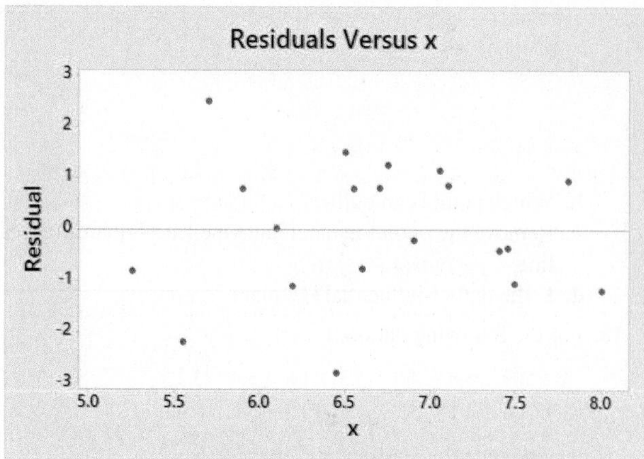

22. **Least-squares OK?** Following is a residual plot produced by MINITAB. Was it appropriate to compute the least-squares regression line? Explain. No

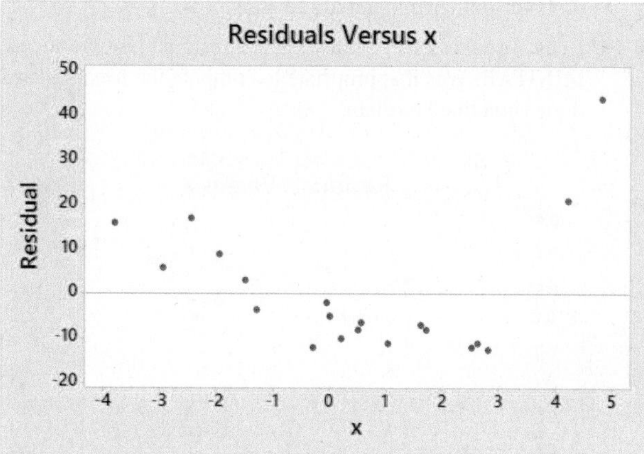

Working with the Concepts

23. **Temperatures:** Following are average temperatures, in degrees Fahrenheit, during the months of January and July for 15 U.S. cities.

City	January Temperature	July Temperature
Albuquerque	35.7	78.5
Atlanta	42.7	80.0
Boston	29.3	73.9
Chicago	22.0	73.3
Cleveland	25.7	71.9
Denver	29.2	73.4
Detroit	24.5	73.5
Houston	51.8	83.6
Indianapolis	26.5	75.4
Minneapolis	13.1	73.2
New Orleans	52.6	82.7
New York	32.1	76.5
Salt Lake City	29.2	77.0
St. Louis	29.6	80.2
Washington, D.C.	34.9	79.2

Source: National Oceanic and Atmospheric Administration

a. Compute the least-squares regression line for predicting July temperature from January temperature. $\hat{y} = 67.0556 + 0.3058x$

b. Construct a residual plot. Does the relationship appear to be approximately linear? Yes

c. Predict the July temperature for a city whose January temperature is 37°F. 78.4

d. The city of Miami, Florida, has an average January temperature of 68.1°F. Use the least-squares line to predict the July temperature for Miami. Is this an appropriate prediction? Why or why not? 87.9; No, the value of 68.1 is outside the range of the data.

24. **Great Pumpkins:** When growing giant pumpkins for competitions, growers need to keep track of the weights of the pumpkins while they are growing. It is difficult to weigh a large pumpkin before it is harvested, so a method has been developed for estimating the weight. The grower measures around the pumpkin both horizontally and vertically, then adds the results. This is called the OTT (over the top) measurement and is used to predict the weight of the pumpkin. Following are the OTT measurements and actual weights of the 20 largest pumpkins entered into official competitions in the year 2018.

OTT (inches)	Weight (pounds)	OTT (inches)	Weight (pounds)
490.0	2528.0	465.0	2138.0
469.0	2469.0	452.0	2136.0
490.0	2433.9	451.0	2114.0
480.0	2416.5	456.0	2091.0
455.0	2283.0	457.0	2079.0
463.0	2170.0	456.0	2077.0
473.0	2166.0	462.0	2070.1
477.0	2157.5	436.0	2027.0
465.0	2152.0	454.0	2020.5
454.0	2138.9	450.0	2017.5

Source: The Great Pumpkin Commonwealth

a. Compute the least-squares regression line for predicting weight (y) from OTT (x). $\hat{y} = -1917.1 + 8.8629x$

b. Construct a residual plot. Does the relationship appear to be approximately linear? Yes

c. Predict the weight of a pumpkin whose OTT measurement is 450 inches. 2071.2 pounds

d. There is much interest among growers as to when a giant pumpkin will break the 3000-pound mark. For what OTT does the least-squares line predict a weight of 3000 pounds? Is this an appropriate prediction? Why or why not? 554.8; No, the value of 3000 is outside the range of the data.

25. **Hot enough for you?** The following table presents the temperature, in degrees Fahrenheit, and barometric pressure, in inches of mercury, on August 15 at 12 noon in Macon, Georgia, over a nine-year period.

Pressure	Temperature
30.16	80.1
30.09	85.5
29.99	87.0
29.83	94.5
29.86	84.0
30.00	83.2
29.98	80.1
29.85	84.9
30.01	84.0

Source: Weather Underground

a. Compute the least-squares regression line for predicting temperature from barometric pressure. $\hat{y} = 740.3765 - 21.8708x$

b. Compute the coefficient of determination. 0.323

c. Construct a scatterplot of the temperature (y) versus the barometric pressure (x).

d. Which point is an outlier? *(29.83, 94.5)*

e. Remove the outlier and compute the least-squares regression line for predicting temperature from barometric pressure. $\hat{y} = 320.5388 - 7.8999x$

f. Is the outlier influential? Explain. *Yes*

g. Compute the coefficient of determination for the data set with the outlier removed. Is the proportion of variation explained by the least-squares regression line greater, less, or about the same without the outlier? Explain.

26. Presidents and first ladies: The following table presents the ages of the last 10 U.S. presidents and their wives on the first day of their presidencies.

Name	Her Age	His Age
Donald and Melania Trump	46	70
Barack and Michelle Obama	45	47
George W. and Laura Bush	54	54
Bill and Hillary Clinton	45	46
George and Barbara Bush	63	64
Ronald and Nancy Reagan	59	69
Jimmy and Rosalynn Carter	49	52
Gerald and Betty Ford	56	61
Richard and Pat Nixon	56	56
Lyndon and Lady Bird Johnson	50	55

a. Compute the least-squares regression line for predicting the president's age from the first lady's age. $\hat{y} = 19.701 + 0.7208x$

b. Compute the coefficient of determination. *0.287*

c. Construct a scatterplot of the presidents' ages (y) versus the first ladies' ages (x).

d. Which point is an outlier? *(46, 70)*

e. Remove the outlier and compute the least-squares regression line for predicting the president's age from the first lady's age. $\hat{y} = -2.8506 + 1.1104x$

f. Is the outlier influential? Explain. *Yes*

g. Compute the coefficient of determination for the data set with the outlier removed. Is the proportion of variation explained by the least-squares regression line greater, less, or about the same without the outlier? Explain. *0.826; greater*

27. Mutant genes: In a study to determine whether the frequency of a certain mutant gene increases with age, the number of mutant genes per microgram of DNA was counted for each of 29 men. The results are presented in the following table.

Age	Number of Mutants	Age	Number of Mutants
44	100	53	49
46	52	80	41
43	93	57	66
46	193	59	109
45	77	65	264
53	35	68	69
58	105	47	29
54	68	61	101
68	71	61	148
55	82	70	78
55	267	70	195
51	43	72	83
52	181	82	449
57	29	83	202
52	58		

Source: *Proceedings of the National Academy of Sciences* 99:14952–14957

a. Compute the least-squares regression line for predicting number of mutants from age. $\hat{y} = -76.2258 + 3.2499x$

b. Construct a scatterplot of number of mutants (y) versus age (x).

c. Which point is an outlier? *(82, 449)*

d. Remove the outlier, and compute the least-squares regression line for predicting number of mutants from age. $\hat{y} = 31.8632 + 1.2282x$

e. Is the outlier influential? Explain. *Yes*

28. Imports and exports: The following table presents the U.S. imports and exports (in billions of dollars) for each of 29 months.

Month	Imports	Exports	Month	Imports	Exports
1	215.9	168.1	16	230.9	184.3
2	211.8	166.6	17	230.5	184.2
3	217.7	174.3	18	227.6	185.2
4	218.1	175.9	19	226.8	183.4
5	223.6	176.2	20	226.1	182.1
6	224.2	173.2	21	228.4	186.8
7	224.9	179.5	22	225.3	182.7
8	224.6	179.9	23	231.6	185.2
9	225.7	181.2	24	227.0	188.7
10	226.6	180.5	25	229.4	186.7
11	226.1	178.3	26	231.0	187.1
12	230.5	179.1	27	222.3	185.2
13	230.9	179.5	28	227.7	187.6
14	225.8	182.1	29	232.1	187.1
15	234.3	186.5			

Source: U.S. Census Bureau

a. Compute the least-squares regression line for predicting exports (y) from imports (x). $\hat{y} = -19.2390 + 0.8868x$

b. Compute the coefficient of determination. *0.6425*

c. The months with the two lowest exports are months 1 and 2, when the exports were 168.1 and 166.6, respectively. Remove these points and compute the least-squares regression line. Is the result noticeably different? $\hat{y} = 23.6335 + 0.6990x$; *Yes*

d. Compute the coefficient of determination for the data set with months 1 and 2 removed. *0.4091*

e. Two economists decide to study the relationship between imports and exports. One uses data from months 1 through 29 and the other uses data from months 3 through 29. For which data set will the proportion of variance explained by the least-squares regression line be greater? *Months 1–29*

29. Energy consumption: The following table presents the average annual energy expenditures (in dollars) for housing units of various sizes (in square feet).

Size	Energy Expenditure
250	1087
750	1228
1250	1583
1750	1798
2250	1939
2750	2138
3250	2172
3750	2315

Source: Energy Information Administration

a. Compute the least-squares regression line for predicting energy consumption from house size. $\hat{y} = 1062.4048 + 0.3600x$

b. Compute the coefficient of determination. 0.9603

c. Construct a residual plot. Does the relationship appear to be linear? No

d. True or false: If the coefficient of determination is large, the relationship must be linear. False

30. Cost of health care: The following table presents the mean cost of a hospital stay, in \$1000s, and the number of hospital stays, in millions, in the United States for each of 15 years.

Cost	Number	Cost	Number
9.8	34.3	15.0	37.2
10.3	34.1	17.3	37.8
10.3	34.2	19.7	38.2
10.6	34.2	20.5	38.7
11.3	34.7	22.3	39.2
11.8	34.9	24.0	39.5
12.5	35.5	26.1	39.5
13.7	36.4		

Source: Agency for Healthcare Research and Quality

a. Compute the least-squares regression line for predicting the number of stays from the cost. $\hat{y} = 30.7147 + 0.3728x$

b. Compute the coefficient of determination. 0.9479

c. Construct a residual plot. Does the relationship appear to be linear? No

d. True or false: If the coefficient of determination is large, the relationship must be linear. False

31. Buy or rent? The following table presents the average sales price (in \$1000s) and the average monthly rental price of homes in selected U.S. cities in a recent month.

City	Sales Price	Rental Price
Charlotte, NC	136.8	1156
Pensacola, FL	118.9	971
San Francisco, CA	590.6	2555
Washington, DC	331.6	2083
Springfield, MA	186.7	1404
Charleston, SC	160.7	1294
Cincinnati, OH	123.5	1110
Jacksonville, FL	127.3	1178
Trenton, NJ	202.0	1777
Warner Robins, GA	101.8	957
Santa Barbara, CA	438.5	2147
Denver, CO	237.9	1566
Tucson, AZ	146.8	1093
Concord, NH	184.1	1481
Providence, RI	211.2	1475

Source: Zillow

a. Compute the least-squares regression line for predicting average rental price (y) from average selling price (x). $\hat{y} = 757.2920 + 3.3009x$

b. Compute the coefficient of determination. 0.9043

c. Construct a residual plot. Does the relationship appear to be linear? No

d. True or false: If the coefficient of determination is large, the relationship must be linear. False

32. Broadway: The following table presents the average ticket price (the average price paid per attendee) in dollars and the gross revenue (in millions of dollars) for Broadway productions for each of 20 seasons.

Ticket Price	Revenue	Ticket Price	Revenue
43.90	356	66.41	771
44.91	406	66.70	769
46.09	436	71.83	862
47.21	499	76.28	939
48.61	558	76.45	938
50.39	588	77.61	943
52.99	603	85.79	1020
56.01	666	86.27	1081
58.72	643	92.38	1139
63.13	721	98.44	1139

Source: Broadway League

a. Compute the least-squares regression line for predicting gross revenue (y) from ticket price (x). $\hat{y} = -179.4279 + 14.2472x$

b. Compute the coefficient of determination. 0.9725

c. Construct a residual plot. Does the relationship appear to be linear? No

d. True or false: If the coefficient of determination is large, the relationship must be linear. False

Extending the Concepts

33. Derive the coefficient of determination: For the house data in Table 4.1, the average selling price is $\bar{y} = 447.0$. Imagine that you do not know the size of any house, so you predict the selling price of each of them to be 447.0. This is equivalent to using the line $y = \bar{y}$ for prediction.

a. Compute the residual for each point using the line $y = \bar{y}$.

b. Compute the sum of squared residuals for the line $y = \bar{y}$. 6168

c. Compute the residual for each point using the least-squares regression line.

d. Compute the sum of squared residuals for the least-squares regression line. 1165.35

e. Determine how much better the least-squares predictions are by computing the difference

Decrease in sum of squared residuals

= (Sum of squared residuals for $y = \bar{y}$)
− (Sum of squared residuals for regression line) 5002.65

f. Compute the ratio

$$\frac{\text{Decrease in sum of squared residuals}}{\text{Sum of squared residuals for } y = \bar{y}}$$ 0.811

g. Show that the ratio obtained in part (f) is equal to the coefficient of determination.

34. Transforming a variable: The following table presents the speed (in mph) and the stopping distance (in feet) for a sample of cars.

Speed	Distance	Speed	Distance
4	12	16	44
7	14	17	48
8	19	18	49
9	20	19	57
10	18	20	59
11	22	22	75
12	25	23	78
13	32	24	87
14	38	25	90
15	38		

a. Compute the least-squares regression line for predicting stopping distance (y) from speed (x). $\hat{y} = -17.1793 + 4.0119x$

b. Construct a residual plot. Explain why the least-squares line is not an appropriate summary of the data. *Obvious pattern*

c. For each data point, square the speed. This is x^2. Compute the least-squares regression line for predicting distance y from x^2. $\hat{y} = 8.3045 + 0.1332x^2$

d. Construct a residual plot. Is this line an appropriate summary? *Yes*

e. Use the equation computed in part (c) to predict the stopping distance for a car whose speed is 15 mph. *38.27 feet*

35. Radon and cancer: Radon is a radioactive element that occurs naturally in soil as a result of the decay of uranium. Long-term exposure to radon, an odorless, invisible gas, is thought to increase the risk of cancer. Radon detectors are often placed in homes to determine whether radon levels are high enough to be dangerous. Two radon detectors were placed in different locations in the basement of a home in the eastern United States. Each provided an hourly measurement of the radon concentration, in units of picocuries/liter. The data are presented in the following table.

R_1	R_2	R_1	R_2	R_1	R_2	R_1	R_2
1.2	1.2	3.4	2.0	4.0	2.6	5.5	3.6
1.3	1.5	3.5	2.0	4.0	2.7	5.8	3.6
1.3	1.6	3.6	2.1	4.3	2.7	5.9	3.9
1.3	1.7	3.6	2.1	4.3	2.8	6.0	4.0
1.5	1.7	3.7	2.1	4.4	2.9	6.0	4.2
1.5	1.7	3.8	2.2	4.4	3.0	6.1	4.4
1.6	1.8	3.8	2.2	4.7	3.1	6.2	4.4
2.0	1.8	3.8	2.3	4.7	3.2	6.5	4.4
2.0	1.9	3.9	2.3	4.8	3.2	6.6	4.4
2.4	1.9	3.9	2.4	4.8	3.5	6.9	4.7
2.9	1.9	3.9	2.4	4.9	3.5	7.0	4.8
3.0	2.0	3.9	2.4	5.4	3.5		

a. Compute the least-squares line for predicting the radon concentration at location 2 from the concentration at location 1. $\hat{y} = 0.4636 + 0.5711x$

b. Construct a residual plot. Does the linear model seem appropriate? *No*

c. Divide the data into two groups: points where $R_1 < 4$ in one group, and points where $R_1 \geq 4$ in the other. Compute the least-squares line and the residual plot for each group. Does the line describe either group well? Which one?

d. Explain why it might be a good idea to fit a linear model to part of these data, and a nonlinear model to the other.

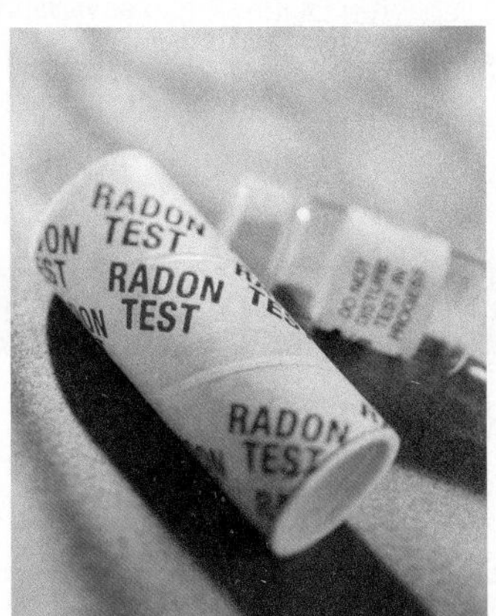

Steve Cole/Getty Images

36. How's your credit? The following table presents the average number of late payments and the average credit score for residents of selected U.S. cities in a recent year.

Late Payments	Credit Score	Late Payments	Credit Score	Late Payments	Credit Score
0.50	729	0.24	777	0.33	762
0.50	727	0.34	776	0.60	733
0.45	758	0.25	771	0.33	769
0.52	731	0.53	724	0.39	760
0.65	755	0.51	740	0.51	737
0.26	778	0.73	702	0.33	772
0.30	772	0.65	729	0.37	767
0.58	729	0.68	714	0.33	762
0.67	739	0.40	739	0.39	761
0.41	761	0.53	746	0.33	776
0.58	717	0.57	711	0.21	784
0.46	752	0.45	766	0.41	767
0.42	767	0.25	787	0.43	752
0.50	742	0.52	731		

Source: Experian

a. Compute the correlation between the number of late payments and the credit score. -0.8702

b. Compute the least-squares regression line for predicting average credit score (y) from average number of late payments (x). $\hat{y} = 814.8943 - 143.3279x$

c. Construct a residual plot. Are the residuals corresponding to $x \geq 0.5$ more spread out than those corresponding to $x < 0.5$? *Yes*

d. Compute the least-squares regression line using only those points with $x < 0.5$. $\hat{y} = 805.8230 - 111.1299x$

e. Compute the least-squares regression line using only those points with $x \geq 0.5$. $\hat{y} = 764.5173 - 60.5335x$

f. Compute the coefficient of determination using only those points with $x < 0.5$. 0.5689

g. Compute the coefficient of determination using only those points with $x \geq 0.5$. 0.1162

h. In which situation is the average number of late payments more strongly related to average credit score, when the average number of late payments is less than 0.5 or when it is greater than or equal to 0.5? *less than 0.5*

Answers to Check Your Understanding Exercises for Section 4.3

1. **a.** \$43,312
   b. It is not possible because 6 is outside the range of the data.
2. **a.** Appropriate **b.** Not appropriate
3. **a.** Not influential **b.** Influential
4. **a.** 36% **b.** 81% **c.** 0% **d.** 100%

Chapter 4 Summary

Section 4.1: Bivariate data are data that consist of ordered pairs. A scatterplot provides a good graphical summary for bivariate data. When large values of one variable are associated with large values of the other, the variables are said to have a positive association. When large values of one variable are associated with small values of the other, the variables are said to have a negative association. When the points on a scatterplot tend to cluster around a straight line, the relationship is said to be linear.

The correlation coefficient r measures the strength of a linear relationship. The value of r is always between -1 and 1. Positive values of r indicate a positive linear association, while negative values of r indicate a negative linear association. Values near 1 or -1 indicate a strong linear association, while values near 0 indicate a weak linear association. The correlation coefficient should not be used when the relationship is not linear.

Correlation is not the same as causation. Even when two variables are highly correlated, it is not necessarily the case that changing the value of one of them will cause a change in the other.

Section 4.2: When two variables have a linear relationship, the points on a scatterplot tend to cluster around a straight line called the least-squares regression line. Given a value of the explanatory variable x, we can predict a value $\hat{y}$ for the outcome variable by substituting the value of x into the equation of the least-squares regression line. The slope of the least-squares regression line predicts the difference between the y-values for two points whose x-values differ by 1. The intercept of the least-squares regression line predicts the y-value of a point whose x-value is 0. The intercept can be interpreted only when the data set contains both positive and negative x-values.

Section 4.3: The least-squares regression line should be used for predictions only for values of x that lie within the range of the data used to compute the equation of the least-squares line. Making predictions outside the range of the data is called extrapolation, and such predictions are generally unreliable.

The difference between the observed value of an outcome variable and the value predicted by the least-squares regression line is called the residual. The sum of the squared residuals is smaller for the least-squares regression line than for any other line. This fact is summarized by saying that the least-squares line satisfies the least-squares property.

A plot of residuals versus values of the explanatory variable is called a residual plot. When a residual plot has no apparent pattern, a linear model is appropriate. The correlation coefficient can be misleading in this regard, because the correlation may be large even when the relationship is not linear.

The least-squares regression line explains some of the variation in the outcome variable. The proportion of variance explained is called the coefficient of determination and is equal to the square of the correlation coefficient.

Vocabulary and Notation

Important Formulas

Correlation coefficient:

$$r = \frac{1}{n-1}\sum\left(\frac{x-\bar{x}}{s_x}\right)\left(\frac{y-\bar{y}}{s_y}\right)$$

Slope of least-squares regression line:

$$b_1 = r\frac{s_y}{s_x}$$

y-intercept of least-squares regression line:

$$b_0 = \bar{y} - b_1\bar{x}$$

Equation of least-squares regression line:

$$\hat{y} = b_0 + b_1 x$$

Chapter Quiz

1. Compute the correlation coefficient for the following data set. *−0.959*

x	2	5	6	7	11
y	15	9	6	4	1

2. The number of theaters showing the movie *Avengers: Endgame* x days after its opening is presented in the following table.

x	Number of Theaters
10	4662
22	4220
29	3810
36	3015
42	2121
49	1450
56	985

Source: Box Office Mojo

Construct a scatterplot with the number of days on the horizontal axis and the number of theaters on the vertical axis.

3. Use the data in Exercise 2 to compute the correlation between the number of days after the opening of the movie and the number of theaters showing the movie. Is the association positive or negative? Weak or strong? *−0.980; strong negative*

4. A scatterplot has a correlation of $r = -1$. Describe the pattern of the points. *Line with negative slope*

5. In a survey of U.S. cities, it is discovered that there is a positive correlation between the number of paved streets in the city and the number of registered cars. Does this mean that paving more streets in the city will result in an increase in the number of registered cars? Explain. *No*

6. The following table presents the average delay in minutes for departures and arrivals of domestic flights at O'Hare Airport in Chicago for selected years.

Average Delay in Departures	Average Delay in Arrivals
63.7	72.4
57.9	64.4
56.7	64.4
59.4	68.5
58.4	67.4
60.7	70.8

Source: Bureau of Transportation Statistics

Compute the least-squares regression line for predicting the delay in arrival time from the delay in departure time.
$\hat{y} = -6.195 + 1.2474x$

7. Use the least-squares regression line computed in Exercise 6 to predict the average delay in arrival time in a year when the average delay in departure time is 58.5 minutes. *66.78*

8. Use the least-squares regression line computed in Exercise 6 to compute the residual for the year when the average delay in departure time was 58.4 minutes and the average delay in arrival time was 67.4 minutes. *0.747*

9. Refer to Exercise 6. If the average delay in departure times differs by 2 minutes from one year to the next, by how much would you predict the average delay in arrival times to change? *2.495*

10. A scatterplot has a least-squares regression line with a slope of 0. What is the correlation coefficient? *0*

11. Compute the least-squares regression line for the following data set. $\hat{y} = 6.0667 - 0.6x$

x	0	1	3	4	7	9
y	7	5	4	3	2	1

12. Two lines are drawn on a scatterplot. The sum of squared residuals for line A is 558.2, and the sum of squared residuals for line B is 723.1. Which of the following is true about the sum of squared residuals for the least-squares regression line? *iii*

i. It will be greater than 723.1.
ii. It will be between 558.2 and 723.1.
iii. It will be less than or equal to 558.2.

13. A sample of students was studied to determine the relationship between sleeping habits and classroom performance. The least-squares regression line for predicting the score on a standardized exam from hours of sleep was computed to be $\hat{y} = 35.6 + 6.8x$, where x is the number of hours of sleep and $\hat{y}$ is the predicted exam score. The number of hours of sleep among the students in the sample ranged from 5 to 8. Should a prediction for the exam score be made for a student who slept for 10 hours? Why or why not? *No*

14. In a scatterplot, the point (−2, 7) is influential. If this point is removed from the scatterplot, which of the following describes the effect on the least-squares regression line? *i*
    i. It will shift its position by a substantial amount.
    ii. It will shift its position slightly.
    iii. It will not shift its position at all.

15. The correlation coefficient for a data set is $r = -0.6$. How much of the variation in the outcome variable is explained by the least-squares regression line? *36%*

Review Exercises

1. **Predicting height:** The heights (y) and lengths of forearms (x) were measured in inches for a sample of men. The following summary statistics were obtained:

$$\bar{x} = 10.1 \quad s_x = 0.8 \quad \bar{y} = 70.1 \quad s_y = 2.5 \quad r = 0.81$$

   a. Compute the least-squares regression line for predicting height from forearm length. *$\hat{y} = 44.534 + 2.5313x$*
   b. Joe's forearm is 1 inch longer than Sam's. How much taller than Sam do you predict Joe to be? *2.5313 inches*
   c. Predict the height of a man whose forearm is 9.5 inches long. *68.581 inches*

2. **How much wood is in that tree?** For a sample of 12 trees, the volume of lumber (y) (in cubic meters) and the diameter (x) (in centimeters) at a fixed height above ground level was measured. The following summary statistics were obtained:

$$\bar{x} = 36.1 \quad s_x = 8.8 \quad \bar{y} = 0.86 \quad s_y = 0.49 \quad r = 0.94$$

   a. Compute the least-squares regression line for predicting volume from diameter. *$\hat{y} = -1.0295 + 0.0523x$*
   b. If the diameters of two trees differ by 8 centimeters, by how much do you predict their volumes to differ? *0.4187*
   c. Predict the volume for a tree whose diameter is 44 centimeters. *1.2735*

3. **How's your mileage?** Weight (in tons) and fuel economy (in mpg) were measured for a sample of seven diesel trucks. The results are presented in the following table.

Weight	8.00	24.50	27.00	14.50	28.50	12.75	21.25
Mileage	7.69	4.97	4.56	6.49	4.34	6.24	4.45

Source: Janet Yanowitz, Ph.D. thesis, Colorado School of Mines

   a. Compute the least-squares regression line for predicting mileage from weight. *$\hat{y} = 8.5593 - 0.1551x$*
   b. Construct a residual plot. Verify that a linear model is appropriate.
   c. If two trucks differ in weight by 5 tons, by how much would you predict their mileages to differ? *0.7755*
   d. Predict the mileage for trucks with a weight of 15 tons. *6.2328*

4. **How's your mileage?** Using the data in Exercise 3:
   a. Compute the correlation coefficient between weight and mileage. *−0.945*
   b. Compute the coefficient of determination. *0.892*
   c. How much of the variation in mileage is explained by the least-squares regression line? *89.2%*

5. **Energy efficiency:** A sample of 10 households was monitored for one year. The household income (in $1000s) and the amount of energy consumed (in 10^{10} joules) were determined. The results follow.

Income	31	40	28	48	195	96	70	100	145	78
Energy	16	40	30	46	185	98	94	77	115	67

   a. Compute the least-squares regression line for predicting energy consumption from income. *$\hat{y} = 2.8827 + 0.8895x$*
   b. Construct a residual plot. Verify that a linear model is appropriate.
   c. If two families differ in income by $12,000, by how much would you predict their energy consumptions to differ? *10.674*
   d. Predict the energy consumption for a family whose income is $50,000. *47.358*

6. **Energy efficiency:** Using the data in Exercise 5:
   a. Compute the correlation coefficient between income and energy consumption. *0.960*
   b. Compute the coefficient of determination. *0.921*
   c. How much of the variation in energy consumption is explained by the least-squares regression line? *92.1%*

7. **Pigskin:** In football, a turnover occurs when a team loses possession of the ball due to a fumble or an interception. Turnovers are bad when they happen to your team, but good when they happen to your opponent. The turnover margin for a team is the difference (Turnovers by opponent − Turnovers by team). The following table presents the turnover margin and the total number of wins for each team in the Southeastern Conference (SEC) in a recent season.

Team	Turnover Margin	Wins	Team	Turnover Margin	Wins
Florida	22	13	S. Carolina	−11	7
Alabama	6	12	Vanderbilt	9	7
Georgia	−3	10	Arkansas	−9	5
Ole Miss	−2	9	Auburn	−8	5
LSU	−1	8	Tennessee	2	5
Kentucky	5	7	Miss. State	−4	4

Source: www.secsports.com

a. Compute the least-squares regression line for predicting team wins from turnover margin. *$\hat{y} = 7.5659 + 0.2015x$*
b. Construct a residual plot. Verify that a linear model is appropriate.
c. Which teams won more games than would be predicted from their turnover margin? *Florida, Alabama, Georgia, Ole Miss, LSU, South Carolina*

8. Pigskin: Using the data in Exercise 7:
a. Compute the correlation coefficient between turnover margin and wins. *0.643*
b. Compute the coefficient of determination. *0.413*
c. How much of the variation in wins for this SEC college football season is explained by the least-squares regression line? *41.3%*

9. SAT scores: The following table presents the number of years of study in English and language arts and the average SAT writing score for students who took the SAT exam.

Years of Study	0.5	1.0	2.0	3.0	4.0
SAT Score	419	444	463	456	498

Source: The College Board

a. Compute the least-squares regression line for predicting mean SAT score from years of study. *$\hat{y} = 417.2012 + 18.476x$*
b. Construct a residual plot. Verify that a linear model is appropriate.
c. Predict the mean SAT score for students with 2.5 years of study. *463*

10. SAT scores: Using the data in Exercise 9:
a. Compute the correlation coefficient between years of study and SAT score. *0.917*
b. Compute the coefficient of determination. *0.842*
c. How much of the variation in SAT score is explained by the least-squares regression line? *84.2%*

11. Baby weights: The average gestational age (time from conception to birth) of a newborn infant is about 40 weeks. The following table presents the gestational age (in weeks) and corresponding mean birth weight (in pounds) for female infants born in Canada.

Gestational Age	36	37	38	39	40	41	42	43
Birth Weight	6.1	6.6	7.0	7.4	7.7	7.9	8.0	8.1

Source: *Pediatrics* Vol. 108, No. 2

a. Compute the least-squares regression line for predicting the birth weight from the gestational age. *$\hat{y} = -3.9357 + 0.2857x$*
b. Compute the coefficient of determination. *0.937*
c. Construct a residual plot. Does the relationship appear to be linear? *No*
d. True or false: If the coefficient of determination is large, the relationship must be linear. *False*

12. Commute times: Every morning, Tania leaves for work a few minutes after 7:00 A.M. For eight days, she keeps track of the time she leaves (the number of minutes after 7:00) and the number of minutes it takes her to get to work. Following are the results.

Time Leaving	13	14	16	30	20	12	9	17	16	10	16
Length of Commute	27	20	23	45	20	21	20	28	27	23	30

a. Construct a scatterplot of the length of commute (y) versus the time leaving (x).
b. Compute the least-squares regression line for predicting the length of commute from the time leaving. *$\hat{y} = 9.7885 + 1.0192x$*
c. Compute the coefficient of determination. *0.635*
d. Which point is an outlier? *(30, 45)*
e. Remove the outlier, and compute the least-squares regression line for predicting the length of commute from the time leaving. *$\hat{y} = 19.5162 + 0.3066x$*
f. Is the outlier influential? Explain. *Yes*
g. Compute the coefficient of determination for the data set with the outlier removed. Is the relationship stronger, weaker, or about equally strong without the outlier? *0.0744; weaker*

13. Interpret technology: The following display from the TI-84 Plus calculator presents the results from computing a least-squares regression line.

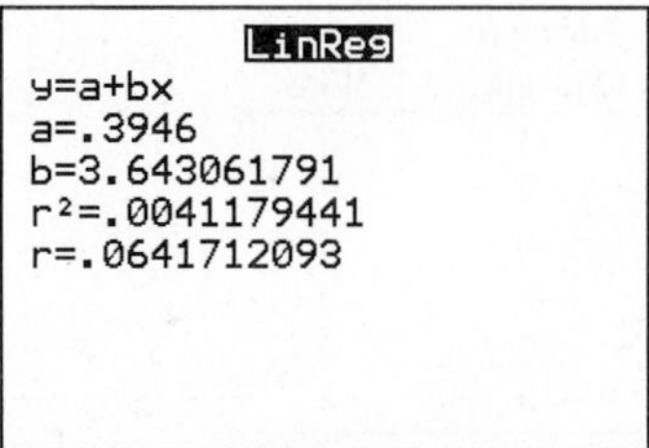

a. Write the equation of the least-squares regression line. *$\hat{y} = 0.3946 + 3.643061791x$*
b. Predict the value of y when the x-value is 10. *36.825*
c. What is the correlation between x and y? *0.064*
d. Is the linear relationship between x and y strong or weak? Explain. *Weak*

14. Interpret technology: The following display from the TI-84 Plus calculator presents the results from computing a least-squares regression line.

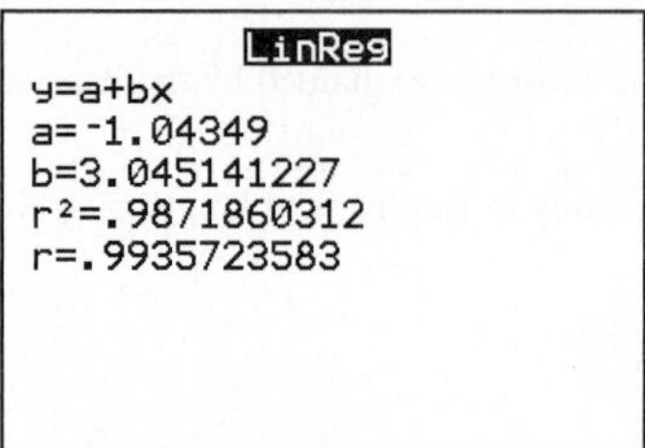

a. Write the equation of the least-squares regression line. *$\hat{y} = -1.04349 + 3.045141227x$*
b. Predict the value of y when the x-value is 50. *151.21*
c. What is the correlation between x and y? *0.994*
d. Is the linear relationship between x and y strong or weak? Explain. *Strong*

15. Interpret technology: The following output from MINITAB presents the results from computing a least-squares regression line.

The regression equation is
Y = 4.99971 + 0.20462 X

Predictor	Coef	SE Coef	T	P
Constant	4.99971	0.02477	201.81	0.000
X	0.20462	0.01115	18.36	0.000

a. Write the equation of the least-squares regression line. *$\hat{y} = 4.99971 + 0.20462x$*
b. Predict the value of y when the x-value is 25. *10.1*

Write About It

1. Describe an example in which two variables are strongly correlated, but changes in one do not cause changes in the other.
2. Two variables x and y have a positive association if large values of x are associated with large values of y. Write an equivalent definition that describes what small values of x are associated with. Then write a definition for negatively associated random variables that describes what small values of y are associated with.
3. Explain why the predicted value $\hat{y}$ is always equal to $\bar{y}$ when $r = 0$.
4. If the slope of the least-squares regression line is negative, can the correlation coefficient be positive? Explain why or why not.
5. Describe conditions under which the slope of the least-squares line will be equal to the correlation coefficient.
6. Describe circumstances under which the sum of the squared residuals will equal zero. What conclusions can be drawn about the least-squares regression line in this case?
7. Explain why extrapolation may lead to unreliable results.
8. Explain how it is possible for a point to be an outlier without being an influential point.
9. Consider the case where there are only two ordered pairs in the data set. What must be true about the residuals?

In-Class Activities

NOTE TO INSTRUCTOR

The activity involving measuring heights and forearm lengths presents an example of confounding. Because men tend to be taller and have longer forearms, the correlation is higher for the combined group than for either one separately. Thus the correlation for the whole group is due in part to variation in gender, which is a confounder.

NOTE TO INSTRUCTOR

Anscombe's quartet illustrates the important point that there is much information about the shape of a data set that is not contained in numerical summaries.

1. **Causation, or just correlation?** Think of some quantities that have been increasing over time (e.g., population) and some that have been decreasing (e.g., the value of a used car). Any two quantities that have been moving in the same direction will be positively correlated, and any two that have been moving in opposite directions will be negatively correlated. Discuss the degree to which causation is present in each case.
2. **Heights and forearms:** For each student, measure both height and forearm length. Compute the correlation between them. Now compute the correlation separately for men and for women. Are the separate correlations higher or lower than the correlation for the whole data set? Why? Draw a scatterplot of height versus forearm length, using different symbols for men and women.
3. **Anscombe's quartet:** Statistician Francis Anscombe developed four data sets that demonstrate the importance of graphing data in addition to computing the correlation and least-squares line. Following are the data sets. Verify that the correlations and least-squares lines are the same for each, to two decimal places. Construct scatterplots for each, and discuss the differences among the data sets, which are not detected by the correlation or least-squares regression line.

Data Set 1		Data Set 2		Data Set 3		Data Set 4	
x	y	x	y	x	y	x	y
10.0	8.04	10.0	9.14	10.0	7.46	8.0	6.58
8.0	6.95	8.0	8.14	8.0	6.77	8.0	5.76
13.0	7.58	13.0	8.74	13.0	12.74	8.0	7.71
9.0	8.81	9.0	8.77	9.0	7.11	8.0	8.84
11.0	8.33	11.0	9.26	11.0	7.81	8.0	8.47
14.0	9.96	14.0	8.10	14.0	8.84	8.0	7.04
6.0	7.24	6.0	6.13	6.0	6.08	8.0	5.25
4.0	4.26	4.0	3.10	4.0	5.39	19.0	12.50
12.0	10.84	12.0	9.13	12.0	8.15	8.0	5.56
7.0	4.82	7.0	7.26	7.0	6.42	8.0	7.91
5.0	5.68	5.0	4.74	5.0	5.73	8.0	6.89

4. **Correlation is definitely not causation:** Following are six data sets, collected during each of the years 2000–2009. Each of the first three is highly correlated with one of the second three. Can you guess which is correlated with which?

   A. German passenger cars sold in the United States (in thousands)
   B. Per capita consumption of turkey (in pounds)
   C. Power generated by nuclear power plants (billions of kilowatt-hours)

   1. Total revenue generated by U.S. skiing facilities (millions of $)
   2. Number of fatalities while boarding a bus
   3. Divorce rate in Connecticut per 1000 people

Year	A	B	C	1	2	3
2000	863	13.7	753.9	1551	1	3.3
2001	837	13.8	768.8	1635	1	3.2
2002	930	14.0	780.1	1801	1	3.3
2003	830	13.7	763.7	1827	2	3.2
2004	810	13.5	788.5	1956	1	3.1
2005	923	13.2	782.0	1989	2	3.0
2006	1154	13.3	787.2	2178	8	3.1
2007	1183	13.8	806.4	2257	7	3.2
2008	1142	13.9	806.2	2476	7	3.4
2009	829	13.3	798.9	2438	1	3.0

Source: http://tylervigen.com/spurious-correlations

Students may guess which are correlated with which (one can only guess), then compute the correlations to see which are in fact highly correlated. Construct scatterplots for the highly correlated pairs of variables, and discuss for which pairs the correlation coefficient is an appropriate summary.

These correlations were discovered by considering thousands of pairs of variables and choosing a few whose correlations were large. In Section 1.1 we described statistics as a process in which the first step is to formulate questions and the second is to collect data. Did this method of finding correlations follow this process? Discuss how this might have led to the discovery of correlations that do not reflect a real relationship.

Case Study: How Are Inflation And Unemployment Related?

The following table, reproduced from the chapter introduction, presents the inflation rate and unemployment rate, both in percent, for the years 1991–2018.

Year	Inflation	Unemployment	Year	Inflation	Unemployment
1991	3.1	6.8	2005	3.4	5.1
1992	2.9	7.5	2006	2.5	4.6
1993	2.7	6.9	2007	4.1	4.6
1994	2.7	6.1	2008	0.1	5.8
1995	2.5	5.6	2009	2.7	9.3
1996	3.3	5.4	2010	1.5	9.6
1997	1.7	4.9	2011	3.0	8.9
1998	1.6	4.5	2012	1.7	8.1
1999	2.7	4.2	2013	1.5	7.4
2000	3.4	4.0	2014	0.8	6.2
2001	1.6	4.7	2015	0.7	5.3
2002	2.4	5.8	2016	2.1	4.9
2003	1.9	6.0	2017	2.1	4.4
2004	3.3	5.5	2018	1.9	3.9

Source: Bureau of Labor Statistics

We will investigate some methods for predicting unemployment. First, we will try to predict the unemployment rate from the inflation rate.

1. Construct a scatterplot of unemployment (y) versus inflation (x). Do you detect any strong nonlinearity? *No*
2. Compute the least-squares line for predicting unemployment from inflation. $\hat{y} = 6.1932 - 0.1160x$
3. Predict the unemployment in a year when inflation is 3.0%. *5.85*
4. Compute the correlation coefficient between inflation and unemployment. *−0.0669*
5. What proportion of the variance in unemployment is explained by inflation? *0.447%*

The relationship between inflation and unemployment is not very strong. However, if we are interested in predicting unemployment, we would probably want to predict next year's unemployment from this year's inflation. We can construct an equation to do this by matching each year's inflation with the next year's unemployment, as shown in the following table.

Year	This Year's Inflation	Next Year's Unemployment	Year	This Year's Inflation	Next Year's Unemployment
1991	3.1	7.5	2005	3.4	4.6
1992	2.9	6.9	2006	2.5	4.6
1993	2.7	6.1	2007	4.1	5.8
1994	2.7	5.6	2008	0.1	9.3
1995	2.5	5.4	2009	2.7	9.6
1996	3.3	4.9	2010	1.5	8.9
1997	1.7	4.5	2011	3.0	8.1
1998	1.6	4.2	2012	1.7	7.4
1999	2.7	4.0	2013	1.5	6.2
2000	3.4	4.7	2014	0.8	5.3
2001	1.6	5.8	2015	0.7	4.9
2002	2.4	6.0	2016	2.1	4.4
2003	1.9	5.5	2017	2.1	3.9
2004	3.3	5.1			

Source: Bureau of Labor Statistics

6. Compute the least-squares line for predicting next year's unemployment from this year's inflation. $\hat{y} = 6.5565 - 0.2875x$

7. Predict next year's unemployment if this year's inflation is 3.0%. 5.69

8. Compute the correlation coefficient between this year's inflation and next year's unemployment. −0.166

9. What proportion of the variance in next year's unemployment is explained by this year's inflation? 2.76%

If we are going to use data from this year to predict unemployment next year, why not use this year's unemployment to predict next year's unemployment? A model like this, in which previous values of a variable are used to predict future values of the same variable, is called an *autoregressive* model. The following table presents the data needed to fit this model.

Year	This Year's Unemployment	Next Year's Unemployment	Year	This Year's Unemployment	Next Year's Unemployment
1991	6.8	7.5	2005	5.1	4.6
1992	7.5	6.9	2006	4.6	4.6
1993	6.9	6.1	2007	4.6	5.8
1994	6.1	5.6	2008	5.8	9.3
1995	5.6	5.4	2009	9.3	9.6
1996	5.4	4.9	2010	9.6	8.9
1997	4.9	4.5	2011	8.9	8.1
1998	4.5	4.2	2012	8.1	7.4
1999	4.2	4.0	2013	7.4	6.2
2000	4.0	4.7	2014	6.2	5.3
2001	4.7	5.8	2015	5.3	4.9
2002	5.8	6.0	2016	4.9	4.4
2003	6.0	5.5	2017	4.4	3.9
2004	5.5	5.1			

Source: Bureau of Labor Statistics

10. Compute the least-squares line for predicting next year's unemployment from this year's unemployment. $\hat{y} = 0.7864 + 0.8511$

11. Predict next year's unemployment if this year's unemployment is 4.0%. 4.19%

12. Compute the correlation coefficient between this year's unemployment and next year's unemployment. 0.829

13. What proportion of the variance in next year's unemployment is explained by this year's unemployment? 68.7%

14. Which of the three models do you think provides the best prediction of unemployment, the one using inflation in the same year, the one using inflation in the previous year, or the one using unemployment in the previous year? Explain.

15. The U.S. economy was in recession during part of 1991, most of 2001, and from December 2007 through June 2009. Compute the residuals for the least-squares line found in Exercise 10. During times of recession, does the least-squares line predict the following year's unemployment well? Does it tend to overpredict or underpredict? *Underpredict*

chapter

5

Probability

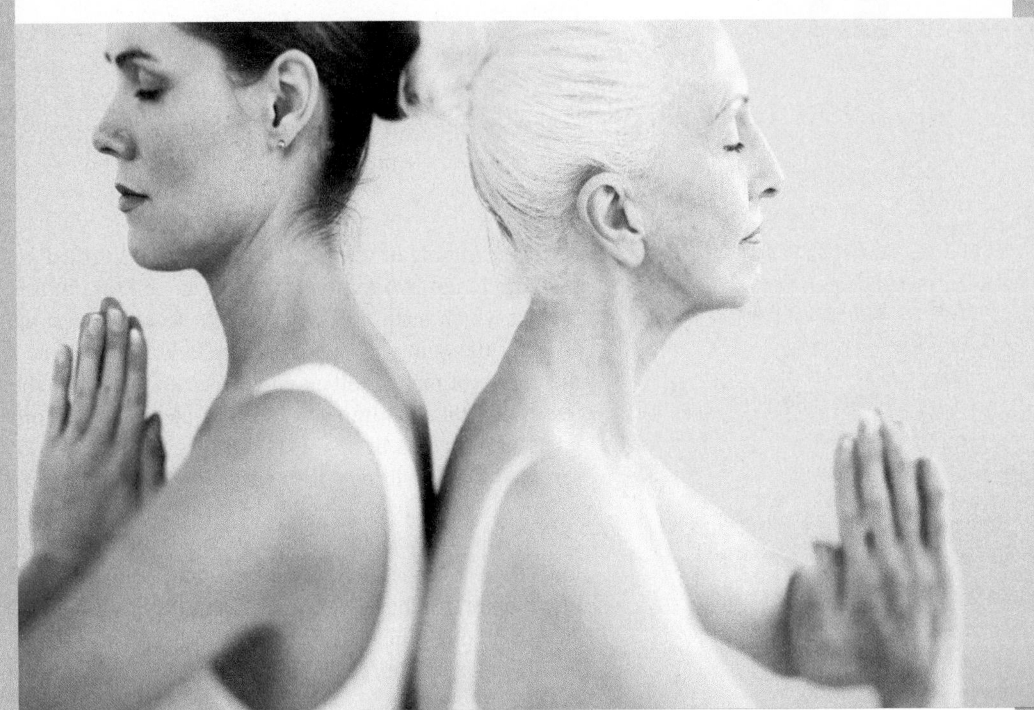

Ryan McVay/Getty Images

Introduction

How likely is it that you will live to be 100 years old? The following table, called a *life table*, can be used to answer this question.

United States Life Table, Total Population

Age Interval	Proportion Surviving	Age Interval	Proportion Surviving
0–10	0.99123	50–60	0.94010
10–20	0.99613	60–70	0.86958
20–30	0.99050	70–80	0.70938
30–40	0.98703	80–90	0.42164
40–50	0.97150	90–100	0.12248

Source: Centers for Disease Control and Prevention

The column labeled "Proportion Surviving" presents the proportion of people alive at the beginning of an age interval who will still be alive at the end of the age interval. For example, among those currently age 20, the proportion who will still be alive at age 30 is 0.99050, or 99.050%. With an understanding of some basic concepts of probability, one can use the life table to compute the probability that a person of a given age will still be alive a given number of years from now. Life insurance companies use this information to determine how much to charge for life insurance policies. In the case study at the end of the chapter, we will use the life table to study some further questions that can be addressed with the methods of probability.

This chapter presents an introduction to probability. Probability is perhaps the only branch of knowledge that owes its existence to gambling. In the seventeenth century, owners of gambling houses hired some of the leading mathematicians of the time to calculate the chances that players would win certain gambling games. Later, people realized that many real-world problems involve chance as well, and since then the methods of probability have been used in almost every area of knowledge.

Section 5.1 Basic Concepts of Probability

Objectives

1. Construct sample spaces
2. Compute and interpret probabilities
3. Approximate probabilities by using the Empirical Method
4. Approximate probabilities by using simulation

NOTE TO INSTRUCTOR

The following topics from the **Statistics Corequisite Workbook** are aligned with the material in this section.

3.1 - Approximating Probabilities with Proportions and Unusual Events

The following topic may also be helpful.

A.2 - Interpreting Scientific Notation

At the beginning of a football game, a coin is tossed to decide which team will get the ball first. There are two reasons for using a coin toss in this situation. First, it is impossible to predict which team will win the coin toss, because there is no way to tell ahead of time whether the coin will land heads or tails. The second reason is that in the long run, over the course of many football games, we know that the home team will win about half of the tosses and the visiting team will win about half. In other words, although we don't know what the outcome of a single coin toss will be, we do know what the outcome of a long series of tosses will be—they will come out about half heads and half tails.

A coin toss is an example of a **probability experiment**. A probability experiment is one in which we do not know what any individual outcome will be, but we do know how a long series of repetitions will come out. Another familiar example of a probability experiment is the rolling of a die. A die has six faces; the faces have from one to six dots. We cannot predict which face will turn up on a single roll of a die, but, assuming the die is evenly balanced (not loaded), we know that in the long run, each face will turn up one-sixth of the time.

The *probability* of an event is the proportion of times that the event occurs in the long run. So, for a "fair" coin, that is, one that is equally likely to come up heads as tails, the probability of heads is $1/2$ and the probability of tails is $1/2$.

DEFINITION

The **probability** of an event is the proportion of times the event occurs in the long run, as a probability experiment is repeated over and over again.

The South African mathematician John Kerrich carried out a famous study that illustrates the idea of the long-run proportion. Kerrich was in Denmark when World War II broke out and spent the war interned in a prisoner-of-war camp. To pass the time, he carried out a series of probability experiments, including one in which he tossed a coin 10,000 times and recorded each toss as a head or a tail.

Figure 5.1 (page 215) summarizes a computer-generated re-creation of Kerrich's study, in which the proportion of heads is plotted against the number of tosses. For example, it turned out that after 5 tosses, 3 heads had appeared, so the proportion of heads was $3/5 = 0.6$. After 100 tosses, 49 heads had appeared, so the proportion of heads was $49/100 = 0.49$. After 10,000 tosses, the proportion of heads was 0.4994, which is very close to the true probability of 0.5. The figure shows that the proportion varies quite a bit within the first few tosses, but the proportion settles down very close to 0.5 as the number of tosses becomes larger.

The fact that the long-run proportion approaches the probability is called the *law of large numbers*.

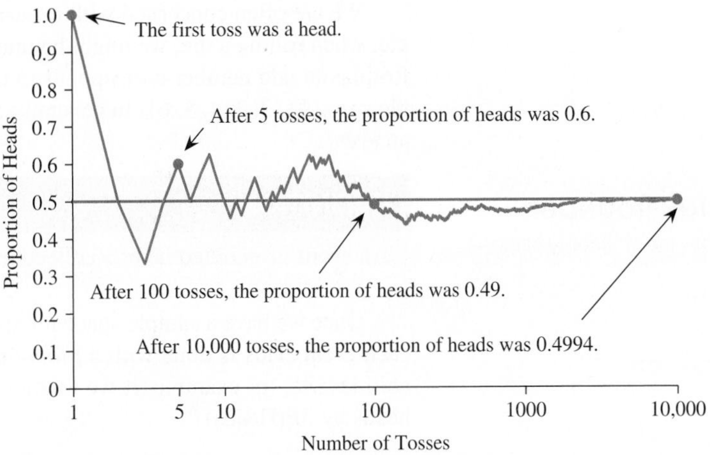

Figure 5.1 As the number of tosses increases, the proportion of heads fluctuates around the true probability of 0.5 and gets closer to 0.5. The horizontal axis is not drawn to scale.

EXPLAIN IT AGAIN

Law of large numbers: The law of large numbers is another way to state our definition of probability.

Law of Large Numbers

The **law of large numbers** says that as a probability experiment is repeated again and again, the proportion of times that a given event occurs will approach its probability.

Objective 1 Construct sample spaces

Probability Models

To study probability formally, we need some basic terminology. The collection of all the possible outcomes of a probability experiment is called a *sample space*.

DEFINITION

A **sample space** contains all the possible outcomes of a probability experiment.

Example 5.1

Describe sample spaces

Describe a sample space for each of the following experiments.

a. The toss of a coin
b. The roll of a die
c. Selecting a student at random from a list of 10,000 students at a large university
d. Selecting a simple random sample of 100 students from a list of 10,000 students

Solution

a. There are two possible outcomes for the toss of a coin: Heads and Tails. So a sample space is {Heads, Tails}.

b. There are six possible outcomes for the roll of a die: the numbers from 1 to 6. So a sample space is {1, 2, 3, 4, 5, 6}.

c. Each of the 10,000 students is a possible outcome for this experiment, so the sample space consists of the 10,000 students.

d. This sample space consists of every group of 100 students that can be chosen from the population of 10,000—in other words, every possible simple random sample of size 100. This is a huge number of outcomes; it can be written approximately as a 6 followed by 241 zeros. This is larger than the number of atoms in the universe.

We are often concerned with occurrences that consist of several outcomes. For example, when rolling a die, we might be interested in the possibility of rolling an odd number. Rolling an odd number corresponds to the collection of outcomes $\{1, 3, 5\}$ from the sample space $\{1, 2, 3, 4, 5, 6\}$. In general, a collection of outcomes of a sample space is called an *event*.

NOTATION ROUNDUP

P(A) denotes the probability of the event *A*.

DEFINITION

An **event** is an outcome or a collection of outcomes from a sample space.

Once we have a sample space for an experiment, we need to specify the probability of each event. This is done with a *probability model*. We use the letter "*P*" to denote probabilities. So, for example, if we toss a coin, we denote the probability that the coin lands heads by "*P*(Heads)."

DEFINITION

A **probability model** for a probability experiment consists of a sample space, along with a probability for each event.

Notation: If A denotes an event, the probability of the event A is denoted $P(A)$.

Objective 2 Compute and interpret probabilities

Probability models with equally likely outcomes

In many situations, the outcomes in a sample space are equally likely. For example, when we toss a coin, we usually assume that the two outcomes "Heads" and "Tails" are equally likely. We call such a coin a *fair* coin. Similarly, a fair die is one in which the numbers from 1 to 6 are equally likely to turn up. When the outcomes in a sample space are equally likely, we can use a simple formula to determine the probability of events.

EXPLAIN IT AGAIN

Fair and unfair: A fair coin or die is one for which all outcomes are equally likely. An unfair coin or die is one for which some outcomes are more likely than others.

Computing Probabilities with Equally Likely Outcomes

If a sample space has n **equally likely outcomes**, and an event A has k outcomes, then

$$P(A) = \frac{\text{Number of outcomes in } A}{\text{Number of outcomes in the sample space}} = \frac{k}{n}$$

Example 5.2

Compute the probability of an event

A fair die is rolled. Find the probability that an odd number comes up.

Solution

The sample space has six equally likely outcomes: $\{1, 2, 3, 4, 5, 6\}$. The event of an odd number has three outcomes: $\{1, 3, 5\}$. The probability is

$$P(\text{odd number}) = \frac{3}{6} = \frac{1}{2}$$

Example 5.3

Compute the probability of an event

In the Georgia Cash-4 Lottery game, a winning number between 0000 and 9999 is chosen at random, with all the possible numbers being equally likely. What is the probability that all four digits of the winning number are the same?

Solution

The outcomes in the sample space are the numbers from 0000 to 9999, so there are 10,000 equally likely outcomes in the sample space. There are 10 outcomes for which all the digits are the same: 0000, 1111, and so on up to 9999. The probability is

$$P(\text{all four digits the same}) = \frac{10}{10{,}000} = 0.001$$

The law of large numbers states that the probability of an event is the long-run proportion of times that the event occurs. An event that never occurs, even in the long run, has a probability of 0. This is the smallest probability an event can have. An event that occurs every time has a probability of 1. This is the largest probability an event can have.

EXPLAIN IT AGAIN

Rules for the value of a probability: A probability can never be negative, and a probability can never be greater than 1.

SUMMARY

The probability of an event is always between 0 and 1. In other words, for any event A, $0 \le P(A) \le 1$.

If A cannot occur, then $P(A) = 0$.

If A is certain to occur, then $P(A) = 1$.

Example 5.4

Computing probabilities

A penny, a nickel, and a dime are tossed. Denoting a head by H and a tail by T, we can denote these three tosses in order. For example, HTH means the penny landed heads, the nickel landed tails, and the dime landed heads. There are eight possible outcomes: HHH, HHT, HTH, HTT, THH, THT, TTH, and TTT. Assume these outcomes are equally likely.

a. What is the probability that there are exactly two heads?

b. What is the probability that all three tosses are the same?

Solution

a. Of the eight equally likely outcomes, the three outcomes HHT, HTH, and THH correspond to having two heads. Therefore

$$P(\text{Two heads}) = \frac{3}{8}$$

b. Of the eight equally likely outcomes, the two outcomes HHH and TTT correspond to having all tosses the same. Therefore

$$P(\text{All three tosses are the same}) = \frac{2}{8} = \frac{1}{4}$$

Check Your Understanding

1. In Example 5.4, what is the probability that the penny comes up heads? 1/2

2. In Example 5.4, what is the probability that the penny and the dime come up the same? 1/2

Answers are on page 225.

Example 5.5

Constructing a sample space

Cystic fibrosis is a disease of the mucous glands whose most common sign is progressive damage to the respiratory system and digestive system. This disease is inherited, as follows. A certain gene may be of type A or type a. Every person has two copies of the gene—one inherited from the person's mother, one from the person's father. If both copies are a, the person will have cystic fibrosis. Assume that a mother and father both have genotype Aa, that is, one gene of each type. Assume that each copy is equally likely to be transmitted to their child. What is the probability that the child will have cystic fibrosis?

Solution

Most of the work in solving this problem is in constructing the sample space. We'll do this in two ways. First, the tree diagram in Figure 5.2 shows that there are four possible outcomes. In the tree diagram, the first two branches indicate the two possible outcomes, A and a, for the mother's gene. Then for each of these outcomes there are two branches indicating the possible outcomes for the father's gene. An alternate method is to construct a table like Table 5.1.

Table 5.1

Mother's Gene	Father's Gene	Child's Genotype
A	A	AA
A	a	Aa
a	A	aA
a	a	aa

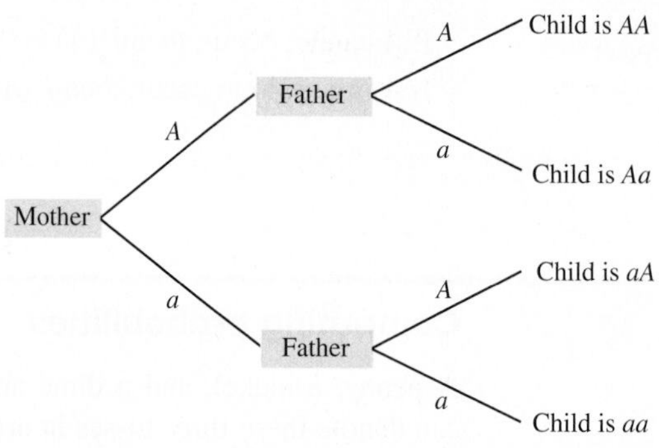

Figure 5.2 Tree diagram illustrating the four outcomes for the child's genotype

We can use either the table or the tree to list the outcomes. Listing the mother's gene first, the four outcomes are AA, Aa, aA, and aa. For one of the four outcomes, aa, the child will have cystic fibrosis. Therefore, the probability of cystic fibrosis is $1/4$.

Check Your Understanding

3. A penny and a nickel are tossed. Each is a fair coin, which means that heads and tails are equally likely.

a. Construct a sample space containing equally likely outcomes. Each outcome should specify the results for both coins. {HH, HT, TH, TT}

b. Find the probability that one coin comes up heads and the other comes up tails. 1/2

Answers are on page 225.

Sampling from a population is a probability experiment

In Section 1.1, we learned that statisticians collect data by drawing samples from populations. Sampling an individual from a population is a probability experiment. The population is the sample space, and the members of the population are equally likely outcomes. For this reason, the ideas of probability are fundamental to statistics.

Example 5.6

Computing probabilities involving sampling

There are 10,000 families in a certain town. They are categorized by their type of housing as follows.

Own a house	4753
Own a condo	1478
Rent a house	912
Rent an apartment	2857

A pollster samples a single family at random from this population.

a. What is the probability that the sampled family owns a house?

b. What is the probability that the sampled family rents?

Solution

a. The sample space consists of the 10,000 households. Of these, 4753 own a house, so the probability that the sampled family owns a house is

$$P(\text{Owns a house}) = \frac{4753}{10{,}000} = 0.4753$$

b. The number of families who rent is $912 + 2857 = 3769$. Therefore, the probability that the sampled family rents is

$$P(\text{Rents}) = \frac{3769}{10{,}000} = 0.3769$$

NOTE TO INSTRUCTOR

The concept of an "unusual event" foreshadows the concept of statistical significance. This can be made explicit by referring to an unusual value as "significantly high" or "significantly low."

In practice, of course, the pollster would sample many people, not just one. In fact, statisticians use the basic ideas of probability to draw conclusions about populations by studying samples drawn from them. In later chapters of this book, we will see how this is done.

Unusual events

As the name implies, an unusual event is one that is not likely to happen—in other words, an event whose probability is small. There are no hard-and-fast rules as to just how small a probability needs to be before an event is considered unusual, but 0.05 is commonly used.

EXPLAIN IT AGAIN

Unusual events: The cutoff value for the probability of an unusual event can be any small value that seems appropriate for a specific situation. The most commonly used value is 0.05.

SUMMARY

An **unusual event** is one whose probability is small.

Sometimes people use the cutoff 0.05; that is, they consider any event whose probability is less than 0.05 to be unusual. But there are no hard-and-fast rules about this.

Example 5.7

Determine whether an event is unusual

In a college of 5000 students, 150 are math majors. A student is selected at random and turns out to be a math major. Is this an unusual event?

NOTE TO INSTRUCTOR

Although there are no hard-and-fast rules as to what value to use as the cutoff for an event to be unusual, the value 0.05 will be used as the default for homework exercises.

Solution

The sample space consists of 5000 students, each of whom is equally likely to be chosen. The event of choosing a math major consists of 150 students. Therefore,

$$P(\text{Math major is chosen}) = \frac{150}{5000} = 0.03$$

Since the probability is less than 0.05, then by the most commonly applied rule, this would be considered an unusual event.

Objective 3 Approximate probabilities by using the Empirical Method

Approximating Probabilities with the Empirical Method

The law of large numbers says that if we repeat a probability experiment a large number of times, then the proportion of times that a particular outcome occurs is likely to be close to the true probability of the outcome. The **Empirical Method** consists of repeating an experiment a large number of times and using the proportion of times an outcome occurs to approximate the probability of the outcome.

Example 5.8

Approximate the probability that a newborn baby is low birth weight

The Centers for Disease Control reports that in a recent year there were 313,752 births in the United States to low birth weight babies (less than 2500 grams), while 3,477,960 were to babies with weight greater than 2500 grams. Approximate the probability that a newborn baby is low birth weight.

EXPLAIN IT AGAIN

The Empirical Method is only approximate: The Empirical Method does not give us the exact probability. But the larger the number of replications of the experiment, the more reliable the approximation will be.

Solution

We compute the number of times the experiment has been repeated:

3,477,960	not low birth weight
+ 313,752	low birth weight
= 3,791,712	births

The proportion of births that are low birth weight is

$$\frac{313{,}752}{3{,}791{,}712} = 0.0827$$

We approximate $P(\text{low birth weight}) \approx 0.0827$.

Example 5.8 is based on a very large number (3,791,712) of replications. The law of large numbers says that the proportion of outcomes approaches the true probability as the number of replications becomes large. For a number this large, we can be virtually certain that the proportion 0.0827 is very close to the true probability. Of course, changes in nutrition and other environmental factors can affect the probability in the future.

Check Your Understanding

4. There are 100,000 voters in a city. A pollster takes a simple random sample of 1000 of them and finds that 513 support a bond issue to support the public library and 487 oppose it. Estimate the probability that a randomly chosen voter in this city supports the bond issue. *0.513*

Answer is on page 225.

Objective 4 Approximate probabilities by using simulation

Simulation

In practice, it can be difficult or impossible to repeat an experiment many times in order to approximate a probability with the Empirical Method. In some cases, we can use technology to repeat an equivalent virtual experiment many times. Conducting a virtual experiment in this way is called **simulation**. Most statistical software packages, and many calculators, will perform simulations. Example 5.9 presents a step-by-step description of a simulation conducted using Excel.

Example 5.9

Approximating a probability with simulation

If three dice are rolled, the smallest possible total is 3 and the largest possible total is 18. We will perform a simulation to estimate the probability that the sum of three dice is equal to 10. If we had no technology, we could roll three dice a large number of times and compute the proportion of times that the sum was equal to 10. With technology, we can use a random number generator in place of dice. For this example, we will show how to use Excel to simulate rolling three dice 1000 times.

Solution

Step 1: Label the first three columns Die 1, Die 2, and Die 3. In the first cell in the Die 1 column, enter the command **=RANDBETWEEN(1,6)** and press **ENTER**. This will produce a random number between 1 and 6 in this cell. Copy this formula for 1000 rows. This will generate 1000 rolls of Die 1. Repeat the process for Die 2 and Die 3 to generate 1000 rolls of all three dice. The first roll of the three dice is 5, 3, 3; the second roll is 6, 1, 1; and so on. Figure 5.3 shows the result.

	A	B	C	D
1	=RANDBETWEEN(1,6)	=RANDBETWEEN(1,6)	=RANDBETWEEN(1,6)	
2				
3	Die 1	Die 2	Die 3	
4	5	3	3	
5	6	1	1	
6	3	5	5	
7	6	1	5	
8	4	4	3	
9	5	2	5	
10	5	1	4	
11	4	6	1	
12	3	1	4	
13	1	6	6	
14	3	4	3	
15	2	2	5	
16	3	6	6	
17	3	4	3	
18	1	1	3	
19	1	3	3	
20	6	3	4	
21	4	4	1	
22	2	1	2	
23	3	2	3	

Figure 5.3 Screenshot of window showing the first 20 rolls of three dice

Step 2: Label the fourth column SUM OF DICE. In the SUM OF DICE column, use the command **=SUM(*data_array*)** to sum the values of the three dice for each roll. Copy this formula for each row. The result is shown in Figure 5.4.

	A	B	C	D	E
1	=RANDBETWEEN(1,6)	=RANDBETWEEN(1,6)	=RANDBETWEEN(1,6)		=SUM(A4:C4)
2					
3	Die 1	Die 2	Die 3	SUM OF DICE	
4	5	3	3	11	
5	6	1	1	8	
6	3	5	5	13	
7	6	1	5	12	
8	4	4	3	11	
9	5	2	5	12	
10	5	1	4	10	
11	4	6	1	11	
12	3	1	4	8	
13	1	6	6	13	
14	3	4	3	10	
15	2	2	5	9	
16	3	6	6	15	
17	3	4	3	10	
18	1	1	3	5	
19	1	3	3	7	
20	6	3	4	13	
21	4	4	1	9	
22	2	1	2	5	
23	3	2	3	8	

Figure 5.4 Screenshot of window showing the sums of the first 20 rolls of three dice

Step 3: To construct the frequency distribution, enter the numbers 3–18 into the rows in an empty column. In the next column, the frequencies will be computed using the **=COUNTIF(*range, condition*)** command. For ***range***, enter the cell range for all of the sums computed in the previous step. The ***condition*** should be the value from the cell range that must be matched in order for it to be counted. Figure 5.5 (page 222) shows the result where the cell range is \$D\$4:\$D\$1003.

=COUNTIF($D$4:$D$1003, 3)

D	E	F	G	H
SUM OF DICE		VALUE	FREQUENCY	
11		3	2	
8		4	18	
13		5	21	
12		6	43	
11		7	70	
12		8	97	
10		9	111	
11		10	129	
8		11	111	
13		12	126	
10		13	102	
9		14	69	
15		15	53	
10		16	27	
5		17	14	
7		18	0	
13				
9				
5				
8				

Figure 5.5 Screenshot of window showing the frequencies

We estimate the probability of obtaining a total of 10 in a roll of three dice to be

$$P(10) = \frac{129}{1000} = 0.129$$

Using Technology

TI-84 PLUS

Simulating 100 rolls of a die

Step 1. Press **MATH**, scroll to the **PRB** menu, and select **5: randInt(**.

Step 2. Enter **1**, comma, **6**, comma, and then the number of rolls you wish to simulate (**100**). Close the parentheses (Figure A).

Step 3. To store the data in **L1**, press **STO**, then **2nd, 1**, then **ENTER** (Figure B).

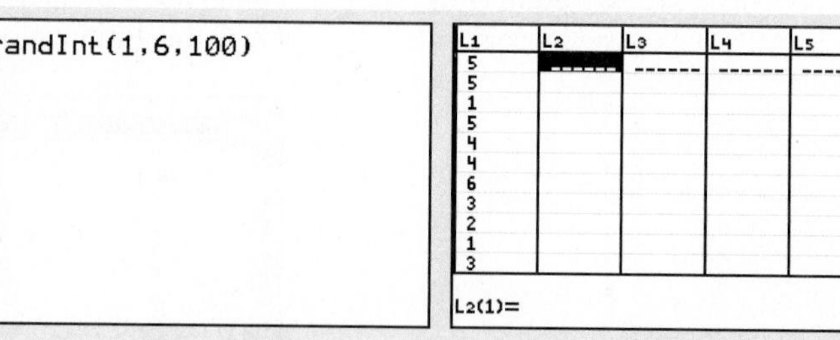

Figure A **Figure B**

EXCEL

Simulating 100 rolls of a die

Step 1. Click on a cell in the worksheet and type **=RANDBETWEEN(1, 6)**. Press **ENTER**.

Step 2. Copy and paste the formula into cells for the number of rolls you wish to simulate (100).

MINITAB

Simulating 100 rolls of a die

Step 1. Click **Calc**, then **Random Data**, then **Integer**.

Step 2. Enter **100** as the number of rows of data.

Step 3. Enter **C1** in the **Store in column(s)** field.

Step 4. Enter **1** as the **Minimum value** and **6** as the **Maximum value**.

Step 5. Click **OK**.

Section 5.1 Exercises

Exercises 1–4 are the Check Your Understanding exercises located within the section.

Understanding the Concepts

In Exercises 5–8, fill in each blank with the appropriate word or phrase.

5. If an event cannot occur, its probability is ________. *0*
6. If an event is certain to occur, its probability is ________. *1*
7. The collection of all possible outcomes of a probability experiment is called a ________. *sample space*
8. An outcome or collection of outcomes from a sample space is called an ________. *event*

In Exercises 9–12, determine whether the statement is true or false. If the statement is false, rewrite it as a true statement.

9. The law of large numbers states that as a probability experiment is repeated, the proportion of times that a given outcome occurs will approach its probability. *True*
10. If A denotes an event, then the sample space is denoted by $P(A)$. *False*
11. The Empirical Method can be used to calculate the exact probability of an event. *False*
12. For any event A, $0 \le P(A) \le 1$. *True*

Practicing the Skills

In Exercises 13–18, assume that a fair die is rolled. The sample space is {1, 2, 3, 4, 5, 6}, and all the outcomes are equally likely.

13. Find $P(2)$. *1/6*
14. Find P(Even number). *1/2*
15. Find P(Less than 3). *1/3*
16. Find P(Greater than 2). *2/3*
17. Find $P(7)$. *0*
18. Find P(Less than 10). *1*
19. A fair coin has probability 0.5 of coming up heads.
    a. If you toss a fair coin twice, are you certain to get one head and one tail? *No*
    b. If you toss a fair coin 100 times, are you certain to get 50 heads and 50 tails? *No*
    c. As you toss the coin more and more times, will the proportion of heads approach 0.5? *Yes*
20. Roulette wheels in Nevada have 38 pockets. They are numbered 0, 00, and 1 through 36. On each spin of the wheel, a ball lands in a pocket, and each pocket is equally likely.
    a. If you spin a roulette wheel 38 times, is it certain that each number will come up once? *No*
    b. If you spin a roulette wheel 3800 times, is it certain that each number will come up 100 times? *No*
    c. As the wheel is spun more and more times, will the proportion of times that each number comes up approach 1/38? *Yes*

In Exercises 21–24, assume that a coin is tossed twice. The coin may not be fair. The sample space consists of the outcomes {HH, HT, TH, TT}.

21. Is the following a probability model for this experiment? Why or why not? *No*

Outcome	HH	HT	TH	TT
Probability	0.55	0.42	0.31	0.25

22. Is the following a probability model for this experiment? Why or why not? *Yes*

Outcome	HH	HT	TH	TT
Probability	0.36	0.24	0.24	0.16

23. Is the following a probability model for this experiment? Why or why not? *Yes*

Outcome	HH	HT	TH	TT
Probability	0.09	0.21	0.21	0.49

24. Is the following a probability model for this experiment? Why or why not? *No*

Outcome	HH	HT	TH	TT
Probability	0.33	0.46	−0.18	0.4

Working with the Concepts

25. **How probable is it?** Someone computes the probabilities of several events. The probabilities are listed on the left, and some verbal descriptions are listed on the right. Match each probability with the best verbal description. Some descriptions may be used more than once.

Probability		Verbal Description
(a) 0.50	*ii*	i. This event is certain to happen.
(b) 0.00	*vi*	ii. This event is as likely to happen as not.
(c) 0.90	*iv*	iii. This event may happen, but it isn't likely.
(d) 1.00	*i*	iv. This event is very likely to happen, but it isn't certain.
(e) 0.10	*iii*	v. It would be unusual for this event to happen.
(f) −0.25	*vii*	vi. This event cannot happen.
(g) 0.01	*v*	vii. Someone made a mistake.
(h) 2.00	*vii*	

26. **Do you know SpongeBob?** According to a survey by Nickelodeon TV, 88% of children under 13 in Germany recognized a picture of the cartoon character SpongeBob SquarePants. What is the probability that a randomly chosen German child recognizes SpongeBob? *0.88*
27. **Who will you vote for?** In a survey of 500 likely voters in a certain city, 275 said that they planned to vote to reelect the incumbent mayor.
    a. What is the probability that a surveyed voter plans to vote to reelect the mayor? *0.55*

b. Interpret this probability by estimating the percentage of all voters in the city who plan to vote to reelect the mayor. 55%

28. Job satisfaction: In a poll conducted by the General Social Survey, 497 out of 1769 people said that their main satisfaction in life comes from their work.

a. What is the probability that a person who was polled finds his or her main satisfaction in life from work? 0.2809

b. Interpret this probability by estimating the percentage of all people whose main satisfaction in life comes from their work. 28.09%

29. True–false exam: A section of an exam contains four true–false questions. A completed exam paper is selected at random, and the four answers are recorded.

a. List all 16 outcomes in the sample space.

b. Assuming the outcomes to be equally likely, find the probability that all the answers are the same. 1/8

c. Assuming the outcomes to be equally likely, find the probability that exactly one of the four answers is "True." 1/4

d. Assuming the outcomes to be equally likely, find the probability that two of the answers are "True" and two of the answers are "False." 3/8

30. A coin flip: A fair coin is tossed three times. The outcomes of the three tosses are recorded.

a. List all eight outcomes in the sample space.

b. Assuming the outcomes to be equally likely, find the probability that all three tosses are "Heads." 1/8

c. Assuming the outcomes to be equally likely, find the probability that the tosses are all the same. 1/4

d. Assuming the outcomes to be equally likely, find the probability that exactly one of the three tosses is "Heads." 3/8

31. Empirical Method: A coin is tossed 400 times and comes up heads 180 times. Use the Empirical Method to approximate the probability that the coin comes up heads. 0.45

32. Empirical Method: A die is rolled 600 times. On 85 of those rolls, the die comes up 6. Use the Empirical Method to approximate the probability that the die comes up 6. 0.1417

33. Pitching: During a recent season, pitcher Clayton Kershaw threw 2515 pitches. Of these, 1303 were fastballs, 12 were changeups, 385 were curveballs, and 815 were sliders.

a. What is the probability that Clayton Kershaw throws a fastball? 0.5181

b. What is the probability that Clayton Kershaw throws a breaking ball (curveball or slider)? 0.4771

34. More pitching: During a recent season, pitcher Jon Lester threw 3727 pitches. Of these, 1449 were thrown with no strikes on the batter, 1168 were thrown with one strike, and 1110 were thrown with two strikes.

a. What is the probability that a Jon Lester pitch is thrown with no strikes? 0.3888

b. What is the probability that a Jon Lester pitch is thrown with fewer than two strikes? 0.7022

35. Risky drivers: An automobile insurance company divides customers into three categories: good risks, medium risks, and poor risks. Assume that of a total of 11,217 customers, 7792 are good risks, 2478 are medium risks, and 947 are poor risks. As part of an audit, one customer is chosen at random.

a. What is the probability that the customer is a good risk? 0.6947

b. What is the probability that the customer is not a poor risk? 0.9156

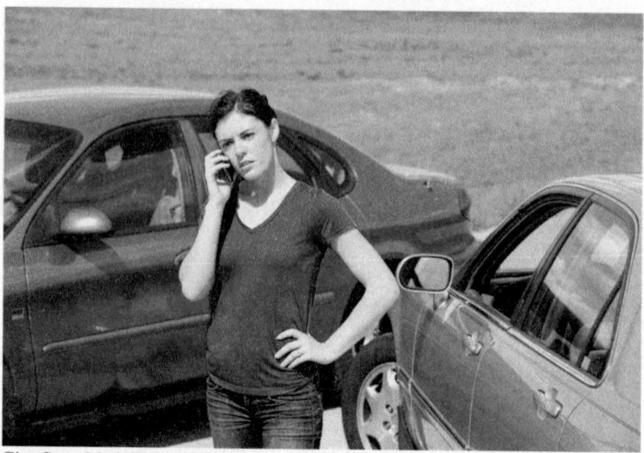

Gino Santa Maria/Shutterstock

36. Pay your bills: A company audit showed that of 875 bills that were sent out, 623 were paid on time, 155 were paid up to 30 days late, 78 were paid between 31 and 90 days late, and 19 were paid after 90 days. One bill is selected at random.

a. What is the probability that the bill was paid on time? 0.712

b. What is the probability that the bill was paid late? 0.288

37. Roulette: A Nevada roulette wheel has 38 pockets. Eighteen of them are red, eighteen are black, and two are green. Each time the wheel is spun, a ball lands in one of the pockets, and each pocket is equally likely.

a. What is the probability that the ball lands in a red pocket? 0.4737

b. If you bet on red on every spin of the wheel, you will lose more than half the time in the long run. Explain why this is so.

38. More roulette: Refer to Exercise 37.

a. What is the probability that the ball lands in a green pocket? 0.0526

b. If you bet on green on every spin of the wheel, you will lose more than 90% of the time in the long run. Explain why this is so.

39. Get an education: The General Social Survey asked 32,201 people how much confidence they had in educational institutions. The results were as follows.

Response	Number
A great deal	10,040
Some	17,890
Hardly any	4,271
Total	32,201

a. What is the probability that a sampled person has either some or a great deal of confidence in educational institutions? 0.8674

b. Assume this is a simple random sample from a population. Use the Empirical Method to estimate the probability that a person has a great deal of confidence in educational institutions. 0.3118

c. If we use a cutoff of 0.05, is it unusual for someone to have hardly any confidence in educational institutions? No

40. How many kids? The General Social Survey asked 46,349 women how many children they had. The results were as follows.

Number of Children	Number of Women
0	12,656
1	7,438
2	11,290
3	7,143
4	3,797
5	1,811
6	916
7	522
8 or more	776
Total	46,349

a. What is the probability that a sampled woman has two children? *0.2436*

b. What is the probability that a sampled woman has fewer than three children? *0.6771*

c. Assume this is a simple random sample of U.S. women. Use the Empirical Method to estimate the probability that a U.S. woman has more than five children. *0.0478*

d. Using a cutoff of 0.05, is it unusual for a woman to have no children? *No*

41. Hospital visits: According to the Agency for Healthcare Research and Quality, there were 409,706 hospital visits for asthma-related illnesses in a recent year. The age distribution was as follows.

Age Range	Number
Less than 1 year	7,866
1–17	103,040
18–44	79,659
45–64	121,728
65–84	80,649
85 and up	16,764
Total	409,706

a. What is the probability that an asthma patient is between 18 and 44 years old? *0.1944*

b. What is the probability that an asthma patient is 65 or older? *0.2378*

c. Using a cutoff of 0.05, is it unusual for an asthma patient to be less than 1 year old? *Yes*

42. Don't smoke: The Centers for Disease Control and Prevention reported that there were 443,000 smoking-related deaths in the United States in a recent year. The numbers of deaths caused by various illnesses attributed to smoking are as follows:

Illness	Number
Lung cancer	128,900
Ischemic heart disease	126,000
Chronic obstructive pulmonary disease	92,900
Other	95,200
Total	443,000

a. What is the probability that a smoking-related death was the result of lung cancer? *0.2910*

b. What is the probability that a smoking-related death was the result of either ischemic heart disease or other? *0.4993*

43. Simulation: If five fair coins are tossed, the possible outcomes for the number of heads obtained are 0, 1, 2, 3, 4, and 5. Use a computer simulation to toss five fair coins 1000 times and estimate the probabilities of these outcomes. (Use 1 to represent a head and 0 to represent a tail.)

44. Simulation: In the game of Dungeons & Dragons, four-sided dice are rolled, with equally likely outcomes 1, 2, 3, and 4. If three four-sided dice are rolled, the possible totals are the numbers 3 through 12. Use a computer simulation to roll three four-sided dice 1000 times and estimate the probability of each outcome.

Extending the Concepts

Two dice are rolled. One is red and one is blue. Each will come up with a number between 1 and 6. There are 36 equally likely outcomes for this experiment. They are ordered pairs of the form (Red die, Blue die).

45. Find a sample space: Construct a sample space for this experiment that contains the 36 equally likely outcomes.

46. Find the probability: What is the probability that the sum of the dice is 5? *1/9*

47. Find the probability: What is the probability that the sum of the dice is 7? *1/6*

48. The red die has been rolled: Now assume that you have rolled the red die, and it has come up 3. How many of the original 36 outcomes are now possible? *6*

49. Find a new sample space: Construct a sample space containing the outcomes that are still possible after the red die has come up 3.

50. New information changes the probability: Given that the red die came up 3, what is the probability that the sum of the dice is 5? Is the probability the same as it was before the red die was observed? *1/6; no*

51. New information doesn't change the probability: Given that the red die came up 3, what is the probability that the sum of the dice is 7? Is the probability the same as it was before the red die was observed? *1/6; yes*

Answers to Check Your Understanding Exercises for Section 5.1

1. 0.5

2. 0.5

3. a.

Penny	Nickel
H	H
H	T
T	H
T	T

b. 0.5

4. 0.513

Section 5.2 The Addition Rule and the Rule of Complements

Objectives

1. Compute probabilities by using the General Addition Rule
2. Compute probabilities by using the Addition Rule for Mutually Exclusive Events
3. Compute probabilities by using the Rule of Complements

NOTE TO INSTRUCTOR

In common English, the phrase *A* or *B* sometimes means that either *A* or *B* occurs, but not both. When discussing probabilities of events, *A* or *B* means that *A* occurs, *B* occurs, or both occur.

If you go out in the evening, you might go to dinner, or to a movie, or to both dinner and a movie. In probability terminology, "go to dinner and a movie" and "go to dinner or a movie" are referred to as *compound events*, because they are composed of combinations of other events—in this case the events "go to dinner" and "go to a movie."

DEFINITION

A **compound event** is an event that is formed by combining two or more events.

In this section, we will focus on compound events of the form "*A* or *B*." We will say that the event "*A* or *B*" occurs whenever *A* occurs, or *B* occurs, or both *A* and *B* occur. We will learn how to compute probabilities of the form $P(A \text{ or } B)$.

NOTE TO INSTRUCTOR

The following topics from the **Statistics Corequisite Workbook** are aligned with the material in this section.

3.2 - Adding and Subtracting Fractions from 1 with Complementary Events

3.3 - Adding and Subtracting Fractions with the Addition Rule

DEFINITION

$P(A \text{ or } B) = P(A \text{ occurs or } B \text{ occurs or both occur})$

Table 5.2 presents the results of a survey in which 1000 adults were asked whether they favored a law that would provide more government support for higher education. In addition, each person was asked whether he or she voted in the last election. Those who had voted were classified as "Likely to vote," and those who had not were classified as "Not likely to vote."

Table 5.2

	Favor	Oppose	Undecided
Likely to vote	372	262	87
Not likely to vote	151	103	25

Table 5.2 is called a **contingency table**. It categorizes people with regard to two variables: whether they are likely to vote, and their opinion on the law. There are six categories, and the numbers in the table present the frequencies for each category. For example, we can see that 372 people are in the row corresponding to "Likely to vote" and the column corresponding to "Favor." Thus, 372 people were likely to vote and favored the law. Similarly, 103 people were not likely to vote and opposed the law.

Example 5.10

Compute probabilities by using equally likely outcomes

Use Table 5.2 to answer the following questions:

a. What is the probability that a randomly selected adult is likely to vote and favors the law?

b. What is the probability that a randomly selected adult is likely to vote?

c. What is the probability that a randomly selected adult favors the law?

Solution

We think of the adults in the survey as outcomes in a sample space. Each adult is equally likely to be the one chosen. We begin by counting the total number of outcomes in the sample space:

$$372 + 262 + 87 + 151 + 103 + 25 = 1000$$

To answer part (a), we observe that there are 372 people who are likely to vote and favor the law. There are 1000 people in the survey. Therefore,

$$P(\text{Likely to vote and Favor}) = \frac{372}{1000} = 0.372$$

To answer part (b), we count the total number of outcomes corresponding to adults who are likely to vote:

$$372 + 262 + 87 = 721$$

There are 1000 people in the survey, and 721 of them are likely to vote. Therefore,

$$P(\text{Likely to vote}) = \frac{721}{1000} = 0.721$$

To answer part (c), we count the total number of outcomes corresponding to adults who favor the law:

$$372 + 151 = 523$$

There are 1000 people in the survey, and 523 of them favor the law. Therefore,

$$P(\text{Favor}) = \frac{523}{1000} = 0.523$$

Objective 1 Compute probabilities by using the General Addition Rule

The General Addition Rule

Example 5.11

Compute a probability of the form *P*(*A* or *B*)

Use the data in Table 5.2 to find the probability that a person is likely to vote or favors the law.

Solution

We will illustrate two approaches to this problem. In the first approach, we will use equally likely outcomes, and in the second, we will develop a method that is especially designed for probabilities of the form $P(A \text{ or } B)$.

Approach 1: To use equally likely outcomes, we reproduce Table 5.2 and circle the numbers that correspond to people who are either likely voters or who favor the law.

	Favor	Oppose	Undecided
Likely to vote	(372)	(262)	(87)
Not likely to vote	(151)	103	25

There are 1000 people altogether. The number of people who either are likely voters or favor the law is

$$372 + 262 + 87 + 151 = 872$$

Therefore,

$$P(\text{Likely to vote or Favor}) = \frac{372 + 262 + 87 + 151}{1000} = \frac{872}{1000} = 0.872$$

Approach 2: In this approach we will begin by computing the probabilities *P*(Likely to vote) and *P*(Favor) separately. We reproduce Table 5.2; this time we circle the numbers that correspond to likely voters and put rectangles around the numbers that correspond to favoring the law. Note that the number 372 has both a circle and a rectangle around it, because these 372 people are both likely to vote and favor the law.

	Favor	**Oppose**	**Undecided**
Likely to vote	372	262	87
Not likely to vote	151	103	25

There are $372 + 262 + 87 = 721$ likely voters and $372 + 151 = 523$ voters who favor the law. If we try to find the number of people who are likely to vote or who favor the law by adding these two numbers, we get $721 + 523 = 1244$, which is too large (there are only 1000 people in total). This happened because there are 372 people who are both likely voters and who favor the law, and these people are counted twice. We can still solve the problem by adding 721 and 523, but we must then subtract 372 to correct for the double counting.

We illustrate this reasoning, using probabilities.

$$P(\text{Likely to vote}) = \frac{721}{1000} = 0.721$$

$$P(\text{Favor}) = \frac{523}{1000} = 0.523$$

$$P(\text{Likely to vote AND Favor}) = \frac{372}{1000} = 0.372$$

$$\begin{aligned} P(\text{Likely to vote OR Favor}) &= P(\text{Likely to vote}) + P(\text{Favor}) \\ &\quad - P(\text{Likely to vote AND Favor}) \\ &= \frac{721}{1000} + \frac{523}{1000} - \frac{372}{1000} \\ &= \frac{872}{1000} = 0.872 \end{aligned}$$

The method of subtracting in order to adjust for double counting is known as the General Addition Rule.

EXPLAIN IT AGAIN

The General Addition Rule: Use the General Addition Rule to compute probabilities of the form $P(A \text{ or } B)$.

The General Addition Rule

For any two events A and B,

$$P(A \text{ or } B) = P(A) + P(B) - P(A \text{ and } B)$$

Example 5.12

Compute a probability by using the General Addition Rule

Refer to Table 5.2. Use the General Addition Rule to find the probability that a randomly selected person is not likely to vote or is undecided.

Solution

Using the General Addition Rule, we compute

$$\begin{aligned} &P(\text{Not likely to vote or Undecided}) \\ &\quad = P(\text{Not likely to vote}) + P(\text{Undecided}) - P(\text{Not likely to vote and Undecided}) \end{aligned}$$

There are $151 + 103 + 25 = 279$ people not likely to vote out of a total of 1000.

Therefore,

$$P(\text{Not likely to vote}) = \frac{279}{1000} = 0.279$$

There are $87 + 25 = 112$ people who are undecided out of a total of 1000. Therefore,

$$P(\text{Undecided}) = \frac{112}{1000} = 0.112$$

Finally, there are 25 people who are both not likely to vote and undecided. Therefore,

$$P(\text{Not likely to vote and Undecided}) = \frac{25}{1000} = 0.025$$

Using the General Addition Rule,

$$P(\text{Not likely to vote or Undecided}) = 0.279 + 0.112 - 0.025 = 0.366$$

Check Your Understanding

1. The following table presents numbers of U.S. workers, in thousands, categorized by type of occupation and educational level.

Type of Occupation	Non-College Graduate	College Graduate
Managers and professionals	17,564	31,103
Service	15,967	2,385
Sales and office	22,352	7,352
Construction and maintenance	12,511	1,033
Production and transportation	14,597	1,308

Source: Bureau of Labor Statistics

a. What is the probability that a randomly selected worker is a college graduate? *0.342*

b. What is the probability that the occupation of a randomly selected worker is categorized either as Sales and office or as Production and transportation? *0.361*

c. What is the probability that a randomly selected worker is either a college graduate or has a service occupation? *0.469*

Answers are on page 235.

Mutually Exclusive Events

Objective 2 Compute probabilities by using the Addition Rule for Mutually Exclusive Events

Sometimes it is impossible for two events both to occur. For example, when a coin is tossed, it is impossible to get both a head and a tail. Two events that cannot both occur are called mutually exclusive. The term *mutually exclusive* means that when one event occurs, it excludes the other.

DEFINITION

Two events are said to be **mutually exclusive** if it is impossible for both events to occur.

EXPLAIN IT AGAIN

Meaning of mutually exclusive events: Two events are mutually exclusive if the occurrence of one makes it impossible for the other to occur.

We can use **Venn diagrams** to illustrate mutually exclusive events. In a Venn diagram, the sample space is represented by a rectangle, and events are represented by circles drawn inside the rectangle. If two circles do not overlap, the two events cannot both occur. If two circles overlap, the overlap area represents the occurrence of both events. Figures 5.6 and 5.7 (page 230) illustrate the idea.

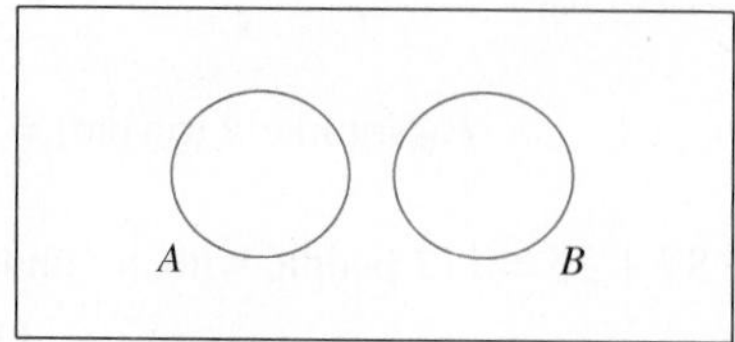

Figure 5.6 Venn diagram illustrating mutually exclusive events

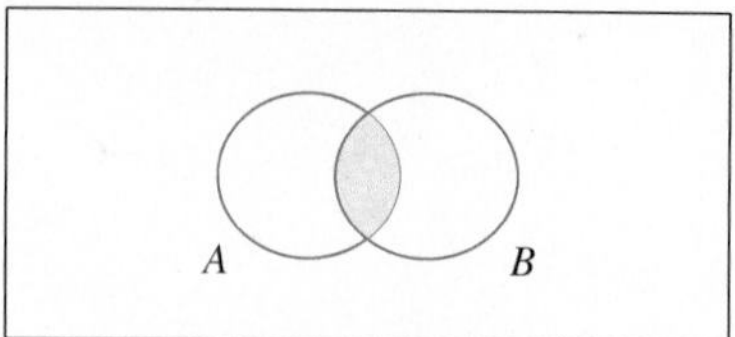

Figure 5.7 Venn diagram illustrating events that are not mutually exclusive

Example 5.13

Determine whether two events are mutually exclusive

In each of the following, determine whether events A and B are mutually exclusive:

a. A die is rolled. Event A is that the die comes up 3, and event B is that the die comes up an even number.

b. A fair coin is tossed twice. Event A is that one of the tosses is a head, and event B is that one of the tosses is a tail.

Solution

a. These events are mutually exclusive. The die cannot both come up 3 and come up an even number.

b. These events are not mutually exclusive. If the two tosses result in HT or TH, then both events occur.

Check Your Understanding

2. A college student is chosen at random. Event A is that the student is older than 21 years, and event B is that the student is taking a statistics class. Are events A and B mutually exclusive? *No*

3. A college student is chosen at random. Event A is that the student is an only child, and event B is that the student has a brother. Are events A and B mutually exclusive? *Yes*

Answers are on page 235.

If events A and B are mutually exclusive, then $P(A \text{ and } B) = 0$. This leads to a simplification of the General Addition Rule.

The Addition Rule for Mutually Exclusive Events

If A and B are mutually exclusive events, then

$$P(A \text{ or } B) = P(A) + P(B)$$

In general, three or more events are mutually exclusive if only one of them can happen. If $A, B, C, \ldots$ are mutually exclusive, then

$$P(A \text{ or } B \text{ or } C \text{ or } \ldots) = P(A) + P(B) + P(C) + \cdots$$

Example 5.14

Compute a probability by using the Addition Rule for Mutually Exclusive Events

In a recent Olympic Games, a total of 11,544 athletes participated. Of these, 554 represented the United States, 314 represented Canada, and 125 represented Mexico.

a. What is the probability that an Olympic athlete chosen at random represents the United States or Canada?

b. What is the probability that an Olympic athlete chosen at random represents the United States, Canada, or Mexico?

Solution

a. These events are mutually exclusive, because it is impossible to compete for both the United States and Canada. We compute P(U.S.) and P(Canada).

$$\begin{aligned} P(\text{U.S. or Canada}) &= P(\text{U.S.}) + P(\text{Canada}) \\ &= \frac{554}{11{,}544} + \frac{314}{11{,}544} \\ &= \frac{868}{11{,}544} \\ &= 0.075191 \end{aligned}$$

b. These events are mutually exclusive, because it is impossible to compete for more than one country. Therefore

$$\begin{aligned} P(\text{U.S. or Canada or Mexico}) &= P(\text{U.S.}) + P(\text{Canada}) + P(\text{Mexico}) \\ &= \frac{554}{11{,}544} + \frac{314}{11{,}544} + \frac{125}{11{,}544} \\ &= \frac{993}{11{,}544} \\ &= 0.086019 \end{aligned}$$

Check Your Understanding

4. In a statistics class of 45 students, 11 got a final grade of A, 22 got a final grade of B, and 8 got a final grade of C.

a. What is the probability that a randomly chosen student got an A or a B? *0.733*

b. What is the probability that a randomly chosen student got an A, a B, or a C? *0.911*

Answers are on page 235.

Objective 3 Compute probabilities by using the Rule of Complements

Complements

If there is a 60% chance of rain today, then there is a 40% chance that it will not rain. The events "Rain" and "No rain" are *complements*. The complement of an event A is the event that A does not occur.

DEFINITION

If A is any event, the **complement** of A is the event that A does not occur. The complement of A is denoted A^c.

Some complements are straightforward. For example, the complement of "the plane was on time" is "the plane was not on time." In other cases, finding the complement requires some thought. Example 5.15 illustrates this.

Example 5.15

Find the complement of an event

Two hundred students were enrolled in a statistics class. Find the complements of the following events.

a. More than 50 of them are business majors.
b. At least 50 of them are business majors.
c. Fewer than 50 of them are business majors.
d. Exactly 50 of them are business majors.

NOTATION ROUNDUP

A^c is the event that A does not occur.

Solution

a. If it is not true that more than 50 are business majors, then the number of business majors must be 50 or less than 50. The complement is that 50 or fewer of the students are business majors.

b. If it is not true that at least 50 are business majors, then the number of business majors must be less than 50. The complement is that fewer than 50 of the students are business majors.

c. If it is not true that fewer than 50 are business majors, then the number of business majors must be 50 or more than 50. Another way of saying this is that at least 50 of the students are business majors. The complement is that at least 50 of the students are business majors.

d. If it is not true that exactly 50 are business majors, then the number of business majors must not equal 50. The complement is that the number of business majors is not equal to 50.

EXPLAIN IT AGAIN

The complement occurs when the event doesn't occur: If an event does not occur, then its complement occurs. If an event occurs, then its complement does not occur.

Two important facts about complements are:

1. Either A or A^c must occur. For example, it must either rain or not rain.
2. A and A^c are mutually exclusive; they cannot both occur. For example, it is impossible for it to both rain and not rain.

In probability notation, fact 1 says that $P(A \text{ or } A^c) = 1$, and fact 2 along with the Addition Rule for Mutually Exclusive Events says that $P(A \text{ or } A^c) = P(A) + P(A^c)$. Putting them together, we get

$$P(A) + P(A^c) = 1$$

Subtracting $P(A)$ from both sides yields

$$P(A^c) = 1 - P(A)$$

This is the Rule of Complements.

The Rule of Complements

$$P(A^c) = 1 - P(A)$$

Example 5.16

Compute a probability by using the Rule of Complements

According to *The Wall Street Journal*, 40% of cars sold in a recent year were small cars. What is the probability that a randomly chosen car sold in that year is not a small car?

Solution

$$P(\text{Not a small car}) = 1 - P(\text{Small car}) = 1 - 0.40 = 0.60$$

Section 5.2 Exercises

Exercises 1–4 are the Check Your Understanding exercises located within the section.

Understanding the Concepts

In Exercises 5–8, fill in each blank with the appropriate word or phrase.

5. The General Addition Rule states that $P(A \text{ or } B) = P(A) + P(B) -$ ________. *P (A and B)*
6. If events A and B are mutually exclusive, then $P(A \text{ and } B) =$ ________. *0*
7. Given an event A, the event that A does not occur is called the ________ of A. *complement*
8. The Rule of Complements states that $P(A^c) =$ ________. *1 − P (A)*

In Exercises 9–12, determine whether the statement is true or false. If the statement is false, rewrite it as a true statement.

9. The General Addition Rule is used for probabilities of the form $P(A \text{ or } B)$. *True*
10. A compound event is formed by combining two or more events. *True*
11. Two events are mutually exclusive if both events can occur. *False*
12. If an event occurs, then its complement also occurs. *False*

Practicing the Skills

13. If $P(A) = 0.75$, $P(B) = 0.4$, and $P(A \text{ and } B) = 0.25$, find $P(A \text{ or } B)$. *0.9*
14. If $P(A) = 0.45$, $P(B) = 0.7$, and $P(A \text{ and } B) = 0.65$, find $P(A \text{ or } B)$. *0.5*
15. If $P(A) = 0.2$, $P(B) = 0.5$, and A and B are mutually exclusive, find $P(A \text{ or } B)$. *0.7*
16. If $P(A) = 0.7$, $P(B) = 0.1$, and A and B are mutually exclusive, find $P(A \text{ or } B)$. *0.8*
17. If $P(A) = 0.3$, $P(B) = 0.4$, and $P(A \text{ or } B) = 0.7$, are A and B mutually exclusive? *Yes*
18. If $P(A) = 0.5$, $P(B) = 0.4$, and $P(A \text{ or } B) = 0.8$, are A and B mutually exclusive? *No*
19. If $P(A) = 0.35$, find $P(A^c)$. *0.65*
20. If $P(B) = 0.6$, find $P(B^c)$. *0.4*
21. If $P(A^c) = 0.27$, find $P(A)$. *0.73*
22. If $P(B^c) = 0.64$, find $P(B)$. *0.36*
23. If $P(A) = 0$, find $P(A^c)$. *1*
24. If $P(A) = P(A^c)$, find $P(A)$. *0.5*

In Exercises 25–30, determine whether events A and B are mutually exclusive.

25. A: Sophie is a member of the debate team; B: Sophie is the president of the theater club. *No*
26. A: Jayden has a math class on Tuesdays at 2:00; B: Jayden has an English class on Tuesdays at 2:00. *Yes*
27. A sample of 20 cars is selected from the inventory of a dealership. A: At least 3 of the cars in the sample are red; B: Fewer than 2 of the cars in the sample are red. *Yes*
28. A sample of 75 books is selected from a library. A: At least 10 of the authors are female; B: At least 10 of the books are fiction. *No*
29. A red die and a blue die are rolled. A: The red die comes up 2; B: The blue die comes up 3. *No*
30. A red die and a blue die are rolled. A: The red die comes up 1; B: The total is 9. *Yes*

In Exercises 31 and 32, find the complements of the events.

31. A sample of 225 internet users was selected.
    - **a.** More than 200 of them use Google as their primary search engine.
    - **b.** At least 200 of them use Google as their primary search engine.
    - **c.** Fewer than 200 of them use Google as their primary search engine.
    - **d.** Exactly 200 of them use Google as their primary search engine.
32. A sample of 700 phone batteries was selected.
    - **a.** Exactly 24 of the batteries were defective.
    - **b.** At least 24 of the batteries were defective.
    - **c.** More than 24 of the batteries were defective.
    - **d.** Fewer than 24 of the batteries were defective.

Working with the Concepts

33. **Traffic lights:** A commuter passes through two traffic lights on the way to work. Each light is either red, yellow, or green. An experiment consists of observing the colors of the two lights.
    - **a.** List the nine outcomes in the sample space.
    - **b.** Let A be the event that both colors are the same. List the outcomes in A.
    - **c.** Let B be the event that the two colors are different. List the outcomes in B.
    - **d.** Let C be the event that at least one of the lights is green. List the outcomes in C.
    - **e.** Are events A and B mutually exclusive? Explain. *Yes*
    - **f.** Are events A and C mutually exclusive? Explain. *No*
34. **Dice:** Two fair dice are rolled. The first die is red and the second is blue. An experiment consists of observing the numbers that come up on the dice.
    - **a.** There are 36 outcomes in the sample space. They are ordered pairs of the form (Red die, Blue die). List the 36 outcomes.
    - **b.** Let A be the event that the same number comes up on both dice. List the outcomes in A.
    - **c.** Let B be the event that the red die comes up 6. List the outcomes in B.
    - **d.** Let C be the event that one die comes up 6 and the other comes up 1. List the outcomes in C.
    - **e.** Are events A and B mutually exclusive? Explain. *No*
    - **f.** Are events A and C mutually exclusive? Explain. *Yes*
35. **Car repairs:** Let E be the event that a new car requires engine work under warranty, and let T be the event that the car requires transmission work under warranty. Suppose that $P(E) = 0.10$, $P(T) = 0.02$, and $P(E \text{ and } T) = 0.01$.

a. Find the probability that the car needs work on either the engine, the transmission, or both. *0.11*

b. Find the probability that the car needs no work on the engine. *0.90*

36. Sick computers: Let V be the event that a computer contains a virus, and let W be the event that a computer contains a worm. Suppose $P(V) = 0.15$, $P(W) = 0.05$, and $P(V \text{ and } W) = 0.03$.

a. Find the probability that the computer contains either a virus or a worm or both. *0.17*

b. Find the probability that the computer does not contain a virus. *0.85*

37. Computer purchases: Out of 800 large purchases made at a computer retailer, 336 were tablets, 398 were laptop computers, and 66 were desktop computers. As part of an audit, one purchase record is sampled at random.

a. What is the probability that it is a tablet? *0.42*

b. What is the probability that it is not a desktop computer? *0.9175*

38. Visit your local library: On a recent Saturday, a total of 1200 people visited a local library. Of these people, 248 were under age 10, 472 were aged 10–18, 175 were aged 19–30, and the rest were more than 30 years old. One person is sampled at random.

a. What is the probability that the person is less than 19 years old? *0.6*

b. What is the probability that the person is more than 30 years old? *0.2542*

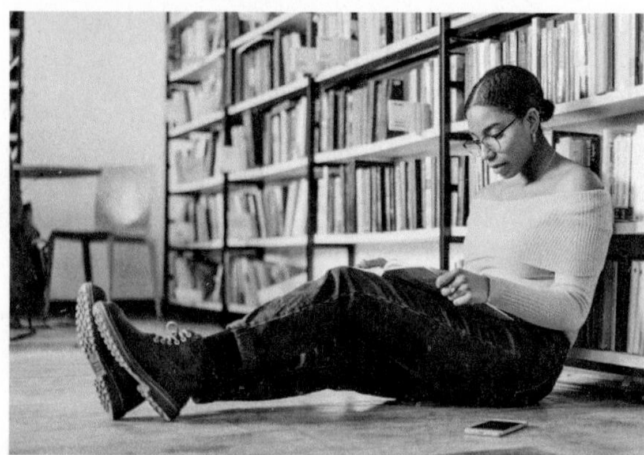

Shutterstock/GaudiLab

39. How are your grades? In a recent semester at a local university, 500 students enrolled in both Statistics I and Psychology I. Of these students, 82 got an A in statistics, 73 got an A in psychology, and 42 got an A in both statistics and psychology.

a. Find the probability that a randomly chosen student got an A in statistics or psychology or both. *0.226*

b. Find the probability that a randomly chosen student did not get an A in psychology. *0.854*

40. Statistics grades: In a statistics class of 30 students, there were 13 men and 17 women. Two of the men and three of the women received an A in the course. A student is chosen at random from the class.

a. Find the probability that the student is a woman. *0.5667*

b. Find the probability that the student received an A. *0.1667*

c. Find the probability that the student is a woman or received an A. *0.6333*

d. Find the probability that the student did not receive an A. *0.8333*

41. Weight and cholesterol: The National Health Examination Survey reported that in a sample of 13,535 adults, 6443 had high cholesterol (total cholesterol above 200 mg/dL), 8841 were overweight (body mass index above 25), and 4629 were both overweight and had high cholesterol. A person is chosen at random from this study.

a. Find the probability that the person is overweight. *0.6532*

b. Find the probability that the person has high cholesterol. *0.4760*

c. Find the probability that the person does not have high cholesterol. *0.5240*

d. Find the probability that the person is overweight or has high cholesterol. *0.7872*

42. Paving stones: Two hundred paving stones were examined for cracks, and 15 were found to be cracked. The same 200 stones were examined for discoloration, and 27 were found to be discolored. A total of 4 stones were both cracked and discolored. One of the 200 stones is selected at random.

a. Find the probability that it is cracked. *0.075*

b. Find the probability that it is discolored. *0.135*

c. Find the probability that it is not cracked. *0.925*

d. Find the probability that it is cracked or discolored. *0.19*

43. Sick children: There are 25 students in Mrs. Bush's sixth-grade class. On a cold winter day in February, many of the students had runny noses and sore throats. After examining each student, the school nurse constructed the following table.

	Sore Throat	No Sore Throat
Runny Nose	6	12
No Runny Nose	4	3

a. Find the probability that a randomly selected student has a runny nose. *0.72*

b. Find the probability that a randomly selected student has a sore throat. *0.4*

c. Find the probability that a randomly selected student has a runny nose or a sore throat. *0.88*

d. Find the probability that a randomly selected student has neither a runny nose nor a sore throat. *0.12*

44. Flawed parts: On a certain day, a foundry manufactured 500 cast aluminum parts. Some of these had major flaws, some had minor flaws, and some had both major and minor flaws. The following table presents the results.

	Minor Flaw	No Minor Flaw
Major Flaw	20	35
No Major Flaw	75	370

a. Find the probability that a randomly chosen part has a major flaw. *0.11*

b. Find the probability that a randomly chosen part has a minor flaw. *0.19*

c. Find the probability that a randomly chosen part has a flaw (major or minor). *0.26*

d. Find the probability that a randomly chosen part has no major flaw. *0.89*

e. Find the probability that a randomly chosen part has no flaw. *0.74*

45. Senators: The following table displays the 100 senators of the 116th U.S. Congress on January 3, 2019, classified by political party affiliation and gender.

	Male	Female	Total
Democrat	28	17	45
Republican	45	8	53
Independent	2	0	2
Total	75	25	100

A senator is selected at random from this group. Compute the following probabilities.

a. The senator is a male Democrat. *0.28*
b. The senator is a Republican or a female. *0.70*
c. The senator is a Republican. *0.53*
d. The senator is not a Republican. *0.47*
e. The senator is a Democrat. *0.45*
f. The senator is an Independent. *0.02*
g. The senator is a Democrat or an Independent. *0.47*

46. Graffiti: The following table presents the number of reports of graffiti in each of New York's five boroughs over a one-year period. These reports were classified as being open, closed, or pending.

Borough	Open Reports	Closed Reports	Pending Reports	Total
Bronx	1,121	1,622	80	2,823
Brooklyn	1,170	2,706	48	3,924
Manhattan	744	3,380	25	4,149
Queens	1,353	2,043	25	3,421
Staten Island	83	118	0	201
Total	4,471	9,869	178	14,518

Source: NYC OpenData

A graffiti report is selected at random. Compute the following probabilities.

a. The report is open and comes from Brooklyn. *0.0806*
b. The report is closed or comes from Queens. *0.7747*
c. The report comes from Manhattan. *0.2858*
d. The report does not come from Manhattan. *0.7142*
e. The report is pending. *0.0123*
f. The report is from the Bronx or Staten Island. *0.2083*

47. Add probabilities? In a certain community, 28% of the houses have fireplaces and 51% have garages. Is the probability that a house has either a fireplace or a garage equal to $0.51 + 0.28 = 0.79$? Explain why or why not. *No*

48. Add probabilities? According to the National Health Statistics Reports, 16% of American women have one child, and 21% have two children. Is the probability that a woman has either one or two children equal to $0.16 + 0.21 = 0.37$? Explain why or why not. *Yes*

Extending the Concepts

49. Mutual exclusivity is not transitive: Give an example of three events A, B, and C, such that A and B are mutually exclusive, B and C are mutually exclusive, but A and C are not mutually exclusive.

50. Complements: Let A and B be events. Express $(A \text{ and } B)^c$ in terms of A^c and B^c. *A^c or B^c*

Answers to Check Your Understanding Exercises for Section 5.2

1. a. 0.342 **b.** 0.361 **c.** 0.469

2. No

3. Yes

4. a. 0.733 **b.** 0.911

Section 5.3 Conditional Probability and the Multiplication Rule

Objectives

1. Compute conditional probabilities
2. Compute probabilities by using the General Multiplication Rule
3. Compute probabilities by using the Multiplication Rule for Independent Events
4. Compute the probability that an event occurs at least once

Objective 1 Compute conditional probabilities

NOTE TO INSTRUCTOR

The following topic from the **Statistics Corequisite Workbook** is aligned with the material in this section.

3.4 - Multiplying Fractions with the General Multiplication Rule

Conditional Probability

Approximately 15% of adult men in the United States are more than six feet tall. Therefore, if a man is selected at random, the probability that he is more than six feet tall is 0.15. Now assume that you learn that the selected man is a professional basketball player. With this extra information, the probability that the man is more than six feet tall becomes much greater than 0.15. A probability that is computed with the knowledge of additional information is called a *conditional probability*; a probability computed without such knowledge is called an *unconditional probability*. As this example shows, the conditional probability of an event can be much different than the unconditional probability.

Example 5.17 Compute an unconditional probability

Joe, Sam, Eliza, and Maria have been elected to the executive committee of their college's student government. They must choose a chairperson and a secretary. They decide to write each name on a piece of paper and draw two names at random. The first name drawn will be the chairperson, and the second name drawn will be the secretary. What is the probability that Joe is the secretary?

Table 5.3 is a sample space for this experiment. The first name in each pair is the chairperson, and the second name is the secretary.

Table 5.3 Twelve Equally Likely Outcomes

(Joe, Sam)	(Sam, Joe)	(Eliza, Joe)	(Maria, Joe)
(Joe, Eliza)	(Sam, Eliza)	(Eliza, Sam)	(Maria, Sam)
(Joe, Maria)	(Sam, Maria)	(Eliza, Maria)	(Maria, Eliza)

There are 12 equally likely outcomes. Three of them, (Sam, Joe), (Eliza, Joe), and (Maria, Joe), correspond to Joe's being secretary. Therefore, $P(\text{Joe is secretary}) = 3/12 = 1/4$.

Example 5.18

Compute a conditional probability

Suppose that Eliza is the first name selected, so she is chairperson. Now what is the probability that Joe is secretary?

Solution

We'll answer this question with intuition first, then show the reasoning. Since Eliza was chosen to be chairperson, she won't be the secretary. That leaves Joe, Sam, and Maria. Each of these three is equally likely to be chosen. Therefore, the probability that Joe is chosen as secretary is 1/3. Note that this probability differs from the probability of 1/4 calculated in Example 5.17.

Now let's look at the reasoning behind this answer. The original sample space, shown in Table 5.3, had 12 outcomes. Once we know that Eliza is chairperson, we know that only three of those outcomes are now possible. Table 5.4 highlights these three outcomes from the original sample space.

Table 5.4

(Joe, Sam)	(Sam, Joe)	(Eliza, Joe)	(Maria, Joe)
(Joe, Eliza)	(Sam, Eliza)	(Eliza, Sam)	(Maria, Sam)
(Joe, Maria)	(Sam, Maria)	(Eliza, Maria)	(Maria, Eliza)

Of the three possible outcomes, only one, (Eliza, Joe), has Joe as secretary. Therefore, given that Eliza is chairperson, the probability that Joe is secretary is $1/3$.

Example 5.18 asked us to compute the probability of an event (that Joe is secretary) after giving us information about another event (that Eliza is chairperson). A probability like this is called a *conditional probability*. The notation for this conditional probability is

$$P(\text{Joe is secretary} \mid \text{Eliza is chairperson})$$

We read this as "the conditional probability that Joe is secretary, given that Eliza is chairperson." It denotes the probability that Joe is secretary, under the assumption that Eliza is chairperson.

NOTATION ROUNDUP

$P(B \mid A)$ is the probability that event B occurs given that event A has occurred.

DEFINITION

The **conditional probability** of an event B, given an event A, is denoted $P(B \mid A)$.

$P(B \mid A)$ is the probability that B occurs, under the assumption that A occurs.

We read $P(B \mid A)$ as "the probability of B, given A."

The General Method for computing conditional probabilities

In Example 5.18, we computed

$$P(\text{Joe is secretary} \mid \text{Eliza is chairperson}) = \frac{1}{3}$$

Let's take a closer look at the answer of $1/3$. The denominator is the number of outcomes that were left in the sample space after it was known that Eliza was chairperson. That is,

$$\text{Number of outcomes where Eliza is chairperson} = 3$$

The numerator is 1, and this corresponds to the one outcome in which Eliza is chairperson and Joe is secretary. That is,

$$\text{Number of outcomes where Eliza is chairperson and Joe is secretary} = 1$$

Therefore, we see that

$$P(\text{Joe is secretary} \mid \text{Eliza is chairperson}) = \frac{\text{Number of outcomes where Eliza is chairperson and Joe is secretary}}{\text{Number of outcomes where Eliza is chairperson}}$$

We can obtain another useful method by recalling that there were 12 outcomes in the original sample space. It follows that

$$P(\text{Eliza is chairperson}) = \frac{3}{12}$$

and

$$P(\text{Eliza is chairperson and Joe is secretary}) = \frac{1}{12}$$

We now see that

$$P(\text{Joe is secretary} \mid \text{Eliza is chairperson}) = \frac{P(\text{Eliza is chairperson and Joe is secretary})}{P(\text{Eliza is chairperson})}$$

This example illustrates the General Method for computing conditional probabilities, which we now state.

The General Method for Computing Conditional Probabilities

The probability of B given A is

$$P(B \mid A) = \frac{P(A \text{ and } B)}{P(A)}$$

Note that we cannot compute $P(B \mid A)$ if $P(A) = 0$.

When the outcomes in the sample space are equally likely, then

$$P(B \mid A) = \frac{\text{Number of outcomes corresponding to } (A \text{ and } B)}{\text{Number of outcomes corresponding to } A}$$

Example 5.19

Use the General Method to compute a conditional probability

Table 5.5 presents the number of U.S. men and women (in millions) 25 years old and older who have attained various levels of education in a recent year.

Table 5.5 Number of Men and Women with Various Levels of Education (in millions)

	Not a high school graduate	High school graduate	Some college, no degree	Associate's degree	Bachelor's degree	Advanced degree
Men	14.0	29.6	15.6	7.2	17.5	10.1
Women	13.7	31.9	17.5	9.6	19.2	9.1

Source: U.S. Census Bureau

A person is selected at random.

a. What is the probability that the person is a man?

b. What is the probability that the person is a man with a bachelor's degree?

c. What is the probability that the person has a bachelor's degree, given that the person is a man?

Solution

a. Each person in the study is an outcome in the sample space. We first compute the total number of people in the study. We'll do this by computing the total number of men, then the total number of women.

$$\text{Total number of men} = 14.0 + 29.6 + 15.6 + 7.2 + 17.5 + 10.1 = 94.0$$

$$\text{Total number of women} = 13.7 + 31.9 + 17.5 + 9.6 + 19.2 + 9.1 = 101.0$$

There are 94.0 million men and 101.0 million women. The total number of people is 94.0 + 101.0 = 195.0 million. We can now compute the probability that a randomly chosen person is a man.

$$P(\text{Man}) = \frac{94.0}{195.0} = 0.4821$$

b. The number of men with bachelor's degrees is found in Table 5.5 to be 17.5 million. The total number of people is 195.0 million. Therefore

$$P(\text{Man with a Bachelor's degree}) = \frac{17.5}{195.0} = 0.08974$$

c. We use the General Method for computing a conditional probability.

$$P(\text{Bachelor's degree}|\text{Man}) = \frac{P(\text{Man with a Bachelor's degree})}{P(\text{Man})} = \frac{17.5/195.0}{94.0/195.0} = 0.1862$$

Check Your Understanding

1. A person is selected at random from the population in Table 5.5.
   a. What is the probability that the person is a woman who is a high school graduate? *0.164*
   b. What is the probability that the person is a high school graduate? *0.315*
   c. What is the probability that the person is a woman, given that the person is a high school graduate? *0.519*

Answers are on page 248.

Objective 2 Compute probabilities by using the General Multiplication Rule

EXPLAIN IT AGAIN

The General Multiplication Rule: Use the General Multiplication Rule to compute probabilities of the form $P(A \text{ and } B)$.

The General Multiplication Rule

The General Method for computing conditional probabilities provides a way to compute probabilities for events of the form "A and B." If we multiply both sides of the equation by $P(A)$, we obtain the General Multiplication Rule.

The General Multiplication Rule

$$P(A \text{ and } B) = P(A)P(B \mid A)$$

or, equivalently,

$$P(A \text{ and } B) = P(B)P(A \mid B)$$

Example 5.20

Use the General Multiplication Rule to compute a probability

Among those who apply for a particular job, the probability of being granted an interview is 0.1. Among those interviewed, the probability of being offered a job is 0.25. Find the probability that an applicant is offered a job.

NOTE TO INSTRUCTOR

It may help to remind students that $P(B \mid A)$ is the probability of event B under the assumption that event A has occurred.

Solution

Being offered a job involves two events. First, a person must be interviewed; then, given that the person has been interviewed, the person must be offered a job. Using the General Multiplication Rule, we obtain

$$\begin{aligned} P(\text{Offered a job}) &= P(\text{Interviewed})P(\text{Offered a job} \mid \text{Interviewed}) \\ &= (0.1)(0.25) \\ &= 0.025 \end{aligned}$$

Check Your Understanding

2. In a certain city, 70% of high school students graduate. Of those who graduate, 40% attend college. Find the probability that a randomly selected high school student will attend college. *0.28*

Answer is on page 248.

Objective 3 Compute probabilities by using the Multiplication Rule for Independent Events

Independence

In some cases, the occurrence of one event has no effect on the probability that another event occurs. For example, if a coin is tossed twice, the occurrence of a head on the first toss does not make it any more or less likely that a head will come up on the second toss. Example 5.21 illustrates this fact.

Example 5.21

Coin tossing probabilities

A fair coin is tossed twice.

a. What is the probability that the second toss is a head?

b. What is the probability that the second toss is a head given that the first toss is a head?

c. Are the answers to parts (a) and (b) different? Does the probability that the second toss is a head change if the first toss is a head?

CAUTION

Do not confuse independent events with mutually exclusive events. Two events are independent if the occurrence of one does not affect the probability of the occurrence of the other. Two events are mutually exclusive if the occurrence of one makes it impossible for the other to occur.

Solution

a. There are four equally likely outcomes for the two tosses. The sample space is {HH, HT, TH, TT}. Of these, there are two outcomes where the second toss is a head. Therefore, $P(\text{Second toss is H}) = 2/4 = 1/2$.

b. We use the General Method for computing conditional probabilities.

$$\begin{aligned} &P(\text{Second toss is H} \mid \text{First toss is H}) \\ &\quad = \frac{\text{Number of outcomes where first toss is H and second is H}}{\text{Number of outcomes where first toss is H}} = \frac{1}{2} \end{aligned}$$

c. The two answers are the same. The probability that the second toss is a head does not change if the first toss is a head. In other words,

$$P(\text{Second toss is H} \mid \text{First toss is H}) = P(\text{Second toss is H})$$

In the case of two coin tosses, the outcome of the first toss does not affect the second toss. Events with this property are said to be *independent*.

DEFINITION

Two events are **independent** if the occurrence of one does not affect the probability that the other event occurs.

If two events are not independent, we say they are **dependent**.

In many situations, we can determine whether events are independent just by understanding the circumstances surrounding the events. Example 5.22 illustrates this.

Example 5.22

Determine whether events are independent

Determine whether the following pairs of events are independent:

a. A college student is chosen at random. The events are "being a freshman" and "being less than 20 years old."

b. A college student is chosen at random. The events are "born on a Sunday" and "taking a statistics class."

Solution

a. These events are not independent. If the student is a freshman, the probability that the student is less than 20 years old is greater than for a student who is not a freshman.

b. These events are independent. If a student was born on a Sunday, this has no effect on the probability that the student takes a statistics class.

When two events, A and B, are independent, then $P(B \mid A) = P(B)$, because knowing that A occurred does not affect the probability that B occurs. This leads to a simplified version of the Multiplication Rule.

EXPLAIN IT AGAIN

The Multiplication Rule for Independent Events: Use the Multiplication Rule for Independent Events to compute probabilities of the form $P(A \text{ and } B)$ when A and B are independent.

The Multiplication Rule for Independent Events

If A and B are independent events, then

$$P(A \text{ and } B) = P(A)P(B)$$

This rule can be extended to the case where there are more than two independent events. If $A, B, C, \ldots$ are independent events, then

$$P(A \text{ and } B \text{ and } C \text{ and } \ldots) = P(A)P(B)P(C)\cdots$$

Example 5.23

Using the Multiplication Rule for Independent Events

According to recent figures from the U.S. Census Bureau, the percentage of people under the age of 18 was 23.5% in New York City, 25.8% in Chicago, and 26.0% in Los Angeles. If one person is selected from each city, what is the probability that all of them are under 18? Is this an unusual event?

Solution

There are three events: person from New York is under 18, person from Chicago is under 18, and person from Los Angeles is under 18. These three events are independent, because the identity of the person chosen from one city does not affect who is chosen in the other cities. We therefore use the Multiplication Rule for Independent Events. Let N denote the event that the person from New York is under 18, and let C and L denote the corresponding events for Chicago and Los Angeles, respectively.

$$P(N \text{ and } C \text{ and } L) = P(N) \cdot P(C) \cdot P(L) = 0.235 \cdot 0.258 \cdot 0.260 = 0.0158$$

The probability is 0.0158. This is an unusual event, if we apply the most commonly used cutoff point of 0.05.

NOTE TO INSTRUCTOR

Stress the distinction between mutually exclusive events and independent events. Students tend to confuse the two concepts.

Distinguishing mutually exclusive from independent

Although the mutually exclusive property and the independence property are quite different, in practice it can be difficult to distinguish them. The following diagram can help you to determine whether two events are mutually exclusive, independent, or neither.

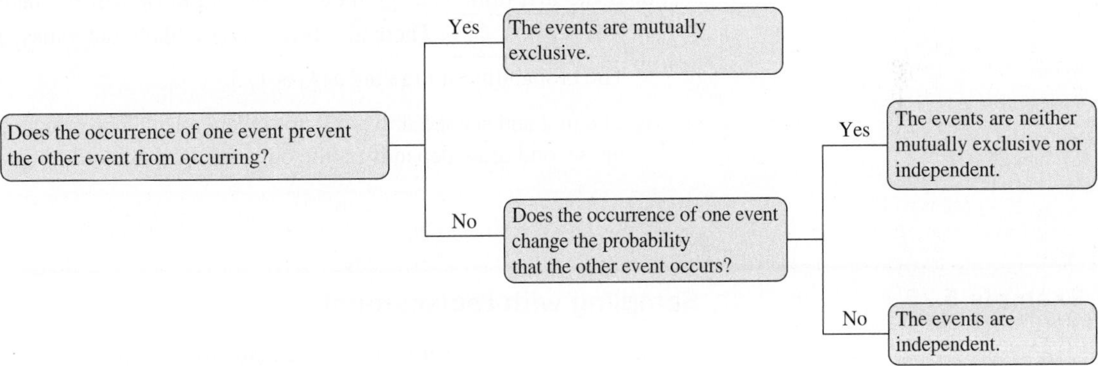

Check Your Understanding

3. Two dice are rolled. Each comes up with a number between 1 and 6. Let A be the event that the number on the first die is even, and let B be the event that the number on the second die is 6.

a. Explain why events A and B are independent.

b. Find $P(A)$, $P(B)$, and $P(A \text{ and } B)$. $P(A) = 1/2$, $P(B) = 1/6$, $P(A \text{ and } B) = 1/12$

Answers are on page 248.

Sampling with and without replacement

When we sample two items from a population, we can proceed in either of two ways. We can replace the first item drawn before sampling the second; this is known as **sampling with replacement**. When sampling with replacement, it is possible to draw the same item more than once. The other option is to leave the first item out when sampling the second one; this is known as **sampling without replacement**. When sampling without replacement, it is impossible to sample an item more than once.

When sampling with replacement, each draw is made from the entire population, so the probability of drawing a particular item on the second draw does not depend on the first draw. In other words, when sampling with replacement, the draws are independent. When sampling without replacement, the draws are not independent. Examples 5.24 and 5.25 illustrate this idea.

Example 5.24

Sampling without replacement

A box contains two cards marked with a "0" and two cards marked with a "1" as shown in the following illustration. Two cards will be sampled without replacement from this population.

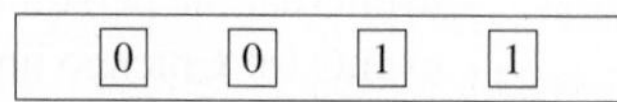

a. What is the probability of drawing a [1] on the second draw given that the first draw is a [0]?

b. What is the probability of drawing a [1] on the second draw given that the first draw is a [1]?

c. Are the first and second draws independent?

Solution

a. If the first draw is a [0], then the second draw will be made from the population [0] [1] [1]. There are three equally likely outcomes, and two of them are [1]. The probability of drawing a [1] is $2/3$.

b. If the first draw is a [1], then the second draw will be made from the population [0] [0] [1]. There are three equally likely outcomes, and one of them is [1]. The probability of drawing a [1] is $1/3$.

c. The first and second draws are not independent. The probability of drawing a [1] on the second draw depends on the outcome of the first draw.

Example 5.25

Sampling with replacement

Two items will be sampled with replacement from the population in Example 5.24. Does the probability of drawing a [1] on the second draw depend on the outcome of the first draw? Are the first and second draws independent?

Solution

Since the sampling is with replacement, then no matter what the first draw is, the second draw will be made from the entire population [0] [0] [1] [1]. Therefore, the probability of drawing a [1] on the second draw is $2/4 = 0.5$ no matter what the first draw is. Since the probability on the second draw does not depend on the outcome of the first draw, the first and second draws are independent.

The population in Examples 5.24 and 5.25 was very small—only four items. When the population is large, the draws will be nearly independent even when sampled without replacement, as illustrated in Example 5.26.

Example 5.26

Sampling without replacement from a large population

A box contains 1000 cards marked with a "0" and 1000 cards marked with a "1," as shown in the following illustration. Two cards will be sampled without replacement from this population.

1000 [0] 's 1000 [1] 's

a. What is the probability of drawing a [1] on the second draw given that the first draw is a [0]?

b. What is the probability of drawing a [1] on the second draw given that the first draw is a [1]?

c. Are the first and second draws independent? Are they approximately independent?

Solution

a. If the first draw is a [0], then the second draw will be made from the population [999 [0]'s 1000 [1]'s]. There are 1999 equally likely outcomes, and 1000 of them are [1]. The probability of drawing a [1] is $1000/1999 = 0.50025$.

b. If the first draw is a [1], then the second draw will be made from the population [1000 [0]'s 999 [1]'s]. There are 1999 equally likely outcomes, and 999 of them are [1]. The probability of drawing a [1] is $999/1999 = 0.49975$.

c. The probability of drawing a [1] on the second draw depends slightly on the outcome of the first draw, so the draws are not independent. However, because the difference in the probabilities is so small (0.50025 versus 0.49975), the draws are approximately independent. In practice, it would be appropriate to treat the two draws as independent.

Example 5.26 shows that when the sample size is small compared to the population size, then items sampled without replacement may be treated as independent. A rule of thumb is that the items may be treated as independent so long as the sample comprises less than 5% of the population.

EXPLAIN IT AGAIN

Replacement doesn't matter when the population is large: When the sample size is less than 5% of the population, it doesn't matter whether the sampling is done with or without replacement. In either case, we will treat the sampled items as independent.

SUMMARY

- When sampling with replacement, the sampled items are independent.
- When sampling without replacement, if the sample size is less than 5% of the population, the sampled items may be treated as independent.
- When sampling without replacement, if the sample size is more than 5% of the population, the sampled items cannot be treated as independent.

Check Your Understanding

4. A pollster plans to sample 1500 voters from a city in which there are 1 million voters. Can the sampled voters be treated as independent? Explain. *Yes*

5. Five hundred students attend a college basketball game. Fifty of them are chosen at random to receive a free T-shirt. Can the sampled students be treated as independent? Explain. *No*

Answers are on page 248.

Objective 4 Compute the probability that an event occurs at least once

Solving "at least once" problems by using complements

Sometimes we need to find the probability that an event occurs **at least once** in several independent trials. We can calculate such probabilities by finding the probability of the complement and subtracting from 1. Examples 5.27 and 5.28 illustrate the method.

Example 5.27

Find the probability that an event occurs at least once

A fair coin is tossed five times. What is the probability that it comes up heads at least once?

EXPLAIN IT AGAIN

Solving "at least once" problems: To compute the probability that an event occurs at least once, find the probability that it does not occur at all, and subtract from 1.

NOTE TO INSTRUCTOR

It is worth taking the time to clearly explain the meaning of "at least once," specifically that it is the complement of "not at all."

Solution

The tosses of a coin are independent, since the outcome of a toss is not affected by the outcomes of other tosses. The complement of coming up heads at least once is coming up tails all five times. We use the Rule of Complements to compute the probability.

$$\begin{aligned}
&P(\text{Comes up heads at least once})\\
&\quad = 1 - P(\text{Does not come up heads at all})\\
&\quad = 1 - P(\text{Comes up tails all five times})\\
&\quad = 1 - P(\text{First toss is T and Second toss is T and} \ldots \text{ and Fifth toss is T})\\
&\quad = 1 - P(\text{First toss is T})P(\text{Second toss is T}) \cdots P(\text{Fifth toss is T})\\
&\quad = 1 - \left(\frac{1}{2}\right)^5\\
&\quad = \frac{31}{32}
\end{aligned}$$

Example 5.28

Find the probability that an event occurs at least once

Items are inspected for flaws by three inspectors. If a flaw is present, each inspector will detect it with probability 0.8. The inspectors work independently. If an item has a flaw, what is the probability that at least one inspector detects it?

Solution

The complement of the event that at least one of the inspectors detects the flaw is that none of the inspectors detects the flaw. We use the Rule of Complements to compute the probability.

We begin by computing the probability that an inspector fails to detect a flaw.

$$P(\text{Inspector fails to detect a flaw}) = 1 - P(\text{Inspector detects flaw}) = 1 - 0.8 = 0.2$$

$$\begin{aligned}
&P(\text{At least one inspector detects the flaw})\\
&\quad = 1 - P(\text{None of the inspectors detects the flaw})\\
&\quad = 1 - P(\text{All three inspectors fail to detect the flaw})\\
&\quad = 1 - P(\text{First fails and second fails and third fails})\\
&\quad = 1 - P(\text{First fails})\,P(\text{Second fails})\,P(\text{Third fails})\\
&\quad = 1 - (0.2)^3\\
&\quad = 0.992
\end{aligned}$$

Check Your Understanding

6. An office has three smoke detectors. In case of fire, each detector has probability 0.9 of detecting it. If a fire occurs, what is the probability that at least one detector detects it? *0.999*

Answer is on page 248.

Determining Which Method to Use

We have studied several methods for finding probabilities of events of the form $P(A \text{ and } B)$, $P(A \text{ or } B)$, and $P(\text{At least one})$. The following diagram can help you to determine the correct method to use for calculating these probabilities.

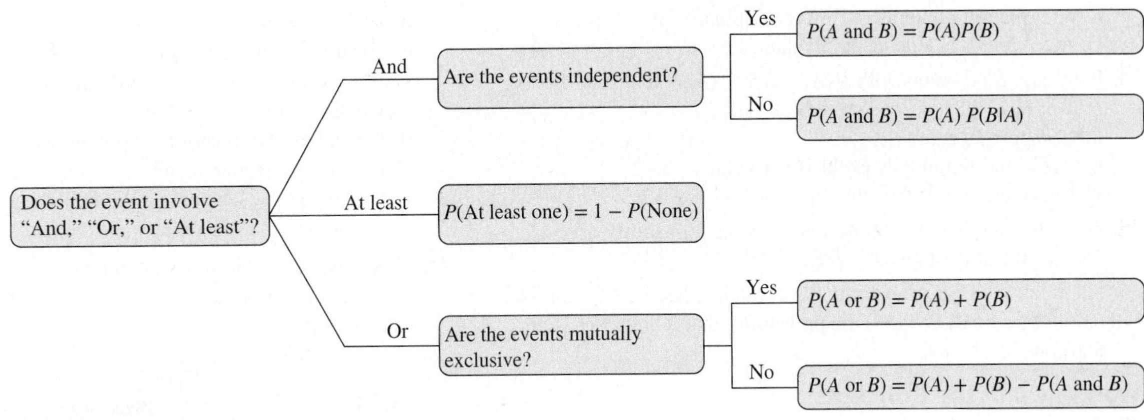

Section 5.3 Exercises

Exercises 1–6 are the Check Your Understanding exercises located within the section.

Understanding the Concepts

In Exercises 7–10, fill in each blank with the appropriate word or phrase.

7. A probability that is computed with the knowledge of additional information is called a ________ probability. *conditional*

8. The General Multiplication Rule states that $P(A \text{ and } B) =$ ________. *$P(A)P(B \mid A)$ or $P(B)P(A \mid B)$*

9. When sampling without replacement, if the sample size is less than ________ % of the population, the sampled items may be treated as independent. *5*

10. Two events are ________ if the occurrence of one does not affect the probability that the other event occurs. *independent*

In Exercises 11–14, determine whether the statement is true or false. If the statement is false, rewrite it as a true statement.

11. $P(B \mid A)$ represents the probability that A occurs under the assumption that B occurs. *False*

12. If A and B are independent events, then $P(A \text{ and } B) = P(A)P(B)$. *True*

13. When sampling without replacement, it is possible to draw the same item from the population more than once. *False*

14. When sampling with replacement, the sampled items are independent. *True*

Practicing the Skills

15. Let A and B be events with $P(A) = 0.4$, $P(B) = 0.7$, and $P(B \mid A) = 0.3$. Find $P(A \text{ and } B)$. *0.12*

16. Let A and B be events with $P(A) = 0.6$, $P(B) = 0.4$, and $P(B \mid A) = 0.4$. Find $P(A \text{ and } B)$. *0.24*

17. Let A and B be events with $P(A) = 0.2$ and $P(B) = 0.9$. Assume that A and B are independent. Find $P(A \text{ and } B)$. *0.18*

18. Let A and B be events with $P(A) = 0.5$ and $P(B) = 0.7$. Assume that A and B are independent. Find $P(A \text{ and } B)$. *0.35*

19. Let A and B be events with $P(A) = 0.8$, $P(B) = 0.1$, and $P(B \mid A) = 0.2$. Find $P(A \text{ and } B)$. *0.16*

20. Let A and B be events with $P(A) = 0.3$, $P(B) = 0.5$, and $P(B \mid A) = 0.7$. Find $P(A \text{ and } B)$. *0.21*

21. Let A, B, and C be independent events with $P(A) = 0.7$, $P(B) = 0.8$, and $P(C) = 0.5$. Find $P(A \text{ and } B \text{ and } C)$. *0.28*

22. Let A, B, and C be independent events with $P(A) = 0.4$, $P(B) = 0.9$, and $P(C) = 0.7$. Find $P(A \text{ and } B \text{ and } C)$. *0.252*

23. A fair coin is tossed four times. What is the probability that all four tosses are heads? *0.0625*

24. A fair coin is tossed four times. What is the probability that the sequence of tosses is HTHT? *0.0625*

25. A fair die is rolled three times. What is the probability that the sequence of rolls is 1, 2, 3? *0.0046*

26. A fair die is rolled three times. What is the probability that all three rolls are 6? *0.0046*

In Exercises 27–30, assume that a student is chosen at random from a class. Determine whether the events A and B are independent, mutually exclusive, or neither.

27. A: The student is a freshman.
B: The student is a sophomore. *Mutually exclusive*

28. A: The student is on the basketball team.
B: The student is more than six feet tall. *Neither*

29. A: The student is a woman.
B: The student belongs to a sorority. *Neither*

30. A: The student is a woman.
B: The student belongs to a fraternity. *Mutually exclusive*

31. Let A and B be events with $P(A) = 0.25$, $P(B) = 0.4$, and $P(A \text{ and } B) = 0.1$.

a. Are A and B independent? Explain. *Yes*
b. Compute $P(A \text{ or } B)$. *0.55*
c. Are A and B mutually exclusive? Explain. *No*

32. Let A and B be events with $P(A) = 0.6$, $P(B) = 0.9$, and $P(A \text{ and } B) = 0.5$.

a. Are A and B independent? Explain. *No*
b. Compute $P(A \text{ or } B)$. *1*
c. Are A and B mutually exclusive? Explain. *No*

33. Let A and B be events with $P(A) = 0.4$, $P(B) = 0.5$, and $P(A \text{ or } B) = 0.6$.

a. Compute $P(A \text{ and } B)$. *0.3*

b. Are A and B mutually exclusive? Explain. No
c. Are A and B independent? Explain. No

34. Let A and B be events with $P(A) = 0.5$, $P(B) = 0.3$, and $P(A \text{ or } B) = 0.8$.
a. Compute $P(A \text{ and } B)$. 0
b. Are A and B mutually exclusive? Explain. Yes
c. Are A and B independent? Explain. No

35. A fair die is rolled three times. What is the probability that it comes up 6 at least once? 0.4213

36. An unfair coin has probability 0.4 of landing heads. The coin is tossed four times. What is the probability that it lands heads at least once? 0.8704

Working with the Concepts

37. **Job interview:** Seven people, named Anna, Bob, Chandra, Darnell, Emma, Francisco, and Gina, will be interviewed for a job. The interviewer will choose two at random to interview on the first day. What is the probability that Anna is interviewed first and Darnell is interviewed second? 0.0238

38. **Shuffle:** Charles has six songs on a playlist. Each song is by a different artist. The artists are Drake, Post Malone, BTS, Ed Sheeran, Taylor Swift, and Cardi B. He programs his player to play the songs in a random order, without repetition. What is the probability that the first song is by Drake and the second song is by Cardi B? 0.0333

39. **Let's eat:** A fast-food restaurant chain has 600 outlets in the United States. The following table categorizes them by city population size and location and presents the number of restaurants in each category. A restaurant is to be chosen at random from the 600 to test market a new menu.

Population	Region			
of city	**NE**	**SE**	**SW**	**NW**
Under 50,000	30	35	15	5
50,000–500,000	60	90	70	30
Over 500,000	150	25	30	60

a. Given that the restaurant is located in a city with a population over 500,000, what is the probability that it is in the Northeast? 0.5660
b. Given that the restaurant is located in the Southeast, what is the probability that it is in a city with a population under 50,000? 0.2333
c. Given that the restaurant is located in the Southwest, what is the probability that it is in a city with a population of 500,000 or less? 0.7391
d. Given that the restaurant is located in a city with a population of 500,000 or less, what is the probability that it is in the Southwest? 0.2537
e. Given that the restaurant is located in the South (either SE or SW), what is the probability that it is in a city with a population of 50,000 or more? 0.8113

40. **Senators:** The following table displays the 100 senators of the 116th U.S. Congress on January 3, 2019, classified by political party affiliation and gender.

	Male	Female	Total
Democrat	28	17	45
Republican	45	8	53
Independent	2	0	2
Total	75	25	100

A senator is selected at random from this group. Compute the following probabilities.

a. What is the probability that the senator is a woman? 0.25
b. What is the probability that the senator is a Republican? 0.53
c. What is the probability that the senator is a Republican and a woman? 0.08
d. Given that the senator is a woman, what is the probability that she is a Republican? 0.32
e. Given that the senator is a Republican, what is the probability that the senator is a woman? 0.1509

41. **Genetics:** A geneticist is studying two genes. Each gene can be either dominant or recessive. A sample of 100 individuals is categorized as follows.

	Gene 2	
Gene 1	**Dominant**	**Recessive**
Dominant	56	24
Recessive	14	6

a. What is the probability that in a randomly sampled individual, gene 1 is dominant? 0.8
b. What is the probability that in a randomly sampled individual, gene 2 is dominant? 0.7
c. Given that gene 1 is dominant, what is the probability that gene 2 is dominant? 0.7
d. Two genes are said to be in linkage equilibrium if the event that gene 1 is dominant is independent of the event that gene 2 is dominant. Are these genes in linkage equilibrium? Yes

42. **Quality control:** A population of 600 semiconductor wafers contains wafers from three lots. The wafers are categorized by lot and by whether they conform to a thickness specification, with the results shown in the following table. A wafer is chosen at random from the population.

Lot	Conforming	Nonconforming
A	88	12
B	165	35
C	260	40

a. What is the probability that a wafer is from Lot A? 0.1667
b. What is the probability that a wafer is conforming? 0.8550
c. What is the probability that a wafer is from Lot A and is conforming? 0.1467
d. Given that the wafer is from Lot A, what is the probability that it is conforming? 0.88
e. Given that the wafer is conforming, what is the probability that it is from Lot A? 0.1715
f. Let E_1 be the event that the wafer comes from Lot A, and let E_2 be the event that the wafer is conforming. Are E_1 and E_2 independent? Explain. No

43. **Stay in school:** In a recent school year in the state of Washington, there were 326,000 high school students. Of these, 167,000 were male. A total of 18,100 students dropped out, and of these, 10,300 were male. A student is chosen at random.

a. What is the probability that the student is male? 0.5123
b. What is the probability that the student dropped out? 0.0555
c. What is the probability that the student is male and dropped out? 0.0316
d. Given that the student is male, what is the probability that he dropped out? 0.0617
e. Given that the student dropped out, what is the probability that the student is male? 0.5691

44. **Management:** The Bureau of Labor Statistics reported that 64.5 million women and 74.6 million men were employed. Of the women, 25.8 million had management jobs, and of the men, 25.0 million had management jobs. An employed person is chosen at random.
   a. What is the probability that the person is a female? 0.4637
   b. What is the probability that the person has a management job? 0.3652
   c. What is the probability that the person is female and has a management job? 0.1855
   d. Given that the person is female, what is the probability that she has a management job? 0.4000
   e. Given that the person has a management job, what is the probability that the person is female? 0.5079

45. **Asthma:** An article in the journal *Risk Analysis* reported that 5.6% of a population has asthma and that on any given day, 2.7% of asthmatics suffer an asthma attack. A person is chosen at random from this population. What is the probability that this person has an asthma attack on that day? 0.0015

46. **Smoking and blood pressure:** According to a recent National Health Examination Survey, the probability that a randomly chosen adult is a smoker is 0.24, and given that a person is a smoker, the probability that the person has high blood pressure (systolic blood pressure above 130 mmHg) is 0.25. What is the probability that a randomly chosen adult is a smoker with high blood pressure? 0.06

47. **GED:** In a certain high school, the probability that a student drops out is 0.05, and the probability that a dropout gets a high-school equivalency diploma (GED) is 0.25. What is the probability that a randomly selected student gets a GED? 0.0125

48. **Working for a living:** The Bureau of Labor Statistics reported that the probability that a randomly chosen employed adult worked in a service occupation was 0.17, and given that a person was in a service occupation, the probability that the person was a woman was 0.57. What is the probability that a randomly chosen employed person was a woman in a service occupation? 0.0969

49. **New car:** At a certain car dealership, the probability that a customer purchases an SUV is 0.20. Given that a customer purchases an SUV, the probability that it is black is 0.25. What is the probability that a customer purchases a black SUV? 0.05

50. **Do you know Squidward?** According to a survey by Nickelodeon TV, 88% of children under 13 in Germany recognized a picture of the cartoon character SpongeBob SquarePants. Assume that among those children, 72% also recognized SpongeBob's cranky neighbor Squidward Tentacles. What is the probability that a German child recognized both SpongeBob and Squidward? 0.6336

51. **Target practice:** Laura and Felipe each fire one shot at a target. Laura has probability 0.5 of hitting the target, and Felipe has probability 0.3. The shots are independent.
   a. Find the probability that both of them hit the target. 0.15
   b. Given that Laura hits the target, the probability is 0.1 that Felipe's shot hits the target closer to the bull's-eye than Laura's. Find the probability that Laura hits the target and that Felipe's shot is closer to the bull's-eye than Laura's shot is. 0.05

52. **Bowling:** Sarah and Thomas are going bowling. The probability that Sarah scores more than 175 is 0.4, and the probability that Thomas scores more than 175 is 0.2. Their scores are independent.
   a. Find the probability that both score more than 175. 0.08
   b. Given that Thomas scores more than 175, the probability that Sarah scores higher than Thomas is 0.3. Find the probability that Thomas scores more than 175 and Sarah scores higher than Thomas. 0.06

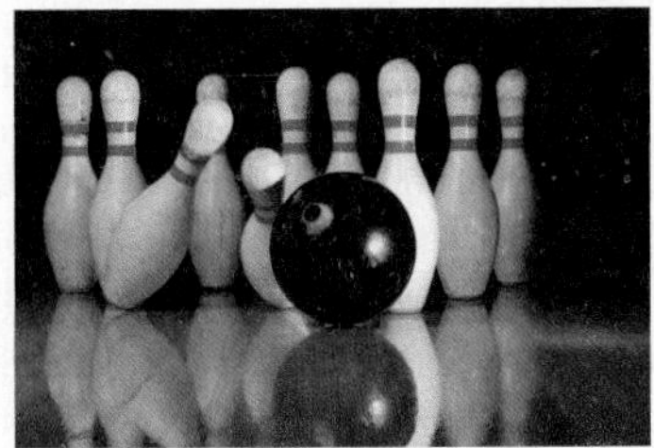

Rim Light/PhotoLink/Getty Images

53. **Defective components:** A lot of 10 components contains 3 that are defective. Two components are drawn at random and tested. Let A be the event that the first component drawn is defective, and let B be the event that the second component drawn is defective.
   a. Find $P(A)$. 0.3
   b. Find $P(B \mid A)$. 0.2222
   c. Find $P(A \text{ and } B)$. 0.0667
   d. Are A and B independent? Explain. No

54. **More defective components:** A lot of 1000 components contains 300 that are defective. Two components are drawn at random and tested. Let A be the event that the first component drawn is defective, and let B be the event that the second component drawn is defective.
   a. Find $P(A)$. 0.3
   b. Find $P(B \mid A)$. 0.2993
   c. Find $P(A \text{ and } B)$. 0.0898
   d. Are A and B independent? Is it reasonable to treat A and B as though they were independent? Explain. No; yes

55. **Multiply probabilities?** In a recent year, 21% of all vehicles in operation were pickup trucks. If someone owns two vehicles, is the probability that they are both pickup trucks equal to $0.21 \times 0.21 = 0.0441$? Explain why or why not. No

56. **Multiply probabilities?** A traffic light at an intersection near Jamal's house is red 50% of the time, green 40% of the time, and yellow 10% of the time. Jamal encounters this light in the morning on his way to work and again in the evening on his way home. Is the probability that the light is green both times equal to $0.4 \times 0.4 = 0.16$? Explain why or why not. Yes

57. **Lottery:** Every day, Jorge buys a lottery ticket. Each ticket has probability 0.2 of winning a prize. After seven days, what is the probability that Jorge has won at least one prize? 0.7903

58. **Car warranty:** The probability that a certain make of car will need repairs in the first six months is 0.3. A dealer sells five such cars. What is the probability that at least one of them will require repairs in the first six months? 0.8319

59. **Tic-tac-toe:** In the game of tic-tac-toe, if all moves are performed randomly the probability that the game will end in a draw is 0.127. Suppose 10 random games of tic-tac-toe are played. What is the probability that at least one of them will end in a draw? 0.743

60. Enter your PIN: The technology consulting company DataGenetics suggests that 17.8% of all four-digit personal identification numbers, or PIN codes, have a repeating digits format such as 2525. Assuming this to be true, if the PIN codes of six people are selected at random, what is the probability that at least one of them will have repeating digits? *0.692*

Extending the Concepts

Exercises 61–64 refer to the following situation:

A medical test is available to determine whether a patient has a certain disease. To determine the accuracy of the test, a total of 10,100 people are tested. Only 100 of these people have the disease, while the other 10,000 are disease free. Of the disease-free people, 9800 get a negative result, and 200 get a positive result. The 100 people with the disease all get positive results.

61. Find the probability: Find the probability that the test gives the correct result for a person who does not have the disease. *0.98*

62. Find the probability: Find the probability that the test gives the correct result for a person who has the disease. *1*

63. Find the probability: Given that a person gets a positive result, what is the probability that the person actually has the disease? *0.3333*

64. Why are medical tests repeated? For many medical tests, if the result comes back positive, the test is repeated. Why do you think this is done?

65. Mutually exclusive and independent? Let A and B be events. Assume that neither A nor B can occur; in other words, $P(A) = 0$ and $P(B) = 0$. Are A and B independent? Are A and B mutually exclusive? Explain. *Yes*

66. Still mutually exclusive and independent? Let A and B be events. Now assume that $P(A) = 0$ but $P(B) > 0$. Are A and B always independent? Are A and B always mutually exclusive? Explain. *Yes*

67. Mutually exclusive and independent again? Let A and B be events. Now assume that $P(A) > 0$ and $P(B) > 0$. Is it possible for A and B to be both independent and mutually exclusive? Explain. *No*

Answers to Check Your Understanding Exercises for Section 5.3

1. a. 0.164 **b.** 0.315 **c.** 0.519

2. 0.28

3. a. The outcome on one die does not influence the outcome on the other die.

b. $P(A) = 1/2$; $P(B) = 1/6$; $P(A \text{ and } B) = 1/12$

4. Yes, because the sample is less than 5% of the population.

5. No, because the sample is more than 5% of the population.

6. 0.999

Section 5.4 Counting

Objectives

1. Count the number of ways a sequence of operations can be performed

2. Count the number of permutations

3. Count the number of combinations

When computing probabilities, it is sometimes necessary to count the number of outcomes in a sample space without being able to list them all. In this section, we will describe several methods for doing this.

Objective 1 Count the number of ways a sequence of operations can be performed

The Fundamental Principle of Counting

The basic rule, which we will call the **Fundamental Principle of Counting**, is presented by means of the following example:

Example 5.29

Using the Fundamental Principle of Counting

A certain make of automobile is available in any of three colors—red, blue, or green—and comes with either a large or small engine. In how many ways can a buyer choose a car?

Solution

There are 3 choices of color and 2 choices of engine. A complete list is shown in the tree diagram in Figure 5.8, and in the form of a table in Table 5.6. The total number of choices is $3 \cdot 2 = 6$.

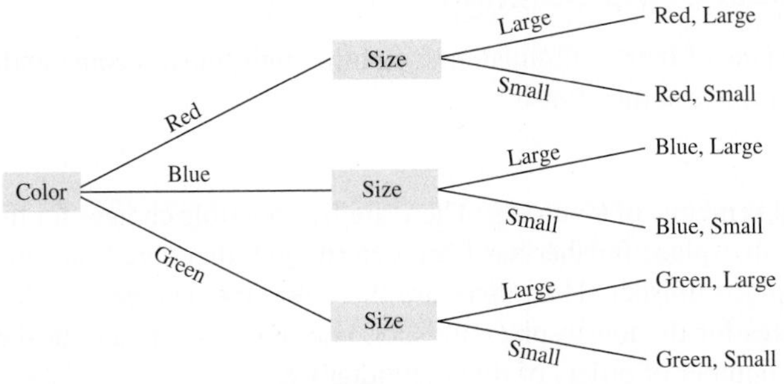

Figure 5.8 Tree diagram illustrating the six choices of color and engine size

Table 5.6 Six Outcomes for the Color and Engine Size

	Large	Small
Red	Red, Large	Red, Small
Blue	Blue, Large	Blue, Small
Green	Green, Large	Green, Small

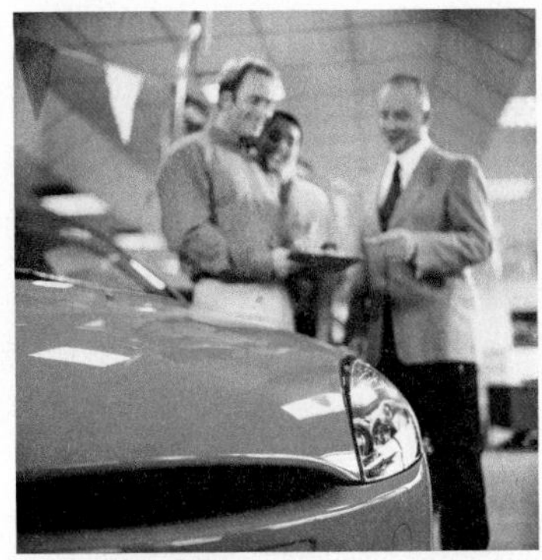

Stockbyte/Getty Images

To generalize Example 5.29, if there are m choices of color and n choices of engine, the total number of choices is mn. This leads to the Fundamental Principle of Counting.

The Fundamental Principle of Counting

If an operation can be performed in m ways, and a second operation can be performed in n ways, then the total number of ways to perform the sequence of two operations is mn.

If a sequence of several operations is to be performed, the number of ways to perform the sequence is found by multiplying together the numbers of ways to perform each of the operations.

Example 5.30 Using the Fundamental Principle of Counting

License plates in a certain state contain three letters followed by three digits. How many different license plates can be made?

Solution

There are six operations in all: choosing three letters and choosing three digits. There are 26 ways to choose each letter and 10 ways to choose each digit. The total number of license plates is therefore

$$26 \cdot 26 \cdot 26 \cdot 10 \cdot 10 \cdot 10 = 17{,}576{,}000$$

NOTE TO INSTRUCTOR

The most effective way to explain the Fundamental Principle of Counting is by presenting examples.

Check Your Understanding

1. When ordering a certain type of computer, there are three choices of hard drive, four choices for the amount of memory, two choices of video card, and three choices of monitor. In how many ways can a computer be ordered? *72*
2. A quiz consists of three true–false questions and two multiple-choice questions with five choices each. How many different sets of answers are there? *200*

Answers are on page 256.

Objective 2 Count the number of permutations

Permutations

The word *permutation* is another word for *ordering*. When we count the number of permutations, we are counting the number of different ways that a group of items can be ordered.

Example 5.31

Counting the number of permutations

Five runners run a race. One of them will finish first, another will finish second, and so on. In how many different orders can they finish?

Solution

We use the Fundamental Principle of Counting. There are five possible choices for the first-place finisher. Once the first-place finisher has been determined, there are four remaining choices for the second-place finisher. Then there are three possible choices for the third-place finisher, two choices for the fourth-place finisher, and only one choice for the fifth-place finisher. The total number of orders of five individuals is

$$\text{Number of orders} = 5 \cdot 4 \cdot 3 \cdot 2 \cdot 1 = 120$$

We say that there are 120 permutations of five individuals.

In Example 5.31, we computed a number of permutations by using the Fundamental Principle of Counting. We can generalize this method, but first we need some notation.

DEFINITION

For any positive integer n, the number $n!$ is pronounced "n factorial" and is equal to the product of all the integers from n down to 1.

$$n! = n(n-1)\cdots(2)(1)$$

By definition, $0! = 1$.

In Example 5.31, we found that the number of permutations of five objects is $5!$. This idea holds in general.

The number of permutations of n objects is $n!$.

Sometimes we want to count the number of permutations of a part of a group. Example 5.32 illustrates the idea.

Example 5.32

Counting the number of permutations

Ten runners enter a race. The first-place finisher will win a gold medal, the second-place finisher will win a silver medal, and the third-place finisher will win a bronze medal. In how many different ways can the medals be awarded?

Solution

We use the Fundamental Principle of Counting. There are 10 possible choices for the gold-medal winner. Once the gold-medal winner is determined, there are nine remaining choices for the silver medal. Finally, there are eight choices for the bronze medal. The total number of ways the medals can be awarded is

$$10 \cdot 9 \cdot 8 = 720$$

In Example 5.32, three runners were chosen from a group of 10, then ordered as first, second, and third. This is referred to as a *permutation* of three items chosen from 10.

DEFINITION

A **permutation** of r items chosen from n items is an ordering of the r items. It is obtained by choosing r items from a group of n items, then choosing an order for the r items.

Notation: The number of permutations of r items chosen from n is denoted ${}_nP_r$.

In Example 5.32, we computed ${}_nP_r$ by using the Fundamental Principle of Counting. We can generalize this method by using factorial notation.

The number of permutations of r objects chosen from n is

$${}_nP_r = n(n-1)\cdots(n-r+1) = \frac{n!}{(n-r)!}$$

Example 5.33

Counting the number of permutations

Five lifeguards are available for duty one Saturday afternoon. There are three lifeguard stations. In how many ways can three lifeguards be chosen and ordered among the stations?

Solution

We are choosing three items from a group of five and ordering them. The number of ways to do this is

$${}_5P_3 = \frac{5!}{(5-3)!} = \frac{5!}{2!} = \frac{5\cdot 4\cdot 3\cdot 2\cdot 1}{2\cdot 1} = 5\cdot 4\cdot 3 = 60$$

In some situations, computing the value of ${}_nP_r$ enables us to determine the number of outcomes in a sample space, and thereby compute a probability. Example 5.34 illustrates the idea.

Example 5.34

Using counting to compute a probability

Refer to Example 5.33. The five lifeguards are named Abby, Bruce, Christopher, Donna, and Esmeralda. Of the three lifeguard stations, one is located at the north end of the beach, one in the middle of the beach, and one at the south end. The lifeguard assignments are made at random. What is the probability that Bruce is assigned to the north station, Donna is assigned to the middle station, and Abby is assigned to the south station?

Solution

The outcomes in the sample space consist of all the choices of three lifeguards chosen from five and ordered. From Example 5.33, we know that there are 60 such outcomes. Only one of the outcomes has Bruce, Donna, Abby, in that order. Thus, the probability is $\frac{1}{60}$.

Example 5.35

Using counting to compute a probability

Refer to Example 5.34. What is the probability that Bruce is assigned to the north station, Abby is assigned to the south station, and either Donna or Esmeralda is assigned to the middle station?

Solution

As in Example 5.34, the sample space consists of the 60 permutations of three lifeguards chosen from five. Two of these permutations satisfy the stated conditions: Bruce, Donna, Abby; and Bruce, Esmeralda, Abby. So the probability is $\frac{2}{60} = \frac{1}{30}$.

Check Your Understanding

3. A committee of eight people must choose a president, a vice president, and a secretary. In how many ways can this be done? *336*

4. Refer to Exercise 3. Two of the committee members are Ellen and Jose. Assume the assignments are made at random.

a. What is the probability that Jose is president and Ellen is vice president? *1/56*

b. What is the probability that either Ellen or Jose is president and the other is vice president? *1/28*

Answers are on page 256.

Objective 3 Count the number of combinations

Combinations

In some cases, when choosing a set of objects from a larger set, we don't care about the ordering of the chosen objects; we care only which objects are chosen. For example, we may not care which lifeguard occupies which station; we might care only which three lifeguards are chosen. Each distinct group of objects that can be selected, without regard to order, is called a **combination**. We will now show how to determine the number of combinations of r objects chosen from a set of n objects. We will illustrate the reasoning with the result of Example 5.33. In that example, we showed that there are 60 permutations of 3 objects chosen from 5. Denoting the objects A, B, C, D, E, Table 5.7 presents a list of all 60 permutations.

NOTE TO INSTRUCTOR

Explain that if we change the order of a list of objects, we have a new permutation, but the same combination.

Table 5.7 The 60 Permutations of 3 Objects Chosen from 5

ABC	ABD	ABE	ACD	ACE	ADE	BCD	BCE	BDE	CDE
ACB	ADB	AEB	ADC	AEC	AED	BDC	BEC	BED	CED
BAC	BAD	BAE	CAD	CAE	DAE	CBD	CBE	DBE	DCE
BCA	BDA	BEA	CDA	CEA	DEA	CDB	CEB	DEB	DEC
CAB	DAB	EAB	DAC	EAC	EAD	DBC	EBC	EBD	ECD
CBA	DBA	EBA	DCA	ECA	EDA	DCB	ECB	EDB	EDC

EXPLAIN IT AGAIN

When to use combinations: Use combinations when the order of the chosen objects doesn't matter. Use permutations when the order does matter.

The 60 permutations in Table 5.7 are arranged in 10 columns of 6 permutations each. Within each column, the three objects are the same, and the column contains the 6 different permutations of those three objects. Therefore, each column represents a distinct combination of 3 objects chosen from 5, and there are 10 such combinations. Table 5.7 thus shows that the number of combinations of 3 objects chosen from 5 can be found by dividing the number of permutations of 3 objects chosen from 5, which is $\frac{5!}{(5-3)!}$, by the number of permutations of 3 objects, which is 3!. In summary:

$$\text{The number of combinations of 3 objects chosen from 5 is } \frac{5!}{3!(5-3)!}$$

The number of combinations of r objects chosen from n is often denoted by the symbol ${}_nC_r$. The reasoning above can be generalized to derive an expression for ${}_nC_r$.

The number of combinations of r objects chosen from a group of n objects is

$${}_nC_r = \frac{n!}{r!(n-r)!}$$

Example 5.36

Counting the number of combinations

Thirty people attend a certain event, and 5 will be chosen at random to receive prizes. The prizes are all the same, so the order in which the people are chosen does not matter. How many different groups of 5 people can be chosen?

Solution

Since the order of the 5 chosen people does not matter, we need to compute the number of combinations of 5 chosen from 30. This is

$$
\begin{aligned}
{}_{30}C_5 &= \frac{30!}{5!(30-5)!} \\
&= \frac{30 \cdot 29 \cdot 28 \cdot 27 \cdot 26}{5 \cdot 4 \cdot 3 \cdot 2 \cdot 1} \\
&= 142{,}506
\end{aligned}
$$

Example 5.37

Using counting to compute a probability

Refer to Example 5.36. Of the 30 people in attendance, 12 are men and 18 are women.

a. What is the probability that all the prize winners are men?

b. What is the probability that at least one prize winner is a woman?

Solution

a. The number of outcomes in the sample space is the number of combinations of 5 chosen from 30. We computed this in Example 5.36 to be ${}_{30}C_5 = 142{,}506$. The number of outcomes in which every prize winner is a man is the number of combinations of 5 men chosen from 12 men. This is

$$
{}_{12}C_5 = \frac{12!}{5!(12-5)!} = \frac{12 \cdot 11 \cdot 10 \cdot 9 \cdot 8}{5 \cdot 4 \cdot 3 \cdot 2 \cdot 1} = 792
$$

The probability that all prize winners are men is

$$
P(\text{All men}) = \frac{792}{142{,}506} = 0.0056
$$

b. This asks for the probability of at least one woman. We therefore find the probability of the complement; that is, we find the probability that none of the prize winners are women. The probability that none of the prize winners are women is the same as the probability that all of the prize winners are men. In part (a), we computed $P(\text{All men}) = 0.0056$. Therefore,

$$
P(\text{At least one woman}) = 1 - P(\text{All men}) = 1 - 0.0056 = 0.9944
$$

Example 5.38

Using counting to compute a probability

A box of lightbulbs contains eight good lightbulbs and two burned-out bulbs. Four bulbs will be selected at random to put into a new lamp. What is the probability that all four bulbs are good?

Solution

The order in which the bulbs are chosen does not matter; all that matters is whether a burned-out bulb is chosen. Therefore, the outcomes in the sample space consist of all the combinations of four bulbs that can be chosen from 10. This number is

$$
{}_{10}C_4 = \frac{10!}{4!(10-4)!} = \frac{3{,}628{,}800}{24 \cdot 720} = 210
$$

To select four good bulbs, we must choose the four bulbs from the eight good bulbs. The number of outcomes that correspond to selecting four good bulbs is therefore the number of combinations of four bulbs that can be chosen from eight. This number is

$$
{}_{8}C_4 = \frac{8!}{4!(8-4)!} = \frac{40{,}320}{24 \cdot 24} = 70
$$

The probability that four good bulbs are selected is therefore

$$P(\text{Four good bulbs are selected}) = \frac{70}{210} = \frac{1}{3}$$

Check Your Understanding

5. Eight college students have applied for internships at a local firm. Three of them will be selected for interviews. In how many ways can this be done? *56*

6. Refer to Exercise 5. Four of the eight students are from Middle Georgia State University. What is the probability that all three of the interviewed students are from Middle Georgia State University? *1/14*

Answers are on page 256.

Using Technology

TI-84 PLUS

Evaluating a factorial

Step 1. To evaluate $n!$, enter n on the home screen.

Step 2. Press **MATH**, scroll to the **PRB** menu, and select **4: !**

Step 3. Press **ENTER**.

Permutations and combinations

Step 1. To evaluate ${}_nP_r$ or ${}_nC_r$, enter n on the home screen.

Step 2. Press **MATH** and scroll to the **PRB** menu.

- For permutations, select **2: nPr** and press **ENTER** (Figure A).
- For combinations, select **3: nCr** and press **ENTER**.

Step 3. Enter the value for r and press **ENTER**.

The results of ${}_{12}P_3$ and ${}_{12}C_3$ are shown in Figure B.

```
MATH NUM CMPLX PROB FRAC
1:rand
2:nPr
3:nCr
4:!
5:randInt(
6:randNorm(
7:randBin(
8:randIntNoRep(
```

Figure A

```
12 nPr 3
                    1320
12 nCr 3
                     220
```

Figure B

EXCEL

Evaluating a factorial

Step 1. To evaluate $n!$, click on a cell in the worksheet and type **=FACT(n)** and press **ENTER**. For example, to compute **12!**, type **=FACT(12)** and press **ENTER**.

Permutations

Step 1. To evaluate ${}_nP_r$, click on a cell in the worksheet and type **=PERMUT(n,r)**. Press **ENTER**.

Combinations

Step 1. To evaluate ${}_nC_r$, click on a cell in the worksheet and type **=COMBIN(n,r)**. Press **ENTER**.

Section 5.4 Exercises

Exercises 1–6 are the Check Your Understanding exercises located within the section.

Understanding the Concepts

In Exercises 7 and 8, fill in the blank with the appropriate word or phrase:

7. If an operation can be performed in m ways, and a second operation can be performed in n ways, then the total number of ways to perform the sequence of two operations is ________. *mn*

8. The number of permutations of six objects is ________. *720*

In Exercises 9 and 10, determine whether the statement is true or false. If the statement is false, rewrite it as a true statement.

9. In a permutation, order is not important. *False*

10. In a combination, order is not important. *True*

Practicing the Skills

In Exercises 11–16, evaluate the factorial.

11. 9! *362,880*
12. 5! *120*
13. 0! *1*
14. 12! *479,001,600*
15. 1! *1*
16. 3! *6*

In Exercises 17–22, evaluate the permutation.

17. $_7P_3$ *210*
18. $_8P_1$ *8*
19. $_{35}P_2$ *1190*
20. $_5P_4$ *120*
21. $_{20}P_0$ *1*
22. $_{45}P_5$ *146,611,080*

In Exercises 23–28, evaluate the combination.

23. $_9C_5$ *126*
24. $_7C_1$ *7*
25. $_{25}C_3$ *2300*
26. $_{10}C_9$ *10*
27. $_{12}C_0$ *1*
28. $_{50}C_{50}$ *1*

Working with the Concepts

29. **Pizza time:** A local pizza parlor is offering a half-price deal on any pizza with one topping. There are eight toppings from which to choose. In addition, there are three different choices for the size of the pizza, and two choices for the type of crust. In how many ways can a pizza be ordered? *48*

30. **Books:** Josephine has six chemistry books, three history books, and eight statistics books. She wants to choose one book of each type to study. In how many ways can she choose the three books? *144*

31. **Playing the horses:** In horse racing, one can make a trifecta bet by specifying which horse will come in first, which will come in second, and which will come in third, in the correct order. One can make a box trifecta bet by specifying which three horses will come in first, second, and third, without specifying the order.
   a. In an eight-horse field, how many different ways can one make a trifecta bet? *336*
   b. In an eight-horse field, how many different ways can one make a box trifecta bet? *56*

32. **Ice cream:** A certain ice cream parlor offers 15 flavors of ice cream. You want an ice cream cone with three scoops of ice cream, all different flavors.
   a. In how many ways can you choose a cone if it matters which flavor is on the top, which is in the middle, and which is on the bottom? *2730*
   b. In how many ways can you choose a cone if the order of the flavors doesn't matter? *455*

Alex Cao/Getty Images

33. **License plates:** In a certain state, license plates consist of four digits from 0 to 9 followed by three letters. Assume the numbers and letters are chosen at random. Replicates are allowed.
   a. How many different license plates can be formed? *175,760,000*
   b. How many different license plates have the letters S-A-M in that order? *10,000*
   c. If your name is Sam, what is the probability that your name is on your license plate? *0.0000569*

34. **Committee:** The Student Council at a certain school has 10 members. Four members will form an executive committee consisting of a president, a vice president, a secretary, and a treasurer.
   a. In how many ways can these four positions be filled? *5040*
   b. In how many ways can four people be chosen for the executive committee if it does not matter who gets which position? *210*
   c. Four of the people on Student Council are Zachary, Yolanda, Xavier, and Walter. What is the probability that Zachary is president, Yolanda is vice president, Xavier is secretary, and Walter is treasurer? *0.000198*
   d. What is the probability that Zachary, Yolanda, Xavier, and Walter are the four committee members? *0.00476*

35. **Day and night shifts:** A company has hired 12 new employees and must assign 8 to the day shift and 4 to the night shift.
   a. In how many ways can the assignment be made? *495*
   b. Assume that the 12 employees consist of six men and six women and that the assignments to day and night shift are made at random. What is the probability that all four of the night-shift employees are men? *0.0303*
   c. What is the probability that at least one of the night-shift employees is a woman? *0.9697*

36. **Keep your password safe:** A computer password consists of eight characters. Replications are allowed.
   a. How many different passwords are possible if each character may be any lowercase letter or digit? *$36^8 = 2.82 \times 10^{12}$*
   b. How many different passwords are possible if each character may be any lowercase letter? *$26^8 = 2.09 \times 10^{11}$*
   c. How many different passwords are possible if each character may be any lowercase letter or digit and at least one character must be a digit? *$36^8 - 26^8 = 2.61 \times 10^{12}$*
   d. A computer is generating passwords. The computer generates eight characters at random, and each is equally likely to be any of the 26 letters or 10 digits. Replications are allowed. What is the probability that the password will contain all letters? *0.0740*
   e. A computer system requires that passwords contain at least one digit. If eight characters are generated at random, what is the probability that they will form a valid password? *0.9260*

37. **It's in your genes:** Human genetic material (DNA) is made up of sequences of the molecules adenosine (A), guanine (G), cytosine (C), and thymine (T), which are called *bases*. A *codon* is a sequence of three bases. Replicates are allowed, so AAA, CGC, and so forth are codons. Codons are important because each codon causes a different protein to be created.
   a. How many different codons are there? *64*
   b. How many different codons are there in which all three bases are different? *24*
   c. The bases A and G are called *purines*, while C and T are called *pyrimidines*. How many different codons are there in which the first base is a purine and the second and third are pyrimidines? *8*
   d. What is the probability that all three bases are different? *0.375*
   e. What is the probability that the first base is a purine and the second and third are pyrimidines? *0.125*

38. **Choosing officers:** A committee consists of 10 women and eight men. Three committee members will be chosen as officers.
   a. How many different choices are possible? *816*
   b. How many different choices are possible if all the officers are to be women? *120*
   c. How many different choices are possible if all the officers are to be men? *56*
   d. What is the probability that all the officers are women? *0.1471*
   e. What is the probability that at least one officer is a man? *0.8529*

39. **Texas hold 'em:** In the game of Texas hold 'em, a player is dealt two cards (called hole cards) from a standard deck of 52 playing cards. The order in which the cards are dealt does not matter.
   a. How many different combinations of hole cards are possible? *1326*
   b. The best hand consists of two aces. There are four aces in the deck. How many combinations are there in which both cards are aces? *6*
   c. What is the probability that a hand consists of two aces? *0.00452*

40. **Blackjack:** In single-deck casino blackjack, the dealer is dealt two cards from a standard deck of 52. The first card is dealt face down, and the second card is dealt face up.
   a. How many dealer hands are possible if it matters which card is face down and which is face up? *2652*
   b. How many dealer hands are possible if it doesn't matter which card is face down and which is face up? *1326*
   c. Of the 52 cards in the deck, four are aces and 16 others (kings, queens, jacks, and tens) are worth 10 points each. The dealer has a blackjack if one card is an ace and the other is worth 10 points; it doesn't matter which card is face up and which card is face down. How many different blackjack hands are there? *64*
   d. What is the probability that a hand is a blackjack? *0.0483*

41. **Lottery:** In the Georgia Fantasy 5 Lottery, balls are numbered from 1 to 42. Five balls are drawn. To win the jackpot, you must mark five numbers from 1 to 42 on a ticket, and your numbers must match the numbers on the five balls. The order does not matter. What is the probability that you win? *$1/{}_{42}C_5 = 0.00000118$*

42. **Lottery:** In the Colorado Lottery Lotto game, balls are numbered from 1 to 42. Six balls are drawn. To win the jackpot, you must mark six numbers from 1 to 42 on a ticket, and your numbers must match the numbers on the six balls. The order does not matter. What is the probability that you win? *$1/{}_{42}C_6 = 0.000000191$*

Extending the Concepts

43. **Sentence completion:** Let A and B be events. Consider the following sentence:

   If A and B are ___(i)___, then to find ___(ii)___, ___(iii)___ $P(A)$ and $P(B)$.

   Each blank in the sentence can be filled in with either of two choices, as follows:
   (i) independent, mutually exclusive
   (ii) $P(A \text{ and } B)$, $P(A \text{ or } B)$
   (iii) multiply, add
   a. In how many ways can the sentence be completed? *8*
   b. If choices are made at random for each of the blanks, what is the probability that the sentence is true? *0.25*

Answers to Check Your Understanding Exercises for Section 5.4

1. 72	**4. a.** 1/56 **b.** 1/28
2. 200	**5.** 56
3. 336	**6.** 1/14

Chapter 5 Summary

Section 5.1: A probability experiment is an experiment that can result in any one of a number of outcomes. The collection of all possible outcomes is a sample space. Sampling from a population is a common type of probability experiment. The population is the sample space, and the individuals in the population are the outcomes. An event is a collection of outcomes from a sample space. The probability of an event is the proportion of times the event occurs in the long run, as the experiment is repeated over and over again. A probability model specifies a probability for every event.

An unusual event is one whose probability is small. There is no hard-and-fast rule about how small a probability has to be for an event to be unusual, but 0.05 is the most commonly used value. The Empirical Method allows us to approximate the probability of an event by repeating a probability experiment many times and computing the proportion of times the event occurs. This is generally done by simulation, in which technology is used to repeat virtual experiments many times.

Section 5.2: A compound event is an event that is formed by combining two or more events. An example of a compound event is one of the form "A or B." The General Addition Rule is used to compute probabilities of the form $P(A \text{ or } B)$. Two events are mutually exclusive if it is impossible for both events to occur. When two events are mutually exclusive, the Addition Rule for Mutually Exclusive Events can be used to find $P(A \text{ or } B)$. The complement of an event A, denoted A^c, is the event that A does not occur. The Rule of Complements states that $P(A^c)$ is found by subtracting $P(A)$ from 1.

Section 5.3: A conditional probability is a probability that is computed with the knowledge of additional information. Conditional probabilities can be computed with the General Method for computing conditional probabilities. Probabilities of the form $P(A \text{ and } B)$ can be computed with the General Multiplication Rule. If A and B are independent, then $P(A \text{ and } B)$ can be computed with the Multiplication Rule for Independent Events. Two events are independent if the occurrence of one does not affect the probability that the other occurs. When sampling from a population, sampled individuals are independent if the sampling is done with replacement, or if the sample size is less than 5% of the population.

Section 5.4: The Fundamental Principle of Counting states that the total number of ways to perform a sequence of operations is found by multiplying together the numbers of ways of performing each operation. We can compute the number of permutations and combinations of r items chosen from a group of n items. The number of ways that a group of r items can be chosen without regard to order is the number of combinations. The number of ways that a group of r items can be chosen and ordered is the number of permutations. Some sample spaces consist of the permutations or combinations of r items chosen from a group of n items. When working with these sample spaces, we can use the counting rules to compute probabilities.

Vocabulary and Notation

Important Formulas

General Addition Rule:

$P(A \text{ or } B) = P(A) + P(B) - P(A \text{ and } B)$

Multiplication Rule for Independent Events:

$P(A \text{ and } B) = P(A)P(B)$

Addition Rule for Mutually Exclusive Events:

$P(A \text{ or } B) = P(A) + P(B)$

Rule of Complements:

$P(A^c) = 1 - P(A)$

General Method for Computing Conditional Probability:

$$P(B \mid A) = \frac{P(A \text{ and } B)}{P(A)}$$

General Multiplication Rule:

$P(A \text{ and } B) = P(A)P(B \mid A) = P(B)P(A \mid B)$

Permutation of r items chosen from n:

$${}_nP_r = \frac{n!}{(n-r)!}$$

Combination of r items chosen from n:

$${}_nC_r = \frac{n!}{r!(n-r)!}$$

Chapter Quiz

1. Fill in the blank: The probability that a fair coin lands heads is 0.5. Therefore, we can be sure that if we toss a coin repeatedly, the proportion of times it lands heads will ________. *i*
  i. approach 0.5
  ii. be equal to 0.5
  iii. be greater than 0.5
  iv. be less than 0.5

2. A pollster will draw a simple random sample of voters from a large city to ask whether they support the construction of a new light rail line. Assume that there are one million voters in the city, and that 560,000 of them support this proposition. One voter is sampled at random.
  a. Identify the sample space.
  b. What is the probability that the sampled voter supports the light rail line? *0.56*

3. State each of the following rules:
  a. General Addition Rule
  b. Addition Rule for Mutually Exclusive Events
  c. Rule of Complements
  d. General Multiplication Rule
  e. Multiplication Rule for Independent Events

4. The following table presents the results of a survey in which 400 college students were asked whether they listen to music while studying.

	Listen	Do Not Listen
Male	121	78
Female	147	54

  a. Find the probability that a randomly selected student does not listen to music while studying. *0.33*
  b. Find the probability that a randomly selected student listens to music or is male. *0.865*

5. Which of the following pairs of events are mutually exclusive? *i*
  i. *A*: A randomly chosen student is 18 years old. *B*: The same student is 20 years old.
  ii. *A*: A randomly chosen student owns a red car. *B*: The same student owns a blue car.

6. In a group of 100 teenagers, 61 received their driver's license on their first attempt on the driver's certification exam and 18 received their driver's license on their second attempt. What is the probability that a randomly selected teenager received their driver's license on their first or second attempt? *0.79*

7. A certain neighborhood has 100 households. Forty-eight households have a dog as a pet. Of these, 32 also have a cat. Given that a household has a dog, what is the probability that it also has a cat? *0.6667*

8. The owner of a bookstore has determined that 80% of people who enter the store will buy a book. Of those who buy a book, 60% will pay with a credit card. Find the probability that a randomly selected person entering the store will buy a book and pay for it using a credit card. *0.48*

9. A jar contains 4 red marbles, 3 blue marbles, and 5 green marbles. Two marbles are drawn from the jar one at a time without replacement. What is the probability that the second marble is red, given that the first was blue? *0.3636*

10. A student is chosen at random. Which of the following pairs of events are independent? *i*
  i. *A*: The student was born on a Monday. *B*: The student's mother was born on a Monday.
  ii. *A*: The student is above average in height. *B*: The student's mother is above average in height.

11. Individual plays on a slot machine are independent. The probability of winning on any play is 0.38. What is the probability of winning 3 plays in a row? *0.0549*

12. Refer to Exercise 11. Suppose that the slot machine is played 5 times in a row. What is the probability of winning at least once? *0.9084*

13. The Roman alphabet (the one used to write English) consists of five vowels (a, e, i, o, u), along with 21 consonants (we are considering y to be a consonant). Gregory needs to make up a computer password containing seven characters. He wants the first six characters to alternate—consonant, vowel, consonant, vowel, consonant, vowel—with repetitions allowed. Then he wants to use a digit for the seventh character.
  a. How many different passwords can he make up? *11,576,250*
  b. If he makes up a password at random, what is the probability that his password is banana7? *1/11,576,250 = 0.0000000864*

14. A caterer offers 24 different types of dessert. In how many ways can 5 of them be chosen for a banquet if the order doesn't matter? 42,504

15. In a standard game of pool, there are 15 balls labeled 1 through 15.
 a. In how many ways can the 15 balls be ordered? 15!
 b. In how many ways can 3 of the 15 balls be chosen and ordered? 2730

Review Exercises

1. **Colored dice:** A six-sided die has one face painted red, two faces painted white, and three faces painted blue. Each face is equally likely to turn up when the die is rolled.
 a. Construct a sample space for the experiment of rolling this die.
 b. Find the probability that a blue face turns up. 1/2

2. **How are your grades?** There were 30 students in last semester's statistics class. Of these, 6 received a grade of A, and 12 received a grade of B. What is the probability that a randomly chosen student received a grade of A or B? 0.6

3. **Statistics, anyone?** Let S be the event that a randomly selected college student has taken a statistics course, and let C be the event that the same student has taken a chemistry course. Suppose $P(S) = 0.4$, $P(C) = 0.3$, and $P(S \text{ and } C) = 0.2$.
 a. Find the probability that a student has taken statistics or chemistry. 0.5
 b. Find the probability that a student has taken statistics given that the student has taken chemistry. 0.6667

4. **Blood types:** Human blood may contain either or both of two antigens, A and B. Blood that contains only the A antigen is called type A, blood that contains only the B antigen is called type B, blood that contains both antigens is called type AB, and blood that contains neither antigen is called type O. A certain blood bank has blood from a total of 1200 donors. Of these, 570 have type O blood, 440 have type A, 125 have type B, and 65 have type AB.
 a. What is the probability that a randomly chosen blood donor is type O? 0.475
 b. A recipient with type A blood may safely receive blood from a donor whose blood does not contain the B antigen. What is the probability that a randomly chosen blood donor may donate to a recipient with type A blood? 0.8417

5. **Start a business:** Suppose that start-up companies in the area of biotechnology have probability 0.2 of becoming profitable, and that those in the area of information technology have probability 0.15 of becoming profitable. A venture capitalist invests in one firm of each type. Assume the companies function independently.
 a. What is the probability that both companies become profitable? 0.03
 b. What is the probability that at least one of the two companies becomes profitable? 0.32

6. **Stop that car:** A drag racer has two parachutes, a main and a backup, that are designed to bring the vehicle to a stop at the end of a run. Suppose that the main chute deploys with probability 0.99, and that if the main fails to deploy, the backup deploys with probability 0.98.
 a. What is the probability that one of the two parachutes deploys? 0.9998
 b. What is the probability that the backup parachute deploys? 0.0098

7. **Defective parts:** A process manufactures microcircuits that are used in computers. Twelve percent of the circuits are defective. Assume that three circuits are installed in a computer. Denote a defective circuit by "D" and a good circuit by "G."
 a. List all eight items in the sample space.
 b. What is the probability that all three circuits are good? 0.6815
 c. The computer will function so long as either two or three of the circuits are good. What is the probability that a computer will function? 0.9603
 d. If we use a cutoff of 0.05, would it be unusual for all three circuits to be defective? Yes

8. **Music to my ears:** Jeri is listening to the songs on a new CD in random order. She will listen to two different songs and will buy the CD if she likes both of them. Assume there are 10 songs on the CD and that she would like five of them.
 a. What is the probability that she likes the first song? 0.5
 b. What is the probability that she likes the second song, given that she liked the first song? 0.4444
 c. What is the probability that she buys the CD? 0.2222

9. **Female business majors:** At a certain university, the probability that a randomly chosen student is female is 0.55, the probability that the student is a business major is 0.20, and the probability that the student is female and a business major is 0.15.
 a. What is the probability that the student is female or a business major? 0.6
 b. What is the probability that the student is female given that the student is a business major? 0.75
 c. What is the probability that the student is a business major given that the student is female? 0.2727
 d. Are the events "female" and "business major" independent? Explain. No
 e. Are the events "female" and "business major" mutually exclusive? Explain. No

10. Heart attack: The following table presents the number of hospitalizations for myocardial infarction (heart attack) for men and women in various age groups.

Age	Male	Female	Total
18–44	26,828	9,265	36,093
45–64	166,340	68,666	235,006
65–84	155,707	124,289	279,996
85 and up	35,524	57,785	93,309
Total	384,399	260,005	644,404

Source: Agency for Healthcare Research and Quality

a. What is the probability that a randomly chosen patient is a woman? *0.4035*
b. What is the probability that a randomly chosen patient is aged 45–64? *0.3647*
c. What is the probability that a randomly chosen patient is a woman and aged 45–64? *0.1066*
d. What is the probability that a randomly chosen patient is a woman or aged 45–64? *0.6616*
e. What is the probability that a randomly chosen patient is a woman given that the patient is aged 45–64? *0.2922*
f. What is the probability that a randomly chosen patient is aged 45–64 given that the patient is a woman? *0.2641*

11. Rainy weekend: Sally is planning to go away for the weekend this coming Saturday and Sunday. At the place she will be going, the probability of rain on any given day is 0.10. Sally says that the probability that it rains on both days is 0.01. She reasons as follows:

$$\begin{aligned} P(\text{Rain Saturday and Rain Sunday}) &= P(\text{Rain Saturday})P(\text{Rain Sunday}) \\ &= (0.1)(0.1) \\ &= 0.01 \end{aligned}$$

a. What assumption is being made in this calculation?
b. Explain why this assumption is probably not justified in the present case.
c. Is the probability of 0.01 likely to be too high or too low? Explain. *Too low*

12. Required courses: A college student must take courses in English, history, mathematics, biology, and physical education. She decides to choose three of these courses to take in her freshman year. In how many ways can this choice be made? *10*

13. Required courses: Refer to Exercise 12. Assume the student chooses three courses at random. What is the probability that she chooses English, mathematics, and biology? *0.1*

14. Bookshelf: Luis has six books: a novel, a biography, a dictionary, a self-help book, a statistics textbook, and a comic book.
a. Luis's bookshelf has room for only three of the books. In how many ways can Luis choose and order three books? *120*
b. In how many ways may the books be chosen and ordered if he does not choose the comic book? *60*

15. Bookshelf: Refer to Exercise 14. Luis chooses three books at random.
a. What is the probability that the books on his shelf are statistics textbook, dictionary, and comic book, in that order? *0.00833*
b. What is the probability that the statistics textbook, dictionary, and comic book are the three books chosen, in any order? *0.05*

Write About It

1. Explain how you could use the law of large numbers to show that a coin is unfair by tossing it many times.

2. When it comes to betting, the chance of winning or losing may be expressed as odds. If there are n equally likely outcomes and m of them result in a win, then the odds of winning are $m{:}(n - m)$, read "m to $n - m$." For example, suppose that a player rolls a die and wins if the number of dots appearing is either 1 or 2. Since there are two winning outcomes out of six equally likely outcomes, the odds of winning are 2:4.

Suppose that a pair of dice is rolled and the player wins if it comes up "doubles," that is, if the same number of dots appears on each die. What are the odds of winning?

3. If the odds of an event occurring are 5:8, what is the probability that the event will occur?

4. Explain why the General Addition Rule $P(A \text{ or } B) = P(A) + P(B) - P(A \text{ and } B)$ may be used even when A and B are mutually exclusive events.

5. Sometimes events are in the form "at least" a given number. For example, if a coin is tossed five times, an event could be getting at least two heads. What would be the complement of the event of getting at least two heads?

6. In practice, one must decide whether to treat two events as independent based on an understanding of the process that creates them. For example, in a manufacturing process that produces electronic circuit boards for calculators, assume that the probability that a board is defective is 0.01. You arrive at the manufacturing plant and sample the next two boards that come off the assembly line. Let A be the event that the first board is defective, and let B be the event that the second board is defective. Describe circumstances under which A and B would not be independent.
7. Describe circumstances under which you would use a permutation.
8. Describe circumstances under which you would use a combination.

In-Class Activities

1. **Law of large numbers:** Each student tosses a coin three times. Compute the proportion of heads among all the students in the class. Repeat a few times, and compute the cumulative proportion of heads each time. Observe how the proportion approaches 1/2, in accordance with the law of large numbers.
2. **Stop after one head:** Each student tosses a coin until a head appears, then stops. Count the total number of tosses. What proportion of tosses were heads? Repeat a few times, and compute the cumulative proportion of heads each time. Does the proportion approach 1/2?

NOTE TO INSTRUCTOR

The activity involving tossing a coin until a head occurs illustrates the fact that the rule for when to stop does not affect the long-run proportion of times that an event occurs.

3. **Monty Hall Problem:** This is based on the old television program *Let's Make a Deal*, hosted by Monty Hall. There are three doors. Behind one of them is a grand prize, and nothing is behind the other two. You select a door. The host then opens one of the doors you didn't select that has nothing behind it. You are offered the opportunity to switch your selection to the other unopened door. Should you switch or stick with your original selection?

   This game can be simulated with playing cards. Use three cards, one of which is an ace, which represents the grand prize. Divide students into pairs. One student puts the three cards face down in front of the other student, who then guesses where the ace is. The first student, who knows where the ace is, turns up one of the cards that is not an ace. The other student can decide whether or not to switch. Compute the proportion of times students win by switching and by not switching. Should you switch?

Case Study: How Likely Are You To Live To Age 100?

The following table is a *life table*, reproduced from the chapter introduction. With an understanding of some basic concepts of probability, one can use the life table to compute the probability that a person of a given age will still be alive a given number of years from now. Life insurance companies use this information to determine how much to charge for life insurance policies.

United States Life Table, Total Population

Age Interval	Proportion Surviving	Age Interval	Proportion Surviving
0–10	0.99123	50–60	0.94010
10–20	0.99613	60–70	0.86958
20–30	0.99050	70–80	0.70938
30–40	0.98703	80–90	0.42164
40–50	0.97150	90–100	0.12248

Source: Centers for Disease Control and Prevention

The column labeled "Proportion Surviving" gives the proportion of people alive at the beginning of an age interval who will still be alive at the end of the age interval. For example, among those currently age 20, the proportion who will still be alive at age 30 is 0.99050, or 99.050%. We will begin by computing the probability that a person lives to any of the ages 10, 20, ... , 100.

The first number in the column is the probability that a person lives to age 10. So

$$P(\text{Alive at age } 10) = 0.99123$$

The key to using the life table is to realize that the rest of the numbers in the "Proportion Surviving" column are conditional probabilities. They are probabilities that a person is alive at the end of the age interval, given that they were alive at the beginning of the age interval. For example, the row labeled "20–30" contains the conditional probability that someone alive at age 20 will be alive at age 30:

$$P(\text{Alive at age 30} \mid \text{Alive at age 20}) = 0.99050$$

In Exercises 1–5, compute the probability that a person lives to a given age.

1. From the table, find the conditional probability $P(\text{Alive at age 20} \mid \text{Alive at age 10})$. *0.99613*
2. Use the result from Exercise 1 along with the result $P(\text{Alive at age 10}) = 0.99123$ to compute $P(\text{Alive at age 20})$. *0.98739*
3. Use the result from Exercise 2 along with the appropriate number from the table to compute $P(\text{Alive at age 30})$. *0.97801*
4. Use the result from Exercise 3 along with the appropriate number from the table to compute $P(\text{Alive at age 40})$. *0.96533*
5. Compute the probability that a person is alive at ages 50, 60, 70, 80, 90, and 100. *50: 0.93782; 60: 0.88164; 70: 0.76666; 80: 0.54385; 90: 0.22931; 100: 0.02809*

In Exercises 1–5, we computed the probability that a newborn lives to a given age. Now let's compute the probability that a person aged x lives to age y. We'll illustrate this with an example to compute the probability that a person aged 20 lives to age 100. This is the conditional probability that a person lives to age 100, given that the person has lived to age 20.

We want to compute the conditional probability

$$P(\text{Alive at age 100} \mid \text{Alive at age 20})$$

Using the definition of conditional probability, we have

$$P(\text{Alive at age 100} \mid \text{Alive at age 20}) = \frac{P(\text{Alive at age 100 and Alive at age 20})}{P(\text{Alive at age 20})}$$

You computed $P(\text{Alive at age 20})$ in Exercise 2. Now we need to compute $P(\text{Alive at age 100 and Alive at age 20})$. The key is to realize that anyone who is alive at age 100 was also alive at age 20. Therefore,

$$P(\text{Alive at age 100 and Alive at age 20}) = P(\text{Alive at age 100})$$

Therefore,

$$P(\text{Alive at age 100} \mid \text{Alive at age 20}) = \frac{P(\text{Alive at age 100})}{P(\text{Alive at age 20})}$$

In general, for $y > x$,

$$P(\text{Alive at age } y \mid \text{Alive at age } x) = \frac{P(\text{Alive at age } y)}{P(\text{Alive at age } x)}$$

6. Find the probability that a person aged 20 is still alive at age 100. *0.028444*
7. Find the probability that a person aged 50 is still alive at age 70. *0.81749*
8. Which is more probable, that a person aged 20 is still alive at age 50, or that a person aged 50 is still alive at age 60? *Aged 20 still alive at age 50*
9. A life insurance company sells term insurance policies. These policies pay \$100,000 if the policyholder dies before age 70, but pay nothing if a person is still alive at age 70. If a person buys a policy at age 40, what is the probability that the insurance company does not have to pay? *0.79419*

chapter

6

Discrete Probability Distributions

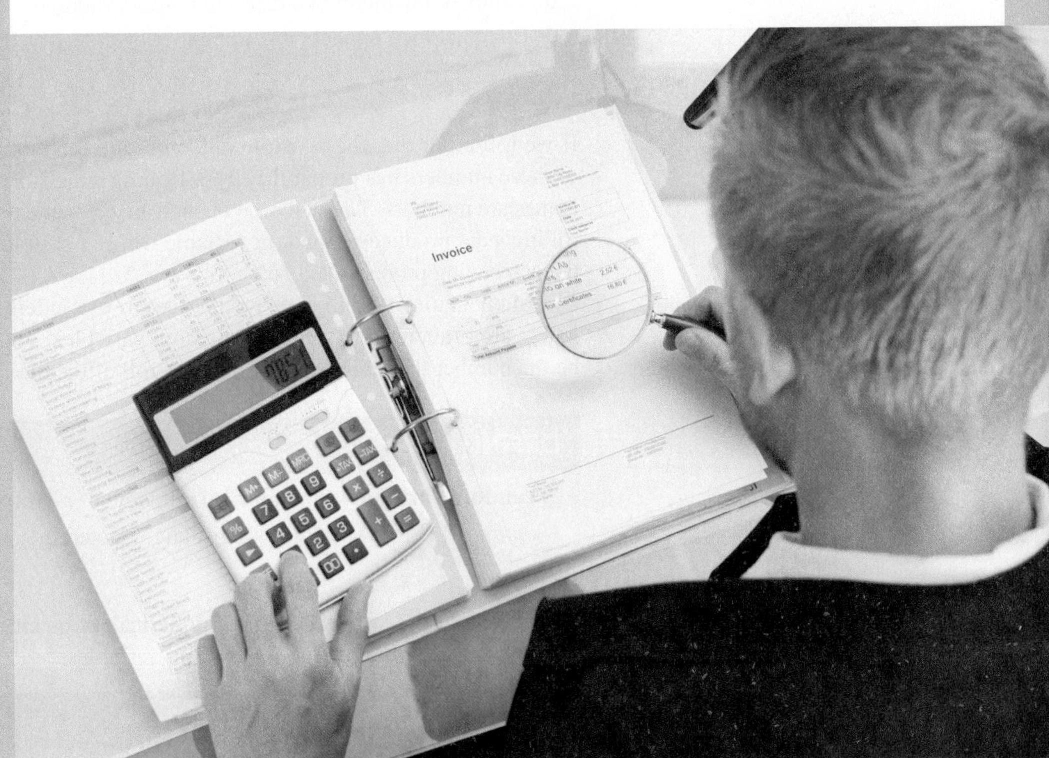
Andriy Popov/123RF

Introduction

How does the Internal Revenue Service detect a fraudulent tax return? One method involves the use of probability in a surprising way. A list of all the amounts claimed as deductions is made. Then a relative frequency distribution is constructed of the first digits of these amounts. This relative frequency distribution is compared to a theoretical distribution, called Benford's law, which gives the probability of each digit. If there is a large discrepancy, the return is suspected of being fraudulent.

The probabilities assigned by Benford's law are given in the following table:

Digit	1	2	3	4	5	6	7	8	9
Probability	0.301	0.176	0.125	0.097	0.079	0.067	0.058	0.051	0.046

Benford's law is an example of a probability distribution. It may be surprising that the first digits of amounts on tax returns are not all equally likely, but in fact the smaller digits occur much more frequently than the larger ones. It turns out that Benford's law describes many data sets that occur naturally. In the case study at the end of the chapter, we will learn more about Benford's law.

Section 6.1 Random Variables

Objectives

1. Distinguish between discrete and continuous random variables
2. Determine a probability distribution for a discrete random variable
3. Describe the connection between probability distributions and populations
4. Construct a probability histogram for a discrete random variable
5. Compute the mean of a discrete random variable
6. Compute the variance and standard deviation of a discrete random variable

NOTE TO INSTRUCTOR

The following topics from the **Statistics Corequisite Workbook** are aligned with the material in this section.

4.1 - Adding and Subtracting Decimals in Discrete Probability Distributions

4.2 - Determining Probability Events from Words for Discrete Random Variables

4.3 - The Mean of a Discrete Random Variable

If we roll a fair die, the possible outcomes are the numbers 1, 2, 3, 4, 5, and 6, and each of these numbers has probability $1/6$. Rolling a die is a probability experiment whose outcomes are numbers. The outcome of such an experiment is called a *random variable*. Thus, rolling a die produces a random variable whose possible values are the numbers 1 through 6, each having probability $1/6$.

Mathematicians and statisticians like to use letters to represent numbers. Uppercase letters are often used to represent random variables. Thus, a statistician might say, "Let X be the number that comes up on the next roll of the die."

DEFINITION

A **random variable** is a numerical outcome of a probability experiment.

Notation: Random variables are usually denoted by uppercase letters.

Objective 1 Distinguish between discrete and continuous random variables

In Section 1.2, we learned that numerical, or quantitative, variables can be discrete or continuous. The same is true for random variables.

DEFINITION

- **Discrete random variables** are random variables whose possible values can be listed. The list may be infinite—for example, the list of all whole numbers.
- **Continuous random variables** are random variables that can take on any value in an interval. The possible values of a continuous variable are not restricted to any list.

Example 6.1

Determining whether a random variable is discrete or continuous

Which of the following random variables are discrete and which are continuous?

a. The number that comes up on the roll of a die

b. The height of a randomly chosen college student

c. The number of siblings a randomly chosen person has

d. Amount of electricity used to light a randomly chosen classroom

Solution

a. The number that comes up on a die is discrete. The possible values are 1, 2, 3, 4, 5, and 6.

b. Height is continuous. A person's height is not restricted to any list of values.

c. The number of siblings is discrete. The possible values are 0, 1, 2, and so forth.

d. The amount of electricity is continuous. It is not restricted to any list of values.

In this chapter, we will focus on discrete random variables. In Chapter 7, we will learn about an important continuous random variable.

Objective 2 Determine a probability distribution for a discrete random variable

DEFINITION

A **probability distribution** for a discrete random variable specifies the probability for each possible value of the random variable.

Example 6.2

Determining a probability distribution

A fair coin is tossed twice. Let X be the number of heads that come up. Find the probability distribution of X.

Solution

There are four equally likely outcomes to this probability experiment, listed in Table 6.1. For each outcome, we count the number of heads, which is the value of the random variable X.

Table 6.1

First Toss	Second Toss	X = Number of Heads
H	H	2
H	T	1
T	H	1
T	T	0

There are three possible values for the number of heads: 0, 1, and 2. One of the four outcomes has the value "0," two of the outcomes have the value "1," and one outcome has the value "2." Therefore, the probabilities are

$$P(0) = \frac{1}{4} = 0.25 \qquad P(1) = \frac{2}{4} = 0.50 \qquad P(2) = \frac{1}{4} = 0.25$$

NOTE TO INSTRUCTOR

An uppercase letter like X represents a random variable. We often use a lowercase letter x to represent a particular value of the random variable.

The probability distribution is presented in Table 6.2.

Table 6.2 Probability Distribution of X

x	0	1	2
$P(x)$	0.25	0.50	0.25

Discrete probability distributions satisfy two properties. First, since the values $P(x)$ are probabilities, they must all be between 0 and 1. Second, since the random variable always takes on one of the values in the list, the sum of the probabilities must equal 1.

EXPLAIN IT AGAIN

Properties of discrete probability distributions: In a probability distribution, each probability must be between 0 and 1, and the sum of all the probabilities must be equal to 1.

SUMMARY

Let $P(x)$ denote the probability that a random variable has the value x. Then

1. $0 \le P(x) \le 1$ for every possible value x

2. $\sum P(x) = 1$

Example 6.3

Identifying probability distributions

Which of the following tables represent probability distributions?

a.

x	$P(x)$
1	0.25
2	0.65
3	−0.30
4	0.11

b.

x	$P(x)$
−1	0.17
−0.5	0.25
0	0.31
0.5	0.22
1	0.05

c.

x	$P(x)$
1	1.02
10	0.31
100	0.90
1000	0.43

d.

x	$P(x)$
0	0.10
1	0.17
2	0.75
3	0.24

Solution

a. This is not a probability distribution. $P(3)$ is not between 0 and 1.

b. This is a probability distribution. All the probabilities are between 0 and 1, and they add up to 1.

c. This is not a probability distribution. $P(1)$ is not between 0 and 1.

d. This is not a probability distribution. Although all the probabilities are between 0 and 1, they do not add up to 1.

When we are given the probability distribution of a random variable, we can use the rules of probability to compute probabilities involving the random variable. Example 6.4 provides an illustration.

Example 6.4

Computing probabilities

Four patients have made appointments to have their blood pressure checked at a clinic. Let X be the number of them who have high blood pressure. Based on data from the National Health and Examination Survey, the probability distribution of X is

x	0	1	2	3	4
$P(x)$	0.23	0.41	0.27	0.08	0.01

a. Find $P(1)$.

b. Find $P(2 \text{ or } 3)$.

c. Find $P(\text{More than } 1)$.

d. Find $P(\text{At least } 1)$.

RECALL

The Addition Rule for Mutually Exclusive Events says that if A and B are mutually exclusive, then $P(A \text{ or } B) = P(A) + P(B)$.

Solution

a. From the probability distribution, we see that the event that exactly one of the patients had high blood pressure is $P(1) = 0.41$.

b. The events "2" and "3" are mutually exclusive, since they cannot both happen. We use the Addition Rule for Mutually Exclusive events:

$$P(2 \text{ or } 3) = P(2) + P(3) = 0.27 + 0.08 = 0.35$$

c. "More than 1" means "2 or 3 or 4." Again we use the Addition Rule for Mutually Exclusive events:

$$P(\text{More than } 1) = P(2 \text{ or } 3 \text{ or } 4) = 0.27 + 0.08 + 0.01 = 0.36$$

RECALL

The Rule of Complements says that $P(A^C) = 1 - P(A)$.

d. We use the Rule of Complements. Recall that the complement of "At least one" is "none."

$$P(\text{At least } 1) = 1 - P(0) = 1 - 0.23 = 0.77$$

Check Your Understanding

1. A school teacher flips a coin and uses the result to determine whether to start the school day with a math lesson or a history lesson. If the first lesson is recorded for three consecutive days, there are eight possible outcomes: MMM, MMH, MHM, MHH, HMM, HMH, HHM, HHH. Assume these outcomes are equally likely. Let X represent the number of days that start with a math lesson. Find the probability distribution of X.

EXPLAIN IT AGAIN

The probability of at least 1: The complement of "at least 1" is "none," or "0." Therefore P (at least 1) $= 1 - P(0)$.

2. Someone says that the following table shows the probability distribution for the number of boys in a family of four children. Is this possible? Explain why or why not.

x	0	1	2	3	4
$P(x)$	0.12	0.37	0.45	0.25	0.18

3. Which of the following tables represent probability distributions? *(a) and (c)*

a.

x	$P(x)$
0	0.45
1	0.15
2	0.30
3	0.10

b.

x	$P(x)$
4	0.27
5	0.15
6	0.11
7	0.34
8	0.25

c.

x	$P(x)$
1	0.02
2	0.41
3	0.24
4	0.33

4. Following is the probability distribution of a random variable that represents the number of extracurricular activities a college freshman participates in.

x	0	1	2	3	4
$P(x)$	0.06	0.14	0.45	0.21	0.14

a. Find the probability that a student participates in exactly two activities. *0.45*
b. Find the probability that a student participates in more than two activities. *0.35*
c. Find the probability that a student participates in at least one activity. *0.94*

Answers are on page 278.

Objective 3 Describe the connection between probability distributions and populations

Connection Between Probability Distributions and Populations

Statisticians are interested in studying samples drawn from populations. Random variables are important because when an item is drawn from a population, the value observed is the value of a random variable. The probability distribution of the random variable tells how frequently we can expect each of the possible values of the random variable to turn up in the sample. Example 6.5 presents the idea.

Example 6.5

Constructing a probability distribution that describes a population

An airport parking facility contains 1000 parking spaces. Of these, 142 are covered long-term spaces that cost \$2.00 per hour, 378 are covered short-term spaces that cost \$4.50 per hour, 423 are uncovered long-term spaces that cost \$1.50 per hour, and 57 are uncovered short-term spaces that cost \$4.00 per hour. A parking space is selected at random. Let X represent the hourly parking fee for the randomly sampled space. Find the probability distribution of X.

EXPLAIN IT AGAIN

A probability distribution describes a population: We can think of a probability distribution as describing a population. The probability of each value represents the proportion of population items that have that value.

Solution

To find the probability distribution, we must list the possible values of X and then find the probability of each of them. The possible values of X are 1.50, 2.00, 4.00, 4.50. We find their probabilities:

$$P(1.50) = \frac{\text{number of spaces costing \$1.50}}{\text{total number of spaces}} = \frac{423}{1000} = 0.423$$

$$P(2.00) = \frac{\text{number of spaces costing \$2.00}}{\text{total number of spaces}} = \frac{142}{1000} = 0.142$$

$$P(4.00) = \frac{\text{number of spaces costing \$4.00}}{\text{total number of spaces}} = \frac{57}{1000} = 0.057$$

$$P(4.50) = \frac{\text{number of spaces costing \$4.50}}{\text{total number of spaces}} = \frac{378}{1000} = 0.378$$

The probability distribution is

x	1.50	2.00	4.00	4.50
$P(x)$	0.423	0.142	0.057	0.378

Sometimes we draw a sample and compute the sample mean $\bar{x}$ and sample standard deviation s. These quantities are also random variables, because their values are different for different samples, but their probability distributions are usually difficult to compute. In Section 7.3 we will learn a way to approximate the probability distribution of the sample mean when the sample size is large.

Check Your Understanding

5. There are 5000 undergraduates registered at a certain college. Of them, 478 are taking one course, 645 are taking two courses, 568 are taking three courses, 1864 are taking four courses, 1357 are taking five courses, and 88 are taking six courses. Let X be the number of courses taken by a student randomly sampled from this population. Find the probability distribution of X.

Answer is on page 278.

Objective 4 Construct a probability histogram for a discrete random variable

Probability histograms

In Section 2.2, we learned to summarize the data in a sample with a histogram. We can represent discrete probability distributions with histograms as well. A histogram that represents a discrete probability distribution is called a **probability histogram**. Constructing a probability histogram from a probability distribution is just like constructing a relative frequency histogram from a relative frequency distribution for discrete data. For each possible value of the random variable, we draw a rectangle whose height is equal to the probability of that value.

EXPLAIN IT AGAIN

A probability histogram is like a histogram for a population: The height of each rectangle in a probability histogram tells us how frequently the value appears in the population.

Table 6.3 presents the probability distribution for the number of heads in a sequence of five tosses, using the assumption that heads and tails are equally likely and that the tosses are independent events. Figure 6.1 presents a probability histogram for this probability distribution.

Table 6.3

x	$P(x)$
0	0.03125
1	0.15625
2	0.31250
3	0.31250
4	0.15625
5	0.03125

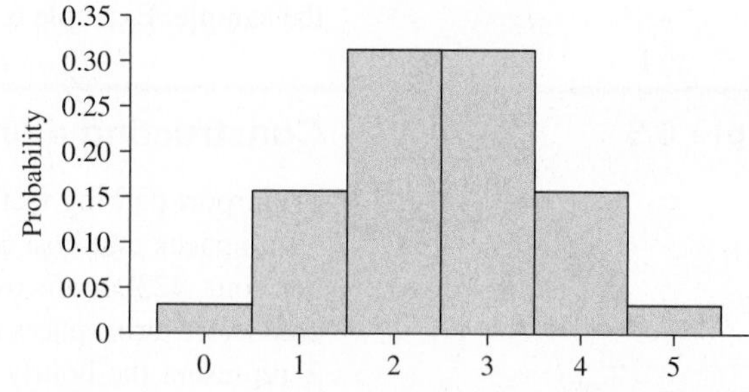

Figure 6.1 Probability histogram for the distribution in Table 6.3

Check Your Understanding

6. Following is a probability histogram for the number of children a woman has. The numbers on the tops of the rectangles are the heights.

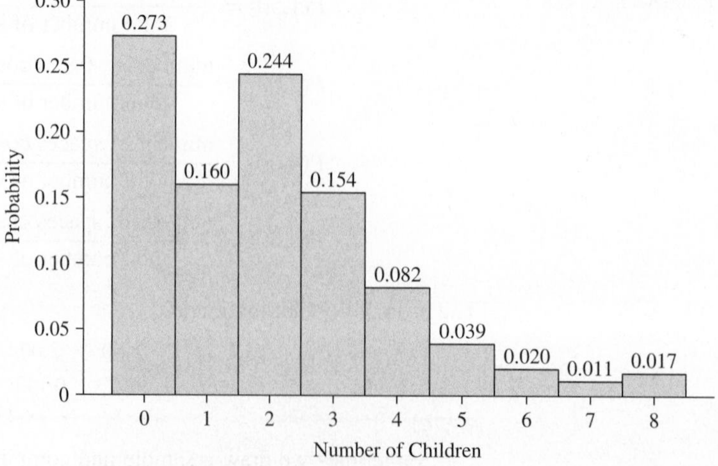

Source: General Social Survey

a. What is the probability that a randomly chosen woman has exactly two children? 0.244

b. What is the probability that a randomly chosen woman has fewer than two children? 0.433

c. What is the probability that a randomly chosen woman has either four or five children? 0.121

Answers are on page 278.

Objective 5 Compute the mean of a discrete random variable

The mean of a random variable

Recall that the mean is a measure of center. The mean of a random variable provides a measure of center for the probability distribution of a random variable.

NOTE TO INSTRUCTOR

It is useful to point out that the mean of a discrete random variable does not have to be one of the possible values of the random variable.

DEFINITION

To find the **mean** of a discrete random variable, multiply each possible value by its probability, then add the products.

In symbols, $\mu_X = \sum [x \cdot P(x)]$.

Another name for the mean of a random variable is the **expected value**.

The notation for the expected value of X is $E(X)$.

Example 6.6

Determining the mean of a discrete random variable

A computer monitor is composed of a very large number of points of light called pixels. It is not uncommon for a few of these pixels to be defective. Let X represent the number of defective pixels on a randomly chosen monitor. The probability distribution of X is as follows:

x	0	1	2	3
$P(x)$	0.2	0.5	0.2	0.1

Find the mean number of defective pixels.

CAUTION

Do not confuse the notation for the mean of random variable (μ_X) with the notation for a population mean (μ).

Solution

The mean is

$$\mu_X = 0(0.2) + 1(0.5) + 2(0.2) + 3(0.1) = 1.2$$

The mean is 1.2.

The mean is a measure of the center of the probability distribution. Figure 6.2 presents a probability histogram for the distribution in Example 6.6 and shows the position of the mean. If we imagine each rectangle to be a weight, the mean is the point at which the histogram would balance.

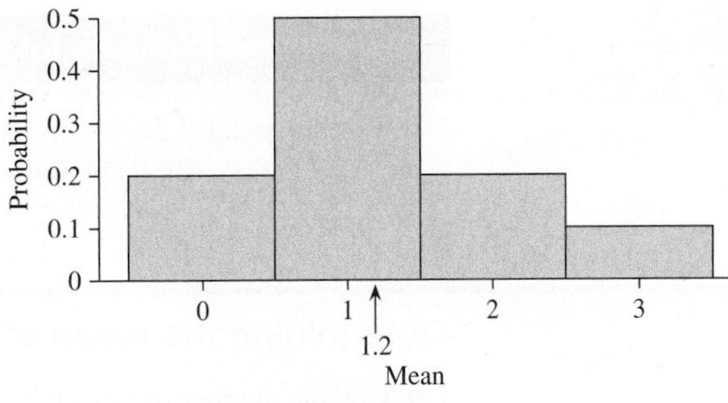

Figure 6.2

The symbol used to represent the mean of a random variable X is μ_X. It is not a coincidence that the mean of a population is also represented by μ. Recall that when we perform a probability experiment to obtain a value of a random variable, it is like sampling an item from a population. The probability distribution of the random variable tells how frequently each of the possible values of the random variable occurs in the population. The mean of the random variable is the same as the mean of the population.

Let's take a closer look at Example 6.6 to see how this is the case. Imagine that we had a population of 10 computer monitors. Figure 6.3 presents a visualization of this population, with each monitor labeled with the number of defective pixels it has.

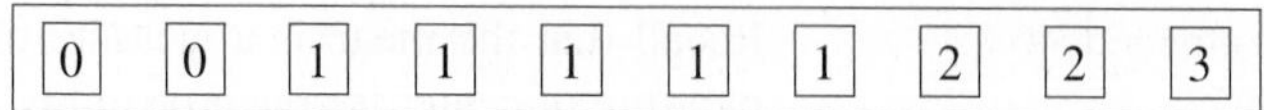

Figure 6.3 Population of 10 computer monitors

The probability distribution in Example 6.6 represents this population. If we sample a monitor from this population, the probability that it will have 0 defective pixels is 0.2, the probability that it will have 1 defective pixel is 0.5, and so forth, just as in the probability distribution. Now we'll compute the mean of this population.

$$\begin{aligned}\mu &= \frac{0+0+1+1+1+1+1+2+2+3}{10} \\ &= \frac{0\cdot 2+1\cdot 5+2\cdot 2+3\cdot 1}{10} \\ &= 0\cdot\frac{2}{10}+1\cdot\frac{5}{10}+2\cdot\frac{2}{10}+3\cdot\frac{1}{10} \\ &= 0(0.2)+1(0.5)+2(0.2)+3(0.1) = 1.2\end{aligned}$$

The mean of the random variable is the same as the mean of the population.

Interpreting the mean of a random variable

In Example 6.6, imagine that we sampled a large number of computer monitors and counted the number of defective pixels on each. We would expect the mean number of defective pixels on our sampled monitors to be close to the population mean of 1.2. This idea provides an interpretation for the mean of a random variable.

Interpretation of the Mean of a Random Variable

If a probability experiment that produces a value of a random variable is repeated over and over again, the average of the values produced will approach the mean of the random variable.

EXPLAIN IT AGAIN

Law of large numbers for means: The law of large numbers for means tells us that for a large sample, the sample mean will almost certainly be close to the population mean.

When the probability experiment is sampling from a population, our interpretation of the mean says that as the sample size increases, the sample mean will approach the population mean. This is known as the **law of large numbers for means**.

Law of Large Numbers for Means

If we sample items from a population, then as the sample size grows larger, the sample mean will approach the population mean.

Example 6.7

Determining the mean of a discrete random variable

Let X be the number of boys in a family of five children. The probability distribution of X is given in Table 6.4 (page 271). Find the mean number of boys and interpret the result.

Table 6.4

x	$P(x)$
0	0.03125
1	0.15625
2	0.31250
3	0.31250
4	0.15625
5	0.03125

Solution

The calculations are presented in Table 6.5. We multiply each x by its corresponding $P(x)$, then add the products to obtain the mean. The mean is 2.5. We interpret this by saying that as we sample more and more families with five children, the mean number of boys in the sampled families will approach 2.5.

Table 6.5

x	$P(x)$	$x \cdot P(x)$
0	0.03125	0.00000
1	0.15625	0.15625
2	0.31250	0.62500
3	0.31250	0.93750
4	0.15625	0.62500
5	0.03125	0.15625
		$\sum[x \cdot P(x)] = 2.5$

Check Your Understanding

7. A true–false quiz with 10 questions was given to a statistics class. Following is the probability distribution for the score of a randomly chosen student. Find the mean score and interpret the result. *7.48*

x	5	6	7	8	9	10
$P(x)$	0.04	0.16	0.36	0.24	0.12	0.08

Answer is on page 278.

Objective 6 Compute the variance and standard deviation of a discrete random variable

The variance and standard deviation of a discrete random variable

The variance and standard deviation are measures of spread. The variance and standard deviation of a random variable measure the spread in the probability distribution of the random variable.

DEFINITION

The **variance** of a discrete random variable X is

$$\sigma_X^2 = \sum[(x - \mu_X)^2 \cdot P(x)]$$

An equivalent expression that is often easier when computing by hand is

$$\sigma_X^2 = \sum[x^2 \cdot P(x)] - \mu_X^2$$

The **standard deviation** of X is the square root of the variance:

$$\sigma_X = \sqrt{\sigma_X^2}$$

Example 6.8

Computing the variance of a discrete random variable

In Example 6.6, we presented the following probability distribution for the number of defective pixels in a computer monitor. Compute the variance and standard deviation of the number of defective pixels.

x	0	1	2	3
$P(x)$	0.2	0.5	0.2	0.1

Solution

In Example 6.6, we computed $\mu_X = 1.2$. The calculations for the variance σ_X^2 are shown in Table 6.6. For each x, we subtract the mean μ_X to obtain $x - \mu_X$. Then we square these quantities and multiply by $P(x)$. We add the products to obtain the variance σ_X^2.

Table 6.6

x	$x - \mu_X$	$(x - \mu_X)^2$	$P(x)$	$(x - \mu_X)^2 \cdot P(x)$
0	−1.2	1.44	0.2	0.288
1	−0.2	0.04	0.5	0.020
2	0.8	0.64	0.2	0.128
3	1.8	3.24	0.1	0.324
				$\sigma_X^2 = \sum[(x - \mu_X)^2 \cdot P(x)] = 0.760$

The variance is $\sigma_X^2 = 0.760$. The standard deviation is

$$\sigma_X = \sqrt{0.760} = 0.872$$

Example 6.9

Computing the mean and standard deviation by using technology

Use technology to compute the mean and standard deviation for the probability distribution in Example 6.8.

```
1-Var Stats
x̄=1.2
Σx=1.2
Σx²=2.2
Sx=
σx=.8717797887
n=1
minX=0
↓Q1=1
```

Figure 6.4

Solution

We use the TI-84 Plus calculator. Enter the possible values of the random variable into **L1** and their probabilities into **L2**. Then use the **1–Var Stats** command. Figure 6.4 presents the calculator display. The mean, μ_X, is denoted by $\bar{x}$. The standard deviation is denoted σx. There is no value for the sample standard deviation Sx because we are not computing the standard deviation of a sample. Step-by-step instructions are presented in the Using Technology section on page 274.

Check Your Understanding

8. Following is the probability distribution for the age of a student at a certain public high school.

x	13	14	15	16	17	18
$P(x)$	0.08	0.24	0.23	0.28	0.14	0.03

a. Find the variance of the ages. *1.6075*

b. Find the standard deviation of the ages. *1.2679*

Answers are on page 278.

Applications of the mean

There are many occasions on which people want to predict how much they are likely to gain or lose if they make a certain decision or take a certain action. Often, this is done by computing the mean of a random variable. In such situations, the mean is sometimes called the "expected value." If the expected value is positive, it is an expected gain, and if it is negative, it is an expected loss. Examples 6.10–6.12 provide some illustrations.

Gambling

Probability was invented by mathematicians who were hired by gamblers to help them create games of chance. For this reason, probability is an extremely useful tool to analyze gambling games—this is what it was designed to do. Example 6.10 analyzes the game of roulette.

Example 6.10

Find the expected loss at roulette

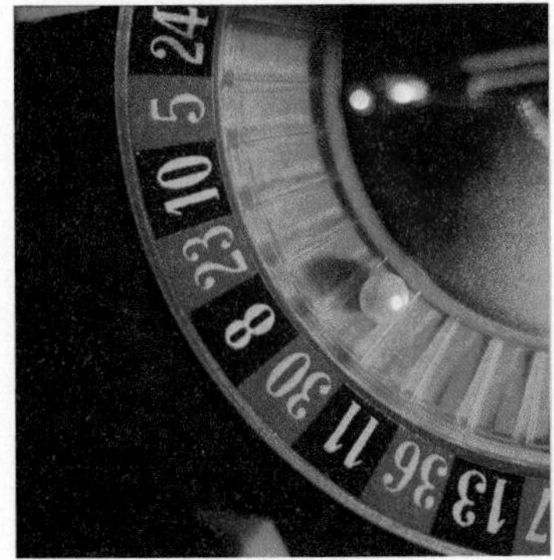

Ingram Publishing/AGE Fotostock

A Nevada roulette wheel contains 38 pockets, numbered 1 to 36 with a zero (0) and a double-zero (00). Eighteen of the numbers are colored red, 18 are colored black, and two (the 0 and the 00) are colored green. Let's say you bet \$1 on red. If a red number comes up, you get your dollar back, along with another dollar, so you win \$1. If a black or green number comes up, you win −\$1 (winning a negative amount is a mathematician's way of saying that you lose). Let X be the amount you win. Find the probability distribution of X and the expected value (mean) of X. Interpret the expected value.

Solution

The possible values of X are 1 and −1. There are 38 outcomes in the sample space, with 18 of them (the red ones) corresponding to 1 and the other 20 corresponding to −1. The probability distribution of X is therefore

x	−1	1
$P(x)$	20/38	18/38

The expected value is the mean of X:

$$E(X) = \mu_X = (-1)\left(\frac{20}{38}\right) + 1\left(\frac{18}{38}\right) = -\frac{2}{38} = -0.0526$$

Since the expected value is negative, this is an expected loss. We can interpret the expected value by saying that, on the average, you can expect to lose 5.26¢ for every dollar you bet.

NOTE TO INSTRUCTOR

You can point out that casinos set their payoffs so that they will have an expected gain and the customer will have an expected loss.

Business projections

When making business decisions, executives often use probability distributions to describe the amount of profit or loss that will result.

Example 6.11

Computing the expected value of a business decision

A mineral economist estimated that a particular mining venture had probability 0.4 of a \$30 million loss, probability 0.5 of a \$20 million profit, and probability 0.1 of a \$40 million profit. Let X represent the profit, in millions of dollars. Find the probability distribution of the profit and the expected value of the profit. Does this venture represent an expected gain or an expected loss?

Source: *Journal of the Australasian Institute of Mining and Metallurgy* 306:18–22

Solution

The probability distribution of the profit X is as follows:

x	−30	20	40
$P(x)$	0.4	0.5	0.1

The expected value is the mean of X:

$$E(X) = \mu_X = (-30)(0.4) + 20(0.5) + 40(0.1) = 2.0$$

The expected value is positive, so this is an expected gain of \$2 million.

Insurance premiums

Insurance companies must determine a price (called a *premium*) to charge for their policies. Computing an expected value is an important part of this process.

Example 6.12

Computing the expected value of an insurance premium

An insurance company sells a one-year term life insurance policy to a 70-year-old man. The man pays a premium of \$400. If he dies within one year, the company will pay \$10,000 to his beneficiary. According to the U.S. Centers for Disease Control and Prevention, the probability that a 70-year-old man is still alive one year later is 0.9715. Let X be the profit made by the insurance company. Find the probability distribution and the expected value of the profit.

Solution

If the man lives, the insurance company keeps the \$400 premium and doesn't have to pay anything. So its profit is \$400. If the man dies, the insurance company still keeps the \$400, but it must also pay \$10,000. So its profit is \$400 − \$10,000 = −\$9600, a loss of \$9600. The probability that the man lives is 0.9715 and the probability that he dies is $1 - 0.9715 = 0.0285$. The probability distribution is

x	−9600	400
$P(x)$	0.0285	0.9715

The expected value is the mean of X:

$$E(X) = \mu_X = (-9600)(0.0285) + 400(0.9715) = 115$$

The expected gain for the insurance company is \$115. We can interpret this by saying that if the insurance company sells many policies like this, it can expect to earn \$115 for each policy, on the average.

In this section, we introduced the concept of a random variable and its probability distribution. In the remaining sections of this chapter, we will discuss some specific probability distributions that are often used in practice, and describe some of their important applications.

Using Technology

We use Example 6.6 to illustrate the technology steps.

TI-84 PLUS

Computing the mean and standard deviation of a discrete random variable

Step 1. Enter the values of the random variable into **L1** in the data editor and the associated probabilities into **L2** (Figure A).

Step 2. Press **STAT** and highlight the **CALC** menu.

Step 3. Select **1–Var Stats** and enter **L1**, comma, **L2**.

Step 4. Press **ENTER** (Figure B).

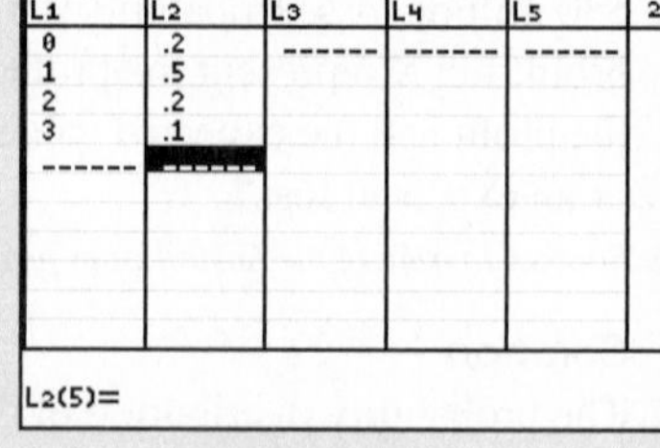

Figure A

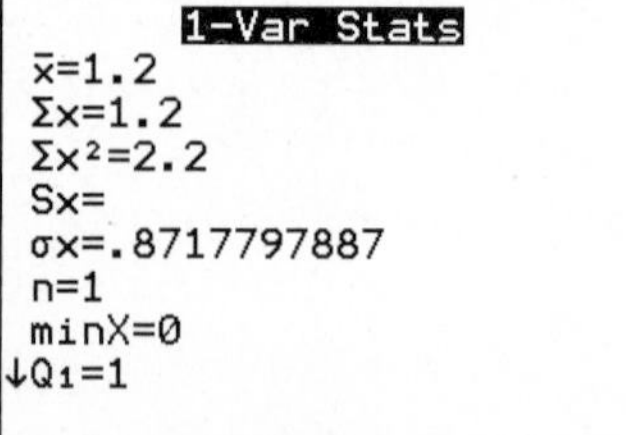

Figure B

Using the TI-84 PLUS Stat Wizards (see Appendix B for more information)

Step 1. Enter the values of the random variable into **L1** in the data editor and the associated probabilities into **L2** (Figure A).

Step 2. Press **STAT** and highlight the **CALC** menu.

Step 3. Select **1–Var Stats**. Enter **L1** into the **List** field and **L2** into the **FreqList** field.

Step 4. Select **Calculate** and press **ENTER** (Figure B).

Section 6.1 Exercises

Exercises 1–8 are the Check Your Understanding exercises located within the section.

Understanding the Concepts

In Exercises 9–12, fill in each blank with the appropriate word or phrase.

9. A numerical outcome of a probability experiment is called a __________. *random variable*
10. The sum of all the probabilities in a discrete probability distribution must be equal to __________. *1*
11. __________ random variables can take on any value in an interval. *Continuous*
12. As the sample size increases, the sample mean approaches the __________ mean. *population*

In Exercises 13–16, determine whether the statement is true or false. If the statement is false, rewrite it as a true statement.

13. To find the mean of a discrete random variable, multiply each possible value of the random variable by its probability, then add the products. *True*
14. The expected value is the mean amount gained or lost. *True*
15. The possible values of a discrete random variable cannot be listed. *False*
16. The standard deviation is found by squaring the variance. *False*

Practicing the Skills

In Exercises 17–28, determine whether the random variable described is discrete or continuous.

17. The number of heads in 100 tosses of a coin *Discrete*
18. The number of people in line at the bank at a randomly chosen time *Discrete*
19. The weight of a randomly chosen student's backpack *Continuous*
20. The amount of rain during the next thunderstorm *Continuous*
21. The number of children absent from school on a randomly chosen day *Discrete*
22. The time it takes to drive to the airport *Continuous*
23. The final exam score of a randomly chosen student from last semester's statistics class *Discrete*
24. The amount of time you wait at a bus stop *Continuous*
25. The height of a randomly chosen college student *Continuous*
26. The number of songs on a randomly chosen playlist *Discrete*
27. The weight of a block of tofu *Continuous*
28. The number of hospital admissions on a given day *Discrete*

In Exercises 29–34, determine whether the table represents a discrete probability distribution. If not, explain why not.

29.

x	P(x)
1	0.4
2	0.2
3	0.1
4	0.3

Yes

30.

x	P(x)
30	0.2
40	0.2
50	0.2
60	0.2
70	0.2

Yes

31.

x	P(x)
2.1	0.1
2.2	0.1
2.3	0.1
2.4	0.1

No

32.

x	P(x)
55	−0.3
65	0.6
75	0.4
85	0.2

No

33.

x	P(x)
100	0.2
200	0.3
300	0.5
400	0.4
500	0.1

No

34.

x	P(x)
−4	0.35
−1	0.25
0	0.15
2	0.25

Yes

In Exercises 35–40, compute the mean and standard deviation of the random variable with the given discrete probability distribution.

35.

x	P(x)
1	0.42
2	0.18
5	0.34
7	0.06

$\mu_X = 2.9$; $\sigma_X = 2.042$

36.

x	P(x)
8	0.15
13	0.23
15	0.25
18	0.27
19	0.10

$\mu_X = 14.7$; $\sigma_X = 3.494$

37.

x	P(x)
4.5	0.33
6	0.11
7	0.21
9.5	0.35

$\mu_X = 6.94$; $\sigma_X = 2.087$

38.

x	P(x)
−3	0.10
0	0.17
1	0.56
3	0.17

$\mu_X = 0.77$; $\sigma_X = 1.548$

39.

x	P(x)
15	0.15
17	0.23
19	0.25
22	0.27
26	0.10

$\mu_X = 19.45$; $\sigma_X = 3.232$

40.

x	P(x)
120	0.30
150	0.30
170	0.15
180	0.25

$\mu_X = 151.5$; $\sigma_X = 23.511$

41. Fill in the missing value so that the following table represents a probability distribution. *0.2*

x	4	5	6	7
P(x)	0.3	?	0.3	0.2

42. Fill in the missing value so that the following table represents a probability distribution. *0.40*

x	15	25	35	45	55
P(x)	0.25	0.15	?	0.05	0.15

Working with the Concepts

43. **Put some air in your tires:** Let X represent the number of tires with low air pressure on a randomly chosen car. The probability distribution of X is as follows.

x	0	1	2	3	4
P(x)	0.1	0.2	0.4	0.2	0.1

   a. Find $P(1)$. *0.2*
   b. Find $P(\text{More than } 2)$. *0.3*
   c. Find the probability that all four tires have low air pressure. *0.1*
   d. Find the probability that no tires have low air pressure. *0.1*
   e. Compute the mean μ_X. *2*
   f. Compute the standard deviation σ_X. *1.095*

44. **Fifteen items or less:** The number of customers in line at a supermarket express checkout counter is a random variable with the following probability distribution.

x	0	1	2	3	4	5
P(x)	0.10	0.25	0.30	0.20	0.10	0.05

a. Find $P(2)$. 0.30
b. Find P(No more than 1). 0.35
c. Find the probability that no one is in line. 0.10
d. Find the probability that at least three people are in line. 0.35
e. Compute the mean μ_X. 2.1
f. Compute the standard deviation σ_X. 1.3
g. If each customer takes 3 minutes to check out, what is the probability that it will take more than 6 minutes for all the customers currently in line to check out? 0.35

45. Defective circuits: The following table presents the probability distribution of the number of defects X in a randomly chosen printed circuit board.

x	0	1	2	3
$P(x)$	0.5	0.3	0.1	0.1

a. Find $P(2)$. 0.1
b. Find P(1 or more). 0.5
c. Find the probability that at least two circuits are defective. 0.2
d. Find the probability that no more than two circuits are defective. 0.9
e. Compute the mean μ_X. 0.8
f. Compute the standard deviation σ_X. 0.980
g. A circuit will function if it has no defects or only one defect. What is the probability that a circuit will function? 0.8

46. Do you carpool? Let X represent the number of occupants in a randomly chosen car on a certain stretch of highway during morning commute hours. A survey of cars showed that the probability distribution of X is as follows.

x	1	2	3	4	5
$P(x)$	0.70	0.15	0.10	0.03	0.02

a. Find $P(2)$. 0.15
b. Find P(More than 3). 0.05
c. Find the probability that a car has only one occupant. 0.7
d. Find the probability that a car has fewer than four occupants. 0.95
e. Compute the mean μ_X. 1.52
f. Compute the standard deviation σ_X. 0.933
g. To save energy, a goal is set to have the mean number of occupants be at least two per car. Has this goal been met? No

47. Dirty air: The federal government has enacted maximum allowable standards for air pollutants such as ozone. Let X be the number of days per year that the level of air pollution exceeds the standard in a certain city. The probability distribution of X is given by

x	0	1	2	3	4
$P(x)$	0.33	0.38	0.19	0.06	0.04

a. Find $P(1)$. 0.38
b. Find P(3 or fewer). 0.96
c. Find the probability that the standard is exceeded on at least one day. 0.67
d. Find the probability that the standard is exceeded on more than two days. 0.10
e. Compute the mean μ_X. 1.1
f. Compute the standard deviation σ_X. 1.054

48. Texting: Five teenagers are selected at random. Let X be the number of them who have sent text messages within the past 30 days. According to a study by the Nielsen Company, the probability distribution of X is as follows:

x	0	1	2	3	4	5
$P(x)$	0.015	0.097	0.258	0.343	0.227	0.060

a. Find $P(2)$. 0.258
b. Find P(More than 1). 0.888
c. Find the probability that three or more of the teenagers sent text messages. 0.630
d. Find the probability that fewer than two of the teenagers sent text messages. 0.112
e. Compute the mean μ_X. 2.85
f. Compute the standard deviation σ_X. 1.107

49. Relax! The General Social Survey asked 1676 people how many hours per day they were able to relax. The results are presented in the following table.

Number of Hours	Frequency
0	114
1	186
2	336
3	251
4	316
5	231
6	149
7	33
8	60
Total	1676

Consider these 1676 people to be a population. Let X be the number of hours of relaxation for a person sampled at random from this population.
a. Construct the probability distribution of X.
b. Find the probability that a person relaxes more than 4 hours per day. 0.282
c. Find the probability that a person doesn't relax at all. 0.068
d. Compute the mean μ_X. 3.36
e. Compute the standard deviation σ_X. 1.97

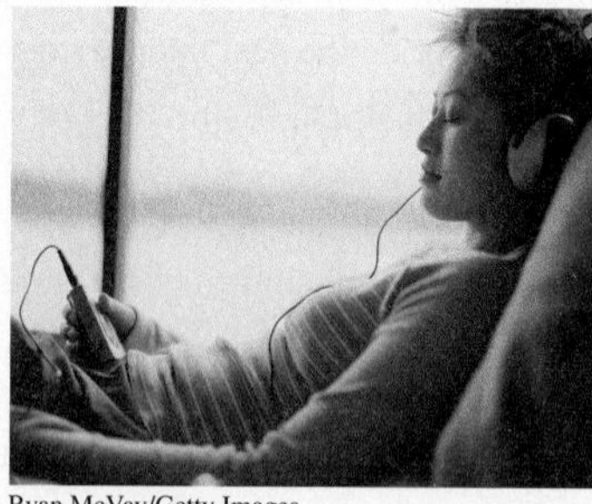
Ryan McVay/Getty Images

50. Pain: The General Social Survey asked 827 people how many days they would wait to seek medical treatment if they were suffering pain that interfered with their ability to work. The results are presented in the following table.

Number of Days	Frequency
0	27
1	436
2	263
3	72
4	19
5	10
Total	827

Consider these 827 people to be a population. Let X be the number of days for a person sampled at random from this population.

a. Construct the probability distribution of X.

b. Find the probability that a person would wait for 3 days. 0.087

c. Find the probability that a person would wait more than 2 days. 0.122

d. Compute the mean μ_X. 1.58

e. Compute the standard deviation σ_X. 0.88

51. School days: The following table presents the numbers of students enrolled in grades 1 through 8 in public schools in the United States.

Grade	Frequency (in 1000s)
1	3750
2	3640
3	3627
4	3585
5	3601
6	3660
7	3715
8	3765
Total	29,343

Source: *Statistical Abstract of the United States*

Consider these students to be a population. Let X be the grade of a student randomly chosen from this population.

a. Construct the probability distribution of X.

b. Find the probability that the student is in fourth grade. 0.122

c. Find the probability that the student is in seventh or eighth grade. 0.255

d. Compute the mean μ_X. 4.51

e. Compute the standard deviation σ_X. 2.31

52. World Cup: The World Cup soccer tournament has been held 29 times since 1930 (21 men's tournaments and 8 women's tournaments). The following table presents the number of goals scored by the winning team in each championship game. Note that there were two games in which no goals were scored; these games were decided on penalty shoot-outs.

Goals	Frequency
0	2
1	4
2	10
3	6
4	5
5	2
Total	29

Consider these 29 games to be a population. Let X be the number of goals scored in a game randomly chosen from this population.

a. Construct the probability distribution of X.

b. Find the probability that three goals were scored. 0.2069

c. Find the probability that fewer than four goals were scored. 0.7586

d. Compute the mean μ_X. 2.4828

e. Compute the standard deviation σ_X. 1.3031

53. Lottery: In the New York State Numbers Lottery, you pay \$1 and pick a number from 000 to 999. If your number comes up, you win \$500, which is a profit of \$499. If you lose, you lose \$1. Your probability of winning is 0.001. What is the expected value of your profit? Is it an expected gain or an expected loss? −\$0.50; loss

54. Lottery: In the New York State Numbers Lottery, you pay \$1 and can bet that the sum of the numbers that come up is 13. The probability of winning is 0.075, and if you win, you win \$6.50, which is a profit of \$5.50. If you lose, you lose \$1. What is the expected value of your profit? Is it an expected gain or an expected loss? −\$0.51; loss

55. Craps: In the game of craps, two dice are rolled, and people bet on the outcome. For example, you can bet \$1 that the dice will total 7. The probability that you win is $1/6$, and if you win, your profit is \$4. If you lose, you lose \$1. What is the expected value of your profit? Is it an expected gain or an expected loss? −\$0.17; loss

56. More craps: Another bet you can make in craps is that the sum of the dice will be 2 (also called "snake eyes"). The probability that you win is $1/36$, and if you win, your profit is \$30. If you lose, you lose \$1. What is the expected value of your profit? Is it an expected gain or an expected loss? −\$0.14; loss

57. Multiple choice: A multiple-choice question has five choices. If you get the question right, you gain one point, and if you get it wrong, you lose $1/4$ point. Assume you have no idea what the right answer is, so you pick one of the choices at random.

a. What is the expected value of the number of points you get? 0

b. If you don't answer a question, you get 0 points. The test makers advise you not to answer a question if you have no idea which answer is correct. Do you think this is good advice? Explain.

58. More multiple choice: Refer to Exercise 57. Assume you can eliminate one of the five choices, and you choose one of the remaining four at random as your answer.

a. What is the expected value of the number of points you get? 0.0625

b. If you don't answer a question, you get 0 points. The test makers advise you to guess if you can eliminate one or more answers. Do you think this is good advice? Explain. Yes

59. Business projection: An investor is considering a \$10,000 investment in a start-up company. She estimates that she has probability 0.25 of a \$20,000 loss, probability 0.20 of a \$10,000 profit, probability 0.15 of a \$50,000 profit, and probability 0.40 of breaking even (a profit of \$0). What is the expected value of the profit? Would you advise the investor to make the investment? \$4500; yes

60. Insurance: An insurance company sells a one-year term life insurance policy to an 80-year-old woman. The woman pays a premium of \$1000. If she dies within one year, the company will pay \$20,000 to her beneficiary. According to the U.S. Centers for Disease Control and Prevention, the probability that an 80-year-old woman will be alive one year later is 0.9516. Let X be the profit made by the insurance company. Find the probability distribution and the expected value of the profit. Expected value is \$32

61. Heads and tails: A fair coin will be tossed until a tail appears or until three tosses have been made, whichever comes first. Let Y be the number of heads that are tossed.

a. Find the probability distribution of Y.
b. Find the mean μ_Y. 0.875
c. Find the standard deviation σ_Y. 1.0533

62. Tails and heads: In Exercise 61, let X be the number of tails that come up.
a. Find the probability distribution of X.
b. Find the mean μ_X. 0.875
c. Find the standard deviation σ_X. 0.3307

Extending the Concepts

63. Success and failure: Three components are randomly sampled, one at a time, from a large lot. As each component is selected, it is tested. If it passes the test, a success (S) occurs; if it fails the test, a failure (F) occurs. Assume that 80% of the components in the lot will succeed in passing the test. Let X represent the number of successes among the three sampled components.
a. What are the possible values for X? 0, 1, 2, 3
b. Find $P(3)$. 0.512
c. The event that the first component fails and the next two succeed is denoted by FSS. Find $P(\text{FSS})$. 0.128
d. Find $P(\text{SFS})$ and $P(\text{SSF})$. 0.128
e. Use the results of parts (c) and (d) to find $P(2)$. 0.384
f. Find $P(1)$. 0.096
g. Find $P(0)$. 0.008
h. Find μ_X. 2.4
i. Find σ_X. 0.6928

Answers to Check Your Understanding Exercises for Section 6.1

1.

x	0	1	2	3
$P(x)$	0.125	0.375	0.375	0.125

2. It is not possible. The probabilities do not add up to 1.

3. (a) and (c)

4. **a.** 0.45 **b.** 0.35 **c.** 0.94

5.

x	1	2	3	4	5	6
$P(x)$	0.0956	0.1290	0.1136	0.3728	0.2714	0.0176

6. **a.** 0.244 **b.** 0.433 **c.** 0.121

7. The mean is 7.48. Interpretation: If we were to give this quiz to more and more students, the mean score for these students would approach 7.48.

8. **a.** 1.6075 **b.** 1.2679

Section 6.2 The Binomial Distribution

Objectives

1. Determine whether a random variable is binomial
2. Determine the probability distribution of a binomial random variable
3. Compute binomial probabilities
4. Compute the mean and variance of a binomial random variable

Your favorite fast-food chain is giving away a coupon with every purchase of a meal. You scratch the coupon to reveal your prize. Twenty percent of the coupons entitle you to a free milkshake, and the rest of them say "better luck next time." You go to this restaurant in a group of 10 people, and everyone orders lunch. What is the probability that three of you win a free milkshake? Let X be the number of people out of 10 who win a free milkshake. What is the probability distribution of X? In this section, we will learn that X has a distribution called the *binomial distribution*, which is one of the most useful probability distributions.

Binomial Random Variables

Objective 1 Determine whether a random variable is binomial

In the situation just described, we are examining 10 coupons. Each time we examine a coupon, we will call it a **trial**, so there are 10 trials. When a coupon is good for a free milkshake, we will call it a "success." The random variable X represents the number of successes in 10 trials.

Under certain conditions, a random variable that represents the number of successes in a series of trials has a probability distribution called the **binomial distribution**. The conditions are:

Conditions for the Binomial Distribution

1. A fixed number of trials are conducted.
2. There are two possible outcomes for each trial. One is labeled "success," and the other is labeled "failure."
3. The probability of success is the same on each trial.
4. The trials are independent. This means that the outcome of one trial does not affect the outcomes of the other trials.
5. The random variable X represents the number of successes that occur.

Notation: The following notation is commonly used:

- The number of trials is denoted by n.
- The probability of success is denoted by p, and the probability of failure is $1 - p$.

It is important to realize that the word *success* does not necessarily refer to a desirable outcome. For example, in medical studies that involve counting the number of people who suffer from a certain disease, the value of p is the probability that someone will come down with the disease. In these studies, disease is a "success," although it is certainly not a desirable outcome.

Example 6.13

Determining whether a random variable is binomial

Determine which of the following are binomial random variables. For those that are binomial, state the two possible outcomes and specify which is a success. Also state the values of n and p.

a. A fair coin is tossed 10 times. Let X be the number of times the coin lands heads.

b. Five basketball players each attempt a free throw. Let X be the number of free throws made.

c. Ten cards are in a box. Five are red and five are green. Three of the cards are drawn at random without replacement. Let X be the number of red cards drawn.

Solution

a. This is a binomial random variable. Each toss of the coin is a trial. There are two possible outcomes—heads and tails. Since X represents the number of heads, a head counts as a success. The trials are independent, because the outcome of one coin toss does not affect the other tosses. The number of trials is $n = 10$, and the success probability is $p = 0.5$.

b. This is not a binomial random variable. The probability of success (making a shot) differs from player to player, because they will not all be equally skilled at making free throws.

c. This is not a binomial random variable because the trials are not independent. If the first card is red, then four of the nine remaining cards will be red, and the probability of a red card on the second draw will be $4/9$. If the first card is not red, then five of the nine remaining cards will be red, and the probability of a red card on the second draw will be $5/9$. Since the probability of success on the second trial depends on the outcome of the first trial, the trials are not independent.

In part (c) of Example 6.13, the sampling was done without replacement, and the trials were not independent. If the sample is less than 5% of the population, however, then in most cases the lack of replacement will have only a negligible effect, and it is appropriate to consider the trials to be independent. (See the discussion in Section 5.3.) In particular, when a simple random sample is drawn from a population, we will consider the sampled individuals to be independent whenever the sample size is less than 5% of the population size.

When a simple random sample comprises less than 5% of the population, we will consider the sampled individuals to be independent.

Check Your Understanding

1. Determine whether X is a binomial random variable.

a. A fair die is rolled 20 times. Let X be the number of times the die comes up 6. *Binomial*

b. A standard deck of 52 cards contains four aces. Four cards are dealt without replacement from this deck. Let X be the number that are aces. *Not binomial*

c. A simple random sample of 50 voters is drawn from the residents in a large city. Let X be the number who plan to vote for a proposition to increase spending on public schools. *Binomial*

Answers are on page 290.

Objective 2 Determine the probability distribution of a binomial random variable

The Binomial Probability Distribution

We will determine the probability distribution of a binomial random variable by considering a simple example. A biased coin has probability 0.6 of coming up heads. The coin is tossed three times. Let X be the number of heads that come up. Since X is the number of heads, coming up heads is a success. Then X is binomial, with $n = 3$ trials and success probability $p = 0.6$. We will compute $P(2)$, the probability that exactly two of the tosses are heads.

There are three arrangements of two heads in three tosses of a coin, HHT, HTH, and THH. We first compute the probability of HHT. The event HHT is a sequence of independent events: H on the first toss, H on the second toss, T on the third toss. We know the probabilities of each of these events separately:

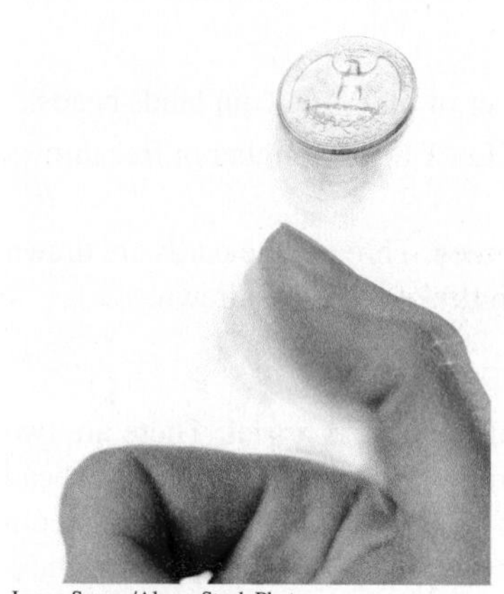

Image Source/Alamy Stock Photo

$P(\text{H on the first toss}) = 0.6, \quad P(\text{H on the second toss}) = 0.6, \quad P(\text{T on the third toss}) = 0.4$

Because the events are independent, the Multiplication Rule for Independent Events tells us that the probability that they all occur is equal to the product of their probabilities. Therefore,

$$P(\text{HHT}) = (0.6)(0.6)(0.4) = (0.6)^2(0.4)^1$$

Similarly, $P(\text{HTH}) = (0.6)(0.4)(0.6) = (0.6)^2(0.4)^1$, and $P(\text{THH}) = (0.4)(0.6)(0.6) = (0.6)^2(0.4)^1$. We can see that all the different arrangements of two heads and one tail have the same probability. Now

$$\begin{aligned} P(2) &= P(\text{HHT or HTH or THH}) \\ &= P(\text{HHT}) + P(\text{HTH}) + P(\text{THH}) \text{ (Addition Rule for Mutually Exclusive Events)} \\ &= (0.6)^2(0.4)^1 + (0.6)^2(0.4)^1 + (0.6)^2(0.4)^1 \\ &= 3(0.6)^2(0.4)^1 \end{aligned}$$

RECALL

$_nC_x = \dfrac{n!}{x!(n-x)!}$, where

$n! = n(n-1)\cdots(2)(1)$

Examining this result, we see that the number 3 represents the number of arrangements of two successes (heads) and one failure (tails). In general, this number will be the number of arrangements of x successes in n trials, which is $_nC_x$. The number 0.6 is the success probability, which in general will be p. The exponent 2 is the number of successes, which in general will be x. The number 0.4 is the failure probability, which is $1 - p$, and the exponent 1 is the number of failures, which is $n - x$.

Formula for Binomial Probabilities

For a binomial random variable X that represents the number of successes in n trials with success probability p, the probability of obtaining x successes is

$$P(x) = {}_nC_x p^x (1-p)^{n-x}$$

The possible values of X are $0, 1, \ldots, n$.

Objective 3 Compute binomial probabilities

Computing Binomial Probabilities

The binomial probability distribution can require tedious calculations. While we can compute simple probabilities by hand, for more involved problems it is better to use a table or technology.

Example 6.14

Calculating probabilities by using the binomial probability distribution

The Pew Research Center recently reported that approximately 30% of internet users in the United States use the image sharing website Pinterest. Suppose a simple random sample of 15 internet users is taken. Use the binomial probability distribution to find the following probabilities.

a. Find the probability that exactly four of the sampled people use Pinterest.

b. Find the probability that fewer than three of the people use Pinterest.

c. Find the probability that more than one person uses Pinterest.

d. Find the probability that the number of people who use Pinterest is between 1 and 4, inclusive.

Solution

a. We use the binomial probability distribution with $n = 15$, $p = 0.3$, and $x = 4$.

$$\begin{aligned} P(4) &= {}_{15}C_4(0.3)^4(1-0.3)^{15-4} \\ &= \frac{15!}{4!(15-4)!}(0.3)^4(0.7)^{11} \\ &= 1365(0.3)^4(0.7)^{11} \\ &= 0.219 \end{aligned}$$

b. The possible numbers of people that are fewer than three are 0, 1, and 2. So we need to find $P(0 \text{ or } 1 \text{ or } 2)$. We use the Addition Rule for Mutually Exclusive Events.

$$\begin{aligned} P(0 \text{ or } 1 \text{ or } 2) &= P(0) + P(1) + P(2) \\ &= {}_{15}C_0(0.3)^0(1-0.3)^{15-0} + {}_{15}C_1(0.3)^1(1-0.3)^{15-1} \\ &\quad + {}_{15}C_2(0.3)^2(1-0.3)^{15-2} \\ &= 0.0047 + 0.0305 + 0.0916 \\ &= 0.127 \end{aligned}$$

c. The possible numbers of people that are more than one are 2, 3, 4, and so forth up to 15. We could find $P(\text{More than } 1)$ by adding $P(2) + P(3) + \cdots + P(15)$, but fortunately there is an easier way. We will use the Rule of Complements. The complement of "more than 1" is "1 or fewer," or, equivalently, 0 or 1. We compute the probability of the complement, and subtract from 1.

$$\begin{aligned} P(0 \text{ or } 1) &= P(0) + P(1) \\ &= {}_{15}C_0(0.3)^0(1-0.3)^{15-0} + {}_{15}C_1(0.3)^1(1-0.3)^{15-1} \\ &= 0.0047 + 0.0305 \\ &= 0.035 \end{aligned}$$

Now we use the Rule of Complements:

$$P(\text{More than } 1) = 1 - P(0 \text{ or } 1) = 1 - 0.035 = 0.965$$

d. Between 1 and 4 inclusive means 1, 2, 3, or 4.

$$P(1 \text{ or } 2 \text{ or } 3 \text{ or } 4) = P(1) + P(2) + P(3) + P(4)$$

$$= {}_{15}C_1(0.3)^1(1-0.3)^{15-1} + {}_{15}C_2(0.3)^2(1-0.3)^{15-2}$$

$$+ {}_{15}C_3(0.3)^3(1-0.3)^{15-3} + {}_{15}C_4(0.3)^4(1-0.3)^{15-4}$$

$$= 0.0305 + 0.0916 + 0.1700 + 0.2186$$

$$= 0.511$$

EXPLAIN IT AGAIN

When can we use a table? Tables contain only a limited selection of values for n and p. Probabilities involving values not in the table must be computed by hand or with technology.

Using a table to compute binomial probabilities

Table A.1 contains probabilities for the binomial distribution. It can be used to compute binomial probabilities for values of n up to 20 and certain values of p.

Example 6.15

Use a table to compute binomial probabilities

The Pew Research Center recently reported that approximately 30% of internet users in the United States use the image sharing website Pinterest. Suppose a simple random sample of 15 internet users is taken. Use the binomial probability distribution to find the following probabilities.

a. Find the probability that exactly five of the sampled people use Pinterest.

b. Find the probability that fewer than four of the people use Pinterest.

c. Find the probability that the number of sampled people who use Pinterest is between 6 and 8, inclusive.

Solution

a. We have $n = 15$, so we go to the section of Table A.1 that corresponds to $n = 15$. This is shown in Figure 6.5. We look at the column corresponding to $p = 0.30$. Now for each value in the column labeled "x," the number in the table is the probability $P(x)$. We therefore look in the row corresponding to $x = 5$. The probability that exactly five people use Pinterest is $P(5) = 0.206$.

							p							
n	x	0.05	0.10	0.20	0.25	0.30	0.40	0.50	0.60	0.70	0.75	0.80	0.90	0.95
15	0	0.463	0.206	0.035	0.013	0.005	0.000+	0.000+	0.000+	0.000+	0.000+	0.000+	0.000+	0.000+
	1	0.366	0.343	0.132	0.067	0.031	0.005	0.000+	0.000+	0.000+	0.000+	0.000+	0.000+	0.000+
	2	0.135	0.267	0.231	0.156	0.092	0.022	0.003	0.000+	0.000+	0.000+	0.000+	0.000+	0.000+
	3	0.031	0.129	0.250	0.225	0.170	0.063	0.014	0.002	0.000+	0.000+	0.000+	0.000+	0.000+
	4	0.005	0.043	0.188	0.225	0.219	0.127	0.042	0.007	0.001	0.000+	0.000+	0.000+	0.000+
	5	0.001	0.010	0.103	0.165	0.206	0.186	0.092	0.024	0.003	0.001	0.000+	0.000+	0.000+
	6	0.000+	0.002	0.043	0.092	0.147	0.207	0.153	0.061	0.012	0.003	0.001	0.000+	0.000+
	7	0.000+	0.000+	0.014	0.039	0.081	0.177	0.196	0.118	0.035	0.013	0.003	0.000+	0.000+
	8	0.000+	0.000+	0.003	0.013	0.035	0.118	0.196	0.177	0.081	0.039	0.014	0.000+	0.000+
	9	0.000+	0.000+	0.001	0.003	0.012	0.061	0.153	0.207	0.147	0.092	0.043	0.002	0.000+
	10	0.000+	0.000+	0.000+	0.001	0.003	0.024	0.092	0.186	0.206	0.165	0.103	0.010	0.001
	11	0.000+	0.000+	0.000+	0.000+	0.001	0.007	0.042	0.127	0.219	0.225	0.188	0.043	0.005
	12	0.000+	0.000+	0.000+	0.000+	0.000+	0.002	0.014	0.063	0.170	0.225	0.250	0.129	0.031
	13	0.000+	0.000+	0.000+	0.000+	0.000+	0.000+	0.003	0.022	0.092	0.156	0.231	0.267	0.135
	14	0.000+	0.000+	0.000+	0.000+	0.000+	0.000+	0.000+	0.005	0.031	0.067	0.132	0.343	0.366
	15	0.000+	0.000+	0.000+	0.000+	0.000+	0.000+	0.000+	0.000+	0.005	0.013	0.035	0.206	0.463

Figure 6.5

EXPLAIN IT AGAIN

Answers using technology may differ: If you find *P*(Fewer than 4) using technology, your answer will be 0.297 rather than 0.298. This difference isn't large enough to matter.

b. $P(\text{Fewer than } 4) = P(0) + P(1) + P(2) + P(3)$. We find these probabilities in Table A.1 and add them. See Figure 6.6.

$$P(\text{Fewer than } 4) = 0.005 + 0.031 + 0.092 + 0.170$$
$$= 0.298$$

							p							
n	*x*	**0.05**	**0.10**	**0.20**	**0.25**	**0.30**	**0.40**	**0.50**	**0.60**	**0.70**	**0.75**	**0.80**	**0.90**	**0.95**
15	0	0.463	0.206	0.035	0.013	0.005	0.000+	0.000+	0.000+	0.000+	0.000+	0.000+	0.000+	0.000+
	1	0.366	0.343	0.132	0.067	0.031	0.005	0.000+	0.000+	0.000+	0.000+	0.000+	0.000+	0.000+
	2	0.135	0.267	0.231	0.156	0.092	0.022	0.003	0.000+	0.000+	0.000+	0.000+	0.000+	0.000+
	3	0.031	0.129	0.250	0.225	0.170	0.063	0.014	0.002	0.000+	0.000+	0.000+	0.000+	0.000+
	4	0.005	0.043	0.188	0.225	0.219	0.127	0.042	0.007	0.001	0.000+	0.000+	0.000+	0.000+
	5	0.001	0.010	0.103	0.165	0.206	0.186	0.092	0.024	0.003	0.001	0.000+	0.000+	0.000+
	6	0.000+	0.002	0.043	0.092	0.147	0.207	0.153	0.061	0.012	0.003	0.001	0.000+	0.000+
	7	0.000+	0.000+	0.014	0.039	0.081	0.177	0.196	0.118	0.035	0.013	0.003	0.000+	0.000+
	8	0.000+	0.000+	0.003	0.013	0.035	0.118	0.196	0.177	0.081	0.039	0.014	0.000+	0.000+
	9	0.000+	0.000+	0.001	0.003	0.012	0.061	0.153	0.207	0.147	0.092	0.043	0.002	0.000+
	10	0.000+	0.000+	0.000+	0.001	0.003	0.024	0.092	0.186	0.206	0.165	0.103	0.010	0.001
	11	0.000+	0.000+	0.000+	0.000+	0.001	0.007	0.042	0.127	0.219	0.225	0.188	0.043	0.005
	12	0.000+	0.000+	0.000+	0.000+	0.000+	0.002	0.014	0.063	0.170	0.225	0.250	0.129	0.031
	13	0.000+	0.000+	0.000+	0.000+	0.000+	0.000+	0.003	0.022	0.092	0.156	0.231	0.267	0.135
	14	0.000+	0.000+	0.000+	0.000+	0.000+	0.000+	0.000+	0.005	0.031	0.067	0.132	0.343	0.366
	15	0.000+	0.000+	0.000+	0.000+	0.000+	0.000+	0.000+	0.000+	0.005	0.013	0.035	0.206	0.463

Figure 6.6

c. $P(\text{Between 6 and 8 inclusive}) = P(6) + P(7) + P(8)$. We find these probabilities in Table A.1 and add them. See Figure 6.7.

$$P(\text{Between 6 and 8 inclusive}) = 0.147 + 0.081 + 0.035$$
$$= 0.263$$

							p							
n	*x*	**0.05**	**0.10**	**0.20**	**0.25**	**0.30**	**0.40**	**0.50**	**0.60**	**0.70**	**0.75**	**0.80**	**0.90**	**0.95**
15	0	0.463	0.206	0.035	0.013	0.005	0.000+	0.000+	0.000+	0.000+	0.000+	0.000+	0.000+	0.000+
	1	0.366	0.343	0.132	0.067	0.031	0.005	0.000+	0.000+	0.000+	0.000+	0.000+	0.000+	0.000+
	2	0.135	0.267	0.231	0.156	0.092	0.022	0.003	0.000+	0.000+	0.000+	0.000+	0.000+	0.000+
	3	0.031	0.129	0.250	0.225	0.170	0.063	0.014	0.002	0.000+	0.000+	0.000+	0.000+	0.000+
	4	0.005	0.043	0.188	0.225	0.219	0.127	0.042	0.007	0.001	0.000+	0.000+	0.000+	0.000+
	5	0.001	0.010	0.103	0.165	0.206	0.186	0.092	0.024	0.003	0.001	0.000+	0.000+	0.000+
	6	0.000+	0.002	0.043	0.092	0.147	0.207	0.153	0.061	0.012	0.003	0.001	0.000+	0.000+
	7	0.000+	0.000+	0.014	0.039	0.081	0.177	0.196	0.118	0.035	0.013	0.003	0.000+	0.000+
	8	0.000+	0.000+	0.003	0.013	0.035	0.118	0.196	0.177	0.081	0.039	0.014	0.000+	0.000+
	9	0.000+	0.000+	0.001	0.003	0.012	0.061	0.153	0.207	0.147	0.092	0.043	0.002	0.000+
	10	0.000+	0.000+	0.000+	0.001	0.003	0.024	0.092	0.186	0.206	0.165	0.103	0.010	0.001
	11	0.000+	0.000+	0.000+	0.000+	0.001	0.007	0.042	0.127	0.219	0.225	0.188	0.043	0.005
	12	0.000+	0.000+	0.000+	0.000+	0.000+	0.002	0.014	0.063	0.170	0.225	0.250	0.129	0.031
	13	0.000+	0.000+	0.000+	0.000+	0.000+	0.000+	0.003	0.022	0.092	0.156	0.231	0.267	0.135
	14	0.000+	0.000+	0.000+	0.000+	0.000+	0.000+	0.000+	0.005	0.031	0.067	0.132	0.343	0.366
	15	0.000+	0.000+	0.000+	0.000+	0.000+	0.000+	0.000+	0.000+	0.005	0.013	0.035	0.206	0.463

Figure 6.7

Example 6.16 Using technology to compute binomial probabilities

The Pew Research Center recently reported that approximately 30% of internet users in the United States use the image sharing website Pinterest. Suppose a simple random sample of

15 internet users is taken. Use the binomial probability distribution to find the following probabilities.

a. Find the probability that exactly four of the sampled people use Pinterest.

b. Find the probability that five or fewer of the people use Pinterest.

c. Find the probability that more than seven of the people use Pinterest.

Solution

a. We will use the TI-84 Plus calculator. We use the **binompdf** command. We input 15 for n, .3 for p, and 4 for x. The following display shows the result. Step-by-step instructions are given in the Using Technology section on page 286.

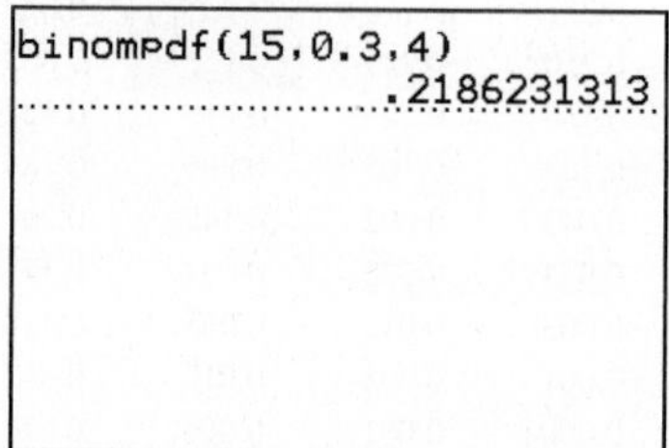

b. We will use the TI-84 Plus calculator. Because we want to find the probability of "five or fewer," which is the same as "less than or equal to five," we will use the **binomcdf** command. We input 15 for n, .3 for p, and 5 for x. The following display shows the result. Step-by-step instructions are given in the Using Technology section on page 286.

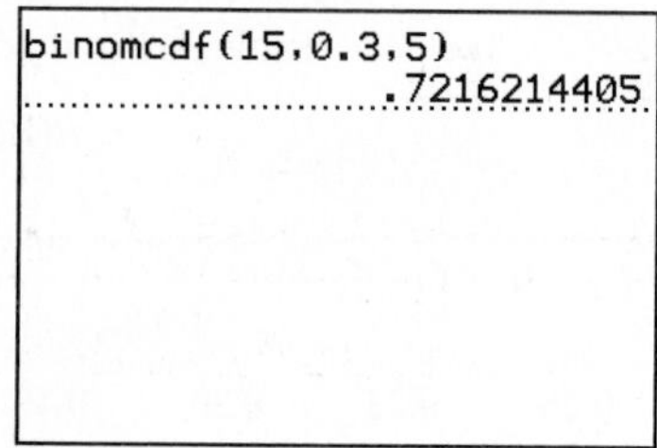

c. We will use Excel. Some versions of Excel are not able to compute probabilities of the form "more than," or "greater than," but they compute probabilities of the form "less than or equal to." We therefore note that the complement of "more than 7" is "less than or equal to 7." We will use Excel to compute P(Less than or equal to 7), and then subtract from 1. The following display shows the result.

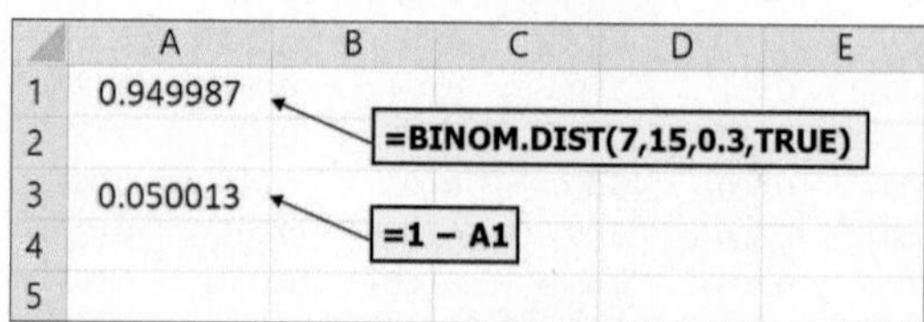

The display shows that P(Less than or equal to 7) = 0.949987. We conclude that P(More than 7) = 1 − 0.949987 = 0.050013. Step-by-step instructions for producing the Excel output are given in the Using Technology section on page 286.

Check Your Understanding

2. In a recent Pew poll, 50% of adults said that they play video games. Assume that 9 adults are randomly sampled.

a. Use the binomial probability distribution to compute the probability that exactly four of them play video games. *0.246*
b. Use Table A.1 to find the probability that fewer than three of them play video games. *0.090*
c. Use any valid method to find the probability that more than three of them play video games. *0.746*
d. Use any valid method to find the probability that the number who play video games is between 4 and 6 inclusive. *0.656*

3. In a recent Pew poll, 10% of adults said that they consider themselves to be "gamers." Assume that 18 adults are randomly sampled.
a. Use the binomial probability distribution to compute the probability that exactly three of them consider themselves to be gamers. *0.168*
b. Use Table A.1 to find the probability that fewer than two of them consider themselves to be gamers. *0.450*
c. Use any valid method to find the probability that more than two of them consider themselves to be gamers. *0.266*
d. Use any valid method to find the probability that the number who consider themselves to be gamers is between 1 and 4 inclusive. *0.822*

Answers are on page 290.

Objective 4 Compute the mean and variance of a binomial random variable

Mean and Variance of a Binomial Random Variable

A fair coin has probability 0.5 of landing heads. If we toss a fair coin 10 times, we expect to get 5 heads, on the average. The reason is that 5 is half of 10, or, in symbols, $5 = 10 \cdot 0.5$. We can see that the mean number of successes was found by multiplying the number of trials by the success probability. This holds true in general.

The variance and standard deviation of a binomial random variable are straightforward to compute as well, although the reasoning behind them is not so obvious.

Mean, Variance, and Standard Deviation of a Binomial Random Variable

Let X be a binomial random variable with n trials and success probability p. Then the mean of X is

$$\mu_X = np$$

The variance of X is

$$\sigma_X^2 = np(1-p)$$

The standard deviation of X is

$$\sigma_X = \sqrt{np(1-p)}$$

NOTE TO INSTRUCTOR

Emphasize that p is the probability of success on a single trial.

Example 6.17

Find the mean and standard deviation of a binomial random variable

The probability that a new car of a certain model will require repairs during the warranty period is 0.15. A particular dealership sells 80 such cars. Let X be the number that will require repairs during the warranty period. Find the mean and standard deviation of X.

Solution

There are $n = 80$ trials, with success probability $p = 0.15$. The mean is

$$\mu_X = np = 80 \cdot 0.15 = 12$$

The standard deviation is

$$\sigma_X = \sqrt{np(1-p)} = \sqrt{80 \cdot 0.15 \cdot (1-0.15)} = 3.194$$

The interpretation of the mean is that in the long run, 12 out of every 80 cars will require repairs. The standard deviation measures the spread in the distribution of the number of cars that will require repairs. We will describe how to use the mean and standard deviation to approximate binomial probabilities in Section 7.5.

Check Your Understanding

4. Gregor Mendel discovered the basic laws of heredity by studying pea plants. In one experiment, he produced plants whose parent plants contained genes for both green and yellow pods. Mendel's theory states that the offspring of two such parents has probability 0.75 of having green pods. Assume that 80 such plants are produced.
   a. Find the mean number of plants that have green pods. *60*
   b. Find the variance of the number of plants that have green pods. *15*
   c. Find the standard deviation of the number of plants that have green pods. *3.8730*

Answers are on page 290.

Using Technology

We use Example 6.14 to illustrate the technology steps.

TI-84 PLUS

Computing binomial probabilities of the form $P(x)$ or P(Less than or equal to x)

Step 1. Press **2nd, VARS** to access the **DISTR** menu.

- To compute $P(x)$, select **binompdf** and enter the values for n, p, and x separated by commas, and press **ENTER**.
- To compute P(Less than or equal to x), select **binomcdf** and enter the values for n, p, and x separated by commas, and press **ENTER**.

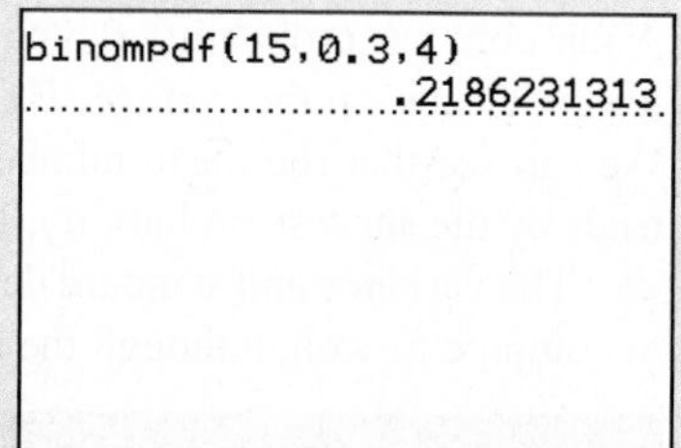

Figure A

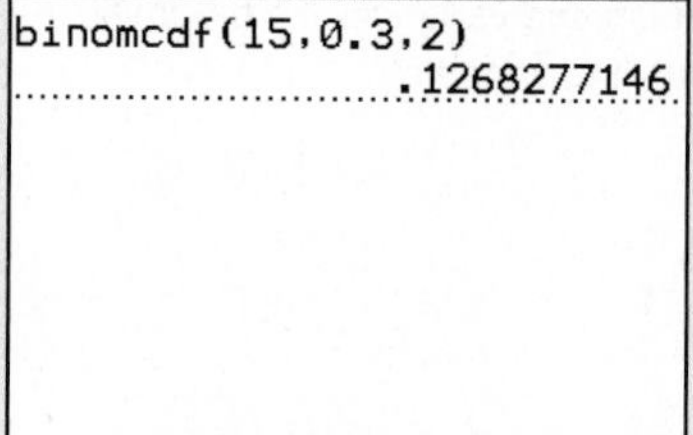

Figure B

Using the TI-84 PLUS Stat Wizards (see Appendix B for more information)

Step 1. Press **2nd, VARS** to access the **DISTR** menu.

- To compute $P(x)$, select **binompdf** and enter the value for n in the **trials** field, the value for p in the **p** field, and the value for x in the **x value** field. Select **Paste** and press **ENTER** to paste the command to the home screen. Press **ENTER** again to run the command.
- To compute $P(x)$, select **binomcdf** and enter the value for n in the **trials** field, the value for p in the **p** field, and the value for x in the **x value** field. Select **Paste** and press **ENTER** to paste the command to the home screen. Press **ENTER** again to run the command.

For Example 6.14, $n = 15$ and $p = 0.3$. Figure A displays the result of part (a), which asks for $P(4)$. Figure B displays the result of part (b), which asks for P(Less than or equal to 2).

EXCEL

Computing binomial probabilities of the form $P(x)$ or P(Less than or equal to x)

The **=BINOM.DIST(*number_s, trials, probability_s, cumulative*)** command computes binomial probabilities where

- ***number_s*** is the number of successes
- ***trials*** is the number of trials
- ***probability_s*** is the probability of a success on each trial
- ***cumulative*** is a true/false value. To compute $P(x)$, enter **FALSE**. To compute P(Less than or equal to x), enter **TRUE**.

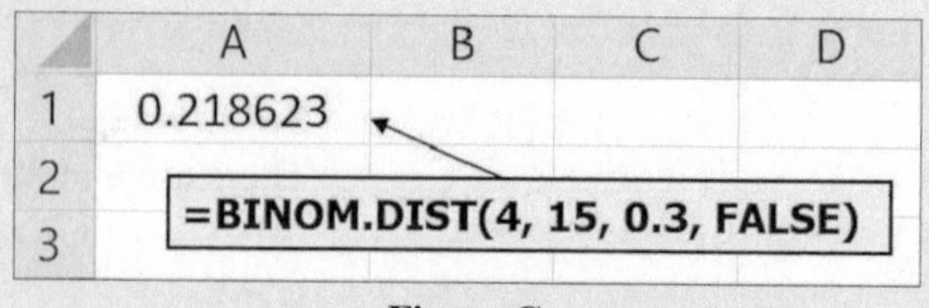

Figure C

Figure C illustrates computing $P(4)$ in part (a) of Example 6.14.

MINITAB

Computing binomial probabilities of the form $P(x)$ or P(Less than or equal to x)

Step 1. Click **Calc**, then **Probability Distributions**, then **Binomial**.

Step 2. Enter the value for n in the **Number of trials** field and the value for p in the **Probability of success** field.

Step 3. To compute $P(x)$, select the **Probability** option and enter the value for x in the **Input constant** field. To compute P(Less than or equal to x), select the **Cumulative probability** option and enter the value for x in the **Input constant** field.

Step 4. Click **OK**.

Note: Binomial probabilities may be computed for a column of values by entering the column name in the **Input column** field.

Section 6.2 Exercises

Exercises 1–4 are the Check Your Understanding exercises located within the section.

Understanding the Concepts

In Exercises 5–7, fill in each blank with the appropriate word or phrase.

5. In a binomial distribution, there are ________ possible outcomes for each trial. *two*

6. To compute a binomial probability, we must know both the success probability and the number n of ________. *trials*

7. If X is a binomial random variable with n trials and success probability p, the standard deviation of X is σ_X = ________. $\sqrt{np(1-p)}$

In Exercises 8–10, determine whether the statement is true or false. If the statement is false, rewrite it as a true statement.

8. The trials in a binomial distribution are independent. *True*

9. A binomial random variable with n trials can sometimes have a value greater than n. *False*

10. The mean of a binomial random variable is found by multiplying the number of trials by the success probability. *True*

Practicing the Skills

In Exercises 11–16, determine whether the random variable X has a binomial distribution. If it does, state the number of trials n. If it does not, explain why not.

11. Ten students are chosen from a statistics class of 25 students. Let X be the number who got an A in the class. *Not binomial*

12. Ten students are chosen from a statistics class of 300 students. Let X be the number who got an A in the class. *Binomial, $n = 10$*

13. A coin is tossed seven times. Let X be the number of heads obtained. *Binomial, $n = 7$*

14. A die is tossed three times. Let X be the sum of the three numbers obtained. *Not binomial*

15. A coin is tossed until a head appears. Let X be the number of tosses. *Not binomial*

16. A random sample of 250 voters is chosen from a list of 10,000 registered voters. Let X be the number who support the incumbent mayor for reelection. *Binomial, $n = 250$*

In Exercises 17–26, determine the indicated probability for a binomial experiment with the given number of trials n and the given success probability p. Then find the mean, variance, and standard deviation.

17. $n = 5,\ p = 0.7,\ P(3)$ *0.3087, mean = 3.5, variance = 1.05, SD = 1.025*

18. $n = 10,\ p = 0.2,\ P(1)$ *0.2684, mean = 2, variance = 1.6, SD = 1.265*

19. $n = 20,\ p = 0.6,\ P(8)$ *0.0355, mean = 12, variance = 4.8, SD = 2.191*

20. $n = 14,\ p = 0.3,\ P(8)$

21. $n = 3,\ p = 0.4,\ P(0)$ *0.2160, mean = 1.2, variance = 0.72, SD = 0.849*

22. $n = 6,\ p = 0.8,\ P(6)$ *0.2621, mean = 4.8, variance = 0.96, SD = 0.980*

23. $n = 8,\ p = 0.2,\ P$(Fewer than 3)

24. $n = 15,\ p = 0.9,\ P$(14 or more)

25. $n = 50,\ p = 0.03,\ P$(2 or fewer)

26. $n = 30,\ p = 0.9,\ P$(More than 27)

27. Match each TI-84 PLUS calculator command with the probability being calculated.
    a. **1 − binomcdf (*n*, *p*, 5)** *ii.*
    b. **binompdf (*n*, *p*, 0)** *iii.*
    c. **binompdf (*n*, *p*, 5)** *i.*
    d. **binomcdf (*n*, *p*, 5) − binompdf (*n*, *p*, 0)** *iv.*
    i. Probability of exactly 5 successes
    ii. Probability of more than 5 successes
    iii. Probability of no successes
    iv. Probability of between 1 and 5 successes, inclusive

28. Match each TI-84 PLUS calculator command with the probability being calculated.
    a. **binomcdf (*n*, *p*, 5)** *iii.*
    b. **1 − binompdf (*n*, *p*, 0)** *ii.*
    c. **binomcdf (*n*, *p*, 5) − binomcdf (*n*, *p*, 2)** *iv.*
    d. **binompdf (*n*, *p*, 1) + binompdf (*n*, *p*, 2)** *i.*
    i. Probability of either 1 or 2 successes
    ii. Probability of at least one success
    iii. Probability of 5 or fewer successes
    iv. Probability of between 3 and 5 successes, inclusive

Working with the Concepts

29. **Take a guess:** A student takes a true–false test that has 10 questions and guesses randomly at each answer. Let X be the number of questions answered correctly.

a. Find $P(4)$. *0.2051*
b. Find $P(\text{Fewer than } 3)$. *0.0547*
c. To pass the test, the student must answer 7 or more questions correctly. Would it be unusual for the student to pass? Explain. *No*

30. Take another guess: A student takes a multiple-choice test that has 10 questions. Each question has four choices. The student guesses randomly at each answer.
a. Find $P(3)$. *0.2503*
b. Find $P(\text{More than } 2)$. *0.4744*
c. To pass the test, the student must answer 7 or more questions correctly. Would it be unusual for the student to pass? Explain. *Yes*

31. Your flight has been delayed: At Denver International Airport, 81% of recent flights have arrived on time. A sample of 12 flights is studied.
a. Find the probability that all 12 of the flights were on time. *0.0798*
b. Find the probability that exactly 10 of the flights were on time. *0.2897*

buurserstraat386/123RF

c. Find the probability that 10 or more of the flights were on time. *0.5940*
d. Would it be unusual for 11 or more of the flights to be on time? *No*
Source: *The Denver Post*

32. Car inspection: Of all the registered automobiles in Colorado, 8% fail the state emissions test. Twelve automobiles are selected at random to undergo an emissions test.
a. Find the probability that exactly three of them fail the test. *0.0532*
b. Find the probability that fewer than three of them fail the test. *0.9348*
c. Find the probability that more than two of them fail the test. *0.0652*
d. Would it be unusual for none of them to fail the test? *No*
Source: *Air Care Colorado*

33. Google it: According to a report of the Nielsen Company, 76% of internet searches used the Google search engine. Assume that a sample of 25 searches is studied.
a. What is the probability that exactly 20 of them used Google? *0.175*
b. What is the probability that 15 or fewer used Google? *0.056*
c. What is the probability that more than 20 of them used Google? *0.248*
d. Would it be unusual if fewer than 12 used Google? *Yes*

34. What should I buy? A study conducted by the Pew Research Center reported that 58% of customers used their phones inside a store for guidance on purchasing decisions. A sample of 15 customers is studied.
a. What is the probability that six or more of them used their phones for guidance on purchasing decisions? *0.9521*
b. What is the probability that fewer than 10 of them used their phones for guidance on purchasing decisions? *0.6570*
c. What is the probability that exactly eight of them used their phones for guidance on purchasing decisions? *0.1900*
d. Would it be unusual if more than 12 of them had used their phones for guidance on purchasing decisions? *Yes*

35. Blood types: The blood type O negative is called the "universal donor" type, because it is the only blood type that may safely be transfused into any person. Therefore, when someone needs a transfusion in an emergency and their blood type cannot be determined, they are given type O negative blood. For this reason, donors with this blood type are crucial to blood banks. Unfortunately, this blood type is fairly rare; according to the Red Cross, only 7% of U.S. residents have type O negative blood. Assume that a blood bank has recruited 20 donors.
a. What is the probability that two or more of them have type O negative blood? *0.4131*
b. What is the probability that fewer than four of them will have type O negative blood? *0.9529*
c. Would it be unusual if none of the donors had type O negative blood? *No*
d. What is the mean number of donors who have type O negative blood? *1.4*
e. What is the standard deviation of the number of donors who have type O negative blood? *1.1411*

36. Coronary bypass surgery: The Agency for Healthcare Research and Quality reported that 53% of people who had coronary bypass surgery in a recent year were over the age of 65. Fifteen coronary bypass patients are sampled.
a. What is the probability that exactly 9 of them are over the age of 65? *0.1780*
b. What is the probability that more than 10 are over the age of 65? *0.0920*
c. What is the probability that fewer than 8 are over the age of 65? *0.4065*
d. Would it be unusual if all of them were over the age of 65? *Yes*
e. What is the mean number of people over the age of 65 in a sample of 15 coronary bypass patients? *7.95*
f. What is the standard deviation of the number of people over the age of 65 in a sample of 15 coronary bypass patients? *1.933*

37. College bound: In a recent year, the *Statistical Abstract of the United States* reported that 66% of students who graduated from high school enrolled in college. Thirty high school graduates are sampled.
a. What is the probability that exactly 18 of them enroll in college? *0.1166*
b. What is the probability that more than 15 enroll in college? *0.9486*
c. What is the probability that fewer than 12 enroll in college? *0.0010*
d. Would it be unusual if more than 25 of them enroll in college? *Yes*
e. What is the mean number who enroll in college in a sample of 30 high school graduates? *19.8*

f. What is the standard deviation of the number who enroll in college in a sample of 30 high school graduates? *2.595*

38. Big babies: The Centers for Disease Control and Prevention reports that 25% of baby boys 6–8 months old in the United States weigh more than 20 pounds. A sample of 16 babies is studied.

a. What is the probability that exactly 5 of them weigh more than 20 pounds? *0.1802*

b. What is the probability that more than 6 weigh more than 20 pounds? *0.0796*

c. What is the probability that fewer than 3 weigh more than 20 pounds? *0.1971*

d. Would it be unusual if more than 8 of them weigh more than 20 pounds? *Yes*

e. What is the mean number who weigh more than 20 pounds in a sample of 16 babies aged 6–8 months? *4*

f. What is the standard deviation of the number who weigh more than 20 pounds in a sample of 16 babies aged 6–8 months? *1.7321*

39. High blood pressure: The National Health and Nutrition Survey reported that 30% of adults in the United States have hypertension (high blood pressure). A sample of 25 adults is studied.

a. What is the probability that exactly 6 of them have hypertension? *0.1472*

b. What is the probability that more than 8 have hypertension? *0.3231*

c. What is the probability that fewer than 4 have hypertension? *0.0332*

d. Would it be unusual if more than 10 of them have hypertension? *No*

e. What is the mean number who have hypertension in a sample of 25 adults? *7.5*

f. What is the standard deviation of the number who have hypertension in a sample of 25 adults? *2.2913*

40. Stress at work: In a poll conducted by the General Social Survey, 81% of respondents said that their jobs were sometimes or always stressful. Ten workers are chosen at random.

a. What is the probability that exactly 7 of them find their jobs stressful? *0.1883*

b. What is the probability that more than 6 find their jobs stressful? *0.8961*

c. What is the probability that fewer than 5 find their jobs stressful? *0.0049*

d. Would it be unusual if fewer than 4 of them find their jobs stressful? *Yes*

e. What is the mean number who find their jobs stressful in a sample of 10 workers? *8.1*

f. What is the standard deviation of the number who find their jobs stressful in a sample of 10 workers? *1.2406*

41. Testing a shipment: A certain large shipment comes with a guarantee that it contains no more than 15% defective items. If the proportion of items in the shipment is greater than 15%, the shipment may be returned. You draw a random sample of 10 items and test each one to determine whether it is defective.

a. If in fact 15% of the items in the shipment are defective (so that the shipment is good, but just barely), what is the probability that 7 or more of the 10 sampled items are defective? *0.000135*

b. Based on the answer to part (a), if 15% of the items in the shipment are defective, would 7 defectives in a sample of size 10 be an unusually large number? *Yes*

c. If you found that 7 of the 10 sample items were defective, would this be convincing evidence that the shipment should be returned? Explain. *Yes*

d. If in fact 15% of the items in the shipment are defective, what is the probability that 2 or more of the 10 sampled items are defective? *0.456*

e. Based on the answer to part (d), if 15% of the items in the shipment are defective, would 2 defectives in a sample of size 10 be an unusually large number? *No*

f. If you found that 2 of the 10 sample items were defective, would this be convincing evidence that the shipment should be returned? Explain. *No*

42. Smoke detectors: An insurance company offers a discount to homeowners who install smoke detectors in their homes. A company representative claims that 80% or more of policy holders have smoke detectors. You draw a random sample of 8 policy holders.

a. If exactly 80% of the policy holders have smoke detectors (so the representative's claim is true, but just barely), what is the probability that at most 2 of the 8 sampled policy holders have smoke detectors? *0.00123*

b. Based on the answer to part (a), if 80% of the policy holders have smoke detectors, would 2 policy holders with smoke detectors in a sample of size 8 be an unusually small number? *Yes*

c. If you found that 2 of the 8 sample policy holders had a smoke detector, would this be convincing evidence that the claim is false? Explain. *Yes*

d. If exactly 80% of the policy holders have smoke detectors, what is the probability that at most 6 of the 8 sampled policy holders have smoke detectors? *0.497*

e. Based on the answer to part (d), if 80% of the policy holders have smoke detectors, would 6 policy holders with smoke detectors in a sample of size 8 be an unusually small number? *No*

f. If you found that 6 of the 8 sample policy holders had smoke detectors, would this be convincing evidence that the claim is false? Explain. *No*

Extending the Concepts

43. Recursive computation of binomial probabilities: Binomial probabilities are often hard to compute by hand, because the computation involves factorials and numbers raised to large powers. It can be shown through algebraic manipulation that if X is a random variable whose distribution is binomial with n trials and success probability p, then

$$P(X = x + 1) = \left(\frac{p}{1-p}\right)\left(\frac{n-x}{x+1}\right)P(X = x)$$

If we know $P(X = x)$, we can use this equation to calculate $P(X = x + 1)$ without computing any factorials or powers.

a. Let X have the binomial distribution with $n = 25$ trials and success probability $p = 0.6$. It can be shown that $P(X = 14) = 0.14651$. Find $P(X = 15)$. *0.16116*

b. Let X have the binomial distribution with $n = 10$ trials and success probability $p = 0.35$. It can be shown that $P(X = 0) = 0.0134627$. Find $P(X = x)$ for $x = 1, 2, ..., 10$.

Answers to Check Your Understanding Exercises for Section 6.2

1. a. X is a binomial random variable.
b. X is not a binomial random variable.
c. X is a binomial random variable.

2. a. 0.246 **b.** 0.090 **c.** 0.746 **d.** 0.656
3. a. 0.168 **b.** 0.450 **c.** 0.266 **d.** 0.822
4. a. 60 **b.** 15 **c.** 3.8730

Section 6.3 The Poisson Distribution

Objectives

1. Compute probabilities for Poisson random variables
2. Compute the mean, variance, and standard deviation of a Poisson random variable
3. Use the Poisson distribution to approximate binomial probabilities

An advertising company has placed an ad on a website. The company's managers are interested to know how many hits the website gets, on the average, during a certain time interval. Assume that between 2:00 P.M. and 3:00 P.M., there are an average of three hits per minute.

If X is the number of hits that occur in a given time interval, then under certain conditions X will have a probability distribution known as the **Poisson distribution.** The Poisson distribution is used to describe certain events that occur in time or space.

EXPLAIN IT AGAIN

Distinguish the Poisson distribution from the binomial: We use the binomial distribution when there is a fixed number n of trials, and we count the number of trials that result in success. We use the Poisson distribution when we count the number of events that occur in a given amount of time or space.

Conditions for the Poisson Distribution

Let X be a random variable that represents the number of events that occur in a time interval of length t. Then X will have the Poisson probability distribution if the following conditions are satisfied.

1. The average rate at which events occur is the same at all times.
2. The numbers of events that occur in nonoverlapping time intervals are independent.
3. For a very short interval of length t:
   a. It is essentially impossible for more than one event to occur within the time interval.
   b. The probability that one event occurs in the interval is approximately equal to λt, where λ is the average rate at which events occur.

Example 6.18 Finding λ and t

RECALL
$x! = x(x-1)\cdots(2)(1)$

The number of hits on a website during a given time interval has a Poisson distribution. If there are an average of three hits per minute, and X is the number of hits between 2:30 P.M. and 2:35 P.M., find λ and t.

Solution

The average number of hits per minute is 3, so $\lambda = 3$ hits per minute. The length of the time interval between 2:30 and 2:35 is 5 minutes, so $t = 5$ minutes.

If the random variable X has a Poisson distribution with rate λ and time t, we can compute probabilities with the following probability distribution.

EXPLAIN IT AGAIN

The number e: The number e is one of the most frequently occurring numbers in nature. Unfortunately, it is an infinitely long decimal, so it is almost impossible to use e in hand computations. Quantities that involve the number e, such as Poisson probabilities, must be computed with technology.

Formula for Poisson Probabilities

For a Poisson random variable X that represents the number of events that occur with average **rate** λ in a time **interval** of length t, the probability that x events occur is

$$P(x) = e^{-\lambda t}\frac{(\lambda t)^x}{x!}$$

The number e is a constant whose value is approximately 2.71828.

The possible values for X are 0, 1, 2, … .

Example 6.19

Computing Poisson probabilities

Objective 1 Compute probabilities for Poisson random variables

The number of hits on a certain website follows a Poisson distribution with $\lambda = 3$ hits per minute. Let X be the number of hits in a 2-minute period.

a. Find $P(5)$.

b. Find $P(\text{Less than } 3)$.

c. Find $P(\text{More than } 2)$.

Solution

a. $P(5) = e^{-3\cdot 2}\frac{(3\cdot 2)^5}{5!} = 0.1606$

b. $P(\text{Less than } 3) = P(0) + P(1) + P(2)$

$$= e^{-3\cdot 2}\frac{(3\cdot 2)^0}{0!} + e^{-3\cdot 2}\frac{(3\cdot 2)^1}{1!} + e^{-3\cdot 2}\frac{(3\cdot 2)^2}{2!}$$

$$= 0.0025 + 0.0149 + 0.0446$$

$$= 0.0620$$

c. We note that the event "More than 2" is the complement of the event "Less than 3." We use the Rule of Complements:

$$P(\text{More than } 2) = 1 - P(\text{Less than } 3) = 1 - 0.0620 = 0.9380$$

RECALL

The Rule of Complements says that $P(A^c) = 1 - P(A)$.

Check Your Understanding

1. The number of traffic accidents at a certain intersection follows a Poisson distribution with $\lambda = 2$ per year. Let X be the number of accidents that occur in a 4-year period.

a. Find $P(7)$. *0.1396*

b. Find $P(\text{Less than } 4)$. *0.0424*

c. Find $P(\text{More than } 2)$. *0.9862*

2. The number of hits on a certain website follows a Poisson distribution with $\lambda = 12$ per minute. Let X be the number of hits that occur in a period of one-half minute.

a. Find $P(5)$. *0.1606*

b. Find $P(\text{Less than } 2)$. *0.0174*

c. Find $P(\text{More than } 3)$. *0.8488*

Answers are on page 296.

In Example 6.19, the events were occurring in time. The Poisson distribution can also be used to compute probabilities involving events that occur in a spatial region. In this situation, the quantity t represents a spatial quantity such as length, area, or volume.

One of the earliest uses of the Poisson distribution involved events occurring in space, and was in fact an application to the brewing of beer. A crucial step in the brewing process is the addition of yeast culture to prepare mash for fermentation. The living yeast cells are kept suspended in a liquid medium. Because the cells are alive, their concentration in the medium changes over time. Therefore, just before the yeast is added, it is necessary to estimate the concentration of yeast cells in the medium, to be sure to add the right amount. Up until the early part of the 20th century, this posed a problem for brewers. The number of yeast cells in a given volume is a random variable, but no one knew how to find the probability distribution.

William Sealy Gosset, a young statistician in his mid-twenties who was employed by the Guinness Brewing Company of Dublin, Ireland, discovered in 1904 that the number of yeast cells in a sampled volume of liquid medium follows a Poisson distribution. Gosset's discovery not only enabled Guinness to produce a more consistent product, it also showed that the Poisson distribution could have important applications in many situations. Gosset wanted to publish his result, but his managers at Guinness considered his discovery to be proprietary information and required him to use the pseudonym "Student."

In Example 6.20, we will solve a problem similar to the one that made Student famous. Before we get to it, though, we will mention that soon after solving this problem, Student made another discovery that made him even more famous. This discovery solved one of the most important outstanding problems in statistics and has profoundly influenced work in virtually all fields of science ever since. We will discuss this result in Section 8.2.

Example 6.20

Computing probability of events in space

Yeast cells are suspended in a liquid medium at a concentration of 4 particles per milliliter. A volume of 2 milliliters is withdrawn. What is the probability that exactly 6 particles are contained in this volume?

Solution

Let X be the number of particles withdrawn. Then X has a Poisson distribution. The rate is $\lambda = 4$ particles per milliliter, and the volume is $t = 2$ milliliters. Therefore

$$P(6) = e^{-4 \cdot 2}\frac{(4 \cdot 2)^6}{6!} = 0.1221$$

Figure 6.8 presents the result in Example 6.20 on the TI-84 Plus calculator. Note that we enter 8, which is the product λt. Step-by-step instructions can be found in the Using Technology section on page 294.

```
poissonpdf(8,6)
                .1221382155
```

Figure 6.8

Check Your Understanding

3. An environmental scientist is counting the number of bacteria in a sample of wastewater. The number of bacteria has a Poisson distribution with $\lambda = 3$ per microliter. What is the probability that exactly 9 bacteria are found in a volume of $t = 4$ microliters? *0.0874*

Answer is on page 296.

Objective 2 Compute the mean, variance, and standard deviation of a Poisson random variable

Mean and Variance of the Poisson Distribution

Imagine that events occur at an average rate of two per minute, and you count the number of events that occur in a 5-minute period. How many events would you expect to see? It is reasonable to think that you would see $5 \cdot 2 = 10$ events, on the average, and this is in fact correct. In general, if X is a Poisson random variable with rate λ and interval length t, the mean of X is λt.

Mean, Variance, and Standard Deviation of a Poisson Random Variable

If X is a Poisson random variable with rate λ and interval length t, then the mean of X is

$$\mu_X = \lambda t$$

The variance of X is

$$\sigma_X^2 = \lambda t \text{ (the variance is equal to the mean)}$$

The standard deviation of X is

$$\sigma_X = \sqrt{\lambda t}$$

Example 6.21

Computing the mean and standard deviation of a Poisson random variable

Yeast cells are suspended in a liquid medium at a concentration of 4 particles per milliliter. A volume of 2 milliliters is withdrawn. Let X be the number of particles contained in this volume. Find the mean and standard deviation of X.

Solution

We have $\lambda = 4$ and $t = 2$. Therefore,

$$\mu_X = 4 \cdot 2 = 8 \qquad \sigma_X = \sqrt{4 \cdot 2} = 2.8284$$

Objective 3 Use the Poisson distribution to approximate binomial probabilities

The Poisson Approximation to the Binomial Distribution

When the number of trials, n, is large, binomial probabilities can be hard to compute, even with technology. The reason is that when n is large, $n!$ is extremely large and will cause most forms of technology to overflow. To be specific, the largest factorial that most forms of technology can handle is 69!. If you try to compute 70!, you will get an error.

Fortunately, when n is large and p is small, the Poisson distribution can be used to approximate binomial probabilities. The **Poisson approximation** may be used for any binomial probability for which $n \geq 100$ and $p \leq 0.1$. To compute $P(x)$ by using the Poisson approximation, use the following steps:

Step 1: Check that $n \geq 100$ and $p \leq 0.1$.

Step 2: Compute $P(x)$ by using the Poisson distribution with p in place of λ and n in place of t.

Example 6.22

Using the Poisson approximation to the binomial

A Nevada roulette wheel has 38 pockets, labeled 0, 00, and 1 through 36. If you bet on 00 (or any other number), the probability that you will win is $1/38 = 0.0263$. Assume you bet on 00 for 200 spins of the wheel. Find the probability that you win exactly six times.

Solution

Step 1: There are 200 trials, and each trial has success probability $1/38 = 0.0263$. The number of successes therefore has a binomial distribution with $n = 200$ and $p = 0.0263$. Since $n \geq 100$ and $p \leq 0.1$, the Poisson approximation is valid.

Step 2: We compute $P(6)$ by using the Poisson distribution with rate $\lambda = 0.0263$ and $t = 200$.

$$P(6) = e^{-0.0263 \cdot 200} \frac{(0.0263 \cdot 200)^6}{6!} = 0.153$$

Check Your Understanding

4. In the Georgia Cash 3 Lottery game, your probability of winning is 0.001. The game is played twice daily, so 730 games are played in a year. If you play twice a day for a year, what is the probability that you win exactly twice? *0.1284*

Answer is on page 296.

Using Technology

We use Example 6.19 to illustrate the technology steps.

TI-84 PLUS

Computing Poisson probabilities of the form *P*(*x*) or *P*(Less than or equal to *x*)

Step 1. Press **2nd, VARS** to access the **DISTR** menu.
- To compute *P*(*x*), select **poissonpdf** and enter the values for λt and x separated by commas and press **ENTER**.
- To compute *P*(Less than or equal to *x*), select **poissoncdf** and enter the values for λt and x separated by commas and press **ENTER**.

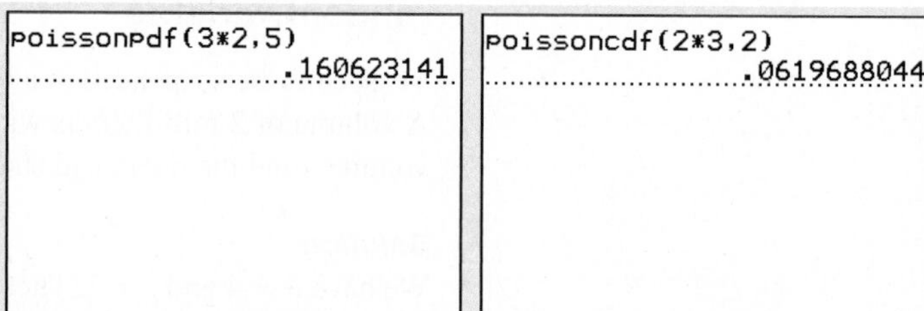

Figure A **Figure B**

Using the TI-84 PLUS Stat Wizards (see Appendix B for more information)

Step 1. Press **2nd, VARS** to access the **DISTR** menu.
- To compute *P*(*x*), select **poissonpdf**, enter the value for λt in the λ field, and enter the value for x in the **x value** field. Select **Paste** and press **ENTER** to paste the command to the home screen. Press **ENTER** again to run the command.
- To compute *P*(Less than or equal to *x*), select **poissoncdf**, enter the value for λt in the λ field, and enter the value for x in the **x value** field. Select **Paste** and press **ENTER** to paste the command to the home screen. Press **ENTER** again to run the command.

For Example 6.19, $\lambda = 3$ and $t = 2$. Figure A displays the result of part (a), which asks for *P*(5). Figure B displays the result of part (b), which asks for *P*(Less than or equal to 2).

EXCEL

Computing Poisson probabilities of the form *P*(*x*) or *P*(Less than or equal to *x*)

Step 1. In an empty cell, select the **Insert Function** icon and highlight **Statistical** in the category field.

Step 2. Click on the **POISSON.DIST** function and click **OK**.

Step 3. Enter the value for x in the **X** field and the value for λt in the **Mean field**.

Step 4. To compute *P*(*x*), enter **FALSE** in the **Cumulative field**. To compute *P*(Less than or equal to *x*), enter **TRUE** in the **Cumulative field**.

Step 5. Click **OK**.

MINITAB

Computing Poisson probabilities of the form *P*(*x*) or *P*(Less than or equal to *x*)

Step 1. Click **Calc**, then **Probability Distributions**, then **Poisson**.

Step 2. Enter the value for λt in the **Mean field**.

Step 3. To compute *P*(*x*), select the **Probability option** and enter the value for x in the **Input constant** field. To compute *P*(Less than or equal to *x*), select the **Cumulative probability** option and enter the value for x in the **Input constant** field.

Step 4. Click **OK**.

Note: Poisson probabilities may be computed for a column of values by entering the column name in the **Input column** field.

Section 6.3 Exercises

Exercises 1–4 are the Check Your Understanding exercises located within the section.

Understanding the Concepts

In Exercises 5 and 6, fill in each blank with the appropriate word or phrase.

5. The Poisson distribution is used to describe events that occur in ________ or ________. *time, space*

6. The mean of the Poisson random variable X with rate λ and interval length t is ________. λt

In Exercises 7 and 8, determine whether the statement is true or false. If the statement is false, rewrite it as a true statement.

7. For a Poisson random variable with rate λ and time interval t, the possible values of x are 0, 1, 2, ..., λ. *False*

8. In a Poisson distribution, the mean and variance are equal. *True*

Practicing the Skills

In Exercises 9–18, determine the indicated probability for a Poisson random variable with the given values of λ and t.

9. $\lambda = 2,\ t = 5,\ P(5)$ *0.0378*
10. $\lambda = 0.5,\ t = 4,\ P(3)$ *0.1804*
11. $\lambda = 0.1,\ t = 10,\ P(2)$ *0.1839*
12. $\lambda = 1,\ t = 2,\ P(0)$ *0.1353*
13. $\lambda = 0.2,\ t = 10,\ P(\text{At least one})$ *0.8647*
14. $\lambda = 0.3,\ t = 8,\ P(\text{At least one})$ *0.9093*
15. $\lambda = 1,\ t = 4,\ P(\text{No more than 5})$ *0.7851*
16. $\lambda = 0.2,\ t = 7,\ P(\text{More than 2})$ *0.1665*
17. $\lambda = 2,\ t = 4,\ P(\text{More than 9})$ *0.2834*
18. $\lambda = 0.3,\ t = 5,\ P(\text{Fewer than 3})$ *0.8088*

Working with the Concepts

19. **Potholes:** In a certain city, the number of potholes on a major street has a Poisson distribution with a rate of 3 per mile. Let X represent the number of potholes in a 2-mile stretch of road. Find
    a. $P(4)$ *0.1339*
    b. $P(\text{More than 1})$ *0.9826*
    c. $P(\text{Between 5 and 7 inclusive})$ *0.4589*
    d. μ_X *6*
    e. σ_X *2.4495*

20. **Flaws in aluminum foil:** The number of flaws in a given area of aluminum foil follows a Poisson distribution with a mean of 3 per square meter. Let X represent the number of flaws in a 1-square-meter sample of foil.
    a. $P(5)$ *0.1008*
    b. $P(0)$ *0.0498*
    c. $P(\text{Less than 2})$ *0.1991*
    d. $P(\text{Greater than 1})$ *0.8009*
    e. μ_X *3*
    f. σ_X *1.7321*

21. **Flaws in lumber:** The number of flaws in a certain type of lumber has a Poisson distribution with a rate of 0.5 per meter of length.
    a. What is the probability that a board 3 meters in length has no flaws? *0.2231*
    b. What is the probability that a board 4 meters in length has at least one flaw? *0.8647*

Comstock Images/Jupiter Images

22. **Flickering screen:** The number of times that screen flicker appears in video recorded on a certain type of digital video recorder (DVR) has a Poisson distribution with a rate of 0.2 per minute.
    a. What is the probability that there are no screen flickers in 8 minutes of video? *0.2019*
    b. What is the probability that there are at least 2 screen flickers in 11 minutes of video? *0.6454*

23. **Computer messages:** The number of tweets received by a certain Twitter user is a Poisson random variable with a mean rate of 8 tweets per hour.
    a. What is the probability that 5 tweets are received in a given hour? *0.0916*
    b. What is the probability that 10 tweets are received in 1.5 hours? *0.1048*
    c. What is the probability that fewer than 3 tweets are received in one-half hour? *0.2381*

24. **Grandma's cookies:** Grandma bakes chocolate chip cookies in batches of 100. She puts 600 chips into each batch, so that the number of chips in a randomly chosen cookie has a Poisson distribution with $\lambda = 6$ chips per cookie.
    a. Grandma gives you one cookie. What is the probability that it contains exactly eight chips? *0.1033*
    b. Because you were extra good, Grandma will give you two cookies. What is the probability that the two cookies will contain a total of exactly 15 chips? *0.0724*

25. **Trees in the forest:** The number of trees of a certain species in a certain forest has a Poisson distribution with mean of 10 trees per acre.
    a. What is the probability that there will be exactly 18 trees in a 2-acre region? *0.0844*
    b. Find the mean number of trees in a 2-acre region. *20*
    c. Find the standard deviation of the number of trees in a 2-acre region. *4.4721*

26. **Bacteria:** An environmental scientist obtains a sample of water from a pond that contains a certain type of bacteria at a concentration of 5 per milliliter.

a. What is the probability that there will be exactly 12 bacteria in a 3-milliliter sample of water? 0.0829

b. Find the mean number of bacteria in a 3-milliliter sample. 15

c. Find the standard deviation of the number of bacteria in a 3-milliliter sample. 3.873

27. Drive safely: In a recent year, there were approximately 250,000,000 registered motor vehicles in the United States. Of these, approximately 4% were involved in an accident during the year. Assume that a simple random sample of $n = 200$ vehicles is drawn. Use the Poisson approximation to the binomial distribution to compute the following:

a. The probability that exactly six of the 200 vehicles were involved in an accident 0.1221

b. The probability that fewer than three of the 200 vehicles were involved in an accident 0.0138

c. The probability that more than 10 of the 200 vehicles were involved in an accident 0.1841

d. The mean number of vehicles that were involved in an accident 8

e. The standard deviation of the number of vehicles that were involved in an accident 2.8284

28. Down Syndrome: A report from the Medical College of Georgia stated that the proportion of infants born with Down Syndrome is 0.0011. Assume a sample of 5000 births is studied. Use the Poisson approximation to the binomial distribution to compute the following:

a. The probability that exactly 5 of the 5000 births involved Down Syndrome 0.1714

b. The probability that fewer than 4 of the 5000 births involved Down Syndrome 0.2017

c. The probability that more than 3 of the 5000 births involved Down Syndrome 0.7983

d. The mean number of births that involved Down Syndrome 5.5

e. The standard deviation of the number of births that involved Down Syndrome 2.3452

Extending the Concepts

The exponential distribution: When the number of events occurring in a given time follows a Poisson distribution, the amount of time that elapses between two events is a random variable, and its distribution is called the exponential distribution.

29. Assume that events occurring in time follow a Poisson distribution with rate λ. Let T be the amount of time, in seconds, that elapses between two events. This exercise will show how to compute probabilities involving T.

a. Let X be the number of events that occur in a 1-second interval. Show that $P(X = 0) = e^{-\lambda}$.

b. Explain why $X = 0$ is the same as $T > 1$.

c. Show that $P(T > 1) = e^{-\lambda}$.

d. Let X be the number of events that occur in a 2-second interval. Show that $P(X = 0) = e^{-2\lambda}$.

e. Explain why $X = 0$ is the same as $T > 2$.

f. Now let t be any amount of time, and let X be the number of events that occur in an interval of length t seconds. Show that $P(X = 0) = e^{-\lambda t}$.

g. Explain why $X = 0$ is the same as $T > t$.

h. Show that $P(T > t) = e^{-\lambda t}$.

Answers to Check Your Understanding Exercises for Section 6.3

1. a. 0.1396 **b.** 0.0424 **c.** 0.9862

2. a. 0.1606 **b.** 0.0174 **c.** 0.8488

3. 0.0874

4. 0.1284

Chapter 6 Summary

Section 6.1: A random variable is a numerical outcome of a probability experiment. Discrete random variables are random variables whose possible values can be listed, whereas continuous random variables can take on any value in some interval. A probability distribution for a discrete random variable specifies the probability for each possible value. A probability histogram is a histogram in which the heights of the rectangles are the probabilities for the possible values of the random variable. A probability histogram can also be thought of as a relative frequency histogram for a population. The law of large numbers for histograms states that as the sample size increases, the relative frequency histogram for the sample approaches the probability histogram.

The mean of a random variable, also called the expected value, measures the center of the distribution. The standard deviation of a random variable measures the spread. The law of large numbers for means states that as the sample size increases, the sample mean approaches the population mean.

Section 6.2: The binomial distribution is an important discrete probability distribution. A random variable has a binomial distribution if it represents the number of successes in a fixed number n of independent trials, all of which have the same success probability p. Binomial probabilities can be found in a table or computed with technology. The mean of a binomial random variable is np, the number of trials multiplied by the success probability. The variance is $np(1 - p)$, and the standard deviation is $\sqrt{np(1 - p)}$.

Section 6.3: The Poisson distribution is used for certain random variables that represent the number of events that occur in time or space. Poisson probabilities are generally computed with technology. The mean and variance of a Poisson random variable are the same: Both are equal to the rate multiplied by the time. The Poisson distribution can also be used to approximate the binomial distribution when the number of trials, n, is large, and the success probability, p, is small.

Vocabulary and Notation

binomial distribution 278
continuous random variable 264
discrete random variable 264
expected value 269
interval length t 290
law of large numbers for means 270
mean 269
Poisson approximation to binomial 293
Poisson distribution 290
probability distribution 265
probability histogram 268
random variable 264
rate λ 290
standard deviation 271
trial 278
variance 271

Important Formulas

Mean of a discrete random variable:
$\mu_X = \sum[x \cdot P(x)]$

Variance of a discrete random variable:
$\sigma_X^2 = \sum[(x - \mu_X)^2 \cdot P(x)] = \sum[x^2 \cdot P(x)] - \mu_X^2$

Standard deviation of a discrete random variable:
$\sigma_X = \sqrt{\sigma_X^2}$

Mean of a binomial random variable:
$\mu_X = np$

Variance of a binomial random variable:
$\sigma_X^2 = np(1 - p)$

Standard deviation of a binomial random variable:
$\sigma_X = \sqrt{np(1 - p)}$

Mean of Poisson random variable:
$\mu_X = \lambda t$

Variance of Poisson random variable:
$\sigma_X^2 = \lambda t$

Standard deviation of Poisson random variable:
$\sigma_X = \sqrt{\lambda t}$

Chapter Quiz

1. Explain why the following is *not* a probability distribution.

x	6	7	8	9	10
$P(x)$	0.32	0.11	0.19	0.28	0.03

2. Find the mean of the random variable X with the following probability distribution. *2*

x	−2	1	4	5
$P(x)$	0.3	0.2	0.1	0.4

3. Refer to Exercise 2.
a. Find the variance of the random variable X. *9*
b. Find the standard deviation of the random variable X. *3*

4. Find the missing value that makes the following a valid probability distribution. *0.19*

x	2	3	5	8	10
$P(x)$	0.23	0.12	0.09	?	0.37

5. The following table presents a probability distribution for the number of pets each family has in a certain neighborhood.

Number of pets	0	1	2	3	4
Probability	0.4	0.2	0.2	0.1	0.1

Construct a probability histogram.

6. Refer to Exercise 5. Find the probability that a randomly selected family has:
a. 1 or 2 pets *0.4*
b. More than 2 pets *0.2*
c. No more than 3 pets *0.9*
d. At least 1 pet *0.6*

7. Refer to Exercise 5. Find the mean number of pets. *1.3*

8. Refer to Exercise 5. Find the standard deviation of the number of pets. *1.3454*

9. At a phone battery plant, 5% of phone batteries produced are defective. A quality control engineer randomly collects a sample of 50 batteries from a large shipment from this plant and inspects them for defects. Find the probability that

a. None of the batteries are defective. *0.0769*
b. At least one of the batteries is defective. *0.9231*
c. No more than 3 of the batteries are defective. *0.7604*

10. Refer to Exercise 9. Find the mean and standard deviation for the number of defective batteries in the sample of size 50. *Mean: 2.5; SD: 1.5411*

11. A meteorologist states that the probability of rain tomorrow is 0.4 and the probability of rain on the next day is 0.6. Assuming these probabilities are accurate, and that the rain events are independent, find the probability distribution for X, the number of days out of the next two that it rains.

12. The number of large packages delivered by a courier service follows a Poisson distribution with a rate of 5 per day. Find the probability that
a. Exactly 4 large packages are delivered on a given day. *0.1755*
b. Fewer than 6 large packages are delivered over a 2-day period. *0.0671*
c. At least one large package is delivered over a half-day period. *0.9179*

13. The number of text messages received on a certain person's phone follows a Poisson distribution with the rate of 10 messages per hour. What is the mean number of messages received in an 8-hour period? *80*

14. Refer to Exercise 13. What are the variance and standard deviation of the number of messages received in an 8-hour period? *Variance: 80; SD: 8.9443*

15. Huntington's disease is a genetically transmitted disease that causes degeneration of nerve cells in the brain. The probability that a person carries a gene for Huntington's disease is 0.00005 (i.e., 1/20,000). In a sample of 100,000 people, what is the probability that exactly four people carry the gene for Huntington's disease? *0.1755*

Review Exercises

1. Which are distributions? Which of the following tables represent probability distributions?

a.

x	$P(x)$
3	0.35
4	0.20
5	0.18
6	0.09
7	0.18

Yes

b.

x	$P(x)$
5	0.27
6	0.45
7	−0.06
8	0.44

No

c.

x	$P(x)$
0	0.02
1	0.34
2	1.02
3	0.01
4	0.43
5	0.14

No

d.

x	$P(x)$
2	0.10
3	0.07
4	0.75
5	0.08

Yes

2. Mean, variance, and standard deviation: A random variable X has the following probability distribution.

x	6	7	8	9	10	11
$P(x)$	0.21	0.12	0.29	0.11	0.01	0.26

a. Find the mean of X. *8.37*
b. Find the variance of X. *3.3131*
c. Find the standard deviation of X. *1.8202*

3. AP tests: Advanced Placement (AP) tests are graded on a scale of 1 (low) through 5 (high). The College Board reported that the distribution of scores on the AP Statistics Exam in a recent year was as follows:

x	1	2	3	4	5
$P(x)$	0.34	0.25	0.18	0.16	0.07

A score of 3 or higher is generally required for college credit. What is the probability that a student scores 3 or higher? *0.41*

4. AP tests again: During a recent academic year, approximately 1.7 million students took one or more AP tests. Following is the frequency distribution of the number of AP tests taken by students who took one or more AP tests.

Number of Tests	Frequency (in 1000s)
1	953
2	423
3	194
4	80
5	29
6	9
7	3
8	1
Total	1692

Source: The College Board

Let X represent the number of exams taken by a student who took one or more.

a. Construct the probability distribution for X.

b. Find the probability that a student took exactly one exam. 0.5632

c. Compute the mean μ_X. 1.731

d. Compute the standard deviation σ_X. 1.049

5. Lottery tickets: Several million lottery tickets are sold, and 60% of the tickets are held by women. Five winning tickets will be drawn at random.

a. What is the probability that three or fewer of the winners will be women? 0.6630

b. What is the probability that three of the winners will be of one gender and two of the winners will be of the other gender? 0.5760

6. Genetic disease: Sickle-cell anemia is a disease that results when a person has two copies of a certain recessive gene. People with one copy of the gene are called carriers. Carriers do not have the disease, but can pass the gene on to their children. A child born to parents who are both carriers has probability 0.25 of having sickle-cell anemia. A medical study samples 18 children in families where both parents are carriers.

a. What is the probability that four or more of the children have sickle-cell anemia? 0.6943

b. What is the probability that fewer than three of the children have sickle-cell anemia? 0.1353

c. Would it be unusual if none of the children had sickle-cell anemia? Yes

7. Craps: In the game of craps, you may bet \$1 that the next roll of the dice will be an 11. If the dice come up 11, your profit is \$15. If the dice don't come up 11, you lose \$1. The probability that the dice come up 11 is 1/18. What is the expected value of your profit? Is it an expected gain or an expected loss? −\$0.11; loss

8. Looking for a job: According to the General Social Survey conducted at the University of Chicago, 59% of employed adults believe that if they lost their job, it would be easy to find another one with a similar salary. Suppose that 10 employed adults are randomly selected.

a. Find the probability that exactly three of them believe it would be easy to find another job. 0.0480

b. Find the probability that more than two of them believe it would be easy to find another job. 0.9854

c. Would it be unusual if all of them believed it would be easy to find another job? Yes

9. Reading tests: According to the National Center for Education Statistics, 66% of fourth graders could read at a basic level in a recent year. Suppose that eight fourth graders are randomly selected.

a. Find the probability that exactly five of them can read at a basic level. 0.2756

b. Find the probability that more than six of them can read at a basic level. 0.1844

c. Would it be unusual if all of them could read at a basic level? Yes

10. Rain, rain, go away: Let X be the number of days during the next month that it rains. Does X have a binomial distribution? Why or why not? No

11. Survey sample: In a college with 5000 students, 100 are randomly chosen to complete a survey in which they rate the quality of the cafeteria food. Let X be the number of freshmen who are chosen. Does X have a binomial distribution? Why or why not? Yes

12. Suspensions: The concentration of particles in a suspension is 2 per milliliter. A volume of 3 milliliters is withdrawn. Find the following probabilities.

a. P(Exactly five particles are withdrawn) 0.1606

b. P(Fewer than two particles are withdrawn) 0.0174

c. P(More than one particle is withdrawn) 0.9826

13. Mean: Refer to Exercise 12. Find the mean number of particles that are withdrawn. 6

14. Standard deviation: Refer to Exercise 12. Find the standard deviation of the number of particles that are withdrawn. 2.4495

15. Congenital disease: The connexin-26 mutation is a genetic mutation that results in deafness. The probability that a person carries a gene with this mutation is 0.0151. Use the Poisson approximation to find the probability that exactly 10 people in a sample of 500 carry this mutation. 0.0873

Write About It

1. Provide an example of a discrete random variable, and explain why it is discrete.
2. Provide an example of a continuous random variable, and explain why it is continuous.
3. If a business decision has an expected gain, is it possible to lose money? Explain.
4. When a population mean is unknown, people will often approximate it with the mean of a large sample. Explain why this is justified.
5. Provide an example of a binomial random variable, and explain how each condition for the binomial distribution is fulfilled.

6. Twenty percent of the men in a certain community are more than 6 feet tall. An anthropologist samples five men from a large family in the community and counts the number X who are more than 6 feet tall. Explain why the binomial distribution is not appropriate in this situation. Is $P(X = 0)$ likely to be greater than or less than the value predicted by the binomial distribution with $n = 10$ and $p = 0.4$? Explain your reasoning.
7. Provide an example of a Poisson random variable, and explain how each condition for the Poisson distribution is fulfilled.

In-Class Activities

1. **Unusual events:** Divide students into pairs. One member of each pair tosses a coin 20 times. The other member tries to guess the outcome of each toss before it is made. Each pair records the number of correct guesses. What is the distribution of the number of correct guesses out of 20? Did any pair get an unusually large or unusually small number of correct guesses? Discuss whether it would be justified to conclude that the person involved has ESP, or whether there is a better conclusion.
2. **Benford's Law:** Ask each student to provide the first digit of their address. Does the distribution of these digits follow Benford's Law, as described in the Case Study for this chapter?

Case Study: Benford's Law: Do The Digits 1–9 Occur Equally Often?

One of the most surprising probability distributions found in practice is given by a rule known as Benford's law. This probability distribution concerns the first digits of numbers. The first digit of a number may be any of the digits 1, 2, 3, 4, 5, 6, 7, 8, or 9. It is reasonable to believe that, for most sets of numbers encountered in practice, these digits would occur equally often. In fact, it has been observed that for many naturally occurring data sets, smaller numbers occur more frequently as the first digit than larger numbers do. Benford's law is named for Frank Benford, an engineer at General Electric, who stated it in 1938.

Following are the populations of the 50 states in July 2015, as estimated by the U.S. Census Bureau. The first digit of each population number is listed separately.

State	Population	First Digit	State	Population	First Digit	State	Population	First Digit
Alabama	4,858,979	4	Louisiana	4,670,724	4	Ohio	11,613,423	1
Alaska	738,432	7	Maine	1,329,328	1	Oklahoma	3,911,338	3
Arizona	6,828,065	6	Maryland	6,006,401	6	Oregon	4,028,977	4
Arkansas	2,978,204	2	Massachusetts	6,794,422	6	Pennsylvania	12,802,503	1
California	39,144,818	3	Michigan	9,922,576	9	Rhode Island	1,056,298	1
Colorado	5,456,574	5	Minnesota	5,489,594	5	South Carolina	4,896,146	4
Connecticut	3,590,886	3	Mississippi	2,992,333	2	South Dakota	858,469	8
Delaware	945,934	9	Missouri	6,083,672	6	Tennessee	6,600,299	6
Florida	20,271,272	2	Montana	1,032,949	1	Texas	27,469,114	2
Georgia	10,214,860	1	Nebraska	1,896,190	1	Utah	2,995,919	2
Hawaii	1,431,603	1	Nevada	2,890,845	2	Vermont	626,042	6
Idaho	1,654,930	1	New Hampshire	1,330,608	1	Virginia	8,382,993	8
Illinois	12,859,995	1	New Jersey	8,958,013	8	Washington	7,170,351	7
Indiana	6,619,680	6	New Mexico	2,085,109	2	Washington, D.C.	672,228	6
Iowa	3,123,899	3	New York	19,795,791	1	West Virginia	1,844,128	1
Kansas	2,911,641	2	North Carolina	10,042,802	1	Wisconsin	5,771,337	5
Kentucky	4,425,092	4	North Dakota	756,927	7	Wyoming	586,107	5

Here is a frequency distribution of the first digits of the state populations:

Digit	Frequency	Digit	Frequency
1	14	6	8
2	8	7	3
3	4	8	3
4	5	9	2
5	4		

For the state populations, the most frequent first digit is 1, with 7, 8, and 9 being the least frequent.

Now here is a table of the closing value of the Dow Jones Industrial Average for each of the years 1975–2016.

Year	Average	First Digit	Year	Average	First Digit	Year	Average	First Digit
1975	852.41	8	1989	2753.20	2	2003	10453.92	1
1976	1004.65	1	1990	2633.66	2	2004	10783.01	1
1977	831.17	8	1991	3168.83	3	2005	10717.50	1
1978	805.01	8	1992	3301.11	3	2006	12463.15	1
1979	838.74	8	1993	3754.09	3	2007	13264.82	1
1980	963.98	9	1994	3834.44	3	2008	8776.39	8
1981	875.00	8	1995	5117.12	5	2009	10428.05	1
1982	1046.55	1	1996	6448.27	6	2010	11557.51	1
1983	1258.64	1	1997	7908.25	7	2011	12217.56	1
1984	1211.57	1	1998	9181.43	9	2012	13104.14	1
1985	1546.67	1	1999	11497.12	1	2013	16576.66	1
1986	1895.95	1	2000	10786.85	1	2014	17823.07	1
1987	1938.83	1	2001	10021.50	1	2015	17425.03	1
1988	2168.57	2	2002	8341.63	8	2016	19762.60	1

Here is a frequency distribution of the first digits of the stock market averages:

Digit	Frequency	Digit	Frequency
1	23	6	1
2	3	7	1
3	4	8	7
4	0	9	2
5	1		

For the stock market averages, the most frequent first digit by far is 1.

The stock market averages give a partial justification for Benford's law. Assume the stock market starts at 1000 and goes up 10% each year. It will take 8 years for the average to exceed 2000. Thus, the first eight averages will begin with the digit 1. Now imagine that the average starts at 5000. If it goes up 10% each year, it would take only 2 years to exceed 6000, so there would be only 2 years starting with the digit 5. In general, Benford's law applies well to data where increments occur as a result of multiplication rather than addition, and where there is a wide range of values. It does not apply to data sets where the range of values is small.

Here is the probability distribution of digits as predicted by Benford's law:

Digit	Frequency	Digit	Frequency
1	0.301	6	0.067
2	0.176	7	0.058
3	0.125	8	0.051
4	0.097	9	0.046
5	0.079		

The surprising nature of Benford's law makes it a useful tool to detect fraud. When people make up numbers, they tend to make the first digits approximately uniformly distributed; in other words, they have approximately equal numbers of 1s, 2s, and so on. Many tax agencies, including the Internal Revenue Service, use software to detect deviations from Benford's law in tax returns.

Following are results from three hypothetical corporate tax returns. Each purports to be a list of expenditures, in dollars, that the corporation is claiming as deductions. Two of the three are genuine, and one is a fraud. Which one is the fraud? *ii*

i.

79,386	17,988
203,374	80,535
11,967	3,037
100,229	132,056
46,428	59,727
7,012	38,354
957,559	137,648
551,284	4,163
97,439	1,279
780,216	91,404
22,443	323,547
1,023	194,288
738,527	24,346
634,814	695,236
850,840	160,546

ii.

1,393	165,648
47,689	601,981
75.854	262,971
5,395	65,407
53,079	6,892
7,791	748,151
93,401	45,054
129,906	83,821
568,823	228,976
4,693	913,337
21,902	252,378
337,122	82,581
162,182	538,342
7,942	99,613
31,121	78,175

iii.

64,888	374,242
1,643	12,338
832,618	14,204
126,811	31,484
13,545	1,818
2,332	104,625
29,288	34,178
81,074	3,684
401,437	11,665
3,040	15,376
244,676	541,894
49,273	65,928
112,111	250,601
56,776	650,316
262,359	90,852

chapter

7

The Normal Distribution

Steve Allen/Getty Images

Introduction

Beverage cans are made from a very thin sheet of aluminum, only 1/80 inch thick. Yet they must withstand pressures of up to 90 pounds per square inch (approximately three times the pressure in an automobile tire). Beverage companies often purchase cans in large shipments. To ensure that can failures are rare, quality control inspectors sample several cans from each shipment and test their strength by placing them in testing machines that apply force until the can fails (is punctured or crushed). The testing process destroys the cans, so the number of cans that can be tested is limited.

Assume that a can is considered defective if it fails at a pressure of less than 90 pounds per square inch. The quality control inspectors want the proportion of defective cans to be no more than 0.001, or 1 in 1000. They test 10 cans, with the following results.

Can	1	2	3	4	5	6	7	8	9	10
Pressure at Failure	95	96	98	99	99	100	101	101	103	104

Although none of the 10 cans were defective, this is not enough by itself to determine whether the proportion of defective cans is less than 1 in 1000. To make this determination, we must know something about the probability distribution of the pressures at which cans fail. In this chapter, we will study the *normal distribution*, which is the most important distribution in statistics. In the case study at the end of this chapter, we will show that if the pressures follow a normal distribution, we can estimate the proportion of defective cans.

Section 7.1 The Standard Normal Curve

Objectives

1. Use a probability density curve to describe a population
2. Use a normal curve to describe a normal population
3. Find areas under the standard normal curve
4. Find z-scores corresponding to areas under the normal curve

Objective 1 Use a probability density curve to describe a population

Figure 7.1, first shown in Section 2.2, presents a relative frequency histogram for the emissions of a sample of 65 vehicles. The amount of emissions is a continuous variable, because its possible values are not limited to some discrete set. The class intervals are chosen so that each rectangle represents a reasonably large number of vehicles. If the sample were larger, we could make the rectangles narrower. In particular, if we had information on the entire population, containing millions of vehicles, we could make the rectangles extremely narrow. The histogram would then look quite smooth and could be approximated by a curve, which might look like Figure 7.2.

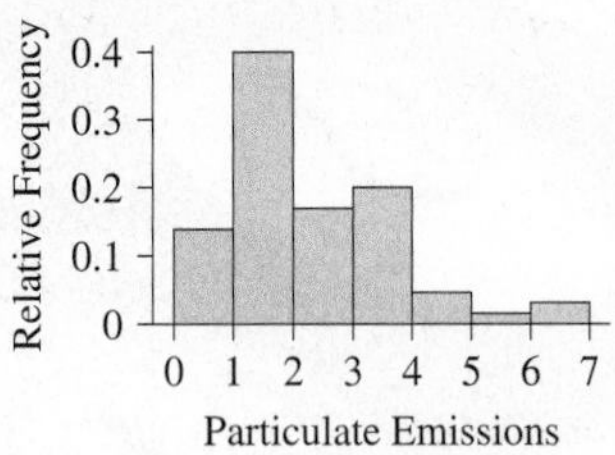

Figure 7.1 Relative frequency histogram for the emissions of a sample of 65 vehicles

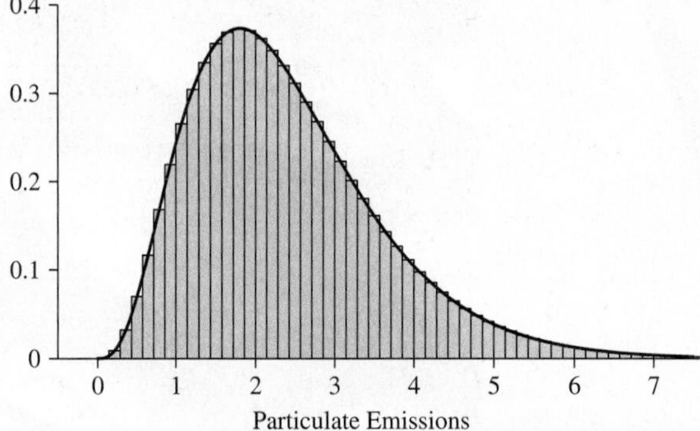

Figure 7.2 The histogram for a large population of vehicles could be drawn with extremely narrow rectangles, and could be represented by a curve.

NOTE TO INSTRUCTOR

The following topics from the **Statistics Corequisite Workbook** are aligned with the material in this section.

5.1 - Inequality Notation in Continuous Distributions

5.2 - Finding Probabilities Using Area of Rectangles

If a vehicle were chosen at random from this population to have its emissions measured, the emissions level would be a continuous random variable. The curve used to describe the distribution of a continuous random variable is called the **probability density curve** of the random variable. The probability density curve tells us what proportion of the population falls within any given interval. For example, Figure 7.3 illustrates the proportion of the population of vehicles whose emissions levels are between 3 and 4. In general, the area under a probability density curve between any two values a and b has two interpretations: It represents the proportion of the population whose values are between a and b, and it also represents the probability that a randomly selected value from the population will be between a and b.

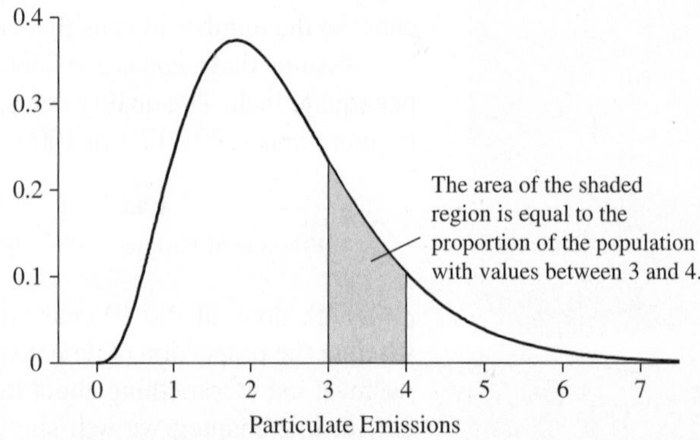

Figure 7.3 The area under a probability density curve between two values is equal to the proportion of the population that falls between the two values.

The region above a single point has zero width, and thus an area of 0. Therefore, when a population is represented with a probability density curve, the probability of obtaining a prespecified value exactly is equal to 0. For this reason, if X is a continuous random variable, then $P(X = a) = 0$ for any number a, and $P(a < X < b) = P(a \leq X \leq b)$ for any numbers a and b. For any probability density curve, the area under the entire curve is equal to 1, because this area represents the entire population.

SUMMARY

- A probability density curve represents the probability distribution of a continuous variable.
- The area under the entire curve is equal to 1.
- The area under the curve between two values a and b has two interpretations:
  1. It is the proportion of the population whose values are between a and b.
  2. It is the probability that a randomly selected individual will have a value between a and b.

Example 7.1

Interpret the area under a probability density curve

Following is a probability density curve for a population.

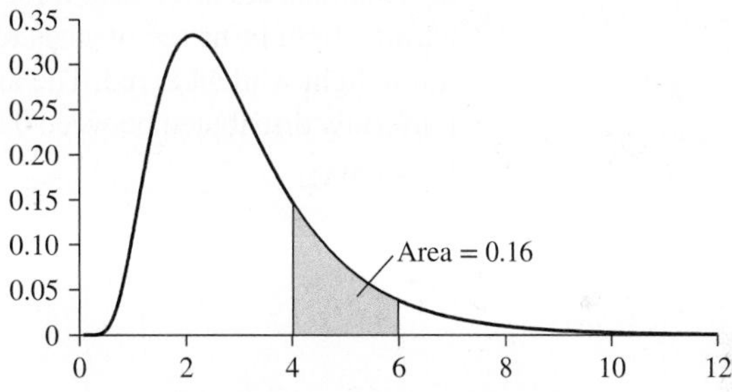

a. What proportion of the population is between 4 and 6?

b. If a value is chosen at random from this population, what is the probability that it will be between 4 and 6?

c. What proportion of the population is not between 4 and 6?

d. If a value is chosen at random from this population, what is the probability that it is not between 4 and 6?

Solution

a. The proportion of the population between 4 and 6 is equal to the area under the curve between 4 and 6, which is 0.16.

b. The probability that a randomly chosen value is between 4 and 6 is equal to the area under the curve between 4 and 6, which is 0.16.

c. The area under the entire curve is equal to 1. Therefore, the proportion that is not between 4 and 6 is equal to $1 - 0.16 = 0.84$.

d. The probability that a randomly chosen value is not between 4 and 6 is equal to the area under the curve that is not between 4 and 6, which is 0.84.

Another way to answer part (d) is to use the Rule of Complements:

$$P(\text{Not between 4 and 6}) = 1 - P(\text{Between 4 and 6}) = 1 - 0.16 = 0.84$$

RECALL

The Rule of Complements says that $P(\text{not } A) = 1 - P(A)$.

Check Your Understanding

1. Following is a probability density curve with the area between 0 and 1 and the area between 1 and 2 indicated.

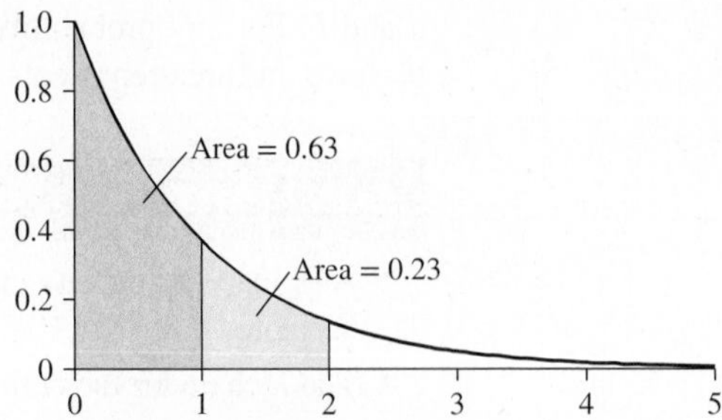

a. What proportion of the population is between 0 and 1? *0.63*
b. What is the probability that a randomly selected value will be between 1 and 2? *0.23*
c. What proportion of the population is between 0 and 2? *0.86*
d. What is the probability that a randomly selected value will be greater than 2? *0.14*

Answers are on page 321.

The Uniform Distribution

A uniform distribution is one in which values in any region are equally likely. Specifically, the probability density curve for a uniform distribution is a horizontal line. Uniform distributions are often useful for modeling waiting times. For example, imagine that when a traffic light turns red, it stays red for 30 seconds before turning green. You pull up to the traffic light while it is red. The amount of time you will wait before the light turns green is uniformly distributed between 0 and 30 seconds. Figure 7.4 presents the probability density function.

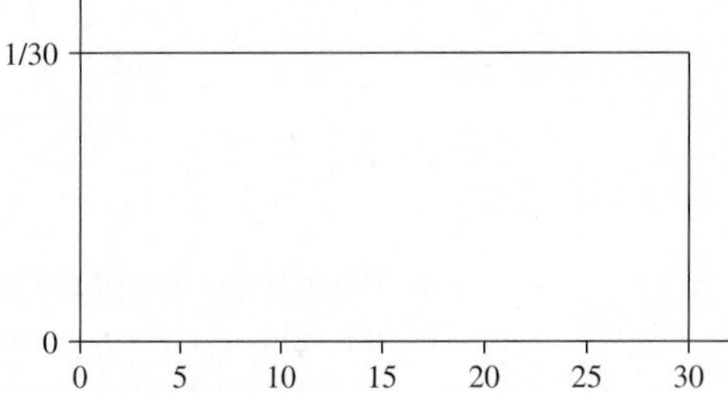

Figure 7.4 Probability density function for the uniform distribution on the interval from 0 to 30.

We can think of the probability density function as a rectangle. The base of the rectangle is the interval of possible values, in this case from 0 to 30. The length of the base is $30 - 0 = 30$. The height of the rectangle is the reciprocal of the length of the base, in this case $1/30$. The reason is that the area of the rectangle must equal 1, because it represents the probability that the waiting time is between 0 and 30 seconds, which is 1.

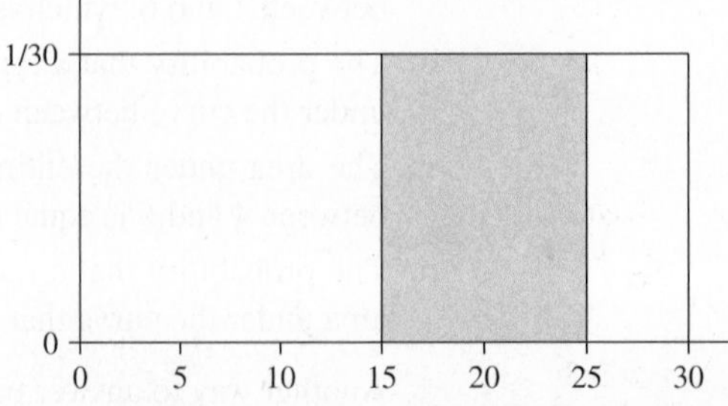

Figure 7.5 The probability that the waiting time is between 15 and 25 seconds is the area of the shaded rectangle.

It is straightforward to compute probabilities for the uniform distribution. For example, let's say we wanted to compute the probability that the time you wait at the light is between 15 and 25 seconds (see Figure 7.5 on page 306). This probability is the area of a rectangle with base $25 - 15 = 10$ and height $1/30$. The probability is $10(1/30) = 1/3$.

Example 7.2

Find probabilities with the uniform distribution

Using the uniform distribution pictured in Figure 7.4, find the following probabilities.

a. The probability that the waiting time is less than 15 seconds.
b. The probability that the waiting time is greater than 25 seconds.
c. The probability that the waiting time is between 5 and 15 seconds.

Solution

a. The probability is the area of a rectangle with base $15 - 0 = 15$ and height $1/30$, which is $15(1/30) = 1/2$.
b. The probability is the area of a rectangle with base $30 - 25 = 5$ and height $1/30$, which is $5(1/30) = 1/6$.
c. The probability is the area of a rectangle with base $15 - 5 = 10$ and height $1/30$, which is $10(1/30) = 1/3$.

Check Your Understanding

2. The waiting time at a bus stop for the next bus to arrive is uniformly distributed between 0 and 10 minutes.
a. Find the probability that the waiting time is less than 3 minutes. 0.3
b. Find the probability that the waiting time is greater than 6 minutes. 0.4
c. Find the probability that the waiting time is between 3 and 8 minutes. 0.5

Answers are on page 321.

The Normal Distribution

Objective 2 Use a normal curve to describe a normal population

Probability density curves come in many varieties, depending on the characteristics of the populations they represent. Remarkably, many important statistical procedures can be carried out using only one type of probability density curve, called a **normal curve**. A population that is represented by a normal curve is said to be *normally distributed*, or to have a **normal distribution**. Figure 7.6 presents some examples of normal curves.

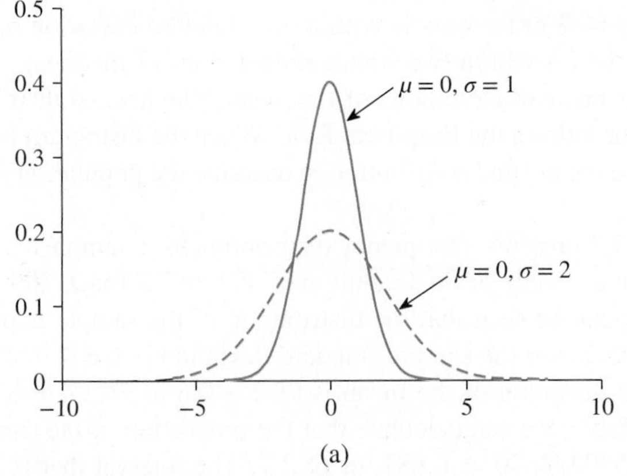

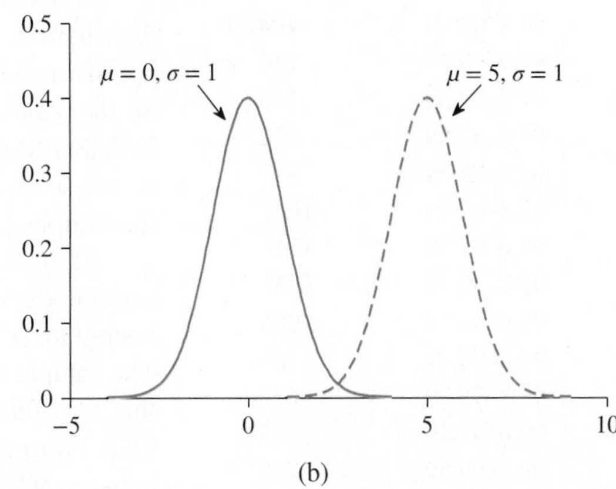

Figure 7.6 (a) Both populations have mean 0. The population with standard deviation 2 is more spread out than the population with standard deviation 1. (b) Both populations have the same spread, because they have the same standard deviation. The curves are centered over the population means.

EXPLAIN IT AGAIN

The mode of a curve: Recall that a peak in a histogram is called a *mode* of the histogram. Similarly, a peak in a probability density curve, such as a normal curve, is called a mode of the curve.

The location and shape of a normal curve reflect the mean and standard deviation of the population. The curve is symmetric around its peak, or mode. Therefore, the mode is equal to the population mean. The population standard deviation measures the spread of the population. Therefore, the normal curve is wide and flat when the population standard deviation is large, and tall and narrow when the population standard deviation is small.

RECALL

The Empirical Rule states that for a unimodal symmetric population, approximately 68% of the population will be within one standard deviation of the mean, approximately 95% will be within two standard deviations, and almost all will be within three standard deviations.

Properties of Normal Distributions

1. Normal distributions have one mode.
2. Normal distributions are symmetric around the mode.
3. The mean and median of a normal distribution are both equal to the mode. In other words, the mean, median, and mode of a normal distribution are all the same.
4. Normal curves extend infinitely far both to the right and to the left.
5. Normal distributions follow the Empirical Rule (see Figure 7.7):
   - Approximately 68% of the population is within one standard deviation of the mean. In other words, approximately 68% of the population is in the interval $\mu - \sigma$ to $\mu + \sigma$.
   - Approximately 95% of the population is within two standard deviations of the mean. In other words, approximately 95% of the population is in the interval $\mu - 2\sigma$ to $\mu + 2\sigma$.
   - Approximately 99.7% of the population is within three standard deviations of the mean. In other words, approximately 99.7% of the population is in the interval $\mu - 3\sigma$ to $\mu + 3\sigma$.

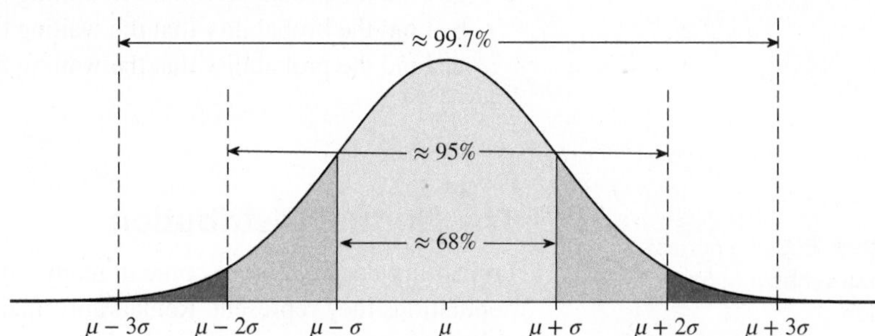

Figure 7.7 Normal curve with mean μ and standard deviation σ

Table 7.1

Temperature (°F)	Frequency
96.4–96.59	24
96.6–96.79	57
96.8–96.99	104
97.0–97.19	210
97.2–97.39	340
97.4–97.59	532
97.6–97.79	817
97.8–97.99	1087
98.0–97.19	1112
98.2–98.39	1035
98.4–98.59	1085
98.6–98.79	768
98.8–98.99	633
99.0–99.19	385
99.2–99.39	245
99.4–99.59	190
99.6–99.79	46
99.8–99.99	0

In Section 3.2 we discussed the Empirical Rule, which states that for a bell-shaped distribution, approximately 68% of the data is within one standard deviation of the mean, approximately 95% of the data is within two standard deviations of the mean, and almost all the data is within three standard deviations of the mean. The normal distribution is a bell-shaped distribution that follows the Empirical Rule. When the histogram for a sample is bell-shaped, we often use the normal distribution to describe the population from which the sample was drawn.

As an example, Table 7.1 presents a frequency distribution for a sample of orally measured body temperatures in a sample of 8670 adult men. Figure 7.8 (page 309) presents a histogram of these data. It can be seen that the distribution of the sample is bell-shaped. The sample mean is $\bar{x} = 98.2$, and the sample standard deviation is $s = 0.6$. The interval that is within one standard deviation of the mean is $98.2 - 0.6$ to $98.2 + 0.6$, or 97.6 to 98.8. From the frequency table, we can calculate that the proportion of the sample that is between 97.6 and 98.8 is $5904/8670 = 0.681$, or 68.1%. The interval that is within two standard deviations of the mean is $98.2 - 2(0.6)$ to $98.2 + 2(0.6)$, or 97.0 to 99.4, and the proportion that is between 97.0 and 99.4 is $8249/8670 = 0.951$, or 95.1%. As expected, the sample follows the Empirical Rule.

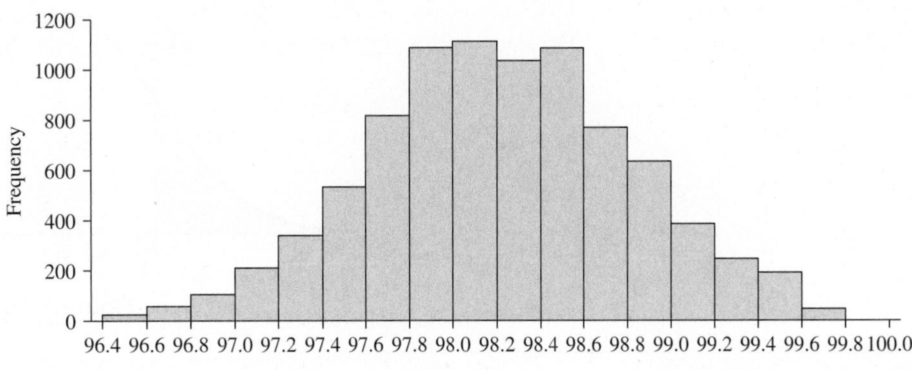

Figure 7.8

We would now like to describe the population of temperatures from which this sample was drawn. We don't know all the values in the population, but because the sample is bell-shaped, we will assume that the population has a normal distribution whose mean and standard deviation are the same as those of the sample: $\mu = 98.2$ and $\sigma = 0.6$. Figure 7.9 presents this normal distribution. Below each value, the number of standard deviations from the mean is indicated.

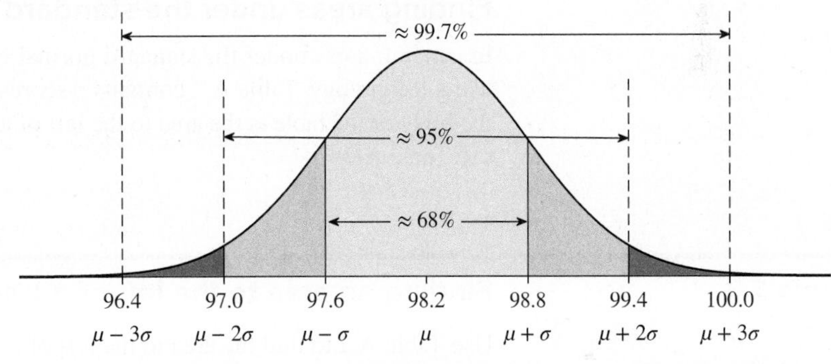

Figure 7.9

Figure 7.9 clearly shows the proportion of the population between values that correspond to whole numbers of standard deviations. For example, we can see that the proportion between 97.6 and 98.8 is approximately 68%, because these numbers correspond to $\mu - \sigma$ and $\mu + \sigma$. The Empirical Rule demonstrates how areas under the probability density curve represent proportions from the population. Sometimes we need to find the proportion of the population corresponding to fractional numbers of standard deviations, such as $\mu + 0.5\sigma$ or $\mu - 0.3\sigma$. To do this we use the standard normal distribution, which is the normal distribution with mean 0 and standard deviation 1.

Objective 3 Find areas under the standard normal curve

Finding Areas Under the Standard Normal Curve

A normal distribution can have any mean and any positive standard deviation, but it is only necessary to work with the normal distribution that has mean 0 and standard deviation 1, which is called the **standard normal distribution**. The probability density function for the standard normal distribution is called the **standard normal curve**.

For any interval, the area under a normal curve over the interval represents the proportion of the population that is contained within the interval. Finding an area under a normal curve is a crucial step in many statistical procedures. When finding an area under the standard normal curve, we use the letter z to indicate a value on the horizontal axis (see Figure 7.10 on page 310). We refer to such a value as a *z-score*. Since the mean of the standard normal distribution, which is located at the mode, is 0, the z-score at the mode of the curve is 0. Points on the horizontal axis to the left of the mode have negative z-scores, and points to the right of the mode have positive z-scores.

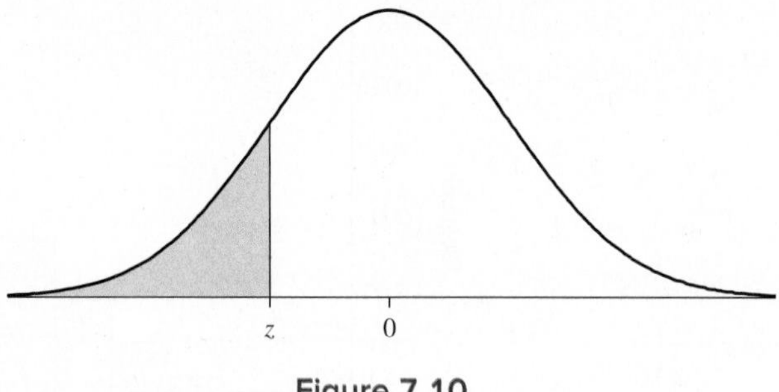

Figure 7.10

SUMMARY

- Points on the horizontal axis to the left of the mode have negative z-scores.
- Points on the horizontal axis to the right of the mode have positive z-scores.
- The mode itself has a z-score of 0.

Finding areas under the standard normal curve by using Table A.2

In general, areas under the standard normal curve can be found by using Table A.2 or by using technology. Table A.2 contains z-scores and areas. Each of the four-digit numbers in the body of the table is the area to the left of a z-score. Examples 7.3–7.5 will show how to use Table A.2.

Example 7.3

Finding an area to the left of a z-score

Use Table A.2 to find the area to the left of $z = 1.26$.

EXPLAIN IT AGAIN

Looking up a z-score: In Table A.2, the units and tenths digits of the z-score correspond to a row and the hundredths digit corresponds to a column. Thus, for the z-score 1.26, we find the row corresponding to 1.2 and the column corresponding to 0.06.

Solution

Step 1: Sketch a normal curve, label the point $z = 1.26$, and shade in the area to the left of it. Note that $z = 1.26$ is located to the right of the mode, since it is positive.

Step 2: Consult Table A.2. To look up $z = 1.26$, find the row containing 1.2 and the column containing 0.06. The value in the intersection of the row and column is 0.8962. This is the area to the left of $z = 1.26$ (see Figure 7.11).

z	0.00	0.01	0.02	0.03	0.04	0.05	0.06	0.07	0.08	0.09
⋮	⋮	⋮	⋮	⋮	⋮	⋮	⋮	⋮	⋮	⋮
1.0	.8413	.8438	.8461	.8485	.8508	.8531	.8554	.8577	.8599	.8621
1.1	.8643	.8665	.8686	.8708	.8729	.8749	.8770	.8790	.8810	.8830
1.2	.8849	.8869	.8888	.8907	.8925	.8944	.8962	.8980	.8997	.9015
1.3	.9032	.9049	.9066	.9082	.9099	.9115	.9131	.9147	.9162	.9177
1.4	.9192	.9207	.9222	.9236	.9251	.9265	.9279	.9292	.9306	.9319
⋮	⋮	⋮	⋮	⋮	⋮	⋮	⋮	⋮	⋮	⋮

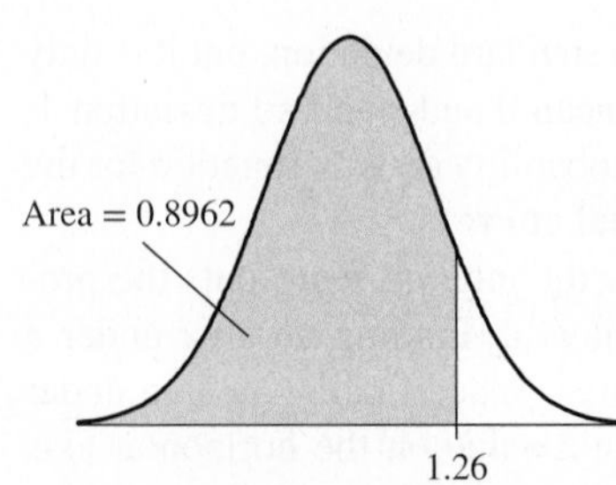

Figure 7.11

An area to the left of a z-score represents the proportion of a population that is less than a given value. Sometimes we need to know the proportion of a population that is greater than a given value. In these cases we need to find the area to the right of a z-score. Since the area under the entire curve is equal to 1, we can find the area to the right of a z-score by finding the area to the left and subtracting from 1.

Example 7.4

Finding an area to the right of a z-score

Use Table A.2 to find the area to the right of $z = -0.58$.

EXPLAIN IT AGAIN

Finding an area to the right of a z-score: To find the area to the right of a z-score, use Table A.2 to find the area to the left, then subtract from 1.

Solution

Step 1: Sketch a normal curve, label the point $z = -0.58$, and shade in the area to the right of it. Note that $z = -0.58$ is located to the left of the mode, since it is negative.

Step 2: Consult Table A.2. To look up $z = -0.58$, find the row containing -0.5 and the column containing 0.08. The value in the intersection of the row and column is 0.2810. This is the area to the *left* of $z = -0.58$. To find the area to the right, we subtract from 1 (see Figure 7.12):

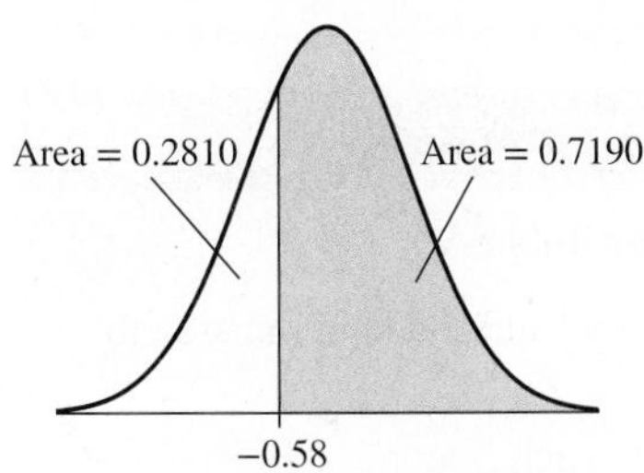

Figure 7.12

$$\begin{aligned}\text{Area to the right of } z = -0.58 &= 1 - \text{Area to the left of } z = -0.58\\ &= 1 - 0.2810\\ &= 0.7190\end{aligned}$$

z	0.00	0.01	0.02	0.03	0.04	0.05	0.06	0.07	0.08	0.09
⋮	⋮	⋮	⋮	⋮	⋮	⋮	⋮	⋮	⋮	⋮
−0.7	.2420	.2389	.2358	.2327	.2296	.2266	.2236	.2206	.2177	.2148
−0.6	.2743	.2709	.2676	.2643	.2611	.2578	.2546	.2514	.2483	.2451
−0.5	.3085	.3050	.3015	.2981	.2946	.2912	.2877	.2843	.2810	.2776
−0.4	.3446	.3409	.3372	.3336	.3300	.3264	.3228	.3192	.3156	.3121
−0.3	.3821	.3783	.3745	.3707	.3669	.3632	.3594	.3557	.3520	.3483
⋮	⋮	⋮	⋮	⋮	⋮	⋮	⋮	⋮	⋮	⋮

Sometimes we need to find the proportion of a population that falls between two values. In these cases, we need to find the area between two z-scores. We can do this using Table A.2 by finding the area to the left of each z-score. The area between the z-scores is found by subtracting the smaller area from the larger area.

Example 7.5

Finding an area between two z-scores

Find the area between $z = -1.45$ and $z = 0.42$.

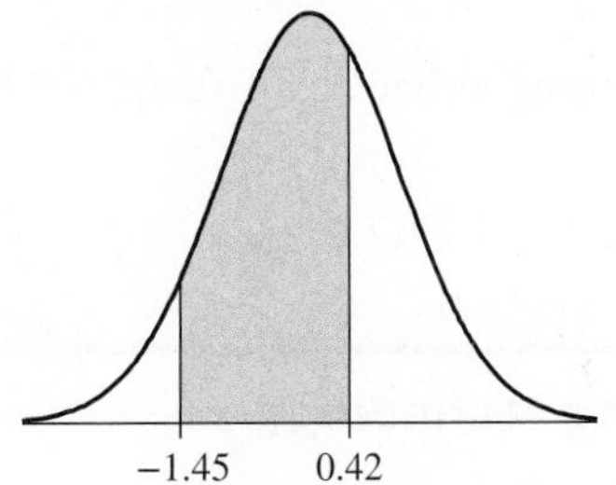

Figure 7.13

Solution

Step 1: Sketch a normal curve, label the points $z = -1.45$ and $z = 0.42$, and shade in the area between them. See Figure 7.13.

Step 2: Use Table A.2 to find the areas to the left of $z = -1.45$ and to the left of $z = 0.42$. The area to the left of $z = -1.45$ is 0.0735, and the area to the left of $z = 0.42$ is 0.6628.

Step 3: Subtract the smaller area from the larger area to find the area between the two z-scores:

$$\begin{aligned}\text{Area between } z = -1.45 \text{ and } z = 0.42 &= (\text{Area left of } 0.42) - (\text{Area left of } -1.45)\\ &= 0.6628 - 0.0735\\ &= 0.5893\end{aligned}$$

The area between $z = -1.45$ and $z = 0.42$ is 0.5893. See Figure 7.14 (page 312).

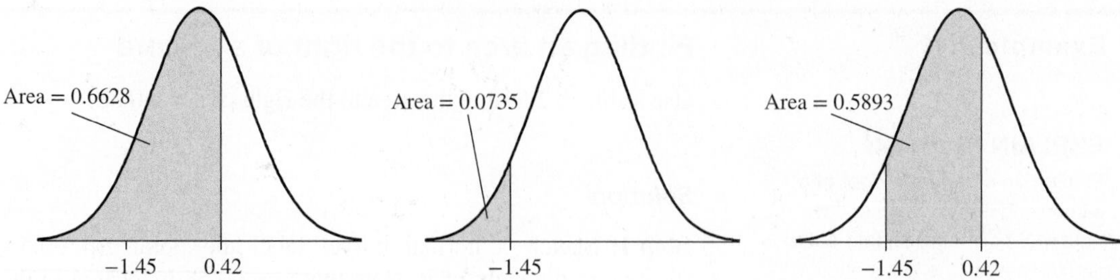

Figure 7.14 We start with the area to the left of $z = 0.42$ and subtract the area to the left of $z = -1.45$. This leaves the area between $z = 0.42$ and $z = -1.45$.

SUMMARY

To find an area under the standard normal curve by using Table A.2:

Step 1: Sketch a normal curve, label the z-score or scores, and shade in the area to be found.

Step 2: Look up the area in Table A.2 corresponding to each z-score.

Step 3:

- To find the area to the left of a z-score, use the area in Table A.2.
- To find the area to the right of a z-score, find the area to the left and subtract from 1.
- To find the area between two z-scores, find the area to the left of each, then subtract the smaller area from the larger area.

Check Your Understanding

3. Use Table A.2 to find the area to the left of $z = 0.25$. *0.5987*

4. Use Table A.2 to find the area to the right of $z = 2.31$. *0.0104*

5. Use Table A.2 to find the area between $z = -1.13$ and $z = 2.02$. *0.8491*

6. Use Table A.2 to find the area between $z = 0.56$ and $z = 1.87$. *0.2570*

Answers are on page 321.

Finding areas under the standard normal curve by using technology

Areas under the standard normal curve can be found by using technology. Examples 7.6 and 7.7 illustrate the method.

Example 7.6

Find the area between two z-scores by using technology

In Example 7.5, we found the area between $z = -1.45$ and $z = 0.42$. Find this area by using technology.

Solution

We present results from the TI-84 Plus calculator. The area is found by using the **normalcdf** command. We enter the left endpoint of the interval (−1.45), the right endpoint (0.42), the mean (0), and the standard deviation (1). Step-by-step instructions are given in the Using Technology section on page 316.

```
normalcdf(-1.45,0.42,0,1)
                .5892279372
```

The TI-84 Plus gives the area as .5892279372. Rounding to four decimal places gives 0.5892.

Example 7.7

Find the area to the right of a z-score by using technology

In Example 7.4, we found the area to the right of $z = -0.58$. Find this area by using technology.

EXPLAIN IT AGAIN

1E99: In Example 7.7, we want to find the area between −0.58 and infinity. Since we can't enter a value of infinity into the calculator, we enter the very large number **1E99** instead.

Solution

We present output from the TI-84 Plus calculator. We use the **normalcdf** command. We enter the left endpoint of the interval (−0.58). Since there is no right endpoint, we enter **1E99**, which represents the very large number that is written with a 1 followed by 99 zeros. Then we enter the mean (0) and the standard deviation (1). Step-by-step instructions are given in the Using Technology section on page 316.

```
normalcdf(-0.58,1E99,0,1)
                .7190427366
```

The TI-84 Plus gives the area as .7190427366. Rounding to four decimal places gives 0.7190.

Objective 4 Find z-scores corresponding to areas under the normal curve

Finding a z-Score Corresponding to a Given Area

Often we are given an area and we need to find the z-score that corresponds to an area under the standard normal curve. Examples 7.8–7.12 show how this is done using either Table A.2 or technology. When using the table, it is useful to remember that the mode, $z = 0$, has an area of 0.5 both to its right and to its left.

Example 7.8

Finding the z-score corresponding to an area to the left

Use Table A.2 to find the z-score that has an area of 0.26 to its left.

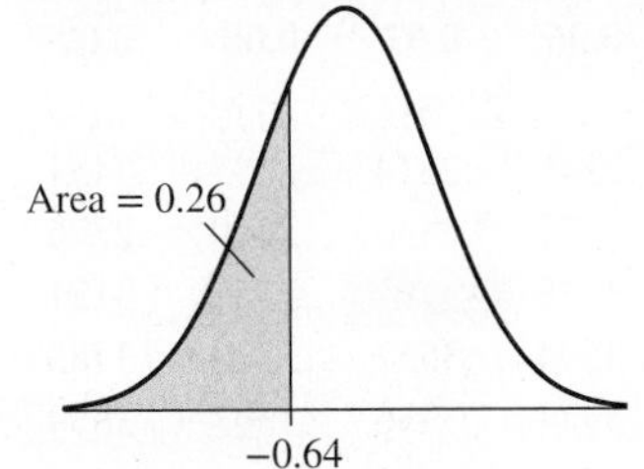

Figure 7.15

Solution

Step 1: Sketch a normal curve and shade in the given area.

Step 2: In Table A.2, look through the body of the table to find the area closest to 0.26. This value is 0.2611. The values in the left-hand column and top row corresponding to 0.2611 are −0.6 and 0.04. Therefore, the z-score is $z = -0.64$. See Figure 7.15.

EXPLAIN IT AGAIN

Finding a z-score corresponding to an area: In Table A.2, the numbers in the body of the table represent *areas* and the numbers down the left-hand column and across the top row represent *z*-scores. When given an area, we find that area in the body of the table and look at the left-hand column and top row to obtain the *z*-score.

z	0.00	0.01	0.02	0.03	0.04	0.05	0.06	0.07	0.08	0.09
⋮	⋮	⋮	⋮	⋮	⋮	⋮	⋮	⋮	⋮	⋮
−0.8	.2119	.2090	.2061	.2033	.2005	.1977	.1949	.1922	.1894	.1867
−0.7	.2420	.2389	.2358	.2327	.2296	.2266	.2236	.2206	.2177	.2148
−0.6	.2743	.2709	.2676	.2643	.2611	.2578	.2546	.2514	.2483	.2451
−0.5	.3085	.3050	.3015	.2981	.2946	.2912	.2877	.2843	.2810	.2776
−0.4	.3446	.3409	.3372	.3336	.3300	.3264	.3228	.3192	.3156	.3121
⋮	⋮	⋮	⋮	⋮	⋮	⋮	⋮	⋮	⋮	⋮

Example 7.9

Using technology to find the *z*-score corresponding to an area to the left

In Example 7.8, we found the z-score that has an area of 0.26 to its left. Use technology to find this z-score.

Solution

We present results from the TI-84 Plus calculator. The z-score is found by using the **invNorm** command. We enter the area to the left (.26), the mean (0), and the standard deviation (1). Step-by-step instructions are given in the Using Technology section on page 316.

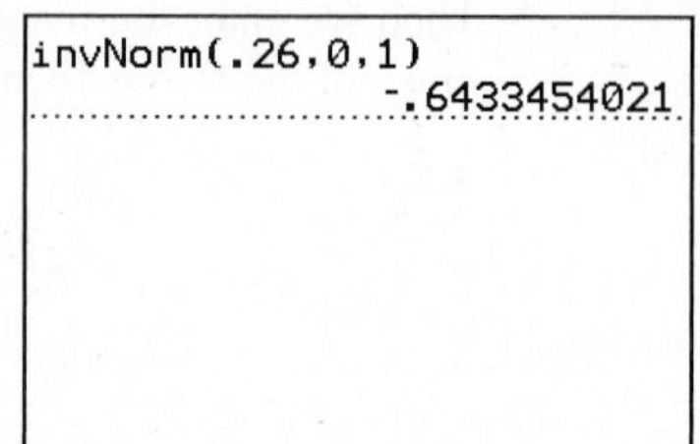

The TI-84 Plus gives $z = -.6433454021$. Rounding to two decimal places gives $z = -0.64$.

Example 7.10

Finding the *z*-score corresponding to an area to the right

Use Table A.2 to find the z-score that has an area of 0.68 to its right.

Solution

Step 1: Sketch a normal curve and shade in the given area.

Step 2: Determine the area to the left of the z-score. Since the area to the right is 0.68, the area to the left is $1 - 0.68 = 0.32$.

Step 3: In Table A.2, look through the body of the table to find the area closest to 0.32. This value is 0.3192. The z-score in Table A.2 corresponding to an area of 0.3192 is $z = -0.47$. See Figure 7.16.

Area = 0.68

−0.47

Figure 7.16

z	0.00	0.01	0.02	0.03	0.04	0.05	0.06	0.07	0.08	0.09
⋮	⋮	⋮	⋮	⋮	⋮	⋮	⋮	⋮	⋮	⋮
−0.6	.2743	.2709	.2676	.2643	.2611	.2578	.2546	.2514	.2483	.2451
−0.5	.3085	.3050	.3015	.2981	.2946	.2912	.2877	.2843	.2810	.2776
−0.4	.3446	.3409	.3372	.3336	.3300	.3264	.3228	.3192	.3156	.3121
−0.3	.3821	.3783	.3745	.3707	.3669	.3632	.3594	.3557	.3520	.3483
−0.2	.4207	.4168	.4129	.4090	.4052	.4013	.3974	.3936	.3897	.3859
⋮	⋮	⋮	⋮	⋮	⋮	⋮	⋮	⋮	⋮	⋮

Example 7.11

Using technology to find the z-score corresponding to an area to the right

In Example 7.10, we found the z-score that has an area of 0.68 to its right. Use technology to find this z-score.

Solution

We use the TI-84 Plus calculator. The first two steps are the same as those for using the table. We must first find the area to the *left* of the desired z-score, by subtracting the area to the right from 1. Since the area to the right is 0.68, the area to the left is $1 - 0.68 = 0.32$.

The results are as follows. Step-by-step instructions are given in the Using Technology section on page 316.

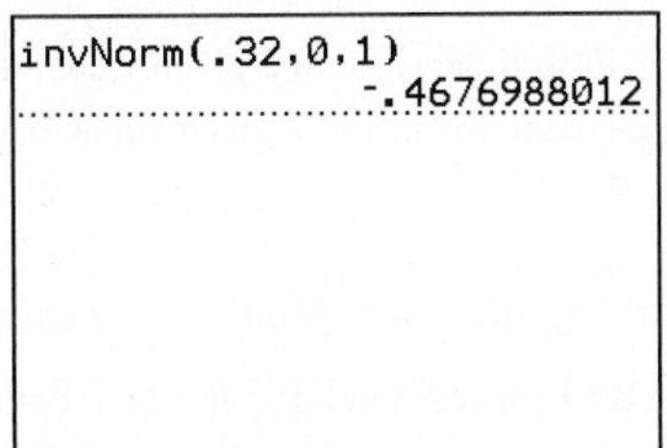

The TI-84 Plus gives $z = -.4676988012$. Rounding to two decimal places gives $z = -0.47$.

Check Your Understanding

7. Use Table A.2 to find the z-score with an area of 0.45 to its left. *−0.13*

8. Use Table A.2 to find the z-score with an area of 0.37 to its right. *0.33*

9. Use any method to find the z-score with an area of 0.74 to its left. *0.64*

10. Use any method to find the z-score with an area of 0.09 to its right. *1.34*

Answers are on page 321.

Example 7.12

Finding the z-score corresponding to an area in the middle

Use Table A.2 to find the z-scores that bound the middle 95% of the area under the standard normal curve.

Solution

Step 1: Sketch a normal curve and shade in the given area. Label the z-score on the left z_1 and the z-score on the right z_2.

Step 2: Find the area to the left of z_1. Since the area in the middle is 0.95, the area in the two tails combined is 0.05. Half of that area, or 0.025, is to the left of z_1. We conclude that z_1 is the z-score that has an area of 0.025 to its left.

Step 3: In Table A.2, an area of 0.025 corresponds to a z-score of -1.96. Therefore, $z_1 = -1.96$.

Step 4: To find z_2, note that the area to the left of z_2 is 0.9750. We can use Table A.2 to determine that $z_2 = 1.96$. Alternatively, note that z_1 and z_2 are equidistant from the mode. Since the normal curve is symmetric, z_1 must be the negative of z_2. Therefore $z_2 = 1.96$. See Figure 7.17.

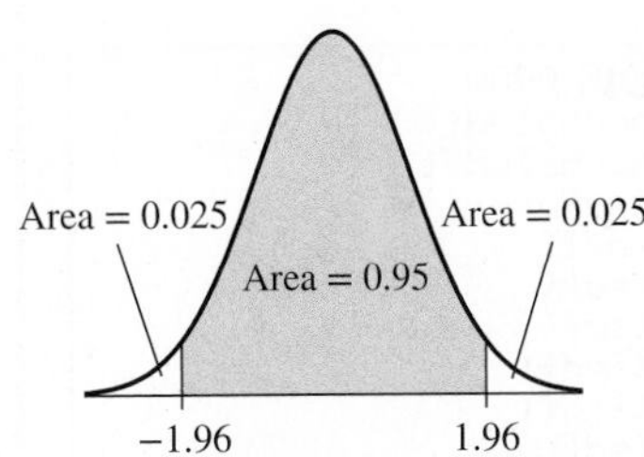

Figure 7.17

Using Technology

TI-84 PLUS

Finding areas under the standard normal curve

The **normalcdf** command is used to calculate area under a normal curve.

Step 1. Press **2nd**, then **VARS** to access the **DISTR** menu. Select **2:normalcdf** (Figure A).

Step 2. Enter the left endpoint, comma, the right endpoint, comma, the mean (**0** for the standard normal), comma, and the standard deviation (**1** for the standard normal).

- When finding the area to the right of a given value, use **1E99** as the right endpoint.
- When finding the area to the left of a given value, use **−1E99** as the left endpoint.

Step 3. Press **ENTER**.

```
DISTR DRAW
1:normalpdf(
2:normalcdf(
3:invNorm(
4:invT(
5:tpdf(
6:tcdf(
7:χ²pdf(
8:χ²cdf(
9↓Fpdf(
```

Figure A

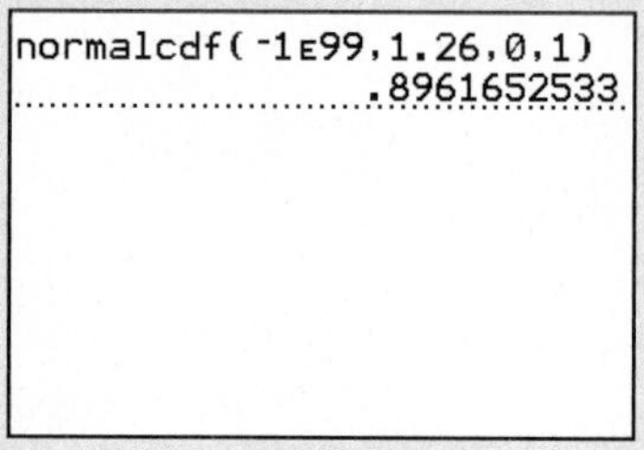

Using the TI-84 PLUS Stat Wizards (see Appendix B for more information)

Step 1. Press **2nd**, then **VARS** to access the **DISTR** menu. Select **2:normalcdf** (Figure A).

Step 2. Enter the left endpoint in the **lower** field, the right endpoint in the **upper** field, the mean in the μ field (**0** for the standard normal), and the standard deviation in the σ field (**1** for the standard normal).

- When finding the area to the right of a given value, use **1E99** as the right endpoint.
- When finding the area to the left of a given value, use **−1E99** as the left endpoint.

Step 3. Select **Paste** and press **ENTER** to paste the command to the home screen. Press **ENTER** again to run the command.

Figure B illustrates finding the area to the left of $z = 1.26$ (Example 7.3).

Figure C illustrates finding the area to the right of $z = -0.58$ (Example 7.4).

Figure D illustrates finding the area between $z = -1.45$ and $z = 0.42$ (Example 7.5).

Note: The quantity 1E99 in the TI-84 PLUS calculator represents a very large number, specifically a 1 followed by 99 zeros.

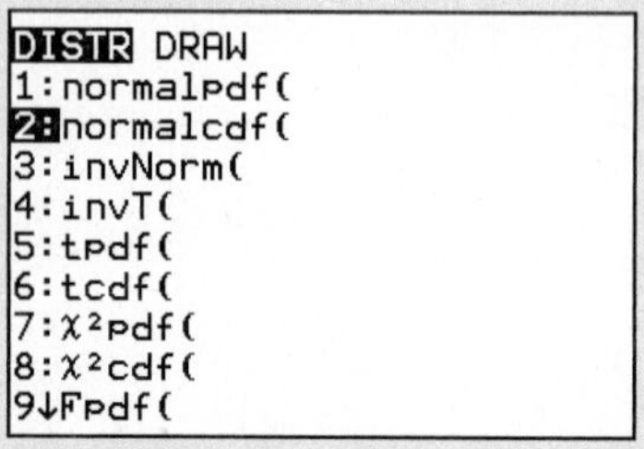

```
normalcdf(-1E99,1.26,0,1)
                .8961652533
```

Figure B

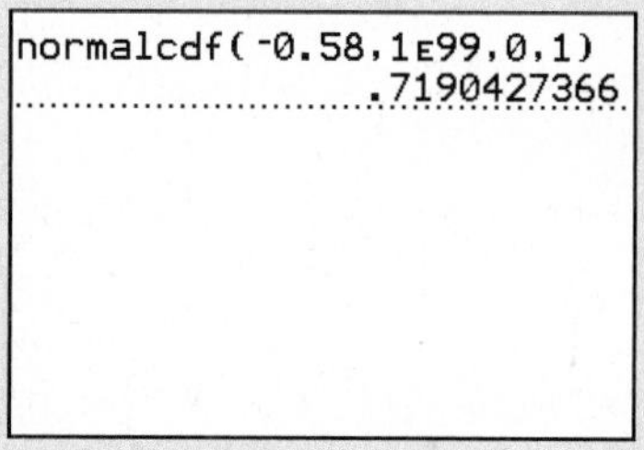

```
normalcdf(-0.58,1E99,0,1)
                .7190427366
```

Figure C

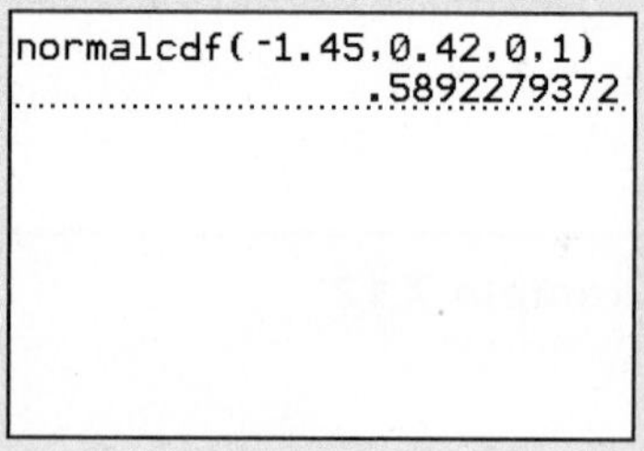

```
normalcdf(-1.45,0.42,0,1)
                .5892279372
```

Figure D

Finding a z-score corresponding to a given area

The **invNorm** command is used to calculate the z-score corresponding to an *area to the left*.

Step 1. Press **2nd**, then **VARS** to access the **DISTR** menu. Select **3:invNorm** (Figure E).

Step 2. Enter the area to the left of the desired z-score, comma, the mean (**0** for the standard normal), comma, and the standard deviation (**1** for the standard normal).

Step 3. Press **ENTER**.

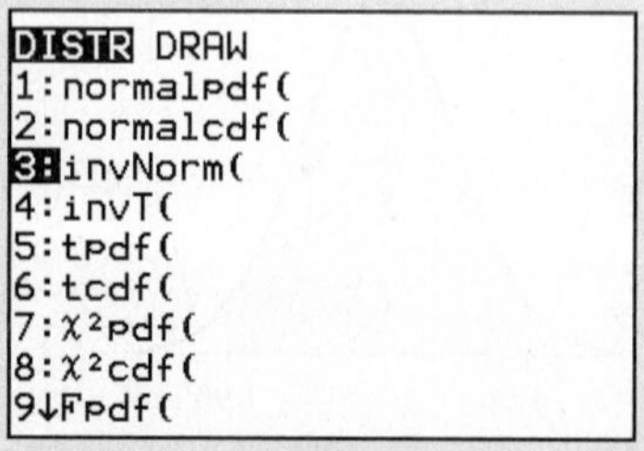

```
DISTR DRAW
1:normalpdf(
2:normalcdf(
3:invNorm(
4:invT(
5:tpdf(
6:tcdf(
7:χ²pdf(
8:χ²cdf(
9↓Fpdf(
```

Figure E

Using the TI-84 PLUS Stat Wizards (see Appendix B for more information)

Step 1. Press **2nd**, then **VARS** to access the **DISTR** menu. Select **3:invNorm** (Figure E).

Step 2. Enter the area to the left of the desired z-score in the **area** field, the mean in the μ field (**0** for the standard normal), and the standard deviation in the σ field (**1** for the standard normal).

Step 3. Select **Paste** and press **ENTER** to paste the command to the home screen. Press **ENTER** again to run the command.

Figure F illustrates finding the z-score that has an area of 0.26 to its left (Example 7.9).

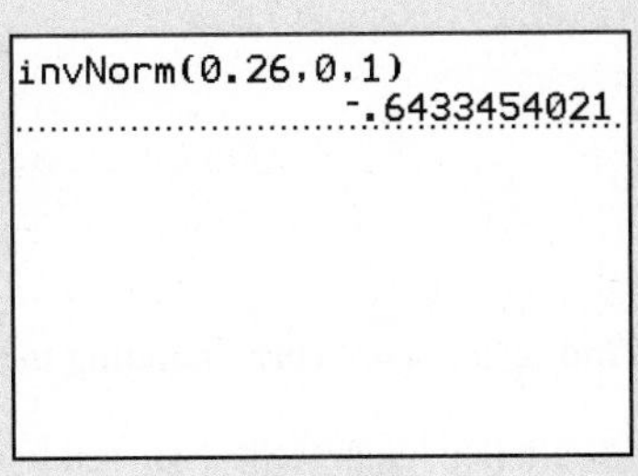

Figure F

EXCEL

Finding areas under the standard normal curve

The following procedure computes the area to the *left* of a given value.

Step 1. In an empty cell, select the **Insert Function** icon and highlight **Statistical** in the category field.

Step 2. Click on the **NORM.S.DIST** function and press **OK**.

Step 3. To compute the area to the left of a given x, enter the value of x in the **X** field.

Step 4. Enter **TRUE** in the **Cumulative** field and click **OK**.

Figure G illustrates finding the area to the left of $z = 1.26$ (Example 7.3).

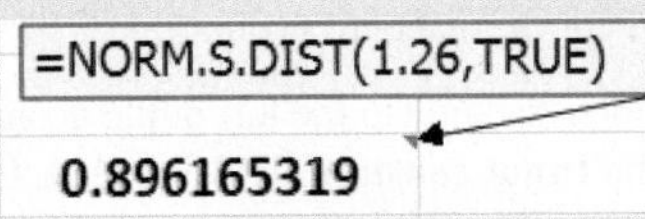

Figure G

Finding a z-score corresponding to a given area

The following procedure is used to calculate the z-score corresponding to an *area to the left*.

Step 1. In an empty cell, select the **Insert Function** icon and highlight **Statistical** in the category field.

Step 2. Click on the **NORM.S.INV** function and press **OK**.

Step 3. Enter the area to the left of the desired z-score in the **Probability** field.

Step 4. Click **OK**.

Figure H illustrates finding the z-score that has an area of 0.26 to its left (Example 7.9).

=NORM.S.INV(0.26)

-0.643345405

Figure H

MINITAB

Finding areas under the standard normal curve

The following procedure computes the area to the *left* of a given value.

Step 1. Click **Calc**, then **Probability Distributions**, then **Normal**.

Step 2. Select the **Cumulative probability** option.

Step 3. Enter the value for the mean in the **Mean** field (**0** for the standard normal) and the value for the standard deviation in the **Standard deviation** field (**1** for the standard normal).

Step 4. To compute the area to the left of a given x, enter the value for x in the **Input constant** field.

Step 5. Click **OK**.

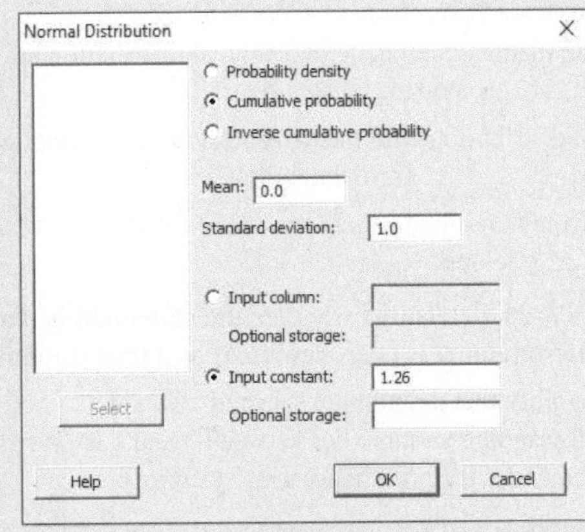

Figure I

Figures I and J illustrate finding the area to the left of $z = 1.26$ (Example 7.3).

Normal with mean = 0 and standard deviation = 1

x	P(X ≤ x)
1.26	0.896165

Figure J

Finding a z-score corresponding to a given area

The following procedure is used to calculate the z-score corresponding to an *area to the left*.

Step 1. Click **Calc**, then **Probability Distributions**, then **Normal**.

Step 2. Select the **Inverse Cumulative Probability** option.

Step 3. Enter the value for the mean in the **Mean** field (**0** for the standard normal) and the value for the standard deviation in the **Standard deviation** field (**1** for the standard normal).

Step 4. Enter the area to the left of the desired z-score in the **Input constant** field and click **OK**.

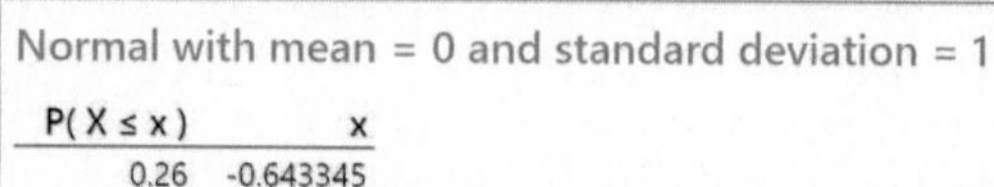

Normal with mean = 0 and standard deviation = 1

P(X ≤ x)	x
0.26	-0.643345

Figure K

Figure K illustrates finding the z-score that has an area of 0.26 to its left (Example 7.9).

Section 7.1 Exercises

Exercises 1–10 are the Check Your Understanding exercises located within the section.

Understanding the Concepts

In Exercises 11–16, fill in each blank with the appropriate word or phrase.

11. The proportion of a population that is contained within an interval corresponds to an area under the probability ________ curve. *density*

12. If X is a continuous random variable, then $P(X = a) =$ ________ for any number a. *0*

13. The area under the entire probability density curve is equal to ________. *1*

14. The mean, median, and mode of a normal distribution are ________ to each other. *equal*

15. A normal distribution with mean 0 and standard deviation 1 is called the ________ normal distribution. *standard*

16. Points on the horizontal axis to the left of the mode have ________ z-scores. *negative*

In Exercises 17–20, determine whether the statement is true or false. If the statement is false, rewrite it as a true statement.

17. The probability that a randomly selected value of a continuous random variable lies between a and b is given by the area under the probability density curve between a and b. *True*

18. A normal curve is symmetric around its mode. *True*

19. A normal curve is wide and flat when the standard deviation is small. *False*

20. The area under the normal curve to the left of the mode is less than 0.5. *False*

Practicing the Skills

21. The following figure is a probability density curve that represents the lifetime, in months, of a certain type of laptop battery.

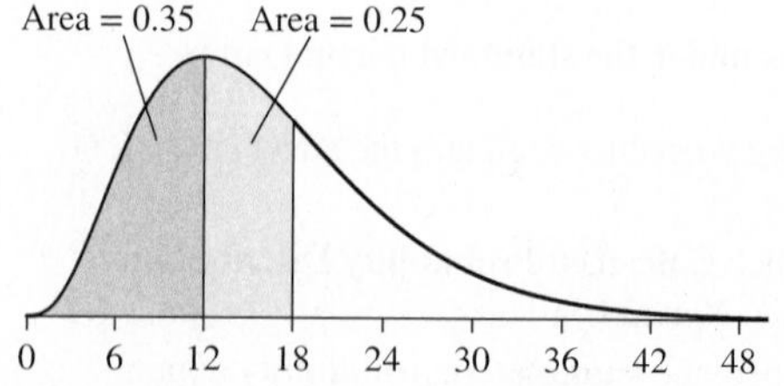

a. Find the proportion of batteries with lifetimes between 12 and 18 months. *0.25*

b. Find the proportion of batteries with lifetimes less than 18 months. *0.6*

c. What is the probability that a randomly chosen battery lasts more than 18 months? *0.4*

22. The following figure is a probability density curve that represents the grade point averages (GPAs) of the graduating seniors at a large university.

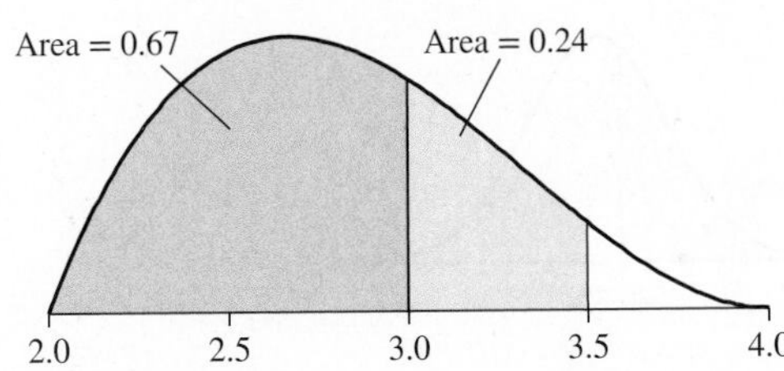

a. Find the proportion of seniors whose GPA is between 3.0 and 3.5. *0.24*

b. What is the probability that a randomly chosen senior will have a GPA greater than 3.5? *0.09*

23. The time waiting for baggage at an airport is uniformly distributed between 0 and 15 minutes.

a. Find the probability that the waiting time is less than 12 minutes. *0.8*

b. Find the probability that the waiting time is greater than 9 minutes. *0.4*

c. Find the probability that the waiting time is between 4 and 13 minutes. *0.6*

24. The waiting time at a certain post office is uniformly distributed between 5 and 25 minutes.

a. Find the probability that the waiting time is less than 15 minutes. *0.5*

b. Find the probability that the waiting time is greater than 11 minutes. *0.7*

c. Find the probability that the waiting time is between 6 and 12 minutes. *0.3*

25. The following figure is a normal curve that represents the heights, in inches, of adult women in the United States.

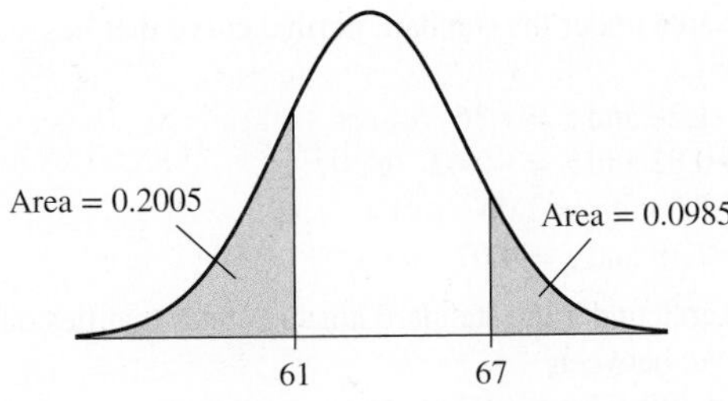

a. Find the proportion of women who are more than 61 inches tall. *0.7995*

b. Find the proportion of women who are less than 67 inches tall. *0.9015*

c. Find the proportion of women who are between 61 and 67 inches tall. *0.7010*

26. The following figure is a normal curve that represents the weights, in pounds, of adult female cats of a certain breed.

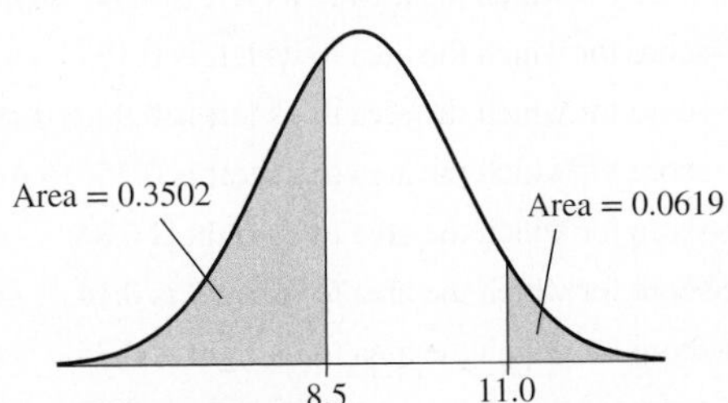

a. Find the proportion of cats who weigh more than 8.5 pounds. *0.6498*

b. Find the proportion of cats who weigh less than 11 pounds. *0.9381*

c. Find the proportion of cats who weigh between 8.5 and 11 pounds. *0.5879*

27. Match each of the normal curves to its mean μ and standard deviation σ.

a. $\mu = -2, \sigma = 2$ *ii* **b.** $\mu = -2, \sigma = 4$ *iv*

c. $\mu = 4, \sigma = 1$ *i* **d.** $\mu = 4, \sigma = 3$ *iii*

i.
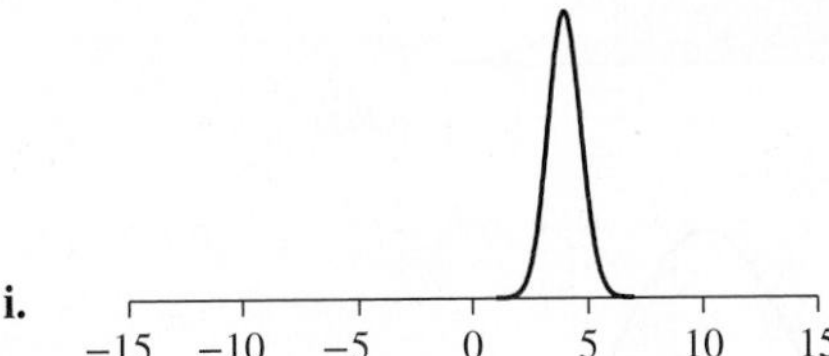

ii.
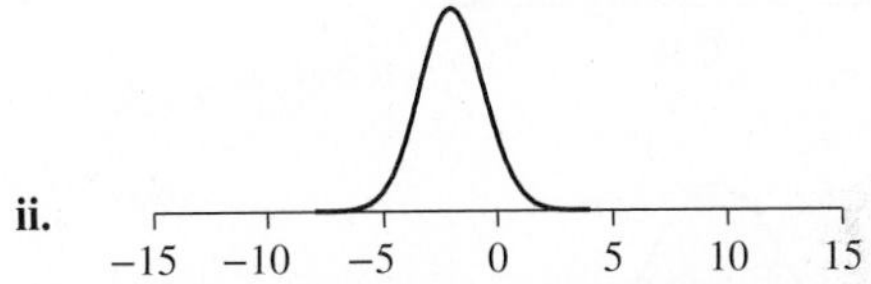

iii.
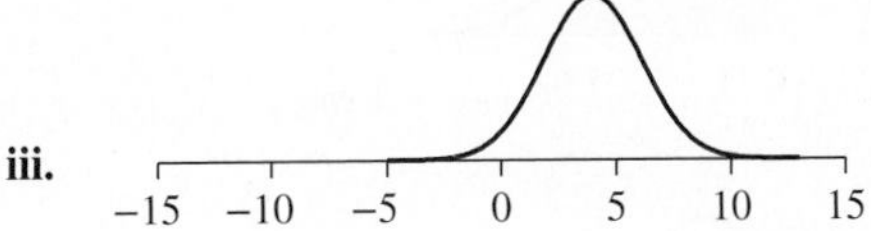

iv.
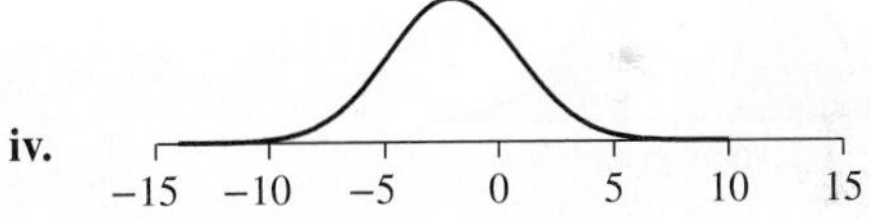

28. Match each of the normal curves to its mean μ and standard deviation σ.

a. $\mu = -3, \sigma = 2$ *iii* **b.** $\mu = 3, \sigma = 6$ *ii*

c. $\mu = -3, \sigma = 4$ *iv* **d.** $\mu = 3, \sigma = 3$ *i*

i.
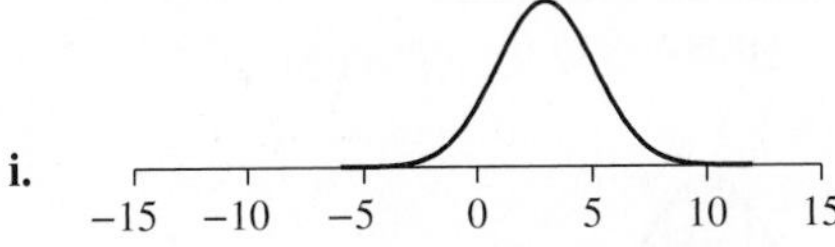

ii.
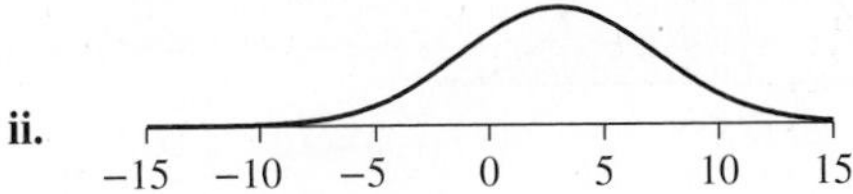

iii.
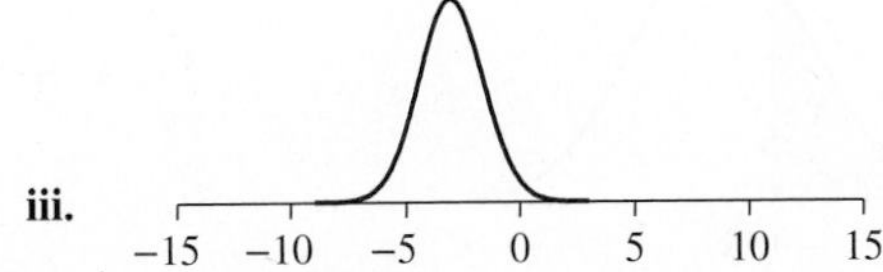

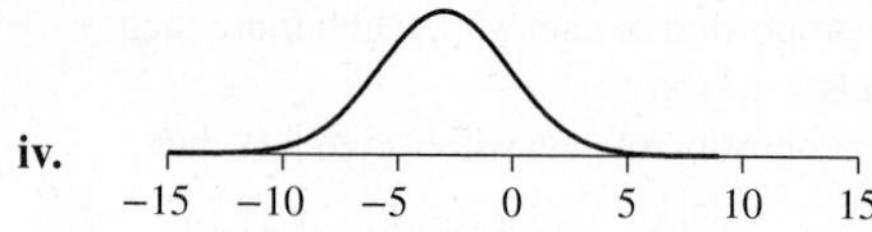

29. Find each of the shaded areas under the standard normal curve.

a.

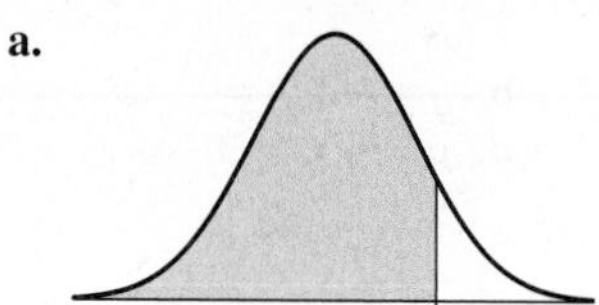

0.8944

b.

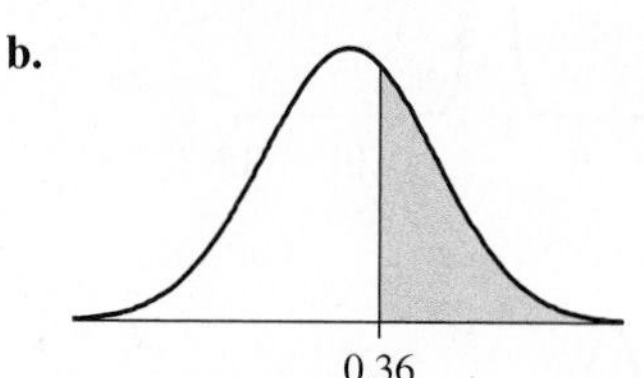

0.3594

c.

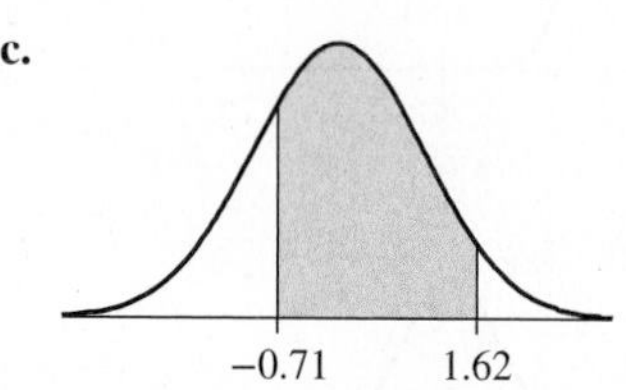

0.7085

d.

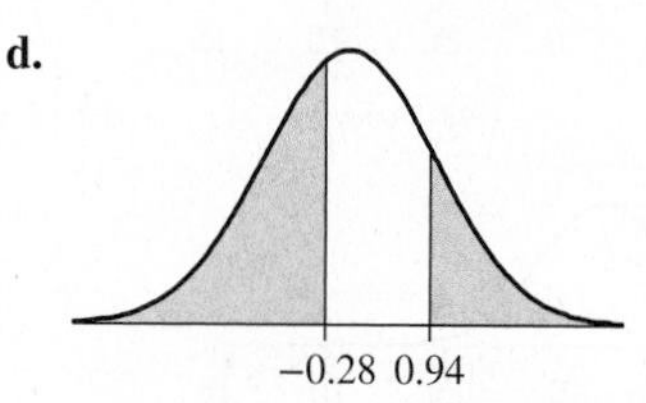

0.5633

30. Find each of the shaded areas under the standard normal curve.

a.

0.3520

b.

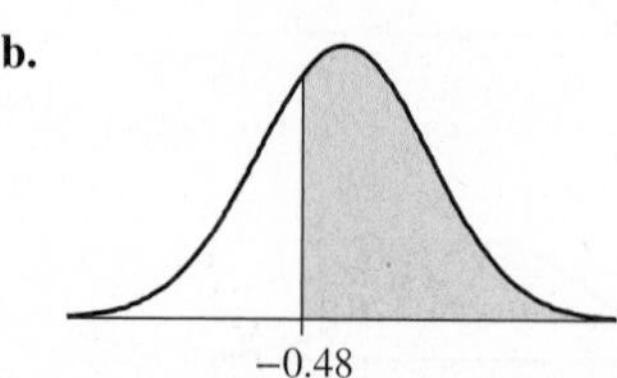

0.6844

c.

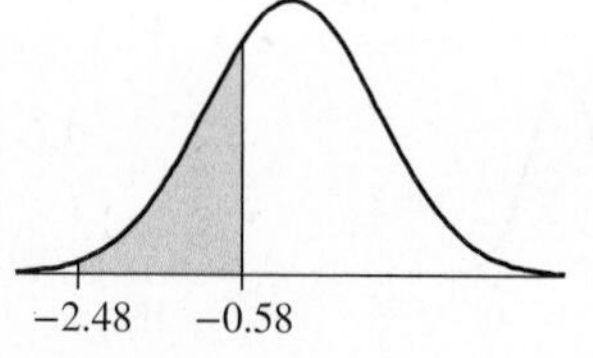

0.2744

d.

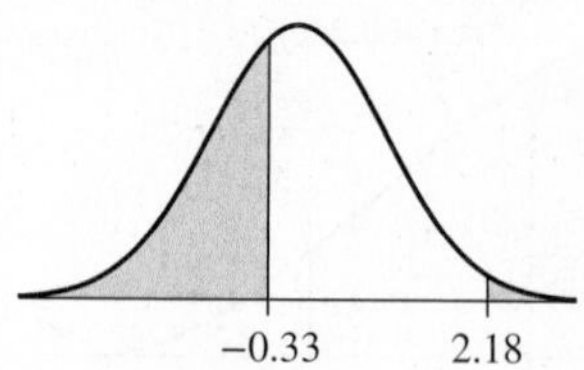

0.3853

31. Find the area under the standard normal curve to the left of
a. $z = 0.74$ *0.7704*
b. $z = -2.16$ *0.0154*
c. $z = 1.02$ *0.8461*
d. $z = -0.15$ *0.4404*

32. Find the area under the standard normal curve to the left of
a. $z = 2.56$ *0.9948*
b. $z = 0.53$ *0.7019*
c. $z = -0.94$ *0.1736*
d. $z = -1.30$ *0.0968*

33. Find the area under the standard normal curve to the right of
a. $z = -1.55$ *0.9394*
b. $z = 0.32$ *0.3745*
c. $z = 3.20$ *0.0007*
d. $z = -2.39$ *0.9916*

34. Find the area under the standard normal curve to the right of
a. $z = 0.47$ *0.3192*
b. $z = -2.91$ *0.9982*
c. $z = 2.04$ *0.0207*
d. $z = 1.09$ *0.1379*

35. Find the area under the standard normal curve that lies between
a. $z = -0.75$ and $z = 1.70$ *0.7288*
b. $z = -2.30$ and $z = 1.08$ *0.8492*
c. $z = -3.27$ and $z = -1.44$ *0.0744*
d. $z = 1.26$ and $z = 2.32$ *0.0936*

36. Find the area under the standard normal curve that lies between
a. $z = -1.28$ and $z = 1.36$ *0.8128*
b. $z = -0.82$ and $z = -0.42$ *0.1311*
c. $z = 1.58$ and $z = 2.06$ *0.0374*
d. $z = -2.19$ and $z = 0.07$ *0.5136*

37. Find the area under the standard normal curve that lies outside the interval between
a. $z = -0.38$ and $z = 1.02$ *0.5059*
b. $z = -1.42$ and $z = 1.78$ *0.1153*
c. $z = 0.01$ and $z = 2.67$ *0.5078*
d. $z = -2.45$ and $z = -0.34$ *0.6402*

38. Find the area under the standard normal curve that lies outside the interval between
a. $z = -1.11$ and $z = 3.21$ *0.1342*
b. $z = -1.93$ and $z = 0.59$ *0.3044*
c. $z = 0.46$ and $z = 1.75$ *0.7173*
d. $z = -2.73$ and $z = -1.39$ *0.9209*

39. Find the z-score for which the area to its left is 0.54. *0.10*

40. Find the z-score for which the area to its left is 0.13. *−1.13*

41. Find the z-score for which the area to its left is 0.93. *1.48*

42. Find the z-score for which the area to its left is 0.25. *−0.67*

43. Find the z-score for which the area to its right is 0.84. *−0.99*

44. Find the z-score for which the area to its right is 0.14. *1.08*

45. Find the z-score for which the area to its right is 0.35. *0.39*

46. Find the z-score for which the area to its right is 0.92. *−1.41*

47. Find the z-scores that bound the middle 50% of the area under the standard normal curve. *−0.67, 0.67*

48. Find the z-scores that bound the middle 70% of the area under the standard normal curve. *−1.04, 1.04*

49. Find the z-scores that bound the middle 80% of the area under the standard normal curve. *−1.28, 1.28*

50. Find the z-scores that bound the middle 98% of the area under the standard normal curve. *−2.33, 2.33*

Working with the Concepts

51. Symmetry: The area under the standard normal curve to the left of $z = -1.75$ is 0.0401. What is the area to the right of $z = 1.75$? *0.0401*

52. Symmetry: The area under the standard normal curve to the right of $z = -0.51$ is 0.6950. What is the area to the left of $z = 0.51$? *0.6950*

53. Symmetry: The area under the standard normal curve between $z = -1.93$ and $z = 0.59$ is 0.6956. What is the area between $z = -0.59$ and $z = 1.93$? *0.6956*

54. Symmetry: The area under the standard normal curve between $z = 1.32$ and $z = 1.82$ is 0.0590. What is the area between $z = -1.82$ and $z = -1.32$? *0.0590*

Extending the Concepts

55. No table, no technology: Let a be the number such that the area to the right of $z = a$ is 0.3. Without using a table or technology, find the area to the left of $z = -a$. *0.3*

56. No table, no technology: Let a be the number such that the area to the right of $z = a$ is 0.21. Without using a table or technology, find the area between $z = -a$ and $z = a$. *0.58*

Answers to Check Your Understanding Exercises for Section 7.1

1. a. 0.63 **b.** 0.23 **c.** 0.86 **d.** 0.14

2. a. 0.3 **b.** 0.4 **c.** 0.5

3. 0.5987

4. 0.0104

5. 0.8491

6. 0.2570

7. −0.13

8. 0.33

9. 0.64

10. 1.34

Section 7.2 Applications of the Normal Distribution

Objectives

1. Convert values from a normal distribution to z-scores

2. Find areas under a normal curve

3. Find the value from a normal distribution corresponding to a given proportion

In Section 7.1, we found areas under a standard normal curve, which has mean 0 and standard deviation 1. In this section, we will learn to find areas under normal curves with any mean and standard deviation.

Converting Normal Values to z-Scores

Objective 1 Convert values from a normal distribution to z-scores

Let x be a value from a normal distribution with mean μ and standard deviation σ. We can convert x to a z-score by using a method known as **standardization**. To standardize a value, subtract the mean and divide by the standard deviation. This produces the z-score.

RECALL

We first described the method for finding the z-score in Section 3.3.

DEFINITION

Let x be a value from a normal distribution with mean μ and standard deviation σ. The **z-score** of x is

$$z = \frac{x - \mu}{\sigma}$$

The z-score satisfies the following properties.

NOTE TO INSTRUCTOR

Emphasize that z-scores are negative for values below the mean and positive for values above the mean.

Properties of the z-Score

1. The z-score follows a standard normal distribution.
2. Values below the mean have negative z-scores, and values above the mean have positive z-scores.
3. The z-score tells how many standard deviations the original value is above or below the mean.

Because the z-score follows a standard normal distribution, we can use the methods of Section 7.1 to find areas under any normal curve, by standardizing to convert the original values to z-scores.

Rounding Off z-Scores

The z-scores in Table A.2 are expressed to two decimal places. For this reason, when converting normal values to z-scores, we will round off the z-scores to two decimal places.

Example 7.13

Finding and interpreting a z-score

Heights in a certain population of women follow a normal distribution with mean $\mu = 64$ inches and standard deviation $\sigma = 3$ inches.

a. A randomly selected woman has a height of $x = 67$ inches. Find and interpret the z-score of this value.

b. Another randomly selected woman has a height of $x = 63$ inches. Find and interpret the z-score of this value.

Solution

EXPLAIN IT AGAIN

Converting x-values to z-scores: After we convert an x-value to a z-score, we use the standard normal curve. This allows us to find areas under the normal curve by using Table A.2.

a. The z-score for $x = 67$ is

$$z = \frac{67 - \mu}{\sigma} = \frac{67 - 64}{3} = 1.00$$

We interpret this by saying that a height of 67 inches is 1 standard deviation above the mean height of 64 inches.

b. The z-score for $x = 63$ is

$$z = \frac{63 - \mu}{\sigma} = \frac{63 - 64}{3} = -0.33$$

We interpret this by saying that a height of 63 inches is 0.33 standard deviation below the mean height of 64 inches.

Figure 7.18 illustrates the results of Example 7.13. Figure 7.18(a) is the normal curve that represents the heights of the population of women. It has a mean of 64. The heights of the

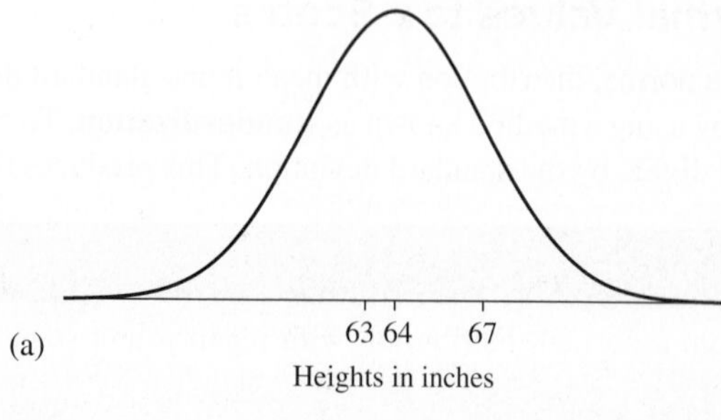

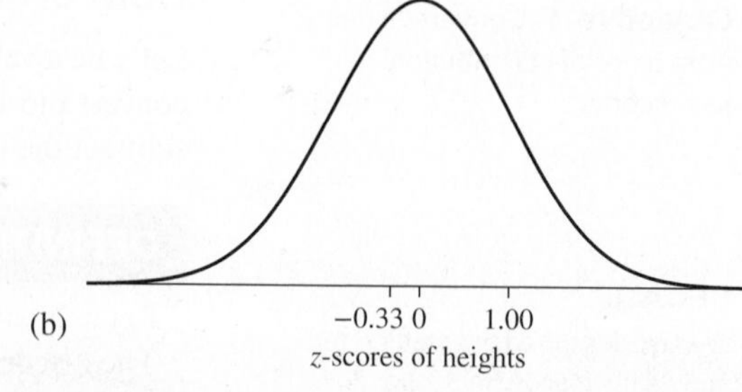

Figure 7.18 (a) This is the normal curve with mean 64 and standard deviation 3. It represents the population of heights of women. The heights of 63 and 67 are shown on the x-axis. (b) This is the standard normal curve. It also represents the population of heights of women, by using the z-scores instead of the actual heights. A height of 63 inches is represented by a z-score of −0.33, and a height of 67 inches is represented by a z-score of 1.00.

two women are indicated at 63 and 67. Figure 7.18(b) is the standard normal curve. The mean is 0, and the heights are represented by their z-scores of −0.33 and 1.00.

Check Your Understanding

1. A normal distribution has mean $\mu = 15$ and standard deviation $\sigma = 4$. Find and interpret the z-score for $x = 11$. $z = -1$
2. A normal distribution has mean $\mu = 60$ and standard deviation $\sigma = 20$. Find and interpret the z-score for $x = 75$. $z = 0.75$
3. Compact fluorescent bulbs are more energy efficient than incandescent bulbs, but they take longer to reach full brightness. The time that it takes for a compact fluorescent bulb to reach full brightness is normally distributed with mean 29.8 seconds and standard deviation 4.5 seconds. A randomly selected bulb takes 28 seconds to reach full brightness. Find and interpret the z-score for $x = 28$. $z = -0.4$

Answers are on page 334.

Objective 2 Find areas under a normal curve

Finding Areas Under a Normal Curve

In Section 7.1, we used z-scores to compute areas under the standard normal curve. By standardizing, we can use z-scores to compute areas under a normal curve with any mean and standard deviation. An area under a normal curve over an interval can be interpreted in two ways: It represents the proportion of the population that is contained within the interval, and it also represents the probability that a randomly selected individual will have a value within the interval.

Example 7.14

Finding an area under a normal curve

A study reported that the length of pregnancy from conception to birth is approximately normally distributed with mean $\mu = 272$ days and standard deviation $\sigma = 9$ days. What proportion of pregnancies last longer than 280 days?

Source: *Singapore Medical Journal* 35:1044–1048

EXPLAIN IT AGAIN

Probabilities and proportions: The probability that a randomly sampled value falls in a given interval is equal to the proportion of the population that is contained in the interval. So the area under a normal curve represents both probabilities and proportions.

Solution

The proportion of pregnancies lasting longer than 280 days is equal to the area under the normal curve corresponding to values of x greater than 280. We find this area as follows.

Step 1: Find the z-score for $x = 280$.

$$z = \frac{x - \mu}{\sigma} = \frac{280 - 272}{9} = 0.89$$

Step 2: Sketch a normal curve, label the mean, x-value, and z-score, and shade in the area to be found. See Figure 7.19.

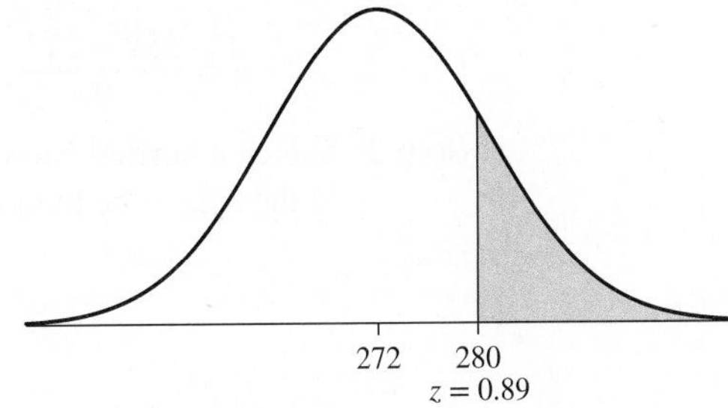

Figure 7.19

NOTE TO INSTRUCTOR

Emphasize that it is helpful to draw a sketch of the normal curve whenever finding a probability or a percentile.

Step 3: Find the area to the right of $z = 0.89$. Using Table A.2, we find the area to the *left* of $z = 0.89$ to be 0.8133. The area to the right is therefore $1 - 0.8133 = 0.1867$. We conclude that the proportion of pregnancies that last longer than 280 days is 0.1867.

Example 7.15

Finding an area under a normal curve by using technology

In Example 7.14, we used Table A.2 to compute the proportion of pregnancies that last longer than 280 days. Find this proportion by using technology.

Solution

We present output from the TI-84 Plus calculator. We use the **normalcdf** command. We enter the left endpoint of the interval (280). Since there is no right endpoint, we enter **1E99**, which represents the very large number that is written with a 1 followed by 99 zeros. Then we enter the mean (272) and the standard deviation (9). Step-by-step instructions are given in the Using Technology section on page 328.

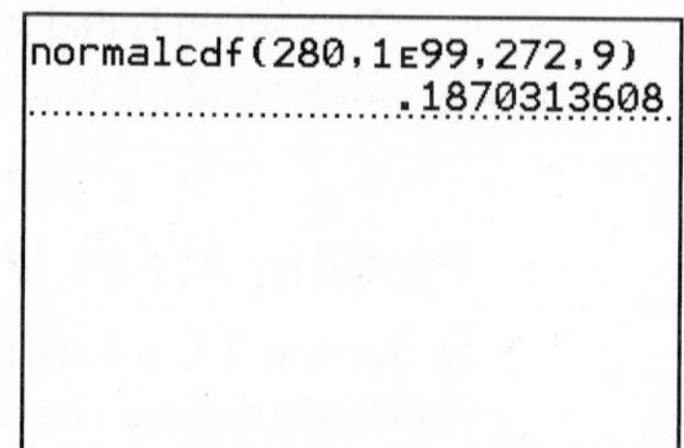

EXPLAIN IT AGAIN

Technology and tables can give slightly different answers: Answers obtained with technology sometimes differ slightly from those obtained by using tables, because the technology is more precise. The differences aren't large enough to matter.

In Example 7.14, we used Table A.2 and found the proportion of pregnancies that last longer than 280 days to be 0.1867. In Example 7.15, the TI-84 Plus calculator found the proportion to be 0.1870. Answers found with technology often differ somewhat from those obtained by using a table. The differences aren't large enough to matter. Whenever the answer obtained from technology differs from the answer obtained by using the table, we will present both answers.

Example 7.16

Finding an area under a normal curve between two values

The length of a pregnancy from conception to birth is approximately normally distributed with mean $\mu = 272$ days and standard deviation $\sigma = 9$ days. A pregnancy is considered full-term if it lasts between 252 days and 298 days. What proportion of pregnancies are full-term?

Solution

Step 1: Find the z-scores for $x = 252$ and $x = 298$.

For $x = 252$:

$$z = \frac{252 - 272}{9} = -2.22$$

For $x = 298$:

$$z = \frac{298 - 272}{9} = 2.89$$

Step 2: Sketch a normal curve; label the mean, the x-values, and the z-scores; and shade in the area to be found. See Figure 7.20.

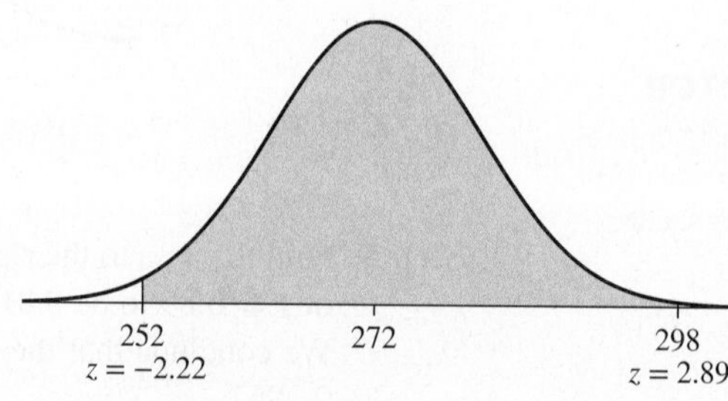

Figure 7.20

Step 3: Using Table A.2, we find that the area to the left of $z = 2.89$ is 0.9981 and the area to the left of $z = -2.22$ is 0.0132. The area between $z = -2.22$ and $z = 2.89$ is therefore $0.9981 - 0.0132 = 0.9849$.

We conclude that the proportion of pregnancies that are full-term is 0.9849.

Check Your Understanding

4. A normal population has mean $\mu = 3$ and standard deviation $\sigma = 1$. Find the proportion of the population that is less than 1. *0.0228*

5. A normal population has mean $\mu = 40$ and standard deviation $\sigma = 10$. Find the probability that a randomly sampled value is greater than 53. *0.0968*

6. A normal population has mean $\mu = 7$ and standard deviation $\sigma = 5$. Find the proportion of the population that is between -2 and 10. *0.6898*

7. A normal population has mean $\mu = 56$ and standard deviation $\sigma = 12$. Find the probability that a randomly sampled value is between 60 and 70. *0.2497 [Tech: 0.2478]*

Answers are on page 334.

Objective 3 Find the value from a normal distribution corresponding to a given proportion

Finding the Value from a Normal Distribution Corresponding to a Given Proportion

Sometimes we want to find the value from a normal distribution that has a given proportion of the population above or below it. The method for doing this is the reverse of the method for finding a proportion for a given value. In particular, we need to find the value from the distribution that has a given z-score.

Recall that the z-score tells how many standard deviations a value is above or below the mean. The value of x that corresponds to a given z-score is given by

$$x = \mu + z\sigma$$

Example 7.17

Finding the value from a normal distribution with a given z-score

Heights in a group of men are normally distributed with mean $\mu = 69$ inches and standard deviation $\sigma = 3$ inches.

a. Find the height whose z-score is 1. Interpret the result.

b. Find the height whose z-score is -2.0. Interpret the result.

c. Find the height whose z-score is 0.6. Interpret the result.

Solution

a. We want the height that is equal to the mean plus one standard deviation. Therefore, $x = \mu + z\sigma = 69 + (1)(3) = 72$. We interpret this by saying that a man 72 inches tall has a height one standard deviation above the mean.

b. We want the height that is equal to the mean minus two standard deviations. Therefore, $x = \mu + z\sigma = 69 + (-2)(3) = 63$. We interpret this by saying that a man 63 inches tall has a height two standard deviations below the mean.

c. We want the height that is equal to the mean plus 0.6 standard deviation. Therefore, $x = \mu + z\sigma = 69 + (0.6)(3) = 70.8$. We interpret this by saying that a man 70.8 inches tall has a height 0.6 standard deviation above the mean.

EXPLAIN IT AGAIN

$x = \mu + z\sigma$: The z-score tells how many standard deviations x is above or below the mean. Therefore, the value of x that corresponds to a given z-score is equal to the mean (μ) plus z times the standard deviation (σ).

To find the value from a normal distribution that has a given proportion above or below it, we can use either Table A.2 or technology. Following are the steps to find the value that has a given proportion above or below it by using Table A.2.

Finding a Normal Value That Has a Given Proportion Above or Below It by Using Table A.2

Step 1: Sketch a normal curve, label the mean, label the value x to be found, and shade in and label the given area.

Step 2: If the given area is on the right, subtract it from 1 to get the area on the left.

Step 3: Look in the body of Table A.2 to find the area closest to the given area. Find the z-score corresponding to that area.

Step 4: Obtain the value from the normal distribution by computing $x = \mu + z\sigma$.

Example 7.18

Finding a normal value corresponding to an area

Scores on an IQ test are normally distributed with mean $\mu = 100$ and standard deviation $\sigma = 15$. What score separates the upper 2% of IQ scores from the lower 98%?

Solution

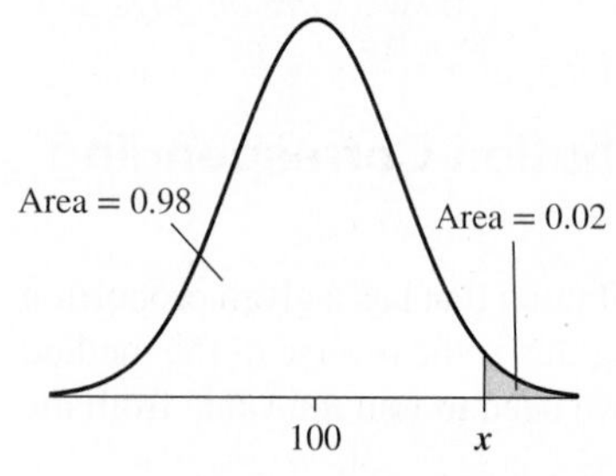

Figure 7.21

Step 1: Figure 7.21 presents a sketch of the normal curve, showing the value x separating the upper 2% from the lower 98%.

Step 2: The area 0.02 is on the right, so we subtract from 1 and work with the area 0.98 on the left.

Step 3: The area closest to 0.98 in Table A.2 is 0.9798, which corresponds to a z-score of 2.05.

Step 4: The IQ score that separates the upper 2% from the lower 98% is

$$x = \mu + z\sigma = 100 + (2.05)(15) = 130.75$$

Since IQ scores are generally whole numbers, we will round this to $x = 131$.

Example 7.19

Find a normal value corresponding to a central area

IQ scores are normally distributed with a mean of 100 and a standard deviation of 15. Find the IQ scores that separate the middle 50% of the scores from the top and bottom 25%.

Solution

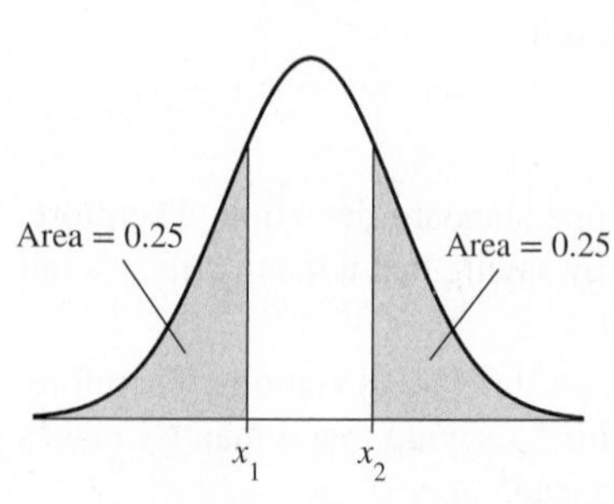

Figure 7.22

Step 1: Figure 7.22 presents a sketch of the normal curve, showing the values x_1 and x_2 separating the lower and upper 25% from the middle 50%.

Step 2: The value x_1 has area 0.25 to the left. The area closest to 0.25 in Table A.2 is 0.2514, which corresponds to a z-score of −0.67.

Step 3: The IQ score separating the lower 25% from the upper 25% is

$$x_1 = \mu + z\sigma = 100 + (-0.67)(15) = 89.95$$

Step 4: The value x_2 has area 0.25 to the right. We subtract from 1 and work with the area 0.75 on the left. The area closest to 0.75 in Table A.2 is 0.7486, which corresponds to a z-score of 0.67.

Step 5: The IQ score separating the upper 25% from the lower 75% is

$$x_1 = \mu + z\sigma = 100 + (0.67)(15) = 110.05$$

The IQ scores separating the middle 50% from the top and bottom 25% are 89.95 and 110.05. Since IQ scores are generally whole numbers, we will round this to 90 and 110.

Finding the normal value corresponding to a given area by using technology

To find the percentile of a normal distribution with technology, follow Steps 1 and 2 of the method for using the table. What is done after that depends on the technology being used. Example 7.20 illustrates the use of the TI-84 Plus calculator.

Example 7.20

Finding the normal value corresponding to an area by using technology

IQ scores are normally distributed with a mean of 100 and a standard deviation of 15. Use technology to find the 90th percentile of IQ scores; in other words, find the IQ score that separates the upper 10% from the lower 90%.

RECALL

The *p*th percentile of a population is the value that separates the lowest *p*% of the population from the highest (100 − *p*)%.

Solution

Step 1: Figure 7.23 presents a sketch of the normal curve, showing the value x separating the upper 10% from the lower 90%.

Step 2: We work with the area 0.90 on the left.

Step 3: For the TI-84 Plus calculator, use the **invNorm** command with area 0.90, mean 100, and standard deviation 15. Step-by-step instructions are given in the Using Technology section on page 329.

Figure 7.24 presents the results from the TI-84 Plus calculator. The IQ score corresponding to the top 10% is 119.

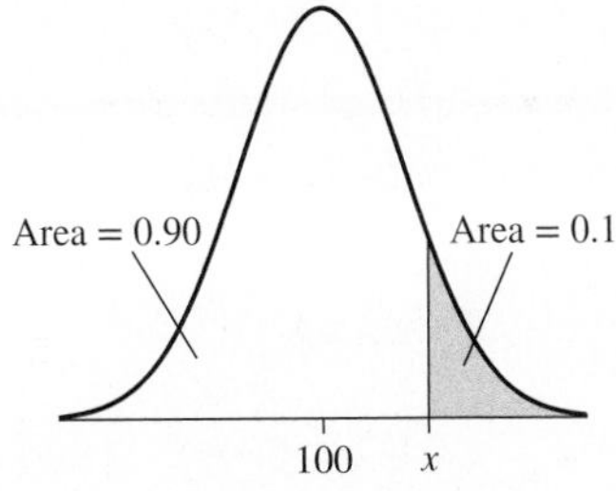

Figure 7.23

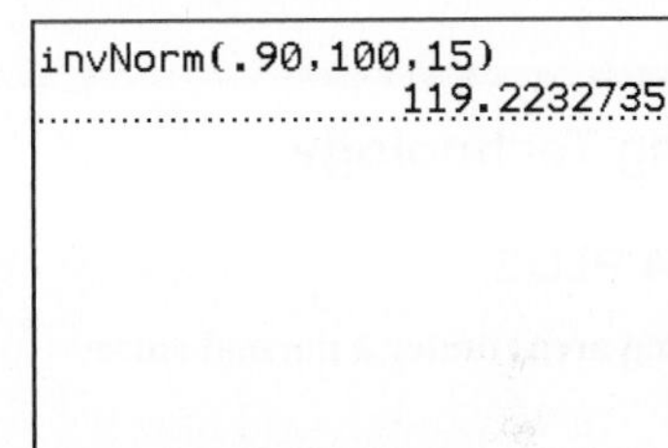

Figure 7.24

Example 7.21

Find a normal value corresponding to a central area by using technology

IQ scores are normally distributed with a mean of 100 and a standard deviation of 15. Find the IQ scores that separate the middle 90% of the scores from the top and bottom 5%.

Solution

Step 1: Figure 7.25 (page 328) presents a sketch of the normal curve, showing the values x_1 and x_2 separating the lower and upper 5% from the middle 90%.

Step 2: The value x_1 has area 0.05 to the left. For the TI-84 Plus calculator, use the **invNorm** command with area 0.05, mean 100, and standard deviation 15. The value x_2 has area 0.95 to the left. Use the **invNorm** command with area 0.95, mean 100, and standard deviation 15. Step by step instructions are given in the Using Technology section on page 328.

Step 3: Figure 7.26 (page 328) presents the results from the TI-84 Plus calculator. The IQ scores separating the middle 90% from the top and bottom 5% are 75 and 125.

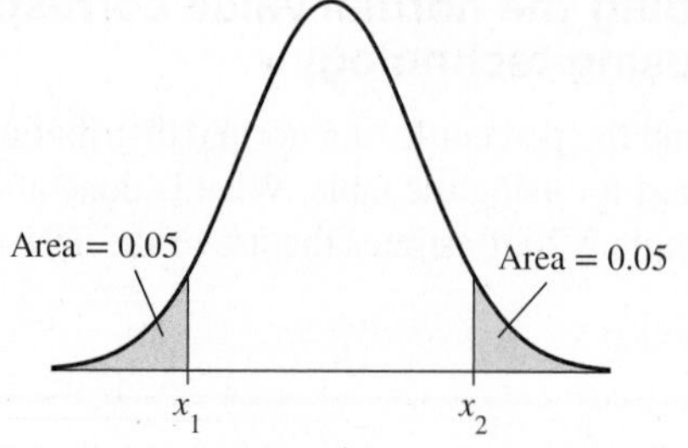

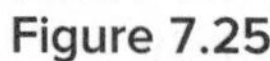

Figure 7.25

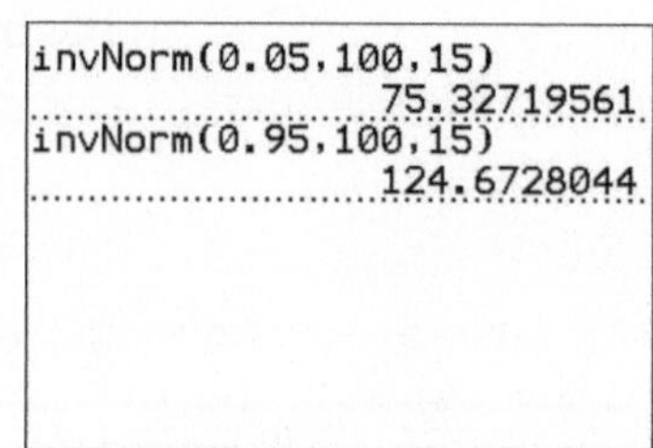

Figure 7.26

Check Your Understanding

8. A normally distributed population has mean $\mu = 6.9$ and standard deviation $\sigma = 2.6$. Find the value that has 80% of the population below it (in other words, the 80th percentile). *9.084 [Tech: 9.088]*

9. A normally distributed population has mean $\mu = 53$ and standard deviation $\sigma = 34$. Find the value that has 35% of the population above it. *66.26 [Tech: 66.10]*

10. A normally distributed population has mean $\mu = 27$ and standard deviation $\sigma = 12$. Find the values that separate the middle 60% of the population from the top and bottom 20%. *16.92, 37.08 [Tech: 16.90, 37.10]*

Answers are on page 334.

Using Technology

TI-84 PLUS

Finding areas under a normal curve

The **normalcdf** command is used to calculate area under a normal curve.

Step 1. Press **2nd**, then **VARS** to access the **DISTR** menu. Select **2:normalcdf** (Figure A).

Step 2. Enter the left endpoint, comma, the right endpoint, comma, the mean, comma, and the standard deviation.

- When finding the area to the right of a given value, use **1E99** as the right endpoint.
- When finding the area to the left of a given value, use **–1E99** as the left endpoint.

Step 3. Press **ENTER**.

Using the TI-84 PLUS Stat Wizards (see Appendix B for more information)

Step 1. Press **2nd**, then **VARS** to access the **DISTR** menu. Select **2:normalcdf** (Figure A).

Step 2. Enter the left endpoint in the **lower** field, the right endpoint in the **upper field**, the mean in the μ field, and the standard deviation in the σ field.

- When finding the area to the right of a given value, use **1E99** as the right endpoint.
- When finding the area to the left of a given value, use **–1E99** as the left endpoint.

Step 3. Select **Paste** and press **ENTER** to paste the command to the home screen. Press **ENTER** again to run the command.

Figure B illustrates finding the area to the right of $x = 280$ with $\mu = 272$ and $\sigma = 9$ (Example 7.14).

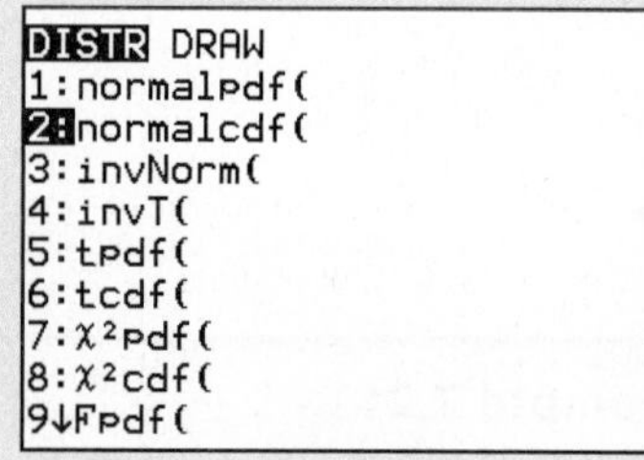

Figure A

```
normalcdf(280,1E99,272,9)
                .1870313608
```

Figure B

Finding a normal value corresponding to a given area

The **invNorm** command is used to calculate the z-score corresponding to an *area to the left.*

Step 1. Press **2nd**, then **VARS** to access the **DISTR** menu. Select **3:invNorm** (Figure C).

Step 2. Enter the area to the left of the desired normal value, comma, the mean, comma, and the standard deviation.

Step 3. Press **ENTER**.

Using the TI-84 PLUS Stat Wizards (see Appendix B for more information)

Step 1. Press **2nd**, then **VARS** to access the **DISTR** menu. Select **3:invNorm** (Figure C).

Step 2. Enter the area to the left of the desired z-score in the **area** field, the mean in the μ field, and the standard deviation in the σ field.

Step 3. Select **Paste** and press **ENTER** to paste the command to the home screen. Press **ENTER** again to run the command.

Figure D illustrates finding the normal value that has an area of 0.98 to its left, where $\mu = 100$ and $\sigma = 15$ (Example 7.18).

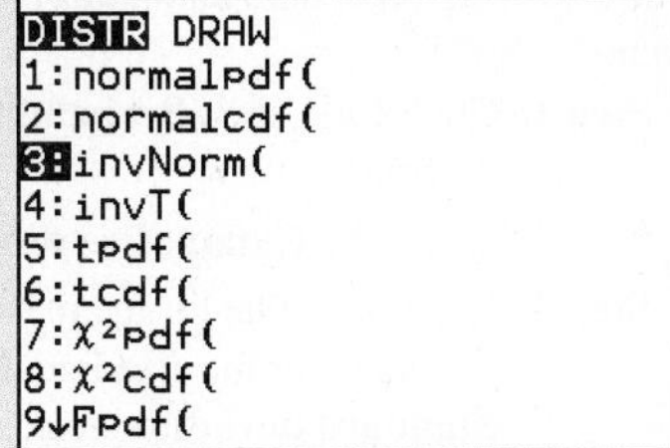

Figure C

```
invNorm(0.98,100,15)
                130.8062337
```

Figure D

EXCEL

Finding areas under a normal curve

The following procedure computes the area to the *left* of a given value.

Step 1. In an empty cell, select the **Insert Function** icon and highlight **Statistical** in the category field.

Step 2. Click on the **NORM.DIST** function and press **OK**.

Step 3. To compute the area to the left of a given x, enter the value of x in the **X** field.

Step 4. Enter the value for the mean in the **Mean** field and the value for the standard deviation in the **Standard deviation** field.

Step 5. Enter **TRUE** in the **Cumulative** field and click **OK**.

Figure E illustrates finding the area to the right of $x = 280$ with $\mu = 272$ and $\sigma = 9$ by subtracting the area on the left from 1 (Example 7.14).

```
=1-NORM.DIST(280,272,9,TRUE)
0.187031399
```

Figure E

Finding a normal value corresponding to a given area

The following procedure is used to calculate the normal value corresponding to an *area to the left*.

Step 1. In an empty cell, select the **Insert Function** icon and highlight **Statistical** in the category field.

Step 2. Click on the **NORM.INV** function and press **OK**.

Step 3. Enter the area to the left of the desired normal value in the **Probability** field.

Step 4. Enter the value for the mean in the **Mean** field and the value for the standard deviation in the **Standard deviation** field.

Step 5. Click **OK**.

Figure F illustrates finding the normal value that has an area of 0.98 to its left, where $\mu = 100$ and $\sigma = 15$ (Example 7.18).

```
=NORM.INV(0.98,100,15)
130.8062337
```

Figure F

MINITAB

Finding areas under a normal curve

The following procedure computes the area to the *left* of a given value.

Step 1. Click **Calc**, then **Probability Distributions**, then **Normal**.

Step 2. Select the **Cumulative probability** option.

Step 3. Enter the value for the mean in the **Mean** field and the value for the standard deviation in the **Standard deviation** field.

Step 4. To compute the area to the left of a given x, enter the value for x in the **Input constant** field.

Step 5. Click **OK**.

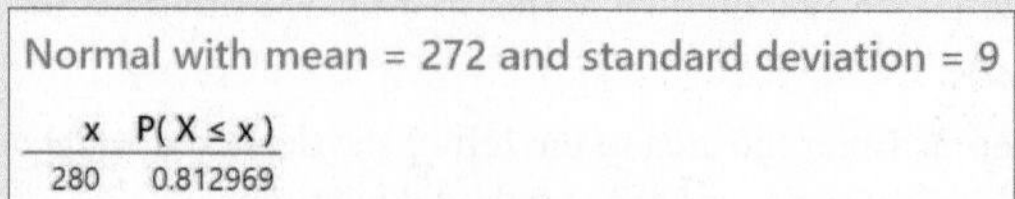

Normal with mean = 272 and standard deviation = 9

x	P(X ≤ x)
280	0.812969

Figure G

Figure G illustrates finding the area to the *left* of $x = 280$ with $\mu = 272$ and $\sigma = 9$. To find the area to the right of $x = 280$, subtract this result from 1 (Example 7.14).

Finding a normal value corresponding to a given area

The following procedure is used to calculate a normal value corresponding to an *area to the left.*

Step 1. Click **Calc**, then **Probability Distributions**, then **Normal**.

Step 2. Select the **Inverse Cumulative Probability** option.

Step 3. Enter the value for the mean in the **Mean** field and the value for the standard deviation in the **Standard deviation** field.

Step 4. Enter the area to the left of the desired normal value and click **OK**.

Normal with mean = 100 and standard deviation = 15

P(X ≤ x)	x
0.98	130.806

Figure H

Figure H illustrates finding the normal value that has an area of 0.98 to its left, where $\mu = 100$ and $\sigma = 15$ (Example 7.18).

Section 7.2 Exercises

Exercises 1–10 are the Check Your Understanding exercises located within the section.

Understanding the Concepts

In Exercises 11 and 12, fill in each blank with the appropriate word or phrase.

11. The process of converting a value x from a normal distribution to a z-score is known as __________. *standardization*

12. A value that is two standard deviations below the mean will have a z-score of __________. *−2*

In Exercises 13–18, determine whether the statement is true or false. If the statement is false, rewrite it as a true statement.

13. z-scores follow a standard normal distribution. *True*

14. A z-score indicates how many standard deviations a value is above or below the mean. *True*

15. If a normal population has a mean of μ and a standard deviation of σ, then the area to the left of μ is less than 0.5. *False*

16. If a normal population has a mean of μ and a standard deviation of σ, then the area to the right of $\mu + \sigma$ is less than 0.5. *True*

17. If a normal population has a mean of μ and a standard deviation of σ, then $P(X = \mu) = 1$. *False*

18. If a normal population has a mean of μ and a standard deviation of σ, then $P(X = \mu) = 0$. *True*

Practicing the Skills

19. A normal population has mean $\mu = 20$ and standard deviation $\sigma = 4$.

a. What proportion of the population is less than 18? *0.3085*

b. What is the probability that a randomly chosen value will be greater than 25? *0.1056*

20. A normal population has mean $\mu = 9$ and standard deviation $\sigma = 6$.
a. What proportion of the population is less than 20?
b. What is the probability that a randomly chosen value will be greater than 5? *0.7486 [Tech: 0.7475]*

21. A normal population has mean $\mu = 25$ and standard deviation $\sigma = 11$.
a. What proportion of the population is greater than 34? *0.2061 [Tech: 0.2066]*
b. What is the probability that a randomly chosen value will be less than 10? *0.0869 [Tech: 0.0863]*

22. A normal population has mean $\mu = 61$ and standard deviation $\sigma = 16$.
a. What proportion of the population is greater than 100?
b. What is the probability that a randomly chosen value will be less than 80? *0.8830 [Tech: 0.8825]*

23. A normal population has mean $\mu = 47$ and standard deviation $\sigma = 3$.
a. What proportion of the population is between 40 and 50?
b. What is the probability that a randomly chosen value will be between 50 and 55? *0.1549 [Tech: 0.1548]*

24. A normal population has mean $\mu = 35$ and standard deviation $\sigma = 8$.
a. What proportion of the population is between 20 and 30?
b. What is the probability that a randomly chosen value will be between 30 and 40? *0.4714 [Tech: 0.4680]*

25. A normal population has mean $\mu = 12$ and standard deviation $\sigma = 3$. What is the 40th percentile of the population? *11.25 [Tech: 11.24]*

26. A normal population has mean $\mu = 56$ and standard deviation $\sigma = 8$. What is the 85th percentile of the population? *64.32 [Tech: 64.29]*

27. A normal population has mean $\mu = 46$ and standard deviation $\sigma = 9$. What is the 19th percentile of the population? *38.08 [Tech: 38.10]*

28. A normal population has mean $\mu = 71$ and standard deviation $\sigma = 33$. What is the 91st percentile of the population? *115.22 [Tech: 115.24]*

Working with the Concepts

29. Check your blood pressure: In a recent study, the Centers for Disease Control and Prevention reported that diastolic blood pressures (in mmHg) of adult women in the United States are approximately normally distributed with mean 80.5 and standard deviation 9.9.
a. What proportion of women have blood pressures lower than 70? *0.1446 [Tech: 0.1444]*
b. What proportion of women have blood pressures between 75 and 90? *0.5438 [Tech: 0.5421]*
c. A diastolic blood pressure greater than 90 is classified as hypertension (high blood pressure). What proportion of women have hypertension? *0.1685 [Tech: 0.1686]*
d. Is it unusual for a woman to have a blood pressure lower than 65? *No*

30. Baby weights: According to a recent National Health Statistics Reports, the weight of male babies less than two months old in the United States is normally distributed with mean 11.5 pounds and standard deviation 2.7 pounds.
a. What proportion of babies weigh more than 13 pounds?
b. What proportion of babies weigh less than 15 pounds?
c. What proportion of babies weigh between 10 and 14 pounds? *0.5361 [Tech: 0.5335]*
d. Is it unusual for a baby to weigh more than 17 pounds? *Yes*

31. Check your blood pressure: The Centers for Disease Control and Prevention reported that diastolic blood pressures (in mmHg) of adult women in the United States are approximately normally distributed with mean 80.5 and standard deviation 9.9.
a. Find the 30th percentile of the blood pressures.
b. Find the 67th percentile of the blood pressures. *84.86*
c. Find the third quartile of the blood pressures.

32. Baby weights: The weight of male babies less than two months old in the United States is normally distributed with mean 11.5 pounds and standard deviation 2.7 pounds.
a. Find the 81st percentile of the baby weights.
b. Find the 10th percentile of the baby weights. *8.04*
c. Find the first quartile of the baby weights. *9.69 [Tech: 9.68]*

33. Vegan Thanksgiving: Tofurkey is a vegan turkey substitute, usually made from tofu. At a certain restaurant, the number of calories in a serving of tofurkey with wild mushroom stuffing and gravy is normally distributed with mean 482 and standard deviation 30.
a. What proportion of servings have more than 440 calories? *0.9192*
b. What proportion of servings have between 450 and 500 calories? *0.5834 [Tech: 0.5827]*
c. Is is unusual for a serving to have less than 420 calories? *Yes*

34. How many apps? According to the website www.appannie.com, the mean number of apps on a smartphone in the United States is 90. Assume the number of apps is normally distributed with mean 90 and standard deviation 25.
a. What proportion of phones have fewer than 110 apps? *0.7881*
b. What proportion of phones have between 70 and 80 apps? *0.1327*
c. Is it unusual for a phone to have more than 120 apps? *No*

35. Vegan Thanksgiving: Tofurkey is a vegan turkey substitute, usually made from tofu. At a certain restaurant, the number of calories in a serving of tofurkey with wild mushroom stuffing and gravy is normally distributed with mean 482 and standard deviation 30.
a. Find the 92nd percentile of the number of calories. *524.15*
b. Find the 30th percentile of the number of calories. *466.27*
c. Find the third quartile of the number of calories. *502.23*

36. How many apps? According to the website www.appannie.com, the mean number of apps on a smartphone in the United States is 90. Assume the number of apps is normally distributed with mean 90 and standard deviation 25.
a. Find the 15th percentile of the number of apps. *64.089*
b. Find the 87th percentile of the number of apps. *118.16*
c. Find the first quartile of the number of apps. *73.138*

37. Fish story: According to a report by the U.S. Fish and Wildlife Service, the mean length of six-year-old rainbow trout in the Arolik River in Alaska is 481 millimeters with a standard deviation of 41 millimeters. Assume these lengths are normally distributed.
a. What proportion of six-year-old rainbow trout are less than 450 millimeters long? *0.2236 [Tech: 0.2248]*
b. What proportion of six-year-old rainbow trout are between 400 and 500 millimeters long? *0.6533 [Tech: 0.6544]*

c. Is it unusual for a six-year-old rainbow trout to be less than 400 millimeters long? *Yes*

38. Big chickens: According to thepoultrysite.com, the weights of broilers (commercially raised chickens) are approximately normally distributed with mean 1387 grams and standard deviation 161 grams.

a. What proportion of broilers weigh between 1100 and 1200 grams? *0.0855 [Tech: 0.0854]*

b. What is the probability that a randomly selected broiler weighs more than 1500 grams? *0.2420 [Tech: 0.2414]*

c. Is it unusual for a broiler to weigh more than 1550 grams? *No*

39. Fish story: The U.S. Fish and Wildlife Service reported that the mean length of six-year-old rainbow trout in the Arolik River in Alaska is 481 millimeters with a standard deviation of 41 millimeters. Assume these lengths are normally distributed.

a. Find the 58th percentile of the lengths. *489.20 [Tech: 489.28]*

b. Find the 76th percentile of the lengths. *510.11 [Tech: 509.96]*

c. Find the first quartile of the lengths. *453.53 [Tech: 453.35]*

d. A size limit is to be put on trout that are caught. What should the size limit be so that 15% of six-year-old trout have lengths shorter than the limit? *438.36 [Tech: 438.51]*

40. Big chickens: A report on thepoultrysite.com stated that the weights of broilers (commercially raised chickens) are approximately normally distributed with mean 1387 grams and standard deviation 161 grams.

a. Find the 22nd percentile of the weights. *1263.03 [Tech: 1262.68]*

b. Find the 93rd percentile of the weights. *1625.28 [Tech: 1624.60]*

c. Find the first quartile of the weights. *1279.13 [Tech: 1278.41]*

d. A chicken farmer wants to provide a money-back guarantee that his broilers will weigh at least a certain amount. What weight should he guarantee so that he will have to give his customers' money back only 1% of the time? *1011.87 [Tech: 1012.46]*

41. Radon: Radon is a naturally occurring radioactive substance that is found in the ground underneath many homes. Radon detectors are often placed in homes to determine whether radon levels are high enough to be dangerous. A radon level less than 4.0 picocuries is considered safe. Because levels fluctuate randomly, the levels measured by detectors are not exactly correct, but are instead normally distributed. It is known from physical theory that when the true level is 4.1 picocuries, the measurement made by a detector over a one-hour period will be normally distributed with mean 4.1 picocuries and standard deviation 0.2 picocurie.

a. If the true level is 4.1, what is the probability that a one-hour measurement will be less than 4.0? *0.3085*

b. If the true level is 4.1, would it be unusual for a one-hour measurement to indicate that the level is safe? *No*

c. If a measurement is made for 24 hours, the mean will still be 4.1 picocuries, but the standard deviation will be only 0.04 picocurie. What is the probability that a 24-hour measurement will be below 4.0? *0.0062*

d. If the true level is 4.1, would it be unusual for a 24-hour measurement to indicate that the level is safe? *Yes*

42. Electric bills: According to the U.S. Energy Information Administration, the mean monthly household electric bill in the United States in a recent year was $110.14. Assume the amounts are normally distributed with standard deviation $20.00.

a. What proportion of bills are greater than $130?

b. What proportion of bills are between $85 and $140?

c. What is the probability that a randomly selected household had a monthly bill less than $120? *0.6879 [Tech: 0.6890]*

43. Radon: Assume that radon measurements are normally distributed with mean 4.1 picocuries and standard deviation of 0.2.

a. Find the 35th percentile of the measurements.

b. Find the 92nd percentile of the measurements.

c. Find the median of the measurements. *4.1*

44. Electric bills: The U.S. Energy Information Agency reported that the mean monthly household electric bill in the United States in a recent year was $110.14. Assume the amounts are normally distributed with standard deviation $20.00.

a. Find the 7th percentile of the bill amounts. *80.54 [Tech: 80.62]*

b. Find the 62nd percentile of the bill amounts.

c. Find the median of the bill amounts. *110.14*

45. Tire lifetimes: The lifetime of a certain type of automobile tire (in thousands of miles) is normally distributed with mean $\mu = 40$ and standard deviation $\sigma = 5$.

a. What is the probability that a randomly chosen tire has a lifetime greater than 48 thousand miles? *0.0548*

b. What proportion of tires have lifetimes between 38 and 43 thousand miles? *0.3811 [Tech: 0.3812]*

c. What proportion of tires have lifetimes less than 46 thousand miles? *0.8849*

46. Tree heights: Cherry trees in a certain orchard have heights that are normally distributed with mean $\mu = 112$ inches and standard deviation $\sigma = 14$ inches.

a. What proportion of trees are more than 120 inches tall? *0.2843 [Tech: 0.2839]*

b. What proportion of trees are less than 100 inches tall? *0.1949 [Tech: 0.1957]*

c. What is the probability that a randomly chosen tree is between 90 and 100 inches tall? *0.1367 [Tech: 0.1376]*

PhotoLink/Getty Images

47. Tire lifetimes: The lifetime of a certain type of automobile tire (in thousands of miles) is normally distributed with mean $\mu = 40$ and standard deviation $\sigma = 5$.

a. Find the 15th percentile of the tire lifetimes. *34.80 [Tech: 34.82]*

b. Find the 68th percentile of the tire lifetimes. *42.35 [Tech: 42.34]*

c. Find the first quartile of the tire lifetimes. *36.65 [Tech: 36.63]*

d. The tire company wants to guarantee that its tires will last at least a certain number of miles. What number of miles (in thousands) should the company guarantee so that only 2% of the tires violate the guarantee? *29.75 [Tech: 29.73]*

48. Tree heights: Cherry trees in a certain orchard have heights that are normally distributed with mean $\mu = 112$ inches and standard deviation $\sigma = 14$ inches.

a. Find the 27th percentile of the tree heights. *103.46 [Tech: 103.42]*

b. Find the 85th percentile of the tree heights. *126.56 [Tech: 126.51]*

c. Find the third quartile of the tree heights. *121.38 [Tech: 121.44]*

d. An agricultural scientist wants to study the tallest 1% of the trees to determine whether they have a certain gene that allows them to grow taller. To do this, she needs to study all the trees above a certain height. What height is this?

49. How much is in that can? The volume of beverage in a 12-ounce can is normally distributed with mean 12.05 ounces and standard deviation 0.02 ounce.

a. What is the probability that a randomly selected can will contain more than 12.06 ounces? *0.3085*

b. What is the probability that a randomly selected can will contain between 12 and 12.03 ounces? *0.1525 [Tech: 0.1524]*

c. Is it unusual for a can to be underfilled (contain less than 12 ounces)? *Yes*

50. How much do you study? A survey among freshmen at a certain university revealed that the number of hours spent studying the week before final exams was normally distributed with mean 25 and standard deviation 7.

a. What proportion of students studied more than 40 hours?

b. What is the probability that a randomly selected student spent between 15 and 30 hours studying? *0.6847 [Tech: 0.6859]*

c. What proportion of students studied less than 30 hours?

51. How much is in that can? The volume of beverage in a 12-ounce can is normally distributed with mean 12.05 ounces and standard deviation 0.02 ounce.

a. Find the 60th percentile of the volumes. *12.055*

b. Find the 4th percentile of the volumes. *12.015*

c. Between what two values are the middle 95% of the volumes? *12.011, 12.089*

52. How much do you study? A survey among freshmen at a certain university revealed that the number of hours spent studying the week before final exams was normally distributed with mean 25 and standard deviation 7.

a. Find the 98th percentile of the number of hours studying.

b. Find the 32nd percentile of the number of hours studying.

c. Between what two values are the middle 80% of the hours spent studying? *16.04 and 33.96 [Tech: 16.03 and 33.97]*

53. Precision manufacturing: A process manufactures ball bearings with diameters that are normally distributed with mean 25.1 millimeters and standard deviation 0.08 millimeter.

a. What proportion of the diameters are less than 25.0 millimeters? *0.1056*

b. What proportion of the diameters are greater than 25.4 millimeters? *0.0001*

c. To meet a certain specification, a ball bearing must have a diameter between 25.0 and 25.3 millimeters. What proportion of the ball bearings meet the specification? *0.8882 [Tech: 0.8881]*

54. Exam grades: Scores on a statistics final in a large class were normally distributed with a mean of 75 and a standard deviation of 8.

a. What proportion of the scores were above 90?

b. What proportion of the scores were below 65? *0.1056*

c. What is the probability that a randomly chosen score is between 70 and 80? *0.4714 [Tech: 0.4680]*

55. Precision manufacturing: A process manufactures ball bearings with diameters that are normally distributed with mean 25.1 millimeters and standard deviation 0.08 millimeter.

a. Find the 60th percentile of the diameters.

b. Find the 32nd percentile of the diameters.

c. A hole is to be designed so that 1% of the ball bearings will fit through it. The bearings that fit through the hole will be melted down and remade. What should the diameter of the hole be? *24.9136 [Tech: 24.9138]*

d. Between what two values are the middle 50% of the diameters? *25.0464 and 25.1536 [Tech: 25.0460 and 25.1540]*

56. Exam grades: Scores on a statistics final in a large class were normally distributed with a mean of 75 and a standard deviation of 8.

a. Find the 40th percentile of the scores. *73.00 [Tech: 72.97]*

b. Find the 65th percentile of the scores. *78.12 [Tech: 78.08]*

c. The instructor wants to give an A to the students whose scores were in the top 10% of the class. What is the minimum score needed to get an A? *85.24 [Tech: 85.25]*

d. Between what two values are the middle 60% of the scores? *68.28 and 81.72 [Tech: 68.27 and 81.73]*

Extending the Concepts

57. Tall men: Heights of men in a certain city are normally distributed with mean 70 inches. Sixteen percent of the men are more than 73 inches tall. What percentage of the men are between 67 and 70 inches tall? *34%*

58. Watch your speed: Speeds of automobiles on a certain stretch of freeway at 11:00 P.M. are normally distributed with mean 65 mph. Twenty percent of the cars are traveling at speeds between 55 and 65 mph. What percentage of the cars are going faster than 75 mph? *30%*

59. Contaminated wells: A study reported that the mean concentration of ammonium in water wells in the state of Iowa was 0.71 milligram per liter, and the standard deviation was 1.09 milligrams per liter. Is it possible to determine whether these concentrations are approximately normally distributed? If so, say whether they are normally distributed, and explain how you know. If not, describe the additional information you would need to determine whether they are normally distributed. *Not normal*

Source: *Water Environment Research* 74:177–186

60. Heights: According to the National Health Statistics Reports, heights of adult women in the United States are normally distributed with mean 64 inches and standard deviation 4 inches. If three women are selected at random, what is the probability that at least one of them is more than 68 inches tall? *0.4045 [Tech: 0.4044]*

61. Exam scores: Scores on an exam were normally distributed. Ten percent of the scores were below 64 and 80% were below 81. Find the mean and standard deviation of the scores. *Mean: 74.26, SD: 8.02 [Tech: 8.01]*

62. Commute to work: Megan drives to work each morning. Her commute time is normally distributed with mean 30 minutes and standard deviation 4 minutes. Her workday begins at 9:00 A.M. At what time should she leave for work so that the probability she is on time is 95%? *8:23 A.M.*

Answers to Check Your Understanding Exercises for Section 7.2

1. $z = -1$. Interpretation: A value of 11 is one standard deviation below the mean.
2. $z = 0.75$. Interpretation: A value of 75 is 0.75 standard deviation above the mean.
3. $z = -0.4$. Interpretation: The length of time for this bulb is 0.4 standard deviation below the mean.
4. 0.0228
5. 0.0968
6. 0.6898
7. 0.2497 [Tech: 0.2478]
8. 9.084 [Tech: 9.088]
9. 66.26 [Tech: 66.10]
10. 16.92, 37.08 [Tech: 16.90, 37.10]

Section 7.3 Sampling Distributions and the Central Limit Theorem

Objectives

1. Construct the sampling distribution of a sample mean
2. Use the Central Limit Theorem to compute probabilities involving sample means

In Section 7.2, we learned to compute probabilities for a randomly sampled individual from a normal population. In practice, statistical studies involve sampling several, perhaps many, individuals. As discussed in Chapter 3, we often compute numerical summaries of samples, and the most commonly used summary is the sample mean $\bar{x}$.

NOTE TO INSTRUCTOR

Remind students that until the sample is selected, $\bar{x}$ is a random variable.

If several samples are drawn from a population, they are likely to have different values for $\bar{x}$. Because the value of $\bar{x}$ varies each time a sample is drawn, $\bar{x}$ is a random variable, and it has a probability distribution. The probability distribution of $\bar{x}$ is called the *sampling distribution* of $\bar{x}$.

Objective 1 Construct the sampling distribution of a sample mean

An Example of a Sampling Distribution

Tetrahedral dice are four-sided dice, used in role-playing games such as Dungeons & Dragons. They are shaped like a pyramid, with four triangular faces. Each face corresponds to a number between 1 and 4, so that when you toss a tetrahedral die, it comes up with one of the numbers 1, 2, 3, or 4. Tossing a tetrahedral die is like sampling a value from the population

Mark Steinmetz/McGraw Hill Education

1	2	3	4

The population mean, variance, and standard deviation are:

$$\text{Population mean: } \mu = \frac{1+2+3+4}{4} = 2.5$$

$$\text{Population variance: } \sigma^2 = \frac{(1-2.5)^2 + (2-2.5)^2 + (3-2.5)^2 + (4-2.5)^2}{4} = 1.25$$

$$\text{Population standard deviation: } \sigma = \sqrt{\sigma^2} = \sqrt{1.25} = 1.118$$

Now imagine tossing a tetrahedral die three times. The sequence of three numbers that is observed is a sample of size 3 drawn with replacement from the population just described. There are 64 possible samples, and they are all equally likely. Table 7.2 (page 335) lists them and provides the value of the sample mean $\bar{x}$ for each.

The columns labeled "$\bar{x}$" contain the values of the sample mean for each of the 64 possible samples. Some of these values appear more than once, because several samples have the same mean. The mean of the sampling distribution is the average of these 64 values. The standard deviation of the sampling distribution is the population standard

Table 7.2 The 64 Possible Samples of Size 3 and Their Sample Means

Sample	$\bar{x}$	Sample	$\bar{x}$	Sample	$\bar{x}$	Sample	$\bar{x}$
1, 1, 1	1.00	2, 1, 1	1.33	3, 1, 1	1.67	4, 1, 1	2.00
1, 1, 2	1.33	2, 1, 2	1.67	3, 1, 2	2.00	4, 1, 2	2.33
1, 1, 3	1.67	2, 1, 3	2.00	3, 1, 3	2.33	4, 1, 3	2.67
1, 1, 4	2.00	2, 1, 4	2.33	3, 1, 4	2.67	4, 1, 4	3.00
1, 2, 1	1.33	2, 2, 1	1.67	3, 2, 1	2.00	4, 2, 1	2.33
1, 2, 2	1.67	2, 2, 2	2.00	3, 2, 2	2.33	4, 2, 2	2.67
1, 2, 3	2.00	2, 2, 3	2.33	3, 2, 3	2.67	4, 2, 3	3.00
1, 2, 4	2.33	2, 2, 4	2.67	3, 2, 4	3.00	4, 2, 4	3.33
1, 3, 1	1.67	2, 3, 1	2.00	3, 3, 1	2.33	4, 3, 1	2.67
1, 3, 2	2.00	2, 3, 2	2.33	3, 3, 2	2.67	4, 3, 2	3.00
1, 3, 3	2.33	2, 3, 3	2.67	3, 3, 3	3.00	4, 3, 3	3.33
1, 3, 4	2.67	2, 3, 4	3.00	3, 3, 4	3.33	4, 3, 4	3.67
1, 4, 1	2.00	2, 4, 1	2.33	3, 4, 1	2.67	4, 4, 1	3.00
1, 4, 2	2.33	2, 4, 2	2.67	3, 4, 2	3.00	4, 4, 2	3.33
1, 4, 3	2.67	2, 4, 3	3.00	3, 4, 3	3.33	4, 4, 3	3.67
1, 4, 4	3.00	2, 4, 4	3.33	3, 4, 4	3.67	4, 4, 4	4.00

NOTATION ROUNDUP

$\mu_{\bar{x}}$ denotes the mean of the sampling distribution of $\bar{x}$.

$\sigma_{\bar{x}}$ denotes the standard deviation of the sampling distribution of $\bar{x}$.

deviation of the 64 sample means, which can be computed by the method presented in Section 3.2. The mean and standard deviation are

$$\text{Mean: } \mu_{\bar{x}} = 2.5 \qquad \text{Standard deviation: } \sigma_{\bar{x}} = 0.6455$$

Comparing the mean and standard deviation of the sampling distribution to the population mean and standard deviation, we see that the mean $\mu_{\bar{x}}$ of the sampling distribution is equal to the population mean μ. The standard deviation $\sigma_{\bar{x}}$ of the sampling distribution is 0.6455, which is less than the population standard deviation $\sigma = 1.118$. It is not immediately obvious how these two quantities are related. Note, however, that

$$\sigma_{\bar{x}} = 0.6455 = \frac{1.118}{\sqrt{3}} = \frac{\sigma}{\sqrt{3}}$$

The sample size is $n = 3$, so $\sigma_{\bar{x}} = \dfrac{\sigma}{\sqrt{n}}$.

These relationships hold in general. Note that the standard deviation $\sigma_{\bar{x}}$ is sometimes called the **standard error** of the mean.

NOTE TO INSTRUCTOR

Emphasize that the sample mean has less variation than do values from the original population.

SUMMARY

Let $\bar{x}$ be the mean of a simple random sample of size n, drawn from a population with mean μ and standard deviation σ.

The mean of the sampling distribution is $\mu_{\bar{x}} = \mu$.

The standard deviation of the sampling distribution is $\sigma_{\bar{x}} = \dfrac{\sigma}{\sqrt{n}}$.

The standard deviation $\sigma_{\bar{x}}$ is sometimes called the standard error of the mean.

Example 7.22

Find the mean and standard deviation of a sampling distribution

Among students at a certain college, the mean number of hours of television watched per week is $\mu = 10.5$, and the standard deviation is $\sigma = 3.6$. A simple random sample of 16 students is chosen for a study of viewing habits. Let $\bar{x}$ be the mean number of hours of TV watched by the sampled students. Find the mean $\mu_{\bar{x}}$ and the standard deviation $\sigma_{\bar{x}}$ of $\bar{x}$.

fstop123/Getty Images

Solution

The mean of $\bar{x}$ is

$$\mu_{\bar{x}} = \mu = 10.5$$

The sample size is $n = 16$. Therefore, the standard deviation of $\bar{x}$ is

$$\sigma_{\bar{x}} = \frac{\sigma}{\sqrt{n}} = \frac{3.6}{\sqrt{16}} = 0.9$$

It makes sense that the standard deviation of $\bar{x}$ is less than the population standard deviation σ. In a sample, it is unusual to get all large values or all small values. Samples usually contain both large and small values that cancel each other out when the sample mean is computed. For this reason, the distribution of $\bar{x}$ is less spread out than the population distribution. Therefore, the standard deviation of $\bar{x}$ is less than the population standard deviation.

Check Your Understanding

1. A population has mean $\mu = 6$ and standard deviation $\sigma = 4$. Find $\mu_{\bar{x}}$ and $\sigma_{\bar{x}}$ for samples of size $n = 25$. $\mu_{\bar{x}} = 6;\ \sigma_{\bar{x}} = 0.8$
2. A population has mean $\mu = 17$ and standard deviation $\sigma = 20$. Find $\mu_{\bar{x}}$ and $\sigma_{\bar{x}}$ for samples of size $n = 100$. $\mu_{\bar{x}} = 17;\ \sigma_{\bar{x}} = 2.0$

Answers are on page 342.

The probability histogram for the sampling distribution of $\bar{x}$

Consider again the example of the tetrahedral die. Let us compare the probability distribution for the population and the sampling distribution. The population consists of the numbers 1, 2, 3, and 4, each of which is equally likely. The sampling distribution for $\bar{x}$ can be determined from Table 7.2. The probability that the sample mean is 1.00 is 1/64, because out of the 64 possible samples, only one has a sample mean equal to 1.00. Similarly, the probability that $\bar{x} = 1.33$ is 3/64, because there are three samples out of 64 whose sample mean is 1.33. Figure 7.27 presents the probability histogram of the population, and Figure 7.28 presents the sampling distribution for $\bar{x}$.

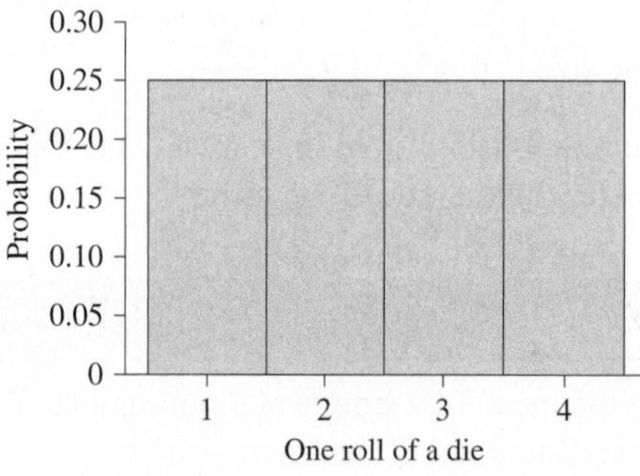

Figure 7.27 Probability histogram for the population

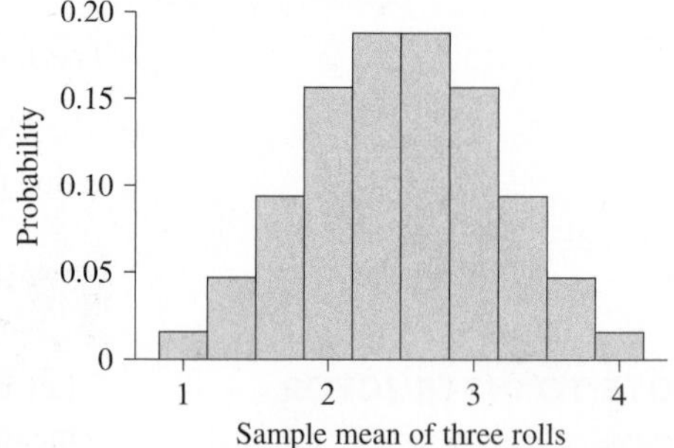

Figure 7.28 Probability histogram for the sampling distribution of $\bar{x}$ for samples of size 3

Note that the probability histogram for the sampling distribution looks a lot like the normal curve, whereas the probability histogram for the population does not. Remarkably, it is true that, for any population, if the sample size is large enough, the sample mean $\bar{x}$ will be approximately normally distributed. For a symmetric population like the one in Figure 7.27, the sample mean is approximately normally distributed even for a small sample size like $n = 3$.

In fact, when a population is normal, the sample mean will also be normal.

For a normal population, the sample mean will be normal for any sample size.

For a skewed population, the sample size must be large for the sample mean to be approximately normal.

Computing the sampling distribution of $\bar{x}$ for a skewed population

For a certain make of car, the number of repairs needed while under warranty has the following probability distribution.

x	$P(x)$
0	0.60
1	0.25
2	0.10
3	0.03
4	0.02

Figure 7.29 presents the probability histogram for this distribution, along with probability histograms for the sampling distribution of $\bar{x}$ for samples of size 3, 10, and 30. The probability histograms for the sampling distributions were created by programming a computer to compute the probability for every possible value of $\bar{x}$.

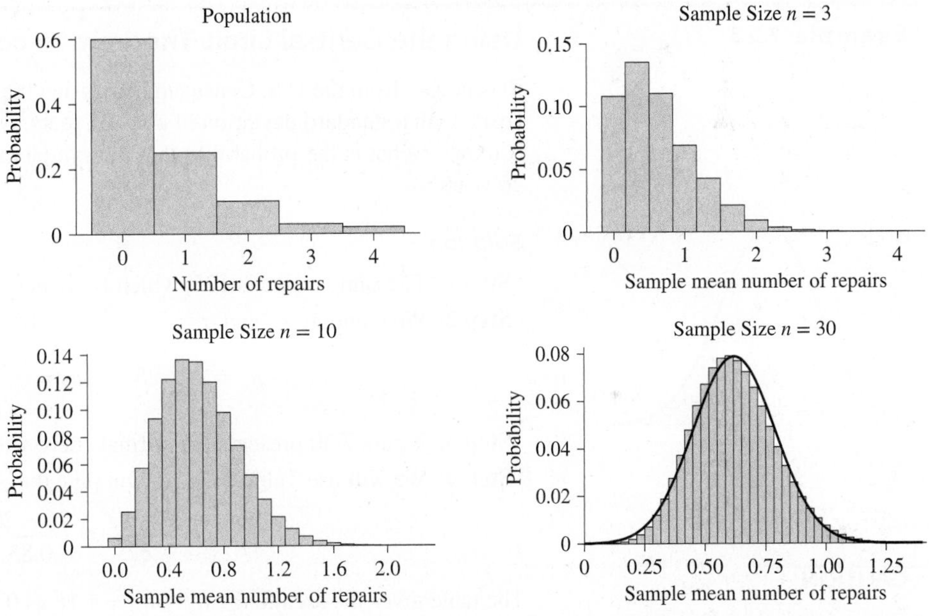

Figure 7.29 The probability histogram for the population distribution is highly skewed. As the sample size increases, the skewness decreases. For a sample size of 30, the probability histogram of the sample mean $\bar{x}$ is reasonably well approximated by a normal curve.

The remarkable fact that the sampling distribution of $\bar{x}$ is approximately normal for a large sample from any distribution is called the **Central Limit Theorem**. The size of the sample needed to obtain approximate normality depends mostly on the skewness of the population. A sample of size $n > 30$ is large enough for most populations encountered in practice. Smaller sample sizes are adequate for distributions that are nearly symmetric.

The Central Limit Theorem

Let $\bar{x}$ be the mean of a large ($n > 30$) simple random sample from a population with mean μ and standard deviation σ.

Then $\bar{x}$ has an approximately normal distribution, with mean $\mu_{\bar{x}} = \mu$ and standard deviation $\sigma_{\bar{x}} = \dfrac{\sigma}{\sqrt{n}}$.

The Central Limit Theorem is the most important result in statistics and forms the basis for much of the work that statisticians do.

Objective 2 Use the Central Limit Theorem to compute probabilities involving sample means

Computing Probabilities with the Central Limit Theorem

To compute probabilities involving a sample mean $\bar{x}$, use the following procedure:

EXPLAIN IT AGAIN

Exactly normal populations: When a population is exactly normal, the sample mean will be normally distributed even if $n < 30$. In practice, however, populations are almost never exactly normal.

Procedure for Computing Probabilities with the Central Limit Theorem

Step 1: Be sure the sample size is greater than 30. If so, it is appropriate to use the normal curve.
Step 2: Find the mean $\mu_{\bar{x}}$ and standard deviation $\sigma_{\bar{x}}$.
Step 3: Sketch a normal curve and shade in the area to be found.
Step 4: Find the area using Table A.2 or technology.

Example 7.23

Using the Central Limit Theorem to compute a probability

Recent data from the U.S. Census indicates that the mean age of college students is $\mu = 25$ years, with a standard deviation of $\sigma = 9.5$ years. A simple random sample of 125 students is drawn. What is the probability that the sample mean age of the students is greater than 26 years?

Solution

Step 1: The sample size is 125, which is greater than 30. We may use the normal curve.
Step 2: We compute $\mu_{\bar{x}}$ and $\sigma_{\bar{x}}$.

$$\mu_{\bar{x}} = \mu = 25 \qquad \sigma_{\bar{x}} = \frac{\sigma}{\sqrt{n}} = \frac{9.5}{\sqrt{125}} = 0.85$$

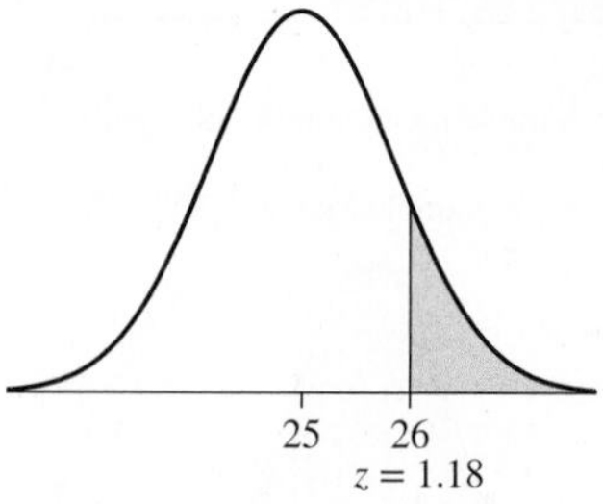

25 26
$z = 1.18$

Figure 7.30

Step 3: Figure 7.30 presents the normal curve with the area of interest shaded.
Step 4: We will use Table A.2. We compute the z-score for 26.

$$z = \frac{x - \mu_{\bar{x}}}{\sigma_{\bar{x}}} = \frac{26 - 25}{0.85} = 1.18$$

The table gives the area to the *left* of $z = 1.18$ as 0.8810. The area to the right of $z = 1.18$ is $1 - 0.8810 = 0.1190$. The probability that the sample mean age of the students is greater than 26 years is 0.1190.

CAUTION

When computing the z-score for the distribution of $\bar{x}$, be sure to use the standard deviation $\sigma_{\bar{x}}$, rather than σ.

Example 7.24

Use the Central Limit Theorem to find a percentile

The mean age of college students is $\mu = 25$ years, with a standard deviation of $\sigma = 9.5$ years. A simple random sample of 125 students is drawn. Find the 30th percentile of the sample mean $\bar{x}$.

Solution

Step 1: The sample size is 125, which is greater than 30. We may use the normal curve.
Step 2: We compute $\mu_{\bar{x}}$ and $\sigma_{\bar{x}}$.

$$\mu_{\bar{x}} = \mu = 25 \qquad \sigma_{\bar{x}} = \frac{\sigma}{\sqrt{n}} = \frac{9.5}{\sqrt{125}} = 0.8497$$

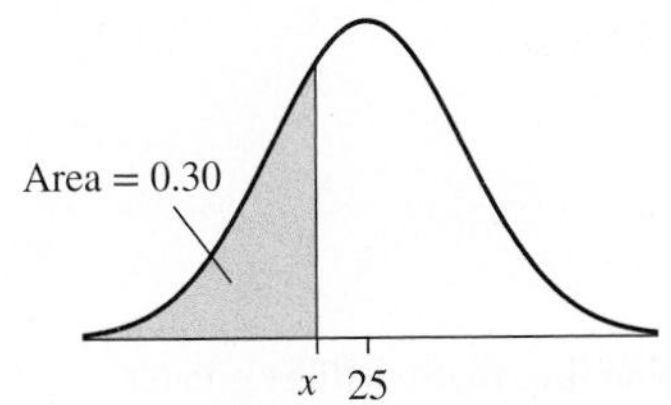

Figure 7.31

Step 3: The 30th percentile is the value that has area 0.30 to its left. Figure 7.31 presents the normal curve, showing the value x as the 30th percentile.

Step 4: The area in Table A.2 closest to 0.30 is 0.3015, which corresponds to a z-score of −0.52.

Step 5: The 30th percentile of $\bar{x}$ is

$$x = \mu_{\bar{x}} + z\sigma_{\bar{x}} = 25 + (-0.52)(0.8497) = 24.67$$

The 30th percentile of $\bar{x}$ is 24.55.

Example 7.25

Using the Central Limit Theorem to determine whether a given value of $\bar{x}$ is unusual

Hereford cattle are one of the most popular breeds of beef cattle. Based on data from the Hereford Cattle Society, the mean weight of a one-year-old Hereford bull is 1135 pounds, with a standard deviation of 97 pounds. Would it be unusual for the mean weight of 100 head of cattle to be less than 1115 pounds?

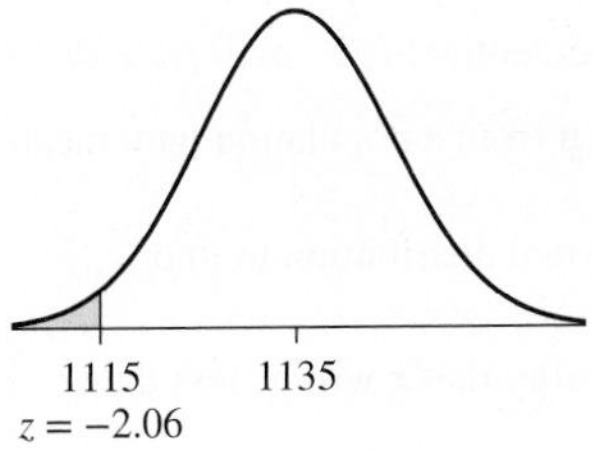

1115 1135
$z = -2.06$

Figure 7.32

Solution

We will compute the probability that the sample mean is less than 1115. We will say that this event is unusual if its probability is less than 0.05.

Step 1: The sample size is 100, which is greater than 30. We may use the normal curve.

Step 2: We compute $\mu_{\bar{x}}$ and $\sigma_{\bar{x}}$.

$$\mu_{\bar{x}} = \mu = 1135 \qquad \sigma_{\bar{x}} = \frac{\sigma}{\sqrt{n}} = \frac{97}{\sqrt{100}} = 9.7$$

Step 3: Figure 7.32 presents the normal curve. We are interested in the area to the left of 1115.

Step 4: We will use Table A.2. We compute the z-score for 1115.

$$z = \frac{x - \mu_{\bar{x}}}{\sigma_{\bar{x}}} = \frac{1115 - 1135}{9.7} = -2.06$$

The area to the left of $z = -2.06$ is 0.0197. The probability that the sample mean weight is less than 1115 is 0.0197. This probability is less than 0.05, so it would be unusual for the sample mean to be less than 1115.

NOTE TO INSTRUCTOR

Using the Central Limit Theorem to determine whether a value is unusual foreshadows the process used in hypothesis testing.

Check Your Understanding

3. A population has mean $\mu = 10$ and standard deviation $\sigma = 8$. A sample of size 50 is drawn.

a. Find the probability that $\bar{x}$ is greater than 11. *0.1894 [Tech: 0.1884]*

b. Would it be unusual for $\bar{x}$ to be less than 8? Explain. *Yes*

4. A population has mean $\mu = 47.5$ and standard deviation $\sigma = 12.6$. A sample of size 112 is drawn.

a. Find the probability that $\bar{x}$ is between 45 and 48. *0.6449*

b. Would it be unusual for $\bar{x}$ to be greater than 48? Explain. *No*

Answers are on page 342.

Section 7.3 Exercises

Exercises 1–4 are the Check Your Understanding exercises located within the section.

Understanding the Concepts

In Exercises 5 and 6, fill in each blank with the appropriate word or phrase.

5. The probability distribution of $\bar{x}$ is called a ________ distribution. *sampling*

6. The ________ states that the sampling distribution of $\bar{x}$ is approximately normal when the sample is large. *Central Limit Theorem*

In Exercises 7 and 8, determine whether the statement is true or false. If the statement is false, rewrite it as a true statement.

7. If $\bar{x}$ is the mean of a large ($n > 30$) simple random sample from a population with mean μ and standard deviation σ, then $\bar{x}$ is approximately normal with $\sigma_{\bar{x}} = \dfrac{\sigma}{\sqrt{n}}$. *True*

8. As the sample size increases, the sampling distribution of $\bar{x}$ becomes more and more skewed. *False*

Practicing the Skills

9. A sample of size 75 will be drawn from a population with mean 10 and standard deviation 12.
   a. Find the probability that $\bar{x}$ will be between 8 and 14.
   b. Find the 15th percentile of $\bar{x}$. *8.56*

10. A sample of size 126 will be drawn from a population with mean 26 and standard deviation 3.
   a. Find the probability that $\bar{x}$ will be between 25 and 27. *0.9998*
   b. Find the 55th percentile of $\bar{x}$. *26.03*

11. A sample of size 68 will be drawn from a population with mean 92 and standard deviation 24.
   a. Find the probability that $\bar{x}$ will be greater than 90.
   b. Find the 90th percentile of $\bar{x}$. *95.73*

12. A sample of size 284 will be drawn from a population with mean 45 and standard deviation 7.
   a. Find the probability that $\bar{x}$ will be greater than 46. *0.0080*
   b. Find the 75th percentile of $\bar{x}$. *45.28*

13. A sample of size 91 will be drawn from a population with mean 33 and standard deviation 17.
   a. Find the probability that $\bar{x}$ will be less than 30.
   b. Find the 25th percentile of $\bar{x}$. *31.81 [Tech: 31.80]*

14. A sample of size 82 will be drawn from a population with mean 24 and standard deviation 9.
   a. Find the probability that $\bar{x}$ will be less than 26.
   b. Find the 10th percentile of $\bar{x}$. *22.73*

15. A sample of size 20 will be drawn from a population with mean 6 and standard deviation 3.
   a. Is it appropriate to use the normal distribution to find probabilities for $\bar{x}$? *No*
   b. If appropriate, find the probability that $\bar{x}$ will be greater than 4. *Not appropriate*
   c. If appropriate, find the 30th percentile of $\bar{x}$. *Not appropriate*

16. A sample of size 42 will be drawn from a population with mean 52 and standard deviation 9.
   a. Is it appropriate to use the normal distribution to find probabilities for $\bar{x}$? *Yes*
   b. If appropriate, find the probability that $\bar{x}$ will be between 53 and 54. *0.1609 [Tech: 0.1608]*
   c. If appropriate, find the 45th percentile of $\bar{x}$. *51.82*

17. A sample of size 5 will be drawn from a normal population with mean 60 and standard deviation 12.
   a. Is it appropriate to use the normal distribution to find probabilities for $\bar{x}$? *Yes*
   b. If appropriate, find the probability that $\bar{x}$ will be between 50 and 70. *0.9372 [Tech: 0.9376]*
   c. If appropriate, find the 80th percentile of $\bar{x}$. *64.51 [Tech: 64.52]*

18. A sample of size 15 will be drawn from a population with mean 125 and standard deviation 28.
   a. Is it appropriate to use the normal distribution to find probabilities for $\bar{x}$? *No*
   b. If appropriate, find the probability that $\bar{x}$ will be less than 120. *Not appropriate*
   c. If appropriate, find the 90th percentile of $\bar{x}$. *Not appropriate*

Working with the Concepts

19. **Summer temperatures:** Following are the temperatures, in degrees Fahrenheit, in Denver for five days in July:

Date	Temperature
July 21	69
July 22	75
July 23	79
July 24	83
July 25	71

   a. Consider this to be a population. Find the population mean μ and the population standard deviation σ.
   b. List all samples of size 2 drawn with replacement. There are $5 \times 5 = 25$ different samples.
   c. Compute the sample mean $\bar{x}$ for each of the 25 samples of size 2. Compute the mean $\mu_{\bar{x}}$ and the standard deviation $\sigma_{\bar{x}}$ of the sample means.
   d. Verify that $\mu_{\bar{x}} = \mu$ and $\sigma_{\bar{x}} = \sigma/\sqrt{2}$.

20. **Age of winners:** Following are the ages of the Grammy award winners for Best New Artist for the years 2015–2019. (Ages are given to the nearest half year, so the five ages are all different.)

Year: Winner	Age
2019: Dua Lipa	23
2018: Alessia Cara	21
2017: Chance the Rapper	23.5
2016: Meghan Trainor	22
2015: Sam Smith	22.5

a. Consider this to be a population. Find the population mean μ and the population standard deviation σ.

b. List all samples of size 2 drawn with replacement. There are $5 \times 5 = 25$ different samples.

c. Compute the sample mean $\bar{x}$ for each of the 25 samples of size 2. Compute the mean $\mu_{\bar{x}}$ and the standard deviation $\sigma_{\bar{x}}$ of the sample means.

d. Verify that $\mu_{\bar{x}} = \mu$ and $\sigma_{\bar{x}} = \sigma/\sqrt{2}$.

21. How's your mileage? The Environmental Protection Agency (EPA) rates the mean highway gas mileage of the 2020 Hyundai Elantra GT to be 28 miles per gallon. Assume the standard deviation is 3 miles per gallon. A rental car company buys 60 of these cars.

a. What is the probability that the average mileage of the fleet is greater than 27.5 miles per gallon? *0.9015 [Tech: 0.9016]*

b. What is the probability that the average mileage of the fleet is between 27 and 27.8 miles per gallon? *0.2966 [Tech: 0.2979]*

c. Would it be unusual if the average mileage of the fleet were less than 27 miles per gallon? *Yes*

22. Watch your cholesterol: The National Health and Nutrition Examination Survey (NHANES) reported that in a recent year, the mean serum cholesterol level for U.S. adults was 202, with a standard deviation of 41 (the units are milligrams per deciliter). A simple random sample of 110 adults is chosen.

a. What is the probability that the sample mean cholesterol level is greater than 210? *0.0202 [Tech: 0.0204]*

b. What is the probability that the sample mean cholesterol level is between than 190 and 200? *0.3039 [Tech: 0.3034]*

c. Would it be unusual for the sample mean to be less than 198? *No*

23. TV sets: A Nielsen Company report states that the mean number of TV sets in a U.S. household is 2.24. Assume the standard deviation is 1.2. A sample of 85 households is drawn.

a. What is the probability that the sample mean number of TV sets is greater than 2? *0.9671 [Tech: 0.9674]*

b. What is the probability that the sample mean number of TV sets is between 2.5 and 3? *0.0228 [Tech: 0.0229]*

c. Find the 30th percentile of the sample mean. *2.17*

d. Would it be unusual for the sample mean to be less than 2? *Yes*

e. Can you tell whether it would be unusual for an individual household to have fewer than 2 TV sets? Explain. *No*

24. SAT scores: The College Board reports that in a recent year, the mean mathematics SAT score was 514, and the standard deviation was 118. A sample of 65 scores is chosen.

a. What is the probability that the sample mean score is less than 500? *0.1685 [Tech: 0.1694]*

b. What is the probability that the sample mean score is between 480 and 520? *0.6489 [Tech: 0.6490]*

c. Find the 80th percentile of the sample mean. *526.3*

d. Would it be unusual if the sample mean were greater than 550? *Yes*

e. Can you tell whether it would be unusual for an individual to get a score greater than 550? Explain. *No*

25. Taxes: The Internal Revenue Service reports that the mean federal income tax paid in a recent year was \$8040. Assume that the standard deviation is \$5000. The IRS plans to draw a sample of 1000 tax returns to study the effect of a new tax law.

a. What is the probability that the sample mean tax is less than \$8000? *0.4013 [Tech: 0.4001]*

b. What is the probability that the sample mean tax is between \$7600 and \$7900? *0.1840 [Tech: 0.1853]*

c. Find the 40th percentile of the sample mean. *8000.5 [Tech: 7999.9]*

d. Would it be unusual if the sample mean were less than \$7500? *Yes*

e. Can you tell whether it would be unusual for an individual to pay a tax of less than \$7500? Explain. *No*

26. High-rent district: The Real Estate Group NY reports that the mean monthly rent for a one-bedroom apartment without a doorman in Manhattan is \$2631. Assume the standard deviation is \$500. A real estate firm samples 100 apartments.

a. What is the probability that the sample mean rent is greater than \$2700? *0.0838*

b. What is the probability that the sample mean rent is between \$2500 and \$2600? *0.2632*

c. Find the 60th percentile of the sample mean.

d. Can you tell whether it would be unusual if the sample mean were greater than \$2800? *Yes*

e. Do you think it would be unusual for an individual apartment to have a rent greater than \$2800? Explain. *No*

27. Roller coaster ride: A roller coaster is being designed that will accommodate 60 riders. The maximum weight the coaster can hold safely is 12,000 pounds. According to the National Health Statistics Reports, the weights of adult U.S. men have mean 194 pounds and standard deviation 68 pounds, and the weights of adult U.S. women have mean 164 pounds and standard deviation 77 pounds.

a. If 60 people are riding the coaster, and their total weight is 12,000 pounds, what is their average weight? *200 pounds*

b. If a random sample of 60 adult men ride the coaster, what is the probability that the maximum safe weight will be exceeded? *0.2483 [Tech: 0.2472]*

c. If a random sample of 60 adult women ride the coaster, what is the probability that the maximum safe weight will be exceeded? *0.0001*

Comstock Images/Getty Images

28. Elevator ride: Engineers are designing a large elevator that will accommodate 40 people. The maximum weight the elevator can hold safely is 8120 pounds. According to the National Health Statistics Reports, the weights of adult U.S. men have mean 194 pounds and standard deviation 68 pounds, and the weights of adult U.S. women have mean 164 pounds and standard deviation 77 pounds.

a. If 40 people are on the elevator, and their total weight is 8120 pounds, what is their average weight? *203 pounds*

b. If a random sample of 40 adult men ride the elevator, what is the probability that the maximum safe weight will be exceeded? *0.2005 [Tech: 0.2013]*

c. If a random sample of 40 adult women ride the elevator, what is the probability that the maximum safe weight will be exceeded? *0.0007*

29. **Annual income:** The mean annual income for people in a certain city (in thousands of dollars) is 42, with a standard deviation of 30. A pollster draws a sample of 90 people to interview.
   a. What is the probability that the sample mean income is less than 38? *0.1038 [Tech: 0.1030]*
   b. What is the probability that the sample mean income is between 40 and 45? *0.5646 [Tech: 0.5651]*
   c. Find the 60th percentile of the sample mean.
   d. Would it be unusual for the sample mean to be less than 35? *Yes*
   e. Can you tell whether it would be unusual for an individual to have an income less than 35? Explain. *No*

30. **Going to work:** An ABC News report stated that the mean distance that commuters in the United States travel each way to work is 16 miles. Assume the standard deviation is 8 miles. A sample of 75 commuters is chosen.
   a. What is the probability that the sample mean commute distance is greater than 13 miles? *0.9994*
   b. What is the probability that the sample mean commute distance is between 18 and 20 miles? *0.0150 [Tech: 0.0152]*
   c. Find the 10th percentile of the sample mean. *14.82*
   d. Would it be unusual for the sample mean distance to be greater than 19 miles? *Yes*
   e. Can you tell whether it would be unusual for an individual to have a commute distance greater than 19 miles? Explain. *No*

31. **Pages in a book:** A 500-page book contains 250 sheets of paper. The thickness of the paper used to manufacture the book has mean 0.08 mm and standard deviation 0.01 mm. Someone wants to know the probability that a randomly chosen page is more than 0.1 mm thick. Is enough information given to compute this probability? If so, compute the probability. If not, explain why not. *No*

32. **An apple a day:** A supermarket sells apples in bags labeled as weighing 10 pounds. A sample of 50 bags had a mean weight of 10.3 pounds with a standard deviation of 0.1 pound. Someone wants to know the probability that a randomly chosen bag weighs more than 10 pounds. Is enough information given to compute this probability? If so, compute the probability. If not, explain why not. *No*

Extending the Concepts

33. **Eat your cereal:** A cereal manufacturer claims that the weight of a box of cereal labeled as weighing 12 ounces has a mean of 12.0 ounces and a standard deviation of 0.1 ounce. You sample 75 boxes and weigh them. Let $\bar{x}$ denote the mean weight of the 75 boxes.
   a. If the claim is true, what is $P(\bar{x} \leq 11.99)$?
   b. Based on the answer to part (a), if the claim is true, is 11.99 ounces an unusually small mean weight for a sample of 75 boxes? *No*
   c. If the mean weight of the boxes were 11.99 ounces, would you be convinced that the claim was false? Explain. *No*
   d. If the claim is true, what is $P(\bar{x} \leq 11.97)$? *0.0047*
   e. Based on the answer to part (d), if the claim is true, is 11.97 ounces an unusually small mean weight for a sample of 75 boxes? *Yes*
   f. If the mean weight of the boxes were 11.97 ounces, would you be convinced that the claim was false? Explain. *Yes*

34. **Battery life:** A battery manufacturer claims that the lifetime of a certain type of battery has a population mean of $\mu = 40$ hours and a standard deviation of $\sigma = 5$ hours. Let $\bar{x}$ represent the mean lifetime of the batteries in a simple random sample of size 100.
   a. If the claim is true, what is $P(\bar{x} \leq 38.5)$? *0.0013*
   b. Based on the answer to part (a), if the claim is true, is a sample mean lifetime of 38.5 hours unusually short? *Yes*
   c. If the sample mean lifetime of the 100 batteries were 38.5 hours, would you find the manufacturer's claim to be plausible? Explain. *No*
   d. If the claim is true, what is $P(\bar{x} \leq 39.8)$? *0.3446*
   e. Based on the answer to part (d), if the claim is true, is a sample mean lifetime of 39.8 hours unusually short? *No*
   f. If the sample mean lifetime of the 100 batteries were 39.8 hours, would you find the manufacturer's claim to be plausible? Explain. *Yes*

35. **Finite population correction:** The mean of a sample of size n has standard deviation $\sigma/\sqrt{n}$, where σ is the population standard deviation. When sampling without replacement, a more accurate expression can be obtained by multiplying by a correction factor. Specifically, if the sample size is more than 5% of the population size, it is better to compute the standard deviation of the sample mean as

$$\frac{\sigma}{\sqrt{n}}\sqrt{\frac{N-n}{N-1}}$$

where N is the population size and n is the sample size. The factor $\sqrt{\frac{N-n}{N-1}}$ is called the finite population correction factor.
   a. One hundred students took an exam. The standard deviation of the 100 scores was 10. Twenty exams were chosen at random as part of a class assessment. Use the finite population correction to compute the standard deviation of the mean of the 20 exams. *2.01*
   b. In general, is the standard deviation computed with the correction smaller or larger than the standard deviation computed without it? *smaller*
   c. Use the finite population correction to show that if all 100 exams are sampled, the standard deviation of the sample mean is 0. Explain why this is so.

Answers to Check Your Understanding Exercises for Section 7.3

1. $\mu_{\bar{x}} = 6, \sigma_{\bar{x}} = 0.8$
2. $\mu_{\bar{x}} = 17, \sigma_{\bar{x}} = 2.0$
3. **a.** 0.1894 [Tech: 0.1884]
   b. The probability that $\bar{x}$ is less than 8 is 0.0384 [Tech: 0.0385]. If we define an event whose probability is less than 0.05 as unusual, then this is unusual.
4. **a.** 0.6449
   b. The probability that $\bar{x}$ is greater than 48 is 0.3372 [Tech: 0.3373]. This event is not unusual.

Section 7.4 The Central Limit Theorem for Proportions

Objectives

1. Construct the sampling distribution for a sample proportion
2. Use the Central Limit Theorem to compute probabilities for sample proportions

NOTE TO INSTRUCTOR

Remind students that until the sample has been selected, $\hat{p}$ is a random variable.

A computer retailer wants to estimate the proportion of people in her city who own laptop computers. She cannot survey everyone in the city, so she draws a sample of 100 people and surveys them. It turns out that 35 out of the 100 people in the sample own laptops. The proportion 35/100 is called the *sample proportion* and is denoted $\hat{p}$. The proportion of people in the entire population who own laptops is called the *population proportion* and is denoted p.

DEFINITION

In a population, the proportion who have a certain characteristic is called the **population proportion**.

In a simple random sample of n individuals, let x be the number in the sample who have the characteristic. The **sample proportion** is

$$\hat{p} = \frac{x}{n}$$

Notation:

- The population proportion is denoted by p.
- The sample proportion is denoted by $\hat{p}$.

If several samples are drawn from a population, they are likely to have different values for $\hat{p}$. Because the value of $\hat{p}$ varies each time a sample is drawn, $\hat{p}$ is a random variable, and it has a probability distribution. The probability distribution of $\hat{p}$ is called the *sampling distribution* of $\hat{p}$.

Objective 1 Construct the sampling distribution for a sample proportion

An Example of a Sampling Distribution

To present an example, consider tossing a fair coin five times. This produces a sample of size $n = 5$, where each item in the sample is either a head or a tail. The proportion of times the coin lands heads will be the sample proportion $\hat{p}$. Because the coin is fair, the probability that it lands heads each time is 0.5. Therefore, the population proportion of heads is $p = 0.5$. There are 32 possible samples. Table 7.3 lists them and presents the sample proportion $\hat{p}$ of heads for each.

Table 7.3 The 32 Possible Samples of Size 5 and Their Sample Proportions of Heads

Sample	$\hat{p}$	Sample	$\hat{p}$	Sample	$\hat{p}$	Sample	$\hat{p}$
TTTTT	0.0	THTTT	0.2	HTTTT	0.2	HHTTT	0.4
TTTTH	0.2	THTTH	0.4	HTTTH	0.4	HHTTH	0.6
TTTHT	0.2	THTHT	0.4	HTTHT	0.4	HHTHT	0.6
TTTHH	0.4	THTHH	0.6	HTTHH	0.6	HHTHH	0.8
TTHTT	0.2	THHTT	0.4	HTHTT	0.4	HHHTT	0.6
TTHTH	0.4	THHTH	0.6	HTHTH	0.6	HHHTH	0.8
TTHHT	0.4	THHHT	0.6	HTHHT	0.6	HHHHT	0.8
TTHHH	0.6	THHHH	0.8	HTHHH	0.8	HHHHH	1.0

The columns labeled "$\hat{p}$" contain the values of the sample proportion for each of the 32 possible samples. Some of these values appear more than once, because several samples have the same proportion. The mean of the sampling distribution is the average of these 32 values. The standard deviation of the sampling distribution is the population standard deviation of these 32 values, which can be computed by the method presented in Section 3.2. The mean and standard deviation are

$$\text{Mean: } \mu_{\hat{p}} = 0.5 \qquad \text{Standard deviation: } \sigma_{\hat{p}} = 0.2236$$

The values of $\mu_{\hat{p}}$ and $\sigma_{\hat{p}}$ are related to the values of the population proportion $p = 0.5$ and the sample size $n = 5$. Specifically,

$$\mu_{\hat{p}} = 0.5 = p$$

The mean of the sample proportion is equal to the population proportion. The relationship among $\sigma_{\hat{p}}$, p, and n is less obvious. However, note that

$$\sigma_{\hat{p}} = 0.2236 = \sqrt{\frac{0.5(1-0.5)}{5}} = \sqrt{\frac{p(1-p)}{n}}$$

These relationships hold in general.

SUMMARY

Let $\hat{p}$ be the sample proportion of a simple random sample of size n, drawn from a population with population proportion p. The mean and standard deviation of the sampling distribution of $\hat{p}$ are

$$\mu_{\hat{p}} = p$$

$$\sigma_{\hat{p}} = \sqrt{\frac{p(1-p)}{n}}$$

Example 7.26

Find the mean and standard deviation of a sampling distribution

The soft-drink cups at a certain fast-food restaurant have tickets attached to them. Customers peel off the tickets to see whether they win a prize. The proportion of tickets that are winners is $p = 0.25$. A total of $n = 70$ people purchase soft drinks between noon and 1:00 P.M. on a certain day. Let $\hat{p}$ be the proportion that win a prize. Find the mean and standard deviation of $\hat{p}$.

Solution

The population proportion is $p = 0.25$, and the sample size is $n = 70$. Therefore,

$$\mu_{\hat{p}} = p = 0.25$$

$$\sigma_{\hat{p}} = \sqrt{\frac{0.25(1-0.25)}{70}} = 0.0518$$

Check Your Understanding

1. Find $\mu_{\hat{p}}$ and $\sigma_{\hat{p}}$ if $n = 20$ and $p = 0.82$. $\mu_{\hat{p}} = 0.82$; $\sigma_{\hat{p}} = 0.08591$
2. Find $\mu_{\hat{p}}$ and $\sigma_{\hat{p}}$ if $n = 217$ and $p = 0.455$. $\mu_{\hat{p}} = 0.455$; $\sigma_{\hat{p}} = 0.03380$

Answers are on page 349.

NOTE TO INSTRUCTOR

Exact binomial probabilities can be calculated with technology, so the normal distribution is not essential for that purpose. However, the normal approximation to the binomial distribution is used when constructing confidence intervals and performing hypothesis tests for proportions.

The probability histogram for the sampling distribution of $\hat{p}$

Figure 7.33 presents the probability histogram for the sampling distribution of $\hat{p}$ for the proportion of heads in five tosses of a fair coin, for which $n = 5$ and $p = 0.5$. The distribution is reasonably well approximated by a normal curve. Figure 7.34 presents the probability histogram for the sampling distribution of $\hat{p}$ for the proportion of heads in 50 tosses of a fair coin, for which $n = 50$ and $p = 0.5$. The distribution is very closely approximated by a normal curve.

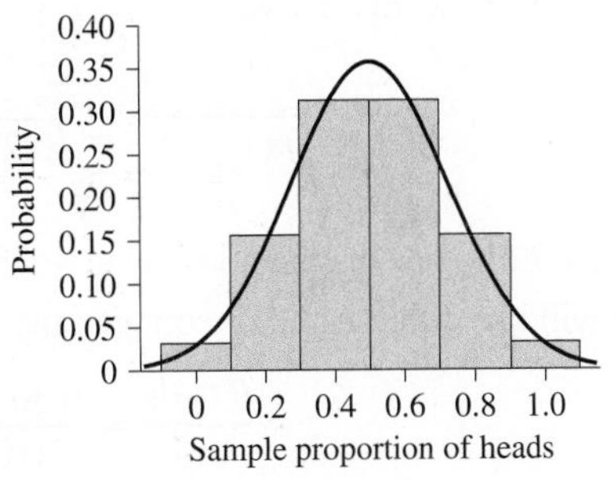

Figure 7.33 The probability histogram for $\hat{p}$ when $n = 5$ and $p = 0.5$. The histogram is reasonably well approximated by a normal curve.

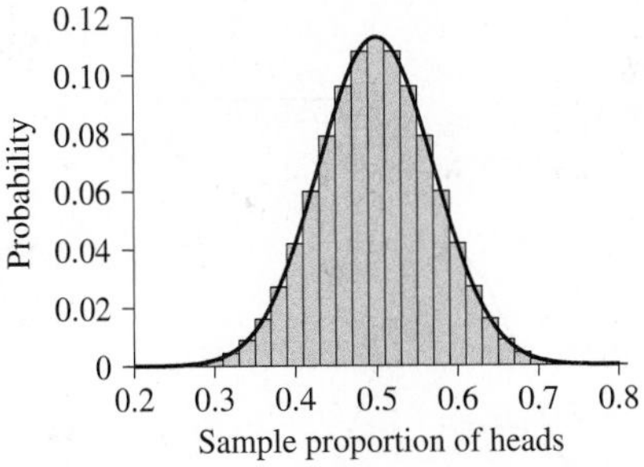

Figure 7.34 The probability histogram for $\hat{p}$ when $n = 50$ and $p = 0.5$. The histogram is very closely approximated by a normal curve.

When $p = 0.5$, the sampling distribution of $\hat{p}$ is somewhat close to normal even for a small sample size like $n = 5$. When p is close to 0 or close to 1, a larger sample size is needed before the distribution of $\hat{p}$ is close to normal. A common rule of thumb is that the distribution may be approximated with a normal curve whenever np and $n(1 - p)$ are both at least 10.

The Central Limit Theorem for Proportions

Let $\hat{p}$ be the sample proportion for a sample size of n and population proportion p. If

$$np \geq 10 \quad \text{and} \quad n(1-p) \geq 10$$

then the distribution of $\hat{p}$ is approximately normal, with mean and standard deviation

$$\mu_{\hat{p}} = p \quad \text{and} \quad \sigma_{\hat{p}} = \sqrt{\frac{p(1-p)}{n}}$$

Objective 2 Use the Central Limit Theorem to compute probabilities for sample proportions

Computing Probabilities with the Central Limit Theorem

To compute probabilities involving a sample proportion $\hat{p}$, use the following procedure:

Procedure for Computing Probabilities with the Central Limit Theorem

Step 1: Check to see that the conditions $np \geq 10$ and $n(1 - p) \geq 10$ are both met. If so, it is appropriate to use the normal curve.

Step 2: Find the mean $\mu_{\hat{p}}$ and standard deviation $\sigma_{\hat{p}}$.

Step 3: Sketch a normal curve and shade in the area to be found.

Step 4: Find the area using Table A.2 or technology.

Example 7.27

Using the Central Limit Theorem to compute a probability

EXPLAIN IT AGAIN

Computing probabilities for sample proportions: Computing probabilities for sample proportions with the Central Limit Theorem is the same as computing probabilities for any normally distributed quantity. Use $\mu_{\hat{p}} = p$ for the mean and $\sigma_{\hat{p}} = \sqrt{p(1-p)/n}$ for the standard deviation.

According to a Harris poll, chocolate is the favorite ice cream flavor for 27% of Americans. If a sample of 100 Americans is taken, what is the probability that the sample proportion of those who prefer chocolate is greater than 0.30?

Solution

Step 1: $np = (100)(0.27) = 27 \geq 10$, and $n(1-p) = (100)(1-0.27) = 73 \geq 10$. We may use the normal curve.

Step 2: $\mu_{\hat{p}} = p = 0.27$.

$$\sigma_{\hat{p}} = \sqrt{\frac{p(1-p)}{n}} = \sqrt{\frac{0.27(1-0.27)}{100}} = 0.0444$$

Step 3: Figure 7.35 presents the normal curve with the area shaded in.

Step 4: We will use Table A.2. We compute the z-score for 0.30.

$$z = \frac{\hat{p} - \mu_{\hat{p}}}{\sigma_{\hat{p}}} = \frac{0.30 - 0.27}{0.0444} = 0.68$$

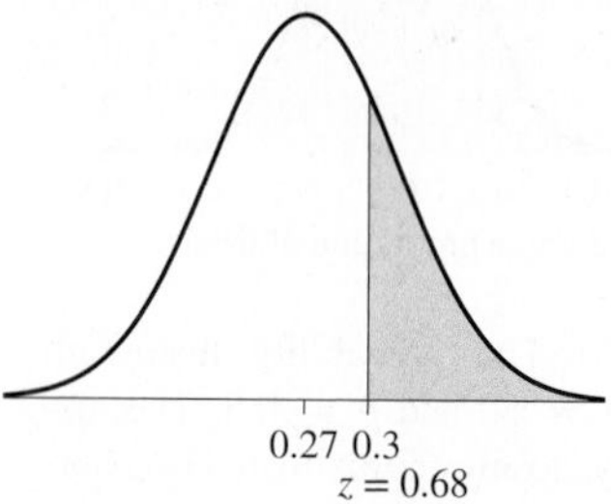

Figure 7.35

The table gives the area to the *left* of $z = 0.68$ as 0.7517. The area to the right of $z = 0.68$ is $1 - 0.7517 = 0.2483$. The probability that the sample proportion of those who prefer chocolate is greater than 0.30 is 0.2483.

Example 7.28

Using the Central Limit Theorem to determine whether a given value of $\hat{p}$ is unusual

Approximately 60% of eligible voters voted in a recent U.S. presidential election. If a sample of 88 eligible voters were polled, would it be unusual if less than half of them had voted?

Solution

We will compute the probability that the sample proportion is less than 0.50. If this probability is less than 0.05, we will say that the event is unusual.

Step 1: $np = (88)(0.6) = 52.8 \geq 10$, and $n(1-p) = (88)(1-0.6) = 35.2 \geq 10$. We may use the normal curve.

Step 2: $\mu_{\hat{p}} = p = 0.6$.

$$\sigma_{\hat{p}} = \sqrt{\frac{p(1-p)}{n}} = \sqrt{\frac{0.6(1-0.6)}{88}} = 0.052223$$

Step 3: Figure 7.36 presents the normal curve with the area shaded in.

Step 4: We will use Table A.2. We compute the z-score for 0.5.

$$z = \frac{\hat{p} - \mu_{\hat{p}}}{\sigma_{\hat{p}}} = \frac{0.5 - 0.6}{0.052223} = -1.91$$

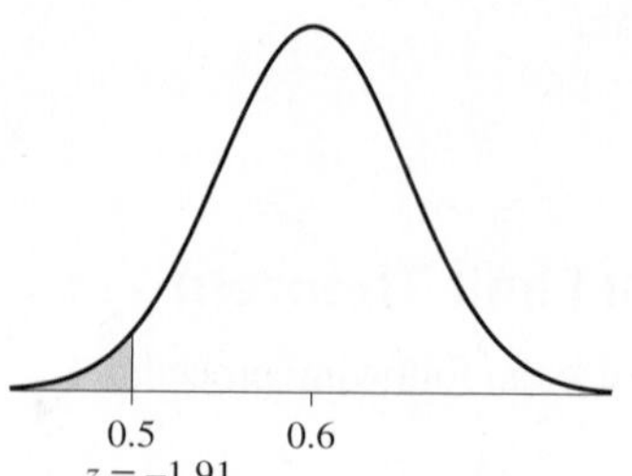

Figure 7.36

The area to the left of $z = -1.91$ is 0.0281. It would be unusual for the sample proportion to be less than 0.5.

Check Your Understanding

3. The General Social Survey reported that 56% of American adults saw a doctor for an illness during the past year. A sample of 65 adults is drawn.
   a. What is the probability that more than 60% of them saw a doctor? *0.2578 [Tech: 0.2580]*
   b. Would it be unusual if more than 70% of them saw a doctor? *Yes*

4. For a certain type of computer chip, the proportion of chips that are defective is 0.10. A computer manufacturer receives a shipment of 200 chips.

a. What is the probability that the proportion of defective chips in the shipment is between 0.08 and 0.15? *0.8173 [Tech: 0.8179]*

b. Would it be unusual for the proportion of defective chips to be less than 0.075? *No*

Answers are on page 349.

Section 7.4 Exercises

Exercises 1–4 are the Check Your Understanding exercises located within the section.

Understanding the Concepts

In Exercises 5 and 6, fill in each blank with the appropriate word or phrase.

5. If n is the sample size and x is the number in the sample who have a certain characteristic, then x/n is called the sample __________. *proportion*

6. The probability distribution of $\hat{p}$ is called a __________ distribution. *sampling*

In Exercises 7 and 8, determine whether the statement is true or false. If the statement is false, rewrite it as a true statement.

7. The distribution of $\hat{p}$ is approximately normal if $np \geq 10$ and $n(1-p) \geq 10$. *True*

8. If n is the sample size, p is the population proportion, and $\hat{p}$ is the sample proportion, then $\sigma_{\hat{p}} = np$. *False*

Practicing the Skills

In Exercises 9–14, n is the sample size, p is the population proportion, and $\hat{p}$ is the sample proportion. If appropriate, use the Central Limit Theorem to find the indicated probability.

9. $n = 147, p = 0.13; P(\hat{p} < 0.11)$ *0.2358 [Tech: 0.2354]*

10. $n = 65, p = 0.86; P(\hat{p} < 0.80)$ *0.0823 [Tech: 0.0816]*

11. $n = 270, p = 0.57; P(\hat{p} > 0.61)$ *0.0918 [Tech: 0.0922]*

12. $n = 103, p = 0.24; P(0.20 < \hat{p} < 0.23)$ *0.2341 [Tech: 0.2352]*

13. $n = 145, p = 0.05; P(0.03 < \hat{p} < 0.08)$ *Not appropriate*

14. $n = 234, p = 0.75; P(0.77 < \hat{p} < 0.81)$ *0.2219 [Tech: 0.2229]*

Working with the Concepts

15. Coffee: The National Coffee Association reported that 63% of U.S. adults drink coffee daily. A random sample of 250 U.S. adults is selected.

a. Find the mean $\mu_{\hat{p}}$. *0.63*

b. Find the standard deviation $\sigma_{\hat{p}}$. *0.0305*

c. Find the probability that more than 67% of the sampled adults drink coffee daily. *0.0951*

d. Find the probability that the proportion of the sampled adults who drink coffee daily is between 0.6 and 0.7. *0.8255 [Tech: 0.8261]*

e. Find the probability that less than 62% of the sampled adults drink coffee daily. *0.3707 [Tech: 0.3716]*

f. Would it be unusual if less than 57% of the sampled adults drink coffee daily? *Yes*

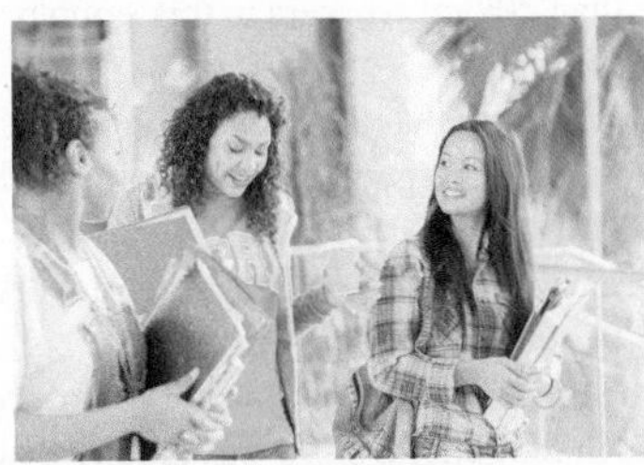

Robert Daly/Caia Image/Glow Images

16. Smartphones: A Pew Research report indicated that 73% of teenagers aged 13–17 own smartphones. A random sample of 150 teenagers is drawn.

a. Find the mean $\mu_{\hat{p}}$. *0.73*

b. Find the standard deviation $\sigma_{\hat{p}}$. *0.0362*

c. Find the probability that more than 70% of the sampled teenagers own a smartphone. *0.7967 [Tech: 0.7961]*

d. Find the probability that the proportion of the sampled teenagers who own a smartphone is between 0.76 and 0.80. *0.1765 [Tech: 0.1772]*

e. Find the probability that less than 75% of the sampled teenagers own smartphones. *0.7088 [Tech: 0.7094]*

f. Would it be unusual if less than 68% of the sampled teenagers owned smartphones? *No*

17. Student loans: The Institute for College Access and Success reported that 68% of college students in a recent year graduated with student loan debt. A random sample of 85 graduates is drawn.

a. Find the mean $\mu_{\hat{p}}$. *0.68*

b. Find the standard deviation $\sigma_{\hat{p}}$. *0.0506*

c. Find the probability that less than 60% of the people in the sample were in debt. *0.0571 [Tech: 0.0569]*

d. Find the probability that between 65% and 80% of the people in the sample were in debt. *0.7135 [Tech: 0.7145]*

e. Find the probability that more than 75% of the people in the sample were in debt. *0.0838 [Tech: 0.0832]*

f. Would it be unusual if less than 65% of the people in the sample were in debt? *No*

18. High school graduates: The National Center for Educational Statistics reported that 82% of freshmen entering public high schools in the U.S. in a recent year graduated with their class four years later. A random sample of 135 freshmen is chosen.

a. Find the mean $\mu_{\hat{p}}$. 0.82

b. Find the standard deviation $\sigma_{\hat{p}}$. 0.0331

c. Find the probability that less than 80% of freshmen in the sample graduated. 0.2743 [Tech: 0.2726]

d. Find the probability that the sample proportion of students who graduated is between 0.75 and 0.85. 0.8016 [Tech: 0.8007]

e. Find the probability that more than 75% of freshmen in the sample graduated. 0.9830 [Tech: 0.9829]

f. Would it be unusual if the sample proportion of students who graduated were more than 0.90? Yes

19. Government workers: The Bureau of Labor Statistics reported that 16% of U.S. nonfarm workers are government employees. A random sample of 50 workers is drawn.

a. Is it appropriate to use the normal approximation to find the probability that less than 20% of the individuals in the sample are government employees? If so, find the probability. If not, explain why not. No

b. A new sample of 90 workers is chosen. Find the probability that more than 20% of workers in this sample are government employees. 0.1492 [Tech: 0.1503]

c. Find the probability that the proportion of workers in the sample of 90 who are government employees is between 0.15 and 0.18. 0.3011 [Tech: 0.2997]

d. Find the probability that less than 25% of workers in the sample of 90 are government employees. 0.9901

e. Would it be unusual if the proportion of government employees in the sample of 90 was greater than 0.25? Yes

20. Working two jobs: The Bureau of Labor Statistics reported in a recent year that 5% of employed adults in the United States held multiple jobs. A random sample of 75 employed adults is chosen.

a. Is it appropriate to use the normal approximation to find the probability that less than 6.5% of the individuals in the sample hold multiple jobs? If so, find the probability. If not, explain why not. No

b. A new sample of 350 employed adults is chosen. Find the probability that less than 6.5% of the individuals in this sample hold multiple jobs. 0.9015 [Tech: 0.9011]

c. Find the probability that more than 6% of the individuals in the sample of 350 hold multiple jobs. 0.1949 [Tech: 0.1953]

d. Find the probability that the proportion of individuals in the sample of 350 who hold multiple jobs is between 0.05 and 0.10. 0.4999 [Tech: 0.5000]

e. Would it be unusual if less than 4% of the individuals in the sample of 350 held multiple jobs? No

21. Future scientists: Education professionals refer to science, technology, engineering, and mathematics as the STEM disciplines. A recent ACT Condition and Career Readiness Report states that 47% of high school graduates have expressed interest in a STEM discipline. A random sample of 85 freshmen is selected.

a. Is it appropriate to use the normal approximation to find the probability that less than 45% of the freshmen in the sample have expressed interest in a STEM discipline? If so, find the probability. If not, explain why not. Yes, 0.3557 [Tech: 0.3559]

b. A new sample of 150 freshmen is selected. Find the probability that less than 45% of the freshmen in this sample have expressed interest in a STEM discipline. 0.3121 [Tech: 0.3118]

c. Find the probability that the proportion of freshmen in the sample of 150 who have expressed interest in a STEM discipline is between 0.40 and 0.45. 0.2694 [Tech: 0.2689]

d. Find the probability that more than 38% of the freshmen in the sample of 150 have expressed interest in a STEM discipline. 0.9864

e. Would it be unusual if less than 42% of the freshmen in the sample of 150 have expressed interest in a STEM discipline? No

22. Blood pressure: High blood pressure has been identified as a risk factor for heart attacks and strokes. The National Health and Nutrition Examination Survey reported that the proportion of U.S. adults with high blood pressure is 0.3. A sample of 38 U.S. adults is chosen.

a. Is it appropriate to use the normal approximation to find the probability that more than 40% of the people in the sample have high blood pressure? If so, find the probability. If not, explain why not. Yes

b. A new sample of 80 adults is drawn. Find the probability that more than 40% of the people in this sample have high blood pressure. 0.0256 [Tech: 0.0255]

c. Find the probability that the proportion of individuals in the sample of 80 who have high blood pressure is between 0.20 and 0.35. 0.8109 [Tech: 0.8100]

d. Find the probability that less than 25% of the people in the sample of 80 have high blood pressure. 0.1635 [Tech: 0.1646]

e. Would it be unusual if more than 45% of the individuals in the sample of 80 had high blood pressure? Yes

23. Pay your taxes: According to the Internal Revenue Service, the proportion of federal tax returns for which no tax was paid was $p = 0.326$. As part of a tax audit, tax officials draw a simple random sample of $n = 120$ tax returns.

a. What is the probability that the sample proportion of tax returns for which no tax was paid is less than 0.30?

b. What is the probability that the sample proportion of tax returns for which no tax was paid is between 0.35 and 0.40? 0.2459 [Tech: 0.2456]

c. What is the probability that the sample proportion of tax returns for which no tax was paid is greater than 0.35? 0.2877 [Tech: 0.2874]

d. Would it be unusual if the sample proportion of tax returns for which no tax was paid was less than 0.25? Yes

24. Weekly paycheck: The Bureau of Labor Statistics reported that in a recent year, the median weekly earnings for people employed full time in the United States was $837.

a. What proportion of full-time employees had weekly earnings of more than $837? 0.5

b. A sample of 150 full-time employees is chosen. What is the probability that more than 55% of them earned more than $837 per week? 0.1112 [Tech: 0.1103]

c. What is the probability that less than 60% of the sample of 150 employees earned more than $837 per week?

d. What is the probability that between 45% and 55% of the sample of 150 employees earned more than $837 per week? 0.7776 [Tech: 0.7793]

e. Would it be unusual if less than 45% of the sample of 150 employees earned more than $755 per week? No

25. Kidney transplants: The Health Resources and Services Administration reported that 5% of people who received kidney transplants were under the age of 18. How large a sample of kidney transplant patients needs to be drawn so that the sample proportion $\hat{p}$ of those under the age of 18 is approximately normally distributed? *200*

26. How's your new car? The General Social Survey reported that 91% of people who bought a car in the past five years were satisfied with their purchase. How large a sample of car buyers needs to be drawn so that the sample proportion $\hat{p}$ who are satisfied is approximately normally distributed? *112*

27. How many heads? Baylee tosses a fair coin 200 times, and her friend Erin tosses a fair coin 50 times.

a. For which of them is it more likely that 55% or more of the tosses will be heads? *Erin*

b. For which of them is it more likely that the percentage of heads will be between 45% and 55%? *Baylee*

28. How many sixes? There are two elementary schools in a town. School A has 500 students and school B has 1000 students. Each student at each school rolls a fair six-sided die.

a. At which school is it more likely that 15% or less of the tosses will be 6s? *School A*

b. At which school is it more likely that the percentage of 6s will be between 15% and 20%? *School B*

Extending the Concepts

29. Flawless tiles: A new process has been designed to make ceramic tiles. The goal is to have no more than 5% of the tiles be nonconforming due to surface defects. A random sample of 1000 tiles is inspected. Let $\hat{p}$ be the proportion of nonconforming tiles in the sample.

a. If 5% of the tiles produced are nonconforming, what is $P(\hat{p} \geq 0.075)$? *0.0001*

b. Based on the answer to part (a), if 5% of the tiles are nonconforming, is a proportion of 0.075 nonconforming tiles in a sample of 1000 unusually large? *Yes*

c. If the sample proportion of nonconforming tiles were 0.075, would it be plausible that the goal had been reached? Explain. *No*

d. If 5% of the tiles produced are nonconforming, what is $P(\hat{p} \geq 0.053)$? *0.3300 [Tech: 0.3317]*

e. Based on the answer to part (d), if 5% of the tiles are nonconforming, is a proportion of 0.053 nonconforming tiles in a sample of 1000 unusually large? *No*

f. If the sample proportion of nonconforming tiles were 0.053, would it be plausible that the goal had been reached? Explain. *Yes*

Answers to Check Your Understanding Exercises for Section 7.4

1. $\mu_{\hat{p}} = 0.82$, $\sigma_{\hat{p}} = 0.08591$

2. $\mu_{\hat{p}} = 0.455$, $\sigma_{\hat{p}} = 0.03380$

3. a. 0.2578 [Tech: 0.2580]

b. The probability that $\hat{p}$ is greater than 0.70 is 0.0116 [Tech: 0.0115]. If we define an event whose probability is less than 0.05 as unusual, then this is unusual.

4. a. 0.8173 [Tech: 0.8179]

b. The probability that $\hat{p}$ is less than 0.075 is 0.1190 [Tech: 0.1193]. This event is not unusual.

Section 7.5 The Normal Approximation to the Binomial Distribution

Objective

1. Use the normal curve to approximate binomial probabilities

Objective 1 Use the normal curve to approximate binomial probabilities

We first introduced binomial random variables in Section 6.2. Recall that a binomial random variable represents the number of successes in a series of independent trials. The sample proportion is found by dividing the number of successes by the number of trials. Since the sample proportion is approximately normally distributed whenever $np \geq 10$ and $n(1 - p) \geq 10$, the number of successes is also approximately normally distributed under these conditions. Therefore, the normal curve can also be used to compute approximate probabilities for the binomial distribution.

We begin by reviewing the conditions under which a random variable has a binomial distribution.

Conditions for the Binomial Distribution

1. A fixed number of trials are conducted.
2. There are two possible outcomes for each trial. One is labeled "success" and the other is labeled "failure."
3. The probability of success is the same on each trial.

4. The trials are independent. This means that the outcome of one trial does not affect the outcomes of the other trials.
5. The random variable X represents the number of successes that occur.

Notation: The following notation is commonly used:

- The number of trials is denoted by n.
- The probability of success is denoted by p, and the probability of failure is $1 - p$.

Mean, Variance, and Standard Deviation of a Binomial Random Variable

Let X be a binomial random variable with n trials and success probability p. Then the mean of X is

$$\mu_X = np$$

The variance of X is

$$\sigma_X^2 = np(1 - p)$$

The standard deviation of X is

$$\sigma_X = \sqrt{np(1 - p)}$$

EXPLAIN IT AGAIN

Calculating binomial probabilities: Binomial probabilities can be computed exactly by using the methods described in Section 6.2. Using these methods by hand is extremely difficult. The normal approximation provides an easier way to approximate these probabilities when computing by hand.

Binomial probabilities can be very difficult to compute exactly by hand, because many terms have to be calculated and added together. For example, imagine trying to compute the probability that the number of heads is between 75 and 125 when a coin is tossed 200 times. To do this, one would need to compute the following sum:

$$P(X = 75) + P(X = 76) + \cdots + P(X = 124) + P(X = 125)$$

This is nearly impossible to do without technology. Fortunately, probabilities like this can be approximated very closely by using the normal curve. In the days before cheap computing became available, use of the normal curve was the only feasible method for doing these calculations. The normal approximation is somewhat less important now but is still useful for quick "back of the envelope" calculations.

Recall from Section 7.4 that a sample proportion $\hat{p}$ is approximately normally distributed whenever $np \geq 10$ and $n(1 - p) \geq 10$. Now if X is a binomial random variable representing the number of successes in n trials, the sample proportion is given by $\hat{p} = X/n$. Since $\hat{p}$ is obtained simply by dividing X by the number of trials, it is reasonable to expect that X will be approximately normal whenever $\hat{p}$ is approximately normal. This is in fact the case.

NOTE TO INSTRUCTOR

Exact binomial probabilities can be calculated with technology, so the normal approximation to the binomial distribution is not essential for that purpose. However, it can provide a foundation for the method used when constructing confidence intervals and performing hypothesis test for proportions.

The Normal Approximation to the Binomial

Let X be a binomial random variable with n trials and success probability p. If $np \geq 10$ and $n(1 - p) \geq 10$, then X is approximately normal with mean $\mu_X = np$ and standard deviation $\sigma_X = \sqrt{np(1 - p)}$.

The continuity correction

The binomial distribution is discrete, whereas the normal distribution is continuous. The **continuity correction** is an adjustment, made when approximating a discrete distribution with a continuous one, that can improve the accuracy of the approximation. To see how it works, imagine that a fair coin is tossed 100 times. Let X represent the number of heads. Then X has the binomial distribution with $n = 100$ trials and success probability $p = 0.5$. Imagine that we want to compute the probability that X is between 45 and 55. This probability will differ depending on whether the endpoints, 45 and 55, are included or excluded. Figure 7.37 illustrates the case where the endpoints are included, that is, where we wish to compute $P(45 \leq X \leq 55)$. The exact probability is given by the total area of the rectangles of the binomial probability histogram corresponding to the integers 45 to 55, inclusive. The approximating normal curve is superimposed. To get the best approximation, we should compute the area under the normal curve between 44.5 and 55.5.

In contrast, Figure 7.38 illustrates the case where we wish to compute $P(45 < X < 55)$. Here the endpoints are excluded. The exact probability is given by the total area of the rectangles of the binomial probability histogram corresponding to the integers 46 to 54. The best normal approximation is found by computing the area under the normal curve between 45.5 and 54.5.

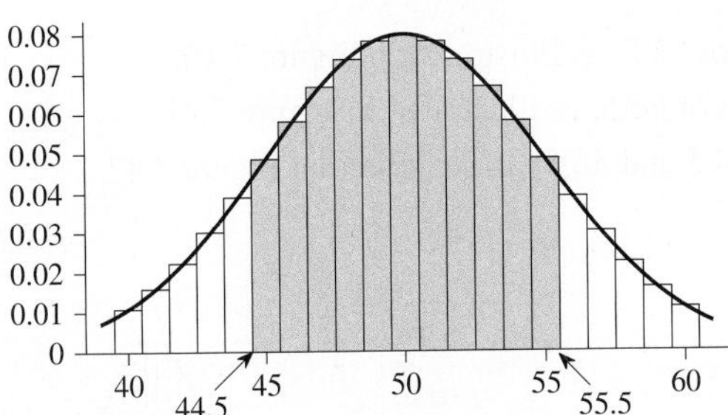

Figure 7.37 To compute $P(45 \leq X \leq 55)$, the areas of the rectangles corresponding to 45 and to 55 should be included. To approximate this probability with the normal curve, compute the area under the curve between 44.5 and 55.5.

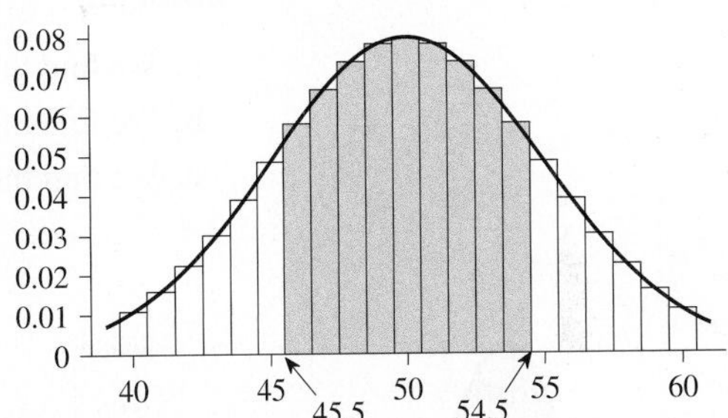

Figure 7.38 To compute $P(45 < X < 55)$, the areas of the rectangles corresponding to 45 and to 55 should be excluded. To approximate this probability with the normal curve, compute the area under the curve between 45.5 and 54.5.

In general, to apply the continuity correction, determine precisely which rectangles of the discrete probability histogram you wish to include, then compute the area under the normal curve corresponding to those rectangles.

Example 7.29

Using the continuity correction to compute a probability

Let X be the number of heads that appear when a fair coin is tossed 100 times. Use the normal curve to find $P(45 \leq X \leq 55)$.

Solution

This situation is illustrated in Figure 7.37.

Step 1: Check the assumptions: The number of trials is $n = 100$. Since the coin is fair, the success probability is $p = 0.5$. Therefore, $np = (100)(0.5) = 50 \geq 10$ and $n(1 - p) = (100)(1 - 0.5) = 50 \geq 10$. We can use the normal approximation.

Step 2: We compute the mean and standard deviation of X:

$$\mu_X = np = (100)(0.5) = 50 \qquad \sigma_X = \sqrt{np(1-p)} = \sqrt{(100)(0.5)(1-0.5)} = 5$$

Step 3: Because the probability is $P(45 \leq X \leq 55)$, we want to *include* both 45 and 55. Therefore, we set the left endpoint to 44.5 and the right endpoint to 55.5.

Step 4: We sketch a normal curve, label the mean of 50, and the endpoints 44.5 and 55.5.

Step 5: We use Table A.2 to find the area. The z-scores for 44.5 and 55.5 are

$$z = \frac{44.5 - 50}{5} = -1.1 \qquad z = \frac{55.5 - 50}{5} = 1.1$$

From Table A.2, we find that the probability is 0.7286. See Figure 7.39.

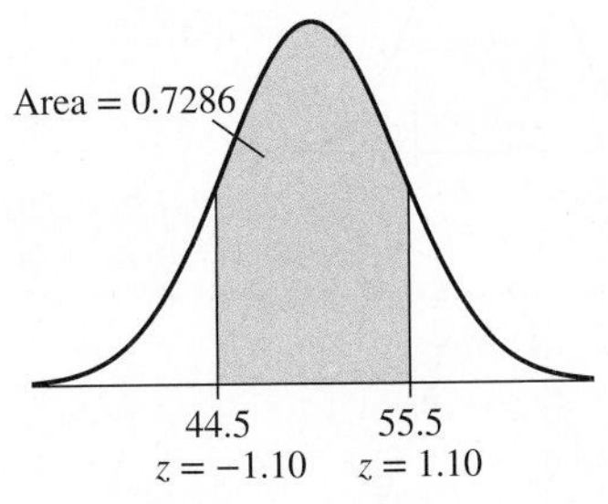

Figure 7.39

In Example 7.29, we used the normal approximation to compute a probability of the form $P(a \leq X \leq b)$. We can also use the normal approximation to compute probabilities of the form $P(X \leq a)$, $P(X \geq a)$, and $P(X = a)$.

Example 7.30

Illustrate areas to be found for the continuity correction

A fair coin is tossed 100 times. Let X be the number of heads that appear. Illustrate the area under the normal curve that represents each of the following probabilities.

a. $P(X \leq 55)$
b. $P(X \geq 55)$
c. $P(X = 55)$

Solution

a. We find the area to the left of 55.5, as illustrated in Figure 7.40.
b. We find the area to the right of 54.5, as illustrated in Figure 7.41.
c. We find the area between 54.5 and 55.5, as illustrated in Figure 7.42.

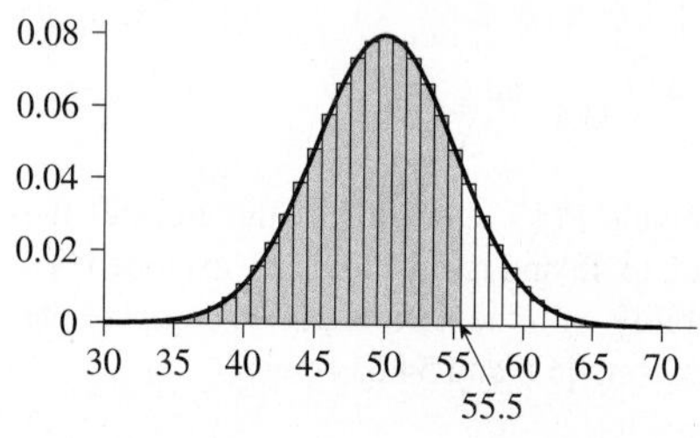

Figure 7.40 To approximate $P(X \leq 55)$, find the area to the left of 55.5.

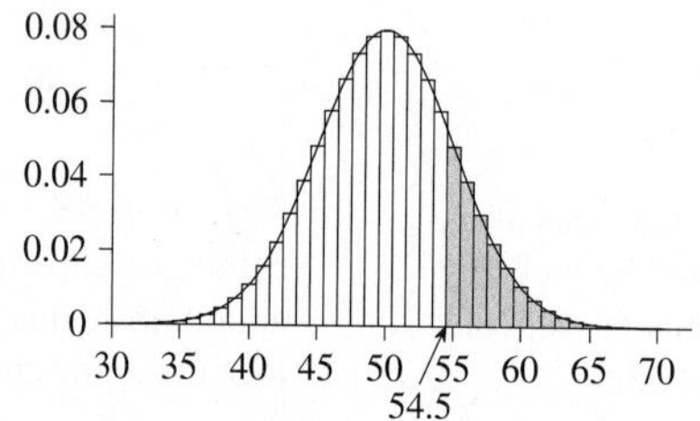

Figure 7.41 To approximate $P(X \geq 55)$, find the area to the right of 54.5.

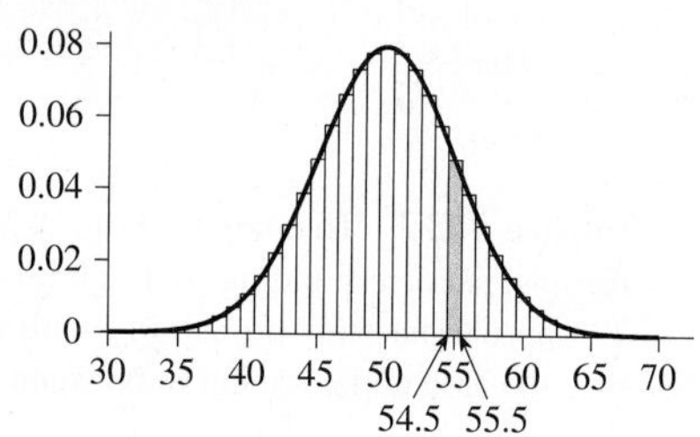

Figure 7.42 To approximate $P(X = 55)$, find the area between 54.5 and 55.5.

SUMMARY

Following are the areas under the normal curve to use when the continuity correction is applied.

$P(a \leq X \leq b)$

Find the area between $a - 0.5$ and $b + 0.5$.

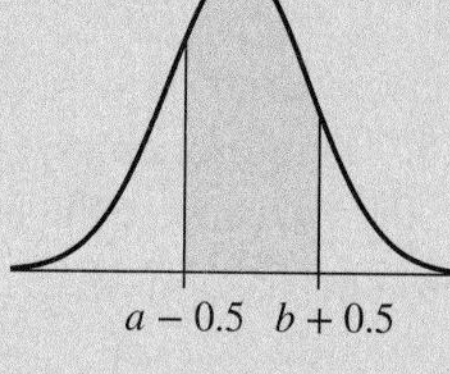

$P(X \leq b)$

Find the area to the left of $b + 0.5$.

$b + 0.5$

$P(X \geq a)$

Find the area to the right of $a - 0.5$.

$a - 0.5$

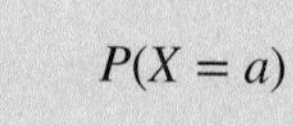

$P(X = a)$

Find the area between $a - 0.5$ and $a + 0.5$.

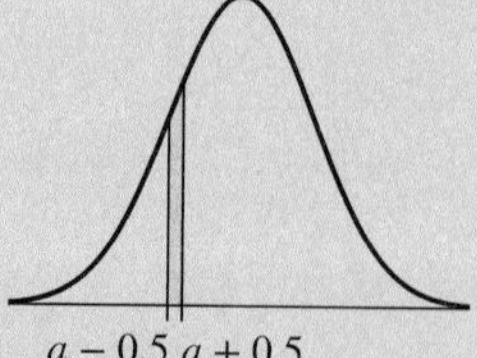

Use the following steps to compute a binomial probability with the normal approximation.

EXPLAIN IT AGAIN

We will use Table A.2 when using the normal approximation to the binomial: The normal approximation is useful when computing by hand. If technology is to be used, the exact probability can be calculated, so the normal approximation is less useful.

Procedure for Computing Binomial Probabilities with the Normal Approximation

Step 1: Check to see that the conditions $np \geq 10$ and $n(1-p) \geq 10$ are both met. If so, it is appropriate to use the normal approximation. If not, the probability must be calculated with the binomial distribution (see Section 6.2).

Step 2: Compute the mean μ_X and the standard deviation σ_X.

Step 3: For each endpoint, determine whether to add 0.5 or subtract 0.5.

Step 4: Sketch a normal curve, label the endpoints, and shade in the area to be found.

Step 5: Find the area using Table A.2 or technology. Note, however, that if you are using technology, you may be able to compute the probability exactly without using the normal approximation.

Example 7.31

Using the continuity correction to compute a probability

The *Statistical Abstract of the United States* reported that 66% of students who graduated from high school in a recent year enrolled in college. One hundred high school graduates are sampled. Let X be the number who enrolled in college. Find $P(X \leq 75)$.

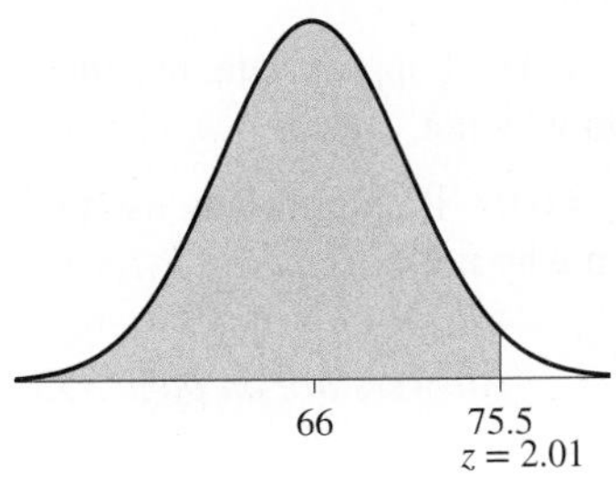

Figure 7.43

Solution

Step 1: Check the assumptions. The number of trials is $n = 100$, and the success probability is $p = 0.66$. Therefore $np = (100)(0.66) = 66 \geq 10$ and $n(1-p) = (100)(1-0.66) = 34 \geq 10$. We can use the normal approximation.

Step 2: We compute the mean and standard deviation of X:

$$\mu_X = np = (100)(0.66) = 66 \qquad \sigma_X = \sqrt{np(1-p)} = \sqrt{(100)(0.66)(1-0.66)} = 4.73709$$

Step 3: Since the probability is $P(X \leq 75)$, we compute the area to the left of 75.5.

Step 4: We sketch a normal curve, and label the mean of 66 and the point 75.5.

Step 5: We use Table A.2 to find the area. The z-score for 75.5 is

$$z = \frac{75.5 - 66}{4.73709} = 2.01$$

From Table A.2 we find that the probability is 0.9778. See Figure 7.43.

Sometimes we want to use the normal approximation to compute a probability of the form $P(X < a)$ or $P(X > a)$. One way to do this is by changing the inequality to a form involving $\leq$ or $\geq$. Example 7.32 provides an illustration.

Example 7.32

Using the continuity correction to compute a probability

The *Statistical Abstract of the United States* reported that 66% of students who graduated from high school in a recent year enrolled in college. One hundred high school graduates are sampled. Approximate the probability that more than 60 enroll in college.

Solution

Let X be the number of students in the sample who enrolled in college. We need to find $P(X > 60)$. We change this to an inequality involving $\geq$ by noting that $P(X > 60) = P(X \geq 61)$. We therefore find $P(X \geq 61)$.

Step 1: Check the assumptions. The number of trials is $n = 100$, and the success probability is $p = 0.66$. Therefore $np = (100)(0.66) = 66 \geq 10$ and $n(1-p) = (100)(1-0.66) = 34 \geq 10$. We can use the normal approximation.

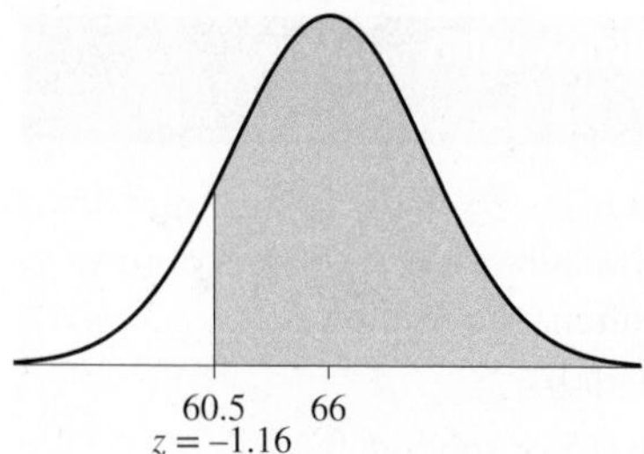

Figure 7.44

Step 2: We compute the mean and standard deviation of X:

$$\mu_X = np = (100)(0.66) = 66$$

$$\sigma_X = \sqrt{np(1-p)} = \sqrt{(100)(0.66)(1-0.66)} = 4.73709$$

Step 3: Since the probability is $P(X \geq 61)$, we compute the area to the right of 60.5.

Step 4: We sketch a normal curve, and label the mean of 66 and the point 60.5.

Step 5: We use Table A.2 to find the area. The z-score for 60.5 is

$$z = \frac{60.5 - 66}{4.73709} = -1.16$$

From Table A.2 we find that the probability is 0.8770. See Figure 7.44.

Check Your Understanding

1. X is a binomial random variable with $n = 50$ and $p = 0.15$. Should the normal approximation be used to find $P(X > 10)$? Why or why not? *No*
2. X is a binomial random variable with $n = 72$ and $p = 0.90$. Should the normal approximation be used to find $P(X \leq 60)$? Why or why not? *No*
3. Let X have a binomial distribution with $n = 64$ and $p = 0.41$. If appropriate, use the normal approximation to find $P(X \leq 20)$. If not, explain why not. *0.0721 [Tech: 0.0723]*
4. Let X have a binomial distribution with $n = 379$ and $p = 0.09$. If appropriate, use the normal approximation to find $P(X > 40)$. If not, explain why not. *0.1251 [Tech: 0.1257]*

Answers are on page 355.

Section 7.5 Exercises

Exercises 1–4 are the Check Your Understanding exercises located within the section.

Understanding the Concepts

In Exercises 5 and 6, fill in each blank with the appropriate word or phrase.

5. If X is a binomial random variable and if $np \geq 10$ and $n(1-p) \geq 10$, then X is approximately normal with $\mu_X =$ ________ and $\sigma_X =$ ________. *np, $\sqrt{np(1-p)}$*
6. The adjustment made when approximating a discrete random distribution with a continuous one is called the ________ correction. *continuity*

In Exercises 7 and 8, determine whether the statement is true or false. If the statement is false, rewrite it as a true statement.

7. If technology is to be used, exact binomial probabilities can be calculated and the normal approximation is not necessary. *True*
8. If X is a binomial random variable with n trials and success probability p, then as n gets larger, the distribution of X becomes more skewed. *False*

Practicing the Skills

In Exercises 9–14, n is the sample size, p is the population proportion of successes, and X is the number of successes in the sample. Use the normal approximation to find the indicated probability.

9. $n = 78, p = 0.43; P(X > 40)$ *0.0559 [Tech: 0.0557]*
10. $n = 538, p = 0.86; P(X \leq 470)$ *0.8340 [Tech: 0.8344]*
11. $n = 99, p = 0.57; P(X \geq 55)$ *0.6517 [Tech: 0.6524]*
12. $n = 442, p = 0.54; P(X < 243)$ *0.6406 [Tech: 0.6423]*
13. $n = 106, p = 0.14; P(14 < X < 18)$ *0.3102 [Tech: 0.3097]*
14. $n = 61, p = 0.34; P(20 \leq X \leq 24)$ *0.4792 [Tech: 0.4765]*

Working with the Concepts

15. **Google it:** According to a report of the Nielsen Company, 76% of internet searches used the Google search engine. A sample of 100 searches is studied.
    a. Approximate the probability that more than 70 of the searches used Google. *0.9015 [Tech: 0.9011]*
    b. Approximate the probability that 75 or fewer of the searches used Google. *0.4522 [Tech: 0.4534]*

c. Approximate the probability that the number of searches that used Google is between 70 and 80 inclusive. *0.7888 [Tech: 0.7900]*

16. Big babies: The Centers for Disease Control and Prevention reports that 25% of baby boys 6–8 months old in the United States weigh more than 20 pounds. A sample of 150 babies is studied.

a. Approximate the probability that more than 40 weigh more than 20 pounds. *0.2843 [Tech: 0.2858]*

b. Approximate the probability that 35 or fewer weigh more than 20 pounds. *0.3520 [Tech: 0.3530]*

c. Approximate the probability that the number who weigh more than 20 pounds is between 30 and 40, exclusive. *0.5546 [Tech: 0.5535]*

17. High blood pressure: The National Health and Nutrition Examination Survey reported that 30% of adults in the United States have hypertension (high blood pressure). A sample of 300 adults is studied.

a. Approximate the probability that 85 or more have hypertension. *0.7549 [Tech: 0.7558]*

b. Approximate the probability that fewer than 80 have hypertension. *0.0934 [Tech: 0.0929]*

c. Approximate the probability that the number who have hypertension is between 75 and 85, exclusive. *0.2115 [Tech: 0.2103]*

18. Stress at work: In a poll conducted by the General Social Survey, 81% of respondents said that their jobs were sometimes or always stressful. Two hundred workers are chosen at random.

a. Approximate the probability that 160 or fewer find their jobs stressful. *0.3936 [Tech: 0.3934]*

b. Approximate the probability that more than 150 find their jobs stressful. *0.9808 [Tech: 0.9809]*

c. Approximate the probability that the number who find their jobs stressful is between 155 and 162, inclusive. *0.4474 [Tech: 0.4477]*

19. What's your opinion? A pollster will interview a sample of 200 voters to ask whether they support a proposal to increase the sales tax to build a new light rail system. Assume that in fact 55% of the voters support the proposal.

a. Approximate the probability that 100 or fewer of the sampled voters support the proposal. *0.0885*

b. Approximate the probability that more than 105 voters support the proposal. *0.7389 [Tech: 0.7388]*

c. Approximate the probability that the number of voters who support the proposal is between 100 and 110, inclusive. *0.4598 [Tech: 0.4605]*

20. Gardening: A gardener buys a package of seeds. Eighty percent of seeds of this type germinate. The gardener plants 90 seeds.

a. Approximate the probability that fewer than 75 seeds germinate. *0.7454 [Tech: 0.7450]*

b. Approximate the probability that 80 or more seeds germinate. *0.0239 [Tech: 0.0241]*

c. Approximate the probability that the number of seeds that germinate is between 67 and 75, exclusive. *0.6284 [Tech: 0.6272]*

21. The car is in the shop: Among automobiles of a certain make, 23% require service during a one-year warranty period. A dealer sells 87 of these vehicles.

a. Approximate the probability that 25 or fewer of these vehicles require repairs. *0.9192 [Tech: 0.9190]*

b. Approximate the probability that more than 17 vehicles require repairs. *0.7389 [Tech: 0.7387]*

c. Approximate the probability that the number of vehicles that require repairs is between 15 and 20, exclusive. *0.3232 [Tech: 0.3230]*

22. Genetics: Pea plants contain two genes for seed color, each of which may be Y (for yellow seeds) or G (for green seeds). Plants that contain one of each type of gene are called heterozygous. According to the Mendelian theory of genetics, if two heterozygous plants are crossed, each of their offspring will have probability 0.75 of having yellow seeds and probability 0.25 of having green seeds. One hundred such offspring are produced.

a. Approximate the probability that more than 30 have green seeds. *0.1020*

b. Approximate the probability that 80 or fewer have yellow seeds. *0.8980*

c. Approximate the probability that the number with green seeds is between 30 and 35, inclusive. *0.1414 [Tech: 0.1417]*

23. Getting bumped: Airlines often sell more tickets for a flight than there are seats, because some ticket holders don't show up for the flight. Assume that an airplane has 100 seats for passengers and that the probability that a person holding a ticket appears for the flight is 0.90. If the airline sells 105 tickets, what is the probability that everyone who appears for the flight will get a seat? *0.9744 [Tech: 0.9745]*

24. College admissions: A small college has enough space to enroll 300 new students in its incoming freshman class. From past experience, the admissions office knows that 65% of students who are accepted actually enroll. If the admissions office accepts 450 students, what is the probability that there will be enough space for all the students who enroll? *0.7852 [Tech: 0.7854]*

Extending the Concepts

25. Probability of a single number: A fair coin is tossed 100 times. Use the normal approximation to approximate the probability that the coin comes up heads exactly 50 times. *0.0796 [Tech: 0.0797]*

Answers to Check Your Understanding Exercises for Section 7.5

1. No, $np = 7.5 < 10$.

2. No, $n(1 - p) = 7.2 < 10$.

3. 0.0721 [Tech: 0.0723]

4. 0.1251 [Tech: 0.1257]

Section 7.6 Assessing Normality

Objectives

1. Use dotplots to assess normality
2. Use boxplots to assess normality
3. Use histograms to assess normality
4. Use stem-and-leaf plots to assess normality
5. Use normal quantile plots to assess normality

NOTE TO INSTRUCTOR

Emphasize that assessing normality is important, because many statistical methods (such as the *t*-test and chi-square test), require that the population be approximately normal.

Many statistical procedures, some of which we will learn about in Chapters 8 and 9, require that we draw a sample from a population whose distribution is approximately normal. Often we don't know whether the population is approximately normal when we draw the sample. So the only way we have to assess whether the population is approximately normal is to examine the sample. In this section, we will describe some ways in which this can be done.

There are three important ideas to remember when assessing normality:

1. We are not trying to determine whether the population is *exactly* normal. No population encountered in practice is *exactly* normal. We are only trying to determine whether the population is *approximately* normal.
2. Assessing normality is more important for small samples than for large samples. When the sample size is large, say $n > 30$, the Central Limit Theorem ensures that $\bar{x}$ is approximately normal. Most statistical procedures designed for large samples rely on the Central Limit Theorem for their validity, so normality of the population is not so important in these cases.
3. Hard-and-fast rules do not work well. They are generally too lenient for very small samples (finding populations to be approximately normal when they are not) or too strict for larger samples (finding populations not to be approximately normal when they are). Informal judgment works as well as or better than hard-and-fast rules.

RECALL

An outlier is a data value that is considerably larger or smaller than most of the rest of the data.

When a sample is very small, it is often impossible to be sure whether it came from an approximately normal population. The best we can do is to examine the sample for signs of nonnormality. If no such signs exist, we will treat the population as approximately normal. Because the normal curve is unimodal and symmetric, samples from normal populations rarely have more than one distinct mode and rarely exhibit a large degree of skewness. In addition, samples from normal populations rarely contain outliers. We summarize the conditions under which we will reject the assumption that a population is approximately normal.

SUMMARY

We will reject the assumption that a population is approximately normal if a sample has *any* of the following features:

1. The sample contains an outlier.
2. The sample exhibits a large degree of skewness.
3. The sample is multimodal; in other words, it has more than one distinct mode.

If the sample has *none* of the preceding features, we will treat the population as being approximately normal.

Many methods have been developed for assessing normality; some of them are quite sophisticated. For our purposes, it will be sufficient to examine dotplots, boxplots, stem-and-leaf plots, and histograms of the sample. We will also describe normal quantile plots, which provide another useful method of assessment.

Objective 1 Use dotplots to assess normality

Dotplots

Dotplots are excellent for detecting outliers and multimodality. They can also be used to detect skewness, although they are not quite as effective as histograms for that purpose.

Example 7.33

Using a dotplot to assess normality

RECALL

Dotplots were introduced in Section 2.3.

The accuracy of an oven thermostat is being tested. The oven is set to 360°F, and the temperature when the thermostat turns off is recorded. A sample of size 7 yields the following results:

358 363 361 355 367 352 368

Is it reasonable to treat this as a sample from an approximately normal population? Explain.

Solution

Figure 7.45 presents a dotplot of the temperatures. The dotplot does not reveal any outliers. The plot does not exhibit a large degree of skewness, and there is no evidence that the population has more than one mode. Therefore, we can treat this as a sample from an approximately normal population.

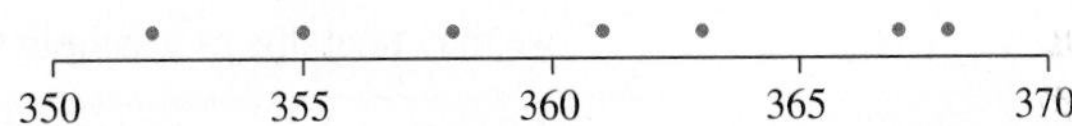

Figure 7.45 The dotplot of the oven temperatures does not reveal any outliers. The plot does not exhibit a large degree of skewness, and there is no evidence that the population has more than one mode. Therefore, we can treat this as a sample from an approximately normal population.

Example 7.34

Using a dotplot to assess normality

At a recent health fair, several hundred people had their pulse rates measured. A simple random sample of six records was drawn, and the pulse rates, in beats per minute, were

68 71 79 98 67 75

Is it reasonable to treat this as a sample from an approximately normal population? Explain.

Solution

Figure 7.46 presents a dotplot of the pulse rates. It is clear that the value 98 is an outlier. Therefore, we should not treat this as a sample from an approximately normal population.

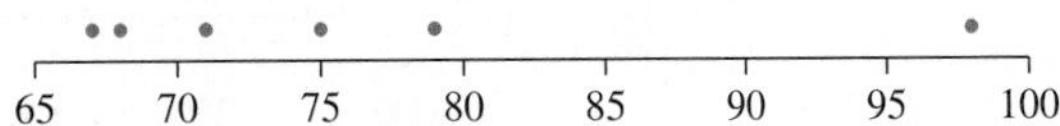

Figure 7.46 The value 98 is an outlier. Therefore, we should not treat this as a sample from an approximately normal population.

Objective 2 Use boxplots to assess normality

Boxplots

Boxplots are very good for detecting outliers and skewness. They work best for data sets that are not too small. For very small samples, it is just as informative to plot all the points with a dotplot. In addition, boxplots do not detect bimodality. When bimodality is a concern, a dotplot is a better choice.

Example 7.35

Using a boxplot to assess normality

RECALL

Boxplots were introduced in Section 3.3.

An insurance adjuster obtains a sample of 20 estimates, in hundreds of dollars, for repairs to cars damaged in collisions. Following are the data.

12.1	15.7	14.2	4.6	8.2	11.6	12.9	11.2	14.9	13.7
6.6	7.2	12.6	9.0	11.9	7.8	9.0	16.2	16.5	12.1

Is it reasonable to treat this as a sample from an approximately normal population? Explain.

Solution

Figure 7.47 presents a boxplot of the repair estimates. There are no outliers. Although the median is not exactly halfway between the quartiles, the skewness is not great. Therefore, we may treat this as a sample from an approximately normal population.

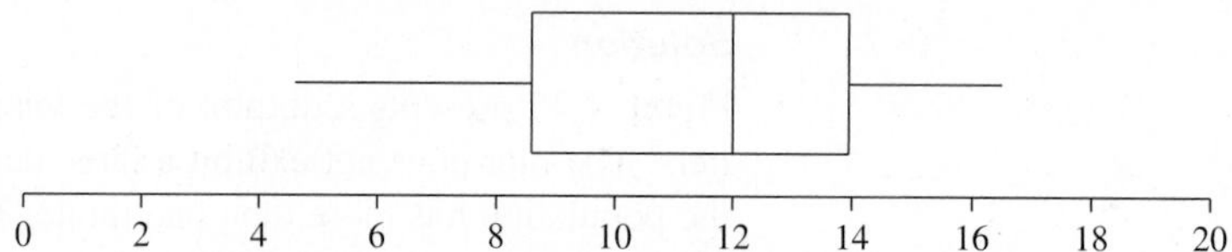

Figure 7.47 There are no outliers, and no evidence of strong skewness. Therefore, we may treat this as a sample from an approximately normal population.

Example 7.36

Using a boxplot to assess normality

A recycler determines the amount of recycled newspaper, in cubic feet, collected each week. Following are the results for a sample of 18 weeks.

2129	2853	2530	2054	2075	2011	2162	2285	2668
3194	4834	2469	2380	2567	4117	2337	3179	3157

Is it reasonable to treat this as a sample from an approximately normal population? Explain.

Solution

Figure 7.48 presents a boxplot of the amount of recycled newspaper. The value 4834 is an outlier. In addition, the upper whisker is much longer than the lower one, which indicates fairly strong skewness. Therefore, we should not treat this as a sample from an approximately normal population.

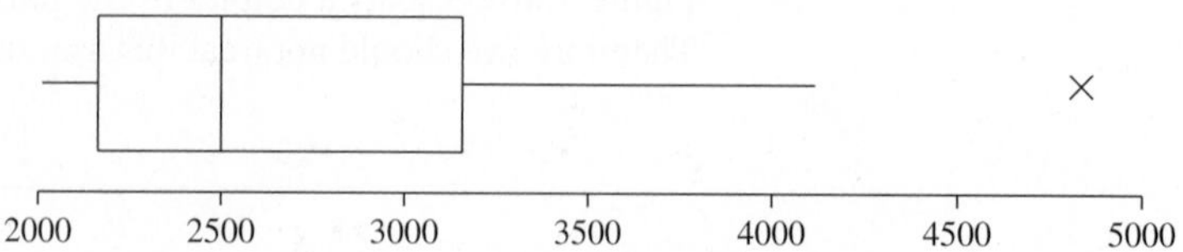

Figure 7.48 The value 4834 is an outlier, and there is evidence of fairly strong skewness as well. Therefore, we should not treat this as a sample from an approximately normal population.

Objective 3 Use histograms to assess normality

Histograms

Histograms are excellent for detecting strong skewness. They are more effective for data sets that are not too small (for very small data sets, a histogram is just like a dotplot with the dots replaced by rectangles).

Example 7.37

Use a histogram to assess normality

RECALL

Histograms were introduced in Section 2.2.

Diameters were measured, in millimeters, for a simple random sample of 20 grade A eggs from a certain farm. The results were

59	60	60	56	59	56	62	58	60	59
61	59	61	61	63	60	56	58	63	58

Construct a histogram for these data. Is it reasonable to treat this as a sample from an approximately normal population? Explain.

Solution

Figure 7.49 presents a relative frequency histogram of the diameters. The histogram does not reveal any outliers, nor does it exhibit a large degree of skewness. There is no evidence that the population has more than one mode. Therefore, we can treat this as a sample from an approximately normal population.

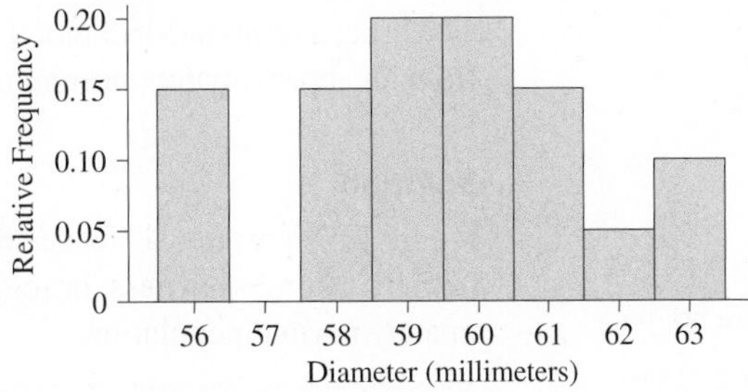

Figure 7.49 The histogram of the egg diameters does not reveal any outliers, nor a large degree of skewness, nor evidence of more than one mode. Therefore, we can treat this as a sample from an approximately normal population.

Example 7.38

Use a histogram to assess normality

A shoe manufacturer is testing a new type of leather sole. A simple random sample of 22 people wore shoes with the new sole for a period of four months. The amount of wear on the right shoe was measured for each person. The results, in thousandths of an inch, were

24.1	2.2	11.8	2.7	4.1	13.9	33.6	2.4	36.2	16.8	5.4
4.6	4.5	4.1	6.1	6.3	22.6	29.1	12.2	4.6	15.8	7.7

Construct a histogram for these data. Is it reasonable to treat this as a sample from an approximately normal population? Explain.

Solution

Figure 7.50 presents a relative frequency histogram of the amounts of wear. The histogram reveals that the sample is strongly skewed to the right. We should not treat this as a sample from an approximately normal population.

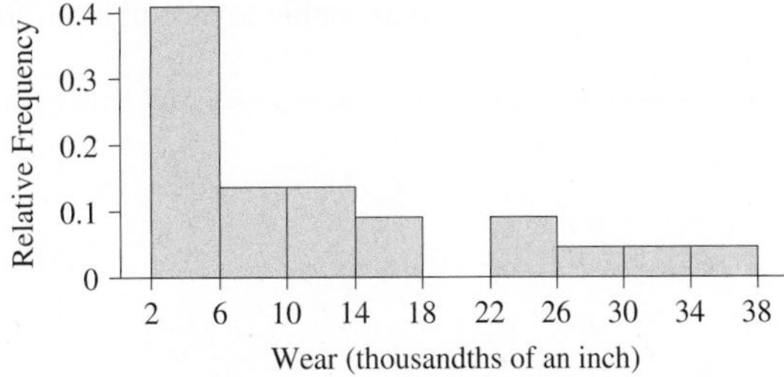

Figure 7.50 The histogram is strongly skewed. Therefore, we should not treat this as a sample from an approximately normal population.

Objective 4 Use stem-and-leaf plots to assess normality

RECALL

Stem-and-leaf plots were introduced in Section 2.3.

Stem-and-Leaf Plots

Stem-and-leaf plots can be used in place of histograms when the number of stems is large enough to provide an idea of the shape of the sample. Like histograms, stem-and-leaf plots are excellent for detecting skewness. They are more useful for data sets that are not too small, so that some of the stems will contain more than one leaf. Stem-and-leaf plots are easier to construct by hand than histograms are, but histograms are sometimes easier to construct with technology. For example, the TI-84 Plus calculator will construct histograms but cannot construct stem-and-leaf plots.

Example 7.39

Use a stem-and-leaf plot to assess normality

A psychologist measures the time it takes for each of 20 rats to run a maze. The times, in seconds, are

54	48	49	54	63	54	66	32	45	52
41	37	56	56	52	53	41	45	48	43

Construct a stem-and-leaf plot for these data. Is it reasonable to treat this as a random sample from an approximately normal population?

Solution

Figure 7.51 presents a stem-and-leaf plot of the times. The stem-and-leaf plot reveals no outliers, strong skewness, or multimodality. We may treat this as a sample from an approximately normal population.

3	2
3	7
4	113
4	55889
5	223444
5	66
6	3
6	6

Figure 7.51 There are no outliers, strong skewness, or multimodality.

Check Your Understanding

1. For each of the following dotplots, determine whether it is reasonable to treat the sample as coming from an approximately normal population.

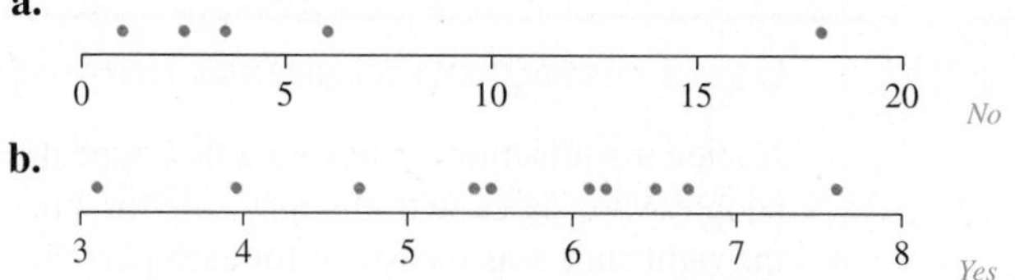

No (a) *Yes* (b)

2. For each of the following histograms, determine whether it is reasonable to treat the sample as coming from an approximately normal population. *(a) Yes (b) No*

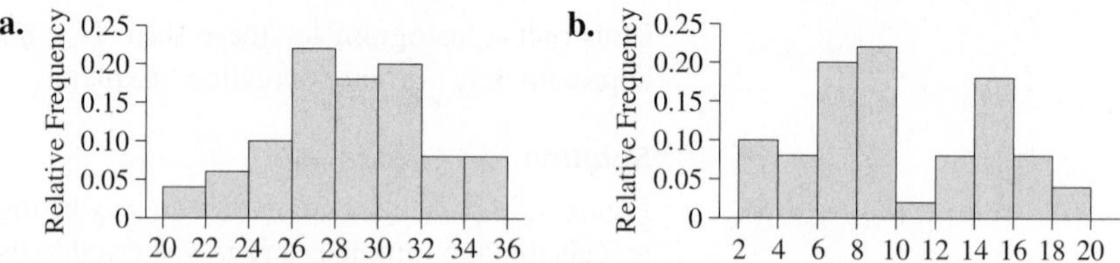

3. The following stem-and-leaf plot represents a sample from a population. Is it reasonable to assume that this population is approximately normal? *No*

1	34579
2	0278
3	25
4	37
5	38
6	4
7	
8	1
9	6

4. The following boxplot represents a sample from a population. Is it reasonable to assume that this population is approximately normal? *Yes*

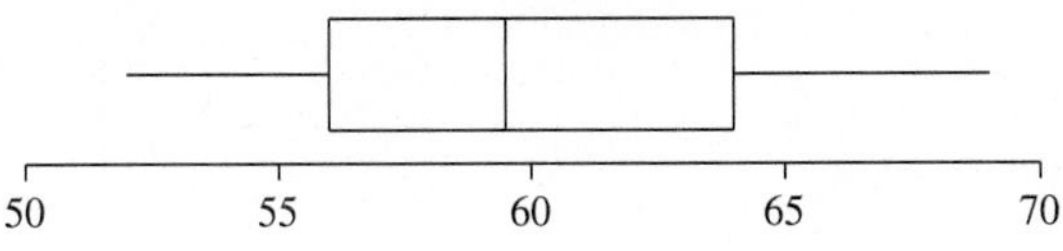

Answers are on page 366.

Objective 5 Use normal quantile plots to assess normality

Normal Quantile Plots

Normal quantile plots are somewhat more complex than dotplots, histograms, and stem-and-leaf plots. We will present the idea behind normal quantile plots with an example. A simple random sample of size $n = 5$ is drawn, and we want to determine whether the population it came from is approximately normal. The five sample values, in increasing order, are

3.0 3.3 4.8 5.9 7.8

NOTE TO INSTRUCTOR

Emphasize the use of technology for constructing normal quantile plots.

We proceed by using the following steps:

Step 1: Let n be the number of values in the data set. Spread the n values evenly over the interval from 0 to 1. This is done by assigning the value $1/(2n)$ to the first sample value, $3/(2n)$ to the second, and so forth. The last sample value will be assigned the value $(2n-1)/(2n)$. These values, denoted a_i, represent areas under the normal curve. For $n = 5$, the values are 0.1, 0.3, 0.5, 0.7, and 0.9.

i	x_i	a_i
1	3.0	0.1
2	3.3	0.3
3	4.8	0.5
4	5.9	0.7
5	7.8	0.9

Step 2: The values assigned in Step 1 represent left-tail areas under the normal curve. We now find the z-scores corresponding to each of these areas. The results are shown in the following table.

i	x_i	a_i	z_i
1	3.0	0.1	−1.28
2	3.3	0.3	−0.52
3	4.8	0.5	0.00
4	5.9	0.7	0.52
5	7.8	0.9	1.28

Step 3: Plot the points (x_i, z_i). The plot is shown in Figure 7.52 (page 362). A straight line has been added to the plot to help in interpreting the results. If the points approximately follow a straight line, then the population may be treated as being approximately normal. If the points deviate substantially from a straight line, the population should not be treated as normal. In this case, the points do approximately follow a straight line, so we may treat this population as approximately normal.

Why do the points on a normal quantile plot tend to follow a straight line when the population is normal? If the population is normal, then, on the average, the values z_i will be close to the actual z-scores of the x_i. Now for any sample, the actual z-scores will follow a straight line when plotted against the x_i. Therefore, if the population is approximately normal, it is likely that the points (x_i, z_i) will approximately follow a straight line.

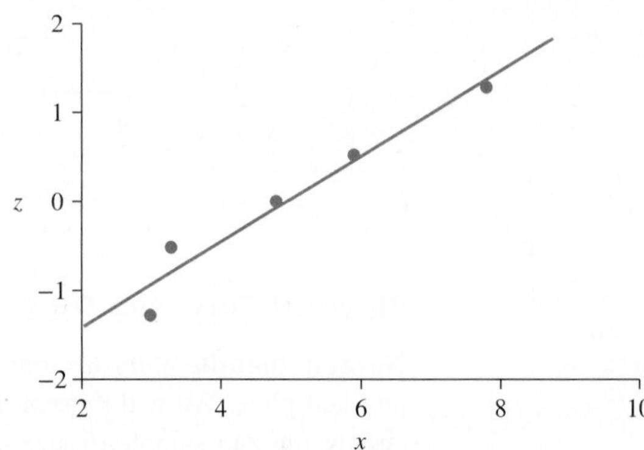

Figure 7.52 The points do not deviate substantially from a straight line. We can treat this population as approximately normal.

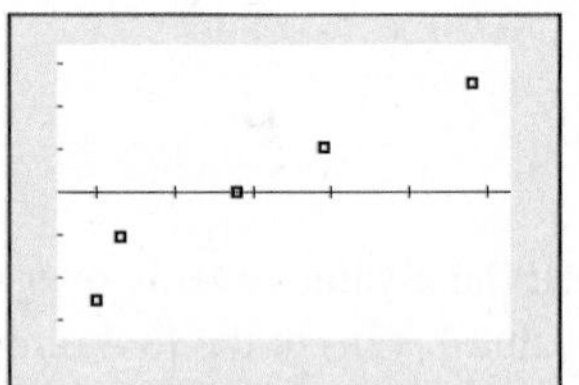

Figure 7.53

In practice, normal quantile plots are always constructed with technology. Figure 7.53 presents the normal quantile plot in Figure 7.52 as drawn by the TI-84 Plus calculator. The calculator does not add a line to the plot. Step-by-step instructions for constructing normal quantile plots with the TI-84 Plus calculator are presented in the Using Technology section on page 363.

Example 7.40

Using a normal quantile plot to assess normality

A placement exam is given to each entering freshman at a large university. A simple random sample of 20 exam scores is drawn, with the following results.

61	60	60	68	63	63	94	66	65	98
61	71	74	63	66	61	61	65	72	85

Construct a normal probability plot. Is the distribution of exam scores approximately normal?

Figure 7.54

Solution

We use the TI-84 Plus calculator. The results are shown in Figure 7.54. The points do not closely follow a straight line. The distribution is not approximately normal.

Check Your Understanding

5. Is it reasonable to treat the sample in the following normal quantile plot as coming from an approximately normal population? Explain.

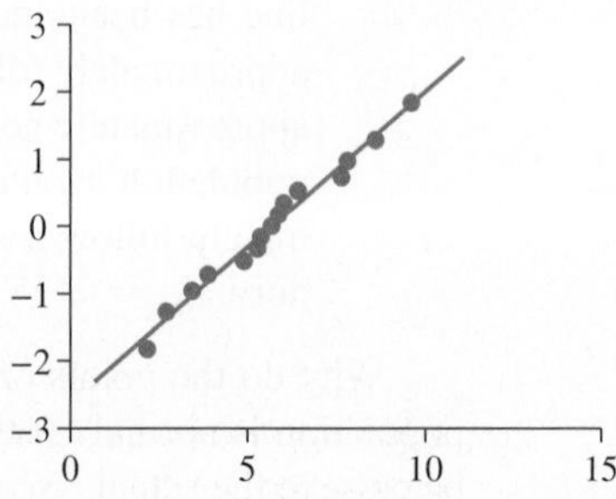

6. Is it reasonable to treat the sample in the following normal quantile plot as coming from an approximately normal population? Explain. *No*

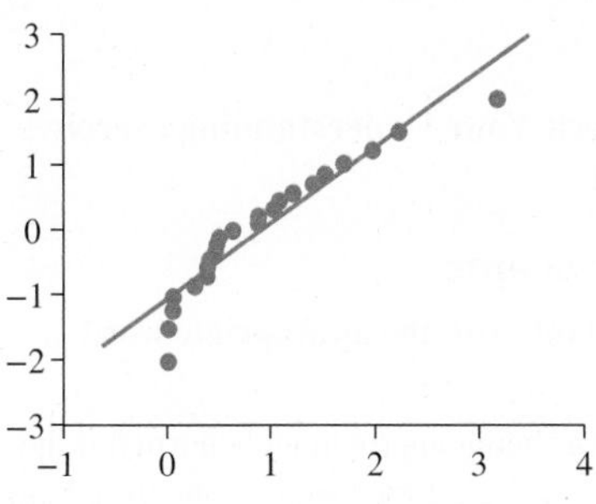

Answers are on page 366.

Using Technology

We use Example 7.40 to illustrate the technology steps.

TI-84 PLUS

Constructing normal quantile plots

Step 1. Enter the data from Example 7.40 into **L1** in the data editor.

Step 2. Press **2nd, Y=** to access the STAT PLOTS menu, and select Plot1 by pressing **1**.

Step 3. Select **On** and the normal quantile plot icon.

Step 4. For **Data List**, select **L1**, and for **Data Axis**, choose the **X** option (Figure A).

Step 5. Press **ZOOM** and then **9: ZoomStat** (Figure B).

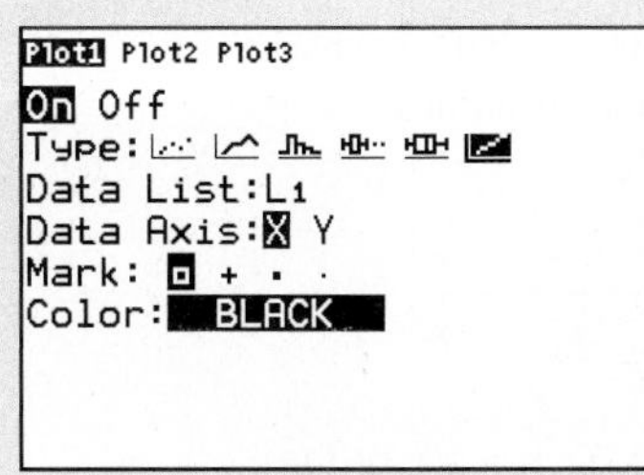

Figure A

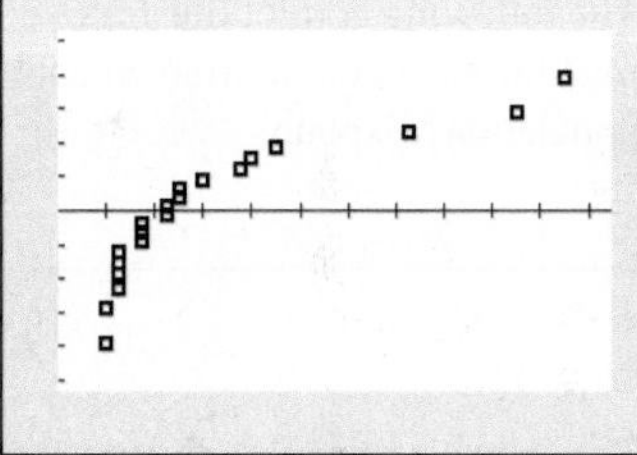

Figure B

MINITAB

Constructing normal quantile plots

Step 1. Enter the data from Example 7.40 into **Column C1**.

Step 2. Click on **Graph** and select **Probability Plot**. Choose the **Single** option and press **OK**.

Step 3. Enter **C1** in the **Graph variables** field.

Step 4. Click **OK** (Figure C).

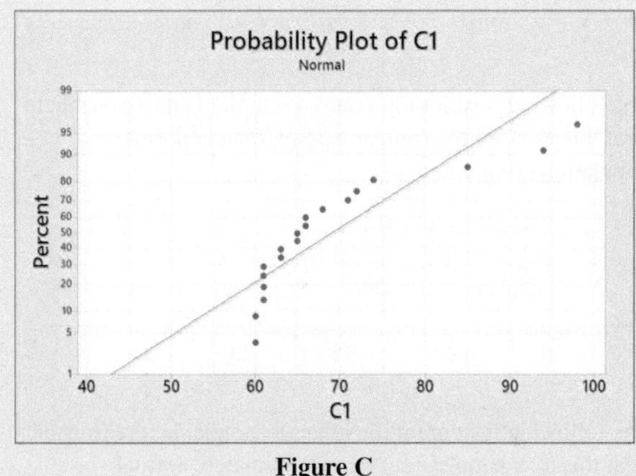

Figure C

Section 7.6 Exercises

Exercises 1–6 are the Check Your Understanding exercises located within the section.

Understanding the Concepts

In Exercise 7, fill in each blank with the appropriate word or phrase.

7. A population is rejected as being approximately normal if the sample contains an ________, if the sample contains a large degree of ________, or if the sample has more than one distinct ________. *outlier, skewness, mode*

In Exercise 8, determine whether the statement is true or false. If the statement is false, rewrite it as a true statement.

8. If the points in a normal quantile plot deviate from a straight line, then the population can be treated as approximately normal. *False*

Practicing the Skills

9. The following dotplot illustrates a sample. Is it reasonable to treat this as a sample from an approximately normal population? Explain. *No*

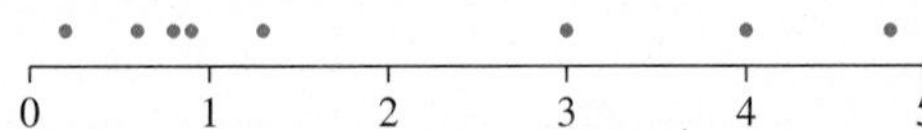

10. The following dotplot illustrates a sample. Is it reasonable to treat this as a sample from an approximately normal population? Explain. *Yes*

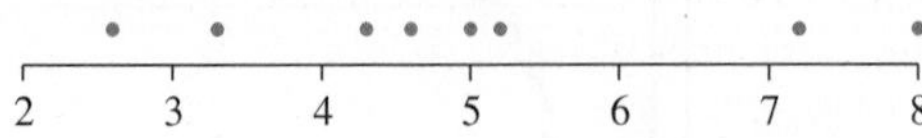

11. The following boxplot illustrates a sample. Is it reasonable to treat this as a sample from an approximately normal population? Explain. *Yes*

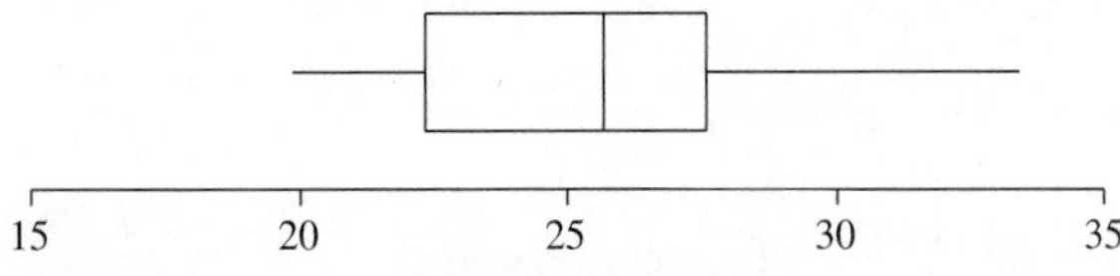

12. The following boxplot illustrates a sample. Is it reasonable to treat this as a sample from an approximately normal population? Explain. *No*

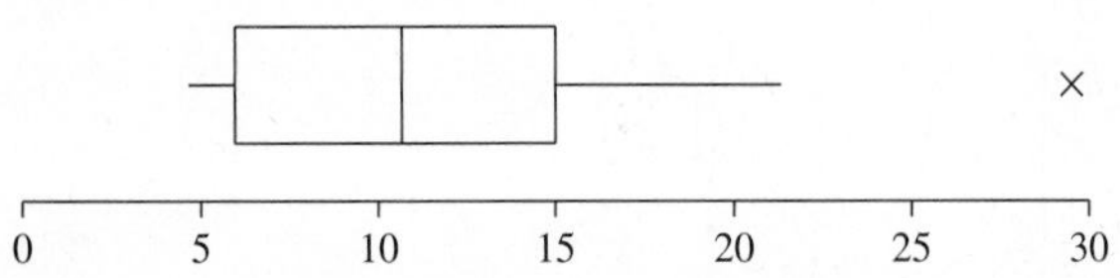

13. The following histogram illustrates a sample. Is it reasonable to treat this as a sample from an approximately normal population? Explain. *Yes*

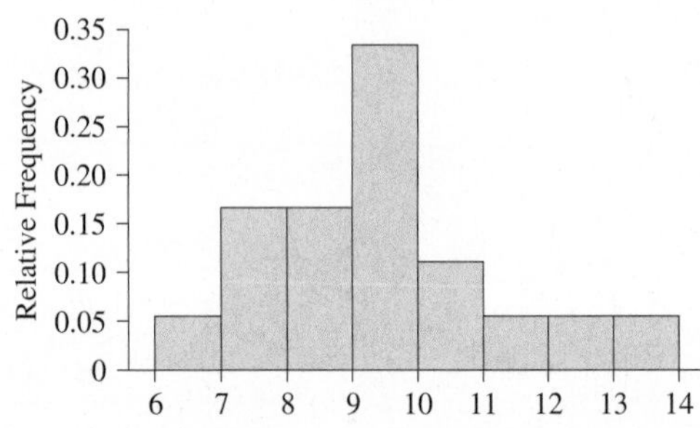

14. The following histogram illustrates a sample. Is it reasonable to treat this as a sample from an approximately normal population? Explain. *No*

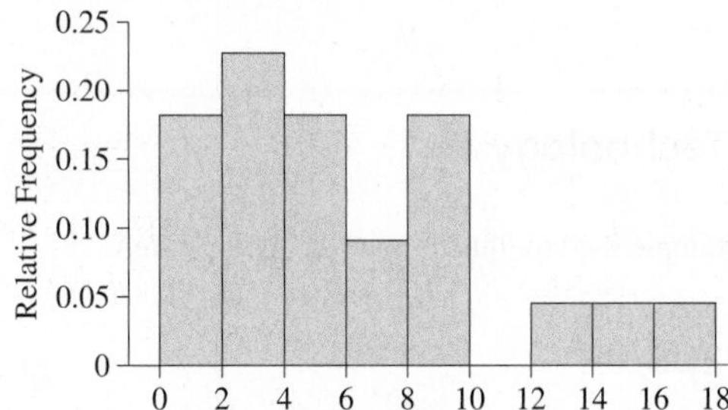

15. The following stem-and-leaf plot illustrates a sample. Is it reasonable to treat this as a sample from an approximately normal population? Explain. *No*

0	35
1	46
2	0022358
3	18
4	3
5	4
6	36
7	19
8	34
9	4
10	
11	8

16. The following stem-and-leaf plot illustrates a sample. Is it reasonable to treat this as a sample from an approximately normal population? Explain. *Yes*

7	35
8	47
9	024
10	379
11	37
12	34
13	0
14	1

17. The following normal quantile plot illustrates a sample. Is it reasonable to treat this as a sample from an approximately normal population? Explain. *Yes*

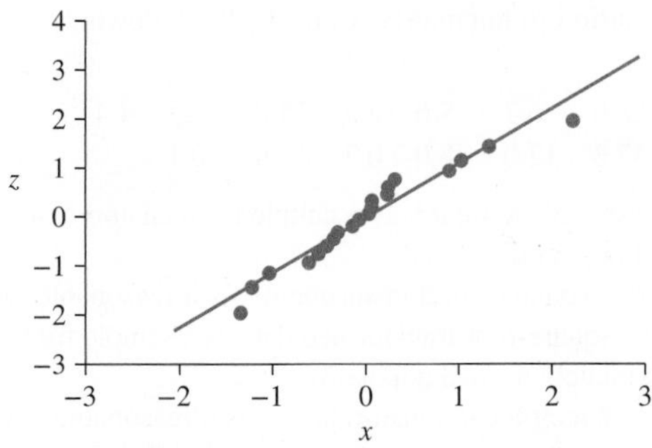

18. The following normal quantile plot illustrates a sample. Is it reasonable to treat this as a sample from an approximately normal population? Explain. No

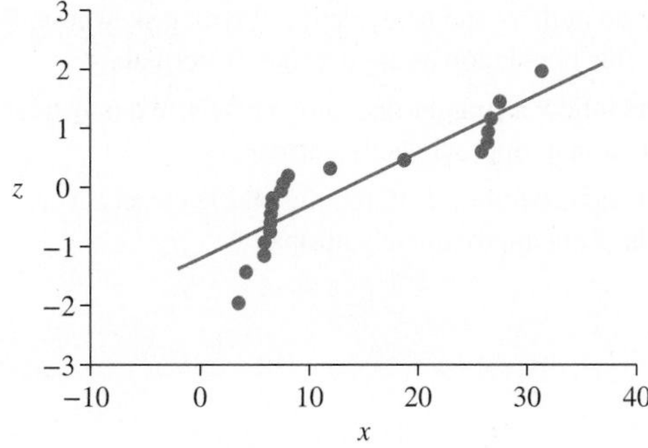

Working with the Concepts

19. **Drug concentrations:** A sample of 10 people ingested a new formulation of a drug. Six hours later, the concentrations in their bloodstreams, in nanograms per milliliter, were as follows.

2.3	1.4	1.8	2.1	1.0	4.1	1.8	2.9	2.5	2.7

Construct a dotplot for this sample. Is it reasonable to treat the sample as coming from an approximately normal population? Explain. No

20. **Reading scores:** A random sample of eight elementary school children were given a standardized reading test. Following are their scores.

72	77	65	85	68	83	73	79

Construct a dotplot for this sample. Is it reasonable to treat the sample as coming from an approximately normal population? Explain. Yes

21. **Timed task:** The number of minutes needed to complete a certain spreadsheet task was measured for 20 clerical workers. The results were as follows.

4.5	5.8	3.7	4.9	4.3	4.7	5.8	3.2	3.0	5.1
3.6	4.3	3.6	5.4	4.7	3.0	3.4	4.3	4.4	3.5

Construct a boxplot for this sample. Is it reasonable to treat the sample as coming from an approximately normal population? Explain. Yes

22. **Impure cans:** A manufacturer of aluminum cans measured the level of impurities in 24 cans. The amounts of impurities, in percent, were as follows.

2.1	1.5	1.9	1.3	2.7	4.9	2.1	1.8
1.3	1.0	3.2	4.4	8.0	4.5	2.5	1.6
2.8	8.2	9.5	3.8	1.9	1.5	2.8	1.6

Construct a boxplot for this sample. Is it reasonable to treat the sample as coming from an approximately normal population? Explain. No

23. **Defective items:** The number of defective items produced on an assembly line during an hour is counted for a random sample of 20 hours. The results are as follows.

21	16	10	10	11	9	13	12	11	29
10	14	27	10	11	11	12	19	11	9

Construct a stem-and-leaf plot for this sample. Is it reasonable to treat the sample as coming from an approximately normal population? Explain. No

24. **Fish weights:** A fish hatchery raises trout to stock streams and lakes. The weights, in ounces, of a sample of 18 trout at their time of release are as follows.

9.9	11.3	11.4	9.0	10.1	8.2	8.9	9.9	10.5
8.6	7.8	10.8	8.4	9.6	9.9	8.4	9.0	9.1

Construct a stem-and-leaf plot for this sample. Is it reasonable to treat the sample as coming from an approximately normal population? Explain. Yes

25. **Timed task:** Construct a histogram for the data in Exercise 21. Explain how the histogram shows whether it is appropriate to treat this sample as coming from an approximately normal population. Approximately normal

26. **Impure cans:** Construct a histogram for the data in Exercise 22. Explain how the histogram shows whether it is appropriate to treat this sample as coming from an approximately normal population. Not normal

27. **Defective items:** Construct a normal quantile plot for the data in Exercise 23. Explain how the plot shows whether it is appropriate to treat this sample as coming from an approximately normal population. Not normal

28. **Fish weights:** Construct a normal quantile plot for the data in Exercise 24. Explain how the plot shows whether it is appropriate to treat this sample as coming from an approximately normal population. Approximately normal

Extending the Concepts

29. **Transformation to normality:** Consider the following data set:

2	37	67	108	148	40	1	9	3	237	12	80

   a. Show that this data set does not come from an approximately normal population.

   b. Take the square root of each value in the data set. This is called a *square-root transformation* of the data. Show that the square roots may be considered to be a sample from an approximately normal population.

30. **Transformation to normality:** Consider the following data set:

−0.5	0.8	1.7	−1.0	−10.0	1.7	0.5	0.3	−5.0

   a. Show that this data set does not come from an approximately normal population.

   b. Take the reciprocal of each value in the data set (the reciprocal of x is $1/x$). This is called a *reciprocal transformation* of the data. Show that the reciprocals may be considered to be a sample from an approximately normal population.

31. Transformation to normality: Consider the following data set:

4.1	2.7	1.2	10.3	0.9	2.4
1.5	1.9	2.1	16.1	1.4	1.0

a. Is it reasonable to treat it as a sample from an approximately normal population? *No*

b. Perform a square-root transformation. Is it reasonable to treat the square-root-transformed data as a sample from an approximately normal population? *No*

c. Perform a reciprocal transformation. Is it reasonable to treat the reciprocal-transformed data as a sample from an approximately normal population? *Yes*

32. Transformation to normality: Consider the following data set:

28.0	6.7	8.6	2.3	25.0	12.5	4.4
37.3	12.0	48.0	0.7	11.6	0.1	

a. Is it reasonable to treat it as a sample from an approximately normal population? *No*

b. Perform a square-root transformation. Is it reasonable to treat the square-root-transformed data as a sample from an approximately normal population? *Yes*

c. Perform a reciprocal transformation. Is it reasonable to treat the reciprocal-transformed data as a sample from an approximately normal population? *No*

Answers to Check Your Understanding Exercises for Section 7.6

1. a. The plot contains an outlier. The population is not approximately normal.
b. We may treat this population as approximately normal.

2. a. We may treat this population as approximately normal.
b. The histogram has more than one mode. The population is not approximately normal.

3. The plot reveals that the sample is strongly skewed. The population is not approximately normal.

4. There are no outliers and no evidence of strong skewness. We may treat this population as approximately normal.

5. The points follow a straight line fairly closely. We may treat this population as approximately normal.

6. The points do not follow a straight line fairly closely. The population is not approximately normal.

Chapter 7 Summary

Section 7.1: Continuous random variables can be described with probability density curves. The area under a probability density curve over an interval can be interpreted in either of two ways. It represents the proportion of the population that is contained in the interval, and it also represents the probability that a randomly chosen value from the population falls within the interval. The normal curve is the most commonly used probability density curve. The standard normal curve represents a normal population with mean 0 and standard deviation 1. We can find areas under the standard normal curve by using Table A.2 or with technology.

Section 7.2: In practice, we need to work with normal distributions with different values for the mean and standard deviation. Technology can be used to compute probabilities for any normal distribution. We can also use Table A.2 to find probabilities for any normal distribution by standardization. Standardization involves computing the z-score by subtracting the mean and dividing by the standard deviation. The z-score has a standard normal distribution, so we can find probabilities by using Table A.2.

Section 7.3: The sampling distribution of a statistic such as a sample mean is the probability distribution of all possible values of the statistic. The Central Limit Theorem states that the sampling distribution of a sample mean is approximately normal so long as the sample size is large enough. Therefore, we can use the normal curve to compute approximate probabilities regarding the sample mean whenever the sample size is sufficiently large. For most populations, samples of size greater than 30 are large enough.

Section 7.4: The Central Limit Theorem can also be used to compute approximate probabilities regarding sample proportions. The sampling distribution of a sample proportion is approximately normal so long as the sample size is large enough. The sample size is large enough if both np and $n(1-p)$ are at least 10.

Section 7.5: A binomial random variable represents the number of successes in a series of independent trials. The number of successes is closely related to the sample proportion, because the sample proportion is found by dividing the number of successes by the number of trials. Since the sample proportion is approximately normally distributed whenever $np \geq 10$ and $n(1-p) \geq 10$, the number of successes is also approximately normally distributed under these conditions. Therefore, the normal curve can also be used to compute approximate probabilities for the binomial distribution. Because the binomial distribution is discrete, the continuity correction can be used to provide more accurate approximations.

Section 7.6: Many statistical procedures require the assumption that a sample is drawn from a population that is approximately normal. Although it is very difficult to determine whether a small sample comes from such a population, we can examine the sample for outliers, multimodality, and large degrees of skewness. If a sample contains no outliers, is not strongly skewed, and has only one distinct mode, we will treat it as though it came from an approximately normal population. Dotplots, boxplots, histograms, stem-and-leaf plots, and normal quantile plots can be used to assess normality.

Vocabulary and Notation

Important Formulas

z-score:

$$z = \frac{x - \mu}{\sigma}$$

Convert z-score to raw score:

$$x = \mu + z\sigma$$

Standard deviation of the sample mean:

$$\sigma_{\bar{x}} = \frac{\sigma}{\sqrt{n}}$$

z-score for a sample mean:

$$z = \frac{\bar{x} - \mu}{\sigma_{\bar{x}}}$$

Standard deviation of the sample proportion:

$$\sigma_{\hat{p}} = \sqrt{\frac{p(1-p)}{n}}$$

z-score for a sample proportion:

$$z = \frac{\hat{p} - p}{\sigma_{\hat{p}}}$$

Chapter Quiz

1. Following is a probability density curve for a population.

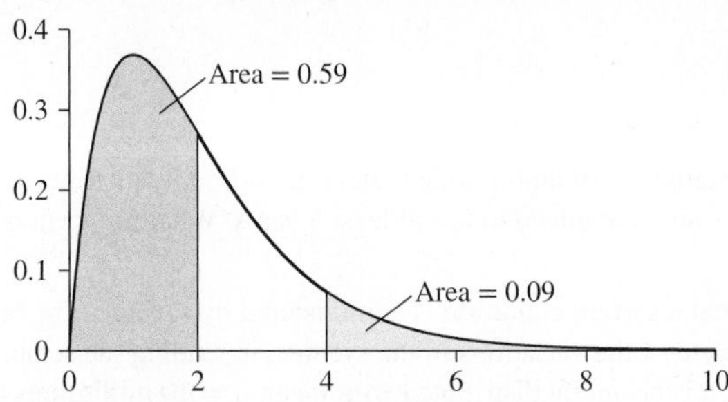

   a. What proportion of the population is between 2 and 4? *0.32*
   b. If a value is chosen at random from this population, what is the probability that it will be greater than 2? *0.41*

2. Find the area under the standard normal curve
   a. To the left of $z = 1.77$ *0.9616*
   b. To the right of $z = 0.41$ *0.3409*
   c. Between $z = -2.12$ and $z = 1.37$ *0.8977*

3. Find the z-score that has
   a. An area of 0.33 to its left *−0.44*
   b. An area of 0.79 to its right *−0.81*

4. Find the z-scores that bound the middle 80% of the area under the normal curve. *$z = -1.28$ and $z = 1.28$*

5. Find $z_{0.15}$. *1.04*

6. Suppose that salaries of recent graduates from a certain college are normally distributed with mean $\mu = \$42{,}650$ and standard deviation $\sigma = \$3800$. What two salaries bound the middle 50%? *$40,104 and $45,196 [Tech: $40,087 and $45,213]*

7. A normal population has mean $\mu = 242$ and standard deviation $\sigma = 31$.
   a. What proportion of the population is greater than 233? *0.6141 [Tech: 0.6142]*
   b. What is the probability that a randomly chosen value will be less than 249? *0.5910 [Tech: 0.5893]*

8. Suppose that in a bowling league, the scores among all bowlers are normally distributed with mean $\mu = 182$ points and standard deviation $\sigma = 14$ points. A trophy is given to each player whose score is at or above the 97th percentile. What is the minimum score needed for a bowler to receive a trophy? *209*

9. State the Central Limit Theorem.

10. A population has mean $\mu = 193$ and standard deviation $\sigma = 42$. Compute $\mu_{\bar{x}}$ and $\sigma_{\bar{x}}$ for samples of size $n = 64$. *$\mu_{\bar{x}} = 193$, $\sigma_{\bar{x}} = 5.25$*

11. The running time for videos submitted to YouTube in a given week is normally distributed with $\mu = 390$ seconds and standard deviation $\sigma = 148$ seconds.

a. If a single video is randomly selected, what is the probability that the running time of the video exceeds 6 minutes (360 seconds)? *0.5793 [Tech: 0.5803]*

b. Suppose that a sample of 40 videos is selected. What is the probability that the mean running time of the sample exceeds 6 minutes? *0.8997 [Tech: 0.9001]*

12. A sample of size $n = 55$ is drawn from a population with proportion $p = 0.34$. Let $\hat{p}$ be the sample proportion.

a. Find $\mu_{\hat{p}}$ and $\sigma_{\hat{p}}$. *$\mu_{\hat{p}} = 0.34$, $\sigma_{\hat{p}} = 0.063875$*

b. Find $P(\hat{p} > 0.21)$. *0.9793 [Tech: 0.9791]*

c. Find $P(\hat{p} < 0.40)$. *0.8264 [Tech: 0.8262]*

13. On a certain television channel, 18% of commercials are local advertisers. A sample of 120 commercials is selected.

a. What is the probability that more than 20% of the commercials in the sample are local advertisers? *0.2843 [Tech: 0.2842]*

b. Would it be unusual for more than 25% of the commercials to be local advertisers? *Yes*

14. Let X have a binomial distribution with $n = 240$ and $p = 0.38$. Use the normal approximation to find:

a. $P(X > 83)$ *0.8461 [Tech: 0.8471]*

b. $P(75 \leq X \leq 95)$ *0.7025 [Tech: 0.7031]*

c. $P(X < 96)$ *0.7157 [Tech: 0.7163]*

15. Is it reasonable to treat the following sample as coming from an approximately normal population? Explain. *Yes*

5.5 8.7 9.3 10.1 15.2 3.5 11.9 7.6 13.7 8.7 14.3 5.8

Review Exercises

1. Find the area: Find the area under the standard normal curve

a. To the left of $z = 0.35$ *0.6368*

b. To the right of $z = -1.56$ *0.9406*

c. Between $z = 0.35$ and $z = 2.47$ *0.3564*

2. Find the z-score: Find the z-score for which the area to its right is 0.89. *−1.23*

3. Your battery is dead: The lifetimes of a certain type of automobile battery are normally distributed with mean 5.9 years and standard deviation 0.4 year. The batteries are guaranteed to last at least 5 years. What proportion of the batteries fail to meet the guarantee? *0.0122*

4. Take your medicine: Medication used to treat a certain condition is administered by syringe. The target dose in a particular application is 10 milligrams. Because of the variations in the syringe, in reading the scale, and in mixing the fluid suspension, the actual dose administered is normally distributed with mean $\mu = 10$ milligrams and standard deviation $\sigma = 1.6$ milligrams.

a. What is the probability that the dose administered is between 9 and 11.5 milligrams? *0.5621 [Tech: 0.5598]*

b. Find the 98th percentile of the administered dose. *13.280 [Tech: 13.286]*

c. If a clinical overdose is defined as a dose larger than 15 milligrams, what is the probability that a patient will receive an overdose? *0.0009*

5. Lightbulbs: The lifetime of lightbulbs has a mean of 1500 hours and a standard deviation of 100 hours. A sample of 50 lightbulbs is tested.

a. What is the probability that the sample mean lifetime is greater than 1520 hours? *0.0793 [Tech: 0.0786]*

b. What is the probability that the sample mean lifetime is less than 1540 hours? *0.9977*

c. What is the probability that the sample mean lifetime is between 1490 and 1550 hours? *0.7609 [Tech: 0.7600]*

6. More lightbulbs: Someone claims to have developed a new lightbulb whose mean lifetime is 1800 hours with a standard deviation of 100 hours. A sample of 100 of these bulbs is tested. The sample mean lifetime is 1770 hours.

a. If the claim is true, what is the probability of obtaining a sample mean that is less than or equal to 1770 hours? *0.0013*

b. If the claim is true, would it be unusual to obtain a sample mean that is less than or equal to 1770 hours? *Yes*

7. Pay your taxes: Among all the state income tax forms filed in a particular state, the mean income tax paid was $\mu = \$2000$ and the standard deviation was $\sigma = \$500$. As part of a study of the impact of a new tax law, a sample of 80 income tax returns is examined. Would it be unusual for the sample mean of these 80 returns to be greater than $2150? *Yes*

8. **Safe delivery:** A certain delivery truck can safely carry a load of 3400 pounds. The cartons that will be loaded onto the truck have a mean weight of 80 pounds with a standard deviation of 20 pounds. Forty cartons are loaded onto the truck.
   a. If the total weight of the 40 cartons is 3400 pounds, what is the sample mean weight? *85 pounds*
   b. What is the probability that the truck can deliver the 40 cartons safely? *0.9429 [Tech: 0.9431]*

9. **Elementary school:** In a certain elementary school, 52% of the students are girls. A sample of 65 students is drawn.
   a. What is the probability that more than 60% of them are girls? *0.0985 [Tech: 0.0984]*
   b. Would it be unusual for more than 70% of them to be girls? *Yes*

10. **Facebook:** Eighty percent of the students at a particular large university have logged on to Facebook at least once in the past week. A sample of 95 students is asked about their internet habits.
    a. What is the probability that less than 75% of the sampled students have logged on to Facebook within the last week? *0.1112 [Tech: 0.1115]*
    b. What is the probability that more than 78% of the sampled students have logged on to Facebook within the last week? *0.6879 [Tech: 0.6870]*
    c. What is the probability that the proportion of the sampled students who have logged on to Facebook within the last week is between 0.82 and 0.85? *0.2009 [Tech: 0.2015]*

11. **It's all politics:** A politician in a close election race claims that 52% of the voters support him. A poll is taken in which 200 voters are sampled, and 44% of them support the politician.
    a. If the claim is true, what is the probability of obtaining a sample proportion that is less than or equal to 0.44? *0.0119 [Tech: 0.0118]*
    b. If the claim is true, would it be unusual to obtain a sample proportion less than or equal to 0.44? *Yes*
    c. If the claim is true, would it be unusual for less than half of the voters in the sample to support the politician? *No*

12. **Side effects:** A new medical procedure produces side effects in 25% of the patients who receive it. In a clinical trial, 60 people undergo the procedure. What is the probability that 20 or fewer experience side effects? *0.9495*

13. **Defective rods:** A grinding machine is used to manufacture steel rods, of which 5% are defective. When a customer orders 1000 rods, a package of 1060 rods is shipped, with a guarantee that at least 1000 of the rods are good. What is the probability that a package of 1060 rods contains 1000 or more that are good? *0.8554 [Tech: 0.8547]*

14. **Is it normal?** Is it reasonable to treat the following sample as though it comes from an approximately normal population? Explain. *No*

    2.6 4.2 1.5 2.0 0.6 0.7 6.6 2.2 9.7 1.8 4.2 4.4 0.6

15. **Is it normal?** Is it reasonable to treat the following sample as though it comes from an approximately normal population? Explain. *Yes*

    8.8 11.2 11.6 6.3 9.3 1.5 14.6 7.5 5.2 9.0 4.3 9.9 7.8 13.1

Write About It

1. Explain why $P(a < X < b)$ is equal to $P(a \le X \le b)$ when X is a continuous random variable.
2. Describe the information you must know to compute the area under the normal curve over a given interval.
3. Describe the information you must know to find the value corresponding to a given proportion of the area under a normal curve.
4. Suppose that in a large class, the instructor announces that the average grade on an exam is 75. Which is more likely to be closer to 75:
   i. The exam grade of a randomly selected student in the class?
   ii. The mean exam grade of a sample of 10 students?
   Explain.
5. Consider the formula for the standard deviation of the sampling distribution of $\hat{p}$ given by $\sigma_{\hat{p}} = \sqrt{\frac{p(1-p)}{n}}$. What happens to the standard deviation as n gets larger and larger? Explain what this means in terms of the spread of the sampling distribution.
6. Explain how to decide when it is appropriate to use the normal approximation to the binomial distribution.
7. Describe the effect, if any, that the size of a sample has in assessing the normality of a population.

In-Class Activities

1. **Converging to normality:** Each student rolls two dice five times. Each time, record both the sum and the product of the numbers on the dice. Compute the mean and standard deviation of the five sums and of the five products. Construct histograms of the means of the sums and the means of the products. Which histogram is closer to the normal distribution? Now construct histograms of the individual sums and products. How do these histograms explain the results observed for the histograms of the means? How would the results differ if you rolled the dice 50 times rather than 5?
2. **Standard error of the mean:** Each student rolls a single die 10 times, records the results, and computes the mean of the 10 values obtained. Construct a histogram of all the individual die rolls and another one of the sample means. Compute the standard deviation of the individual die rolls and the standard deviation of the sample means. Discuss how these results illustrate the ideas presented in this chapter.

Case Study: Testing The Strength Of Cans

In the chapter opener, we discussed a method used to determine whether shipments of aluminum cans are strong enough to withstand the pressure of containing a carbonated beverage. Several cans are sampled from a shipment and tested to determine the pressure they can withstand. Based on this small sample, quality inspectors must estimate the proportion of cans that will fail at or below a certain threshold, which we will take to be 90 pounds per square inch.

The quality control inspectors want the proportion of defective cans to be no more than 0.001, or 1 in 1000. They test 10 cans, with the following results.

Can	1	2	3	4	5	6	7	8	9	10
Pressure at failure	95	96	98	99	99	100	101	101	103	104

None of the cans in the sample were defective; in other words, none of them failed at a pressure of 90 or less. The quality control inspectors want to use these data to estimate the proportion of defective cans in the shipment. If the estimate is to be no more than 0.001, or 1 in 1000, they will accept the shipment; otherwise, they will return it for a refund.

The following exercises will lead you through the process used by the quality control inspectors. Assume the failure pressures are normally distributed.

1. Compute the sample mean $\bar{x}$ and the sample standard deviation s. $\bar{x} = 99.6$, $s = 2.8363$
2. Estimate the population mean μ with $\bar{x}$ and the population standard deviation σ with s. In other words, assume that the data are a sample from a normal population with mean $\mu = \bar{x}$ and standard deviation $\sigma = s$. Under this assumption, what proportion of cans will fail at a pressure of 90 or less? 0.0004
3. The shipment will be accepted if we estimate that the proportion of cans that fail at a pressure of 90 or less is less than 0.001. Will this shipment be accepted? Yes
4. A second shipment of cans is received. Ten randomly sampled cans are tested with the following results.

Can	1	2	3	4	5	6	7	8	9	10
Pressure at failure	96	97	99	100	100	100	101	103	103	120

   Explain why the second sample of cans is stronger than the first sample.
5. Compute the sample mean $\bar{x}$ and the sample standard deviation s for the second sample. $\bar{x} = 101.9$, $s = 6.7404$
6. Using the same method as for the first sample, estimate the proportion of cans that will fail at a pressure of 90 or less. 0.0384 [Tech: 0.0387]
7. The shipment will be accepted if we estimate that the proportion of cans that fail at a pressure of 90 or less is less than 0.001. Will this shipment be accepted? No
8. Make a boxplot of the pressures for the second sample. Is the method appropriate for the second shipment? No

Confidence Intervals

chapter

8

Schroptschop/E+/Getty Images

Introduction

Air pollution is a serious problem in many places. Particulate matter (PM), which consists of tiny particles in the air, is a form of air pollution that is suspected of causing respiratory illness. PM can come from many sources, such as ash from burning, and tiny particles of rubber from automobile and truck tires.

The town of Libby, Montana, has experienced high levels of PM, especially in the winter. Many houses in Libby are heated by wood stoves, which produce a lot of particulate pollution. In an attempt to reduce the winter levels of air pollution in Libby, a program was undertaken in which almost every wood stove in the town was replaced with a newer, cleaner-burning model. Measurements of several air pollutants, including PM, were taken both before and after the stove replacement. Following are PM levels for 20 days in the winter after the stoves were replaced. The units are micrograms per cubic meter.

21.7	27.8	24.7	15.3	18.4	14.4	19.0	23.7	22.4	25.6
15.0	17.0	23.2	17.7	11.1	29.8	20.0	21.6	14.8	21.0

Clearly, the amount varies from day to day. The sample mean is 20.21 and the sample standard deviation is 4.86. We would like to use this information to estimate the population mean, which is the mean over all days that winter. What is the best way to do this? If we had to pick a single number to estimate the population mean, the best choice would be the sample mean, which is 20.21. The estimate 20.21 is called a *point estimate*,

NOTE TO INSTRUCTOR

Section 8.1 covers confidence intervals for a mean with standard deviation known, which is the situation where the computations are simplest. Concepts such as the relationship between confidence and margin of error, calculating sample sizes, and distinguishing between probability and confidence are covered in Section 8.2, so as to present them in a setting more commonly encountered in practice.

because it is a single number. The problem with point estimates is that they are almost never exactly equal to the true values they are estimating. They are almost always off—sometimes by a little, sometimes by a lot. It is unlikely that the mean level of PM for the entire winter is exactly equal to 20.21; it is somewhat more or less than that. In order for a point estimate to be useful, it is necessary to describe just how close to the true value it is likely to be. To do this, statisticians construct *confidence intervals*. A confidence interval gives a range of values that is likely to contain the true value being estimated.

To construct a confidence interval, we put a plus-or-minus number on the point estimate. So, for example, we might estimate that the population mean is 20.21 ± 2.0, or equivalently, that the population mean is between 18.21 and 22.21. The interval 20.21 ± 2.0, or equivalently, (18.21, 22.21), is a confidence interval for the population mean.

One of the benefits of confidence intervals is that they come with a measure of the level of confidence we can have that they actually cover the true value being estimated. For example, we will show that we can be 95% confident that the population mean PM level during the winter is in the interval 20.21 ± 2.27, or equivalently, between 17.94 and 22.48. If we want more confidence, we can widen the interval. For example, we will learn how to show that we can be 99% confident that the population mean PM level is in the interval 20.21 ± 3.11, or equivalently, between 17.10 and 23.32. In the case study at the end of the chapter, we will use confidence intervals to further study the effects of the Libby stove replacement program.

We begin with the construction of confidence intervals for a population mean. The fundamental principle is that the smaller the population standard deviation, the closer one can expect the sample mean to be to the population mean, so the narrower the confidence interval can be. When the population standard deviation is known, it is straightforward to illustrate how this principle is used to determine the appropriate width of the confidence interval. For this reason, we begin with the case where the population standard deviation is known. In practice, the population standard deviation is generally not known. In Section 8.2, we describe how to construct confidence intervals in that case.

Section 8.1 Confidence Intervals for a Population Mean, Standard Deviation Known

Objectives

1. Construct and interpret confidence intervals for a population mean when the population standard deviation is known
2. Find critical values for confidence intervals

The Investigative Process of Statistics

In Chapter 1, we described statistical analysis as a four-step process, whose steps are:

- Formulate questions.
- Collect data needed to answer the questions.
- Describe the data.
- Draw conclusions, using appropriate methods.

We have covered the collection of data (Chapter 1), and the description of data (Chapters 2–4). Having laid the groundwork for drawing conclusions (Chapters 5–7), we now discuss the construction of confidence intervals, which are one of the most commonly used methods for drawing conclusions.

NOTE TO INSTRUCTOR

The following topics from the **Statistics Corequisite Workbook** are aligned with the material in Sections 8.2 and 8.3 of this chapter.

6.1 – Operations in Confidence Intervals

6.2 – Finding the Necessary Sample Size for a Confidence Interval

Objective 1 Construct and interpret confidence intervals for a population mean when the population standard deviation is known

Estimating a Population Mean

How can we measure the reading ability of elementary school students? In a certain school district, administrators are trying out a new experimental approach to teach reading to fourth graders. A simple random sample of 100 fourth graders is selected to take part in the program. At the end of the program, the students are given a standardized reading test. On the basis of past results, it is known that scores on this test have a population standard deviation of $\sigma = 15$.

RECALL

A parameter is a numerical summary of a population, such as a population mean μ or a population proportion p.

The sample mean score for the 100 students was $\bar{x} = 67.30$. The administrators want to estimate what the mean score μ would be if the entire population of fourth graders in the district had enrolled in the program. The best estimate for the population mean is the sample mean, $\bar{x} = 67.30$. The sample mean is a *point estimate*, because it is a single number.

DEFINITION

A **point estimate** is a single number that is used to estimate the value of an unknown parameter.

EXPLAIN IT AGAIN

The symbol ±: The symbol ± means to form an interval by adding and subtracting. For example, 67.30 ± 1 means the interval from 67.30 − 1 to 67.30 + 1, or, in other words, from 66.30 to 68.30.

It is very unlikely that the point estimate $\bar{x}$ is exactly equal to the population mean μ. Therefore, in order for the estimate to be useful, we must describe how close it is likely to be. For example, if we think that $\bar{x} = 67.30$ is likely to be within 1 point of the population mean, we would estimate μ with the interval $66.30 < \mu < 68.30$. This could also be written as 67.30 ± 1. If we think that $\bar{x} = 67.30$ could be off by as much as 10 points from the population mean, we would estimate μ with the interval $57.30 < \mu < 77.30$, which could also be written as 67.30 ± 10. The plus-or-minus number is called the **margin of error**. We need to determine how large to make the margin of error so that the interval is likely to contain the population mean. To do this, we use the sampling distribution of $\bar{x}$.

Because the sample size is large ($n > 30$), the Central Limit Theorem tells us that the sampling distribution of $\bar{x}$ is approximately normal with mean μ and standard deviation (also called the **standard error**) given by

$$\text{Standard error} = \frac{\sigma}{\sqrt{n}} = \frac{15}{\sqrt{100}} = 1.5$$

RECALL

The quantity $\sigma/\sqrt{n}$ is called the standard error of the mean.

We will now construct a *95% confidence interval* for μ. We begin with a normal curve, and using the method described in Section 7.1 (Example 7.12), we find the z-scores that bound the middle 95% of the area under the curve. These z-scores are 1.96 and −1.96 (see Figure 8.1). The value 1.96 is called the **critical value**. To obtain the margin of error, we multiply the critical value by the standard error.

$$\text{Margin of error} = \text{Critical value} \cdot \text{Standard error} = (1.96)(1.5) = 2.94$$

A 95% confidence interval for μ is therefore

$$\bar{x} - 2.94 < \mu < \bar{x} + 2.94$$

$$67.30 - 2.94 < \mu < 67.30 + 2.94$$

$$64.36 < \mu < 70.24$$

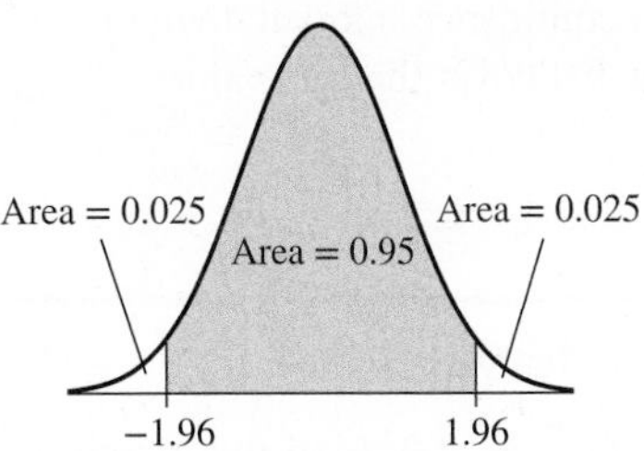

Figure 8.1 95% of the area under the standard normal curve is between $z = -1.96$ and $z = 1.96$.

There are several ways to express this confidence interval. We can write $64.36 < \mu < 70.24$, 67.30 ± 2.94, or (64.36, 70.24). In words, we would say, "We are 95% confident that the population mean is between 64.36 and 70.24."

Figures 8.2 and 8.3 help explain why this interval is called a 95% confidence interval. Figure 8.2 illustrates a sample whose mean $\bar{x}$ is in the middle 95% of its distribution. The 95% confidence interval constructed from this value of $\bar{x}$ covers the true population mean μ.

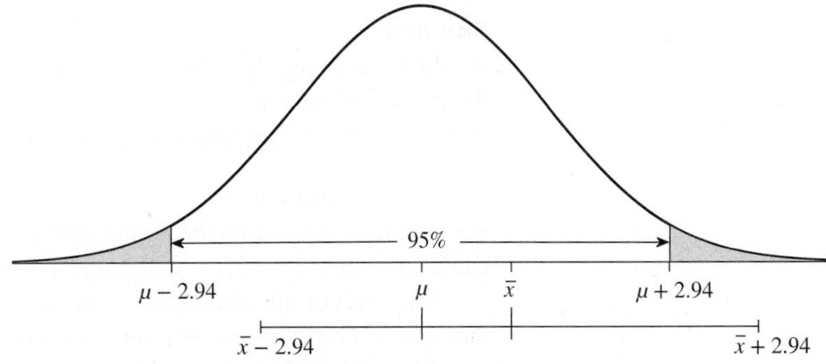

Figure 8.2 The sample mean $\bar{x}$ comes from the middle 95% of the distribution, so the 95% confidence interval $\bar{x} \pm 2.94$ succeeds in covering the population mean μ.

Figure 8.3 illustrates a sample whose mean $\bar{x}$ is in one of the tails of the distribution, outside the middle 95%. The 95% confidence interval constructed from this value of $\bar{x}$ does *not* cover the true population mean μ.

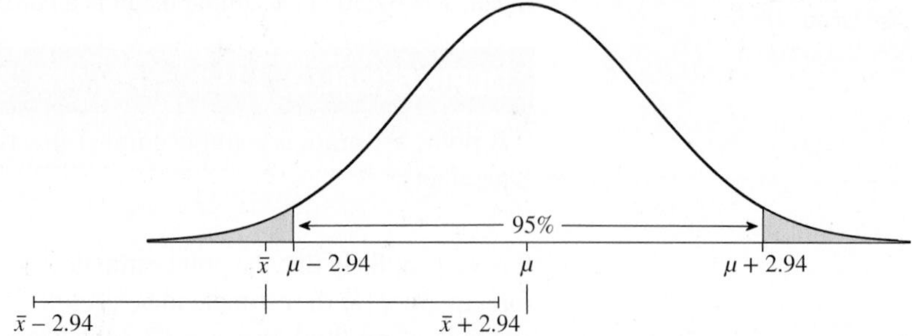

Figure 8.3 The sample mean $\bar{x}$ comes from the outer 5% of the distribution, so the 95% confidence interval $\bar{x} \pm 2.94$ fails to cover the population mean μ.

Now in the long run, 95% of the sample means we observe will be like the one in Figure 8.2. They will be in the middle 95% of the distribution and their confidence intervals will cover the population mean. Therefore, in the long run, 95% of the confidence intervals we construct will cover the population mean. Only 5% of the sample means we observe will be outside the middle 95% of the distribution like the one in Figure 8.3. So in the long run, only 5% of the confidence intervals we construct will fail to cover the population mean.

To summarize, the confidence interval we just constructed is a 95% confidence interval, because the method we used to construct it will cover the population mean μ for 95% of all the possible samples that might be drawn. We can also say that the interval has a *confidence level* of 95%.

EXPLAIN IT AGAIN

Confidence levels: A confidence level may be any percentage between 0% and 100%. However, in practice we use large values such as 90%, 95%, or 99%, because we want to have high confidence that our intervals contain the true value.

DEFINITION

- A **confidence interval** is an interval that is used to estimate the value of a parameter.
- The **confidence level** is a percentage between 0% and 100% that measures the success rate of the method used to construct the confidence interval. If we were to draw many samples and use each one to construct a confidence interval, then in the long run, the percentage of confidence intervals that cover the true value would be equal to the confidence level.

Example 8.1

Construct and interpret a 95% confidence interval

A large sample has mean $\bar{x} = 7.1$ and standard error $\sigma/\sqrt{n} = 2.3$. Find the margin of error for a 95% confidence interval. Construct a 95% confidence interval for the population mean μ, and explain what it means to say that the confidence level is 95%.

Solution

As shown in Figure 8.1, the critical value for a 95% confidence interval is 1.96. Therefore, the margin of error is

$$\text{Margin of error} = \text{Critical value} \cdot \text{Standard error} = (1.96)(2.3) = 4.5$$

The point estimate of μ is $\bar{x} = 7.1$. To construct a confidence interval, we add and subtract the margin of error from the point estimate. So the 95% confidence interval is 7.1 ± 4.5. We can also write this as $7.1 - 4.5 < \mu < 7.1 + 4.5$, or $2.6 < \mu < 11.6$.

The level of this confidence interval is 95% because if we were to draw many samples and use this method to construct the corresponding confidence intervals, then in the long run, 95% of the intervals would cover the true value of the population mean μ. Unless we were unlucky in the sample we drew, the population mean μ will be between 2.6 and 11.6.

Objective 2 Find critical values for confidence intervals

Finding the critical value for a given confidence level

Although 95% is the most commonly used confidence level, sometimes we will want to construct a confidence interval with a different level. We can construct a confidence interval with any confidence level between 0% and 100% by finding the appropriate critical value for that level. We have seen that the critical value for a 95% confidence interval is $z = 1.96$, because 95% of the area under a normal curve is between $z = -1.96$ and $z = 1.96$. Similarly, the critical value for a 99% confidence interval is the z-score for which the area between z and $-z$ is 0.99, the critical value for a 98% confidence interval is the z-score for which the area between z and $-z$ is 0.98, and so on.

The row of Table A.3 labeled "z" presents critical values for several confidence levels. Following is part of that row, which presents critical values for four of the most commonly used confidence levels.

z	···	1.645	1.96	2.326	2.576	···
	···	90%	95%	98%	99%	···
	Confidence level					

Example 8.2

Construct confidence intervals of various levels

A large sample has mean $\bar{x} = 7.1$ and standard error $\sigma/\sqrt{n} = 2.3$. Construct confidence intervals for the population mean μ with the following levels:

a. 90% **b.** 98% **c.** 99%

NOTE TO INSTRUCTOR

The intuition behind the fact that higher levels of confidence correspond to wider confidence intervals is straightforward, but worth explaining. The wider we make the confidence interval, the more confident we are that it reaches far enough to contain the true value.

Solution

The point estimate of μ is $\bar{x} = 7.1$.

a. From the bottom row of Table A.3, we see that the critical value for a 90% confidence interval is 1.645, so the margin of error is

$$\text{Margin of error} = \text{Critical value} \cdot \text{Standard error} = (1.645)(2.3) = 3.8$$

The 90% confidence interval is 7.1 ± 3.8. We can also write this as

$$7.1 - 3.8 < \mu < 7.1 + 3.8, \text{ or } 3.3 < \mu < 10.9.$$

b. From the bottom row of Table A.3, we see that the critical value for a 98% confidence interval is 2.326, so the margin of error is

$$\text{Margin of error} = \text{Critical value} \cdot \text{Standard error} = (2.326)(2.3) = 5.3$$

The 98% confidence interval is 7.1 ± 5.3. We can also write this as

$$7.1 - 5.3 < \mu < 7.1 + 5.3, \text{ or } 1.8 < \mu < 12.4.$$

c. From the bottom row of Table A.3, we see that the critical value for a 99% confidence interval is 2.576, so the margin of error is

$$\text{Margin of error} = \text{Critical value} \cdot \text{Standard error} = (2.576)(2.3) = 5.9$$

The 99% confidence interval is 7.1 ± 5.9. We can also write this as

$$7.1 - 5.9 < \mu < 7.1 + 5.9, \text{ or } 1.2 < \mu < 13.0.$$

The notation z_α

Sometimes we may need to find a critical value for a confidence level not given in the last row of Table A.3. To do this, it is useful to learn a notation for a z-score with a given area to its right.

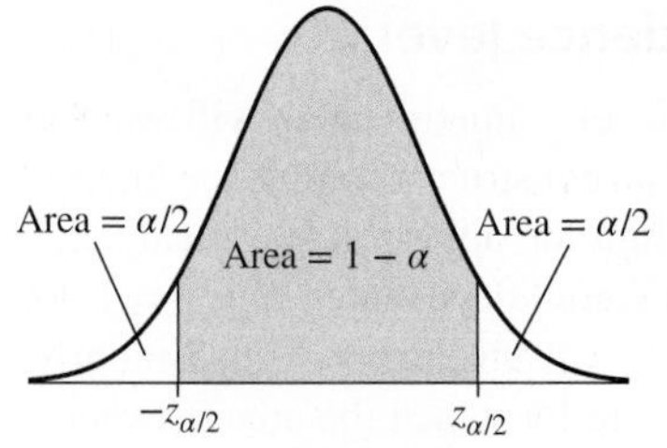

Figure 8.4

DEFINITION

Let α be any number between 0 and 1; in other words, $0 < \alpha < 1$.

- The notation z_α refers to the z-score with an area of α to its right.
- The notation $z_{\alpha/2}$ refers to the z-score with an area of $\alpha/2$ to its right.

To find the critical value for a confidence interval with a given level, let $1 - \alpha$ be the confidence level expressed as a decimal. The critical value is then $z_{\alpha/2}$, because the area under the standard normal curve between $-z_{\alpha/2}$ and $z_{\alpha/2}$ is $1 - \alpha$. See Figure 8.4.

Example 8.3

Find a critical value

Find the critical value $z_{\alpha/2}$ for a 92% confidence interval.

NOTE TO INSTRUCTOR

We define the confidence level as $1 - \alpha$, because, as we will discuss in Chapter 9, α is the significance level for a hypothesis test. We reject a two-tailed hypothesis at level α if a level $1 - \alpha$ confidence interval does not contain the null value of the parameter.

Solution

The level is 92%, so we have $1 - \alpha = 0.92$. It follows that $\alpha = 1 - 0.92 = 0.08$, so $\alpha/2 = 0.04$. The critical value is $z_{0.04}$. We now must find the value of $z_{0.04}$. To do this using Table A.2, we find the area to the left of $z_{0.04}$. Since the area to the right of $z_{0.04}$ is 0.04, the area to the left is $1 - 0.04 = 0.96$. See Figure 8.5.

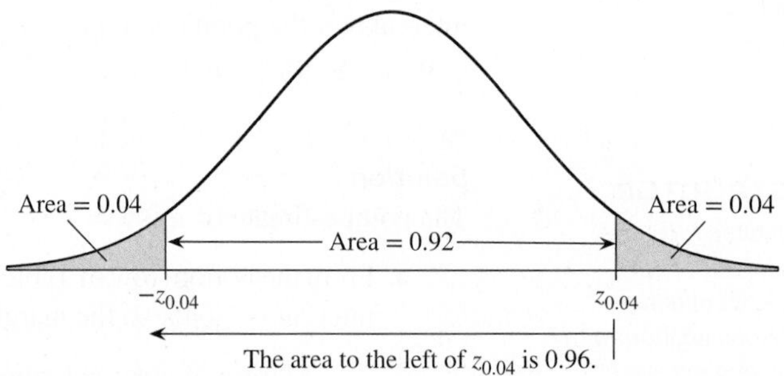

Figure 8.5 The critical value $z_{0.04}$ contains an area of 0.92 under the standard normal curve between $-z_{0.04}$ and $z_{0.04}$.

We now look in the body of Table A.2 (a portion of which is shown in Figure 8.6) to find the closest value to 0.96. This value is 0.9599, and it corresponds to a z-score of 1.75. Therefore, the critical value is $z_{0.04} = 1.75$.

z	0.00	0.01	0.02	0.03	0.04	0.05	0.06	0.07	0.08	0.09
⋮	⋮	⋮	⋮	⋮	⋮	⋮	⋮	⋮	⋮	⋮
1.2	.8849	.8869	.8888	.8907	.8925	.8944	.8962	.8980	.8997	.9015
1.3	.9032	.9049	.9066	.9082	.9099	.9115	.9131	.9147	.9162	.9177
1.4	.9192	.9207	.9222	.9236	.9251	.9265	.9279	.9292	.9306	.9319
1.5	.9332	.9345	.9357	.9370	.9382	.9394	.9406	.9418	.9429	.9441
1.6	.9452	.9463	.9474	.9484	.9495	.9505	.9515	.9525	.9535	.9545
1.7	.9554	.9564	.9573	.9582	.9591	.9599	.9608	.9616	.9625	.9633
1.8	.9641	.9649	.9656	.9664	.9671	.9678	.9686	.9693	.9699	.9706
1.9	.9713	.9719	.9726	.9732	.9738	.9744	.9750	.9756	.9761	.9767
2.0	.9772	.9778	.9783	.9788	.9793	.9798	.9803	.9808	.9812	.9817
2.1	.9821	.9826	.9830	.9834	.9838	.9842	.9846	.9850	.9854	.9857
2.2	.9861	.9864	.9868	.9871	.9875	.9878	.9881	.9884	.9887	.9890
⋮	⋮	⋮	⋮	⋮	⋮	⋮	⋮	⋮	⋮	⋮

Figure 8.6

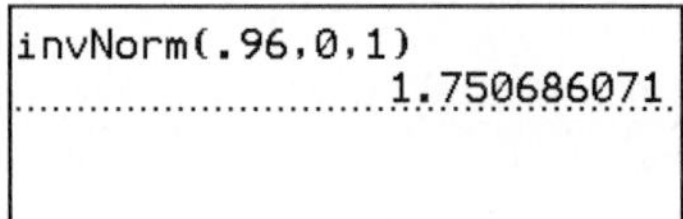

Figure 8.7

As an alternative to Table A.2, we can find $z_{0.04}$ with technology, for example, by using the TI-84 Plus calculator. Figure 8.5 shows that the area to the left of $z_{0.04}$ is 0.96. Therefore, we can find $z_{0.04}$ with the command **invNorm(.96,0,1)**. See Figure 8.7.

Check Your Understanding

1. Find the critical value $z_{\alpha/2}$ to construct a confidence interval with level

a. 90% 1.645 **b.** 98% 2.326 **c.** 99.5% 2.81 **d.** 80% 1.28

2. Find the levels of the confidence intervals that have the following critical values.

a. $z_{\alpha/2} = 1.96$ 95% **b.** $z_{\alpha/2} = 2.17$ 97% **c.** $z_{\alpha/2} = 1.28$ 80%

d. $z_{\alpha/2} = 3.28$ 99.9%

3. Find the margin of error given the standard error and the confidence level.

a. Standard error = 1.2, confidence level 95% 2.352

b. Standard error = 0.4, confidence level 99% 1.030

c. Standard error = 3.5, confidence level 90% 5.758 [Tech: 5.757]

d. Standard error = 2.75, confidence level 98% 6.397

Answers are on page 385.

Assumptions

The method we have described for constructing a confidence interval requires us to assume that the population standard deviation σ is known. In practice, it is more common that σ is not known. The advantage of first learning the method that assumes σ known is that it allows us to use the familiar normal distribution, so we can focus on the basic ideas of confidence intervals. We will learn how to construct confidence intervals when σ is unknown in Section 8.2.

Following are the assumptions for the method we describe.

EXPLAIN IT AGAIN

$\bar{x}$ must be approximately normal: We need assumption 2 to be sure that the sampling distribution of $\bar{x}$ is approximately normal. This allows us to use $z_{\alpha/2}$ as the critical value.

Assumptions for Constructing a Confidence Interval for μ When σ Is Known

1. We have a simple random sample.
2. The sample size is large ($n > 30$), or the population is approximately normal.

When these assumptions are met, we can use the following steps to construct a confidence interval for μ when σ is known.

Procedure for Constructing a Confidence Interval for μ When σ Is Known

Check to be sure the assumptions are satisfied. If they are, then proceed with the following steps.

Step 1: Find the value of the point estimate $\bar{x}$, if it isn't given.

Step 2: Find the critical value $z_{\alpha/2}$ corresponding to the desired confidence level from the last row of Table A.3, from Table A.2, or with technology.

Step 3: Find the standard error $\sigma/\sqrt{n}$, and multiply it by the critical value to obtain the margin of error $z_{\alpha/2}\dfrac{\sigma}{\sqrt{n}}$.

Step 4: Use the point estimate and the margin of error to construct the confidence interval:

$$\text{Point estimate} \pm \text{Margin of error}$$

$$\bar{x} \pm z_{\alpha/2}\frac{\sigma}{\sqrt{n}}$$

$$\bar{x} - z_{\alpha/2}\frac{\sigma}{\sqrt{n}} < \mu < \bar{x} + z_{\alpha/2}\frac{\sigma}{\sqrt{n}}$$

Step 5: Interpret the result.

Rounding off the final result

When constructing a confidence interval for a population mean, you may be given a value for $\bar{x}$, or you may be given the data and have to compute $\bar{x}$ yourself. If you are given the value of $\bar{x}$, round the final result to the same number of decimal places as $\bar{x}$. If you are given data, then round $\bar{x}$ and the final result to one more decimal place than is given in the data.

Although you should round off your final answer, do not round off the calculations you make along the way. Doing so may affect the accuracy of your final answer.

Example 8.4

Construct a confidence interval

The mean test score for a simple random sample of $n = 100$ students was $\bar{x} = 67.30$. The population standard deviation of test scores is $\sigma = 15$. Construct a 98% confidence interval for the population mean test score μ.

Solution

First we check the assumptions. The sample is a simple random sample, and the sample size is large ($n > 30$). The assumptions are met, so we may proceed.

Step 1: Find the point estimate. The point estimate is the sample mean $\bar{x} = 67.30$.

Step 2: Find the critical value $z_{\alpha/2}$. The desired confidence level is 98%. We look on the last line of Table A.3 and find that the critical value is $z_{\alpha/2} = 2.326$.

Step 3: Find the standard error and the margin of error. The standard error is

$$\frac{\sigma}{\sqrt{n}} = \frac{15}{\sqrt{100}} = 1.5$$

We multiply the standard error by the critical value to obtain the margin of error:

$$\text{Margin of error} = z_{\alpha/2}\frac{\sigma}{\sqrt{n}} = 2.326(1.5) = 3.489$$

Step 4: Construct the confidence interval. The 98% confidence interval is

$$\bar{x} - z_{\alpha/2}\frac{\sigma}{\sqrt{n}} < \mu < \bar{x} + z_{\alpha/2}\frac{\sigma}{\sqrt{n}}$$

$$67.30 - 3.489 < \mu < 67.30 + 3.489$$

$$63.81 < \mu < 70.79 \quad \text{(rounded to two decimal places, like } \bar{x}\text{)}$$

Step 5: Interpret the result. We are 98% confident that the population mean score μ is between 63.81 and 70.79. Another way to say this is that we are 98% confident that μ is in the interval 67.30 ± 3.49. If we were to draw many different samples and use this method to construct the corresponding confidence intervals, then in the long run, 98% of the intervals would cover the true population mean μ. So unless we were quite unlucky in the sample we drew, the population mean μ will be between 63.81 and 70.79.

Check Your Understanding

4. An IQ test was given to a simple random sample of 75 students at a certain college. The sample mean score was 105.2. Scores on this test are known to have a standard deviation of $\sigma = 10$. It is desired to construct a 90% confidence interval for the mean IQ score of students at this college.

- **a.** What is the point estimate? *105.2*
- **b.** Find the critical value. *1.645*
- **c.** Find the standard error. *1.1547*
- **d.** Find the margin of error. *1.90*
- **e.** Construct the 90% confidence interval. *(103.3, 107.1)*
- **f.** Is it likely that the population mean μ is greater than 100? Explain. *Yes*

5. The lifetime of a certain type of battery is known to be normally distributed with standard deviation $\sigma = 20$ hours. A sample of 50 batteries had a mean lifetime of 120.1 hours. It is desired to construct a 95% confidence interval for the mean lifetime for this type of battery.

a. What is the point estimate? *120.1*

b. Find the critical value. *1.96*

c. Find the standard error. *2.8284*

d. Find the margin of error. *5.54*

e. Construct the 95% confidence interval. *(114.6, 125.6)*

f. Is it likely that the population mean μ is greater than 130? Explain. *No*

6. In a survey of a simple random sample of students at a certain college, the sample mean time per week spent watching television was 18.3 hours and the margin of error for a 95% confidence interval was 1.2 hours. True or false:

a. A 95% confidence interval for the mean number of hours per week spent watching television by students at this college is $17.1 < \mu < 19.5$. *True*

b. Approximately 95% of the students at this university watch between 17.1 and 19.5 hours of television per week. *False*

7. Use the data in Exercise 4 to construct a 95% confidence interval for the mean IQ score. *(102.9, 107.5)*

8. Use the data in Exercise 5 to construct a 98% confidence interval for the mean lifetime for this type of battery. *(113.5, 126.7)*

Answers are on page 385.

Constructing confidence intervals with technology

In Example 8.4, we found a 98% confidence interval for the mean test score, based on a sample size of $n = 100$, a sample mean of $\bar{x} = 67.30$, and a population standard deviation $\sigma = 15$. The following TI-84 Plus display presents the results.

```
ZInterval
(63.81,70.79)
x̄=67.3
n=100
```

The display presents the confidence interval (63.81, 70.79), along with the values of $\bar{x}$ and n. Note that the confidence level (98%) is not given.

CAUTION

Confidence intervals constructed using technology may differ from those constructed by hand due to rounding. The differences are never large enough to matter.

Following is Excel output for this example. The Excel function **=CONFIDENCE .NORM** returns the margin of error. The inputs are the value of α, the population standard deviation σ, and the sample size n.

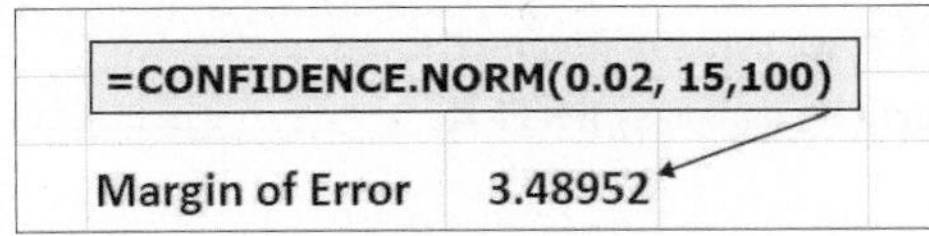

Finally, we present MINITAB output for the same example.

```
The assumed sigma = 15.0000
Variable     N      Mean      StDev    SE Mean          98% CI
Score      100   67.3000    15.0000    1.50000    (63.8105, 70.7895)
```

The output is fairly straightforward. Going from left to right, "N" represents the sample size, "Mean" is the sample mean $\bar{x}$, "StDev" is the population standard deviation σ, and "SE Mean" is the standard error $\sigma/\sqrt{n}$. The lower and upper confidence limits of the 98% confidence interval are given on the right. Note that neither the critical value nor the margin of error is given explicitly in the output.

Step-by-step instructions for constructing confidence intervals with technology are given in the Using Technology section on page 380.

Check Your Understanding

9. To estimate the accuracy of a laboratory scale, a weight known to have a mass of 100 grams is weighed 32 times. The reading of the scale is recorded each time. The following MINITAB output presents a 95% confidence interval for the mean reading of the scale.

```
The assumed sigma = 2.5000
Variable       N      Mean     StDev   SE Mean           95% CI
Scale Reading  32  102.3527   2.5000   0.44194   (101.4865, 103.2189)
```

A scientist claims that the mean reading μ is actually 100 grams. Is it likely that this claim is true? *No*

10. Using the output in Exercise 9:

a. Find the critical value $z_{\alpha/2}$ for a 99% confidence interval. *2.576*

b. Use the critical value along with the information in the output to construct a 99% confidence interval for the mean reading of the scale. *(101.2143, 103.4911)*

Answers are on page 385.

Using Technology

We use Example 8.4 to illustrate the technology steps.

TI-84 PLUS

Constructing a confidence interval for the mean when σ is known

Step 1. Press **STAT** and highlight the **TESTS** menu.

Step 2. Select **ZInterval** and press **ENTER** (Figure A). The **ZInterval** menu appears.

Step 3. For **Inpt**, select the **Stats** option and enter the values of σ, $\bar{x}$, and n. For Example 8.4, we use $\sigma = 15$, $\bar{x} = 67.30$, and $n = 100$.

Step 4. In the **C-Level** field, enter the confidence level. For Example 8.4, we use 0.98 (Figure B).

Step 5. Highlight **Calculate** and press **ENTER** (Figure C).

Note that if the raw data are given, the **ZInterval** command may be used by selecting **Data** as the **Inpt** option and entering the location of the data as the **List** option (Figure D).

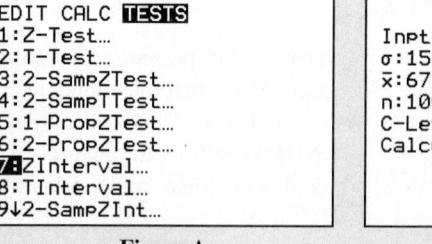

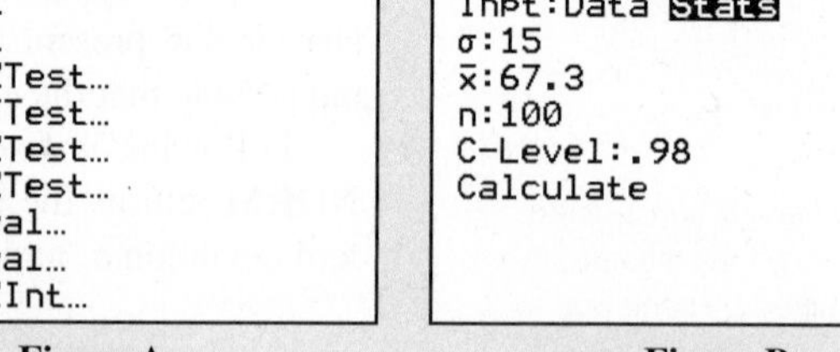

Figure A

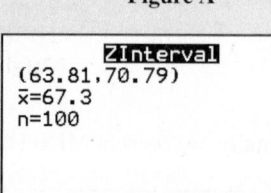

Figure B

```
ZInterval
(63.81,70.79)
x̄=67.3
n=100
```

Figure C

```
ZInterval
Inpt:Data Stats
σ:15
List:L1
Freq:1
C-Level:.98
Calculate
```

Figure D

EXCEL

Constructing a confidence interval for the mean when σ is known

The =**CONFIDENCE.NORM**(*alpha, standard_dev, size*) command returns the margin of error when the population standard deviation is known.

To obtain the margin of error m for Example 8.4, we use $\alpha = 0.02$ in the *alpha* field, the population standard deviation $\sigma = 15$ in the *standard_dev* field, and the sample size $n = 100$ in the *size* field (Figure E).

The confidence interval is given by $\bar{x} - m < \mu < \bar{x} + m$.

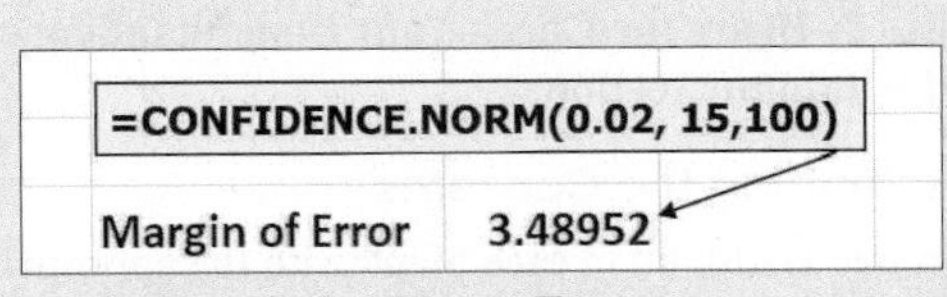

Figure E

Finding a critical value from a normal distribution

The =**NORM.S.INV**(*probability*) command returns a z-score where the *probability* input corresponds to the area on the left.

The critical value for confidence level $1 - \alpha$ is the absolute value of $z_{\alpha/2}$. Figure F shows the command for finding the critical value of a 95% confidence interval where $\frac{\alpha}{2} = 0.025$.

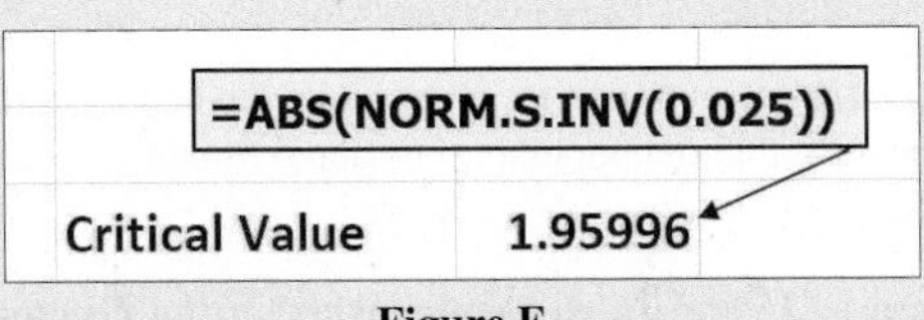

Figure F

MINITAB

Constructing a confidence interval for the mean when σ is known

Step 1. Click on **Stat**, then **Basic Statistics**, then **1-Sample Z.**

Step 2. Choose one of the following:

- If the summary statistics are given, select **Summarized Data** and enter the **Sample Size** (100), the **Sample Mean** (67.30), and the **Standard Deviation** (15) (Figure G).
- If the raw data are given, select **One or more samples, each in a column** and select the column that contains the data. Enter the **Standard Deviation**.

Step 3. Click **Options**, and enter the confidence level in the **Confidence Level** (98) field. Click **OK**.

Step 4. Click **OK** (Figure H).

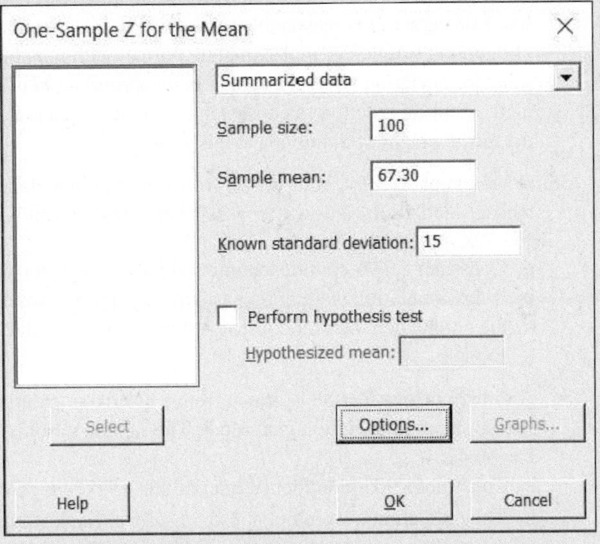

Figure G

N	Mean	SE Mean	98% CI for μ
100	67.30	1.50	(63.81, 70.79)

μ: mean of Sample
Known standard deviation = 15

Figure H

Section 8.1 Exercises

Exercises 1–10 are the Check Your Understanding exercises located within the section.

Understanding the Concepts

In Exercises 11–14, fill in each blank with the appropriate word or phrase.

11. A single number that estimates the value of an unknown parameter is called a ________ estimate. *point*

12. The margin of error is the product of the standard error and the ________. *critical value*

13. In the confidence interval 24.3 ± 1.2, the quantity 1.2 is called the ________. *margin of error*

14. To construct a confidence interval for μ when $n \leq 30$, it is necessary that the population be approximately ________. *normal*

In Exercises 15 and 16, determine whether the statement is true or false. If the statement is false, rewrite it as a true statement.

15. The confidence level is the proportion of all possible samples for which the confidence interval will cover the true value. *True*

16. To construct a confidence interval for a population mean, we add and subtract the critical value from the point estimate. *False*

17. A simple random sample of 15 cars on a used car lot had a mean odometer reading of $\bar{x} = 42{,}000$ miles. The sample is skewed. Assume the population standard deviation is $\sigma = 10{,}000$ miles. Is it appropriate to use the methods of this section to construct a confidence interval for the mean odometer reading for cars on this lot? If not, explain why not. *No*

18. Credit scores range from 300 to 800, and are used by lenders to determine whether someone is a good credit risk. A loan officer at a bank samples 50 personal loans that were made within the past year and finds that the mean credit score of the borrowers is $\bar{x} = 730$. Assume the population standard deviation of credit scores for those receiving loans is $\sigma = 40$. Is it appropriate to use the methods of this section to construct a confidence interval for the mean credit score for those receiving loans at this bank in the past year? If not, explain why not. *Yes*

19. A manager at a supermarket sampled 100 receipts from customers who have visited the store within the past month. The mean amount spent was $\bar{x} = \$87$. Assume the population standard deviation is $\sigma = \$35$. Is it appropriate to use the methods of this section to construct a confidence interval for the mean amount spent by customers in the past month? If not, explain why not. *Yes*

20. The registrar at a certain college examines the records of 75 graduating seniors to determine how many courses each is taking. The mean is $\bar{x} = 3.8$. Assume the population standard deviation is $\sigma = 1.5$. Is it appropriate to use the methods of this section to construct a confidence interval for the mean number of classes taken by all students at this college? If not, explain why not. *No*

Practicing the Skills

In Exercises 21–24, find the critical value $z_{\alpha/2}$ needed to construct a confidence interval with the given level.

21. Level 95% *1.96* **22.** Level 85% *1.44*

23. Level 96% *2.05* **24.** Level 99.7% *2.97*

In Exercises 25–28, find the levels of the confidence intervals that have the given critical values.

25. 2.326 *98%* **26.** 2.576 *99%* **27.** 2.81 *99.5%* **28.** 1.04 *70%*

29. A sample of size $n = 49$ is drawn from a population whose standard deviation is $\sigma = 4.8$. Find the margin of error for a 95% confidence interval for μ. *1.344*

30. A sample of size $n = 50$ is drawn from a population whose standard deviation is $\sigma = 26$. Find the margin of error for a 90% confidence interval for μ. *6.049 [Tech: 6.048]*

31. A sample of size $n = 32$ is drawn from a population whose standard deviation is $\sigma = 12.1$. Find the margin of error for a 95% confidence interval for μ. *4.192*

32. A sample of size $n = 64$ is drawn from a population whose standard deviation is $\sigma = 24.18$. Find the margin of error for a 95% confidence interval for μ. *5.924*

33. A simple random sample of 20 bicycle helmets had a mean price of $\bar{x} = \$38.12$. It is reasonable to assume that the population of helmet prices is approximately normal with population standard deviation $\sigma = \$7.79$. Is it appropriate to use the methods of this section to construct a confidence interval for the mean helmet price? *Yes*

34. The weight of a diamond is measured in carats. A random sample of 18 diamonds in a retail store had a mean weight of $\bar{x} = 1.08$ carats. It is reasonable to assume that the population of diamond weights is approximately normal with population standard deviation $\sigma = 0.12$ carats. Is it appropriate to use the methods of this section to construct a confidence interval for the mean weight of diamonds at this store? *Yes*

35. A sample of size $n = 10$ is drawn from a normal population whose standard deviation is $\sigma = 2.5$. The sample mean is $\bar{x} = 7.92$.

a. Construct a 95% confidence interval for μ. *(6.37, 9.47)*

b. If the population were not approximately normal, would the confidence interval constructed in part (a) be valid? Explain. *No*

36. A sample of size $n = 80$ is drawn from a normal population whose standard deviation is $\sigma = 6.8$. The sample mean is $\bar{x} = 40.41$.

a. Construct a 90% confidence interval for μ. *(39.16, 41.66)*

b. If the population were not approximately normal, would the confidence interval constructed in part (a) be valid? Explain. *Yes*

Working with the Concepts

37. SAT scores: A college admissions officer takes a simple random sample of 100 entering freshmen and computes their mean mathematics SAT score to be 458. Assume the population standard deviation is $\sigma = 116$.

a. Construct a 99% confidence interval for the mean mathematics SAT score for the entering freshman class. *(428, 488)*

b. If the sample size were 75 rather than 100, would the margin of error be larger or smaller than the result in part (a)? Explain. *Larger*

c. If the confidence level were 95% rather than 99%, would the margin of error be larger or smaller than the result in part (a)? Explain. *Smaller*

d. Based on the confidence interval constructed in part (a), is it likely that the mean mathematics SAT score for the entering freshman class is greater than 500? *No*

38. How many computers? In a simple random sample of 150 households, the sample mean number of personal computers was 1.32. Assume the population standard deviation is $\sigma = 0.41$.

a. Construct a 95% confidence interval for the mean number of personal computers. *(1.25, 1.39)*

b. If the sample size were 100 rather than 150, would the margin of error be larger or smaller than the result in part (a)? Explain. *Larger*

c. If the confidence level were 98% rather than 95%, would the margin of error be larger or smaller than the result in part (a)? Explain. *Larger*

d. Based on the confidence interval constructed in part (a), is it likely that the mean number of personal computers is greater than 1.25? *Yes*

39. Babies: According to the National Health Statistics Reports, a sample of 360 one-year-old baby boys in the United States had a mean weight of 25.5 pounds. Assume the population standard deviation is $\sigma = 5.3$ pounds.

a. Construct a 95% confidence interval for the mean weight of all one-year-old baby boys in the United States *(25.0, 26.0)*

b. Should this confidence interval be used to estimate the mean weight of all one-year-old babies in the United States? Explain. *No*

c. Based on the confidence interval constructed in part (a), is it likely that the mean weight of all one-year-old boys is less than 28 pounds? *Yes*

40. Watch your cholesterol: A sample of 314 patients between the ages of 38 and 82 were given a combination of the drugs ezetimibe and simvastatin. They achieved a mean reduction in total cholesterol of 0.94 millimole per liter. Assume the population standard deviation is $\sigma = 0.18$.

a. Construct a 98% confidence interval for the mean reduction in total cholesterol in patients who take this combination of drugs. *(0.92, 0.96)*

b. Should this confidence interval be used to estimate the mean reduction in total cholesterol for patients over the age of 85? Explain. *No*

c. Based on the confidence interval constructed in part (a), is it likely that the mean reduction in cholesterol level is less than 0.90? *No*

Source: *International Journal of Clinical Practice* 58:653–658

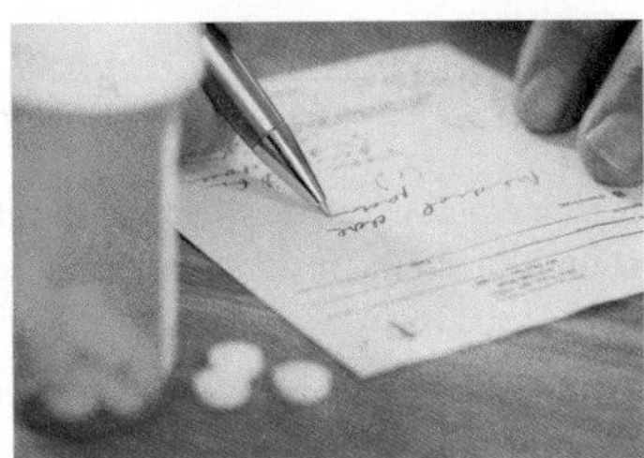

JGI/Blend Images LLC

41. How smart is your phone? A random sample of 11 smartphones sold over the internet had the following prices, in dollars:

199	169	385	329	269	149
135	249	349	299	249	

Assume the population standard deviation is $\sigma = 85$.

a. Explain why it is necessary to check whether the population is approximately normal before constructing a confidence interval. *$n \leq 30$*

b. Following is a dotplot of these data. Is it reasonable to assume that the population is approximately normal? *Yes*

100 150 200 250 300 350 400

c. If appropriate, construct a 95% confidence interval for the mean price for all phones of this type being sold on the internet. If not appropriate, explain why not. *(202.6, 303.1)*

42. Stock prices: The Standard and Poor's (S&P) 500 is a group of 500 large companies traded on the New York Stock Exchange. Following are prices, in dollars, for a random sample of ten stocks on a recent day.

84.86	8.11	74.23	35.25	13.19
53.55	84.25	201.94	24.68	53.47

Assume the population standard deviation is $\sigma = 50$.

a. Explain why it is necessary to check whether the population is approximately normal before constructing a confidence interval. *$n \leq 30$*

b. Following is a dotplot of these data. Is it reasonable to assume that the population is approximately normal? *No*

0 50 100 150 200 250

c. If appropriate, construct a 95% confidence interval for the mean price for all S&P 500 stocks on this day. If not appropriate, explain why not. *Not appropriate*

43. High energy: A random sample of energy drinks had the following amounts of caffeine per fluid ounce.

14.2	8.3	80.8	6.7	3.6
13.7	12.9	11.5	24.7	9.5

Assume the population standard deviation is $\sigma = 24$.

a. Explain why it is necessary to check whether the population is approximately normal before constructing a confidence interval. *$n \leq 30$*

b. Following is a dotplot of these data. Is it reasonable to assume that the population is approximately normal? *No*

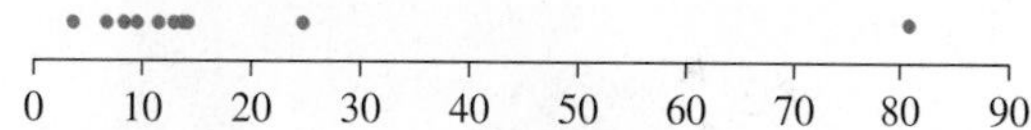

c. If appropriate, construct a 95% confidence interval for the mean amount of caffeine in all energy drinks. If not appropriate, explain why not. *Not appropriate*

44. Let's shake on it: A random sample of 12-ounce milkshakes from 14 fast-food restaurants had the following number of calories.

504	399	580	476	450	591	510
700	608	472	642	613	473	375

Assume the population standard deviation is $\sigma = 90$.

a. Explain why it is necessary to check whether the population is approximately normal before constructing a confidence interval. *$n \leq 30$*

b. Following is a dotplot of these data. Is it reasonable to assume that the population is approximately normal? *yes*

c. If appropriate, construct a 95% confidence interval for the mean number of calories for all 12-ounce milkshakes sold at fast-food restaurants. If not appropriate, explain why not. *(480.93, 575.22)*

45. Don't construct a confidence interval: A psychology professor at a certain college gave a test to the students in her class. The test was designed to measure students' attitudes toward school, with higher scores indicating a more positive attitude. There were 30 students in the class, and their mean score was 78. Scores on this test are known to be normally distributed with a standard deviation of 10. Explain why these data should not be used to construct a confidence interval for the mean score for all the students in the college.

46. Don't construct a confidence interval: A college alumni organization sent a survey to all recent graduates to ask their annual income. Twenty percent of the alumni responded, and their mean annual income was $40,000. Assume the population standard deviation is $\sigma = \$10{,}000$. Explain why these data should not be used to construct a confidence interval for the mean annual income of all recent graduates.

47. Interpret a confidence interval: A dean at a certain college looked up the GPA for a random sample of 85 students. The sample mean GPA was 2.82, and a 95% confidence interval for the mean GPA of all students in the college was $2.76 < \mu < 2.88$. True or false, and explain:

a. We are 95% confident that the mean GPA of all students in the college is between 2.76 and 2.88. *True*

b. We are 95% confident that the mean GPA of all students in the sample is between 2.76 and 2.88. *False*

c. The probability is 0.95 that the mean GPA of all students in the college is between 2.76 and 2.88. *False*

d. 95% of the students in the sample had a GPA between 2.76 and 2.88. *False*

48. Interpret a confidence interval: A survey organization drew a simple random sample of 625 households from a city of 100,000 households. The sample mean number of people in the 625 households was 2.30, and a 95% confidence interval for the mean number of people in the 100,000 households was $2.16 < \mu < 2.44$. True or false, and explain:

a. We are 95% confident that the mean number of people in the 625 households is between 2.16 and 2.44. *False*

b. We are 95% confident that the mean number of people in the 100,000 households is between 2.16 and 2.44. *True*

c. The probability is 0.95 that the mean number of people in the 100,000 households is between 2.16 and 2.44. *False*

d. 95% of the households in the sample contain between 2.16 and 2.44 people. *False*

49. Interpret calculator display: The number of words typed per minute was measured for a sample of typists. The following display from a TI-84 Plus calculator presents a 95% confidence interval for the population mean typing speed.

```
ZInterval
(56.019,60.881)
x̄=58.45
n=65
```

a. Fill in the blanks: We are ________ confident that the population mean number of words per minute is between ________ and ________. *95%, 56.019, 60.881*

b. Assume the population is not normally distributed. Is the confidence interval still valid? Explain. *Yes*

50. Interpret calculator display: Flight times, in minutes, were measured for a sample of drones delivering packages. The following display from a TI-84 Plus calculator presents a 99% confidence interval for the population mean flight time.

```
ZInterval
(17.012,20.048)
x̄=18.53
n=15
```

a. Fill in the blanks: We are ________ confident that the population mean flight time is between ________ and ________. *99%, 17.012, 20.048*

b. Assume the population is not normally distributed. Is the confidence interval still valid? Explain. *No*

51. Interpret computer output: A sample of patients was put on a special diet. After six months, the number of pounds lost was measured for each of them. The following MINITAB output presents a confidence interval for the population mean weight loss.

```
The assumed sigma = 6.5000
Variable    N     Mean    SE Mean          95% CI
X          23   12.352     1.3553   (9.6956, 15.0084)
```

a. Fill in the blanks: We are ________ confident that the population mean weight loss is between ________ and ________. *95%, 9.6956, 15.0084*

b. Use the appropriate critical value along with the information in the computer output to construct a 99% confidence interval. *(8.861, 15.843)*

52. Interpret computer output: The number of years employed was determined for a sample of employees at a large company. The following MINITAB output presents a confidence interval for the population mean length of employment.

```
The assumed sigma = 8.0000
Variable    N     Mean    SE Mean        98% CI
X          58    2.657     1.0505   (0.1233, 5.1007)
```

a. Fill in the blanks: We are ________ confident that the population mean number of years employed is between ________ and ________. *98%, 0.1233, 5.1007*

b. Use the appropriate critical value along with the information in the computer output to construct a 95% confidence interval. *(0.598, 4.716)*

Answers to Check Your Understanding Exercises for Section 8.1

1. a. 1.645 **b.** 2.326 **c.** 2.81 **d.** 1.28
2. a. 95% **b.** 97% **c.** 80% **d.** 99.9%
3. a. 2.352 **b.** 1.030 **c.** 5.758 [Tech: 5.757] **d.** 6.397
4. a. 105.2 **b.** 1.645 **c.** 1.1547 **d.** 1.90
e. $103.3 < \mu < 107.1$
f. Yes. We are 90% confident that μ is between 103.3 and 107.1, so it is likely that $\mu > 100$.
5. a. 120.1 **b.** 1.96 **c.** 2.8284 **d.** 5.54
e. $114.6 < \mu < 125.6$
f. No. We are 95% confident that μ is between 114.6 and 125.6, so it is not likely that $\mu > 130$.
6. a. True **b.** False
7. $102.9 < \mu < 107.5$
8. $113.5 < \mu < 126.7$
9. The confidence interval does not contain the value 100. Therefore, it is not likely that the claim that $\mu = 100$ is true.
10. a. 2.576 **b.** $101.2143 < \mu < 103.4911$

Section 8.2 Confidence Intervals for a Population Mean, Standard Deviation Unknown

Objectives

1. Describe the properties of the Student's t distribution
2. Construct confidence intervals for a population mean when the population standard deviation is unknown
3. Describe the relationship between the confidence level and the margin of error
4. Find the sample size necessary to obtain a confidence interval of a given width
5. Distinguish between confidence and probability

Objective 1 Describe the properties of the Student's t distribution

The Student's *t* Distribution

In Section 8.1, we showed how to construct a confidence interval for the mean μ of a normal population when the population standard deviation σ is known. The confidence interval is

$$\bar{x} \pm z_{\alpha/2}\frac{\sigma}{\sqrt{n}}$$

The critical value is $z_{\alpha/2}$ because the quantity

$$\frac{\bar{x} - \mu}{\sigma/\sqrt{n}}$$

has a normal distribution.

In practice, it is more common that σ is unknown. When we don't know the value of σ, we replace it with the sample standard deviation s. However, we cannot then use $z_{\alpha/2}$ as the critical value, because the quantity

$$\frac{\bar{x} - \mu}{s/\sqrt{n}}$$

does not have a normal distribution. One reason is that s is, on the average, a bit smaller than σ, so replacing σ with s tends to increase the magnitude. Another reason is that s is random whereas σ is constant, so replacing σ with s increases the spread.

The distribution of this quantity is called the **Student's *t* distribution**. It was discovered in 1908 by William Sealy Gosset, a statistician who worked for the Guinness Brewing Company in Dublin, Ireland. The management at Guinness considered the discovery to be proprietary information and required him to use the pseudonym "Student." This had happened before; see Section 6.3.

RECALL

Degrees of freedom were introduced in Section 3.2.

In fact, there are many different Student's *t* distributions; they are distinguished by a quantity called the **degrees of freedom**. When using the Student's *t* distribution to construct a confidence interval for a population mean, the number of degrees of freedom is 1 less than the sample size.

Degrees of Freedom for the Student's *t* Distribution

When constructing a confidence interval for a population mean, the number of degrees of freedom for the Student's *t* distribution is 1 less than the sample size n.

$$\text{Number of degrees of freedom} = n - 1$$

Figure 8.8 presents *t* distributions for several different degrees of freedom, along with a standard normal distribution for comparison. The *t* distributions are symmetric and unimodal, just like the normal distribution. The *t* distribution is more spread out than the standard normal distribution, because the sample standard deviation s is, on the average, a bit less than σ. When the number of degrees of freedom is small, this tendency is more pronounced, so the *t* distributions are much more spread out than the normal. When the number of degrees of freedom is large, s tends to be very close to σ, so the *t* distribution is very close to the normal. Figure 8.8 shows that with 10 degrees of freedom, the difference between the *t* distribution and the normal is not great. If a *t* distribution with 30 degrees of freedom were plotted in Figure 8.8, it would be indistinguishable from the normal distribution.

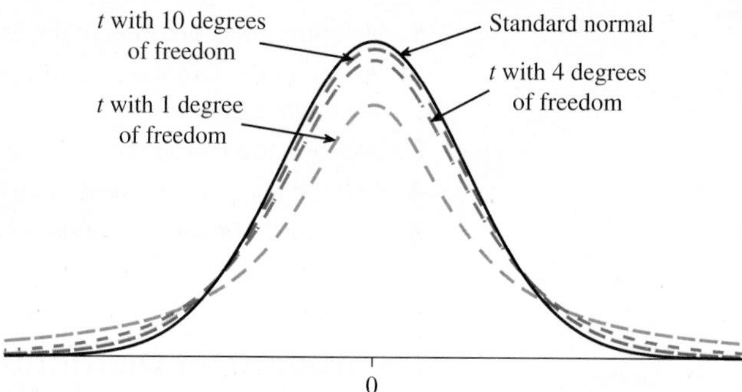

Figure 8.8 Plots of the Student's *t* distribution for 1, 4, and 10 degrees of freedom. The standard normal distribution is plotted for comparison. The *t* distributions are more spread out than the normal, but the amount of extra spread decreases as the number of degrees of freedom increases.

SUMMARY

The Student's *t* distribution has the following properties:

- It is symmetric and unimodal.
- It is more spread out than the standard normal distribution.
- If we increase the number of degrees of freedom, the Student's *t* curve becomes closer to the standard normal curve.

Finding the critical value

We use the Student's *t* distribution to construct confidence intervals for μ when σ is unknown. The idea behind the critical value is the same as for the normal distribution. To find

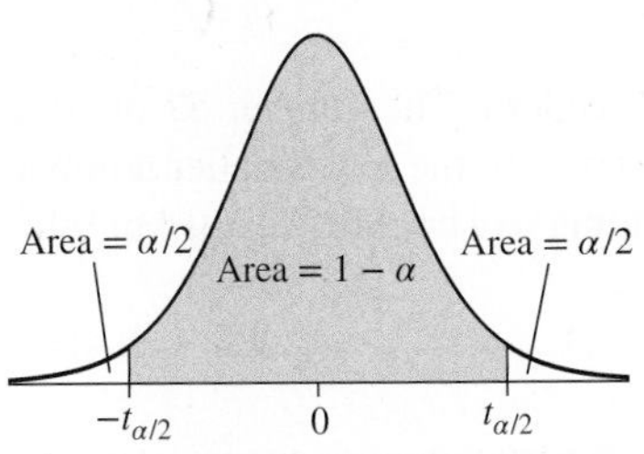

Figure 8.9

the critical value for a confidence interval with a given level, let $1 - \alpha$ be the confidence level expressed as a decimal. The critical value is then $t_{\alpha/2}$, because the area under the Student's t curve between $-t_{\alpha/2}$ and $t_{\alpha/2}$ is $1 - \alpha$. See Figure 8.9. The critical value $t_{\alpha/2}$ can be found in Table A.3, in the row corresponding to the number of degrees of freedom and the column corresponding to the desired confidence level. Example 8.5 shows how to find a critical value.

Example 8.5

Finding a critical value

A simple random sample of size 10 is drawn from a normal population. Find the critical value $t_{\alpha/2}$ for a 95% confidence interval.

Solution

The sample size is $n = 10$, so the number of degrees of freedom is $n - 1 = 9$. We consult Table A.3, looking in the row corresponding to 9 degrees of freedom, and in the column with confidence level 95% (the confidence levels are listed along the bottom of the table). The critical value is $t_{\alpha/2} = 2.262$. See Figure 8.10.

Area = 0.025
Area = 0.95
Area = 0.025
−2.262
0
2.262

Figure 8.10 95% of the area under the Student's t curve with 9 degrees of freedom is between $t = -2.262$ and $t = 2.262$.

⋮	⋮	⋮	⋮	⋮	⋮	⋮	⋮	⋮	⋮	⋮
4	0.271	0.741	1.533	2.132	2.776	3.747	4.604	5.598	7.173	8.610
5	0.267	0.727	1.476	2.015	2.571	3.365	4.032	4.773	5.893	6.869
6	0.265	0.718	1.440	1.943	2.447	3.143	3.707	4.317	5.208	5.959
7	0.263	0.711	1.415	1.895	2.365	2.998	3.499	4.029	4.785	5.408
8	0.262	0.706	1.397	1.860	2.306	2.896	3.355	3.833	4.501	5.041
9	0.261	0.703	1.383	1.833	2.262	2.821	3.250	3.690	4.297	4.781
10	0.260	0.700	1.372	1.812	2.228	2.764	3.169	3.581	4.144	4.587
11	0.260	0.697	1.363	1.796	2.201	2.718	3.106	3.497	4.025	4.437
12	0.259	0.695	1.356	1.782	2.179	2.681	3.055	3.428	3.930	4.318
13	0.259	0.694	1.350	1.771	2.160	2.650	3.012	3.372	3.852	4.221
14	0.258	0.692	1.345	1.761	2.145	2.624	2.977	3.326	3.787	4.140
⋮	⋮	⋮	⋮	⋮	⋮	⋮	⋮	⋮	⋮	⋮
	20%	50%	80%	90%	95%	98%	99%	99.5%	99.8%	99.9%
					Confidence Level					

```
invT(0.025,9)
                -2.262157158
```

What if the number of degrees of freedom isn't in the table?

The largest number of degrees of freedom shown in Table A.3 is 200. If the desired number of degrees of freedom is less than 200 and is not shown in Table A.3, use the next smaller number of degrees of freedom in the table. If the desired number of degrees of freedom is more than 200, use the z-value found in the last row of Table A.3, or use Table A.2. This problem will not arise if a calculator or computer is used to construct a confidence interval, because technology can compute critical values for any number of degrees of freedom.

SUMMARY

If the desired number of degrees of freedom isn't listed in Table A.3, then

- If the desired number is less than 200, use the next smaller number that is in the table.
- If the desired number is greater than 200, use the z-value found in the last row of Table A.3, or use Table A.2.

Example 8.6

Finding a critical value

Use Table A.3 to find the critical value for a 99% confidence interval for a sample of size 58.

Solution

Because the sample size is 58, there are 57 degrees of freedom. The number 57 doesn't appear in the degrees of freedom column in Table A.3, so we use the next smaller number that does appear, which is 50. The value of $t_{\alpha/2}$ corresponding to a confidence level of 99% is $t_{\alpha/2} = 2.678$.

EXPLAIN IT AGAIN

$\bar{x}$ must be approximately normal: We need assumption 2 to be sure that the sampling distribution of $\bar{x}$ is approximately normal.

Following are the assumptions that are necessary to construct confidence intervals by using the Student's t distribution.

Assumptions for Constructing a Confidence Interval for μ When σ Is Unknown

1. We have a simple random sample.
2. Either the sample size is large ($n > 30$), *or* the population is approximately normal.

Checking the assumptions

When the sample size is small ($n \leq 30$), we must check to determine whether the sample comes from a population that is approximately normal. We will focus on detecting skewness and outliers, since these departures from normality have the greatest effect on the performance of the confidence intervals we discuss here. This can be done using the methods described in Section 7.6. A simple method is to draw a dotplot or boxplot of the sample. If there are no outliers, and if the sample is not strongly skewed, then it is reasonable to construct a confidence interval using the Student's t distribution.

Check Your Understanding

1. Use Table A.3 to find the critical value $t_{\alpha/2}$ needed to construct a confidence interval of the given level with the given sample size:
   - **a.** Level 95%, sample size 15 *2.145*
   - **b.** Level 99%, sample size 22 *2.831*
   - **c.** Level 90%, sample size 63 *1.671*
   - **d.** Level 95%, sample size 2 *12.706*
2. In each of the following situations, state whether the methods of this section should be used to construct a confidence interval for the population mean. Assume that σ is unknown.
   - **a.** A simple random sample of size 8 is drawn from a distribution that is approximately normal. *Yes*
   - **b.** A simple random sample of size 15 is drawn from a distribution that is not close to normal. *No*
   - **c.** A simple random sample of size 150 is drawn from a distribution that is not close to normal. *Yes*
   - **d.** A nonrandom sample is drawn. *No*

Answers are on page 404.

Objective 2 Construct confidence intervals for a population mean when the population standard deviation is unknown

Constructing a Confidence Interval for μ When σ Is Unknown

The ingredients for a confidence interval for a population mean μ when σ is unknown are the point estimate $\bar{x}$, the critical value $t_{\alpha/2}$, and the standard error $s/\sqrt{n}$. The margin of error is $t_{\alpha/2}s/\sqrt{n}$. When the assumptions for the Student's t distribution are met, we can use the following step-by-step procedure for constructing a confidence interval for a population mean.

Procedure for Constructing a Confidence Interval for μ When σ Is Unknown

Check to be sure the assumptions are satisfied. If they are, then proceed with the following steps.

Step 1: Compute the sample mean $\bar{x}$ and sample standard deviation s, if they are not given.

Step 2: Find the number of degrees of freedom $n - 1$ and the critical value $t_{\alpha/2}$.

Step 3: Compute the standard error $s/\sqrt{n}$, and multiply it by the critical value to obtain the margin of error $t_{\alpha/2}\dfrac{s}{\sqrt{n}}$.

Step 4: Use the point estimate and the margin of error to construct the confidence interval:

$$\text{Point estimate} \pm \text{Margin of error}$$

$$\bar{x} \pm t_{\alpha/2}\frac{s}{\sqrt{n}}$$

$$\bar{x} - t_{\alpha/2}\frac{s}{\sqrt{n}} < \mu < \bar{x} + t_{\alpha/2}\frac{s}{\sqrt{n}}$$

Step 5: Interpret the result.

Example 8.7

Constructing a confidence interval

A food chemist analyzed the calorie content for a popular type of chocolate cookie. Following are the numbers of calories in a sample of eight cookies.

113 114 111 116 115 120 118 116

Find a 98% confidence interval for the mean number of calories in this type of cookie.

Solution

We check the assumptions. We have a simple random sample. Because the sample size is small, the population must be approximately normal. We check this with a dotplot of the data.

110 111 112 113 114 115 116 117 118 119 120 121

There is no evidence of strong skewness, and no outliers. Therefore, we may proceed.

Step 1: Find the sample mean and sample standard deviation. We compute the mean and standard deviation of the sample values. We obtain

$$\bar{x} = 115.375 \qquad s = 2.8253$$

Step 2: Find the number of degrees of freedom and the critical value $t_{\alpha/2}$. The number of degrees of freedom is $n - 1 = 8 - 1 = 7$. Using Table A.3, we find that the critical value corresponding to a level of 98% is $t_{\alpha/2} = 2.998$.

Step 3: Compute the margin of error. The margin of error is

$$t_{\alpha/2}\frac{s}{\sqrt{n}}$$

We substitute $t_{\alpha/2} = 2.998$, $s = 2.8253$, and $n = 8$ to obtain

$$t_{\alpha/2}\frac{s}{\sqrt{n}} = 2.998\frac{2.8253}{\sqrt{8}} = 2.9947$$

Step 4: Construct the confidence interval. The 98% confidence interval is given by

$$\bar{x} - t_{\alpha/2}\frac{s}{\sqrt{n}} < \mu < \bar{x} + t_{\alpha/2}\frac{s}{\sqrt{n}}$$

$$115.375 - 2.9947 < \mu < 115.375 + 2.9947$$

$$112.4 < \mu < 118.4$$

Note that we round the final result to one decimal place, because the sample values were whole numbers.

Step 5: Interpret the result. We are 98% confident that the mean number of calories per cookie is between 112.4 and 118.4.

Example 8.8

Constructing a confidence interval

dolgachov/123RF

The General Social Survey is a survey of opinions and lifestyles of U.S. adults, conducted by the National Opinion Research Center at the University of Chicago. A sample of 123 people aged 18–22 reported the number of hours they spent on the internet in an average week. The sample mean was 8.20 hours, with a sample standard deviation of 9.84 hours. Assume this is a simple random sample from the population of people aged 18–22 in the United States. Construct a 95% confidence interval for μ, the population mean number of hours per week spent on the internet by people aged 18–22 in the United States.

Solution

We check the assumptions. We have a simple random sample. Now either the sample size must be greater than 30, or the population must be approximately normal. Since the sample size is $n = 123$, the assumptions are met.

Step 1: Find the sample mean and sample standard deviation. These are given as $\bar{x} = 8.20$ and $s = 9.84$.

Step 2: Find the number of degrees of freedom and the critical value $t_{\alpha/2}$. The number of degrees of freedom is $n - 1 = 123 - 1 = 122$. Since this number of degrees of freedom does not appear in Table A.3, we use the next smaller value in the table, which is 100. The critical value corresponding to a level of 95% is $t_{\alpha/2} = 1.984$.

Step 3: Compute the margin of error. The margin of error is

$$t_{\alpha/2}\frac{s}{\sqrt{n}}$$

We substitute $t_{\alpha/2} = 1.984$, $s = 9.84$, and $n = 123$ to obtain

$$t_{\alpha/2}\frac{s}{\sqrt{n}} = 1.984\frac{9.84}{\sqrt{123}} = 1.7603$$

Step 4: Construct the confidence interval. The 95% confidence interval is given by

$$\bar{x} - t_{\alpha/2}\frac{s}{\sqrt{n}} < \mu < \bar{x} + t_{\alpha/2}\frac{s}{\sqrt{n}}$$

$$8.20 - 1.7603 < \mu < 8.20 + 1.7603$$

$$6.44 < \mu < 9.96$$

Note that we round the final result to two decimal places, because the value of $\bar{x}$ was given to two decimal places.

Step 5: Interpret the result. We are 95% confident that the mean number of hours per week spent on the internet by people 18–22 years old is between 6.44 and 9.96.

Note that in Example 8.8, the sample standard deviation of 9.84 is larger than the sample mean of 8.20. Since the minimum possible time to spend on the internet is 0, the smallest sample value is less than one standard deviation below the mean. This indicates that the

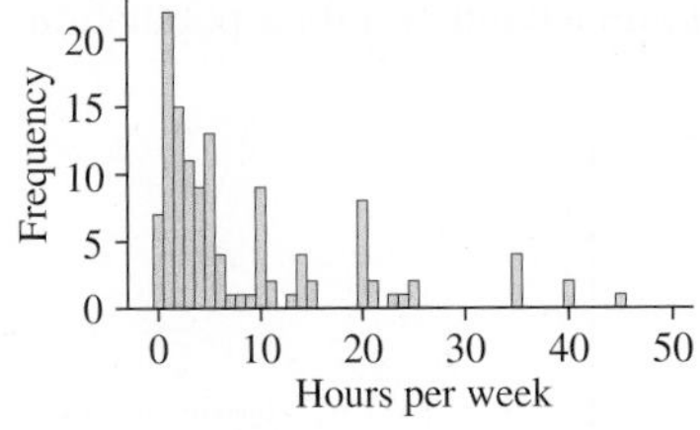

Figure 8.11

CAUTION

Confidence intervals constructed using technology may differ from those constructed by hand due to rounding. The differences are never large enough to matter.

sample is fairly skewed. Figure 8.11 confirms this. Even though the sample is skewed, the t statistic is still appropriate, because the sample size of 123 is large.

Constructing confidence intervals with technology

The following TI-84 Plus display presents the results of Example 8.8.

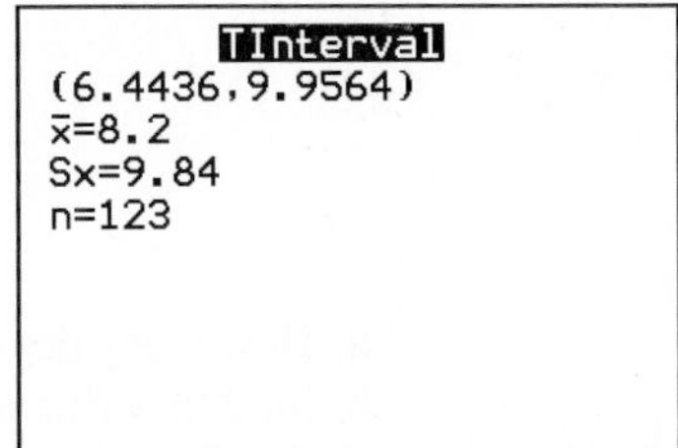

The display is fairly straightforward. The quantity Sx is the sample standard deviation s. The TI-84 Plus uses the exact number of degrees of freedom, 122, rather than 100 as we did in the solution to Example 8.8. This does not make a difference when the answer is rounded to two decimal places. Note that the confidence level (95%) is not given in the display.

Following is Excel output for this example. The Excel function **=CONFIDENCE.T** returns the margin of error. The inputs are the value of α, the population standard deviation σ, and the sample size n.

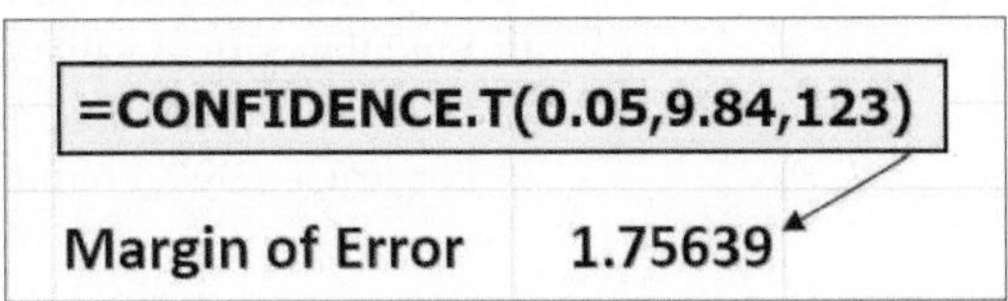

Finally, we present MINITAB output for this example.

Variable	N	Mean	StDev	SE Mean	95% CI
Hours	123	8.20000	9.84000	0.88724	(6.44361, 9.95639)

The output is fairly straightforward. Going from left to right, "N" represents the sample size, "Mean" is the sample mean $\bar{x}$, and "StDev" is the sample standard deviation s. The quantity labeled "SE Mean" is the standard error $s/\sqrt{n}$. Note that neither the critical value nor the margin of error is given explicitly in the output. Finally, the lower and upper limits of the 95% confidence interval are given on the right. Like the TI-84 Plus, MINITAB uses 122 degrees of freedom rather than 100 as we did in the solution to the example. This does not make a difference when the answer is rounded to two decimal places.

Step-by-step instructions for constructing confidence intervals with technology are given in the Using Technology section on page 398.

Check Your Understanding

3. A potato chip company wants to evaluate the accuracy of its potato chip bag-filling machine. Bags are labeled as containing 8 ounces of potato chips. A simple random sample of 12 bags had mean weight 8.12 ounces with a sample standard deviation of 0.1 ounce. Assume the weights are approximately normally distributed. Construct a 99% confidence interval for the population mean weight of bags of potato chips. *(8.03, 8.21)*

4. A company has developed a new type of lightbulb and wants to estimate its mean lifetime. A simple random sample of 100 bulbs had a sample mean lifetime of 750.2 hours with a sample standard deviation of 30 hours. Construct a 95% confidence interval for the population mean lifetime of all bulbs manufactured by this new process. *(744.2, 756.2)*

5. The following TI-84 Plus display presents a 95% confidence interval for a population mean.

```
TInterval
(41.772,49.348)
x̄=45.56
Sx=6.84
n=15
```

a. How many degrees of freedom are there? 14
b. Find the critical value $t_{\alpha/2}$ for a 99% confidence interval. 2.977
c. Use the critical value along with the information in the display to construct a 99% confidence interval for the population mean. (40.30, 50.82)

6. The following MINITAB output presents a confidence interval for a population mean.

Variable	N	Mean	StDev	SE Mean	95% CI
X	10	8.5963	0.11213	0.03546	(8.5161, 8.6765)

a. How many degrees of freedom are there? 9
b. Find the critical value $t_{\alpha/2}$ for a 99% confidence interval. 3.250
c. Use the critical value along with the information in the computer output to construct a 99% confidence interval for the population mean. (8.4811, 8.7115)

Answers are on page 404.

Objective 3 Describe the relationship between the confidence level and the margin of error

More Confidence Means a Bigger Margin of Error

Other things being equal, it is better to have more confidence than less. We would also rather have a smaller margin of error than a larger one. However, when it comes to confidence intervals, there is a trade-off. If we increase the level of confidence, we must increase the critical value, which in turn increases the margin of error. Examples 8.9 and 8.10 help explain this idea.

Example 8.9

Construct a confidence interval

A machine that fills cereal boxes is supposed to put 20 ounces of cereal in each box. A simple random sample of 6 boxes is found to contain a sample mean of 20.25 ounces of cereal, with a standard deviation of $s = 0.2$ ounce. It is known from past experience that the fill weights are normally distributed. Construct a 90% confidence interval for the mean fill weight.

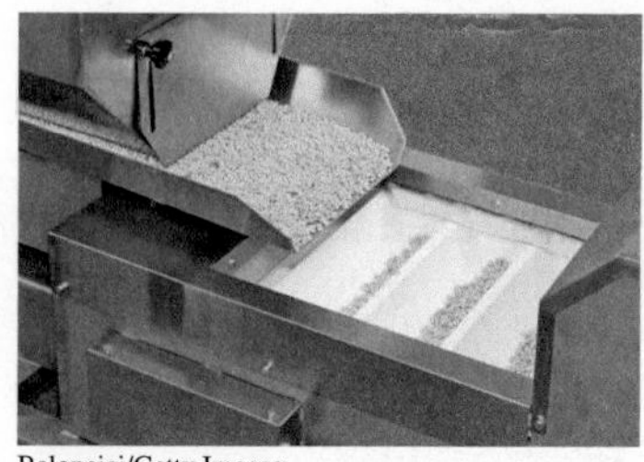
Baloncici/Getty Images

Solution

We check the assumptions. The sample is a simple random sample, and the population is known to be normal. The assumptions are met, so we may proceed.

Step 1: Find the sample mean and the sample standard deviation. These are given as $\bar{x} = 20.25$ and $s = 0.2$.

Step 2: Find the number of degrees of freedom and the critical value $t_{\alpha/2}$. The number of degrees of freedom is $n - 1 = 6 - 1 = 5$. Using Table A.3, we find that the critical value corresponding to a level of 90% is $t_{\alpha/2} = 2.015$.

Step 3: Compute the margin of error. The margin of error is

$$t_{\alpha/2}\frac{s}{\sqrt{n}}$$

```
TInterval
(20.085,20.415)
x̄=20.25
Sx=.2
n=6
```

Figure 8.12

We substitute $t_{\alpha/2} = 2.015$, $s = 0.2$, and $n = 6$ to obtain

$$t_{\alpha/2}\frac{s}{\sqrt{n}} = 2.015\frac{0.2}{\sqrt{6}} = 0.1645$$

Step 4: Construct the confidence interval. The 90% confidence interval is given by

$$\bar{x} - t_{\alpha/2}\frac{s}{\sqrt{n}} < \mu < \bar{x} + t_{\alpha/2}\frac{s}{\sqrt{n}}$$

$$20.25 - 0.1645 < \mu < 20.25 + 0.1645$$

$$20.09 < \mu < 20.41 \qquad \text{(rounded to two decimal places, like } \bar{x}\text{)}$$

Figure 8.12 presents the results from the TI-84 Plus calculator.

Step 5: Interpret the result. We are 90% confident that the mean weight μ is between 20.09 and 20.41. Another way to say this is that we are 90% confident that the mean weight μ is in the interval 20.25 ± 0.16. If we were to draw many different samples and use this method to construct the corresponding confidence intervals, then in the long run, 90% of them would cover the true population mean μ. So unless we were somewhat unlucky in the sample we drew, the true mean weight is between 20.09 and 20.41 ounces.

A confidence level of 90% is the lowest level commonly used in practice. In Example 8.10, we will construct a 99% confidence interval.

Example 8.10

Construct a confidence interval

Use the data in Example 8.9 to construct a 99% confidence interval for the mean fill weight. Compare the margin of error of this confidence interval to the 90% confidence interval constructed in Example 8.9.

EXPLAIN IT AGAIN

The relationship between confidence and the margin of error: If we want to increase our confidence that an interval contains the true value, we must increase the critical value. This increases the margin of error, which makes the confidence interval wider.

Solution

As in Example 8.9, the assumptions are met, so we may proceed.

Step 1: Find the sample mean and the sample standard deviation. These are given as $\bar{x} = 20.25$ and $s = 0.2$.

Step 2: Find the number of degrees of freedom and the critical value $t_{\alpha/2}$. The number of degrees of freedom is $n - 1 = 6 - 1 = 5$. Using Table A.3, we find that the critical value corresponding to a level of 99% is $t_{\alpha/2} = 4.032$.

Step 3: Compute the margin of error. The margin of error is

$$t_{\alpha/2}\frac{s}{\sqrt{n}}$$

We substitute $t_{\alpha/2} = 4.032$, $s = 0.2$, and $n = 6$ to obtain

$$t_{\alpha/2}\frac{s}{\sqrt{n}} = 4.032\frac{0.2}{\sqrt{6}} = 0.3292$$

Step 4: Construct the confidence interval. The 99% confidence interval is given by

$$\bar{x} - t_{\alpha/2}\frac{s}{\sqrt{n}} < \mu < \bar{x} + t_{\alpha/2}\frac{s}{\sqrt{n}}$$

$$20.25 - 0.3292 < \mu < 20.25 + 0.3292$$

$$19.92 < \mu < 20.58 \qquad \text{(rounded to two decimal places, like } \bar{x}\text{)}$$

Figure 8.13 presents the results from the TI-84 Plus calculator.

```
TInterval
(19.921,20.579)
x̄=20.25
Sx=.2
n=6
```

Figure 8.13

Step 5: Interpret the result. We are 99% confident that the mean weight μ is between 19.92 and 20.58. Another way to say this is that we are 99% confident that the mean weight μ is in the interval 20.25 ± 0.33. If we were to draw many different samples and use this method to construct the corresponding confidence intervals, then in the long run, 99% of them would cover the true population mean μ. So unless we were very unlucky in the sample we drew, the true mean is between 19.92 and 20.58 ounces.

For this 99% confidence interval, the margin of error is 0.3292. For the 90% confidence interval in Example 8.9, the margin of error was only 0.1645. The reason is that for a 90% confidence interval, we used a critical value of 2.015, and for the 99% confidence interval, we must use a larger critical value of 4.032.

We can see that if we want to be more confident that our interval contains the true value, we must increase the critical value, which increases the margin of error. There is a trade-off. We would rather have a higher level of confidence than a lower level, but we would also rather have a smaller margin of error than a larger one. So we have to choose a level of confidence that strikes a good balance. The most common choice is 95%. In some cases where high confidence is very important, a larger confidence level such as 99% may be chosen. In general, intervals with confidence levels less than 90% are not considered to be reliable enough to be used in practical situations.

Figure 8.14 illustrates the trade-off between confidence level and margin of error. One hundred samples were drawn from a population with mean μ. The center diagram presents one hundred 95% confidence intervals, each based on one of these samples. The confidence intervals are all different, because each sample has a different mean $\bar{x}$. The diagram on the left presents 70% confidence intervals based on the same samples. These intervals are narrower because they have a smaller margin of error, but many of them fail to cover the population mean. These intervals are too unreliable to be of any practical value. The figure on the right presents 99.7% confidence intervals. These intervals are very reliable. In the long run, only 3 in 1000 of these intervals will fail to cover the population mean. However, they are wider due to the larger margin of error, so they do not convey as much information.

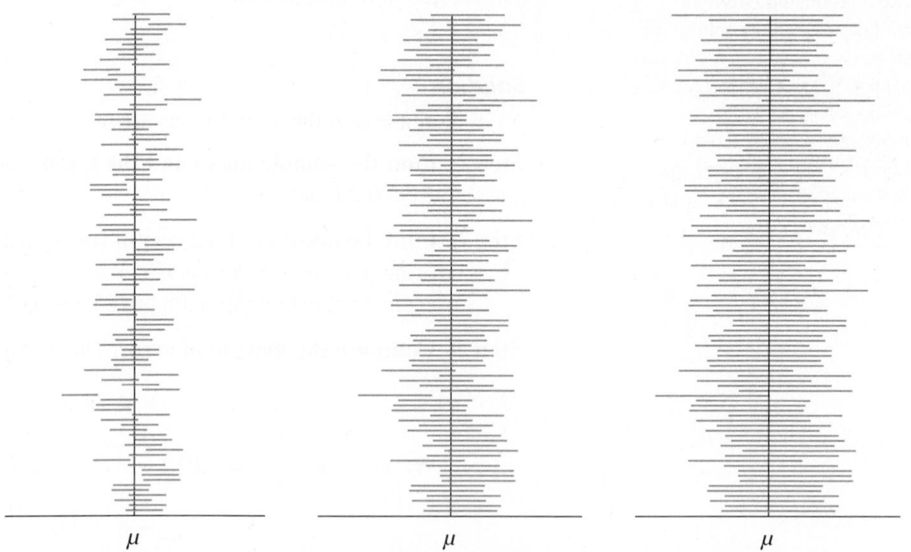

Figure 8.14 *Left:* One hundred 70% confidence intervals for a population mean, each constructed from a different sample. Although their margin of error is small, they cover the population mean only 70% of the time. This low success rate makes the 70% confidence interval unacceptable for practical purposes. *Center:* One hundred 95% confidence intervals constructed from these samples. This represents a good compromise between reliability and margin of error for many purposes. *Right:* One hundred 99.7% confidence intervals constructed from these samples. These intervals cover the population mean 997 times out of 1000. They almost always succeed in covering the population mean, but their margin of error is large.

The confidence level measures the success rate of the method used to construct the confidence interval

The center diagram in Figure 8.14 presents 100 different 95% confidence intervals. When we construct a confidence interval with level 95%, we are essentially getting a look at one of these confidence intervals. However, we don't get to see any of the other confidence intervals, nor do we get to see the vertical line that indicates where the true value μ is. Therefore, we cannot be sure whether we got one of the confidence intervals that covers μ, or whether we were unlucky enough to get one of the unsuccessful ones. What we do know is that our confidence interval was constructed by a method that succeeds 95% of the time. The confidence level describes the success rate of the *method* used to construct a confidence interval, not the success of any particular interval.

Check Your Understanding

7. To determine how well a new method of teaching vocabulary is working in a certain elementary school, education researchers plan to give a vocabulary test to a sample of 40 sixth graders. Assume the sample standard deviation is $s = 8$. The researchers plan to compute the sample mean $\bar{x}$, then construct a 95% confidence interval for the population mean test score.

a. What is the critical value $t_{\alpha/2}$ for this confidence interval? 2.023

b. Find the margin of error for this confidence interval. 2.559

c. Let m represent the margin of error for this confidence interval. For what percentage of all samples will the confidence interval $\bar{x} \pm m$ cover the true population mean? 95%

8. The researchers now plan to construct a 99% confidence interval for the test scores described in Exercise 7.

a. What is the critical value $t_{\alpha/2}$ for this confidence interval? 2.708

b. Find the margin of error for this confidence interval. 3.425

c. Let m represent the margin of error for this confidence interval. For what percentage of all samples will the confidence interval $\bar{x} \pm m$ cover the true population mean? 99%

Answers are on page 404.

Objective 4 Find the sample size necessary to obtain a confidence interval of a given width

Finding the Necessary Sample Size

We have seen that we can make the margin of error smaller if we are willing to reduce our level of confidence. We can also reduce the margin of error by increasing the sample size. We will demonstrate this first in the case where the population standard deviation σ is known, then extend it to the case where σ is unknown. If σ is known, the margin of error is

$$m = z_{\alpha/2}\frac{\sigma}{\sqrt{n}}$$

Since the sample size n appears in the denominator, making it larger will make the value of m smaller. We will show how we can manipulate this formula using algebra to express the sample size n in terms of the margin of error m.

$$m = z_{\alpha/2}\frac{\sigma}{\sqrt{n}}$$

$$m\sqrt{n} = \frac{z_{\alpha/2} \cdot \sigma}{\cancel{\sqrt{n}}}\cancel{\sqrt{n}} \qquad \text{(Multiply both sides by } \sqrt{n}\text{)}$$

$$\frac{\cancel{m}\sqrt{n}}{\cancel{m}} = \frac{z_{\alpha/2} \cdot \sigma}{m} \qquad \text{(Divide both sides by } m\text{)}$$

$$n = \left(\frac{z_{\alpha/2} \cdot \sigma}{m}\right)^2 \qquad \text{(Square both sides)}$$

With this formula, if we know how small we want the margin of error to be, we can compute the sample size needed to achieve the desired margin of error.

NOTE TO INSTRUCTOR

The sample size computed using the sample standard deviation will tend to be a slight underestimate, because $z_{\alpha/2}$ is slightly less than $t_{\alpha/2}$. Although it is not necessary, one could adjust for this by doing the calculation twice: The first calculation gives an initial estimate n, then the second calculation is done using the $t_{\alpha/2}$ critical value with $n - 1$ degrees of freedom.

When σ is unknown, the margin of error is

$$m = t_{\alpha/2}\frac{s}{\sqrt{n}}$$

There are two difficulties in using this equation to determine a sample size n. First, the value $t_{\alpha/2}$ depends on n. Second, the sample standard deviation s is not equal to σ. If we have a reasonably large sample from the population (sample size greater than 30), these difficulties are less serious. In this case, the sample standard deviation s will be reasonably close to σ with high probability. In addition, since we already have a sample of size greater than 30, the value $t_{\alpha/2}$ will be close to the value $z_{\alpha/2}$. We can therefore use the formula that assumes σ is known to estimate a sample size n, by replacing σ with s.

CAUTION

Always round the sample size *up*. For example, if the value of n given by the formula is 84.01, round it *up* to 85.

SUMMARY

Let m be the desired margin of error. Let σ be the population standard deviation, and let $z_{\alpha/2}$ be the critical value for a confidence interval. The sample size n needed so that the confidence interval will have margin of error m is given by

$$n = \left(\frac{z_{\alpha/2} \cdot \sigma}{m}\right)^2$$

If a sample standard deviation s has been calculated from a sample of size greater than 30, an estimate of the sample size needed so that the confidence interval will have a margin of error m is

$$n = \left(\frac{z_{\alpha/2} \cdot s}{m}\right)^2$$

If the value of n given by the formula is not a whole number, round it *up* to the nearest whole number.

Example 8.11

Finding the necessary sample size

Scientists want to estimate the mean weight of mice after they have been fed a special diet. A sample of 60 mice have been weighed, and the standard deviation was $s = 3$ grams. Estimate the number of mice that must be weighed for a 95% confidence interval to have a margin of error of 0.5 grams.

Solution

Since we want a 95% confidence interval, we use $z_{\alpha/2} = 1.96$. We are also given $s = 3$ and $m = 0.5$. We therefore use the formula as follows:

$$n = \left(\frac{z_{\alpha/2} \cdot s}{m}\right)^2 = \left(\frac{1.96 \cdot 3}{0.5}\right)^2 = 138.30; \text{ round up to } 139$$

We must weigh 139 mice in order to obtain a 95% confidence interval with a margin of error of 0.5 grams.

Factors that limit sample size

Since larger sample sizes result in narrower confidence intervals, it is natural to wonder why we don't always collect a large sample when we want to construct a confidence interval. In practice, the size of the sample that is feasible to obtain is often limited. In some cases, an expensive experimental procedure must be repeated each time an observation is made. For example, studies of automobile safety that require the crashing of new cars are not likely to have large sample sizes. Sometimes ethical considerations restrict the sample size. For example, when a new drug is being tested, there is a risk of adverse health effects to the subjects who take the drug. It is important that the sample size not be larger than necessary, to limit the health risk to as few people as possible.

Check Your Understanding

9. A machine used to fill beverage cans is supposed to put exactly 12 ounces of beverage in each can, but the actual amount varies randomly from can to can. In a sample of 50 cans, the standard deviation of the amount was $s = 0.05$ ounces. A simple random sample of filled cans will have their volumes measured, and a 95% confidence interval for the mean fill volume will be constructed. Estimate the number of cans that must be sampled for the margin of error to be equal to 0.01 ounces. *97*

10. An IQ test is designed to have scores that have a standard deviation of $\sigma = 15$. A simple random sample of students at a large university will be given the test in order to construct a 98% confidence interval for the mean IQ of all students at the university. How many students must be tested so that the margin of error will be equal to 3 points? *136*

Answers are on page 404.

Objective 5 Distinguish between confidence and probability

Distinguish Between Confidence and Probability

In Example 8.10, a 99% confidence interval for the population mean weight μ was computed to be $19.92 < \mu < 20.58$. It is tempting to say that the probability is 99% that μ is between 19.92 and 20.58. This, however, is not correct. The term *probability* refers to random events, which can come out differently when experiments are repeated. The numbers 19.92 and 20.58 are fixed, not random. The population mean is also fixed. The population mean weight is either between 19.92 and 20.58 or it is not. There is no randomness involved. Therefore, we say that we have 99% *confidence* (not probability) that the population mean is in this interval.

On the other hand, let's say that we are discussing a *method* used to construct a 99% confidence interval. The method will succeed in covering the population mean 99% of the time and fail the other 1% of the time. In this case, whether the population mean is covered or not is a random event, because it can vary from experiment to experiment. Therefore, it *is* correct to say that a *method* for constructing a 99% confidence interval has probability 99% of covering the population mean.

Example 8.12

Interpreting a confidence level

A hospital administrator plans to draw a simple random sample of 100 records of patients who were admitted for cardiac bypass surgery. She will compute the sample mean number of days spent in the hospital and construct a 95% confidence interval for the population mean, using an appropriate method. She claims that the probability is 0.95 that the confidence interval will cover the population mean. Is she right?

Solution

Yes, she is right. The probability that a 95% confidence interval constructed by an appropriate method will cover the true value is 0.95.

Example 8.13

Interpreting a confidence interval

Refer to Example 8.12. After drawing the sample, the hospital administrator constructs the 95% confidence interval, and it turns out to be $7.1 < \mu < 7.5$. The administrator claims that the probability is 0.95 that the population mean is between 7.1 and 7.5. Is she right?

Solution

No, she is not right. Once a specific confidence interval has been constructed, there is no more probability. She should say that she is 95% confident that the population mean is between 7.1 and 7.5.

Check Your Understanding

11. A scientist plans to construct a 95% confidence interval for the mean length of steel rods that are manufactured by a certain process. She will draw a simple random sample of rods and compute the confidence interval using the methods described in this section. She says, "The probability is 95% that the population mean length will be covered by the confidence interval." Is she right? Explain. *Yes*

12. The scientist in Exercise 11 constructs the 95% confidence interval for the mean length in centimeters, and it turns out to be $25.1 < \mu < 27.2$. She says, "The probability is 95% that the population mean length is between 25.1 and 27.2 centimeters." Is she right? Explain. *No*

Answers are on page 404.

Using Technology

We use Example 8.8 to illustrate the technology steps.

TI-84 PLUS

Constructing a confidence interval for the mean when σ is unknown

Step 1. Press **STAT** and highlight the **TESTS** menu.

Step 2. Select **TInterval** and press **ENTER** (Figure A). The **TInterval** menu appears.

Step 3. For **Inpt**, select the **Stats** option and enter the values of $\bar{x}$, s, and n. For Example 8.8, we use $\bar{x} = 8.20$, $s = 9.84$, and $n = 123$.

Step 4. In the **C-Level** field, enter the confidence level. For Example 8.8, we use 0.95 (Figure B).

Step 5. Highlight **Calculate** and press **ENTER** (Figure C).

Note that if the raw data are given, the **TInterval** command may be used by selecting **Data** as the **Inpt** option and entering the location of the data as the **List** option (Figure D).

```
EDIT CALC TESTS
1:Z-Test…
2:T-Test…
3:2-SampZTest…
4:2-SampTTest…
5:1-PropZTest…
6:2-PropZTest…
7:ZInterval…
8:TInterval…
9↓2-SampZInt…
```

Figure A

```
TInterval
Inpt:Data Stats
x̄:8.2
Sx:9.84
n:123
C-Level:.95
Calculate
```

Figure B

```
TInterval
(6.4436,9.9564)
x̄=8.2
Sx=9.84
n=123
```

Figure C

```
TInterval
Inpt:Data Stats
List:L1
Freq:1
C-Level:.95
Calculate
```

Figure D

EXCEL

Constructing a confidence interval for the mean when σ is unknown

The **=CONFIDENCE.T**(*alpha, standard_dev, size*) command returns the margin of error when the population standard deviation is not known.

To obtain the margin of error m for Example 8.12, we use $\alpha = 0.05$ in the *alpha* field, the sample standard deviation $s = 9.84$ in the *standard_dev* field, and the sample size $n = 123$ in the *size* field (Figure E).

The confidence interval is given by $\bar{x} - m < \mu < \bar{x} + m$.

=CONFIDENCE.T(0.05,9.84,123)	
Margin of Error	1.75639

Figure E

Finding a critical value from a Student's t distribution

The **=T.INV**(*probability, deg_freedom*) command returns the value from the Student's t distribution, where the *probability* input corresponds to the area on the left and the *deg_freedom* input is the number of degrees of freedom.

The critical value for confidence level $1-\alpha$ is the absolute value of $t_{\alpha/2}$. Figure F shows the command for finding the critical value of a 95% confidence interval where $\frac{\alpha}{2} = 0.025$ and with 122 degrees of freedom.

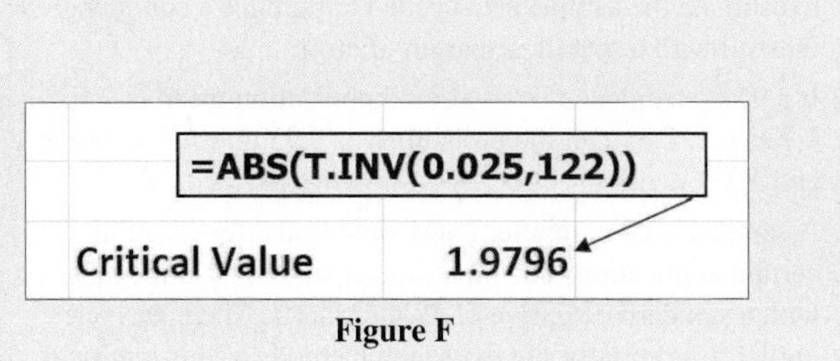

Figure F

MINITAB

Constructing a confidence interval for the mean when σ is unknown

Step 1. Click on **Stat**, then **Basic Statistics**, then **1-Sample t**.

Step 2. Choose one of the following:

- If the summary statistics are given, select **Summarized Data** and enter the **Sample Size** (123), the **Mean** (8.20), and the **Standard Deviation** (9.84) (Figure G).
- If the raw data are given, select **One or more samples, each in a column** and select the column that contains the data.

Step 3. Click **Options**, and enter the confidence level in the **Confidence Level** field. For Example 8.8, enter **95%**. Click **OK**.

Step 4. Click **OK** (Figure H).

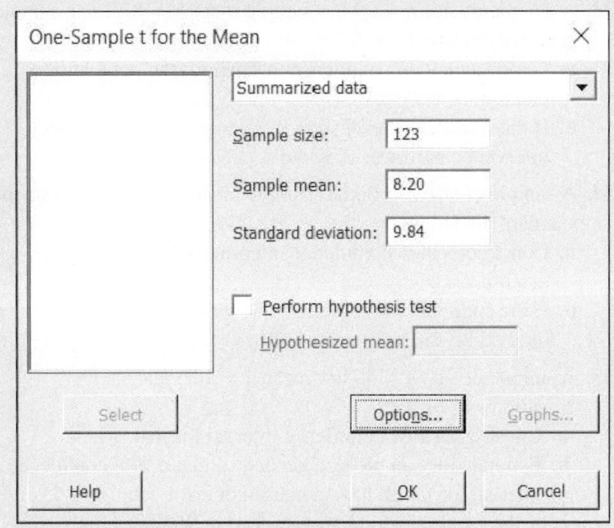

Figure G

N	Mean	StDev	SE Mean	95% CI for μ
123	8.200	9.840	0.887	(6.444, 9.956)

Figure H

Section 8.2 Exercises

Exercises 1–12 are the Check Your Understanding exercises located within the section.

Understanding the Concepts

In Exercises 13–16, fill in each blank with the appropriate word or phrase.

13. When constructing a confidence interval for a population mean μ from a sample of size 12, the number of degrees of freedom for the critical value $t_{\alpha/2}$ is ________. *11*

14. When the number of degrees of freedom is large, the Student's t distribution is close to the ________ distribution. *normal*

15. If we increase the sample size and keep the confidence level the same, we ________ the margin of error. *decrease*

16. If we increase the confidence level and keep the sample size the same, we ________ the margin of error. *increase*

In Exercises 17–20, determine whether the statement is true or false. If the statement is false, rewrite it as a true statement.

17. The Student's t curve is less spread out than the standard normal curve. *False*

18. The Student's t distribution should not be used to find a confidence interval for μ if outliers are present in a small sample. *True*

19. We may use the standard deviation from a sample of size $n > 30$ to estimate the sample size needed to produce a confidence interval with a specified margin of error. *True*

20. If a 95% confidence interval for a population mean is $1.7 < \mu < 2.3$, then the probability is 0.95 that μ is between 1.7 and 2.3. *False*

21. A sample of 20 computer chips were tested for speed in a certain application. The mean speed was $\bar{x} = 495$ megahertz, with a standard deviation of 19 megahertz. There was one outlier. Is it appropriate to use the methods of this section to construct a confidence interval for the mean speed of this chip? If not, explain why not. *No*

22. To estimate the mean lifetime of a certain type of truck tire, a mechanic put 18 of the tires on an 18-wheeler. The mean lifetime of the tires is $\bar{x} = 40{,}000$ miles, with a standard deviation of $s = 5000$ miles. The sample had no outliers and no strong skewness. Is it appropriate to use the methods of this section to construct a confidence interval for the mean lifetime of this type of tire? If not, explain why not. *Yes*

23. A simple random sample of 10 students at a certain college reported the number of minutes per week they typically spent exercising. The results were

60 120 0 150 450 0 30 75 300 90

Is it appropriate to use the methods of this section to construct a confidence interval for the mean number of minutes spent exercising? If not, explain why not. *No*

24. A pollster sampled 14 people aged 18–22 and asked them how many hours per day they spent on social media. The results were as follows:

2.8	1.4	1.9	1.7	3.7	3.1	2.7
2.7	1.7	3.5	2.8	2.2	2.3	2.4

Is it appropriate to use the methods of this section to construct a confidence interval for the mean number of hours spent on social media? If not, explain why not. *Yes*

Practicing the Skills

25. Find the critical value $t_{\alpha/2}$ needed to construct a confidence interval of the given level with the given sample size.

a. Level 95%, sample size 23 *2.074*
b. Level 90%, sample size 3 *2.920*
c. Level 98%, sample size 18 *2.567*
d. Level 99%, sample size 29 *2.763*

26. Find the critical value $t_{\alpha/2}$ needed to construct a confidence interval of the given level with the given sample size.

a. Level 90%, sample size 6 *2.015*
b. Level 98%, sample size 12 *2.718*
c. Level 95%, sample size 32 *2.040*
d. Level 99%, sample size 10 *3.250*

27. A sample of size $n = 18$ is drawn from a normal population, and a 95% confidence interval is constructed, using a critical value of $t_{\alpha/2} = 2.110$. If the sample size were $n = 25$, would the critical value be smaller or larger? *Smaller*

28. A sample of size $n = 22$ is drawn from a normal population, and a 90% confidence interval is constructed, using a critical value of $t_{\alpha/2} = 1.721$. If the sample size were $n = 15$, would the critical value be smaller or larger? *Larger*

29. A sample of size $n = 5$ is drawn from a normal population. The sample mean is $\bar{x} = 4.31$, and the sample standard deviation is $s = 2.7$.

a. Construct a 95% confidence interval for μ. *(0.96, 7.66)*
b. If the population were not approximately normal, would the confidence interval constructed in part (a) be valid? Explain. *No*

30. A sample of size $n = 92$ is drawn from a normal population. The sample mean is $\bar{x} = 12.7$, and the sample standard deviation is $s = 8.6$.

a. Construct a 90% confidence interval for μ. *(11.2, 14.2)*
b. If the population were not approximately normal, would the confidence interval constructed in part (a) be valid? Explain. *Yes*

31. A sample of size $n = 15$ has sample mean $\bar{x} = 2.1$ and sample standard deviation $s = 1.7$.

a. Construct a 95% confidence interval for the population mean μ. *(1.2, 3.0)*
b. If the sample size were $n = 25$, would the confidence interval be narrower or wider? *Narrower*

32. A sample of size $n = 44$ has sample mean $\bar{x} = 56.9$ and sample standard deviation $s = 9.1$.

a. Construct a 98% confidence interval for the population mean μ. *(53.6, 60.2)*
b. If the sample size were $n = 30$, would the confidence interval be narrower or wider? *Wider*

33. A sample of size $n = 89$ has sample mean $\bar{x} = 87.2$ and sample standard deviation $s = 5.3$.

a. Construct a 95% confidence interval for the population mean μ. *(86.1, 88.3)*
b. If the confidence level were 99%, would the confidence interval be narrower or wider? *Wider*

34. A sample of size $n = 35$ has sample mean $\bar{x} = 34.85$ and sample standard deviation $s = 17.9$.

a. Construct a 98% confidence interval for the population mean μ. *(27.46, 42.24)*
b. If the confidence level were 95%, would the confidence interval be narrower or wider? *Narrower*

35. A sample of size $n = 45$ has mean $\bar{x} = 10.5$ and standard deviation $s = 14.3$.

a. Construct a 99% confidence interval for μ.
b. Estimate the sample size needed so that a 99% confidence interval for μ will have a margin of error equal to 2.5. *218*
c. If the required confidence level were 95%, would the necessary sample size be larger or smaller? *Smaller*

36. A sample of size $n = 56$ has mean $\bar{x} = 25.2$ and standard deviation $s = 17.1$.

a. Construct a 95% confidence interval for μ. *(20.6, 29.8)*
b. Estimate the sample size needed so that a 95% confidence interval for μ will have a margin of error equal to 3.6. *87*
c. If the required confidence level were 98%, would the necessary sample size be larger or smaller? *Larger*

37. A sample of size $n = 61$ has mean $\bar{x} = 18.3$ and standard deviation $s = 2.7$.

a. Construct a 90% confidence interval for μ. *(17.7, 18.9)*
b. Estimate the sample size needed so that a 90% confidence interval for μ will have a margin of error equal to 0.5. *79*
c. If the required margin of error were 0.2, would the necessary sample size be larger or smaller? *Larger*

38. A sample of size $n = 72$ has mean $\bar{x} = 45.9$ and standard deviation $s = 9.2$.

a. Construct a 95% confidence interval for μ. *(43.7, 48.1)*
b. Estimate the sample size needed so that a 95% confidence interval for μ will have a margin of error equal to 1.1. *269*

c. If the required margin of error were 1.5, would the necessary sample size be larger or smaller? *Larger*

Working with the Concepts

39. Online courses: A sample of 263 students who were taking online courses were asked to describe their overall impression of online learning on a scale of 1–7, with 7 representing the most favorable impression. The average score was 5.53, and the standard deviation was 0.92.

a. Construct a 95% confidence interval for the mean score. *(5.42, 5.64)*

b. Assume that the mean score for students taking traditional courses is 5.55. A college that offers online courses claims that the mean scores for online courses and traditional courses are the same. Does the confidence interval contradict this claim? Explain. *No*

Source: *Innovations in Education and Teaching International* 45:115–126

40. Get an education: The General Social Survey asked 1972 adults how many years of education they had. The sample mean was 13.37 years with a standard deviation of 3.13 years.

a. Construct a 98% confidence interval for the mean number of years of education. *(13.21, 13.53)*

b. Data collected in an earlier study suggests that the mean 10 years ago was 13.26 years. A sociologist believes that the mean now is the same. Does the confidence interval contradict this belief? Explain. *No*

41. Fake Twitter followers: Many celebrities and public figures have Twitter accounts with large numbers of followers. However, some of these followers are fake, resulting from accounts generated by spamming computers. In a sample of 46 twitter audits, the mean percentage of fake followers was 14.1 with a standard deviation of 9.6.

a. Construct a 90% confidence interval for the mean percentage of fake Twitter followers. *(11.7, 16.5)*

b. Based on the confidence interval, is it reasonable that the mean percentage of fake Twitter followers is less than 10? *No*

Source: www.twitteraudit.com

42. Let's go to the movies: A random sample of 35 Hollywood movies made in the last 10 years had a mean length of 125.2 minutes, with a standard deviation of 20.4 minutes.

a. Construct a 95% confidence interval for the true mean length of all Hollywood movies made in the last 10 years. *(118.19, 132.21)*

b. To date, there have been five movies released in the Bourne film series, and their mean length is 119.8 minutes. Someone claims that the mean length of Bourne movies is actually greater than the mean length of all Hollywood movies. Does the confidence interval contradict this claim? Explain. *No*

image100/Alamy

43. Hip surgery: In a sample of 123 hip surgeries of a certain type, the average surgery time was 136.9 minutes with a standard deviation of 22.6 minutes.

a. Construct a 95% confidence interval for the mean surgery time for this procedure. *(132.9, 140.9)*

b. If a 99% confidence interval were constructed with these data, would it be wider or narrower than the interval constructed in part (a)? Explain. *Wider*

Source: *Journal of Engineering in Medicine* 221:699–712

44. Sound it out: Phonics is an instructional method in which children are taught to connect sounds with letters or groups of letters. A sample of 134 first graders who were learning English were asked to identify as many letter sounds as possible in a period of one minute. The average number of letter sounds identified was 34.06 with a standard deviation of 23.83.

a. Construct a 98% confidence interval for the mean number of letter sounds identified in one minute. *(29.19, 38.93) [Tech: (29.21, 38.91)]*

b. If a 95% confidence interval were constructed with these data, would it be wider or narrower than the interval constructed in part (a)? Explain. *Narrower*

Source: *School Psychology Review* 37:5–17

45. Software instruction: A hybrid course is one that contains both online and classroom instruction. In a study performed at Middle Georgia State University, a software package was used as the main source of instruction in a hybrid college algebra course. The software tracked the number of hours it took for each student to meet the objectives of the course. In a sample of 45 students, the mean number of hours was 80.5, with a standard deviation of 51.2.

a. Construct a 95% confidence interval for the mean number of hours it takes for a student to meet the course objectives. *(65.1, 95.9)*

b. If a sample of 90 students had been studied, would you expect the confidence interval to be wider or narrower than the interval constructed in part (a)? Explain. *Narrower*

46. Baby talk: In a sample of 77 children, the mean age at which they first began to combine words was 16.51 months, with a standard deviation of 9.59 months.

a. Construct a 95% confidence interval for the mean age at which children first begin to combine words. *(14.32, 18.70) [Tech: (14.33, 18.69)]*

b. If a sample of 50 children had been studied, would you expect the confidence interval to be wider or narrower than the interval constructed in part (a)? Explain. *Wider*

Source: *Proceedings of the 4th International Symposium on Bilingualism*, pp. 58–77

47. Baby weights: Following are weights, in pounds, of 12 two-month-old baby girls. It is reasonable to assume that the population is approximately normal.

12.23	12.32	11.87	12.34	11.48	12.66
8.51	14.13	12.95	10.30	9.34	8.63

a. Construct a 98% confidence interval for the mean weight of two-month-old baby girls. *(9.986, 12.808)*

b. According to the National Health Statistics Reports, the mean weight of two-month-old baby boys is 11.5 pounds. Based on the confidence interval, is it reasonable to believe that the mean weight of two-month-old baby girls may be the same as that of two-month-old baby boys? Explain. *Yes*

48. Eat your cereal: Boxes of cereal are labeled as containing 14 ounces. Following are the weights, in ounces, of a sample of 12 boxes: It is reasonable to assume that the population is approximately normal.

14.02	13.97	14.11	14.12	14.10	14.02
14.15	13.97	14.05	14.04	14.11	14.12

a. Construct a 98% confidence interval for the mean weight. *(14.017, 14.113)*

b. The quality control manager is concerned that the mean weight is actually less than 14 ounces. Based on the confidence interval, is there a reason to be concerned? Explain. *No*

49. Eat your spinach: Six measurements were made of the mineral content (in percent) of spinach, with the following results. It is reasonable to assume that the population is approximately normal.

19.1 20.8 20.8 21.4 20.5 19.7

a. Construct a 95% confidence interval for the mean mineral content. *(19.50, 21.26)*

b. Based on the confidence interval, is it reasonable to believe that the mean mineral content of spinach may be greater than 21%? Explain. *Yes*

Source: *Journal of Nutrition* 66:55–66

50. Mortgage rates: Following are interest rates (annual percentage rates) for a 30-year fixed-rate mortgage from a sample of lenders in Macon, Georgia, on a recent day. It is reasonable to assume that the population is approximately normal.

4.750	4.375	4.176	4.679	4.426	4.227
4.125	4.250	3.950	4.191	4.299	4.415

Source: www.bankrate.com

a. Construct a 99% confidence interval for the mean rate. *(4.1190, 4.5248)*

b. One week earlier, the mean rate was 4.050%. A mortgage broker claims that the mean rate is now higher. Based on the confidence interval, is this a reasonable claim? Explain. *Yes*

51. Hi-def: Following are prices of a random sample of 18 smart TVs sold on shopper.cnet.com with screen sizes between 46 and 50 inches, along with a dotplot of the data.

548	598	697	699	749	799
829	849	928	1050	1098	1169
1198	1269	1299	1399	1455	1599

400 800 1200 1600

a. Is it reasonable to assume that the conditions for constructing a confidence interval for the mean price are satisfied? Explain. *Yes*

b. If appropriate, construct a 95% confidence interval for the mean price of all smart TVs in this size range. *(856.9, 1168.9)*

52. Big salary for the boss: Following is the total compensation, in millions of dollars, for Chief Executive Officers at 20 large U.S. corporations, along with a boxplot of the data.

2.75	9.15	2.32	23.84	13.12	9.91	6.39
2.19	4.28	0.70	3.18	8.20	1.68	4.64
6.05	6.59	11.31	1.35	25.84	2.85	

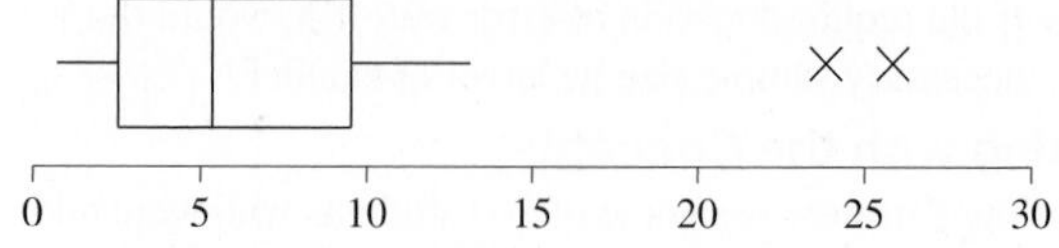

a. Is it reasonable to assume that the conditions for constructing a confidence interval for the mean compensation are satisfied? Explain. *No*

b. If appropriate, construct a 95% confidence interval for the mean compensation of a Chief Executive Officer. *Not appropriate*

53. Pain relief: One of the ways in which doctors try to determine how long a single dose of pain reliever will provide relief is to measure the drug's half-life, which is the length of time it takes for one-half of the dose to be eliminated from the body. Following are half-lives (in hours) of the pain reliever oxycodone for a sample of 18 individuals, along with a boxplot of the data, based on a report of the National Institutes of Health.

3.3	1.7	2.0	5.0	1.2	2.8	3.7	3.5	4.8
4.7	4.9	2.5	5.1	6.0	3.9	4.3	2.1	3.0

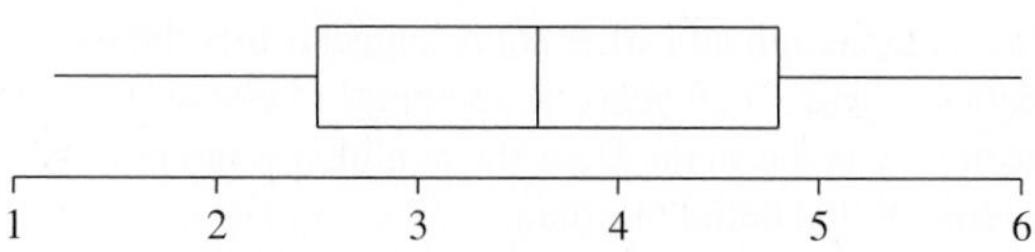

a. Is it reasonable to assume that the conditions for constructing a confidence interval for the mean half-life are satisfied? Explain. *Yes*

b. If appropriate, construct a 95% confidence interval for the mean half-life. *(2.91, 4.26)*

c. The National Institutes of Health report states that the mean half-life is 3.51 hours. If appropriate, explain whether this confidence interval contradicts this claim. *Does not contradict*

54. How's your mileage? Gas mileage, in miles per gallon, was measured for five Ford Explorer SUVs. Following are the results, along with a dotplot of the data.

20.4 19.6 17.1 17.9 18.7

17 17.5 18 18.5 19 19.5 20 20.5 21

a. Is it reasonable to assume that the conditions for constructing a confidence interval for the mean gas mileage are satisfied? Explain. *Yes*

b. If appropriate, construct a 95% confidence interval for the mean gas mileage. *(17.11, 20.37)*

c. The EPA gas mileage rating for a Ford Explorer is 19 miles per gallon. If appropriate, explain whether the rated mileage may be equal to the mean mileage, based on the confidence interval. *May be equal*

Source: TrueDelta.com

55. Different levels: Joe and Arya are going to construct confidence intervals from the same simple random sample. Joe's confidence interval will have level 90%, and Arya's will have level 95%.

a. Which confidence interval will have the larger margin of error? Or will they both be the same? *Arya's*

b. Which confidence interval is more likely to cover the population mean? Or are they both equally likely to do so? *Arya's*

56. Different levels: Bertha and Nico are going to construct confidence intervals from the same simple random sample. Bertha's confidence interval will have level 98%, and Nico's will have level 95%.

a. Which confidence interval will have the larger margin of error? Or will they both be the same? *Bertha's*

b. Which confidence interval is more likely to cover the population mean? Or are they both equally likely to do so? *Bertha's*

57. Different sample sizes: Javier and Kai are going to construct confidence intervals from different simple random samples. Both confidence intervals will have level 95%. Javier will use a sample of size 50, and Kai will use a sample of size 100. The samples will be drawn from the same population.

a. Which confidence interval will have the larger margin of error? Or will they both be the same? *Javier's*

b. Which confidence interval is more likely to cover the population mean? Or are they both equally likely to do so? *Both equally likely*

58. Different sample sizes: Emily and Julio are going to construct confidence intervals from different simple random samples. Both confidence intervals will have level 99%. Emily will use a sample of size 40, and Julio will use a sample of size 75. The samples will be drawn from the same population.

a. Which confidence interval will have the larger margin of error? Or will they both be the same? *Emily's*

b. Which confidence interval is more likely to cover the population mean? Or are they both equally likely to do so? *Both equally likely*

59. Different standard deviations: Maria and Bob are going to construct confidence intervals from different simple random samples. Both confidence intervals will have level 95%. Maria's sample has standard deviation $s = 1$, and Bob's sample has standard deviation $s = 2$. Both sample sizes are the same.

a. Which confidence interval will have the larger margin of error? Or will they both be the same? *Bob's*

b. Which confidence interval is more likely to cover the population mean? Or are they both equally likely to do so? *Both equally likely*

60. Different standard deviations: Martin and Bianca are going to construct confidence intervals from different simple random samples. Both confidence intervals will have level 99%. Martin's sample has standard deviation $s = 25$, and Bianca's sample has standard deviation $s = 18$. Both sample sizes are the same.

a. Which confidence interval will have the larger margin of error? Or will they both be the same? *Martin's*

b. Which confidence interval is more likely to cover the population mean? Or are they both equally likely to do so? *Both equally likely*

61. Which interval is which? Alessandro constructed three confidence intervals, all from the same random sample. The confidence levels are 90%, 95%, and 99%. The confidence intervals are $5.6 < \mu < 14.4$, $7.2 < \mu < 12.8$, and $6.6 < \mu < 13.4$. Unfortunately, Alessandro has forgotten which confidence interval has which level. Match each confidence interval with its level. *90%: (7.2, 12.8), 95%: (6.6, 13.4), 99%: (5.6, 14.4)*

62. Which interval is which? Matilda has constructed three confidence intervals, all from the same random sample. The confidence levels are 95%, 98%, and 99.9%. The confidence intervals are $6.4 < \mu < 12.3$, $5.1 < \mu < 13.6$, and $6.8 < \mu < 11.9$. Unfortunately, Matilda has forgotten which confidence interval has which level. Match each confidence interval with its level. *95%: (6.8, 11.9), 98%: (6.4, 12.3), 99.9%: (5.1, 13.6)*

63. Eat your kale: Kale is a type of cabbage commonly found in salad and used in cooking in many parts of the world. Six measurements were made of the mineral content (in percent) of kale, with the following results.

26.1 17.5 15.4 16.4 15.1 12.8

It turns out that the value 26.1 came from a specimen that the investigator forgot to wash before measuring.

a. The data contain an outlier that is clearly a mistake. Eliminate the outlier, then construct a 95% confidence interval for the mean mineral content from the remaining values. *(13.27, 17.61)*

b. Leave the outlier in and construct the 95% confidence interval. Are the results noticeably different? Explain why it is important to check data for outliers. *(12.36, 22.07)*

Source: *Journal of Nutrition* 66:55–66

64. Sleeping outlier: A simple random sample of eight college freshmen were asked how many hours of sleep they typically got per night. The results were

7.5 8.0 6.5 24 8.5 6.5 7.0 7.5

Notice that one joker said that he sleeps 24 hours a day.

a. The data contain an outlier that is clearly a mistake. Eliminate the outlier, then construct a 95% confidence interval for the mean amount of sleep from the remaining values. *(6.67, 8.05)*

b. Leave the outlier in and construct the 95% confidence interval. Are the results noticeably different? Explain why it is important to check data for outliers. *(4.48, 14.39)*

65. How much confidence? In a sample of 100 U.S. adults aged 18–24 who celebrate Halloween, the mean amount spent on a costume was \$37.51 with a standard deviation of \$16.44. A retail specialist claims that the mean amount spent on Halloween costumes for all U.S. adults aged 18–24 is between \$35.07 and \$39.95. With what level of confidence can this claim be made? *85.9%*

66. How much confidence? In a survey of 200 adult women in a certain city, the mean number of children they had was 2.3 with a standard deviation of 1.2. A sociologist states that the mean number of children per woman in this city is between 2.13 and 2.47. With what level of confidence can this claim be made? *95.4%*

67. Don't construct a confidence interval: There have been 44 presidents of the United States. Their mean height is 70.8 inches, with a standard deviation of 2.7 inches. Explain why these data should not be used to construct a confidence interval for the mean height of the presidents.

68. Don't construct a confidence interval: As of July 1, 2012, the mean population of the 50 states of the United States was 6.3 million, with a standard deviation of 7.0 million. Explain why these data should not be used to construct a confidence interval for the mean population of the states.

69. Interpret calculator display: Weights, in pounds, of a sample of male college soccer players were measured. The following display from a TI-84 Plus calculator presents a 98% confidence interval for the population mean weight.

```
TInterval
(178.08,181.58)
x̄=179.83
Sx=6.86
n=86
```

a. Fill in the blanks: We are ________ confident that the population mean weight is between ________ and ________. *98%, 178.08, 181.58*

b. Assume the population is not normally distributed. Is the confidence interval still valid? Explain. *Yes*

70. Interpret calculator display: A sample of college students was asked how many hours per week they spent reading. The following display from a TI-84 Plus calculator presents a 95% confidence interval for the population mean time spent reading.

```
TInterval
(.85638,2.2836)
x̄=1.57
Sx=.68
n=6
```

a. Fill in the blanks: We are ________ confident that the population mean number of hours is between ________ and ________. *95%, 0.85638, 2.2836*

b. Assume the population is not normally distributed. Is the confidence interval still valid? Explain. *No*

71. Interpret computer output: A sample of used cars on a lot was taken, and the age of each car was recorded. The following MINITAB output presents a confidence interval for the population mean age.

```
Variable  N   Mean   StDev  SE Mean       95% CI
X        15  5.9373  2.0387  0.5264  (4.8083, 7.0663)
```

a. How many degrees of freedom are there? *14*

b. If the population were not approximately normal, would this confidence interval be valid? Explain. *No*

c. Find the critical value $t_{\alpha/2}$ for a 98% confidence interval. *2.624*

d. Use the critical value and the information in the output to construct a 98% confidence interval. *(4.5561, 7.3185) [Tech: (4.5558, 7.3188)]*

72. Interpret computer output: A sample of college students was taken, and their ages were recorded. The following MINITAB output presents a confidence interval for the population mean age.

```
Variable  N    Mean    StDev  SE Mean       99% CI
X        71  23.8760  3.9385   0.4674  (22.638, 25.114)
```

a. How many degrees of freedom are there? *70*

b. If the population were not approximately normal, would this confidence interval be valid? Explain. *Yes*

c. Find the critical value $t_{\alpha/2}$ for a 95% confidence interval. *2.000 [Tech: 1.994]*

d. Use the critical value and the information in the output to construct a 95% confidence interval. *(22.9412, 24.8108) [Tech: (22.9438, 24.8082)]*

Extending the Concepts

73. Sample of size 1: The concentration of carbon monoxide in parts per million is believed to be normally distributed with a standard deviation $\sigma = 8$. A single measurement of the concentration is made, and its value is 85.

a. Use the methods of Section 8.1 to construct a 95% confidence interval for the mean concentration. *(69.3, 100.7)*

b. Would it be possible to construct a confidence interval using the methods of this section if the population standard deviation were unknown? Explain. *No*

A confidence interval provides likely minimum and maximum values for a parameter. In some cases, we are interested only in a maximum or only in a minimum. In these cases, we construct a ***one-sided confidence bound****. A one-sided confidence bound can be an* ***upper confidence bound****, which has the form $\bar{x} + t_\alpha s/\sqrt{n}$, or a* ***lower confidence bound****, which has the form $\bar{x} - t_\alpha s/\sqrt{n}$. Note that the critical value is t_α rather than $t_{\alpha/2}$.*

74. One-sided confidence bound: A simple random sample of 50 middle-school children participated in an experimental class designed to introduce them to computer programming. At the end of the class, the students took a final exam to assess their learning. The sample mean score was 78 points, and the sample standard deviation was 8 points. Compute a lower 99% confidence bound for the mean score. *75.3*

75. One-sided confidence bound: A random sample of 65 charges on a credit card had a mean of $56.85, and the sample standard deviation was $20.08. Compute an upper 95% confidence bound for the mean amount charged. *$61.01*

Answers to Check Your Understanding Exercises for Section 8.2

1. a. 2.145 **b.** 2.831 **c.** 1.671 **d.** 12.706

2. a. Yes **b.** No **c.** Yes **d.** No

3. $8.03 < \mu < 8.21$

4. $744.2 < \mu < 756.2$

5. a. 14 **b.** 2.977 **c.** $40.30 < \mu < 50.82$

6. a. 9 **b.** 3.250 **c.** $8.4811 < \mu < 8.7115$

7. a. 2.023 **b.** 2.559 **c.** 95%

8. a. 2.708 **b.** 3.425 **c.** 99%

9. 97

10. 136

11. Yes

12. No

Section 8.3 Confidence Intervals for a Population Proportion

Objectives

1. Construct a confidence interval for a population proportion
2. Find the sample size necessary to obtain a confidence interval of a given width
3. Describe a method for constructing confidence intervals with small samples

Objective 1 Construct a confidence interval for a population proportion

Construct a Confidence Interval for a Population Proportion

Does learning music improve your grades? The National Association of Music Merchants surveyed 800 parents of children 5–18, and 632 of them said that music education has a positive overall effect on academic performance.

This is an example of a population whose items fall into two categories. In this example, the categories are those parents who believe that music education has a positive effect on academic performance, and those who don't. We are interested in the population proportion of those who believe there is a positive effect. We will use the following notation.

NOTATION

- p is the population proportion of individuals who are in a specified category.
- x is the number of individuals in the sample who are in the specified category.
- n is the sample size.
- $\hat{p}$ is the sample proportion of individuals who are in the specified category. $\hat{p} = x/n$

To construct a confidence interval, we need a point estimate and a margin of error. The point estimate we use for the population proportion p is the sample proportion

$$\hat{p} = \frac{x}{n}$$

EXPLAIN IT AGAIN

The population proportion and the sample proportion: The population proportion p is unknown. The sample proportion $\hat{p}$ is known, and we use the value of $\hat{p}$ to estimate the unknown value p.

To compute the margin of error, we multiply the standard error of the point estimate by the critical value. The standard error and the critical value are determined by the sampling distribution of $\hat{p}$. In Section 7.4 we found that when the sample size n is large enough, the sample proportion $\hat{p}$ is approximately normal with standard deviation

$$\sqrt{\frac{p(1-p)}{n}}$$

In practice, we don't know the value of p, so we substitute $\hat{p}$ instead to obtain the standard error we use for the confidence interval:

$$\text{Standard error} = \sqrt{\frac{\hat{p}(1-\hat{p})}{n}}$$

NOTE TO INSTRUCTOR

It is interesting to note that the margin of error reported in political polls is the margin of error for a 95% confidence interval for a proportion.

Since the point estimate $\hat{p}$ is approximately normal with standard error $\sqrt{\hat{p}(1-\hat{p})/n}$, the appropriate margin of error is

$$\text{Margin of error} = z_{\alpha/2}\sqrt{\frac{\hat{p}(1-\hat{p})}{n}}$$

The confidence interval is

$$\text{Point estimate} \pm \text{Margin of error}$$

$$\hat{p} \pm z_{\alpha/2}\sqrt{\frac{\hat{p}(1-\hat{p})}{n}}$$

The method we have just described requires certain assumptions, which we now state.

EXPLAIN IT AGAIN

Reasons for the assumptions: The population must be much larger than the sample (at least 20 times as large), so that the sampled items are independent. The assumption that there are at least 10 items in each category is an approximate check on the assumption that both np and $n(1-p)$ are at least 10, which ensures that the sampling distribution of $\hat{p}$ is approximately normal.

Assumptions for Constructing a Confidence Interval for p

1. We have a simple random sample.
2. The population is at least 20 times as large as the sample.
3. The items in the population are divided into two categories.
4. The sample must contain at least 10 individuals in each category.

Following is a step-by-step description of the procedure for constructing a confidence interval for a population proportion p.

Procedure for Constructing a Confidence Interval for p

Check to be sure the assumptions are satisfied. If they are, then proceed with the following steps.

Step 1: Compute the value of the point estimate $\hat{p}$.

Step 2: Find the critical value $z_{\alpha/2}$ corresponding to the desired confidence level, either from the last line of Table A.3, from Table A.2, or with technology.

Step 3: Compute the standard error $\sqrt{\hat{p}(1-\hat{p})/n}$, and multiply it by the critical value to obtain the margin of error $z_{\alpha/2}\sqrt{\hat{p}(1-\hat{p})/n}$.

Step 4: Use the point estimate and the margin of error to construct the confidence interval:

$$\text{Point estimate} \pm \text{Margin of error}$$

$$\hat{p} \pm z_{\alpha/2}\sqrt{\frac{\hat{p}(1-\hat{p})}{n}}$$

$$\hat{p} - z_{\alpha/2}\sqrt{\frac{\hat{p}(1-\hat{p})}{n}} < p < \hat{p} + z_{\alpha/2}\sqrt{\frac{\hat{p}(1-\hat{p})}{n}}$$

Step 5: Interpret the result.

EXPLAIN IT AGAIN

Round-off rule: When constructing a confidence interval for a proportion, round the final result to three decimal places.

Rounding off the final result

When constructing confidence intervals for a proportion, we will round the final result to three decimal places. Note that you should round only the final result, and not the calculations you have made along the way.

Example 8.14

Construct a confidence interval for a proportion

In a survey of 800 parents, 632 said that music education has a positive effect on academic performance.

Tim Pannell/Fuse/Getty Images

a. Construct a 95% confidence interval for the proportion of parents who believe that music education has a positive effect.

b. A music teacher claims that 80% of parents believe that music education has a positive effect. Does the confidence interval contradict this claim?

Solution

a. We begin by checking the assumptions. We have a simple random sample. It is certainly true that the population of parents in the United States is at least 20 times as large as the sample. The items in the population can be divided into two categories:

those who believe that the music education has a positive effect, and those who do not. There are 632 parents who believe that music education has a positive effect, and $800 - 632 = 168$ who do not, so there are 10 or more items in each category. The assumptions are met.

Step 1: Compute the point estimate $\hat{p}$. The sample size is $n = 800$, and the number who believe that music education has a positive effect is $x = 632$. The point estimate is

$$\hat{p} = \frac{632}{800} = 0.79$$

Step 2: Find the critical value. The critical value for a 95% confidence interval is $z_{\alpha/2} = 1.96$.

Step 3: Compute the margin of error. The margin of error is

$$z_{\alpha/2}\sqrt{\frac{\hat{p}(1-\hat{p})}{n}}$$

We substitute $z_{\alpha/2} = 1.96$, $\hat{p} = 0.79$, and $n = 800$ to obtain

$$z_{\alpha/2}\sqrt{\frac{\hat{p}(1-\hat{p})}{n}} = 1.96\sqrt{\frac{0.79(1-0.79)}{800}} = 0.028225$$

Step 4: Construct the 95% confidence interval. The point estimate is 0.790, and the margin of error is 0.028225. The 95% confidence interval is

$$0.79 - 0.028225 < p < 0.79 + 0.028225$$

$$0.762 < p < 0.818 \text{ (rounded to three decimal places)}$$

Step 5: Interpret the result. We are 95% confident that the proportion of parents who believe that music education has a positive effect is between 0.762 and 0.818.

b. Since the value 0.80 is within the confidence interval, the confidence interval does not contradict the claim.

Constructing confidence intervals with technology

Example 8.14 presented a 95% confidence interval for the proportion of parents who believe that music education has a positive effect on academic performance. The confidence interval obtained was $0.762 < p < 0.818$. Following are the results from a TI-84 Plus calculator.

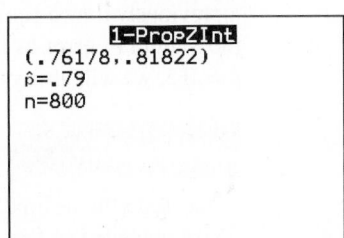

```
1-PropZInt
(.76178,.81822)
p̂=.79
n=800
```

The display presents the confidence interval, followed by the point estimate $\hat{p}$ and the sample size n. Note that the level (95%) is not displayed.

Following is MINITAB output for the same example.

Sample	X	N	Sample p	95% CI
1	632	800	0.790000	(0.761775, 0.818225)

Step-by-step instructions for constructing confidence intervals with technology are presented in the Using Technology section on page 412.

Check Your Understanding

1. A simple random sample of 200 third graders in a large school district was chosen to participate in an after-school program to improve reading skills. After completing the program, the children were tested, and 142 of them showed improvement in their reading skills.
   a. Find a point estimate for the proportion of third graders in the school district whose reading scores would improve after completing the program. *0.710*
   b. Construct a 95% confidence interval for the proportion of third graders in the school district whose reading scores would improve after completing the program. *(0.647, 0.773)*
   c. Is it reasonable to conclude that more than 60% of the students would improve their reading scores after completing the program? Explain. *Yes*
   d. The district superintendent wants to construct a confidence interval for the proportion of all elementary schoolchildren in the district whose scores would improve. Should the sample of 200 third graders be used for this purpose? Explain. *No*

Answers are on page 417.

Objective 2 Find the sample size necessary to obtain a confidence interval of a given width

Finding the Necessary Sample Size

If we wish to make the margin of error of a confidence interval smaller while keeping the confidence level the same, we can do this by making the sample size larger. Sometimes we have a specific value m that we would like the margin of error to attain, and we wish to compute a sample size n that is likely to give us a margin of error of that size.

The method for computing the sample size is based on the formula for the margin of error:

$$m = z_{\alpha/2}\sqrt{\frac{\hat{p}(1-\hat{p})}{n}}$$

By manipulating this formula algebraically, we can solve for n:

$$n = \hat{p}(1-\hat{p})\left(\frac{z_{\alpha/2}}{m}\right)^2$$

NOTE TO INSTRUCTOR

Emphasize that the expression that assumes $p = 0.5$ is conservative, in that the margin of error produced will always be at least as large as desired.

This formula shows that the sample size n depends not only on the margin of error m but also on the sample proportion $\hat{p}$. Therefore, in order to compute the sample size n, we need a value for $\hat{p}$ as well as a value for m. Now of course we don't know ahead of time what $\hat{p}$ is going to be. The approach, therefore, is to determine a value for $\hat{p}$.

There are two ways to determine a value for $\hat{p}$. One is to use a value that is available from a previously drawn sample. The other is to assume that $\hat{p} = 0.5$. The value $\hat{p} = 0.5$ makes the margin of error as large as possible for any sample size. Therefore, if we assume that $\hat{p} = 0.5$, we will always get a margin of error that is less than or equal to the desired value.

EXPLAIN IT AGAIN

Computing the sample size: When computing the necessary sample size, use a value of $\hat{p}$ from a previously drawn sample if one is available. Otherwise, use $\hat{p} = 0.5$.

SUMMARY

Let m be the desired margin of error, and let $z_{\alpha/2}$ be the critical value. The sample size n needed so that a confidence interval for a proportion will have margin of error approximately equal to m is

$$n = \hat{p}(1-\hat{p})\left(\frac{z_{\alpha/2}}{m}\right)^2 \quad \text{if a value for } \hat{p} \text{ is available}$$

$$n = 0.25\left(\frac{z_{\alpha/2}}{m}\right)^2 \quad \text{if no value for } \hat{p} \text{ is available}$$

(This is equivalent to assuming that $\hat{p} = 0.5$.)

If the value of n given by the formula is not a whole number, round it *up* to the nearest whole number. By rounding up, we can be sure that the margin of error is no greater than the desired value m.

Example 8.15

Find the necessary sample size

Example 8.14 described a sample of 800 parents, 632 of whom believe that music education has a positive effect on academic performance. Estimate the sample size needed so that a 95% confidence interval will have a margin of error of 0.025.

Solution

The desired level is 95%. The critical value is therefore $z_{\alpha/2} = 1.96$. We now compute $\hat{p}$:

$$\hat{p} = \frac{632}{800} = 0.79$$

The desired margin of error is $m = 0.025$. Since we have a value for $\hat{p}$, we substitute $\hat{p} = 0.79$, $z_{\alpha/2} = 1.96$, and $m = 0.025$ into the formula

$$n = \hat{p}(1 - \hat{p})\left(\frac{z_{\alpha/2}}{m}\right)^2$$

and obtain

$$n = (0.79)(1 - 0.79)\left(\frac{1.96}{0.025}\right)^2 = 1019.71 \quad \text{(round up to 1020)}$$

We estimate that we need to sample 1020 parents to obtain a 95% confidence interval with a margin of error of 0.025.

Example 8.16

Find the necessary sample size

We plan to sample parents in order to construct a 95% confidence interval for the proportion who believe that physical education has a positive effect on academic performance. We have no value of $\hat{p}$ available. Estimate the sample size needed so that a 95% confidence interval will have a margin of error of 0.025. Explain why the estimated sample size in this example is larger than the one in Example 8.15.

Solution

The desired level is 95%. The critical value is therefore $z_{\alpha/2} = 1.96$. The desired margin of error is $m = 0.025$. Since we have no value of $\hat{p}$, we substitute the values $z_{\alpha/2} = 1.96$ and $m = 0.025$ into the formula

$$n = 0.25\left(\frac{z_{\alpha/2}}{m}\right)^2$$

and obtain

$$n = 0.25\left(\frac{1.96}{0.025}\right)^2 = 1536.64 \quad \text{(round up to 1537)}$$

We estimate that we need to sample 1537 parents to obtain a 95% confidence interval with a margin of error of 0.025. This estimate is larger than the one in Example 8.15 because we used a value of 0.5 for $\hat{p}$, which provides a sample size large enough to guarantee that the margin of error will be no greater than 0.025 no matter what the true value of p is.

Check Your Understanding

2. In a preliminary study, a simple random sample of 100 computer chips was tested, and 17 of them were found to be defective. Now another sample will be drawn in order to construct a 95% confidence interval for the proportion of chips that are defective. Use the results of the preliminary study to estimate the sample size needed so that the confidence interval will have a margin of error of 0.06. *151*

3. A pollster is going to sample a number of voters in a large city and construct a 98% confidence interval for the proportion who support the incumbent candidate for mayor. Find a sample size so that the margin of error will be no larger than 0.05. *542*

Answers are on page 417.

The margin of error does not depend on the population size

At the time of the last presidential election, there were about 18 million registered voters in the state of California, and about 0.24 million registered voters in the state of Wyoming. A simple random sample of 1000 Wyoming voters is selected to estimate the proportion of voters who favor the Democratic candidate for president. Another simple random sample of 1000 California voters is selected to determine the proportion of Democratic voters in that state. Which estimate has the smaller standard error? Because California has a much larger population of registered voters than does Wyoming, it might seem that a larger sample would be needed in California to produce the same standard error. Surprisingly enough, this is not the case. In fact, the standard errors for the two estimates will be about the same. This is clear from the formula for the standard error: The population size does not enter into the calculation. Since the standard errors are about the same, the margins of error will be about the same if confidence intervals of the same level are constructed for both population proportions.

Intuitively, we can see that population size doesn't matter by considering an analogy with testing the water in a swimming pool. To determine whether the chemical balance is correct, one withdraws a few drops of water to test. As long as the contents of the pool are well mixed, so that the water removed constitutes a simple random sample of molecules from the pool, it doesn't matter how large the pool is. One doesn't need to sample more water from a bigger pool.

Example 8.17

The margin of error does not depend on the population size

A pollster has conducted a poll using a sample of 500 drawn from a town with population 25,000. He now wants to conduct the poll in a larger town with population 250,000 and to obtain approximately the same margin of error as in the smaller town. How large a sample must he draw?

Solution

He should draw a sample of 500, just as in the small town. The population size does not affect the margin of error.

Check Your Understanding

4. A pollster is planning to draw a simple random sample of 500 people in Colorado (population 5.5 million). He will then conduct a similar poll in Texas (population 27.9 million). He wants to have approximately the same standard error in both polls. True or false:
   a. The pollster needs a sample in Texas that is about five times as large as the one in Colorado. *False*
   b. The pollster needs a sample in Texas that is about the same size as the one in Colorado. *True*
5. A marketing firm in New York City (population 8.6 million) plans to draw a simple random sample of 1000 people to estimate the proportion who have heard about a new product. The firm then plans to take a simple random sample of 500 in Denver (population 683,000) for the same purpose. True or false:

a. The margin of error for a 95% confidence interval will be larger in New York. *False*

b. The margin of error for a 95% confidence interval will be larger in Denver. *True*

c. The margin of error for a 95% confidence interval will be about the same in both cities. *False*

Answers are on page 417.

Objective 3 Describe a method for constructing confidence intervals with small samples

A Method for Constructing Confidence Intervals with Small Samples

The method that we have presented for constructing a confidence interval for a proportion requires that we have at least 10 individuals in each category. When this condition is not met, we can still construct a confidence interval by adjusting the sample proportion a bit. We increase the number of individuals in each category by 2, so that the sample size increases by 4. Thus, instead of using the sample proportion $\hat{p} = x/n$, we use the *adjusted sample proportion*

$$\tilde{p} = \frac{x+2}{n+4}$$

NOTE TO INSTRUCTOR

The material on confidence intervals for proportions with small samples may be omitted without loss of continuity.

The standard error and critical value are calculated in the same way as in the traditional method, except that we use the adjusted sample proportion $\tilde{p}$ in place of $\hat{p}$, and $n + 4$ in place of n.

Constructing Confidence Intervals for a Proportion with Small Samples

If x is the number of individuals in a sample of size n who have a certain characteristic, and p is the population proportion, then:

The adjusted sample proportion is

$$\tilde{p} = \frac{x+2}{n+4}$$

A confidence interval for p is

$$\tilde{p} - z_{\alpha/2}\sqrt{\frac{\tilde{p}(1-\tilde{p})}{n+4}} < p < \tilde{p} + z_{\alpha/2}\sqrt{\frac{\tilde{p}(1-\tilde{p})}{n+4}}$$

Another way to write this is

$$\tilde{p} \pm z_{\alpha/2}\sqrt{\frac{\tilde{p}(1-\tilde{p})}{n+4}}$$

Example 8.18

Construct a confidence interval with a small sample

In a random sample of 10 businesses in a certain city, 6 of them had more than 15 employees. Use the small-sample method to construct a 95% confidence interval for the proportion of businesses in this city that have more than 15 employees.

Solution

The adjusted sample proportion is

$$\tilde{p} = \frac{x+2}{n+4} = \frac{6+2}{10+4} = 0.5714$$

The critical value is $z_{\alpha/2} = 1.96$. The 95% confidence interval is therefore

$$\tilde{p} - z_{\alpha/2}\sqrt{\frac{\tilde{p}(1-\tilde{p})}{n+4}} < p < \tilde{p} + z_{\alpha/2}\sqrt{\frac{\tilde{p}(1-\tilde{p})}{n+4}}$$

$$0.5714 - 1.96\sqrt{\frac{0.5714(1-0.5714)}{10+4}} < p < 0.5714 + 1.96\sqrt{\frac{0.5714(1-0.5714)}{10+4}}$$

$$0.312 < p < 0.831$$

Check Your Understanding

6. In a simple random sample of 15 seniors from a certain college, 8 of them had found jobs. Use the small-sample method to construct a 95% confidence interval for the proportion of seniors at that college who have found jobs. *(0.302, 0.751)*

Answer is on page 417.

Using technology to implement the small-sample method

Because the only difference between the small-sample method and the traditional method is the use of $\tilde{p}$ rather than $\hat{p}$, a software package or calculator such as the TI-84 Plus that uses the traditional method can be made to produce a confidence interval using the small-sample method. Simply input $x + 2$ for the number of individuals in the category of interest, and $n + 4$ for the sample size.

The small-sample method is better overall

The small-sample method can be used for any sample size, and recent research has shown that it has two advantages over the traditional method. First, the margin of error is smaller, because we divide by $n + 4$ rather than n. Second, the actual probability that the small-sample confidence interval covers the true population proportion is almost always at least as great as, or greater than, that of the traditional method. This holds for confidence levels of 90% or more, which are the levels commonly used in practice. For more information on this method, see the article "Approximate is Better Than 'Exact' for Interval Estimation of Binomial Proportions" (A. Agresti and B. Coull, *The American Statistician*, 52:119–126).

Using Technology

We use Example 8.14 to illustrate the technology steps.

TI-84 PLUS

Constructing a confidence interval for a proportion

Step 1. Press **STAT** and highlight the **TESTS** menu.

Step 2. Select **1–PropZInt** and press **ENTER** (Figure A). The **1–PropZInt** menu appears.

Step 3. Enter the values of x and n. For Example 8.14, we use $x = 632$ and $n = 800$.

Step 4. In the **C-Level** field, enter the confidence level. For Example 8.14, we use 0.95 (Figure B).

Step 5. Highlight **Calculate** and press **ENTER** (Figure C).

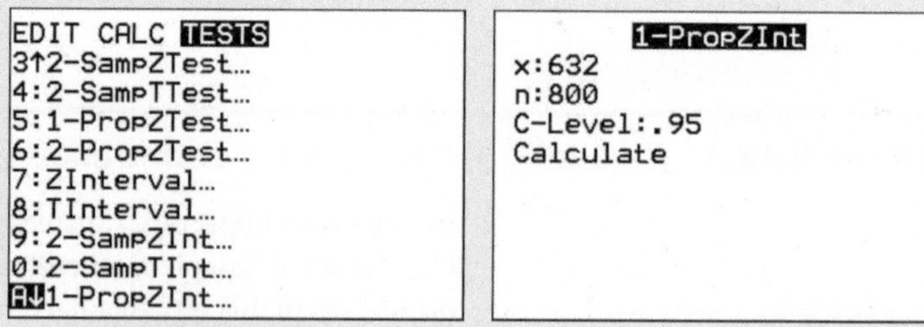

Figure A **Figure B**

Note: The preceding steps produce the traditional confidence interval. To produce the small-sample interval, enter the value of $x + 2$ for x and the value of $n + 4$ for n.

```
1-PropZInt
(.76178,.81822)
p̂=.79
n=800
```

Figure C

EXCEL

Constructing a confidence interval for a proportion

Figure D demonstrates the calculations involved for the margin of error of a confidence interval for a proportion. In Example 8.14, $x = 632$, $n = 800$, and the confidence level is 95%.

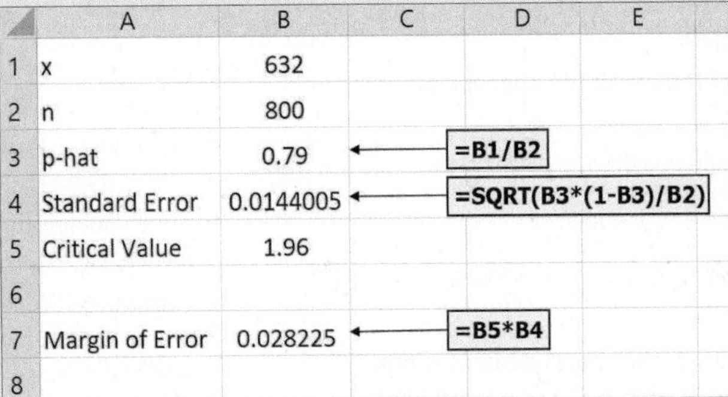

	A	B	C	D	E
1	x	632			
2	n	800			
3	p-hat	0.79		=B1/B2	
4	Standard Error	0.0144005		=SQRT(B3*(1-B3)/B2)	
5	Critical Value	1.96			
6					
7	Margin of Error	0.028225		=B5*B4	
8					

Figure D

MINITAB

Constructing a confidence interval for a proportion

Step 1. Click on **Stat**, then **Basic Statistics**, then **1 Proportion**.

Step 2. Select **Summarized Data**, and enter the value of n in the **Number of trials** field and the value of x in the **Number of events** field. For Example 8.14, we use $x = 632$ and $n = 800$ (Figure E).

Step 3. Click **Options**, and enter the confidence level in the **Confidence Level** (95) field. Click **OK**.

Step 4. Click **OK** (Figure F).

Note: The preceding steps produce the traditional confidence interval. To produce the small-sample interval, enter the value of $x + 2$ for x and the value of $n + 4$ for n.

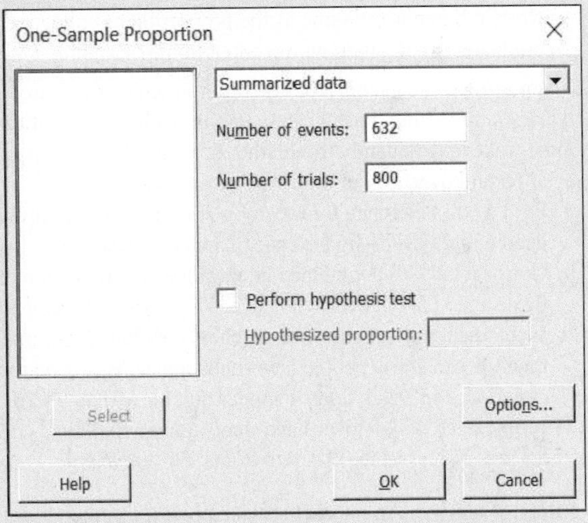

Figure E

N	Event	Sample p	95% CI for p
800	632	0.790000	(0.760106, 0.817738)

Figure F

Section 8.3 Exercises

Exercises 1–6 are the Check Your Understanding exercises located within the section.

Understanding the Concepts

In Exercises 7 and 8, fill in each blank with the appropriate word or phrase.

7. If $\hat{p}$ is the sample proportion and n is the sample size, then $\sqrt{\frac{\hat{p}(1-\hat{p})}{n}}$ is the ________. *standard error*

8. To estimate the necessary sample size when no value of $\hat{p}$ is available, we use $\hat{p}$ = ________. *0.5*

In Exercises 9 and 10, determine whether the statement is true or false. If the statement is false, rewrite it as a true statement.

9. If we estimate the necessary sample size and no value for $\hat{p}$ is available, the estimated sample size will be larger than if a value for $\hat{p}$ were available. *True*

10. The margin of error does not depend on the sample size. *False*

11. A firm sent out a random sample of 1000 shipments in a new type of heavy-duty packaging. Six of them were damaged upon arrival at their destinations. Is it appropriate to use the methods of this section to construct a confidence interval for the proportion of shipments in heavy-duty packaging that are damaged upon arrival? If not, explain why not. *No*

12. Last year, the air quality in a certain city was unhealthy on 25 of 365 days. Is it appropriate to use the methods of this section to construct a confidence interval for the proportion of days on which the air quality is unhealthy? If not, explain why not. *No*

13. In a simple random sample of 200 students at a certain large university, 25 said they were left-handed. Is it appropriate to use the methods of this section to construct a confidence interval for the proportion of students at this university who are left-handed? If not, explain why not. *Yes*

14. In a simple random sample of 100 employees at a large company, 27 of them reported that they owned stock in the company. Is it appropriate to use the methods of this section to construct a confidence interval for the proportion of employees at this company who own stock? If not, explain why not. *Yes*

Practicing the Skills

In Exercises 15–18, find the point estimate, the standard error, and the margin of error for the given confidence levels and values of x and n.

15. $x = 146$, $n = 762$, confidence level 95% *0.1916, 0.01426, 0.02794*

16. $x = 46$, $n = 97$, confidence level 99% *0.4742, 0.0507, 0.1306*

17. $x = 236$, $n = 474$, confidence level 90% *0.4979, 0.02297, 0.0378*

18. $x = 29$, $n = 80$, confidence level 92% *0.3625, 0.05375, 0.0941*

In Exercises 19–22, use the given data to construct a confidence interval of the requested level.

19. $x = 28$, $n = 64$, confidence level 93% *(0.325, 0.550)*

20. $x = 52$, $n = 71$, confidence level 97% *(0.618, 0.846)*

21. $x = 125$, $n = 317$, confidence level 95% *(0.341, 0.448)*

22. $x = 178$, $n = 531$, confidence level 90% *(0.302, 0.369)*

Working with the Concepts

23. **Smartphone:** Among 238 smartphone owners aged 18–24 surveyed by the Pew Research Center, 102 said their phone was an Android phone.
    a. Find a point estimate for the proportion of smartphone owners aged 18–24 who have an Android phone. *0.429*
    b. Construct a 95% confidence interval for the proportion of smartphone owners aged 18–24 who have an Android phone. *(0.366, 0.491)*
    c. Assume that an advertisement claimed that 45% of smartphone owners aged 18–24 have an Android phone. Does the confidence interval contradict this claim? *No*

Leszek Kobusinski/Shutterstock

24. **Working at home:** According to the U.S. Census Bureau, 43% of men who worked at home were college graduates. In a sample of 500 women who worked at home, 162 were college graduates.
    a. Find a point estimate for the proportion of college graduates among women who work at home. *0.324*
    b. Construct a 98% confidence interval for the proportion of women who work at home who are college graduates. *(0.275, 0.373)*
    c. Based on the confidence interval, is it reasonable to believe that the proportion of college graduates among women who work at home is the same as the proportion of college graduates among men who work at home? Explain. *No*

25. **Sleep apnea:** Sleep apnea is a disorder in which there are pauses in breathing during sleep. People with this condition must wake up frequently to breathe. In a sample of 427 people aged 65 and over, 104 of them had sleep apnea.
    a. Find a point estimate for the population proportion of those aged 65 and over who have sleep apnea. *0.244*
    b. Construct a 99% confidence interval for the proportion of those aged 65 and over who have sleep apnea. *(0.190, 0.297)*
    c. In another study, medical researchers concluded that more than 9% of elderly people have sleep apnea. Based on the confidence interval, does it appear that more than 9% of people aged 65 and over have sleep apnea? Explain. *Yes*

    Sources: *Sleep* 14:486–495; *Mayo Clinic Proceedings* 76:897–905

26. **Internet service:** An internet service provider sampled 540 customers and found that 75 of them experienced an interruption in high-speed service during the previous month.

a. Find a point estimate for the population proportion of all customers who experienced an interruption. 0.139
b. Construct a 90% confidence interval for the proportion of all customers who experienced an interruption. (0.114, 0.163)
c. The company's quality control manager claims that no more than 10% of its customers experienced an interruption during the previous month. Does the confidence interval contradict this claim? Explain. Yes

27. **Volunteering:** The General Social Survey asked 1294 people whether they performed any volunteer work during the past year. A total of 517 people said they did.
a. Find a point estimate for the proportion of people who performed volunteer work during the past year. 0.400
b. Construct a 95% confidence interval for the proportion of people who performed volunteer work during the past year. (0.373, 0.426)
c. A sociologist states that 50% of Americans perform volunteer work in a given year. Does the confidence interval contradict this statement? Explain. Yes

28. **SAT scores:** A college admissions officer sampled 120 entering freshmen and found that 42 of them scored more than 550 on the math SAT.
a. Find a point estimate for the proportion of all entering freshmen at this college who scored more than 550 on the math SAT. 0.350
b. Construct a 98% confidence interval for the proportion of all entering freshmen at this college who scored more than 550 on the math SAT. (0.249, 0.451)
c. According to the College Board, 39% of all students who took the math SAT scored more than 550. The admissions officer believes that the proportion at her university is also 39%. Does the confidence interval contradict this belief? Explain. No

29. **LoL:** In the computer game League of Legends, some of the strikes are critical strikes, which do more damage. Assume that the probability of a critical strike is the same for every attack and that attacks are independent. Assume that a character has 242 critical strikes out of 595 attacks.
a. Construct a 95% confidence interval for the proportion of strikes that are critical strikes. (0.367, 0.446)
b. Construct a 98% confidence interval for the proportion of strikes that are critical strikes. (0.360, 0.454)
c. What is the effect of increasing the level of confidence on the width of the interval? Makes it wider

30. **Contaminated water:** In a sample of 42 water specimens taken from a construction site, 26 contained detectable levels of lead.
a. Construct a 90% confidence interval for the proportion of water specimens that contain detectable levels of lead. (0.496, 0.742)
b. Construct a 95% confidence interval for the proportion of water specimens that contain detectable levels of lead. (0.472, 0.766)
c. What is the effect of increasing the level of confidence on the width of the interval? Makes it wider

Source: *Journal of Environmental Engineering* 128:237–245

31. **Vegetarians:** In a recent poll of 1033 adults in the United States, 51 said they were vegetarians.
a. Construct a 95% confidence interval for the proportion of adults in the United States who are vegetarians. (0.036, 0.063)
b. Using an estimate of 0.05 for the proportion of vegetarians, what sample size is needed so that a 95% confidence interval will have a margin of error of 0.01? 1825
c. A dietitian claims that 10% of U.S. adults are vegetarians. Does the confidence interval contradict this claim? Yes

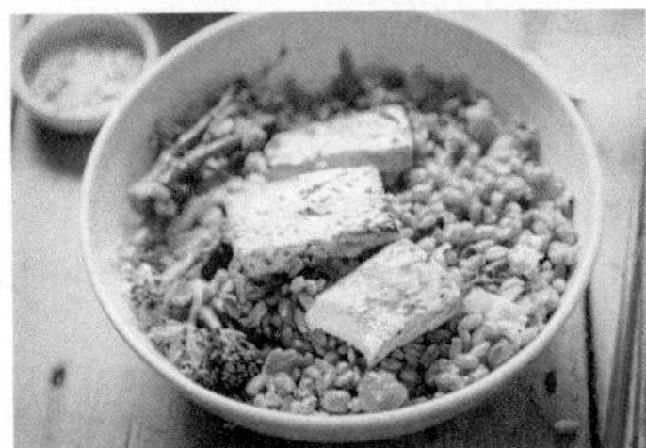
Bartosz Luczak/Getty Images

32. **Quit smoking:** The National Center for Health Studies interviewed 5409 adult smokers and found that 2636 of them had attempted to quit smoking within the past year.
a. Construct a 99% confidence interval for the proportion of adult smokers who attempted to quit smoking within the past year. (0.470, 0.505)
b. Assume that no estimate of p is available. Estimate the sample size needed so that a 99% confidence interval will have a margin of error of 0.015. 7396
c. A public health official claims that 50% of adult smokers attempted to quit within the past year. Does the confidence interval contradict this claim? No

33. **Call me:** A sociologist wants to construct a 95% confidence interval for the proportion of children aged 8–12 living in New York who own a smartphone.
a. A survey by the National Consumers League estimated the nationwide proportion to be 0.56. Using this estimate, what sample size is needed so that the confidence interval will have a margin of error of 0.02? 2367
b. Estimate the sample size needed if no estimate of p is available. 2401
c. If the sociologist wanted to estimate the proportion in the entire United States rather than in New York, would the necessary sample size be larger, smaller, or about the same? Explain. About the same

34. **Reading proficiency:** An educator wants to construct a 98% confidence interval for the proportion of elementary schoolchildren in Colorado who are proficient in reading.
a. The results of a recent statewide test suggested that the proportion is 0.70. Using this estimate, what sample size is needed so that the confidence interval will have a margin of error of 0.05? 455
b. Estimate the sample size needed if no estimate of p is available. 542
c. If the educator wanted to estimate the proportion in the entire United States rather than in Colorado, would the necessary sample size be larger, smaller, or about the same? Explain. About the same

35. **Surgical complications:** A medical researcher wants to construct a 99% confidence interval for the proportion of knee replacement surgeries that result in complications.
a. An article in the *Journal of Bone and Joint Surgery* suggested that approximately 8% of such operations result in complications. Using this estimate, what sample size is needed so that the confidence interval will have a margin of error of 0.04? 306

b. Estimate the sample size needed if no estimate of p is available. *1037*

Source: *Journal of Bone and Joint Surgery* 87:1719–1724

36. How's the economy? A pollster wants to construct a 95% confidence interval for the proportion of adults who believe that economic conditions are getting better.

a. A Gallup poll estimates this proportion to be 0.34. Using this estimate, what sample size is needed so that the confidence interval will have a margin of error of 0.03? *958*

b. Estimate the sample size needed if no estimate of p is available. *1068*

37. Changing jobs: A sociologist sampled 200 people who work in computer-related jobs and found that 42 of them have changed jobs in the past six months.

a. Construct a 95% confidence interval for the proportion of those who work in computer-related jobs who have changed jobs in the past six months. *(0.154, 0.266)*

b. Among the 200 people, 120 of them are under the age of 35. These constitute a simple random sample of workers under the age of 35. If this sample were used to construct a 95% confidence interval for the proportion of workers under the age of 35 who have changed jobs in the past six months, is it likely that the margin of error would be larger, smaller, or about the same as the one in part (a)? *Larger*

38. Political polling: A simple random sample of 300 voters was polled several months before a presidential election. One of the questions asked was: "Are you satisfied with the choice of candidates for president?" A total of 123 of them said that they were not satisfied.

a. Construct a 99% confidence interval for the proportion of voters who are not satisfied with the choice of candidates. *(0.337, 0.483)*

b. Among the 300 voters were 158 women. These constitute a simple random sample of women voters. If this sample were used to construct a 99% confidence interval for the proportion of women voters who are satisfied with the choice of candidates for president, is it likely that the margin of error would be larger, smaller, or about the same as the one in part (a)? *Larger*

39. Small sample: Eighteen concrete blocks were sampled and tested for crushing strength in order to estimate the proportion that were sufficiently strong for a certain application. Sixteen of the 18 blocks were sufficiently strong. Use the small-sample method to construct a 95% confidence interval for the proportion of blocks that are sufficiently strong. *(0.657, 0.979)*

40. Small sample: During an economic downturn, 20 companies were sampled and asked whether they were planning to increase their workforce. Only 3 of the 20 companies were planning to increase their workforce. Use the small-sample method to construct a 98% confidence interval for the proportion of companies that are planning to increase their workforce. *(0.015, 0.401)*

41. Interpret calculator display: A sample of voters in a certain city was asked whether they planned to vote for the incumbent mayor in the next election. The following display from a TI-84 Plus calculator presents a 99% confidence interval for the population proportion that plan to vote for the incumbent mayor.

```
1-PropZInt
(.41911,.73714)
p̂=.578125
n=64
```

a. Fill in the blanks. We are ________ confident that the population proportion is between ________ and ________. *99%, 0.41911, 0.73714*

b. Use the information in the display to construct a 95% confidence interval for p. *(0.457, 0.699)*

42. Interpret calculator display: A sample of employed people was asked whether they had changed jobs in the past two years. The following display from a TI-84 Plus calculator presents a 95% confidence interval for the population proportion who had changed jobs during that time.

```
1-PropZInt
(.19525,.38253)
p̂=.2888888889
n=90
```

a. Fill in the blanks. We are ________ confident that the population proportion is between ________ and ________. *95%, 0.19525, 0.38253*

b. Use the information in the display to construct a 98% confidence interval for p. *(0.178, 0.400)*

43. Interpret computer output: A sample of drivers was asked whether they regularly use seat belts. The following MINITAB output presents a confidence interval for the population proportion who regularly use seat belts.

Sample	X	N	Sample p	98% CI
1	145	181	0.801105	(0.732082, 0.870128)

a. Fill in the blanks. We are ________ confident that the population proportion is between ________ and ________. *98%, 0.732082, 0.870128*

b. Use the information in the display to construct a 90% confidence interval for p. *(0.752, 0.850)*

44. Interpret computer output: A football game starts with the toss of a coin to determine which team will get the ball first. Before the game, the referee tosses the coin a number of times and records the number of times the coin lands heads. The following MINITAB output presents a confidence interval for the probability that the coin lands heads.

Sample	X	N	Sample p	95% CI
1	31	58	0.534483	(0.406111, 0.662854)

a. Fill in the blanks. We are ________ confident that the population proportion is between ________ and ________. *95%, 0.406111, 0.662854*

b. Use the information in the display to construct a 99% confidence interval for p. (0.366, 0.703)

45. Don't construct a confidence interval: The United States Senate consists of 100 senators. In January 2017, 21 of them were women. Explain why these data should not be used to construct a 95% confidence interval for the proportion of senators who are women.

46. Don't construct a confidence interval: At the end of a television documentary on the nature of government, viewers are invited to tweet an answer to the question, "Do you believe that women are more effective at governing than men are?" A total of 2348 viewers answer the question, and 1247 of them answer "Yes." Explain why these data should not be used to construct a confidence interval for the proportion of people who believe that women are more effective at governing than men are.

Extending the Concepts

Wilson's interval: The small-sample method for constructing a confidence interval is a simple approximation of a more complicated interval known as Wilson's interval. Let $\hat{p} = x/n$. Wilson's confidence interval for p is given by

$$\frac{\hat{p} + \dfrac{z_{\alpha/2}^2}{2n} \pm z_{\alpha/2}\sqrt{\dfrac{\hat{p}(1-\hat{p})}{n} + \dfrac{z_{\alpha/2}^2}{4n^2}}}{1 + \dfrac{z_{\alpha/2}^2}{n}}$$

47. College-bound: In a certain high school, 9 out of 15 tenth graders said they planned to go to college after graduating. Construct a 95% confidence interval for the proportion of tenth graders who plan to attend college:

a. Using Wilson's method (0.357, 0.802)

b. Using the small-sample method (0.357, 0.801)

c. Using the traditional method (0.352, 0.848)

48. Comparing the methods: Refer to Exercise 47.

a. Which of the three confidence intervals is the narrowest? Small-sample

b. Does the small-sample method provide a good approximation to Wilson's interval in this case? Yes

c. Explain why the traditional interval is the widest of the three.

49. Approximation depends on the level: The small-sample method is a good approximation to Wilson's method for all confidence levels commonly used in practice but is best when $z_{\alpha/2}$ is close to 2. Refer to Exercise 47.

a. Use Wilson's method to construct a 90% confidence interval, a 95% confidence interval, and a 99% confidence interval for the proportion of tenth graders who plan to attend college. 90%: (0.393, 0.777); 95%: (0.357, 0.802); 99%: (0.296, 0.842)

b. Use the small-sample method to construct a 90% confidence interval, a 95% confidence interval, and a 99% confidence interval for the proportion of tenth graders who plan to attend college. 90%: (0.393, 0.765); 95%: (0.357, 0.801); 99%: (0.287, 0.871)

c. For which level is the small-sample method the closest to Wilson's method? Explain why this is the case. 95%

Answers to Check Your Understanding Exercises for Section 8.3

1. a. 0.710 **b.** $0.647 < p < 0.773$

c. Yes. We are 95% confident that the proportion who would improve their scores is between 0.647 and 0.773. Therefore, it is reasonable to conclude that the proportion is greater than 0.60.

d. No. Because the sample contains only third graders, it should not be used to construct a confidence interval for all elementary schoolchildren.

2. 151

3. 542

4. a. False **b.** True

5. a. False **b.** True **c.** False

6. $0.302 < p < 0.751$

Section 8.4 Confidence Intervals for a Standard Deviation

Objectives

1. Find critical values of the chi-square distribution
2. Construct confidence intervals for the variance and standard deviation of a normal distribution

Objective 1 Find critical values of the chi-square distribution

The Chi-Square Distribution

Most confidence intervals constructed in practice are for means and proportions. However, when the population is normal, it is possible to construct confidence intervals for the standard deviation or variance. These confidence intervals are based on a distribution known as the **chi-square distribution**, denoted χ^2. The symbol χ is the Greek letter chi (pronounced "kigh"; rhymes with sky). We will begin by describing this distribution.

CAUTION

The methods of this section apply only for samples drawn from a normal distribution. If the distribution differs even slightly from normal, these methods should not be used.

There are actually many different chi-square distributions, each with a different number of degrees of freedom. Figure 8.15 presents chi-square distributions for several different degrees of freedom. There are two important points to notice:

- The chi-square distributions are not symmetric. They are skewed to the right.
- Values of the χ^2 statistic are always greater than or equal to 0. They are never negative.

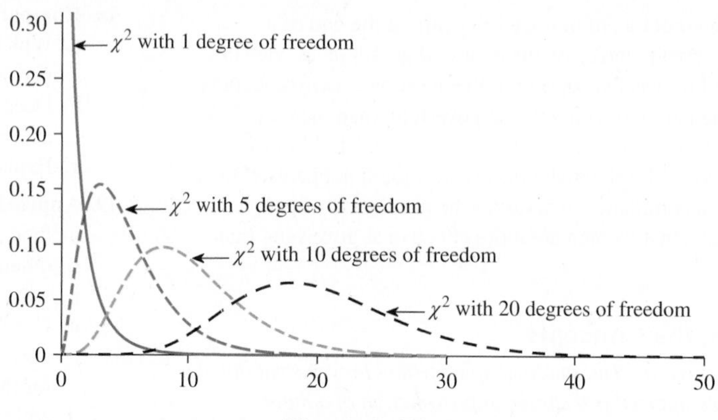

Figure 8.15

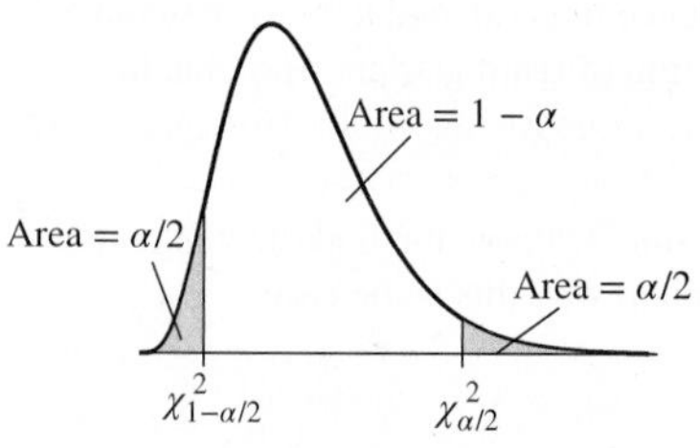

Figure 8.16

Confidence intervals for variances and standard deviations follow a somewhat different pattern than those for means and proportions. Whereas confidence intervals for means and proportions consist of a point estimate, a critical value, and a standard error, confidence intervals for variances and standard deviations consist of a point estimate and two critical values. The point estimate for the population variance σ^2 is s^2, the sample variance. The critical values come from the chi-square distribution.

The critical values for a level $100(1 - \alpha)\%$ confidence interval are the values that contain the middle $100(1 - \alpha)\%$ of the area under the curve between them. The notation for the critical values tells how much area is to the *right* of the critical value. Thus, for a level $1 - \alpha$ confidence interval, the critical values are denoted $\chi^2_{1-\alpha/2}$ and $\chi^2_{\alpha/2}$. See Figure 8.16.

Example 8.19 shows how to use Table A.4 to find critical values.

Example 8.19

Find critical values

Find the critical values for a 95% confidence interval using the chi-square distribution with 10 degrees of freedom.

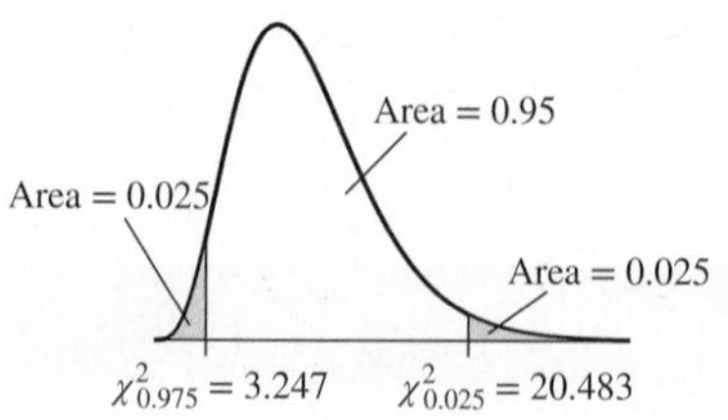

Figure 8.17

Solution

Figure 8.17 presents the chi-square distribution with 10 degrees of freedom and shows the locations of the critical values. The confidence level is 95%, so the critical values are the values that contain the middle 95% of the area under the curve between them. The lower critical value, denoted $\chi^2_{0.975}$, has an area of 0.975 to its right, and the upper critical value, denoted $\chi^2_{0.025}$, has an area of 0.025 to its right. These critical values are found in Table A.4, at the intersection of the row corresponding to 10 degrees of freedom and the columns corresponding to 0.975 and 0.025. The critical values are $\chi^2_{0.975} = 3.247$ and $\chi^2_{0.025} = 20.483$.

=CHISQ.INV(0.025,10)	
Lower Crit Value	3.24697
=CHISQ.INV.RT(0.025,10)	
Upper Crit Value	20.4832

Degrees of Freedom	Area in Right Tail									
	0.995	0.99	0.975	0.95	0.90	0.10	0.05	0.025	0.01	0.005
⋮	⋮	⋮	⋮	⋮	⋮	⋮	⋮	⋮	⋮	⋮
8	1.344	1.646	2.180	2.733	3.490	13.362	15.507	17.535	20.090	21.955
9	1.735	2.088	2.700	3.325	4.168	14.684	16.919	19.023	21.666	23.589
10	2.156	2.558	3.247	3.940	4.865	15.987	18.307	20.483	23.209	25.188
11	2.603	3.053	3.816	4.575	5.578	17.275	19.675	21.920	24.725	26.757
12	3.074	3.571	4.404	5.226	6.304	18.549	21.026	23.337	26.217	28.300
⋮	⋮	⋮	⋮	⋮	⋮	⋮	⋮	⋮	⋮	⋮

Check Your Understanding

1. Find the critical values for a 95% confidence interval using the chi-square distribution with 18 degrees of freedom. *8.231, 31.526*
2. Find the critical values for a 99% confidence interval using the chi-square distribution with 25 degrees of freedom. *10.520, 46.928*

Answers are on page 422.

Objective 2 Construct confidence intervals for the variance and standard deviation of a normal distribution

Confidence Intervals for the Variance and Standard Deviation

We will present the method for constructing confidence intervals for the variance and the standard deviation of a normal distribution. We will follow with an example, then present a justification for the method.

Confidence Intervals for the Variance and Standard Deviation

Let s^2 be the sample variance from a simple random sample of size n from a normal population. A level $100(1 - \alpha)\%$ confidence interval for the population variance σ^2 is

$$\frac{(n-1)s^2}{\chi^2_{\alpha/2}} < \sigma^2 < \frac{(n-1)s^2}{\chi^2_{1-\alpha/2}}$$

A level $100(1 - \alpha)\%$ confidence interval for the population standard deviation σ is

$$\sqrt{\frac{(n-1)s^2}{\chi^2_{\alpha/2}}} < \sigma < \sqrt{\frac{(n-1)s^2}{\chi^2_{1-\alpha/2}}}$$

The critical values are taken from a chi-square distribution with $n - 1$ degrees of freedom.

EXPLAIN IT AGAIN

Degrees of freedom: When constructing a confidence interval for a variance or standard deviation, the number of degrees of freedom is always 1 less than the sample size.

Step-by-step procedure for constructing confidence intervals

We summarize the procedure for constructing confidence intervals for the variance and standard deviation of a normal population.

Procedure for Constructing Confidence Intervals for the Variance and Standard Deviation of a Normal Distribution

Step 1: Compute the sample variance s^2, if it isn't given.

Step 2: Find the critical values $\chi^2_{1-\alpha/2}$ and $\chi^2_{\alpha/2}$, using the chi-square distribution with $n - 1$ degrees of freedom.

Step 3: Compute the lower and upper confidence bounds:

$$\text{Lower bound} = \frac{(n-1)s^2}{\chi^2_{\alpha/2}} \qquad \text{Upper bound} = \frac{(n-1)s^2}{\chi^2_{1-\alpha/2}}$$

Step 4: The level $100(1 - \alpha)\%$ confidence interval for σ^2 is

$$\frac{(n-1)s^2}{\chi^2_{\alpha/2}} < \sigma^2 < \frac{(n-1)s^2}{\chi^2_{1-\alpha/2}}$$

To find a level $100(1 - \alpha)\%$ confidence interval for the population standard deviation σ, take the square roots of the confidence bounds for the variance:

$$\sqrt{\frac{(n-1)s^2}{\chi^2_{\alpha/2}}} < \sigma < \sqrt{\frac{(n-1)s^2}{\chi^2_{1-\alpha/2}}}$$

Step 5: Interpret the result.

Example 8.20

Constructing a confidence interval

The compressive strengths of seven concrete blocks, in pounds per square inch, are measured, with the following results.

1989.9 1993.8 2074.5 2070.5 2070.9 2033.6 1939.6

Assume these values are a simple random sample from a normal population. Construct a 95% confidence interval for the population standard deviation σ.

Solution

Step 1: Find s^2.

$$s^2 = \frac{\sum(x - \bar{x})^2}{7 - 1} = 2699.8648$$

Step 2: Find the critical values. We have $7 - 1 = 6$ degrees of freedom. Since the confidence level is 95%, the critical values are $\chi^2_{0.975}$ and $\chi^2_{0.025}$. From Table A.4, we find

$$\chi^2_{0.975} = 1.237 \qquad \chi^2_{0.025} = 14.449$$

Step 3: Compute the lower and upper confidence bounds.

$$\text{Lower bound} = \frac{(n-1)s^2}{\chi^2_{\alpha/2}} = \frac{(7-1)(2699.8648)}{14.449} = 1121.129$$

$$\text{Upper bound} = \frac{(n-1)s^2}{\chi^2_{1-\alpha/2}} = \frac{(7-1)(2699.8648)}{1.237} = 13{,}095.545$$

Step 4: The 95% confidence interval for σ^2 is

$$1121.129 < \sigma^2 < 13{,}095.545$$

To find the confidence interval for σ, we take square roots. We find that $\sqrt{1121.129} = 33.48$ and $\sqrt{13{,}095.545} = 114.44$. The 95% confidence interval for σ is

$$33.48 < \sigma < 114.44$$

Step 5: Interpret the result. We are 95% confident that the population standard deviation of the strengths of the concrete blocks is between 33.48 and 114.44.

Check Your Understanding

3. Construct a 95% confidence interval for the population standard deviation σ if a sample of size 10 has standard deviation $s = 6$. *(4.13, 10.95)*
4. Construct a 99% confidence interval for the population standard deviation σ if a sample of size 23 has standard deviation $s = 12$. *(8.60, 19.15)*

Answers are on page 422.

Justification for the method

Confidence intervals for the variance of a normal distribution are based on the fact that when a sample of size n is drawn from a normal distribution, the quantity $(n - 1)s^2/\sigma^2$ follows a chi-square distribution with $n - 1$ degrees of freedom. Therefore, for a proportion $1 - \alpha$ of all possible samples,

$$\chi^2_{1-\alpha/2} < \frac{(n-1)s^2}{\sigma^2} < \chi^2_{\alpha/2}$$

Through algebraic manipulation, we can solve for σ^2, to obtain a level $100(1 - \alpha)\%$ confidence interval for σ^2:

$$\frac{(n-1)s^2}{\chi^2_{\alpha/2}} < \sigma^2 < \frac{(n-1)s^2}{\chi^2_{1-\alpha/2}}$$

Using Technology

We use Example 8.20 to illustrate the technology steps.

MINITAB

Constructing a confidence interval for σ

Step 1. Enter the data in **Column C1**. For Example 8.20, we use **1989.9, 1993.8, 2074.5, 2070.5, 2070.9, 2033.6, 1939.6**.

Step 2. Click on **Stat**, then **Basic Statistics**, then **Graphical Summary**.

Step 3. Enter **C1** in the **Variables** field, and enter the confidence level in the **Confidence Level** field. For Example 8.20, we use **95**.

Step 4. Click **OK** (Figure A).

95% Confidence Interval for Mean	
1976.6	2072.7
95% Confidence Interval for Median	
1976.5	2071.9
95% Confidence Interval for StDev	
33.5	114.4

Figure A

Section 8.4 Exercises

Exercises 1–4 are the Check Your Understanding exercises located within the section.

Understanding the Concepts

In Exercises 5 and 6, fill in each blank with the appropriate word or phrase.

5. To find a confidence interval for a standard deviation from a sample of size 15, we use a chi-square distribution with ________ degrees of freedom. *14*

6. The method described for finding confidence intervals should be used only when the distribution of the population is almost exactly ________ . *normal*

In Exercises 7 and 8, determine whether the statement is true or false. If the statement is false, rewrite it as a true statement.

7. When constructing a confidence interval for a standard deviation, we must find two critical values. *True*

8. If we have a confidence interval for a variance, we can obtain a confidence interval for the standard deviation by squaring the confidence bounds. *False*

Practicing the Skills

9. Find the critical values for a 95% confidence interval using the chi-square distribution with 15 degrees of freedom. *6.262, 27.488*

10. Find the critical values for a 99% confidence interval using the chi-square distribution with 5 degrees of freedom. *0.412, 16.750*

11. Construct a 95% confidence interval for the population standard deviation σ if a sample of size 25 has standard deviation $s = 15$. *(11.71, 20.87)*

12. Construct a 99% confidence interval for the population standard deviation σ if a sample of size 8 has standard deviation $s = 7.5$. *(4.41, 19.95)*

Working with the Concepts

13. SAT scores: Scores on the math SAT are normally distributed. A sample of 20 SAT scores had standard deviation $s = 87$.

a. Construct a 98% confidence interval for the population standard deviation σ. *(63.04, 137.26)*

b. Someone says that the scoring system for the SAT is designed so that the population standard deviation will be $\sigma = 100$. Does this confidence interval contradict this claim? Explain. *No*

14. IQ scores: Scores on an IQ test are normally distributed. A sample of 25 IQ scores had standard deviation $s = 8$.

a. Construct a 95% confidence interval for the population standard deviation σ. *(6.25, 11.13)*

b. The developer of the test claims that the population standard deviation is $\sigma = 15$. Does this confidence interval contradict this claim? Explain. *Yes*

15. Baby weights: Following are weights of 12 two-month-old baby girls. Assume that the population is normally distributed.

12.23	12.32	11.87	12.34	11.48	12.66
8.51	14.13	12.95	10.30	9.34	8.63

a. Find the sample standard deviation s. *1.798*

b. Construct a 95% confidence interval for the population standard deviation σ. *(1.27, 3.05)*

c. According to the National Health Statistics Reports, the standard deviation of the weight of two-month-old baby boys is 2.7 pounds. Based on the confidence interval, is it reasonable to believe that the standard deviation of the weights of two-month-old baby girls is the same as that of two-month-old baby boys? Explain. Yes

16. **Eat your cereal:** Boxes of cereal are labeled as containing 14 ounces. Following are the weights of a sample of 12 boxes. Assume that the population is normally distributed.

14.02	13.97	14.11	14.12	14.10	14.02
14.15	13.97	14.05	14.04	14.11	14.12

a. Find the sample standard deviation s. 0.0614
b. Construct a 98% confidence interval for the population standard deviation σ. (0.04, 0.12)
c. The goal of the quality control manager is for the population standard deviation of the weights to be less than 0.03. Based on the confidence interval, is it reasonable to believe that the goal has been met? No

17. **Eat your spinach:** Six measurements were made of the mineral content (in percent) of spinach, with the following results. Assume that the population is normally distributed.

19.1 20.8 20.8 21.4 20.5 19.7

a. Find the sample standard deviation s. 0.838
b. Construct a 99% confidence interval for the population standard deviation σ. (0.46, 2.92)
c. Based on the confidence interval, is it reasonable to believe that the population standard deviation might be less than 1.5? Explain. Yes

Source: *Journal of Nutrition* 66:55–66

18. **Mortgage rates:** Following are interest rates (annual percentage rates) for a 30-year fixed-rate mortgage from a sample of lenders in Macon, Georgia, on a recent day. It is reasonable to assume that the population is approximately normal.

4.750	4.375	4.176	4.679	4.426	4.227
4.125	4.250	3.950	4.191	4.299	4.415

Source: www.bankrate.com

a. Find the sample standard deviation s. 0.2263
b. Construct a 95% confidence interval for population standard deviation σ. (0.16, 0.38)
c. Based on the confidence interval, is it reasonable to believe that the population standard deviation is greater than 0.5? Explain. No

Extending the Concepts

The chi-square distribution is skewed, but as the number of degrees of freedom becomes large, the skewness diminishes. If the number of degrees of freedom, k, is large enough, the chi-square distribution is reasonably well approximated by a normal distribution with mean k and variance 2k.

19. **Exact confidence interval:** A sample of size 101 from a normal population has sample standard deviation $s = 40$. Use Table A.4 to find the exact critical values $\chi^2_{0.025}$ and $\chi^2_{0.975}$ for a 95% confidence interval, and construct a 95% confidence interval for σ. $\chi^2_{0.975} = 74.22$, $\chi^2_{0.025} = 129.56$; 95% CI: (35.14, 46.43)

20. **Using the normal approximation:** Refer to Exercise 19. Use the normal approximation to estimate the critical values $\chi^2_{0.025}$ and $\chi^2_{0.975}$ for a 95% confidence interval, and construct a 95% confidence interval for σ. $\chi^2_{0.975} = 72.28$, $\chi^2_{0.025} = 127.72$; 95% CI: (35.39, 47.05)

21. **Comparing results:** How close is the confidence interval based on the normal approximation constructed in Exercise 20 to the exact confidence interval constructed in Exercise 19?

A more accurate normal approximation to χ^2_α is given by $\chi^2_\alpha \approx 0.5\left(z_\alpha + \sqrt{2k-1}\right)^2$, where z_α is the z-score that has area α to its right.

22. **More accuracy:** Refer to Exercise 19. Use the more accurate normal approximation to estimate the critical values $\chi^2_{0.025}$ and $\chi^2_{0.975}$ for a 95% confidence interval, and construct a 95% confidence interval for σ. $\chi^2_{0.975} = 73.77$, $\chi^2_{0.025} = 129.07$; 95% CI: (35.21, 46.57)

23. **Comparing results:** How close is the confidence interval based on the more accurate normal approximation to the exact one?

Answers to Check Your Understanding Exercises for Section 8.4

1. $\chi^2_{0.975} = 8.231$, $\chi^2_{0.025} = 31.526$
2. $\chi^2_{0.995} = 10.520$, $\chi^2_{0.005} = 46.928$
3. $4.13 < \sigma < 10.95$
4. $8.60 < \sigma < 19.15$

Section 8.5 Determining Which Method to Use

Objective

1. Determine which method to use when constructing a confidence interval

Objective 1 Determine which method to use when constructing a confidence interval

One of the challenges in constructing a confidence interval is to determine which method to use. The first step is to determine which type of parameter we are estimating. There are three types of parameters for which we have learned to construct confidence intervals:

- Population mean μ
- Population proportion p
- Population standard deviation σ or variance σ^2

Once you have determined which type of parameter you are estimating, proceed as follows:

- **Population mean:** There are two methods for constructing a confidence interval for a population mean, the z method (Section 8.1) and the t method (Section 8.2). To determine which method to use, we must determine whether the population standard deviation is known, whether the population is approximately normal, and whether the sample size is large ($n > 30$). The following diagram can help you make the correct choice.

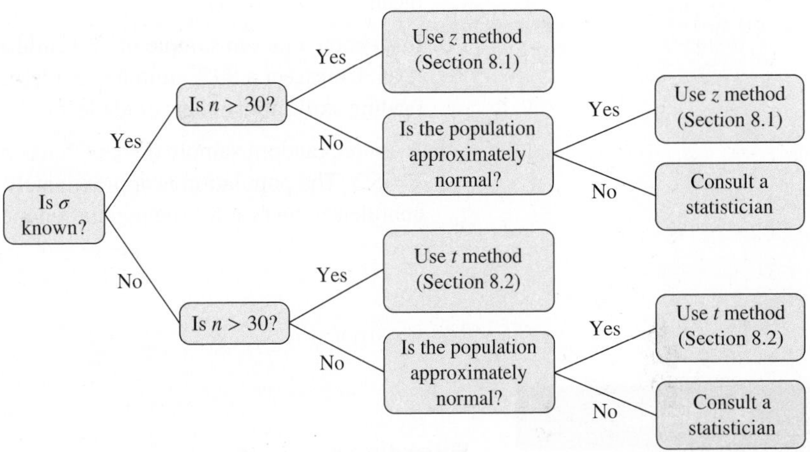

- **Population proportion:** To construct a confidence interval for a population proportion, use the method described in Section 8.3.
- **Population standard deviation or variance:** To construct a confidence interval for a population standard deviation or variance, use the method described in Section 8.4.

Example 8.21

Determining which method to use

A random sample of 41 pumpkins harvested from a pumpkin patch has a mean weight of $\bar{x} = 8.53$ pounds with a sample standard deviation of $s = 1.32$ pounds. Construct a 95% confidence interval for the mean weight of pumpkins from this patch. Determine the type of parameter that is to be estimated, and construct the confidence interval.

Solution

We are asked to find a confidence interval for the mean weight; this is a population mean. We consult the diagram to determine the correct method. We must first determine whether σ is known. There is no information given about σ, so σ is unknown. We follow the "No" path. Next we must determine whether $n > 30$. There are 41 pumpkins in the sample, so $n > 30$. We follow the "Yes" path, and find that we should use the t method described in Section 8.2.

To construct the confidence interval, there are $41 - 1 = 40$ degrees of freedom. Because the confidence level is 95%, the critical value is $t_{0.025} = 2.021$. The 95% confidence interval is

$$8.53 \pm 2.021 \frac{1.32}{\sqrt{41}}$$

$$8.11 < \mu < 8.95$$

Check Your Understanding

In Exercises 1–4, state which type of parameter is to be estimated, then construct the confidence interval.

1. A simple random sample of size 15 has mean $\bar{x} = 10.34$ and standard deviation $s = 3.48$. The population is normally distributed. Construct a 95% confidence interval for the population standard deviation. *Standard deviation; (2.55, 5.49)*
2. A simple random sample of size 80 has mean $\bar{x} = 7.31$. The population standard deviation is $\sigma = 6.26$. Construct a 99% confidence interval for the population mean. *Mean; (5.51, 9.11)*
3. In a simple random sample of 100 children, 22 had reading skills above their grade level. Construct a 99% confidence interval for the proportion of children who have reading skills above their grade level. *Proportion; (0.113, 0.327)*
4. A simple random sample of size 25 has mean $\bar{x} = 17.4$ and standard deviation $s = 5.3$. The population is approximately normally distributed. Construct a 95% confidence interval for the population mean. *Mean; (15.2, 19.6)*

Answers are on page 425.

Section 8.5 Exercises

Exercises 1–4 are the Check Your Understanding exercises located within the section.

Practicing the Skills

In Exercises 5–12, state which type of parameter is to be estimated, then construct the confidence interval.

5. A simple random sample of size 18 has mean $\bar{x} = 71.32$ and standard deviation $s = 15.78$. The population is approximately normally distributed. Construct a 95% confidence interval for the population mean. *Mean; (63.47, 79.17)*
6. In a simple random sample of 400 voters, 220 said that they were planning to vote for the incumbent mayor in the next election. Construct a 99% confidence interval for the proportion of voters who plan to vote for the incumbent mayor in the next election. *Proportion; (0.486, 0.614)*
7. A simple random sample of size 8 has mean $\bar{x} = 3.21$ and standard deviation $s = 1.69$. The population is normally distributed. Construct a 99% confidence interval for the population standard deviation. *Standard deviation; (0.99, 4.50)*
8. A simple random sample of size 12 has mean $\bar{x} = 3.37$. The population standard deviation is $\sigma = 1.62$. The population is approximately normally distributed. Construct a 95% confidence interval for the population mean. *Mean; (2.45, 4.29)*
9. In a survey of 250 employed adults, 185 said that they had missed one or more days of work in the past six months. Construct a 95% confidence interval for the proportion of employed adults who missed one or more days of work in the past six months. *Proportion; (0.686, 0.794)*
10. A simple random sample of size 17 has mean $\bar{x} = 8.44$ and standard deviation $s = 5.38$. The population is normally distributed. Construct a 95% confidence interval for the population standard deviation. *Standard deviation; (4.01, 8.19)*
11. A simple random sample of size 120 has mean $\bar{x} = 8.45$. The population standard deviation is $\sigma = 4.81$. Construct a 99% confidence interval for the population mean. *Mean; (7.32, 9.58)*
12. A simple random sample of size 23 has mean $\bar{x} = 1.48$ and standard deviation $s = 1.32$. The population is approximately normally distributed. Construct a 99% confidence interval for the population mean. *Mean; (0.70, 2.26)*

Working with the Concepts

13. **Football players:** The weights of 52 randomly selected NFL football players are presented below. The sample mean is $\bar{x} = 248.38$, and the sample standard deviation is $s = 46.68$.

305	265	287	285	290	235	300	230	195
236	244	194	190	307	218	315	265	210
194	216	255	300	315	190	185	183	313
246	212	201	308	270	241	242	306	237
315	215	200	295	187	204	257	185	255
318	230	316	200	324	245	185		

Source: Chicago Tribune

Construct a 95% confidence interval for the mean weight of NFL football players. *(235.38, 261.38)*

14. **Ages of students:** A simple random sample of 100 U.S. college students had a mean age of 22.68 years. Assume the population standard deviation is $\sigma = 4.74$ years. Construct a 99% confidence interval for the mean age of U.S. college students. *(21.46, 23.90)*

15. Calories in bread: Following are the numbers of calories in a random sample of 10 slices of bread. Assume the population is normally distributed.

55 51 49 48 68 52 62 70 67 70

Construct a 95% confidence interval for the standard deviation of the number of calories. (6.26, 16.62)

16. Credit card debt: In a survey of 1118 U.S. adults conducted by the Financial Industry Regulatory Authority, 626 said they always pay their credit cards in full each month. Construct a 95% confidence interval for the proportion of U.S. adults who pay their credit cards in full each month. (0.531, 0.589)

17. Windy place: Mt. Washington, New Hampshire, is one of the windiest places in the United States. Wind speed measurements on a simple random sample of 50 days had a sample mean of 45.01 mph. Assume the population standard deviation is $\sigma = 25.6$ mph. Construct a 95% confidence interval for the mean wind speed on Mt. Washington. (37.91, 52.11)

18. An apple a day: Following are the numbers of grams of sugar per 100 grams of apple in a random sample of six Red Delicious apples. Assume the population is normally distributed.

12.0 12.6 13.1 13.5 12.1 10.5

Construct a 95% confidence interval for the standard deviation of the number of grams of sugar. (0.66, 2.58)

19. Pneumonia: In a simple random sample of 1500 patients admitted to the hospital with pneumonia, 145 were under the age of 18. Construct a 99% confidence interval for the proportion of pneumonia patients who are under the age of 18. (0.077, 0.116)

20. College tuition: A simple random sample of 35 colleges and universities in the United States had a mean tuition of \$18,702 with a standard deviation of \$10,653. Construct a 95% confidence interval for the mean tuition for all colleges and universities in the United States. (15,043, 22,361)

Answers to Check Your Understanding Exercises for Section 8.5

1. The parameter is the population standard deviation. The 95% confidence interval is $2.55 < \sigma < 5.49$.
2. The parameter is the population mean. The 99% confidence interval is $5.51 < \mu < 9.11$.
3. The parameter is the population proportion. The 99% confidence interval is $0.113 < p < 0.327$.
4. The parameter is the population mean. The 95% confidence interval is $15.2 < \mu < 19.6$.

Chapter 8 Summary

Section 8.1: In this section, we presented the basic ideas behind confidence intervals. We learned that a point estimate is a single number that is used to estimate the value of an unknown parameter. For example, the sample mean $\bar{x}$ is a point estimate of the population mean μ. The standard error of a point estimate tells us roughly how far from the true value the point estimate is likely to be. We multiply the standard error by a critical value to obtain the margin of error. By adding and subtracting the margin of error from the point estimate, we obtain a confidence interval. The level of a confidence interval is the proportion of samples for which the confidence interval will contain the true value. If the sample size is large ($n > 30$) or if the population is approximately normal, then the confidence interval for the population mean is $\bar{x} \pm z_{\alpha/2}\sigma/\sqrt{n}$ if the population mean σ is known.

Section 8.2: When the population is approximately normal, or the sample size is large ($n > 30$), we can use the Student's t distribution to construct a confidence interval for a population mean μ when the population standard deviation is unknown. Let s be the sample standard deviation. The confidence interval is $\bar{x} \pm t_{\alpha/2}s/\sqrt{n}$ if σ is unknown.

Section 8.3: In this section, we learned to construct confidence intervals for population proportions. The assumptions are that the individuals in the population can be divided into two categories, that the sample contains at least 10 individuals in each category, and that the population is at least 20 times as large as the sample. We denote the sample size by n and the number of individuals in the sample who fall into the specified category by x. When the assumptions are met, the confidence interval for the population proportion p is $\hat{p} \pm z_{\alpha/2}\sqrt{\hat{p}(1-\hat{p})/n}$, where $\hat{p} = x/n$. A small-sample method can also be used. In the small-sample method, we define $\tilde{p} = \dfrac{x+2}{n+4}$. The confidence interval is $\tilde{p} \pm z_{\alpha/2}\sqrt{\dfrac{\tilde{p}(1-\tilde{p})}{n+4}}$. The small-sample confidence interval is actually valid for any sample size.

Section 8.4: When the population is almost exactly normal, it is possible to find a confidence interval for the variance or standard deviation of the population using the chi-square distribution. This method is very sensitive to the assumption of normality, and should not be used unless it is certain that the population is almost exactly normal. The lower bound of the confidence interval is $(n-1)s^2/\chi^2_{\alpha/2}$ and the upper bound is $(n-1)s^2/\chi^2_{1-\alpha/2}$.

Section 8.5: We have learned to construct confidence intervals for a population mean, a population proportion, and a population standard deviation or variance. There are two methods for constructing a confidence interval for a population mean, the z method and the t method. The method to use depends on whether the population standard deviation σ is known.

Vocabulary and Notation

Important Formulas

Confidence interval for a mean, standard deviation known

$$\bar{x} - z_{\alpha/2}\frac{\sigma}{\sqrt{n}} < \mu < \bar{x} + z_{\alpha/2}\frac{\sigma}{\sqrt{n}}$$

Sample size to construct an interval for μ with margin of error m, standard deviation known:

$$n = \left(\frac{z_{\alpha/2} \cdot \sigma}{m}\right)^2$$

Confidence interval for a mean, standard deviation unknown:

$$\bar{x} - t_{\alpha/2}\frac{s}{\sqrt{n}} < \mu < \bar{x} + t_{\alpha/2}\frac{s}{\sqrt{n}}$$

Sample size to construct an interval for μ with margin of error m, standard deviation unknown:

$$n = \left(\frac{z_{\alpha/2} \cdot s}{m}\right)^2$$

Confidence interval for a proportion:

$$\hat{p} - z_{\alpha/2}\sqrt{\frac{\hat{p}(1-\hat{p})}{n}} < p < \hat{p} + z_{\alpha/2}\sqrt{\frac{\hat{p}(1-\hat{p})}{n}}$$

Sample size to construct an interval for p with margin of error m:

$$n = \hat{p}(1-\hat{p})\left(\frac{z_{\alpha/2}}{m}\right)^2 \quad \text{if a value for } \hat{p} \text{ is available}$$

$$n = 0.25\left(\frac{z_{\alpha/2}}{m}\right)^2 \quad \text{if no value for } \hat{p} \text{ is available}$$

Confidence interval for the variance of a normal distribution:

$$\frac{(n-1)s^2}{\chi^2_{\alpha/2}} < \sigma^2 < \frac{(n-1)s^2}{\chi^2_{1-\alpha/2}}$$

Confidence interval for the standard deviation of a normal distribution:

$$\sqrt{\frac{(n-1)s^2}{\chi^2_{\alpha/2}}} < \sigma < \sqrt{\frac{(n-1)s^2}{\chi^2_{1-\alpha/2}}}$$

Chapter Quiz

1. Define the following terms:
   a. Point estimate
   b. Confidence interval
   c. Confidence level

2. Find the critical value $t_{\alpha/2}$ needed to construct a 90% confidence interval for a population mean with sample size 27. *1.706*

3. An owner of a fleet of taxis wants to estimate the mean gas mileage, in miles per gallon, of the cars in the fleet. A random sample of 40 cars is followed for one month, and the sample mean gas mileage is 23.2 with a standard deviation of 5.8. Construct a 90% confidence interval for the mean gas mileage in the fleet. *(21.7, 24.7)*

4. Construct a 95% confidence interval for the population standard deviation σ if a sample of size 20 has standard deviation $s = 10$. *(7.60, 14.61)*

5. A cookie manufacturer wants to estimate the length of time that her boxes of cookies spend in the store before they are bought. She visits a sample of 15 supermarkets and determines the number of days since manufacture of the oldest box of cookies in the store. The mean is 54.8 days with a standard deviation of 11.3 days. A dotplot of the data indicates that the assumptions for constructing a confidence interval for the mean are satisfied. Construct a 99% confidence interval for the mean number of days. *(46.1, 63.5)*

6. A person selects a random sample of 15 credit cards and determines the annual interest rate, in percent, of each. The sample mean is 12.42 with a sample standard deviation of 1.3. Construct a 95% confidence interval for the mean credit card annual interest rate, assuming that the rates are approximately normally distributed. *(11.70, 13.14)*

7. Construct a 90% confidence interval for the population standard deviation σ if a sample of size 6 has standard deviation $s = 22$. *(14.79, 45.97) [Tech: (14.79, 45.96)]*

8. Find the critical value $z_{\alpha/2}$ needed to construct a confidence interval for a population proportion with confidence level 92%. *1.75 [Tech: 1.751]*

9. Find the critical values for a 98% confidence interval using the chi-square distribution with 18 degrees of freedom. *7.015, 34.805*

10. The amount of time that a certain phone will keep a charge is known to be normally distributed with standard deviation $\sigma = 16$ hours. A sample of 40 phones had a mean time of 141 hours. Let μ represent the population mean time that a phone will keep a charge.
    a. What is the point estimate of μ? *141*
    b. What is the standard error of the point estimate? *2.53*
11. Refer to Exercise 10. Suppose that a 95% confidence interval is to be constructed for the mean time.
    a. What is the critical value? *1.96*
    b. What is the margin of error? *4.958*
    c. Construct the 95% confidence interval. *(136, 146)*
12. Refer to Exercise 10. What sample size is necessary so that a 95% confidence interval will have a margin of error of 1 hour? *984*
13. In a survey of 802 U.S. adult drivers, 265 state that traffic is getting worse in their community. Construct a 99% confidence interval for the proportion of adult drivers who think that traffic is getting worse. *(0.288, 0.373)*
14. Refer to Exercise 13. How large a sample is needed so that a 99% confidence interval will have margin of error of 0.08, using the sample proportion for $\hat{p}$? *230*
15. Refer to Exercise 13. How large a sample is needed so that a 99% confidence interval will have margin of error of 0.08, assuming no estimate of $\hat{p}$ is available? *260*

Review Exercises

1. **Build more parking?** A survey is to be conducted in which a random sample of residents in a certain city will be asked whether they favor or oppose the building of a new parking structure downtown. How many residents should be polled to be sure that a 90% confidence interval for the proportion who favor the construction will have a margin of error no greater than 0.05? *271*
2. **Drill lifetime:** A sample of 50 drills had a mean lifetime of 12.68 holes drilled when drilling a low-carbon steel. Assume the population standard deviation is 6.83.
    a. Construct a 95% confidence interval for the mean lifetime of this type of drill. *(10.79, 14.57)*
    b. The manufacturer of the drills claims that the mean lifetime is greater than 13. Does this confidence interval contradict this claim? Explain. *No*
    c. How large would the sample need to be so that a 95% confidence interval would have a margin of error of 1.0? *180*
    Source: *Journal of Engineering Manufacture* 216:301–305
3. **Cost of environmental restoration:** In a survey of 189 Scottish voters, 61 said they would be willing to pay additional taxes in order to restore the Affric forest.
    a. Assuming that the 189 voters who responded constitute a random sample, construct a 99% confidence interval for the proportion of voters who would be willing to pay to restore the Affric forest. *(0.235, 0.410)*
    b. Use the results from the sample of size 189 to estimate the sample size needed so that the 99% confidence interval will have a margin of error of 0.03. *1612*
    c. Another survey is planned, in which voters will be asked whether they would be willing to pay in order to restore the Strathspey forest. At this point, no estimate of this proportion is available. Find an estimate of the sample size needed so that the margin of error of a 99% confidence interval will be 0.03. *1844*
    Source: *Environmental and Resource Economics* 18:391–410
4. **More repairs:** A sample of six records for repairs of a component showed the following costs:

    93 97 27 79 81 87

    a. Construct a 90% confidence interval for the mean cost of a repair for this type of component. *(56.3, 98.4)*
    b. Is there any evidence to suggest that this confidence interval may not be reliable? Explain. *Yes*
5. **More repairs:** Refer to Exercise 4. Would it be appropriate to use these data to construct a confidence interval for σ using the methods of Section 8.4? Explain. *No*
6. **Contaminated water:** Polychlorinated biphenyls (PCBs) are a group of synthetic oil-like chemicals that were at one time widely used as insulation in electrical equipment and were discharged into rivers. They were discovered to be a health hazard and were banned in the 1970s. Assume that water samples are being drawn from a river in order to estimate the PCB concentration. Suppose that a random sample of size 60 has a sample mean of 1.96 parts per billion (ppb). Assume the population standard deviation is $\sigma = 0.35$ ppb.
    a. Construct a 98% confidence interval for the PCB concentration. *(1.85, 2.07)*
    b. EPA standards require that the PCB concentration in drinking water be no more than 0.5 ppb. Based on the confidence interval, is it reasonable to believe that this water meets the EPA standard for drinking water? Explain. *No*
    c. Estimate the sample size needed so that a 98% confidence interval will have a margin of error of 0.03. *737*

7. **Defective electronics:** A simple random sample of 200 electronic components was tested, and 17 of them were found to be defective.
   a. Construct a 99% confidence interval for the proportion of components that are defective. *(0.034, 0.136)*
   b. Use the results from the sample of 200 to estimate the sample size needed so that the 99% confidence interval will have a margin of error equal to 0.04. *323*
   c. A simple random sample of a different type of component will be tested. At this point, there is no estimate of the proportion defective. Find a sample size so that the 99% confidence interval will have a margin of error no greater than 0.04. *1037*

8. **Cost of repairs:** A sample of eight repair records for a certain fiber-optic component was drawn, and the cost of each repair, in dollars, was recorded, with the following results:
   30 35 19 23 27 22 26 16
   a. Construct a dotplot for these data. Are the assumptions for constructing a confidence interval for the mean satisfied? Explain. *Yes*
   b. If appropriate, construct a 98% confidence interval for the mean cost of a repair. *(18.3, 31.2)*

9. **Cost of repairs:** Refer to Exercise 8. Assume the population is normal. Construct a 95% confidence interval for the population standard deviation σ. *(4.03, 12.39)*

10. **Super Bowl:** A simple random sample of 140 residents in a certain town was polled the week after the Super Bowl, and 75 of them said they had watched the game on television.
    a. Construct a 95% confidence interval for the proportion of people in the town who watched the Super Bowl on television.
    b. Ratings for the 2019 Super Bowl between the New England Patriots and the Los Angeles Rams indicate that 41% of television sets in the United States were tuned to the game. Someone claims that the percentage of people who watched the game in this town was less than 41%. Does the confidence interval contradict this claim? Explain. *Yes*
    c. Use the results from the sample of 140 to estimate the sample size necessary for a 95% confidence interval to have a margin of error of 0.025. *1529*

11. **Testing math skills:** In order to test the effectiveness of a program to improve mathematical skills, a simple random sample of 45 fifth graders was chosen to participate in the program. The students were given an exam at the beginning of the program and again at the end. The sample mean increase in the exam score was 12.2 points, with a sample standard deviation of 4.7 points.
    a. Construct a 99% confidence interval for the mean increase in score. *(10.3, 14.1)*
    b. The developers of the program claim that the program will produce a mean increase of more than 15 points. Does the confidence interval contradict this claim? Explain. *Yes*

12. **Sleep time:** In a sample of 87 young adults, the average time per day spent in bed asleep was 7.06 hours. Assume the population standard deviation is 1.11 hours.
    a. Construct a 99% confidence interval for the mean time spent in bed asleep. *(6.75, 7.37)*
    b. Some health experts recommend that people get 8 hours or more of sleep per night. Based on the confidence interval, is it reasonable to believe that the mean number of hours of sleep for young adults is 8 or more? Explain. *No*
    c. How large would the sample have to be so that a 99% confidence interval would have a margin of error of 0.1? *818*

    Source: *Behavioral Medicine* 27:71–76

13. **Leaking tanks:** Leakage from underground fuel tanks has been a source of water pollution. In a random sample of 107 gasoline stations, 18 were found to have at least one leaking underground tank.
    a. Find a point estimate for the proportion of gasoline stations with at least one leaking underground tank. *0.168*
    b. Construct a 95% confidence interval for the proportion of gasoline stations with at least one leaking underground tank.
    c. Use the point estimate computed in part (a) to determine the number of stations that must be sampled so that a 95% confidence interval will have a margin of error of 0.03. *598*

14. **Waist size:** According to the National Health Statistics Reports, a sample of 783 men aged 20–29 years had a mean waist size of 36.9 inches with a standard deviation of 8.8 inches.
    a. Construct a 95% confidence interval for the mean waist size. *(36.3, 37.5)*
    b. The results of another study suggest that the mean waist size for men aged 30–39 is 38.7 inches. Based on the confidence interval, is it reasonable to believe that the mean waist size for men aged 20–29 may be 38.7 inches? Explain. *No*

15. **Don't construct a confidence interval:** A meteorology student examines precipitation records for a certain city and discovers that of the last 365 days, it rained on 46 of them. Explain why these data cannot be used to construct a confidence interval for the proportion of days in this city that are rainy.

Write About It

1. When constructing a confidence interval for μ when σ is known, we assume that we have a simple random sample, that σ is known, and that either the sample size is large or the population is approximately normal. Why is it necessary for these assumptions to be met?
2. What factors can you think of that may affect the width of a confidence interval? In what way does each factor affect the width?
3. Explain the difference between confidence and probability.

In Exercises 4 and 5, express the following survey results in terms of confidence intervals for p:

4. According to a survey of 1000 American adults, 55% of Americans do not have a will specifying the handling of their estate. The survey's margin of error was plus or minus 3%.
Source: FindLaw.com
5. In a survey of 5050 U.S. adults, 29% would consider traveling abroad for medical care because of medical costs. The survey's margin of error was plus or minus 2%.
Source: The Gallup Poll
6. When constructing a confidence interval for μ, how do you decide whether to use the t distribution or the normal distribution? Are there any circumstances when it is acceptable to use either distribution?
7. It is stated in the text that there are many different t distributions. Explain how this is so.

In-Class Activities

1. **Calculating a sample size:** Each student tosses a coin 10 times and calculates $\hat{p}$, the proportion of heads observed. Use that value of $\hat{p}$ to estimate the sample size needed (call it n_1) so that a 90% confidence interval for p will have a margin of error of 0.05. Make additional tosses to bring the total to n_1, and compute a 90% confidence interval for p. What proportion of confidence intervals cover the true value of 0.5? Is it close to 90%? If someone's confidence interval doesn't contain the value 0.5, should we conclude that the person's coin is unfair? Or is there a better explanation?
2. **Construct confidence intervals:** Ask each student how far they traveled to get to school that day. Choose five students at random, and make a dotplot of their results to check for skewness and outliers. Regardless of the results, use these five results to construct a confidence interval for the mean distance traveled. Repeat for several samples of students.

Case Study: Do Newer Wood Stoves Produce Less Pollution?

The town of Libby, Montana, has experienced high levels of air pollution in the winter because many of the houses in Libby are heated by wood stoves that produce a lot of pollution. In an attempt to reduce the level of air pollution in Libby, a program was undertaken in which almost every wood stove in the town was replaced with a newer, cleaner-burning model. Measurements of several air pollutants were taken both before and after the stove replacement. They included particulate matter (PM); total carbon (TC); organic carbon (OC), which is carbon bound in organic molecules; and levoglucosan (LE), which is a compound found in charcoal and is thus an indicator of the amount of wood smoke in the atmosphere.

In order to determine how much the pollution levels were reduced, scientists measured the levels of these pollutants for three winters prior to the replacement. The mean levels over this period of time are referred to as the *baseline* levels. Following are the baseline levels for these pollutants. The units are micrograms per cubic meter.

PM: 27.08 OC: 17.41 TC: 18.87 LE: 2.57

The following table presents values measured on samples of winter days during the two years following replacement.

Year 1				Year 2			
PM	**OC**	**TC**	**LE**	**PM**	**OC**	**TC**	**LE**
21.7	15.6	17.73	1.78	27.0	15.79	19.46	2.06
27.8	15.6	17.87	2.25	24.7	13.61	15.98	3.10
24.7	17.2	18.75	1.98	21.8	12.94	15.79	2.68
15.3	8.3	9.21	0.67	23.2	12.97	16.32	2.80
18.4	11.3	12.46	0.86	23.3	11.19	13.49	2.07
14.4	8.4	9.66	1.93	16.2	9.61	12.44	2.14
19.0	13.2	14.73	1.51	13.4	6.97	8.40	2.32
23.7	11.4	13.23	1.98	13.0	7.96	10.02	2.18
22.4	13.8	17.08	1.69	16.9	8.43	11.08	2.06
25.6	13.2	15.86	2.30	26.3	14.92	21.46	1.94
15.0	15.7	17.27	1.24	31.4	17.15	20.57	1.85
17.0	9.3	10.21	1.44	40.1	15.13	19.64	2.11
23.2	10.5	11.47	1.43	28.0	8.66	10.75	2.50
17.7	14.2	15.64	1.07	4.2	15.95	20.36	2.27
11.1	11.6	13.48	0.59	15.9	11.73	14.59	2.17
29.8	7.0	7.795	2.10	20.5	14.34	17.64	2.74
20.0	19.9	21.20	1.73	23.8	8.99	11.75	2.45
21.6	14.8	15.65	1.56	14.6	10.63	13.12	
14.8	12.6	13.51	1.10	17.8			
21.0	9.1	9.94					

1. For each of the four pollutants, construct a boxplot for the values for Year 1 to verify that the assumptions for constructing a confidence interval are satisfied.
2. Construct a 95% confidence interval for the mean level of each pollutant for the Year 1. *PM: (17.94, 22.48); OC: (11.08, 14.19); TC: (12.428, 15.847); LE: (1.291, 1.784)*
3. Is it reasonable to conclude that the mean levels in Year 1 were lower than the baseline levels for some or all of the pollutants? Which ones, if any? *All of them*

The investigators were concerned that the reduction in pollution levels might be only temporary. Specifically, they were concerned that people might use their new stoves carefully at first, thus obtaining the full advantage of their cleaner burning, but then become more casual in their operation, leading to an increase in pollution levels. We will investigate this issue by constructing confidence intervals for the mean levels in Year 2.

4. Repeat Exercises 1 and 2 for the Year 2 data. *PM: (17.33, 25.00); OC: (10.493, 13.615); TC: (13.149, 17.169); LE: (2.144, 2.496)*
5. Is it reasonable to conclude that the mean levels in Year 2 were lower than the baseline levels for some or all of the pollutants? Which ones, if any? *All of them*

chapter

9

Hypothesis Testing

Murat Taner/Getty Images

Introduction

How serious is global warming? There is much evidence to suggest that temperatures have been increasing since the early part of the 20th century. Studies of global warming involve analyzing temperature records. The following table presents the record high and low temperatures in degrees Fahrenheit in Washington, D.C., along with the year in which the record was set, for a selection of days in April through September. The column labeled "More Recent" tells which record, high or low, occurred more recently.

If there were no temperature trend, we would expect that the proportion of days on which the high temperature record was more recent to be about one-half. It would not be surprising if this proportion were somewhat different from one-half, because we would expect to see some difference just by chance. However, we would not expect the proportion to be much different from one-half. If the proportion were much different from one-half, we would conclude that there was a temperature trend.

There are 26 days in the table. Half of 26 is 13. If the record high had been more recent on 13 of the days, there would be no reason to believe there was a warming trend. If the record high had been more recent on 14 or 15 days, this would be slightly more than one-half, but it would seem reasonable to believe that a difference this small was just due to chance.

In fact, the record high temperature was more recent for 18 of the 26 days. The question we need to address is whether this difference from one-half is too large for us to believe that it is simply due to chance. This is the sort of question that hypothesis tests are designed to answer. In this chapter, we will learn to perform hypothesis tests in a variety of commonly occurring situations. In the case study at the end of the chapter, we will study a data set that includes the data in the preceding table and investigate the possibility of a warming trend in Washington, D.C.

Date	High	Year	Low	Year	More Recent	Date	High	Year	Low	Year	More Recent	Date	High	Year	Low	Year	More Recent
Apr 2	89	1963	23	1907	High	Jun 4	99	1925	46	1929*	Low	Aug 6	106	1918	53	1912	High
9	90	1959	28	1972*	Low	11	101	1911	45	1913	Low	13	101*	2016	55	1930*	High
16	92	2002	29	1928	High	18	97	1944	51	1965*	Low	20	101	1983	50	1896	High
23	95	1960	33	1933*	High	25	100	1997	53	1902	High	27	100	1987	51	1885	High
30	92	1942*	34	1874	High	Jul 2	101	1898	55	1940	Low	Sep 3	98	1953	48	1909*	High
May 7	95	1930	38	1970	Low	9	104	1936	55	1891	High	10	98	1983	44	1883	High
14	93	1956	41	1928	High	16	104	1988	56	1930*	High	17	96	1991	44	1923	High
21	95	1934	41	1907	High	23	102	2011	56	1890	High	24	99	2010	39	1963	High
28	97	1941	42	1961	Low	30	99	1953	56	1914	High						

*Indicates that the record occurred more than once; only the most recent year is given.
Source: National Weather Service

Section 9.1 Basic Principles of Hypothesis Testing

Objectives

1. Define the null and alternate hypotheses
2. Describe the reasoning used in hypothesis testing
3. State conclusions to hypothesis tests
4. Distinguish between Type I and Type II errors

Objective 1 Define the null and alternate hypotheses

The Null Hypothesis and the Alternate Hypothesis

Air pollution has become a serious health problem in many cities. One of the forms of air pollution that health officials are most concerned about is particulate matter (PM), which refers to fine particles that can be trapped in the lungs, increasing the risk of respiratory disease. Some of the PM in the atmosphere comes from car exhaust, so one important way to reduce PM pollution is to design automobile engines that produce less PM. The following example will show how hypothesis testing can play a part in this effort.

A study published in the *Journal of the Air and Waste Management Association* reported that the mean amount of PM produced by cars and light trucks in an urban setting is 35 milligrams of PM per mile of travel. Suppose that a new engine design is proposed that is intended to reduce this level. Now there are two possibilities: either the new design will reduce the level, or it will not. These possibilities are called *hypotheses*. To be specific,

1. The *null hypothesis* says that the new design will not reduce the level, so the mean for the new engines will be $\mu = 35$.
2. The *alternate hypothesis* says that the new design will reduce the level, so $\mu < 35$.

In general, the null hypothesis says that a parameter is equal to a certain value, while the alternate hypothesis says that the parameter differs from this value. Often the null hypothesis is a statement of no change or no difference, while the alternate hypothesis states that a change or difference has occurred.

NOTE TO INSTRUCTOR

The following topic from the **Statistics Corequisite Workbook** is aligned with the material in this section.

6.3 - Translating Between Symbolic Notation and Words for Hypothesis Tests

EXPLAIN IT AGAIN

A statistical hypothesis is often a statement about one or more population parameters.

DEFINITION

- The **null hypothesis** about a parameter states that the parameter is equal to a specific value, for example, H_0: $\mu = 35$. The null hypothesis is denoted H_0.
- The **alternate hypothesis** about a parameter states that the value of the parameter differs from the value specified by the null hypothesis. The alternate hypothesis is denoted H_1.

There are three types of alternate hypothesis, which we now define.

DEFINITION

- **A left-tailed** alternate hypothesis states that the parameter is less than the value specified by the null hypothesis, for example, H_1: $\mu < 35$.
- **A right-tailed** alternate hypothesis states that the parameter is greater than the value specified by the null hypothesis, for example, H_1: $\mu > 35$.
- **A two-tailed** alternate hypothesis states that the parameter is not equal to the value specified by the null hypothesis, for example, H_1: $\mu \neq 35$.

Left-tailed and right-tailed hypotheses are called **one-tailed** hypotheses.

Example 9.1

State the null and alternate hypotheses

Boxes of a certain kind of cereal are labeled as containing 20 ounces. An inspector thinks that the mean weight may be less than this. State the appropriate null and alternate hypotheses.

Solution

The null hypothesis says that there is no difference, so the null hypothesis is H_0: $\mu = 20$. The inspector thinks that the mean weight may be less than 20, so the alternate hypothesis is H_1: $\mu < 20$.

Example 9.2

State the null and alternate hypotheses

Last year, the mean monthly rent for an apartment in a certain city was $800. A real estate agent believes that the mean rent is higher this year. State the appropriate null and alternate hypotheses.

Solution

The null hypothesis says that there is no change, so the null hypothesis is H_0: $\mu = 800$. The real estate agent wants to know whether the mean is higher, so the alternate hypothesis is H_1: $\mu > 800$.

Example 9.3

State the null and alternate hypotheses

Scores on a standardized test have a mean of 70. Some modifications are made to the test, and an educator believes that the mean may have changed. State the appropriate null and alternate hypotheses.

Solution

The null hypothesis says that there is no change, so the null hypothesis is H_0: $\mu = 70$. The educator wants to know whether the mean has changed, without specifying whether it has increased or decreased. Therefore, the alternate hypothesis is H_1: $\mu \neq 70$.

Check Your Understanding

1. Last year, the mean amount spent by customers at a certain restaurant was $35. The restaurant owner believes that the mean may be higher this year. State the appropriate null and alternate hypotheses. H_0: $\mu = 35$, H_1: $\mu > 35$
2. In a recent year, the mean weight of newborn boys in a certain country was 6.6 pounds. A doctor wants to know whether the mean weight of newborn girls differs from this. State the appropriate null and alternate hypotheses. H_0: $\mu = 6.6$, H_1: $\mu \neq 6.6$
3. A certain model of car can be ordered with either a large or small engine. The mean number of miles per gallon for cars with a small engine is 25.5. An automotive engineer thinks that the mean for cars with the larger engine will be less than this. State the appropriate null and alternate hypotheses. H_0: $\mu = 25.5$, H_1: $\mu < 25.5$

Answers are on page 438.

A hypothesis test is like a trial

The purpose of a **hypothesis test** is to determine how plausible the null hypothesis is. The idea behind hypothesis testing is the same as the idea behind a criminal trial. At the start of a trial, the defendant is assumed to be innocent. Then the evidence is presented. If the evidence strongly indicates that the defendant is guilty, we abandon the assumption of innocence and find the defendant guilty. In a hypothesis test, the null hypothesis plays the role of the defendant. At the start of a hypothesis test, we assume that the null hypothesis is true. Then we look at the evidence, which comes from data that have been collected. If the data strongly indicate that the null hypothesis is false, we abandon our assumption that it is true and believe the alternate hypothesis instead. This is referred to as **rejecting the null hypothesis**.

SUMMARY

- We begin a hypothesis test by assuming the null hypothesis to be true.
- If the data provide strong evidence against the null hypothesis, we reject it, and believe the alternate hypothesis.

Objective 2 Describe the reasoning used in hypothesis testing

The Reasoning Used in Hypothesis Testing

We reject the null hypothesis when the data provide strong evidence against it. The evidence is in the form of a statistic calculated from the data. In many cases, we make a decision by computing the difference between this statistic and the parameter value specified by the null hypothesis. When this difference is sufficiently large, we reject H_0.

For example, imagine that an exam has been given to a large class. Someone claims that the mean score on the exam was 80. This is the null hypothesis: H_0: $\mu = 80$. Assume the alternate hypothesis is H_1: $\mu \neq 80$. Now you sample some students and find that the sample mean exam score is $\bar{x} = 78$. This is not exactly equal to the null mean $\mu = 80$. However, we expect the sample mean to differ somewhat from the population mean, just by chance. Because 78 is fairly close to 80, it is plausible that the null hypothesis is true: the population mean is $\mu = 80$. Now imagine that the sample mean had been $\bar{x} = 50$. This value is far from the null mean of 80, and it is unlikely that a difference this large is due just to chance. Because 50 is far from 80, it does not seem plausible that the population mean is 80. Therefore, we would probably reject H_0. See Figures 9.1 and 9.2 on page 435.

We see that the farther the test statistic is from the parameter value specified by H_0, the less likely the difference is due to chance, and the less plausible H_0 becomes. The question then is: How big does the difference have to be before we reject H_0? To answer this question, we need methods that enable us to calculate just how plausible H_0 is. Hypothesis tests provide these methods taking into account things such as the size of the sample and the amount of spread in the distribution. In the remaining sections of this chapter, we will learn ways to calculate a value that tells us how plausible H_0 is, so we can decide whether to reject it.

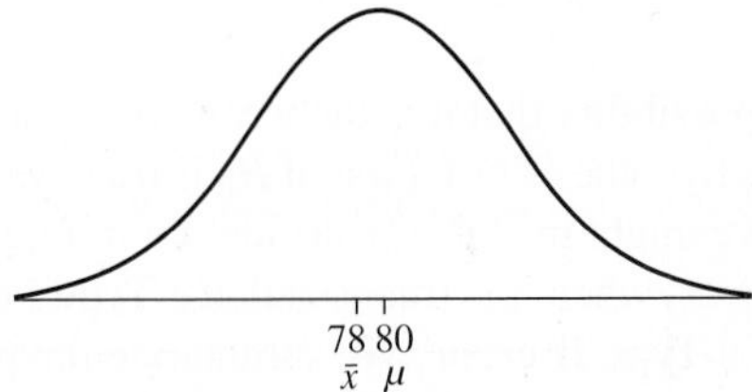

Figure 9.1 The null hypothesis is H_0: $\mu = 80$. The sample mean is $\bar{x} = 78$. The sample mean is close to the value specified by the null hypothesis, so we would not reject H_0.

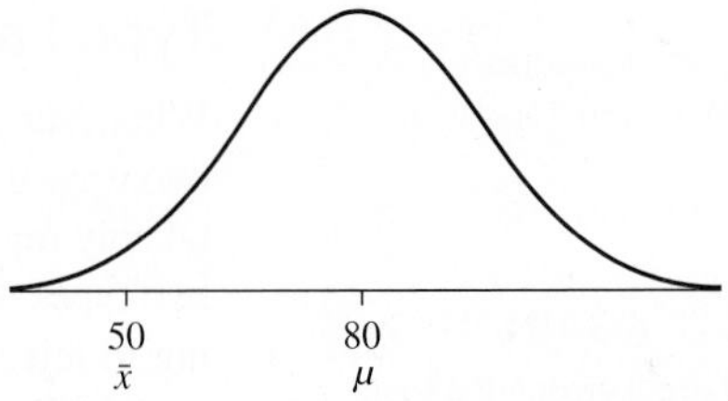

Figure 9.2 The null hypothesis is H_0: $\mu = 80$. The sample mean is $\bar{x} = 50$. The sample mean is far from the value specified by the null hypothesis, so we would reject H_0.

Objective 3 State conclusions to hypothesis tests

Stating Conclusions

If the null hypothesis is rejected, we conclude that H_1 is true. We can state this conclusion by expressing H_1 in words. We should not simply say "we reject the null hypothesis."

Example 9.4

State a conclusion when the null hypothesis is rejected

Boxes of a certain kind of cereal are labeled as containing 20 ounces. An inspector thinks that the mean weight may be less than this, so he performs a test of H_0: $\mu = 20$ versus H_1: $\mu < 20$. He rejects the null hypothesis. State an appropriate conclusion.

NOTE TO INSTRUCTOR

It is important that students express the conclusion of a hypothesis test in words relevant to the problem. It is not enough simply to state whether or not the null hypothesis is rejected.

Solution

Because the null hypothesis is rejected, we conclude that the alternate hypothesis is true. We express the alternate hypothesis in words: "We conclude that the mean weight of cereal boxes is less than 20 ounces."

If the null hypothesis is rejected, the conclusion is straightforward: We conclude that the null hypothesis is false and the alternate hypothesis is true. However, if the null hypothesis is not rejected, we do *not* conclude that the null hypothesis is true. In our formulation, the null hypothesis says that a parameter, such as μ, is equal to a certain value. Now we can never be sure that a parameter is *exactly* equal to a particular value. Therefore, we can never be sure that the null hypothesis is true. When we do not reject the null hypothesis, this just means that the evidence wasn't strong enough to reject it. An appropriate way to state a conclusion when the null hypothesis is not rejected is to state that there is not sufficient evidence to conclude that H_1 is true.

EXPLAIN IT AGAIN

The conclusion of a hypothesis test is like the verdict of a jury: Not rejecting H_0 is like a jury verdict of not guilty. A not guilty verdict doesn't mean that the defendant is innocent; it just means that the evidence wasn't strong enough to be sure of guilt. Not rejecting H_0 does not mean that H_0 is true; it just means that the evidence wasn't strong enough to reject it.

SUMMARY

- If there is sufficient evidence to reject the null hypothesis, we conclude that the alternate hypothesis is true.
- If there is not sufficient evidence to reject the null hypothesis, we conclude that the null hypothesis *might* be true, but we never conclude that the null hypothesis *is* true.

Example 9.5

State a conclusion when the null hypothesis is not rejected

Boxes of a certain kind of cereal are labeled as containing 20 ounces. An inspector thinks that the mean weight may be less than this, so he performs a test of H_0: $\mu = 20$ versus H_1: $\mu < 20$. He does not reject the null hypothesis. State an appropriate conclusion.

Solution

The null hypothesis is not rejected, so we do not have sufficient evidence to conclude that the alternate hypothesis is true. We can express this as follows: "There is not enough evidence to conclude that the mean weight of cereal boxes is less than 20 ounces." Another way to state this is: "The mean weight of cereal boxes may be equal to 20 ounces."

Objective 4 Distinguish between Type I and Type II errors

Type I and Type II Errors

Whenever a decision is made, there is a possibility that it is the wrong decision. There are two ways to make a wrong decision with a hypothesis test. First, if H_0 is true, we might mistakenly reject it. Second, if H_0 is false, we might mistakenly decide not to reject it. These two types of errors have names. Rejecting H_0 when it is true is called a **Type I error**. Failing to reject H_0 when it is false is called a **Type II error**. We summarize the possibilities in the following table.

EXPLAIN IT AGAIN

Type I and Type II errors in a trial: In a trial, the null hypothesis is that the defendant is innocent. A Type I error occurs if an innocent defendant is found guilty. A Type II error occurs if a guilty defendant is found not guilty.

	Reality	
Decision	**H_0 True**	**H_0 False**
Reject H_0	Type I error	Correct decision
Don't reject H_0	Correct decision	Type II error

In general, a Type I error is more serious than a Type II error, because a Type I error results in a false conclusion, while a Type II error results only in no conclusion being made. For example, in a trial a Type I error occurs when an innocent defendant is found guilty, and a Type II error occurs when a guilty defendant is found not guilty.

Ideally, we would like to minimize the probability of both errors. Unfortunately, the only way to do this is to increase the sample size, which in practice is often not possible. With a fixed sample size, decreasing the probability of a Type I error increases the probability of a Type II error, and vice versa. To see this, note that if we want to reduce the probability of a Type I error, we must increase the strength of the evidence needed to reject the null hypothesis, making it less likely that we reject H_0 when it is true. However, it will then also be less likely that we reject H_0 when it is false, which will increase the probability of a Type II error. Similarly, if we reduce the strength of evidence needed to reject H_0, we decease the probability of a Type II error but increase the probability of a Type I error.

Because a Type I error is usually more serious than a Type II error, hypothesis tests are often designed so that the probability of a Type I error will be acceptably small, often 0.05 or 0.01. This value is called the **significance level** of the test, which we will discuss further in Section 9.2.

Example 9.6

Determining which type of error has been made

The dean of a business school wants to determine whether the mean starting salary of graduates of her school is greater than \$50,000. She will perform a hypothesis test with the following null and alternate hypotheses:

$$H_0\colon \mu = \$50{,}000 \qquad H_1\colon \mu > \$50{,}000$$

NOTE TO INSTRUCTOR

This is a good place to emphasize that rejecting the null hypothesis does not *prove* that it is false. It indicates that it is unlikely to be true, but it is possible to make an error.

a. Suppose that the true mean is $\mu = \$50{,}000$, and the dean rejects H_0. Is this a Type I error, a Type II error, or a correct decision?

b. Suppose that the true mean is $\mu = \$55{,}000$, and the dean rejects H_0. Is this a Type I error, a Type II error, or a correct decision?

c. Suppose that the true mean is $\mu = \$55{,}000$, and the dean does not reject H_0. Is this a Type I error, a Type II error, or a correct decision?

Solution

a. The true mean is $\mu = \$50{,}000$, so H_0 is true. Because the dean rejects H_0, this is a Type I error.

b. The true mean is $\mu = \$55{,}000$, so H_0 is false. Because the dean rejects H_0, this is a correct decision.

c. The true mean is $\mu = \$55{,}000$, so H_0 is false. Because the dean does not reject H_0, this is a Type II error.

Check Your Understanding

4. A test is made of H_0: $\mu = 100$ versus H_1: $\mu \neq 100$. The true value of μ is 150, and H_0 is rejected. Is this a Type I error, a Type II error, or a correct decision? *Correct decision*

5. A test is made of H_0: $\mu = 18$ versus H_1: $\mu > 18$. The true value of μ is 20, and H_0 is not rejected. Is this a Type I error, a Type II error, or a correct decision? *Type II error*

6. A test is made of H_0: $\mu = 3$ versus H_1: $\mu < 3$. The true value of μ is 3, and H_0 is rejected. Is this a Type I error, a Type II error, or a correct decision? *Type I error*

Answers are on page 438.

Section 9.1 Exercises

Exercises 1–6 are the Check Your Understanding exercises located within the section.

Understanding the Concepts

In Exercises 7 and 8, fill in each blank with the appropriate word or phrase.

7. The ________ hypothesis states that a parameter is equal to a certain value, while the ________ hypothesis states that the parameter differs from this value. *null, alternate*

8. Rejecting H_0 when it is true is called a ________ error, and failing to reject H_0 when it is false is called a ________ error. *Type I, Type II*

In Exercises 9–12, determine whether the statement is true or false. If the statement is false, rewrite it as a true statement.

9. H_1: $\mu > 50$ is an example of a left-tailed alternate hypothesis. *False*

10. If we reject H_0, we conclude that H_0 is false. *True*

11. If we do not reject H_0, then we conclude that H_1 is false. *False*

12. If we do not reject H_0, we conclude that H_0 is true. *False*

Practicing the Skills

In Exercises 13–16, determine whether the alternate hypothesis is left-tailed, right-tailed, or two-tailed.

13. H_0: $\mu = 5$ H_1: $\mu < 5$ *Left-tailed*

14. H_0: $\mu = 10$ H_1: $\mu > 10$ *Right-tailed*

15. H_0: $\mu = 1$ H_1: $\mu \neq 1$ *Two-tailed*

16. H_0: $\mu = 26$ H_1: $\mu \neq 26$ *Two-tailed*

In Exercises 17–20, determine whether the outcome is a Type I error, a Type II error, or a correct decision.

17. A test is made of H_0: $\mu = 20$ versus H_1: $\mu \neq 20$. The true value of μ is 25, and H_0 is rejected. *Correct decision*

18. A test is made of H_0: $\mu = 5$ versus H_1: $\mu < 5$. The true value of μ is 5, and H_0 is rejected. *Type I error*

19. A test is made of H_0: $\mu = 63$ versus H_1: $\mu > 63$. The true value of μ is 75, and H_0 is not rejected. *Type II error*

20. A test is made of H_0: $\mu = 45$ versus H_1: $\mu < 45$. The true value of μ is 40, and H_0 is rejected. *Correct decision*

21. A medical research team wants to know whether more than 75% of people who take a certain allergy medicine experience relief. Let p be the proportion who experience relief. A test will be made of H_0: $p = 0.75$ versus H_1: $p > 0.75$. Let $\hat{p}$ be the sample proportion of those who experience relief. Which value of $\hat{p}$ provides stronger evidence against H_0, $\hat{p} = 0.8$ or $\hat{p} = 0.9$? *0.9*

22. A machine that fills cereal boxes is supposed to produce a mean fill weight of 12 ounces. Let μ denote the true mean fill weight. A test is made of H_0: $\mu = 12$ versus H_1: $\mu \neq 12$. Let $\bar{x}$ be the mean weight of a sample of filled boxes. Which value of $\bar{x}$ provides greater evidence against H_0, $\bar{x} = 11$ or $\bar{x} = 14$? *14*

Working with the Concepts

23. **Fertilizer:** A new type of fertilizer is being tested on a plot of land in an orange grove, to see whether it increases the amount of fruit produced. The mean number of pounds of fruit on this plot of land with the old fertilizer was 400 pounds. Agriculture scientists believe that the new fertilizer may increase the yield. State the appropriate null and alternate hypotheses. *H_0: $\mu = 400$, H_1: $\mu > 400$*

24. **Big fish:** A sample of 100 flounder of a certain species have sample mean weight 21.5 grams. Scientists want to perform a hypothesis test to determine how strong the evidence is that the mean weight differs from 20 grams. State the appropriate null and alternate hypotheses. *H_0: $\mu = 20$, H_1: $\mu \neq 20$*

25. **Check, please:** A restaurant owner claims that the mean amount spent by diners at his restaurant is more than \$30. A test is made of H_0: $\mu = 30$ versus H_1: $\mu > 30$. The null hypothesis is rejected. State an appropriate conclusion.

26. **Coffee:** The mean caffeine content per cup of regular coffee served at a certain coffee shop is supposed to be 100 milligrams.

A test is made of H_0: $\mu = 100$ versus H_1: $\mu \neq 100$. The null hypothesis is rejected. State an appropriate conclusion.

27. Big dogs: A veterinarian claims that the mean weight of adult German shepherd dogs is 75 pounds. A test is made of H_0: $\mu = 75$ versus H_1: $\mu \neq 75$. The null hypothesis is not rejected. State an appropriate conclusion.

28. Business trips: A sales manager believes that the mean number of days per year her company's sales representatives spend traveling is less than 50. A test is made of H_0: $\mu = 50$ versus H_1: $\mu < 50$. The null hypothesis is not rejected. State an appropriate conclusion.

29. Which type of error? Batteries used in a certain heart pacemaker have a mean life of 7 years. A new type of battery is being tested and will be used in place of the old battery if it can be shown to have a mean lifetime of more than 7 years. A test is made of H_0: $\mu = 7$ versus H_1: $\mu > 7$. There are two possible errors:
i. The true mean is 7 or less, but the new batteries are used.
ii. The true mean is greater than 7, but the new batteries are not used.
a. Which is a Type I error? *ii*
b. Which is a Type II error? *i*

30. Which type of error? Water pipe to be used in a certain application is required to have a mean breaking strength of more than 2000 pounds per foot. A test is made of H_0: $\mu = 2000$ versus H_1: $\mu > 2000$. There are two possible errors:
i. The true mean is greater than 2000, but the pipe is not used.
ii. The true mean is 2000 or less, but the pipe is used.
a. Which is a Type I error? *i*
b. Which is a Type II error? *ii*

31. Type I error: A company that manufactures steel wires guarantees that the mean breaking strength (in kilonewtons) of the wires is greater than 50. They measure the strengths for a sample of wires and test H_0: $\mu = 50$ versus H_1: $\mu > 50$.
a. If a Type I error is made, what conclusion will be drawn regarding the mean breaking strength?
b. If a Type II error is made, what conclusion will be drawn regarding the mean breaking strength?
c. This test uses a one-tailed alternate hypothesis. Explain why a one-tailed hypothesis is more appropriate than a two-tailed hypothesis in this situation.

32. Type I error: Washers used in a certain application are supposed to have a thickness of 2 millimeters. A quality control engineer measures the thicknesses for a sample of washers and tests H_0: $\mu = 2$ versus H_1: $\mu \neq 2$.
a. If a Type I error is made, what conclusion will be drawn regarding the mean washer thickness?
b. If a Type II error is made, what conclusion will be drawn regarding the mean washer thickness?
c. This test uses a two-tailed alternate hypothesis. Explain why a two-tailed hypothesis is more appropriate than a one-tailed hypothesis in this situation.

33. Scales: It is desired to check the calibration of a scale by weighing a standard 10-gram weight 100 times. Let μ be the population mean reading on the scale, so that the scale is in calibration if $\mu = 10$ and out of calibration if $\mu \neq 10$. A test is made of the hypotheses H_0: $\mu = 10$ versus H_1: $\mu \neq 10$. Consider three possible conclusions: (i) The scale is in calibration. (ii) The scale is not in calibration. (iii) The scale might be in calibration.
a. Which of the three conclusions is best if H_0 is rejected? *(ii)*
b. Which of the three conclusions is best if H_0 is not rejected? *(iii)*
c. Assume that the scale is in calibration, but the conclusion is reached that the scale is not in calibration. Which type of error is this? *Type I*
d. Assume that the scale is not in calibration. Is it possible to make a Type I error? Explain. *No*
e. Assume that the scale is not in calibration. Is it possible to make a Type II error? Explain. *Yes*

34. IQ: Scores on a certain IQ test are known to have a mean of 100. A random sample of 60 students attend a series of coaching classes before taking the test. Let μ be the population mean IQ score that would occur if every student took the coaching classes. The classes are successful if $\mu > 100$. A test is made of the hypotheses H_0: $\mu = 100$ versus H_1: $\mu > 100$. Consider three possible conclusions: (i) The classes are successful. (ii) The classes are not successful. (iii) The classes might not be successful.
a. Which of the three conclusions is best if H_0 is rejected? *(i)*
b. Which of the three conclusions is best if H_0 is not rejected? *(iii)*
c. Assume that the classes are successful but the conclusion is reached that the classes might not be successful. Which type of error is this? *Type II*
d. Assume that the classes are not successful. Is it possible to make a Type I error? Explain. *Yes*
e. Assume that the classes are not successful. Is it possible to make a Type II error? Explain. *No*

35. Probability of error: A coin has probability p of landing heads when tossed. A test will be made of the hypotheses H_0: $p = 0.1$ versus H_1: $p > 0.1$, as follows. The coin will be tossed once. If it comes up heads, H_0 will be rejected. If it comes up tails, H_0 will not be rejected.
a. If the true value of p is 0.1, what is the probability that the test results in a Type I error? *0.1*
b. If the true value of p is 0.4, what is the probability that the test results in a Type II error? *0.6*

Answers to Check Your Understanding Exercises for Section 9.1

1. H_0: $\mu = 35$, H_1: $\mu > 35$
2. H_0: $\mu = 6.6$, H_1: $\mu \neq 6.6$
3. H_0: $\mu = 25.5$, H_1: $\mu < 25.5$
4. Correct decision
5. Type II error
6. Type I error

Section 9.2 Hypothesis Tests for a Population Mean, Standard Deviation Known

Objectives

1. Perform hypothesis tests with the critical value method
2. Perform hypothesis tests with the P-value method

NOTE TO INSTRUCTOR

Section 9.2 covers hypothesis tests for a mean with standard deviation known, which is the situation where the computations are simplest. Concepts such as the relationship between hypothesis tests and confidence intervals, and the difference between statistical and practical significance, are covered in Section 9.3, so as to present them in a setting more commonly encountered in practice.

Does coaching improve SAT scores? The College Board reported that the mean math SAT score in a recent year was 515, with a standard deviation of 116. Results of an earlier study (*Preparing for the SAT—An Update*, College Board Report 98–5) suggest that coached students should have a mean SAT score of approximately 530. A teacher who runs an online coaching program thinks that students coached by his method have a higher mean score than this. We will see how to perform a hypothesis test to determine whether the teacher is right.

There are two ways to perform hypothesis tests; both methods produce the same results. The first one we will discuss is called the **critical value method**. Then we will discuss the second method, known as the *P-value method*.

Objective 1 Perform hypothesis tests with the critical value method

The Critical Value Method

In the SAT example, the teacher believes that the mean score for his students is greater than 530. Therefore, the null hypothesis says that the mean μ is equal to 530, and the alternate hypothesis says that μ is greater than 530. In symbols,

$$H_0: \mu = 530 \qquad H_1: \mu > 530$$

Image100 Ltd

Now assume that the teacher draws a random sample of 100 students who are planning to take the SAT and enrolls them in the online coaching program. After completing the program, their sample mean SAT score is $\bar{x} = 562$. This is higher than 530. Can he reject H_0 and conclude that the mean SAT math score for his students is greater than 530?

The sample mean, $\bar{x} = 562$, differs somewhat from the null hypothesis value for the population mean, $\mu = 530$. The key idea behind a hypothesis test is to measure how large this difference is. If the sample mean differs from H_0 only slightly, then H_0 may well be true, because slight differences can easily be due to chance. However, if the difference is larger, it is less likely to be due to chance, and H_0 is less likely to be true.

RECALL

The z-score tells us how many standard deviations $\bar{x}$ is from μ.

We must now determine how strong the disagreement is between the sample mean $\bar{x} = 562$ and the null hypothesis $\mu = 530$. We do this by calculating the value of a **test statistic**. In this example, the test statistic is the z-score of the sample mean $\bar{x}$. We now show how to compute the z-score.

RECALL

When the sample size is large ($n > 30$), the sample mean $\bar{x}$ is approximately normally distributed with mean μ and standard deviation $\sigma/\sqrt{n}$.

Recall that in a hypothesis test, we begin by assuming that H_0 is true. We therefore assume that the mean of $\bar{x}$ is $\mu = 530$. Because the sample size is large ($n = 100$), we know that $\bar{x}$ is approximately normally distributed. Suppose the population standard deviation is known to be $\sigma = 116$. The standard deviation of $\bar{x}$ is

$$\frac{\sigma}{\sqrt{n}} = \frac{116}{\sqrt{100}} = 11.6$$

The z-score for $\bar{x}$ is

$$z = \frac{\bar{x} - \mu}{\sigma/\sqrt{n}} = \frac{562 - 530}{116/\sqrt{100}} = 2.76$$

We have found that the value of the test statistic is $z = 2.76$. Does this present strong evidence against H_0? Figure 9.3 (page 440) presents the distribution of the sample mean under the assumption that H_0 is true. The value $\bar{x} = 562$ that we observed has a z-score of 2.76, which means that our observed mean is 2.76 standard deviations away from the assumed mean of 530. Visually, we can see from Figure 9.3 that our observed value $\bar{x} = 562$ is pretty far out in the tail of the distribution—far from the null hypothesis value $\mu = 530$. Intuitively, therefore, it appears that the evidence against H_0 is fairly strong.

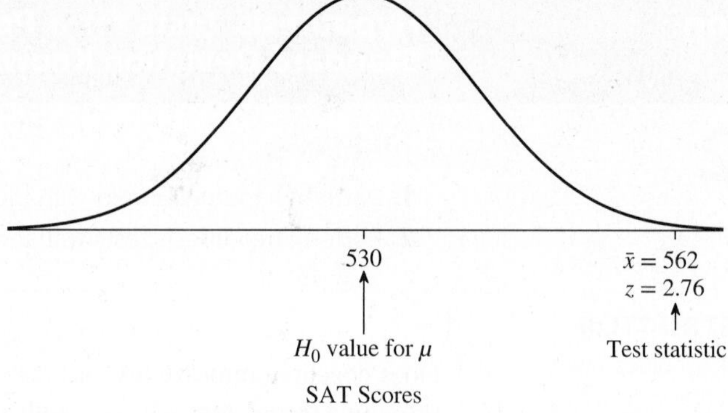

Figure 9.3 If H_0 is true, then the value of $\bar{x}$ is in the tail of the distribution and far from the null hypothesis mean $\mu = 530$. Visually, it appears that the evidence against H_0 is fairly strong.

The critical value method is based on the idea that we should reject H_0 if the value of the test statistic is unusual when we assume H_0 to be true. In this method, we choose a **critical value**, which forms a boundary between values that are considered unusual and values that are not. The region that contains the unusual values is called the **critical region**. If the value of the test statistic is in the critical region, we reject H_0.

The critical value we choose depends on how small we believe a probability should be for an event to be considered unusual. Let's say that an event with a probability of 0.05 or less is unusual. Figure 9.4 illustrates a critical value of 1.645 and a critical region consisting of z-scores greater than or equal to 1.645. The probability that a z-score is in the critical region is 0.05, so the critical region contains the z-scores that are considered unusual. We have observed a z-score of 2.76, which is in the critical region. Therefore, we reject H_0. We conclude that the mean SAT math score for students completing the online coaching program is greater than 530.

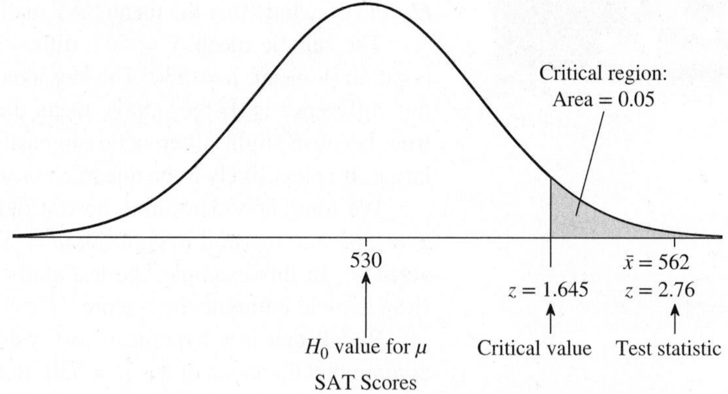

Figure 9.4 The critical value is 1.645. The critical region contains all z-scores greater than or equal to 1.645. The value of our test statistic is $z = 2.76$. This value is in the critical region, so we reject H_0.

The probability that we use to determine whether an event is unusual is called the **significance level** of the test and is denoted with the letter α. In Figure 9.4, we used $\alpha = 0.05$. This is the most commonly used value for α, but other values are sometimes used as well. Next to $\alpha = 0.05$, the most commonly used value is $\alpha = 0.01$.

Note that the significance level α is the probability of a Type I error. If H_0 is true, the distribution shown in Figure 9.4 is the true distribution of the sample mean $\bar{x}$. A Type I error occurs whenever $\bar{x}$ is in the critical region. The probability of a Type I error is therefore equal to the area of the critical region, which is the significance level.

The choice of α is determined by how strong we require the evidence against H_0 to be in order to reject it. The smaller the value of α, the stronger we require the evidence to be. For example, if we choose $\alpha = 0.05$, we will reject H_0 if the test statistic is in the most extreme 5% of its distribution. However, if we choose $\alpha = 0.01$, we will not reject H_0 unless the test statistic is in the most extreme 1% of its distribution.

DEFINITION

If we reject H_0 after choosing a significance level α, we say that the result is **statistically significant** at the α level.

We also say that H_0 *is rejected at the* α *level.*

In our SAT example, we rejected H_0 at the $\alpha = 0.05$ level, and the result was statistically significant at the $\alpha = 0.05$ level.

Our alternate hypothesis of $\mu > 530$ was a right-tailed alternative. For this reason, the critical region was in the right tail of the distribution. The location of the critical region depends on whether the alternate hypothesis is left-tailed, right-tailed, or two-tailed.

Critical Values for Hypothesis Tests

Let α denote the chosen significance level. The critical value depends on whether the alternate hypothesis is left-tailed, right-tailed, or two-tailed.

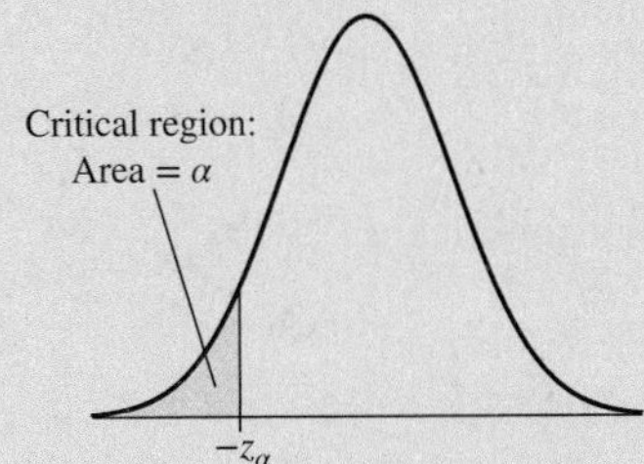

For left-tailed H_1: The critical value is $-z_\alpha$, which has area α to its left. Reject H_0 if $z \le -z_\alpha$.

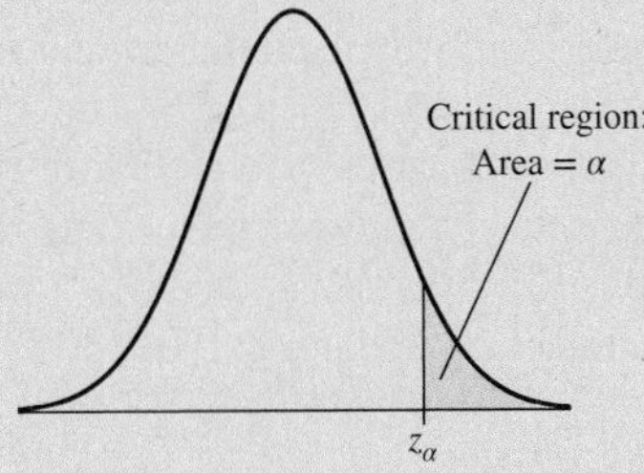

For right-tailed H_1: The critical value is z_α, which has area α to its right. Reject H_0 if $z \ge z_\alpha$.

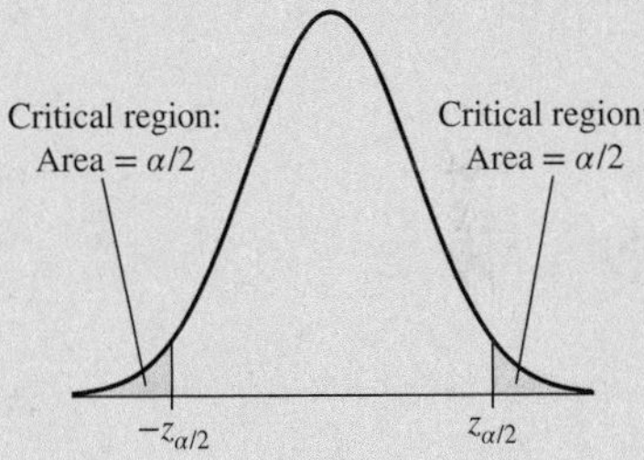

For two-tailed H_1: The critical values are $z_{\alpha/2}$, which has area $\alpha/2$ to its right, and $-z_{\alpha/2}$, which has area $\alpha/2$ to its left. Reject H_0 if $z \ge z_{\alpha/2}$ or $z \le -z_{\alpha/2}$.

Table 9.1 presents critical values for some commonly used significance levels α.

Table 9.1 Table of Critical Values

	Significance Level α			
H_1	0.10	0.05	0.02	0.01
Left-tailed	−1.282	−1.645	−2.054	−2.326
Right-tailed	1.282	1.645	2.054	2.326
Two-tailed	±1.645	±1.96	±2.326	±2.576

Example 9.7

Find the critical region for a right-tailed alternate hypothesis

A test is made of H_0: $\mu = 1$ versus H_1: $\mu > 1$. The value of the test statistic is $z = 1.85$.

a. Is H_0 rejected at the $\alpha = 0.05$ level?

b. Is H_0 rejected at the $\alpha = 0.01$ level?

Solution

The alternate hypothesis is H_1: $\mu > 1$, so this is a right-tailed test.

a. From Table 9.1, we see that the critical value for $\alpha = 0.05$ is $z_\alpha = 1.645$. For a right-tailed test, we reject H_0 if $z \ge z_\alpha$. Because $1.85 > 1.645$, we reject H_0 at the $\alpha = 0.05$ level.

b. The critical value for $\alpha = 0.01$ is 2.326. Because $1.85 < 2.326$, we do not reject H_0 at the $\alpha = 0.01$ level.

Check Your Understanding

1. A test is made of H_0: $\mu = 25$ versus H_1: $\mu < 25$. The value of the test statistic is $z = -1.84$.
   a. Find the critical value and the critical region for a significance level of $\alpha = 0.05$. *-1.645, $z \le -1.645$*
   b. Do you reject H_0 at the $\alpha = 0.05$ level? *Yes*
   c. Find the critical value and the critical region for a significance level of $\alpha = 0.01$. *-2.326, $z \le -2.326$*
   d. Do you reject H_0 at the $\alpha = 0.01$ level? *No*
2. A test is made of H_0: $\mu = 7.5$ versus H_1: $\mu > 7.5$. The value of the test statistic is $z = 2.71$.
   a. Find the critical value and the critical region for a significance level of $\alpha = 0.05$. *1.645, $z \ge 1.645$*
   b. Do you reject H_0 at the $\alpha = 0.05$ level? *Yes*
   c. Find the critical value and the critical region for a significance level of $\alpha = 0.01$. *2.326, $z \ge 2.326$*
   d. Do you reject H_0 at the $\alpha = 0.01$ level? *Yes*
3. A test is made of H_0: $\mu = 12$ versus H_1: $\mu \neq 12$. The value of the test statistic is $z = 1.78$.
   a. Find the critical value and the critical region for a significance level of $\alpha = 0.05$. *$1.96, -1.96$, $z \le -1.96$ or $z \ge 1.96$*
   b. Do you reject H_0 at the $\alpha = 0.05$ level? *No*
   c. Find the critical value and the critical region for a significance level of $\alpha = 0.01$. *$2.576, -2.576$, $z \le -2.576$ or $z \ge 2.576$*
   d. Do you reject H_0 at the $\alpha = 0.01$ level? *No*

Answers are on page 457.

The method we have described requires certain assumptions, which we now state.

Assumptions for Performing a Hypothesis Test About μ When σ Is Known

1. We have a simple random sample.
2. The sample size is large ($n > 30$), or the population is approximately normal.

When these assumptions are met, a hypothesis test can be performed using the following steps.

Performing a Hypothesis Test for a Population Mean with σ Known Using the Critical Value Method

Check to be sure the assumptions are satisfied. If they are, then proceed with the following steps.

Step 1: State the null and alternate hypotheses. The null hypothesis specifies a value for the population mean μ. We will call this value μ_0. So the null hypothesis is of the form H_0: $\mu = \mu_0$. The alternate hypothesis can be stated in one of three ways:

Left-tailed: H_1: $\mu < \mu_0$
Right-tailed: H_1: $\mu > \mu_0$
Two-tailed: H_1: $\mu \neq \mu_0$

Step 2: Choose a significance level α, and find the critical value or values.

Step 3: Compute the test statistic $z = \dfrac{\bar{x} - \mu_0}{\sigma/\sqrt{n}}$.

Step 4: Determine whether to reject H_0, as follows:

Left-tailed: H_1: $\mu < \mu_0$ Reject if $z \le -z_\alpha$.
Right-tailed: H_1: $\mu > \mu_0$ Reject if $z \ge z_\alpha$.
Two-tailed: H_1: $\mu \neq \mu_0$ Reject if $z \ge z_{\alpha/2}$ or $z \le -z_{\alpha/2}$.

Step 5: State a conclusion.

Example 9.8

Performing a hypothesis test with the critical value method

The American Automobile Association reported that the mean price of a gallon of regular grade gasoline in the city of Los Angeles in July 2019 was \$4.07. A recently taken simple random sample of 50 gas stations in Los Angeles had an average price of \$4.02 for a gallon of regular grade gasoline. Assume that the standard deviation of prices is \$0.15. An economist is interested in determining whether the mean price for all Los Angeles gas stations is less than \$4.07. Use the critical value method to perform a hypothesis test at the $\alpha = 0.05$ level of significance.

Solution

We first check the assumptions. We have a simple random sample, the sample size is large ($n > 30$), and the population standard deviation σ is known. The assumptions are satisfied.

Step 1: State H_0 and H_1. The null hypothesis says that the mean price is \$4.07. Therefore, we have

$$H_0: \mu = 4.07$$

We are interested in knowing whether the mean price is less than \$4.07. Therefore, the alternate hypothesis is

$$H_1: \mu < 4.07$$

At this point, we assume H_0 to be true.

Step 2: Choose a significance level and find the critical value. The significance level is $\alpha = 0.05$. Since the alternate hypothesis is $\mu < 4.07$, this is a left-tailed test. The critical value corresponding to $\alpha = 0.05$ is -1.645.

Step 3: Compute the test statistic. The test statistic is the z-score of the sample mean $\bar{x}$. The population standard deviation is $\sigma = 0.15$. Since we assume H_0 to be true, the population mean is $\mu_0 = 4.07$. The sample size is $n = 50$. Therefore, the test statistic is

$$z = \frac{\bar{x} - \mu_0}{\sigma/\sqrt{n}} = \frac{4.02 - 4.07}{0.15/\sqrt{50}} = -2.36$$

Step 4: Determine whether to reject H_0. This is a left-tailed test, so we reject H_0 if $z < -1.645$. Since $-2.36 < -1.645$, we reject H_0 at the $\alpha = 0.05$ level. See Figure 9.5.

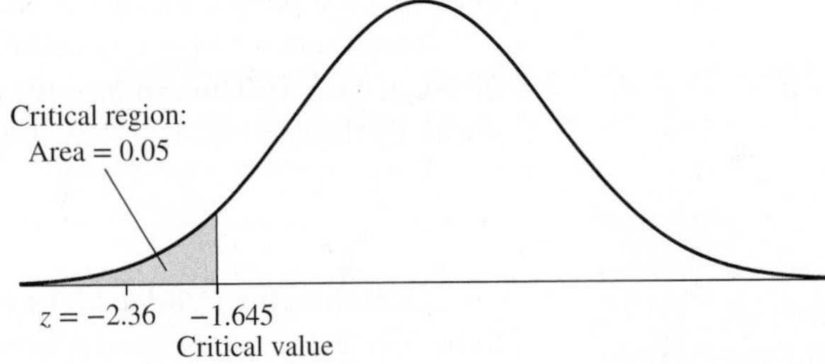

Figure 9.5 The value of the test statistic, $z = -2.36$, is in the level $\alpha = 0.05$ critical region. Therefore, we reject H_0 at the $\alpha = 0.05$ level.

Step 5: State a conclusion. We conclude that the mean price of a gallon of regular gasoline in Los Angeles is less than \$4.07.

Check Your Understanding

4. A test is made of H_0: $\mu = 15$ versus H_1: $\mu > 15$. The sample mean is $\bar{x} = 16.5$, the sample size is $n = 50$, and the population standard deviation is $\sigma = 5$.

a. Find the value of the test statistic z. *2.12*

b. Find the critical region for a level $\alpha = 0.05$ test. *$z \geq 1.645$*

c. Do you reject H_0 at the $\alpha = 0.05$ level? *Yes*

5. A test is made of H_0: $\mu = 125$ versus H_1: $\mu < 125$. The sample mean is $\bar{x} = 123$, the sample size is $n = 100$, and the population standard deviation is $\sigma = 20$.
a. Find the value of the test statistic z. *−1.00*
b. Find the critical region for a level $\alpha = 0.02$ test. $z \le -2.054$
c. Do you reject H_0 at the $\alpha = 0.02$ level? *No*

6. A test is made of H_0: $\mu = 100$ versus H_1: $\mu \ne 100$. The sample mean is $\bar{x} = 97$, the sample size is $n = 75$, and the population standard deviation is $\sigma = 8$.
a. Find the value of the test statistic z. *−3.25*
b. Find the critical region for a level $\alpha = 0.01$ test. $z \le -2.576$ or $z \ge 2.576$
c. Do you reject H_0 at the $\alpha = 0.01$ level? *Yes*

Answers are on page 457.

With the critical value method, the value of the test statistic is considered to be unusual if it is in the critical region and not unusual if it is not in the critical region. We will now describe the ***P*-value method**, which provides more information than the critical value method. Whereas the critical value method tells us only whether the test statistic was unusual or not, the *P*-value method tells us exactly how unusual the test statistic is. For this reason, the *P*-value method is the one more often used in practice. In particular, almost all forms of technology use the *P*-value method.

Objective 2 Perform hypothesis tests with the *P*-value method

The *P*-Value Method

We will introduce the *P*-value method with our SAT example. An online coaching program is supposed to increase the mean SAT math score to a value greater than 530. The null and alternate hypotheses are

$$H_0\colon \mu = 530 \qquad H_1\colon \mu > 530$$

Now assume that 100 students are randomly chosen to participate in the program, and their sample mean score is $\bar{x} = 562$. Suppose that the population standard deviation for SAT math scores is known to be $\sigma = 116$. Does this provide strong evidence against the null hypothesis $\mu = 530$?

To measure just how strong the evidence against H_0 is, we compute a quantity called the ***P*-value**. The *P*-value is the probability that a number drawn from the distribution of the sample mean would be as extreme as or more extreme than our observed value of 562. The more extreme the value, the stronger is the evidence against H_0. Because our alternate hypothesis is H_1: $\mu > 530$, this is a right-tailed test, so values of $\bar{x}$ greater than 562 are more extreme than our observed value is. We find the *P*-value by computing the z-score of our observed sample mean $\bar{x} = 562$. We now explain how to do this.

Recall that we begin by assuming that H_0 is true. We therefore assume that the mean of $\bar{x}$ is $\mu = 530$. The sample size is large ($n = 100$), so we know that $\bar{x}$ is approximately normally distributed. The standard deviation of $\bar{x}$ is

$$\frac{\sigma}{\sqrt{n}} = \frac{116}{\sqrt{100}} = 11.6$$

RECALL

When the sample size is large ($n > 30$), the sample mean $\bar{x}$ is approximately normally distributed with mean μ and standard deviation $\sigma/\sqrt{n}$.

Therefore, the *P*-value is the probability that $\bar{x}$ is greater than 562 when μ is assumed to be 530 and the standard deviation is 11.6.

The z-score for $\bar{x}$ is

$$z = \frac{\bar{x} - \mu}{\sigma/\sqrt{n}} = \frac{562 - 530}{116/\sqrt{100}} = 2.76$$

The *P*-value is therefore the area under the normal curve to the right of $z = 2.76$. Using Table A.2, we see that the area to the *left* of $z = 2.76$ is 0.9971. Therefore, the area to the right of $z = 2.76$ is $1 - 0.9971 = 0.0029$ (see Figure 9.6 on page 445). Therefore, the *P*-value for this test is 0.0029.

This *P*-value tells us that if H_0 were true, the probability of observing a value of $\bar{x}$ as large as 562 is only 0.0029. Therefore, there are only two possibilities:

- H_0 is false.
- H_0 is true, and we got an unusual sample, whose mean lies in the most extreme 0.0029 of its distribution.

EXPLAIN IT AGAIN

Using technology: The *P*-value is the area to the right of $\bar{x} = 562$ when the mean is 530 and the standard deviation is 11.6. This area can be found with technology. The following display illustrates the **normalcdf** command on the TI-84 Plus calculator.

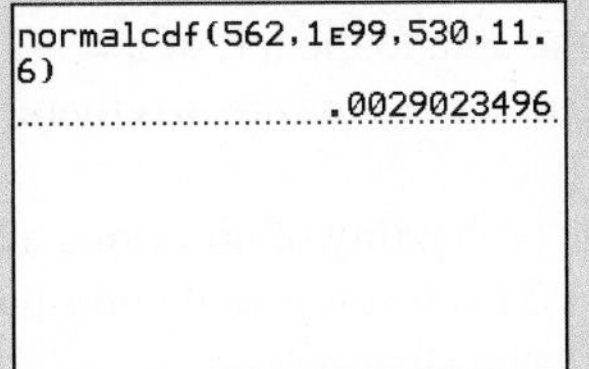

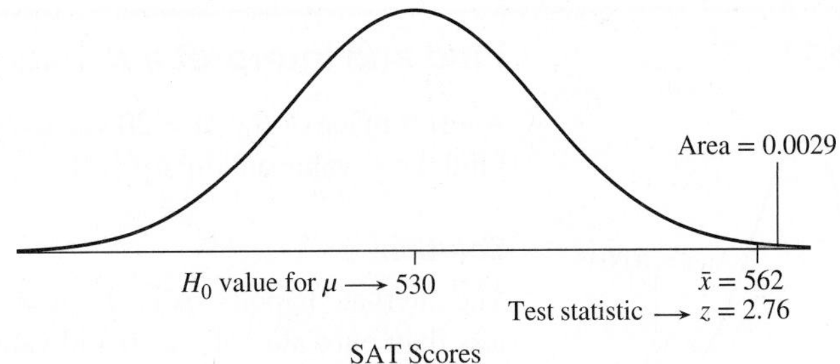

Figure 9.6 If H_0 is true, the probability that $\bar{x}$ takes on a value as extreme as or more extreme than the observed value of 562 is 0.0029. This is the *P*-value.

In practice, events in the most extreme 0.0029 of their distributions are very unusual. This means that a *P*-value as small as 0.0029 is very unlikely to occur if H_0 is true. A *P*-value of 0.0029 is very strong evidence against H_0.

SUMMARY

- The *P*-value is the probability, assuming that H_0 is true, of observing a value for the test statistic that is as extreme as or more extreme than the value actually observed.
- The smaller the *P*-value, the stronger the evidence against H_0.

Example 9.9

Find and interpret a *P*-value

A test is made of H_0: $\mu = 10$ versus H_1: $\mu > 10$. The value of the test statistic is $z = 2.25$. Find the *P*-value and interpret it.

Solution

The alternate hypothesis is H_1: $\mu > 10$, so this is a right-tailed test. Therefore, values of z greater than our observed value of 2.25 are more extreme than our value is. The *P*-value is the area under the normal curve to the right of the test statistic $z = 2.25$. Using Table A.2, we see that the area to the *left* of $z = 2.25$ is 0.9878. Therefore, the area to the right of $z = 2.25$ is $1 - 0.9878 = 0.0122$. See Figure 9.7.

The *P*-value of 0.0122 tells us that if H_0 is true, then the probability of observing a test statistic of 2.25 or more is only 0.0122. This result is fairly unusual if we assume H_0 to be true. Therefore, this is fairly strong evidence against H_0.

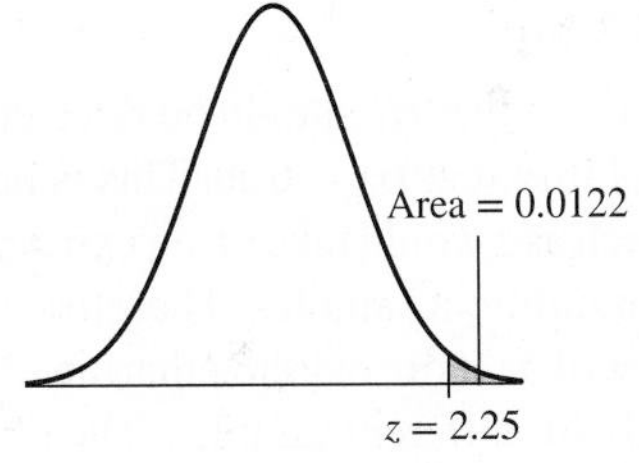

Figure 9.7

Example 9.10

Find and interpret a *P*-value

A test is made of H_0: $\mu = 5$ versus H_1: $\mu < 5$. The value of the test statistic is $z = -0.63$. Find the *P*-value and interpret it.

Solution

The alternate hypothesis is H_1: $\mu < 5$, so this is a left-tailed test. Therefore, values of z less than our value of -0.63 are more extreme than our value is. The *P*-value is the area under the normal curve to the left of the test statistic $z = -0.63$. Using Table A.2, we see that the area to the left of $z = -0.63$ is 0.2643. See Figure 9.8.

The *P*-value of 0.2643 tells us that if H_0 is true, then the probability of observing a test statistic of -0.63 or less is 0.2643. This is not particularly unusual, so this is not strong evidence against H_0.

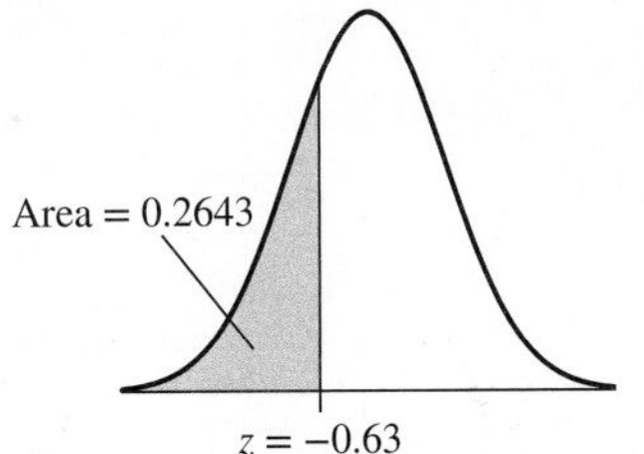

Figure 9.8

Example 9.11

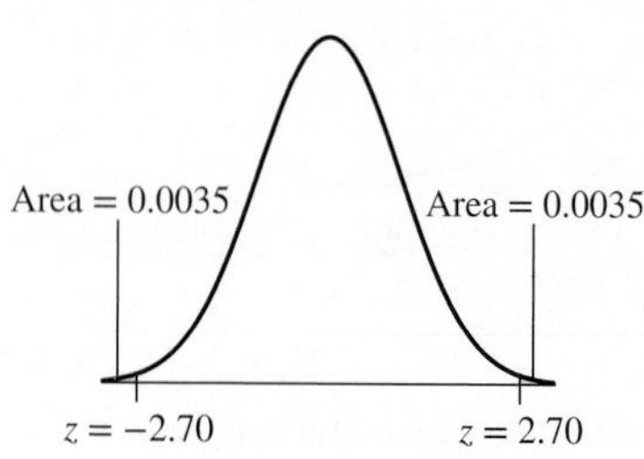

Figure 9.9

Find and interpret a *P*-value

A test is made of H_0: $\mu = 20$ versus H_1: $\mu \neq 20$. The value of the test statistic is $z = -2.70$. Find the *P*-value and interpret it.

Solution

The alternate hypothesis is H_1: $\mu \neq 20$, so this is a two-tailed test. Therefore, values of z less than our value of -2.70 and values greater than 2.70 are both more extreme than our value is. The *P*-value is the sum of the areas under the normal curve to the right of $z = 2.70$ and to the left of $z = -2.70$. Using Table A.2, we see that the area to the left of $z = -2.70$ is 0.0035. The area to the right of $z = 2.70$ is also 0.0035. The sum of the areas is therefore $0.0035 + 0.0035 = 0.0070$. See Figure 9.9.

The *P*-value of 0.0070 tells us that if H_0 is true, then the probability of observing a test statistic greater than 2.70 or less than -2.70 is only 0.0070. This result is quite unusual if we assume H_0 to be true. Therefore, this is very strong evidence against H_0.

Check Your Understanding

7. Which provides stronger evidence against H_0: a *P*-value of 0.05 or a *P*-value of 0.50? *0.05*

8. A test is made of H_0: $\mu = 30$ versus H_1: $\mu < 30$. The test statistic is $z = -1.28$. Find and interpret the *P*-value. $P = 0.1003$

9. A test is made of H_0: $\mu = 6$ versus H_1: $\mu \neq 6$.

a. The test statistic is $z = 0.75$. Find and interpret the *P*-value. $P = 0.4532$

b. The test statistic is $z = -2.20$. Find and interpret the *P*-value. $P = 0.0278$

c. Which provides stronger evidence against H_0: $z = 0.75$ or $z = -2.20$? $z = -2.20$

Answers are on page 457.

The *P*-value is not the probability that H_0 is true

Because the *P*-value is a probability and small *P*-values indicate that H_0 should be rejected, it is tempting to think that the *P*-value represents the probability that H_0 is true. This is not the case. The *P*-value is the probability that a test statistic such as z would take on an extreme value. Probability is used for events that can be different for different samples. Therefore, it makes sense to talk about the probability that the value of z will be more extreme than an observed value, because the value of z can come out differently for different samples. The null hypothesis, however, is either true or not true. The truth of H_0 does not change from sample to sample. For this reason, it does not make sense to talk about the probability that H_0 is true.

SUMMARY

The *P*-value is the probability, under the assumption that H_0 is true, that the test statistic takes on a value as extreme as or more extreme than the value actually observed.

The *P*-value is not the probability that the null hypothesis is true.

Check Your Understanding

10. If $P = 0.02$, which is the best conclusion? *ii*

i. The probability that H_0 is true is 0.02.

ii. If H_0 is true, the probability of obtaining a test statistic more extreme than the one actually observed is 0.02.

iii. The probability that H_1 is true is 0.02.

iv. If H_1 is true, the probability of obtaining a test statistic more extreme than the one actually observed is 0.02.

Answer is on page 458.

Choosing a significance level

We have seen that the smaller the P-value, the stronger the evidence against H_0. In practice, people often do not choose a significance level. They simply report the P-value and let the reader decide whether the evidence is strong enough to reject H_0. Sometimes, however, we need to make a firm decision whether to reject H_0. We then choose a significance level α between 0 and 1 before performing the test and reject H_0 if the P-value is less than or equal to α. The most commonly used value is $\alpha = 0.05$, but other values are sometimes used as well. Next to $\alpha = 0.05$, the most commonly used value is $\alpha = 0.01$.

SUMMARY

To make a decision whether to reject H_0 when using the P-value method:

- Choose a significance level α between 0 and 1.
- Compute the P-value.
- If $P \leq \alpha$, reject H_0. If $P > \alpha$, do not reject H_0.

If $P \leq \alpha$, we say that H_0 is rejected at the α level, or that the result is statistically significant at the α level.

Example 9.12 Find the *P*-value

In Example 9.9, the P-value was $P = 0.0122$.

a. Do you reject H_0 at the $\alpha = 0.05$ level?
b. Do you reject H_0 at the $\alpha = 0.01$ level?
c. Is the result statistically significant at the $\alpha = 0.05$ level?
d. Is the result statistically significant at the $\alpha = 0.01$ level?

Solution

a. Because $P \leq 0.05$, we reject H_0 at the $\alpha = 0.05$ level.
b. Because $P > 0.01$, we do not reject H_0 at the $\alpha = 0.01$ level.
c. We reject H_0 at the $\alpha = 0.05$ level, so the result is statistically significant at the $\alpha = 0.05$ level.
d. We do not reject H_0 at the $\alpha = 0.01$ level, so the result is not statistically significant at the $\alpha = 0.01$ level.

Check Your Understanding

11. Ellie and Brenna each plan to perform a hypothesis test, using the same sample. Ellie will use a significance level of $\alpha = 0.05$. Brenna wants to require the evidence to be stronger in order to reject the null hypothesis. Should she use a significance level that is greater than or less than Ellie's? *Less than*

12. Ellie and Brenna each plan to perform a hypothesis test, using the same sample. Ellie will use a significance level of $\alpha = 0.01$, and Brenna will use a significance level of $\alpha = 0.05$. Which of them requires the evidence to be stronger in order to reject the null hypothesis? *Ellie*

13. A hypothesis test is performed with a significance level of $\alpha = 0.05$.
a. If the P-value is 0.08, is H_0 rejected? *No*
b. If the P-value is 0.08, are the results statistically significant at the 0.05 level? *No*
c. If the P-value is 0.03, is H_0 rejected? *Yes*
d. If the P-value is 0.03, are the results statistically significant at the 0.05 level? *Yes*

14. For each of the following P-values, state whether H_0 will be rejected at the 0.10 level.

a. $P = 0.12$ *No*
b. $P = 0.07$ *Yes*
c. $P = 0.05$ *Yes*
d. $P = 0.20$ *No*

15. For each of the following P-values, state whether the result is statistically significant at the 0.10 level.

a. $P = 0.08$ *Yes*
b. $P = 0.15$ *No*
c. $P = 0.01$ *Yes*
d. $P = 0.50$ *No*

Answers are on page 458.

The assumptions for using the P-value method are the same as for the critical value method. We repeat these assumptions here.

Assumptions for Performing a Hypothesis Test About μ When σ Is Known

1. We have a simple random sample.
2. The sample size is large ($n > 30$), or the population is approximately normal.

We now summarize the steps in testing a hypothesis with the P-value method.

Performing a Hypothesis Test for a Population Mean with σ Known Using the *P*-Value Method

Check to be sure the assumptions are satisfied. If they are, then proceed with the following steps.

Step 1: State the null and alternate hypotheses. The null hypothesis specifies a value for the population mean μ. We will call this value μ_0. So the null hypothesis is of the form H_0: $\mu = \mu_0$. The alternate hypothesis can be stated in one of three ways:

Left-tailed: H_1: $\mu < \mu_0$
Right-tailed: H_1: $\mu > \mu_0$
Two-tailed: H_1: $\mu \neq \mu_0$

Step 2: If making a decision, choose a significance level α.

Step 3: Compute the test statistic $z = \dfrac{\bar{x} - \mu_0}{\sigma/\sqrt{n}}$.

Step 4: Compute the P-value of the test statistic. The P-value is the probability, assuming that H_0 is true, of observing a value for the test statistic that is as extreme or more extreme than the value actually observed. The P-value is an area under the standard normal curve; it depends on the type of alternate hypothesis. Note that the inequality in the alternate hypothesis points in the direction of the tail that contains the area for the P-value.

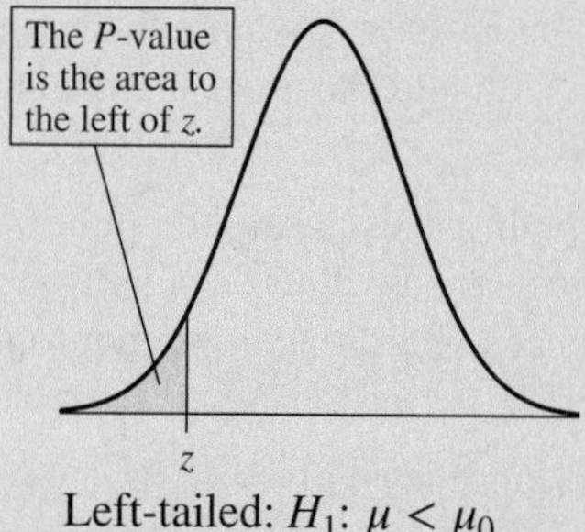
The *P*-value is the area to the left of *z*.
z
Left-tailed: H_1: $\mu < \mu_0$

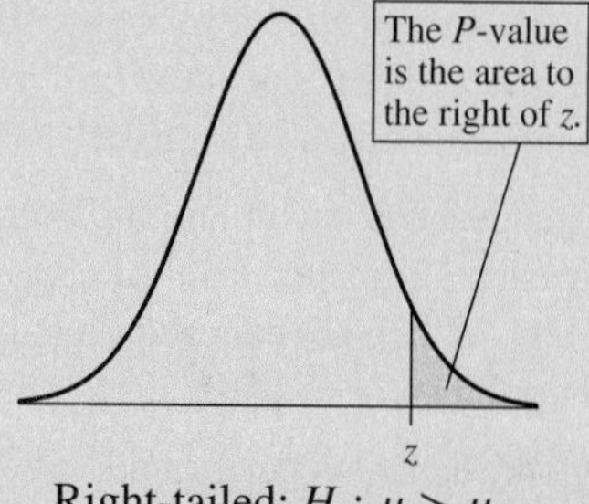
The *P*-value is the area to the right of *z*.
z
Right-tailed: H_1: $\mu > \mu_0$

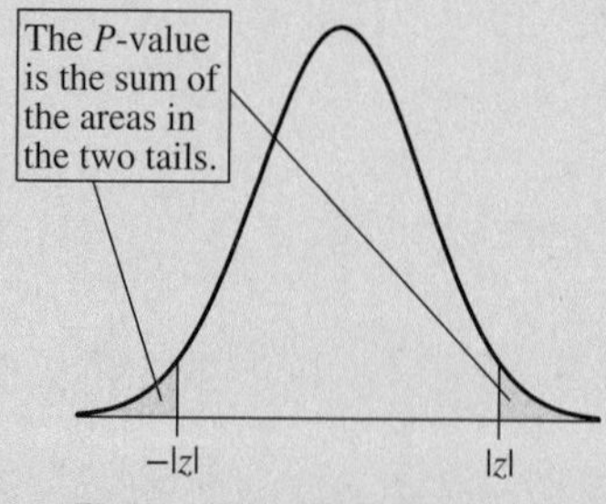
The *P*-value is the sum of the areas in the two tails.
−|z|
|z|
Two-tailed: H_1: $\mu \neq \mu_0$

Step 5: Interpret the P-value. If making a decision, reject H_0 if the P-value is less than or equal to the significance level α.

Step 6: State a conclusion.

Example 9.13

Perform a hypothesis test

The National Health and Nutrition Examination Surveys (NHANES) are designed to assess the health and nutritional status of adults and children in the United States. According to a recent NHANES survey, the mean height of adult men in the United States is 69.7 inches, with a standard deviation of 3 inches. A sociologist believes that taller men may be more likely to be promoted to positions of leadership, so the mean height μ of male business executives may be greater than the mean height of the entire male population. A simple random sample of 100 male business executives has a mean height of 69.9 in. Assume that the standard deviation of male executive heights is $\sigma = 3$ inches. Can we conclude that male business executives are taller, on the average, than the general male population at the $\alpha = 0.05$ level?

Solution

We first check the assumptions. We have a simple random sample, the sample size is large ($n > 30$), and the population standard deviation is known. The assumptions are satisfied.

Step 1: State H_0 and H_1. The null hypothesis, H_0, says that there is no difference between the mean heights of executives and others. Therefore, we have

$$H_0: \mu = 69.7$$

We are interested in determining whether the mean height of executives is greater than 69.7. Therefore, we have

$$H_1: \mu > 69.7$$

At this point, we assume that H_0 is true.

Step 2: Choose a level of significance. The level of significance is $\alpha = 0.05$.

Step 3: Compute the test statistic. Because the sample size is large ($n = 100$), the sample mean $\bar{x}$ is approximately normally distributed. The test statistic is the z-score for $\bar{x}$. To find the z-score, we first need to find the mean and standard deviation of $\bar{x}$. Because we are assuming that H_0 is true, we assume that the mean of $\bar{x}$ is $\mu = 69.7$. We know that the population standard deviation is $\sigma = 3$. The standard deviation of $\bar{x}$ is therefore

$$\frac{\sigma}{\sqrt{n}} = \frac{3}{\sqrt{100}} = 0.3$$

It follows that $\bar{x}$ is normally distributed with mean 69.7 and standard deviation 0.3. We observed a value of $\bar{x} = 69.9$. The z-score is

$$z = \frac{\bar{x} - \mu_0}{\sigma/\sqrt{n}} = \frac{69.9 - 69.7}{0.3} = 0.67$$

Step 4: Compute the *P*-value. Since the alternate hypothesis is $\mu > 69.7$, this is a right-tailed test. The P-value is the area under the curve to the right of $z = 0.67$. Using Table A.2, we see that the area to the *left* of $z = 0.67$ is 0.7486. Therefore, the area to the right of $z = 0.67$ is $1 - 0.7486 = 0.2514$. The P-value is 0.2514. See Figure 9.10.

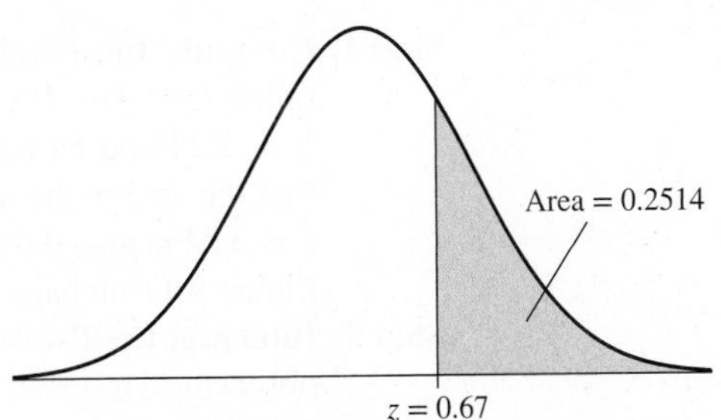

Figure 9.10

EXPLAIN IT AGAIN

Using technology: In Example 9.13, the P-value is the area to the right of $z = 0.67$. This area can be found with technology. The following display illustrates the **normalcdf** command on the TI-84 Plus calculator. We enter the values $\bar{x} = 69.9$, $\mu_0 = 69.7$, and $\sigma/\sqrt{n} = 0.3$. The result given by the calculator differs slightly from the result found by using Table A.2.

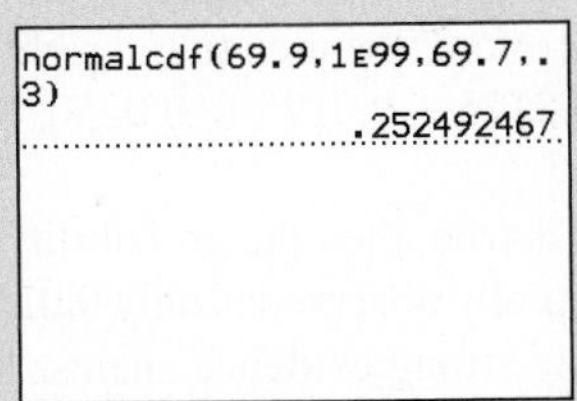

Step 5: Interpret the *P*-value. The *P*-value of 0.2514 says that if H_0 is true, the probability of observing a test statistic as large or larger than 0.67 is 0.2514. This is not unusual; it will happen for about one out of every four samples. Therefore, this sample does not strongly disagree with H_0. In particular, $P > 0.05$, so we do not reject the null hypothesis at the $\alpha = 0.05$ level.

Step 6: State a conclusion. There is not enough evidence to conclude that male executives have a greater mean height than adult males in general. The mean height of male executives may be the same as the mean height of adult males in general.

Example 9.14

Perform a two-tailed hypothesis test

At a large company, the attitudes of workers are regularly measured with a standardized test. The scores on the test range from 0 to 100, with higher scores indicating greater satisfaction with their jobs. The mean score over all of the company's employees was 74, with a standard deviation of $\sigma = 8$. Some time ago, the company adopted a policy of telecommuting. Under this policy, workers could spend one day per week working from home. After the policy had been in place for some time, a random sample of 80 workers was given the test to determine whether their mean level of satisfaction had changed since the policy was put into effect. The sample mean was 76. Assume the standard deviation is still $\sigma = 8$. Can we conclude that the mean level of satisfaction is different since the policy change at the $\alpha = 0.05$ level?

Solution

We first check the assumptions. We have a simple random sample, the sample size is large ($n > 30$), and the population standard deviation is known. The assumptions are satisfied.

Step 1: State H_0 and H_1. The null hypothesis, H_0, says that there is no difference between the mean level of satisfaction before and after telecommuting. Therefore, we have

$$H_0\colon \mu = 74$$

We are interested in knowing whether the mean level has changed. We are not specifically interested in whether it went up or down. Therefore, the alternate hypothesis is

$$H_1\colon \mu \neq 74$$

At this point, we assume that H_0 is true.

Step 2: Choose a level of significance. The level of significance is $\alpha = 0.05$.

Step 3: Compute the test statistic. Since the sample size, $n = 80$, is large, $\bar{x}$ is approximately normally distributed. The test statistic is the z-score for the sample mean $\bar{x}$. The population standard deviation is $\sigma = 8$. Because we assume H_0 to be true, the population mean is $\mu = 74$. Therefore, $\bar{x}$ is normally distributed with mean 74 and standard error

$$\frac{\sigma}{\sqrt{n}} = \frac{8}{\sqrt{80}} = 0.8944$$

We observed a value of $\bar{x} = 76$. The z-score is

$$z = \frac{76 - 74}{0.8944} = 2.24$$

Step 4: Compute the *P*-value. The alternate hypothesis is $\mu \neq 74$, so this is a two-tailed test. The *P*-value is thus the sum of two areas: the area to the right of $z = 2.24$ and an equal area to the left of $z = -2.24$. Using Table A.2, we see that the area to the left of $z = -2.24$ is 0.0125. Therefore, the area to the right of $z = 2.24$ is also 0.0125. The *P*-value is therefore $0.0125 + 0.0125 = 0.0250$. See Figure 9.11 on page 451.

Step 5: Interpret the *P*-value. The *P*-value says that if H_0 is true, then the probability of observing a test statistic as extreme as the one we actually observed is only 0.0250. In practice, this would generally be considered fairly strong evidence against H_0. In particular, $P < 0.05$, so we reject H_0 at the $\alpha = 0.05$ level.

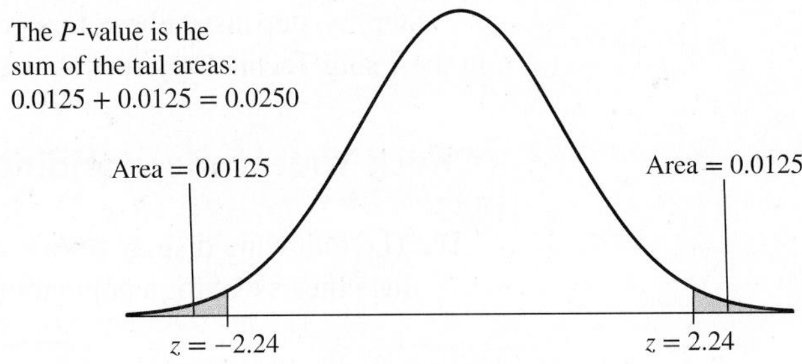
The P-value is the sum of the tail areas: $0.0125 + 0.0125 = 0.0250$
Area = 0.0125
Area = 0.0125
$z = -2.24$
$z = 2.24$

Figure 9.11

Step 6: State a conclusion. We conclude that the mean score among employees has changed since the adoption of telecommuting.

Check Your Understanding

16. A social scientist suspects that the mean number of years of education μ for adults in a certain large city is greater than 12 years. She will test the null hypothesis H_0: $\mu = 12$ against the alternate hypothesis H_1: $\mu > 12$. She surveys a random sample of 100 adults and finds that the sample mean number of years is $\bar{x} = 12.98$. Assume that the standard deviation for the number of years of education is $\sigma = 3$ years.

a. Compute the value of the test statistic. *3.27*

b. Compute the P-value. *0.0005*

c. Interpret the P-value.

d. Is H_0 rejected at the $\alpha = 0.05$ level? State a conclusion. *Yes*

e. Is H_0 rejected at the $\alpha = 0.01$ level? State a conclusion. *Yes*

Answers are on page 458.

Performing hypothesis tests with technology

Following are the results of Example 9.14, as presented by the TI-84 Plus calculator.

```
Z-Test
μ≠74
z=2.236067977
p=.0253472347
x̄=76
n=80
```

CAUTION

P-values computed using technology may differ slightly from those computed with tables due to rounding. The differences are never large enough to matter.

The first line presents the alternate hypothesis, $\mu \neq 74$. Following that are the test statistic (z), the P-value (p), the sample mean ($\bar{x}$), and the sample size (n). Note that the P-value differs slightly from the value obtained in Example 9.14 by using Table A.2. This is common. Results given by technology are more precise and therefore often differ slightly from results obtained from tables. The differences are never large enough to matter.

Following are the results of Example 9.14 as presented by MINITAB.

```
Test of mu = 74.0 vs not = 74.0
The assumed standard deviation = 8.0

 N    Mean   SE Mean        95% CI           Z       P
80   76.00    0.8944   (74.247, 77.753)   2.236   0.025
```

The second line of the output presents both the null and alternate hypotheses. The quantity labeled "SE Mean" is the standard error of the mean, $\sigma/\sqrt{n}$, which is the standard deviation of $\bar{x}$. Notice that MINITAB provides a 95% confidence interval for μ along with the hypothesis test.

Step-by-step instructions for performing hypothesis tests with technology are presented in the Using Technology section that follows.

Check Your Understanding

17. The following display from a TI-84 Plus calculator presents the results of a hypothesis test for a population mean.

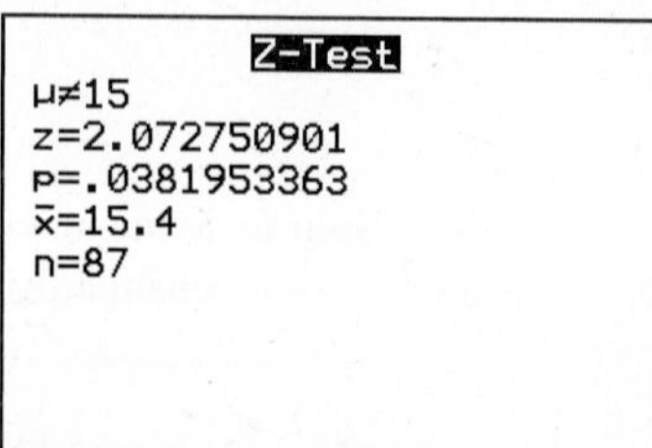

a. What are the null and alternate hypotheses? *H_0: $\mu = 15$, H_1: $\mu \neq 15$*

b. What is the value of the test statistic? *2.072750901*

c. What is the *P*-value? *0.0381953363*

d. Do you reject H_0 at the $\alpha = 0.05$ level? *Yes*

18. The following output from MINITAB presents the results of a hypothesis test for a population mean.

```
Test of mu = 53.5 vs not = 53.5
The assumed standard deviation = 2.3634

  N    Mean   SE Mean         95% CI           Z      P
145  53.246    0.1962   (52.861, 53.631)   -1.29  0.196
```

a. What are the null and alternate hypotheses? *H_0: $\mu = 53.5$, H_1: $\mu \neq 53.5$*

b. What is the value of the test statistic? *−1.29*

c. What is the *P*-value? *0.196*

d. Do you reject H_0 at the $\alpha = 0.05$ level? *No*

Answers are on page 458.

Using Technology

We use Example 9.14 to illustrate the technology steps.

TI-84 PLUS

Testing a hypothesis about the population mean when σ is known

Step 1. Press **STAT** and highlight the **TESTS** menu.

Step 2. Select **Z–Test** and press **ENTER** (Figure A). The **Z–Test** menu appears.

Step 3. For **Inpt**, select the **Stats** option and enter the values of μ_0, σ, $\bar{x}$, and n. For Example 9.14, we use $\mu_0 = 74$, $\sigma = 8$, $\bar{x} = 76$, and $n = 80$.

```
EDIT CALC TESTS
1:Z-Test...
2:T-Test...
3:2-SampZTest...
4:2-SampTTest...
5:1-PropZTest...
6:2-PropZTest...
7:ZInterval...
8:TInterval...
9↓2-SampZInt...
```

Figure A

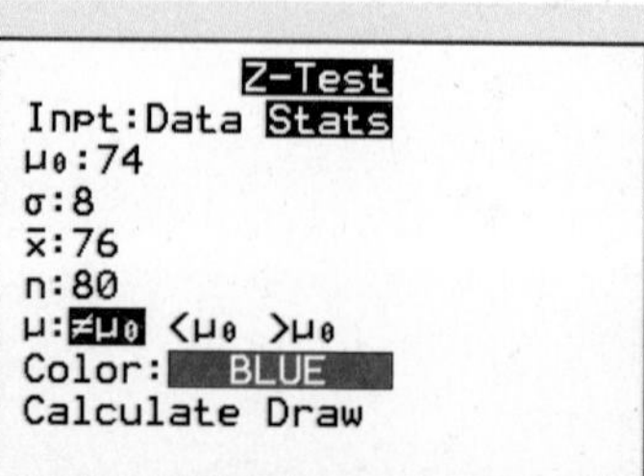

Figure B

Step 4. Select the form of the alternate hypothesis. For Example 9.14, the alternate hypothesis has the form $\mu \neq \mu_0$ (Figure B).

Step 5. Highlight **Calculate** and press **ENTER** (Figure C).

Note that if the raw data are given, the **Z–Test** command can be used by selecting **Data** as the **Inpt** option and entering the location of the data as the **List** option (Figure D).

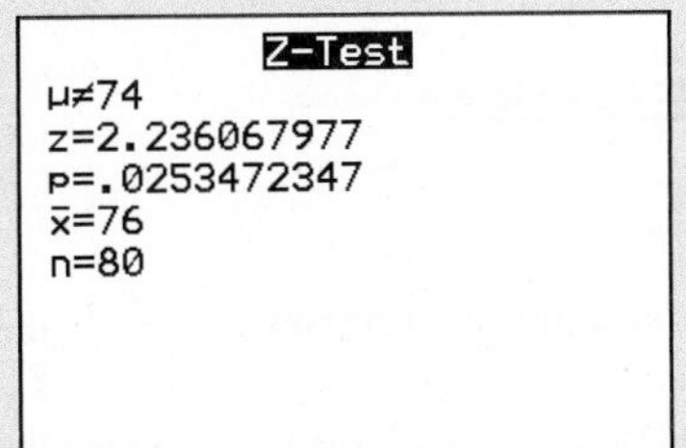

Figure C

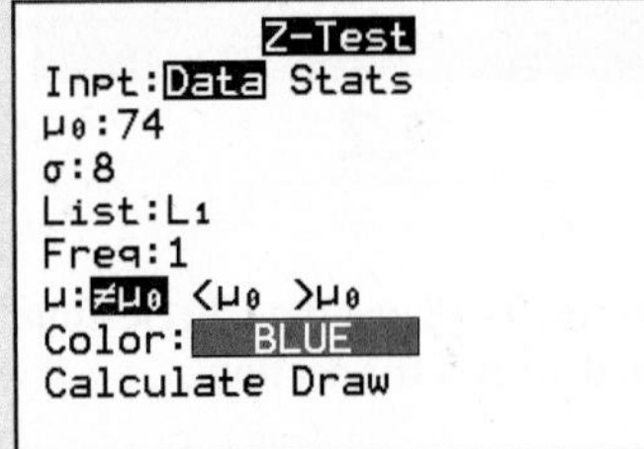

Figure D

EXCEL

Finding the test statistic for a hypothesis test about the mean when σ is known

Figure E demonstrates the calculations involved for finding the test statistic for a hypothesis test about the mean when σ is known. In Example 9.14, we test H_0: $\mu = 74$ versus H_1: $\mu \neq 74$, where $\bar{x} = 76$, $\sigma = 8$, and $n = 80$.

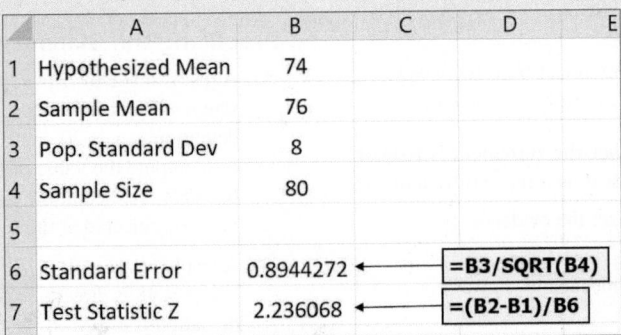

	A	B	C	D	E
1	Hypothesized Mean	74			
2	Sample Mean	76			
3	Pop. Standard Dev	8			
4	Sample Size	80			
5					
6	Standard Error	0.8944272 ←	=B3/SQRT(B4)		
7	Test Statistic Z	2.236068 ←	=(B2-B1)/B6		

Figure E

MINITAB

Testing a hypothesis about the population mean when σ is known

Step 1. Click on **Stat**, then **Basic Statistics**, then **1-Sample Z.**

Step 2. Choose one of the following:

- If the summary statistics are given, select **Summarized Data** and enter the **Sample Size** (80), the **Sample Mean** (76), and the **Standard Deviation** (8).
- If the raw data are given, select **One or more samples, each in a column** and select the column that contains the data. Enter the **Standard Deviation.**

Step 3. Select the **Perform hypothesis test** option, and enter the **Hypothesized Mean** (74) (Figure F).

Step 4. Click **Options**, and select the form of the alternate hypothesis. For Example 9.14, we select **Mean ≠ hypothesized mean**. Given significance level α, enter $100(1 - \alpha)$ as the **Confidence Level**. For Example 9.14, $\alpha = 0.05$ and the confidence level is $100(1 - 0.05) = 95$. Click **OK.**

Step 5. Click **OK** (Figure G).

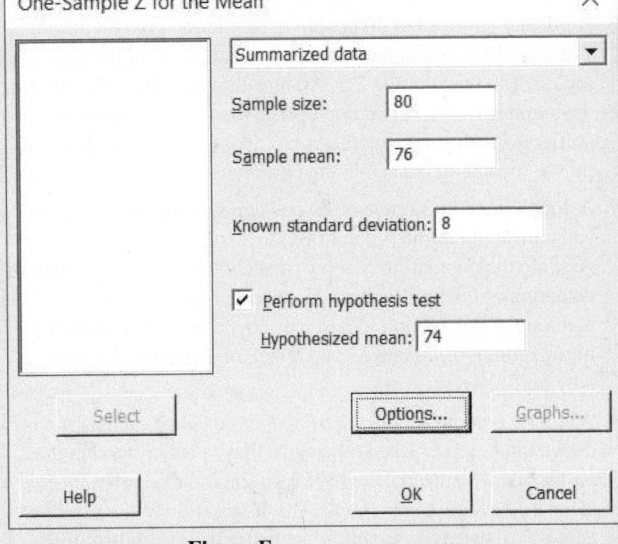

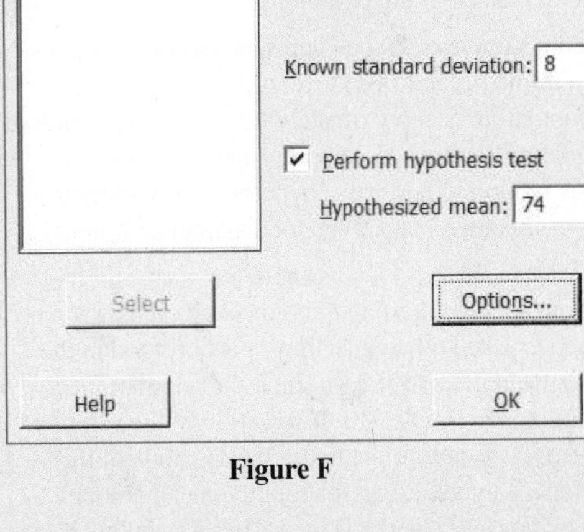

Figure F

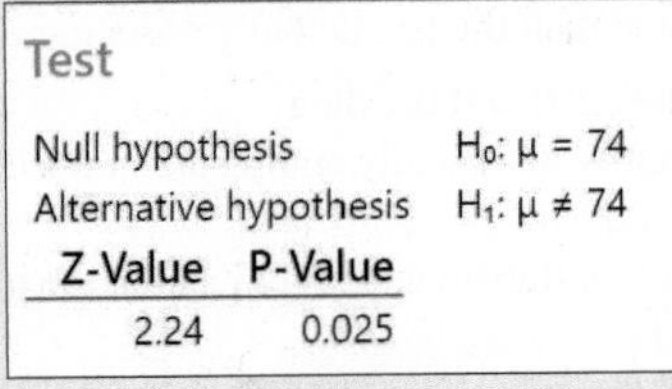

Test

Null hypothesis	H_0: $\mu = 74$
Alternative hypothesis	H_1: $\mu \neq 74$

Z-Value	P-Value
2.24	0.025

Figure G

Section 9.2 Exercises

Exercises 1–18 are the Check Your Understanding exercises located within the section.

Understanding the Concepts

In Exercises 19–22, fill in each blank with the appropriate word or phrase.

19. The ________ is the probability, assuming H_0 is true, of observing a value for the test statistic that is as extreme as or more extreme than the value actually observed. *P-value*

20. The smaller the *P*-value is, the stronger the evidence against the ________ hypothesis becomes. *null*

21. When using the critical value method, the region that contains the unusual values is called the ________ region. *critical*

22. When using the *P*-value method, we reject H_0 if the *P*-value is less than or equal to the ________ level. *significance*

In Exercises 23–26, determine whether the statement is true or false. If the statement is false, rewrite it as a true statement.

23. The smaller the *P*-value, the stronger the evidence against H_0. *True*

24. If the *P*-value is less than the significance level, we reject H_0. *True*

25. The *P*-value represents the probability that H_0 is true. *False*

26. If we reject H_0 at the $\alpha = 0.01$ level, we will also reject H_0 at the $\alpha = 0.05$ level. *True*

27. A simple random sample of 50 students at a certain school was drawn, and each was asked how many credits they are currently taking. The mean was $\bar{x} = 14$. Assume the population standard deviation is $\sigma = 1.1$. Is it appropriate to use the methods of this section to perform a hypothesis test about the mean number of credits students are currently taking? If not, why not? *Yes*

28. A pollster approaches 40 people as they leave a movie theater and asks them how many movies they have seen in the past month. The mean is $\bar{x} = 2.3$. Assume the population standard deviation is $\sigma = 1.2$. Is it appropriate to use the methods of this section to perform a hypothesis test about the mean number of movies people have seen in the past month? If not, why not? *No*

29. A simple random sample of 10 one-cup servings of asparagus had a mean thiamine (vitamin B1) level of 0.29 milligrams. Assume the population is approximately normal with a standard deviation of 0.03 milligrams. Is it appropriate to use the methods of this section to perform a hypothesis test about the mean amount of thiamine in a serving of asparagus? If not, why not? *Yes*

30. A simple random sample of 10 households in a certain town is drawn, and each is asked how much they spent on food in the past week. The mean is $\bar{x} = \$150$. Assume the population standard deviation is $\sigma = \$30$. The distribution of the sample values is skewed. Is it appropriate to use the methods of this section to perform a hypothesis test about the mean amount spent on food by households in this town? If not, why not? *No*

31. During a 30-day period, there were 130 recorded earthquakes in Los Angeles County. The mean magnitude was $\bar{x} = 2.0$. Assume the population standard deviation of earthquake magnitudes in Los Angeles is $\sigma = 0.4$. Is it appropriate to use the methods of this section to perform a hypothesis test about the mean magnitude of earthquakes in Los Angeles? If not, why not? *No*

32. A simple random sample of 60 employees at a large company was drawn, and each was asked how many sick days they have taken in the past year. The mean was $\bar{x} = 5.1$. Assume the population standard deviation is $\sigma = 1.4$. Is it appropriate to use the methods of this section to perform a hypothesis test about the mean number of sick days taken by employees at this company in the past year? If not, why not? *Yes*

Practicing the Skills

33. A test is made of H_0: $\mu = 50$ versus H_1: $\mu > 50$. A sample of size $n = 75$ is drawn, and $\bar{x} = 56$. The population standard deviation is $\sigma = 20$.
    a. Compute the value of the test statistic z. *2.60*
    b. Is H_0 rejected at the $\alpha = 0.05$ level? *Yes*
    c. Is H_0 rejected at the $\alpha = 0.01$ level? *Yes*

34. A test is made of H_0: $\mu = 14$ versus H_1: $\mu \neq 14$. A sample of size $n = 48$ is drawn, and $\bar{x} = 12$. The population standard deviation is $\sigma = 6$.
    a. Compute the value of the test statistic z. *−2.31*
    b. Is H_0 rejected at the $\alpha = 0.05$ level? *Yes*
    c. Is H_0 rejected at the $\alpha = 0.01$ level? *No*

35. A test is made of H_0: $\mu = 130$ versus H_1: $\mu \neq 130$. A sample of size $n = 63$ is drawn, and $\bar{x} = 135$. The population standard deviation is $\sigma = 40$.
    a. Compute the value of the test statistic z. *0.99*
    b. Is H_0 rejected at the $\alpha = 0.05$ level? *No*
    c. Is H_0 rejected at the $\alpha = 0.01$ level? *No*

36. A test is made of H_0: $\mu = 5$ versus H_1: $\mu < 5$. A sample of size $n = 87$ is drawn, and $\bar{x} = 4.5$. The population standard deviation is $\sigma = 25$.
    a. Compute the value of the test statistic z. *−0.19*
    b. Is H_0 rejected at the $\alpha = 0.05$ level? *No*
    c. Is H_0 rejected at the $\alpha = 0.01$ level? *No*

37. A test of the hypothesis H_0: $\mu = 65$ versus H_1: $\mu \neq 65$ was performed. The *P*-value was 0.035. Fill in the blank: If $\mu = 65$, then the probability of observing a test statistic as extreme as or more extreme than the one actually observed is ________. *0.035*

38. A test of the hypothesis H_0: $\mu = 150$ versus H_1: $\mu < 150$ was performed. The *P*-value was 0.28. Fill in the blank: If $\mu = 150$, then the probability of observing a test statistic as extreme as or more extreme than the one actually observed is ________. *0.28*

39. True or false: If $P = 0.02$, then
    a. The result is statistically significant at the $\alpha = 0.05$ level. *True*
    b. The result is statistically significant at the $\alpha = 0.01$ level. *False*
    c. The null hypothesis is rejected at the $\alpha = 0.05$ level. *True*
    d. The null hypothesis is rejected at the $\alpha = 0.01$ level. *False*

40. True or false: If $P = 0.08$, then
  a. The result is statistically significant at the $\alpha = 0.05$ level. *False*
  b. The result is statistically significant at the $\alpha = 0.10$ level. *True*
  c. The null hypothesis is rejected at the $\alpha = 0.05$ level. *False*
  d. The null hypothesis is rejected at the $\alpha = 0.10$ level. *True*

41. A test of H_0: $\mu = 17$ versus H_1: $\mu < 17$ is performed using a significance level of $\alpha = 0.01$. The value of the test statistic is $z = -2.68$.
  a. Is H_0 rejected? *Yes*
  b. If the true value of μ is 17, is the result a Type I error, a Type II error, or a correct decision? *Type I error*
  c. If the true value of μ is 10, is the result a Type I error, a Type II error, or a correct decision? *Correct decision*

42. A test of H_0: $\mu = 50$ versus H_1: $\mu \neq 50$ is performed using a significance level of $\alpha = 0.01$. The value of the test statistic is $z = 1.23$.
  a. Is H_0 rejected? *No*
  b. If the true value of μ is 50, is the result a Type I error, a Type II error, or a correct decision? *Correct decision*
  c. If the true value of μ is 65, is the result a Type I error, a Type II error, or a correct decision? *Type II error*

43. A test of H_0: $\mu = 0$ versus H_1: $\mu \neq 0$ is performed using a significance level of $\alpha = 0.05$. The *P*-value is 0.15.
  a. Is H_0 rejected? *No*
  b. If the true value of μ is 1, is the result a Type I error, a Type II error, or a correct decision? *Type II error*
  c. If the true value of μ is 0, is the result a Type I error, a Type II error, or a correct decision? *Correct decision*

44. A test of H_0: $\mu = 6$ versus H_1: $\mu > 6$ is performed using a significance level of $\alpha = 0.01$. The *P*-value is 0.002.
  a. Is H_0 rejected? *Yes*
  b. If the true value of μ is 8, is the result a Type I error, a Type II error, or a correct decision? *Correct decision*
  c. If the true value of μ is 6, is the result a Type I error, a Type II error, or a correct decision? *Type I error*

45. If H_0 is rejected at the $\alpha = 0.05$ level, which of the following is the best conclusion? *iii*
  i. H_0 is also rejected at the $\alpha = 0.01$ level.
  ii. H_0 is not rejected at the $\alpha = 0.01$ level.
  iii. We cannot determine whether H_0 is rejected at the $\alpha = 0.01$ level.

46. If H_0 is rejected at the $\alpha = 0.01$ level, which of the following is the best conclusion? *i*
  i. H_0 is also rejected at the $\alpha = 0.05$ level.
  ii. H_0 is not rejected at the $\alpha = 0.05$ level.
  iii. We cannot determine whether H_0 is rejected at the $\alpha = 0.05$ level.

47. If $P = 0.03$, which of the following is the best conclusion? *i*
  i. If H_0 is true, the probability of obtaining a test statistic as extreme as or more extreme than the one actually observed is 0.03.
  ii. The probability that H_0 is true is 0.03.
  iii. The probability that H_0 is false is 0.03.
  iv. If H_0 is false, the probability of obtaining a test statistic as extreme as or more extreme than the one actually observed is 0.03.

48. If $P = 0.25$, which of the following is the best conclusion? *iii*
  i. The probability that H_0 is true is 0.25.
  ii. If H_0 is false, the probability of obtaining a test statistic as extreme as or more extreme than the one actually observed is 0.25.
  iii. If H_0 is true, the probability of obtaining a test statistic as extreme as or more extreme than the one actually observed is 0.25.
  iv. The probability that H_0 is false is 0.25.

Working with the Concepts

49. Netflix: A study conducted by the technology company TiVo showed that the mean time spent per day browsing the video streaming service Netflix for something to watch was 19.3 minutes. Assume the standard deviation is $\sigma = 8$. Suppose a simple random sample of 100 visits taken this year has a sample mean of $\bar{x} = 21.5$ minutes. A social scientist is interested to know whether the mean time browsing Netflix has increased.
  a. State the appropriate null and alternate hypotheses. *H_0: $\mu = 19.3$ H_1: $\mu > 19.3$*
  b. Compute the value of the test statistic. *2.75*
  c. State a conclusion. Use the $\alpha = 0.05$ level of significance. *Reject H_0.*

50. Are you smarter than a second grader? A random sample of 60 second graders in a certain school district are given a standardized mathematics skills test. The sample mean score is $\bar{x} = 52$. Assume the standard deviation of test scores is $\sigma = 15$. The nationwide average score on this test is 50. The school superintendent wants to know whether the second graders in her school district have greater math skills than the nationwide average.
  a. State the appropriate null and alternate hypotheses. *H_0: $\mu = 50$, H_1: $\mu > 50$*
  b. Compute the value of the test statistic. *1.03*
  c. State a conclusion. Use the $\alpha = 0.01$ level of significance. *Do not reject H_0.*

51. Height and age: Are older men shorter than younger men? According to the National Health Statistics Reports, the mean height for U.S. men is 69.4 inches. In a sample of 300 men between the ages of 60 and 69, the mean height was $\bar{x} = 69.0$ inches. Public health officials want to determine whether the mean height μ for older men is less than the mean height of all adult men.
  a. State the appropriate null and alternate hypotheses. *H_0: $\mu = 69.4$, H_1: $\mu < 69.4$*
  b. Assume the population standard deviation to be $\sigma = 2.84$ inches. Compute the value of the test statistic. *−2.44*
  c. State a conclusion. Use the $\alpha = 0.01$ level of significance. *Reject H_0.*

52. Calibrating a scale: Making sure that the scales used by businesses in the United States are accurate is the responsibility of the National Institute for Standards and Technology (NIST) in Washington, D.C. Suppose that NIST technicians are testing a scale by using a weight known to weigh exactly 1000 grams. They weigh this weight on the scale 50 times and read the result each time. The 50 scale readings have a sample mean of $\bar{x} = 1000.6$ grams. The scale is out of calibration if the mean scale reading differs from 1000 grams. The technicians want to perform a hypothesis test to determine whether the scale is out of calibration.

a. State the appropriate null and alternate hypotheses. H_0: $\mu = 1000$, H_1: $\mu \neq 1000$

b. The standard deviation of scale reading is known to be $\sigma = 2$. Compute the value of the test statistic. 2.12

c. State a conclusion. Use the $\alpha = 0.05$ level of significance. Reject H_0.

53. Measuring lung function: One of the measurements used to determine the health of a person's lungs is the amount of air a person can exhale under force in one second. This is called the forced expiratory volume in one second and is abbreviated FEV_1. Assume the mean FEV_1 for 10-year-old boys is 2.1 liters and that the population standard deviation is $\sigma = 0.3$. A random sample of 100 10-year-old boys who live in a community with high levels of ozone pollution are found to have a sample mean FEV_1 of 1.95 liters. Can you conclude that the mean FEV_1 in the high-pollution community is less than 2.1 liters? Use the $\alpha = 0.05$ level of significance. Yes, reject H_0.

54. Watch your weight: Are men heavier now than in the past? The National Health Examination and Nutrition Survey (NHANES) published in 2008 reported that the mean weight of men aged 40–59 was 199.9 pounds. Another NHANES survey, published in 2016, reported that a sample of 835 men aged 40–59 had an average weight of 200.9 pounds. Assume the population standard deviation is $\sigma = 35$ pounds. Can you conclude that the mean weight of men aged 40–59 is higher in 2016 than in 2008? Use the $\alpha = 0.01$ level of significance. No, do not reject H_0.

55. House prices: Data from the Denver Metro Association of Realtors indicates that the mean price of a home in Denver, Colorado, in December 2018 was 522.8 thousand dollars. A random sample of 50 homes sold in 2019 had a mean price of 553.3 thousand dollars.

a. Assume the population standard deviation is $\sigma = 150$. Can you conclude that the mean price in 2019 differs from the mean price in December 2018? Use the $\alpha = 0.05$ level of significance. No

b. Following is a boxplot of the data. Explain why it is not reasonable to assume that the population is approximately normally distributed. Outliers

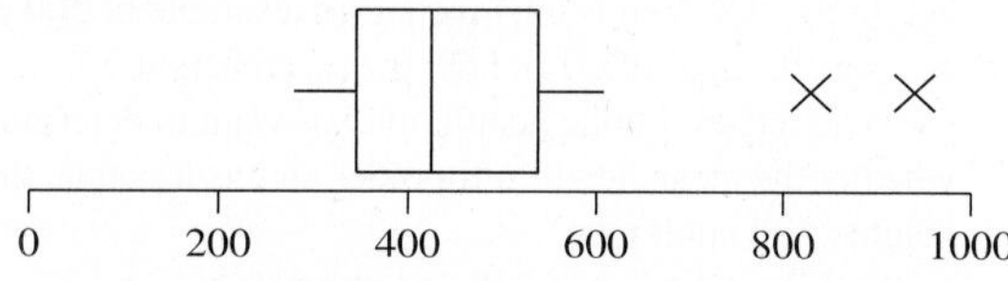

c. Explain why the assumptions for the hypothesis test are satisfied even though the population is not normal. $n > 30$

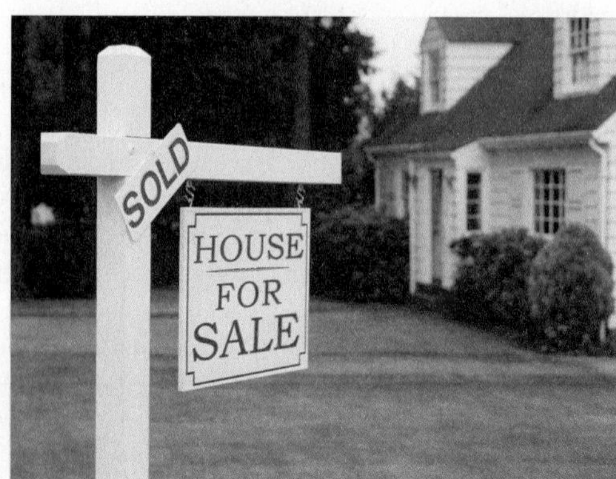

Ryan McVay/Getty Images

56. SAT scores: The College Board reports that in 2018 the mean score on the math SAT was 527 and the population standard deviation was $\sigma = 107$. A random sample of 20 students who took the test in 2019 had a mean score of 532. Following is a dotplot of the 20 scores. Following is a dotplot of the data.

a. Are the assumptions for a hypothesis test satisfied? Explain. Yes

b. If appropriate, perform a hypothesis test to determine whether you can conclude that the mean score in 2019 differs from the mean score in 2018. Assume the population standard deviation is $\sigma = 107$. Use the $\alpha = 0.05$ level of significance. Do not reject H_0.

57. What are you drinking? Environmental Protection Agency standards require that the amount of lead in drinking water be less than 15 micrograms per liter. Twelve samples of water from a particular source have the following concentrations, in units of micrograms per liter:

11.4	13.9	11.2	14.5	15.2	8.1
12.4	8.6	10.5	17.1	9.8	15.9

a. Explain why it is necessary to check that the population is approximately normal before performing a hypothesis test. $n \leq 30$

b. Following is a dotplot of the data. Is it reasonable to assume that the population is approximately normal? Yes

c. Assume that the population standard deviation is $\sigma = 3$. If appropriate, perform a hypothesis test at the $\alpha = 0.01$ level to determine whether you can conclude that the mean concentration of lead meets the EPA standard. What do you conclude? Reject H_0.

58. GPA: The mean GPA at a certain university is 2.80. Following are GPAs for a random sample of 16 business students from this university.

2.27	3.05	2.57	3.36	3.10	3.03	3.19	3.08
2.60	2.92	2.77	3.55	2.63	2.79	2.70	2.92

a. Following is a boxplot of the data. Is it reasonable to assume that the population is approximately normal? Yes

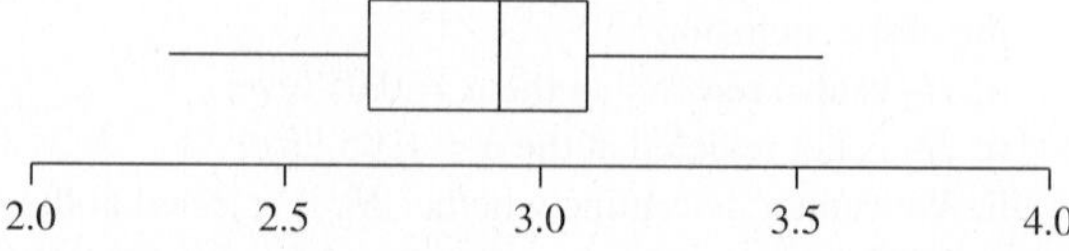

b. Assume that the population standard deviation is $\sigma = 0.3$. If appropriate, perform a hypothesis test at the $\alpha = 0.05$ level to determine whether the mean GPA for business students differs from the mean GPA at the whole university. What do you conclude? Do not reject H_0.

59. Interpret calculator display: The age in years was recorded for a sample of books in a college library. The following display from a TI-84 Plus calculator presents the results of a hypothesis test regarding the mean age of books in this library.

```
Z-Test
μ≠45
z=3.094063348
p=.0019744896
x̄=48.78
n=67
```

a. What are the null and alternate hypotheses? *H_0: $\mu = 45$, H_1: $\mu \neq 45$*

b. What is the value of the test statistic? *3.094063348*

c. What is the *P*-value? *0.0019744896*

d. Do you reject H_0 at the $\alpha = 0.05$ level? State a conclusion. *Yes*

e. Do you reject H_0 at the $\alpha = 0.01$ level? State a conclusion. *Yes*

60. Interpret calculator display: The number of characters was determined for a sample of text messages sent by a certain student. The following display from a TI-84 Plus calculator presents the results of a hypothesis test regarding the mean number of characters in a text message.

```
Z-Test
μ≠125
z=1.73116377
p=.0834224813
x̄=131.6
n=43
```

a. What are the null and alternate hypotheses? *H_0: $\mu = 125$, H_1: $\mu \neq 125$*

b. What is the value of the test statistic? *1.73116377*

c. What is the *P*-value? *0.0834224813*

d. Do you reject H_0 at the $\alpha = 0.05$ level? State a conclusion. *No*

e. Do you reject H_0 at the $\alpha = 0.01$ level? State a conclusion. *No*

61. Interpret computer output: A zoologist recorded the weight, in grams, for a sample of white-fronted Amazon parrots. The following MINITAB output presents the results of a hypothesis test regarding the mean weight of this variety of parrot.

```
Test of mu = 225.0 vs not = 225.0
The assumed standard deviation = 35.0

 N    Mean   SE Mean          95% CI            Z       P
50  235.32    4.9497  (225.619, 245.021)  2.085   0.037
```

a. What are the null and alternate hypotheses? *H_0: $\mu = 225$, H_1: $\mu \neq 225$*

b. What is the value of the test statistic? *2.085*

c. What is the *P*-value? *0.037*

d. Do you reject H_0 at the $\alpha = 0.05$ level? State a conclusion. *Yes*

e. Do you reject H_0 at the $\alpha = 0.01$ level? State a conclusion. *No*

f. Use the results of the output to compute the value of the test statistic z for a test of H_0: $\mu = 230$ versus H_1: $\mu > 230$. *1.07*

g. Find the *P*-value. *0.1423 [Tech: 0.1412]*

h. Do you reject the null hypothesis in part (f) at the $\alpha = 0.05$ level? State a conclusion. *No*

62. Interpret computer output: A sample of students at a certain college was drawn, and the age of each, in years, was determined. The following MINITAB output presents the results of a hypothesis test regarding the mean age of students at this college.

```
Test of mu = 20.0 vs > 20.0
The assumed standard deviation = 6.5

 N    Mean   SE Mean        95% CI           Z       P
45  21.324    0.9690  (19.425, 23.223)  1.366   0.086
```

a. What are the null and alternate hypotheses? *H_0: $\mu = 20$, H_1: $\mu > 20$*

b. What is the value of the test statistic? *1.366*

c. What is the *P*-value? *0.086*

d. Do you reject H_0 at the $\alpha = 0.05$ level? State a conclusion. *No*

e. Do you reject H_0 at the $\alpha = 0.01$ level? State a conclusion. *No*

f. Use the results of the output to compute the value of the test statistic z for a test of H_0: $\mu = 24$ versus H_0: $\mu \neq 24$. *−2.76*

g. Find the *P*-value. *0.0058 [Tech: 0.0057]*

h. Do you reject the null hypothesis in part (f) at the $\alpha = 0.05$ level? State a conclusion. *Yes*

Answers to Check Your Understanding Exercises for Section 9.2

1. a. Critical value is −1.645, critical region is $z \leq -1.645$.
b. Yes
c. Critical value is −2.326, critical region is $z \leq -2.326$.
d. No

2. a. Critical value is 1.645, critical region is $z \geq 1.645$.
b. Yes
c. Critical value is 2.326, critical region is $z \geq 2.326$.
d. Yes

3. a. Critical values are 1.96 and −1.96, critical region is $z \leq -1.96$ or $z \geq 1.96$.
b. No
c. Critical values are 2.576 and −2.576, critical region is $z \leq -2.576$ or $z \geq 2.576$.
d. No

4. a. $z = 2.12$ **b.** $z \geq 1.645$ **c.** Yes

5. a. $z = -1.00$ **b.** $z \leq -2.054$ **c.** No

6. a. $z = -3.25$ **b.** $z \leq -2.576$ or $z \geq 2.576$ **c.** Yes

7. $P = 0.05$

8. $P = 0.1003$. If H_0 is true, then the probability of observing a test statistic less than or equal to the value we actually observed is 0.1003. This result is not very unusual, so the evidence against H_0 is not strong.

9. a. $P = 0.4532$. If H_0 is true, then the probability of observing a test statistic as extreme as or more extreme than the value we actually observed is 0.4532. This result is not unusual, so the evidence against H_0 is not strong.

b. $P = 0.0278$. If H_0 is true, then the probability of observing a test statistic as extreme as or more extreme than the value we actually observed is 0.0278. This result is fairly unusual, so the evidence against H_0 is fairly strong.
c. $z = -2.20$

10. ii

11. Less than

12. Ellie

13. a. No **b.** No **c.** Yes **d.** Yes

14. a. No **b.** Yes **c.** Yes **d.** No

15. a. Yes **b.** No **c.** Yes **d.** No

16. a. $z = 3.27$ **b.** $P = 0.0005$
c. If H_0 is true, then the probability of observing a test statistic greater than or equal to the value we actually observed is 0.0005. This result is very unusual, so the evidence against H_0 is very strong.
d. Yes, we conclude that the mean number of years of education is greater than 12.
e. Yes, we conclude that the mean number of years of education is greater than 12.

17. a. H_0: $\mu = 15$, H_1: $\mu \neq 15$ **b.** $z = 2.072750901$
c. $P = 0.0381953363$ **d.** Yes

18. a. H_0: $\mu = 53.5$, H_1: $\mu \neq 53.5$ **b.** $z = -1.29$
c. $P = 0.196$ **d.** No

Section 9.3 Hypothesis Tests for a Population Mean, Standard Deviation Unknown

Objectives

1. Test a hypothesis about a mean using the P-value method
2. Test a hypothesis about a mean using the critical value method
3. Describe the relationship between hypothesis tests and confidence intervals
4. Describe the relationship between α and the probability of error
5. Report the P-value or the test statistic value
6. Distinguish between statistical significance and practical significance

Objective 1 Test a hypothesis about a mean using the P-value method

NOTE TO INSTRUCTOR

The following topic from the **Statistics Corequisite Workbook** is aligned with the material in Sections 9.3 and 9.4 of this chapter.
6.4 - Operations in Hypothesis Tests

Do low-fat diets work? The following study was reported in the *Journal of the American Medical Association* (297:969–977). A total of 76 subjects were placed on a low-fat diet. After 12 months, their sample mean weight loss was $\bar{x} = 2.2$ kilograms, with a sample standard deviation of $s = 6.1$ kilograms. How strong is the evidence that people who adhere to this diet will lose weight, on the average?

To answer this question, we need to perform a hypothesis test on a population mean. Assume that the subjects in the study constitute a simple random sample from a population of interest. We are interested in their population mean weight loss μ. We know the sample mean $\bar{x} = 2.2$. We do not know the population standard deviation σ, but we know that the sample standard deviation is $s = 6.1$.

Because we do not know σ, we cannot use the z-score

$$z = \frac{\bar{x} - \mu}{\sigma/\sqrt{n}}$$

as our test statistic. Instead, we replace σ with the sample standard deviation s and use the t statistic

$$t = \frac{\bar{x} - \mu}{s/\sqrt{n}}$$

RECALL

When $\bar{x}$ is the mean of a sample from a normal population, the quantity $\frac{\bar{x} - \mu}{s/\sqrt{n}}$ has a Student's t distribution with $n - 1$ degrees of freedom.

When the null hypothesis is true, the t statistic has a Student's t distribution with $n - 1$ degrees of freedom. We described the Student's t distribution in Section 8.2. When we perform a test using the t statistic, we call the test a ***t*-test**. We can perform a t-test for a population mean whenever the following assumptions are satisfied.

Assumptions for a Test of a Population Mean μ When σ Is Unknown

1. We have a simple random sample.
2. The sample size is large ($n > 30$), or the population is approximately normal.

When these assumptions are met, a hypothesis test can be performed. Either the critical value method or the *P*-value method may be used. Following are the steps for the *P*-value method.

Performing a Hypothesis Test on a Population Mean with σ Unknown Using the *P*-Value Method

Check to determine whether the assumptions are satisfied. If they are, then proceed with the following steps.

Step 1: State the null and alternate hypotheses. The null hypothesis specifies a value for the population mean μ. We will call this value μ_0. So the null hypothesis is of the form H_0: $\mu = \mu_0$. The alternate hypothesis can be stated in one of three ways:

Left-tailed: H_1: $\mu < \mu_0$
Right-tailed: H_1: $\mu > \mu_0$
Two-tailed: H_1: $\mu \neq \mu_0$

Step 2: If making a decision, choose a significance level α.

Step 3: Compute the test statistic $t = \dfrac{\bar{x} - \mu_0}{s/\sqrt{n}}$.

Step 4: Compute the *P*-value of the test statistic. The *P*-value is the probability, assuming that H_0 is true, of observing a value for the test statistic that disagrees as strongly as or more strongly with H_0 than the value actually observed. The *P*-value is an area under the Student's *t* curve with $n - 1$ degrees of freedom. The area is in the left tail, the right tail, or in both tails, depending on the type of alternate hypothesis. Note that the inequality points in the direction of the tail that contains the area for the *P*-value.

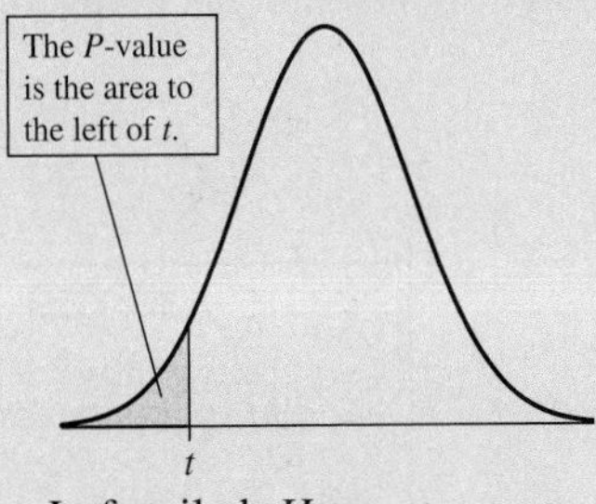
The *P*-value is the area to the left of *t*.
t
Left-tailed: H_1: $\mu < \mu_0$

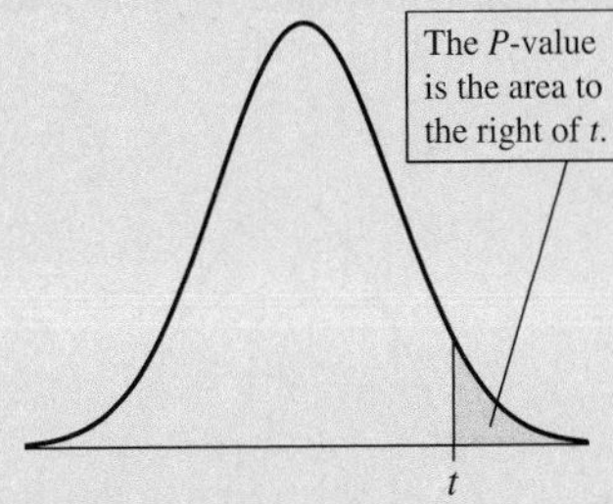
The *P*-value is the area to the right of *t*.
t
Right-tailed: H_1: $\mu > \mu_0$

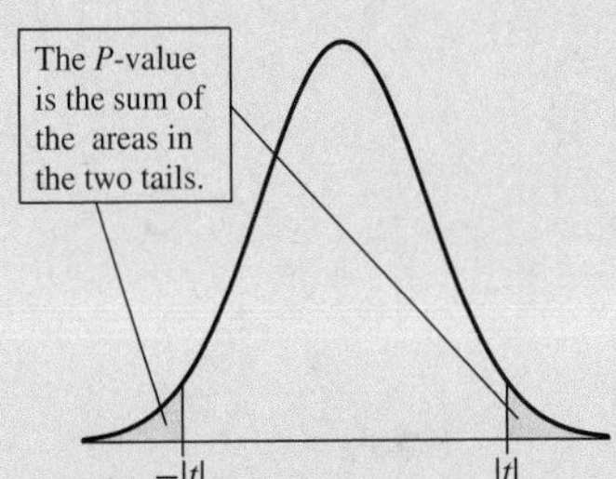
The *P*-value is the sum of the areas in the two tails.
−|*t*|
|*t*|
Two-tailed: H_1: $\mu \neq \mu_0$

Step 5: Interpret the *P*-value. If making a decision, reject H_0 if the *P*-value is less than or equal to the significance level α.

Step 6: State a conclusion.

Example 9.15

Perform a hypothesis test

In a recent medical study, 76 subjects were placed on a low-fat diet. After 12 months, their sample mean weight loss was $\bar{x}$ = 2.2 kilograms, with a sample standard deviation of s = 6.1 kilograms. Can we conclude that the mean weight loss is greater than 0? Use the $\alpha = 0.05$ level of significance.

Source: *Journal of the American Medical Association* 297:969–977

Solution

We first check the assumptions. We have a simple random sample. The sample size is 76, so $n > 30$. The assumptions are satisfied.

Step 1: State H_0 and H_1. The issue is whether the mean weight loss μ is greater than 0. So the null and alternate hypotheses are

$$H_0: \mu = 0 \qquad H_1: \mu > 0$$

Note that we have a right-tailed test, because we are particularly interested in whether the diet results in a weight loss.

Step 2: Choose a level of significance. The level of significance is $\alpha = 0.05$.

Step 3: Compute the test statistic. The test statistic is

$$t = \frac{\bar{x} - \mu_0}{s/\sqrt{n}}$$

To compute its value, we note that $\bar{x} = 2.2$, $s = 6.1$, and $n = 76$. We set $\mu_0 = 0$, the value for μ specified by H_0. The value of the test statistic is

$$t = \frac{2.2 - 0}{6.1/\sqrt{76}} = 3.144$$

Step 4: Compute the *P*-value. When H_0 is true, the test statistic t has the Student's t distribution with $n - 1$ degrees of freedom. In this case, the sample size is $n = 76$, so there are $n - 1 = 75$ degrees of freedom. To obtain the *P*-value, note that the alternate hypothesis is H_1: $\mu > 0$. Therefore, values of the t statistic in the right tail of the Student's t distribution provide evidence against H_0. The *P*-value is the probability that a value as extreme as or more extreme than the observed value of 3.144 is observed from a t distribution with 75 degrees of freedom. To find the *P*-value exactly, it is necessary to use technology. The *P*-value is 0.0012. Figure 9.12 illustrates the *P*-value as an area under the Student's t curve, and presents the results from the TI-84 Plus calculator. Step-by-step instructions for performing hypothesis tests with technology are given in the Using Technology section on page 471.

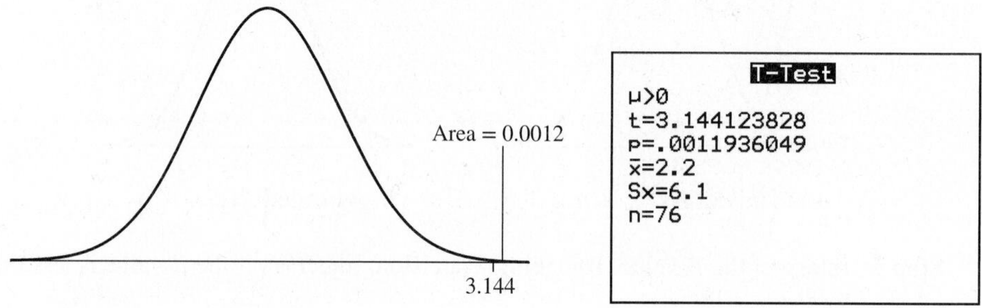

Figure 9.12 The *P*-value is the area to the right of the observed value of the test statistic, 3.144. The TI-84 Plus display shows that $P = 0.0012$, rounded to four decimal places.

Step 5: Interpret the *P*-value. The *P*-value is 0.0012. Because $P < 0.05$, we reject H_0.

Step 6: State a conclusion. We conclude that the mean weight loss of people who adhered to this diet for 12 months is greater than 0.

Estimating the *P*-Value from a Table

If no technology is available to compute the *P*-value, the t table (Table A.3) can be used to provide an approximation. When using a t table, we cannot find the *P*-value exactly. Instead, we can only specify that P is between two values. We now show how to use Table A.3 to bracket P between two values.

In Example 9.15, there are 75 degrees of freedom. We consult Table A.3 and find that the number 75 does not appear in the degrees of freedom column. We therefore use the next smallest number, which is 60. Now look across the row for two numbers that bracket the observed value 3.144. These are 2.915 and 3.232. The upper-tail probabilities are 0.0025 for 2.915 and 0.001 for 3.232. The P-value must therefore be between 0.001 and 0.0025 (see Figure 9.13). We can conclude that the P-value is small enough to reject H_0.

Degrees of	Area in the Right Tail									
Freedom	**0.40**	**0.25**	**0.10**	**0.05**	**0.025**	**0.01**	**0.005**	**0.0025**	**0.001**	**0.0005**
1	0.325	1.000	3.078	6.314	12.706	31.821	63.657	127.321	318.309	636.619
2	0.289	0.816	1.886	2.920	4.303	6.965	9.925	14.089	22.327	31.599
3	0.277	0.765	1.638	2.353	3.182	4.541	5.841	7.453	10.215	12.924
⋮	⋮	⋮	⋮	⋮	⋮	⋮	⋮	⋮	⋮	⋮
38	0.255	0.681	1.304	1.686	2.024	2.429	2.712	2.980	3.319	3.566
39	0.255	0.681	1.304	1.685	2.023	2.426	2.708	2.976	3.313	3.558
40	0.255	0.681	1.303	1.684	2.021	2.423	2.704	2.971	3.307	3.551
50	0.255	0.679	1.299	1.676	2.009	2.403	2.678	2.937	3.261	3.496
60	0.254	0.679	1.296	1.671	2.000	2.390	2.660	2.915	3.232	3.460
80	0.254	0.678	1.292	1.664	1.990	2.374	2.639	2.887	3.195	3.416
100	0.254	0.677	1.290	1.660	1.984	2.364	2.626	2.871	3.174	3.390
200	0.254	0.676	1.289	1.653	1.972	2.345	2.601	2.839	3.131	3.340

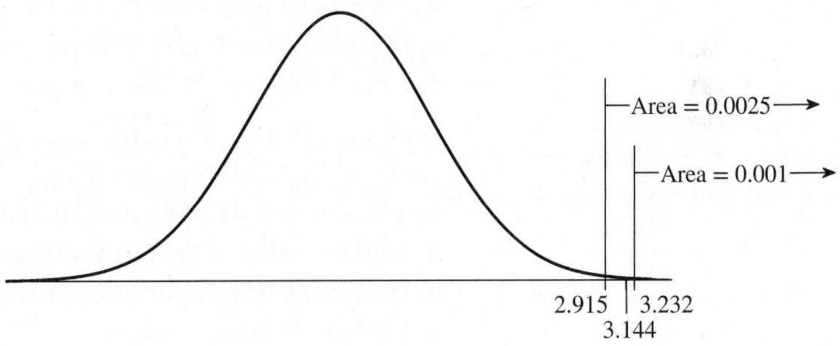

Figure 9.13 The P-value is the area to the right of the observed value of the test statistic, 3.144. The P-value is between 0.001 and 0.0025.

Finding the *P*-value for a two-tailed test from a table

In the previous example, what if the alternate hypothesis were H_1: $\mu \neq 0$? The P-value would be the sum of the areas in two tails. We know that the area in the right tail is 0.0012 (see Figure 9.12). Since the t distribution is symmetric, the sum of the areas in two tails is twice as much: $0.0012 + 0.0012 = 0.0024$. This is shown in Figure 9.14.

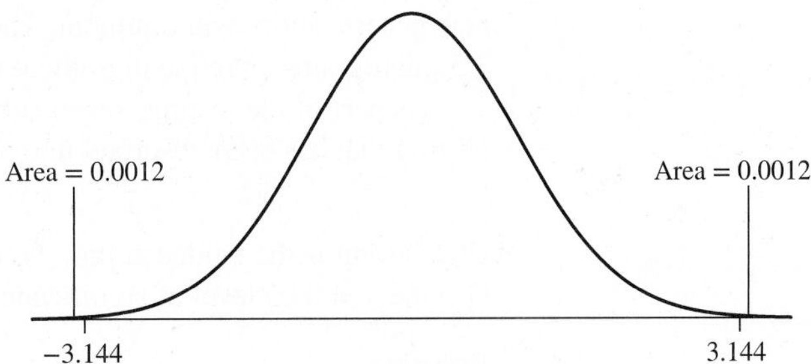

Figure 9.14 The P-value for a two-tailed test is the sum of the areas in the two tails. Each tail has area 0.0012. The P-value is $0.0012 + 0.0012 = 0.0024$.

If we are using Table A.3, we can only specify that P is between two values. We know that the area in one tail is between 0.001 and 0.0025. Therefore, the area in both tails is between $2(0.001) = 0.002$ and $2(0.0025) = 0.005$. This is shown in Figure 9.15 (page 462).

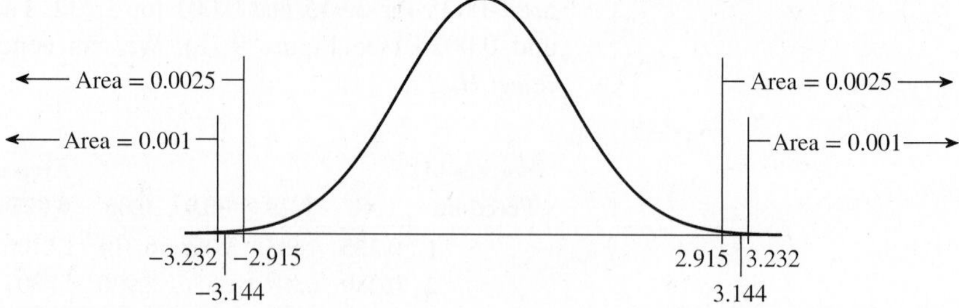

Figure 9.15 The P-value for a two-tailed test is the sum of the areas in the two tails. The area in each tail is between 0.001 and 0.0025. The sum of the areas in both tails is therefore between 2(0.001) = 0.002 and 2(0.0025) = 0.005.

Check Your Understanding

1. Find the P-value for the following values of the test statistic t, sample size n, and alternate hypothesis H_1. If you use Table A.3, you may specify that P is between two values.

a. $t = 2.584, n = 12, H_1: \mu > \mu_0$ *0.0127*

b. $t = -1.741, n = 21, H_1: \mu < \mu_0$ *0.0485*

c. $t = 3.031, n = 14, H_1: \mu \neq \mu_0$ *0.0096*

d. $t = -2.584, n = 31, H_1: \mu \neq \mu_0$ *0.0148*

2. In Example 9.15, the sample size was $n = 76$, and we observed $\bar{x} = 2.2$ and $s = 6.1$. We tested $H_0: \mu = 0$ versus $H_1: \mu > 0$, and the P-value was 0.0012. Assume that the sample size was 41 instead of 76, but that the values of $\bar{x}$ and s were the same.

a. Find the value of the test statistic t. *2.309*

b. How many degrees of freedom are there? *40*

c. Find the P-value. *0.0131*

d. Is the evidence against H_0 stronger or weaker than the evidence from the sample of 76? Explain. *Weaker*

Answers are on page 478.

Example 9.16

Perform a hypothesis test

Generic drugs are lower-cost substitutes for brand-name drugs. Before a generic drug can be sold in the United States, it must be tested and found to perform equivalently to the brand-name product. The U.S. Food and Drug Administration is now supervising the testing of a new generic antifungal ointment. The brand-name ointment is known to deliver a mean of 3.5 micrograms of active ingredient to each square centimeter of skin.

As part of the testing, seven subjects apply the ointment. Six hours later, the amount of drug that has been absorbed into the skin is measured. The amounts, in micrograms, are

2.6 3.2 2.1 3.0 3.1 2.9 3.7

How strong is the evidence that the mean amount absorbed differs from 3.5 micrograms? Use the $\alpha = 0.05$ level of significance.

Solution

We first check the assumptions. Because the sample is small, the population must be approximately normal. We check this with a dotplot of the data.

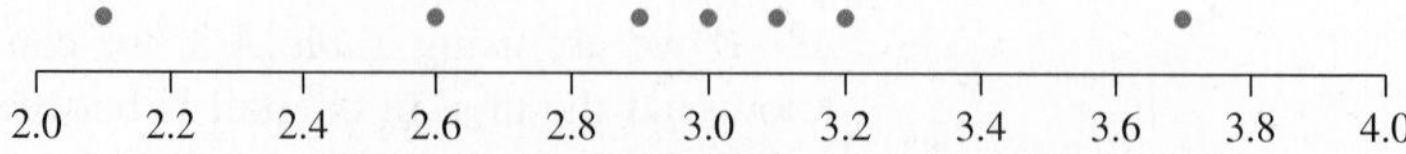

There is no evidence of strong skewness, and no outliers. Therefore, we can proceed.

Step 1: State the null and alternate hypotheses. The issue is whether the mean μ differs from 3.5. Therefore, the null and alternate hypotheses are

$$H_0\colon \mu = 3.5 \qquad H_1\colon \mu \neq 3.5$$

Step 2: Choose a significance level α. The significance level is $\alpha = 0.05$.

Step 3: Compute the value of the test statistic t. To compute t, we need to know the sample mean $\bar{x}$, the sample standard deviation s, the null hypothesis mean μ_0, and the sample size n. We compute $\bar{x}$ and s from the sample. The values are

$$\bar{x} = 2.9429 \qquad s = 0.4995$$

The null hypothesis mean is $\mu_0 = 3.5$. The sample size is $n = 7$. The value of the t statistic is

$$t = \frac{\bar{x} - \mu_0}{s/\sqrt{n}} = \frac{2.9429 - 3.5}{0.4995/\sqrt{7}} = -2.951$$

Step 4: Compute the P-value. The number of degrees of freedom is $n - 1 = 7 - 1 = 6$. The alternate hypothesis is two-tailed, so the P-value is the sum of the area to the left of the observed t statistic -2.951 and the area to the right of 2.951, in a t distribution with 6 degrees of freedom. We can use technology to find that $P = 0.0256$. The following TI-84 Plus display presents the results. Step-by-step instructions for performing hypothesis tests with technology are given in the Using Technology section on page 471.

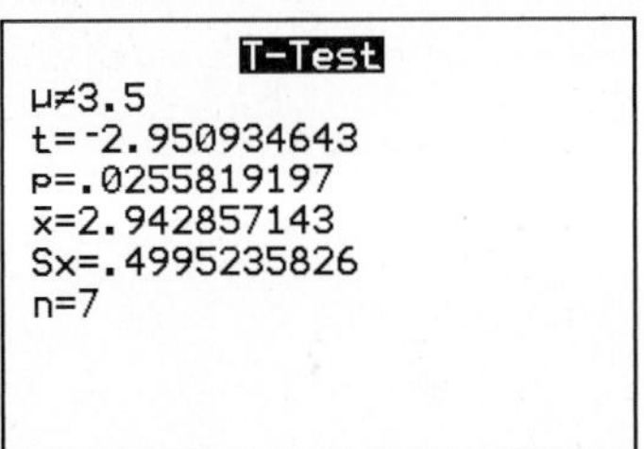

The P-value is given on the third line of the display. Rounding off to four decimal places, we see that $P = 0.0256$.

Alternatively, we can use Table A.3 to specify that the P-value is between two numbers. In the row corresponding to 6 degrees of freedom, the two values closest to 2.951 are 2.447 and 3.143. The area to the right of 2.447 is 0.025, and the area to the right of 3.143 is 0.01. Therefore, the area in the right tail is between 0.01 and 0.025. The P-value is twice the area in the right tail, so we conclude that P is between $2(0.01) = 0.02$ and $2(0.025) = 0.05$. See Figure 9.16.

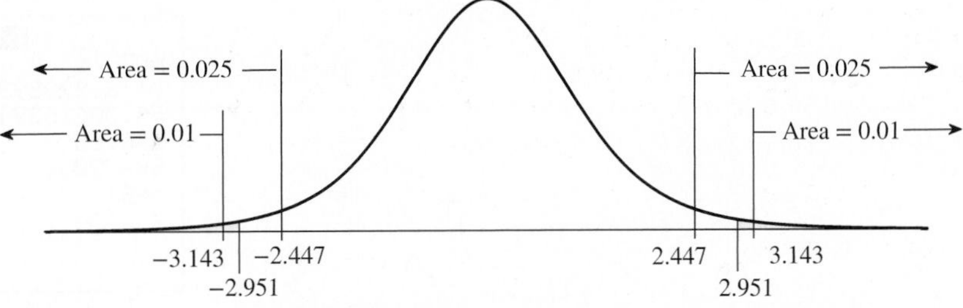

Figure 9.16 The P-value for a two-tailed test is the sum of the areas in the two tails. The area in each tail is between 0.01 and 0.025. The sum of the areas in both tails is therefore between $2(0.01) = 0.02$ and $2(0.025) = 0.05$.

Step 5: Interpret the *P*-value. The *P*-value of 0.0256 tells us that if H_0 is true, the probability of observing a value of the test statistic as extreme as or more extreme than the value of −2.951 that we observed is 0.0256. The *P*-value is small enough to give us doubt about the truth of H_0. Because $P < 0.05$, we reject H_0 at the 0.05 level.

Step 6: State a conclusion. We conclude that the mean amount of drug absorbed differs from 3.5 micrograms.

Performing hypothesis tests with technology

The following output (from MINITAB) presents the results of Example 9.16.

```
Test of mu = 3.5 vs not = 3.5

N      Mean    StDev   SE Mean          95% CI             T       P
7    2.9429   0.4995    0.1888   (2.4809, 3.4048)    −2.95   0.026
```

Most of the output is straightforward. The first line specifies the null and alternate hypotheses. The sample size, sample mean, and sample standard deviation are given as "N," "Mean," and "StDev," respectively. The quantity labeled "SE Mean" is the standard error of the mean, which is the quantity $s/\sqrt{n}$ that appears in the denominator of the *t* statistic. Next, MINITAB provides a 99% confidence interval for μ. Finally, the value of the *t* statistic and the *P*-value are given.

The following TI-84 Plus display presents the results of Example 9.16. This display was also shown in the solution to Example 9.16.

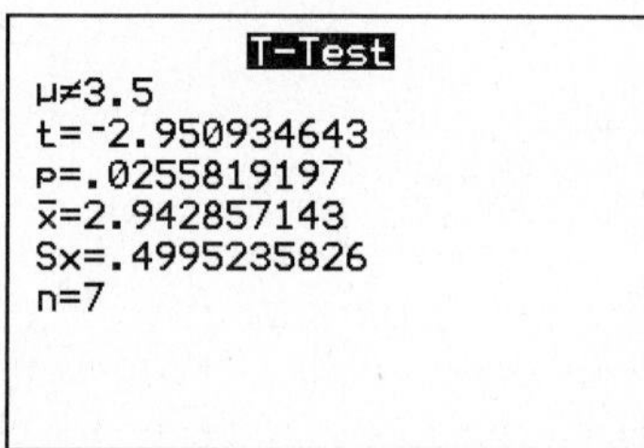

The first line states the alternate hypothesis. The quantity labeled "Sx" is the sample standard deviation *s*. Step-by-step instructions for performing hypothesis tests with technology are given in the Using Technology section on page 471.

Check Your Understanding

3. In Example 9.16, the alternate hypothesis was H_1: $\mu \neq 3.5$ and the *P*-value for the two-tailed test was $P = 0.0256$. What would the *P*-value be for the alternate hypothesis H_1: $\mu < 3.5$? *0.0128*

4. The following TI-84 Plus display presents the results of a *t*-test.

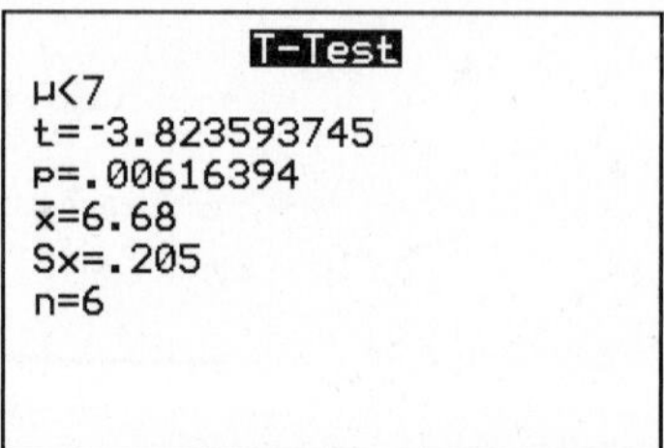

a. What are the null and alternate hypotheses? *H_0: $\mu = 7$, H_1: $\mu < 7$*

b. What is the sample size? *6*

c. How many degrees of freedom are there? *5*

d. What is the value of $\bar{x}$? *6.68*
e. What is the value of s? *0.205*
f. What is the value of the test statistic? *−3.823593745*
g. What is the P-value? *0.00616394*
h. Do you reject H_0 at the $\alpha = 0.01$ level? *Yes*

5. Refer to the display in Exercise 4. If the sample mean were 6.80 instead of 6.68, what would the P-value be? If you use Table A.3, you may specify that P is between two values. *0.0312*

Answers are on page 478.

Objective 2 Test a hypothesis about a mean using the critical value method

Testing a Hypothesis About a Population Mean Using the Critical Value Method

The critical value method when σ is unknown is the same as that when σ is known, except that we use the Student's t distribution rather than the normal distribution. The critical value can be found in Table A.3 or with technology. The procedure depends on whether the alternate hypothesis is left-tailed, right-tailed, or two-tailed.

Critical Values for the *t* Statistic

Let α denote the chosen significance level, and let n denote the sample size. The critical value depends on whether the alternate hypothesis is left-tailed, right-tailed, or two-tailed. We use the Student's t distribution with $n - 1$ degrees of freedom.

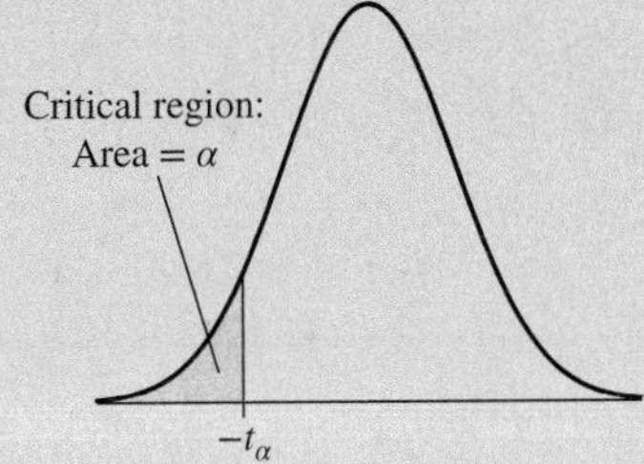

For left-tailed H_1: The critical value is $-t_\alpha$, which has area α to its left. Reject H_0 if $t \le -t_\alpha$.

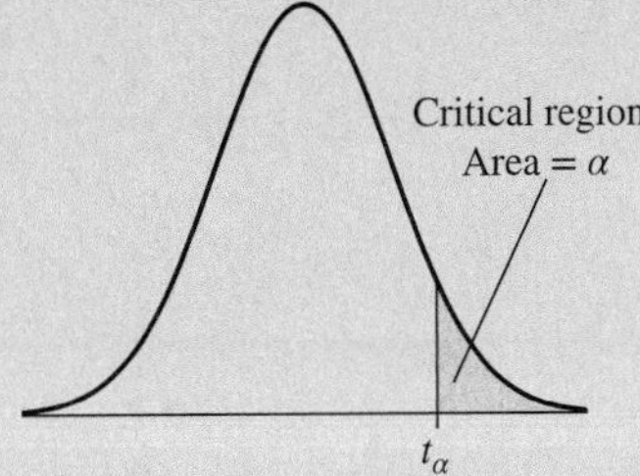

For right-tailed H_1: The critical value is t_α, which has area α to its right. Reject H_0 if $t \ge t_\alpha$.

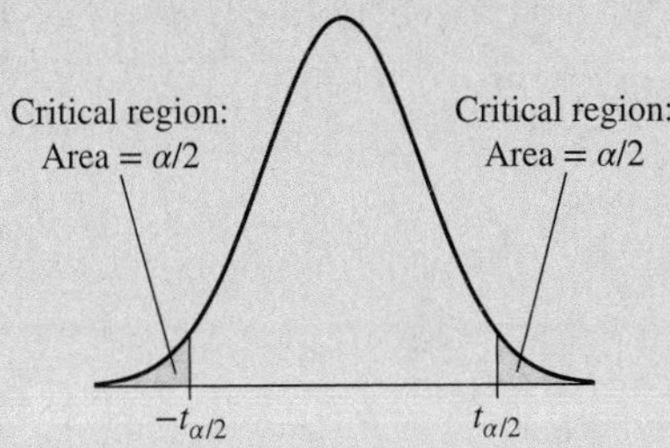

For two-tailed H_1: The critical values are $t_{\alpha/2}$, which has area $\alpha/2$ to its right, and $-t_{\alpha/2}$, which has area $\alpha/2$ to its left. Reject H_0 if $t \ge t_{\alpha/2}$ or $t \le -t_{\alpha/2}$.

Check Your Understanding

6. Find the critical value or values for the following values of the significance level α, sample size n, and alternate hypothesis H_1.
a. $\alpha = 0.05,\quad n = 3,\quad H_1\colon \mu > \mu_0$ *2.920*
b. $\alpha = 0.01,\quad n = 26,\quad H_1\colon \mu \ne \mu_0$ *−2.787, 2.787*
c. $\alpha = 0.10,\quad n = 81,\quad H_1\colon \mu < \mu_0$ *−1.292*
d. $\alpha = 0.05,\quad n = 14,\quad H_1\colon \mu \ne \mu_0$ *−2.160, 2.160*

Answers are on page 478.

The assumptions for the critical value method are the same as those for the P-value method. We repeat them here.

Assumptions for a Test of a Population Mean μ When σ Is Unknown

1. We have a simple random sample.
2. The sample size is large ($n > 30$), or the population is approximately normal.

When these assumptions are satisfied, a hypothesis test can be performed using the following steps.

Performing a Hypothesis Test on a Population Mean with σ Unknown Using the Critical Value Method

Check to be sure that the assumptions are satisfied. If they are, then proceed with the following steps:

Step 1: State the null and alternate hypotheses. The null hypothesis specifies a value for the population mean μ. We will call this value μ_0, so the null hypothesis is of the form H_0: $\mu = \mu_0$. The alternate hypothesis can be stated in one of three ways:

Left-tailed: H_1: $\mu < \mu_0$
Right-tailed: H_1: $\mu > \mu_0$
Two-tailed: H_1: $\mu \neq \mu_0$

Step 2: Choose a significance level α, and find the critical value or values. Use $n - 1$ degrees of freedom, where n is the sample size.

Step 3: Compute the test statistic $t = \dfrac{\bar{x} - \mu_0}{s/\sqrt{n}}$.

Step 4: Determine whether to reject H_0, as follows:

Left-tailed: H_1: $\mu < \mu_0$ Reject if $t \leq -t_\alpha$.
Right-tailed: H_1: $\mu > \mu_0$ Reject if $t \geq t_\alpha$.
Two-tailed: H_1: $\mu \neq \mu_0$ Reject if $t \geq t_{\alpha/2}$ or $t \leq -t_{\alpha/2}$.

Step 5: State a conclusion.

Example 9.17

Test a hypothesis using the critical value method

A computer software vendor claims that a new version of its operating system will crash fewer than six times per year on average. A system administrator installs the operating system on a random sample of 41 computers. At the end of a year, the sample mean number of crashes is 7.1, with a standard deviation of 3.6. Can you conclude that the vendor's claim is false? Use the $\alpha = 0.05$ significance level.

Solution

We first check the assumptions. We have a large ($n > 30$) random sample, so the assumptions are satisfied.

Step 1: State the null and alternate hypotheses. To conclude that the vendor's claim is false, we must conclude that $\mu > 6$. This is H_1. The hypotheses are H_0: $\mu = 6$ versus H_1: $\mu > 6$.

Step 2: Choose a significance level α, and find the critical value. We will use a significance level of $\alpha = 0.05$. We use Table A.3. The number of degrees of freedom is $41 - 1 = 40$. This is a right-tailed test, so the critical value is the t-value with area 0.05 above it in the right tail. Thus, the critical value is $t_\alpha = 1.684$.

Step 3: Compute the test statistic. We have $\bar{x} = 7.1$, $\mu_0 = 6$, $s = 3.6$, and $n = 41$. The test statistic is

$$t = \frac{7.1 - 6}{3.6/\sqrt{41}} = 1.957$$

Step 4: Determine whether to reject H_0. Because this is a right-tailed test, we reject H_0 if $t \geq t_\alpha$. Because $t = 1.957$ and $t_\alpha = 1.684$, we reject H_0. Figure 9.17 (page 467) illustrates the critical region and the test statistic.

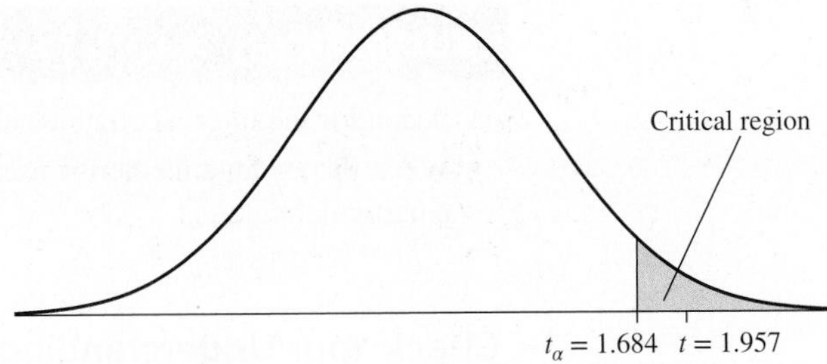

Figure 9.17

Step 5: State a conclusion. We conclude that the mean number of crashes is greater than six per year.

Objective 3 Describe the relationship between hypothesis tests and confidence intervals

The Relationship Between Hypothesis Tests and Confidence Intervals

In Example 9.16, we tested the hypotheses H_0: $\mu = 3.5$ versus H_1: $\mu \neq 3.5$ and obtained a P-value of 0.0256. Because $P < 0.05$, H_0 is rejected at the 0.05 level. Informally, this says that the value 3.5 is not plausible for μ.

Another way to express information about μ is through a confidence interval. A 95% confidence interval for μ is $2.4809 < \mu < 3.4048$. (This confidence interval is displayed in the MINITAB output following Example 9.16.) Note that the 95% confidence interval does not contain the null hypothesis value of 3.5. In this way, the 95% confidence interval agrees with the results of the hypothesis test. Informally, a confidence interval for μ contains all the values that are plausible for μ. Because 3.5 is not in the confidence interval, 3.5 is not a plausible value for μ.

This relationship holds for any confidence interval for a population mean, and any two-tailed hypothesis test.

If we test H_0: $\mu = \mu_0$ versus H_1: $\mu \neq \mu_0$, then

- If the 95% confidence interval contains μ_0, then H_0 will not be rejected at the 0.05 level.
- If the 95% confidence interval does not contain μ_0, then H_0 will be rejected at the 0.05 level.

This relationship between hypothesis tests and confidence intervals holds exactly for population means, but only approximately for other parameters such as population proportions. The reason is that the standard error that is used in a hypothesis test for a proportion differs somewhat from the standard error that is used in a confidence interval for a proportion.

Although hypothesis tests are closely related to confidence intervals, the two address different questions. A confidence interval provides all of the values that are plausible at a specified level. A hypothesis test tells us about only one value, but when the P-value method is used, it tells us much more precisely how plausible that one value is.

For example, consider the 95% confidence interval $2.4809 < \mu < 3.4048$ previously mentioned for the mean amount of drug absorbed in Example 9.16. The value $\mu = 3.5$ is not in the confidence interval, so we can conclude that the hypothesis H_0: $\mu = 3.5$ will be rejected at the $\alpha = 0.05$ level with a two-tailed test. However, this tells us only that $P < 0.05$. It does not tell us exactly how much less than 0.05 the P-value is. By performing the hypothesis test, we find that $P = 0.025$. This tells us much more precisely just how plausible or implausible the value of 3.5 is for μ.

SUMMARY

- A confidence interval contains all the values that are plausible at a particular level.
- When the *P*-value method is used, a hypothesis test tells us precisely how plausible a particular value is.

Check Your Understanding

7. A 95% confidence interval for μ is computed to be (1.75, 3.25). For each of the following hypotheses, state whether H_0 will be rejected at the 0.05 level.

a. H_0: $\mu = 3$ versus H_1: $\mu \neq 3$ *No*

b. H_0: $\mu = 4$ versus H_1: $\mu \neq 4$ *Yes*

c. H_0: $\mu = 1.7$ versus H_1: $\mu \neq 1.7$ *Yes*

d. H_0: $\mu = 3.5$ versus H_1: $\mu \neq 3.5$ *Yes*

8. You want to test H_0: $\mu = 4$ versus H_1: $\mu \neq 4$, so you compute a 95% confidence interval for μ. The 95% confidence interval is $5.1 < \mu < 7.2$.

a. Do you reject H_0 at the $\alpha = 0.05$ level? *Yes*

b. Your friend thinks that $\alpha = 0.01$ is a more appropriate significance level. Can you tell from the confidence interval whether to reject at this level? *No*

Answers are on page 478.

Objective 4 Describe the relationship between α and the probability of error

The Relationship Between α and the Probability of an Error

Recall that a Type I error occurs if we reject H_0 when it is true, and a Type II error occurs if we do not reject H_0 when it is false (see Table 9.2). When designing a hypothesis test, we would like to make the probabilities of these two errors small. In order to do this, we need to know how to calculate the probabilities of these errors. It is straightforward to find the probability of a Type I error: It is equal to the significance level. So, for example, if we perform a test at a significance level of $\alpha = 0.05$, the probability of a Type I error is 0.05.

Table 9.2

	H_0 true	H_0 false
Reject H_0	Type I error	Correct
Don't reject H_0	Correct	Type II error

SUMMARY

When a test is performed with a significance level α, the probability of a Type I error is α.

The probability of a Type II error is denoted by the letter β. Computing the probability of a Type II error is more difficult than finding the probability of a Type I error. A Type II error occurs when H_0 is false, and a decision is made not to reject. The probability of a Type II error depends on the true value of the parameter being tested. We will learn how to compute these probabilities in Section 9.7.

Because α is the probability of a Type I error, why don't we always choose a very small value for α? The reason is that the smaller a value we choose for α, the larger the value of β, the probability of making a Type II error, becomes (unless we increase the sample size).

SUMMARY

The smaller a value we choose for the significance level α:

- The smaller the probability of a Type I error becomes.
- The larger the probability of a Type II error becomes.

In general, making a Type I error is more serious than making a Type II error. When a Type I error is much more serious, a smaller value of α is appropriate. When a Type I error is only slightly more serious, a larger value of α can be justified.

Check Your Understanding

9. A hypothesis test is performed at a significance level $\alpha = 0.05$. What is the probability of a Type I error? *0.05*

10. Charlie will perform a hypothesis test at the $\alpha = 0.05$ level. Felice will perform the same test at the $\alpha = 0.01$ level.

a. If H_0 is true, who has a greater probability of making a Type I error? *Charlie*

b. If H_0 is false, who has a greater probability of making a Type II error? *Felice*

Answers are on page 478.

Objective 5 Report the *P*-value or the test statistic value

Report the *P*-Value or the Test Statistic Value

Sometimes people report only that a test result was statistically significant at a certain level, without giving the *P*-value. It is common, for example, to read that a result was "statistically significant at the 0.05 level" or "statistically significant ($P \leq 0.05$)." It is much better to report the *P*-value along with the decision whether to reject. There are three reasons for this.

The first reason is that there is a big difference between a *P*-value that is just barely small enough to reject, say $P = 0.049$, and a *P*-value that is extremely small, say $P = 0.0001$. If $P = 0.049$, the evidence is just barely strong enough to reject H_0 at the $\alpha = 0.05$ level, whereas if $P = 0.0001$, the evidence against H_0 is overwhelming. Thus, reporting the *P*-value describes exactly how strong the evidence against H_0 is.

The second reason is that there is a big difference between a *P*-value that is barely large enough not to reject, say $P = 0.051$, and a truly large value, say $P = 0.4$. Reporting the *P*-value distinguishes between a situation in which there is almost enough evidence to reject H_0 and one in which there is very little evidence against H_0.

The third reason is that not everyone may agree with your choice of α. For example, let's say you have chosen a significance level of $\alpha = 0.05$. You obtain a *P*-value of $P = 0.03$. Since $P < 0.05$, you reject H_0. Let's say that you report only that H_0 is rejected at the $\alpha = 0.05$ level, without stating the *P*-value. Now imagine that the person reading your report believes that a Type I error would be very serious, so that a significance level of $\alpha = 0.01$ would be more appropriate. This reader cannot tell whether to reject H_0 at the $\alpha = 0.01$ level, because you have not reported the *P*-value. It is much more helpful to report that $P = 0.03$, so that people can decide for themselves whether or not to reject H_0.

When using the critical value method, you should report the value of the test statistic, rather than simply stating whether the test statistic was in the critical region. In this way, the reader can tell whether the value of the test statistic was just barely inside the critical region, or well inside. In addition, reporting the value of the test statistic gives the reader the opportunity to choose a different critical value and determine whether H_0 can be rejected at a different level.

SUMMARY

When presenting the results of a hypothesis test, state the *P*-value or the value of the test statistic. Don't just state whether or not H_0 was rejected.

Check Your Understanding

11. A test was made of the hypotheses H_0: $\mu = 15$ versus H_1: $\mu > 15$. Four statisticians wrote summaries of the results. For each summary, state whether it contains enough information. If there is not enough information, indicate what needs to be added.

a. The *P*-value was 0.02, so we reject H_0 at the $\alpha = 0.05$ level. *Enough information*

b. The critical value was 1.645. Because $t > 1.645$, we reject H_0 at the $\alpha = 0.05$ level. *State value of test statistic*

c. The critical value was 1.645. Because $t = 2.05$, we reject H_0 at the $\alpha = 0.05$ level. *Enough information*

d. Because $P < 0.05$, we reject H_0 at the $\alpha = 0.05$ level. *State P-value*

Answers are on page 478.

Objective 6 Distinguish between statistical significance and practical significance

Statistical Significance Is Not the Same as Practical Significance

When a result has a small P-value, we say that it is "statistically significant." In common usage, the word *significant* means "important." It is therefore tempting to think that statistically significant results must always be important. This is not the case. Sometimes statistically significant results do not have any practical importance. Example 9.18 illustrates the idea.

Example 9.18

Determining practical significance

At a large company, employee satisfaction is measured with a standardized test for which scores range from 0 to 100. The mean score on this test was 74. The company then implemented a new policy that allowed telecommuting, so that employees could work from home. After the policy change, the mean score for a sample of employees was 76. In order to determine whether the mean score for all employees, μ, had changed after the new policy was implemented, a hypothesis test was performed of

$$H_0: \mu = 74 \qquad H_1: \mu \neq 74$$

We performed this test in Example 9.14 in Section 9.2. The standard error of $\bar{x}$ was 0.8944 and the P-value was 0.0250, so we rejected H_0 at the $\alpha = 0.05$ level. We concluded that the mean satisfaction level changed after the new policy was implemented. The human resources manager now writes a report stating that the new policy resulted in a large change in employee satisfaction. Explain why the human resources manager is not interpreting the result correctly.

NOTE TO INSTRUCTOR

It is important for students to appreciate the distinction between statistical significance and practical significance.

Solution

The increase in mean score was from 74 to 76. Although this is statistically significant, it is only two points out of 100. It is unlikely that this difference is large enough to matter. The lesson here is that a result can be statistically significant without being large enough to be of practical importance. How can this happen? A difference is statistically significant when it is large compared to its standard error. In the example, a difference of two points was statistically significant because the standard error of $\bar{x}$ was small—only 0.8944. When the standard error is small, even a small difference can be statistically significant.

NOTE TO INSTRUCTOR

In the medical field, practical significance is known as "clinical significance."

SUMMARY

When a result is statistically significant, we can only conclude that the true value of the parameter is different from the value specified by H_0. We cannot conclude that the difference is large enough to be important.

Check Your Understanding

12. A certain type of calculator battery has a mean lifetime of 100 hours and a standard deviation of $s = 10$ hours. A company has developed a new battery and claims it has a longer mean life. A random sample of 1000 batteries is tested, and their sample mean lifetime is $\bar{x} = 101$ hours. A test was made of the hypotheses

$$H_0: \mu = 100 \qquad H_1: \mu > 100$$

a. Show that H_0 is rejected at the $\alpha = 0.01$ level.

b. The battery manufacturer says that because the evidence is strong that $\mu > 100$, you should be willing to pay a much higher price for its battery than for the old type of battery. Do you agree? Why or why not? *Do not agree*

Answers are on page 478.

Using Technology

We use Example 9.16 to illustrate the technology steps.

TI-84 PLUS

Testing a hypothesis about a population mean when σ is unknown

Step 1. Press **STAT** and highlight the **TESTS** menu.

Step 2. Select **T–Test** and press **ENTER** (Figure A). The **T–Test** menu appears.

Step 3. Choose one of the following:

- If the summary statistics are given, select **Stats** as the **Inpt** option and enter μ_0, $\bar{x}$, s, and n.
- If the raw data are given, select **Data** as the **Inpt** option and enter the location of the data as the **List** option. For Example 9.16, the sample has been entered in list **L1**.

Step 4. Select the form of the alternate hypothesis. For Example 9.16, the alternate hypothesis has the form $\neq \mu_0$ (Figure B).

Step 5. Highlight **Calculate** and press **ENTER** (Figure C).

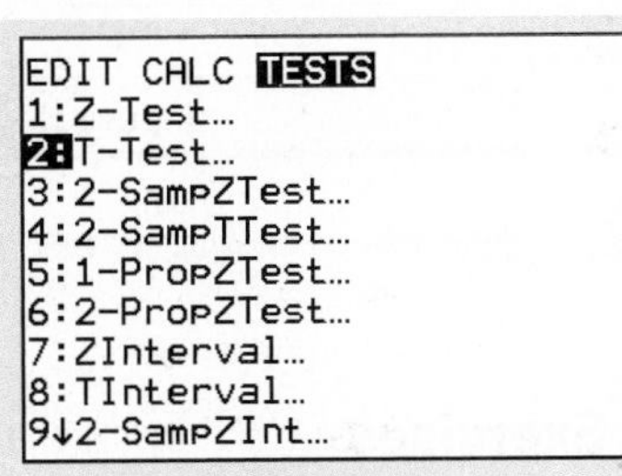

Figure A

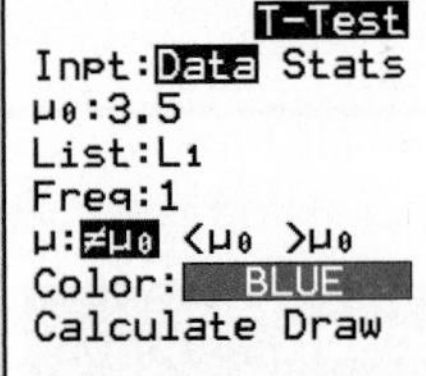

Figure B

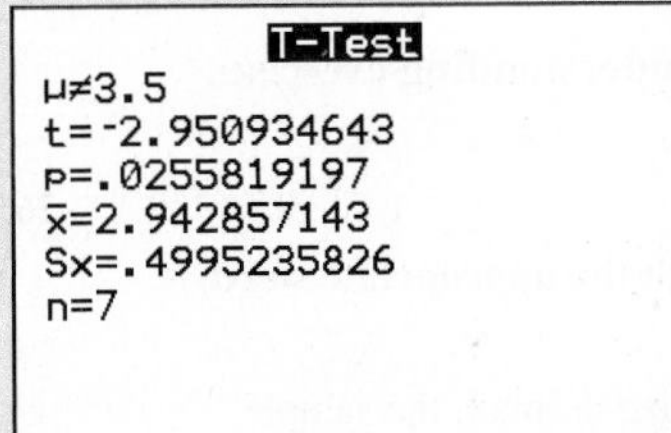

Figure C

EXCEL

Finding the test statistic for a hypothesis test about the mean when σ is unknown

Figure D demonstrates the calculations involved for finding the test statistic for a hypothesis test about the mean when σ is unknown. In Example 9.16, we test H_0: $\mu = 3.5$ versus H_1: $\mu \neq 3.5$ where $\bar{x} = 2.94286$, $\sigma = 0.49952$, and $n = 7$.

	A	B	C	D	E
1	Hypothesized Mean	3.5			
2	Sample Mean	2.94286			
3	Sample Standard Dev	0.49952			
4	Sample Size	7			
5					
6	Standard Error	0.1888008 ← =B3/SQRT(B4)			
7	Test Statistic T	-2.950941 ← =(B2-B1)/B6			

Figure D

MINITAB

Testing a hypothesis about a population mean when σ is unknown

Step 1. Click on **Stat**, then **Basic Statistics**, then **1-Sample t**.

Step 2. Choose one of the following:

- If the summary statistics are given, select **Summarized Data** and enter the **Sample Size**, the **Mean**, and the **Standard Deviation** for the sample.
- If the raw data are given, select **One or more samples, each in a column** and select the column that contains the data. For Example 9.16, the sample has been entered in column **C1**.

Step 3. Select the **Perform hypothesis test** option, and enter the **Hypothesized Mean** (3.5).

Step 4. Click **Options**, and select the form of the alternate hypothesis. For Example 9.16, we select **Mean ≠ hypothesized mean**. Given significance level α, enter $100(1 - \alpha)$ as the **Confidence Level**. For Example 9.16, since $\alpha = 0.05$, the confidence level is $100(1 - 0.05) = 95$. Click **OK**.

Step 5. Click **OK** (Figure E).

Test

Null hypothesis $H_0: \mu = 3.5$
Alternative hypothesis $H_1: \mu \neq 3.5$

T-Value	P-Value
-2.95	0.026

Figure E

Section 9.3 Exercises

Exercises 1–12 are the Check Your Understanding exercises located within the section.

Understanding the Concepts

In Exercises 13–18, fill in each blank with the appropriate word or phrase.

13. To perform a t-test when the sample size is small, the sample must show no evidence of strong ________ and must contain no ________. *skewness, outliers*

14. The number of degrees of freedom for the Student's t-test of a population mean is always 1 less than the ________. *sample size*

15. When testing H_0: $\mu = \mu_0$ versus H_1: $\mu \neq \mu_0$, if a 95% confidence interval does not contain μ_0, we reject H_0 at the ________ level. *0.05*

16. If we decrease the value of the significance level α, we ________ the probability of a Type I error. *decrease*

17. If we decrease the value of the significance level α, we ________ the probability of a Type II error. *increase*

18. When results are statistically significant, they do not necessarily have ________ significance. *practical*

In Exercises 19–24, determine whether the statement is true or false. If the statement is false, rewrite it as a true statement.

19. A t-test is used when the population standard deviation is unknown. *True*

20. A t-test is used when the number of degrees of freedom is unknown. *False*

21. The probability of a Type II error is α, the significance level. *False*

22. If the P-value is small enough, we can be sure the results have practical significance. *False*

23. If the P-value is small enough, we can be sure the results are statistically significant. *True*

24. When presenting the results of a hypothesis test, one should report the P-value or the value of the test statistic. *True*

25. A simple random sample of six classes offered at a certain university was drawn, and the numbers of students in the classes were

27 35 252 41 78 31

Is it appropriate to use the methods of this section to perform a hypothesis test about the mean number of students in a class at this university? If not, why not? *No*

26. A simple random sample of ten employees at a large company was asked how far (in miles) they commute to work each day. Following are the results:

10.1 11.7 9.7 11.1 11.4 12.1 12.8 9.4 10.6 13.2

Is it appropriate to use the methods of this section to perform a hypothesis test about the mean distance commuted by employees at this company? If not, why not? *Yes*

27. A simple random sample of 50 steel canisters has a mean wall thickness of $\bar{x} = 1.2$ millimeters with a standard deviation of $s = 0.05$ millimeters. Is it appropriate to use the methods of this section to perform a hypothesis test about the mean wall thickness of this type of steel canister? If not, why not? *Yes*

28. The mean and standard deviation of salaries of players on the Denver Broncos were recorded. Is it appropriate to use the methods of this section to perform a hypothesis test about the mean salary of players in the National Football League? If not, why not? *No*

29. The weight of a diamond is measured in carats. A random sample of 12 diamonds in a retail store had a mean weight of $\bar{x} = 1.15$ carats, with a standard deviation of $s = 0.15$ carats. Following is a dotplot of the data. Is it appropriate to use the methods of this section to perform a hypothesis test for the mean weight of diamonds at this store? *Yes*

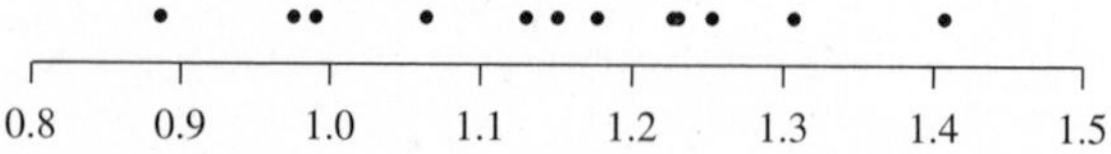

30. Voltages were measured for a random sample of 18 lithium-ion phone batteries that had been in use for one year. The mean was $\bar{x} = 3.78$ volts, with a standard deviation of $s = 0.2$ volts. Following is a boxplot of the data. Is it appropriate to use the methods of this section to perform a hypothesis test for the mean voltage of lithium-ion batteries after one year of use? *No*

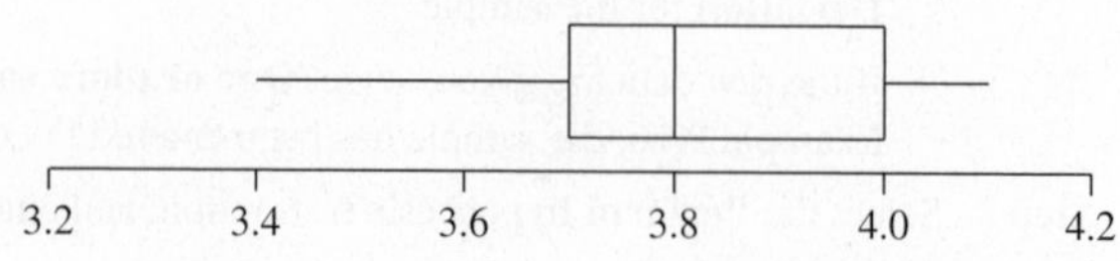

Practicing the Skills

31. Find the P-value for the following values of the test statistic t, sample size n, and alternate hypothesis H_1. If you use Table A.3, you may specify that P is between two values.

a. $t = 2.336, n = 5, H_1\colon \mu > \mu_0$ *0.0399*
b. $t = 1.307, n = 18, H_1\colon \mu \neq \mu_0$ *0.2086*
c. $t = -2.864, n = 51, H_1\colon \mu < \mu_0$ *0.0030*
d. $t = -2.031, n = 3, H_1\colon \mu \neq \mu_0$ *0.1793*

32. Find the P-value for the following values of the test statistic t, sample size n, and alternate hypothesis H_1. If you use Table A.3, you may specify that P is between two values.

a. $t = -1.584, n = 19, H_1\colon \mu \neq \mu_0$ *0.1306*
b. $t = -2.473, n = 41, H_1\colon \mu < \mu_0$ *0.0089*
c. $t = 1.491, n = 30, H_1\colon \mu \neq \mu_0$ *0.1468*
d. $t = 3.635, n = 4, H_1\colon \mu > \mu_0$ *0.0179*

33. Find the critical value or values for the following values of the significance level α, sample size n, and alternate hypothesis H_1.

a. $\alpha = 0.05, n = 27, H_1\colon \mu \neq \mu_0$ *−2.056, 2.056*
b. $\alpha = 0.01, n = 61, H_1\colon \mu > \mu_0$ *2.390*
c. $\alpha = 0.10, n = 16, H_1\colon \mu \neq \mu_0$ *−1.753, 1.753*
d. $\alpha = 0.05, n = 11, H_1\colon \mu < \mu_0$ *−1.812*

34. Find the critical value or values for the following values of the significance level α, sample size n, and alternate hypothesis H_1.

a. $\alpha = 0.05, n = 39, H_1\colon \mu > \mu_0$ *1.686*
b. $\alpha = 0.01, n = 34, H_1\colon \mu < \mu_0$ *−2.445*
c. $\alpha = 0.10, n = 6, H_1\colon \mu \neq \mu_0$ *−2.015, 2.015*
d. $\alpha = 0.05, n = 25, H_1\colon \mu \neq \mu_0$ *−2.064, 2.064*

35. A test is made of $H_0\colon \mu = 40$ versus $H_1\colon \mu > 40$. A sample of size 26 is drawn. The sample mean and standard deviation are $\bar{x} = 46$ and $s = 10$.

a. Compute the value of the test statistic t. *3.06*
b. How many degrees of freedom are there? *25*
c. Is H_0 rejected at the $\alpha = 0.05$ level? *Yes*
d. Is H_0 rejected at the $\alpha = 0.01$ level? *Yes*

36. A test is made of $H_0\colon \mu = 22$ versus $H_1\colon \mu \neq 22$. A sample of size 51 is drawn. The sample mean and standard deviation are $\bar{x} = 20$ and $s = 7$.

a. Compute the value of the test statistic t. *−2.04*
b. How many degrees of freedom are there? *50*
c. Is H_0 rejected at the $\alpha = 0.05$ level? *Yes*
d. Is H_0 rejected at the $\alpha = 0.01$ level? *No*

37. A test is made of $H_0\colon \mu = 105$ versus $H_1\colon \mu \neq 105$. A sample of size 75 is drawn. The sample mean and standard deviation are $\bar{x} = 107$ and $s = 8$.

a. Compute the value of the test statistic t. *2.17*
b. How many degrees of freedom are there? *74*
c. Is H_0 rejected at the $\alpha = 0.05$ level? *Yes*
d. Is H_0 rejected at the $\alpha = 0.01$ level? *No*

38. A test is made of $H_0\colon \mu = 48$ versus $H_1\colon \mu \neq 48$. A sample of size 162 is drawn. The sample mean and standard deviation are $\bar{x} = 49$ and $s = 28$.

a. Compute the value of the test statistic t. *0.45*
b. How many degrees of freedom are there? *161*
c. Is H_0 rejected at the $\alpha = 0.05$ level? *No*
d. Is H_0 rejected at the $\alpha = 0.01$ level? *No*

39. A test was made of $H_0\colon \mu = 37$ versus $H_1\colon \mu > 37$. The P-value was 0.10. Fill in the blank: If $\mu = 37$, then the probability of observing a test statistic as extreme or more extreme than the one actually observed is ________.

40. A test was made of $H_0\colon \mu = 55$ versus $H_1\colon \mu \neq 55$. The P-value was 0.02. Fill in the blank: If $\mu = 55$, then the probability of observing a test statistic as extreme or more extreme than the one actually observed is ________.

41. True or false: If $P = 0.03$, then

a. The result is statistically significant at the $\alpha = 0.05$ level. *True*
b. The result is statistically significant at the $\alpha = 0.01$ level. *False*
c. The null hypothesis is rejected at the $\alpha = 0.05$ level. *True*
d. The null hypothesis is rejected at the $\alpha = 0.01$ level. *False*

42. True or false: If $P = 0.07$, then

a. The result is statistically significant at the $\alpha = 0.05$ level. *False*
b. The result is statistically significant at the $\alpha = 0.10$ level. *True*
c. The null hypothesis is rejected at the $\alpha = 0.05$ level. *False*
d. The null hypothesis is rejected at the $\alpha = 0.10$ level. *True*

43. A test of $H_0\colon \mu = 4$ versus $H_1\colon \mu \neq 4$ is performed using a significance level of $\alpha = 0.01$. The sample size is $n = 12$, and the value of the test statistic is $t = 1.59$.

a. Is H_0 rejected? *No*
b. If the true value of μ is 4, is the result a Type I error, a Type II error, or a correct decision? *Correct decision*
c. If the true value of μ is 6, is the result a Type I error, a Type II error, or a correct decision? *Type II error*

44. A test of $H_0\colon \mu = 97$ versus $H_1\colon \mu < 97$ is performed using a significance level of $\alpha = 0.05$. The sample size is $n = 37$, and the value of the test statistic is $t = -2.86$.

a. Is H_0 rejected? *Yes*
b. If the true value of μ is 97, is the result a Type I error, a Type II error, or a correct decision? *Type I error*
c. If the true value of μ is 92, is the result a Type I error, a Type II error, or a correct decision? *Correct decision*

45. A test of $H_0\colon \mu = 46$ versus $H_1\colon \mu \neq 46$ is performed using a significance level of $\alpha = 0.05$. The P-value is 0.12.

a. Is H_0 rejected? *No*
b. If the true value of μ is 50, is the result a Type I error, a Type II error, or a correct decision? *Type II error*
c. If the true value of μ is 46, is the result a Type I error, a Type II error, or a correct decision? *Correct decision*

46. A test of $H_0\colon \mu = 80$ versus $H_1\colon \mu > 80$ is performed using a significance level of $\alpha = 0.01$. The P-value is 0.0078.

a. Is H_0 rejected? *Yes*
b. If the true value of μ is 90, is the result a Type I error, a Type II error, or a correct decision? *Correct decision*
c. If the true value of μ is 80, is the result a Type I error, a Type II error, or a correct decision? *Type I error*

47. If $P = 0.30$, which of the following is the best conclusion? *ii*

i. The probability that H_0 is true is 0.30.
ii. If H_0 is true, the probability of obtaining a test statistic as extreme or more extreme than the one actually observed is 0.30.
iii. If H_0 is false, the probability of obtaining a test statistic as extreme or more extreme than the one actually observed is 0.30.
iv. The probability that H_0 is false is 0.30.

48. If $P = 0.02$, which of the following is the best conclusion? *iii*

i. The probability that H_0 is false is 0.02.
ii. The probability that H_0 is true is 0.02.

iii. If H_0 is true, the probability of obtaining a test statistic as extreme or more extreme than the one actually observed is 0.02.

iv. If H_0 is false, the probability of obtaining a test statistic as extreme or more extreme than the one actually observed is 0.02.

Working with the Concepts

49. Is there a doctor in the house? The market research firm Salary.com reported that the mean annual earnings of all family practitioners in the United States was $178,258. A random sample of 55 family practitioners in Los Angeles that month had mean earnings of $\bar{x}$ = $192,340 with a standard deviation of $42,387. Do the data provide sufficient evidence to conclude that the mean salary for family practitioners in Los Angeles is greater than the national average?

a. State the null and alternate hypotheses. *H_0: $\mu = 178{,}258$, H_1: $\mu > 178{,}258$*

b. Compute the value of the t statistic. How many degrees of freedom are there? *2.464; 54 degrees of freedom*

c. State your conclusion. Use the $\alpha = 0.05$ level of significance. *Reject H_0.*

50. College tuition: The mean annual tuition and fees for a sample of 14 private colleges in California was $37,900 with a standard deviation of $7200. A dotplot shows that it is reasonable to assume that the population is approximately normal. Can you conclude that the mean tuition and fees for private institutions in California differs from $35,000?

a. State the null and alternate hypotheses. *H_0: $\mu = 35{,}000$, H_1: $\mu \neq 35{,}000$*

b. Compute the value of the t statistic. How many degrees of freedom are there? *1.507; 13 degrees of freedom*

c. State your conclusion. Use the $\alpha = 0.01$ level of significance. *Do not reject H_0.*

Based on data from collegeprowler.com

51. Big babies: The National Health Statistics Reports described a study in which a sample of 360 one-year-old baby boys were weighed. Their mean weight was 25.5 pounds with standard deviation 5.3 pounds. A pediatrician claims that the mean weight of one-year-old boys is greater than 25 pounds. Do the data provide convincing evidence that the pediatrician's claim is true? Use the $\alpha = 0.01$ level of significance. *Do not reject H_0.*

52. Good credit: The Fair Isaac Corporation (FICO) credit score is used by banks and other lenders to determine whether someone is a good credit risk. Scores range from 300 to 850, with a score of 720 or more indicating that a person is a very good credit risk. An economist wants to determine whether the mean FICO score is lower than the cutoff of 720. She finds that a random sample of 100 people had a mean FICO score of 703 with a standard deviation of 92. Can the economist conclude that the mean FICO score is less than 720? Use the $\alpha = 0.05$ level of significance. *Reject H_0.*

53. Commuting to work: The American Community Survey sampled 1923 people in Colorado and asked them how long it took them to commute to work each day. The sample mean one-way commute time was 24.5 minutes with a standard deviation of 13.0 minutes. A transportation engineer claims that the mean commute time is less than 25 minutes. Do the data provide convincing evidence that the engineer's claim is true? Use the $\alpha = 0.05$ level of significance. *Reject H_0.*

54. Watching TV: The General Social Survey asked a sample of 1298 people how much time they spent watching TV each day. The mean number of hours was 3.09 with a standard deviation of 2.87. A sociologist claims that people watch a mean of three hours of TV per day. Do the data provide sufficient evidence to disprove the claim? Use the $\alpha = 0.01$ level of significance. *Do not reject H_0.*

55. Weight loss: In a study to determine whether counseling could help people lose weight, a sample of people experienced a group-based behavioral intervention, which involved weekly meetings with a trained interventionist for a period of six months. The following data are the numbers of pounds lost for 14 people, based on means and standard deviations given in the article.

18.2	24.8	3.9	20.0	17.1	8.8	13.4
17.3	33.8	29.7	8.5	31.2	19.3	15.1

Source: *Journal of the American Medical Association* 299:1139–1148

a. Following is a boxplot for these data. Is it reasonable to assume that the conditions for performing a hypothesis test are satisfied? Explain. *Yes*

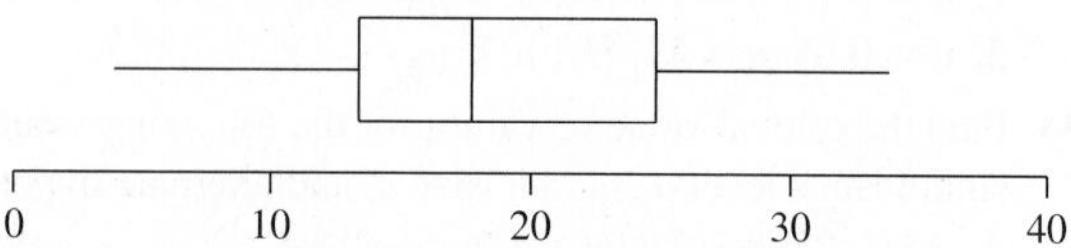

b. If appropriate, perform a hypothesis test to determine whether the mean weight loss is greater than 10 pounds. Use the $\alpha = 0.05$ level of significance. What do you conclude? *Reject H_0.*

Comstock Images

56. How much is in that can? A machine that fills beverage cans is supposed to put 12 ounces of beverage in each can. Following are the amounts measured in a simple random sample of eight cans.

11.96 12.10 12.04 12.13 11.98 12.05 11.91 12.03

a. Following is a dotplot for these data. Is it reasonable to assume that the conditions for performing a hypothesis test are satisfied? Explain. *Yes*

11.90 11.95 12.00 12.05 12.10 12.15

b. If appropriate, perform a hypothesis test to determine whether the mean volume differs from 12 ounces. Use the $\alpha = 0.05$ level of significance. What do you conclude? *Do not reject H_0.*

57. Credit card debt: Following are outstanding credit card balances for a sample of 16 college seniors at a large university.

870	419	1021	723	152	387	335	334
2618	529	593	769	502	485	1213	347

a. Following is a dotplot for these data. Is it reasonable to assume that the conditions for performing a hypothesis test are satisfied? Explain. *No*

0 500 1000 1500 2000 2500 3000

b. According to the report *How America Pays for College*, by Sallie Mae, the mean outstanding balance for college seniors in 2012 was \$515. If appropriate, perform a hypothesis test to determine whether the mean debt for seniors at this university differs from \$515. *Not appropriate*

58. Rats: A psychologist is designing an experiment in which rats will navigate a maze. Ten rats run the maze, and the time it takes for each to complete the maze is recorded. The results are as follows.

66.3	68.1	52.5	68.3	62.6
55.6	42.1	60.9	69.2	69.3

a. Following is a boxplot for these data. Is it reasonable to assume that the conditions for performing a hypothesis test are satisfied? Explain. *No*

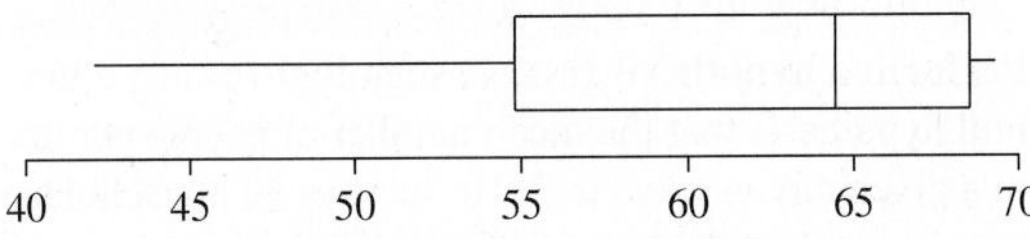

b. The psychologist hopes that the mean time for a rat to run the maze will be greater than 60 seconds. If appropriate, perform a hypothesis test to determine whether the mean time is greater than 60 seconds. *Not appropriate*

59. Keep cool: Following are prices, in dollars, of a random sample of ten 7.5-cubic-foot refrigerators.

314	377	330	285	319
274	332	350	299	306

a. Following is a dotplot for these data. Is it reasonable to assume that the conditions for performing a hypothesis test are satisfied? Explain. *Yes*

260 280 300 320 340 360 380

b. A consumer organization reports that the mean price of 7.5-cubic-foot refrigerators is greater than \$300. Do the data provide convincing evidence of this claim? Use the $\alpha = 0.01$ level of significance. *Do not reject H_0.*

60. Free dessert: In an attempt to increase business on Monday nights, a restaurant offers a free dessert with every dinner order. Before the offer, the mean number of dinner customers on Monday was 150. Following are the numbers of diners on a random sample of 12 days while the offer was in effect.

206	169	191	152	212	139
142	151	174	220	192	153

a. Following is a boxplot for these data. Is it reasonable to assume that the conditions for performing a hypothesis test are satisfied? Explain. *Yes*

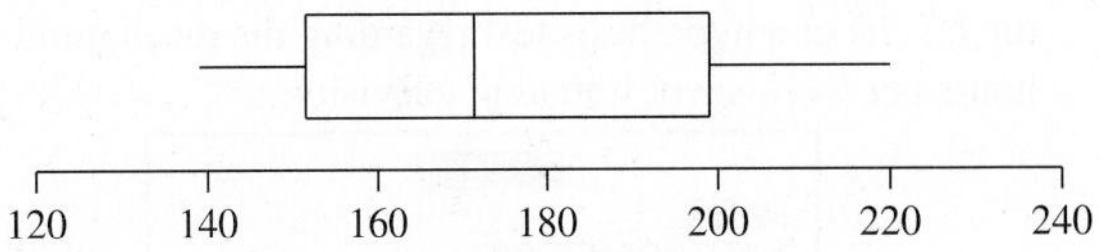

b. Can you conclude that the mean number of diners increased while the free dessert offer was in effect? Use the $\alpha = 0.01$ level of significance. *Reject H_0.*

61. Effective drugs: When testing a new drug, scientists measure the amount of the active ingredient that is absorbed by the body. In a study done at the Colorado School of Mines, a new antifungal medication that was designed to be applied to the skin was tested. The medication was applied to the skin of eight adult subjects. One hour later, the amount of active ingredient that had been absorbed into the skin was measured for each subject. The results, in micrograms, were

1.28 1.81 2.71 3.13 1.55 2.55 3.36 3.86

a. Construct a boxplot for these data. Is it appropriate to perform a hypothesis test? *Yes*

b. If appropriate, perform a hypothesis test to determine whether the mean amount absorbed is greater than 2 micrograms. Use the $\alpha = 0.05$ level of significance. What do you conclude? *Do not reject H_0.*

62. More effective drugs: An antifungal medication was applied to the skin of eight adult subjects. One hour later, the amount of active ingredient that had been absorbed into the skin was measured for each subject. The results, in micrograms, were

2.13 1.88 2.07 1.19 2.51 5.61 2.81 3.05

a. Construct a boxplot for these data. Is it appropriate to perform a hypothesis test? *No*

b. If appropriate, perform a hypothesis test to determine whether the mean amount absorbed is less than 3 micrograms. Use the $\alpha = 0.05$ level of significance. What do you conclude? *Not appropriate*

63. Interpret calculator display: A sample of adults was asked how many hours per day they spend on social media. The following display from a TI-84 Plus calculator presents the results of a hypothesis test regarding the mean number of hours per day spent social media.

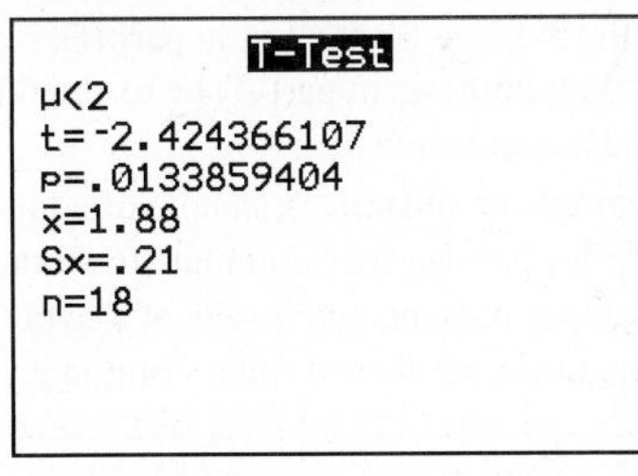

a. State the null and alternate hypotheses. *H_0: $\mu = 2$, H_1: $\mu < 2$*

b. What is the value of $\bar{x}$? *1.88*

c. What is the value of s? *0.21*

d. How many degrees of freedom are there? *17*

e. Do you reject H_0 at the 0.05 level? State a conclusion. *Yes*

f. Someone wants to test the hypothesis H_0: $\mu = 1.8$ versus H_1: $\mu > 1.8$. Use the information in the display to compute the t statistic for this test. *1.616*

g. Compute the P-value for the test in part (f). *0.0622*

h. Can the null hypothesis in part (f) be rejected at the 0.05 level? State a conclusion. *No*

64. Interpret calculator display: A sample of adults was asked how many hours per week they spend watching television. The following display from a TI-84 Plus calculator presents the results of a hypothesis test regarding the mean number of hours per week spent watching television.

```
T-Test
μ≠20
t=1.206976206
p=.2347112421
x̄=20.75
Sx=3.93
n=40
```

a. State the null and alternate hypotheses. H_0: $\mu = 20$, H_1: $\mu \neq 20$
b. What is the value of $\bar{x}$? 20.75
c. What is the value of s? 3.93
d. How many degrees of freedom are there? 39
e. Do you reject H_0 at the 0.05 level? State a conclusion. No
f. Someone wants to test the hypothesis H_0: $\mu = 22.5$ versus H_1: $\mu \neq 22.5$. Use the information in the display to compute the t statistic for this test. −2.816
g. Compute the P-value for the test in part (f).
h. Can the null hypothesis in part (f) be rejected at the 0.05 level? State a conclusion. Yes

65. Interpret computer output: A veterinarian recorded the weights, in grams, for a sample of hamsters. The following MINITAB output presents the results of a hypothesis test regarding the mean weight of hamsters.

```
Test of mu = 5.5 vs > 5.5

                                           95%
                                          Lower
N     Mean     StDev    SE Mean   Bound      T      P
5   5.92563  0.15755   0.07046   5.77542   6.04   0.002
```

a. State the null and alternate hypotheses. H_0: $\mu = 5.5$, H_1: $\mu > 5.5$
b. What is the value of $\bar{x}$? 5.92563
c. What is the value of s? 0.15755
d. How many degrees of freedom are there? 4
e. Do you reject H_0 at the 0.05 level? State a conclusion. Yes
f. Someone wants to test the hypothesis H_0: $\mu = 6.5$ versus H_1: $\mu < 6.5$. Use the information in the output to compute the t statistic for this test. −8.152
g. Compute the P-value for the test in part (f). 0.00062
h. Can the null hypothesis in part (f) be rejected at the 0.05 level? State a conclusion. Yes

66. Interpret computer output: A sample of adults was asked how many miles per day they commute to work. The following MINITAB output presents the results of a hypothesis test regarding the mean number of miles commuted to work.

```
Test of mu = 16 vs not = 16

 N    Mean   StDev  SE Mean        95% CI             T      P
11 13.2874  6.0989   1.8389  (9.1901, 17.3847)  -1.48  0.171
```

a. State the null and alternate hypotheses. H_0: $\mu = 16$, H_1: $\mu \neq 16$
b. What is the value of $\bar{x}$? 13.2874
c. What is the value of s? 6.0989
d. How many degrees of freedom are there? 10
e. Do you reject H_0 at the 0.05 level? State a conclusion. No
f. Someone wants to test the hypothesis H_0: $\mu = 9$ versus H_1: $\mu > 9$. Use the information in the output to compute the t statistic for this test. 2.332
g. Compute the P-value for the test in part (f). 0.0210
h. Can the null hypothesis in part (f) be rejected at the 0.05 level? State a conclusion. Yes

67. Does this diet work? In a study of the effectiveness of a certain diet, 100 subjects went on the diet for a period of six months. The sample mean weight loss was 0.5 pound, with a sample standard deviation of 4 pounds.
a. Find the t statistic for testing H_0: $\mu = 0$ versus H_1: $\mu > 0$. 1.25
b. Find the P-value for testing H_0: $\mu = 0$ versus H_1: $\mu > 0$. 0.1071
c. Can you conclude that the diet produces a mean weight loss that is greater than 0? Use the $\alpha = 0.05$ level of significance. No

68. Effect of larger sample size: The study described in Exercise 67 is repeated with a larger sample of 1000 subjects. Assume that the sample mean is once again 0.5 pound and the sample standard deviation is once again 4 pounds.
a. Find the t statistic for testing H_0: $\mu = 0$ versus H_1: $\mu > 0$. Is the value of the t statistic greater than or less than the value obtained with a smaller sample of 100? 3.953; larger
b. Find the P-value for testing H_0: $\mu = 0$ versus H_1: $\mu > 0$. 0.000041
c. Can you conclude that the diet produces a mean weight loss that is greater than 0? Use the $\alpha = 0.05$ level of significance. Yes
d. Explain why the mean weight loss is not of practical significance, even though the results are statistically significant at the 0.05 level.

69. Perform a hypothesis test? A sociologist wants to test the null hypothesis that the mean number of people per household in a given city is equal to 3. He surveys 50 households on a certain block in the city and finds that the sample mean number of people is 3.4 with a standard deviation of 1.2. Should these data be used to perform a hypothesis test? Explain why or why not. No

70. Perform a hypothesis test? A health professional wants to test the null hypothesis that the mean length of hospital stay for a certain surgical procedure is 4 days. She obtains records for all the patients who have undergone the procedure at a certain hospital during a given year and finds that the mean length of stay is 4.7 days with a standard deviation of 1.1 days. Should these data be used to perform a hypothesis test? Explain why or why not. No

71. Larger or smaller *P*-value? In a study of sleeping habits, a researcher wants to test the null hypothesis that adults in a certain community get a mean of 8 hours of sleep versus the alternative that the mean is not equal to 8. In a sample of 250 adults, the mean number of hours of sleep was 8.2. A second researcher repeated the study with a different sample of 250 and obtained a sample mean of 7.5. Both researchers obtained the same standard deviation. Will the P-value of the second researcher be greater than or less than that of the first researcher? Explain. less

72. Larger or smaller *P*-value? Juan and Mary want to test the null hypothesis that the mean length of text messages sent by students at their school is 10 characters versus the alternative that it is less. Juan samples 100 text messages and finds the mean length to be 8.4 characters. Mary samples 100 messages and finds the mean length to be 7.3 characters. Both Juan and Mary obtained the same standard deviation. Will Juan's P-value be greater than, less than, or the same as Mary's P-value? Explain. greater

73. Statistical or practical significance: A new method of teaching arithmetic to elementary school students was evaluated. The students who were taught by the new method were given a standardized test with a maximum score of 100 points. They scored an average of one point higher than students taught by the old method. A hypothesis test was performed in which the null hypothesis stated that there was no difference between the two groups, and the alternate hypothesis stated that the mean score for the new method was higher. The *P*-value was 0.001. True or false:

a. Because the *P*-value is very small, we can conclude that the mean score for students taught by the new method is higher than for students taught by the old method. *True*

b. Because the *P*-value is very small, we can conclude that the new method represents an important improvement over the old method. *False*

74. Statistical or practical significance: A new method of postoperative treatment was evaluated for patients undergoing a certain surgical procedure. Under the old method, the mean length of hospital stay was 6.3 days. The sample mean for the new method was 6.1 days. A hypothesis test was performed in which the null hypothesis stated that the mean length of stay was the same for both methods, and the alternate hypothesis stated that the mean stay was lower for the new method. The *P*-value was 0.002. True or false:

a. Because the *P*-value is very small, we can conclude that the new method provides an important reduction in the mean length of hospital stay. *False*

b. Because the *P*-value is very small, we can conclude that the mean length of hospital stay is less for patients treated by the new method than for patients treated by the old method. *True*

75. Interpret a *P*-value: A real estate agent believes that the mean size of houses in a certain city is greater than 1500 square feet. He samples 100 houses and performs a test of H_0: $\mu = 1500$ versus H_1: $\mu > 1500$. He obtains a *P*-value of 0.0002.

a. The real estate agent concludes that because the *P*-value is very small, the mean house size must be much greater than 1500. Is this conclusion justified? *No*

b. Another real estate agent says that because the *P*-value is very small, we can be fairly certain that the mean size is greater than 1500, but we cannot conclude that it is a lot greater. Is this conclusion justified? *Yes*

76. Interpret a *P*-value: The manufacturer of a medication designed to lower blood pressure claims that the mean systolic blood pressure for people taking their medication is less than 135. To test this claim, blood pressure is measured for a sample of 500 people who are taking the medication. The *P*-value for testing H_0: $\mu = 135$ versus H_1: $\mu < 135$ is $P = 0.001$.

a. The manufacturer concludes that because the *P*-value is very small, we can be fairly certain that the mean pressure is less than 135, but we cannot conclude that it is a lot smaller. Is this conclusion justified? *Yes*

b. Someone else says that because the *P*-value is very small, we can conclude that the mean pressure is a lot less than 135. Is this conclusion justified? *No*

77. Test scores: A math teacher has developed a new program to help high school students prepare for the math SAT. A sample of 100 students enroll in the program. They take a math SAT exam before the program starts and again at the end to measure their improvement. The mean number of points improved was $\bar{x} = 2.5$. The sample standard deviation was $s = 10$. Let μ be the population mean number of points improved. To determine whether the program is effective, a test is made of the hypotheses H_0: $\mu = 0$ versus H_1: $\mu > 0$.

a. Compute the value of the test statistic. *2.50*

b. Do you reject H_0 at the $\alpha = 0.05$ level? *Yes*

c. Is the result of practical significance? Explain. *No*

78. Weight loss: A doctor has developed a new diet to help people lose weight. A random sample of 500 people went on the diet for six weeks. The mean number of pounds lost was $\bar{x} = 0.5$. The sample standard deviation is $s = 5$. Let μ be the population mean number of pounds lost. To determine whether the diet is effective, a test is made of the hypotheses H_0: $\mu = 0$ versus H_1: $\mu > 0$.

a. Compute the value of the test statistic. *2.24*

b. Do you reject H_0 at the $\alpha = 0.05$ level? *Yes*

c. Is the result of practical significance? Explain. *No*

79. Enough information? A test was made of the hypotheses H_0: $\mu = 70$ versus H_1: $\mu \neq 70$. A report of the results stated: "$P < 0.05$, so we reject H_0 at the $\alpha = 0.05$ level." Is there any additional information that should have been included in the report? If so, what is it? *Yes; P-value*

80. Enough information? A test was made of the hypotheses H_0: $\mu = 10$ versus H_1: $\mu > 10$. A report of the results stated: The critical value was 1.645. Since $z > 1.645$, we reject H_0 at the $\alpha = 0.05$ level. Is there any additional information that should have been included in the report? If so, what is it? *Yes; value of test statistic*

Extending the Concepts

81. Somebody's wrong: Cindy computes a 95% confidence interval for μ and obtains (94.6, 98.3). Luis performs a test of the hypotheses H_0: $\mu = 100$ versus H_1: $\mu \neq 100$ and obtains a *P*-value of 0.12. Explain why they can't both be right.

82. Large samples and practical significance: A sample of size $n = 100$ is used to test H_0: $\mu = 20$ versus H_1: $\mu > 20$. The value of μ will not have practical significance unless $\mu > 25$. The sample standard deviation is $s = 10$. The value of $\bar{x}$ is 21.

a. Assume the sample size is $n = 100$. Compute the *P*-value. Show that you do not reject H_0 at the $\alpha = 0.05$ level. *$0.10 < P < 0.25$ [Tech: $P = 0.1599$]*

b. Assume the sample size is $n = 1000$. Compute the *P*-value. Show that you reject H_0 at the $\alpha = 0.05$ level. *$0.00005 < P < 0.001$ [Tech: $P = 0.0008$]*

c. Is the difference of practical significance? Explain. *No*

d. Explain why a larger sample can be more likely to produce a statistically significant result that is not practically significant.

83. Using z instead of t: When the sample size is large, some people treat the sample standard deviation s as if it were the population standard deviation σ, and use the standard normal distribution rather than the Student's t distribution, to find a critical value. Assume that a right-tailed test will be made with a sample of size 100 from a normal population, using the $\alpha = 0.05$ significance level.

a. Find the critical value under the assumption that σ is known. *1.645*

b. In fact, σ is unknown. How many degrees of freedom should be used for the Student's t distribution? *99*

c. What is the probability of rejecting H_0 when it is true if the critical value in part (a) is used? You will need technology to find the answer. *0.0516*

Answers to Check Your Understanding Exercises for Section 9.3

1. **a.** P-value is between 0.01 and 0.025 [Tech: 0.0127]
   b. P-value is between 0.025 and 0.05 [Tech: 0.0485]
   c. P-value is between 0.005 and 0.01 [Tech: 0.0096]
   d. P-value is between 0.01 and 0.02 [Tech: 0.0148]
2. **a.** $t = 2.309$ **b.** 40
   c. Between 0.01 and 0.025 [Tech: 0.0131]
   d. Weaker; the P-value is larger.
3. 0.0128
4. **a.** H_0: $\mu = 7$, H_1: $\mu < 7$ **b.** 6 **c.** 5 **d.** 6.68
   e. 0.205 **f.** −3.823593745 **g.** 0.00616394 **h.** Yes
5. $0.025 < P < 0.05$ [Tech: 0.0312]
6. **a.** 2.920 **b.** −2.787, 2.787 **c.** 1.292 **d.** −2.160, 2.160
7. **a.** No **b.** Yes **c.** Yes **d.** Yes
8. **a.** Yes **b.** No
9. 0.05
10. **a.** Charlie **b.** Felice
11. **a.** Contains enough information
    b. The value of the test statistic needs to be added.
    c. Contains enough information
    d. The P-value needs to be added.
12. **a.** $t = 3.16$. $0.0005 < P < 0.001$ [Tech: $P = 0.0008$], so H_0 is rejected at the $\alpha = 0.01$ level.
    b. No. The difference between 100 and 101 is not of practical significance.

Section 9.4 Hypothesis Tests for Proportions

Objectives

1. Test a hypothesis about a proportion using the P-value method
2. Test a hypothesis about a proportion using the critical value method

Objective 1 Test a hypothesis about a proportion using the P-value method

Can virtual reality be used to enhance education? In a recent survey of teachers conducted by Samsung, 85% of them stated that using virtual reality in the classroom would have a positive effect on education. One educational technology specialist believes that the percentage has now increased to more than 90% since virtual reality equipment has become more available. She samples 500 teachers and finds that 471 of them believe that virtual reality would have a positive effect. Can she conclude that the proportion of teachers who believe that virtual reality would have a positive effect is greater than 0.90?

This is an example of a problem that calls for a hypothesis test about a population proportion. There are two categories, those who believe that virtual reality would have a positive effect, and those who do not. The quantity 0.90 represents the proportion who believe that virtual reality would have a positive effect. To perform the test we will need some notation, which we summarize as follows.

NOTATION

- p is the population proportion of individuals who are in a specified category.
- p_0 is the population proportion specified by H_0.
- x is the number of individuals in the sample who are in the specified category.
- n is the sample size.
- $\hat{p}$ is the sample proportion of individuals who are in the specified category. $\hat{p} = x/n$.

We can perform a test whenever the sample proportion $\hat{p}$ is approximately normally distributed. This will occur when the following assumptions are met.

EXPLAIN IT AGAIN

Reasons for the assumptions: The population must be much larger than the sample (at least 20 times as large), so that the sampled items are independent. The assumption that both np_0 and $n(1 - p_0)$ are at least 10 ensures that the sampling distribution of $\hat{p}$ is approximately normal when we assume that H_0 is true.

Assumptions for Performing a Hypothesis Test for a Population Proportion

1. We have a simple random sample.
2. The population is at least 20 times as large as the sample.
3. The individuals in the population are divided into two categories.
4. The values np_0 and $n(1 - p_0)$ are both at least 10.

Either the critical value method or the *P*-value method may be used to perform a hypothesis test for a population proportion. We will present the steps for the *P*-value method first.

Performing a Hypothesis Test for a Population Proportion Using the *P*-Value Method

Check to be sure the assumptions are satisfied. If they are, then proceed with the following steps:

Step 1: State the null and alternate hypotheses. The null hypothesis will have the form H_0: $p = p_0$. The alternate hypothesis will be $p < p_0$, $p > p_0$, or $p \neq p_0$.

Step 2: If making a decision, choose a significance level α.

Step 3: Compute the test statistic $z = \dfrac{\hat{p} - p_0}{\sqrt{\dfrac{p_0(1 - p_0)}{n}}}$.

Step 4: Compute the *P*-value. The *P*-value is an area under the standard normal curve; it depends on the alternate hypothesis as follows:

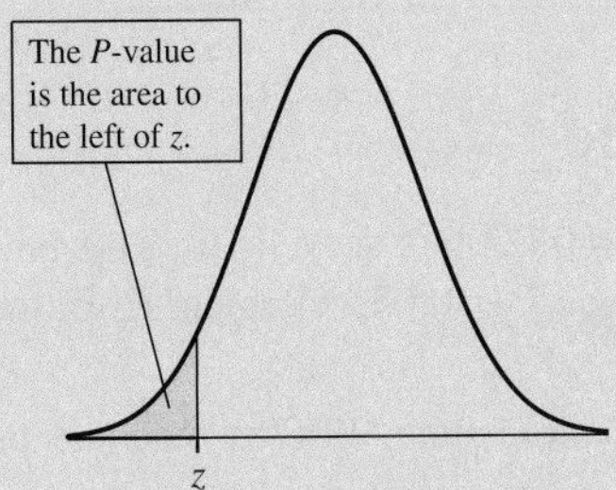
The P-value is the area to the left of z.
z
Left-tailed: H_1: $p < p_0$

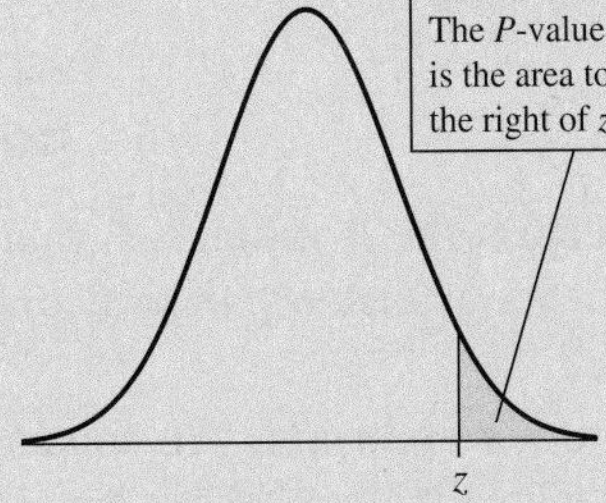
The P-value is the area to the right of z.
z
Right-tailed: H_1: $p > p_0$

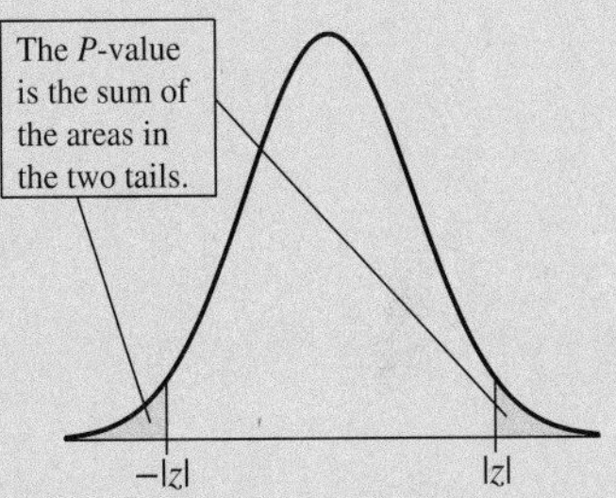
The P-value is the sum of the areas in the two tails.
−|z|
|z|
Two-tailed: H_1: $p \neq p_0$

Step 5: Interpret the *P*-value. If making a decision, reject H_0 if the *P*-value is less than or equal to the significance level α.

Step 6: State a conclusion.

Example 9.19

Perform a hypothesis test

An educational technology specialist is studying attitudes of teachers about the use of virtual reality in the classroom. She samples 500 teachers and finds that 471 of them believe that virtual reality would have a positive effect. Can she conclude that the proportion of teachers who believe that virtual reality would have a positive effect is greater than 0.90? Use the $\alpha = 0.05$ level of significance.

Solution

We first check the assumptions. We have a simple random sample of teachers. The members of the population fall into two categories: those who believe that virtual reality would have a positive effect, and those who do not. The population of teachers is more than 20 times the sample size of $n = 500$. The proportion specified by the null hypothesis is $p_0 = 0.90$. Now $np_0 = (500)(0.90) = 450 > 10$ and $n(1 - p_0) = (500)(1 - 0.90) = 50 > 10$. The assumptions are satisfied.

Step 1: State H_0 and H_1: We are asked whether we can conclude that the population proportion p is greater than 0.90. The null and alternate hypotheses are therefore

$$H_0: p = 0.90 \quad H_1: p > 0.90$$

Step 2: Choose a significance level: The significance level is $\alpha = 0.05$.

Step 3: Compute the test statistic: The sample proportion $\hat{p}$ is

$$\hat{p} = \frac{471}{500} = 0.942$$

The value of p specified by the null hypothesis is $p_0 = 0.90$. The test statistic is the z-score of $\hat{p}$:

$$z = \frac{\hat{p} - p_0}{\sqrt{\frac{p_0(1-p_0)}{n}}} = \frac{0.942 - 0.90}{\sqrt{\frac{0.90(1-0.90)}{500}}} = 3.13$$

Step 4: Compute the *P*-value. The alternate hypothesis is H_1: $\mu > 0.90$, which is right-tailed. The *P*-value is therefore the area to the right of $z = 3.13$. Using Table A.2, we see that the area to the left of $z = 3.13$ is 0.9991. The area to the right of $z = 3.13$ is therefore $1 - 0.9991 = 0.0009$. The *P*-value is $P = 0.0009$. See Figure 9.18.

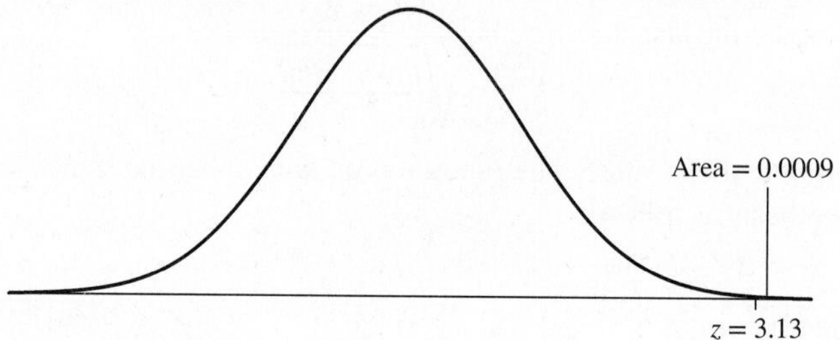

Figure 9.18

Step 5: Interpret the *P*-value. A *P*-value of $P = 0.0009$ is very small. This is very strong evidence against H_0. In particular, because $P < 0.05$, we reject H_0 at the $\alpha = 0.05$ level.

Step 6: State a conclusion. We conclude that more than 90% of teachers believe that virtual reality would have a positive effect on education.

Check Your Understanding

1. The Pew Research Center reported that only 15% of 18- to 24-year-olds read a daily newspaper. The publisher of a local newspaper wants to know whether the percentage of newspaper readers among students at a nearby large university differs from the percentage among 18- to 24-year-olds in general. She surveys a simple random sample of 200 students at the university and finds that 40 of them, or 20%, read a newspaper each day. Can she conclude that the proportion of students who read a daily newspaper differs from 0.15? Use the $\alpha = 0.05$ level of significance.

a. State the null and alternate hypotheses. *H_0: $p = 0.15$, H_1: $p \neq 0.15$*

b. Compute the test statistic. *$z = 1.98$*

c. Compute the *P*-value. *0.0478 [Tech: 0.0477]*

d. State a conclusion.

Answers are on page 488.

Performing a hypothesis test with technology

The following computer output (from MINITAB) presents the results of Example 9.19.

```
Test of p = 0.9 vs p > 0.9
                                         95%
                                        Lower
Sample      X     N    Sample p         Bound    Z-Value    P-Value
1         471   500    0.942000      0.921757       3.13      0.001
```

Most of the output is straightforward. The first line specifies the null and alternate hypotheses. The quantity labeled "X" is the number of teachers in the sample who believe that virtual reality would have a positive effect, and N is the sample size. The quantity labeled

"Sample p" is the sample proportion $\hat{p}$. The quantity labeled "95% Lower Bound" is a 95% lower confidence bound for the population proportion p. The interpretation of this quantity is that we are 95% confident that the population proportion p is greater than or equal to 0.921757. Next is the value of the test statistic z, labeled "Z-value," and finally at the end of the row is the P-value.

The following display from a TI-84 Plus calculator presents the results of Example 9.19.

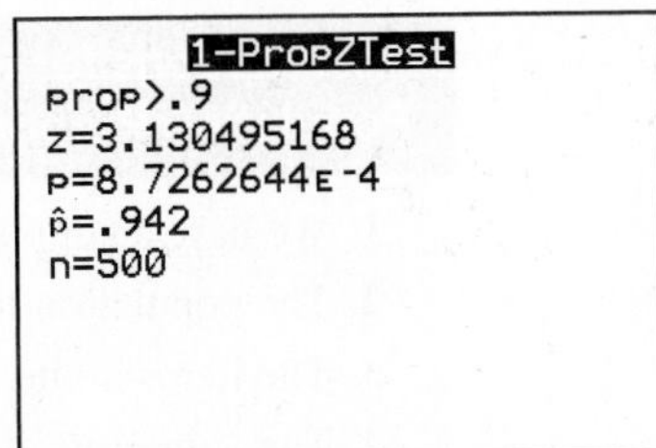

The first line in the display presents the alternate hypothesis. The word "prop" refers to the population proportion p. Note that the letter "p" in the third line is the P-value, not the population proportion. This number is written in scientific notation as 8.7262644E-4. This indicates that we should move the decimal point four places to the left, so $P = 0.00087262644$.

Step-by-step instructions for performing hypothesis tests with technology are presented in the Using Technology section on page 484.

Check Your Understanding

2. The following output from MINITAB presents the results of a hypothesis test.

Test of p = 0.60 vs p < 0.60

Sample	X	N	Sample p	95% Lower Bound	Z-Value	P-Value
1	72	150	0.480000	0.547097	−3.00	0.001

a. What are the null and alternate hypotheses? H_0: $p = 0.6$, H_1: $p < 0.6$
b. What is the sample size? 150
c. What is the value of $\hat{p}$? 0.48
d. What is the value of the test statistic? −3.00
e. Do you reject H_0 at the 0.05 level? Yes
f. Do you reject H_0 at the 0.01 level? Yes

3. The following display from a TI-84 Plus calculator presents the results of a hypothesis test.

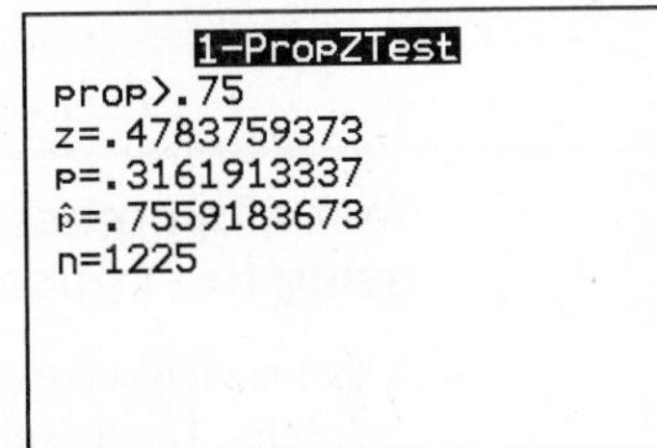

a. What are the null and alternate hypotheses? H_0: $p = 0.75$, H_1: $p > 0.75$
b. What is the sample size? 1225
c. What is the value of $\hat{p}$? 0.7559183673
d. What is the value of the test statistic? 0.4783759373
e. Do you reject H_0 at the 0.05 level? No
f. Do you reject H_0 at the 0.01 level? No

Answers are on page 488.

Objective 2 Test a hypothesis about a proportion using the critical value method

Testing Hypotheses for a Proportion Using the Critical Value Method

To use the critical value method, compute the test statistic as before. Because the test statistic is a z-score, critical values can be found in Table A.2, in the last line of Table A.3, or with technology. The assumptions for the critical value method are the same as for the P-value method.

Assumptions for Performing a Hypothesis Test for a Population Proportion

1. We have a simple random sample.
2. The population is at least 20 times as large as the sample.
3. The items in the population are divided into two categories.
4. The values np_0 and $n(1 - p_0)$ are both at least 10.

Following are the steps for the critical value method.

Performing a Hypothesis Test for a Proportion Using the Critical Value Method

Check to be sure the assumptions are satisfied. If they are, then proceed with the following steps:

Step 1: State the null and alternate hypotheses. The null hypothesis will have the form $H_0: p = p_0$. The alternate hypothesis will be $p < p_0$, $p > p_0$, or $p \neq p_0$.

Step 2: Choose a significance level α and find the critical value or values.

Step 3: Compute the test statistic $z = \dfrac{\hat{p} - p_0}{\sqrt{\dfrac{p_0(1 - p_0)}{n}}}$.

Step 4: Determine whether to reject H_0, as follows:

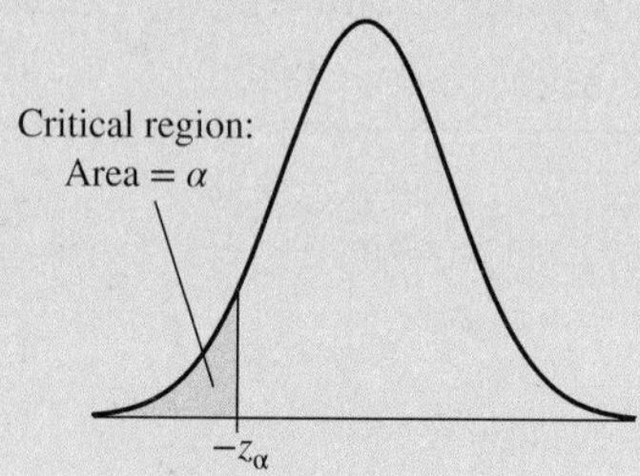

Left-tailed: $H_1: p < p_0$
Reject if $z \leq -z_\alpha$.

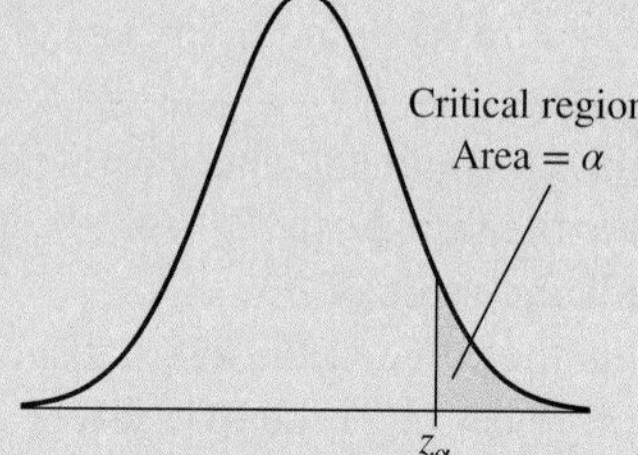

Right-tailed: $H_1: p > p_0$
Reject if $z \geq z_\alpha$.

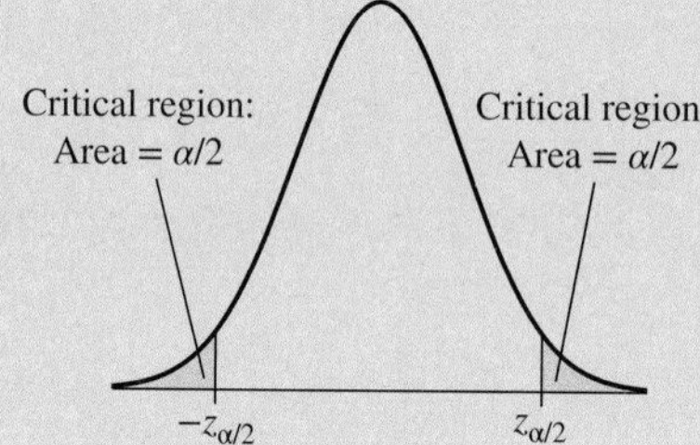

Two-tailed: $H_1: p \neq p_0$
Reject if $z \geq z_{\alpha/2}$ or $z \leq -z_{\alpha/2}$.

Step 5: State a conclusion.

Example 9.20

Test a hypothesis about a population proportion using the critical value method

A nationwide survey of working adults indicates that only 50% of them are satisfied with their jobs. The president of a large company believes that more than 50% of employees at his company are satisfied with their jobs. To test his belief, he surveys a random sample of 100 employees, and 54 of them report that they are satisfied with their jobs. Can he conclude that more than 50% of employees at the company are satisfied with their jobs? Use the $\alpha = 0.05$ level of significance.

Solution

We first check the assumptions. We have a simple random sample from the population of employees. Each employee is categorized as being satisfied or not satisfied. The sample size is $n = 100$ and the proportion p_0 specified by H_0 is 0.5. Therefore, we calculate that

$np_0 = 100(0.5) = 50 > 10$, and $n(1 - p_0) = 100(1 - 0.5) = 50 > 10$. If the total number of employees in the company is more than 2000, as we shall assume, then the population is more than 20 times as large as the sample. All the assumptions are therefore satisfied.

Step 1: State the null and alternate hypotheses. The issue is whether the proportion of employees that are satisfied with their jobs is more than 0.5. Therefore, the null and alternate hypotheses are

$$H_0: p = 0.5 \qquad H_1: p > 0.5$$

Step 2: Choose a significance level and find the critical value. The significance level is $\alpha = 0.05$. The alternate hypothesis is $p > 0.5$, so this is a right-tailed test. The critical value corresponding to $\alpha = 0.05$ is $z_\alpha = 1.645$.

Step 3: Compute the test statistic. The test statistic is

$$z = \frac{\hat{p} - p_0}{\sqrt{\dfrac{p_0(1 - p_0)}{n}}}$$

The value of the sample proportion $\hat{p}$ is

$$\hat{p} = \frac{\text{Number of satisfied employees}}{\text{Sample size}} = \frac{54}{100} = 0.54$$

The quantity p_0 is the value of p specified by H_0, so $p_0 = 0.5$. The sample size is $n = 100$. Therefore, the value of the test statistic is

$$z = \frac{0.54 - 0.5}{\sqrt{\dfrac{0.5(1 - 0.5)}{100}}} = 0.80$$

Step 4: Determine whether to reject H_0. Because this is a right-tailed test, we reject H_0 if $z \geq 1.645$. Because $0.80 < 1.645$, we do not reject H_0. See Figure 9.19.

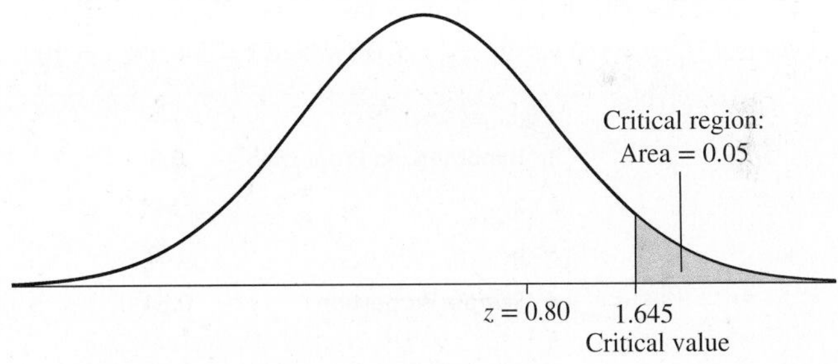

Figure 9.19

Step 5: State a conclusion. There is not enough evidence to conclude that the company president is correct in his belief that the proportion of employees who are satisfied with their jobs is greater than 0.5. The proportion may be equal to 0.5.

Check Your Understanding

4. A Gallup poll sampled 1000 adults in the United States. Of these people, 770 said they enjoyed situations in which they competed with other people. Can you conclude that less than 80% of U.S. adults like to compete? Use the critical value method with significance level $\alpha = 0.05$.

a. State the null and alternate hypotheses. H_0: $p = 0.8$, H_1: $p < 0.8$

b. Compute the test statistic. −2.37

c. Find the critical value. −1.645

d. State a conclusion.

Answers are on page 488.

Using Technology

We use Example 9.20 to illustrate the technology steps.

TI-84 PLUS

Testing a hypothesis about a proportion

Step 1. Press **STAT** and highlight the **TESTS** menu.

Step 2. Select **1–PropZTest** and press **ENTER** (Figure A). The **1–PropZTest** menu appears.

Step 3. Enter the values of p_0, x, and n. For Example 9.20, we use $p_0 = 0.5$, $x = 54$, and $n = 100$.

Step 4. Select the form of the alternate hypothesis. For Example 9.20, the alternate hypothesis has the form $> p_0$ (Figure B).

Step 5. Highlight **Calculate** and press **ENTER** (Figure C).

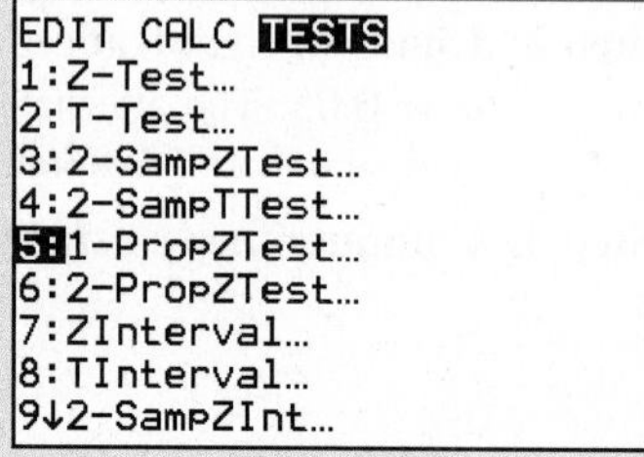

Figure A

1-PropZTest
p0:.5
x:54
n:100
prop:≠p0 <p0 >p0
Color: BLUE
Calculate Draw

Figure B

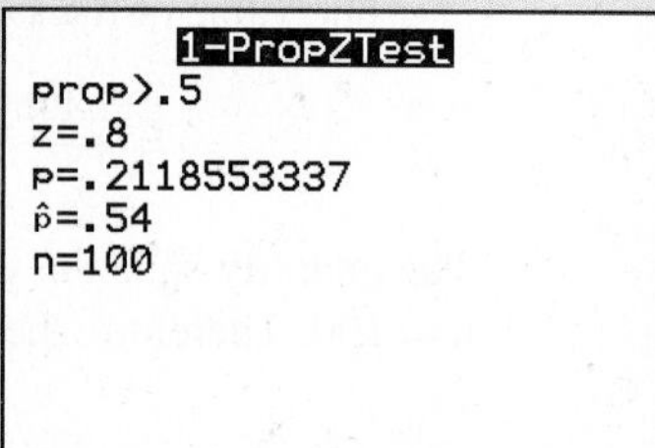

Figure C

EXCEL

Finding the test statistic for a hypothesis test about a population proportion

Figure D demonstrates the calculations involved for finding the test statistic for a hypothesis test about a population proportion. In Example 9.20, we test $H_0: p = 0.5$ versus $H_1: p > 0.5$ where $x = 54$ and $n = 100$.

	A	B	C	D	E
1	Hypothesized Proportion	0.5			
2	x	54			
3	n	100			
4	Sample Proportion	0.54			
5					
6	Standard Error	0.05	← =SQRT(B1*(1-B1)/B3)		
7	Test Statistic Z	0.8	← =(B4-B1)/B6		

Figure D

MINITAB

Testing a hypothesis about a proportion

Step 1. Click on **Stat**, then **Basic Statistics**, then **1 Proportion**.

Step 2. Select **Summarized Data**, and enter the value of n in the **Number of Trials** field and the value of x in the **Number of Events** field. For Example 9.20, we use $x = 54$ and $n = 100$ (Figure E).

Step 3. Select the **Perform hypothesis test** option and enter the value of p_0 in the **Hypothesized proportion** field. For Example 9.20, we enter 0.5 for p_0. Click on **Options**, and select the form of the alternate hypothesis. We use **Proportion > hypothesized proportion**.

Step 4. Given significance level α, enter $100(1 - \alpha)$ as the **Confidence Level**. For Example 9.20, $\alpha = 0.05$, so the confidence level is $100(1 - 0.05) = 95$ (Figure F).

Step 5. In the **Method** field, select the **Normal approximation** option and click **OK**.

Step 6. Click **OK** (Figure G).

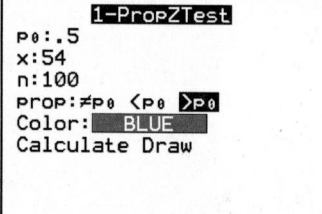

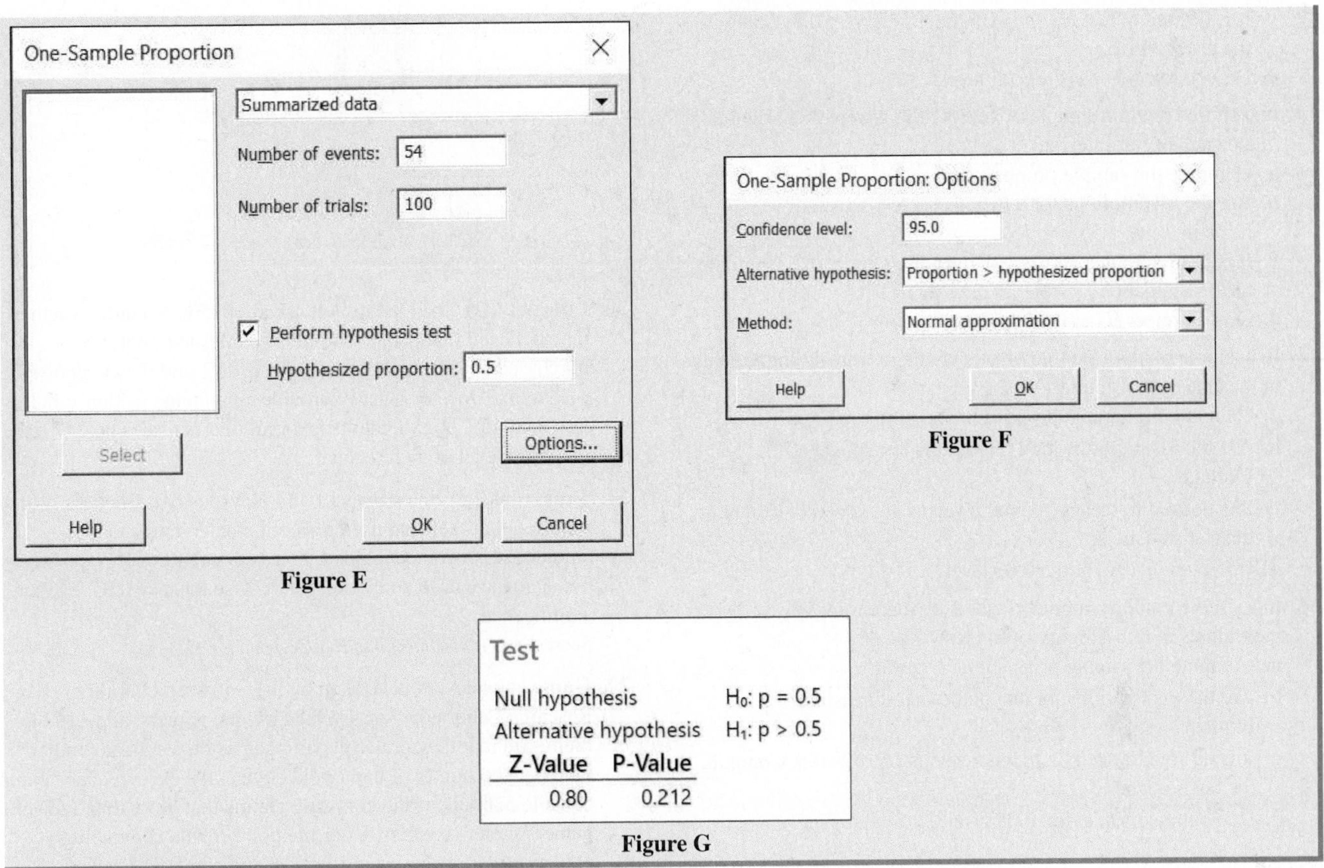

Figure E

Figure F

Figure G

Section 9.4 Exercises

Exercises 1–4 are the Check Your Understanding exercises located within the section.

Understanding the Concepts

In Exercises 5 and 6, fill in each blank with the appropriate word or phrase.

5. To test H_0: $p = p_0$ with the methods in this section, the values np_0 and $n(1 - p_0)$ must both be at least ________. *10*

6. To test H_0: $p = p_0$ with the methods in this section, the population size must be at least ________ times as large as the sample size. *20*

In Exercises 7 and 8, determine whether the statement is true or false. If the statement is false, rewrite it as a true statement.

7. When testing a hypothesis for a proportion, we assume that the items in the population are divided into two categories. *True*

8. When testing a hypothesis for a proportion, the alternate hypothesis is always two-tailed. *False*

9. In a simple random sample of 100 adults aged 18–25, 20 of them said they had purchased a phone within the past year. Is it appropriate to use the methods of this section to perform a hypothesis test about the proportion of adults aged 18–25 who have purchased a phone in the past year? If not, why not? *Yes*

10. In a simple random sample of 200 allergy sufferers, 125 of them reported obtaining relief from a new allergy medication. Is it appropriate to use the methods of this section to perform a hypothesis test about the proportion of allergy sufferers who experience relief from this medication? If not, why not? *Yes*

11. A simple random sample of 50 households is drawn from a neighborhood that contains a total of 600 households. Twenty of the households report that they own an SUV. Is it appropriate to use the methods of this section to perform a hypothesis test about the proportion of households in this neighborhood who own an SUV? If not, why not? *No*

12. Several thousand people participate in a Twitter poll in which they state which candidate they plan to vote for in an upcoming election. Fifty-two percent of the participants state that they plan to vote for the incumbent candidate. Is it appropriate to use the methods of this section to perform a hypothesis test about the proportion of voters who plan to vote for the incumbent candidate? If not, why not? *No*

Practicing the Skills

13. In a simple random sample of size 80, there were 54 individuals in the category of interest.
    a. Compute the sample proportion $\hat{p}$. *0.675*
    b. Are the assumptions for a hypothesis test satisfied? Explain. *Yes*

c. It is desired to test H_0: $p = 0.8$ versus H_1: $p < 0.8$. Compute the test statistic z. -2.80

d. Do you reject H_0 at the 0.05 level? Yes

14. In a simple random sample of size 60, there were 38 individuals in the category of interest.

a. Compute the sample proportion $\hat{p}$. 0.633

b. Are the assumptions for a hypothesis test satisfied? Explain. Yes

c. It is desired to test H_0: $p = 0.7$ versus H_1: $p \neq 0.7$. Compute the test statistic z. -1.13

d. Do you reject H_0 at the 0.05 level? No

15. In a simple random sample of size 75, there were 42 individuals in the category of interest.

a. Compute the sample proportion $\hat{p}$. 0.560

b. Are the assumptions for a hypothesis test satisfied? Explain. Yes

c. It is desired to test H_0: $p = 0.6$ versus H_1: $p \neq 0.6$. Compute the test statistic z. -0.71

d. Do you reject H_0 at the 0.05 level? No

16. In a simple random sample of size 150, there were 90 individuals in the category of interest.

a. Compute the sample proportion $\hat{p}$. 0.600

b. Are the assumptions for a hypothesis test satisfied? Explain. Yes

c. It is desired to test H_0: $p = 0.5$ versus H_1: $p > 0.5$. Compute the test statistic z. 2.45

d. Do you reject H_0 at the 0.05 level? Yes

Working with the Concepts

17. Spam: According to SecureList, 71.8% of all email sent is spam. A system manager at a large corporation believes that the percentage at his company may be 80%. He examines a random sample of 500 emails received at an email server, and finds that 382 of the messages are spam.

a. State the appropriate null and alternate hypotheses. H_0: $p = 0.8$, H_1: $p \neq 0.8$

b. Compute the test statistic z. -2.01

c. Using $\alpha = 0.05$, can you conclude that the percentage of emails that are spam differs from 80%? Yes

d. Using $\alpha = 0.01$, can you conclude that the percentage of emails that are spam differs from 80%? No

18. Confidence in banks: A poll conducted by the General Social Survey asked a random sample of 1325 adults in the United States how much confidence they had in banks and other financial institutions. A total of 149 adults said that they had a great deal of confidence. An economist claims that less than 15% of U.S. adults have a great deal of confidence in banks.

a. State the appropriate null and alternate hypotheses. H_0: $p = 0.15$, H_1: $p < 0.15$

b. Compute the test statistic z. -3.83

c. Using $\alpha = 0.05$, can you conclude that the economist's claim is true? Yes

d. Using $\alpha = 0.01$, can you conclude that the economist's claim is true? Yes

19. Kids with phones: A marketing manager for a phone company claims that more than 55% of children aged 8–12 have their own phones. In a survey of 802 children aged 8–12 by the National Consumers League, 449 of them had phones. Can you conclude that the manager's claim is true? Use the $\alpha = 0.01$ level of significance. Do not reject H_0.

locrifa/Shutterstock

20. Internet tax: The Gallup Poll asked 1015 U.S. adults whether they believed that people should pay sales tax on items purchased over the internet. Of these, 437 said they supported such a tax. Does the survey provide convincing evidence that less than 45% of U.S. adults favor an internet sales tax? Use the $\alpha = 0.05$ level of significance. Do not reject H_0.

21. Quit smoking: In a survey of 444 HIV-positive smokers, 170 reported that they had used a nicotine patch to try to quit smoking. Can you conclude that less than half of HIV-positive smokers have used a nicotine patch? Use the $\alpha = 0.05$ level of significance. Reject H_0.

Source: *American Journal of Health Behavior* 32:3–15

22. Game consoles: A poll taken by the Software Usability Research Laboratory surveyed 341 video gamers, and 110 of them said that they prefer playing games on a console, rather than on a computer or hand-held device. An executive at a game console manufacturing company claims that more than 25% of gamers prefer consoles. Does the poll provide convincing evidence that the claim is true? Use the $\alpha = 0.01$ level of significance. Reject H_0.

23. Tattoo: A Harris poll taken surveyed 2016 adults and found that 423 of them had one or more tattoos. Can you conclude that the percentage of adults who have a tattoo is less than 25%? Use the $\alpha = 0.01$ level of significance. Reject H_0.

24. Curing diabetes: Vertical banded gastroplasty is a surgical procedure that reduces the volume of the stomach in order to produce weight loss. In a recent study, 82 patients with Type 2 diabetes underwent this procedure, and 59 of them experienced a recovery from diabetes. Does this study provide convincing evidence that more than 60% of those with diabetes who undergo this surgery will recover from diabetes? Use the $\alpha = 0.05$ level of significance. Reject H_0.

Source: *New England Journal of Medicine* 357:753–761

25. Tweet tweet: An article in *Forbes* magazine reported that 73% of Fortune 500 companies have Twitter accounts. An economist thinks the percentage is higher at technology companies. She samples 70 technology companies and finds that 55 of them have Twitter accounts. Can she conclude that more than 73% of technology companies have Twitter accounts? Use the $\alpha = 0.05$ level of significance. Do not reject H_0.

26. Online photos: A Pew poll surveyed 1802 internet users and found that 829 of them had posted a photo or video online. Can you conclude that less than half of internet users have posted photos or videos online? Use the $\alpha = 0.05$ level of significance. Reject H_0.

27. Choosing a doctor: Which do patients value more when choosing a doctor: interpersonal skills or technical ability? In a recent study, 304 people were asked to choose a physician based on two hypothetical descriptions. One physician was described as having high technical skills and average

interpersonal skills, and the other was described as having average technical skills and high interpersonal skills. The physician with high interpersonal skills was chosen by 116 of the people. Can you conclude that less than half of patients prefer a physician with high interpersonal skills? Use the $\alpha = 0.01$ level of significance. *Reject H_0.*
Source: *Health Services Research* 40:957–977

28. Streaming video: A streaming service is considering raising its fees. A marketing executive at the company surveys a sample of 400 subscribers and asks them whether they would continue subscribing if fees were raised by 20%. A total of 215 of the 400 replied that they would continue subscribing. The marketing executive claims that more than 50% of its subscribers would continue subscribing. Can you conclude that the executive's claim is true? Use the $\alpha = 0.01$ level of significance. *Do not reject H_0.*

29. Blood sugar: High levels of blood sugar are associated with an increased risk of type 2 diabetes. A blood sugar level above normal is referred to as "impaired fasting glucose." An article in the journal *Environmental Health Perspectives* reports that in a sample of 155 people with impaired fasting glucose, 47 had type 2 diabetes. Can you conclude that less than 35% of people with impaired fasting glucose have type 2 diabetes? Use the $\alpha = 0.05$ level of significance. *Do not reject H_0.*
Source: *Environmental Health Perspectives* https:doi.org/10.1289/EHP2566

30. Knee injury: The anterior cruciate ligament (ACL) runs diagonally in the middle of the knee. An article in the *Journal of Bone and Joint Surgery* reported results for 85 people who had suffered ACL injuries. Of the 85 injuries, 51 were to the left knee and 34 were to the right knee. Can you conclude that more than half of ACL injuries are to the left knee? Use the $\alpha = 0.05$ level of significance. *Reject H_0.*
Source: *Journal of Bone and Joint Surgery* 99:897–904

31. Interpret calculator display: In a recent poll, people were asked whether they supported an increase in the sales tax. The following display from a TI-84 Plus calculator presents the results of a hypothesis test regarding the proportion of individuals who said they supported an increase in the sales tax.

```
1-PropZTest
prop<.5
z=-2.667891875
p=.0038164856
p̂=.3382352941
n=68
```

a. What are the null and alternate hypotheses?
b. What is the value of the sample proportion $\hat{p}$? *0.3382352941*
c. Can H_0 be rejected at the 0.05 level? State a conclusion. *Yes*
d. Someone wants to use these data to test $H_0\colon p = 0.25$ versus $H_1\colon p \neq 0.25$. Find the test statistic z and use the method of this section to find the P-value. Do you reject H_0 at the $\alpha = 0.05$ level? State a conclusion. *1.68; no*

32. Interpret calculator display: A sample of mothers was asked how old they were when their first child was born. The following display from a TI-84 Plus calculator presents the results of a hypothesis test regarding the proportion who had their first child before the age of 24.

```
1-PropZTest
prop≠.4
z=-.7071067812
p=.4794999735
p̂=.36
n=75
```

a. What are the null and alternate hypotheses? *H_0: $p = 0.4$, H_1: $p \neq 0.4$*
b. What is the value of the sample proportion $\hat{p}$? *0.36*
c. Can H_0 be rejected at the 0.05 level? State a conclusion. *No*
d. Someone wants to use these data to test $H_0\colon p = 0.5$ versus $H_1\colon p < 0.5$. Find the test statistic z, and use the method of this section to find the P-value. Do you reject H_0 at the $\alpha = 0.05$ level? State a conclusion. *−2.42; Yes*

33. Interpret computer output: A sample of college students was asked whether they had a job outside of school. The following MINITAB output presents the results of a hypothesis test regarding the proportion of college students who have a job outside of school.

```
Test of p = 0.6 vs p > 0.6
                               95%
                              Lower
  X     N   Sample p          Bound   Z-Value   P-Value
539   871   0.618829       0.591760      1.13     0.129
```

a. What are the null and alternate hypotheses? *H_0: $p = 0.6$, H_1: $p > 0.6$*
b. What is the value of the sample proportion $\hat{p}$? *0.618829*
c. Can H_0 be rejected at the 0.05 level? State a conclusion. *No*
d. Someone wants to use these data to test $H_0\colon p = 0.65$ versus $H_1\colon p < 0.65$. Find the test statistic z, and use the method of this section to find the P-value. Do you reject H_0 at the $\alpha = 0.05$ level? State a conclusion. *−1.93; yes*

34. Interpret computer output: A sample of adults was asked whether they were interested in economic issues. The following MINITAB output presents the results of a hypothesis test regarding the proportion who said they were interested in economic issues.

```
Test of p = 0.7 vs p not equal 0.7

 X  N Sample p          95% CI          Z-Value P-Value
27 52 0.519231 (0.383432, 0.655029)     -2.84    0.004
```

a. What are the null and alternate hypotheses? *H_0: $p = 0.7$, H_1: $p \neq 0.7$*
b. What is the value of the sample proportion $\hat{p}$? *0.519231*
c. Can H_0 be rejected at the 0.05 level? State a conclusion. *Yes*
d. Someone wants to use these data to test $H_0\colon p = 0.6$ versus $H_1\colon p \neq 0.6$. Find the test statistic z, and use the method of this section to find the P-value. Do you reject H_0 at the $\alpha = 0.05$ level? State a conclusion. *−1.19; no*

35. Satisfied with college? A simple random sample of 500 students at a certain college were surveyed and asked whether they were satisfied with college life. Two hundred eighty of them replied that they were satisfied. The Dean of Students claims that more than half of the students at the college are satisfied. To test this claim, a test of the hypotheses $H_0: p = 0.5$ versus $H_1: p > 0.5$ is performed.

a. Show that the P-value is 0.004.

b. The P-value is very small, so H_0 is rejected. Someone claims that because P is very small, the population proportion p must be much greater than 0.5. Is this a correct interpretation of the P-value? *No*

c. Someone else claims that because the P-value is very small, we can be fairly certain that the population proportion p is greater than 0.5, but we cannot be certain that it is a lot greater. Is this a correct interpretation of the P-value? *Yes*

36. Who will you vote for? A simple random sample of 1500 voters were surveyed and asked whether they were planning to vote for the incumbent mayor for re-election. Seven hundred ninety-eight of them replied that they were planning to vote for the mayor. The mayor claims that more than half of all voters are planning to vote for her. To test this claim, a test of the hypotheses $H_0: p = 0.5$ versus $H_1: p > 0.5$ is performed.

a. Show that the P-value is 0.007.

b. The P-value is very small, so H_0 is rejected. A pollster claims that because the P-value is very small, we can be fairly certain that the population proportion p is greater than 0.5, but we cannot be certain that it is a lot greater. Is this a correct interpretation of the P-value? *Yes*

c. The mayor's campaign manager claims that because the P-value is very small, the population proportion of voters who plan to vote for the mayor must be much greater than 0.5. Is this a correct interpretation of the P-value? *No*

37. Don't perform a test: A few weeks before election day, a TV station broadcast a debate between the two leading candidates for governor. Viewers were invited to send a tweet to indicate which candidate they plan to vote for. A total of 3125 people tweeted, and 1800 of them said that they planned to vote for candidate A. Explain why these data should not be used to test the claim that more than half of the voters plan to vote for candidate A.

38. Don't perform a test: Over the past 100 days, the price of a certain stock went up on 60 days and went down on 40 days. Explain why these data should not be used to test the claim that this stock price goes down on less than half of the days.

Extending the Concepts

39. Exact test: When $np_0 < 10$ or $n(1 - p_0) < 10$, we cannot use the normal approximation, but we can use the binomial distribution to perform what is known as an *exact test*. Let p be the probability that a given coin lands heads. The coin is tossed 10 times and comes up heads 9 times. Test $H_0: p = 0.5$ versus $H_1: p > 0.5$, as follows.

a. Let n be the number of tosses and let X denote the number of heads. Find the values of n and X in this example. *$n = 10, X = 9$*

b. The distribution of X is binomial. Assuming H_0 is true, find n and p. *$n = 10, p = 0.5$*

c. Because the alternate hypothesis is $p > 0.5$, large values of X support H_1. Find the probability of observing a value of X as extreme as or more extreme than the value actually observed, assuming H_0 to be true. This is the P-value. *0.0107*

d. Do you reject H_0 at the $\alpha = 0.05$ level? *Yes*

Answers to Check Your Understanding Exercises for Section 9.4

1. a. $H_0: p = 0.15$, $H_1: p \neq 0.15$ **b.** $z = 1.98$
c. $P = 0.0478$ [Tech: 0.0477]
d. We conclude that the proportion of students who read a newspaper differs from 0.15.

2. a. $H_0: p = 0.6$, $H_1: p < 0.6$ **b.** 150 **c.** 0.48
d. −3.00 **e.** Yes **f.** Yes

3. a. $H_0: p = 0.75$, $H_1: p > 0.75$ **b.** 1225
c. 0.7559183673 **d.** 0.4783759373 **e.** No **f.** No

4. a. $H_0: p = 0.80$, $H_1: p < 0.80$ **b.** $z = -2.37$
c. −1.645
d. We conclude that less than 80% of U.S. adults enjoy competing with others.

Section 9.5 Hypothesis Tests for a Standard Deviation

Objectives

1. Find critical values of the chi-square distribution

2. Test hypotheses about the standard deviation of a normal distribution

Objective 1 Find critical values of the chi-square distribution

The chi-square distribution was introduced in Section 8.4. The critical value χ^2_α represents the value that has area α to its right. We review the method for finding critical values for the chi-square distribution from Table A.4.

Example 9.21 Find a critical value

Find the critical value $\chi^2_{0.05}$ for a chi-square distribution with 10 degrees of freedom.

Solution

We consult Table A.4. The critical value is located at the intersection of the row corresponding to 10 degrees of freedom and the column corresponding to $\alpha = 0.05$. The critical value is $\chi^2_{0.05} = 18.307$.

CAUTION

The methods of this section apply only for samples drawn from a normal distribution. If the distribution differs even slightly from normal, these methods should not be used.

NOTE TO INSTRUCTOR

The material in this section may be omitted without loss of continuity.

Degrees of Freedom	Area in Right Tail 0.995	0.99	0.975	0.95	0.90	0.10	0.05	0.025	0.01	0.005
⋮	⋮	⋮	⋮	⋮	⋮	⋮	⋮	⋮	⋮	⋮
8	1.344	1.646	2.180	2.733	3.490	13.362	15.507	17.535	20.090	21.955
9	1.735	2.088	2.700	3.325	4.168	14.684	16.919	19.023	21.666	23.589
10	2.156	2.558	3.247	3.940	4.865	15.987	18.307	20.483	23.209	25.188
11	2.603	3.053	3.816	4.575	5.578	17.275	19.675	21.920	24.725	26.757
12	3.074	3.571	4.404	5.226	6.304	18.549	21.026	23.337	26.217	28.300
⋮	⋮	⋮	⋮	⋮	⋮	⋮	⋮	⋮	⋮	⋮

Objective 2 Test hypotheses about the standard deviation of a normal distribution

Hypothesis Tests for the Standard Deviation

The null hypothesis for a standard deviation σ is of the form H_0: $\sigma = \sigma_0$. The test is based on the fact that if H_0 is true, then the test statistic

$$\chi^2 = \frac{(n-1) \cdot s^2}{\sigma_0^2}$$

has a chi-square distribution with $n-1$ degrees of freedom. We will focus on how to perform a test using the critical value method and Table A.4.

Performing a Hypothesis Test for a Standard Deviation

Check to be sure that the assumptions are satisfied. We must have a simple random sample from a normal population.

Step 1: State the null and alternate hypotheses. The null hypothesis will be of the form H_0: $\sigma = \sigma_0$. The alternate hypothesis can be stated in one of three ways:

Left-tailed: H_1: $\sigma < \sigma_0$
Right-tailed: H_1: $\sigma > \sigma_0$
Two-tailed: H_1: $\sigma \neq \sigma_0$

Step 2: Choose a significance level α, and find the critical value, using Table A.4 with $n-1$ degrees of freedom.

Left-tailed: H_1: $\sigma < \sigma_0$: The critical value is $\chi^2_{1-\alpha}$.
Right-tailed: H_1: $\sigma > \sigma_0$: The critical value is χ^2_{α}.
Two-tailed: H_1: $\sigma \neq \sigma_0$: The critical values are $\chi^2_{\alpha/2}$ and $\chi^2_{1-\alpha/2}$.

Step 3: Compute the test statistic $\chi^2 = \dfrac{(n-1) \cdot s^2}{\sigma_0^2}$.

Step 4: Determine whether to reject H_0, as follows:

Left-tailed: H_1: $\sigma < \sigma_0$: Reject if $\chi^2 \leq \chi^2_{1-\alpha}$.
Right-tailed: H_1: $\sigma > \sigma_0$: Reject if $\chi^2 \geq \chi^2_{\alpha}$.
Two-tailed: H_1: $\sigma \neq \sigma_0$: Reject if $\chi^2 \geq \chi^2_{\alpha/2}$ or $\chi^2 \leq \chi^2_{1-\alpha/2}$.

Step 5: State a conclusion.

EXPLAIN IT AGAIN

Degrees of freedom: When a hypothesis test for a standard deviation is performed, the degrees of freedom is always 1 less than the sample size.

Example 9.22

Performing a hypothesis test

To check the reliability of a scale in a butcher shop, a test weight known to weigh 400 grams was weighed 16 times. For the scale to be considered reliable, the standard deviation of repeated measurements must be less than 1 gram. The standard deviation of the 16 measured weights was $s = 0.8$ gram. Assume that the measured weights are independent and follow a normal distribution. Can we conclude that the population standard deviation of the measurements is less than 1 gram? Use the $\alpha = 0.05$ level of significance.

Solution

We first check the assumptions. We have a random sample from a normal population, so the assumptions are satisfied.

Step 1: State the null and alternate hypotheses. Because we are interested in determining whether $\sigma < 1$, the hypotheses are

$$H_0: \sigma = 1 \qquad H_1: \sigma < 1$$

Step 2: Choose a significance level and find the critical value. We will use a significance level of $\alpha = 0.05$. We have $16 - 1 = 15$ degrees of freedom. This is a left-tailed test, so the critical value is $\chi^2_{1-\alpha} = \chi^2_{0.95} = 7.261$.

Step 3: Compute the test statistic. The sample size is $n = 16$. The sample variance is $s^2 = (0.8)^2 = 0.64$, and the value specified by H_0 is $\sigma_0 = 1$. The value of the test statistic is

$$\chi^2 = \frac{(n-1)s^2}{\sigma_0^2} = \frac{(16-1)(0.64)}{1^2} = 9.6$$

Step 4: Determine whether to reject H_0. The value of the test statistic is $\chi^2 = 9.6$. The critical value is $\chi^2_{0.95} = 7.261$. This is a left-tailed test, so we reject H_0 if $\chi^2 < \chi^2_{0.95}$. Because $9.6 > 7.261$, we do not reject H_0 at the $\alpha = 0.05$ level.

Step 5: State a conclusion. There is not enough evidence to conclude that the population standard deviation σ is less than 1 gram. We cannot consider the scale to be reliable.

Check Your Understanding

1. A random sample of size 20 from a normal distribution has standard deviation $s = 50$. Test $H_0: \sigma = 45$ versus $H_1: \sigma > 45$ at the $\alpha = 0.05$ level. Do you reject H_0? *No*

2. A random sample of size 12 from a normal distribution has standard deviation $s = 7$. Test $H_0: \sigma = 15$ versus $H_1: \sigma < 15$ at the $\alpha = 0.01$ level. Do you reject H_0? *Yes*

3. A random sample of size 28 from a normal distribution has standard deviation $s = 5$. Test $H_0: \sigma = 3$ versus $H_1: \sigma \neq 3$ at the $\alpha = 0.01$ level. Do you reject H_0? *Yes*

4. A random sample of size 5 from a normal distribution has standard deviation $s = 28$. Test $H_0: \sigma = 40$ versus $H_1: \sigma \neq 40$ at the $\alpha = 0.05$ level. Do you reject H_0? *No*

Answers are on page 491.

Section 9.5 Exercises

Exercises 1–4 are the Check Your Understanding exercises located within the section.

Understanding the Concepts

In Exercises 5 and 6, fill in each blank with the appropriate word or phrase.

5. To test a hypothesis about a standard deviation using a sample of size 15, we use a chi-square distribution with __________ degrees of freedom. *14*

6. The method described for testing hypotheses about standard deviations should be used only when the distribution of the population is almost exactly __________. *normal*

In Exercises 7 and 8, determine whether the statement is true or false. If the statement is false, rewrite it as a true statement.

7. When a test for a standard deviation is performed, it does not matter whether the population is normal, so long as the sample is large. *False*

8. Hypothesis tests for a standard deviation may be either one- or two-tailed. *True*

Practicing the Skills

9. A random sample of size 11 from a normal distribution has standard deviation $s = 98$. Test $H_0: \sigma = 70$ versus $H_1: \sigma > 70$. Use the $\alpha = 0.05$ level of significance. *Reject H_0.*

10. A random sample of size 29 from a normal distribution has standard deviation $s = 49$. Test H_0: $\sigma = 55$ versus H_1: $\sigma < 55$. Use the $\alpha = 0.01$ level of significance. *Do not reject H_0.*

11. A random sample of size 24 from a normal distribution has standard deviation $s = 29$. Test H_0: $\sigma = 35$ versus H_1: $\sigma < 35$. Use the $\alpha = 0.01$ level of significance. *Do not reject H_0.*

12. A random sample of size 13 from a normal distribution has standard deviation $s = 83$. Test H_0: $\sigma = 60$ versus H_1: $\sigma > 60$. Use the $\alpha = 0.05$ level of significance. *Reject H_0.*

13. A random sample of size 25 from a normal distribution has standard deviation $s = 51$. Test H_0: $\sigma = 30$ versus H_1: $\sigma \neq 30$. Use the $\alpha = 0.05$ level of significance. *Reject H_0.*

14. A random sample of size 8 from a normal distribution has standard deviation $s = 75$. Test H_0: $\sigma = 50$ versus H_1: $\sigma \neq 50$. Use the $\alpha = 0.01$ level of significance. *Do not reject H_0.*

Working with the Concepts

15. **Babies:** A sample of 25 one-year-old girls had a mean weight of 24.1 pounds with a standard deviation of 4.3 pounds. Assume that the population of weights is normally distributed. A pediatrician claims that the standard deviation of the weights of one-year-old girls is less than 5 pounds. Do the data provide convincing evidence that the pediatrician's claim is true? Use the $\alpha = 0.05$ level of significance. *Do not reject H_0.*

    Based on data from the National Health Statistics Reports

16. **Watching TV:** The General Social Survey asked a large number of people how much time they spent watching TV each day. The mean number of hours was 3.09 with a standard deviation of 2.87. Assume that in a sample of 40 teenagers, the sample standard deviation of daily TV time is 2.0 hours, and that the population of TV watching times is normally distributed. Can you conclude that the population standard deviation of TV watching times for teenagers is less than 2.87? Use the $\alpha = 0.01$ level of significance. *Reject H_0.*

17. **IQ scores:** Scores on an IQ test are normally distributed. A sample of 25 IQ scores had standard deviation $s = 8$. The developer of the test claims that the population standard deviation is $\sigma = 15$. Do these data provide sufficient evidence to contradict this claim? Use the $\alpha = 0.05$ level of significance. *Reject H_0.*

18. **SAT scores:** Scores on the math SAT are normally distributed. A sample of 20 SAT scores had standard deviation $s = 87$. Someone says that the scoring system for the SAT is designed so that the population standard deviation will be $\sigma = 100$. Do these data provide sufficient evidence to contradict this claim? Use the $\alpha = 0.05$ level of significance. *Do not reject H_0.*

19. **How much is in that can?** A machine that fills beverage cans is supposed to put 12 ounces of beverage in each can. The standard deviation of the amount in each can is 0.1 ounce. The machine is moved to a new location. To determine whether the standard deviation has changed, ten cans are filled. Following are the amounts in the ten cans. Assume them to be a random sample from a normal population.

12.18	11.77	12.09	12.03	11.87
11.96	12.03	12.36	12.28	11.85

    Perform a hypothesis test to determine whether the standard deviation differs from 0.1 ounce. Use the $\alpha = 0.05$ level of significance. What do you conclude? *Reject H_0.*

20. **Long-lasting drugs:** One of the ways in which doctors try to determine how long a single dose of pain reliever will provide relief is to measure the drug's half-life, which is the length of time it takes for one-half of the dose to be eliminated from the body. A report of the National Institutes of Health states that the standard deviation of the half-life of the pain reliever oxycodone is $\sigma = 1.43$ hours. Assume that a sample of 25 patients is given the drug, and the sample standard deviation of the half-lives was $s = 1.5$ hours. Assume the population is normally distributed. Can you conclude that the true standard deviation is greater than the value reported by the National Institutes of Health? Use the $\alpha = 0.01$ level of significance. *Do not reject H_0.*

Extending the Concepts

The chi-square distribution is skewed, but as the number of degrees of freedom becomes large, the skewness diminishes. If the number of degrees of freedom, k, is large enough, the chi-square distribution is reasonably well approximated by a normal distribution with mean k and variance 2k.

21. **Exact test:** A sample of size 101 from a normal population has sample standard deviation $s = 40$. Use Table A.4 to find the exact critical values $\chi^2_{0.025}$ and $\chi^2_{0.975}$ to test H_0: $\sigma = 30$ versus H_1: $\sigma \neq 30$. Can you reject H_0 at the $\alpha = 0.05$ level? *Yes*

22. **Using the normal approximation:** Refer to Exercise 21. Use the normal approximation to estimate the critical values $\chi^2_{0.025}$ and $\chi^2_{0.975}$. Using these critical values, can you reject H_0 at the $\alpha = 0.05$ level? *Yes*

    A more accurate normal approximation to χ^2_α is given by $\chi^2_\alpha \approx 0.5\left(z_\alpha + \sqrt{2k-1}\right)^2$, where z_α is the z-score that has area α to its right.

23. **More accuracy:** Refer to Exercise 21. Use the more accurate normal approximation to estimate the critical values $\chi^2_{0.025}$ and $\chi^2_{0.975}$. Using these critical values, can you reject H_0 at the $\alpha = 0.05$ level? *Yes*

Answers to Check Your Understanding Exercises for Section 9.5

1. $\chi^2 = 23.457$, critical value is 30.144. Do not reject H_0.
2. $\chi^2 = 2.396$, critical value is 3.053. Reject H_0.
3. $\chi^2 = 75.000$, critical values are 11.808 and 49.645. Reject H_0.
4. $\chi^2 = 1.960$, critical values are 0.484 and 11.143. Do not reject H_0.

Section 9.6 Determining Which Method to Use

Objective

1. Determine which method to use when performing a hypothesis test

Objective 1 Determine which method to use when performing a hypothesis test

One of the challenges in performing a hypothesis test is to determine which method to use. The first step is to determine which type of parameter we are testing. There are three types of parameters about which we have learned to perform hypothesis tests:

- Population mean μ
- Population proportion p
- Population standard deviation σ or variance σ^2

Once you have determined which type of parameter you are testing, proceed as follows:

- **Population mean:** There are two methods for performing a hypothesis test for a population mean, the z-test (Section 9.2) and the t-test (Section 9.3). To determine which method to use, we must determine whether the population is approximately normal, and whether the sample size is large ($n > 30$). The following diagram can help you make the correct choice.

CAUTION

Be sure to check the assumptions before performing a hypothesis test.

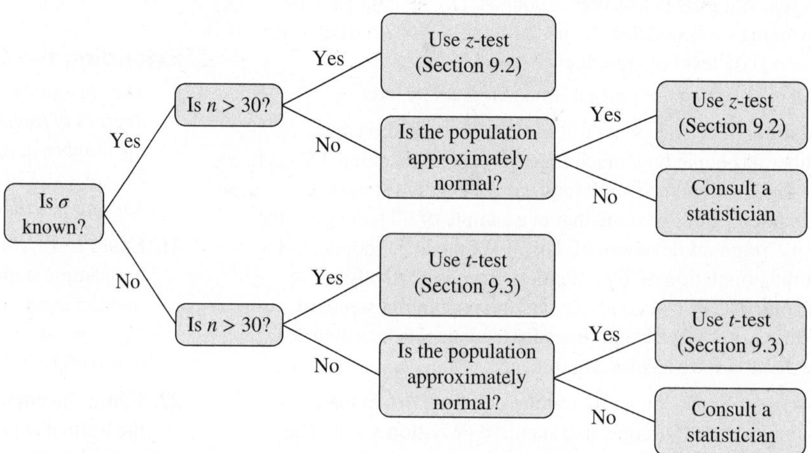

- **Population proportion:** To perform a hypothesis test for a population proportion, use the method described in Section 9.4.
- **Population standard deviation or variance:** To perform a hypothesis test for a population standard deviation or variance, use the method described in Section 9.5.

Example 9.23

Determining which method to use

Starting salaries for a random sample of 51 physicians had a mean of \$103,000. Assume that the population standard deviation is \$10,500. Can you conclude that the mean starting salary for physicians is greater than \$100,000? Determine the type of parameter that is to be tested, and perform the hypothesis test. Use the $\alpha = 0.05$ level of significance.

Solution

We are asked to perform a hypothesis test for the mean salary; this is a population mean. We consult the diagram to determine the correct method. We must first determine whether σ is known. We are told that the population standard deviation is \$10,500. Therefore, $\sigma = 10{,}500$. We follow the "Yes" path. Next we must determine whether $n > 30$. The sample size is 51, so $n > 30$. We follow the "Yes" path, and find that we should use the z-test described in Section 9.2. To perform the test, we compute the value of the test statistic:

$$z = \frac{103{,}000 - 100{,}000}{10{,}500/\sqrt{51}} = 2.04$$

Because the alternate hypothesis is H_1: $\mu > 100{,}000$, this is a right-tailed test. The P-value is the area under the normal curve to the right of $z = 2.04$. Using Table A.2 or technology, we find that the area to the left of $z = 2.04$ is 0.9793. The area to the right is therefore $1 - 0.9793 = 0.0207$. The P-value is 0.0207. Because $P < 0.05$, we conclude that the mean starting salary is greater than \$100,000.

Check Your Understanding

In Exercises 1–4, state which type of parameter is to be tested; then perform the hypothesis test.

1. In a simple random sample of 150 cars undergoing emissions testing, 23 failed the test. Can you conclude that the proportion of cars that fail the test is less than 20%? Use the $\alpha = 0.05$ level of significance. *Proportion; do not reject H_0.*
2. A simple random sample of size 15 has mean $\bar{x} = 27.72$ and standard deviation $s = 8.21$. The population is approximately normally distributed. Can you conclude that the population mean differs from 35? Use the $\alpha = 0.01$ level of significance. *Mean; reject H_0.*
3. A simple random sample of size 7 has mean $\bar{x} = 32.5$ and standard deviation $s = 8.27$. The population is normally distributed. Can you conclude that the population standard deviation is greater than 6? Use the $\alpha = 0.05$ level of significance. *Standard deviation; do not reject H_0.*
4. A simple random sample of size 65 has mean $\bar{x} = 38.16$. The population standard deviation is $\sigma = 5.95$. Can you conclude that the mean is less than 40? Use the $\alpha = 0.01$ level of significance. *Mean; reject H_0.*

Answers are on page 494.

Section 9.6 Exercises

Exercises 1–4 are the Check Your Understanding exercises located within the section.

Practicing the Skills

In Exercises 5–12, state which type of parameter is to be tested; then perform the hypothesis test.

5. A simple random sample of size 13 has mean $\bar{x} = 7.26$ and standard deviation $s = 2.45$. The population is approximately normally distributed. Can you conclude that the population mean differs from 9? Use the $\alpha = 0.01$ level of significance. *Mean; do not reject H_0.*
6. A simple random sample of size 65 has mean $\bar{x} = 57.3$. The population standard deviation is $\sigma = 12.6$. Can you conclude that the population mean is less than 60? Use the $\alpha = 0.05$ level of significance. *Mean; reject H_0.*
7. A simple random sample of size 16 has mean $\bar{x} = 4.1$ and standard deviation $s = 6.2$. The population is normally distributed. Can you conclude that the population standard deviation is greater than 5? Use the $\alpha = 0.05$ level of significance. *Standard deviation; do not reject H_0.*
8. In a simple random sample of 120 law students, 54 were women. Can you conclude that less than half of law students are women? Use the $\alpha = 0.01$ level of significance. *Proportion; do not reject H_0.*
9. A simple random sample of size 23 has mean $\bar{x} = 41.8$. The population standard deviation is $\sigma = 3.72$. Can you conclude that the population mean differs from 40? Use the $\alpha = 0.05$ level of significance. *Mean; reject H_0.*
10. A simple random sample of size 6 has mean $\bar{x} = 5.49$ and standard deviation $s = 2.37$. The population is approximately normally distributed. Can you conclude that the population mean is greater than 4? Use the $\alpha = 0.05$ level of significance. *Mean; do not reject H_0.*
11. In a simple random sample of 95 families, 70 had one or more pets at home. Can you conclude that the proportion of families with one or more pets differs from 0.6? Use the $\alpha = 0.01$ level of significance. *Proportion; reject H_0.*

12. A simple random sample of size 22 has mean $\bar{x} = 91.7$ and standard deviation $s = 19.3$. The population is normally distributed. Can you conclude that the population standard deviation is greater than 15? Use the $\alpha = 0.05$ level of significance. *Standard deviation; reject H_0.*

Working with the Concepts

13. Saving for college: In a survey of 909 U.S. adults with children conducted by the Financial Industry Regulatory Authority, 309 said that they had saved money for their children's college education. Can you conclude that more than 30% of U.S. adults with children have saved money for college? Use the $\alpha = 0.05$ level of significance. *Reject H_0.*

14. Big houses: The U.S. Census Bureau reported that the mean area of U.S. homes built in 2018 was 2559 square feet. Assume that a simple random sample of 20 homes built in 2020 had a mean area of 2635 square feet, with a standard deviation of 225 square feet. Assume the population of areas is normally distributed. Can you conclude that the mean area of homes built in 2020 is greater than that of homes built in 2018? Use the $\alpha = 0.01$ level of significance. *Do not reject H_0.*

15. Cookies: Following are the weights of 8 boxes of cookies, each of which is labeled as containing 16 ounces. Assume that the population of weights is normally distributed.

15.91 16.06 16.08 15.97 16.02 15.88 15.89 16.01

Can you conclude that the population standard deviation is less than 0.1? Use the $\alpha = 0.01$ level of significance. *Do not reject H_0.*

16. Careers in science: The General Social Survey asked 514 people whether they had ever considered a career in science, and 178 said that they had. Can you conclude that more than 30% of people have considered a career in science? Use the $\alpha = 0.05$ level of significance. *Reject H_0.*

17. Teacher salaries: A random sample of 50 public school teachers in Georgia had a mean annual salary of $48,300. Assume the population standard deviation is $\sigma = \$8{,}000$. Can you conclude that the mean salary of public school teachers in Georgia differs from $50,000? Use the $\alpha = 0.01$ level of significance. *Do not reject H_0.*

18. Highway speeds: Speeds for a sample of nine cars were measured by radar along a stretch of highway. The results, in miles per hour, were as follows. Assume that the population of speeds is normally distributed.

56 60 53 55 54 51 54 51 56

Can you conclude that the population standard deviation is greater than 2? Use the $\alpha = 0.05$ level of significance. *Reject H_0.*

19. Mercury pollution: Mercury is a toxic metal that is used in many industrial applications. Seven measurements, in milligrams per cubic meter, were taken of the mercury concentration in a lake, with the following results. Assume that the population of measurements is approximately normally distributed.

1.02 1.23 0.91 1.29 1.01 1.35 1.43

Can you conclude that the mean concentration is greater than 1 milligram per cubic meter? Use the $\alpha = 0.05$ level of significance. *Reject H_0.*

20. Ladies' shoes: A random sample of 100 pairs of ladies' shoes had a mean size of 8.3. Assume the population standard deviation is $\sigma = 1.5$. Can you conclude that the mean size of ladies' shoes differs from 8? Use the $\alpha = 0.01$ level of significance. *Do not reject H_0.*

Answers to Check Your Understanding Exercises for Section 9.6

1. The parameter is the population proportion. The test statistic is $z = -1.43$. The P-value is 0.0764 [Tech: 0.0765]. Do not reject H_0. There is not enough evidence to conclude that the proportion of cars that fail the test is less than 0.20.
2. The parameter is the population mean. The test statistic is $t = -3.43$. The P-value is $0.002 < P < 0.005$ [Tech: 0.004]. Reject H_0. We conclude that the mean differs from 35.
3. The parameter is the population standard deviation. The test statistic is $\chi^2 = 11.399$. The critical value is $\chi^2_{0.975} = 12.59$. Do not reject H_0. There is not enough evidence to conclude that the population standard deviation is greater than 6.
4. The parameter is the population mean. The test statistic is $z = -2.49$. The P-value is 0.0064 [Tech: 0.0063]. Reject H_0. We conclude that the population mean is less than 40.

Section 9.7 Power

Objective

1. Compute the power of a test

Objective 1 Compute the power of a test

Recall that there are two types of errors that can be made with a hypothesis test. A Type I error occurs when H_0 is true and we reject it. A Type II error occurs when H_0 is false and we fail to reject it. Table 9.3 (page 495) summarizes the possibilities. The probability of a Type I error is α, the significance level of the test. The probability of a Type II error is denoted β.

The *power* of a test is the probability that we do not make a Type II error. In other words, the power of a test is the probability that we reject H_0 when it is false.

Table 9.3

	H_0 true	H_0 false
Reject H_0	Type I error	Correct
Don't reject H_0	Correct	Type II error

DEFINITION

The **power** of a test is the probability of rejecting H_0 when it is false.

$$\text{Power} = 1 - \beta$$

where $\beta = P(\text{Type II error})$.

We would like to have a small probability of rejecting H_0 when it is true, and a large probability of rejecting H_0 when it is false. In other words, we would like the probability of a Type I error to be small, and the power to be large. We can make the probability of a Type I error small by choosing a small value for the significance level α. As we have seen, the values $\alpha = 0.05$ and $\alpha = 0.01$ are considered to be appropriate in many situations. Once the value of α has been chosen, we can increase the power only by increasing the sample size n. What is done in practice is to choose values for α and n and then compute the power. Values of 0.8 or 0.9 are considered to be acceptably large for the power. If the power is not large enough, then if sufficient time and resources are available, the sample size can be increased.

The power of a test about a population mean depends on the true value of the population mean. To compute the power, we specify a value μ_1 for the population mean that satisfies the alternate hypothesis. The power is the probability that the test statistic falls in the critical region when μ_1 is the true value of the population mean. The following steps can be used to compute the power of a hypothesis test for the population mean when the population standard deviation σ is known.

Computing the Power of a Test

Let μ_0 be the value of μ specified by H_0, let α be the significance level, let n be the sample size, and let σ be the population standard deviation.

Step 1: Find the critical value: z_α for a one-tailed test or $z_{\alpha/2}$ for a two-tailed test.

Step 2:

- For a one-tailed test, find the value of $\bar{x}$ whose z-score is equal to the critical value. We call this value $\bar{x}^*$. Find the value of $\bar{x}^*$ as follows:

  Left-tailed: H_1: $\mu < \mu_0$ $\qquad \bar{x}^* = \mu_0 - z_\alpha \cdot \dfrac{\sigma}{\sqrt{n}}$

  Right-tailed: H_1: $\mu > \mu_0$ $\qquad \bar{x}^* = \mu_0 + z_\alpha \cdot \dfrac{\sigma}{\sqrt{n}}$

- For a two-tailed test, there are two values of $\bar{x}^*$. We call them $\bar{x}^*_{\text{left}}$ and $\bar{x}^*_{\text{right}}$. They are computed as follows:

$$\bar{x}^*_{\text{left}} = \mu_0 - z_{\alpha/2} \cdot \frac{\sigma}{\sqrt{n}} \qquad \bar{x}^*_{\text{right}} = \mu_0 + z_{\alpha/2} \cdot \frac{\sigma}{\sqrt{n}}$$

Step 3: Let μ_1 be a specific value that satisfies the alternate hypothesis. Sketch a normal curve with mean μ_1.

Step 4: The power is an area under the normal curve sketched in Step 3. The area depends on the form of the alternate hypothesis, as follows:

Left-tailed: H_1: $\mu < \mu_0$	Area to the left of $\bar{x}^*$
Right-tailed: H_1: $\mu > \mu_0$	Area to the right of $\bar{x}^*$
Two-tailed: H_1: $\mu \neq \mu_0$	Sum of the area to the left of $\bar{x}^*_{\text{left}}$ and the area to the right of $\bar{x}^*_{\text{right}}$

Example 9.24

Compute the power of a test

The General Social Survey indicates that Americans watch an average of 2.98 hours of television per day, with a standard deviation of $\sigma = 2.66$ hours. A sociologist believes that the mean viewing time for college students is less, because students spend more time on the internet and playing video games. The sociologist will sample 75 college students and test the hypotheses

$$H_0: \mu = 2.98 \qquad H_1: \mu < 2.98$$

at the $\alpha = 0.05$ level. Assume the population standard deviation for college students is also $\sigma = 2.66$. Find the power of the test against the alternative $\mu_1 = 2$.

Solution

Step 1: This is a one-tailed test, with significance level $\alpha = 0.05$. Therefore, we use the critical value $z_\alpha = 1.645$.

Step 2: We have $\mu_0 = 2.98$, $z_\alpha = 1.645$, $\sigma = 2.66$, and $n = 75$. This is a left-tailed test. Therefore,

$$\bar{x}^* = 2.98 - (1.645)\frac{2.66}{\sqrt{75}} = 2.475$$

Step 3: The following figure presents a normal curve with mean 2. The value of $\bar{x}^* = 2.475$ is indicated as well.

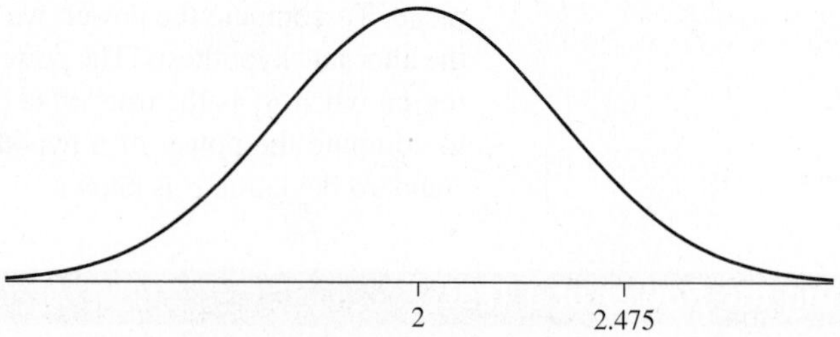

Step 4: This is a left-tailed test, so the power is the area to the left of $\bar{x}^* = 2.475$. To find this area, we find the z-score for 2.475, using the value $\mu_1 = 2$.

$$z = \frac{2.475 - 2}{2.66/\sqrt{75}} = 1.55$$

The power is the area under the normal curve to the left of $z = 1.55$. This area is 0.9394, as the following figure shows.

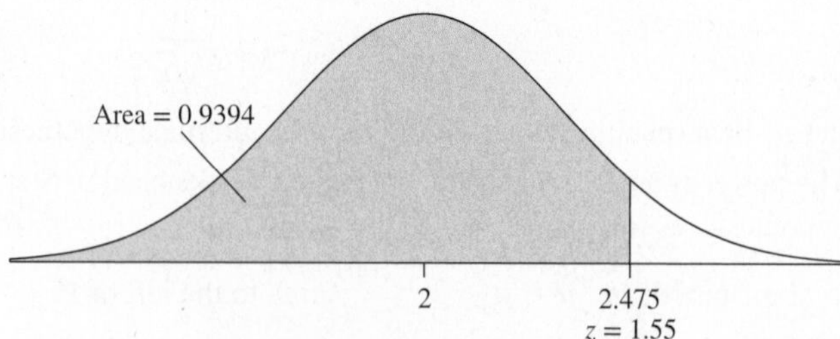

Figure 9.20 (page 497) presents two distributions for $\bar{x}$. The curve on the right is the distribution under the assumption that H_0 is true. This distribution has mean $\mu_0 = 2.98$. The curve on the left is the distribution under the assumption that the mean is equal to the alternate value $\mu_1 = 2$. The critical region is the region to the left of $\bar{x}^* = 2.475$. The area of the critical region under the null hypothesis distribution is the significance level $\alpha = 0.05$. The area under the alternate distribution is the power, 0.9394.

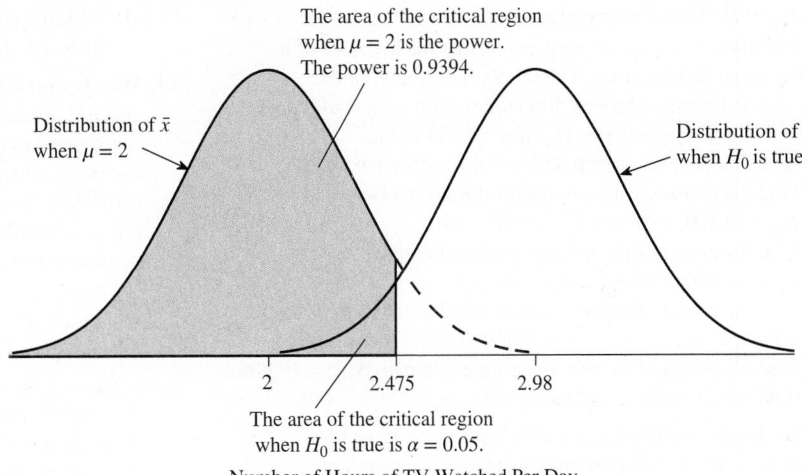

Figure 9.20 The critical region is shaded. The area of the critical region is $\alpha = 0.05$ when the curve for H_0 is used and is equal to the power, 0.9394, when the curve with the alternate mean of 2 is used.

Check Your Understanding

1. For testing the hypotheses H_0: $\mu = 10$ versus H_1: $\mu < 10$ at the $\alpha = 0.05$ level, assume the population standard deviation is $\sigma = 3$ and the sample size is $n = 50$. Find the power against the alternative $\mu_1 = 9$. *0.7611 [Tech: 0.7618]*

2. For testing the hypotheses H_0: $\mu = 35$ versus H_1: $\mu > 35$ at the $\alpha = 0.01$ level, assume the population standard deviation is $\sigma = 10$ and the sample size is $n = 100$. Find the power against the alternative $\mu_1 = 40$. *0.9962*

Answers are on page 499.

Section 9.7 Exercises

Exercises 1 and 2 are the Check Your Understanding exercises located within the section.

Understanding the Concepts

In Exercises 3–5, fill in each blank with the appropriate word or phrase.

3. The greater the power, the less likely we are to make a __________ error. *Type II*

4. The power is the probability of rejecting H_0 when it is __________. *false*

5. If the sample size is increased, the power will __________. *increase*

In Exercises 6–8, determine whether the statement is true or false. If the statement is false, rewrite it as a true statement.

6. Once the significance level α has been chosen, the only way to increase the power of a test is to decrease the sample size. *False*

7. The power is the probability of making a Type I error. *False*

8. For a test about the population mean, the power of a test depends on a specified value that satisfies the alternate hypothesis. *True*

Practicing the Skills

9. A test has power 0.90 when $\mu_1 = 15$. True or false:
   a. The probability of rejecting H_0 when $\mu_1 = 15$ is 0.90. *True*
   b. The probability of making a correct decision when $\mu_1 = 15$ is 0.90. *True*
   c. The probability of making a correct decision when $\mu_1 = 15$ is 0.10. *False*
   d. The probability that H_0 is true when $\mu_1 = 15$ is 0.10. *False*

10. A test has power 0.80 when $\mu_1 = 3.5$. True or false:
   a. The probability of rejecting H_0 when $\mu_1 = 3.5$ is 0.80. *True*
   b. The probability of making a Type I error when $\mu_1 = 3.5$ is 0.80. *False*
   c. The probability of making a Type I error when $\mu_1 = 3.5$ is 0.20. *False*
   d. The probability of making a Type II error when $\mu_1 = 3.5$ is 0.80. *False*
   e. The probability of making a Type II error when $\mu_1 = 3.5$ is 0.20. *True*
   f. The probability that H_0 is false when $\mu_1 = 3.5$ is 0.80. *False*

Working with the Concepts

11. Tire lifetimes: A tire company claims that the lifetimes of its tires average 50,000 miles. The standard deviation of tire lifetimes is known to be $\sigma = 5000$ miles. You sample 100 tires and will test the hypotheses H_0: $\mu = 50{,}000$ versus H_1: $\mu < 50{,}000$ at the $\alpha = 0.05$ level of significance.

a. Find the power of the test against the alternative $\mu_1 = 49{,}500$. 0.2578 [Tech: 0.2595]

b. Find the power of the test against the alternative $\mu_1 = 49{,}000$. 0.6406 [Tech: 0.6388]

c. Find the power of the test against the alternative $\mu_1 = 49{,}500$ if the test is made at level $\alpha = 0.01$. 0.0918 [Tech: 0.0924]

d. Find the power of the test against the alternative $\mu_1 = 49{,}000$ if the test is made at level $\alpha = 0.01$. 0.3707 [Tech: 0.3721]

12. Coffee beans: Shipments of coffee beans are checked for moisture content. A high moisture content indicates water contamination and will result in the shipment being rejected. Let μ represent the mean water content (in percent by weight) in a shipment. Fifty moisture measurements will be made on beans chosen at random from the shipment. A test of the hypotheses H_0: $\mu = 10$ versus H_1: $\mu > 10$ will be made at the $\alpha = 0.05$ level of significance. Assume the standard deviation of moisture content is $\sigma = 5.0$.

a. Find the power of the test against the alternative $\mu_1 = 11$. 0.4090 [Tech: 0.4088]

b. Find the power of the test against the alternative $\mu_1 = 12$. 0.8810 [Tech: 0.8817]

c. Find the power of the test against the alternative $\mu_1 = 11$ if the test is made at level $\alpha = 0.01$. 0.1814 [Tech: 0.1808]

d. Find the power of the test against the alternative $\mu_1 = 12$ if the test is made at level $\alpha = 0.01$. 0.6915 [Tech: 0.6922]

Ingram Publishing/SuperStock

13. SAT scores: A college admissions officer will draw a simple random sample of 100 mathematics SAT scores from the entering freshman class. The admissions officer will perform a test of the hypotheses H_0: $\mu = 500$ versus H_1: $\mu > 500$ at the $\alpha = 0.05$ level of significance. Assume the population standard deviation is $\sigma = 116$.

a. Find the power of the test against the alternative $\mu_1 = 520$. 0.5319 [Tech: 0.5316]

b. Find the power of the test against the alternative $\mu_1 = 550$. 0.9962

c. Find the power of the test against the alternative $\mu_1 = 520$ if the test is made at level $\alpha = 0.01$. 0.2743 [Tech: 0.2735]

d. Find the power of the test against the alternative $\mu_1 = 550$ if the test is made at level $\alpha = 0.01$. 0.9761 [Tech: 0.9764]

14. Watch your cholesterol: An article in the *International Journal of Clinical Practice* described a study in which a sample of 314 patients took a combination of the drugs ezetimibe and simvastatin in order to reduce their total blood cholesterol levels. A test of the hypotheses H_0: $\mu = 1$ versus H_1: $\mu > 1$ will be made at the $\alpha = 0.05$ level of significance. Assume the population standard deviation is $\sigma = 0.2$.

a. Find the power of the test against the alternative $\mu_1 = 1.02$. 0.5517 [Tech: 0.5506]

b. Find the power of the test against the alternative $\mu_1 = 1.05$. 0.9974 [Tech: 0.9973]

c. Find the power of the test against the alternative $\mu_1 = 1.02$ if the test is made at level $\alpha = 0.01$. 0.2912 [Tech: 0.2897]

d. Find the power of the test against the alternative $\mu_1 = 1.05$ if the test is made at level $\alpha = 0.01$. 0.9821 [Tech: 0.9823]

15. Tire lifetimes: Refer to Exercise 11. A test of the hypotheses H_0: $\mu = 50{,}000$ versus H_1: $\mu \neq 50{,}000$ will be made at the $\alpha = 0.05$ level of significance. Assume that the standard deviation of tire lifetimes is $\sigma = 5000$ and that the sample size is $n = 100$. Find the power of the test against the alternative $\mu_1 = 49{,}000$. 0.5160

16. Coffee beans: Refer to Exercise 12. A test of the hypotheses H_0: $\mu = 10$ versus H_1: $\mu \neq 10$ will be made at the $\alpha = 0.05$ level of significance. Assume that the standard deviation of moisture content is $\sigma = 5.0$ and that the sample size is $n = 50$. Find the power of the test against the alternative $\mu_1 = 12$. 0.8079 [Tech: 0.8074]

17. SAT scores: Refer to Exercise 13. The admissions officer will perform a test of the hypotheses H_0: $\mu = 500$ versus H_1: $\mu \neq 500$ at the $\alpha = 0.01$ level of significance. Assume that the population standard deviation is $\sigma = 116$ and that the sample size is $n = 100$. Find the power of the test against the alternative $\mu_1 = 550$. 0.9582 [Tech: 0.9586]

18. Watch your cholesterol: Refer to Exercise 14. A test of the hypotheses H_0: $\mu = 1$ versus H_1: $\mu \neq 1$ will be made at the $\alpha = 0.01$ level of significance. Assume that the population standard deviation is $\sigma = 0.2$ and that the sample size is $n = 314$. Find the power of the test against the alternative $\mu_1 = 0.98$. 0.2119 [Tech: 0.2107]

Extending the Concepts

The power for a test of a population proportion can be found by a method similar to that for a population mean.

19. Power for test of a proportion: A marketing firm samples 150 residents of a certain town to determine the sample proportion $\hat{p}$ that have seen a new advertisement. A test of the hypotheses H_0: $p = 0.4$ versus H_1: $p < 0.4$ will be performed, using a significance level of $\alpha = 0.05$.

a. Find the critical value z_α. 1.645

b. Find the value of $\hat{p}$ whose z-score is $-z_\alpha$. Call this value p^*. Note that the population standard deviation is $\sigma = \sqrt{p_0(1 - p_0)}$. 0.3342

c. We want to find the power against the alternative $p_1 = 0.3$. The power is the area to the left of p^* under the assumption that the true proportion is $p_1 = 0.3$. Find the power. [Note that the standard deviation is now $\sigma = \sqrt{p_1(1 - p_1)}$.] 0.8186 [Tech: 0.8197]

Answers to Check Your Understanding Exercises for Section 9.7

1. 0.7611 [Tech: 0.7618] **2.** 0.9962

Chapter 9 Summary

Section 9.1: A hypothesis test involves a null hypothesis, H_0, which makes a statement about one or more population parameters, and an alternate hypothesis, which contradicts H_0. We begin by assuming that H_0 is true. If the data provide strong evidence against H_0, we then reject H_0 and believe H_1. A Type I error occurs when a true null hypothesis is rejected. A Type II error occurs when a false H_0 is not rejected.

Section 9.2: We follow one of two methods in performing a hypothesis test. In the critical value method, we choose a significance level α, then find a critical region. We reject H_0 if the test statistic falls inside the critical region. The probability of a Type I error is α, the significance level of the test. In the *P*-value method, we compute a *P*-value, which is the probability of observing a value for the test statistic that is as extreme as or more extreme than the value actually observed, under the assumption that H_0 is true. The smaller the *P*-value, the stronger the evidence against H_0. If we want to make a firm decision about the truth of H_0, we choose a significance level α and reject H_0 if $P \le \alpha$.

When testing a hypothesis about a population mean with the population standard deviation σ known, the test statistic, z, has a standard normal distribution. If the sample size is not large, the population must be approximately normal. We can check normality with a boxplot or dotplot.

Section 9.3: When testing a hypothesis about a population mean with the population standard deviation σ unknown, the test statistic has a Student's t distribution. The number of degrees of freedom is 1 less than the sample size. The population must be approximately normal, or the sample size must be large ($n > 30$). We can check normality with a boxplot or dotplot.

Statistical significance is not the same as practical significance. When a result is statistically significant, we can conclude only that the true value of the parameter is different from the value specified by H_0. We cannot conclude that the difference is large enough to be important.

When presenting the results of a hypothesis test, it is important to state the *P*-value or the value of the test statistic, so that others can decide for themselves whether to reject H_0. It isn't enough simply to state whether or not H_0 was rejected.

Section 9.4: When testing a hypothesis about a population proportion, the test statistic is z. The sample proportion must be approximately normal. We check this by requiring that both np_0 and $n(1 - p_0)$ are at least 10.

Section 9.5: When a population is almost exactly normal, we can test a hypothesis about a population standard deviation. We use the chi-square distribution, with degrees of freedom 1 less than the sample size.

Section 9.6: We have learned to perform hypothesis tests for a population mean, a population proportion, and a population standard deviation or variance. There are two tests for a population mean, the z-test and the t-test. The test to use depends on whether the population standard deviation σ is known.

Section 9.7: The power of a test is the probability that a false H_0 is rejected. It is desirable for a test to have a high degree of power. If the sample size remains the same, however, increasing the power also increases the probability of a Type I error. In order to increase the power without increasing the probability of a Type I error, it is necessary to increase the sample size.

Vocabulary and Notation

Important Formulas

Test statistic for a mean, standard deviation known:

$$z = \frac{\bar{x} - \mu_0}{\sigma / \sqrt{n}}$$

Test statistic for a mean, standard deviation unknown:

$$t = \frac{\bar{x} - \mu_0}{s / \sqrt{n}}$$

Test statistic for a proportion:

$$z = \frac{\hat{p} - p_0}{\sqrt{\dfrac{p_0(1 - p_0)}{n}}}$$

Test statistic for a standard deviation:

$$\chi^2 = \frac{(n - 1) \cdot s^2}{\sigma_0^2}$$

Chapter Quiz

1. Fill in the blank: A test of the hypotheses H_0: $\mu = 65$ versus H_1: $\mu \neq 65$ was performed. The P-value was 0.035. Fill in the blank: If $\mu = 65$, then the probability of observing a test statistic as extreme as or more extreme than the one actually observed is ________. *0.035*
2. A hypothesis test results in a P-value of 0.008. Which is the best conclusion? *iv*
   i. H_0 is definitely false.
   ii. H_0 is definitely true.
   iii. H_0 is plausible.
   iv. H_0 might be true, but it's very unlikely.
   v. H_0 might be false, but it's very unlikely.
3. True or false: If $P = 0.03$, then
   a. The result is statistically significant at the $\alpha = 0.05$ level. *True*
   b. The result is statistically significant at the $\alpha = 0.01$ level. *False*
   c. The null hypothesis is rejected at the $\alpha = 0.05$ level. *True*
   d. The null hypothesis is rejected at the $\alpha = 0.01$ level. *False*
4. A null hypothesis is rejected at the $\alpha = 0.05$ level. True or false:
   a. The P-value is greater than 0.05. *False*
   b. The P-value is less than or equal to 0.05. *True*
   c. The result is statistically significant at the $\alpha = 0.05$ level. *True*
   d. The result is statistically significant at the $\alpha = 0.10$ level. *True*
5. A sample of size 8 is drawn from a normal population with mean μ, and the population standard deviation is unknown.
   a. Is it appropriate to perform a z-test? Explain. *No*
   b. Is it appropriate to perform a t-test? Explain. *Yes*
6. A test will be made of H_0: $\mu = 4$ versus H_1: $\mu > 4$, using a sample of size 25. The population standard deviation is unknown. Find the critical value of the test statistic if the significance level is $\alpha = 0.05$. *1.711*
7. True or false: We never conclude that H_0 is true. *True*
8. In a random sample of 500 people who took their driver's test, 445 passed. Let p be the population proportion who pass. A test will be made of H_0: $p = 0.85$ versus H_1: $p > 0.85$.
   a. Compute the value of the test statistic. *2.50*
   b. Do you reject H_0 at the $\alpha = 0.05$ level? *Yes*
   c. State a conclusion.
9. For testing H_0: $\mu = 3$ versus H_1: $\mu < 3$, a P-value of 0.024 is obtained.
   a. If the significance level is $\alpha = 0.05$, would you conclude that $\mu < 3$? Explain. *Yes*
   b. If the significance level is $\alpha = 0.01$, would you conclude that $\mu < 3$? Explain. *No*
10. True or false: When we reject H_0, we are certain that H_1 is true. *False*
11. The result of a hypothesis test is reported as follows: "We reject H_0 at the $\alpha = 0.05$ level." What additional information should be included? *P-value or value of test statistic*
12. In a test of H_0: $\mu = 5$ versus H_1: $\mu > 5$, the value of the test statistic is $t = 2.96$. There are 17 degrees of freedom. Do you reject H_0 at the $\alpha = 0.05$ level? *Yes*
13. True or false: We can perform a test for a standard deviation only when the population is almost exactly normal. *True*
14. A random sample of size 20 from a normal population has sample standard deviation $s = 10$. Test H_0: $\sigma = 15$ versus H_1: $\sigma < 15$. Use the $\alpha = 0.05$ level. *Reject H_0.*
15. A test of H_0: $\mu = 50$ versus H_1: $\mu > 50$ will be made at a significance level of $\alpha = 0.05$. The population standard deviation is $\sigma = 10$ and the sample size is $n = 60$. Find the power of the test against the alternative $\mu_1 = 55$. *0.9871*

Review Exercises

1. **What's the conclusion?** A hypothesis test is performed, and $P = 0.02$. Which of the following is the best conclusion? *i*
   i. H_0 is rejected at the 0.05 level.
   ii. H_0 is rejected at the 0.01 level.
   iii. H_1 is rejected at the 0.05 level.
   iv. H_1 is rejected at the 0.01 level.
2. **Scoring runs:** In 2018, the mean number of runs scored by both teams in a Major League Baseball game was 8.87. Following are the numbers of runs scored in a sample of 24 games in 2019.

2	10	3	9	15	10	7	4	3	7	5	9
5	9	15	15	4	5	13	6	14	11	6	12

a. Construct a boxplot of the data. Is it appropriate to perform a hypothesis test? *Yes*

b. If appropriate, perform a hypothesis test to determine whether the mean number of runs in 2018 is less than it was in 2019. Use the $\alpha = 0.05$ level. *Do not reject H_0.*

3. Facebook: A popular blog reports that 60% of college students log in to Facebook on a daily basis. The Dean of Students at a certain university thinks that the proportion may be different at her university. She polls a simple random sample of 200 students, and 134 of them report that they log in to Facebook daily. Can you conclude that the proportion of students who log in to Facebook daily differs from 0.60?

a. State the null and alternate hypotheses. $H_0: p = 0.6,\ H_1: p \neq 0.6$

b. Compute the value of the test statistic. *2.02*

c. Do you reject H_0? Use the $\alpha = 0.05$ level. *Yes*

d. State a conclusion.

4. Playing the market: The Russell 2000 is a group of 2000 small-company stocks. On a recent day, a random sample of 35 of these stocks had a mean price of \$26.89, with a standard deviation of \$23.41. A stock market analyst predicted that the mean price of all 2000 stocks would be \$25.00. Can you conclude that the mean price differs from \$25.00?

a. State the null and alternate hypotheses. $H_0: \mu = 25,\ H_1: \mu \neq 25$

b. Should we perform a z-test or a t-test? Explain. *t-test*

c. Compute the value of the test statistic. *0.478*

d. Do you reject H_0? Use the $\alpha = 0.05$ level. *No*

e. State a conclusion.

5. Power: A test of $H_0: \mu = 100$ versus $H_1: \mu > 100$ will be made at a significance level of $\alpha = 0.01$. The population standard deviation is $\sigma = 50$, and the sample size is $n = 75$. Find the power of the test against the alternative $\mu_1 = 110$. *0.2776 [Tech: 0.2762]*

6. More power: Refer to Exercise 5. If the test is made at the $\alpha = 0.05$ level with the same sample size, would the power be greater than or less than in Exercise 5? Explain. *Greater*

7. Household size: For the past several years, the mean number of people in a household has been declining. A social scientist believes that in a certain large city, the mean number of people per household is less than 2.5. To investigate this, she takes a simple random sample of 150 households in the city, and finds that the sample mean number of people is 2.3 with a sample standard deviation of 1.5. Can you conclude that the mean number of people per household is less than 2.5?

a. State the null and alternate hypotheses. $H_0: \mu = 2.5,\ H_1: \mu < 2.5$

b. Should we perform a z-test or a t-test? Explain. *t-test*

c. Compute the value of the test statistic. *−1.633*

d. Do you reject H_0? Use the $\alpha = 0.01$ level. *No*

e. State a conclusion.

8. Job satisfaction: The General Social Survey sampled 762 employed people and asked them how satisfied they were with their jobs. Of the 762 people sampled, 386 said that they were completely satisfied or very satisfied with their jobs. Can you conclude that more than 45% of employed people in the United States are completely or very satisfied with their jobs?

a. State the null and alternate hypotheses. $H_0: p = 0.45,\ H_1: p > 0.45$

b. Compute the value of the test statistic. *3.14*

c. Do you reject H_0? Use the $\alpha = 0.01$ level. *Yes*

d. State a conclusion.

9. Sugar content: The sugar content in milligrams of a syrup used to pack canned fruit is measured for 8 batches of syrup. The measurements are normally distributed, and the sample standard deviation is $s = 3$. Can you conclude that the population standard deviation is greater than 2? Use the $\alpha = 0.01$ level of significance. *No, do not reject H_0.*

10. Interpret computer output: The following output from MINITAB presents the results of a hypothesis test.

```
Test of mu = 4.7 vs not = 4.7
The assumed standard deviation = 2.0

 N    Mean    SE Mean       95% CI            Z       P
35   5.401     0.3381   (4.738, 6.064)    2.074   0.038
```

a. What are the null and alternate hypotheses? $H_0: \mu = 4.7,\ H_1: \mu \neq 4.7$

b. What is the value of the test statistic? *2.074*

c. What is the P-value? *0.038*

d. Do you reject H_0 at the $\alpha = 0.05$ level? *Yes*

e. Do you reject H_0 at the $\alpha = 0.01$ level? *No*

11. Interpret calculator display: The following TI-84 Plus display presents the results of a hypothesis test.

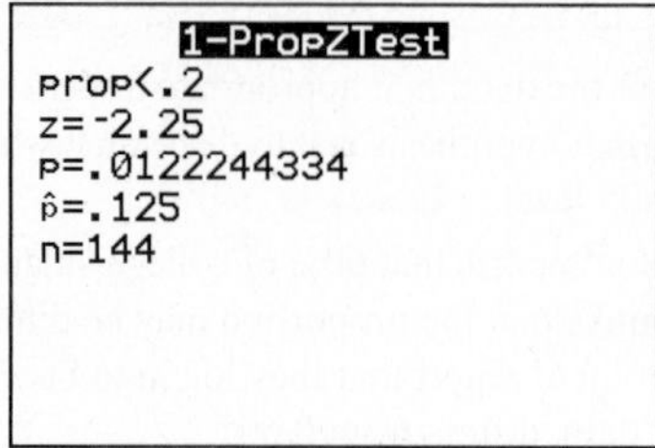

a. Is this a test for a mean, a proportion, or a standard deviation? *Proportion*
b. What are the null and alternate hypotheses? H_0: $p = 0.2$, H_1: $p < 0.2$
c. What is the value of the test statistic? *−2.25*
d. What is the *P*-value? *0.0122244334*
e. Do you reject H_0 at the $\alpha = 0.05$ level? *Yes*
f. Do you reject H_0 at the $\alpha = 0.01$ level? *No*

12. How many TV sets? A survey organization sampled 60 households in a community and found that the sample mean number of TV sets per household was 3.1. The population standard deviation is $\sigma = 1.5$. Can you conclude that the mean number of TV sets per household is greater than 3?
a. State the null and alternate hypotheses. H_0: $\mu = 3$, H_1: $\mu > 3$
b. Should we perform a *z*-test or a *t*-test? Explain. *z-test*
c. Compute the value of the test statistic. *0.52*
d. Do you reject H_0? Use the $\alpha = 0.01$ level. *No*
e. State a conclusion.

13. Crackers: Boxes of crackers are filled by a machine. The weights of a random sample of 25 boxes of crackers have standard deviation $s = 5$ grams. Assume the weights are normally distributed. Can you conclude that the population standard deviation is less than 10 grams? Use the $\alpha = 0.05$ level of significance. *Yes, reject* H_0.

14. How much rent? A housing official in a certain city claims that the mean monthly rent for apartments in the city is more than $1000. To verify this claim, a simple random sample of 40 renters in the city was taken, and the sample mean rent paid was $1100 with a sample standard deviation of $300. Can you conclude that the mean monthly rent in the city is greater than $1000?
a. State the null and alternate hypotheses. H_0: $\mu = 1000$, H_1: $\mu > 1000$
b. Should we perform a *z*-test or a *t*-test? Explain. *t-test*
c. Compute the value of the test statistic. *2.108*
d. Do you reject H_0? Use the $\alpha = 0.05$ level. *Yes*
e. State a conclusion.

15. What's the news? The Pew Research Center reported that 23% of 18- to 29-year-olds watch a cable news channel regularly. The director of media relations at a large university wants to know whether the population proportion of cable news viewers among students at her university is greater than the proportion among 18- to 29-year-olds in general. She surveys a simple random sample of 200 students at the university and finds that 66 of them watch cable news regularly. Can she conclude that the proportion of students at her university who watch cable news regularly is greater than 0.23?
a. State the null and alternate hypotheses. H_0: $p = 0.23$, H_1: $p > 0.23$
b. Compute the value of the test statistic. *3.36*
c. Do you reject H_0? Use the $\alpha = 0.01$ level. *Yes*
d. State a conclusion.

Write About It

1. A result is significant at the 0.01 level. Explain why it must also be significant at the 0.05 level.
2. What does the *P*-value represent?
3. Why is it important to report the *P*-value or the test statistic when presenting the results of a hypothesis test?
4. Why don't we need to know the population standard deviation when performing a test about a population proportion?
5. In what ways are hypothesis tests for a population mean different from hypothesis tests for a proportion? In what ways are they similar?
6. A test will be made of H_0: $\mu = 12$ versus H_1: $\mu > 12$. Explain why the power will be greater when the true value of μ is 20 than when the true value of μ is 15.

In-Class Activities

1. **Multiple testing problem:** Each student tosses a coin 50 times, then performs a test of the hypothesis $H_0: p = 0.5$ versus $H_1: p \neq 0.5$. If some students reject H_0 at the $\alpha = 0.05$ level, can we conclude their coins were unfair? Or is there a better explanation?

   Construct a histogram of the *P*-values. Is it approximately uniform? How does the uniform distribution of the *P*-value explain the relationship between the significance level and the probability of a Type I error?

2. **Choosing H_0 and H_1:** In a manufacturing process, each item is tested as it is manufactured to determine whether it has sufficient strength for the use to which it will be put. There are two choices: The item will be used, or the process will be shut down for recalibration at some expense. Should H_0 be that the item has sufficient strength, or that it does not? Which of the two choices will be made if H_0 is rejected in each case? If H_0 is not rejected? Discuss the implications of Type I and Type II errors for each choice of H_0.

   In a clinical trial, a new drug is tested to see how its effectiveness compares with that of a currently used drug. Should H_1 be that the new drug is more effective, or that it is less effective? Discuss the decisions that would be made and the implications of Type I and Type II errors in each case.

3. **One or two tails:** Discuss some situations in which a one-tailed alternate hypothesis is appropriate and some in which a two-tailed alternate is appropriate.

Case Study: Is It Getting Warmer In Washington, D.C.?

There is substantial evidence to indicate that temperatures on the surface of the Earth have been increasing for the past 100 years or so. We will investigate the possibility of warming trends in one location: Washington, D.C. Table 9.4 presents the record high and low temperatures, along with the year they occurred, for every seventh day at the site of Reagan National Airport in Washington, D.C. The data span the years 1871–2018.

Table 9.4 Dates of Record Temperatures in Washington, D.C.

Date	High	Year	Low	Year	More Recent	Date	High	Year	Low	Year	More Recent	Date	High	Year	Low	Year	More Recent
Jan 1	69	2005	−14	1881	High	May 7	95	1930	38	1970	Low	Sep 3	98	1953	48	1909*	High
8	73	2008	0	1878	High	14	93	1956	41	1928	High	10	98	1983	44	1883	High
15	77	1932	4	1886	High	21	95	1934	41	1907	High	17	96	1991	44	1923	High
22	76	1927	1	1893	High	28	97	1941	42	1961	Low	24	99	2010	39	1963	High
29	76	1975	2	1873	High	Jun 4	99	1925	46	1929*	Low	Oct 1	93	1941*	36	1899	High
Feb 5	70	1991*	−2	1918*	High	11	101	1911	45	1913	Low	8	91	2007	36	1964*	High
12	74	1999	4	1899	High	18	97	1944	51	1965*	Low	15	87	1975	32	1874	High
19	74	1939	4	1903	High	25	100	1997	53	1902	High	22	84	1979*	29	1895	High
26	74	1932	12	1970	Low	Jul 2	101	1898	55	1940	Low	29	82	1918	30	1976*	Low
Mar 5	83	1976	6	1872	High	9	104	1936	55	1891	High	Nov 5	81	2003*	20	1879	High
12	89	1990	11	1900	High	16	104	1988	56	1930*	High	12	77	1912*	24	1926	Low
19	87	1945	12	1876	High	23	102	2011	56	1890	High	19	77	1928	18	1891	High
26	87	1921	23	1955	Low	30	99	1953	56	1914	High	26	74	1979	17	1950	High
Apr 2	89	1963	23	1907	High	Aug 6	106	1918	53	1912	High	Dec 3	71	2012	15	1976	High
9	90	1959	28	1972*	Low	13	101	2016*	55	1930*	High	10	67	1966*	4	1876	High
16	92	2002	29	1928	High	20	101	1983	50	1896	High	17	64	1984*	10	1876	High
23	95	1960	33	1933*	High	27	100	1987	51	1885	High	24	71	2015	5	1983	High
30	92	1942*	34	1874	High												

*Indicates that the record occurred more than once; only the most recent year is given.
Source: National Weather Service

1. If there have been no temperature trends over the years, then it will be equally likely for the record high or the record low to be more recent. If there has been a warming trend, it might be more likely for the record high to be more recent. Let p be the probability that the record high occurred more recently than the record low. Use the sample proportion of dates where the high occurred more recently to test $H_0: p = 0.5$ versus $H_1: p > 0.5$. What do you conclude?
2. The following table presents the records for every day in June. The data show that it is common for records to be set on two or more consecutive days in the same year. This is due to hot spells and cold spells in the weather. For example, five consecutive record highs, from June 2 through June 6, occurred in 1925. Explain why using data for every day may violate the assumption, used in Exercise 1, that the data are a simple random sample.

Date	High	Year	Low	Year	Date	High	Year	Low	Year	Date	High	Year	Low	Year
Jun 1	98	2011	45	1938*	Jun 11	101	1911	45	1913	Jun 21	99	2012	51	1940
2	97	1925*	43	1897	12	95	2002*	50	1907*	22	101	1988	51	1992*
3	99	1925	45	1910	13	96	1954	51	1887	23	98	1988*	51	1918
4	99	1925	46	1929*	14	98	1994	49	1933	24	100	2010	46	1902
5	100	1925	48	1926	15	101	1994	47	1933	25	100	1997	53	1902
6	97	1925	46	1945*	16	99	1994	50	1917	26	101	1952	56	1979
7	98	2008*	47	1894	17	97	2014	50	1926	27	99	2010	57	1927*
8	99	2011	49	1977*	18	97	1944	51	1965*	28	100	1969	54	1927
9	102	2011*	45	1913*	19	99	1994	51	1909	29	104	2012	54	1888
10	100	1964	46	1913	20	99	1931	54	1926*	30	100	1959	50	1919

*Indicates that the record occurred more than once; only the most recent year is given.

3. We will perform another test to determine whether record highs are more likely to have occurred recently. If a record high is equally likely to occur in any year of observation, the mean year in which a record is observed would occur at the midpoint of the observation period, which is $(1871 + 2018)/2 = 1944.5$. Use the data in Table 9.4 to test the hypothesis that the mean year in which a record high occurred is 1944.5 against the alternative that it is greater. What do you conclude?
4. For some records, marked with a *, the record temperature occurred more than once. In these cases, only the most recent year is listed. Explain how this might cause the mean to be greater than the midpoint of 1944.5, even if records are equally likely to occur in any year.
5. Using the data in Table 9.4, drop the dates in which the record high occurred more than once, and test the hypothesis in Exercise 3 again. Does your conclusion change?
6. Perform a hypothesis test on the record lows, after dropping dates on which the record low occurred more than once, in which the alternate hypothesis is that the mean year is less than 1944.5. What do you conclude?
7. Using the analyses you have performed, write a summary of your findings. Describe how strong you believe the evidence to be that record highs have tended to occur more recently than record lows.

chapter

10

Two-Sample Confidence Intervals

Ridofranz/Getty Images

Introduction

When a new medical treatment is proposed, a clinical trial is conducted to determine whether the treatment is safe and effective. In a clinical trial, patients are assigned to receive either the new treatment or an existing treatment. If the patients receiving the new treatment tend to have better outcomes, this is evidence that the new treatment represents an improvement over the old one. When assigning patients to treatments, it is important that the two groups be approximately equal with regard to prior health status. If one group is much healthier than the other, this can bias the results of the trial.

An article in the *New England Journal of Medicine* (361:1329–1338) reported the results of a clinical trial to compare the effectiveness of a new type of heart pacemaker in preventing cardiac failure in patients with heart disease. A total of 1820 patients participated, with 1089 receiving the new treatment and 731 receiving the standard treatment. The assignment to treatments was not made by simple random sampling, but instead by an algorithm that was designed to balance the two groups. The following tables present some of the important characteristics of the two groups.

Characteristic	Standard Treatment Mean	Standard Treatment Standard Deviation	New Treatment Mean	New Treatment Standard Deviation
Age	64	11	65	11
Systolic blood pressure	121	18	124	17
Diastolic blood pressure	71	10	72	10

Characteristic	Standard Treatment Percent with the Characteristic	New Treatment Percent with the Characteristic
Treatment for hypertension	63.2	63.7
Atrial fibrillation	12.6	11.1
Diabetes	30.6	30.2
Cigarette smoking	12.8	11.4
Coronary bypass surgery	28.5	29.1

There are differences between the two groups in all of these characteristics. This is not surprising, because we would expect to see some differences just by chance. The question is whether the differences are large enough to suggest that they may be due to the assignment procedure, and if so, whether the differences may be large enough to be of concern when evaluating the results of the trial.

In this chapter, we will learn to construct confidence intervals that will address questions like this. In the case study at the end of the chapter, we will investigate the differences in the table, to determine whether any differences that result from the assignment procedure might be large enough to be of concern.

Section 10.1 Confidence Intervals for the Difference Between Two Means: Independent Samples

Objectives

1. Distinguish between independent and paired samples
2. Construct confidence intervals for the difference between two population means
3. Describe the pooled standard deviation and the known standard deviation methods

Objective 1 Distinguish between independent and paired samples

How can we tell whether a new drug reduces blood pressure better than an old one? A drug company has developed a new drug that is designed to reduce high blood pressure. The researchers wish to design a study to compare the effectiveness of the new drug to that of the old drug. Here are two ways in which the study can be designed.

Design 1: Two samples of individuals are chosen. One sample is given the old drug, and the other sample is given the new drug. After several months, blood pressures of the members of both samples are measured. We compare the blood pressures in the first sample to the blood pressures in the second sample to determine which drug is more effective.

In design 1, we have **independent samples**. This means that the observations in one sample do not influence the observations in the other.

Design 2: A single group of individuals is chosen. They are given the old drug for a month, then their blood pressures are measured. They then switch to the new drug for a month, after which their blood pressures are measured again. This produces two samples of measurements, the first one from the old drug and the second one from the new drug. We compare the blood pressures in the first sample to the blood pressures in the second sample to determine which drug is more effective.

In design 2, we have **paired samples**. Each observation in one sample can be paired with an observation in the second.

SUMMARY

- Two samples are independent if the observations in one sample do not influence the observations in the other.
- Two samples are paired if each observation in one sample can be paired with an observation in the other. Typically the samples consist of pairs of measurements on the same individual, or on pairs of individuals who are related, such as spouses and siblings.

In this section, we will learn how to construct confidence intervals from independent samples. In Section 10.2, we will learn how to construct confidence intervals from paired samples.

Check Your Understanding

1. A sample of students is enrolled in a speed-reading class. Each takes a reading test before and again after the class. The two samples of scores are compared to determine how large an improvement in reading speed occurred. Are these samples independent or paired? *Paired*
2. A sample of students is enrolled in an online statistics class, and another sample is enrolled in a traditional statistics class. At the end of the semester, the students are given a test. The scores from each sample are compared to determine which class was more effective. Are these samples independent or paired? *Independent*

Answers are on page 517.

Objective 2 Construct confidence intervals for the difference between two population means

Construct Confidence Intervals for the Difference Between Two Population Means

Imagine that we will compare the effectiveness of a new drug designed to reduce blood pressure to the effectiveness of a standard drug. We will draw two independent samples. The first sample will get the new drug, and the second sample will get the standard drug. We can imagine these samples as coming from two populations—a population of patients who take the new drug, and a population of patients who take the standard drug. For each population, there is a mean reduction in blood pressure. We are interested in estimating the difference between the population means.

Let μ_1 be the population mean reduction for the new drug, and let μ_2 be the population mean reduction for the standard drug. We wish to construct a confidence interval for the difference $\mu_1 - \mu_2$. We need a point estimate, a standard error, and a critical value.

The point estimate for the population mean μ_1 is the sample mean $\bar{x}_1$, and the point estimate for the population mean μ_2 is the sample mean $\bar{x}_2$. It follows that the point estimate for the difference $\mu_1 - \mu_2$ between the population means is the difference between the sample means:

$$\text{Point estimate for } \mu_1 - \mu_2 \text{ is } \bar{x}_1 - \bar{x}_2$$

The sample means $\bar{x}_1$ and $\bar{x}_2$ have variances σ_1^2/n_1 and σ_2^2/n_2, respectively, where σ_1^2 and σ_2^2 are the population variances. It is a fact that when samples are independent, the variance of the difference $\bar{x}_1 - \bar{x}_2$ is the *sum* of the variances, so

$$\text{Variance of } \bar{x}_1 - \bar{x}_2 = \frac{\sigma_1^2}{n_1} + \frac{\sigma_2^2}{n_2}$$

The standard error of $\bar{x}_1 - \bar{x}_2$ is the square root of the variance. We don't know the values of σ_1^2 and σ_2^2, so we approximate them with the sample variances s_1^2 and s_2^2. The standard error is

$$\text{Standard error of } \bar{x}_1 - \bar{x}_2 = \sqrt{\frac{s_1^2}{n_1} + \frac{s_2^2}{n_2}}$$

Now we need a critical value. Let $1 - \alpha$ be the confidence level, expressed as a decimal. It can be shown by advanced methods that the appropriate critical value for a confidence interval is $t_{\alpha/2}$, based on the Student's t distribution. We need to determine the number of degrees of freedom for $t_{\alpha/2}$. There are two ways to do this: a simple method that is easier when computing by hand, and a more complicated method that is used by software packages. The simple method is:

$$\text{Degrees of freedom} = \text{Smaller of } n_1 - 1 \text{ and } n_2 - 1$$

The margin of error is obtained by multiplying the critical value by the standard error:

$$\text{Margin of error} = t_{\alpha/2}\sqrt{\frac{s_1^2}{n_1} + \frac{s_2^2}{n_2}}$$

Finally, we construct the confidence interval by adding and subtracting the margin of error from the point estimate:

$$\bar{x}_1 - \bar{x}_2 - t_{\alpha/2}\sqrt{\frac{s_1^2}{n_1} + \frac{s_2^2}{n_2}} < \mu_1 - \mu_2 < \bar{x}_1 - \bar{x}_2 + t_{\alpha/2}\sqrt{\frac{s_1^2}{n_1} + \frac{s_2^2}{n_2}}$$

This method is often referred to as Welch's t method, or more simply, as **Welch's method**.

The method we have just described requires some assumptions, which we now state.

EXPLAIN IT AGAIN

Reason for assumption 3: Assumption 3 is necessary to ensure that the sampling distributions of $\bar{x}_1$ and $\bar{x}_2$ are approximately normal. This justifies the use of the Student's t distribution.

Assumptions for Constructing a Confidence Interval for $\mu_1 - \mu_2$ with Independent Samples (Welch's method)

1. We have simple random samples from two populations.
2. The samples are independent of one another.
3. Each sample size is large ($n > 30$), *or* its population is approximately normal.

When these assumptions are satisfied, we can construct a confidence interval by using the following steps.

Procedure for Constructing a Confidence Interval for $\mu_1 - \mu_2$ with Independent Samples (Welch's method)

Check to be sure that the assumptions are satisfied. If they are, then proceed with the following steps.

Step 1: Compute the sample means $\bar{x}_1$ and $\bar{x}_2$ if they are not given; then compute the point estimate $\bar{x}_1 - \bar{x}_2$.

Step 2: Find the number of degrees of freedom, which is the smaller of $n_1 - 1$ and $n_2 - 1$, and the critical value $t_{\alpha/2}$.

Step 3: Compute the sample standard deviations s_1 and s_2 if they are not given, and compute the standard error $\sqrt{\frac{s_1^2}{n_1} + \frac{s_2^2}{n_2}}$. Multiply the standard error by the critical value to obtain the margin of error: $t_{\alpha/2}\sqrt{\frac{s_1^2}{n_1} + \frac{s_2^2}{n_2}}$

Step 4: Use the point estimate and the margin of error to construct the confidence interval:

Point estimate ± Margin of error

$$\bar{x}_1 - \bar{x}_2 \pm t_{\alpha/2}\sqrt{\frac{s_1^2}{n_1} + \frac{s_2^2}{n_2}}$$

$$\bar{x}_1 - \bar{x}_2 - t_{\alpha/2}\sqrt{\frac{s_1^2}{n_1} + \frac{s_2^2}{n_2}} < \mu_1 - \mu_2 < \bar{x}_1 - \bar{x}_2 + t_{\alpha/2}\sqrt{\frac{s_1^2}{n_1} + \frac{s_2^2}{n_2}}$$

Step 5: Interpret the result.

Example 10.1

Constructing a confidence interval

A drug company has developed a new drug that is designed to reduce high blood pressure. To test the drug, a sample of 15 patients is recruited to take the drug. Their systolic blood pressures are reduced by an average of 28.3 millimeters of mercury (mmHg), with a standard deviation of 12.0 mmHg. In addition, another sample of 20 patients takes a standard drug. The blood pressures in this group are reduced by an average of 17.1 mmHg with a standard deviation of 9.0 mmHg. Assume that blood pressure reductions are approximately normally distributed. Find a 95% confidence interval for the difference between the population mean reduction for the new drug and that of the standard drug.

Solution

To help us keep track of the relevant information, we present it in the following table:

	New Drug	Standard Drug
Sample mean	$\bar{x}_1 = 28.3$	$\bar{x}_2 = 17.1$
Sample standard deviation	$s_1 = 12$	$s_2 = 9$
Sample size	$n_1 = 15$	$n_2 = 20$
Population mean	μ_1 (unknown)	μ_2 (unknown)

We check the assumptions. We have two independent random samples, and the populations are approximately normally distributed. The assumptions are satisfied.

Step 1: Compute the point estimate.

$$\bar{x}_1 - \bar{x}_2 = 28.3 - 17.1 = 11.2$$

Step 2: Find the critical value. In this example, $n_1 = 15$ and $n_2 = 20$, so the degrees of freedom is the smaller of $n_1 - 1 = 14$ and $n_2 - 1 = 19$, which is 14. We look up the critical value $t_{\alpha/2}$ in Table A.3. The value corresponding to 14 degrees of freedom with a confidence level of 95% is $t_{\alpha/2} = 2.145$.

Step 3: Compute the standard error and the margin of error. The standard deviations are $s_1 = 12.0$ and $s_2 = 9.0$. The sample sizes are $n_1 = 15$ and $n_2 = 20$. The standard error is

$$\sqrt{\frac{s_1^2}{n_1} + \frac{s_2^2}{n_2}} = \sqrt{\frac{12.0^2}{15} + \frac{9.0^2}{20}} = 3.6946$$

The margin of error is obtained by multiplying the standard error by the critical value.

$$\text{Margin of error} = t_{\alpha/2}\sqrt{\frac{s_1^2}{n_1} + \frac{s_2^2}{n_2}}$$

In this example, the margin of error is

$$t_{\alpha/2}\sqrt{\frac{s_1^2}{n_1} + \frac{s_2^2}{n_2}} = 2.145\sqrt{\frac{12.0^2}{15} + \frac{9.0^2}{20}} = 7.925$$

Step 4: Construct the confidence interval. The 95% confidence interval is

$$11.2 - 7.925 < \mu_1 - \mu_2 < 11.2 + 7.925$$

$$3.3 < \mu_1 - \mu_2 < 19.1$$

Note that we have rounded the final result to one decimal place, because each of the sample means (28.3 and 17.1) was given to one decimal place.

Step 5: Interpret the result. We are 95% confident that the new drug provides a greater reduction in systolic blood pressure, and that the improvement due to the new drug is between 3.3 and 19.1 mmHg.

Check Your Understanding

3. **Big fish:** A sample of 87 one-year-old spotted flounder had a mean length of 126.31 millimeters with a sample standard deviation of 18.10 millimeters, and a sample of 132 two-year-old spotted flounder had a mean length of 162.41 millimeters with a sample standard deviation of 28.49 millimeters. Construct a 95% confidence interval for the mean length difference between two-year-old flounder and one-year-old flounder. *(29.83, 42.37) [Tech: (29.89, 42.31)]*
Source: *Turkish Journal of Veterinary and Animal Science* 29:1013–1018

4. **Traffic speed:** The mean speed for a sample of 39 cars at a certain intersection was 26.50 kilometers per hour with a standard deviation of 2.37 kilometers per hour, and the mean speed for a sample of 142 motorcycles was 37.14 kilometers per hour with a standard deviation of 3.66 kilometers per hour. Construct a 99% confidence interval for the difference between the mean speeds of motorcycles and cars at this intersection. *(9.32, 11.96) [Tech: (9.36, 11.92)]*
Source: *Journal of Transportation Engineering* 121:317–323

Answers are on page 517.

Technology calculates the degrees of freedom differently

If you construct a confidence interval for the difference between two means with technology, you will usually get a somewhat different answer than you will get using the method we have presented here. The reason is that computers and calculators compute the number of degrees of freedom differently, using a more accurate but rather complicated formula. We present this formula, but you don't need to use it when computing by hand. When computing by hand, it is acceptable just to use the smaller of $n_1 - 1$ and $n_2 - 1$ for the degrees of freedom.

More Accurate Formula for the Degrees of Freedom

Most computer packages compute the degrees of freedom as follows:

$$\text{Degrees of freedom} = \frac{\left[\dfrac{s_1^2}{n_1} + \dfrac{s_2^2}{n_2}\right]^2}{\dfrac{(s_1^2/n_1)^2}{n_1 - 1} + \dfrac{(s_2^2/n_2)^2}{n_2 - 1}}$$

When computing by hand, it is acceptable, and simpler, just to use the smaller of $n_1 - 1$ and $n_2 - 1$ for the degrees of freedom.

Constructing confidence intervals with technology

The following TI-84 Plus display presents results for Example 10.1.

```
2-SampTInt
(3.5912,18.809)
df=25.02267409
x̄1=28.3
x̄2=17.1
Sx1=12
Sx2=9
n1=15
n2=20
```

The results differ from those we obtained, because the degrees of freedom (labeled "df") has been calculated by the more accurate formula. Note that the degrees of freedom is not a whole number.

The following MINITAB output presents results for Example 10.1.

```
        N    Mean    StDev   SE Mean
A      15    28.3     12.0       3.1
B      20    17.1      9.0       2.0

Difference = mu (A) - mu (B)
Estimate for difference: 11.2
95% CI for difference: (3.5908, 18.8092)  DF = 25
```

Most of the output is straightforward. The quantities labeled "SE Mean" are the values of $s_1/\sqrt{n_1}$ and $s_2/\sqrt{n_2}$. MINITAB uses the more accurate formula for the degrees of freedom but rounds the degrees of freedom down to the nearest whole number, which in this case is 25. For this reason, the MINITAB confidence interval differs slightly from the one calculated by the TI-84 Plus.

Step-by-step instructions for constructing confidence intervals with technology are presented in the Using Technology section on page 512.

In most situations in practice, Welch's method is the method of choice for constructing confidence intervals for the difference between two means with independent samples. There are two other methods that have sometimes been used. We describe them here, because they are often offered as options with statistical software. They are generally not the best to use in practice, however, so we will always use Welch's method.

Objective 3 Describe the pooled standard deviation and the known standard deviation methods

Constructing confidence intervals by using the pooled standard deviation

When the two population variances, σ_1^2 and σ_2^2, are known to be equal, there is an alternate method for computing a confidence interval. This alternate method was widely used in the past and is still an option in many forms of technology. We recommend against using it, for reasons that we will discuss.

NOTE TO INSTRUCTOR

The material on pooled standard deviations and known standard deviations can be skipped without loss of continuity.

A Method for Constructing a Confidence Interval When $\sigma_1 = \sigma_2$ (Not Recommended)

Step 1: Compute the **pooled standard deviation**, s_p, as follows:

$$s_p = \sqrt{\frac{(n_1 - 1)s_1^2 + (n_2 - 1)s_2^2}{n_1 + n_2 - 2}}$$

Step 2: Compute the degrees of freedom:

$$\text{Degrees of freedom} = n_1 + n_2 - 2$$

A level $100(1 - \alpha)\%$ confidence interval is

$$\bar{x}_1 - \bar{x}_2 - t_{\alpha/2}s_p\sqrt{\frac{1}{n_1} + \frac{1}{n_2}} < \mu_1 - \mu_2 < \bar{x}_1 - \bar{x}_2 + t_{\alpha/2}s_p\sqrt{\frac{1}{n_1} + \frac{1}{n_2}}$$

The major problem with this method is that the assumption that the population variances are equal is very strict. The method can be quite unreliable if it is used when the population variances are not equal. Now in practice, it is rarely possible to be sure that the variances are equal. There is a test for equality of variances (the F-test, Section 11.4), for which the null hypothesis is that the variances are equal. One could perform this test and assume that the variances are equal if the null hypothesis is not rejected. We recommend against this for two reasons. First, the F-test is unreliable unless the populations are almost exactly normal. Second, failure to reject the null hypothesis does not allow one to assume that the null hypothesis is true. Finally, even when the variances are equal, this method is usually only slightly better than Welch's method. In summary, the best practice is not to use the method that assumes the population variances are equal unless you are very sure that they are.

Constructing confidence intervals when the population standard deviations are known

When the two population variances, σ_1^2 and σ_2^2, are known, we can use $z_{\alpha/2}$, rather than $t_{\alpha/2}$, as the critical value. In practice, σ_1^2 and σ_2^2 are rarely known, so this method is not often applicable. We present it here because it is often offered as an option in statistical calculators and software. The assumptions for this method are the same as for Welch's method, with the additional assumption that the population standard deviations are known.

A Method for Constructing a Confidence Interval When σ_1 and σ_2 Are Known

A level $100(1-\alpha)\%$ confidence interval when σ_1 and σ_2 are known is given by

$$\bar{x}_1 - \bar{x}_2 \pm z_{\alpha/2}\sqrt{\frac{\sigma_1^2}{n_1} + \frac{\sigma_2^2}{n_2}}$$

$$\bar{x}_1 - \bar{x}_2 - z_{\alpha/2}\sqrt{\frac{\sigma_1^2}{n_1} + \frac{\sigma_2^2}{n_2}} < \mu_1 - \mu_2 < \bar{x}_1 - \bar{x}_2 + z_{\alpha/2}\sqrt{\frac{\sigma_1^2}{n_1} + \frac{\sigma_2^2}{n_2}}$$

Note that this method is the same as Welch's except that the sample standard deviations s_1 and s_2 are replaced with the population standard deviations σ_1 and σ_2, and $t_{\alpha/2}$ is replaced with $z_{\alpha/2}$.

Using Technology

We use Example 10.1 to illustrate the technology steps.

TI-84 PLUS

Constructing a confidence interval for the difference between two means

Step 1. Press **STAT** and highlight the **TESTS** menu.

Step 2. Select **2–SampTInt** and press **ENTER** (Figure A). The **2–SampTInt** menu appears.

Step 3. Choose one of the following:

- If the summary statistics are given, select **Stats** as the **Inpt** option and enter $\bar{x}_1$, s_1, n_1, $\bar{x}_2$, s_2, and n_2. For Example 10.1, we use $\bar{x}_1 = 28.3$, $s_1 = 12$, $n_1 = 15$, $\bar{x}_2 = 17.1$, $s_2 = 9$, $n_2 = 20$ (Figure B).
- If the raw data are given, select **Data** as the **Inpt** option and enter the location of the data as the **List1** and **List2** options.

Step 4. In the **C-Level** field, enter the confidence level. For Example 10.1, we use 0.95.

Step 5. Select **No** for the **Pooled** option.

Step 6. Highlight **Calculate** and press **ENTER** (Figure C).

```
EDIT CALC TESTS
2↑T-Test...
3:2-SampZTest...
4:2-SampTTest...
5:1-PropZTest...
6:2-PropZTest...
7:ZInterval...
8:TInterval...
9:2-SampZInt...
0↓2-SampTInt...
```

Figure A

```
2-SampTInt
Inpt:Data Stats
x̄1:28.3
Sx1:12
n1:15
x̄2:17.1
Sx2:9
n2:20
C-Level:.95
↓Pooled:No Yes
```

Figure B

```
2-SampTInt
(3.5912,18.809)
df=25.02267409
x̄1=28.3
x̄2=17.1
Sx1=12
Sx2=9
n1=15
n2=20
```

Figure C

EXCEL

Constructing a confidence interval for the difference between two means

Figure D demonstrates the calculations involved for constructing the margin of error of a confidence interval for the difference between two means. In Example 10.1, $x_1 = 28.3$, $x_2 = 17.1$, $s_1 = 12$, $s_2 = 9$, $n_1 = 15$, and $n_2 = 20$. The confidence level is 95%.

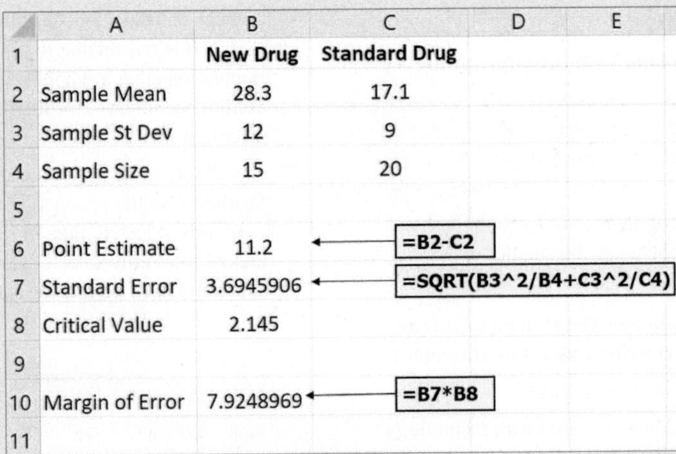

	A	B	C	D	E
1		New Drug	Standard Drug		
2	Sample Mean	28.3	17.1		
3	Sample St Dev	12	9		
4	Sample Size	15	20		
5					
6	Point Estimate	11.2	← =B2-C2		
7	Standard Error	3.6945906	← =SQRT(B3^2/B4+C3^2/C4)		
8	Critical Value	2.145			
9					
10	Margin of Error	7.9248969	← =B7*B8		
11					

Figure D

MINITAB

Constructing a confidence interval for the difference between two means

Step 1. Click on **Stat**, then **Basic Statistics**, then **2-Sample t**.

Step 2. Choose one of the following:

- If the summary statistics are given, select **Summarized Data** and enter the **Sample Size**, the **Mean**, and the **Standard Deviation** for each sample. For Example 10.1, we use $\bar{x}_1 = 28.3$, $s_1 = 12$, $n_1 = 15$, $\bar{x}_2 = 17.1$, $s_2 = 9$, $n_2 = 20$.
- If the raw data are given, select **Each sample is in its own column** and select the columns that contain the data.

Step 3. Click **Options**, and enter the confidence level in the **Confidence Level** field (95) and choose **Difference ≠ hypothesized difference** in the **Alternative field**. Click **OK**.

Step 4. Click **OK** (Figure E).

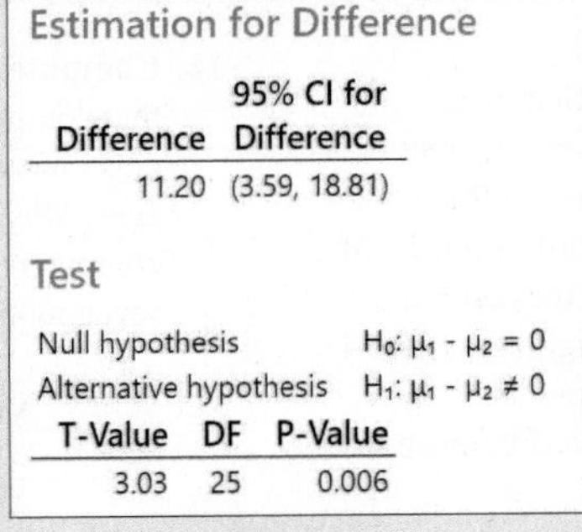

Estimation for Difference

Difference	95% CI for Difference
11.20	(3.59, 18.81)

Test

Null hypothesis H_0: $\mu_1 - \mu_2 = 0$

Alternative hypothesis H_1: $\mu_1 - \mu_2 \neq 0$

T-Value	DF	P-Value
3.03	25	0.006

Figure E

Section 10.1 Exercises

Exercises 1–4 are the Check Your Understanding exercises located within the section.

Understanding the Concepts

In Exercises 5 and 6, fill in each blank with the appropriate word or phrase.

5. If observations in one sample do not influence the observations in another sample, the samples are said to be ________. *independent*

6. When determining the number of degrees of freedom by hand with sample sizes n_1 and n_2, we choose the smaller of ________ and ________. $n_1 - 1, n_2 - 1$

In Exercises 7 and 8, determine whether the statement is true or false. If the statement is false, rewrite it as a true statement.

7. The point estimate for $\mu_1 - \mu_2$ is $\bar{x}_1 + \bar{x}_2$. *False*

8. The number of degrees of freedom calculated with technology is generally different from the number calculated by hand. *True*

Practicing the Skills

In Exercises 9–14, construct the confidence interval for the difference $\mu_1 - \mu_2$ for the given level and values of $\bar{x}_1, \bar{x}_2, s_1, s_2, n_1$, and n_2.

9. Level 90%: $\bar{x}_1 = 104.6, \bar{x}_2 = 92.9, s_1 = 4.8, s_2 = 6.9, n_1 = 26, n_2 = 19$ *(8.5, 14.9) [Tech: (8.6, 14.8)]*

10. Level 95%: $\bar{x}_1 = 478.81, \bar{x}_2 = 322.49, s_1 = 42.84, s_2 = 25.17, n_1 = 14, n_2 = 16$ *(128.10, 184.54) [TI-84 Plus: (129.10, 183.54)] [MINITAB: (129.07, 183.57)]*

11. Level 99%: $\bar{x}_1 = 603.55, \bar{x}_2 = 516.63, s_1 = 54.7, s_2 = 45.2, n_1 = 15, n_2 = 24$ *(36.70, 137.14) [TI-84 Plus: (39.99, 133.85)] [MINITAB: (39.90, 133.94)]*

12. Level 98%: $\bar{x}_1 = 77.3, \bar{x}_2 = 72.6, s_1 = 9.1, s_2 = 8.8, n_1 = 12, n_2 = 16$ *(−4.6, 14.0) [Tech: (−3.9, 13.3)]*

13. Level 95%: $\bar{x}_1 = 47.7, \bar{x}_2 = 42.6, s_1 = 33.9, s_2 = 17.6, n_1 = 13, n_2 = 19$ *(−17.2, 27.4) [TI-84 Plus: (−16.5, 26.7)] [MINITAB: (−16.6, 26.8)]*

14. Level 99.5%: $\bar{x}_1 = 82.9, \bar{x}_2 = 64.1, s_1 = 9.8, s_2 = 6.2, n_1 = 19, n_2 = 10$ *(7.8, 29.9) [Tech: (9.6, 28.0)]*

Working with the Concepts

15. **Does this diet help?** A group of 78 people enrolled in a weight-loss program that involved adhering to a special diet and to a daily exercise program. After six months, their mean weight loss was 25 pounds, with a sample standard deviation of 9 pounds. A second group of 43 people went on the diet but didn't exercise. After six months, their mean weight loss was 14 pounds, with a sample standard deviation of 7 pounds. Construct a 95% confidence interval for the mean difference in weight losses. *(8.0, 14.0) [Tech: (8.1, 13.9)]*

16. **Contaminated water:** The concentration of benzene was measured in units of milligrams per liter for a simple random sample of five specimens of untreated wastewater produced at a gas field. The sample mean was 7.8 with a sample standard deviation of 1.4. Seven specimens of treated wastewater had an average benzene concentration of 3.2 with a standard deviation of 1.7. It is reasonable to assume that both samples come from populations that are approximately normal. Construct a 99% confidence interval for the reduction in benzene concentration after treatment. *(0.5, 8.7) [Tech: (1.7, 7.5)]*

17. **Fertilizer:** In an agricultural experiment, the effects of two fertilizers on the production of oranges were measured. Sixteen randomly selected plots of land were treated with fertilizer A, and 12 randomly selected plots were treated with fertilizer B. The number of pounds of harvested fruit was measured from each plot. Following are the results.

Fertilizer A							
445	523	464	483	441	491	403	466
448	457	437	516	417	420	400	506

Fertilizer B					
362	414	408	398	382	368
393	437	387	373	424	384

a. Explain why it is necessary to check whether the populations are approximately normal before constructing a confidence interval. *Small samples*

b. Following are boxplots of these data. Is it reasonable to assume that the populations are approximately normal? *Yes*

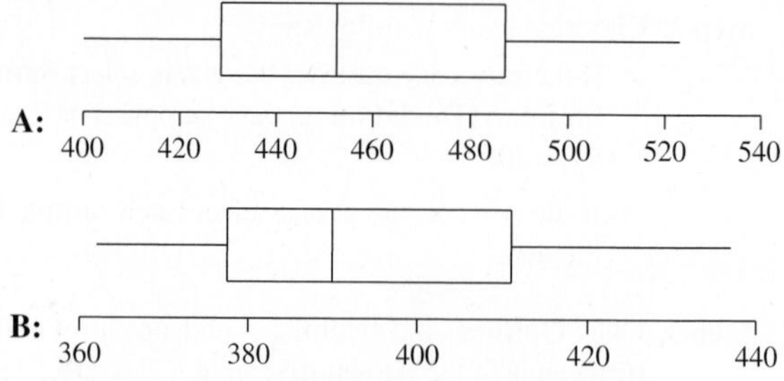

c. If appropriate, construct a 98% confidence interval for the difference between the mean yields for the two types of fertilizer. If not appropriate, explain why not. *(31.4, 94.9) [TI-84 Plus: (34.1, 92.2)] [MINITAB: (34.0, 92.3)]*

18. **Computer crashes:** A computer system administrator notices that computers running a particular operating system seem to crash more often as the installation of the operating system ages. She measures the time (in minutes) before crash for seven computers one month after installation, and for nine computers seven months after installation. The results are as follows:

One month after installation						
209	230	217	230	221	243	247

Seven months after installation								
85	59	129	201	176	240	149	154	105

a. Explain why it is necessary to check whether the populations are approximately normal before constructing a confidence interval. *Small samples*

b. Following are dotplots of these data. Is it reasonable to assume that the populations are approximately normal? *Yes*

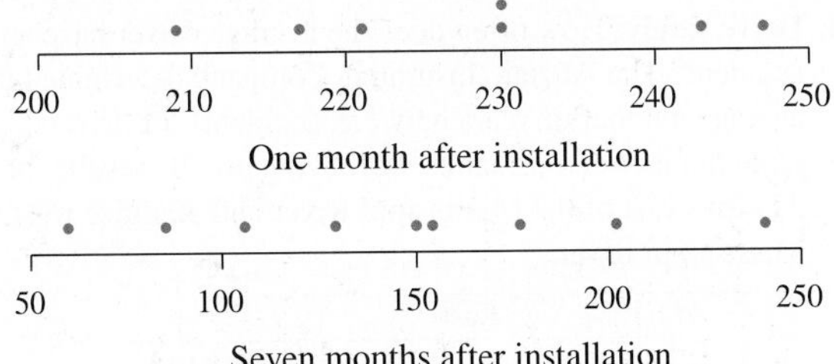

c. If appropriate, construct a 95% confidence interval for the mean difference in time to crash between the first month after installation and the seventh. If not appropriate, explain why not. *(35.7, 132.2) [TI-84 Plus: (39.4, 128.4)] [MINITAB: (39.3, 128.5)]*

19. Are you smarter than your older brother? In a study of birth order and intelligence, IQ tests were given to 18- and 19-year-old men to estimate the size of the difference, if any, between the mean IQs of firstborn sons and of secondborn sons. The following data for 10 firstborn sons and 10 secondborn sons are consistent with the means and standard deviations reported in the article. It is reasonable to assume that the samples come from populations that are approximately normal.

Firstborn				
104	82	102	96	129
89	114	107	89	103

Secondborn				
103	103	91	113	102
103	92	90	114	113

a. Construct a 95% confidence interval for the difference in mean IQ between firstborn and secondborn sons. *(−12.6, 10.8) [TI-84 Plus: (−11.9, 10.1)] [MINITAB: (−12.0, 10.2)]*

b. Based on the confidence interval, is it reasonable to believe that firstborn sons and secondborn sons may have the same mean IQ? *Yes*

Source: Based on data in *Science* 316:1717

20. Effectiveness of distance learning: A study was done to compare the effectiveness of distance learning with traditional classroom instruction. Twelve students took a business administration course online, while 14 students took it in a classroom. The final exam scores were as follows. It is reasonable to assume that the samples come from populations that are approximately normal.

Online					
66	75	85	64	88	77
74	91	72	69	77	83

Classroom						
80	83	64	81	75	80	86
81	91	64	99	85	74	77

a. Construct a 95% confidence interval for the difference between the mean scores for the two types of instruction. *(−11.0, 4.5) [Tech: (−10.5, 4.0)]*

b. An educator claims that both methods of instruction are equally effective. Does the confidence interval contradict this claim? *No*

21. Boys and girls: The National Health Statistics Reports stated that a sample of 318 one-year-old boys had a mean weight of 25.0 pounds with a standard deviation of 3.6 pounds. In addition, a sample of 297 one-year-old girls had a mean weight of 24.1 pounds with a standard deviation of 3.8 pounds.

a. Construct a 95% confidence interval for the difference between the mean weights. *(0.3, 1.5)*

b. A magazine article states that the mean weight of one-year-old boys is the same as that of one-year-old girls. Does the confidence interval contradict this statement? *Yes*

22. Body mass index: In a survey of adults with diabetes, the average body mass index (BMI) in a sample of 1924 women was 31.1 with a standard deviation of 0.2. The BMI in a sample of 1559 men was 30.4, with a standard deviation of 0.6.

a. Construct a 99% confidence interval for the difference in the mean BMI between women and men with diabetes. *(0.66, 0.74)*

b. Does the confidence interval contradict the claim that the mean BMI is the same for both men and women with diabetes? *Yes*

Source: *Journal of Women's Health* 16:1421–1428

23. Energy drinks: A survey of college students reported that in a sample of 413 male college students, the average number of energy drinks consumed per month was 2.49 with a standard deviation of 4.87, and in a sample of 382 female college students, the average was 1.22 with a standard deviation of 3.24.

a. Construct a 99% confidence interval for the difference between men and women in the mean number of energy drinks consumed. *(0.52, 2.02)*

b. Based on the confidence interval, is it reasonable to believe that the mean number of energy drinks consumed may be the same for both male and female college students? *No*

Source: *Journal of American College Health* 56:481–489

24. Low-fat or low-carb? Are low-fat diets or low-carb diets more effective for weight loss? A sample of 77 subjects went on a low-carbohydrate diet for six months. At the end of that time, the sample mean weight loss was 4.7 kilograms with a sample standard deviation of 7.2 kilograms. A second sample of 79 subjects went on a low-fat diet. Their sample mean weight loss was 2.6 kilograms with a standard deviation of 5.9 kilograms.

a. Construct a 98% confidence interval for the difference in mean weight loss between the low-fat and low-carb diets. *(−0.4, 4.6)*

b. A dietitian claims that both diets are equally effective. Does the confidence interval contradict this claim? *No*

Source: *Journal of the American Medical Association* 297:969–977

25. Heart disease and blood pressure: A study of women with heart disease compared blood pressure measurements of women who also had symptoms of coronary artery disease to measurements of women who did not have these symptoms. In a sample of 110 women with symptoms of coronary artery disease, the mean peak systolic blood pressure was 169.9 mmHg, with a standard deviation of 24.8 mmHg. In a sample of 225 women without symptoms, the mean peak systolic blood pressure was 163.3 mmHg, with a standard deviation of 25.8 mmHg.

a. Construct a 95% confidence interval for the difference in mean systolic blood pressure between these two groups of women. *(0.799, 12.401) [Tech: (0.838, 12.362)]*

b. Does the confidence interval contradict the claim that the mean systolic blood pressure is the same in both groups? *Yes*

Source: *Journal of Women's Health* 24:617–622

26. Treadmill exercise: A study was done to investigate whether treadmill exercise could improve the walking ability of patients suffering from claudication, which is pain caused by insufficient blood flow to the legs. A sample of 48 patients exercised on a treadmill every day. After six months, the mean distance walked in six minutes was 348 meters, with a standard deviation of 80 meters. For a control group of 46 patients who did not exercise on a treadmill, the mean distance was 309 meters with a standard deviation of 89 meters.

a. Construct a 99% confidence interval for the difference in mean distance walked between the two groups of patients. *(−8.264, 86.264) [Tech: (−6.998, 84.998)]*

b. Does the confidence interval contradict the claim that treadmill exercise does not increase walking ability? *No*

Source: *Journal of the American Medical Association* 302:165–174

27. Paint manufacture: Two methods are being considered to increase production of a paint manufacturing process. In a random sample of 50 days, the mean daily production using the first method was 645 tons with a standard deviation of 50 tons. In a random sample of 64 days, the mean daily production using the second method was 630 tons with a standard deviation of 40 tons.

a. Construct a 95% confidence interval for the difference in mean daily production between the two methods. *(−2.50, 32.50) [Tech: (−2.20, 32.20)]*

b. Based on the confidence interval, is it reasonable to believe that the mean daily production is the same for both methods? *Yes*

28. Eggs: In a sample of 85 large grade AA eggs, the mean number of calories was 80 with a standard deviation of 2.7. In a sample of 90 large grade A eggs, the mean number of calories was 70 with a standard deviation of 2.5.

a. Construct a 99% confidence interval for the difference in the mean number of calories between large grade A and grade AA eggs. *(8.960, 11.040) [Tech: (8.974, 11.026)]*

b. Based on the confidence interval, is it reasonable to believe that the mean number of calories may be the same for both types of eggs? *No*

29. Online testing: Do you prefer taking tests on paper or online? A college instructor gave identical tests to two randomly sampled groups of 35 students. One group took the test on paper, and the other took it online. Following are the test scores.

Paper						
79	75	49	78	73	81	70
63	79	65	60	74	64	64
74	69	57	65	61	58	92
71	69	66	71	68	67	81
78	80	76	67	49	56	45

Online						
79	75	71	81	56	72	49
75	63	81	74	72	71	73
83	59	78	65	53	47	63
82	81	76	65	82	78	76
65	72	85	84	81	50	79

a. Construct a 95% confidence interval for the difference in mean scores between paper and online tests. *(−8.0, 2.2) [Tech: (−7.9, 2.1)]*

b. The instructor claims that the mean scores are the same for both the paper and the online versions of the test. Does the confidence interval contradict this claim? *No*

30. Drive safely: How often does the average driver have an accident? The Allstate Insurance Company determined the average number of years between accidents for drivers in a large number of U.S. cities. Following are the results for 32 cities east of the Mississippi River and 32 cities west of the Mississippi River.

East							
10.2	9.5	11.2	11.4	9.8	10.0	11.6	10.7
9.5	9.0	7.2	11.5	7.7	9.0	8.9	8.4
6.8	5.3	8.0	8.8	9.0	10.0	9.7	11.6
7.9	9.9	10.5	11.9	10.9	8.6	6.7	9.6

West							
9.5	7.9	13.4	12.6	9.7	8.2	14.0	7.1
10.2	9.4	10.8	9.5	7.5	9.1	8.4	7.6
7.4	7.7	7.1	11.9	10.2	11.9	7.7	7.0
10.0	8.1	11.5	8.4	9.6	9.5	9.9	11.5

a. Construct a 95% confidence interval for the difference in mean scores between western and eastern cities. *(−0.79, 1.01) [Tech: (−0.77, 0.99)]*

b. An insurance company executive claims that the mean number of years between accidents for western cities is 1.5 years greater than the mean for eastern cities. Does the confidence interval contradict this claim? *Yes*

31. Interpret calculator display: The following TI-84 Plus calculator display presents a 95% confidence interval for the difference between two means. The sample sizes are $n_1 = 7$ and $n_2 = 10$.

```
2-SampTInt
(20.904,74.134)
df=12.28537157
x̄1=150.375
x̄2=102.856
Sx1=25.724
Sx2=23.548
n1=7
n2=10
```

a. Compute the point estimate of $\mu_1 - \mu_2$. *47.519*

b. How many degrees of freedom did the calculator use? *12.28537157*

c. Fill in the blanks: We are 95% confident that the difference between the means is between ________ and ________. *20.904, 74.134*

32. Interpret calculator display: The following TI-84 Plus calculator display presents a 99% confidence interval for the difference between two means. The sample sizes are $n_1 = 50$ and $n_2 = 42$.

```
2-SampTInt
(3.0101,4.6114)
df=86.51750655
x̄1=6.83562
x̄2=3.02487
Sx1=1.72541
Sx2=1.17482
n1=50
n2=42
```

a. Compute the point estimate of $\mu_1 - \mu_2$. *3.81075*

b. How many degrees of freedom did the calculator use? *86.51750655*

c. Fill in the blanks: We are 99% confident that the difference between the means is between ______ and ______. *3.0101, 4.6114*

33. Interpret computer output: The following MINITAB output display presents a 98% confidence interval for the difference between two means.

```
     N     Mean     StDev   SE Mean
A   17   72.9172   10.7134   2.5984
B   25   52.1743    9.1237   1.8247

Difference = mu (A) - mu (B)
Estimate for difference: 20.7429
98% CI for difference: (12.9408, 28.5450 )  DF = 30
```

a. What is the point estimate of $\mu_1 - \mu_2$? *20.7429*

b. How many degrees of freedom did MINITAB use? *30*

c. Fill in the blanks: We are ______ confident that the difference between the means is between ______ and ______. *98%, 12.9408, 28.5450*

34. Interpret computer output: The following MINITAB output display presents a 95% confidence interval for the difference between two means.

```
     N     Mean     StDev   SE Mean
A   48   33.827    8.423    1.2157
B   57   10.372    9.314    1.2337

Difference = mu (A) - mu (B)
Estimate for difference: 23.455
95% CI for difference: (20.019, 29.891 )  DF = 102
```

a. What is the point estimate of $\mu_1 - \mu_2$? *23.455*

b. How many degrees of freedom did MINITAB use? *102*

c. Fill in the blanks: We are ______ confident that the difference between the means is between ______ and ______. *95%, 20.019, 26.891*

Extending the Concepts

35. Calculator display: The following TI-84 Plus display presents a 95% confidence interval, using the more accurate formula for the degrees of freedom. The sample sizes are $n_1 = 12$ and $n_2 = 10$.

```
2-SampTInt
(40.739,83.861)
df=16.46306189
x̄1=109.5
x̄2=47.2
Sx1=31.2
Sx2=15.1
n1=12
n2=10
```

a. Use the simpler method to compute the degrees of freedom as the smaller of $n_1 - 1$ and $n_2 - 1$. *9*

b. Find the critical value $t_{\alpha/2}$ for a 95% confidence interval using this value for the degrees of freedom. *2.262*

c. Use this critical value along with the results on the TI-84 Plus display to construct the 95% confidence interval using this value for the degrees of freedom. *(39.241, 85.359)*

Answers to Check Your Understanding Exercises for Section 10.1

1. Paired
2. Independent
3. $29.83 < \mu_1 - \mu_2 < 42.37$ [Tech: $29.89 < \mu_1 - \mu_2 < 42.31$]
4. $9.32 < \mu_1 - \mu_2 < 11.96$ [Tech: $9.36 < \mu_1 - \mu_2 < 11.92$]

Section 10.2 Confidence Intervals for the Difference Between Two Means: Paired Samples

Objective

1. Construct confidence intervals with paired samples

Objective 1 Construct confidence intervals with paired samples

Construct Confidence Intervals with Paired Samples

Does drinking a small amount of alcohol reduce reaction time noticeably? Sixteen volunteers were given a test in which they had to push a button in response to the appearance of an image on a screen. Their reaction times were measured. Then the subjects consumed enough alcohol to raise their blood alcohol level to 0.05%. (In most states, a person is not considered to be "under the influence" until the blood alcohol level reaches 0.08%.) They then took the reaction time test again. Their reaction times, in milliseconds, are presented in Table 10.1 on page 518. The row labeled "Difference" is the increase in reaction time after consuming alcohol. A negative difference occurs when the reaction time after consuming alcohol is less.

Table 10.1 Reaction Times Before and After Consuming Alcohol

	1	2	3	4	5	6	7	8	9	10	11	12	13	14	15	16	Sample Mean
Blood alcohol 0.05%	102	100	77	61	85	50	95	115	64	98	107	44	47	92	70	94	81.3
Blood alcohol 0	103	99	69	50	96	26	71	109	53	89	103	27	50	100	66	86	74.8
Difference	−1	1	8	11	−11	24	24	6	11	9	4	17	−3	−8	4	8	6.5

We have two samples, a sample of reaction times before alcohol consumption and a sample after alcohol consumption. These are paired samples, because each value in one sample can be paired with the value from the same person in the other sample. For example, the first pair is (102, 103), which are the two values from subject 1. These pairs are called **matched pairs**. The bottom row of Table 10.1 contains the differences between the values in each matched pair. These differences are a sample from the population of differences. We can compute the means of the two original samples, along with the mean of the sample of differences. Denote the means of the original samples by $\bar{x}_1$ and $\bar{x}_2$. Denote the mean of the sample of differences by $\bar{d}$. The sample means are presented in the rightmost column of Table 10.1. They are

$$\bar{x}_1 = 81.3 \qquad \bar{x}_2 = 74.8 \qquad \bar{d} = 6.5$$

Simple arithmetic shows that the mean of the differences is the same as the difference between the sample means. In other words, $\bar{d} = \bar{x}_1 - \bar{x}_2$. The same relationship holds for the population means. If we denote the population means by μ_1 and μ_2, and denote the population mean of the differences by μ_d, then $\mu_d = \mu_1 - \mu_2$. This is a very useful fact. It means that a confidence interval for the mean μ_d is also a confidence interval for the difference $\mu_1 - \mu_2$. The matched pairs reduce the two-sample problem to a one-sample problem.

The data show that the sample mean reaction time increased after the subjects consumed alcohol. We would like to construct a 95% confidence interval for the population mean increase μ_d. The method for computing a confidence interval for μ_d is the usual method for computing a confidence interval for a population mean. This method was presented in Section 8.2. We now list the assumptions for this method, when applied to matched pairs.

Assumptions for Constructing a Confidence Interval Using Matched Pairs

1. We have two paired random samples.
2. Either the sample size is large ($n > 30$), *or* the differences between the matched pairs come from a population that is approximately normal.

Notation:

- $\bar{d}$ is the sample mean of the differences between the values in the matched pairs.
- s_d is the sample standard deviation of the differences between the values in the matched pairs.
- μ_d is the population mean difference for the matched pairs.

Constructing a Confidence Interval Using Matched Pairs

Let $\bar{d}$ be the sample mean of the differences between matched pairs, and let s_d be the sample standard deviation. Let μ_d be the population mean difference between matched pairs.

A level $100(1 - \alpha)\%$ confidence interval for μ_d is

$$\bar{d} - t_{\alpha/2}\frac{s_d}{\sqrt{n}} < \mu_d < \bar{d} + t_{\alpha/2}\frac{s_d}{\sqrt{n}}$$

Another way to write this is

$$\bar{d} \pm t_{\alpha/2}\frac{s_d}{\sqrt{n}}$$

Example 10.2

Construct a confidence interval

Use the data in Table 10.1 to construct a 95% confidence for μ_d, the mean difference in reaction times.

We check the assumptions. Because the sample size is small ($n = 16$), we construct a boxplot for the differences to check for outliers or strong skewness.

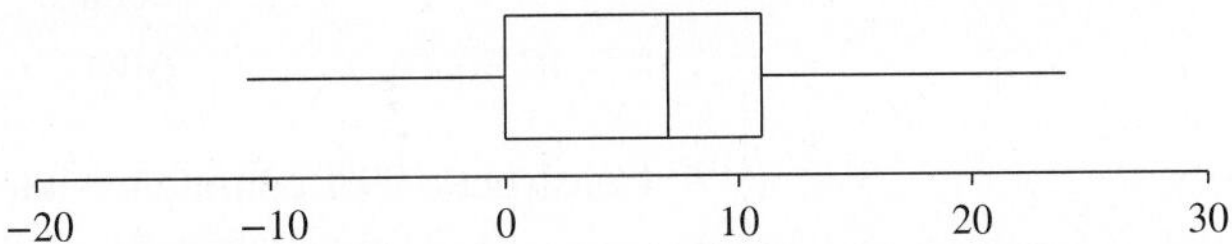

There are no outliers and no evidence of strong skewness, so we can proceed.

Step 1: Compute the sample mean difference $\bar{d}$ and the sample standard deviation of the differences s_d. The sample mean and standard deviation are

$$\bar{d} = 6.500 \qquad s_d = 9.93311$$

Step 2: Compute the critical value. We use the t statistic. The sample size is $n = 16$, so the number of degrees of freedom is $16 - 1 = 15$. The confidence level is 95%. From Table A.3, we find the critical value to be

$$t_{\alpha/2} = 2.131$$

Step 3: Compute the standard error and the margin of error. The standard error is

$$\frac{s_d}{\sqrt{n}} = \frac{9.93311}{\sqrt{16}} = 2.48328$$

The margin of error is

$$t_{\alpha/2}\frac{s_d}{\sqrt{n}} = 2.131(2.48328) = 5.292$$

Step 4: Construct the confidence interval. The 95% confidence interval is

$$\text{Point estimate} \pm \text{Margin of error}$$

$$6.5 - 5.292 < \mu_d < 6.5 + 5.292$$

$$1.2 < \mu_d < 11.8$$

Note that we have rounded the final result to one decimal place, because the original data (the differences) were given as whole numbers (no places after the decimal point).

Step 5: Interpret the result. We are 95% confident that the mean difference is between 1.2 and 11.8. In particular, the confidence interval does not contain 0, and all the values in the confidence interval are positive. We can be fairly certain that the mean reaction time is greater when the blood alcohol level is 0.05%.

Check Your Understanding

1. **High blood pressure:** A group of five individuals with high blood pressure were given a new drug that was designed to lower blood pressure. Systolic blood pressure was measured before and after treatment for each individual. The results follow. Construct a 95% confidence interval for the mean reduction in systolic blood pressure. *(22.3, 40.5)*

	Individual				
	1	2	3	4	5
Before	170	164	168	158	183
After	145	132	129	135	145

2. **Extra help:** The statistics department at a large university instituted a program in which students could get extra help with statistics in the evening. The following table presents scores for tests taken before and after the program for a random sample of six students.

	Student					
	1	2	3	4	5	6
Before	67	58	78	61	75	80
After	73	66	85	69	80	82

Construct a 99% confidence interval for the mean increase in test score. (2.2, 9.8)

Answers are on page 525.

Constructing confidence intervals with technology

The following TI-84 Plus display presents results for Example 10.2.

```
TInterval
(1.207,11.793)
x̄=6.5
Sx=9.933109617
n=16
```

The following output from MINITAB presents the results for Example 10.2.

```
                N     Mean    StDev   SE Mean
Difference     16  6.50000  9.93311   2.48328

95% CI for mean difference: (1.20702, 11.79298)
```

Most of the output is straightforward. The quantity labeled "StDev" is the standard deviation of the differences, s_d, and the quantity labeled "SE Mean" is $s_d/\sqrt{n}$, which is the standard error.

Step-by-step instructions for constructing confidence intervals with technology are given in the Using Technology section on page 521.

Matched pairs usually have a smaller margin of error than independent samples

NOTE TO INSTRUCTOR

It is worth emphasizing that experiments should be designed using matched pairs rather than independent samples when possible, because of the smaller margin of error.

When designing a study to estimate the difference between two means, there is an advantage to using matched pairs when possible. The reason is that the margin of error with matched pairs is much less in most cases than the margin of error for two independent samples. To see this, we will compute the sample standard deviations for the reaction time data in Table 10.1. Let s_1 denote the sample standard deviation for the blood level 0.05% sample, and let s_2 denote the sample standard deviation for the blood level 0 sample. The values of s_1 and s_2 are

$$s_1 = 22.71 \qquad s_2 = 27.39$$

If the samples had been independent, the standard error would have been

$$\sqrt{\frac{s_1^2}{16}+\frac{s_2^2}{16}} = \sqrt{\frac{22.71^2}{16}+\frac{27.39^2}{16}} = 8.90$$

Because we were able to use the sample of differences, the standard error was only 2.48.

The number of degrees of freedom for matched pairs is usually somewhat less than for independent samples. This makes the critical value a bit larger for matched pairs. However, the slight increase in the critical value is in most cases more than made up for by a larger reduction in the standard error.

Using Technology

We use Example 10.2 to illustrate the technology steps.

TI-84 PLUS

Constructing a confidence interval for the difference using matched pairs

Step 1. Enter the data into **L1** and **L2** in the data editor. On the home screen, enter (**L1–L2**) STO **L3** to assign the differences in list **L3** (Figure A).

Step 2. Press **STAT** and highlight the **TESTS** menu.

Step 3. Select **TInterval** and press **ENTER** (Figure B). The **TInterval** menu appears.

Step 4. For **Inpt**, select the **Data** option and enter **L3** as the **List** option.

Step 5. In the **C-Level** field, enter the confidence level. For Example 10.2, we use 0.95 (Figure C).

Step 6. Highlight **Calculate** and press **ENTER** (Figure D).

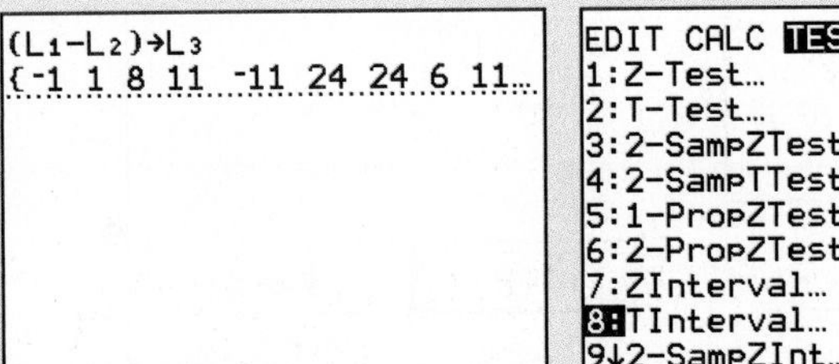

Figure A

Figure B

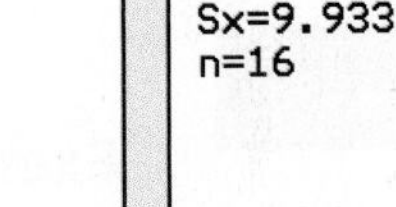

Figure C

Figure D

EXCEL

Constructing a confidence interval for the difference using matched pairs

We use the **=CONFIDENCE.T**(*alpha, standard_dev, size*) command to compute the margin of error for a confidence interval for the difference between the mean of two paired samples. To obtain the margin of error m for Example 10.2, we use $\alpha = 0.05$ in the *alpha* field, the sample standard deviation of the differences $s_d = 9.93311$ in the *standard_dev* field, and the sample size $n = 16$ in the *size* field (Figure E). The confidence interval is given by $\bar{d} - m < \mu < \bar{d} + m$.

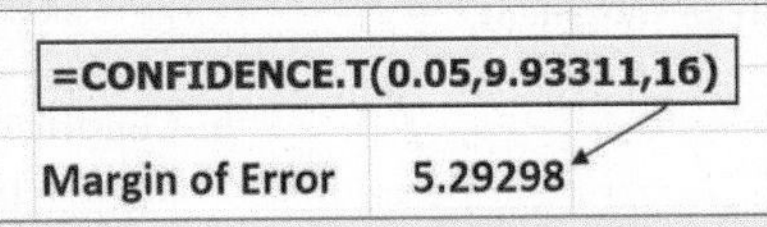

Figure E

MINITAB

Constructing a confidence interval for the difference using matched pairs

Step 1. Enter the data from Example 10.2 into **Columns C1** and **C2**.

Step 2. Click on **Stat**, then **Basic Statistics**, then **Paired t**.

Step 3. Select **Each sample is in a column**, and enter **C1** in the **Sample 1** field and **C2** in the **Sample 2** field.

Step 4. Click **Options**, and enter the confidence level in the **Confidence Level** (95) field. Enter **0** in the **Hypothesized Mean** field and choose **Difference ≠ hypothesized difference** in the **Alternative** field. Click **OK** (Figure F).

Step 5. Click **OK** (Figure G).

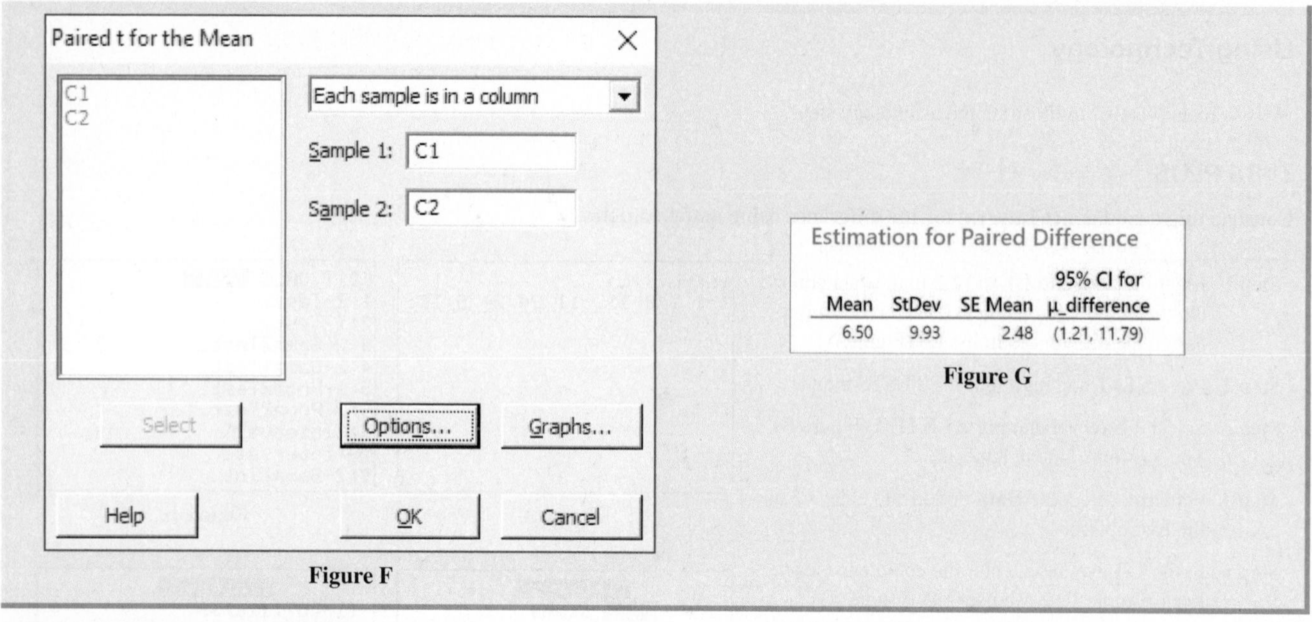

Figure F

Figure G

Section 10.2 Exercises

Exercises 1 and 2 are the Check Your Understanding exercises located within the section.

Understanding the Concepts

In Exercises 3 and 4, fill in each blank with the appropriate word or phrase.

3. With ________ samples, each value in one sample can be matched with a corresponding value from another sample. *paired*
4. The margin of error for the mean difference between matched pairs is usually ________ than the margin of error for independent samples. *smaller*

In Exercises 5 and 6, determine whether the statement is true or false. If the statement is false, rewrite it as a true statement.

5. Paired data reduce a two-sample problem to a one-sample problem. *True*
6. To construct a confidence interval using matched pairs, we must compute the standard deviation of each sample. *False*

Working with the Concepts

7. **Fast computer:** Two microprocessors are compared on a sample of 6 benchmark codes to determine whether there is a difference in speed. The times (in seconds) used by each processor on each code are as follows:

	Code					
	1	2	3	4	5	6
Processor A	27.2	18.1	27.2	19.7	24.5	22.1
Processor B	24.1	19.3	26.8	20.1	27.6	29.8

   a. Find a 95% confidence interval for the difference between the mean speeds. *(−5.33, 2.36)*

   b. A computer scientist claims that the mean speed is the same for both processors. Does the confidence interval contradict this claim? *No*

8. **Brake wear:** For a sample of 9 automobiles, the mileage (in 1000s of miles) at which the original front brake pads were worn to 10% of their original thickness was measured, as was the mileage at which the original rear brake pads were worn to 10% of their original thickness. The results were as follows:

Car	Rear	Front
1	41.2	32.5
2	35.8	26.5
3	46.6	35.6
4	46.9	36.2
5	39.2	29.8
6	51.5	40.9
7	51.0	40.7
8	46.0	34.5
9	47.3	36.5

   a. Construct a 99% confidence interval for the difference in mean lifetime between the front and rear brake pads. *(9.23, 11.29)*

   b. An automotive engineer claims that the mean lifetime for rear brake pads is more than 10,000 miles more than the mean lifetime for front brake pads. Does the confidence interval contradict this claim? *No*

9. **Strength of concrete:** The compressive strength, in kilopascals, was measured for concrete blocks from five different batches of concrete, both three and six days after pouring. The data are as follows:

	Batch				
	1	2	3	4	5
After 6 days	1376	1373	1366	1384	1358
After 3 days	1341	1316	1352	1355	1327

   a. Construct a 95% confidence interval for the difference between the mean strengths of blocks cured three days and blocks cured six days. *(14.0, 52.4)*

b. A civil engineer claims that the mean strength of blocks cured three days is the same as that of blocks cured six days. Does the confidence interval contradict this claim? *Yes*

10. Truck pollution: In an experiment to determine the effect of ambient temperature on the emissions of oxides of nitrogen (NO_x) of diesel trucks, 10 trucks were run at temperatures of 40°F and 80°F. The emissions, in parts per billion, are presented in the following table.

Truck	40°F	80°F
1	834.7	815.2
2	753.2	765.2
3	855.7	842.6
4	901.2	797.1
5	785.4	764.3
6	862.9	819.5
7	882.7	783.6
8	740.3	694.5
9	748.0	772.9
10	848.6	794.7

a. Construct a 99% confidence interval for the difference in mean emissions between trucks running at 40°F and trucks running at 80°F. *(−7.19, 79.81) [Tech: (−7.18, 79.80)]*

b. Based on the confidence interval, is it reasonable to believe that the mean emissions at 40°F may be the same as the mean at 80°F? *Yes*

11. Water loss: Human skin loses water through evaporation, and, in general, damaged skin loses water at a faster rate than undamaged skin. In a study in which the outer layer of skin on a small section of the forearm was partially removed, water loss (in g/m^2 per hour) was measured both before and after skin removal. The results for ten individuals were as follows:

Subject	Before	After
1	18	27
2	12	19
3	14	19
4	11	20
5	12	22
6	17	26
7	16	18
8	18	26
9	14	22
10	14	24

a. Construct a 95% confidence interval for the mean increase in water loss. *(5.914, 9.486)*

b. A dermatologist claims that the difference in water loss is 8 g/m^2 per hour. Does the confidence interval contradict this claim? *No*

12. Breathe deeply: Breathing rates, in breaths per minute, were measured for a group of ten subjects at rest, and then during moderate exercise. The results were as follows:

Subject	Rest	Exercise
1	15	30
2	16	37
3	21	39
4	17	37
5	18	40
6	15	39
7	19	34
8	21	40
9	18	38
10	14	34

a. Construct a 99% confidence interval for the mean increase in breathing rate due to exercise. *(16.49, 22.31)*

b. Based on the confidence interval, is it reasonable to believe that the mean increase is 20 breaths per minute? *Yes*

13. Lower blood pressure: Five individuals with high blood pressure were given a new drug that was designed to lower blood pressure. Systolic blood pressure (in mmHg) was measured before and after treatment for each individual, with the following results:

Subject	Before	After
1	170	145
2	164	132
3	168	129
4	158	135
5	183	145

a. Construct a 98% confidence interval for the mean reduction in systolic blood pressure. *(19.166, 43.634)*

b. Based on the confidence interval, is it reasonable to believe that the mean reduction may be 50 mmHg? *No*

14. Absorption rates: In a study to compare the absorption rates of two antifungal ointments (labeled "A" and "B"), equal amounts of the two drugs were applied to the skin of 14 volunteers. After six hours, the amounts absorbed into the skin (in μg/cm^2) were measured. The results were as follows:

Subject	A	B
1	2.56	2.34
2	2.08	1.49
3	2.46	1.86
4	2.61	2.17
5	1.94	1.29
6	4.50	4.07
7	2.27	1.83
8	3.19	2.93
9	3.16	2.58
10	3.02	2.73
11	3.46	2.84
12	2.57	2.37
13	2.64	2.42
14	2.63	2.44

a. Construct a 95% confidence interval for the mean difference between the amounts absorbed. *(0.3076, 0.5109)*

b. It is claimed that the absorption rates of the two ointments are the same. Does the confidence interval contradict this claim? *Yes*

15. High cholesterol: A group of eight individuals with high cholesterol levels were given a new drug that was designed to lower cholesterol levels. Cholesterol levels, in milligrams per deciliter, were measured before and after treatment for each individual, with the following results:

Individual	Before	After
1	283	215
2	299	206
3	274	187
4	284	212
5	248	178
6	275	212
7	293	192
8	277	196

a. Construct a 90% confidence interval for the mean reduction in cholesterol level. *(70.4, 88.3)*

b. A physician claims that the mean reduction in cholesterol level is more than 80 milligrams per deciliter. Does the confidence interval contradict this claim? *No*

16. Tires and fuel economy: A tire manufacturer is interested in testing the fuel economy for two different tread patterns. Tires of each tread type were driven for 1000 miles on each of 9 different cars. The mileages, in miles per gallon, were as follows:

Car	Tread A	Tread B
1	24.7	20.3
2	22.5	19.0
3	24.0	22.5
4	26.9	23.1
5	22.5	20.9
6	23.5	23.6
7	22.7	21.4
8	19.7	18.2
9	27.5	25.9

a. Construct a 95% confidence interval for the mean difference in fuel economy. *(1.01, 3.24)*

b. Based on the confidence interval, is it reasonable to believe that the mean mileage may be the same for both types of tread? *No*

17. Growth spurt: It is generally known that boys grow at an unusually fast rate between the ages of about 12 and 14. Following are heights, in inches of 40 boys, measured at age 12 and again at age 14.

Age 12	Age 14	Age 12	Age 14
57.9	62.9	55.4	61.8
61.1	65.9	58.7	64.2
62.7	67.5	64.3	69.8
67.5	73.7	58.1	63.3
59.2	64.9	63.3	69.6
61.4	67.0	61.2	66.4
60.7	65.5	64.5	69.2
55.9	62.1	55.9	62.0
59.7	65.4	60.4	65.7
56.3	61.5	57.8	62.8
63.0	68.5	68.3	73.6
58.6	63.9	63.0	67.7
61.1	65.8	64.4	69.2
59.5	64.5	58.2	64.6
61.6	66.3	59.7	66.1
59.3	64.2	60.2	65.9
62.1	67.1	63.7	68.3
62.8	68.1	60.2	65.2
65.3	69.9	62.7	67.9
60.4	66.7	55.6	61.7

a. Construct a 95% confidence interval for the mean increase in height for boys between the ages of 12 and 14. *5.18, 5.55*

b. A pediatrician claims that the mean increase in height is 5.5 inches. Does the confidence interval contradict this claim? *No*

18. SAT coaching: A sample of 32 students took a class designed to improve their SAT math scores. Following are their scores before and after the class:

Before	After	Before	After
383	420	394	430
334	368	513	525
378	396	483	482
467	488	447	482
470	489	440	479
473	473	439	451
443	448	435	431
459	473	451	454
426	428	453	463
493	525	491	511
382	382	526	529
473	474	473	493
408	407	440	466
433	434	481	482
478	490	459	455
502	508	399	404

a. Construct a 95% confidence interval for the mean increase in scores after the class. *8.0, 17.8*

b. The class instructor claims that the mean increase is greater than 20 points. Does the confidence interval contradict this claim? *Yes*

19. Interpret calculator display: The following TI-84 Plus calculator display presents a 95% confidence interval for the mean difference between matched pairs.

```
TInterval
(6.5788,10.898)
x̄=8.7385
Sx=3.7405
n=14
```

a. What is the point estimate of μ_d? *8.7385*

b. How many degrees of freedom are there? *13*

c. Fill in the blanks: We are 95% confident that the mean difference is between ________ and ________. *6.5788, 10.898*

20. Interpret calculator display: The following TI-84 Plus calculator display presents a 99% confidence interval for the mean difference between matched pairs.

```
TInterval
(-4.798,9.9348)
x̄=2.56842
Sx=5.2569
n=7
```

a. What is the point estimate of μ_d? *2.56842*

b. How many degrees of freedom are there? *6*

c. Fill in the blanks: We are 99% confident that the mean difference is between ________ and ________. *−4.798, 9.9348*

21. Interpret computer output: The following output from MINITAB presents a confidence interval for the mean difference between matched pairs.

```
             N     Mean    StDev    SE Mean
Difference   6   2.5324   3.6108    1.47410

99% CI for mean difference: (−3.411394, 8.476194)
```

a. What is the point estimate of μ_d? *2.5324*

b. How many degrees of freedom are there? *5*

c. Fill in the blanks: We are ________ confident that the mean difference is between ________ and ________. *99%, −3.411394, 8.476194*

22. Interpret computer output: The following output from MINITAB presents a confidence interval for the mean difference between matched pairs.

```
              N     Mean    StDev    SE Mean
Difference   12   16.412    3.626    1.04674

95% CI for mean difference: (14.10815, 18.71585)
```

a. What is the point estimate of μ_d? *16.412*

b. How many degrees of freedom are there? *11*

c. Fill in the blanks: We are ________ confident that the mean difference is between ________ and ________. *95%, 14.10815, 18.71585*

Extending the Concepts

23. Advantage of matched pairs: Refer to Exercise 16.

a. Assume that the measurements for tread A and for tread B come from independent samples, and construct a 95% confidence interval for the difference in mean tread wear using the methods of Section 10.1.

b. Which confidence interval is narrower, the one using matched pairs or the one constructed under the assumption that the samples are independent? Explain. *Matched pairs*

24. Paired or independent? To construct a confidence interval for each of the following quantities, say whether it would be better to use paired samples or independent samples, and explain why.

a. The mean difference in height between identical twins. *paired*

b. The mean difference in test scores between students taught by different methods. *independent*

c. The mean difference in height between men and women. *independent*

d. The mean difference in apartment rents in a certain town between this year and last year. *paired*

Answers to Check Your Understanding Exercises for Section 10.2

1. $22.3 < \mu_d < 40.5$

2. $2.2 < \mu_d < 9.8$

Section 10.3 Confidence Intervals for the Difference Between Two Proportions

Objective

1. Construct confidence intervals for the difference between two proportions

Objective 1 Construct confidence intervals for the difference between two proportions

Construct Confidence Intervals for the Difference Between Two Proportions

Does breathing polluted air reduce lung capacity? In a hypothetical study of the effect of air pollution on lung function, a sample of 50 children living in a community with a high level of ozone pollution had their lung functions tested, and 14 of them had lung capacities that were below normal for their size. A second sample of 80 children was drawn from a community with a low level of ozone pollution, and 12 of them had lung capacities that were below normal for their size.

These samples come from two different populations—the populations of children living in the two communities. Each population is one whose individuals fall into two categories: those with reduced lung function, and those with normal lung function. The samples are independent, because the choice of the individuals to be in one sample did not affect the choice of individuals to be in the other.

If we compute the sample proportions, we see that the sample proportion of children in the high-pollution community with reduced lung function is $14/50 = 0.28$, while the proportion in the low-pollution community is only $12/80 = 0.15$. We would like to estimate the difference between the population proportions. Specifically, we would like to construct a 95% confidence interval for the difference between the population proportions.

We will need some notation for the population proportions, the numbers of individuals in each category, and the sample sizes:

NOTATION

- p_1 and p_2 are the population proportions of the category of interest in the two populations.
- x_1 and x_2 are the numbers of individuals in the category of interest in the two samples.
- n_1 and n_2 are the two sample sizes.

We wish to construct a 95% confidence interval for the difference $p_1 - p_2$. As usual, we need three ingredients: a point estimate, a critical value, and a standard error.

We begin with the point estimate. The natural point estimates for the population proportions are the sample proportions

$$\hat{p}_1 = \frac{x_1}{n_1} \qquad \hat{p}_2 = \frac{x_2}{n_2}$$

The point estimate of $p_1 - p_2$ is $\hat{p}_1 - \hat{p}_2$.

Now we compute the standard error. The variances of $\hat{p}_1$ and $\hat{p}_2$ are

$$\text{Variance of } \hat{p}_1 = \frac{p_1(1-p_1)}{n_1} \qquad \text{Variance of } \hat{p}_2 = \frac{p_2(1-p_2)}{n_2}$$

The samples are independent, so the variance of $\hat{p}_1 - \hat{p}_2$ is the sum of the variances of $\hat{p}_1$ and $\hat{p}_2$.

$$\text{Variance of } \hat{p}_1 - \hat{p}_2 = \frac{p_1(1-p_1)}{n_1} + \frac{p_2(1-p_2)}{n_2}$$

The standard error is the square root of the variance. In practice, we don't know the values of p_1 and p_2, so we approximate them with $\hat{p}_1$ and $\hat{p}_2$. The standard error is

$$\text{Standard error of } \hat{p}_1 - \hat{p}_2 = \sqrt{\frac{\hat{p}_1(1-\hat{p}_1)}{n_1} + \frac{\hat{p}_2(1-\hat{p}_2)}{n_2}}$$

Now when the sample sizes are large enough, $\hat{p}_1$ and $\hat{p}_2$ are approximately normally distributed, so the critical value is $z_{\alpha/2}$, found in Table A.2. We multiply the standard error by the critical value to obtain the margin of error

$$\text{Margin of error} = z_{\alpha/2}\sqrt{\frac{\hat{p}_1(1-\hat{p}_1)}{n_1} + \frac{\hat{p}_2(1-\hat{p}_2)}{n_2}}$$

The level $100(1-\alpha)\%$ confidence interval is

$$\text{Point estimate} \pm \text{Margin of error}$$

$$\hat{p}_1 - \hat{p}_2 \pm z_{\alpha/2}\sqrt{\frac{\hat{p}_1(1-\hat{p}_1)}{n_1} + \frac{\hat{p}_2(1-\hat{p}_2)}{n_2}}$$

The method we have just described requires some assumptions, which we now list.

Assumptions for Constructing a Confidence Interval for the Difference Between Proportions

1. We have two independent simple random samples.
2. Each population is at least 20 times as large as the sample drawn from it.
3. The individuals in each sample are divided into two categories.
4. Both samples contain at least 10 individuals in each category.

When these assumptions are met, we can construct a confidence interval for the difference between two proportions by using the following steps.

Procedure for Constructing a Confidence Interval for $p_1 - p_2$

Check to be sure that the assumptions are satisfied. If they are, then proceed with the following steps:

Step 1: Compute the value of the point estimate $\hat{p}_1 - \hat{p}_2$.

Step 2: Find the critical value $z_{\alpha/2}$ corresponding to the desired confidence level from the last line of Table A.3, from Table A.2, or with technology.

Step 3: Compute the standard error

$$\sqrt{\frac{\hat{p}_1(1-\hat{p}_1)}{n_1} + \frac{\hat{p}_2(1-\hat{p}_2)}{n_2}}$$

and multiply it by the critical value to obtain the margin of error

$$z_{\alpha/2}\sqrt{\frac{\hat{p}_1(1-\hat{p}_1)}{n_1} + \frac{\hat{p}_2(1-\hat{p}_2)}{n_2}}$$

Step 4: Use the point estimate and the margin of error to construct the confidence interval:

$$\text{Point estimate} \pm \text{Margin of error}$$

$$\hat{p}_1 - \hat{p}_2 \pm z_{\alpha/2}\sqrt{\frac{\hat{p}_1(1-\hat{p}_1)}{n_1} + \frac{\hat{p}_2(1-\hat{p}_2)}{n_2}}$$

$$\hat{p}_1 - \hat{p}_2 - z_{\alpha/2}\sqrt{\frac{\hat{p}_1(1-\hat{p}_1)}{n_1} + \frac{\hat{p}_2(1-\hat{p}_2)}{n_2}} < p_1 - p_2 < \hat{p}_1 - \hat{p}_2 + z_{\alpha/2}\sqrt{\frac{\hat{p}_1(1-\hat{p}_1)}{n_1} + \frac{\hat{p}_2(1-\hat{p}_2)}{n_2}}$$

Step 5: Interpret the results.

Example 10.3

Constructing a confidence interval

In a study of the effect of air pollution on lung function, a sample of 50 children living in a community with a high level of ozone pollution had their lung capacities measured, and 14 of them had capacities that were below normal for their size. A second sample of 80 children was drawn from a community with a low level of ozone pollution, and 12 of them had lung capacities that were below normal for their size. Construct a 95% confidence interval for the difference between the proportions of children with lung capacities below normal in the two communities.

Solution

We begin by summarizing the available information in a table:

	High Pollution	Low Pollution
Sample size	$n_1 = 50$	$n_2 = 80$
Number with below-normal lung capacity	$x_1 = 14$	$x_2 = 12$
Population proportion	p_1 (unknown)	p_2 (unknown)

We check the assumptions: We have two independent random samples. The populations of children are more than 20 times as large as the samples. The individuals are divided into two categories. In the first sample, there are 14 children with lung capacity below normal, and $50 - 14 = 36$ whose lung capacity is not below normal. In the second sample, the corresponding numbers are 12 and 68. Therefore, each sample contains at least 10 individuals in each category.

Step 1: Compute the value of the point estimate. The sample proportions are

$$\hat{p}_1 = \frac{14}{50} = 0.280 \qquad \hat{p}_2 = \frac{12}{80} = 0.150$$

The point estimate is

$$\hat{p}_1 - \hat{p}_2 = 0.280 - 0.150 = 0.130$$

RECALL

Some commonly used critical values are shown in the following table:

Level	Critical Value
95%	1.96
98%	2.326
99%	2.576

Step 2: Find the critical value. The desired confidence level is 95%, so the critical value is $z_{\alpha/2} = 1.96$.

Step 3: Compute the standard error and the margin of error. The standard error is

$$\sqrt{\frac{\hat{p}_1(1-\hat{p}_1)}{n_1} + \frac{\hat{p}_2(1-\hat{p}_2)}{n_2}} = \sqrt{\frac{0.280(1-0.280)}{50} + \frac{0.150(1-0.150)}{80}}$$
$$= 0.075005$$

The margin of error is

$$z_{\alpha/2}\sqrt{\frac{\hat{p}_1(1-\hat{p}_1)}{n_1} + \frac{\hat{p}_2(1-\hat{p}_2)}{n_2}} =$$

$$1.96\sqrt{\frac{0.280(1-0.280)}{50} + \frac{0.150(1-0.150)}{80}} = 0.14701$$

Step 4: Construct the confidence interval. The 95% confidence interval is

$$\text{Point estimate} \pm \text{Margin of error}$$

$$0.130 \pm 0.14701$$

$$0.130 - 0.14701 < p_1 - p_2 < 0.130 + 0.14701$$

$$-0.017 < p_1 - p_2 < 0.277$$

EXPLAIN IT AGAIN

Round-off rule: When computing a confidence interval for the difference between proportions, round the final result to three decimal places.

Step 5: Interpret the results. We are 95% confident that the difference between the proportions is between −0.017 and 0.277. This confidence interval contains 0. Therefore, we cannot be sure that the proportions of children with diminished lung capacity differ between the two communities.

In Example 10.3, we rounded the final result to three decimal places. We will follow this rule in general.

Constructing confidence intervals with technology

Example 10.3 presented a 95% confidence interval for the difference between the proportions of children with diminished lung capacity in two communities. We now present the results from a TI-84 Plus calculator and the software package MINITAB.

Following is the TI-84 Plus display.

```
2-PropZInt
(-.017,.27701)
p̂1=.28
p̂2=.15
n1=50
n2=80
```

In addition to the confidence interval, the display presents the two sample proportions, $\hat{p}_1$ and $\hat{p}_2$, along with the sample sizes n_1 and n_2.

Following is the MINITAB output.

```
Sample    X    N   Sample p
1        14   50   0.280000
2        12   80   0.150000

Difference = p (1) − p (2)
Estimate for difference: 0.130000
95% CI for difference: (−0.01701, 0.27701)
```

The MINITAB output is mostly straightforward. The quantities listed under the heading "Sample p" are $\hat{p}_1$ and $\hat{p}_2$. The quantity "Estimate for difference" is $\hat{p}_1 - \hat{p}_2$.

Step-by-step instructions for constructing confidence intervals with technology are given in the Using Technology section that follows.

Check Your Understanding

1. **Teaching methods:** A class of 30 computer science students were taught introductory computer programming class with an innovative teaching method that used a graphical interface and drag-and-drop methods of creating computer programs. At the end of the class, 23 of these students said that they felt confident in their ability to write computer programs. Another class of 40 students were taught the same material using a standard method. At the end of class, 25 of these students said they felt confident. Assume that each class contained a simple random sample of students. Construct a 95% confidence interval for the difference between the proportions of students who felt confident. *(−0.071, 0.355)*
2. **Damp electrical connections:** In a test of the effect of dampness on electrical connections, 80 electrical connections were tested under damp conditions and 130 were tested under dry conditions. Twenty of the damp connections failed, and only 8 of the dry ones failed. If possible, construct a 90% confidence interval for the difference between the proportions of connections that fail when damp as opposed to dry. If not possible, explain why. *Not possible*

Answers are on page 533.

Using Technology

We use Example 10.3 to illustrate the technology steps.

TI-84 PLUS

Constructing a confidence interval for the difference between two proportions

Step 1. Press **STAT** and highlight the **TESTS** menu.

Step 2. Select **2–PropZInt** and press **ENTER** (Figure A). The **2–PropZInt** menu appears.

Step 3. Enter x_1, n_1, x_2, and n_2. For Example 10.3, we use $x_1 = 14$, $n_1 = 50$, $x_2 = 12$, and $n_2 = 80$.

Step 4. In the **C-Level** field, enter the confidence level. For Example 10.3, we use 0.95 (Figure B).

```
EDIT CALC TESTS
4↑2-SampTTest…
5:1-PropZTest…
6:2-PropZTest…
7:ZInterval…
8:TInterval…
9:2-SampZInt…
0:2-SampTInt…
A:1-PropZInt…
B↓2-PropZInt…
```

Figure A

```
2-PropZInt
x1:14
n1:50
x2:12
n2:80
C-Level:.95
Calculate
```

Figure B

Step 5. Highlight **Calculate** and press **ENTER** (Figure C).

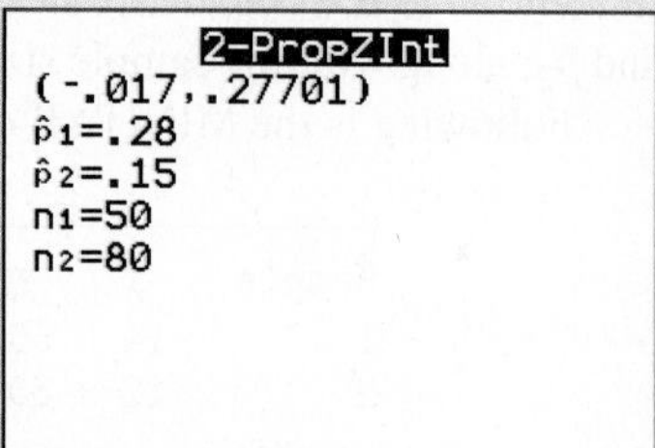

Figure C

EXCEL

Constructing a confidence interval for the difference between two proportions

Figure D demonstrates the calculations involved for constructing the margin of error of a confidence interval for the difference between two proportions. In Example 10.3, $x_1 = 14$, $n_1 = 50$, $x_2 = 12$, and $n_2 = 80$.

	A	B	C	D	E	F
1		High Pollution	Low Pollution			
2	Sample Size	50	80			
3	Number Below Normal	14	12			
4	Sample Proportion	0.28	0.15			
5						
6	Point Estimate	0.13	← =B4-C4			
7	Standard Error	0.075005	← =SQRT(B4*(1-B4)/B2+C4*(1-C4)/C2)			
8	Critical Value	1.96				
9						
10	Margin of Error	0.1470098	← =B7*B8			
11						

Figure D

MINITAB

Constructing a confidence interval for the difference between two proportions

Step 1. Click on **Stat**, then **Basic Statistics**, then **2-Proportions**.

Step 2. Choose one of the following:

- If the summary statistics are given, click **Summarized Data** and enter the values of x_1 and n_1 for the **Number of events** and the **Number of trials** for sample 1. Enter the values of x_2 and n_2 for the **Number of events** and the **Number of trials** for sample 2. For Example 10.3, we use $x_1 = 14$, $n_1 = 50$, $x_2 = 12$, and $n_2 = 80$.
- If the raw data are given, select **Each sample in its own column** and select the columns that contain the data.

Step 3. Click **Options**, and enter the confidence level in the **Confidence Level** field (95) and choose **Difference ≠ hypothesized difference** in the **Alternative field**. Click **OK**.

Step 4. Click **OK** (Figure E).

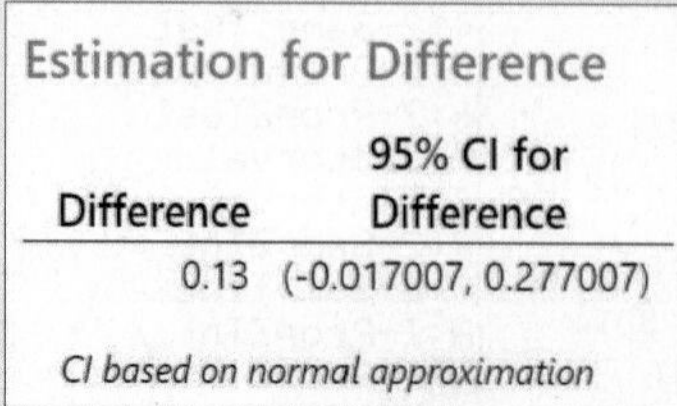

Figure E

Section 10.3 Exercises

Exercises 1 and 2 are the Check Your Understanding exercises located within the section.

Understanding the Concepts

In Exercises 3 and 4, fill in each blank with the appropriate word or phrase.

3. To construct a confidence interval for $p_1 - p_2$, we must have two _______ samples. *independent*
4. When constructing a confidence interval for $p_1 - p_2$, we assume that items in each sample are divided into _______ categories and that there are at least _______ items in each category. *two, ten*

In Exercises 5 and 6, determine whether the statement is true or false. If the statement is false, rewrite it as a true statement.

5. The point estimate for $p_1 - p_2$ is $\hat{p}_1 - \hat{p}_2$, where $\hat{p}_1 = x_1/n_1$ and $\hat{p}_2 = x_2/n_2$. *True*
6. The margin of error for $\hat{p}_1 - \hat{p}_2$ is $\sqrt{\frac{\hat{p}_1(1-\hat{p}_1)}{n_1} + \frac{\hat{p}_2(1-\hat{p}_2)}{n_2}}$. *False*

Practicing the Skills

In Exercises 7–12, construct the confidence interval for the difference $p_1 - p_2$ for the given level and values of x_1, n_1, x_2, and n_2.

7. Level 95%: $x_1 = 42, n_1 = 80, x_2 = 18, n_2 = 60$ *(0.066, 0.384)*
8. Level 90%: $x_1 = 14, n_1 = 25, x_2 = 12, n_2 = 40$ *(0.058, 0.462)*
9. Level 99%: $x_1 = 57, n_1 = 147, x_2 = 86, n_2 = 118$ *(−0.489, −0.193)*
10. Level 95%: $x_1 = 63, n_1 = 106, x_2 = 70, n_2 = 126$ *(−0.089, 0.166)*
11. Level 98%: $x_1 = 49, n_1 = 74, x_2 = 62, n_2 = 153$ *(0.099, 0.415)*
12. Level 99%: $x_1 = 24, n_1 = 53, x_2 = 17, n_2 = 41$ *(−0.227, 0.303)*

Working with the Concepts

13. **Traffic accidents:** Traffic engineers compared rates of traffic accidents at intersections with raised medians with rates at intersections with two-way left-turn lanes. They found that out of 4644 accidents at intersections with raised medians, 2280 were rear-end accidents, and out of 4584 accidents at two-way left-turn lanes, 1982 were rear-end accidents.
    a. Assuming these to be random samples of accidents from the two types of intersection, construct a 95% confidence interval for the difference between the proportions of accidents that are of the rear-end type at the two types of intersection. *(0.038, 0.079)*
    b. Does the confidence interval contradict the claim that the proportion of rear-end accidents is the same at both types of intersection? *Yes*

    Source: *Journal of Transportation Engineering* 121:317–323

14. **Computers in the classroom:** In a new experimental teaching method, when students have questions in class, they send them to the instructor via a laptop computer. From time to time, the instructor pauses to read the questions and to provide the answers, without revealing the identities of the students who are asking the questions. This new method is supposed to eliminate the reluctance of students to ask questions for fear of revealing their lack of understanding in public. In a study of this new method, 67 male students and 72 female students participated. Treat these as if they were simple random samples. Of the male students, 38 said that they liked the new method better than the traditional one in which students raise their hands to ask questions. Of the female students, only 22 said they liked the new method better.
    a. Construct a 95% confidence interval for the difference between the proportions of male and female students who like the new method better. *(0.102, 0.421)*
    b. An educator claims that the proportion of students who like the new method better is the same for males and females. Does the confidence interval contradict this claim? *Yes*

Ariel Skelley/Getty Images

15. **Pain after surgery:** In a random sample of 50 patients undergoing a standard surgical procedure, 15 required medication for postoperative pain. In a random sample of 90 patients undergoing a new procedure, only 16 required pain medication.
    a. Construct a 95% confidence interval for the difference in the proportions of patients needing pain medication between the old and new procedures. *(−0.027, 0.272)*
    b. A physician claims that the proportion of patients who need pain medication is the same for both procedures. Does the confidence interval contradict this claim? *No*
16. **Pretzels:** In order to judge the effectiveness of an advertising campaign for a certain brand of pretzel, a company obtained a simple random sample of 90 convenience store receipts the week before the ad campaign began, and found that 21 of them showed a purchase of the pretzels. Another simple random sample of 70 receipts was taken the week after the ad campaign, and 39 of them showed a pretzel purchase.
    a. Construct a 98% confidence interval for the difference between the proportions of customers purchasing pretzels before and after the ad campaign. *(−0.497, −0.151)*
    b. A marketing manager claims that the proportion of customers who purchased pretzels did not change after the ad

campaign. Does the confidence interval contradict this claim? *Yes*

17. Angioplasty: Angioplasty is a medical procedure in which an obstructed blood vessel is widened. In some cases, a wire mesh tube, called a stent, is placed in the vessel to help it remain open. A study was done to compare the effectiveness of a bare metal stent to one that was coated with a drug designed to prevent reblocking of the vessel. A total of 5320 patients received bare metal stents, and of these, 841 needed treatment for reblocking within a year. A total of 1120 patients received drug coated stents, and 134 of them required treatment within a year.

a. Construct a 95% confidence interval for the difference between the proportions for bare metal and drug coated stents. *(0.0171, 0.0598)*

b. A physician says that the difference in proportions is 0.02. Does the confidence interval contradict this claim? *No*

Source: *Canadian Medical Association Journal* 180:167–174

18. Drinking and smoking: A study of soft drink and smoking habits was conducted. Of 11,805 people who said they consume soft drinks weekly, 2527 said they were smokers. Of 14,985 who said they almost never consume soft drinks, 2503 were smokers.

a. Construct a 98% confidence interval for the difference between the proportions of smokers in the two groups. *(0.0357, 0.0583)*

b. Based on the confidence interval, is it reasonable to believe that the proportion of smokers is the same in both groups? *No*

Source: *European Journal of Nutrition* 57:2113–2121

19. Mushroom compost: An experiment was done to determine whether adding mushroom compost to soil improves plant germination by removing petroleum contamination from soil. Of 150 seeds planted in soil containing 3% mushroom compost by weight, 74 germinated. Of 150 seeds planted in soil containing 5% mushroom compost by weight, 86 germinated.

a. Construct a 99% confidence interval for the difference between the proportions of seeds that germinate. *(−0.0679, 0.2279)*

b. Based on the confidence interval, is it reasonable to believe that the proportion of seeds that germinate is the same for both concentrations of mushroom compost? *Yes*

Source: *Journal of Environmental Health Science and Engineering* 15:8

20. Pain relief: In a clinical trial to study the effectiveness of a new pain reliever, a sample of 250 patients were given a new drug, and another sample of 100 patients were given a placebo. Of the patients given the new drug, 186 reported feeling relief, and of the patients given the placebo, 56 reported feeling relief.

a. Construct a 95% confidence interval for the difference between the proportions feeling relief. *(0.073, 0.295)*

b. Does the confidence interval contradict the claim that the proportions feeling relief are the same for both groups? *Yes*

21. Defective electronics: A team of designers was given the task of reducing the defect rate in the manufacture of a certain printed circuit board. The team decided to reconfigure the cooling system. A total of 973 boards were produced the week before the reconfiguration was implemented, and 254 of these were defective. A total of 847 boards were produced the week after reconfiguration, and 95 of these were defective.

a. Construct a 90% confidence interval for the decrease in the defective rate after the reconfiguration. *(0.120, 0.178)*

b. A quality control engineer claims that the reconfiguration has decreased the proportion of defective parts by more than 0.15. Does the confidence interval contradict this claim? *No*

Source: *The American Statistician* 56:312–315

22. Satisfied? A poll taken by the General Social Survey in 2014 asked people in the United States whether they were satisfied with their financial situation. A total of 698 out of 2532 people said they were satisfied. The same question was asked in 2018, and 737 out of 2338 people said they were satisfied.

a. Construct a 95% confidence interval for the difference between the proportion of adults who said they were satisfied in 2018 and the proportion in 2014. *(0.014, 0.065)*

b. A sociologist claims that the proportion of people who are satisfied increased from 2014 to 2018 by more than 0.02. Does the confidence interval contradict this claim? *No*

23. Cancer prevention: Colonoscopy is a medical procedure that is designed to find and remove precancerous lesions in the colon before they become cancerous. In a sample of 51,460 people without colorectal cancer, 5043 had previously had a colonoscopy, and in a sample of 10,292 people diagnosed with colorectal cancer, 720 had previously had a colonoscopy.

a. Construct a 95% confidence interval for the difference in the proportions of people who had colonoscopies between those who were diagnosed with colorectal cancer and those who were not. *(0.022, 0.034)*

b. Does the confidence interval contradict the claim that the proportion of people who have had colonoscopies is the same among those with colorectal cancer and those without? *Yes*

Source: *Annals of Internal Medicine* 150:1–8

24. Social media: A Pew poll found that in a sample of 1002 adults, 621 had a social media account. The study was repeated a year later, and 631 out of 971 adults had social media accounts.

a. Construct a 95% confidence interval for the increase in the proportion of adults with social media accounts during the year between the surveys. *(−0.0124, 0.0726)*

b. An executive at a social media company claims that the proportion of adults with social media accounts increased by more than 0.10 during the year. Does the confidence interval contradict this claim? *Yes*

25. Interpret calculator display: The following TI-84 Plus calculator display presents a 95% confidence interval for the difference between two proportions.

```
2-PropZInt
(-.0815,.25484)
p̂1=.3866666667
p̂2=.3
n1=75
n2=50
```

a. Compute the point estimate of $p_1 - p_2$. *0.086666667*

b. Fill in the blanks: We are 95% confident that the difference between the proportions is between ________ and ________. *−0.0815, 0.25484*

26. Interpret calculator display: The following TI-84 Plus calculator display presents a 99% confidence interval for the difference between two proportions.

```
2-PropZInt
(.04811,.33732)
p̂1=.8620689655
p̂2=.6693548387
n1=87
n2=124
```

a. Compute the point estimate of $p_1 - p_2$. *0.19271*

b. Fill in the blanks: We are 99% confident that the difference between the proportions is between ________ and ________. *0.04811, 0.33732*

27. Interpret computer output: The following MINITAB output presents a confidence interval for the difference between two proportions.

```
Sample    X    N   Sample p
1        32   59   0.542373
2        23   63   0.365079

Difference = p(1) - p(2)
Estimate for difference: 0.177294
99% CI for difference: (-0.051451, 0.406038)
```

a. What is the point estimate of $p_1 - p_2$? *0.177294*

b. Fill in the blanks: We are ________ confident that the difference between the proportions is between ________ and ________. *99%, −0.051451, 0.406038*

28. Interpret computer output: The following MINITAB output presents a confidence interval for the difference between two proportions.

```
Sample    X    N   Sample p
1        16  546   0.029304
2        18  935   0.019251

Difference = p(1) - p(2)
Estimate for difference: 0.010053
95% CI for difference: (-0.006612, 0.026717)
```

a. What is the point estimate of $p_1 - p_2$? *0.010053*

b. Fill in the blanks: We are ________ confident that the difference between the proportions is between ________ and ________. *95%, −0.006612, 0.026717*

Extending the Concepts

29. Finding the sample size: Polls are to be conducted in two cities to determine the difference in the proportions of residents who believe that the economy will improve over the next year. A 95% confidence interval will be constructed for the difference between the proportions. If the sample sizes are to be equal in the two cities, how large should each sample be so that the margin of error will be no more than 0.08? *301*

Answers to Check Your Understanding Exercises for Section 10.3

1. $-0.071 < p_1 - p_2 < 0.355$

2. It is not possible to construct a confidence interval, because the sample of dry connections contains fewer than 10 that failed.

Chapter 10 Summary

Section 10.1: When we want to estimate the difference between two population means, we may use either independent samples or paired samples. Two samples are independent if the observations in one sample do not influence the observations in the other. Two samples are paired if each observation in one sample can be paired with an observation in the other. When the samples are independent, Welch's method can be used to construct a confidence interval for the difference between the population means. If both sample sizes are large or both populations are approximately normal, a level $100(1 - \alpha)\%$ confidence interval for the difference between two population means is $\bar{x}_1 - \bar{x}_2 \pm t_{\alpha/2}\sqrt{s_1^2/n_1 + s_2^2/n_2}$.

Section 10.2: When the samples are paired, we can construct a confidence interval for the mean difference by computing the difference for each pair and constructing a confidence interval for the mean. The assumptions for the population of differences are the same as the assumptions for constructing a confidence interval for a population mean. A level $100(1 - \alpha)\%$ confidence interval is $\bar{d} \pm t_{\alpha/2} s_d/\sqrt{n}$, where $\bar{d}$ is the average difference and s_d is the standard deviation of the sample of differences.

Section 10.3: Confidence intervals can also be constructed for the difference between two proportions. The assumptions necessary to construct a confidence interval for a single population proportion must hold for both populations. A level $100(1 - \alpha)\%$ confidence interval for the difference between two proportions is given by $\hat{p}_1 - \hat{p}_2 \pm z_{\alpha/2}\sqrt{\hat{p}_1(1 - \hat{p}_1)/n_1 + \hat{p}_2(1 - \hat{p}_2)/n_2}$.

Vocabulary and Notation

Important Formulas

Confidence interval for the difference between two means, independent samples:

$$\bar{x}_1 - \bar{x}_2 - t_{\alpha/2}\sqrt{\frac{s_1^2}{n_1} + \frac{s_2^2}{n_2}} < \mu_1 - \mu_2 < \bar{x}_1 - \bar{x}_2 + t_{\alpha/2}\sqrt{\frac{s_1^2}{n_1} + \frac{s_2^2}{n_2}}$$

Confidence interval for the difference between two proportions:

$$\hat{p}_1 - \hat{p}_2 - z_{\alpha/2}\sqrt{\frac{\hat{p}_1(1-\hat{p}_1)}{n_1} + \frac{\hat{p}_2(1-\hat{p}_2)}{n_2}} < p_1 - p_2 < \hat{p}_1 - \hat{p}_2 + z_{\alpha/2}\sqrt{\frac{\hat{p}_1(1-\hat{p}_1)}{n_1} + \frac{\hat{p}_2(1-\hat{p}_2)}{n_2}}$$

Confidence interval for the difference between two means, matched pairs:

$$\bar{d} - t_{\alpha/2}\frac{s_d}{\sqrt{n}} < \mu_d < \bar{d} + t_{\alpha/2}\frac{s_d}{\sqrt{n}}$$

Chapter Quiz

In Exercises 1 and 2, determine whether the samples described are paired or independent.

1. A sample of 15 weight lifters is tested to see how much weight they can bench press. They then follow a special training program for three weeks, after which they are tested again. The samples are the amounts of weights that were lifted before and after the training program. *Paired*

2. A sample of 20 weight lifters is tested to see how much weight they can bench press. Ten of them are chosen at random as the treatment group. They participate in a special training program for three weeks. The remaining 10 are the control group. They follow their usual program. At the end of three weeks, all 20 weight lifters are tested again and the increases in the amounts they can lift are recorded. The two samples are the increases in the amounts lifted by the treatment group and the control group. *Independent*

3. In a survey of 300 randomly selected female and 240 male holiday shoppers, 87 of the females and 98 of the males stated that they will wait until the last week before Christmas to finish buying gifts. Let p_1 be the population proportion of males who will wait until the last week, and let p_2 be the population proportion of females. Compute a point estimate for the difference $p_1 - p_2$. *0.1183*

4. Refer to Exercise 3. Find the critical value for a 95% confidence interval for the difference $p_1 - p_2$. *1.96*

5. Refer to Exercise 3. Compute the margin of error for a 95% confidence interval for the difference $p_1 - p_2$. *0.0806*

6. Refer to Exercise 3. Construct a 95% confidence interval for the difference $p_1 - p_2$. *(0.038, 0.199)*

7. Eight students in a particular college course are given a pre-test at the beginning of the semester and are then given the same exam at the end to test what they have learned. The exam scores at the beginning and at the end are given in the following table.

	Student							
	1	2	3	4	5	6	7	8
End	83	71	79	95	84	72	69	78
Beginning	72	58	76	81	69	63	71	77

Let μ_d be the mean difference End – Beginning. Compute a point estimate for μ_d. *8*

8. Refer to Exercise 7. Find the critical value for a 99% confidence interval for μ_d. *3.499*

9. Refer to Exercise 7. Compute the margin of error for a 99% confidence interval for the difference μ_d. *8.0172 [Tech: 8.0183]*

10. Refer to Exercise 7. Construct a 99% confidence interval for the difference μ_d. *(−0.02, 16.02)*

11. A random sample of 76 residents in a small town had a mean annual income of $34,214, with a sample standard deviation of $2171. In a neighboring town, a random sample of 88 residents had a mean annual income of $31,671 with a sample standard deviation of $3279. Let μ_1 be the population mean annual income in the first town, and let μ_2 be the population mean annual income in the neighboring town. Compute a point estimate for the difference $\mu_1 - \mu_2$. *2543*

12. Refer to Exercise 11. Find the critical value for a 90% confidence interval for the difference $\mu_1 - \mu_2$. *1.671 [Tech: 1.655]*

13. Refer to Exercise 11. Compute the margin of error for a 90% confidence interval for the difference $\mu_1 - \mu_2$. *717.161 [TI-84 Plus: 710.262] [MINITAB: 710.269]*

14. Refer to Exercise 11. Construct a 90% confidence interval for the difference $\mu_1 - \mu_2$. *(1825.8, 3260.2) [Tech: (1832.7, 3253.3)]*

15. In a poll of 100 voters, 57 said they were planning to vote for the incumbent governor, and 48 said they were planning to vote for the incumbent mayor. Explain why these data should not be used to construct a confidence interval for the difference between the proportions of voters who plan to vote for the governor and those who plan to vote for the mayor. *Not independent samples*

Review Exercises

1. **Hatching shrimp eggs:** A study of the effect of water pollution on the habitat of shrimp reported that out of 1985 eggs produced by shrimp at the Diesel Creek site in Charleston, South Carolina, 1919 hatched, and at the Shipyard Creek site, also in Charleston, 4561 out of 4988 eggs hatched. Construct a 95% confidence interval for the difference between the proportions of eggs that hatch at the two sites. *(0.041, 0.063)*

2. **Don't construct a confidence interval:** An investor is trying to decide in which of two stocks to invest. He examines records for the past 365 days and finds that the price of stock A increased on 197 of them and the price of stock B increased on 158 of them. Explain why these data should not be used to construct a confidence interval for the difference between the proportions of days that the stock prices increase. *Not independent samples*

3. **Exercise and heart rate:** A simple random sample of seven people embarked on a program of regular aerobic exercise. Their heart rates, in beats per minute, were measured before and after, with the following results:

	Person						
	1	2	3	4	5	6	7
Before	81	84	79	85	79	84	87
After	73	77	73	78	71	75	80

Construct a 95% confidence interval for the mean reduction in heart rate. *(6.5, 8.3)*

4. **College tuition:** Independent random samples of private college and university tuition and fees for a recent academic year from California and New York yielded the following data, in dollars.

California:	29,940	35,650	27,231	21,000	32,610	37,850	36,992	34,584	22,850	25,788
New York:	29,283	40,870	28,460	33,835	24,410	30,868	26,400	37,882	22,656	38,980

Based on data from www.collegeprowler.com.

Construct a 95% confidence interval for the difference between the mean amounts of tuition in the two states. *(−7153.4, 5323.6) [TI-84 Plus: (−6710.1, 4880.3)] [MINITAB: (−6733.7, 4903.9)]*

5. **Recovery time from surgery:** A new postsurgical treatment is being compared with a standard treatment. Seven subjects receive the new treatment, while seven others (the controls) receive the standard treatment. The recovery times, in days, are given below.

Control: 18 23 24 30 32 35 39 Treatment: 12 13 15 19 20 21 24

Construct a 98% confidence interval for the reduction in the mean recovery times associated with treatment. *(0.7, 21.3) [TI-84 Plus: (2.0, 20.0)] [MINITAB: (1.8, 20.2)]*

6. **Polling results:** A simple random sample of 400 voters in the town of East Overshoe was polled, and 242 said they planned to vote in favor of a bond issue to raise money for elementary schools. A simple random sample of 300 voters in West Overshoe was polled, and 161 said they were in favor. Construct a 98% confidence interval for the difference between the proportions of voters in the two towns who favor the bond issue. *(−0.020, 0.156)*

7. **Treating bean plants:** In a study to measure the effect of an herbicide on the phosphate content of bean plants, a sample of 75 plants treated with the herbicide had a mean phosphate concentration (in percent) of 3.52 with a standard deviation of 0.41, and 100 untreated plants had a mean phosphate concentration of 5.82 with a standard deviation of 0.52. Construct a 95% confidence interval for the difference in mean phosphate concentration between treated and untreated plants. *(−2.44, −2.16)*

8. **Dueling scales:** In an experiment to determine whether there is a systematic difference between the weights obtained with two different scales, 10 rock specimens were weighed, in grams, on each scale. The following data were obtained:

	Specimen									
	1	2	3	4	5	6	7	8	9	10
Weight on scale 1	11.24	14.36	8.31	10.50	23.44	9.12	13.46	6.44	12.41	19.35
Weight on scale 2	11.27	14.40	8.35	10.53	23.41	9.16	13.50	6.43	12.44	19.34

Construct a 98% confidence interval for mean difference in weight between the two scales. *(−0.043, 0.003)*

9. **Automobile pollution:** In a random sample of 340 cars driven at low altitudes, 46 exceeded a threshold for the amount of particulate pollution. In a random sample of 85 cars driven at high altitudes, 21 exceeded a threshold for the amount of particulate pollution. Construct a 99% confidence interval for the difference between the proportions for high-altitude and low-altitude vehicles. *(−0.241, 0.018)*

10. **Interpret calculator display:** The following TI-84 Plus calculator display presents a 95% confidence interval for the difference between two means. The sample sizes are $n_1 = 85$ and $n_2 = 71$.

```
2-SampTInt
(9.8059,12.998)
df=113.2701584
x̄1=49.81472
x̄2=38.41269
Sx1=3.69057
Sx2=5.89133
n1=85
n2=71
```

a. Find the point estimate of $\mu_1 - \mu_2$. *11.402*
b. How many degrees of freedom did the calculator use? *113.2701584*
c. Fill in the blanks: We are 95% confident that the difference between the means is between ______ and ______. *9.8059, 12.998*

11. Interpret computer output: The following MINITAB output display presents a 95% confidence interval for the difference between two means.

```
     N    Mean    StDev   SE Mean
A   12   9.5713   1.025    0.2959
B    8   7.2198   5.173    1.8289

Difference = mu (A) − mu (B)
Estimate for difference: 2.3515
95% CI for difference: (−2.02947, 6.73247)  DF = 7
```

a. Find the point estimate of $\mu_1 - \mu_2$. *2.3515*
b. How many degrees of freedom did MINITAB use? *7*
c. Fill in the blanks: We are ______ confident that the difference between the means is between ______ and ______. *95%, −2.02947, 6.73247*

12. Interpret calculator display: The following TI-84 Plus calculator display presents a 95% confidence interval for the difference between two proportions.

```
2-PropZInt
(.12307,.5092)
p̂1=.7307692308
p̂2=.4146341463
n1=52
n2=41
```

a. Find the point estimate of $p_1 - p_2$. *0.31614*
b. Fill in the blanks: We are 95% confident that the difference between the proportions is between ______ and ______. *0.12307, 0.5092*

13. Interpret computer output: The following MINITAB output presents a confidence interval for the difference between two proportions.

```
Sample    X     N    Sample p
1        85   103    0.825243
2        72   125    0.576000

Difference = p(1) − p(2)
Estimate for difference: 0.249243
95% CI for difference: (0.135735, 0.362751)
```

a. Find the point estimate of $p_1 - p_2$. *0.249243*
b. Fill in the blanks: We are ______ confident that the difference between the proportions is between ______ and ______. *95%, 0.135735, 0.362751*

14. Interpret calculator display: The following TI-84 Plus calculator display presents a 95% confidence interval for the mean difference between matched pairs.

```
TInterval
(7.7632,9.3028)
x̄=8.533
Sx=1.548
n=18
```

a. Find the point estimate of μ_d. 8.533

b. How many degrees of freedom are there? 17

c. Fill in the blanks: We are 95% confident that the mean difference is between ________ and ________. 7.7632, 9.3028

15. Interpret computer output: The following output from MINITAB presents a confidence interval for the mean difference between matched pairs.

```
              N      Mean      StDev    SE Mean
Difference   15    9.8612     3.7149    0.95918

95% CI for mean difference: (7.803957, 11.918443)
```

a. Find the point estimate of μ_d. 9.8612

b. How many degrees of freedom are there? 14

c. Fill in the blanks: We are ________ confident that the mean difference is between ________ and ________. 95%, 7.803957, 11.918443

Write About It

1. Why is it necessary for all values in the confidence interval to be positive to conclude that $\mu_1 > \mu_2$? What would have to be true to conclude that $\mu_1 < \mu_2$?
2. In what ways is the procedure for constructing a confidence interval for the difference between two proportions similar to constructing a confidence interval for one proportion? In what ways is it different?
3. Provide an example of two samples that are independent. Explain why these samples are independent.
4. Provide an example of two samples that are paired. Explain why these samples are paired.

In-Class Activities

1. **Confidence level:** Each student tosses both a penny and a nickel 50 times each and constructs a 90%, a 95%, and a 99% confidence interval for the difference in the proportion of heads. How many of the 90% confidence intervals would you expect to contain the true value of 0? How many do? Repeat for the 95% and 99% confidence intervals.
2. **Paired versus independent samples:** For each student, measure the distance from the knee to the ankle and the distance from the elbow to the wrist. Construct a 95% confidence interval for the difference in two ways: first using paired differences, and second treating the samples as independent. Which confidence interval is wider? Why? Why does the paired method give a different result?

Case Study: Evaluating The Assignment Of Subjects In A Clinical Trial

In the chapter opener, we described a study in which patients were assigned to receive either a new treatment or a standard treatment for the prevention of heart failure. A total of 1820 patients participated, with 1089 receiving the new treatment and 731 receiving the standard treatment. The assignment was not made by simple random sampling; instead, an algorithm was used that was designed to produce balance between the groups. The following table presents summary statistics describing several health characteristics of the people in each group.

Characteristic	Standard Treatment		New Treatment	
	Mean	Standard Deviation	Mean	Standard Deviation
Age	64	11	65	11
Systolic blood pressure	121	18	124	17
Diastolic blood pressure	71	10	72	10

Characteristic	Standard Treatment Percentage with the Characteristic	New Treatment Percentage with the Characteristic
Treatment for hypertension	63.2	63.7
Atrial fibrillation	12.6	11.1
Diabetes	30.6	30.2
Cigarette smoking	12.8	11.4
Coronary bypass surgery	28.5	29.1

1. For each health characteristic, construct a 95% confidence interval for the difference between the group means or group proportions.
2. Based on the confidence intervals in Exercise 1, for which health characteristics does it appear that the assignment to groups is not balanced? *Systolic blood pressure and Diastolic blood pressure*
3. Based on the confidence intervals in Exercise 1, does it appear that the imbalance is large enough to be of concern? Explain. *No*

chapter

11

Two-Sample Hypothesis Tests

Malcolm Fife/age fotostock

Introduction

Does exposure to electrical fields cause cancer? This question was addressed in a study carried out at the University of Southern California. Three hundred children who suffered from brain cancer (referred to as "cases") were sampled, along with an equal number of children who did not have cancer (referred to as "controls"). The investigators counted the number of cases and controls who lived in neighborhoods in which the electrical wiring was underground, rather than overhead. In addition, they collected information about ethnicity and income. The following table presents some of the information collected.

	Cases	Controls
Proportion with high income	0.09	0.13
Proportion owning a home	0.72	0.76
Proportion with underground wiring	0.14	0.07
Proportion with father white non-Hispanic	0.43	0.36
Proportion with father Hispanic	0.38	0.47

Cases and controls differ in their proportions for all of these categories. This is not surprising. After all, if we were to sample two groups of people at random, we would expect to see some differences, just by chance. The question we need to address is whether the differences between the cases and controls are just due to chance, or whether they reflect a real association. The basic principle is simple: The smaller the difference, the more likely it is just due to chance. Larger differences are more likely to reflect a real association. Now the question becomes: How do we determine when a difference is large enough to justify concluding that it reflects a real association? That is the question that hypothesis tests are designed to answer. In this chapter, we will learn to perform hypothesis tests in a variety of commonly occurring situations. In the case study at the end of the chapter, we will investigate the differences in the preceding table and determine which factors seem to be associated with cancer.

Section 11.1 Hypothesis Tests for the Difference Between Two Means: Independent Samples

Objectives

1. Perform a hypothesis test for the difference between two means using the *P*-value method
2. Perform a hypothesis test for the difference between two means using the critical value method

Objective 1 Perform a hypothesis test for the difference between two means using the *P*-value method

Perform a Hypothesis Test for the Difference Between Two Means Using the *P*-Value Method

Do computers help high school students to learn math? One way to address this question is to give a test to two samples of students—one from the population of students who used computers in their math classes, and another from the population of students who did not. If the difference between the sample mean scores is large enough, we can conclude that there is a real difference between the two populations.

The National Assessment of Educational Progress (NAEP) has been testing students for the past 30 years. Scores on the NAEP mathematics test range from 0 to 500. In a recent year, the sample mean score for students using a computer was 309, with a sample standard deviation of 29. For students not using a computer, the sample mean was 303, with a sample standard deviation of 32. Assume there were 60 students in the computer sample, and 40 students in the sample that didn't use a computer. We can see that the sample mean scores differ by 6 points: $309 - 303 = 6$. Now, we are interested in the difference between the population means, which will not be exactly the same as the difference between the sample means. Is it plausible that the difference between the population means could be 0? How strong is the evidence that the population mean scores are different?

RECALL

Two samples are independent if the observations in one sample do not influence the observations in the other.

This is an example of a situation in which the data consist of two **independent samples**. We will describe a method for performing a hypothesis test to determine whether the population means are equal. We will need some notation for the population means, the sample means, the sample standard deviations, and the sample sizes:

NOTATION

- μ_1 and μ_2 are the population means.
- $\bar{x}_1$ and $\bar{x}_2$ are the sample means.
- s_1 and s_2 are the sample standard deviations.
- n_1 and n_2 are the sample sizes.

We will now describe how to perform the hypothesis test.

The null and alternate hypotheses

The issue is whether the population means μ_1 and μ_2 are equal. The null hypothesis says that the population means are equal:

$$H_0: \mu_1 = \mu_2$$

There are three possibilities for the alternate hypothesis.

$$H_1: \mu_1 < \mu_2 \qquad H_1: \mu_1 > \mu_2 \qquad H_1: \mu_1 \neq \mu_2$$

The test statistic

The test statistic is based on the difference between the two sample means $\bar{x}_1 - \bar{x}_2$. The mean of $\bar{x}_1 - \bar{x}_2$ is $\mu_1 - \mu_2$. We approximate the standard deviation of $\bar{x}_1 - \bar{x}_2$ with the standard error, which we derived in Section 10.1.

$$\text{Standard error of } \bar{x}_1 - \bar{x}_2 = \sqrt{\frac{s_1^2}{n_1} + \frac{s_2^2}{n_2}}$$

The test statistic is

$$t = \frac{(\bar{x}_1 - \bar{x}_2) - (\mu_1 - \mu_2)}{\sqrt{\frac{s_1^2}{n_1} + \frac{s_2^2}{n_2}}}$$

Under the assumption that H_0 is true, the test statistic has approximately a Student's t distribution. The number of degrees of freedom can be calculated with a rather complicated formula. (The formula is presented in Section 10.1, although you don't need to know it.) A simpler method, which is acceptable, is

$$\text{Degrees of freedom} = \text{Smaller of } n_1 - 1 \text{ and } n_2 - 1$$

The method just described requires certain assumptions, which we now state:

EXPLAIN IT AGAIN

Reason for assumption 3: Assumption 3 is necessary to ensure that the sampling distributions of $\bar{x}_1$ and $\bar{x}_2$ are approximately normal. This justifies the use of the Student's t distribution.

Assumptions for Performing a Hypothesis Test for the Difference Between Two Means with Independent Samples

1. We have simple random samples from two populations.
2. The samples are independent of one another.
3. Each sample size is large ($n > 30$), *or* its population is approximately normal.

We now summarize the steps in testing a hypothesis about the difference between two means with independent samples, using the *P*-value method. Later we will describe the critical value method.

Performing a Hypothesis Test for the Difference Between Two Means Using the *P*-Value Method

Check to be sure the assumptions are satisfied. If they are, then proceed with the following steps.

Step 1: State the null and alternate hypotheses. The null hypothesis specifies that the population means are equal: H_0: $\mu_1 = \mu_2$. The alternate hypothesis will be $\mu_1 < \mu_2$, $\mu_1 > \mu_2$, or $\mu_1 \neq \mu_2$.

Step 2: If making a decision, choose a significance level α.

Step 3: Compute the test statistic $t = \dfrac{(\bar{x}_1 - \bar{x}_2) - (\mu_1 - \mu_2)}{\sqrt{\frac{s_1^2}{n_1} + \frac{s_2^2}{n_2}}}$.

Step 4: Compute the *P*-value. The *P*-value is an area under the Student's t curve. If using Table A.3, approximate the number of degrees of freedom with the smaller of $n_1 - 1$ and $n_2 - 1$. The *P*-value depends on the alternate hypothesis as follows:

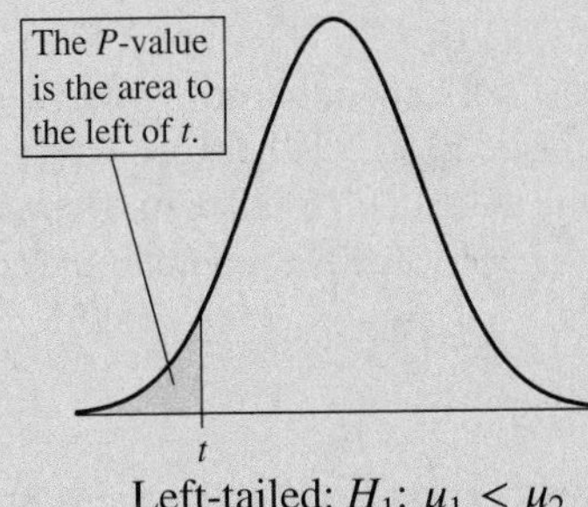
The P-value is the area to the left of t.
t
Left-tailed: H_1: $\mu_1 < \mu_2$

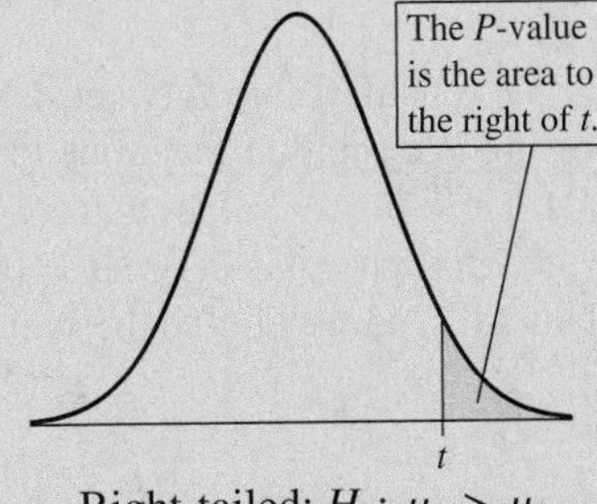
The P-value is the area to the right of t.
t
Right-tailed: H_1: $\mu_1 > \mu_2$

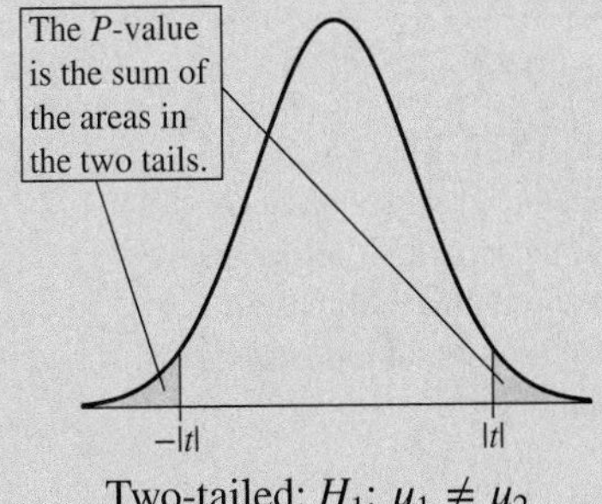
The P-value is the sum of the areas in the two tails.
−|t|
|t|
Two-tailed: H_1: $\mu_1 \neq \mu_2$

Step 5: Interpret the *P*-value. If making a decision, reject H_0 if the *P*-value is less than or equal to the significance level α.

Step 6: State a conclusion.

Example 11.1

Perform a hypothesis test

The National Assessment of Educational Progress (NAEP) tested a sample of students who had used a computer in their mathematics classes, and another sample of students who had not used a computer. The sample mean score for students using a computer was 309, with a sample standard deviation of 29. For students not using a computer, the sample mean was 303, with a sample standard deviation of 32. Assume there were 60 students in the computer sample and 40 students in the sample that hadn't used a computer. Can you conclude that the population mean scores differ? Use the $\alpha = 0.05$ level.

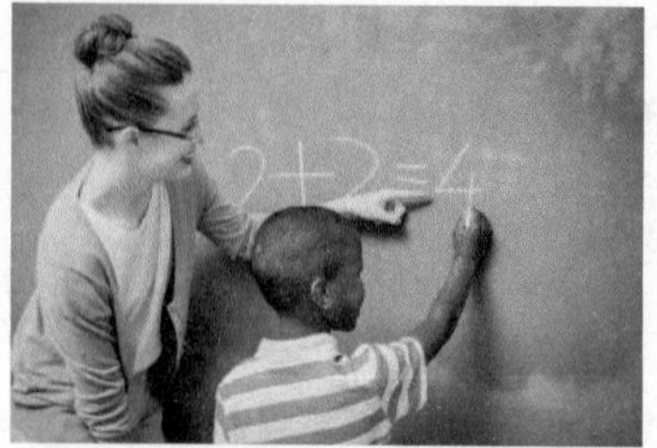

ESB Professional/Shutterstock

Solution

We first check the assumptions. We have two independent random samples. Both sample sizes are larger than 30. The assumptions are satisfied.

Step 1: State the null and alternate hypotheses. We are asked whether we can conclude that the two means differ. Therefore, this is a two-tailed test. The null and alternate hypotheses are

$$H_0\colon \mu_1 = \mu_2 \qquad H_1\colon \mu_1 \neq \mu_2$$

Step 2: Choose a significance level. The significance level is $\alpha = 0.05$.

Step 3: Compute the test statistic. The test statistic is

$$t = \frac{(\bar{x}_1 - \bar{x}_2) - (\mu_1 - \mu_2)}{\sqrt{\dfrac{s_1^2}{n_1} + \dfrac{s_2^2}{n_2}}}$$

To help keep track of things, we'll begin by organizing the relevant information in the following table:

	With Computer	Without Computer
Sample mean	$\bar{x}_1 = 309$	$\bar{x}_2 = 303$
Sample standard deviation	$s_1 = 29$	$s_2 = 32$
Sample size	$n_1 = 60$	$n_2 = 40$
Population mean	μ_1 (unknown)	μ_2 (unknown)

Under the assumption that H_0 is true, $\mu_1 - \mu_2 = 0$. The value of the test statistic is

$$t = \frac{(309 - 303) - (0)}{\sqrt{\dfrac{29^2}{60} + \dfrac{32^2}{40}}} = 0.953$$

Step 4: Compute the *P*-value. We will approximate the *P*-value by using Table A.3. We begin by finding the number of degrees of freedom. The sample sizes are $n_1 = 60$ and $n_2 = 40$, so the number of degrees of freedom is $40 - 1 = 39$. The value of the test statistic is $t = 0.953$. This is a two-tailed test, so the *P*-value is the sum of the areas to the right of 0.953 and to the left of -0.953. Figure 11.1 illustrates the *P*-value.

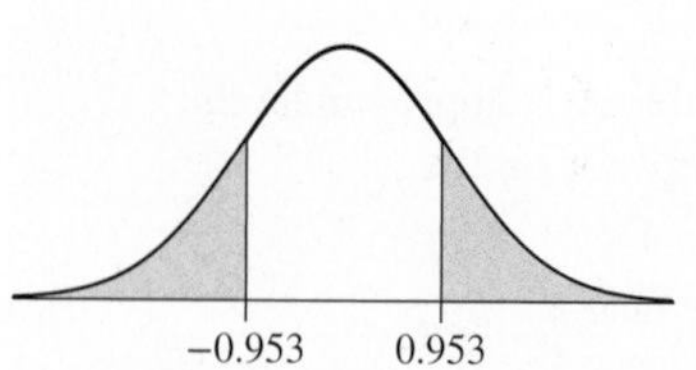

Figure 11.1 The *P*-value for a two-tailed test is the sum of the areas in the two tails. These areas can be approximated by using Table A.3, or found more precisely with technology.

We consult Table A.3 and look at the row corresponding to 39 degrees of freedom. We see that the value of the test statistic, 0.953, is between 0.681 and 1.304. These are the values that correspond to tail areas of 0.25 and 0.10. Therefore, the area in each tail is between 0.10 and 0.25. The *P*-value is the sum of the areas in both tails, so we double these numbers:

$$0.20 < P\text{-value} < 0.50$$

Step 5: Interpret the *P*-value. If H_0 were true, we would expect to observe a value of the test statistic as extreme as or more extreme than our value of 0.953 between 20% and 50% of the time. This is not unusual, so there is no strong evidence against H_0. In particular, $P > 0.05$, so we do not reject H_0 at the $\alpha = 0.05$ level.

Step 6: State a conclusion. There is not enough evidence to conclude that the mean scores differ between those students who use a computer and those who do not. The mean scores may be the same.

Performing a hypothesis test with technology

The following computer output (from MINITAB) presents the results of Example 11.1.

```
Two-sample T for Computer vs No Computer

                N     Mean     StDev    SE Mean
Computer       60    309.0      29.0    3.74388
No computer    40    303.0      32.0    5.05964

Difference = mu (Treatment1) − mu (Treatment2)
Estimate for difference: 6.000
95% CI for difference: (−6.5333, 18.5333 )
T-Test of difference = 0 (vs not =):   T-Value = 0.95
                                       P-Value = 0.343    DF = 77
```

EXPLAIN IT AGAIN

Results from technology will differ: Results found with technology will differ from those obtained by hand, because computers and calculators use the more complicated formula for the degrees of freedom that we presented in Section 10.1.

The output presents the sample sizes (N), the sample means (Mean), and sample standard deviations (StDev). The column labeled "SE Mean" presents the standard errors of $\bar{x}_1$ and $\bar{x}_2$, which are $s_1/\sqrt{n_1}$ and $s_2/\sqrt{n_2}$, respectively. The row labeled "Estimate for difference" presents the difference between the sample means: $\bar{x}_1 - \bar{x}_2$. The next row contains a 95% confidence interval for $\mu_1 - \mu_2$. The row after that specifies that the alternate hypothesis is two-tailed, then presents the value of the test statistic (T-Value), the *P*-value, and the number of degrees of freedom. The number of degrees of freedom (DF) is 77, which differs from the value of 39 that we used in the solution to Example 11.1. The reason is that MINITAB uses a more complicated formula to compute the number of degrees of freedom, then rounds the value down to the nearest whole number.

The following TI-84 Plus display presents results for Example 11.1.

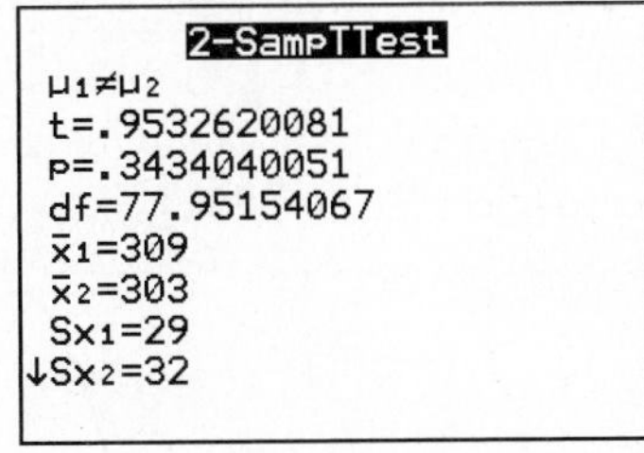

As with the MINITAB output, the results differ somewhat from those we obtained, because the degrees of freedom (labeled "df") have been calculated by the more complicated formula. Note that the degrees of freedom is not a whole number. Unlike MINITAB, the TI-84 Plus does not round the degrees of freedom.

Step-by-step instructions for performing hypothesis tests with technology are presented in the Using Technology section on page 547.

Check Your Understanding

1. A test was made of H_0: $\mu_1 = \mu_2$ versus H_1: $\mu_1 > \mu_2$. Independent random samples were drawn from approximately normal populations. The sample means were $\bar{x}_1 = 6.8$ and $\bar{x}_2 = 4.9$. The sample standard deviations were $s_1 = 1.6$ and $s_2 = 1.3$. The sample sizes were $n_1 = 12$ and $n_2 = 10$.
   - **a.** Find the value of the test statistic t. *3.073*
   - **b.** Find the number of degrees of freedom for t. *9*
   - **c.** Find the *P*-value. *0.0066 [TI-84 Plus: 0.0030] [MINITAB: 0.0031]*
   - **d.** Interpret the *P*-value. Do you reject H_0 at the $\alpha = 0.01$ level? *Yes*

2. A test was made of H_0: $\mu_1 = \mu_2$ versus H_1: $\mu_1 \neq \mu_2$. Independent random samples were drawn from approximately normal populations. The sample means were $\bar{x}_1 = 73.9$ and $\bar{x}_2 = 71.8$. The sample standard deviations were $s_1 = 4.2$ and $s_2 = 3.8$. The sample sizes were $n_1 = 23$ and $n_2 = 17$.

a. Find the value of the test statistic t. *1.652*

b. Find the number of degrees of freedom for t. *16*

c. Find the P-value. *0.1180 [TI-84 Plus: 0.1072] [MINITAB: 0.1073]*

d. Interpret the P-value. Do you reject H_0 at the $\alpha = 0.05$ level? *No*

Answers are on page 551.

Objective 2 Perform a hypothesis test for the difference between two means using the critical value method

Performing Hypothesis Tests Using the Critical Value Method

To use the critical value method, compute the test statistic as before. The procedure for finding the critical value is the same as for hypothesis tests for a mean with σ unknown. The critical value can be found in Table A.3 or with technology. The procedure depends on whether the alternate hypothesis is left-tailed, right-tailed, or two-tailed. The assumptions for the critical value method are the same as for the P-value method.

Following are the steps for the critical value method.

Performing a Hypothesis Test for the Difference Between Two Means Using the Critical Value Method

Check to be sure the assumptions are satisfied. If they are, then proceed with the following steps.

Step 1: State the null and alternate hypotheses. The null hypothesis will have the form H_0: $\mu_1 = \mu_2$. The alternate hypothesis will be $\mu_1 < \mu_2$, $\mu_1 > \mu_2$, or $\mu_1 \neq \mu_2$.

Step 2: Choose a significance level α, and find the critical value or values.

Step 3: Compute the test statistic

$$t = \frac{(\bar{x}_1 - \bar{x}_2) - (\mu_1 - \mu_2)}{\sqrt{\dfrac{s_1^2}{n_1} + \dfrac{s_2^2}{n_2}}}.$$

Step 4: Determine whether to reject H_0, as follows:

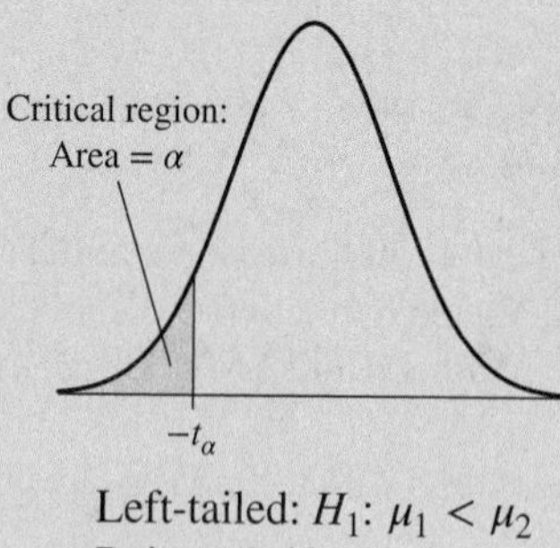

Left-tailed: H_1: $\mu_1 < \mu_2$
Reject H_0 if $t \leq -t_\alpha$.

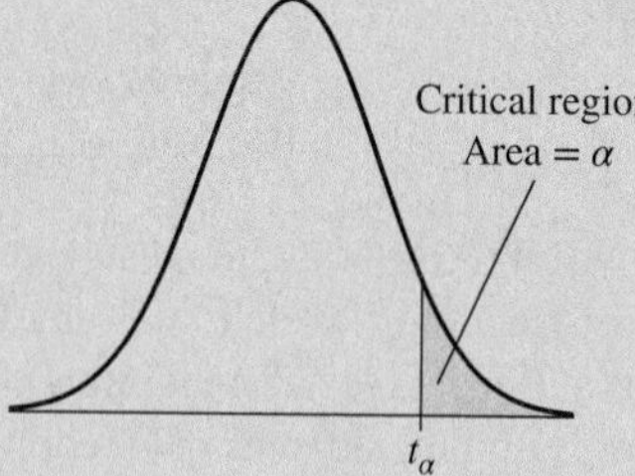

Right-tailed: H_1: $\mu_1 > \mu_2$
Reject H_0 if $t \geq t_\alpha$.

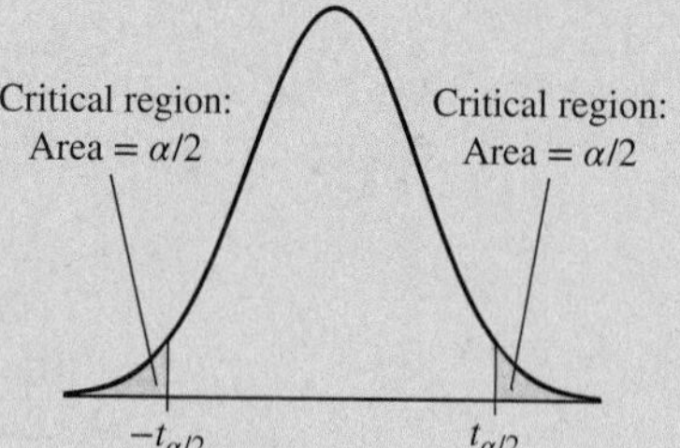

Two-tailed: H_1: $\mu_1 \neq \mu_2$
Reject H_0 if $t \geq t_{\alpha/2}$ or $t \leq -t_{\alpha/2}$.

Step 5: State a conclusion.

Example 11.2

Test a hypothesis using the critical value method

Treatment of wastewater is important to reduce the concentration of undesirable pollutants. One such substance is benzene, which is used as an industrial solvent. Two methods of water treatment are being compared. Treatment 1 is applied to five specimens of wastewater, and treatment 2 is applied to seven specimens. The benzene concentrations, in units of milligrams per liter, for each specimen are as follows:

Treatment 1:	7.8	7.6	5.6	6.8	6.4		
Treatment 2:	4.1	6.5	3.7	7.7	7.3	4.7	5.9

How strong is the evidence that the mean concentration is less for treatment 2 than for treatment 1? We will test at the $\alpha = 0.05$ significance level.

Solution

We first check the assumptions. Because the samples are small, we must check for strong skewness and outliers. We construct dotplots for each sample.

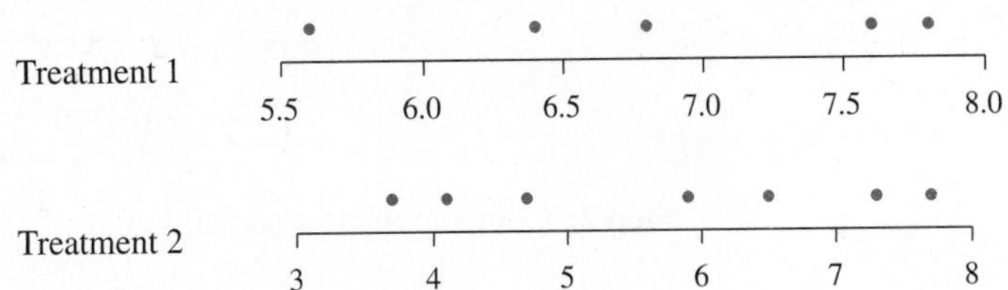

There are no outliers, and no evidence of strong skewness, in either sample. Therefore, we can proceed.

Step 1: State the null and alternate hypotheses. The issue is whether the mean for treatment 2 is less than the mean for treatment 1. Let μ_1 denote the mean for treatment 1 and μ_2 denote the mean for treatment 2. The hypotheses are

$$H_0\colon \mu_1 = \mu_2 \qquad H_1\colon \mu_1 > \mu_2$$

Step 2: Choose a significance level and find the critical value. The significance level is $\alpha = 0.05$. We will find the critical value in Table A.3. The sample sizes are $n_1 = 5$ and $n_2 = 7$. For the number of degrees of freedom, we use the smaller of $5-1=4$ and $7-1=6$, which is 4. Because the alternate hypothesis, $\mu_1 - \mu_2 > 0$, is right-tailed, the critical value is the value with area 0.05 to its right. We consult Table A.3 with 4 degrees of freedom and find that $t_\alpha = 2.132$.

Step 3: Compute the test statistic. To compute the test statistic, we first compute the sample means and standard deviations. These are

$$\bar{x}_1 = 6.84 \qquad \bar{x}_2 = 5.70 \qquad s_1 = 0.8989 \qquad s_2 = 1.5706$$

The sample sizes are $n_1 = 5$ and $n_2 = 7$. The test statistic is

$$t = \frac{(\bar{x}_1 - \bar{x}_2) - (\mu_1 - \mu_2)}{\sqrt{\dfrac{s_1^2}{n_1} + \dfrac{s_2^2}{n_2}}}$$

Under the assumption that H_0 is true, $\mu_1 - \mu_2 = 0$. The value of the test statistic is

$$t = \frac{(6.84 - 5.70) - 0}{\sqrt{\dfrac{0.8989^2}{5} + \dfrac{1.5706^2}{7}}} = 1.590$$

Step 4: Determine whether to reject H_0. This is a right-tailed test, so we reject H_0 if $t \geq t_\alpha$. Because $t = 1.590$ and $t_\alpha = 2.132$, we do not reject H_0.

Step 5: State a conclusion. There is not enough evidence to conclude that the mean benzene concentration with treatment 1 is greater than that with treatment 2. The concentrations may be the same.

NOTE TO INSTRUCTOR

The material on equal variances and known variances can be skipped without loss of continuity.

An alternate method when the population variances are equal

When the two population standard deviations, σ_1 and σ_2, are known to be equal, there is an alternate method for testing hypotheses about $\mu_1 - \mu_2$. This alternate method was widely used in the past and is still an option in many computer packages. We will describe the method here, because it is still sometimes used. However, the method is rarely appropriate, for the same reasons that the pooled method for constructing confidence intervals is rarely appropriate. These reasons were discussed in Section 10.1.

A Method for Testing a Hypothesis About $\mu_1 - \mu_2$ When $\sigma_1 = \sigma_2$ (Not Recommended)

Step 1: Compute the **pooled standard deviation,** s_p, as follows:

$$s_p = \sqrt{\frac{(n_1 - 1)s_1^2 + (n_2 - 1)s_2^2}{n_1 + n_2 - 2}}$$

Step 2: Compute the test statistic:

$$t = \frac{(\bar{x}_1 - \bar{x}_2) - (\mu_1 - \mu_2)}{s_p\sqrt{\dfrac{1}{n_1} + \dfrac{1}{n_2}}}$$

Step 3: Compute the degrees of freedom:

$$\text{Degrees of freedom} = n_1 + n_2 - 2$$

Step 4: Compute the P-value using a Student's t distribution with $n_1 + n_2 - 2$ degrees of freedom.

When using technology, you may be offered a choice of whether to assume that the variances are equal. The best practice is not to assume that the variances are equal.

Performing hypothesis tests when the population standard deviations are known

When the two population standard deviations, σ_1 and σ_2, are known, we can modify the test statistic presented here by replacing s_1 and s_2 with σ_1 and σ_2, and using the standard normal distribution rather than the Student's t distribution to find the P-value or critical values. In practice, σ_1 and σ_2 are rarely known, so this method is not often applicable. We present it here because it is sometimes offered as an option in statistical calculators and software. The assumptions for this method are the same as for the method using the Student's t distribution, with the additional assumption that the population standard deviations are known.

SUMMARY

When the population standard deviations σ_1 and σ_2 are known, tests for the difference between two means can be carried out by using the following test statistic:

$$z = \frac{(\bar{x}_1 - \bar{x}_2) - (\mu_1 - \mu_2)}{\sqrt{\dfrac{\sigma_1^2}{n_1} + \dfrac{\sigma_2^2}{n_2}}}$$

The test statistic has a standard normal distribution when H_0 is true.

Using Technology

We use Example 11.2 to illustrate the technology steps.

TI-84 PLUS

Testing a hypothesis about the difference between two means

Step 1. Press **STAT** and highlight the **TESTS** menu.

Step 2. Select **2–SampTTest** and press **ENTER** (Figure A). The **2–SampTTest** menu appears.

Step 3. Choose one of the following:

- If the summary statistics are given, select **Stats** as the **Inpt** option and enter $\bar{x}_1$, s_1, n_1, $\bar{x}_2$, s_2, n_2.
- If the raw data are given, select **Data** as the **Inpt** option and enter the location of the data as the **List1** and **List2** options. For Example 11.2, the sample has been entered in lists **L1** and **L2**.

Step 4. Select the form of the alternate hypothesis. For Example 11.2, the alternate hypothesis has the form $>\mu_2$.

Step 5. Select **No** for the **Pooled** option (Figure B).

Step 6. Highlight **Calculate** and press **ENTER** (Figure C).

```
EDIT CALC TESTS
1:Z-Test...
2:T-Test...
3:2-SampZTest...
4:2-SampTTest...
5:1-PropZTest...
6:2-PropZTest...
7:ZInterval...
8:TInterval...
9↓2-SampZInt...
```

Figure A

```
2-SampTTest
Inpt:Data Stats
List1:L1
List2:L2
Freq1:1
Freq2:1
μ1:≠μ2 <μ2 >μ2
Pooled:No Yes
Color: BLUE
Calculate Draw
```

Figure B

```
2-SampTTest
μ1>μ2
t=1.590125268
p=.0719103361
df=9.703795747
x̄1=6.84
x̄2=5.7
Sx1=.898888202
↓Sx2=1.57056253
```

Figure C

EXCEL

Finding the test statistic for a hypothesis test about the difference between two means

Figure D demonstrates the calculations involved for finding the test statistic for a hypothesis test about the difference between two means. In Example 11.2, $\bar{x}_1 = 6.84$, $\bar{x}_2 = 5.70$, $s_1 = 0.8989$, $s_2 = 1.5706$, $n_1 = 5$, and $n_2 = 7$.

	A	B	C	D	E	F
1		Treatment 1	Treatment 2			
2	Sample Mean	6.84	5.70			
3	Sample St Dev	0.8989	1.5706			
4	Sample Size	5	7			
5	$\mu_1 - \mu_2$	0				
6						
7	$\bar{x}_1 - \bar{x}_2$	1.14	← =B2-C2			
8	Standard Error	0.71693933	← =SQRT(B3^2/B4+C3^2/C4)			
9						
10	Test Statistic T	1.5900927	← =(B7-B5)/B8			
11						

Figure D

MINITAB

Testing a hypothesis about the difference between two means

Step 1. Click on **Stat**, then **Basic Statistics**, then **2-Sample t**.

Step 2. Choose one of the following:

- If the summary statistics are given, select **Summarized Data** and enter the **Sample Size**, the **Mean**, and the **Standard Deviation** for each sample.
- If the raw data are given, select **Each sample is in its own column** and select the columns that contain the data. For Example 11.2, the two samples have been entered in columns **C1** and **C2**.

Step 3. Click **Options**, and enter the difference between the means in the **Test difference** field and select the form of the alternate hypothesis. Given significance level α, enter $100(1 - \alpha)$ as the **Confidence Level.** For Example 11.2, we use **95** as the **Confidence Level, 0** as the **Hypothesized difference**, and **Difference > hypothesized difference** as the **Alternative**. Click **OK.**

Step 4. Click **OK** (Figure E).

Descriptive Statistics

Sample	N	Mean	StDev	SE Mean
C1	5	6.840	0.899	0.40
C2	7	5.70	1.57	0.59

Test

Null hypothesis $H_0: \mu_1 - \mu_2 = 0$

Alternative hypothesis $H_1: \mu_1 - \mu_2 > 0$

T-Value	DF	P-Value
1.59	9	0.073

Figure E

Section 11.1 Exercises

Exercises 1 and 2 are the Check Your Understanding exercises located within the section.

Understanding the Concepts

In Exercises 3 and 4, fill in each blank with the appropriate word or phrase.

3. To use the methods of this section to test a hypothesis about the difference between two means when the samples are small, the samples must show no evidence of strong ________ and must contain no ________. *skewness, outliers*

4. If the sample sizes are n_1 and n_2, the simplest way to compute the degrees of freedom is to use the smaller of ________ and ________. $n_1 - 1, n_2 - 1$

In Exercises 5 and 6, determine whether the statement is true or false. If the statement is false, rewrite it as a true statement.

5. To use the methods of this section to test a hypothesis about the difference between two means, the population standard deviations must be known. *False*

6. In general, it is not recommended to use a pooled standard deviation when testing a hypothesis for the difference between means. *True*

Practicing the Skills

7. A test was made of H_0: $\mu_1 = \mu_2$ versus H_1: $\mu_1 < \mu_2$. The sample means were $\bar{x}_1 = 6$ and $\bar{x}_2 = 11$, the sample standard deviations were $s_1 = 3$ and $s_2 = 5$, and the sample sizes were $n_1 = 10$ and $n_2 = 20$.
   - **a.** How many degrees of freedom are there for the test statistic, using the simple method? *9*
   - **b.** Compute the value of the test statistic. *−3.410*
   - **c.** Is H_0 rejected at the 0.05 level? Explain. *Yes*

8. A test was made of H_0: $\mu_1 = \mu_2$ versus H_1: $\mu_1 \neq \mu_2$. The sample means were $\bar{x}_1 = 10$ and $\bar{x}_2 = 8$, the sample standard deviations were $s_1 = 4$ and $s_2 = 7$, and the sample sizes were $n_1 = 15$ and $n_2 = 27$.
   - **a.** How many degrees of freedom are there for the test statistic, using the simple method? *14*
   - **b.** Compute the value of the test statistic. *1.178*
   - **c.** Is H_0 rejected at the 0.05 level? Explain. *No*

Working with the Concepts

9. **More time on the internet:** The General Social Survey polled a sample of 1399 adults in the year 2014, asking them how many hours per week they spent on the internet. The sample mean was 11.62 with a standard deviation of 15.02. A second sample of 1361 adults was taken in the year 2018. For this sample, the mean was 13.84 with a standard deviation of 16.96. Assume these are simple random samples from populations of adults. Can you conclude that the mean number of hours per week spent on the internet increased between 2014 and 2018? Use the $\alpha = 0.05$ level.
   - **a.** State the appropriate null and alternate hypotheses. $H_0: \mu_1 = \mu_2, H_1: \mu_1 < \mu_2$
   - **b.** Compute the test statistic. *−3.637*
   - **c.** How many degrees of freedom are there, using the simple method? *1360*
   - **d.** Do you reject H_0? State a conclusion. *Yes*

10. **Low-fat or low-carb?** Are low-fat diets or low-carb diets more effective for weight loss? A sample of 77 subjects went on a low-carbohydrate diet for six months. At the end of that time, the sample mean weight loss was 4.7 kilograms with a sample standard deviation of 7.16 kilograms. A second sample of 79 subjects went on a low-fat diet. Their sample mean weight loss was 2.6 kilograms with a standard deviation of 5.90 kilograms. Can you conclude that the mean weight loss differs between the two diets? Use the $\alpha = 0.01$ level.
    - **a.** State the appropriate null and alternate hypotheses. $H_0: \mu_1 = \mu_2, H_1: \mu_1 \neq \mu_2$
    - **b.** Compute the test statistic. *1.996*
    - **c.** How many degrees of freedom are there, using the simple method? *76*

d. Do you reject H_0? State a conclusion. No
Source: *Journal of the American Medical Association* 297:969–977

11. Are you smarter than your older brother? In a study of birth order and intelligence, IQ tests were given to 18- and 19-year-old men to estimate the size of the difference, if any, between the mean IQs of firstborn sons and secondborn sons. The following data for 10 firstborn sons and 10 secondborn sons are consistent with the means and standard deviations reported in the article. It is reasonable to assume that the samples come from populations that are approximately normal.

Firstborn				
104	82	102	96	129
89	114	107	89	103

Secondborn				
103	103	91	113	102
103	92	90	114	113

Can you conclude that there is a difference in mean IQ between firstborn and secondborn sons? Use the $\alpha = 0.01$ level. Do not reject H_0.
Based on data in *Science* 316:1717

12. Recovering from surgery: A new postsurgical treatment was compared with a standard treatment. Seven subjects received the new treatment, while seven others (the controls) received the standard treatment. The recovery times, in days, are given below.

Treatment:	12	13	15	19	20	21	24
Control:	18	23	24	30	32	35	39

Can you conclude that the mean recovery time for those receiving the new treatment is less than the mean for those receiving the standard treatment? Use the $\alpha = 0.05$ level. Reject H_0.

13. Contaminated water: The concentration of benzene was measured in units of milligrams per liter for a simple random sample of five specimens of untreated wastewater produced at a gas field. The sample mean was 7.8 with a sample standard deviation of 1.4. Seven specimens of treated wastewater had an average benzene concentration of 3.2 with a standard deviation of 1.7. It is reasonable to assume that both samples come from populations that are approximately normal. Can you conclude that the mean benzene concentration is less in treated water than in untreated water? Use the $\alpha = 0.05$ level of significance. Reject H_0.

14. Exercise: Medical researchers conducted a study to determine whether treadmill exercise could improve the walking ability of patients suffering from claudication, which is pain caused by insufficient blood flow to the muscles of the legs. A sample of 48 patients walked on a treadmill for six minutes every day. After six months, the mean distance walked in six minutes was 348 meters, with a standard deviation of 80 meters. For a control group of 46 patients who did not walk on a treadmill, the mean distance was 309 meters with a standard deviation of 89 meters. Can you conclude that the mean distance walked for patients using a treadmill is greater than the mean for the controls? Use the $\alpha = 0.05$ level of significance. Reject H_0.
Source: *Journal of the American Medical Association* 301:165–174

15. Mummy's curse: King Tut was an ancient Egyptian ruler whose tomb was discovered and opened in 1923. Legend has it that the archaeologists who opened the tomb were subject to a "mummy's curse," which would shorten their life spans. A team of scientists conducted an investigation of the mummy's curse. They reported that the 25 people exposed to the curse had a mean life span of 70.0 years with a standard deviation of 12.4 years, while a sample of 11 Westerners in Egypt at the time who were not exposed to the curse had a mean life span of 75.0 years with a standard deviation of 13.6 years. Assume that the populations are approximately normal. Can you conclude that the mean life span of those exposed to the mummy's curse is less than the mean of those not exposed? Use the $\alpha = 0.05$ level. Do not reject H_0.
Source: *British Medical Journal* 325:1482

16. Baby weights: Following are weights in pounds for random samples of 25 newborn baby boys and baby girls born in Denver. Boxplots indicate that the samples come from populations that are approximately normal.

Boys								
6.6	5.9	6.4	7.6	6.4	8.1	7.9	8.3	7.3
6.4	8.4	8.5	6.9	6.3	7.4	7.8	7.5	6.9
7.8	8.6	7.7	7.4	7.7	8.1	6.4		

Girls								
7.7	7.0	8.2	7.4	6.0	6.7	8.2	7.5	5.7
6.6	6.4	8.5	7.2	6.9	8.2	6.5	6.7	7.2
6.3	5.9	8.1	8.2	6.7	6.2	7.7		

Can you conclude that the mean weights differ between boys and girls? Use the $\alpha = 0.05$ level of significance. Do not reject H_0.

17. Empathy: The Interpersonal Reactivity Index is a survey designed to assess four different types of empathy. One type of empathy, called Empathetic Concern, measures the tendency to feel sympathy and compassion for people who are less fortunate. The index ranges from 0 (less empathetic) to 28 (more empathetic). The following data, representing random samples of 16 males and 16 females, are consistent with results reported in psychological studies. Boxplots show that it is reasonable to assume that the populations are approximately normal.

Males							
13	20	12	16	13	26	21	23
8	15	18	25	15	23	17	22

Females							
22	20	26	25	28	24	16	19
20	23	21	23	15	26	19	25

Can you conclude that there is a difference in mean empathy score between men and women? Use the $\alpha = 0.05$ level. Reject H_0.

18. Text me: The Pew Internet and American Life Project surveyed 2227 adults, asking them about their texting habits. Among men, the mean number of text messages sent per day was 40.9, and among women it was 42.0. Assume that 1085 men and 1142 women were sampled, and that the standard deviations were 56.5 for the men and 51.4 for the women. Can you conclude that the mean number of text messages differs between men and women? Use the $\alpha = 0.05$ level of significance. Do not reject H_0.

19. Online testing: Do you prefer taking tests on paper or online? A college instructor gave identical tests to two randomly sampled groups of 35 students. One group took the test on paper, and the other took it online. Following are the test scores.

Paper						
79	75	49	78	73	81	70
63	79	65	60	74	64	64
74	69	57	65	61	58	92
71	69	66	71	68	67	81
78	80	76	67	49	56	45

Online						
79	75	71	81	56	72	49
75	63	81	74	72	71	73
83	59	78	65	53	47	63
82	81	76	65	82	78	76
65	72	85	84	81	50	79

Can you conclude that there is a difference in the mean score between paper and online testing? Use the $\alpha = 0.05$ level of significance. *No*

20. Drive safely: How often does the average driver have an accident? The Allstate Insurance Company determined the average number of years between accidents for drivers in a large number of U.S. cities. Following are the results for 32 cities east of the Mississippi River and 32 cities west of the Mississippi River.

East							
10.2	9.5	11.2	11.4	9.8	10.0	11.6	10.7
9.5	9.0	7.2	11.5	7.7	9.0	8.9	8.4
6.8	5.3	8.0	8.8	9.0	10.0	9.7	11.6
7.9	9.9	10.5	11.9	10.9	8.6	6.7	9.6

West							
9.5	7.9	13.4	12.6	9.7	8.2	14.0	7.1
10.2	9.4	10.8	9.5	7.5	9.1	8.4	7.6
7.4	7.7	7.1	11.9	10.2	11.9	7.7	7.0
10.0	8.1	11.5	8.4	9.6	9.5	9.9	11.5

Can you conclude that the mean time between accidents is greater for western cities than for eastern cities? Use the $\alpha = 0.05$ level of significance. *No*

21. Interpret calculator display: The following TI-84 Plus calculator display presents the results of a hypothesis test for the difference between two means. The sample sizes are $n_1 = 12$ and $n_2 = 15$.

```
2-SampTTest
μ1>μ2
t=1.308978076
p=.101223442
df=24.99965945
x̄1=6.8526
x̄2=5.5758
Sx1=2.23666745
↓Sx2=2.83164884
```

a. Is this a left-tailed test, a right-tailed test, or a two-tailed test? *Right-tailed*

b. How many degrees of freedom did the calculator use? *24.99965945*

c. What is the P-value? *0.101223442*

d. Can you reject H_0 at the $\alpha = 0.05$ level? *No*

22. Interpret calculator display: The following TI-84 Plus calculator display presents the results of a hypothesis test for the difference between two means. The sample sizes are $n_1 = 25$ and $n_2 = 28$.

```
2-SampTTest
μ1≠μ2
t=2.39806437
p=.0209030051
df=42.91558207
x̄1=26.80805
x̄2=24.66451
Sx1=3.70502
↓Sx2=2.64524
```

a. Is this a left-tailed test, a right-tailed test, or a two-tailed test? *Two-tailed*

b. How many degrees of freedom did the calculator use? *42.91558207*

c. What is the P-value? *0.0209030051*

d. Can you reject H_0 at the $\alpha = 0.05$ level? *Yes*

23. Interpret computer output: The following computer output (from MINITAB) presents the results of a hypothesis test for the difference $\mu_1 - \mu_2$ between two population means:

```
Two-sample T for X1 vs X2

      N     Mean     StDev    SE Mean
X1   58   35.848    10.233     1.3437
X2   36   26.851    15.329     2.5548

Difference = mu (X1) − mu (X2)
Estimate for difference: 8.997
95% CI for difference: (4.16605, 13.827951)
T-Test of difference = 0 (vs not =):  T-Value = 3.12
                                      P-Value = 0.003   DF = 54
```

a. What is the alternate hypothesis? $\mu_1 \neq \mu_2$

b. Can H_0 be rejected at the $\alpha = 0.05$ level? Explain. *Yes*

c. How many degrees of freedom are there for the test statistic? *54*

d. Use the simpler method to compute the degrees of freedom as the smaller of $n_1 - 1$ and $n_2 - 1$. *35*

e. Find the P-value using this value for the degrees of freedom. *0.0036*

24. Interpret computer output: The following computer output (from MINITAB) presents the results of a hypothesis test for the difference $\mu_1 - \mu_2$ between two population means:

```
Two-sample T for X1 vs X2

      N     Mean     StDev    SE Mean
X1   20     3.44      2.65       0.23
X2   25     4.43      2.38       0.18

Difference = mu (X1) − mu (X2)
Estimate for difference: −0.99
95% upper bound for difference: 0.291437
T-Test of difference = 0 (vs <):  T-Value = −1.30
                                  P-Value = 0.100   DF = 38
```

a. What is the alternate hypothesis? $\mu_1 < \mu_2$

b. Can H_0 be rejected at the $\alpha = 0.05$ level? Explain. *No*

c. How many degrees of freedom are there for the test statistic? *38*

d. Use the simpler method to compute the degrees of freedom as the smaller of $n_1 - 1$ and $n_2 - 1$. *19*

e. Find the P-value using this value for the degrees of freedom. *0.1046*

Extending the Concepts

25. More accurate degrees of freedom: A test will be made of $H_0\colon \mu_1 = \mu_2$ versus $H_1\colon \mu_1 > \mu_2$. The sample sizes were $n_1 = 10$ and $n_2 = 20$. The sample standard deviations were $s_1 = 3$ and $s_2 = 8$.

a. Compute the critical value for a level $\alpha = 0.05$ test, using 1 less than the smaller of the two sample sizes for the degrees of freedom. *1.833*

b. Use the expression in Section 10.1 to compute a more accurate number of degrees of freedom. *26.727*

c. Use the more accurate number of degrees of freedom to compute the probability of rejecting H_0 when it is true when the critical value found in part (a) is used. You will need technology to find the answer. *0.039*

Answers to Check Your Understanding Exercises for Section 11.1

1. a. 3.073 **b.** 9

c. The P-value is between 0.005 and 0.01. [Tech: 0.0066] [TI-84 Plus: 0.0030] [MINITAB: 0.0031]

d. If H_0 is true, the probability of observing a value for the test statistic as extreme as or more extreme than the value actually observed is 0.0066. This is unusual, so the evidence against H_0 is strong. Because $P < 0.01$, we reject H_0 at the $\alpha = 0.01$ level.

2. a. 1.652 **b.** 16

c. The P-value is between 0.10 and 0.20. [Tech: 0.1180] [TI-84 Plus: 0.1072] [MINITAB: 0.1073]

d. If H_0 is true, the probability of observing a value for the test statistic as extreme as or more extreme than the value actually observed is 0.1073. This is not very unusual, so the evidence against H_0 is not strong. Because $P > 0.05$, we do not reject H_0 at the $\alpha = 0.05$ level.

Section 11.2 Hypothesis Tests for the Difference Between Two Means: Paired Samples

Objectives

1. Perform a hypothesis test with matched pairs using the P-value method
2. Perform a hypothesis test with matched pairs using the critical value method

Objective 1 Perform a hypothesis test with matched pairs using the P-value method

Hypothesis Tests with Matched Pairs Using the P-Value Method

Does tuning a car engine improve the gas mileage? A sample of eight automobiles were run to determine their mileage, in miles per gallon. Then each car was given a tune-up and run again to measure the mileage a second time. The results are presented in Table 11.1.

Table 11.1 Gas Mileage Before and After Tune-up for Eight Automobiles

	Automobile							
	1	2	3	4	5	6	7	8
After Tune-up	35.44	35.17	31.07	31.57	26.48	23.11	25.18	32.39
Before Tune-up	33.76	34.30	29.55	30.90	24.92	21.78	24.30	31.25
Difference	1.68	0.87	1.52	0.67	1.56	1.33	0.88	1.14

The sample mean mileage was higher after tune-up. We would like to determine how strong the evidence is that the population mean mileage is higher after tune-up.

RECALL
When we have paired samples, the pairs are called matched pairs.

These are paired samples, because each value before tune-up is paired with the value from the same car after tune-up. By computing the difference between the values in each matched pair, we construct a sample of differences, shown in the last row of Table 11.1.

If we denote the population mean mileage before tune-up by μ_1, and the population mean mileage after tune-up by μ_2, then we are interested in the difference $\mu_1 - \mu_2$. Because these are paired samples, the population mean of the differences, μ_d, is the same as $\mu_1 - \mu_2$. Therefore, performing a hypothesis test on μ_d is the same as performing a hypothesis test on the difference of the population means $\mu_1 - \mu_2$.

NOTE TO INSTRUCTOR

It is worth emphasizing that experiments should be designed using matched pairs rather than independent samples when possible, because of the smaller margin of error.

We summarize the notation we will use in this section.

NOTATION

- $\bar{d}$ is the sample mean of the differences between the values in the matched pairs.
- s_d is the sample standard deviation of the differences between the values in the matched pairs.
- μ_d is the population mean difference for the matched pairs.

The method for testing a hypothesis about μ_d is the usual method for testing a hypothesis about a population mean, as presented in Section 9.3. We now list the assumptions for this method, when applied to paired samples.

Assumptions for Performing a Hypothesis Test for the Difference Between Two Means with Matched Pairs

1. We have a simple random sample of matched pairs.
2. Either the sample size is large ($n > 30$), *or* the differences between items in the matched pairs show no evidence of strong skewness and no outliers. This is required to be sure that $\bar{d}$ will be approximately normally distributed.

When these assumptions are satisfied, a hypothesis test can be performed in the same way as for a test for a population mean, using the following steps.

Performing a Hypothesis Test with Matched-Pair Data Using the *P*-Value Method

Check to be sure the assumptions are satisfied. If they are, then proceed with the following steps.

Step 1: State the null and alternate hypotheses. The null hypothesis will have the form H_0: $\mu_d = \mu_0$. The alternate hypothesis will be of the form $\mu_d < \mu_0$, $\mu_d > \mu_0$, or $\mu_d \neq \mu_0$.

Step 2: If making a decision, choose a significance level α.

Step 3: Compute the test statistic $t = \dfrac{\bar{d} - \mu_0}{s_d / \sqrt{n}}$.

Step 4: Compute the *P*-value. The *P*-value is an area under the t curve with $n - 1$ degrees of freedom. The *P*-value depends on the alternate hypothesis as follows:

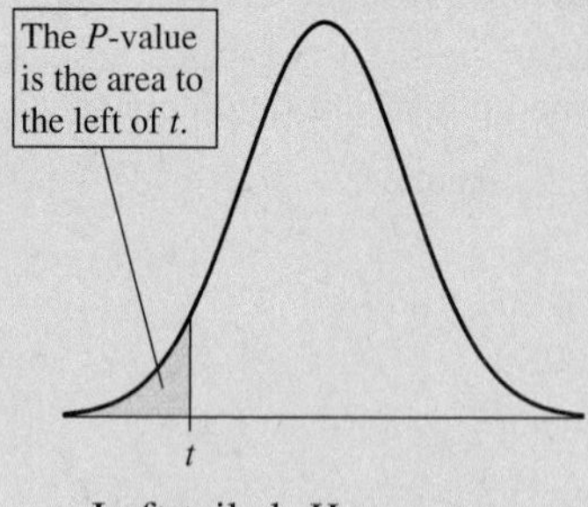
The P-value is the area to the left of t.
t
Left-tailed: H_1: $\mu_d < \mu_0$

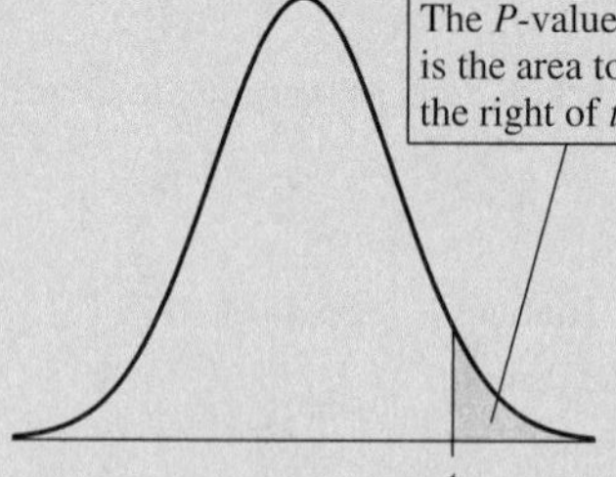
The P-value is the area to the right of t.
t
Right-tailed: H_1: $\mu_d > \mu_0$

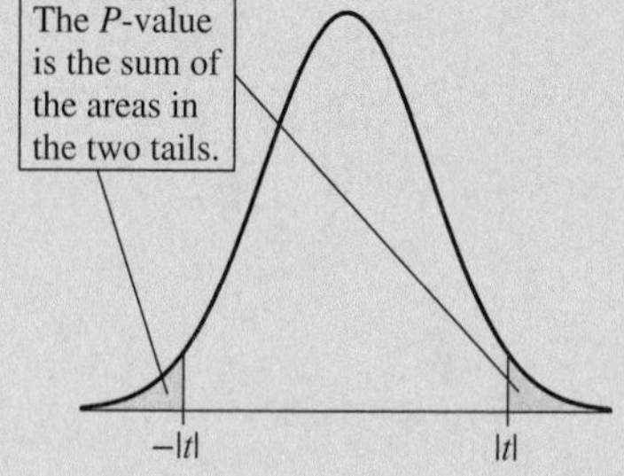
The P-value is the sum of the areas in the two tails.
−|t|
|t|
Two-tailed: H_1: $\mu_d \neq \mu_0$

Step 5: Interpret the *P*-value. If making a decision, reject H_0 if the *P*-value is less than or equal to the significance level α.

Step 6: State a conclusion.

Example 11.3

Test a hypothesis with matched-pair data

Test H_0: $\mu_d = 0$ versus H_1: $\mu_d > 0$, using the data in Table 11.1. Use the $\alpha = 0.01$ significance level.

Solution

We first check the assumptions. We have a simple random sample of differences. Because the sample size is small ($n = 8$), we must check for signs of strong skewness or outliers. Following is a dotplot of the differences.

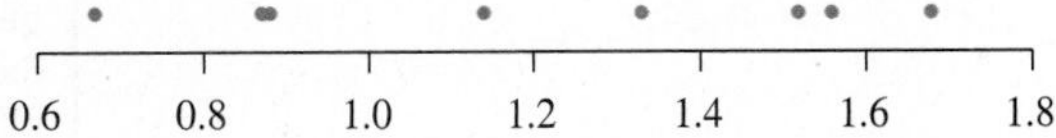

The dotplot does not reveal any outliers or strong skewness. Therefore, we can proceed.

Step 1: State H_0 and H_1. The issue is whether the mileage is greater after tune-up, so the null and alternate hypotheses are

$$H_0: \mu_d = 0 \qquad H_1: \mu_d > 0$$

Step 2: Choose a significance level. We will use $\alpha = 0.01$.

Step 3: Compute the test statistic. First we compute the sample mean and sample standard deviation of the differences. These are

$$\bar{d} = 1.20625 \qquad s_d = 0.37317$$

The test statistic is

$$t = \frac{\bar{d} - \mu_0}{s_d/\sqrt{n}}$$

Under the assumption that H_0 is true, $\mu_d = \mu_0 = 0$. The value of the test statistic is therefore

$$t = \frac{\bar{d} - \mu_0}{s_d/\sqrt{n}} = \frac{1.20625 - 0}{0.37317/\sqrt{8}} = 9.1427$$

Step 4: Compute the P-value. Under the assumption that H_0 is true, the test statistic has a t distribution. The number of degrees of freedom is $n - 1 = 8 - 1 = 7$. The alternate hypothesis is $\mu_d > 0$, so the P-value is the area to the right of the observed value of 9.1427. Technology gives $P = 0.0000193$.

Step 5: Interpret the P-value. The P-value is nearly 0. If H_0 were true, there would be virtually no chance of observing a test statistic as extreme as the value of 9.1427 that we observed. Because $P < 0.01$, we reject H_0 at the $\alpha = 0.01$ level.

Step 6: State a conclusion. We conclude that the gas mileage increased after a tune-up.

Performing hypothesis tests with technology

The following computer output (from MINITAB) presents the results of the hypothesis test performed in Example 11.3.

```
Paired T for After - Before

              N       Mean      StDev    SE Mean
After         8   30.05125    4.60928    1.62963
Before        8   28.84500    4.63519    1.63879
Difference    8    1.20625    0.37317    0.13194

99% lower bound for mean difference: 0.81071
T-Test of mean difference = 0 (vs > 0):  T-Value = 9.14  P-Value = 0.000
```

The column labeled "SE Mean" presents the standard errors for the means of the two samples and for the mean of the differences. The standard error is the standard deviation divided by the square root of the sample size. Note that the sample standard deviation of the differences is much smaller than the sample standard deviations of the original samples. This is the case for most matched-pair data and is the reason that tests based on matched pairs have more power than tests based on independent samples.

Following are the results of Example 11.3 as displayed on a TI-84 Plus calculator.

```
T-Test
μ>0
t=9.142739886
p=1.9251423E-5
x̄=1.20625
Sx=.3731693411
n=8
```

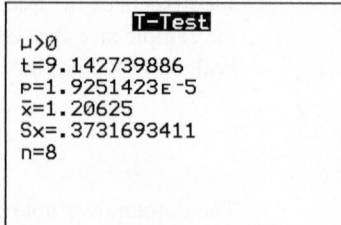

The second line states the alternate hypothesis. This is followed by the value of the test statistic t and the P-value. The quantity x̄ is the sample mean of the differences $\bar{d}$, and the quantity Sx is the sample standard deviation of the differences s_d.

Step-by-step instructions for performing hypothesis tests with technology are presented in the Using Technology section on page 556.

Check Your Understanding

Comstock/PictureQuest

1. A sample of five third graders took a reading test. They then participated in a reading improvement program and took the test again to determine whether their reading ability had improved. Following are the test scores for each of the students both before and after the program. Can you conclude that the mean reading score increased after the program?

	1	2	3	4	5
After	67	68	78	75	84
Before	59	63	81	74	78

a. Let μ_d denote the population mean difference After − Before. State the appropriate null and alternate hypotheses about μ_d. *H_0: $\mu_d = 0$, H_1: $\mu_d > 0$*

b. Compute the differences After − Before. *8, 5, −3, 1, 6*

c. Compute the value of the test statistic. *1.731*

d. Compute the P-value. *0.0793*

e. Do you reject H_0 at the $\alpha = 0.05$ level? *No*

f. State a conclusion.

2. Following are the annual amounts of rainfall, in inches, in six randomly chosen cities for two consecutive years. Can you conclude that the mean rainfall was greater in year 2 than in year 1?

	1	2	3	4	5	6
Year 2	34.6	18.7	42.6	41.3	60.6	29.9
Year 1	25.1	15.3	46.4	31.2	51.7	24.2

a. Let μ_d denote the population mean difference Year 2 − Year 1. State the appropriate null and alternate hypotheses about μ_d. *H_0: $\mu_d = 0$, H_1: $\mu_d > 0$*

b. Compute the differences Year 2 − Year 1. *9.5, 3.4, −3.8, 10.1, 8.9, 5.7*

c. Compute the value of the test statistic. *2.612*

d. Compute the P-value. *0.0238*

e. Do you reject H_0 at the $\alpha = 0.05$ level? *Yes*

f. State a conclusion.

Answers are on page 560.

Objective 2 Perform a hypothesis test with matched pairs using the critical value method

Testing a Hypothesis with Matched-Pair Data Using the Critical Value Method

The critical value method for matched-pair data is essentially the same as that for a population mean with σ unknown. We can find the critical value in Table A.3 or with technology. The assumptions for the critical value method are the same as for the P-value method. When the assumptions are satisfied, a hypothesis test can be performed using the following steps.

Performing a Hypothesis Test with Matched-Pair Data Using the Critical Value Method

Check to be sure the assumptions are satisfied. If they are, then proceed with the following steps.

Step 1: State the null and alternate hypotheses. The null hypothesis will have the form H_0: $\mu_d = \mu_0$. The alternate hypothesis will be of the form $\mu_d < \mu_0$, $\mu_d > \mu_0$, or $\mu_d \neq \mu_0$.

Step 2: Choose a significance level α, and find the critical value or values.

Step 3: Compute the test statistic $t = \dfrac{\bar{d} - \mu_0}{s_d/\sqrt{n}}$.

Step 4: Determine whether to reject H_0, as follows:

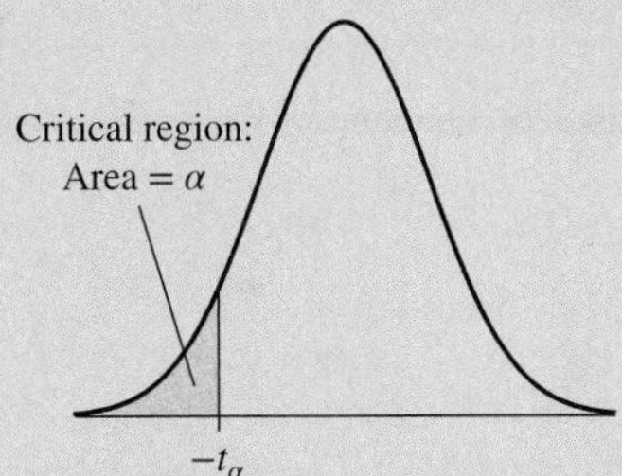

Left-tailed: H_1: $\mu_d < \mu_0$

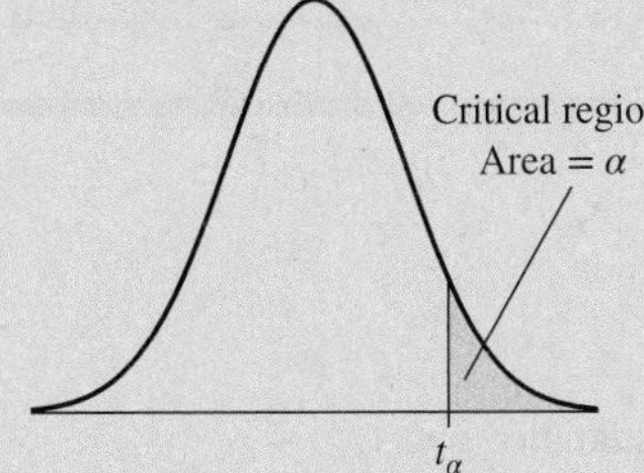

Right-tailed: H_1: $\mu_d > \mu_0$

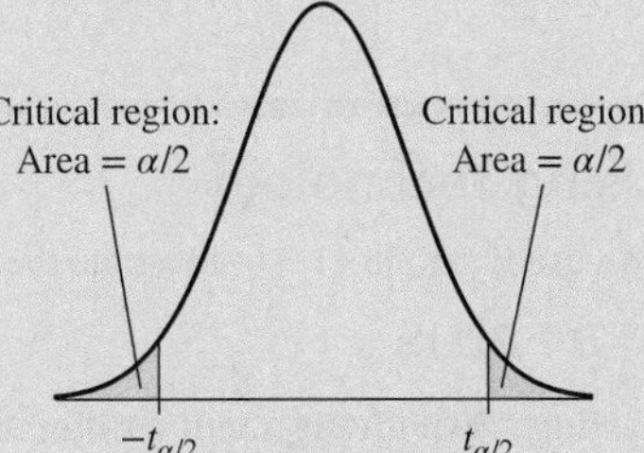

Two-tailed: H_1: $\mu_d \neq \mu_0$

Step 5: State a conclusion.

Example 11.4

Testing hypotheses with the critical value method

For a sample of nine automobiles, the mileage (in 1000s of miles) at which the original front brake pads were worn to 10% of their original thickness was measured, as was the mileage at which the original rear brake pads were worn to 10% of their original thickness. The results are given below.

	Automobile								
	1	2	3	4	5	6	7	8	9
Rear	42.7	36.7	46.1	46.0	39.9	51.7	51.6	46.1	47.3
Front	32.8	26.6	35.6	36.4	29.2	40.9	40.9	34.8	36.6
Difference	9.9	10.1	10.5	9.6	10.7	10.8	10.7	11.3	10.7

The differences in the last line of the table are Rear − Front. Can you conclude that the mean time for the rear brake pads to wear out is longer than the mean time for the front pads? Use the $\alpha = 0.05$ significance level.

Solution

We first check the assumptions. Because the sample size is small, we will construct a dotplot.

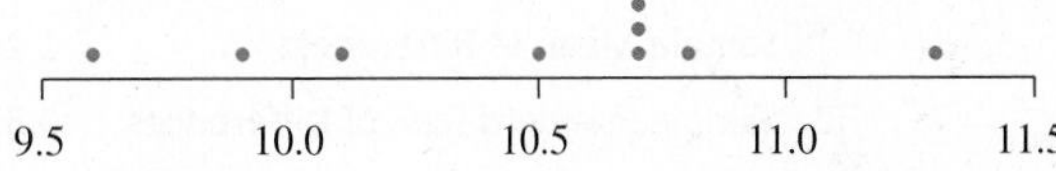

The dotplot shows no evidence of outliers or extreme skewness, so we can proceed.

Step 1: State the null and alternate hypotheses. We are interested in determining whether the mean time for the rear pads is longer than for the front. Therefore, the hypotheses are

$$H_0\colon \mu_d = 0 \qquad H_1\colon \mu_d > 0$$

Step 2: Choose a significance level α, and find the critical value. We will use $\alpha = 0.05$. Because this is a right-tailed test, the critical value is the value for which the area to the right is 0.05. The sample size is $n = 9$, so there are $9 - 1 = 8$ degrees of freedom. The critical value is $t_\alpha = 1.860$.

Step 3: Compute the test statistic. The sample size is $n = 9$. We compute the sample mean and standard deviation of the differences:

$$\bar{d} = 10.478 \qquad s_d = 0.5215$$

The test statistic is

$$t = \frac{\bar{d} - 0}{s_d/\sqrt{n}} = \frac{10.478 - 0}{0.5215/\sqrt{9}} = 60.28$$

Step 4: Determine whether to reject H_0. This is a right-tailed test, so we reject H_0 if $t \geq t_\alpha$. Because $t = 60.28$ and $t_\alpha = 1.860$, we reject H_0 at the $\alpha = 0.05$ level.

Step 5: State a conclusion. We conclude that the mean time for rear brake pads to wear out is longer than the mean time for front brake pads.

Using Technology

We use Example 11.3 to illustrate the technology steps.

TI-84 PLUS

Testing a hypothesis about a difference using matched pairs

Step 1. Enter the data into **L1** and **L2** in the data editor. On the home screen, enter **(L1 – L2) STO L3** to assign the differences in list **L3** (Figure A).

Step 2. Press **STAT** and highlight the **TESTS** menu.

Step 3. Select **T–Test** and press **ENTER** (Figure B). The **T–Test** menu appears.

Step 4. For **Inpt**, select the **Data** option and enter **L3** as the **List** option.

Step 5. Enter the null hypothesis mean for μ_0, and select the form of the alternate hypothesis. For Example 11.3, we have $\mu_0 = 0$ and the alternate hypothesis has the form $>\mu_0$ (Figure C).

Step 6. Highlight **Calculate** and press **ENTER** (Figure D).

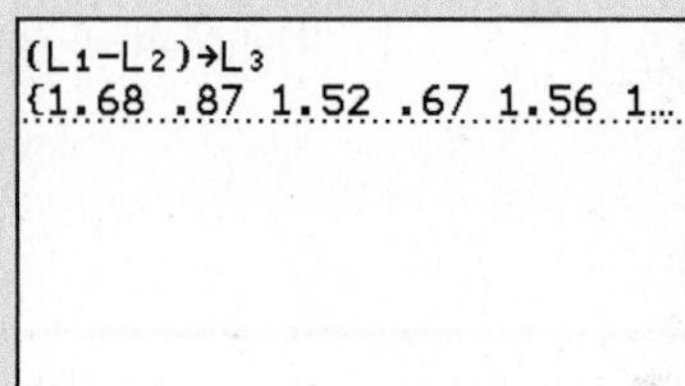

Figure A

```
EDIT CALC TESTS
1:Z-Test…
2:T-Test…
3:2-SampZTest…
4:2-SampTTest…
5:1-PropZTest…
6:2-PropZTest…
7:ZInterval…
8:TInterval…
9↓2-SampZInt…
```

Figure B

```
T-Test
Inpt:Data Stats
μ0:0
List:L3
Freq:1
μ:≠μ0 <μ0 >μ0
Color: BLUE
Calculate Draw
```

Figure C

```
T-Test
μ>0
t=9.142739886
p=1.9251423E-5
x̄=1.20625
Sx=.3731693411
n=8
```

Figure D

EXCEL

Finding the test statistic for a hypothesis test about the difference between two means using matched pairs

Figure E demonstrates the calculations involved for finding the test statistic for a hypothesis test about the difference between two means using matched pairs. In Example 11.3, we test H_0: $\mu_d = 0$ versus H_1: $\mu_d > 0$ where $\bar{d} = 1.20625$, $s_d = 0.37317$, and $n = 8$.

	A	B	C	D
1	Sample Mean of Differences	1.20625		
2	Sample Standard Dev. of Differences	0.37317		
3	Sample Size	8		
4	μ_0	0		
5				
6	Standard Error	0.1319355 ←	=B2/SQRT(B3)	
7				
8	Test Statistic T	9.1427237 ←	=(B1-B4)/B6	
9				

Figure E

MINITAB

Testing a hypothesis about a difference using matched pairs

Step 1. Enter the data from Example 11.3 into **Columns C1** and **C2**.

Step 2. Click on **Stat**, then **Basic Statistics**, then **Paired t**.

Step 3. Select **Each sample is in a column**, and enter **C1** in the **Sample 1** field and **C2** in the **Sample 2** field.

Step 4. Click **Options**, and enter the null hypothesis difference between the means in the **Hypothesized difference** field and select the form of the alternate hypothesis. Given significance level α, enter $100(1-\alpha)$ as the **Confidence Level**. For Example 11.3, we use **99** as the **Confidence Level**, **0** as the **Hypothesized difference**, and **Difference > hypothesized difference** as the **Alternative**. Click **OK**.

Step 5. Click **OK** (Figure F).

Descriptive Statistics

Sample	N	Mean	StDev	SE Mean
C1	8	30.05	4.61	1.63
C2	8	28.85	4.64	1.64

Test

Null hypothesis H_0: μ_difference = 0
Alternative hypothesis H_1: μ_difference > 0

T-Value	P-Value
9.14	0.000

Figure F

Section 11.2 Exercises

Exercises 1 and 2 are the Check Your Understanding exercises located within the section.

Understanding the Concepts

In Exercises 3 and 4, fill in each blank with the appropriate word or phrase.

3. If the sample size is small, the differences between the items in the matched pairs must show no evidence of strong _________ and must contain no _________. *skewness, outliers*

4. With matched pairs, the test for the difference between population means is the same as the test for a single population _________. *mean*

In Exercises 5 and 6, determine whether the statement is true or false. If the statement is false, rewrite it as a true statement.

5. Paired data are data for which each value in one sample can be matched with a corresponding value in another sample. *True*

6. To compute the test statistic for a test with matched pairs, we must compute the standard deviations of the samples. *False*

Practicing the Skills

7. Following is a sample of five matched pairs.

Sample 1	19	15	16	23	24
Sample 2	18	19	10	14	17

Let μ_1 and μ_2 represent the population means and let $\mu_d = \mu_1 - \mu_2$. A test will be made of the hypotheses $H_0\colon \mu_d = 0$ versus $H_1\colon \mu_d > 0$.

a. Compute the differences. *1, −4, 6, 9, 7*
b. Compute the test statistic. *1.614*
c. Can you reject H_0 at the $\alpha = 0.05$ level of significance? *No*
d. Can you reject H_0 at the $\alpha = 0.01$ level of significance? *No*

8. Following is a sample of 10 matched pairs.

Sample 1	28	29	22	25	26	29	27	24	27	28
Sample 2	34	30	31	26	31	30	31	32	29	37

Let μ_1 and μ_2 represent the population means and let $\mu_d = \mu_1 - \mu_2$. A test will be made of the hypotheses $H_0\colon \mu_d = 0$ versus $H_1\colon \mu_d \neq 0$.

a. Compute the differences. *−6, −1, −9, −1, −5, −1, −4, −8, −2, −9*
b. Compute the test statistic. *−4.399*
c. Can you reject H_0 at the $\alpha = 0.05$ level of significance? *Yes*
d. Can you reject H_0 at the $\alpha = 0.01$ level of significance? *Yes*

Working with the Concepts

9. **Crossover trial:** A crossover trial is a type of experiment used to compare two drugs. Subjects take one drug for a period of time, then switch to the other. The responses of the subjects are then compared using matched-pair methods. In an experiment to compare two pain relievers, seven subjects took one pain reliever for two weeks, then switched to the other. They rated their pain level from 1 to 10, with larger numbers representing higher levels of pain. The results were:

	Subject						
	1	2	3	4	5	6	7
Drug A	6	3	4	5	7	1	4
Drug B	5	1	5	5	5	2	2

Can you conclude that the mean pain level is less with drug B?

a. State the null and alternate hypotheses. *$H_0\colon \mu_d = 0$, $H_1\colon \mu_d > 0$*
b. Compute the test statistic. *1.369*
c. State a conclusion. Use the $\alpha = 0.05$ level of significance. *Do not reject H_0.*

10. Comparing scales: In an experiment to determine whether there is a systematic difference between the weights obtained with two different scales, 10 rock specimens were weighed, in grams, on each scale. The following data were obtained:

Specimen	Weight on Scale 1	Weight on Scale 2
1	11.23	11.27
2	14.36	14.41
3	8.33	8.35
4	10.50	10.52
5	23.42	23.41
6	9.15	9.17
7	13.47	13.52
8	6.47	6.46
9	12.40	12.45
10	19.38	19.35

Can you conclude that the mean weight differs between the scales?

a. State the null and alternate hypotheses. *H_0: $\mu_d = 0$, H_1: $\mu_d \neq 0$*

b. Compute the test statistic. *−2.206*

c. State a conclusion. Use the $\alpha = 0.01$ level of significance.

11. Shoot or root? Six bean plants had their carbohydrate concentrations (in percent by weight) measured both in the shoot and in the root. The following results were obtained:

Plant	Shoot	Root
1	4.42	3.66
2	5.81	5.51
3	4.65	3.91
4	4.77	4.47
5	5.25	4.69
6	4.75	3.93

Can you conclude that there is a difference in mean concentration between the shoot and the root?

a. State the null and alternate hypotheses. *H_0: $\mu_d = 0$, H_1: $\mu_d \neq 0$*

b. Compute the test statistic. *6.082*

c. State a conclusion. Use the $\alpha = 0.05$ level of significance. *Reject H_0.*

12. Tire wear: An automobile manufacturer wants to compare the lifetimes of two brands of tire. She obtains samples of seven tires of each brand. On each of seven cars, she mounts one tire of each brand on each front wheel. The cars are driven until only 20% of the original tread remains. The distances, in thousands of miles, for each tire are presented in the following table.

Car	Brand A	Brand B
1	36.9	34.3
2	45.3	42.2
3	36.2	35.5
4	32.1	31.9
5	37.2	38.1
6	48.3	47.8
7	38.2	33.2

Brand A is more expensive than Brand B. Can you conclude that the mean lifetime of Brand A tires is greater than that of Brand B?

a. State the null and alternate hypotheses. *H_0: $\mu_d = 0$, H_1: $\mu_d > 0$*

b. Compute the test statistic. *2.072*

c. State a conclusion. Use the $\alpha = 0.01$ level of significance. *Do not reject H_0.*

13. Drug comparison: A study was done of a generic and a brand name antifungal ointment to determine whether there was a difference in the amounts of active ingredient absorbed into the skin. Both the brand name and generic products were applied to the arms of 14 subjects, and the amounts absorbed, in $\mu g/cm^2$, were measured. The results are presented in the following table.

Subject	Brand Name	Generic
1	2.23	1.42
2	1.68	1.95
3	1.96	2.58
4	2.81	2.25
5	1.14	1.21
6	3.20	3.01
7	2.33	2.76
8	4.06	3.65
9	2.92	2.89
10	2.92	2.85
11	2.83	2.44
12	3.45	3.11
13	2.72	2.64
14	3.74	2.82

Source: *Pharmaceutical Research* 26:316–328

Can you conclude that the mean amount absorbed differs between the brand name and the generic drug?

a. State the null and alternate hypotheses. *H_0: $\mu_d = 0$, H_1: $\mu_d \neq 0$*

b. Compute the test statistic. *1.459*

c. State a conclusion. Use the $\alpha = 0.01$ level of significance. *Do not reject H_0.*

14. Lowering cholesterol: A group of eight individuals with high cholesterol levels were given a new drug that was designed to lower cholesterol levels. Total cholesterol levels, in mg/dL, were measured before and after treatment, with the following results.

Subject	Before	After
1	283	215
2	299	206
3	274	187
4	284	212
5	248	178
6	275	212
7	293	192
8	277	196

Can you conclude that the mean cholesterol level is reduced by more than 65 mg/dL after treatment?

a. State the null and alternate hypotheses. *H_0: $\mu_d = 65$, H_1: $\mu_d > 65$*

b. Compute the test statistic. *3.038*

c. State a conclusion. Use the $\alpha = 0.05$ level of significance. *Reject H_0.*

15. Strength of concrete: The compressive strength, in kilopascals, was measured for concrete blocks from five different batches of concrete, both three and six days after pouring. The data are as follows:

	Block				
	1	2	3	4	5
After 3 days	1341	1316	1352	1355	1327
After 6 days	1376	1373	1366	1384	1358

Can you conclude that the mean strength after three days differs from the mean strength after six days?

a. State the null and alternate hypotheses. *H_0: $\mu_d = 0$, H_1: $\mu_d \neq 0$*

b. Compute the test statistic. *−4.790*

c. State a conclusion. Use the $\alpha = 0.05$ level of significance. *Reject H_0.*

16. Truck pollution: In an experiment to determine the effect of ambient temperature on the emissions of oxides of nitrogen (NO_x) of diesel trucks, 10 trucks were run at temperatures of 40°F and 80°F. The emissions, in parts per billion, are presented in the following table.

Truck	40°F	80°F
1	834.7	815.2
2	753.2	765.2
3	855.7	842.6
4	901.2	797.1
5	785.4	764.3
6	862.9	819.5
7	882.7	783.6
8	740.3	694.5
9	748.0	772.9
10	848.6	794.7

Can you conclude that the mean emissions are higher at 40°F?

a. State the null and alternate hypotheses. *H_0: $\mu_d = 0$, H_1: $\mu_d > 0$*

b. Compute the test statistic. *2.713*

c. State a conclusion. Use the $\alpha = 0.05$ level of significance. *Reject H_0.*

17. Growth spurt: It is generally known that boys grow at an unusually fast rate between the ages of about 12 and 14. Following are heights, in inches, of 40 boys measured at age 12 and again at age 14.

Age 12	Age 14	Age 12	Age 14
57.9	62.9	55.4	61.8
61.1	65.9	58.7	64.2
62.7	67.5	64.3	69.8
67.5	73.7	58.1	63.3
59.2	64.9	63.3	69.6
61.4	67.0	61.2	66.4
60.7	65.5	64.5	69.2
55.9	62.1	55.9	62.0
59.7	65.4	60.4	65.7
56.3	61.5	57.8	62.8
63.0	68.5	68.3	73.6
58.6	63.9	63.0	67.7
61.1	65.8	64.4	69.2
59.5	64.5	58.2	64.6
61.6	66.3	59.7	66.1
59.3	64.2	60.2	65.9
62.1	67.1	63.7	68.3
62.8	68.1	60.2	65.2
65.3	69.9	62.7	67.9
60.4	66.7	55.6	61.7

Can you conclude that the mean increase in height is greater than 5 inches?

a. State the null and alternate hypotheses.

b. Compute the test statistic. *4.009*

c. State a conclusion. Use the $\alpha = 0.05$ level of significance. *Reject H_0.*

18. SAT coaching: A sample of 32 students took a class designed to improve their SAT math scores. Following are their scores before and after the class.

Before	After	Before	After
383	420	394	430
334	368	513	525
378	396	483	482
467	488	447	482
470	489	440	479
473	473	439	451
443	448	435	431
459	473	451	454
426	428	453	463
493	525	491	511
382	382	526	529
473	474	473	493
408	407	440	466
433	434	481	482
478	490	459	455
502	508	399	404

Can you conclude that the mean increase in score is less than 15 points?

a. State the null and alternate hypotheses.

b. Compute the test statistic. *−0.860*

c. State a conclusion. Use the $\alpha = 0.05$ level of significance. *Do not reject H_0.*

19. Interpret calculator display: The following TI-84 Plus calculator display presents the results of a hypothesis test for the mean difference between matched pairs.

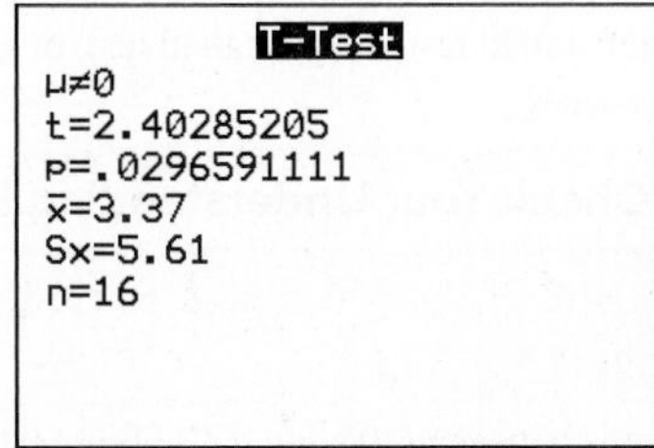

a. Is this a left-tailed test, a right-tailed test, or a two-tailed test? *Two-tailed*

b. How many degrees of freedom are there? *15*

c. What is the P-value? *0.0296591111*

d. Can H_0 be rejected at the 0.05 level? Explain. *Yes*

20. Interpret calculator display: The following TI-84 Plus calculator display presents the results of a hypothesis test for the mean difference between matched pairs.

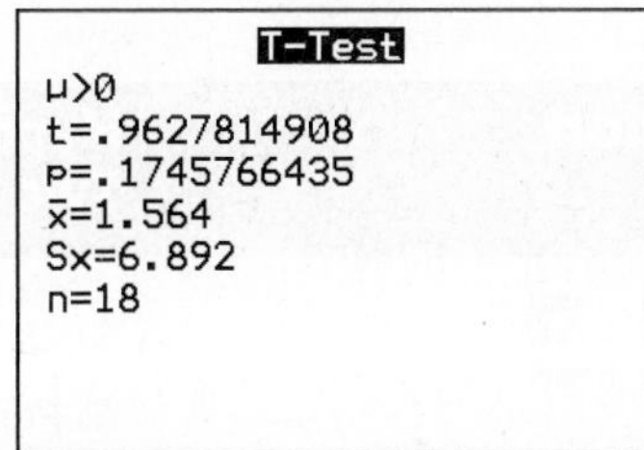

a. Is this a left-tailed test, a right-tailed test, or a two-tailed test? *Right-tailed*

b. How many degrees of freedom are there? *17*

c. What is the P-value? *0.1745766435*

d. Can H_0 be rejected at the 0.05 level? Explain. *No*

21. Interpret computer output: The following MINITAB output presents the results of a hypothesis test for a mean difference.

```
Paired T for X - Y

              N      Mean     StDev   SE Mean
X            12   134.233    68.376    19.739
Y            12   100.601    94.583    27.304
Difference   12   33.6316    59.511    17.179

95% lower bound for mean difference: 2.7793
T-Test of mean difference = 0 (vs > 0 ):  T-Value = 1.96
                                            P-Value = 0.038
```

a. Is this a left-tailed test, a right-tailed test, or a two-tailed test? *Right-tailed*

b. Can H_0 be rejected at the 0.05 level? Explain. *Yes*

c. Can H_0 be rejected at the 0.01 level? Explain. *No*

22. Interpret computer output: The following MINITAB output presents the results of a hypothesis test for a mean difference.

```
Paired T for X - Y

              N     Mean      StDev    SE Mean
X             7   21.4236    10.145    3.8344
Y             7   19.2587     8.4049   3.1767
Difference    7    2.16485    5.3707   2.0299

95% CI for mean difference: (-2.802239, 7.131933)
T-Test of mean difference = 0 (vs not = 0):  T-Value = 1.07
                                             P-Value = 0.327
```

a. Is this a left-tailed test, a right-tailed test, or a two-tailed test? *Two-tailed*

b. Can H_0 be rejected at the 0.05 level? Explain. *No*

c. Can H_0 be rejected at the 0.01 level? Explain. *No*

Extending the Concepts

23. Advantage of matched pairs: Refer to Exercise 16. Assume you did not know that the two samples were paired, so you used the methods of Section 11.1 to perform the test.

a. What is the P-value? *0.0726 [TI-84 Plus: 0.0653] [MINITAB: 0.0658]*

b. Explain why the P-value is greater when the methods of Section 11.1 are used. *Standard error is larger.*

Answers to Check Your Understanding Exercises for Section 11.2

1. a. H_0: $\mu_d = 0$, H_1: $\mu_d > 0$ **b.** 8, 5, −3, 1, 6

c. $t = 1.731$

d. P-value is between 0.05 and 0.10 [Tech: 0.0793].

e. If H_0 is true, the probability of observing a value for the test statistic as extreme as or more extreme than the value actually observed is 0.0793. This is somewhat unusual, so there is some evidence against H_0. However, because $P > 0.05$, we do not reject H_0 at the $\alpha = 0.05$ level.

f. We conclude that the mean reading score may have remained the same after the reading program.

2. a. H_0: $\mu_d = 0$, H_1: $\mu_d > 0$

b. 9.5, 3.4, −3.8, 10.1, 8.9, 5.7 **c.** $t = 2.612$

d. P-value is between 0.01 and 0.025 [Tech: 0.0238].

e. If H_0 is true, the probability of observing a value for the test statistic as extreme as or more extreme than the value actually observed is 0.0238. This is fairly unusual, so there is fairly strong evidence against H_0. In particular, because $P < 0.05$, we reject H_0 at the $\alpha = 0.05$ level.

f. We conclude that the mean rainfall was greater in year 2 than in year 1.

Section 11.3 Hypothesis Tests for the Difference Between Two Proportions

Objectives

1. Perform a hypothesis test for the difference between two proportions using the P-value method
2. Perform a hypothesis test for the difference between two proportions using the critical value method

Objective 1 Perform a hypothesis test for the difference between two proportions using the P-value method

Perform a Hypothesis Test for the Difference Between Two Proportions Using the *P*-Value Method

Are older, more experienced workers more likely to use a computer at work than younger workers? The General Social Survey took a poll to address this question. They asked 350 employed people aged 25–40 whether they used a computer at work, and 259 said they did. They also asked the same question of 500 employed people aged 41–65, and 384 of them said that they used a computer at work.

We can compute the sample proportions of people who used a computer at work in each of these age groups. Among those 25–40, the sample proportion was $259/350 = 0.740$, and among those aged 41–65 the sample proportion was $384/500 = 0.768$. So the sample proportion is larger among older workers. The question of interest, however, involves the population proportions. There are two populations involved; the population of all employed people aged 25–40, and the population of all employed people aged 41–65. The question is whether the population proportion of people aged 41–65 who use a computer at work is greater than the population proportion among those aged 25–40.

This is an example of a situation in which we have two independent samples, with the sample proportion computed for each one. We will describe a method for performing a hypothesis test to determine whether the two population proportions are equal. We will need some notation for the population proportions, the sample proportions, the numbers of individuals in each category, and the sample sizes.

NOTATION

- p_1 and p_2 are the proportions of the category of interest in the two populations.
- $\hat{p}_1$ and $\hat{p}_2$ are the proportions of the category of interest in the two samples.
- x_1 and x_2 are the numbers of individuals in the category of interest in the two samples.
- n_1 and n_2 are the two sample sizes.

We will now describe how to perform the hypothesis test.

The null and alternate hypotheses

The issue is whether the population proportions p_1 and p_2 are equal. The null hypothesis says that they are equal:

$$H_0: p_1 = p_2$$

There are three possibilities for the alternate hypothesis:

$$H_1: p_1 < p_2 \qquad H_1: p_1 > p_2 \qquad H_1: p_1 \neq p_2$$

The test statistic

The test statistic is based on the difference between the sample proportions, $\hat{p}_1 - \hat{p}_2$. When the sample size is large, this difference is approximately normally distributed.

The mean and standard deviation of this distribution are

RECALL

Because the samples are independent, the variance of $\hat{p}_1 - \hat{p}_2$ is the sum of the variances of $\hat{p}_1$ and $\hat{p}_2$. The standard deviation is the square root of the variance.

$$\text{Mean} = p_1 - p_2 \qquad \text{Standard deviation} = \sqrt{\frac{p_1(1-p_1)}{n_1} + \frac{p_2(1-p_2)}{n_2}}$$

To compute the test statistic, we must find values for the mean and standard deviation. The mean is straightforward: Under the assumption that H_0 is true, $p_1 - p_2 = 0$. The standard deviation is a bit more involved. The standard deviation depends on the population proportions p_1 and p_2, which are unknown. We need to estimate p_1 and p_2. Under H_0, we assume that $p_1 = p_2$. Therefore, we need to estimate p_1 and p_2 with the same value. The value to use is the **pooled proportion**, which we will denote by $\hat{p}$. The pooled proportion is found by treating the two samples as though they were one big sample. We divide the total number of individuals in the category of interest in the two samples by the sum of the two sample sizes:

$$\hat{p} = \frac{x_1 + x_2}{n_1 + n_2}$$

The standard deviation is estimated with the standard error:

$$\text{Standard error} = \sqrt{\frac{\hat{p}(1-\hat{p})}{n_1} + \frac{\hat{p}(1-\hat{p})}{n_2}} = \sqrt{\hat{p}(1-\hat{p})\left(\frac{1}{n_1} + \frac{1}{n_2}\right)}$$

The test statistic is the z-score for $\hat{p}_1 - \hat{p}_2$:

$$z = \frac{(\hat{p}_1 - \hat{p}_2) - (p_1 - p_2)}{\sqrt{\hat{p}(1-\hat{p})\left(\frac{1}{n_1} + \frac{1}{n_2}\right)}} = \frac{(\hat{p}_1 - \hat{p}_2) - 0}{\sqrt{\hat{p}(1-\hat{p})\left(\frac{1}{n_1} + \frac{1}{n_2}\right)}} = \frac{\hat{p}_1 - \hat{p}_2}{\sqrt{\hat{p}(1-\hat{p})\left(\frac{1}{n_1} + \frac{1}{n_2}\right)}}$$

The method just described requires certain assumptions, which we now state:

EXPLAIN IT AGAIN

Reason for assumption 4: We require that both samples contain at least 10 individuals in each category in order to be sure that $\hat{p}_1$ and $\hat{p}_2$ are both approximately normally distributed.

Assumptions for Performing a Hypothesis Test for the Difference Between Two Proportions

1. There are two simple random samples that are independent of one another.
2. Each population is at least 20 times as large as the sample drawn from it.
3. The individuals in each sample are divided into two categories.
4. Both samples contain at least 10 individuals in each category.

We summarize the steps for testing a hypothesis about the difference between two proportions using the P-value method. Later we will describe the critical value method.

Performing a Hypothesis Test for the Difference Between Two Proportions Using the *P*-Value Method

Check to be sure the assumptions are satisfied. If they are, then proceed with the following steps.

Step 1: State the null and alternate hypotheses. The null hypothesis will have the form H_0: $p_1 = p_2$. The alternate hypothesis will be $p_1 < p_2$, $p_1 > p_2$, or $p_1 \neq p_2$.

Step 2: If making a decision, choose a significance level α.

Step 3: Compute the test statistic $z = \dfrac{\hat{p}_1 - \hat{p}_2}{\sqrt{\hat{p}(1-\hat{p})\left(\dfrac{1}{n_1} + \dfrac{1}{n_2}\right)}}$

where $\hat{p}$ is the pooled proportion: $\hat{p} = \dfrac{x_1 + x_2}{n_1 + n_2}$

Step 4: Compute the P-value. The P-value is an area under the normal curve. The P-value depends on the alternate hypothesis as follows:

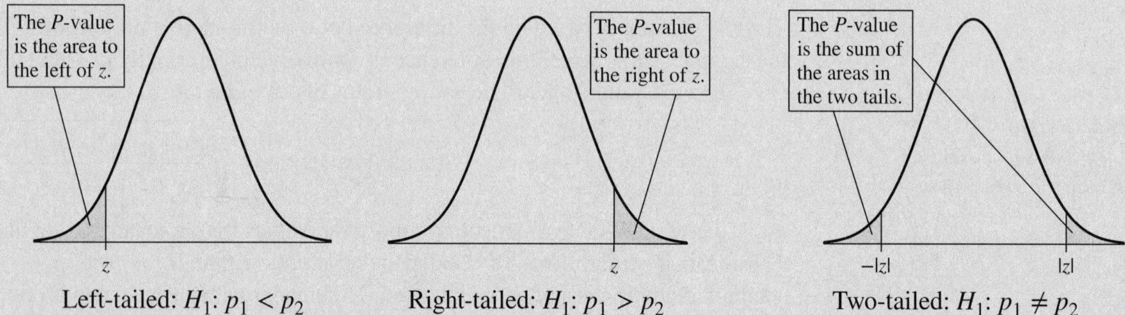
The P-value is the area to the left of z.
z
The P-value is the area to the right of z.
z
The P-value is the sum of the areas in the two tails.
−|z|
|z|

Left-tailed: H_1: $p_1 < p_2$ Right-tailed: H_1: $p_1 > p_2$ Two-tailed: H_1: $p_1 \neq p_2$

Step 5: Interpret the P-value. If making a decision, reject H_0 if the P-value is less than or equal to the significance level α.

Step 6: State a conclusion.

Example 11.5

Perform a hypothesis test

Are older, more experienced workers more likely to use a computer at work than younger workers? The General Social Survey took a poll to address this question. They asked 350 employed people aged 25–40 whether they used a computer at work, and 259 said they did. They also asked the same question of 500 employed people aged 41–65, and 384 of them said that they used a computer at work. Can you conclude that the proportion of people who use a computer at work is greater among those aged 41–65 than among those aged 25–40? Use the $\alpha = 0.05$ level.

Solution

We first check the assumptions. We have two independent random samples, and the populations are more than 20 times as large as the samples. The individuals in each sample are divided into two categories: those who use a computer at work and those who do not. Finally, each sample contains more than 10 individuals in each category. The assumptions are satisfied.

Step 1: State the null and alternate hypotheses: We'll let p_1 be the population proportion of people aged 25–40 who used a computer at work, and p_2 be the proportion among those aged 41–65. The issue is whether the proportions are the same, or whether the proportion of those aged 41–65, p_2, is greater than the proportion of those aged 25–40, p_1. Therefore the null and alternate hypotheses are

$$H_0: p_1 = p_2 \qquad H_1: p_1 < p_2$$

Step 2: Choose a significance level: The significance level is $\alpha = 0.05$.

Step 3: Compute the test statistic: We'll begin by summarizing the necessary information in a table.

	Ages 25–40	Ages 41–65
Sample size	$n_1 = 350$	$n_2 = 500$
Number of individuals	$x_1 = 259$	$x_2 = 384$
Sample proportion	$\hat{p}_1 = 259/350 = 0.740$	$\hat{p}_2 = 384/500 = 0.768$
Population proportion	p_1 (unknown)	p_2 (unknown)

Next, we compute the pooled proportion $\hat{p}$.

$$\hat{p} = \frac{x_1 + x_2}{n_1 + n_2} = \frac{259 + 384}{350 + 500} = 0.756471$$

The value of the test statistic is

$$z = \frac{\hat{p}_1 - \hat{p}_2}{\sqrt{\hat{p}(1-\hat{p})\left(\frac{1}{n_1} + \frac{1}{n_2}\right)}} = \frac{0.740 - 0.768}{\sqrt{0.756471(1 - 0.756471)\left(\frac{1}{350} + \frac{1}{500}\right)}}$$
$$= -0.94$$

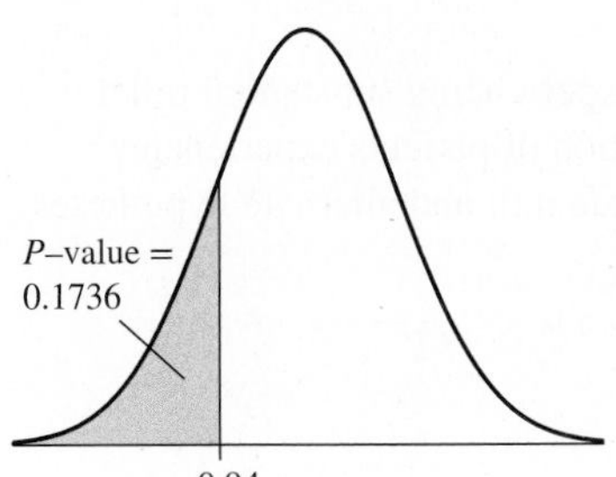

Figure 11.2

Step 4: Compute the *P*-value: The alternate hypothesis, $p_1 < p_2$, is left-tailed. Therefore the *P*-value is the area to the left of $z = -0.94$. Using Table A.2, we find this area to be 0.1736. Figure 11.2 illustrates the *P*-value.

Step 5: Interpret the *P*-value: The *P*-value of 0.1736 is greater than the significance level $\alpha = 0.05$. Therefore, we do not reject H_0.

Step 6: State a conclusion: We cannot conclude that the proportion of workers aged 41–65 who use a computer at work is greater than the proportion among those aged 25–40.

Performing a hypothesis test with technology

The following computer output (from MINITAB) presents the results of Example 11.5.

```
Test and CI for Two Proportions: 25-40, 41-65

Sample        X      N    Sample p
25-40       259    350    0.740000
41-65       384    500    0.768000

Difference = p (25-40) − p (41-65)
Estimate for difference: −0.028000
95% upper bound for difference: 0.021516
Test of difference = 0 (vs < 0): Z = −0.94 P-Value = 0.175
```

The numbers of individuals in the category of interest (X), the sample sizes (N), and the sample proportions (Sample p) are given. The quantity "Estimate for difference" is the difference between the sample proportions, $\hat{p}_1 - \hat{p}_2$. Next is the 95% upper bound for the difference. We can be 95% confident that the difference $p_1 - p_2$ is less than or equal to this value of 0.021516. The last line in the output presents the alternate hypothesis, the value of the test statistic (Z), and the *P*-value.

Following are the results as presented by the TI-84 Plus calculator.

```
2-PropZTest
p1<p2
z=-.9360431283
p=.1746254715
p̂1=.74
p̂2=.768
p̂=.7564705882
n1=350
n2=500
```

The letter "p" in the fourth line is the P-value.

Step-by-step instructions for performing hypothesis tests with technology are presented in the Using Technology section on page 566.

Check Your Understanding

1. In a clinical trial to compare the effectiveness of two pain relievers, a sample of 100 patients was given drug 1 and an independent sample of 200 patients was given drug 2. Of the patients on drug 1, 76 experienced substantial relief, while of the patients on drug 2, 128 experienced substantial relief. Investigators want to know whether the proportion of patients experiencing substantial relief is greater for drug 1. They will use the $\alpha = 0.05$ level of significance.

a. Let p_1 be the population proportion of patients experiencing substantial relief from drug 1, and let p_2 be the population proportion of patients experiencing substantial relief from drug 2. State the appropriate null and alternate hypotheses about p_1 and p_2. *H_0: $p_1 = p_2$, H_1: $p_1 > p_2$*

b. Compute the sample proportions $\hat{p}_1$ and $\hat{p}_2$. *$\hat{p}_1 = 0.76$, $\hat{p}_2 = 0.64$*

c. Compute the value of the test statistic. *2.10*

d. Compute the P-value. *0.0179*

e. Interpret the P-value. *Reject H_0 at the $\alpha = 0.05$ level.*

f. State a conclusion.

2. A sample of 200 voters over the age of 60 were asked whether they thought Social Security benefits should be increased for people over the age of 65. A total of 95 of them answered yes. A sample of 150 voters aged 18–25 were asked the same question, and 63 of them answered yes. A pollster wants to know whether the proportion of voters who support an increase in Social Security benefits is greater among older voters. He will use the $\alpha = 0.05$ level of significance.

a. Let p_1 be the population proportion of older voters expressing support, and let p_2 be the population proportion of younger voters expressing support. State the appropriate null and alternate hypotheses about p_1 and p_2. *H_0: $p_1 = p_2$, H_1: $p_1 > p_2$*

b. Compute the sample proportions $\hat{p}_1$ and $\hat{p}_2$. *$\hat{p}_1 = 0.475$, $\hat{p}_2 = 0.420$*

c. Compute the value of the test statistic. *1.02*

d. Compute the P-value. *0.1539 [Tech: 0.1531]*

e. Interpret the P-value. *Do not reject at the $\alpha = 0.05$ level.*

f. State a conclusion.

Answers are on page 570.

Objective 2 Perform a hypothesis test for the difference between two proportions using the critical value method

Using the Critical Value Method

To use the critical value method, compute the test statistic as before. Because the test statistic is a z-score, critical values can be found in Table A.2, in Table 9.1 in Section 9.2, or with technology. The assumptions for the critical value method are the same as for the P-value method.

Following are the steps for the critical value method.

Performing a Hypothesis Test for the Difference Between Two Proportions Using the Critical Value Method

Check to be sure that the assumptions are satisfied. If they are, then proceed with the following steps.

Step 1: State the null and alternate hypotheses. The null hypothesis will have the form H_0: $p_1 = p_2$. The alternate hypothesis will be $p_1 < p_2$, $p_1 > p_2$, or $p_1 \neq p_2$.

Step 2: Choose a significance level α, and find the critical value or values.

Step 3: Compute the test statistic $z = \dfrac{\hat{p}_1 - \hat{p}_2}{\sqrt{\hat{p}(1-\hat{p})\left(\dfrac{1}{n_1}+\dfrac{1}{n_2}\right)}}$

where $\hat{p}$ is the pooled proportion: $\hat{p} = \dfrac{x_1 + x_2}{n_1 + n_2}$

Step 4: Determine whether to reject H_0, as follows:

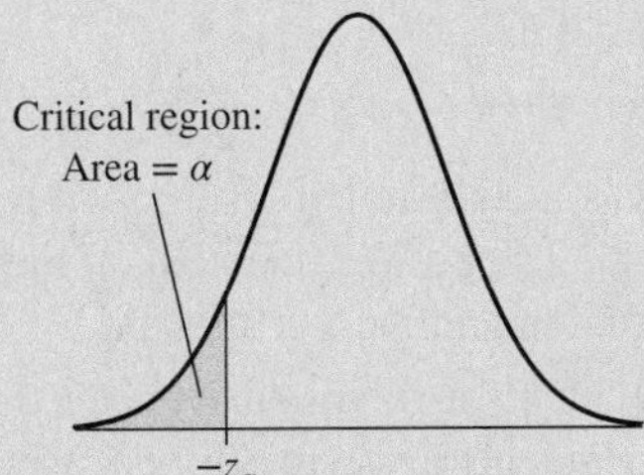

Left-tailed: H_1: $p_1 < p_2$
Reject if $z \leq -z_\alpha$.

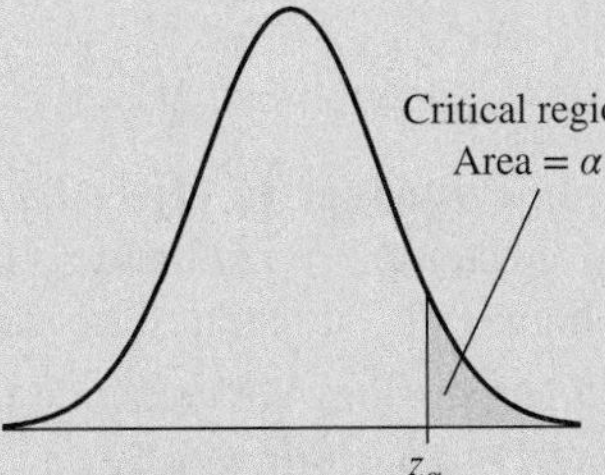

Right-tailed: H_1: $p_1 > p_2$
Reject if $z \geq z_\alpha$.

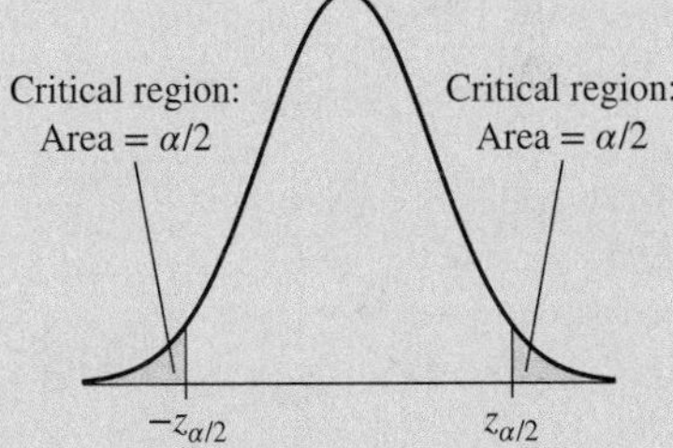

Two-tailed: H_1: $p_1 \neq p_2$
Reject if $z \geq z_{\alpha/2}$ or $z \leq -z_{\alpha/2}$.

Step 5: State a conclusion.

Example 11.6

Testing a hypothesis using the critical value method

Are younger drivers more likely to have accidents in their driveways? Traffic engineers tabulated types of car accidents by drivers of various ages. Out of a total of 82,486 accidents involving drivers aged 15–24 years, 4243 of them, or 5.1%, occurred in a driveway. Out of a total of 219,170 accidents involving drivers aged 25–64 years, 10,701 of them, or 4.9%, occurred in a driveway. Can you conclude that accidents involving drivers aged 15–24 are more likely to occur in driveways than accidents involving drivers aged 25–64? Use the $\alpha = 0.05$ significance level.

Source: *Journal of Transportation Engineering* 125:502–507

Solution

We first check the assumptions. We have two independent samples, and the individuals in each sample fall into two categories. Each sample contains at least 10 individuals in each category. The assumptions are satisfied.

Step 1: State H_0 and H_1. We are interested in whether the proportion of accidents in driveways is greater among younger drivers. Let p_1 be the population proportion for younger drivers, and let p_2 be the population proportion for older drivers. Then the null and alternate hypotheses are:

$$H_0: p_1 = p_2 \qquad H_1: p_1 > p_2$$

Step 2: Choose a significance level α, and find the critical value. We will use $\alpha = 0.05$. Because this is a right-tailed test, the critical value is the value for which the area to the right is 0.05. This value is $z_\alpha = 1.645$.

Step 3: Compute the test statistic. We begin by organizing the available information in a table:

	Ages 15–24	Ages 25–64
Sample size	$n_1 = 82{,}486$	$n_2 = 219{,}170$
Number of individuals	$x_1 = 4{,}243$	$x_2 = 10{,}701$
Sample proportion	$\hat{p}_1 = 4{,}243/82{,}486 = 0.051439$	$\hat{p}_2 = 10{,}701/219{,}170 = 0.048825$
Population proportion	p_1 (unknown)	p_2 (unknown)

Next, we compute the pooled proportion, $\hat{p}$.

$$\hat{p} = \frac{x_1 + x_2}{n_1 + n_2} = \frac{4{,}243 + 10{,}701}{82{,}486 + 219{,}170} = 0.049540$$

The test statistic is

$$z = \frac{\hat{p}_1 - \hat{p}_2}{\sqrt{\hat{p}(1-\hat{p})\left(\dfrac{1}{n_1} + \dfrac{1}{n_2}\right)}} = 2.95$$

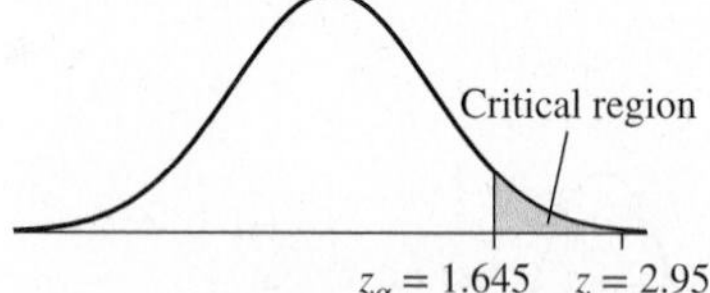

Figure 11.3 The value of the test statistic, $z = 2.95$, is inside the critical region. Therefore, we reject H_0.

Step 4: Determine whether to reject H_0. This is a right-tailed test, so we reject H_0 if $z > z_\alpha$. Because $z = 2.95$ and $z_\alpha = 1.645$, $z > z_\alpha$. We reject H_0 at the $\alpha = 0.05$ level. Figure 11.3 illustrates the critical region and the test statistic.

Step 5: State a conclusion. We conclude that accidents involving drivers aged 15–24 are more likely to occur in a driveway than accidents involving drivers aged 25–64.

Using Technology

We use Example 11.6 to illustrate the technology steps.

TI-84 PLUS

Testing a hypothesis about the difference between two proportions

Step 1. Press **STAT** and highlight the **TESTS** menu.

Step 2. Select **2–PropZTest** and press **ENTER** (Figure A). The **2–PropZTest** menu appears.

Step 3. Enter the values of x_1, n_1, x_2, and n_2. For Example 11.6, we use $x_1 = 4243$, $n_1 = 82486$, $x_2 = 10701$, and $n_2 = 219170$.

Step 4. Select the form of the alternate hypothesis. For Example 11.6, the alternate hypothesis has the form $>\mathbf{p_2}$ (Figure B).

Step 5. Highlight **Calculate** and press **ENTER** (Figure C).

```
EDIT CALC TESTS
1:Z-Test...
2:T-Test...
3:2-SampZTest...
4:2-SampTTest...
5:1-PropZTest...
6:2-PropZTest...
7:ZInterval...
8:TInterval...
9↓2-SampZInt...
```

Figure A

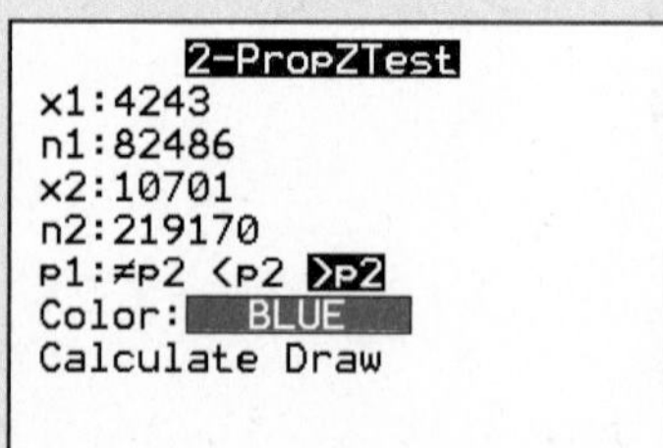

Figure B

```
2-PropZTest
p1>p2
z=2.948984202
p=.0015941698
p̂1=.0514390321
p̂2=.0488251129
p̂=.0495398732
n1=82486
n2=219170
```

Figure C

EXCEL

Finding the test statistic for a hypothesis test about the difference between two proportions

Figure D demonstrates the calculations involved for finding the test statistic for a hypothesis test about the difference between two proportions. In Example 11.6, $x_1 = 4243$, $n_1 = 82{,}486$, $x_2 = 10{,}701$, and $n_2 = 219{,}170$.

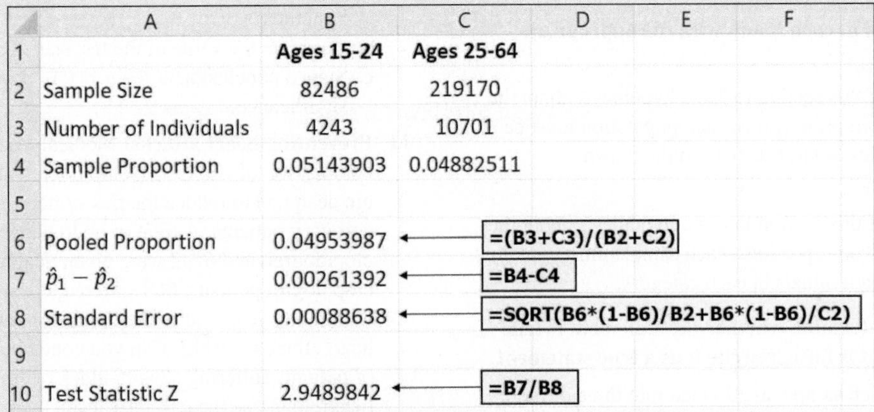

	A	B	C	D	E	F
1		**Ages 15-24**	**Ages 25-64**			
2	Sample Size	82486	219170			
3	Number of Individuals	4243	10701			
4	Sample Proportion	0.05143903	0.04882511			
5						
6	Pooled Proportion	0.04953987				
7	$\hat{p}_1 - \hat{p}_2$	0.00261392				
8	Standard Error	0.00088638				
9						
10	Test Statistic Z	2.9489842				

Figure D

MINITAB

Testing a hypothesis about the difference between two proportions

Step 1. Click on **Stat**, then **Basic Statistics**, then **2 Proportions**.

Step 2. Click **Summarized Data**, and enter the value of x_1 and n_1 for the **Number of events** and the **Number of trials** for sample 1. Enter the values x_2 and n_2 for the **Number of events** and the **Number of trials** for sample 2. For Example 11.6, we use $x_1 = 4243$, $n_1 = 82486$, $x_2 = 10701$, and $n_2 = 219170$ (Figure E).

Step 3. Click **Options**, and enter **0** in the **Hypothesized difference** field and select the form of the alternate hypothesis. Given significance level α, enter $100(1 - \alpha)$ as the **Confidence Level**. For Example 11.6, we use **95** as the **Confidence Level**, **0** as the **Test difference**, and **Difference > hypothesized difference** as the **Alternative**. Select the **Use the pooled estimate of the proportion** option. Click **OK**.

Step 4. Click **OK** (Figure F).

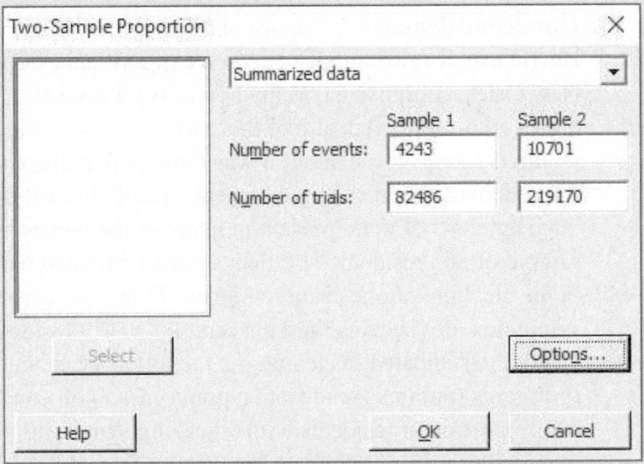

Figure E

Descriptive Statistics

Sample	N	Event	Sample p
Sample 1	82486	4243	0.051439
Sample 2	219170	10701	0.048825

Test

Null hypothesis H_0: $p_1 - p_2 = 0$
Alternative hypothesis H_1: $p_1 - p_2 > 0$

Method	Z-Value	P-Value
Normal approximation	2.95	0.002
Fisher's exact		0.002

Figure F

Section 11.3 Exercises

Exercises 1 and 2 are the Check Your Understanding exercises located within the section.

Understanding the Concepts

In Exercises 3 and 4, fill in each blank with the appropriate word or phrase.

3. To use the method of this section to test a hypothesis about the difference between two proportions, each population must be at least ________ times as large as the sample drawn from it. *20*

4. To use the method of this section to test a hypothesis about the difference between two proportions, each sample must contain at least ________ individuals in each category. *10*

In Exercises 5 and 6, determine whether the statement is true or false. If the statement is false, rewrite it as a true statement.

5. The individuals in each sample are divided into three or more categories. *False*

6. To compute the test statistic, it is necessary to compute the pooled proportion. *True*

Practicing the Skills

7. In a test for the difference between two proportions, the sample sizes were $n_1 = 120$ and $n_2 = 85$, and the numbers of events were $x_1 = 55$ and $x_2 = 45$. A test is made of the hypotheses H_0: $p_1 = p_2$ versus H_1: $p_1 \neq p_2$.
   a. Compute the value of the test statistic. *−1.00*
   b. Can you reject H_0 at the $\alpha = 0.05$ level of significance? *No*
   c. Can you reject H_0 at the $\alpha = 0.01$ level of significance? *No*

8. In a test for the difference between two proportions, the sample sizes were $n_1 = 68$ and $n_2 = 76$, and the numbers of events were $x_1 = 41$ and $x_2 = 25$. A test is made of the hypotheses H_0: $p_1 = p_2$ versus H_1: $p_1 > p_2$.
   a. Compute the value of the test statistic. *3.29*
   b. Can you reject H_0 at the $\alpha = 0.05$ level of significance? *Yes*
   c. Can you reject H_0 at the $\alpha = 0.01$ level of significance? *Yes*

Working with the Concepts

9. **Childhood obesity:** The National Health and Nutrition Examination Survey (NHANES) weighed a sample of 546 boys aged 6–11 and found that 87 of them were overweight. They weighed a sample of 508 girls aged 6–11 and found that 74 of them were overweight. Can you conclude that the proportion of boys who are overweight differs from the proportion of girls who are overweight?
   a. State the appropriate null and alternate hypotheses. H_0: $p_1 = p_2$, H_1: $p_1 \neq p_2$
   b. Compute the value of the test statistic. *0.62*
   c. State a conclusion at the $\alpha = 0.05$ level of significance. *Do not reject* H_0.

10. **Pollution and altitude:** In a random sample of 340 cars driven at low altitudes, 46 of them exceeded a standard of 10 grams of particulate pollution per gallon of fuel consumed. In an independent random sample of 85 cars driven at high altitudes, 21 of them exceeded the standard. Can you conclude that the proportion of high-altitude vehicles exceeding the standard is greater than the proportion of low-altitude vehicles exceeding the standard?
   a. State the appropriate null and alternate hypotheses. H_0: $p_1 = p_2$, H_1: $p_1 < p_2$
   b. Compute the value of the test statistic. *−2.53*
   c. State a conclusion at the $\alpha = 0.01$ level of significance. *Reject* H_0.

11. **Preventing heart attacks:** Medical researchers performed a comparison of two drugs, clopidogrel and ticagrelor, which are designed to reduce the risk of heart attack or stroke in coronary patients. A total of 6676 patients were given clopidogrel, and 6732 were given ticagrelor. Of the clopidogrel patients, 668 suffered a heart attack or stroke within one year, and of the ticagrelor patients, 569 suffered a heart attack or stroke. Can you conclude that the proportion of patients suffering a heart attack or stroke is less for ticagrelor? Use the $\alpha = 0.01$ level. *Reject* H_0.
   Source: *Lancet* 375:283–293

12. **Cholesterol:** An article in the *Archives of Internal Medicine* reported that in a sample of 244 men, 73 had elevated total cholesterol levels (more than 200 milligrams per deciliter). In a sample of 232 women, 44 had elevated cholesterol levels. Can you conclude that the proportion of people with elevated cholesterol levels differs between men and women? Use the $\alpha = 0.05$ level. *Reject* H_0.

13. **Treating circulatory disease:** Angioplasty is a medical procedure in which an obstructed blood vessel is widened. In some cases, a wire mesh tube, called a stent, is placed in the vessel to help it remain open. A study was conducted to compare the effectiveness of a bare metal stent with one that has been coated with a drug designed to prevent reblocking of the vessel. A total of 5320 patients received bare metal stents, and of these, 841 needed treatment for reblocking within a year. A total of 1120 received drug-coated stents, and 134 of them required treatment within a year. Can you conclude that the proportion of patients who needed retreatment is less for those who received drug-coated stents? Use the $\alpha = 0.05$ level. *Reject* H_0.
   Source: *Canadian Medical Association Journal* 180:167–174

14. **Hurricane damage:** In August and September 2005, Hurricanes Katrina and Rita caused extraordinary flooding in New Orleans, Louisiana. Many homes were severely damaged or destroyed, and of those that survived, many required extensive cleaning. It was thought that cleaning flood-damaged homes might present a health hazard due to the large amounts of mold present in many of the homes. In a sample of 365 residents of Orleans Parish who had participated in the cleaning of one or more homes, 77 had experienced symptoms of wheezing, and in a sample of 179 residents who had not participated in cleaning, 23 reported wheezing symptoms (numbers read from a graph). Can you conclude that the proportion of residents with wheezing symptoms is greater among those who participated in the cleaning of flood-damaged homes? Use the $\alpha = 0.05$ level. *Reject* H_0.
   Source: *American Journal of Public Health* 98:869–875

Jocelyn Augustino/FEMA

15. What's the key? A method of analyzing digital music files to determine the key in which the music was written was tested to determine its accuracy. In a sample of 307 pop music selections, the key was correctly identified in 245 of them. In a sample of 347 new age selections, the key was correctly identified in 304 of them. Can you conclude that the method is more accurate for new age music than for pop music? Use the $\alpha = 0.05$ level of significance. *Reject H_0.*

Source: *International Journal of Knowledge-based and Intelligent Engineering Systems* 15:165–175

16. Yellow light: When the light turns yellow, should you stop or go through it? A recent study of driver behavior defined the "indecision zone" as the period when a vehicle is between 2.5 and 5.5 seconds away from an intersection. At the intersection of Route 7 and North Shrewsbury in Clarendon, Vermont, 154 vehicles were observed to encounter a yellow light in the indecision zone, and 21 of them ran the red light. At the intersection of Route 62 and Paine Turnpike in Berlin, Vermont, 183 vehicles entered the intersection in the indecision zone, and 20 ran the red light. Can you conclude that the proportion of red light runners differs between the two intersections? Use the $\alpha = 0.01$ level of significance. *Do not reject H_0.*

Source: *Journal of Transportation Engineering* 137:277–286

17. Quit smoking: A recent study reported that in a sample of 230 European-American HIV-positive smokers, 102 of them had used a nicotine patch to try to quit smoking, and in a sample of 72 Hispanic-American HIV-positive smokers, 20 had used a nicotine patch. Can you conclude that the proportion of patch users is greater among European-Americans? Use the $\alpha = 0.01$ level of significance. *Reject H_0.*

Source: *American Journal of Health Behavior* 32:3–15

18. Parkinson's disease: A survey was conducted of patients with Parkinson's disease. Of 164 patients who said they exercised regularly, 76 reported falling within the previous six months. Of 96 patients who said they did not exercise regularly, 48 reported falling within the previous six months. Can you conclude that the proportion of patients who fall is less for those who exercise than for those who do not? Use the $\alpha = 0.05$ level of significance. *Do not reject H_0.*

Source: *Physical Therapy* 91:1838–1848

19. Don't perform a hypothesis test: In a certain year, there was measurable snowfall on 80 out of 365 days in Denver, and 63 out of 365 days in Chicago. A meteorologist proposes to perform a test of the hypothesis that the proportions of days with snow are equal in the two cities. Explain why this cannot be done using the method presented in this section. *Samples are not independent.*

20. Don't perform a hypothesis test: A new reading program is being tested. Parents are asked whether they would like to enroll their children, and 50 children are enrolled in the program. There are 45 children whose parents do not choose to enroll their children. At the end of the school year, the children are tested. Of the 50 children who participated in the program, 38 are found to be reading at grade level. Of the 45 children who did not participate, 24 were reading at grade level. Explain why these data should not be used to test the hypothesis that the proportion of children reading at grade level is higher for those who participate in the program. *Samples are not random samples.*

21. Interpret calculator display: The following TI-84 Plus calculator display presents the results of a hypothesis test for the difference between two proportions. The sample sizes are $n_1 = 165$ and $n_2 = 152$.

```
2-PropZTest
p1>p2
z=2.673676852
p=.0037512809
p̂1=.3757575758
p̂2=.2368421053
p̂=.309148265
n1=165
n2=152
```

a. Is this a left-tailed test, a right-tailed test, or a two-tailed test? *Right-tailed*

b. What is the P-value? *0.0037512809*

c. Can you reject H_0 at the $\alpha = 0.05$ level? *Yes*

22. Interpret calculator display: The following TI-84 Plus calculator display presents the results of a hypothesis test for the difference between two proportions. The sample sizes are $n_1 = 71$ and $n_2 = 62$.

```
2-PropZTest
p1≠p2
z=.8141749055
p=.4155446361
p̂1=.6338028169
p̂2=.564516129
p̂=.6015037594
n1=71
n2=62
```

a. Is this a left-tailed test, a right-tailed test, or a two-tailed test? *Two-tailed*

b. What is the P-value? *0.4155446361*

c. Can you reject H_0 at the $\alpha = 0.05$ level? *No*

23. Interpret computer output: The following computer output (from MINITAB) presents the results of a hypothesis test on the difference between two proportions.

```
Test and CI for Two Proportions: P1, P2

Variable    X    N    Sample p
P1         22   43    0.511628
P2         55   93    0.591398

Difference = p (P1) - p (P2)
Estimate for difference: -0.079770
95% upper bound for difference: 0.071079
Test for difference = 0 (vs < 0):  Z = -0.87  P-Value = 0.192
```

a. Is this a left-tailed test, a right-tailed test, or a two-tailed test? *Left-tailed*

b. What is the P-value? 0.192

c. Can H_0 be rejected at the 0.05 level? Explain. No

24. Interpret computer output: The following computer output (from MINITAB) presents the results of a hypothesis test on the difference between two proportions.

```
Test and CI for Two Proportions: P1, P2

Variable    X     N    Sample p
P1         405   577   0.701906
P2         363   578   0.628028

Difference = p (P1) − p (P2)
Estimate for difference: 0.073879
95% CI for difference: (0.0194191, 0.127827)
Test of difference = 0 (vs not = 0):   Z = −2.66  P-Value = 0.008
```

a. Is this a left-tailed test, a right-tailed test, or a two-tailed test? Two-tailed

b. What is the P-value? 0.008

c. Can H_0 be rejected at the 0.05 level? Explain. Yes

Extending the Concepts

Null difference other than 0: Occasionally someone may wish to test a hypothesis of the form H_0: $p_1 - p_2 = p_d$, where $p_d \neq 0$. In this situation, the null hypothesis says that the population proportions are unequal, so we do not compute the pooled proportion, which assumes the population proportions are equal. One approach to testing this hypothesis is to use the test statistic

$$z = \frac{(\hat{p}_1 - \hat{p}_2) - p_d}{\sqrt{\dfrac{\hat{p}_1(1-\hat{p}_1)}{n_1} + \dfrac{\hat{p}_2(1-\hat{p}_2)}{n_2}}}$$

When the assumptions of this section are met, this statistic has approximately a standard normal distribution when H_0 is true.

25. Computer chips: A computer manufacturer has a choice of two machines, a less expensive one and a more expensive one, to manufacture a particular computer chip. Out of 500 chips manufactured on the less expensive machine, 70 were defective. Out of 400 chips manufactured on the more expensive machine, only 20 were defective. The manufacturer will buy the more expensive machine if he is convinced that the proportion of defectives is more than 5% less than on the less expensive machine. Let p_1 represent the proportion of defectives produced by the less expensive machine, and let p_2 represent the proportion of defectives produced by the more expensive machine.

a. State appropriate null and alternate hypotheses. H_0: $p_1 - p_2 = 0.05$, H_1: $p_1 - p_2 > 0.05$

b. Compute the value of the test statistic. 2.11

c. Can you reject H_0 at the $\alpha = 0.05$ level? Yes

d. Which machine should the manufacturer buy? More expensive

Answers to Check Your Understanding Exercises for Section 11.3

1. a. H_0: $p_1 = p_2$, H_1: $p_1 > p_2$
   b. $\hat{p}_1 = 0.76$, $\hat{p}_2 = 0.64$ c. $z = 2.10$ d. $P = 0.0179$
   e. If H_0 is true, the probability of observing a value for the test statistic as extreme as or more extreme than the value actually observed is 0.0179. This is unusual, so the evidence against H_0 is strong. Because $P < 0.05$, we reject H_0 at the $\alpha = 0.05$ level.
   f. We conclude that the proportion of patients experiencing substantial relief is greater for drug 1.
2. a. H_0: $p_1 = p_2$, H_1: $p_1 > p_2$
   b. $\hat{p}_1 = 0.475$, $\hat{p}_2 = 0.420$ c. $z = 1.02$
   d. $P = 0.1539$ [Tech: 0.1531]
   e. If H_0 is true, the probability of observing a value for the test statistic as extreme as or more extreme than the value actually observed is 0.1539. This is not unusual, so the evidence against H_0 is not strong. Because $P > 0.05$, we do not reject H_0 at the $\alpha = 0.05$ level.
   f. We conclude that the proportions of younger and older voters that support an increase in Social Security benefits may be the same.

Section 11.4 Hypothesis Tests for Two Population Standard Deviations

Objectives

1. Find critical values of the F distribution
2. Perform a hypothesis test for two population standard deviations

Objective 1 Find critical values of the F distribution

Find Critical Values of the F Distribution

The tests we have studied so far have involved means or proportions. Occasionally it may be desirable to test a null hypothesis that two populations have equal standard deviations. In general, there is no good way to do this. In the special case where both populations are normal, however, a method is available.

CAUTION

The test for two standard deviations is sensitive to the assumption of normality. If the populations differ even slightly from normal, the test should not be used.

Assumptions for Performing a Hypothesis Test for Two Population Standard Deviations

1. We have independent random samples from two populations.
2. Both populations are normal.

In what follows, σ_1 and σ_2 will denote the standard deviations of the two populations, s_1 and s_2 will denote the sample standard deviations, and n_1 and n_2 will denote the sample sizes.

Any of three hypotheses may be tested. They are

$$H_0: \sigma_1 = \sigma_2 \qquad H_1: \sigma_1 > \sigma_2$$
$$H_0: \sigma_1 = \sigma_2 \qquad H_1: \sigma_1 < \sigma_2$$
$$H_0: \sigma_1 = \sigma_2 \qquad H_1: \sigma_1 \neq \sigma_2$$

The test statistic used in this test is one that we have not seen before; it is denoted by the letter F. We compute F using the sample variances s_1^2 and s_2^2. We divide the larger sample variance by the smaller sample variance:

$$F = \frac{\text{Larger of } s_1^2 \text{ and } s_2^2}{\text{Smaller of } s_1^2 \text{ and } s_2^2}$$

When H_0 is true, $\sigma_1 = \sigma_2$. Therefore, the sample standard deviations s_1 and s_2 are, on average, approximately the same size, so the test statistic F is likely to be near 1. If the test statistic is much greater than 1, we will reject H_0. In order to use F as a test statistic, we must know its distribution when H_0 is true. This distribution is called an ***F* distribution**, which we now describe.

The *F* distribution

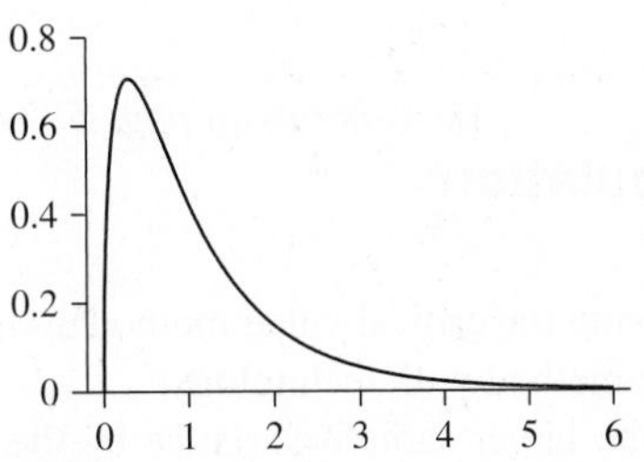

Figure 11.4 The F distribution is skewed to the right.

Statistics that have an F distribution are ratios of quantities, such as the ratio of two sample variances. The F distribution, therefore, has two values for the degrees of freedom: one associated with the numerator, and one associated with the denominator. The degrees of freedom are indicated with subscripts under the letter F. For example, the symbol $F_{3,16}$ denotes the F distribution with 3 degrees of freedom for the numerator and 16 degrees of freedom for the denominator. Note that the degrees of freedom for the numerator are always listed first. Unlike the normal and Student's t distributions, which are symmetric, the F distribution is skewed to the right. Figure 11.4 presents an F curve.

The level α critical value of the F distribution is the value for which the area in the tail to the right of the critical value is α. We will denote the level α critical value by f_α. We will not need to find critical values for left tails.

Example 11.7

Find a critical value of an *F* distribution

Use Table A.5 to find $f_{0.05}$, the $\alpha = 0.05$ critical value, for the $F_{3,16}$ distribution.

Solution

In Table A.5, we find the column corresponding to 3 degrees of freedom for the numerator, and the rows corresponding to 16 degrees of freedom for the denominator. We choose the row for which the tail area is 0.05. The critical value is $f_{0.05} = 3.24$.

Denominator Degrees of Freedom	Area	Numerator Degrees of Freedom 1	2	3	4	5	6	7	8	9
15	0.100	3.07	2.70	2.49	2.36	2.27	2.21	2.16	2.12	2.09
15	0.050	4.54	3.68	3.29	3.06	2.90	2.79	2.71	2.64	2.59
15	0.025	6.20	4.77	4.15	3.80	3.58	3.41	3.29	3.20	3.12
15	0.010	8.68	6.36	5.42	4.89	4.56	4.32	4.14	4.00	3.89
15	0.001	16.59	11.34	9.34	8.25	7.57	7.09	6.74	6.47	6.26
16	0.100	3.05	2.67	2.46	2.33	2.24	2.18	2.13	2.09	2.06
16	0.050	4.49	3.63	3.24	3.01	2.85	2.74	2.66	2.59	2.54
16	0.025	6.12	4.69	4.08	3.73	3.50	3.34	3.22	3.12	3.05
16	0.010	8.53	6.23	5.29	4.77	4.44	4.20	4.03	3.89	3.78
16	0.001	16.12	10.97	9.01	7.94	7.27	6.80	6.46	6.19	5.98
17	0.100	3.03	2.64	2.44	2.31	2.22	2.15	2.10	2.06	2.03
17	0.050	4.45	3.59	3.20	2.96	2.81	2.70	2.61	2.55	2.49
17	0.025	6.04	4.62	4.01	3.66	3.44	3.28	3.16	3.06	2.98
17	0.010	8.40	6.11	5.18	4.67	4.34	4.10	3.93	3.79	3.68
17	0.001	15.72	10.66	8.73	7.68	7.02	6.56	6.22	5.96	5.75

Figure 11.5 presents the probability density curve of the $F_{3,16}$ distribution with the $\alpha = 0.05$ critical region shaded. The $\alpha = 0.05$ critical value is $f_{0.05} = 3.24$.

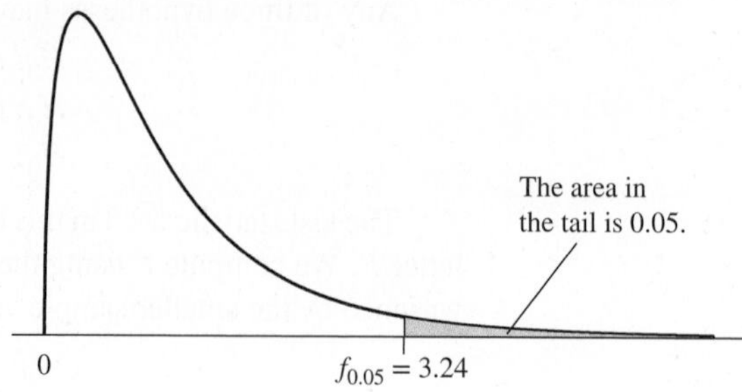

Figure 11.5 Probability density curve of the $F_{3,16}$ distribution. The $\alpha = 0.05$ critical value is $f_{0.05} = 3.24$.

Check Your Understanding

1. Find $f_{0.05}$ for $F_{8,18}$. *2.51*

2. Find $f_{0.01}$ for $F_{12,8}$. *5.67*

Answers are on page 577.

Objective 2 Perform a hypothesis test for two population standard deviations

Perform a Hypothesis Test for Two Population Standard Deviations

We will first describe how to perform a hypothesis test using the critical value method and Table A.5. Then we will describe how to use the P-value method with technology.

Because we compute the test statistic by dividing the larger sample variance by the smaller sample variance, the value of the test statistic is never less than 1. Therefore, the test statistic cannot fall into the left tail of the distribution. The critical region for a one-tailed test (either right- or left-tailed) is $F \geq f_\alpha$. The critical region for a two-tailed test differs from that of other tests. For most other two-tailed tests, the critical region consists of an area of $\alpha/2$ in both left and right tails. For the F-test, however, the test statistic never falls into the left tail, so the critical region for a two-tailed test consists of only the right tail. We reject H_0 when $F > f_{\alpha/2}$.

We can perform a hypothesis test for two population standard deviations using the following steps.

Performing a Hypothesis Test for Two Standard Deviations

Check to be sure the assumptions are satisfied. If they are, then proceed with the following steps.

Step 1: State the null and alternate hypotheses. The null hypothesis is $H_0\colon \sigma_1 = \sigma_2$. The alternate hypothesis can be stated in one of three ways:

Left-tailed: $H_1\colon \sigma_1 < \sigma_2$
Right-tailed: $H_1\colon \sigma_1 > \sigma_2$
Two-tailed: $H_1\colon \sigma_1 \neq \sigma_2$

Step 2: Choose a significance level α and find the critical value, as follows:

One-tailed:	$H_1\colon \sigma_1 < \sigma_2$ or $H_1\colon \sigma_1 > \sigma_2$	The critical value is f_α.
Two-tailed:	$H_1\colon \sigma_1 \neq \sigma_2$	The critical value is $f_{\alpha/2}$.

Step 3: Compute the test statistic $F = \dfrac{\text{Larger of } s_1^2 \text{ and } s_2^2}{\text{Smaller of } s_1^2 \text{ and } s_2^2}$.

Step 4: Determine whether to reject H_0, as follows:

One-tailed:	$H_1\colon \sigma_1 < \sigma_2$ or $H_1\colon \sigma_1 > \sigma_2$	Reject if $F \geq f_\alpha$.
Two-tailed:	$H_1\colon \sigma_1 \neq \sigma_2$	Reject if $F \geq f_{\alpha/2}$.

Step 5: State a conclusion.

Example 11.8

Perform a hypothesis test

In a series of experiments to determine the absorption rate of certain pesticides into skin, two pesticides were applied to several skin specimens. After a time, the amounts absorbed (in micrograms) were measured. For pesticide 1, the variance of the amounts absorbed in 6 specimens was 2.3, while for pesticide 2, the variance of the amounts absorbed in 10 specimens was 0.6. Assume that for each pesticide, the amounts absorbed are a simple random sample from a normal population. Can we conclude that the standard deviation of the amount absorbed is greater for pesticide 1 than for pesticide 2? Use the $\alpha = 0.05$ significance level.

Solution

We first check the assumptions. We have two independent samples from normal populations, so the assumptions are satisfied.

Step 1: State the null and alternate hypotheses. Because we are interested in determining whether $\sigma_1 > \sigma_2$, the hypotheses are

$$H_0\colon \sigma_1 = \sigma_2 \qquad H_1\colon \sigma_1 > \sigma_2$$

Step 2: Choose a significance level and find the critical value. We will use a significance level of $\alpha = 0.05$. To find the critical value, we first find the degrees of freedom. The value 2.3 in the numerator of the test statistic came from a sample of size 6. Therefore, there are $6 - 1 = 5$ degrees of freedom for the numerator. The value 0.6 in the denominator of the test statistic came from a sample of size 10. Therefore, there are $10 - 1 = 9$ degrees of freedom for the denominator. This is a one-tailed test, so the critical value is $f_{0.05}$. We consult Table A.5 with 5 and 9 degrees of freedom and a tail area of 0.05. The critical value is $f_{0.05} = 3.48$.

Denominator Degrees of Freedom	Area	Numerator Degrees of Freedom 1	2	3	4	5	6	7	8	9
8	0.100	3.46	3.11	2.92	2.81	2.73	2.67	2.62	2.59	2.56
8	0.050	5.32	4.46	4.07	3.84	3.69	3.58	3.50	3.44	3.39
8	0.025	7.57	6.06	5.42	5.05	4.82	4.65	4.53	4.43	4.36
8	0.010	11.26	8.65	7.59	7.01	6.63	6.37	6.18	6.03	5.91
8	0.001	25.41	18.49	15.83	14.39	13.48	12.86	12.40	12.05	11.77
9	0.100	3.36	3.01	2.81	2.69	2.61	2.55	2.51	2.47	2.44
9	0.050	5.12	4.26	3.86	3.63	3.48	3.37	3.29	3.23	3.18
9	0.025	7.21	5.71	5.08	4.72	4.48	4.32	4.20	4.10	4.03
9	0.010	10.56	8.02	6.99	6.42	6.06	5.80	5.61	5.47	5.35
9	0.001	22.86	16.39	13.90	12.56	11.71	11.13	10.70	10.37	10.11
10	0.100	3.29	2.92	2.73	2.61	2.52	2.46	2.41	2.38	2.35
10	0.050	4.96	4.10	3.71	3.48	3.33	3.22	3.14	3.07	3.02
10	0.025	6.94	5.46	4.83	4.47	4.24	4.07	3.95	3.85	3.78
10	0.010	10.04	7.56	6.55	5.99	5.64	5.39	5.20	5.06	4.94
10	0.001	21.04	14.91	12.55	11.28	10.48	9.93	9.52	9.20	8.96

Step 3: Compute the test statistic. The larger sample variance is 2.3 and the smaller one is 0.6. The test statistic is

$$F = \frac{2.3}{0.6} = 3.83$$

Step 4: Determine whether to reject H_0. The value of the test statistic is $F = 3.83$. The critical value is $f_{0.05} = 3.48$. Because $F > f_{0.05}$, we reject H_0 at the $\alpha = 0.05$ level.

Step 5: State a conclusion. We conclude that the standard deviation of the amount absorbed is greater for pesticide 1 than for pesticide 2.

Check Your Understanding

3. For testing H_0: $\sigma_1 = \sigma_2$ versus H_1: $\sigma_1 < \sigma_2$, we observe $s_1 = 0.9$, $s_2 = 1.5$, $n_1 = 16$, and $n_2 = 11$. Test at the significance level $\alpha = 0.01$.

a. Find the critical value. *3.80*

b. Do you reject H_0? *No*

4. For testing H_0: $\sigma_1 = \sigma_2$ versus H_1: $\sigma_1 \neq \sigma_2$, we observe $s_1 = 9.3$, $s_2 = 2.1$, $n_1 = 6$, and $n_2 = 13$. Test at the significance level $\alpha = 0.05$.

a. Find the critical value. *3.89*

b. Do you reject H_0? *Yes*

Answers are on page 577.

Performing hypothesis tests using the *P*-value method

We show how to perform tests for two population standard deviations using the *P*-value method. We calculate the *P*-value with technology.

The following TI-84 Plus display presents the results for the test described in Example 11.8.

```
2-SampFTest
σ1>σ2
F=3.833426299
p=.0387586595
Sx1=1.5166
Sx2=.7746
n1=6
n2=10
```

The first line specifies the alternate hypothesis. Then follows the test statistic *F*, the *P*-value, and the sample standard deviations ($1.5166 = \sqrt{2.3}$ and $0.7746 = \sqrt{0.6}$). The *P*-value is 0.039, rounded to three decimal places. We reject H_0 at the $\alpha = 0.05$ level.

The *F*-test is sensitive to the normality assumption

The *F*-test, like the *t*-test, requires that the samples come from normal populations. Unlike the *t*-test, the *F*-test for comparing variances is very sensitive to this assumption. If the shapes of the populations differ much from the normal curve, the *F*-test may give misleading results. For this reason, the *F*-test for comparing variances must be used with caution.

Using Technology

We use Example 11.8 to illustrate the technology steps. The TI-84 Plus calculator and EXCEL require as input the standard deviations. In Example 11.8, the variances $s_1^2 = 2.3$ and $s_2^2 = 0.6$ are given. We find the standard deviations by taking the square root: $s_1 = \sqrt{2.3} = 1.5166$; $s_2 = \sqrt{0.6} = 0.7746$.

TI-84 PLUS

Testing a hypothesis about two population standard deviations

Step 1. Press **STAT** and highlight the **TESTS** menu.

Step 2. Select **2–SampFTest** and press **ENTER** (Figure A). The **2–SampFTest** menu appears.

Step 3. Choose one of the following:

- If the summary statistics are given, select **Stats** as the **Inpt** option and enter s_1, n_1, s_2, n_2. For Example 11.8, we use $s_1 = 1.5166$, $n_1 = 6$, $s_2 = 0.7746$, $n_2 = 10$.
- If the raw data are given, select **Data** as the **Inpt** option and enter the location of the data as the **List1** and **List2** options.

```
EDIT CALC TESTS
7↑ZInterval…
8:TInterval…
9:2-SampZInt…
0:2-SampTInt…
A:1-PropZInt…
B:2-PropZInt…
C:χ²-Test…
D:χ²GOF-Test…
E↓2-SampFTest…
```

Figure A

```
2-SampFTest
Inpt:Data Stats
Sx1:1.5166
n1:6
Sx2:.7746
n2:10
σ1:≠σ2 <σ2 >σ2
Color: BLUE
Calculate Draw
```

Figure B

Step 4. Select the form of the alternate hypothesis. For Example 11.8, the alternate hypothesis has the form **>σ2** (Figure B).

Step 5. Highlight **Calculate** and press **ENTER** (Figure C).

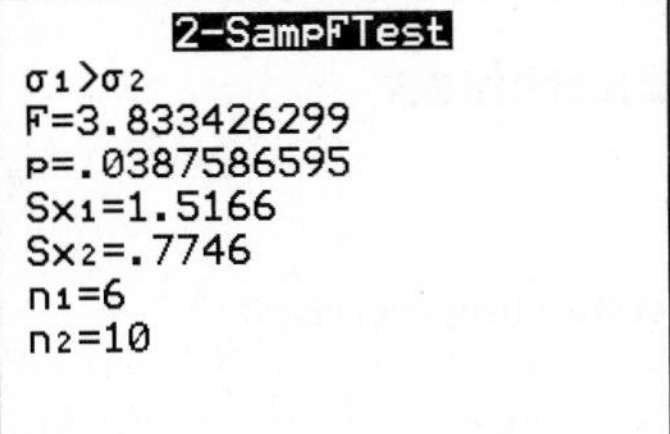

Figure C

EXCEL

Testing a hypothesis about two standard deviations

Figure D demonstrates the calculations involved for finding the test statistic for a hypothesis test about two standard deviations. In Example 11.8, we test H_0: $\sigma_1 = \sigma_2$ versus H_1: $\sigma_1 > \sigma_2$, where $s_1^2 = 2.3$, $n_1 = 6$, $s_2^2 = 0.6$, $n_2 = 10$.

	A	B	C	D	E
1		**Pesticide 1**	**Pesticide 2**		
2	Variance	2.3	0.6		
3	Sample Size	6	10		
4					
5	Test Statistic F	3.833333333 ←	**=B2/C2**		
6					
7	P-Value (one-tailed)	0.038761201 ←	**=F.DIST.RT(B5,B3-1,C3-1)**		
8					

Figure D

MINITAB

Testing a hypothesis about two population standard deviations

Step 1. Click on **Stat**, then **Basic Statistics**, then **2 Variances**.

Step 2. Choose one of the following:

- If the summary statistics are given, select **Sample Variances**, and enter the **Sample Size** and the **Variance** for each sample. For Example 11.8, we use $s_1^2 = 2.3$, $n_1 = 6$, $s_2^2 = 0.6$, $n_2 = 10$.
- If the raw data are given, select **Each sample is in its own column** and select the columns that contain the data.

Select Options, and select the form of the alternate hypothesis. For Example 11.8, we use **Ratio >** hypothesized value.

Step 3. Click **OK** (Figure E).

Test

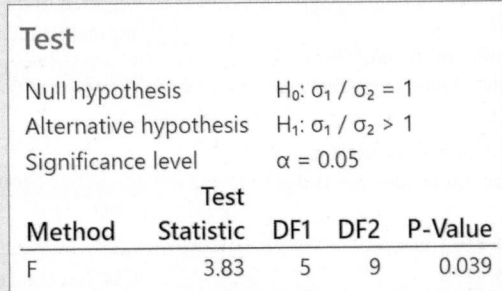

Null hypothesis	H_0: $\sigma_1 / \sigma_2 = 1$
Alternative hypothesis	H_1: $\sigma_1 / \sigma_2 > 1$
Significance level	$\alpha = 0.05$

Method	Test Statistic	DF1	DF2	P-Value
F	3.83	5	9	0.039

Figure E

Section 11.4 Exercises

Exercises 1–4 are the Check Your Understanding exercises located within the section.

Understanding the Concepts

In Exercises 5 and 6, fill in each blank with the appropriate word or phrase.

5. To use an F-test, both populations must have a __________ distribution. *normal*
6. The F statistic is the quotient of the sample __________. *variances*

In Exercises 7 and 8, determine whether the statement is true or false. If the statement is false, rewrite it as a true statement.

7. The test statistic F is never greater than 1. *False*
8. The F-test for two standard deviations is very sensitive to the normality assumption. *True*

Practicing the Skills

9. Find the critical value $f_{0.05}$ for $F_{7,20}$. *2.51*
10. Find the critical value $f_{0.01}$ for $F_{2,5}$. *13.27*
11. An F-test with 12 degrees of freedom in the numerator and 6 degrees of freedom in the denominator produced a test statistic whose value was 3.42. The null and alternate hypotheses were H_0: $\sigma_1 = \sigma_2$ versus H_1: $\sigma_1 < \sigma_2$.
    a. Do you reject H_0 at the $\alpha = 0.05$ level? *No*
    b. Do you reject H_0 at the $\alpha = 0.01$ level? *No*
12. An F-test with 5 degrees of freedom in the numerator and 7 degrees of freedom in the denominator produced a test statistic whose value was 5.31. The null and alternate hypotheses were H_0: $\sigma_1 = \sigma_2$ versus H_1: $\sigma_1 > \sigma_2$.
    a. Do you reject H_0 at the $\alpha = 0.05$ level? *Yes*
    b. Do you reject H_0 at the $\alpha = 0.01$ level? *No*

Working with the Concepts

13. **Sugar content:** A broth used to manufacture a pharmaceutical product has its sugar content, in milligrams per milliliter, measured several times on each of three successive days.

    Day 1: 5.0 4.8 5.1 5.1 4.8 5.1 4.8 4.8 5.0
    5.2 4.9 4.9 5.0
    Day 2: 5.8 4.7 4.7 4.9 5.1 4.9 5.4 5.3 5.3
    4.8 5.7 5.1 5.7
    Day 3: 6.3 4.7 5.1 5.9 5.1 5.9 4.7 6.0 5.3
    4.9 5.7 5.3 5.6

    a. Can you conclude that the variability of the process is greater on the second day than on the first day? Use the $\alpha = 0.05$ level of significance. *Yes*
    b. Can you conclude that the variability of the process is greater on the third day than on the second day? Use the $\alpha = 0.01$ level of significance. *No*
14. **Hockey sticks:** The breaking strength of hockey stick shafts made of two different graphite-kevlar composites yields the following results (in newtons).

    Composite 1: 487.3 444.5 467.7 456.3 449.7
    459.2 478.9 461.5 477.2
    Composite 2: 488.5 501.2 475.3 467.2 462.5
    499.7 470.0 469.5 481.5 485.2
    509.3 479.3 498.3

    Can you conclude that the standard deviation of the breaking strength differs between the two composites? Use the $\alpha = 0.05$ level of significance. *No*
15. **Thread strength:** Two types of thread are being considered for use in personal flotation devices. Breaking strengths, in newtons, of ten threads of each type were measured, with the following results.

    Type A: 43 52 52 60 49 52 39 52 56 54
    Type B: 49 45 49 56 52 45 47 56 56 41

    Can you conclude that the standard deviations of the breaking strength differ between the two types? Use the $\alpha = 0.05$ level of significance. *No*
16. **Salty chocolate:** Measurements of the sodium content in samples of two brands of chocolate yielded the following results (in grams).

    Brand A: 34 31 37 28 33 32 34 34 30
    Brand B: 46 40 42 40 36 38 34 42 32
    32 34 22 44

    Can you conclude that the sodium content is more variable in Brand B? Use the $\alpha = 0.01$ level of significance. *Yes*
17. **Frozen computer:** A computer system administrator notices that computers running a particular operating system seem to freeze up more often as the installation of the operating system ages. She measures the time (in minutes) before freeze-up for 7 computers one month after installation, and for 9 computers seven months after installation. The results are as follows.

    One month: 207.4 233.1 215.9 235.1 225.6
    244.4 245.3
    Seven months: 84.3 53.2 127.3 201.3 174.2
    246.2 149.4 156.4 103.3

    Can you conclude that the time to freeze-up is more variable in the seventh month than in the first month after installation? Use the $\alpha = 0.01$ level of significance. *Yes*
18. **Are you smarter than your older brother?** In a study of birth order and intelligence, IQ tests were given to 18- and 19-year-old men to estimate the size of the difference, if any, between the mean IQs of firstborn sons and secondborn sons. The following data for 10 firstborn sons and 10 secondborn sons are consistent with the means and standard deviations reported in the article. Assume that the samples come from normal populations.

Firstborn				
104	82	102	96	129
89	114	107	89	103

Secondborn				
103	103	91	113	102
103	92	90	114	113

Can you conclude that the standard deviation of IQ differs between firstborn and secondborn sons? Use the $\alpha = 0.05$ level. *No*

Source: *Science* 316:1717

Extending the Concepts

19. Left tail: It is desired to find the value $f_{0.99}$ for an $F_{10,5}$ distribution. Use Table A.5 and the $F_{5,10}$ distribution to find this value. *0.177*

Answers to Check Your Understanding Exercises for Section 11.4

1. 2.51
2. 5.67
3. a. 3.80 **b.** No
4. a. 3.89 **b.** Yes

Section 11.5 The Multiple Testing Problem

Objectives

1. Describe the multiple testing problem
2. Use the Bonferroni method to adjust the P-value

Objective 1 Describe the multiple testing problem

NOTE TO INSTRUCTOR

The material in this section is highly relevant to statistical practice and is worth emphasizing. There are many situations in practice in which several hypotheses are tested, and it is important to know how to take this into account when interpreting the results.

The Multiple Testing Problem

Rates of diseases such as cancer are monitored by the Centers for Disease Control and Prevention (CDC). According to a recent CDC study, the nationwide rate of new cancer cases in a recent year was 487 per 100,000 people. Many scientists believe that some of these cancers are caused by exposure to substances in the environment. How can we determine which environmental substances increase the risk of cancer? One way is to measure cancer rates in various local regions. Some regions turn out to have cancer rates that are significantly larger than the nationwide rate. These regions are sometimes referred to as cancer "hot spots." When a hot spot is discovered, scientists can investigate the environmental conditions there to try to find possible cancer-causing agents.

A natural way to discover hot spots is through hypothesis testing. A region whose cancer rate is statistically significant and higher than the national average is a candidate for a hot spot. It would seem, then, that a good way to proceed would be to perform hypothesis tests on a large number of regions, then look for cancer-causing agents in those whose rates are significantly higher than the nationwide rate. Unfortunately, this approach doesn't work very well, because of a phenomenon known as the **multiple testing problem.**

The multiple testing problem is this: *As more hypothesis tests are performed, small P-values become less meaningful.* We illustrate this point with the cancer hot spot example. Suppose that 20 cities are chosen, and the cancer rates are measured in each. Because we are interested in determining the cities in which the rate per 100,000 people is above the national average of 487, we perform, in each city, a test of the hypotheses

$$H_0\colon \mu = 487 \qquad H_1\colon \mu > 487$$

We will use a significance level of 0.05, so we will reject H_0 whenever $P \leq 0.05$.

Now suppose that for 19 of the 20 cities, the P-value is greater than 0.05, so H_0 is not rejected. For one city, $P \leq 0.05$, so H_0 is rejected. It might seem reasonable to conclude that this city really does have a cancer rate that is above the national average. However, this conclusion is not justified. Examples 11.9 and 11.10 show why.

Example 11.9

Find the probability of a false rejection

If only one city were tested, and the true mean for this city was equal to the national average of 487, what is the probability that H_0 would be rejected, leading to a wrong conclusion?

Solution

If the true mean is 487, H_0 is true. Rejecting H_0 is then a Type I error. The question is therefore asking for the probability of a Type I error. This probability is equal to the significance level of the test, which in this case is 0.05. The probability is 0.05 that H_0 will be rejected.

Example 11.10

Find the probability of at least one false rejection with multiple tests

If 20 cities are tested, and H_0 is rejected for one of the 20 cities, is it plausible that the mean rate for this city is actually equal to 487?

Solution

Yes. It is quite plausible that all of the cities, including the one for which H_0 was rejected, have mean rates equal to 487. There were 20 hypothesis tests made. For each test, there was a probability of 0.05 (that is, 1 chance in 20) of a Type I error. Therefore, we expect, on the average, that out of every 20 true null hypotheses, one will be rejected. So rejecting H_0 in one out of the 20 tests is exactly what one would expect in the case that all of the cities had a mean equal to 487.

Examples 11.9 and 11.10 illustrate the multiple testing problem. Put simply, the multiple testing problem is this: When H_0 is rejected, we have strong evidence that it is false. But strong evidence is not certainty. Occasionally a true null hypothesis will be rejected. When many tests are performed, it is more likely that some true null hypotheses will be rejected, resulting in Type I errors. Thus, when many tests are performed, it is difficult to tell which of the rejected null hypotheses are really false and which correspond to Type I errors.

Objective 2 Use the Bonferroni method to adjust the *P*-value

The Bonferroni Method

The **Bonferroni method** provides a way to adjust *P*-values upward when several hypothesis tests are performed. If a *P*-value remains small after the adjustment, the null hypothesis may be rejected. To make the Bonferroni adjustment, simply multiply the *P*-value by the number of tests performed. Here are two examples.

Example 11.11

Use the Bonferroni method to adjust the *P*-value

Four different cities are tested to determine whether their cancer rates are above 487 per 100,000 people. The null hypothesis H_0: $\mu = 487$ versus H_1: $\mu > 487$ is tested for each city, and the results are

City A: $P = 0.37$
City B: $P = 0.41$
City C: $P = 0.005$
City D: $P = 0.21$

Scientists suspect that city C may be a hot spot, but they know that the *P*-value of 0.005 is unreliable, because several tests have been performed. Use the Bonferroni adjustment to produce a reliable *P*-value.

Solution

Four tests were performed, so the Bonferroni adjustment yields $P = (4)(0.005) = 0.02$ for city C. So the evidence is reasonably strong that city C is in fact a hot spot.

Example 11.12

Interpret an adjusted *P*-value

In Example 11.11, assume the *P*-value for city C had been 0.03 instead of 0.005. What conclusion would you reach in that case?

Solution

The Bonferroni adjustment would yield $P = (4)(0.03) = 0.12$. This is probably not strong enough evidence to conclude that city C is in fact a hot spot. Because the original *P*-value was small, however, it is likely that one would still be suspicious about city C.

The Bonferroni adjustment is conservative; in other words, the P-value it produces is never smaller than the true P-value. So when the Bonferroni-adjusted P-value is small, the null hypothesis can be rejected conclusively. Unfortunately, as Example 11.12 shows, there are many occasions in which the original P-value is small enough to arouse a strong suspicion that a null hypothesis may be false, but the Bonferroni adjustment does not allow the hypothesis to be rejected.

When the Bonferroni-adjusted P-value is too large to reject a null hypothesis, yet the original P-value is small enough to lead one to suspect that the hypothesis is in fact false, often the best thing to do is to retest the hypothesis that appears to be false, using data from a new experiment. For a suspected cancer hot spot, an appropriate procedure would be to collect new data from that city at a future time and test the hypothesis again. If the P-value is again small, this time without multiple tests, this provides real evidence against the null hypothesis.

Check Your Understanding

1. Five tests are performed, and the P-values are 0.02, 0.11, 0.23, 0.38, and 0.45. Use the Bonferroni adjustment to adjust the P-value of 0.02. *0.10*

2. Six tests are performed, and the smallest P-value is 0.03. Which is the best conclusion? *iii*

i. Reject H_0 because $P < 0.05$.

ii. Do not reject H_0 because the Bonferroni adjustment yields $P = 6(0.03) = 0.18$, which is greater than 0.05.

iii. Do not reject H_0, but collect new data and repeat the test.

Answers are on page 580.

Report all *P*-values

The multiple testing problem shows us that to properly interpret a P-value we must know how many tests were performed. For example, imagine that a test produces a P-value of 0.005. If this is the only test performed, it is reasonable to reject the null hypothesis. However, if this is only one of 50 tests performed, it is not clear that we should reject H_0. When performing several tests, it is tempting to report only those with small P-values.

Reporting only small P-values, without disclosing the number of tests performed, is known as "*p*-hacking," and is misleading. For the results of a hypothesis test to be useful, it is important to report the number of tests conducted, and how the values to be reported were chosen.

Section 11.5 Exercises

Exercises 1 and 2 are the Check Your Understanding exercises located within the section.

Understanding the Concepts

In Exercises 3 and 4, fill in each blank with the appropriate word or phrase.

3. The multiple testing problem states that as more hypothesis tests are performed, small P-values become __________ meaningful. *less*

4. The Bonferroni adjustment is made by multiplying the P-value by the number of __________. *tests*

In Exercises 5 and 6, determine whether the statement is true or false. If the statement is false, rewrite it as a true statement.

5. The Bonferroni-adjusted P-value is always greater than the uncorrected P-value. *True*

6. When testing several hypotheses, it is reasonable to reject every hypothesis for which $P < 0.05$. *False*

Practicing the Skills

7. Five null hypotheses were tested, and the P-values were:

Hypothesis	1	2	3	4	5
P-value	0.35	0.019	0.012	0.008	0.042

a. Which hypotheses, if any, can be rejected at the $\alpha = 0.05$ level? *4*

b. Which hypotheses, if any, can be rejected at the $\alpha = 0.01$ level? *None*

8. Six null hypotheses were tested, and the P-values were:

Hypothesis	1	2	3	4	5	6
P-value	0.003	0.012	0.002	0.001	0.024	0.032

a. Which hypotheses, if any, can be rejected at the $\alpha = 0.05$ level? *1, 3, 4*

b. Which hypotheses, if any, can be rejected at the $\alpha = 0.01$ level? *4*

Working with the Concepts

9. Gender differences: A sociologist surveys a sample of college students to determine whether there are differences in the attitudes and behaviors of male and female students. The survey contains 20 questions. For one question, which asks how much time students spend studying each week, the difference between males and females is statistically significant with a P-value of 0.01. On all the other questions, the differences are not statistically significant. The sociologist concludes at the $\alpha = 0.05$ level that the time spent studying differs between males and females. Explain why this conclusion is not justified.

BananaStock/JupiterImages

10. Defective parts: Six different settings are tried on a machine to determine whether any of them will reduce the proportion of defective parts. For each setting, an appropriate null hypothesis is tested to determine whether the proportion of defective parts has been reduced. The six P-values are 0.34, 0.27, 0.002, 0.45, 0.03, and 0.19. A quality engineer concludes at the $\alpha = 0.05$ level that the method whose P-value is 0.002 reduces the proportion of defective parts. Explain why this conclusion is justified.

11. More gender differences: Refer to Exercise 9. For the result that was statistically significant, the P-value was 0.01. What P-value would be needed to conclude at the $\alpha = 0.05$ level that the time spent studying differs between males and females after applying the Bonferroni adjustment? *0.0025*

12. More defective parts: Refer to Exercise 10. The quality engineer suspects that the setting with a P-value of 0.03 may actually reduce the proportion of defective parts. What P-value would be needed to make this conclusion at the $\alpha = 0.05$ level after applying the Bonferroni adjustment? *0.008333*

13. Fertilizer: An agricultural scientist tests six types of fertilizer, labeled A, B, C, D, E, and F, to determine whether any of them produces an increase in crop yield over that obtained with the current fertilizer. For fertilizer C, the increase in yield was statistically significant at the $\alpha = 0.05$ level. For the other five, the increase was not statistically significant at the $\alpha = 0.05$ level. The scientist concludes that the yield obtained with fertilizer C is greater than that of the current fertilizer. Is this conclusion justified? Explain. *No*

14. More fertilizer: Refer to Exercise 13. The P-value for fertilizer C was 0.03. Use the Bonferroni adjustment to produce a reliable P-value for this fertilizer. Can you reject H_0 at the $\alpha = 0.05$ level? *0.15; No*

15. Coatings: Twenty formulations of a coating for gears are being tested to see if any of them reduce wear. For the Bonferroni-adjusted P-value for a formulation to be 0.05, what must the original P-value be? *0.0025*

16. Paint drying: Five new paint additives, labeled A, B, C, D, and E, have been tested to see if any of them can reduce the mean drying time from the current value of 12 minutes. Ten specimens have been painted with each of the new types of paint, and the drying times (in minutes) have been measured. The results are given below.

Additive	A	B	C	D	E
1	13.6	10.4	15.5	10.4	11.3
2	11.0	10.4	9.2	7.3	10.8
3	12.4	11.4	11.4	10.3	11.5
4	12.9	9.7	10.8	11.6	10.5
5	12.1	11.0	11.0	10.7	13.4
6	12.3	11.7	10.6	12.2	10.2
7	11.8	11.1	15.1	10.2	11.0
8	9.9	9.9	12.0	9.3	13.2
9	12.7	13.7	13.4	10.8	12.3
10	10.6	9.5	8.3	11.8	11.1

For each additive, perform a hypothesis test of the null hypothesis H_0: $\mu = 12$ against the alternate H_1: $\mu < 12$. You may assume that each population is approximately normal.

a. What are the P-values for the five tests?

b. On the basis of the results, which of the three following conclusions seems most appropriate? Explain your answer. *(i)*

(i) At least one of the new additives results in an improvement.

(ii) None of the new additives result in an improvement.

(iii) Some of the new additives may result in an improvement, but the evidence is inconclusive.

Extending the Concepts

17. Many multiple tests: Five hundred null hypotheses are tested at the $\alpha = 0.05$ level, and 40 of them are rejected. Assume the tests are independent,

a. Assuming that all 500 null hypotheses are true, what is the probability that a given hypothesis is rejected? *0.05*

b. Let X be the number of hypotheses that are rejected, so that the observed value of X is 40. Under the assumption that all 500 null hypotheses are true, what is the distribution of X? *Binomial with $n = 500$, $p = 0.05$*

c. Under the assumption that all 500 null hypotheses are true, what is the probability of observing a value of X as extreme as or more extreme than the value actually observed? *0.002701*

d. Can you conclude that some of the null hypotheses are false? Explain. *Yes*

Answers to Check Your Understanding Exercises for Section 11.5

1. 0.10

2. iii

Chapter 11 Summary

Section 11.1: We can test hypotheses about the difference between two population means. The test statistic has a Student's t distribution. The number of degrees of freedom can be taken to be 1 less than the smaller sample size. A more complicated formula, used by technology, provides a greater number of degrees of freedom. The assumptions that are necessary for one-sample tests must hold for both populations.

Section 11.2: When the data consist of matched pairs, we can test hypotheses about the difference between the population means by computing the difference between the values in each pair, then following the procedure for hypotheses about a single population mean. The assumptions required for a test of a population mean must hold for the population of differences.

Section 11.3: We can test hypotheses about the difference between two proportions. The test statistic has a standard normal distribution. The assumptions that are necessary for tests involving a single proportion must hold for both populations.

Section 11.4: We can use the F distribution to test whether two population standard deviations are equal. The validity of this test is sensitive to the assumption that both populations are normal. Therefore, this test should not be used unless it is certain that both populations are very close to normally distributed.

Section 11.5: When several tests are made, we have several chances to make an error. The Bonferroni adjustment can be used to correct P-values for the fact that several tests have been made. To make the adjustment, we multiply each P-value by the number of tests that are made.

Vocabulary and Notation

Bonferroni method 578
F distribution 571
independent samples 540
multiple testing problem 577
pooled proportion 561
pooled standard deviation 546

Important Formulas

Test statistic for the difference between two means, independent samples:

$$t = \frac{(\bar{x}_1 - \bar{x}_2) - (\mu_1 - \mu_2)}{\sqrt{\dfrac{s_1^2}{n_1} + \dfrac{s_2^2}{n_2}}}$$

Test statistic for the difference between two means, matched pairs:

$$t = \frac{\bar{d} - \mu_0}{s_d/\sqrt{n}}$$

Test statistic for the difference between two proportions:

$$z = \frac{\hat{p}_1 - \hat{p}_2}{\sqrt{\hat{p}(1-\hat{p})\left(\dfrac{1}{n_1} + \dfrac{1}{n_2}\right)}}$$

where $\hat{p}$ is the pooled proportion $\hat{p} = \dfrac{x_1 + x_2}{n_1 + n_2}$

Test statistic for two standard deviations:

$$F = \frac{\text{Larger of } s_1^2 \text{ and } s_2^2}{\text{Smaller of } s_1^2 \text{ and } s_2^2}$$

Chapter Quiz

1. A simple random sample of 75 people are given a new drug that is designed to relieve pain. A second sample of 50 people are given a standard drug. The question of interest is whether the proportion of people experiencing relief is greater among those taking the new drug. To address this question, which of the following is the most appropriate type of hypothesis test?

i. A test for the difference between two population means using independent samples
ii. A test for the difference between two population proportions
iii. A test for the difference between two population means using matched pairs
iv. A test for the difference between two population standard deviations

2. Two machines are used to fill cans. The machines are supposed to fill each can with 12 ounces of liquid. The amounts actually filled follow a normal distribution. A simple random sample of 10 cans is filled by each machine. The question of interest is whether the variability in the fill volume differs between the two machines. To address this question, which of the following is the most appropriate type of hypothesis test?

i. A test for the difference between two population means using independent samples
ii. A test for the difference between two population proportions
iii. A test for the difference between two population means using matched pairs
iv. A test for the difference between two population standard deviations

3. A fleet of 100 taxis is divided into two groups of 50 cars each to determine whether premium gasoline reduces maintenance costs. Premium unleaded fuel is used in group A, while regular unleaded fuel is used in group B. The total maintenance cost for each vehicle during a one-year period is recorded. The question of interest is whether the mean maintenance cost is less for vehicles using premium fuel. To address this question, which of the following is the most appropriate type of hypothesis test? *i*
   i. A test for the difference between two population means using independent samples
   ii. A test for the difference between two population proportions
   iii. A test for the difference between two population means using matched pairs
   iv. A test for the difference between two population standard deviations

4. A simple random sample of 75 people are given a new drug that is designed to relieve pain. After taking this drug for a month, they switch to a standard drug. The question of interest is whether the proportion of people who experienced relief is greater when taking the new drug. To address this question, which of the following is the most appropriate type of hypothesis test? *iii*
   i. A test for the difference between two population means using independent samples
   ii. A test for the difference between two population proportions
   iii. A test for the difference between two population means using matched pairs
   iv. A test for the difference between two population standard deviations

5. In a test of H_0: $p_1 = p_2$ versus H_1: $p_1 \neq p_2$, the value of the test statistic is $z = -1.21$. What do you conclude about the difference $p_1 - p_2$ at the $\alpha = 0.05$ level of significance? $p_1 - p_2$ *may equal 0.*

6. A sample of size 15 is drawn from a normal population. The sample standard deviation is $s_1 = 5.2$. A sample of size 10 is drawn from another normal population. The sample standard deviation is $s_2 = 9.3$. Can you conclude that the variances of these two populations are different? Use the $\alpha = 0.05$ level of significance. *No*

7. For a test of H_0: $\mu_1 = \mu_2$ versus H_1: $\mu_1 \neq \mu_2$, the sample sizes were $n_1 = 15$ and $n_2 = 25$. How many degrees of freedom are there for the test statistic? Use the simple method. *14*

8. In a set of 12 matched pairs, the mean difference was $\bar{d} = 18$ and the standard deviation of the differences was $s_d = 4$. Find the value of the test statistic for testing H_0: $\mu_d = 15$ versus H_1: $\mu_d > 15$. Can you reject H_0 at the $\alpha = 0.05$ level? *Yes*

9. Two suppliers of machine parts delivered large shipments. A simple random sample of 150 parts was chosen from each shipment.
   For supplier A, 12 of the 150 parts were defective. For supplier B, 28 of the 150 parts were defective. The question of interest is whether the proportion of defective parts is greater for supplier B than for supplier A. Let p_1 be the population proportion of defective parts for supplier A, and let p_2 be the population proportion of defective parts for supplier B. State appropriate null and alternate hypotheses. H_0: $p_1 = p_2$, H_1: $p_1 < p_2$

10. Refer to Exercise 9. Compute the value of the test statistic. *−2.72*

11. Refer to Exercise 9. Can you reject H_0 at the $\alpha = 0.01$ level? State a conclusion. *Reject* H_0.

12. A simple random sample of 17 business majors from a certain university had a mean GPA of 2.81 with a standard deviation of 0.27. A simple random sample of 23 psychology majors was selected from the same university, and their mean GPA was 2.97 with a standard deviation of 0.23. Boxplots show that it is reasonable to assume that the populations are approximately normal. The question of interest is whether the mean GPAs differ between business majors and psychology majors. Let μ_1 be the population mean GPA for business majors, and let μ_2 be the population mean GPA for psychology majors. State the null and alternate hypotheses. H_0: $\mu_1 = \mu_2$, H_1: $\mu_1 \neq \mu_2$

13. Refer to Exercise 12. Compute the value of the test statistic. *−1.971*

14. Refer to Exercise 12. Can you reject H_0 at the $\alpha = 0.05$ level? State a conclusion. *Do not reject* H_0.

15. Five null hypotheses were tested, and the *P*-values were 0.24, 0.17, 0.03, 0.002, and 0.02. How many of the hypotheses are rejected at the $\alpha = 0.05$ level if the Bonferroni adjustment is made? *One*

Review Exercises

1. **Sick days:** A large company is considering a policy of flextime, in which employees can choose their own work schedules within broad limits. The company is interested to determine whether this policy would reduce the number of sick days taken. They chose two simple random samples of 100 employees each. The employees in one sample were allowed to choose their own schedules. The other sample was a control group. Employees in that sample were required to come to work according to a schedule set by management. At the end of one year, the employees in the flextime group had a sample mean of 4.7 days missed, with a sample standard deviation of 3.1 days. The employees in the control group had a sample mean of 5.9 days missed, with a sample standard deviation of 3.9 days. Perform a hypothesis test to measure the strength of the evidence that the mean number of days missed is less in the flextime group. State the null and alternate hypotheses, find the *P*-value, and state your conclusion. Use the $\alpha = 0.01$ level of significance. *Reject* H_0.

2. **Political polling:** In a certain state, a referendum is being held to determine whether the transportation authority should issue additional highway bonds. A sample of 500 voters is taken in county A, and 285 say that they favor the bond proposal. A sample of 600 voters is taken in county B, and 305 say that they favor the bond issue. Perform a hypothesis test to measure the strength of the evidence that the proportion of voters who favor the proposal is greater in county A than in county B. State the null and alternate hypotheses, find the *P*-value, and state your conclusion. Use the $\alpha = 0.05$ level of significance. *Reject H_0.*

3. **Contaminated water:** The concentration of benzene was measured in units of milligrams per liter for a simple random sample of five specimens of untreated wastewater produced at a gas field. The sample mean was 7.8 with a sample standard deviation of 1.4. Seven specimens of treated wastewater had an average benzene concentration of 3.2 with a standard deviation of 1.7. Assume that both samples come from normal populations. Can you conclude that the standard deviation of benzene concentration differs between treated water and untreated water? Use the $\alpha = 0.05$ level of significance. *No*

4. **Sales commissions:** A company studied two programs for compensating its sales staff. Nine salespeople participated in the study. In program A, salespeople were paid a higher salary, plus a small commission for each item they sold. In program B, they were paid a lower salary with a larger commission. Following are the amounts sold, in thousands of dollars, for each salesperson on each program.

	Salesperson								
Program	**1**	**2**	**3**	**4**	**5**	**6**	**7**	**8**	**9**
A	55	22	34	22	25	61	55	36	68
B	53	24	36	28	31	61	58	38	72

Can you conclude that the mean sales differ between the two programs? Use the $\alpha = 0.05$ level of significance. *Yes*

5. **Exercise:** Medical researchers conducted a study to determine whether treadmill exercise could improve the walking ability of patients suffering from claudication, which is pain caused by insufficient blood flow to the muscles of the legs. A sample of 8 patients walked on a treadmill for six minutes every day. After six months, the mean distance walked in six minutes was 348 meters, with a standard deviation of 80 meters. For a control group of 7 patients who did not walk on a treadmill, the mean distance was 305 meters with a standard deviation of 93 meters. Can you conclude that the standard deviations of the distances walked differ between treatment and control? Use the $\alpha = 0.05$ level of significance. *No*

Based on data in the *Journal of the American Medical Association* 301:165–174

6. **Watching television:** The General Social Survey reported that in a sample of 68 men aged 18–25, the mean number of hours of television watched per day was 2.76 with a standard deviation of 2.21. In a sample of 72 women aged 18–25, the mean number of hours of television watched per day was 2.88 with a standard deviation of 2.43. Can you conclude that the mean number of hours of television watched differs between men and women? Use the $\alpha = 0.05$ level of significance. *No*

7. **Spreadsheets:** An accounting firm tested two spreadsheet applications to determine whether there is a difference between them in the mean speed with which a standard accounting problem can be solved. The times needed for each of twelve accountants to solve the problem are as follows:

	Accountant											
Application	**1**	**2**	**3**	**4**	**5**	**6**	**7**	**8**	**9**	**10**	**11**	**12**
A	42	34	45	35	40	44	48	49	41	33	31	38
B	50	43	48	31	39	56	45	60	46	47	33	50

Can you conclude that the mean time differs between the two applications? Use the $\alpha = 0.05$ level of significance. *Yes*

8. **Meeting specifications:** Two extrusion machines that manufacture steel rods are being compared. In a sample of 1000 rods taken from machine A, 960 met specifications regarding length and diameter. In a sample of 600 rods taken from machine B, 582 met the specifications. Machine B is more expensive to run, so it is decided that machine A will be used unless it can be convincingly shown that machine B produces a larger proportion of rods meeting specifications.
   a. State the appropriate null and alternate hypotheses for making the decision as to which machine to use. H_0: $p_1 = p_2$, H_1: $p_1 < p_2$
   b. Perform a hypothesis test at the $\alpha = 0.05$ level to determine which machine to use. State your conclusion. *Machine A should be used.*

9. **Interpret calculator display:** The following TI-84 Plus calculator display presents the results of a hypothesis test for the difference between two means. The sample sizes are $n_1 = 18$ and $n_2 = 16$.

```
2-SampTTest
μ1>μ2
t=3.414595477
p=.0012913382
df=21.18819537
x̄1=34.3569
x̄2=23.5185
Sx1=12.6882
↓Sx2=4.2544
```

a. Is this a left-tailed test, a right-tailed test, or a two-tailed test? *Right-tailed*
b. How many degrees of freedom did the calculator use? *21.18819537*
c. What is the P-value? *0.0012913382*
d. Can you reject H_0 at the $\alpha = 0.05$ level? *Yes*

10. Interpret computer output: The following MINITAB output presents a 95% confidence interval for the difference between two means.

```
Two-sample T for Population1 vs Population2

                N      Mean     StDev  SE Mean
Population1    55  16.48435  10.23430  1.52564
Population2    47  18.32197   8.38450  1.22301

Difference = mu (Treatment1) − mu (Treatment2)
Estimate for difference: −1.83762
95% upper bound for difference: 2.22314
T-Test of difference = 0 (vs < 0): T-Value = −0.997 P-Value = 0.161  DF = 99
```

a. Is this a left-tailed test, a right-tailed test, or a two-tailed test? *Left-tailed*
b. How many degrees of freedom did MINITAB use? *99*
c. What is the P-value? *0.161*
d. Can you reject H_0 at the $\alpha = 0.05$ level? *No*

11. Interpret calculator display: The following TI-84 Plus calculator display presents the results of a hypothesis test for the difference between two proportions. The sample sizes are $n_1 = 125$ and $n_2 = 150$.

```
2-PropZTest
p1≠p2
z=1.269800442
p=.2041558545
p̂1=.272
p̂2=.2066666667
p̂=.2363636364
n1=125
n2=150
```

a. Is this a left-tailed test, a right-tailed test, or a two-tailed test? *Two-tailed*
b. What is the P-value? *0.2041558545*
c. Can you reject H_0 at the $\alpha = 0.05$ level? *No*

12. Interpret computer output: The following MINITAB output presents the results of a hypothesis test for the difference between two proportions.

```
Test and CI for Two Proportions: A, B

Variable    X    N   Sample p
A          31   45   0.688889
B          23   58   0.396552

Difference = p (A) − p (B)
Estimate for difference: 0.292337
95% Lower Bound for difference: 0.128108
Test for difference = 0 (vs > 0): Z = 2.95  P-Value = 0.003
```

a. Is this a left-tailed test, a right-tailed test, or a two-tailed test? *Right-tailed*
b. What is the P-value? *0.003*
c. Can you reject H_0 at the $\alpha = 0.05$ level? *Yes*

13. **Interpret calculator display:** The following TI-84 Plus calculator display presents the results of a hypothesis test for the mean difference between matched pairs.

```
T-Test
μ>0
t=3.405554322
p=.0021327513
x̄=1.02
Sx=1.16
n=15
```

a. Is this a left-tailed test, a right-tailed test, or a two-tailed test? *Right-tailed*
b. How many degrees of freedom are there? *14*
c. What is the P-value? *0.0021327513*
d. Can you reject H_0 at the $\alpha = 0.05$ level? *Yes*

14. **Interpret computer output:** The following output from MINITAB presents a confidence interval for the mean difference between matched pairs.

```
Paired T-Test and CI: Before, After

Paired T for Before - After

              N       Mean     StDev   SE Mean
Before       12   20.50337   9.69467   2.79861
After        12   17.82398   6.36154   1.83642
Difference   12    2.67399   3.09085   0.89225
95% confidence interval for mean difference: (0.71016, 4.63782)
T-Test of mean difference = 0 (vs ne 0):   T-Value = 3.00  P-Value = 0.012
```

a. Is this a left-tailed test, a right-tailed test, or a two-tailed test? *Two-tailed*
b. How many degrees of freedom are there? *11*
c. What is the P-value? *0.012*
d. Can you reject H_0 at the $\alpha = 0.05$ level? *Yes*

15. **Strong bolts:** Five different variations of a bolt-making process are run to determine whether any of them can increase the mean breaking strength of the bolts over that of the current process. The P-values are 0.13, 0.34, 0.03, 0.28, and 0.38. Of the following choices, which is the best thing to do next? *iii*
   i. Implement the process whose P-value was 0.03, because it performed the best.
   ii. Because none of the processes had Bonferroni-adjusted P-values less than 0.05, we should continue with the current process.
   iii. Rerun the process whose P-value was 0.03 to determine whether it remains small in the absence of multiple testing.
   iv. Rerun all five variations again, to determine whether any of them produce a small P-value the second time around.

Write About It

1. Provide an example, real or imagined, of a hypothesis test for the difference between two means.
2. Describe under what circumstances a hypothesis test for the difference between two proportions would be performed. Provide an example.
3. Describe the differences between performing a hypothesis test for $\mu_1 - \mu_2$ with paired samples and performing the same test for independent samples.
4. How do hypothesis tests for two population standard deviations differ from other tests presented in this chapter? In what ways are they similar?
5. Several null hypotheses are tested, and one of them is rejected after the Bonferroni adjustment. Explain why this null hypothesis would also have been rejected if it had been the only one tested.

In-Class Activities

1. **Multiple testing problem:** Each student tosses both a penny and a nickel 50 times each, and performs a test of the hypothesis $H_0: p_1 = p_2$ versus $H_1: p_1 \neq p_2$, where p_1 is the probability of heads for the penny and p_2 is the probability of heads for the nickel. If some students reject H_0 at the $\alpha = 0.05$ level, can we conclude the probabilities for coins were different? Or is there a better explanation? Construct a histogram of the *P*-values. Is it approximately uniform? How does the uniform distribution of the *P*-value explain the relationship between the significance level and the probability of a Type I error?
2. **Paired versus independent samples:** Each student draws two lines, by eye, on a piece of paper. Students should try to make the first line as close to 3 inches long as possible, and the second line as close to 6 inches long as possible. Then students measure the lengths of the lines and calculate the errors between the true lengths and the attempted lengths. Each student reports the error in the 3-inch line, the error in the 6-inch line, and the difference between the errors. Test the hypothesis that the mean error is greater for the 6-inch line with a paired test. Then, pretending the samples are independent, test the same hypothesis using an independent samples test. For which test is the *P*-value smaller? Why?

Case Study: Does Exposure To Electrical Fields Cause Brain Cancer?

In the chapter opener, we presented the results of a study that was designed to determine whether exposure to electrical fields could cause cancer. Three hundred children with brain cancer (cases) and three hundred healthy children (controls) were studied. The proportions with underground wiring in each group were computed, along with proportions in certain income and ethnic groups. The results for five factors are presented in the following table.

	Cases	Controls
Proportion with high income	0.09	0.13
Proportion owning a home	0.72	0.76
Proportion with underground wiring	0.14	0.07
Proportion with father white non-Hispanic	0.43	0.36
Proportion with father Hispanic	0.38	0.47

1. For each of the five factors, test the null hypothesis that the population proportions are equal against the alternate hypothesis that they are not equal. Find the *P*-value for each test.
2. Which factors are significant at the $\alpha = 0.05$ level?
3. For the *P*-values that are less than 0.05, use the Bonferroni adjustment to adjust them for the fact that several tests have been performed.
4. For each factor, describe how strong the evidence is that the proportions differ for that factor.

Tests with Qualitative Data

chapter

12

Mark Scott/Getty Images

Introduction

Do graduate schools discriminate against women? This issue was addressed in a famous study carried out at the University of California at Berkeley. The following table presents the numbers of male and female applicants to six of the most popular departments at the University of California at Berkeley. Out of 2691 male applicants, 1198, or 44.5%, were accepted. Out of 1835 female applicants, only 557, or 30.4%, were accepted. Is this difference due to discrimination against women? In the case study at the end of the chapter, we will use the methods presented in this chapter to determine the real reason for this difference.

Gender	Accept	Reject	Total
Male	1198	1493	2691
Female	557	1278	1835
Total	1755	2771	4526

Section 12.1 Testing Goodness of Fit

Objectives

1. Find critical values of the chi-square distribution
2. Perform goodness-of-fit tests

Objective 1 Find critical values of the chi-square distribution

The Chi-Square Distribution

In this chapter, we will introduce hypothesis tests for qualitative data, also called categorical data. These tests are based on the **chi-square distribution**, which we first introduced in Section 8.4. The test statistic used for these tests is called the **chi-square statistic**, denoted χ^2. The symbol χ is the Greek letter chi (pronounced "kigh"; rhymes with sky). We find critical values for this statistic by using the chi-square distribution. We will begin by reviewing the features of this distribution.

RECALL
Qualitative data classify individuals into categories.

NOTE TO INSTRUCTOR
The chi-square distribution was first introduced in Section 8.4. We repeat the introduction here.

There are actually many different chi-square distributions, each with a different number of degrees of freedom. Figure 12.1 presents chi-square distributions for several different degrees of freedom. There are two important points to notice.

- The chi-square distribution is not symmetric. It is skewed to the right.
- Values of the χ^2 statistic are always greater than or equal to 0. They are never negative.

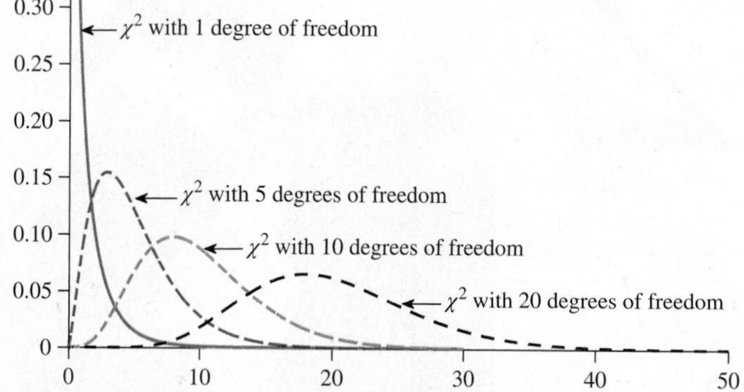

Figure 12.1 Chi-square distributions with various degrees of freedom

Finding critical values for the chi-square distribution

We find right-tail critical values for the chi-square distribution. These values can be found using Table A.4.

Example 12.1

Find a critical value

Find the $\alpha = 0.05$ critical value for the chi-square distribution with 12 degrees of freedom.

Solution

The critical value is found at the intersection of the row corresponding to 12 degrees of freedom and the column corresponding to $\alpha = 0.05$. The critical value is 21.026.

Degrees of Freedom	Area in Right Tail									
	0.995	0.99	0.975	0.95	0.90	0.10	0.05	0.025	0.01	0.005
⋮	⋮	⋮	⋮	⋮	⋮	⋮	⋮	⋮	⋮	⋮
10	2.156	2.558	3.247	3.940	4.865	15.987	18.307	20.483	23.209	25.188
11	2.603	3.053	3.816	4.575	5.578	17.275	19.675	21.920	24.725	26.757
12	3.074	3.571	4.404	5.226	6.304	18.549	21.026	23.337	26.217	28.300
13	3.565	4.107	5.009	5.892	7.042	19.812	22.362	24.736	27.688	29.819
14	4.075	4.660	5.629	6.571	7.790	21.064	23.685	26.119	29.141	31.319
15	4.601	5.229	6.262	7.261	8.547	22.307	24.996	27.488	30.578	32.801
⋮	⋮	⋮	⋮	⋮	⋮	⋮	⋮	⋮	⋮	⋮

Figure 12.2 presents the chi-square distribution with 12 degrees of freedom, with the $\alpha = 0.05$ critical value labeled.

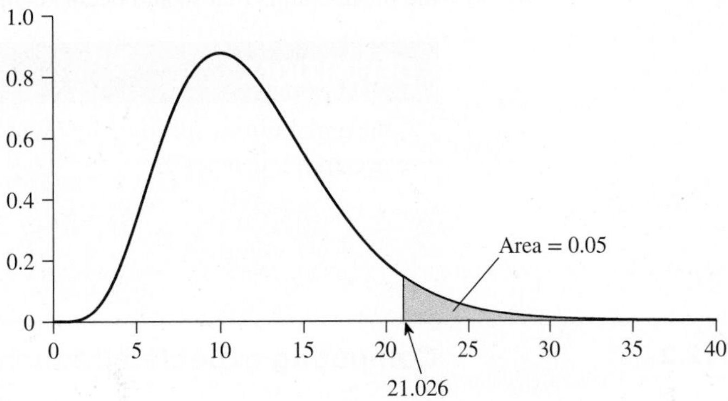

Figure 12.2 The area to the right of 21.026 is 0.05, so 21.026 is the $\alpha = 0.05$ critical value.

Occasionally we may be given a number of degrees of freedom that is not found in Table A.4. When this occurs, simply use the next larger number of degrees of freedom in the table.

Check Your Understanding

1. Find the $\alpha = 0.05$ critical value for the chi-square distribution with 18 degrees of freedom. *28.869*
2. Find the $\alpha = 0.10$ critical value for the chi-square distribution with 4 degrees of freedom. *7.779*
3. Find the area to the right of 29.141 under the chi-square distribution with 14 degrees of freedom. *0.01*
4. Find the area to the right of 46.979 under the chi-square distribution with 30 degrees of freedom. *0.025*

Answers are on page 597.

Objective 2 Perform goodness-of-fit tests

Goodness-of-Fit Tests

Imagine that you want to determine whether a coin is fair. You could toss the coin a number of times and compute the sample proportion $\hat{p}$ of heads. You could then use a test for a population proportion (Section 9.4) to test the hypotheses $H_0: p = 0.5$ versus $H_1: p \neq 0.5$. The test for a population proportion is designed for an experiment with two possible outcomes, such as the toss of a coin. Sometimes we work with experiments that have more than two possible outcomes. For example, imagine that a gambler wants to test a die to determine whether it is fair. The roll of a die has six possible outcomes: 1, 2, 3, 4, 5, and 6; and the die is fair if each of these outcomes is equally likely. The gambler rolls the die 60 times and counts the number of times each number comes up. These counts, which are called the **observed frequencies**, are presented in Table 12.1.

Table 12.1 Observed Frequencies for 60 Rolls of a Die

Outcome	1	2	3	4	5	6
Observed	12	7	14	15	4	8

EXPLAIN IT AGAIN

Hypotheses for goodness-of-fit tests: The null hypothesis always specifies a probability for each category. The alternate hypothesis says that some or all of these probabilities differ from the true probabilities of the categories.

The gambler wants to perform a hypothesis test to determine whether the die is fair. The null hypothesis for this test says that the die is fair; in other words, it says that each of the six outcomes has probability 1/6 of occurring. Let p_1 be the probability of rolling a 1, p_2 be the probability of rolling a 2, and so on. Then the null hypothesis is

$$H_0: p_1 = p_2 = p_3 = p_4 = p_5 = p_6 = 1/6$$

The alternate hypothesis says that the roll of a die does not follow the distribution specified by H_0; in other words, it states that not all of the p_i are equal to 1/6.

Computing expected frequencies

To test H_0, we begin by computing **expected frequencies**. The expected frequencies are the mean counts that would occur if H_0 were true.

DEFINITION

If the probabilities specified by H_0 are $p_1, p_2, ...,$ and the total number of trials is n, the expected frequencies are

$$E_1 = np_1, \quad E_2 = np_2, \quad \text{and so on}$$

Example 12.2

Computing expected frequencies

Compute the expected frequencies for the die example.

Solution

The probabilities specified by H_0 are $p_1 = p_2 = \cdots = p_6 = 1/6$. The total number of trials is $n = 60$. Therefore, the expected frequencies are

$$E_1 = E_2 = E_3 = E_4 = E_5 = E_6 = (60)(1/6) = 10$$

Table 12.2 presents both observed and expected frequencies for the die example.

EXPLAIN IT AGAIN

The expected frequency: The expected value of a binomial random variable is $E = np$. Here n is the total number of trials, and p is the probability of a particular outcome.

Table 12.2 Observed and Expected Frequencies

Outcome	1	2	3	4	5	6
Observed	12	7	14	15	4	8
Expected	10	10	10	10	10	10

Check Your Understanding

5. A researcher wants to determine whether children are more likely to be born on certain days of the week. She will sample 350 births and record the day of the week for each. The null hypothesis is that a birth is equally likely to occur on any day of the week. Compute the expected frequencies.
6. A researcher wants to test the hypothesis that births are more likely to occur on weekdays. The null hypothesis is that 17% of births occur on each of the days Monday through Friday, 10% occur on Saturday, and 5% occur on Sunday. If 350 births are sampled, find the expected frequencies.

Answers are on page 597.

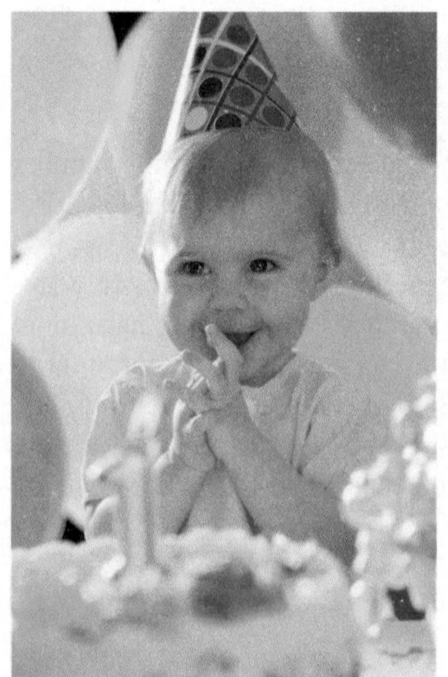

Elyse Lewin/Getty Images

If H_0 is true, the observed and expected frequencies should be fairly close. The larger the differences are between the observed and expected frequencies, the stronger the evidence is against H_0. We compute a test statistic that measures how large these differences are. As mentioned previously, the statistic is called the chi-square statistic, denoted χ^2.

DEFINITION

Let k be the number of categories, let $O_1, ..., O_k$ be the observed frequencies, and let $E_1, ..., E_k$ be the expected frequencies. The chi-square statistic is

$$\chi^2 = \sum \frac{(O - E)^2}{E}$$

When H_0 is true, the chi-square statistic has approximately a chi-square distribution, provided that all the expected frequencies are 5 or more.

EXPLAIN IT AGAIN

Degrees of freedom and the number of categories: For a goodness-of-fit test, the number of degrees of freedom for the chi-square statistic is always 1 less than the number of categories.

When H_0 is true, the statistic

$$\chi^2 = \sum \frac{(O-E)^2}{E}$$

has a chi-square distribution with $k-1$ degrees of freedom, where k is the number of categories, provided that all the expected frequencies are greater than or equal to 5.

We will first describe how to perform a hypothesis test using the critical value method and Table A.4. Then we will describe how to use the P-value method with technology.

CAUTION

The assumption that the expected frequencies are at least 5 does not apply to the observed frequencies. There are no assumptions on the observed frequencies.

Performing a Goodness-of-Fit Test

Step 1: State the null and alternate hypotheses. The null hypothesis specifies a probability for each category. The alternate hypothesis says that some or all of the actual probabilities differ from those specified by H_0.

Step 2: Compute the expected frequencies, and check to be sure that all of them are 5 or more. If they are, then proceed.

Step 3: Choose a significance level α.

Step 4: Compute the test statistic $\chi^2 = \sum \frac{(O-E)^2}{E}$.

Step 5: Find the critical value from Table A.4, using $k-1$ degrees of freedom, where k is the number of categories. If χ^2 is greater than or equal to the critical value, reject H_0. Otherwise, do not reject H_0.

Step 6: State a conclusion.

NOTE TO INSTRUCTOR

You may wish to point out that chi-square tests in this chapter are all one-tailed. Only large values of the test statistic provide evidence against the null hypothesis.

Example 12.3

Perform a goodness-of-fit test

Table 12.3 presents the observed frequencies for the die example. Can you conclude at the $\alpha = 0.05$ level that the die is not fair?

Table 12.3

Category	Observed
1	12
2	7
3	14
4	15
5	4
6	8

Solution

Step 1: State the null and alternate hypotheses. The null hypothesis says that the die is fair, and the alternate hypothesis says that the die is not fair, so we have

H_0: $p_1 = p_2 = p_3 = p_4 = p_5 = p_6 = 1/6$ $\quad$ H_1: Some or all of the p_i differ from $1/6$

Step 2: Compute the expected frequencies. The expected frequencies were computed in Example 12.2. We present them in Table 12.4. They are all greater than 5, so we proceed.

Table 12.4

Category	Observed	Expected
1	12	10
2	7	10
3	14	10
4	15	10
5	4	10
6	8	10

Step 3: Choose a level of significance. We will use $\alpha = 0.05$.

Step 4: Compute the value of the test statistic. The following table presents the calculations.

Category	O	E	$O-E$	$(O-E)^2$	$\frac{(O-E)^2}{E}$
1	12	10	2	4	0.4
2	7	10	−3	9	0.9
3	14	10	4	16	1.6
4	15	10	5	25	2.5
5	4	10	−6	36	3.6
6	8	10	−2	4	0.4

$$\chi^2 = \sum \frac{(O-E)^2}{E} = 9.4$$

The value of the test statistic is $\chi^2 = 9.4$.

Step 5: Find the critical value. There are six categories, so there are $6 - 1 = 5$ degrees of freedom. From Table A.4, we find that the $\alpha = 0.05$ critical value for 5 degrees of freedom is 11.070. The value of the test statistic is $\chi^2 = 9.4$. Because $9.4 < 11.070$, we do not reject H_0.

Step 6: State a conclusion. There is not enough evidence to conclude that the die is unfair.

Example 12.4 Perform a goodness-of-fit test using technology

The following TI-84 Plus display presents the results of Example 12.3.

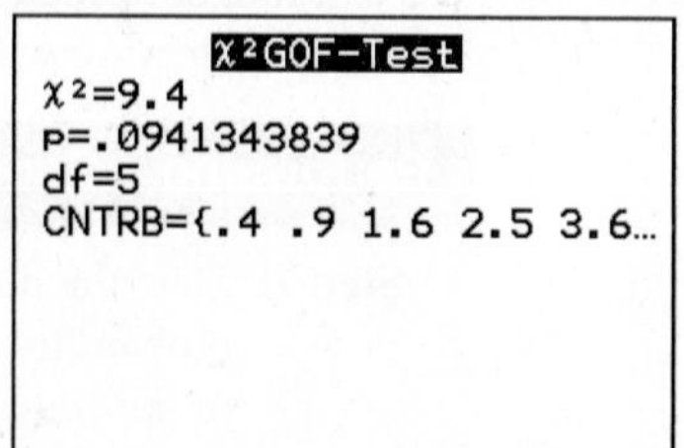

Most of the output is straightforward. The first, second, and third lines present the value of the chi-square statistic, the *P*-value, and the degrees of freedom, respectively. The last line, labeled "CNTRB," presents the quantities $\frac{(O-E)^2}{E}$ for the first two categories.

The *P*-value is 0.0941. Because $P > 0.05$, we do not reject H_0 at the $\alpha = 0.05$ level. This conclusion is the same as the one we reached when we used the critical value method in Example 12.3.

In EXCEL, the function **CHISQ.TEST** returns the *P*-value for a goodness-of-fit test. The inputs are the range of cells that contain the observed frequencies and the range of cells that contain the expected frequencies. Following is the EXCEL output. The *P*-value is shown to be 0.0941.

	A	B
1	Observed	Expected
2	12	10
3	7	10
4	14	10
5	15	10
6	4	10
7	8	10

=CHISQ.TEST(A2:A7, B2:B7)

0.0941

Step-by-step instructions for using technology are presented in the Using Technology section on page 594.

Example 12.5 Perform a goodness-of-fit test

A poll conducted by the General Social Survey asked 1155 people whether they thought that people with high incomes should pay a greater or smaller percentage of their income in tax than low-income people. The results are presented in the following table.

Category	Observed
Pay much more	218
Pay somewhat more	497
Pay the same	425
Pay less	15

Five years earlier, it was determined that 18.5% believed that the rich should pay much more, 39.2% believed they should pay somewhat more, 41.2% believed they should pay the same, and 1.1% believed they should pay less. Can we conclude that the current percentages differ from these? Use the $\alpha = 0.05$ level of significance.

Solution

Step 1: State the null and alternate hypotheses. The null hypothesis is $H_0\colon p_1 = 0.185$, $p_2 = 0.392, p_3 = 0.412, p_4 = 0.011$. The alternate hypothesis states that some of the probabilities are not equal to the values specified by H_0.

Step 2: Compute the expected frequencies. The number of trials is $n = 1155$. The expected frequencies are

$$E_1 = np_1 = (1155)(0.185) = 213.675 \qquad E_2 = np_2 = (1155)(0.392) = 452.76$$

$$E_3 = np_3 = (1155)(0.412) = 475.86 \qquad E_4 = np_4 = (1155)(0.011) = 12.705$$

All the expected frequencies are 5 or more, so we proceed.

Step 3: Choose a significance level. We will use $\alpha = 0.05$.

Step 4: Compute the test statistic. Using the observed and expected frequencies, we compute the value of the test statistic to be

$$\chi^2 = \frac{(218 - 213.675)^2}{213.675} + \frac{(497 - 452.76)^2}{452.76} + \frac{(425 - 475.86)^2}{475.86} + \frac{(15 - 12.705)^2}{12.705} = 10.261$$

Step 5: Find the critical value. There are four categories, so there are $4 - 1 = 3$ degrees of freedom. From Table A.4, we find that the $\alpha = 0.05$ critical value for 3 degrees of freedom is 7.815. The value of the test statistic is $\chi^2 = 10.261$. Because $10.261 > 7.815$, we reject H_0.

Step 6: State a conclusion. We conclude that the distribution of opinions on this issue changed during the 5 years prior to the survey.

Check Your Understanding

7. Following are observed frequencies for five categories:

Category	1	2	3	4	5
Observed	25	14	23	6	2

a. Compute the expected frequencies for testing $H_0\colon p_1 = 0.3,\ p_2 = 0.25, p_3 = 0.2,\ p_4 = 0.15,\ p_5 = 0.1$.

b. One of the observed frequencies is less than 5. Is the chi-square test appropriate? *Yes*

c. Compute the value of χ^2. *12.748*

d. How many degrees of freedom are there? *4*

e. Find the level $\alpha = 0.05$ critical value. *9.488*

f. Do you reject H_0 at the 0.05 level? *Yes*

g. Find the level $\alpha = 0.01$ critical value. *13.277*

h. Do you reject H_0 at the 0.01 level? *No*

8. For the data in Exercise 7:

a. Compute the expected frequencies for testing $H_0\colon p_1 = 0.4, p_2 = 0.3, p_3 = 0.1, p_4 = 0.15, p_5 = 0.05$.

b. Is it appropriate to perform a chi-square test for the hypothesis in part (a)? Explain. *No*

Answers are on page 597.

Using Technology

We use Example 12.3 to illustrate the technology steps.

TI-84 PLUS

Testing goodness-of-fit

Step 1. Enter observed frequencies into **L1** in the data editor and expected frequencies into **L2**. Figure A illustrates this for Example 12.3.

Step 2. Press **STAT** and highlight the **TESTS** menu. Select χ^2**GOF–Test** and press **ENTER** (Figure B). The χ^2**GOF–Test** menu appears.

Step 3. In the **Observed** field, enter **L1**, and in the **Expected** field, enter **L2**.

Step 4. Enter the degrees of freedom in the **df** field. For Example 12.3, there are 5 degrees of freedom (Figure C).

Step 5. Highlight **Calculate** and press **ENTER** (Figure D).

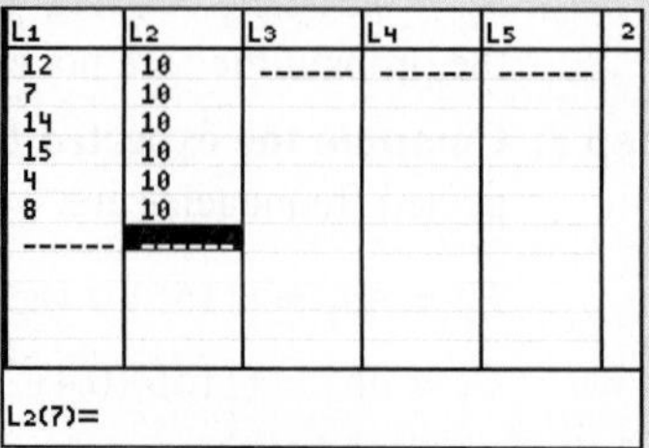

L1 L2 L3 L4 L5 2
12 10
7 10
14 10
15 10
4 10
8 10
L2(7)=

Figure A

```
EDIT CALC TESTS
6↑2-PropZTest…
7:ZInterval…
8:TInterval…
9:2-SampZInt…
0:2-SampTInt…
A:1-PropZInt…
B:2-PropZInt…
C:χ²-Test…
D↓χ²GOF-Test…
```

Figure B

```
χ²GOF-Test
Observed:L1
Expected:L2
df:5
Color: BLUE
Calculate Draw
```

Figure C

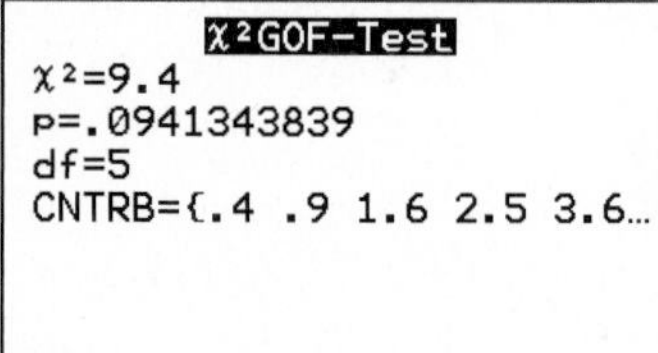

Figure D

EXCEL

Testing goodness-of-fit

The **CHISQ.TEST** command returns the *P*-value for a goodness-of-fit test given the observed and expected frequencies

Step 1. Enter the observed and expected frequencies into EXCEL. For Example 12.3, the observed frequencies are contained in A2:A7 and the expected frequencies are in B2:B7 (Figure E).

Step 2. In an empty cell, select the **Insert Function** icon and highlight **Statistical** in the category field.

Step 3. Click on the **CHISQ.TEST** function and press **OK**.

Step 4. In the **Actual_range** field, enter or select the range of cells containing the observed frequencies. In the **Expected_range** field, enter or select the range of cells containing the expected frequencies.

Step 5. Click **OK** to obtain the *P*-value (Figure F).

	A	B
1	Observed	Expected
2	12	10
3	7	10
4	14	10
5	15	10
6	4	10
7	8	10

Figure E

```
=CHISQ.TEST(A2:A7, B2:B7)
0.0941
```

Figure F

Section 12.1 Exercises

Exercises 1–8 are the Check Your Understanding exercises located within the section.

Understanding the Concepts

In Exercises 9 and 10, fill in each blank with the appropriate word or phrase.

9. For the goodness-of-fit test to be valid, each of the __________ frequencies must be at least 5. *expected*

10. In a goodness-of-fit test, we reject H_0 if the __________ frequencies are much different from the expected frequencies. *observed*

In Exercises 11 and 12, determine whether the statement is true or false. If the statement is false, rewrite it as a true statement.

11. The chi-square distribution is symmetric. *False*

12. The alternate hypothesis for a goodness-of-fit test says that some of the probabilities differ from those specified by the null hypothesis. *True*

Practicing the Skills

13. Find the $\alpha = 0.05$ critical value for the chi-square statistic with 14 degrees of freedom. *23.685*

14. Find the $\alpha = 0.01$ critical value for the chi-square statistic with 5 degrees of freedom. *15.086*

15. Find the area to the right of 24.725 under the chi-square distribution with 11 degrees of freedom. *0.01*

16. Find the area to the right of 40.256 under the chi-square distribution with 30 degrees of freedom. *0.10*

17. For the following observed and expected frequencies:

Observed	9	22	53	9	7
Expected	15	20	40	15	10

a. Compute the value of χ^2. *10.125*
b. How many degrees of freedom are there? *4*
c. Can you conclude that the distribution of observed frequencies differs from that given by the expected frequencies? Use the $\alpha = 0.05$ level of significance. *Reject H_0.*

18. For the following observed and expected frequencies:

Observed	43	42	31	19	34	32
Expected	44	44	33	15	36	29

a. Compute the value of χ^2. *1.723*
b. How many degrees of freedom are there? *5*
c. Can you conclude that the distribution of observed frequencies differs from that given by the expected frequencies? Use the $\alpha = 0.01$ level of significance. *Do not reject H_0.*

19. Following are observed frequencies. The null hypothesis is $H_0\colon p_1 = 0.5, p_2 = 0.3, p_3 = 0.15, p_4 = 0.05$.

Category	1	2	3	4
Observed	106	64	24	6

a. Compute the expected frequencies.
b. Compute the value of χ^2. *3.427*
c. How many degrees of freedom are there? *3*
d. Can you conclude that the distribution of observed frequencies differs from that given by the null hypothesis? Use the $\alpha = 0.01$ level of significance. *Do not reject H_0.*

20. Following are observed frequencies. The null hypothesis is $H_0\colon p_1 = 0.4, p_2 = 0.25, p_3 = 0.05, p_4 = 0.1, p_5 = 0.2$.

Category	1	2	3	4	5
Observed	50	51	14	12	23

a. Compute the expected frequencies.
b. Compute the value of χ^2. *14.393*
c. How many degrees of freedom are there? *4*
d. Can you conclude that the distribution of observed frequencies differs from that given by the null hypothesis? Use the $\alpha = 0.05$ level of significance. *Reject H_0.*

21. For the following observed and expected frequencies:

Observed	16	18	3	17	14
Expected	14	16	8	18	12

a. Is a goodness-of-fit test valid for these data? Why or why not? *Yes*
b. If the test is valid, determine whether you can conclude that the distribution of the observed frequencies differs from that given by the expected frequencies. Use the $\alpha = 0.05$ level of significance. *Do not reject H_0.*

22. For the following observed and expected frequencies:

Observed	17	12	14	13	7	12
Expected	15	14	15	12	3	15

a. Is a goodness-of-fit test valid for these data? Why or why not? *No*
b. If the test is valid, determine whether you can conclude that the distribution of the observed frequencies differs from that given by the expected frequencies. Use the $\alpha = 0.05$ level of significance.

23. Following are observed frequencies. The null hypothesis is $H_0\colon p_1 = 0.05, p_2 = 0.25, p_3 = 0.3, p_4 = 0.2, p_5 = 0.2$.

Category	1	2	3	4	5
Observed	6	10	14	11	9

a. Is a goodness-of-fit test valid for these data? Why or why not? *No*
b. If the test is valid, determine whether you can conclude that the distribution of the observed frequencies differs from that given by the expected frequencies. Use the $\alpha = 0.01$ level of significance.

24. Following are observed frequencies. The null hypothesis is $H_0\colon p_1 = 0.1, p_2 = 0.2, p_3 = 0.2, p_4 = 0.3, p_5 = 0.1, p_6 = 0.1$.

Category	1	2	3	4	5	6
Observed	12	18	22	28	12	8

a. Is a goodness-of-fit test valid for these data? Why or why not? *Yes*
b. If the test is valid, determine whether you can conclude that the distribution of the observed frequencies differs from that given by the expected frequencies. Use the $\alpha = 0.01$ level of significance. *Do not reject H_0.*

Working with the Concepts

25. Is the lottery fair? Mega Millions is a multistate lottery in which players try to guess the numbers that will turn up in a drawing of numbered balls. One of the balls drawn is the Mega Ball. Matching the number drawn on the Mega Ball increases one's winnings. During a recent three-year period, the Mega Ball was drawn from a collection of 15 balls numbered 1 through 15, and a total of 344 drawings were made. For the purposes of this exercise, we grouped the numbers into five categories: 1–3, 4–6, and so on. If the lottery is fair, then the winning number is equally likely to occur in any category. Following are the observed frequencies.

Category	1–3	4–6	7–9	10–12	13–15
Observed	73	64	74	62	71

Source: www.usamega.com

a. Compute the expected frequencies. *All are 68.8*
b. Compute the value of χ^2 *1.7267*
c. How many degrees of freedom are there? *4*
d. Can you conclude that the categories are not equally likely? Use the $\alpha = 0.05$ level of significance. *Do not reject H_0.*

26. Grade distribution: A statistics teacher claims that, on the average, 20% of her students get a grade of A, 35% get a B, 25% get a C, 10% get a D, and 10% get an F. The grades of a random sample of 100 students were recorded. The following table presents the results.

Grade	A	B	C	D	F
Observed	29	42	20	5	4

a. How many of the students in the sample got an A? How many got an F? *29; 4*
b. Compute the expected frequencies.
c. Which grades were given more often than expected? Which grades were given less often than expected? *A and B more; C, D, and F less*
d. What is the value of χ^2? *12.550*
e. How many degrees of freedom are there? *4*

f. Can you conclude that the distribution of the grades differs from the teacher's claim? Use the $\alpha = 0.05$ level of significance. *Reject H_0.*

27. False alarm: The numbers of false fire alarms were counted each month at a number of sites. The results are given in the following table.

Month	Number of Alarms
January	32
February	15
March	37
April	38
May	45
June	48
July	46
August	42
September	34
October	36
November	28
December	26

Source: *Journal of Architectural Engineering* 5:62–65

Can you conclude that false alarms are not equally likely to occur in any month? Use the $\alpha = 0.01$ level of significance. *Reject H_0.*

28. Crime rates: The FBI computed the proportion of violent crimes in the United States falling into each of four categories. A simple random sample of 500 violent crimes committed in California were categorized in the same way. The following table presents the results.

Category	U.S. Proportion	California Frequency
Murder	0.013	5
Forcible Rape	0.051	23
Robbery	0.360	206
Aggravated Assault	0.576	266

Can you conclude that the proportions of crimes in the various categories in California differ from the United States as a whole? Use the $\alpha = 0.05$ level of significance. *Do not reject H_0.*

29. Where do you live? The U.S. Census Bureau computed the proportion of U.S. residents who lived in each of four geographic regions in 2015. Then a simple random sample was drawn of 1000 people living in the United States in 2018. The following table presents the results:

	2015 Proportion	2018 Frequency
Northeast	0.175	172
Midwest	0.212	209
South	0.236	238
West	0.377	381

Can you conclude that the proportions of people living in the various regions changed between 2015 and 2018? Use the $\alpha = 0.05$ level of significance. *Do not reject H_0*

30. Abortion policy: A Gallup poll taken in September 2019 asked 1025 adult Americans to state their opinion on the availability of abortions. The following table presents the results, along with the proportions of people who held these views in 2018.

Opinion	2018 Proportion	2019 Frequency
Legal in all cases	0.24	297
Legal in some cases	0.54	513
Never legal	0.19	195
No opinion	0.03	20

Can you conclude that the proportions of people giving the various responses changed between 2018 and 2019? Use the $\alpha = 0.01$ level of significance. *Reject H_0*

31. Economic future: A Gallup poll taken in May 2019 obtained responses from 1522 adult Americans to the question, "How would you rate the current state of the economy? Is it excellent, good, fair, or poor?" The following table presents the results, along with the proportions of people who gave these responses in 2018.

Opinion	2018 Proportion	2019 Frequency
Excellent	0.08	167
Good	0.41	609
Fair	0.39	563
Poor	0.12	183

Can you conclude that the proportions of people giving the various responses changed between 2018 and 2019? Use the $\alpha = 0.01$ level of significance. *Reject H_0*

32. Guess the answer: A statistics instructor gave a four-question true–false quiz to his class of 150 students. The results were as follows.

Number correct	0	1	2	3	4
Observed	2	13	29	71	35

The instructor thinks that the students may have answered the questions by guessing, so that the probability that any given answer is correct is 0.5. Under this null hypothesis, the number of correct answers has a binomial distribution with 4 trials and success probability 0.5. Perform a chi-square test of this hypothesis. Can you reject H_0 at the $\alpha = 0.05$ level? *Reject H_0.*

Extending the Concepts

33. Fair die? A gambler rolls a die 600 times to determine whether or not it is fair. Following are the results.

Outcome	1	2	3	4	5	6
Observed	113	101	106	81	108	91

a. Let p_1 be the probability that the die comes up 1, let p_2 be the probability that the die comes up 2, and so on. Use the chi-square distribution to test the null hypothesis, at the $\alpha = 0.05$ level of significance, that the die is fair. *Do not reject H_0.*

b. The gambler decides to use the test for proportions (discussed in Section 9.4) to test H_0: $p_i = 1/6$ for each p_i. Find the P-values for each of these tests.

c. Show that the hypothesis H_0: $p_4 = 1/6$ is rejected at level 0.05.

d. The gambler now reasons as follows: "I reject H_0: $p_4 = 1/6$, and I conclude that $p_4 \neq 1/6$. Therefore, I can reject the null hypothesis that the die is fair." Explain why this reasoning is incorrect. (*Hint:* See Section 11.5 on the multiple testing problem.)

e. Use the Bonferroni correction (Section 11.5) to adjust the P-value for the test of $H_0: p_4 = 1/6$. Can you now conclude that the die is not fair? Explain. *Do not reject H_0.*

Answers to Check Your Understanding Exercises for Section 12.1

1. 28.869

2. 7.779

3. 0.01

4. 0.025

5.

Sunday	Monday	Tuesday	Wednesday	Thursday	Friday	Saturday
50	50	50	50	50	50	50

6.

Sunday	Monday	Tuesday	Wednesday	Thursday	Friday	Saturday
17.5	59.5	59.5	59.5	59.5	59.5	35.0

7. a.

1	2	3	4	5
21.0	17.5	14.0	10.5	7.0

b. Yes, because all the expected frequencies are 5 or more.
c. 12.748 **d.** 4 **e.** 9.488 **f.** Yes
g. 13.277 **h.** No

8. a.

1	2	3	4	5
28.0	21.0	7.0	10.5	3.5

b. No, one of the expected frequencies is less than 5.

Section 12.2 Tests for Independence and Homogeneity

Objectives

1. Interpret contingency tables
2. Perform tests of independence
3. Perform tests of homogeneity

Objective 1 Interpret contingency tables

Contingency Tables

Do some college majors require more studying than others? The National Survey of Student Engagement asked a number of college freshmen what their major was and how many hours per week they spent studying, on average. A sample of 1000 of these students was chosen, and the numbers of students in each category are presented in Table 12.5.

Andersen Ross/Getty Images

Table 12.5 Observed Frequencies

Hours Studying Per Week	Major			
	Humanities	Social Science	Business	Engineering
0–10	68	106	131	40
11–20	119	103	127	81
More Than 20	70	52	51	52

Table 12.5 is called a **contingency table**. A contingency table relates two qualitative variables. One of the variables, called the **row variable**, has one category for each row of the table. The other variable, called the **column variable**, has one category for each column of the table. In Table 12.5, hours studying is the row variable and major is the column variable. In general, it does not matter which variable is the row variable and which is the column variable. We could just as well have made major the row variable and hours studying the column variable. The intersection of a row and a column is called a **cell**. For example, the number 68 appears in the upper left cell, which tells us that 68 students were humanities majors who study 0–10 hours per week.

Objective 2 Perform tests of independence

Performing a Test of Independence

We are interested in determining whether the distribution of one variable differs, depending on the value of the other variable. If so, the variables are *dependent*. If the distribution of one variable is the same for all the values of the other variable, the variables are *independent*. For Table 12.5, the null and alternate hypotheses are

H_0: Hours studying and major are independent.

H_1: Hours studying and major are not independent.

EXPLAIN IT AGAIN

Independence: If hours studied and major are independent, then the distribution of hours studied will be the same for all majors, and the distribution of majors will be the same for all categories of hours studied.

We will use the chi-square statistic to test the null hypothesis that major and hours studying are independent. If we reject H_0, we will conclude that the variables are dependent. The values in Table 12.5 are the observed frequencies. To compute the value of χ^2, we must compute the expected frequencies.

Computing the expected frequencies

The first step in computing the expected frequencies is to compute the row and column totals. For example, the total in the first row is

$$\text{Total number of students studying 0–10 hours} = 68 + 106 + 131 + 40 = 345$$

The total in the first column is

$$\text{Total number of humanities majors} = 68 + 119 + 70 = 257$$

Table 12.6 presents Table 12.5 with the row and column totals included. The total number of individuals in the table, 1000, is included as well. This total is often called the **grand total**.

Table 12.6 Observed Frequencies with Row and Column Totals

	Major				
Hours Studying Per Week	**Humanities**	**Social Science**	**Business**	**Engineering**	**Row Total**
0–10	68	106	131	40	345
11–20	119	103	127	81	430
More Than 20	70	52	51	52	225
Column Total	257	261	309	173	1000

RECALL

The Multiplication Rule for Independent Events says that if events *E* and *F* are independent, then $P(E \text{ and } F) = P(E) \cdot P(F)$.

As with any hypothesis test, we begin by assuming the null hypothesis to be true. The null hypothesis says that hours studying and major are independent. We can now use the Multiplication Rule for Independent Events to compute the expected frequencies. For example,

$$P(\text{Study 0–10 hours and Humanities major}) = P(\text{Study 0–10 hours})P(\text{Humanities major})$$

Now out of a total of 1000 students, 345 studied 0–10 hours per week. Therefore,

$$P(\text{Study 0–10 hours}) = \frac{345}{1000}$$

Out of a total of 1000 students, 257 were humanities majors. Therefore,

$$P(\text{Humanities major}) = \frac{257}{1000}$$

The Multiplication Rule for Independent Events tells us that if H_0 is true, then

$$P(\text{Study 0–10 hours and Humanities major}) = \left(\frac{345}{1000}\right)\left(\frac{257}{1000}\right)$$

We obtain the expected frequency for those who study 0–10 hours and are humanities majors by multiplying this probability by the grand total, which is 1000.

$$\text{Expected frequency} = 1000\left(\frac{345}{1000}\right)\left(\frac{257}{1000}\right)$$

Before calculating this quantity, simplify it as follows:

$$\text{Expected frequency} = \cancel{1000}\left(\frac{345}{\cancel{1000}}\right)\left(\frac{257}{1000}\right) = \frac{345 \cdot 257}{1000} = 88.665$$

We see that the expected frequency can be computed as

$$\text{Expected frequency} = \frac{\text{Row total} \cdot \text{Column total}}{\text{Grand total}}$$

SUMMARY

To find the expected frequency for a cell, multiply the row total by the column total, then divide by the grand total.

$$E = \frac{\text{Row total} \cdot \text{Column total}}{\text{Grand total}}$$

The expected frequency for a cell represents the number of individuals we would expect to find in that cell under the assumption that the two variables are independent. If the differences between the observed and expected frequencies tend to be large, we will reject the null hypothesis of independence.

Check Your Understanding

1. The following contingency table presents observed frequencies. Compute the expected frequencies.

	1	2	3
A	13	8	27
B	18	21	35
C	19	13	15
D	20	17	27

Answer is on page 607.

Once the expected frequencies are computed, we check to determine whether all of them are at least 5. If so, we use the chi-square statistic as a test statistic. The number of degrees of freedom is $(r-1)(c-1)$, where r is the number of rows and c is the number of columns. We will first describe how to perform a hypothesis test using the critical value method and Table A.4. Then we will describe how to use the P-value method with technology.

RECALL

We can use the chi-square statistic whenever the expected frequencies are all at least 5.

Performing a Test of Independence

Step 1: State the null and alternate hypotheses. The null hypothesis says that the row and column variables are independent. The alternate hypothesis says that they are not independent.

Step 2: Compute the row and column totals.

Step 3: Compute the expected frequencies:

$$E = \frac{\text{Row total} \cdot \text{Column total}}{\text{Grand total}}$$

Check to be sure that all the expected frequencies are at least 5.

Step 4: Choose a level of significance α, and compute the test statistic:

$$\chi^2 = \sum \frac{(O-E)^2}{E}$$

Step 5: Find the critical value from Table A.4, using $(r-1)(c-1)$ degrees of freedom, where r is the number of rows and c is the number of columns. If χ^2 is greater than or equal to the critical value, reject H_0. Otherwise, do not reject H_0.

Step 6: State a conclusion.

Example 12.6

Perform a test of independence

Refer to Table 12.6. Can you conclude that major and hours studying are not independent? Use the $\alpha = 0.01$ level of significance.

Solution

Step 1: State the null and alternate hypotheses. The hypotheses are

H_0: Major and hours studying are independent.

H_1: Major and hours studying are not independent.

Step 2: Compute the row and column totals. These are shown in Table 12.6.

Step 3: Compute the expected frequencies. As an example, we compute the expected frequency for the cell corresponding to business major, studying 11–20 hours. The row total is 430, the column total is 309, and the grand total is 1000. The expected frequency is

$$E = \frac{430 \cdot 309}{1000} = 132.87$$

Table 12.7 presents the expected frequencies. All the expected frequencies are at least 5, so we can proceed.

Table 12.7 Expected Frequencies

	Major			
Hours Studying Per Week	**Humanities**	**Social Science**	**Business**	**Engineering**
0–10	88.665	90.045	106.605	59.685
11–20	110.510	112.230	132.870	74.390
More Than 20	57.825	58.725	69.525	38.925

Step 4: Choose a level of significance, and compute the test statistic. We will use the $\alpha = 0.01$ level of significance. To compute the test statistic, we use the observed frequencies in Table 12.6 and the expected frequencies in Table 12.7:

$$\chi^2 = \frac{(68 - 88.665)^2}{88.665} + \cdots + \frac{(52 - 38.925)^2}{38.925} = 34.638$$

Step 5: Find the critical value. There are $r = 3$ rows and $c = 4$ columns, so the number of degrees of freedom is $(3 - 1)(4 - 1) = 6$. From Table A.4, we find that the critical value corresponding to 6 degrees of freedom and $\alpha = 0.01$ is 16.812. The value of the test statistic is 34.638. Because $34.638 > 16.812$, we reject H_0.

Step 6: State a conclusion. We conclude that the choice of major and the number of hours spent studying are not independent. The numbers of hours that students study varies among majors.

Check Your Understanding

2. The following contingency table presents observed frequencies.

	Observed		
	1	**2**	**3**
A	8	13	14
B	18	1	15
C	16	17	15
D	19	19	8

a. Compute the expected frequencies.

b. One of the observed frequencies is less than 5. Is it appropriate to perform a test of independence? Explain. *Yes*

c. Compute the value of the test statistic. *20.973*

d. How many degrees of freedom are there? *6*

e. Do you reject H_0 at the $\alpha = 0.05$ level? *Yes*

Answers are on page 607.

Example 12.7

Perform a test of independence with technology

Use technology to test the hypothesis of independence of major and hours studied.

Solution

We present the results from MINITAB.

In the MINITAB output, each cell (intersection of row and column) contains three numbers. The top number is the observed frequency, the middle number is the expected frequency, and the bottom number is the contribution $\frac{(O-E)^2}{E}$ to the chi-square statistic from that cell. The *P*-value is given as 0.000. This means that the *P*-value is less than 0.0005, so when rounded to three decimal places, the value is 0.000.

Because $P < 0.01$, we reject H_0 at the $\alpha = 0.01$ level and conclude that the choice of major and the number of hours studying are not independent. This conclusion is the same as the one we reached in Example 12.6 using the critical value method.

Step-by-step instructions for using technology are presented in the Using Technology section on page 603.

```
Chi-Square Test: Humanities, Science, Business, Engineering

Expected counts are printed below observed counts
Chi-Square contributions are printed below expected counts

       Humanities   Science   Business   Engineering   Total
  1            68       106        131            40     345
            88.67     90.05     106.61         59.69
            4.816     2.827      5.582         6.492

  2           119       103        127            81     430
           110.51    112.23     132.87         74.39
            0.652     0.759      0.259         0.587

  3            70        52         51            52     225
            57.83     58.73      69.53         38.93
            2.563     0.770      4.936         4.392

Total         257       261        309           173    1000

Chi-Sq = 34.638, DF = 6, P-Value = 0.000
```

Objective 3 Perform tests of homogeneity

Tests of Homogeneity

In the contingency tables we have seen so far, the individuals in the table were sampled from a single population. For each individual, the values of both the row and column variables were random. In some cases, values of one of the variables (say, the row variable) are assigned by the investigator and are not random. In these cases, we consider the rows as representing separate populations, and we are interested in testing the hypothesis that the distribution of the column variable is the same for each row. This is known as a **test of homogeneity**. Following is an example.

The drugs telmisartan and ramipril are designed to reduce high blood pressure. In a clinical trial to compare the effectiveness of these drugs in preventing heart attacks, 25,620 patients were divided into three groups. One group took one telmisartan tablet each day, another took one ramipril tablet each day, and the third group took one tablet of each drug each day. The patients were followed for 56 months, and the numbers who suffered fatal and nonfatal heart attacks were counted. Table 12.8 on page 602 presents the results.

In this table, the patients were assigned a row category, so only the column variable is random. We are interested in performing a test of homogeneity, to test the hypothesis that the distribution of outcomes is the same for each row. We have already seen the method for performing a test of homogeneity. It is the same as the method for performing a test of independence.

Table 12.8

	Fatal Heart Attack	Nonfatal Heart Attack	No Heart Attack
Telmisartan only	598	431	7513
Ramipril only	603	400	7573
Both drugs	620	424	7458

Source: www.clinicaltrials.gov

The method for performing a test of homogeneity is identical to the method for performing a test of independence.

Example 12.8

Perform a test of homogeneity

Refer to Table 12.8. Can you conclude that the distribution of outcomes differs among the treatment groups? Use the $\alpha = 0.05$ level.

EXPLAIN IT AGAIN

Interpretation of a test of homogeneity: If we reject the null hypothesis, we conclude that the distributions are not all the same, but we cannot tell which ones are different.

Solution

We follow the same steps as for a test of independence.

Step 1: State the null and alternate hypotheses. The null hypothesis says that the distribution of outcomes (fatal heart attack, nonfatal heart attack, no heart attack) is the same for all the drug treatments. The alternate hypothesis says that the distribution of outcomes is not the same for all the treatments.

Step 2: Compute the row and column totals. These are shown in the following table.

	Fatal Heart Attack	Nonfatal Heart Attack	No Heart Attack	Row Total
Telmisartan only	598	431	7,513	8,542
Ramipril only	603	400	7,573	8,576
Both drugs	620	424	7,458	8,502
Column Total	1,821	1,255	22,544	25,620

Step 3: Compute the expected frequencies. As an example, we compute the expected frequency for the cell corresponding to Telmisartan, Fatal heart attack. The row total is 8542, the column total is 1821, and the grand total is 25,620. The expected frequency is

$$\frac{8542 \cdot 1821}{25{,}620} = 607.14$$

The following table presents the expected frequencies. All the expected frequencies are at least 5, so we can proceed.

	Fatal Heart Attack	Nonfatal Heart Attack	No Heart Attack
Telmisartan only	607.14	418.43	7516.43
Ramipril only	609.56	420.10	7546.34
Both drugs	604.30	416.47	7481.23

Step 4: Choose a level of significance, and compute the test statistic. We will use the $\alpha = 0.05$ level of significance. The test statistic is

$$\chi^2 = \frac{(598 - 607.14)^2}{607.14} + \cdots + \frac{(8502 - 7481.23)^2}{7481.23} = 2.259$$

Step 5: Find the critical value. There are $r = 3$ rows and $c = 3$ columns, so the number of degrees of freedom is $(3 - 1)(3 - 1) = 4$. From Table A.4, we find that the critical value corresponding to 4 degrees of freedom and $\alpha = 0.05$ is 9.488. The value of the test statistic is 2.259. Since $2.259 < 9.488$, we do not reject H_0.

Step 6: State a conclusion. There is not enough evidence to conclude that the distribution of outcomes is different for different drug treatments.

Using Technology

We use Table 12.5 to illustrate the technology steps.

TI-84 PLUS

Testing for independence

Step 1. Press **2nd**, then **MATRIX** to access the Matrix menu. Highlight **EDIT** and press **ENTER**. Select **1:[A]**.

Step 2. To input the data from Table 12.5, enter the size of the matrix as **3 × 4**. Enter each of the data values (Figure A; note that the last column does not show).

Step 3. Press **STAT** and highlight the **TESTS** menu. Select **χ^2–Test** and press **ENTER** (Figure B). The **χ^2–Test** menu appears.

Step 4. Enter **[A]** in the **Observed** field. The default value for the **Expected** field is **[B]** (Figure C).

Step 5. Highlight **Calculate** and press **ENTER** (Figure D).

Note: The expected frequencies can be viewed by accessing matrix **[B]** from the Matrix menu.

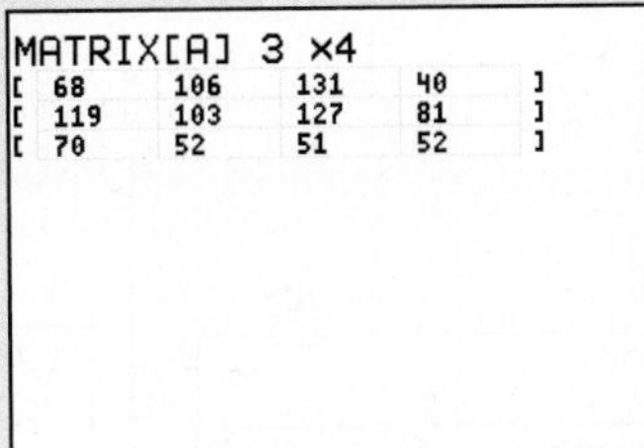

MATRIX[A] 3 ×4
[68 106 131 40]
[119 103 127 81]
[70 52 51 52]

Figure A

```
EDIT CALC TESTS
5↑1-PropZTest...
6:2-PropZTest...
7:ZInterval...
8:TInterval...
9:2-SampZInt...
0:2-SampTInt...
A:1-PropZInt...
B:2-PropZInt...
C↓χ²-Test...
```

Figure B

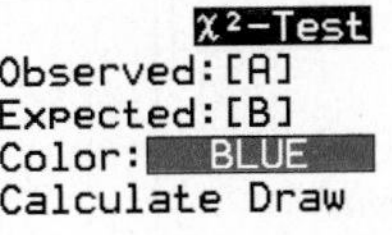

Figure C

```
χ²-Test
χ²=34.63775048
p=5.0648653E-6
df=6
```

Figure D

EXCEL

Testing independence

The **CHISQ.TEST** command returns the *P*-value for a test of independence given the observed and expected frequencies.

Step 1. Enter the observed and expected frequencies into EXCEL. For Table 12.5, the observed frequencies are contained in B3:E5 and the expected frequencies are in B9:E11 (Figure E).

Step 2. In an empty cell, select the **Insert Function** icon and highlight **Statistical** in the category field.

Step 3. Click on the **CHISQ.TEST** function, and press **OK**.

Step 4. In the **Actual_range** field, enter or select the range of cells containing the observed frequencies. In the **Expected_range** field, enter or select the range of cells containing the expected frequencies.

Step 5. Click **OK** to obtain the *P*-value (Figure F).

	A	B	C	D	E
1			OBSERVED		
2	Hours	Humanities	Social Science	Business	Engineering
3	0 - 10	68	106	131	40
4	11 - 20	119	103	127	81
5	More than 20	70	52	51	52
6					
7			EXPECTED		
8	Hours	Humanities	Social Science	Business	Engineering
9	0 - 10	88.665	90.045	106.605	59.685
10	11 - 20	110.51	112.23	132.87	74.39
11	More than 20	57.825	58.725	69.525	38.925

Figure E

Figure F

MINITAB

Testing for independence

Step 1. Enter the data from Table 12.5 as shown in Figure G.

Step 2. Click on **Stat**, then **Tables**, then **Chi-Square Test for Association.**

Step 3. Select **Summarized data in a two-way table** and then select all columns **(Humanities–Engineering)** in the **Columns containing the table** field (Figure H).

Step 4. Click **OK** (Figure I).

↓	C1	C2	C3	C4
	Humanities	Social Science	Business	Engineering
1	68	106	131	40
2	119	103	127	81
3	70	52	51	52

Figure G

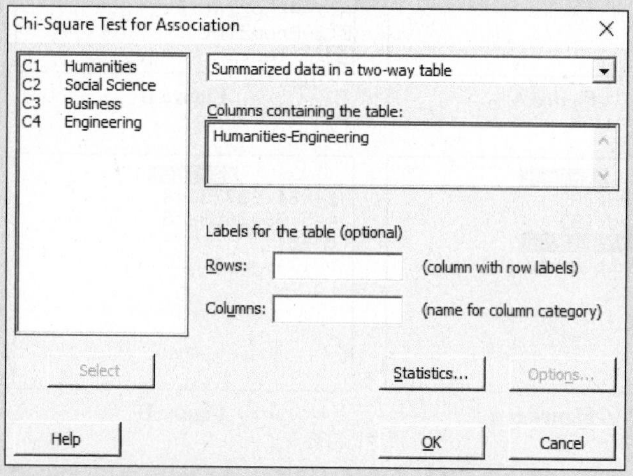

Chi-Square Test for Association

C1 Humanities
C2 Social Science
C3 Business
C4 Engineering

Summarized data in a two-way table

Columns containing the table:
Humanities-Engineering

Labels for the table (optional)

Rows: (column with row labels)

Columns: (name for column category)

Select | Statistics... | Options...

Help | OK | Cancel

Figure H

Rows: Worksheet rows Columns: Worksheet columns

	Humanities	Social Science	Business	Engineering	All
1	68	106	131	40	345
	88.67	90.05	106.61	59.69	
2	119	103	127	81	430
	110.51	112.23	132.87	74.39	
3	70	52	51	52	225
	57.83	58.73	69.53	38.92	
All	257	261	309	173	1000

Cell Contents
Count
Expected count

Chi-Square Test

	Chi-Square	DF	P-Value
Pearson	34.638	6	0.000
Likelihood Ratio	35.334	6	0.000

Figure I

Section 12.2 Exercises

Exercises 1 and 2 are the Check Your Understanding exercises located within the section.

Understanding the Concepts

In Exercises 3–5, fill in each blank with the appropriate word or phrase.

3. To calculate the expected frequencies, we must know the row totals, the column totals, and the ________ total. *grand*

4. We reject H_0 if the value of the test statistic is ________ the critical value. *greater than or equal to*

5. In the test for ________, the null hypothesis is that the distribution of the column variable is the same in each row. *homogeneity*

In Exercises 6–8, determine whether the statement is true or false. If the statement is false, rewrite it as a true statement.

6. A contingency table containing observed values has three rows and four columns. The number of degrees of freedom for the chi-square statistic is 7. *False*

7. In a test of homogeneity, the alternate hypothesis says that the distributions in the rows are the same. *False*

8. The procedure for testing homogeneity is the same as the procedure for testing independence. *True*

Practicing the Skills

9. For the given table of observed frequencies:

	1	2	3
A	15	10	12
B	3	11	11
C	9	14	12

a. Compute the row totals, the column totals, and the grand total.

b. Construct the corresponding table of expected frequencies.

c. Compute the value of the chi-square statistic. *6.481*

d. How many degrees of freedom are there? *4*

e. If appropriate, perform a test of independence, using the $\alpha = 0.05$ level of significance. If not appropriate, explain why. *Do not reject H_0.*

10. For the given table of observed frequencies:

	1	2	3
A	25	4	11
B	3	3	4
C	42	3	5

a. Compute the row totals, the column totals, and the grand total.
b. Construct the corresponding table of expected frequencies.
c. Compute the value of the chi-square statistic. *14.432*
d. How many degrees of freedom are there? *4*
e. If appropriate, perform a test of independence, using the $\alpha = 0.05$ level of significance. If not appropriate, explain why. *Not appropriate*

Working with the Concepts

11. Carbon monoxide: A recent study examined the effects of carbon monoxide exposure on a group of construction workers. The following table presents the numbers of workers who reported various symptoms, along with the shift (morning, evening, or night) that they worked.

	Shift		
	Morning	Evening	Night
Influenza	16	13	18
Headache	24	33	6
Weakness	11	16	5
Shortness of Breath	7	9	9

Source: *Journal of Environmental Science and Health* A39:1129–1139

a. Compute the expected frequencies under the null hypothesis.
b. Compute the value of the chi-square statistic. *17.572*
c. How many degrees of freedom are there? *6*
d. Can you conclude that symptoms and shift are not independent? Use the $\alpha = 0.05$ level of significance. *Reject H_0.*

12. Beryllium disease: Beryllium is an extremely lightweight metal that is used in many industries, such as aerospace and electronics. Long-term exposure to beryllium can cause people to become sensitized. Once an individual is sensitized, continued exposure can result in chronic beryllium disease, which involves scarring of the lungs. In a study of the effects of exposure to beryllium, workers were categorized by their duration of exposure (in years) and by their disease status (diseased, sensitized, or normal). The results were as follows:

	Duration of Exposure		
	< 1	1 to < 5	≥ 5
Diseased	10	8	23
Sensitized	9	19	11
Normal	70	136	206

Source: *Environmental Health Perspectives* 113:1366–1372

a. Compute the expected frequencies under the null hypothesis.
b. Compute the value of the chi-square statistic. *10.829*
c. How many degrees of freedom are there? *4*
d. Can you conclude that disease status and duration of exposure are not independent? Use the $\alpha = 0.01$ level of significance. *Do not reject H_0.*

13. No smoking: The General Social Survey conducted a poll of 668 adults in which the subjects were asked whether they agree that the government should prohibit smoking in public places. In addition, each was asked how many people lived in his or her household. The results are summarized in the following contingency table.

	Household Size				
	1	2	3	4	5
Agree	73	109	48	37	37
No Opinion	31	52	29	20	18
Disagree	42	71	38	40	23

a. Compute the expected frequencies.
b. Compute the value of the test statistic. *6.377*
c. How many degrees of freedom are there? *8*
d. Can you conclude that opinion and household size are not independent? Use the $\alpha = 0.01$ level of significance. *Do not reject H_0.*

14. How big is your family? The General Social Survey asked a sample of 2780 adults how many children they had, and also how many siblings (brothers and sisters) they had. The results are summarized in the following contingency table.

	Number of Children					
Siblings	0	1	2	3	4	More Than 4
0	60	19	34	15	10	6
1	189	97	163	63	29	12
2	179	98	137	81	41	20
3	124	85	132	83	31	16
4	85	53	88	61	32	22
More Than 4	128	121	175	156	77	58

a. Compute the expected frequencies.
b. Compute the value of the test statistic. *126.444*
c. How many degrees of freedom are there? *25*
d. Can you conclude that the number of siblings and the number of children are not independent? Use the $\alpha = 0.05$ level of significance. *Reject H_0.*

15. Age discrimination: The following table presents the numbers of employees, by age group, who were promoted, or not promoted, in a sample drawn from a certain industry during the past year.

	Age			
	Under 30	30–39	40–49	50 and Over
Promoted	16	22	26	15
Not Promoted	38	39	45	38

Can you conclude that the people in some age groups are more likely to be promoted than those in other age groups? Use the $\alpha = 0.05$ level of significance. *Do not reject H_0.*

16. Museums: Do people who are interested in environmental issues visit museums more often than people who are not? The General Social Survey asked 990 people how interested they were in environmental issues and how often they had visited a science museum in the past year. The following table presents the results.

	Number of Visits		
	None	One	More Than One
Very interested	331	96	40
Moderately interested	347	57	25
Not interested	70	15	9

Can you conclude that the number of visits to a science museum is related to interest in environmental issues? Use the $\alpha = 0.01$ level of significance. *Do not reject H_0.*

17. Genes: At a certain genetic locus on a chromosome, each individual has one of three different DNA sequences (alleles). The three alleles are denoted A, B, C. At another genetic locus on the same chromosome, each organism has one of three alleles, denoted 1, 2, 3. Each individual therefore has one of nine possible allele pairs: A1, A2, A3, B1, B2, B3, C1, C2, or C3. These allele pairs are called *haplotypes*. The loci are said to be in *linkage equilibrium* if the two alleles in an individual's haplotype are independent. Haplotypes were determined for 316 individuals. The following MINITAB output presents the results of a chi-square test for independence.

```
Chi-Square Test: A, B, C

Expected counts are printed below observed counts
Chi-Square contributions are printed below expected counts

          A       B       C   Total
  1      66      44      34     144
      61.06   47.39   35.54
      0.399   0.243   0.067

  2      36      38      20      94
      39.86   30.94   23.20
      0.374   1.613   0.442

  3      32      22      24      78
      33.08   25.67   19.25
      0.035   0.525   1.170

Total   134     104      78     316

Chi-Sq= 4.868, DF = 4, P-Value = 0.301
```

a. How many individuals were observed to have the haplotype B3? *22*

b. What is the expected number of individuals with the haplotype A2? *39.86*

c. Which of the nine haplotypes was least frequently observed? *C2*

d. Which of the nine haplotypes has the smallest expected count? *C3*

e. What is the *P*-value? *0.301*

f. Can you conclude at the $\alpha = 0.05$ level that the loci are not in linkage equilibrium (i.e., not independent)? Explain. *No*

g. Can you conclude that the loci are in linkage equilibrium (i.e., independent)? Explain. *No*

18. Product rating: A firm that is planning to market a new cleaning product surveyed 1268 users of the leading competitor's product. Each person rated the product as fair, good, or excellent. In addition, each person stated whether they use the product rarely (1), occasionally (2), or frequently (3). The firm is interested in determining whether the rating given to the product is independent of the frequency with which the product is used. The following MINITAB output presents the results of a chi-square test for independence.

```
Chi-Square Test: Fair, Good, Excellent

Expected counts are printed below observed counts
Chi-Square contributions are printed below expected counts

         Fair     Good   Excellent   Total
  1        97      136          92     325
        79.97   141.23      103.81
        3.627    0.193       1.343

  2       128      234         155     517
       127.21   224.66      165.13
        0.005    0.388       0.621

  3        87      181         158     426
       104.82   185.12      136.06
        3.030    0.091       3.536

Total     312      551         405    1268

Chi-Sq= 12.835, DF = 4, P-Value = 0.012
```

a. How many individuals who rarely use the product rated it good? *136*

b. What is the expected number of individuals who used the product occasionally and rated it excellent? *165.13*

c. Which of the nine combinations was least frequently observed? *Frequent and fair*

d. Which of the nine combinations has the smallest expected count? *Rarely and fair*

e. What is the *P*-value? *0.012*

f. Can you conclude at the $\alpha = 0.05$ level that rating and frequency of use are not independent? Explain. *Yes*

Extending the Concepts

19. Degrees of freedom: In the following contingency table, the row and column totals are presented, but the data for one row and one column have been omitted. Thus, there are data for only $r - 1$ rows and $c - 1$ columns. Show that you can calculate the missing values.

	1	2	3	4	Row Total
A	10	52	29	—	98
B	25	38	10	—	92
C	—	—	—	—	147
Column Total	65	117	83	72	

Conclude that if $(r - 1)(c - 1)$ entries in the table are known, the remaining ones are automatically determined, so there are no degrees of freedom left.

20. Are you an optimist? The General Social Survey asked 1373 men and 993 women in the United States whether they agreed that they were generally optimistic about the future. The results are presented in the following table.

	Male	Female
Optimistic	1148	815
Pessimistic	225	178

a. Compute the value of the χ^2 statistic for testing the hypothesis that the opinion on optimism is independent of gender. *0.96455*

b. Compute the proportion of men who were optimistic. 0.836

c. Compute the proportion of women who were optimistic. 0.821

d. Compute the test statistic z for testing the null hypothesis that the two proportions are equal versus the alternative that they are not equal. 0.982115

e. Show that $\chi^2 = z^2$.

f. Use technology to find the P-value for each test. Show that the P-values are equal. Both are 0.3260.

g. Conclude that when a contingency table has two rows and two columns, the chi-square test is equivalent to the test for the difference between proportions.

Answers to Check Your Understanding Exercises for Section 12.2

1.

	1	2	3
A	14.421	12.155	21.425
B	22.232	18.738	33.030
C	14.120	11.901	20.979
D	19.227	16.206	28.567

2. a.

	1	2	3
A	13.098	10.736	11.166
B	12.724	10.429	10.847
C	17.963	14.724	15.313
D	17.215	14.110	14.675

b. Yes, because all the expected frequencies are 5 or more.

c. 20.973 d. 6 e. Yes

Chapter 12 Summary

Section 12.1: In this section, we introduced the chi-square test for goodness of fit. This test is based on the chi-square statistic. The data consist of observed frequencies in several categories. The null hypothesis specifies a distribution, which consists of a probability for each category. The chi-square statistic is used to test this null hypothesis. The degrees of freedom is 1 less than the number of categories. We reject H_0 if the value of the test statistic is greater than or equal to the critical value.

Section 12.2: In this section, we studied contingency tables. A contingency table relates the values of two qualitative variables. When both row and column totals are random, we test the hypothesis that the two variables are independent. When one of the variables is assigned, and the other is random, we test the hypothesis of homogeneity. The procedure is the same for both tests. The test statistic is the chi-square statistic with $(r - 1)(c - 1)$ degrees of freedom, where r is the number of rows and c is the number of columns. We reject H_0 if the value of the test statistic is greater than or equal to the critical value.

Vocabulary and Notation

chi-square distribution 588
chi-square statistic 588
cell 597
column variable 597
contingency table 597
expected frequency 590
goodness-of-fit test 589
grand total 598
observed frequency 589
row variable 597
test of homogeneity 601
test of independence 597

Important Formulas

Chi-square statistic:

$$\chi^2 = \Sigma \frac{(O - E)^2}{E}$$

Expected frequency for goodness-of-fit:

$E = np$

Expected frequency for independence or homogeneity:

$$E = \frac{\text{Row total} \cdot \text{Column total}}{\text{Grand total}}$$

Chapter Quiz

1. A contingency table containing observed values has four rows and five columns. The value of the chi-square statistic for testing independence is 22.87. Is H_0 rejected at the $\alpha = 0.05$ level? Yes

2. A goodness-of-fit test is performed to test the null hypothesis that each of the six faces on a die has probability 1/6 of coming up. The null hypothesis is rejected. True or false: We can conclude that the probability that a 6 comes up is not equal to 1/6. False

3. A test of homogeneity is performed, and the null hypothesis is rejected. True or false: We can conclude that the distributions are not the same in every row. *True*

Exercises 4–9 refer to the following data:

Electric motors are assembled on four different production lines. Random samples of motors are taken from each line and inspected. The numbers that pass and that fail the inspection are counted for each line, with the following results:

	Line			
	1	2	3	4
Pass	482	467	458	404
Fail	57	59	37	47

4. State the appropriate null and alternate hypotheses for determining whether to conclude that the failure rates differ among the four lines.

5. Compute the expected frequencies.

6. Compute the value of the chi-square statistic. *4.676*

7. How many degrees of freedom are there? *3*

8. Find the critical value for the $\alpha = 0.05$ level of significance. *7.815*

9. State a conclusion. *Do not reject H_0.*

Exercises 10–15 refer to the following data:

Anthropologists can estimate the birthrate of an ancient society by studying the age distribution of skeletons found in ancient cemeteries. An article in the journal *Current Anthropology* presented the following numbers of skeletons of various ages found at two such sites.

	Ages of Skeletons		
Site	**0–4 Years**	**5–19 Years**	**20 Years or More**
Casa da Moura	27	61	126
Wandersleben	38	60	118

Source: *Current Anthropology* 43:637–650

10. State the appropriate null and alternate hypotheses for determining whether to conclude that the age distributions differ between the two sites.

11. Compute the expected frequencies.

12. Compute the value of the chi-square statistic. *2.123*

13. How many degrees of freedom are there? *2*

14. Find the critical value for the $\alpha = 0.01$ level of significance. *9.210*

15. State a conclusion. *Do not reject H_0.*

Review Exercises

Exercises 1–3 refer to the following data:

Two hundred people each tossed four coins, and the number of heads was recorded for each. The results are presented in the following table.

Number of heads	0	1	2	3	4
Observed	10	57	80	38	15

1. Null hypothesis: Let p_0 be the probability that no heads appear in a sequence of four coin tosses, p_1 be the probability that one head appears, and so on. The null hypothesis is that the number of heads in a sequence of four coin tosses follows a binomial distribution with $n = 4$ and $p = 0.5$. State the null hypothesis in terms of p_0, p_1, p_2, p_3, and p_4.

2. Expected frequencies: Compute the expected frequencies.

3. State a conclusion: Can you conclude that the distribution of the number of heads differs from the binomial with $p = 0.5$? Use the $\alpha = 0.05$ level of significance. *Do not reject H_0.*

Exercises 4–6 refer to the following data:

The General Social Survey polled 1280 men and 1531 women to determine their level of education. The results are presented in the following table.

	Educational Level				
	No High School Diploma	**High School Diploma**	**Associate's Degree**	**Bachelor's Degree**	**Graduate Degree**
Men	178	608	96	248	150
Women	186	827	128	259	131

4. **Expected frequencies:** Compute the expected frequencies under the null hypothesis of independence.
5. **Test statistic:** Compute the value of the chi-square statistic. *17.419*
6. **State a conclusion:** Can you conclude that education level is not independent of gender? Use the $\alpha = 0.01$ level of significance. *Reject H_0.*

Exercises 7–9 refer to the following data:

At an assembly plant for light trucks, routine monitoring of the quality of welds yielded the following data.

	Number of Welds		
	High Quality	**Moderate Quality**	**Low Quality**
Day Shift	467	191	42
Evening Shift	445	171	34
Night Shift	254	129	17

7. **Expected frequencies:** Compute the expected frequencies under the null hypothesis of homogeneity.
8. **Test statistic:** Compute the value of the chi-square statistic. *5.760*
9. **State a conclusion:** Can you conclude that the quality varies among shifts? Use the $\alpha = 0.01$ level of significance. *Do not reject H_0.*

Exercises 10–12 refer to the following data:

In a hypothetical study, four hospitals were compared with regard to the outcome of a particular type of surgery. For each patient at each hospital, the outcome was classified as Substantial improvement, Some improvement, or No improvement. The results are presented in the following contingency table.

	Outcome		
Hospital	**Substantial Improvement**	**Some Improvement**	**No Improvement**
A	114	64	22
B	132	52	16
C	100	64	36
D	126	56	18

10. **Expected frequencies:** Compute the expected frequencies under the null hypothesis of homogeneity.
11. **Test statistic:** Compute the value of the chi-square statistic. *17.524*
12. **State a conclusion:** Can you conclude that the distribution of outcomes differs among the hospitals? Use the $\alpha = 0.05$ level of significance. *Reject H_0.*
13. **Can't read the numbers:** Because of printer failure, none of the observed frequencies in the following table were printed, but some of the row and column totals were. Is it possible to construct the corresponding table of expected frequencies from the information given? If so, construct it. If not, describe the additional information you would need. *Yes*

	Observed			
	1	**2**	**3**	**Total**
A	—	—	—	25
B	—	—	—	—
C	—	—	—	40
D	—	—	—	75
Total	50	20	—	150

14. Lottery: Mega Millions is a multistate lottery in which players try to guess the numbers that will turn up in a drawing of numbered balls. During a recent 39-month period, a total of 1690 balls numbered 1 through 75 were drawn. Following are the observed frequencies.

Number	Frequency	Number	Frequency	Number	Frequency	Number	Frequency	Number	Frequency
1	18	16	21	31	31	46	27	61	16
2	28	17	25	32	17	47	23	62	22
3	23	18	22	33	22	48	21	63	20
4	15	19	22	34	21	49	32	64	17
5	17	20	30	35	32	50	27	65	21
6	20	21	24	36	23	51	29	66	26
7	25	22	21	37	21	52	19	67	13
8	25	23	19	38	23	53	22	68	26
9	21	24	22	39	23	54	24	69	21
10	23	25	33	40	18	55	14	70	19
11	28	26	25	41	32	56	23	71	19
12	22	27	22	42	17	57	24	72	16
13	18	28	25	43	15	58	23	73	20
14	25	29	33	44	30	59	21	74	25
15	21	30	21	45	28	60	16	75	17

Source: www.usamega.com

Can you conclude that some numbers are more likely to come up than others? Use the $\alpha = 0.05$ level of significance. *Do not reject H_0*

15. Absent from school: Following are the total numbers of absences on each day of the week for a recent academic year at a Montana elementary school.

Monday	Tuesday	Wednesday	Thursday	Friday
844	909	781	795	837

Can you conclude that absences are more likely on some days of the week than on others? Use the $\alpha = 0.05$ level of significance. *Reject H_0.*

Write About It

1. Why do large values of χ^2 provide evidence against H_0? Why don't small values of χ^2 provide evidence against H_0?
2. Explain what the expected frequencies represent in a goodness-of-fit test.
3. If the row variable and column variable are interchanged, how are the expected frequencies affected? Is the value of the chi-square statistic affected? Is the number of degrees of freedom affected?
4. Explain what the expected frequencies represent in a test of independence.

In-Class Activities

1. **Addresses:** Each student provides the last digit of his or her address. Test whether all 10 digits are equally likely.
2. **Fair dice:** Each student tosses a die 60 times and performs a test of the null hypothesis that the numbers 1 through 6 are equally likely. How many students reject the null hypothesis at the $\alpha = 0.05$ level? Can you conclude that their dice are unfair?

Case Study: Gender Bias In Graduate Admissions

The chapter opener described a famous study carried out at the University of California at Berkeley in the 1970s, regarding what appeared to be a case of gender discrimination in admissions to graduate school. The information in this case study was taken from the article "Sex Bias in Graduate Admissions Data from Berkeley," which appeared in *Science* magazine.

Table 12.9 presents the numbers of male and female applicants to six of the most popular departments at the University of California at Berkeley. Out of 2691 male applicants, 1198, or 44.5%, were accepted. Out of 1835 female applicants, only 557, or 30.4%, were accepted.

Table 12.9

Gender	Accept	Reject	Total
Male	1198	1493	2691
Female	557	1278	1835
Total	1755	2771	4526

In graduate school, each department conducts its own admissions process. Table 12.10 presents the numbers of applicants accepted and rejected by each of the six departments.

Table 12.10

Department	A	B	C	D	E	F	Total
Accept	601	370	322	269	147	46	1755
Reject	332	215	596	523	437	668	2771
Total	933	585	918	792	584	714	4526

University policy does not allow these departments to be identified by name.

Table 12.11 presents the numbers of men and women who applied to each of the six departments.

Table 12.11

Department	A	B	C	D	E	F	Total
Male	825	560	325	417	191	373	2691
Female	108	25	593	375	393	341	1835
Total	933	585	918	792	584	714	4526

University policy does not allow these departments to be identified by name.

1. Use the data in Table 12.9 to test the null hypothesis that acceptance to graduate school is independent of gender. Show that this hypothesis is rejected at the $\alpha = 0.01$ level.
2. Use the data in Table 12.10 to test the null hypothesis that acceptance to graduate school is independent of department. Show that this hypothesis is rejected at the $\alpha = 0.01$ level.
3. Use the data in Table 12.11 to test the null hypothesis that gender is independent of department. Show that this hypothesis is rejected at the $\alpha = 0.01$ level.

We conclude that department is associated with both gender and admissions. Therefore, department is a confounder for the relationship between gender and admissions. This means that it is possible that the differences in admissions rates between men and women may be due to differences in the departments they apply to, rather than to gender. To determine whether this is the case, we must look at the admissions rates for men and women in each department separately. The following six tables present these data.

Department A			
Gender	Accept	Reject	Total
Male	512	313	825
Female	89	19	108
Total	601	332	933

Department B			
Gender	Accept	Reject	Total
Male	353	207	560
Female	17	8	25
Total	370	215	585

Department C			
Gender	Accept	Reject	Total
Male	120	205	325
Female	202	391	593
Total	322	596	918

Department D			
Gender	Accept	Reject	Total
Male	138	279	417
Female	131	244	375
Total	269	523	792

Department E			
Gender	Accept	Reject	Total
Male	53	138	191
Female	94	299	393
Total	147	437	584

Department F			
Gender	Accept	Reject	Total
Male	22	351	373
Female	24	317	341
Total	46	668	714

4. For each department, test the hypothesis that gender and admissions are independent. Use the $\alpha = 0.01$ level of significance. Show that the only department for which this hypothesis is rejected is department A. Show that, in fact, the admissions rate for women in department A is significantly higher than the rate for men.
5. For each department, compute the proportion of all applicants who are accepted.
6. For each department, compute the proportion of all applicants who are men.
7. In general, do the departments with the higher acceptance proportions tend to have the higher proportions of male applicants? *Yes*
8. Which of the following is the best explanation for the fact that the graduate admissions rate is lower for women than for men? *ii*
   i. The admissions process discriminates against women.
   ii. Women tend to apply to departments that are harder to get into.

The data set in this case study provides an example of a result known as *Simpson's Paradox*. Simpson's Paradox occurs when two variables in a contingency table are dependent, but the dependence vanishes or reverses when separate tables are made for each level of a confounder. The lesson to be learned is that lack of independence is not the same as causation. When we reject the null hypothesis of independence in an observational study, it is possible that the result may be due to confounding. To determine whether a certain variable is a confounder, it is necessary to make a separate table for each value of the confounder.

chapter

13

Inference in Linear Models

Vincent Ting/Getty Images

Introduction

Renewable energy sources have the potential to produce clean energy indefinitely. One form of renewable energy that has been extensively studied is wind. Windmills are used to produce electric power from the energy in the wind. Of course, stronger winds produce more power. The following data set, based on a model relating wind speed and generated power, represents measurements collected at a particular site on 42 different days. Wind speed is measured in miles per hour and power is measured in watts.

Wind Speed	Power	Wind Speed	Power	Wind Speed	Power	Wind Speed	Power
4.2	10.3	4.2	5.3	2.6	9.1	4.6	5.0
1.4	1.4	3.7	1.1	7.7	46.1	8.0	65.1
6.6	39.4	5.9	29.2	6.1	26.9	7.7	54.6
4.7	12.3	6.0	24.4	5.5	19.7	1.6	0.6
2.6	9.0	5.3	13.9	4.7	5.4	5.1	17.9
5.8	20.6	5.1	9.5	4.0	12.4	6.6	32.5
1.8	3.8	4.9	2.9	2.3	5.0	2.3	9.4
5.8	21.1	8.3	63.7	8.6	67.4	5.8	10.8
7.3	41.3	7.1	37.1	5.6	15.2	6.9	29.3
7.1	45.2	9.2	87.5	4.2	11.0	6.2	21.4
6.4	28.4	4.4	0.5				

It is important to be able to predict the amount of power that will be obtained from a given wind speed. In the case study at the end of the chapter, we will use the methods of the chapter to determine how precisely we can predict power from wind speed.

Section 13.1 Inference on the Slope of the Regression Line

Objectives

1. State the assumptions of the linear model
2. Check the assumptions of the linear model
3. Construct confidence intervals for the slope
4. Test hypotheses about the slope

Objective 1 State the assumptions of the linear model

The Linear Model

In Chapter 4, we discussed bivariate data—data that are in the form of ordered pairs. We learned to construct scatterplots, and we found that, in many cases, the points on a scatterplot tend to cluster around a straight line. We learned to summarize a scatterplot by computing the least-squares regression line.

Before continuing, we will review these ideas with a data set based on a nutritional analysis of more than 700 candy products conducted by the U.S. Department of Agriculture. Table 13.1 presents the number of calories and the number of grams of fat per 100 grams of product for a sample of 18 candy products.

NOTE TO INSTRUCTOR

The methods in this chapter are computationally intensive. It is best to focus on the interpretation of results and on the use of technology to produce them.

Table 13.1 Grams of Fat and Number of Calories per 100 Grams of Product

Product	Fat (x)	Calories (y)	Product	Fat (x)	Calories (y)
3 Musketeers	12.75	436	Mr. Goodbar	33.21	538
Kit Kat	25.99	518	100 Grand	19.33	468
M&M Plain	21.13	492	Baby Ruth	21.60	459
M&M Peanut	26.13	515	Bit O' Honey	7.50	375
Milky Way	17.23	456	Butterfinger	18.90	459
Skittles	4.37	405	Oh Henry!	23.00	462
Snickers	23.85	491	Reese's Pieces	24.77	497
Starburst	8.36	408	Tootsie Roll	3.31	387
Twix	24.85	502	Twizzlers	2.32	350

Source: USDA National Nutrient Database

The data consist of 18 points (x, y). To summarize the data, we compute the least-squares regression line, which was introduced in Section 4.2. We repeat the definition here.

The Least-Squares Regression Line

Let x be an explanatory variable and y an outcome variable, with means $\bar{x}$ and $\bar{y}$, standard deviations s_x and s_y, and correlation coefficient r. The equation of the least-squares regression line for predicting y from x is

$$\hat{y} = b_0 + b_1 x$$

Formulas for b_1 and b_0 are

$$b_1 = r\frac{s_y}{s_x}, \text{ where } r \text{ is the correlation coefficient}$$

$$b_0 = \bar{y} - b_1\bar{x}$$

An alternate formula for b_1 is

$$b_1 = \frac{\sum(x-\bar{x})(y-\bar{y})}{\sum(x-\bar{x})^2}$$

Example 13.1

Compute the least-squares regression line

Use technology to find the least-squares regression line for the data in Table 13.1.

Solution

The following TI-84 Plus display shows that the equation of the least-squares regression line is $\hat{y} = 356.7392193 + 5.639341034x$.

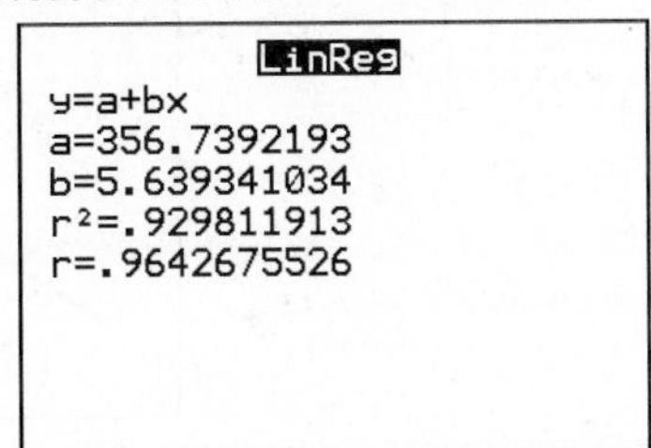

Note that the display labels the y-intercept as "a" and the slope as "b," whereas in our notation, we refer to the y-intercept as b_0 and the slope as b_1. In our notation, the intercept of the line is $b_0 = 356.7392193$ and the slope is $b_1 = 5.639341034$.

Figure 13.1 presents a scatterplot of the data, from MINITAB, with the least-squares regression line superimposed.

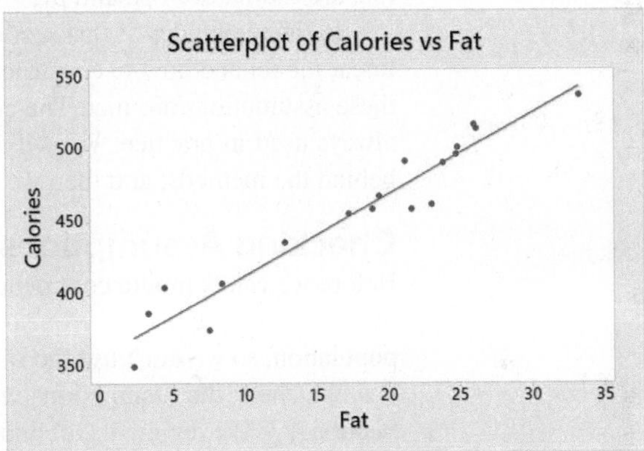

Figure 13.1 A scatterplot of the data in Table 13.1. The least-squares regression line is superimposed.

When the points on a scatterplot are a random sample from a population, we can imagine plotting every point in the population on a scatterplot. Then, if certain assumptions are met, we say that the population follows a **linear model**. The intercept b_0 and the slope b_1 of the least-squares regression line are then estimates of a population intercept β_0 and a population slope β_1. We cannot determine the exact values of β_0 and β_1, because we cannot observe the entire population. However, we can use the sample points to construct confidence intervals and test hypotheses about β_0 and β_1. We will focus on β_1.

The assumptions for a linear model are illustrated in Figure 13.2 on page 616, which presents a hypothetical scatterplot of an entire population. The scatterplot is divided up into narrow vertical strips, so that the points within each strip have approximately the same x-value. The following three conditions hold:

Assumptions for the Linear Model

1. The mean of the y-values within a strip is denoted $\mu_{y|x}$. As x varies, the values of $\mu_{y|x}$ follow a straight line: $\mu_{y|x} = \beta_0 + \beta_1 x$.
2. The amount of vertical spread is approximately the same in each strip, except perhaps near the ends.
3. The y-values within a strip are approximately normally distributed. (This is not obvious from the scatterplot.) This assumption is not necessary if the sample size is large ($n > 30$).

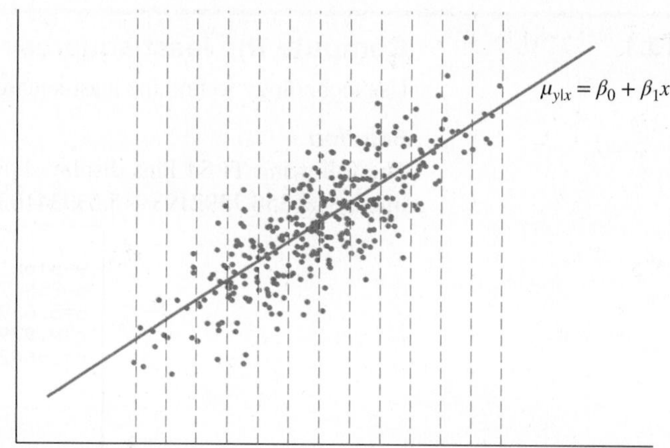

Figure 13.2 A scatterplot of a population that follows the linear model

When the assumptions of the linear model hold, the points (x, y) satisfy the following linear model equation:

$$y = \beta_0 + \beta_1 x + \varepsilon$$

In this equation, β_0 is the y-intercept, β_1 is the slope of the line that summarizes the entire population, and ε is a random error. The y-intercept b_0 and the slope b_1 of the least-squares line are estimates of β_0 and β_1.

In the remainder of this section, we will describe how to check the assumptions of the linear model and how to construct confidence intervals and perform hypothesis tests when these assumptions are met. The computations are rather lengthy, and technology is almost always used in practice. We will present the hand calculations first, to illustrate the ideas behind the methods, and then we will present examples using technology.

Objective 2 Check the assumptions of the linear model

Checking Assumptions

Before we can compute confidence intervals and perform hypothesis tests, we must check that the assumptions of the linear model are satisfied. In practice, we do not see the entire population, so we must use the sample to do the checking.

We check the assumptions with a residual plot. Residual plots were first discussed in Section 4.3. We review the definition here.

DEFINITION

Given a point (x, y) and the least-squares regression line $\hat{y} = b_0 + b_1 x$, the **residual** for the point (x, y) is the difference between the observed value y and the predicted value $\hat{y}$:

$$\text{Residual} = y - \hat{y}$$

A **residual plot** is a plot in which the residuals are plotted against the values of the explanatory variable x. In other words, the points on the residual plot are $(x,\ y - \hat{y})$.

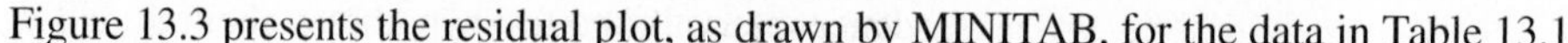

Figure 13.3 presents the residual plot, as drawn by MINITAB, for the data in Table 13.1.

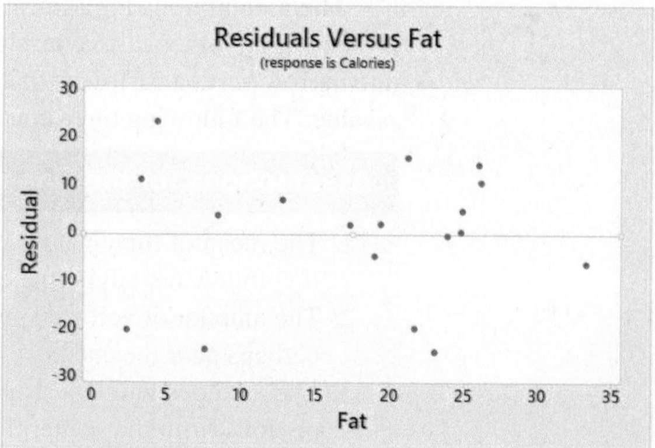

Figure 13.3 Residual plot for the data in Table 13.1

EXPLAIN IT AGAIN

Reasons for the conditions: Condition 1 checks that $\mu_{y|x}$ follows a straight line. Condition 2 checks that the variance of the *y*-values is the same for every *x*-value. Condition 3 checks that the *y*-values corresponding to a given *x* are approximately normally distributed.

When the linear model assumptions are satisfied, the residual plot will not show any obvious pattern. In particular, the residual plot must satisfy the following conditions.

Conditions for the Residual Plot

1. The residual plot must not exhibit an obvious pattern.
2. The vertical spread of the points in the residual plot must be roughly the same across the plot.
3. There must be no outliers.

Figure 13.3 satisfies all three conditions, so we can assume that the linear model is valid, and we can construct confidence intervals and test hypotheses.

Example 13.2

Checking the assumptions with a residual plot

For each of the following residual plots, determine whether the assumptions of the linear model are satisfied. If they are not, specify which assumptions are violated.

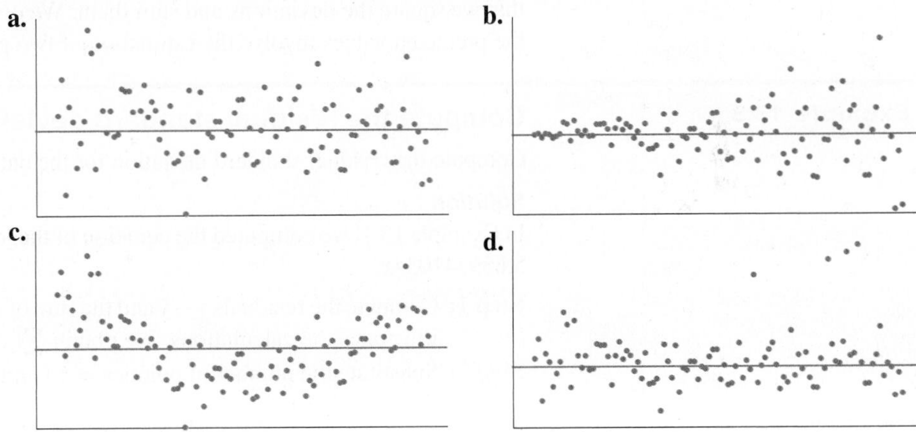

Solution

a. The plot shows no obvious pattern. The assumptions of the linear model are satisfied.
b. The vertical spread varies. The linear model is not valid.
c. The plot shows an obvious curved pattern. The linear model is not valid.
d. The plot contains outliers. The linear model is not valid.

Check Your Understanding

1. For each of the following residual plots, determine whether the assumptions of the linear model are satisfied. If they are not, specify which assumptions are violated. *Satisfied for (c), violated for (a), (b), (d)*

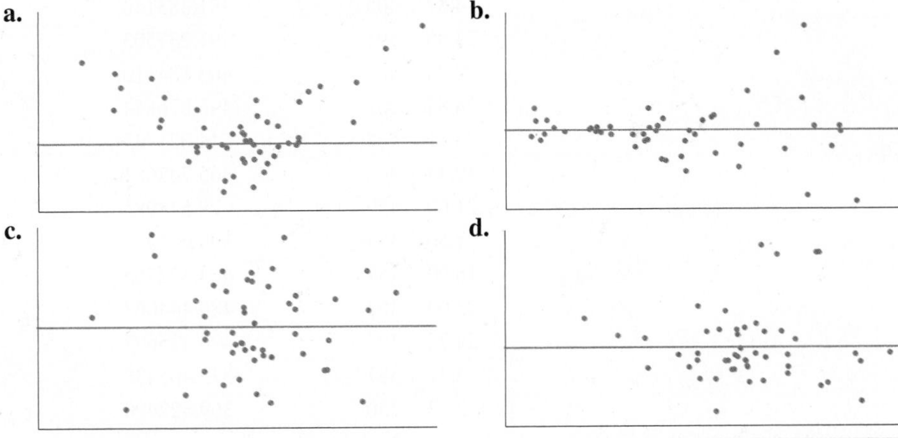

Answers are on page 629.

Objective 3 Construct confidence intervals for the slope

Constructing Confidence Intervals for the Slope

The slope b_1 of the least-squares regression line is a point estimate of the population slope β_1. When the assumptions of the linear model are satisfied, we can construct a confidence interval for β_1. To form a confidence interval, we need a point estimate, a standard error, and a critical value. The point estimate for β_1 is b_1.

The standard error of b_1

To compute the standard error of b_1, we first compute a quantity called the **residual standard deviation**. This quantity, denoted s_e, measures the spread of the points on the scatterplot around the least-squares regression line. The formula for s_e is

$$s_e = \sqrt{\frac{\sum(y - \hat{y})^2}{n - 2}}$$

To understand this formula, note that the predicted value $\hat{y}$ is an estimate of the mean of y. The residual $y - \hat{y}$ estimates the deviation between the observed data value y and its mean. The formula for s_e is similar to the formula for the sample standard deviation, in that we square the deviations and sum them. We divide by $n - 2$ rather than $n - 1$ because the predicted values involve the estimation of two parameters, β_0 and β_1.

Example 13.3

Compute the residual standard deviation

Compute the residual standard deviation for the data in Table 13.1.

Solution

In Example 13.1, we computed the equation of the least-squares line as $\hat{y} = 356.7392193 + 5.639341034x$.

Step 1: Compute the residuals $y - \hat{y}$ and the sum of squared residuals $\sum(y - \hat{y})^2$. Table 13.2 illustrates the calculations. We obtain $\sum(y - \hat{y})^2 = 3398.6787$.

Step 2: Substitute the number of points $n = 18$ and the value $\sum(y - \hat{y})^2 = 3398.6787$ into the formula for s_e:

$$s_e = \sqrt{\frac{\sum(y - \hat{y})^2}{n - 2}} = \sqrt{\frac{3398.6787}{18 - 2}} = 14.574547$$

Table 13.2

x	y	Predicted Value $\hat{y} = 356.73921925 + 5.63934103x$	Residual $y - \hat{y}$	Residual² $(y - \hat{y})^2$
12.75	436	428.640817	7.359183	54.15756880
25.99	518	503.305693	14.694307	215.92266939
21.13	492	475.898495	16.101505	259.25845638
26.13	515	504.095200	10.904800	118.91465510
17.23	456	453.905065	2.094935	4.38875183
4.37	405	381.383140	23.616860	557.75609746
23.85	491	491.237503	−0.237503	0.05640759
8.36	408	403.884110	4.115890	16.94054835
24.85	502	496.876844	5.123156	26.24672898
33.21	538	544.021735	−6.021735	36.26129068
19.33	468	465.747681	2.252319	5.07293926
21.60	459	478.548985	−19.548985	382.16283400
7.50	375	399.034277	−24.034277	577.64646971
18.90	459	463.322765	−4.322765	18.68629480
23.00	462	486.444063	−24.444063	597.51221301
24.77	497	496.425697	0.574303	0.32982444
3.31	387	375.405438	11.594562	134.43386660
2.32	350	369.822490	−19.822490	392.93112723

$\sum(y - \hat{y})^2 = 3398.6787$

EXPLAIN IT AGAIN

Relationship between $\sum(x-\bar{x})^2$ and the sample variance: We can compute $\sum(x-\bar{x})^2$ as $\sum(x-\bar{x})^2 = (n-1)s_x^2$, where s_x^2 is the sample variance of the x-values.

Once we have computed s_e, we divide it by the quantity $\sqrt{\sum(x-\bar{x})^2}$ to obtain the standard error of b_1. The quantity $\sum(x-\bar{x})^2$ is called the **sum of squares for x**. The standard error of b_1 is

$$s_b = \frac{s_e}{\sqrt{\sum(x-\bar{x})^2}}$$

Example 13.4

Compute the standard error of b_1

Compute the standard error s_b for the data in Table 13.1.

Solution

Step 1: Compute s_e, the residual standard deviation. We did this in Example 13.3 and obtained $s_e = 14.574547$.

Step 2: Compute $\sum(x-\bar{x})^2$. The calculations are presented in Table 13.3. We obtain $\sum(x-\bar{x})^2 = 1415.7452$.

Table 13.3

x	$x-\bar{x}$	$(x-\bar{x})^2$
12.75	−4.95	24.5025
25.99	8.29	68.7241
21.13	3.43	11.7649
26.13	8.43	71.0649
17.23	−0.47	0.2209
4.37	−13.33	177.6889
23.85	6.15	37.8225
8.36	−9.34	87.2356
24.85	7.15	51.1225
33.21	15.51	240.5601
19.33	1.63	2.6569
21.60	3.90	15.2100
7.50	−10.20	104.0400
18.90	1.20	1.4400
23.00	5.30	28.0900
24.77	7.07	49.9849
3.31	−14.39	207.0721
2.32	−15.38	236.5444
$\bar{x} = \frac{\sum x}{n} = 17.7$		$\sum(x-\bar{x})^2 = 1415.7452$

We now compute the standard error s_b:

$$s_b = \frac{s_e}{\sqrt{\sum(x-\bar{x})^2}} = \frac{14.574547}{\sqrt{1415.7452}} = 0.387349$$

The critical value and the margin of error

Under the assumptions of the linear model, the quantity

$$\frac{b_1 - \beta_1}{s_b}$$

has a Student's t distribution with $n-2$ degrees of freedom. Therefore, the critical value for a level $100(1-\alpha)\%$ confidence interval is the value $t_{\alpha/2}$ for which the area under the t curve with $n-2$ degrees of freedom between $-t_{\alpha/2}$ and $t_{\alpha/2}$ is $1-\alpha$.

The margin of error for a level $100(1-\alpha)\%$ confidence interval is

$$\text{Margin of error} = t_{\alpha/2} \cdot s_b$$

A level $100(1-\alpha)\%$ confidence interval for β_1 is

$$b_1 \pm t_{\alpha/2} \cdot s_b$$

$$b_1 - t_{\alpha/2} \cdot s_b < \beta_1 < b_1 + t_{\alpha/2} \cdot s_b$$

The steps for constructing a confidence interval for β_1 are as follows:

Constructing a Confidence Interval for β_1

Step 1: Compute the least-squares regression line $\hat{y} = b_0 + b_1 x$.

Step 2: Compute the residuals and construct a residual plot to be sure the assumptions of the linear model are satisfied.

Step 3: Compute the residual standard deviation: $s_e = \sqrt{\dfrac{\Sigma(y-\hat{y})^2}{n-2}}$.

Step 4: Compute the standard error of b_1: $s_b = \dfrac{s_e}{\sqrt{\Sigma(x-\bar{x})^2}}$.

Step 5: Find the critical value $t_{\alpha/2}$ from the Student's t curve with $n-2$ degrees of freedom, and multiply it by the standard error to obtain the margin of error $t_{\alpha/2} \cdot s_b$.

Step 6: Use the point estimate b_1 and the margin of error $t_{\alpha/2} \cdot s_b$ to compute the confidence interval.

$$\text{Point estimate} \pm \text{Margin of error}$$

$$b_1 \pm t_{\alpha/2} \cdot s_b$$

$$b_1 - t_{\alpha/2} \cdot s_b < \beta_1 < b_1 + t_{\alpha/2} \cdot s_b$$

Step 7: Interpret the result.

Example 13.5

Construct a confidence interval

Construct a 95% confidence interval for the slope β_1 for the candy data.

Solution

Step 1: Compute the least-squares regression line. We computed the equation of the least-squares regression line in Example 13.1 and obtained $\hat{y} = 356.7392193 + 5.639341034x$.

Step 2: Compute the residuals and construct a residual plot to be sure the assumptions of the linear model are satisfied. The residual plot was presented in Figure 13.3. The assumptions of the linear model are satisfied.

Step 3: Compute the residual standard deviation. We computed the residual standard deviation in Example 13.3 and obtained $s_e = 14.574547$.

Step 4: Compute the standard error of b_1. We computed the standard error of b_1 in Example 13.4 and obtained $s_b = 0.387349$.

Step 5: Find the critical value $t_{\alpha/2}$ from the Student's t curve with $n-2$ degrees of freedom, and multiply it by the standard error to obtain the margin of error $t_{\alpha/2} \cdot s_b$. There are $n - 2 = 18 - 2 = 16$ degrees of freedom for the Student's t distribution. From Table A.3, we find that the critical value for a 95% confidence interval is $t_{\alpha/2} = 2.120$. The margin of error is therefore

$$\text{Margin of error} = t_{\alpha/2} \cdot s_b = 2.120 \cdot 0.387349 = 0.82118$$

Step 6: Use the point estimate b_1 and the margin of error to construct the confidence interval. The point estimate is $b_1 = 5.63934103$. The margin of error is 0.82118. The 95% confidence interval is

$$\text{Point estimate} \pm \text{Margin of error}$$

$$5.63934103 \pm 0.82118$$

$$4.8182 < \beta_1 < 6.4605$$

Step 7: Interpret the result. We are 95% confident that the mean difference in calories for items that differ by 1 gram in fat content is between 4.8182 and 6.4605.

Example 13.6

Interpreting a confidence interval

A nutritionist believes that if two candies differ in their fat content by 1 gram, that their calorie count will differ, on average, by 5 calories. Is the confidence interval constructed in Example 13.5 consistent with this belief?

Solution

The parameter β_1 represents the mean difference in calories corresponding to a difference of 1 gram of fat. The confidence interval constructed in Example 13.5 is $4.8182 < \beta_1 < 6.4605$. This interval contains 5, so it is consistent with the belief.

Construct confidence intervals with technology

Example 13.7

Construct a confidence interval with technology

Use the TI-84 Plus calculator to construct the confidence interval in Example 13.5.

Solution

The TI-84 Plus display is as follows:

```
LinRegTInt
y=a+bx
(4.8182,6.4605)
b=5.639341034
df=16
s=14.57454704
a=356.7392193
r²=.929811913
r=.9642675526
```

The confidence interval is shown on the second line of the display. Step-by-step instructions for constructing confidence intervals with technology are presented in the Using Technology section on page 625.

Check Your Understanding

2. A certain data set contains 27 points. The least-squares regression line is computed, with the following results: $b_1 = 5.78$, $s_e = 1.35$, and $\sum(x - \bar{x})^2 = 3.4$. Construct a 95% confidence interval for β_1. *(4.27, 7.29)*

3. Following is a TI-84 Plus display showing a 95% confidence interval for β_1.

```
LinRegTInt
y=a+bx
(.17354,.2208)
b=.1971681416
df=10
s=.3254404775
a=1.077522124
r²=.9749242655
r=.9873825325
```

a. What is the slope of the least-squares regression line? *0.1971681416*

b. How many degrees of freedom are there? *10*

c. How many points are in the data set? *12*

d. What is the 95% confidence interval for β_1? *(0.17354, 0.2208)*

Answers are on page 629.

Objective 4 Test hypotheses about the slope

EXPLAIN IT AGAIN

$\beta_1 = 0$ means no linear relationship: If $\beta_1 = 0$, then the mean of the outcome variable is $\mu_{y|x} = \beta_0$ for every value of x. Therefore, no matter what the value of x, the mean of y stays the same.

Testing Hypotheses About the Slope

We can use the values of b_1 and s_b to test hypotheses about the population slope β_1. If $\beta_1 = 0$, then there is no linear relationship between the explanatory variable x and the outcome variable y. For this reason, the null hypothesis most often tested is $H_0\colon \beta_1 = 0$. If this null hypothesis is rejected, we conclude that there is a linear relationship between x and y and that the explanatory variable x is useful in predicting the outcome variable y.

Recall that the quantity

$$\frac{b_1 - \beta_1}{s_b}$$

has a Student's t distribution with $n - 2$ degrees of freedom. We construct the test statistic for testing $H_0\colon \beta_1 = 0$ by setting $\beta_1 = 0$. The test statistic is

$$t = \frac{b_1}{s_b}$$

When H_0 is true, the test statistic has a Student's t distribution with $n - 2$ degrees of freedom. If the assumptions of the linear model are satisfied, a test of the hypothesis $\beta_1 = 0$ can be performed. The steps are as follows:

Testing $H_0\colon \beta_1 = 0$

Step 1: Compute the least-squares regression line. Verify that the assumptions of the linear model are satisfied.

Step 2: State the null and alternate hypotheses. The null hypothesis is $H_0\colon \beta_1 = 0$. The alternate hypothesis may be stated in any of three ways:

Left-tailed: $H_1\colon \beta_1 < 0$
Right-tailed: $H_1\colon \beta_1 > 0$
Two-tailed: $H_1\colon \beta_1 \neq 0$

Step 3: If making a decision, choose a significance level α.

Step 4: Compute the standard error of the slope s_b.

Step 5: Compute the value of the test statistic $t = \dfrac{b_1}{s_b}$ and the number of degrees of freedom $n - 2$.

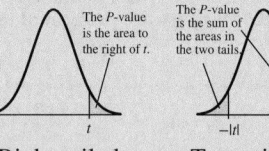

The *P*-Value Method

Step 6: Compute the *P*-value of the test statistic.

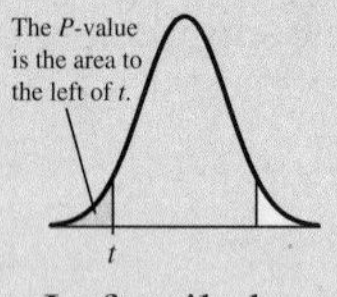
The P-value is the area to the left of t.
t

Left-tailed

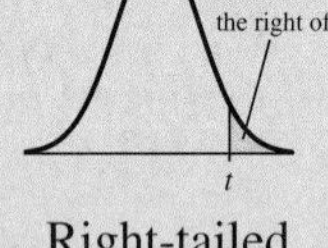
The P-value is the area to the right of t.
t

Right-tailed

The P-value is the sum of the areas in the two tails.
$-|t|$ $|t|$

Two-tailed

Step 7: Interpret the *P*-value. If making a decision, reject H_0 if the *P*-value is less than or equal to the significance level α.

Step 8: State a conclusion.

The Critical Value Method

Step 6: Find the critical value.

Step 7: Determine whether to reject H_0, as follows:

Left-tailed: $H_1\colon \beta_1 < 0$ Reject if $t \le -t_\alpha$.
Right-tailed: $H_1\colon \beta_1 > 0$ Reject if $t \ge t_\alpha$.
Two-tailed: $H_1\colon \beta_1 \neq 0$ Reject if $t \ge t_{\alpha/2}$ or $t \le -t_{\alpha/2}$.

Step 8: State a conclusion.

Example 13.8

Test a hypothesis about the slope

For the data in Table 13.1, perform a test of $H_0\colon \beta_1 = 0$ versus $H_1\colon \beta_1 > 0$. Use the $\alpha = 0.05$ level of significance.

Solution

Step 1: Compute the least-squares regression line, and verify that the assumptions of the linear model are satisfied. The least-squares regression line was computed in Example 13.1. We obtained $b_1 = 5.639341034$. The assumptions were verified in Figure 13.3.

Step 2: State the null and alternate hypotheses. The hypotheses are

$$H_0\colon \beta_1 = 0 \qquad H_1\colon \beta_1 > 0$$

Step 3: Choose a significance level. We will choose $\alpha = 0.05$.

Step 4: Compute s_b. In Example 13.4, we computed $s_b = 0.387349$.

Step 5: Compute the value of the test statistic. The value of s_b is 0.387349. The point estimate is $b_1 = 5.639341034$. We compute

$$t = \frac{b_1}{s_b} = \frac{5.639341034}{0.387349} = 14.56$$

Because the sample size is $n = 18$, the number of degrees of freedom is $n-2 = 16$.

P-Value Method

Step 6: Compute the P-value of the test statistic. We use technology to compute the P-value. The following TI-84 Plus calculator display presents the P-value in scientific notation. The notation **E-11** indicates that the decimal point should be moved 11 places to the left. Thus, the P-value is $P = 0.0000000000597$.

```
LinRegTTest
y=a+bx
β>0 and ρ>0
t=14.55880877
p=5.972739E-11
df=16
a=356.7392193
b=5.639341034
↓s=14.57454704
```

Step 7: Interpret the P-value. Because $P < 0.05$, we reject H_0. We conclude that $\beta_1 > 0$.

Step 8: State a conclusion. There is a linear relationship between the amount of fat and the number of calories in candy products. Because we conclude that $\beta_1 > 0$, we conclude that products with more fat tend to have more calories.

Critical Value Method

Step 6: Find the critical value. This is a right-tailed test, so the critical value is the value t_α for which the area to the right is $\alpha = 0.05$. We use Table A.3 with 16 degrees of freedom. The critical value is $t_\alpha = 1.746$.

Step 7: Determine whether to reject H_0. The value of the test statistic is $t = 14.56$. Because $t > t_\alpha$, we reject H_0. We conclude that $\beta_1 > 0$.

Step 8: State a conclusion. There is a linear relationship between the amount of fat and the number of calories in candy products. Because we conclude that $\beta_1 > 0$, we conclude that products with more fat tend to have more calories.

Testing the correlation

In Section 4.1, we introduced the sample correlation coefficient r, which is computed from a sample. If we knew the entire population and computed the correlation from it, we would obtain the **population correlation**, which is denoted with the Greek letter ρ (rho). The correlation measures the strength of the linear relationship between two variables. The population correlation ρ and the population slope β_1 always have the same sign. In particular, whenever one of them is equal to 0, the other is equal to 0 as well. For this reason, a test of the hypothesis $\beta_1 = 0$ is also a test of the hypothesis $\rho = 0$.

We now present a specialized test of H_0: $\rho = 0$. It always produces the same result as the test of H_0: $\beta_1 = 0$ that was illustrated in Example 13.8. The test statistic for testing H_0: $\rho = 0$ is

$$U = \frac{r\sqrt{n-2}}{\sqrt{1-r^2}}$$

where r is the correlation coefficient and n is the number of points. When H_0 is true, U has a Student's t-distribution with $n-2$ degrees of freedom. The value of U is always the same as the value of the test statistic used to test H_0: $\beta_1 = 0$.

NOTE TO INSTRUCTOR

The statistic U for testing H_0: $\rho = 0$ is equal to the statistic b_1/s_b for testing H_0: $\beta_1 = 0$. Therefore, these tests always produce the same P-value.

Example 13.9

Test a hypothesis about the correlation

For the data in Table 13.1, test H_0: $\rho = 0$ versus H_1: $\rho > 0$. Use the $\alpha = 0.05$ level of significance. Compare the result with the result in Example 13.8.

Solution

From Table 13.1, we compute $\sum(x-\bar{x})(y-\bar{y}) = 7983.87$, $\sum(x-\bar{x})^2 = 1415.745$, and $\sum(y-\bar{y})^2 = 48{,}422.44$.

The correlation is

$$r = \frac{\sum(x-\bar{x})(y-\bar{y})}{\sqrt{\sum(x-\bar{x})^2}\sqrt{\sum(y-\bar{y})^2}} = \frac{7983.87}{\sqrt{1415.745}\sqrt{48{,}422.44}} = 0.964268$$

The number of points is $n = 18$. The value of the test statistic is

$$U = \frac{r\sqrt{n-2}}{\sqrt{1-r^2}} = \frac{0.964268\sqrt{18-2}}{\sqrt{1-0.964268^2}} = 14.56$$

Note that the value of the test statistic is the same as in Example 13.8. The *P*-value is also the same: 0.0000000000597. We reject H_0.

Example 13.10

Test a hypothesis with technology

For the data in Table 13.1, perform a test of $H_0\colon \beta_1 = 0$ versus $H_1\colon \beta_1 > 0$. Use the $\alpha = 0.05$ level of significance.

Solution

The following TI-84 Plus display was presented in Example 13.8. We explain it in more detail.

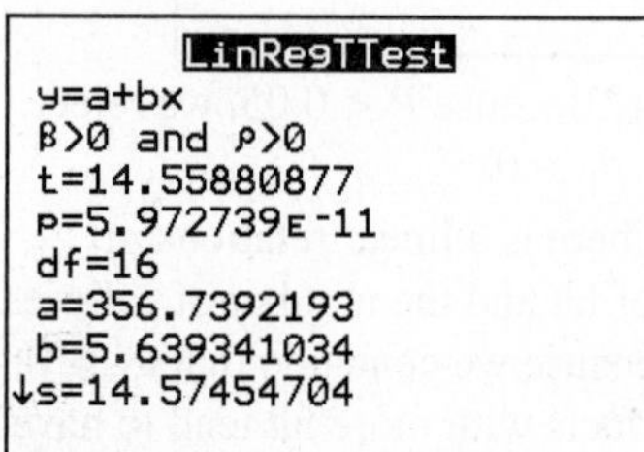

The second line specifies the alternate hypothesis. The third line presents the value of the test statistic, and the fourth line shows the *P*-value. As explained in Example 13.8, the symbol **E-11** means that the decimal point should be moved over 11 places to the left.

Now we will present the results from MINITAB.

```
Regression Analysis: Calories versus Fat

The regression equation is
Calories = 356.7 + 5.6 Fat

Predictor      Coef   SE Coef       T       P
Constant     356.74     7.669   46.52   0.000
Fat          5.6393     0.387   14.56   0.000
```

The *P*-value is given in the column headed "P," in the row labeled "Fat."

Check Your Understanding

4. For a given data set containing 18 points, the assumptions of the linear model are satisfied. The following values are computed: $b_1 = 5.58$ and $s_b = 4.42$. Perform a test of the hypothesis $H_0\colon \beta_1 = 0$ versus $H_1\colon \beta_1 \neq 0$. Use the $\alpha = 0.05$ level of significance. Can you conclude that the explanatory variable is useful in predicting the outcome variable? Explain. *No*

5. For a given data set containing 26 points, the assumptions of the linear model are satisfied. The following values are computed: $b_1 = 46.8$ and $s_b = 15.2$. Perform a test of the hypothesis $H_0\colon \beta_1 = 0$ versus $H_1\colon \beta_1 > 0$. Use the $\alpha = 0.01$ level of significance. Can you conclude that the explanatory variable is useful in predicting the outcome variable? Explain. *Yes*

6. The following TI-84 Plus display presents the results of a test of the null hypothesis $H_0: \beta_1 = 0$.

```
LinRegTTest
y=a+bx
β≠0 and ρ≠0
t=2.656594953
p=.0240393815
df=10
a=-21.87181303
b=8.973087819
↓s=73.27794742
```

a. What is the alternate hypothesis? *$\beta_1 \neq 0$*

b. What is the value of the test statistic? *2.656594953*

c. How many degrees of freedom are there? *10*

d. What is the *P*-value? *0.0240393815*

e. Can you conclude that the explanatory variable is useful in predicting the outcome variable? Answer this question using the $\alpha = 0.05$ significance level. *Yes*

Answers are on page 629.

Using Technology

We use the data in Table 13.1 and Examples 13.5 and 13.8 to illustrate the technology steps.

TI-84 PLUS

Constructing a confidence interval for the slope of the least-squares regression line

Step 1. Enter the *x*-values from Table 13.1 into **L1** and the *y*-values into **L2** (Figure A).

Step 2. Press **STAT** and highlight the **TESTS** menu. Select **LinRegTInt** and press **ENTER** (Figure B). The **LinRegTInt** menu appears.

Step 3. Enter **L1** in the **Xlist** field and **L2** in the **Ylist** field.

Step 4. Enter the confidence level in the **C-Level** field. For Example 13.5, we enter **.95** (Figure C).

Step 5. Highlight **Calculate** and press **ENTER** (Figure D).

L1	L2	L3	L4	L5
12.75	436	------	------	------
25.99	518			
21.13	492			
26.13	515			
17.23	456			
4.37	405			
23.85	491			
8.36	408			
24.85	502			
33.21	538			
19.33	468			

L1(1)=12.75

Figure A

```
EDIT CALC TESTS
9↑2-SampZInt…
0:2-SampTInt…
A:1-PropZInt…
B:2-PropZInt…
C:χ²-Test…
D:χ²GOF-Test…
E:2-SampFTest…
F:LinRegTTest…
G↓LinRegTInt…
```

Figure B

```
LinRegTInt
Xlist:L1
Ylist:L2
Freq:1
C-Level:.95
RegEQ:
Calculate
```

Figure C

```
LinRegTInt
y=a+bx
(4.8182,6.4605)
b=5.639341034
df=16
s=14.57454704
a=356.7392193
r²=.929811913
r=.9642675526
```

Figure D

TI-84 PLUS

Testing a hypothesis about the slope of the least-squares regression line

Step 1. Enter the *x*-values from Table 13.1 into **L1** and the *y*-values into **L2** (Figure E).

Step 2. Press **STAT** and highlight the **TESTS** menu. Select **LinRegTTest** and press **ENTER** (Figure F). The **LinRegTTest** menu appears.

Step 3. Enter **L1** in the **Xlist** field and **L2** in the **Ylist** field.

L1	L2	L3	L4	L5
12.75	436	------	------	------
25.99	518			
21.13	492			
26.13	515			
17.23	456			
4.37	405			
23.85	491			
8.36	408			
24.85	502			
33.21	538			
19.33	468			

L1(1)=12.75

Figure E

```
EDIT CALC TESTS
8↑TInterval…
9:2-SampZInt…
0:2-SampTInt…
A:1-PropZInt…
B:2-PropZInt…
C:χ²-Test…
D:χ²GOF-Test…
E:2-SampFTest…
F↓LinRegTTest…
```

Figure F

Step 4. Select the form of the alternate hypothesis. For Example 13.8, the alternate hypothesis has the form **>0** (Figure G).

Step 5. Highlight **Calculate** and press **ENTER** (Figure H).

```
LinRegTTest
Xlist:L1
Ylist:L2
Freq:1
β & ρ:≠0 <0 >0
RegEQ:
Calculate
```

Figure G

```
LinRegTTest
y=a+bx
β>0 and ρ>0
t=14.55880877
p=5.972739E-11
df=16
a=356.7392193
b=5.639341034
↓s=14.57454704
```

Figure H

EXCEL

Constructing a confidence interval and testing a hypothesis for the slope of the least-squares regression line

Step 1. Enter the values for Fat (x) from Table 13.1 into **Column A** and the values for Calories (y) into **Column B** (Figure I).

Step 2. Select **Data**, then **Data Analysis**. Highlight **Regression** and click **OK**.

Step 3. Enter the range of cells that contain the x-values in the **Input X Range** field and the range of cells that contain the y-values in the **Input Y Range** field. Select the **Confidence Level** option, and enter **95%**.

Step 4. Click **OK** (Figure J).

	A	B
1	**Fat**	**Calories**
2	12.75	436
3	25.99	518
4	21.13	492
5	26.13	515
6	17.23	456
7	⋮	⋮

Figure I

	Coefficients	*Standard Error*	*t Stat*	*P-value*	*Lower 95%*	*Upper 95%*
Intercept	356.7392193	7.668557932	46.51973	1.66E-18	340.4826027	372.9958359
Fat	5.639341034	0.38734907	14.55881	1.19E-10	4.818197688	6.46048438

Figure J

MINITAB

Testing a hypothesis about the slope of the least-squares regression line

Step 1. Label **Column C1** as **Fat** and **Column C2** as **Calories**. Enter the x-values from Table 13.1 into the **Fat** column and the y-values into the **Calories** column.

Step 2. Click on **Stat**, then **Regression**, then **Regression**, and then **Fit Regression Model.**

Step 3. Select the y-variable **(Calories)** as the **Response variable** and the x-variable as the **Continuous predictors (Fat)**.

Step 4. Click **OK** (Figure K).

Note: MINITAB presents the P-value for a two-tailed test by default. For a one-tailed test, divide this value by 2.

Regression Equation

Calories = 356.74 + 5.639 Fat

Coefficients

Term	Coef	SE Coef	T-Value	P-Value	VIF
Constant	356.74	7.67	46.52	0.000	
Fat	5.639	0.387	14.56	0.000	1.00

Figure K

Section 13.1 Exercises

Exercises 1–6 are the Check Your Understanding exercises located within the section.

Understanding the Concepts

In Exercises 7 and 8, fill in each blank with the appropriate word or phrase.

7. If there are 20 pairs (x, y) in a data set, then the number of degrees of freedom for the critical value is ________. *18*

8. Under the assumptions of the linear model, the values of $\mu_{y|x}$ follow a ________. *straight line*

In Exercises 9 and 10, determine whether the statement is true or false. If the statement is false, rewrite it as a true statement.

9. Under the assumptions of the linear model, the residual plot will exhibit a linear pattern. *False*

10. Under the assumptions of the linear model, the vertical spread in a residual plot will be about the same across the plot. *True*

Practicing the Skills

11. The summary statistics for a certain set of points are: $n = 30$, $s_e = 3.975$, $\sum(x - \bar{x})^2 = 15.425$, and $b_1 = 1.212$. Assume the conditions of the linear model hold. A 95% confidence interval for β_1 will be constructed.

a. How many degrees of freedom are there for the critical value? *28*

b. What is the critical value? *2.048*

c. What is the margin of error? *2.073*

d. Construct the 95% confidence interval. *(−0.861, 3.285)*

12. The summary statistics for a certain set of points are: $n = 20$, $s_e = 4.65$, $\sum(x - \bar{x})^2 = 118.26$, and $b_1 = 1.62$. Assume the conditions of the linear model hold. A 99% confidence interval for β_1 will be constructed.

a. How many degrees of freedom are there for the critical value? *18*

b. What is the critical value? *2.878*

c. What is the margin of error? *1.231*

d. Construct the 99% confidence interval. *(0.389, 2.851)*

13. Use the summary statistics in Exercise 11 to test the null hypothesis H_0: $\beta_1 = 0$ versus H_1: $\beta_1 \neq 0$. Use the $\alpha = 0.01$ level of significance. *Do not reject H_0.*

14. Use the summary statistics in Exercise 12 to test the null hypothesis H_0: $\beta_1 = 0$ versus H_1: $\beta_1 > 0$. Use the $\alpha = 0.05$ level of significance. *Reject H_0.*

In Exercises 15–18, use the given set of points to

a. Compute b_1.

b. Compute the residual standard deviation s_e.

c. Compute the sum of squares for x, $\sum(x - \bar{x})^2$.

d. Compute the standard error of b_1, s_b.

e. Find the critical value for a 95% confidence interval for β_1.

f. Compute the margin of error for a 95% confidence interval for β_1.

g. Construct a 95% confidence interval for β_1.

h. Test the null hypothesis H_0: $\beta_1 = 0$ versus H_1: $\beta_1 \neq 0$. Use the $\alpha = 0.05$ level of significance.

15.

x	12	21	27	27	10	15
y	52	90	113	111	45	65

16.

x	12	17	3	17	16	11	14	9
y	13	14	16	13	14	14	13	14

17.

x	18	20	17	12	10
y	71	77	68	52	46

18.

x	12	13	15	13	12	14	13
y	18	19	18	16	16	15	20

Working with the Concepts

19. Calories and protein: The following table presents the number of grams of protein and the number of calories per 100 grams for each of 18 fast-food products.

Protein	Calories	Protein	Calories
17.25	289	5.62	367
13.73	376	3.58	188
12.96	315	11.97	275
21.24	268	5.84	408
22.54	221	12.22	460
16.23	309	8.05	189
10.64	226	5.15	366
6.16	344	15.93	334
3.69	163	16.52	241

Source: United States Department of Agriculture

a. Compute the least-squares regression line for predicting calories (y) from protein (x). *$\hat{y} = 302.5212 - 0.5082x$*

b. Construct a 95% confidence interval for the slope. *(−7.8848, 6.8684) [Tech: (−7.8845, 6.8680)]*

c. Test H_0: $\beta_1 = 0$ versus H_1: $\beta_1 \neq 0$. Can you conclude that the amount of protein is useful in predicting the number of calories? Use the $\alpha = 0.05$ level of significance. *No*

20. Like father, like son: In 1906, the statistician Karl Pearson measured the heights of 1078 pairs of fathers and sons. The following table presents a sample of 16 pairs, with height measured in inches, simulated from the distribution specified by Pearson.

Father's Height	Son's Height	Father's Height	Son's Height
70.8	69.8	72.4	69.1
65.4	66.0	65.7	65.3
65.7	70.9	69.1	71.8
69.0	69.1	70.7	71.0
73.6	74.9	72.3	71.9
66.7	68.8	73.6	76.5
70.1	73.3	69.3	71.4
68.3	68.3	64.5	68.5

a. Compute the least-squares regression line for predicting son's height (y) from father's height (x). *$\hat{y} = 19.4093 + 0.7370x$*

b. Construct a 95% confidence interval for the slope. *(0.3527, 1.1214)*

c. Test H_0: $\beta_1 = 0$ versus H_1: $\beta_1 \neq 0$. Can you conclude that father's height is useful in predicting son's height? Use the $\alpha = 0.05$ level of significance. *Yes*

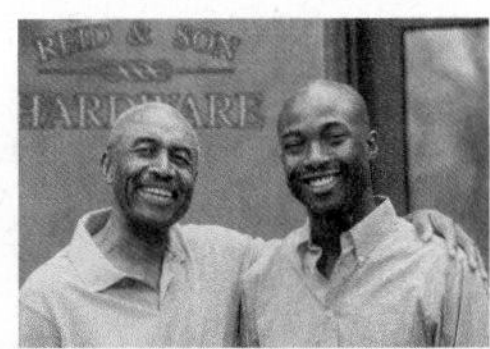

Ariel Skelley/Blend Images/Getty Images

21. Butterfly wings: Do larger butterflies live longer? The wingspan (in millimeters) and the lifespan in the adult state (in days) were measured for 22 species of butterfly. Following are the results.

Wingspan	Lifespan	Wingspan	Lifespan
35.5	19.8	25.9	32.5
30.6	17.3	31.3	27.5
30.0	27.5	23.0	31.0
32.3	22.4	26.3	37.4
23.9	40.7	23.7	22.6
27.7	18.3	27.1	23.1
28.8	25.9	28.1	18.5
35.9	23.1	25.9	32.3
25.4	24.0	28.8	29.1
24.6	38.8	31.4	37.0
28.1	36.5	28.5	33.7

Source: *Oikos Journal of Ecology* 105:41–54

a. Compute the least-squares regression line for predicting lifespan (y) from wingspan (x). *$\hat{y} = 52.0434 - 0.8445x$*

b. Construct a 99% confidence interval for the slope. *(−2.0322, 0.3432) [Tech: (−2.0324, 0.3434)]*

c. Test H_0: $\beta_1 = 0$ versus H_1: $\beta_1 < 0$. Can you conclude that wingspan is useful in predicting lifespan? Use the $\alpha = 0.05$ level of significance. *Yes*

d. Do larger butterflies tend to live for a longer or shorter time than smaller butterflies? Explain. *Shorter*

22. Blood pressure: A blood pressure measurement consists of two numbers: the systolic pressure, which is the maximum pressure taken when the heart is contracting, and the diastolic pressure, which is the minimum pressure taken at the beginning of the heartbeat. Blood pressures were measured, in millimeters, for a sample of 16 adults. The following table presents the results.

Systolic	Diastolic	Systolic	Diastolic
134	87	133	91
115	83	112	75
113	77	107	71
123	77	110	74
119	69	108	69
118	88	105	66
130	76	157	103
116	70	154	94

Based on results published in the *Journal of Hypertension* 26:199–209

a. Compute the least-squares regression line for predicting diastolic pressure (y) from systolic pressure (x). $\hat{y} = 9.1828 + 0.5748x$

b. Construct a 99% confidence interval for the slope. (0.2995, 0.8500)

c. Test H_0: $\beta_1 = 0$ versus H_1: $\beta_1 > 0$. Can you conclude that systolic blood pressure is useful in predicting diastolic blood pressure? Use the $\alpha = 0.01$ level of significance. *Yes*

d. Do people with higher diastolic pressure tend to have higher or lower systolic pressures? Explain. *Higher*

23. Noisy streets: How much noisier are streets where cars travel faster? The following table presents noise levels in decibels and average speed in kilometers per hour for a sample of roads.

Speed	Noise
28.26	78.1
36.22	79.6
38.73	81.0
29.07	78.7
30.28	78.6
30.25	78.5
29.03	78.4
33.17	79.6

Source: *Journal of Transportation Engineering* 125:152–159

a. Compute the least-squares regression line for predicting noise level (y) from speed (x). $\hat{y} = 71.4363 + 0.2392x$

b. Construct a 95% confidence interval for the slope. (0.1669, 0.3116)

c. Test H_0: $\beta_1 = 0$ versus H_1: $\beta_1 \neq 0$. Can you conclude that speed is useful in predicting noise level? Use the $\alpha = 0.01$ level of significance. *Yes*

24. Fast reactions: In a study of reaction times, the time to respond to a visual stimulus (x) and the time to respond to an auditory stimulus (y) were recorded for each of 10 subjects. Times were measured in thousandths of a second. The results are presented in the following table.

Visual	Auditory
161	159
203	206
235	241
176	163
201	197
188	193
228	209
211	189
191	169
178	201

a. Compute the least-squares regression line for predicting auditory response time (y) from visual response time (x). $\hat{y} = 22.3600 + 0.8638x$

b. Construct a 95% confidence interval for the slope. (0.3647, 1.3629)

c. Test H_0: $\beta_1 = 0$ versus H_1: $\beta_1 \neq 0$. Can you conclude that visual response is useful in predicting auditory response? Use the $\alpha = 0.01$ level of significance. *Yes*

25. Getting bigger: Concrete expands both horizontally and vertically over time. Measurements of horizontal and vertical expansion (in units of parts per hundred thousand) were made at several locations on a bridge in Quebec City in Canada. The results are presented in the following table.

Horizontal	Vertical
20	58
15	58
43	55
5	80
18	58
24	68
32	57
10	69
21	63

Source: *Canadian Journal of Civil Engineering* 32:463–479

a. Compute the least-squares line for predicting vertical expansion (y) from horizontal expansion (x). $\hat{y} = 73.2662 - 0.4968x$

b. Construct a 95% confidence interval for the slope β_1. (−0.9521, −0.0415) [Tech: (−0.9520, −0.0416)]

c. Test H_0: $\beta_1 = 0$ versus H_1: $\beta_1 \neq 0$. Can you conclude that horizontal expansion is useful in predicting vertical expansion? Use the $\alpha = 0.05$ level of significance. *Yes*

26. Dry up: In a study to determine the relationship between ambient outdoor temperature and the rate of evaporation of water from soil, measurements of average daytime temperature in °C and evaporation in millimeters per day were taken for 10 days. The results are shown in the following table.

Temperature	Evaporation
11.8	2.4
21.5	4.4
16.5	5.0
23.6	4.1
19.1	6.0
21.6	5.9
31.0	4.8
18.9	3.0
24.2	7.1
19.1	1.6

a. Compute the least-squares line for predicting evaporation (y) from temperature (x). $\hat{y} = 1.4314 + 0.1446x$

b. Construct a 99% confidence interval for β_1. $(-0.2163, 0.5056)$ [Tech: $(-0.2164, 0.5057)$]

c. Test H_0: $\beta_1 = 0$ versus H_1: $\beta_1 \neq 0$. Can you conclude that temperature is useful in predicting evaporation? Use the $\alpha = 0.05$ level of significance. No

27. Calculator display: The following TI-84 Plus display presents the results of a test of the null hypothesis H_0: $\beta_1 = 0$.

```
LinRegTTest
y=a+bx
β≠0 and ρ≠0
t=2.60388259
p=.0404509768
df=6
a=16.59028418
b=3.132227343
↓s=21.70238057
```

a. What is the alternate hypothesis? $\beta_1 \neq 0$

b. What is the value of the test statistic? 2.60388259

c. How many degrees of freedom are there? 6

d. What is the P-value? 0.0404509768

e. Can you conclude that the explanatory variable is useful in predicting the outcome variable? Answer this question using the $\alpha = 0.05$ level of significance. Yes

28. Calculator display: The following TI-84 Plus display presents the results of a test of the null hypothesis H_0: $\beta_1 = 0$.

```
LinRegTTest
y=a+bx
β<0 and ρ<0
t=-1.194401785
p=.138695956
df=6
a=11.86643045
b=-1.564032211
↓s=27.05487986
```

a. What is the alternate hypothesis? $\beta_1 < 0$

b. What is the value of the test statistic? -1.194401785

c. How many degrees of freedom are there? 6

d. What is the P-value? 0.138695956

e. Can you conclude that the explanatory variable is useful in predicting the outcome variable? Answer this question using the $\alpha = 0.05$ level of significance. No

29. Air pollution: Ozone is a major component of air pollution in many cities. Atmospheric ozone levels are influenced by many factors, including weather. In one study, the mean percent relative humidity (x) and the ozone levels (y) were measured for 120 days in a western city. Ozone levels were measured in parts per billion. The following MINITAB output describes the fit of a linear model to these data. Assume that the assumptions of the linear model are satisfied.

```
The regression equation is
Ozone = 88.761 - 0.7524 Humidity

Predictor     Coef     SE Coef       T       P
Constant    88.761       7.288   12.18   0.000
Humidity   -0.7524     0.13024   -5.78   0.000
```

a. What are the slope and intercept of the least-squares regression line? -0.7524, 88.761

b. Can you conclude that relative humidity is useful in predicting ozone levels? Answer this question using the $\alpha = 0.05$ level of significance. Yes

30. Cholesterol: Serum cholesterol levels (y) and age in years (x) were recorded for several men in a medical center. Cholesterol levels were measured in milligrams per deciliter. The following MINITAB output describes the fit of a linear model to these data. Assume that the assumptions of the linear model are satisfied.

```
The regression equation is
Cholesterol = 162.15 + 1.2499 Age

Predictor     Coef     SE Coef       T       P
Constant    162.15      16.439   9.863   0.000
Age         1.2499     0.38708   3.772   0.007
```

a. What are the slope and intercept of the least-squares regression line? 1.2499, 162.15

b. Can you conclude that age is useful in predicting cholesterol levels? Answer this question using the $\alpha = 0.05$ level of significance. Yes

Extending the Concepts

31. Confidence interval for the conditional mean: In Example 13.5, we constructed a 95% confidence interval for the slope β_1 in the model to predict the number of calories from the number of grams of fat. The 95% confidence interval is $4.8182 < \beta_1 < 6.4605$. Let $\mu_{y|15}$ be the mean number of calories for food products containing 15 grams of fat, and let $\mu_{y|20}$ be the mean number of calories for food products containing 20 grams of fat. Construct a 95% confidence interval for the difference $\mu_{y|20} - \mu_{y|15}$. (24.091, 32.3025)

Answers to Check Your Understanding Exercises for Section 13.1

1. a. Not satisfied; plot exhibits a pattern.

b. Not satisfied; vertical spread varies.

c. Satisfied

d. Not satisfied; plot contains outliers.

2. $4.27 < \beta_1 < 7.29$

3. a. 0.1971681416 **b.** 10 **c.** 12

d. $0.17354 < \beta_1 < 0.2208$

4. No. $P > 0.05$, so we do not reject H_0. It is possible that $\beta_1 = 0$ so that the explanatory variable is not helpful in predicting the outcome variable.

5. Yes. $P < 0.01$, so we reject H_0. We conclude that $\beta_1 > 0$, so the explanatory variable is helpful in predicting the outcome variable.

6. a. $\beta_1 \neq 0$ **b.** $t = 2.906470757$ **c.** 10

d. 0.0156590864 **e.** Yes

Section 13.2 Inference About the Response

Objectives

1. Construct confidence intervals for the mean response

2. Construct prediction intervals for an individual response

Objective 1 Construct confidence intervals for the mean response

Confidence Intervals for the Mean Response

In Section 13.1, we learned how to construct confidence intervals for the slope in a linear model that was used to predict the number of calories in a candy product from the number of grams of fat. The least-squares regression line was $\hat{y} = 356.7392193 + 5.639341034x$.

In this section, we will consider two further problems.

1. Given that the number of grams of fat is x, estimate the mean number of calories for all candy products whose fat content is x.
2. Given that the number of grams of fat is x, predict the number of calories for a particular candy product whose fat content is x.

Figure 13.4 presents an intuitive picture of these two problems. Imagine that this figure represents the entire population of candy products. Each point represents a particular product. The x-value of the point represents the fat content, and the y-value of the point represents the number of calories. The vertical strip contains all the points for which the fat content is x. We can visualize the two problems by looking at this figure.

1. The mean number of calories for all the products whose fat content is x is the value $\mu_{y|x}$. This is the y-value of the line $\mu_{y|x} = \beta_0 + \beta_1 x$ through the middle of the vertical strip.
2. The number of calories in a particular candy product whose fat content is x is the y-value of a randomly chosen point in the vertical strip.

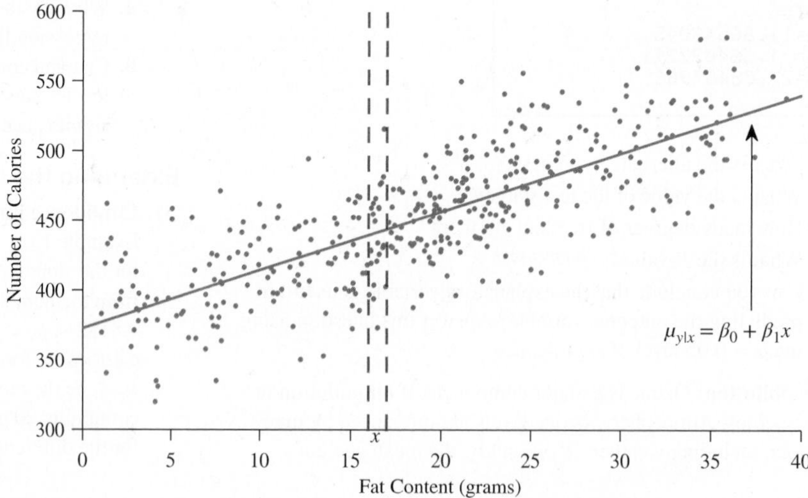

Figure 13.4 Imagine that this scatterplot represents the entire population of candy products. The vertical strip contains all the products for which the fat content is x.

The point estimate is the same for both of these problems. The point estimate is the y-value on the least-squares regression line: $\hat{y} = 356.7392193 + 5.639341034x$. However, if we want to construct intervals around these point estimates, the interval for the predicted number of calories in a particular product will have a larger margin of error than the interval for the estimated mean number of calories of all the products. The reason is that there is less variation in the mean of all the values in a vertical strip than in the distribution of the individual points.

Constructing a confidence interval for the mean response

The mean y-value corresponding to a given x-value is called the **mean response**. We show how to construct a confidence interval for the mean response.

CAUTION

Be sure that the assumptions of the linear model are satisfied before constructing a confidence interval for the mean response.

Constructing a Confidence Interval for the Mean Response

Let x^* be a value of the explanatory variable x, let $\hat{y} = b_0 + b_1x^*$ be the predicted value corresponding to x^*, and let n be the sample size. A level $100(1-\alpha)\%$ confidence interval for the mean response is

$$\hat{y} \pm t_{\alpha/2} \cdot s_e \sqrt{\frac{1}{n} + \frac{(x^* - \bar{x})^2}{\sum(x - \bar{x})^2}}$$

The critical value $t_{\alpha/2}$ is based on $n-2$ degrees of freedom.

Example 13.11

Constructing a confidence interval for the mean response

Construct a 95% confidence interval for the mean number of calories for candy products containing 18 grams of fat.

Solution

The least-squares regression line is $\hat{y} = 356.7392193 + 5.639341034x$. To obtain the point estimate $\hat{y}$, we replace x with $x^* = 18$ to obtain

$$\hat{y} = 356.7392193 + 5.639341034(18) = 458.24736$$

The sample size is $n = 18$, so there are 16 degrees of freedom. The critical value is $t_{\alpha/2} = 2.120$.

To obtain the margin of error, recall that we computed $\bar{x} = 17.7$, $s_e = 14.574547$, and $\sum(x - \bar{x})^2 = 1415.7452$ in Example 13.4 in Section 13.1. The margin of error is therefore

RECALL

The quantity s_e is the residual standard deviation:

$$s_e = \sqrt{\frac{\sum(y-\hat{y})^2}{n-2}}$$

$$t_{\alpha/2} \cdot s_e \sqrt{\frac{1}{n} + \frac{(x^* - \bar{x})^2}{\sum(x - \bar{x})^2}} = 2.120 \cdot 14.574547 \sqrt{\frac{1}{18} + \frac{(18 - 17.7)^2}{1415.7452}} = 7.286903$$

The 95% confidence interval is

$$458.24736 \pm 7.286903$$

$$450.960 < \text{Mean response} < 465.534$$

We are 95% confident that the mean number of calories for candy products containing 18 grams of fat is between 450.960 and 465.534.

Example 13.12

Interpreting a confidence interval

In Example 13.11, we found that we are 95% confident that the mean number of calories for candy products containing 18 grams of fat is between 450.960 and 465.534. You are planning to purchase a particular product that contains 18 grams of fat. Can you be 95% confident that the number of calories in your particular product will be between 450.960 and 465.534? Explain why or why not.

Solution

No, you cannot be 95% confident that the number of calories in your particular product will be between 450.960 and 465.534. The confidence interval for the mean response provides information about the mean number of calories for all fast-food products with 18 grams of fat. To estimate the number of calories in a particular product, we need a prediction interval for an **individual response**. The prediction interval will have a larger margin of error than the confidence interval for the mean response. We will learn to construct prediction intervals later in this section.

Check Your Understanding

1. For a sample of size $n = 20$, the following values were obtained: $b_0 = 1.05$, $b_1 = 4.50$, $s_e = 0.54$, $\sum(x - \bar{x})^2 = 10.9$, $\bar{x} = 8.52$. Construct a 95% confidence interval for the mean response when $x = 10$. (45.482, 46.618)

Answer is on page 635.

Objective 2 Construct prediction intervals for an individual response

Prediction Intervals for an Individual Response

Following is the method for constructing a **prediction interval**.

Constructing a Prediction Interval

Let x^* be a value of the explanatory variable x, let $\hat{y} = b_0 + b_1x^*$ be the predicted value corresponding to x^*, and let n be the sample size. A level $100(1 - \alpha)\%$ prediction interval for an individual response is

$$\hat{y} \pm t_{\alpha/2} \cdot s_e \sqrt{1 + \frac{1}{n} + \frac{(x^* - \bar{x})^2}{\sum(x - \bar{x})^2}}$$

The critical value $t_{\alpha/2}$ is based on $n - 2$ degrees of freedom.

CAUTION

Be sure that the assumptions of the linear model are satisfied before constructing a prediction interval.

Note that the standard error for the prediction interval is similar to the one for the confidence interval for a mean response. The difference is that the standard error for prediction has a "1" added to the quantity under the square root. This reflects the extra variability in the prediction interval.

Example 13.13

Constructing a prediction interval

A particular candy product has a fat content of 18 grams. Construct a 95% prediction interval for the number of calories in this product.

Solution

The least-squares regression line is $\hat{y} = 356.7392193 + 5.639341034x$. To obtain the point estimate $\hat{y}$, we replace x with $x^* = 18$ to obtain

$$\hat{y} = 356.7392193 + 5.639341034(18) = 458.24736$$

The sample size is $n = 18$, so there are 16 degrees of freedom. The critical value is $t_{\alpha/2} = 2.120$.

To obtain the margin of error, recall that we computed $\bar{x} = 17.7$, $s_e = 14.574547$, and $\sum(x - \bar{x})^2 = 1415.7452$ in Example 13.4 in Section 13.1. The margin of error is therefore

$$t_{\alpha/2} \cdot s_e \sqrt{1 + \frac{1}{n} + \frac{(x^* - \bar{x})^2}{\sum(x - \bar{x})^2}} = 2.120 \cdot 14.574547 \sqrt{1 + \frac{1}{18} + \frac{(18 - 17.7)^2}{1415.7452}}$$
$$= 31.744256$$

The 95% prediction interval is

$$458.24736 \pm 31.744256$$

$$426.503 < \text{Number of calories} < 489.992$$

We are 95% confident that a particular candy product with a fat content of 18 grams will have between 426.503 and 489.992 calories.

Example 13.14

Interpret a prediction interval

Refer to Example 13.13. You are planning to eat a candy bar, and you want to consume less than 500 calories. If you choose an item that contains 18 grams of fat, can you be reasonably sure that it will contain less than 500 calories?

Solution

Yes. We are 95% confident that a particular candy bar with a fat content of 18 grams will have between 426.503 and 489.992 calories. Therefore, we can be reasonably sure that it will contain less than 500 calories.

Check Your Understanding

2. For a sample of size $n = 15$, the following values were obtained: $b_0 = 3.71$, $b_1 = 8.38$, $s_e = 1.13$, $\sum(x - \bar{x})^2 = 7.71$, $\bar{x} = 13.16$. Construct a 95% prediction interval for an individual response when $x = 8$. *(65.561, 75.939)*

Answer is on page 635.

Example 13.15

Constructing intervals with technology

Use technology to construct a 95% confidence interval for the mean number of calories for candy products containing 18 grams of fat and to construct a 95% prediction interval for the number of calories in a particular candy product that contains 18 grams of fat.

Solution

We will use MINITAB. The output is as follows.

Fit	StDev Fit	95.0% CI	95.0% PI
458.247	3.437	(450.960, 465.534)	(426.503, 489.992)

The 95% confidence interval and the 95% prediction interval are shown. Step-by-step instructions for constructing confidence intervals for the mean response and prediction intervals are given in the Using Technology section that follows.

Using Technology

We use the data in Table 13.1 and Example 13.13 to illustrate the technology steps.

MINITAB

Constructing confidence and prediction intervals

Step 1. Click on **Regression**, then **Regression**, and then **Predict**.

Step 2. Select the y-variable (**Calories**) as the **Response**. Select the **Enter individual values** option. For Example 13.13, we use **18**.

Step 3. Click **Options**, and enter the confidence level in the **Confidence level** field. We use **95**. Click **OK**.

Step 4. Click **OK** (Figure A).

Fit	SE Fit	95% CI	95% PI
458.247	3.43722	(450.961, 465.534)	(426.503, 489.992)

Figure A

Section 13.2 Exercises

Exercises 1 and 2 are the Check Your Understanding exercises located within the section.

Understanding the Concepts

In Exercises 3 and 4, fill in each blank with the appropriate word or phrase.

3. A ________ interval estimates the mean y-value for all individuals with a given x-value. *confidence*

4. A ________ interval estimates the y-value for a particular individual with a given x-value. *prediction*

In Exercises 5 and 6, determine whether the statement is true or false. If the statement is false, rewrite it as a true statement.

5. For a given x-value, the 95% confidence interval for the mean response will always be wider than the 95% prediction interval. *False*

6. For a given x-value, the point estimate for a 95% confidence interval for the mean response is the same as the one for the 95% prediction interval. *True*

Practicing the Skills

7. For a sample of size 25, the following values were obtained: $b_0 = 3.25$, $b_1 = 2.32$, $s_e = 3.53$, $\sum(x - \bar{x})^2 = 224.05$, and $\bar{x} = 0.98$.
   a. Construct a 95% confidence interval for the mean response when $x = 2$. *(6.35, 9.43)*
   b. Construct a 95% prediction interval for an individual response when $x = 2$. *(0.43, 15.35)*

8. For a sample of size 18, the following values were obtained: $b_0 = 2.27$, $b_1 = -1.46$, $s_e = 5.72$, $\sum(x - \bar{x})^2 = 360.26$, and $\bar{x} = 1.95$.
   a. Construct a 99% confidence interval for the mean response when $x = 2$. *(−4.59, 3.29)*
   b. Construct a 99% prediction interval for an individual response when $x = 2$. *(−17.82, 16.52)*

In Exercises 9 and 10, use the given set of points to
   a. Compute b_0 and b_1.
   b. Compute the predicted value $\hat{y}$ for the given value of x.
   c. Compute the residual standard deviation s_e.
   d. Compute the sum of squares for x, $\sum(x - \bar{x})^2$.
   e. Find the critical value for a 95% confidence or prediction interval.
   f. Construct a 95% confidence interval for the mean response for the given value of x.
   g. Construct a 95% prediction interval for an individual response for the given value of x.

9.

x	15	11	17	15	11	16
y	30	15	33	27	22	37

$x = 12$

10.

x	23	16	17	19	30	19	18	27
y	51	22	56	34	67	59	55	25

$x = 25$

Working with the Concepts

11. **Calories and protein:** Use the data in Exercise 19 in Section 13.1 for the following:
    a. Compute a point estimate for the mean number of calories in fast-food products that contain 15 grams of protein. *294.9*
    b. Construct a 95% confidence interval for the mean number of calories in fast-food products that contain 15 grams of protein. *(245.69, 344.1)*
    c. Predict the number of calories in a particular product that contains 15 grams of protein. *294.9*
    d. Construct a 95% prediction interval for the number of calories in a particular product that contains 15 grams of protein. *(108.17, 481.63)*

12. **Like father, like son:** Use the data in Exercise 20 in Section 13.1 for the following.
    a. Compute a point estimate of the mean height of sons whose fathers are 70 inches tall. *71.002*
    b. Construct a 95% confidence interval for the mean height of sons whose fathers are 70 inches tall. *(69.855. 72.149)*
    c. Predict the height of a particular son whose father is 70 inches tall. *71.002*
    d. Construct a 95% prediction interval for the height of a particular son whose father is 70 inches tall. *(66.435, 75.57)*

13. **Butterfly wings:** Use the data in Exercise 21 in Section 13.1 for the following.
    a. Compute a point estimate of the mean lifespan of butterflies with a wingspan of 30 millimeters. *26.708*
    b. Construct a 95% confidence interval for the mean lifespan of butterflies with a wingspan of 30 millimeters. *(23.367, 30.05)*
    c. Predict the lifespan of a particular butterfly whose wingspan is 30 millimeters. *26.708*
    d. Construct a 95% prediction interval for the lifespan of a particular butterfly whose wingspan is 30 millimeters. *(12.247, 41.17)*

14. **Blood pressure:** Use the data in Exercise 22 in Section 13.1 for the following.
    a. Compute a point estimate of the mean diastolic pressure for people whose systolic pressure is 120. *78.15*
    b. Construct a 95% confidence interval for the mean diastolic pressure for people whose systolic pressure is 120. *(75.099, 81.208)*
    c. Predict the diastolic pressure of a particular person whose systolic pressure is 120. *78.15*
    d. Construct a 95% prediction interval for the diastolic pressure of a particular person whose systolic pressure is 120. *(65.674, 90.634)*

15. **Noisy streets:** Use the data in Exercise 23 in Section 13.1 for the following.
    a. Compute a point estimate for the mean noise level for streets with a mean speed of 35 kilometers per hour. *79.81*
    b. Construct a 99% confidence interval for the mean noise level for streets with a mean speed of 35 kilometers per hour. *(79.29, 80.33)*
    c. Predict the noise level for a particular street with a mean speed of 35 kilometers per hour. *79.81*
    d. Construct a 99% prediction interval for the noise level of a particular street with a mean speed of 35 kilometers per hour. *(78.588, 81.032)*

16. **Fast reactions:** Use the data in Exercise 24 in Section 13.1 for the following.
   a. Compute a point estimate for the mean auditory response time for subjects with a visual response time of 200. *195.12*
   b. Construct a 99% confidence interval for the mean auditory response time for subjects with a visual response time of 200. *(178.97, 211.27)*
   c. Predict the auditory response time for a particular subject whose visual response time is 200. *195.12*
   d. Construct a 99% prediction interval for the auditory response time for a particular subject whose visual response time is 200. *(141.94, 248.3)*

17. **Getting bigger:** Use the data in Exercise 25 in Section 13.1 for the following.
   a. Compute a point estimate for the mean vertical expansion at locations where the horizontal expansion is 25. *60.847*
   b. Construct a 99% confidence interval for the mean vertical expansion at locations where the horizontal expansion is 25. *(53.103, 68.59)*
   c. Predict the vertical expansion at a particular location where the horizontal expansion is 25. *60.847*
   d. Construct a 99% prediction interval for the vertical expansion at a particular location where the horizontal expansion is 25. *(37.813, 83.88)*

18. **Dry up:** Use the data in Exercise 26 in Section 13.1 for the following.
   a. Compute a point estimate for the mean evaporation rate when the temperature is 20°C. *4.3244*
   b. Construct a 99% confidence interval for the mean evaporation rate for all days with a temperature of 20°C. *(2.5604, 6.0884)*
   c. Predict the evaporation rate when the temperature is 20°C. *4.3244*
   d. Construct a 99% prediction interval for the evaporation rate on a given day with a temperature of 20°C. *(−1.4665, 10.115)*

19. **Air pollution:** The following MINITAB output presents a 95% confidence interval for the mean ozone level on days when the relative humidity is 60%, and a 95% prediction interval for the ozone level on a particular day when the relative humidity is 60%. The units of ozone are parts per billion.

```
Predicted Values for New Observations

New Obs    Fit    SE Fit      95.0% CI         95.0% PI
1         43.62     1.20   (41.23, 46.00)   (20.86, 66.37)

Values of Predictors for New Observations
New Obs    Humidity
1              60.0
```

   a. What is the point estimate for the mean ozone level for days when the relative humidity is 60%? *43.62*
   b. What is the 95% confidence interval for the mean ozone level for days when the relative humidity is 60%? *(41.23, 46.00)*
   c. Predict the ozone level for a day when the relative humidity is 60%. *43.62*
   d. Upon learning that the relative humidity on a certain day is 60%, someone predicts that the ozone level that day will be 80 parts per billion. Is this a reasonable prediction? If so, explain why. If not, give a reasonable range of predicted values. *No*

20. **Cholesterol:** The following MINITAB output presents a 95% confidence interval for the mean cholesterol levels for men aged 50 years, and a 95% prediction interval for an individual man aged 50. The units of cholesterol are milligrams per deciliter.

```
Predicted Values for New Observations

New Obs     Fit    SE Fit       95.0% CI           95.0% PI
1         224.64     6.08   (211.40, 237.89)   (182.66, 266.42)

Values of Predictors for New Observations
New Obs     Age
1           50.0
```

   a. What is the point estimate for the mean cholesterol level for men aged 50? *224.64*
   b. What is the 95% confidence interval for the mean cholesterol level for men aged 50? *(211.40, 237.89)*
   c. Predict the cholesterol level for a man aged 50. *224.64*
   d. Upon learning that a man is 50 years old, someone predicts that his cholesterol level is 160. Is this a reasonable prediction? If so, explain why. If not, give a reasonable range of predicted values. *No*

Extending the Concepts

21. **Margin of error:** Several 95% confidence intervals for the mean response will be constructed, based on a data set for which the sample mean value for the explanatory variable is $\bar{x} = 10$. The values of x^* for which the confidence intervals will be constructed are $x^* = 9$, $x^* = 12$, and $x^* = 14$.
   a. For which of these values of x^* will the margin of error be the smallest? *9*
   b. For which of these values of x^* will the margin of error be the largest? *14*
   c. If one wanted to construct a 95% confidence interval with the smallest possible margin of error, which value of x^* would one use? *10*

Answers to Check Your Understanding Exercises for Section 13.2

1. 45.482 < Mean response < 46.618

2. 65.561 < Individual response < 75.939

Section 13.3 Multiple Regression

Objectives

1. Use technology to compute the multiple regression equation
2. Interpret the multiple regression coefficients
3. Test the multiple regression coefficients
4. Construct confidence intervals and prediction intervals
5. Test goodness-of-fit
6. Eliminate unnecessary variables from a multiple regression model

Objective 1 Use technology to compute the multiple regression equation

The Multiple Regression Equation

In Sections 13.1 and 13.2 we studied the least-squares regression line, which is used to predict a value of an outcome variable y from the value of a single explanatory variable x. Sometimes we need several explanatory variables to predict the value of the response variable. In these situations we use **multiple regression**. We refer to the explanatory variables as x_1, x_2, and so on. If there are p explanatory variables, the multiple regression equation is

$$\hat{y} = b_0 + b_1x_1 + \cdots + b_px_p$$

Table 13.4 presents the results of a hypothetical study in which the lung capacities of 16 children were measured. Lung capacity is measured by having a person take as deep a breath as possible, then blowing into a tube connected to a machine called a spirometer, which measures the volume of air exhaled. Variables that might be useful to predict a child's lung capacity are height, weight, and age, along with the temperature and barometric pressure at the time the lung capacity was measured. These variables were measured as well.

Table 13.4 Lung Capacity, Height, Weight, Age, Barometric Pressure, and Temperature for Measurements on Sixteen Children

Lung Capacity (y)	Height (x_1)	Weight (x_2)	Age (x_3)	Barometric Pressure (x_4)	Temperature (x_5)
2.68	61	105	13	30.0	70.4
2.22	49	50	9	30.3	71.6
2.29	61	128	15	29.9	73.6
2.16	54	81	10	29.6	70.8
3.19	68	120	13	29.6	69.1
1.97	56	92	10	30.0	70.1
2.98	68	140	16	30.2	67.7
1.96	57	86	10	30.7	56.3
3.05	62	120	16	30.5	63.8
2.48	60	114	13	30.3	69.9
2.09	57	94	12	30.1	73.0
3.27	67	145	16	30.6	70.3
2.26	52	66	10	28.8	75.5
3.20	72	159	16	30.0	72.1
2.10	53	65	10	30.5	67.5
1.79	51	56	9	29.7	65.6

In Table 13.4, lung capacity (y) is measured in liters, height (x_1) in inches, weight (x_2) in pounds, age (x_3) in years, barometric pressure (x_4) in inches, and temperature (x_5) in degrees Fahrenheit. To predict the lung capacity for a person of a given height, weight, age, and given values of the pressure and temperature, we must compute the multiple regression equation

$$\hat{y} = b_0 + b_1x_1 + b_2x_2 + b_3x_3 + b_4x_4 + b_5x_5$$

The calculations involved in computing the multiple regression equation are quite complex and almost impossible to do by hand. Technology is always used in practice. Figure 13.5 presents MINITAB output for the computation of the multiple regression equation. Step-by-step instructions are given in the Using Technology section on page 645.

```
The regression equation is
Lung Capacity = −7.82 + 0.119 Height − 0.0246 Weight + 0.164 Age
          + 0.076 Pressure + 0.0202 Temperature

Predictor          Coef     SE Coef      T       P
Constant         −7.817       6.288  −1.24   0.242
Height          0.11927     0.03292   3.62   0.005
Weight        −0.024627    0.008965  −2.75   0.021
Age             0.16411     0.06252   2.62   0.025
Pressure         0.0764      0.1645   0.46   0.652
Temperature     0.02022     0.01716   1.18   0.266

S = 0.214478     R−Sq = 88.0%     R−Sq(adj) = 81.9%

Analysis of Variance

Source           DF        SS        MS       F      P
Regression        5   3.35908   0.67182   14.60  0.000
Residual Error   10   0.46001   0.04600
Total            15   3.81909
```

Figure 13.5 Multiple regression equation and related output for data in Table 13.4, from MINITAB.

The multiple regression equation is presented at the top of the output. The column labeled "Coef" presents the coefficients of the multiple regression equation. The rest of the information in the output describes how well the equation fits the data; we will explain this as we go along.

The residual plot

For each individual, the residual is the difference $y - \hat{y}$ between their outcome variable y and the predicted value $\hat{y}$. As with the least-squares regression line, we determine whether the multiple regression equation is appropriate by constructing a residual plot. This is a plot of residuals against the fitted values $\hat{y}$. When the multiple regression is appropriate, the residual plot will not exhibit any noticeable pattern. If the residual plot does exhibit a pattern, such as a curved pattern, then the multiple regression equation should not be used.

Figure 13.6 (page 638) presents a residual plot, produced by MINITAB, for the data in Table 13.4. The residual plot does not exhibit any particular pattern, which suggests that the multiple regression equation is appropriate. Step-by-step instructions for constructing residual plots with MINITAB are presented in the Using Technology section on page 645.

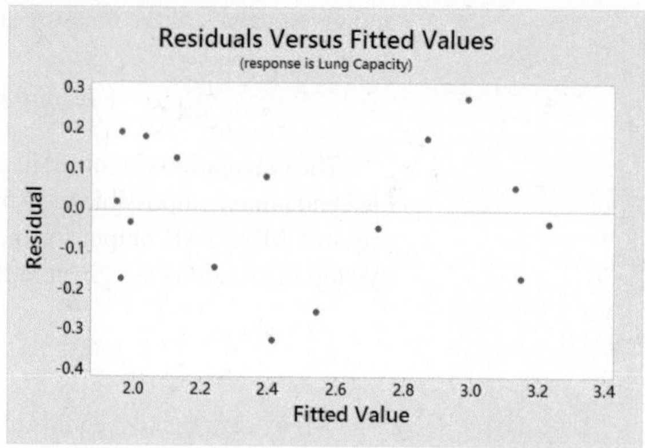

Figure 13.6 Residual plot for the data in Table 13.4. The plot exhibits no noticeable pattern, so use of the multiple regression equation is appropriate.

Example 13.16

Predict the value of an outcome variable

Predict the lung capacity for a 12-year-old who is 61 inches tall and weighs 105 pounds, at a pressure of 30.1 inches and a temperature of 70 degrees.

Solution

We use the coefficients of the multiple regression equation, presented in Figure 13.5, to compute

$$\begin{aligned}\hat{y} &= -7.817 + 0.11927(61) - 0.024627(105) + 0.16411(12) \\ &\quad + 0.0764(30.1) + 0.02022(70) \\ &= 2.557\end{aligned}$$

The predicted value is 2.557 liters.

Objective 2 Interpret the multiple regression coefficients

Interpret the Multiple Regression Coefficients

The interpretation of the coefficients in multiple regression is similar to the interpretation of the slope of the least-squares regression line. If the values of an explanatory variable x_i for two individuals differ by 1, and we assume that the values of the other explanatory values are the same, then the predicted values will differ by b_i. If the values of the explanatory variable differ by an amount d, then their predicted values will differ by $b_i \cdot d$.

Example 13.17

Compute the predicted difference in outcomes

Two people differ in height by 2 inches. Their weights and ages are the same, and their lung capacities are measured at the same pressure and temperature. By how much should we predict their lung capacities to differ?

Solution

The coefficient of height, shown in Figure 13.5, is 0.11927. We predict their lung capacities to differ by (0.11927)(2) = 0.23854 liters.

Check Your Understanding

1. Use the coefficients of the multiple regression equation in Figure 13.5 to predict the lung capacity for a 10-year-old who is 57 inches tall and weighs 90 pounds, at a pressure of 30.2 inches and a temperature of 65 degrees. 2.028

2. Two people differ in age by 1.5 years. Their heights and weights are the same, and their lung capacities are measured at the same pressure and temperature. By how much should we predict their lung capacities to differ? *0.2462*

Answers are on page 652.

Objective 3 Test the multiple regression coefficients

Test the Multiple Regression Coefficients

The 16 individuals described in Table 13.4 are a sample from a large population. When we compute the multiple regression equation for this sample, it is an estimate of what the multiple regression equation would be for the whole population. The coefficients of the multiple regression equation for the sample are denoted b_0, b_1, and so forth. The coefficients of the multiple regression equation for the population are denoted β_0, β_1, and so forth. Thus, b_0 is an estimate of β_0, b_1 is an estimate of β_1, and so forth.

When a coefficient β_i is equal to 0, the variable x_i has no linear relation to the outcome variable and can be omitted from the multiple regression equation. The most important hypothesis test to perform on β_i is therefore H_0: $\beta_i = 0$. This test is automatically performed for each β_i by most computer software packages. For each coefficient, the column labeled "T" in the MINITAB output presents test statistic, and the column labeled "P" presents the P-value for the test of H_0: $\beta_i = 0$ versus H_1: $\beta_i \neq 0$.

Example 13.18

Test hypotheses about the multiple regression coefficients

Following is part of the MINITAB output shown in Figure 13.5, which presents the coefficients, test statistics and P-values for the multiple regression equation for the data in Table 13.4.

Predictor	Coef	SE Coef	T	P
Constant	-7.817	6.288	-1.24	0.242
Height	0.11927	0.03292	3.62	0.005
Weight	-0.024627	0.008965	-2.75	0.021
Age	0.16411	0.06252	2.62	0.025
Pressure	0.0764	0.1645	0.46	0.652
Temperature	0.02022	0.01716	1.18	0.266

a. For the data in Table 13.4, β_1 is the coefficient of height. Test the hypothesis H_0: $\beta_1 = 0$ versus H_1: $\beta_1 \neq 0$ at the $\alpha = 0.05$ level. What do you conclude?

b. For the data in Table 13.4, β_5 is the coefficient of temperature. Test the hypothesis H_0: $\beta_5 = 0$ versus H_1: $\beta_5 \neq 0$ at the $\alpha = 0.05$ level. What do you conclude?

Solution

a. In Figure 13.5, the P-value for height is 0.005. This value is less than 0.05, so we reject H_0 and conclude that $\beta_1 \neq 0$. The explanatory variable height has a linear relationship with the outcome variable, lung capacity.

b. In Figure 13.5, the P-value for temperature is 0.266. This value is greater than 0.05, so we do not reject H_0. There is not enough evidence to conclude that $\beta_5 \neq 0$. We cannot conclude that the explanatory variable temperature may not have a linear relationship with the outcome variable, lung capacity.

Objective 4 Construct confidence intervals and prediction intervals

Construct Confidence Intervals and Prediction Intervals

In Section 13.2 we learned to construct confidence intervals for a mean response and prediction intervals for an individual response based on the least-squares regression line. With technology we can do the same with the multiple regression equation.

Recall the difference between a confidence interval and a prediction interval. A confidence interval presents plausible values for the mean value of the response for all individuals with a specified set of values for their explanatory variables. A prediction interval presents a plausible range of values for a particular individual with a specified set of values for the explanatory variables.

Example 13.19

Find a confidence interval and a prediction interval

a. Use the data in Table 13.4 to construct a 95% confidence interval for the mean lung capacity for a 12-year-old who is 65 inches tall and weighs 110 pounds, if the measurement is taken at a pressure of 30 inches and a temperature of 70 degrees.

b. Use the data in Table 13.4 to construct a 95% prediction interval for the lung capacity of a particular 12-year-old who is 65 inches tall and weighs 110 pounds, if the measurement is taken at a pressure of 30 inches and a temperature of 70 degrees.

Solution

The MINITAB output follows:

```
Predicted Values for New Observations

New
Obs        Fit      SE Fit          95% CI               95% PI

  1     2.9032      0.1510     (2.5667, 3.2397)     (2.3187, 3.4876)

Values of Predictors for New Observations

New
Obs  Height  Weight   Age  Pressure  Temperature
  1    65.0     110  12.0      30.0         70.0
```

The values of the explanatory variables are listed under the heading "Values of Predictors for New Observations."

a. The 95% confidence interval is given under the heading "95% CI." The confidence interval is (2.5667, 3.2397). We are 95% confident that the mean lung function over all individuals with the given values for the explanatory variables is between 2.5667 and 3.2397 liters.

b. The 95% prediction interval is given under the heading "95% PI." The prediction interval is (2.3187, 3.4876). We are 95% confident that the lung function for a particular individual with the given values for the explanatory variables will be between 2.3187 and 3.4876 liters.

Check Your Understanding

3. The following MINITAB output presents a multiple regression equation $\hat{y} = b_0 + b_1x_1 + b_2x_2 + b_3x_3$. Test H_0: $\beta_i = 0$ versus H_1: $\beta_i \neq 0$ for $i = 1, 2, 3$. Use the $\alpha = 0.05$ level. *Reject H_0 for β_1 and β_3. Do not reject H_0 for β_2.*

```
The regression equation is
Y = 3.4105 + 0.9543 X1 + 0.2970 X2 + 2.6126 X3

Predictor      Coef     SE Coef       T        P
Constant     3.4105      0.5832  5.8475    0.000
X1           0.9543      0.2975  3.2074    0.008
X2           0.2970      0.5796  0.5124    0.619
X3           2.6136      0.4252  6.1461    0.000

S = 1.8421      R-Sq = 81.2%     R-Sq (adj) = 76.0%

Analysis of Variance

Source          DF        SS        MS        F       P
Regression       3    160.71    53.570   15.787   0.000
Residual Error  11    37.327    3.3934
Total           14    198.04
```

4. The following MINITAB output presents a confidence interval for a mean response and a prediction interval for an individual response.

```
New Obs      Fit     SE Fit        95.0% CI            95.0% PI
1         10.523     1.3996   (7.443, 13.604)   (5.4315, 15.615)

Values of Predictors for New Observations

New Obs       X1       X2       X3
1           1.32     1.58     2.06
```

a. Predict the value of y when $x_1 = 1.32$, $x_2 = 1.58$, and $x_3 = 2.06$. *10.523*

b. We are 95% confident that an individual whose values are $x_1 = 1.32$, $x_2 = 1.58$, and $x_3 = 2.06$ will have a response between ________ and ________ . *5.4315, 15.615*

c. We are 95% confident that the mean response when $x_1 = 1.32$, $x_2 = 1.58$, and $x_3 = 2.06$ is between ________ and ________ . *7.443, 13.604*

Answers are on page 652.

Objective 5 Test goodness-of-fit

Testing Goodness-of-Fit

The coefficient of determination

For each individual, we can compute the difference $y - \bar{y}$ between that individual's observed value y and the average observation $\bar{y}$. In multiple regression, as with the least-squares regression line, we can divide the difference into two parts. The difference $\hat{y} - \bar{y}$ between $\hat{y}$, the predicted value for the individual, and $\bar{y}$ is called the explained difference, and the difference $y - \hat{y}$ is called the unexplained difference. The quantity $\sum(\hat{y} - \bar{y})^2$ is called the **regression sum of squares**, and the quantity $\sum(y - \hat{y})^2$ is called the **error sum of squares**. We add the regression sum of squares and the error sum of squares to obtain the **total sum of squares**.

The **coefficient of determination**, denoted R^2, is defined by

$$R^2 = \frac{\text{regression sum of squares}}{\text{total sum of squares}}$$

The coefficient of determination represents the proportion of variation in the response variable explained by the multiple regression equation.

EXPLAIN IT AGAIN

Proportion of variation explained: The regression sum of squares is the same as the explained variation in the least-squares regression line, and the total sum of squares is the total variation. For this reason, R^2 is the proportion of variation explained by the multiple regression equation.

The coefficient of determination has a disadvantage as a measure of goodness-of-fit. If more explanatory variables are added to the model, the value of R^2 will not decrease and will almost certainly increase, even if the explanatory variables are unrelated to the outcome variable. To compensate for this, a quantity known as **adjusted R^2** is used. This quantity adjusts the value of R^2 by using the number of observations n and the number of explanatory variables p. The value of adjusted R^2 is

$$\text{Adjusted } R^2 = R^2 - \frac{p}{n-p-1}(1-R^2)$$

Because the value of adjusted R^2 is found by subtracting a positive quantity from R^2, its value is always somewhat less than the value of R^2. The values of R^2 and adjusted R^2 are routinely provided in the output from computer software packages.

Example 13.20

Find R^2 and adjusted R^2

Find the values of R^2 and adjusted R^2 for the data in Table 13.4.

Solution

Figure 13.5 presents the MINITAB output. The value of R^2 is labeled "R-sq." The value of R^2 is 88.0%. The value of adjusted R^2 is labeled "R-sq(adj)." The value of adjusted R^2 is 81.9%.

The *F*-test for goodness-of-fit

To determine whether a multiple regression equation is useful for prediction, we test the null hypothesis

$$H_0\colon \beta_1 = \beta_2 = \cdots = \beta_p = 0 \quad \text{versus} \quad H_1\colon \text{at least one of the } \beta_i \neq 0$$

If H_0 is false, then one or more of the β_i is not zero, and the model is useful for prediction. If H_0 is true, the model is not useful for prediction. The test statistic is based on the regression and error sums of squares, as follows.

Each sum of squares has a number of degrees of freedom associated with it. The degrees of freedom for the regression sum of squares is p, the number of explanatory variables in the model. The degrees of freedom for the error sum of squares is $n - p - 1$. When a sum of squares is divided by its degrees of freedom, the result is known as a mean square. The regression mean square is denoted *MSR*, and the error mean square is denoted *MSE*. The statistic for testing goodness-of-fit is

$$F = \frac{MSR}{MSE}$$

When H_0 is true, the test statistic has an F distribution with p and $n - p - 1$ degrees of freedom. We could use the F-table to determine whether to reject H_0, but computer packages routinely provide the P-value, so in practice this is not necessary.

Example 13.21

Test the goodness-of-fit of a multiple regression equation

For the data in Table 13.4, perform an F-test. Is the model useful for prediction? Use the $\alpha = 0.01$ level.

Solution

Figure 13.5 presents the MINITAB output. Information regarding the F-test is found under the heading "Analysis of Variance." The value of the test statistic, labeled "F," is 14.60.

The P-value, labeled "P," is given as 0.000. This means that the P-value is 0 when rounded to three decimal places. Since $P < 0.01$, we reject H_0. We conclude that at least one of the coefficients is not equal to 0, and the model is useful for prediction.

Check Your Understanding

5. The following MINITAB output presents a multiple regression equation.

```
The regression equation is
Y = 5.3296 + 3.6204 X1 + 1.5721 X2 + 2.3328 X3

Predictor        Coef      SE Coef         T        P
Constant       5.3296       1.5051    3.5410    0.000
X1             3.6204       1.2845    2.8186    0.010
X2             1.5721       1.3634    1.1531    0.261
X3             2.3328       1.4494    1.6095    0.122

S = 7.4857         R-Sq = 32.6%         R-Sq (adj) = 23.3%

Analysis of Variance

Source            DF       SS        MS         F        P
Regression         3   595.14    198.38    3.5402    0.031
Residual Error    22   1232.8    56.036
Total             25   1827.9
```

a. What percentage of the variation in the response is explained by the multiple regression equation? *32.6%*

b. What percentage of the variation in the response is explained by the multiple regression equation, if we account for the number of explanatory variables in the model? *23.3%*

6. Refer to Exercise 5.

a. Is the multiple regression equation useful for prediction? Explain. Use the $\alpha = 0.05$ level. *Yes*

b. Is the multiple regression equation useful for prediction? Explain. Use the $\alpha = 0.01$ level. *No*

Answers are on page 652.

Objective 6 Eliminate unnecessary variables from a multiple regression model

Eliminating Unnecessary Variables from a Multiple Regression Model

In practice, there are often many explanatory variables that could potentially be included in a multiple regression equation. When using multiple regression, it is important to build a model that includes only as many explanatory variables as necessary, to avoid needless complexity. There are many methods available for choosing variables for multiple regression models; entire books have been written on the subject. Here, we will show how to use one basic principle to eliminate unnecessary variables.

Principle for Determining Whether a Variable Should Be Removed

If removing a variable increases the adjusted R^2, then the variable should be removed.

Note: You cannot remove the variable labeled "Constant." This is the intercept b_0, which should always be part of the equation.

We show how to use this principle to eliminate variables from the equation shown in Figure 13.5. The variable pressure has the largest P-value, which means that it explains less of the variation in the response variable than any of the other variables. We will remove pressure to see what happens to the adjusted R^2. Figure 13.7 presents the results of fitting the multiple regression equation using only the variables height, weight, age, and temperature.

```
The regression equation is
Lung Capacity = -5.01 + 0.114 Height - 0.0237 Weight + 0.169 Age
                + 0.0149 Temperature

Predictor          Coef    SE Coef      T      P
Constant         -5.009      1.662  -3.01  0.012
Height          0.11423    0.02994   3.81  0.003
Weight        -0.023688   0.008418  -2.81  0.017
Age             0.16915    0.05934   2.85  0.016
Temperature     0.01485    0.01223   1.21  0.250

S = 0.206692    R-Sq = 87.7%    R-Sq (adj) = 83.2%

Analysis of Variance

Source          DF       SS       MS      F      P
Regression       4  3.34916  0.83729  19.60  0.000
Residual Error  11  0.46994  0.04272
Total           15  3.81909
```

Figure 13.7 Multiple regression equation and related output using explanatory variables height, weight, age, and temperature, for data in Table 13.4, from MINITAB

Comparing the adjusted R^2 value in Figure 13.7 to the value in Figure 13.5, we see that eliminating pressure has increased the value of adjusted R^2 from 81.9% to 83.2%. This tells us that including pressure in the model does not significantly improve the fit, so we should leave it out. Note that removing pressure from the model caused the values of the coefficients for the remaining variables to change somewhat. This is typical.

We now see that temperature has the largest P-value of the remaining variables. We therefore fit the model using only the variables height, weight, and age. Figure 13.8 (page 645) presents the results.

Comparing the adjusted R^2 value in Figure 13.8 to the value in Figure 13.7, we see that eliminating temperature has decreased the value of adjusted R^2 from 83.2% to 82.6%. This tells us that we should leave temperature in the model. We would therefore use as our final model the equation that contains the explanatory variables height, weight, age, and temperature, shown in Figure 13.7.

```
The regression equation is
Lung Capacity = −3.65 + 0.105 Height − 0.0217 Weight + 0.169 Age

Predictor        Coef      SE Coef      T       P
Constant       -3.647      1.252     -2.91   0.013
Height          0.10519    0.02957    3.56   0.004
Weight         -0.021737   0.008425  -2.58   0.024
Age             0.16950    0.06050    2.80   0.016

S = 0.210748     R-Sq = 86.0%     R-Sq (adj) = 82.6%

Analysis of Variance

Source            DF       SS        MS       F       P
Regression         3    3.2861    1.0954   24.66   0.000
Residual Error    12    0.5330    0.0444
Total             15    3.8191
```

Figure 13.8 Multiple regression equation and related output using explanatory variables height, weight, and age, for data in Table 13.4, from MINITAB

Using Technology

We use Table 13.4 to illustrate the technology steps.

EXCEL

Constructing the multiple regression equation and residual plot

Step 1. Enter the values for Lung Capacity (y) from Table 13.4 into **Column A** and the values for Height, Weight, Age, Barometric Pressure, and Temperature, into **Columns B** through **F** (Figure A)

Step 2. Select **Data**, then **Data Analysis**. Highlight **Regression** and click **OK**.

Step 3. Enter the range of cells that contain the x-values in the **Input X Range** field and the range of cells that contain the y-values in the **Input Y Range** field. The residuals may be generated by selecting the **Residuals** option.

Step 4. Click **OK** (Figure B).

A residual plot may be generated using the procedure presented in Section 4.3 on page 192.

	A	B	C	D	E	F
1	Lung Cap	Height	Weight	Age	Pressure	Temp
2	2.68	61	105	13	30	70.4
3	2.22	49	50	9	30.3	71.6
4	2.29	61	128	15	29.9	73.6
5	2.16	54	81	10	29.6	70.8
6	3.19	68	120	13	29.6	69.1
7	1.97	56	92	10	30	70.1
8	2.98	68	140	16	30.2	67.7
9	1.96	57	86	10	30.7	56.3
10	3.05	62	120	16	30.5	63.8
11	2.48	60	114	13	30.3	69.9
12	2.09	57	94	12	30.1	73
13	3.27	67	145	16	30.6	70.3
14	2.26	52	66	10	28.8	75.5
15	3.2	72	159	16	30	72.1
16	2.1	53	65	10	30.5	67.5
17	1.79	51	56	9	29.7	65.6

Figure A

Regression Statistics	
Multiple R	0.93784
R Square	0.87955
Adjusted R Square	0.81933
Standard Error	0.21448
Observations	16

ANOVA

	df	*SS*	*MS*	*F*	*Significance F*
Regression	5	3.35908	0.67182	14.60440	0.00025
Residual	10	0.46001	0.04600		
Total	15	3.81909			

	Coefficients	*Standard Error*	*t Stat*	*P-value*
Intercept	-7.81721	6.28758	-1.24328	0.24212
Height	0.11927	0.03292	3.62364	0.00466
Weight	-0.02463	0.00897	-2.74683	0.02059
Age	0.16411	0.06252	2.62480	0.02539
Pressure	0.07640	0.16447	0.46451	0.65223
Temp	0.02022	0.01716	1.17827	0.26597

Figure B

MINITAB

Constructing the multiple regression equation and residual plot

Step 1. Enter the values for Lung Capacity (y) from Table 13.4 into column **C1** and the values for Height, Weight, Age, Barometric Pressure, and Temperature into columns **C2** through **C6**.

Step 2. Select **Stat**, then **Regression**, then **Regression**, and then **Fit Regression Model**.

Step 3. Select the y-variable (**C1**) as the **Response** and the x-variables (**C2, C3, C4, C5, C6**) as the **Continuous predictors**. To generate the residual plot, select **Graphs...** and select **Residuals versus fits** and click **OK**.

Step 4. Click **OK** (Figures C and D).

Regression Equation

Lung Capacity = -7.82 + 0.1193 Height - 0.02463 Weight + 0.1641 Age + 0.076 Barometric Presure + 0.0202 Temperature

Coefficients

Term	Coef	SE Coef	T-Value	P-Value	VIF
Constant	-7.82	6.29	-1.24	0.242	
Height	0.1193	0.0329	3.62	0.005	16.56
Weight	-0.02463	0.00897	-2.75	0.021	28.58
Age	0.1641	0.0625	2.62	0.025	9.50
Barometric Presure	0.076	0.164	0.46	0.652	2.00
Temperature	0.0202	0.0172	1.18	0.266	1.97

Figure C

Model Summary

S	R-sq	R-sq(adj)	R-sq(pred)
0.214478	87.96%	81.93%	68.50%

Analysis of Variance

Source	DF	Adj SS	Adj MS	F-Value	P-Value
Regression	5	3.35908	0.671817	14.60	0.000
Height	1	0.60403	0.604027	13.13	0.005
Weight	1	0.34708	0.347081	7.55	0.021
Age	1	0.31693	0.316928	6.89	0.025
Barometric Presure	1	0.00993	0.009926	0.22	0.652
Temperature	1	0.06386	0.063864	1.39	0.266
Error	10	0.46001	0.046001		
Total	15	3.81909			

Figure D

Section 13.3 Exercises

Exercises 1–6 are the Check Your Understanding exercises located within the section.

Understanding the Concepts

In Exercises 7 and 8, fill in each blank with the appropriate word or phrase.

7. A ________ plot can be used to determine whether a multiple regression equation is appropriate. *residual*

8. We should leave a variable out of a multiple regression equation when removing it ________ the value of adjusted R^2. *increases*

In Exercises 9 and 10, determine whether the statement is true or false. If the statement is false, rewrite it as a true statement.

9. The coefficient of determination R^2 measures the percentage of variation in the outcome that is explained by the model. *True*

10. If the value of the F-statistic is large, the multiple regression equation is not useful for making predictions. *False*

Practicing the Skills

11. For the following data set:

y	x_1	x_2	x_3
13.9	4.1	0.6	0.1
47.2	8.4	4.4	9.3
21.9	6.9	2.8	1.1
27.9	6.0	0.0	7.2
48.1	6.2	2.9	9.1
44.1	0.9	8.0	7.7
49.7	6.9	7.6	9.0
32.7	7.3	2.2	4.7
30.6	9.7	4.3	3.7
41.1	6.9	1.5	7.4

a. Construct the multiple regression equation $\hat{y} = b_0 + b_1x_1 + b_2x_2 + b_3x_3$.

b. Predict the value of y when $x_1 = 1, x_2 = 4.5, x_3 = 6.2$. *35.26*

c. What percentage of the variation in y is explained by the model? 94.1%

d. Is the model useful for prediction? Why or why not? Use the $\alpha = 0.05$ level. Yes

e. Test H_0: $\beta_1 = 0$ versus H_1: $\beta_1 \neq 0$ at the $\alpha = 0.05$ level. Can you reject H_0? Repeat for β_2 and β_3. Reject H_0 for β_2, β_3

12. For the following data set:

y	x_1	x_2	x_3
49.4	35.1	11.7	36.0
33.2	16.3	6.2	32.8
31.9	38.7	2.3	32.3
23.3	0.6	5.7	16.5
48.0	23.7	8.9	37.7
49.1	35.7	32.9	27.7
31.7	34.8	17.7	20.7
28.3	11.1	25.1	36.1
34.5	21.9	4.4	25.9
17.1	12.9	21.8	3.9
36.6	37.5	29.4	7.4
22.9	27.1	0.4	12.1

a. Construct the multiple regression equation $\hat{y} = b_0 + b_1x_1 + b_2x_2 + b_3x_3$.
$\hat{y} = 7.502 + 0.3838x_1 + 0.2417x_2 + 0.5617x_3$

b. Predict the value of y when $x_1 = 10.1$, $x_2 = 8.5$, $x_3 = 26.2$. 28.15

c. What percentage of the variation in y is explained by the model? 69.6%

d. Is the model useful for prediction? Why or why not? Use the $\alpha = 0.05$ level. Yes

e. Test H_0: $\beta_1 = 0$ versus H_1: $\beta_1 \neq 0$ at the $\alpha = 0.05$ level. Can you reject H_0? Repeat for β_2 and β_3. Reject H_0 for β_3

13. For the following data set:

y	x_1	x_2	x_3	x_4
27.4	12.3	4.7	3.4	4.2
55.2	9.7	12.9	10.7	2.1
35.3	12.4	2.8	9.1	9.7
36.3	5.2	13.6	5.0	0.7
42.7	1.0	9.3	2.6	13.5
58.5	14.8	9.1	3.2	10.1
58.9	5.9	11.1	3.7	4.2
54.0	11.7	11.8	8.6	0.8
64.2	3.7	14.7	1.8	14.6
61.1	13.3	12.1	0.4	2.3
53.3	14.5	9.4	6.3	10.0
59.5	5.0	9.6	13.4	14.7
43.3	4.1	7.0	10.4	9.4
40.3	2.4	9.0	8.3	6.2
43.3	3.8	6.9	7.8	11.3

a. Construct the multiple regression equation $\hat{y} = b_0 + b_1x_1 + b_2x_2 + b_3x_3 + b_4x_4$.
$\hat{y} = -1.9446 + 1.1325x_1 + 3.0963x_2 + 0.4224x_3 + 1.2384x_4$

b. Predict the value of y when $x_1 = 5.2$, $x_2 = 9.1$, $x_3 = 8.7$, $x_4 = 2.8$. 39.263

c. What percentage of the variation in y is explained by the model? 72.0%

d. Is the model useful for prediction? Why or why not? Use the $\alpha = 0.01$ level. Yes

e. Test H_0: $\beta_1 = 0$ versus H_1: $\beta_1 \neq 0$ at the $\alpha = 0.05$ level. Can you reject H_0? Repeat for β_2 and β_3. Reject H_0 for β_1, β_2, β_4

14. For the following data set:

y	x_1	x_2	x_3	x_4
163.4	2.9	24.2	4.3	4.4
120.9	22.2	12.9	10.6	13.7
116.7	5.5	6.8	17.3	13.9
63.9	8.8	12.0	0.9	1.2
86.9	5.1	18.3	2.5	13.4
74.2	14.7	8.0	8.5	5.6
72.8	7.1	17.2	0.2	24.7
84.1	1.6	5.6	0.5	4.3
136.4	21.6	2.0	3.9	15.7
122.3	14.5	13.3	14.7	13.1
78.9	17.7	12.7	3.2	8.0
108.0	16.4	15.0	1.6	23.3
124.5	24.3	20.6	9.1	11.3
90.9	21.7	0.6	5.7	5.8
143.5	18.6	11.2	13.9	14.7

a. Construct the multiple regression equation $\hat{y} = b_0 + b_1x_1 + b_2x_2 + b_3x_3 + b_4x_4$.
$\hat{y} = 63.14 + 0.532x_1 + 1.3975x_2 + 2.5277x_3 + 0.2046x_4$

b. Predict the value of y when $x_1 = 15.3$, $x_2 = 4.7$, $x_3 = 0.6$, $x_4 = 8.2$. 81.042

c. What percentage of the variation in y is explained by the model? 33.7%

d. Is the model useful for prediction? Why or why not? Use the $\alpha = 0.05$ level. No

e. Test H_0: $\beta_1 = 0$ versus H_1: $\beta_1 \neq 0$ at the $\alpha = 0.05$ level. Can you reject H_0? Repeat for β_2, β_3 and β_4. Do not reject H_0 for β_1, β_2, β_3, β_4

Working with the Concepts

15. Engine emissions: In a laboratory test of a new automobile engine design carried out at the Colorado School of Mines, the emission rate (in milligrams per second) of oxides of nitrogen was measured for 32 engines at various engine speeds (in rpm) and engine torque (in foot-pounds). The following MINITAB output presents the multiple regression equation Emission rate = $b_0 + b_1$ Speed + b_2 Torque.

```
The regression equation is
Emission rate = −321 + 0.378 Speed − 0.160 Torque

Predictor        Coef        StDev         T         P
Constant      −320.59        98.14     −3.27     0.003
Speed         0.37820      0.06861      5.51     0.000
Torque       −0.16047      0.06082     −2.64     0.013

S = 67.13         R−Sq = 51.6%        R−Sq (adj) = 48.3%

Analysis of Variance
Source              DF          SS          MS         F         P
Regression           2    139419.6     69079.8    15.469     0.000
Residual Error      29    130685.2      4506.4
Total               31    270104.8
```

a. Predict the emissions rate for an engine with speed 1500 and torque 500. *166.475*

b. Two engines are run at the same torque, but their speeds differ by 300 rpm. By how much should we predict their emissions rates to differ? *113.46*

c. Is the model useful for prediction? Explain. Use the $\alpha = 0.05$ level. *Yes*

d. How much of the variation in the emission rate is explained by the model? *51.6%*

e. Let β_1 be the coefficient of speed. Test H_0: $\beta_1 = 0$ versus H_1: $\beta_1 \neq 0$ at the $\alpha = 0.05$ level. What do you conclude? *Reject H_0.*

f. Let β_2 be the coefficient of torque. Test H_0: $\beta_2 = 0$ versus H_1: $\beta_2 \neq 0$ at the $\alpha = 0.01$ level. What do you conclude? *Do not reject H_0.*

16. Fracking: Natural gas is found in rock formations underground. In order to extract the gas, a procedure known as hydraulic fracturing, or "fracking," is often used. In this procedure, fluid mixed with sand is pumped into the gas well and forced through cracks in the rock. The sand holds open the cracks, allowing the gas to flow through. A study was performed at the Colorado School of Mines to determine whether increasing the amounts of fluid or sand would increase the production of a well. Varying amounts of fluid and sand were pumped into 255 gas wells, and the amount of gas recovered was measured. The amount of gas produced is measured in units of 100 cubic feet per each foot of well depth, fluid is measured in thousands of gallons per foot, and sand is measured in thousands of pounds per foot. The following MINITAB output presents the multiple regression equation Production = $b_0 + b_1$ Fluid + b_2 Sand.

```
The regression equation is
Production = 1.0729 + 1.1752 Fluid − 0.024886 Sand

Predictor         Coef        StDev          T         P
Constant        1.0729      0.23348     4.5954     0.000
Fluid           1.1752      0.24609     4.7753     0.000
Sand         −0.024886     0.096792   −0.25711     0.797

S = 2.0197         R−Sq = 29.1%        R−Sq (adj) = 28.5%

Analysis of Variance
Source              DF         SS          MS         F         P
Regression           2     421.70      210.85    51.689     0.000
Residual Error     252     1028.0      4.0792
Total              254     1449.7
```

a. Predict the gas production for a well treated with 3 thousand gallons of fluid and 7 thousand pounds of sand. *4.4242*

b. Two wells have the same amount of sand pumped in, and the amount of fluid differs by 2 thousand pounds. By how much should we predict their productions to differ? *2.3504*

c. Is the model useful for prediction? Explain. Use the $\alpha = 0.01$ level. *Yes*

d. How much of the variation in the gas production is explained by the model? *29.1%*

e. Let β_1 be the coefficient of fluid. Test H_0: $\beta_1 = 0$ versus H_1: $\beta_1 \neq 0$ at the $\alpha = 0.01$ level. What do you conclude? *Reject H_0.*
f. Let β_2 be the coefficient of sand. Test H_0: $\beta_2 = 0$ versus H_1: $\beta_2 \neq 0$ at the $\alpha = 0.05$ level. What do you conclude? *Do not reject H_0.*

17. Drop that variable: The following MINITAB output presents a multiple regression equation: $\hat{y} = b_0 + b_1x_1 + b_2x_2 + b_3x_3 + b_4x_4$.

```
The regression equation is
Y = 0.742 + 3.58 X1 - 2.18 X2 - 0.818 X3 - 1.58 X4

Predictor      Coef      StDev        T        P
Constant     0.7424     0.7557   0.9824    0.333
X1           3.5829     1.0602   3.3794    0.002
X2          -2.1772     0.6896  -3.1574    0.003
X3          -0.8177     0.8606  -0.9501    0.349
X4          -1.5840     0.8851  -1.7897    0.082

S = 4.6846       R-sq = 46.0%       R-sq (adj) = 39.8%

Analysis of Variance
Source           DF       SS       MS         F       P
Regression        4   654.38   163.59    7.4547   0.000
Residual Error   35   768.08   21.945
Total            39   1422.5
```

It is desired to drop one of the explanatory variables. Which of the following is the most appropriate action?
i. Drop x_3, then see whether R^2 increases.
ii. Drop x_3, then see whether adjusted R^2 increases.
iii. Drop x_1, then see whether R^2 increases.
iv. Drop x_1, then see whether adjusted R^2 increases.

18. Drop that variable: The following MINITAB output presents a multiple regression equation: $\hat{y} = b_0 + b_1x_1 + b_2x_2 + b_3x_3 + b_4x_4 + b_5x_5$.

```
The regression equation is
Y = 1.38 + 1.13 X1 + 4.39 X2 - 1.45 X3 + 0.741 X4 + 1.71 X5

Predictor      Coef      StDev        T        P
Constant     1.3774     0.7317   1.8823    0.066
X1           1.1307     0.6932   1.6311    0.110
X2           4.3897     0.6755   6.4987    0.000
X3          -1.4479     0.7978  -1.8149    0.076
X4           0.7414     0.6447   1.1500    0.256
X5           1.7095     0.6266   2.7283    0.009

S = 4.5160       R-sq = 63.7%       R-sq (adj) = 59.6%

Analysis of Variance
Source           DF       SS        MS        F       P
Regression        5   1573.38   314.79   15.435   0.000
Residual Error   44   897.35    20.394
Total            49   2471.3
```

It is desired to drop one of the explanatory variables. Which of the following is the most appropriate action?
i. Drop x_2, then see whether R^2 increases.
ii. Drop x_4, then see whether R^2 increases.
iii. Drop x_4, then see whether adjusted R^2 increases.
iv. Drop x_2, then see whether adjusted R^2 increases.

19. Predicting GPA: Twenty college students were sampled after their freshman year. Following are their freshman GPAs, their high school GPAs, their SAT reading scores, and their SAT math scores.

Freshman GPA	High School GPA	SAT Reading	SAT Math
2.98	2.90	506	497
3.59	3.58	572	521
2.26	2.18	402	498
2.43	2.44	490	474
2.14	2.69	440	398
3.77	3.61	645	625
3.26	3.61	541	481
3.18	3.40	569	490
2.80	3.05	447	626
3.20	3.37	477	594
3.02	2.41	486	508
2.84	2.95	555	466
2.66	3.30	486	425
3.46	3.47	535	523
2.52	2.47	397	453
2.01	1.89	482	388
2.79	2.41	459	294
3.01	2.62	421	319
3.31	3.54	525	549
3.17	3.10	514	542

a. Let y represent freshman GPA, x_1 represent high school GPA, x_2 represent SAT reading score, and x_3 represent SAT math score. Construct the multiple regression equation $\hat{y} = b_0 + b_1x_1 + b_2x_2 + b_3x_3$. *$\hat{y} = 0.1210 + 0.5408x_1 + 0.002146x_2 + 0.0002816x_3$*

b. An applicant for next year's freshman class has a high school GPA of 3.05, an SAT reading score of 510, and an SAT math score of 515. Predict the freshman GPA for this student. *3.01*

c. Refer to part (b). Construct a 95% confidence interval for the freshman GPA. *(2.8720, 3.1483)*

d. Refer to part (b). Construct a 95% prediction interval for the freshman GPA. *(2.4147, 3.6056)*

e. What percentage of the variation in freshman GPA is explained by the model? *72.5%*

f. Is the model useful for prediction? Why or why not? Use the $\alpha = 0.01$ level. *Yes*

g. Test H_0: $\beta_1 = 0$ versus H_1: $\beta_1 \neq 0$ at the $\alpha = 0.05$ level. Can you reject H_0? Repeat for β_2 and β_3. *Reject H_0 for β_1. Do not reject H_0 for β_2 and β_3.*

20. Paint lifetime: A paint company collected data on the lifetime (in years) of its paint in eleven U.S. cities. The data are in the following table.

City	Paint Lifetime	Average January Temperature	Average July Temperature	Average Annual Precipitation (inches)
Atlanta, GA	11.5	41.9	78.6	48.6
Boston, MA	11.7	29.6	73.5	43.8
Kansas City, KS	12.3	28.4	80.9	29.3
Minneapolis, MN	10.5	11.2	73.1	26.4
Dallas, TX	11.2	45.0	86.3	34.2
Denver, CO	15.2	29.5	73.3	15.3
Miami, FL	8.7	67.1	82.4	57.5
Phoenix, AZ	11.1	52.3	92.3	7.1
San Francisco, CA	16.7	48.5	62.2	19.7
Seattle, WA	14.2	40.6	65.3	38.9
Washington, DC	12.6	35.2	78.9	39.0

a. Let y represent paint lifetime, x_1 represent January temperature, x_2 represent July temperature, and x_3 represent annual precipitation. Construct the multiple regression equation $\hat{y} = b_0 + b_1x_1 + b_2x_2 + b_3x_3$. *$\hat{y} = 29.292 + 0.031527x_1 - 0.20046x_2 - 0.084195x_3$*

b. In Cheyenne, Wyoming, the average January temperature is 26.1, the average July temperature is 68.9, and the average annual precipitation is 13.3. Predict the lifetime of this paint in Cheyenne. *15.183 years*

c. Refer to part (b). Construct a 95% confidence interval for the paint lifetime in Cheyenne. *(13.343, 17.023)*

d. Refer to part (b). Construct a 95% prediction interval for the paint lifetime in Cheyenne. *(11.609, 18.757)*

e. What percentage of the variation in the paint lifetime in Cheyenne is explained by the model? *77.1%*

f. Is the model useful for prediction? Why or why not? Use the $\alpha = 0.05$ level. *Yes*

g. Test H_0: $\beta_1 = 0$ versus H_1: $\beta_1 \neq 0$ at the $\alpha = 0.05$ level. Can you reject H_0? Repeat for β_2 and β_3. *Reject H_0 for β_2 and β_3. Do not reject H_0 for β_1.*

21. Chemical reaction: A chemical reaction was run 48 times. In each run, different values were chosen for the temperature in degrees Celsius (x_1), the concentration of the primary reactant (x_2), and the number of hours the reaction was allowed to run (x_3). The outcome variable (y) is the amount of product.

y	x_1	x_2	x_3	y	x_1	x_2	x_3	y	x_1	x_2	x_3	y	x_1	x_2	x_3
31.3	50	19	4.0	37.4	60	30	5.0	24.0	60	24	4.0	51.0	80	34	7.5
56.9	90	38	8.0	43.3	60	26	7.0	36.2	50	21	7.0	34.3	60	22	2.5
43.1	70	28	6.5	36.3	70	25	7.5	26.5	50	17	2.0	31.5	60	24	5.0
41.5	70	25	5.5	38.4	70	31	5.5	47.1	80	34	8.5	33.2	60	23	4.0
39.0	60	26	6.5	41.5	60	27	7.5	38.1	70	27	5.5	39.2	60	29	6.5
40.9	70	29	5.0	36.1	60	23	6.0	33.5	60	24	2.5	46.7	70	27	7.5
35.9	60	23	5.5	38.5	60	23	6.0	43.6	70	27	10.0	30.4	70	32	4.0
43.5	70	28	5.5	42.4	60	24	9.0	41.0	80	32	6.5	43.2	60	25	5.5
47.9	80	34	6.5	46.5	70	31	5.5	50.2	50	26	9.0	30.6	60	26	3.5
33.8	70	26	4.5	43.1	80	32	6.0	34.4	50	22	4.0	43.3	70	28	7.5
41.1	70	26	8.0	50.8	60	26	10.0	47.9	80	34	6.5	32.4	60	22	5.5
38.7	70	26	8.0	44.2	70	28	4.5	46.2	50	21	10.0	35.5	60	22	4.5

a. Construct the multiple regression equation $\hat{y} = b_0 + b_1x_1 + b_2x_2 + b_3x_3$. $\hat{y} = 8.8721 - 0.006795x_1 + 0.71632x_2 + 2.030x_3$

b. A reaction is run at a temperature of 50°C, with the concentration of the primary reactant set to 30, and the reaction is allowed to run for 6 hours. Predict the amount of product produced. *42.202*

c. Refer to part (b). Construct a 95% confidence interval for the amount produced. *(36.934, 47.470)*

d. Refer to part (b). Construct a 95% prediction interval for the amount produced. *(32.888, 51.517)*

e. What percentage of the variation in the amount of product is explained by the model? *70.2%*

f. Is the model useful for prediction? Why or why not? Use the $\alpha = 0.05$ level. *Yes*

g. Test H_0: $\beta_1 = 0$ versus H_1: $\beta_1 \neq 0$ at the $\alpha = 0.05$ level. Can you reject H_0? Repeat for β_2 and β_3. *Reject H_0 for β_2 and β_3. Do not reject H_0 for β_1.*

22. Old Faithful: The following table lists values measured for 60 consecutive eruptions of the geyser Old Faithful in Yellowstone National Park. They are the duration of the eruption (x_1), the duration of the dormant period immediately before the eruption (x_2), and the duration of the dormant period immediately after the eruption (y). All the times are in minutes.

x_1	x_2	y	x_1	x_2	y	x_1	x_2	y	x_1	x_2	y
3.5	80	84	1.8	42	91	4.7	88	51	4.1	70	79
4.1	84	50	4.1	91	51	1.8	51	80	3.7	79	60
2.3	50	93	1.8	51	79	4.6	80	49	3.8	60	86
4.7	93	55	3.2	79	53	1.9	49	82	3.4	86	71
1.7	55	76	1.9	53	82	3.5	82	75	4.0	71	67
4.9	76	58	4.6	82	51	4.0	75	73	2.3	67	81
1.7	58	74	2.0	51	76	3.7	73	67	4.4	81	76
4.6	74	75	4.5	76	82	3.7	67	68	4.1	76	83
3.4	75	80	3.9	82	84	4.3	68	86	4.3	83	76
4.3	80	56	4.3	84	53	3.6	86	72	3.3	76	55
1.7	56	80	2.3	53	86	3.8	72	75	2.0	55	73
3.9	80	69	3.8	86	51	3.8	75	75	4.3	73	56
3.7	69	57	1.9	51	85	3.8	75	66	2.9	56	83
3.1	57	90	4.6	85	45	2.5	66	84	4.6	83	57
4.0	90	42	1.8	45	88	4.5	84	70	1.9	57	71

a. Construct the multiple regression equation $\hat{y} = b_0 + b_1x_1 + b_2x_2$. $\hat{y} = 120.15 + 0.89597x_1 - 0.74088x_2$

b. Predict the duration of dormant period immediately after an eruption, if the duration of the eruption is 3.0 minutes and the duration of the dormant period immediately before the eruption is 70 minutes. *70.979*

c. Refer to part (b). Construct a 95% confidence interval for the duration of the dormant period immediately following the eruption. *(67.875, 74.084)*

d. Refer to part (b). Construct a 95% prediction interval for the duration of the dormant period immediately following the eruption. *(50.880, 91.079)*

e. What percentage of the variation in the duration is explained by the model? *47.2%*

f. Is the model useful for prediction? Why or why not? Use the $\alpha = 0.01$ level. *Yes*

g. Test H_0: $\beta_1 = 0$ versus H_1: $\beta_1 \neq 0$ at the $\alpha = 0.05$ level. Can you reject H_0? Repeat for β_2. *Reject H_0 for β_2. Do not reject H_0 for β_1.*

23. How's your credit? Credit data were collected on a random sample of 25 U.S. cities in a recent year. Following are the average credit scores, the average debt, the average number of late payments, the average percentage of credit available, and the average number of open credit cards.

Credit Score	Debt	Late Payments	% of Credit Available	Credit Cards	Credit Score	Debt	Late Payments	% of Credit Available	Credit Cards
748	24,713	0.40	68.16	1.85	759	23,602	0.36	73.03	2.10
755	23,393	0.36	71.63	1.73	779	21,113	0.30	76.53	1.91
713	25,054	0.66	65.88	1.65	727	25,733	0.50	66.41	1.86
764	24,166	0.35	74.21	1.85	767	26,555	0.41	69.79	2.14
751	25,076	0.49	69.68	1.91	747	23,762	0.48	69.98	1.77
710	26,074	0.61	66.55	1.60	756	24,376	0.45	72.51	1.97
766	24,854	0.35	71.44	2.13	770	24,884	0.33	74.85	1.83
739	23,104	0.40	68.74	2.05	724	26,242	0.53	69.35	2.01
761	24,982	0.41	71.08	2.14	767	24,955	0.37	70.69	2.00
755	24,705	0.39	68.26	2.02	729	23,602	0.58	71.98	1.55
729	26,055	0.65	66.36	1.72	781	23,622	0.33	75.58	1.86
742	23,767	0.48	71.61	1.73	768	24,842	0.33	72.06	1.92
702	26,414	0.73	65.76	1.36					

a. Let y represent average credit score, x_1 represent average debt, x_2 represent average number of late payments, x_3 represent average percentage of credit available, and x_4 represent average number of open credit cards. Construct the multiple regression equation $\hat{y} = b_0 + b_1x_1 + b_2x_2 + b_3x_3 + b_4x_4$. *$\hat{y} = 524.7948 + 0.0022154x_1 - 113.3986x_2 + 2.8074x_3 + 11.8733x_4$*

b. In Denver, Colorado, the average debt was $26,775, the average number of late payments was 0.42, the average percentage of available credit was 70.24, and the average number of open credit cards was 2.18. Predict the average credit score in Denver. *759.55*

c. Refer to part (b). Construct a 95% confidence interval for the average credit score. *(750.368, 768.751)*

d. Refer to part (b). Construct a 95% prediction interval for the average credit score. *(741.385, 777.734)*

e. What percentage of the variation in the average credit score is explained by the model? *90%*

f. Is the model useful for prediction? Why or why not? Use the $\alpha = 0.01$ level. *Yes*

g. Test $H_0\colon \beta_1 = 0$ versus $H_1\colon \beta_1 \neq 0$ at the $\alpha = 0.05$ level. Can you reject H_0? Repeat for β_2, β_3, and β_4. *Reject H_0 for β_2 and β_3. Do not reject H_0 for β_1 and β_4.*

Answers to Check Your Understanding Exercises for Section 13.3

1. The predicted value is 2.0276 liters.
2. By 0.2462 liters.
3. The P-value for β_1 is 0.008; reject H_0. The P-value for β_2 is 0.619; do not reject H_0. The P-value for β_3 is 0.000; reject H_0.
4. (a) The predicted value is 10.523. (b) 5.4315, 15.615 (c) 7.443, 13.604
5. (a) 32.6% (b) 23.3%
6. (a) The P-value is 0.031. We reject $H_0\colon \beta_1 = \beta_2 = \beta_3$ at the 0.05 level. We conclude that the model is useful for prediction. (b) We do not reject $H_0\colon \beta_1 = \beta_2 = \beta_3$ at the 0.01 level. There is not enough evidence to conclude at the $\alpha = 0.01$ level that the model is useful for prediction.

Chapter 13 Summary

Section 13.1: When the assumptions of the linear model are satisfied, the intercept b_0 and slope b_1 of the least-squares regression line are estimates of a true intercept β_0 and a true slope β_1. The linear model assumptions can be checked by constructing a residual plot. When the residual plot exhibits no obvious pattern, the vertical spread is approximately the same across the plot, and there are no outliers, we may conclude that the assumptions of the linear model are satisfied. We may then compute confidence intervals and test hypotheses about β_1. If we reject $H_0\colon \beta_1 = 0$, we may conclude that the explanatory variable is useful to help predict the value of the outcome variable.

Section 13.2: When the assumptions of the linear model are satisfied, we may construct confidence intervals for the mean response and prediction intervals for an individual response. A confidence interval for the mean response is an interval that is likely to contain the mean value of the response variable y for a given value of the explanatory variable x. A prediction interval for an individual response is an interval that is likely to contain the value of the response variable y for a randomly chosen individual whose value of the explanatory variable is x.

Section 13.3: Sometimes we need several explanatory variables to predict the value of the response variable. In these situations we use multiple regression. As with the least-squares regression line, we can construct confidence intervals for the mean response and prediction intervals for an individual response. The coefficient of determination, R^2, measures the proportion of variation in the response variable that is explained by the multiple regression equation. The adjusted R^2 value can be used to determine which explanatory variables can be dropped from the model. If removing a variable increases the adjusted R^2, then the variable should be removed.

Vocabulary and Notation

Important Formulas

Residual standard deviation:

$$s_e = \sqrt{\frac{\sum(y - \hat{y})^2}{n - 2}}$$

Standard error for b_1:

$$s_b = \frac{s_e}{\sqrt{\sum(x - \bar{x})^2}}$$

Confidence interval for slope:

$$b_1 - t_{\alpha/2} \cdot s_b < \beta_1 < b_1 + t_{\alpha/2} \cdot s_b$$

Confidence interval for the mean response:

$$\hat{y} \pm t_{\alpha/2} \cdot s_e \sqrt{\frac{1}{n} + \frac{(x^* - \bar{x})^2}{\sum(x - \bar{x})^2}}$$

Test statistic for slope b_1:

$$t = \frac{b_1}{s_b}$$

Test statistic for correlation:

$$U = \frac{r\sqrt{n-2}}{\sqrt{1 - r^2}}$$

Prediction interval for an individual response:

$$\hat{y} \pm t_{\alpha/2} \cdot s_e \sqrt{1 + \frac{1}{n} + \frac{(x^* - \bar{x})^2}{\sum(x - \bar{x})^2}}$$

Chapter Quiz

1. A confidence interval for β_1 is to be constructed from a sample of 20 points. How many degrees of freedom are there for the critical value? *18*
2. A confidence interval for a mean response and a prediction interval for an individual response are to be constructed from the same data. True or false: The number of degrees of freedom for the critical value is the same for both intervals. *True*
3. True or false: If we fail to reject the null hypothesis H_0: $\beta_1 = 0$, we can conclude that there is no linear relationship between the explanatory variable and the outcome variable. *False*
4. True or false: When the sample size is large, confidence intervals and hypothesis tests for β_1 are valid even when the assumptions of the linear model are not met. *False*
5. A statistics student has constructed a confidence interval for the mean height of daughters whose mothers are 66 inches tall, and a prediction interval for the height of a particular daughter whose mother is 66 inches tall. One of the intervals is (65.3, 68.2) and the other is (63.8, 69.7). Unfortunately, the student has forgotten which interval is which. Can you tell which is the confidence interval and which is the prediction interval? Explain. *(65.3, 68.2) is the confidence interval.*

Exercises 6–10 refer to the following data set:

x	25	13	16	19	29	19	16	30
y	40	20	33	30	50	37	34	37

6. Compute the point estimates b_0 and b_1. *$b_0 = 13.0508$, $b_1 = 1.0574$*
7. Construct a 95% confidence interval for β_1. *(0.2247, 1.8902)*
8. Test the hypotheses H_0: $\beta_1 = 0$ versus H_1: $\beta_1 \neq 0$. Use the $\alpha = 0.01$ level of significance. *Do not reject H_0.*
9. Construct a 95% confidence interval for the mean response when $x = 20$. *(29.195, 39.205)*
10. Construct a 95% prediction interval for an individual response when $x = 20$. *(19.327, 49.073)*

Exercises 11–15 refer to the following data set:

y	x_1	x_2	x_3
69.8	7.9	37.3	62.4
32.3	9.3	20.2	40.7
66.9	13.3	30.5	48.7
87.5	27.4	38.8	35.8
93.5	9.3	40.1	28.9
65.1	23.3	29.2	46.9
26.3	24.1	11.5	50.4
48.7	21.5	19.0	58.2
52.8	25.5	32.2	34.7
58.0	8.3	37.9	23.7
65.6	30.1	34.6	56.2
42.3	21.1	23.9	42.6

11. Construct the multiple regression equation $\hat{y} = b_0 + b_1x_1 + b_2x_2 + b_3x_3$. *$\hat{y} = -13.542 + 0.204x_1 + 2.056x_2 + 0.181x_3$*

12. Predict the value of y when $x_1 = 20$, $x_2 = 20$, and $x_3 = 30$. *37.09*

13. What percentage of the variation in y is explained by the model? *78.2%*

14. Is this model useful for prediction? Why or why not? Use the $\alpha = 0.05$ level. *Yes*

15. Test H_0: $\beta_1 = 0$ versus H_1: $\beta_1 \neq 0$ at the $\alpha = 0.05$ level. Repeat for β_2 and β_3. *Reject H_0 for β_2. Do not reject H_0 for β_1 and β_3.*

Review Exercises

1. How's your mileage? Weight (in tons) and fuel economy (in mpg) were measured for a sample of seven diesel trucks. The results are presented in the following table.

Weight	Mileage
8.00	7.69
24.50	4.97
27.00	4.56
14.50	6.49
28.50	4.34
12.75	6.24
21.25	4.45

Source: J. Yanowitz, Ph.D. thesis, Colorado School of Mines

a. Compute the least-squares regression line for predicting mileage (y) from weight (x). *$\hat{y} = 8.5593 - 0.1551x$*

b. Construct a 95% confidence interval for the slope. *(−0.217, −0.093)*

c. Test H_0: $\beta_1 = 0$ versus H_1: $\beta_1 < 0$. Can you conclude that weight is useful in predicting mileage? Use the $\alpha = 0.05$ level of significance. *Yes*

2. How's your mileage? Use the data in Exercise 1 for the following.

a. Compute a point estimate for the mean mileage for trucks that weigh 20 tons. *5.4567*

b. Construct a 95% confidence interval for the mean mileage of trucks that weigh 20 tons. *(5.0064, 5.907)*

c. Predict the mileage of a particular truck that weighs 20 tons. *5.4567*

d. Construct a 95% prediction interval for the mileage of a particular truck that weighs 20 tons. *(4.1857, 6.7278)*

3. How much wood is in that tree? For a sample of 12 trees, the volume of lumber (in cubic meters) and the diameter (in centimeters) at a fixed height above ground level were measured. The results were as follows.

Diameter	Volume	Diameter	Volume
35.1	0.81	33.8	0.80
48.4	1.39	45.3	1.69
47.9	1.31	25.2	0.30
35.3	0.67	28.5	0.19
47.3	1.46	30.1	0.63
26.4	0.47	30.0	0.64

a. Compute the least-squares regression line for predicting volume (y) from diameter (x). *$\hat{y} = -1.0123 + 0.0519x$*

b. Construct a 95% confidence interval for the slope. *(0.0389, 0.0650)*

c. Test H_0: $\beta_1 = 0$ versus H_1: $\beta_1 > 0$. Can you conclude that diameter is useful in predicting volume? Use the $\alpha = 0.01$ level of significance. *Yes*

4. How much wood is in that tree? Use the data in Exercise 3 for the following.

a. Compute a point estimate for the mean volume for trees with a diameter of 44 centimeters. *1.2733*

b. Construct a 95% confidence interval for the mean volume of trees with a diameter of 44 centimeters. *(1.1224, 1.4241)*

c. Predict the volume for a particular tree whose diameter is 44 centimeters. *1.2733*

d. Construct a 95% prediction interval for the volume of a particular tree whose diameter is 44 centimeters. *(0.8631, 1.6834)*

5. Watching paint dry: In tests designed to measure the effect of the concentration (in percent) of a certain additive on the drying time (in hours) of paint, the following data were obtained.

Concentration of Additive	Drying Time
4.0	8.7
4.2	8.8
4.4	8.3
4.6	8.7
4.8	8.1
5.0	8.0
5.2	8.1
5.4	7.7
5.6	7.5
5.8	7.2

a. Compute the least-squares regression line for predicting drying time (y) from concentration (x). $\hat{y} = 12.1933 - 0.8333x$

b. Construct a 95% confidence interval for the slope. *(−1.078, −0.589)*

c. Test H_0: $\beta_1 = 0$ versus H_1: $\beta_1 < 0$. Can you conclude that concentration is useful in predicting drying time? Use the $\alpha = 0.05$ level of significance. *Yes*

6. Watching paint dry: Use the data in Exercise 5 for the following.

a. Compute a point estimate for the mean drying time for paint with a concentration of 5.1. *7.9433*

b. Construct a 95% confidence interval for the mean drying time for paint whose concentration is 5.1. *(7.7945, 8.0922)*

c. Predict the mean drying time for a particular can of paint with a concentration of 5.1. *7.9433*

d. Construct a 95% prediction interval for the drying time of a particular can of paint whose concentration is 5.1. *(7.4745, 8.4122)*

7. Energy use: A sample of 10 households was monitored for one year. The household income (in $1000s) and the amount of energy consumed (in 10^{10} joules) were determined. The results follow.

Income	Energy	Income	Energy
31	16.0	96	98.3
40	40.2	70	93.8
28	29.8	100	77.1
48	45.6	145	114.8
195	184.6	78	67.0

a. Compute the least-squares line for predicting energy consumption (y) from income (x). $\hat{y} = 2.9073 + 0.8882x$

b. Compute a 95% confidence interval for the slope. *(0.677, 1.100)*

c. Test H_0: $\beta_1 = 0$ versus H_1: $\beta_1 > 0$. Can you conclude that income is useful in predicting energy consumption? Use the $\alpha = 0.01$ level of significance. *Yes*

8. Energy use: Use the data in Exercise 7 for the following.

a. Compute a point estimate for the mean energy use for families with an income of $50,000. *47.319*

b. Construct a 95% confidence interval for the mean energy use for families with an income of $50,000. *(34.48, 60.158)*

c. Predict the energy use for a particular family with an income of $50,000. *47.319*

d. Construct a 95% prediction interval for the energy use for a particular family with an income of $50,000. *(10.954, 83.685)*

9. Interpret calculator display: The following TI-84 Plus display presents the results of a test of the null hypothesis H_0: $\beta_1 = 0$.

```
LinRegTTest
y=a+bx
β≠0 and ρ≠0
t=3.461106178
p=.0085538598
df=8
a=-2.051191527
b=2.122998676
↓s=18.46693196
```

a. What is the alternate hypothesis? $\beta_1 \neq 0$
b. What is the value of the test statistic? 3.461106178
c. How many degrees of freedom are there? 8
d. What is the *P*-value? 0.0085538598
e. Can you conclude that the explanatory variable is useful in predicting the outcome variable? Answer this question using the $\alpha = 0.01$ level of significance. Yes

10. Interpret computer output: The following MINITAB output presents the results of a test of the null hypothesis H_0: $\beta_1 = 0$.

```
Regression Analysis :Y versus X

The regression equation is
Y = 1.92 + 5.56 X

Predictor      Coef     SE Coef       T        P
Constant      1.9167     0.1721    11.14    0.000
X             5.5582     2.1158     2.63    0.006
```

a. What are the slope and intercept of the least-squares regression line? 5.56, 1.92
b. Can you conclude that *x* is useful for predicting *y*? Use the $\alpha = 0.05$ level of significance. Yes

11. Air pollution: Following are measurements of particulate matter (PM) concentration (in micrograms per cubic meter), temperature, in degrees Fahrenheit, wind speed in miles per hour, and humidity in percent for 38 days in Denver, Colorado.

PM	Temperature	Wind Speed	Humidity	PM	Temperature	Wind Speed	Humidity
24.6	35.8	4.8	37.8	12.6	36.0	3.0	48.6
71.7	41.5	13.3	32.6	7.3	47.9	2.5	19.7
13.7	38.1	9.5	39.2	17.4	33.5	3.5	50.9
19.5	35.6	6.0	37.2	7.7	25.1	4.5	55.1
79.6	35.1	12.7	37.5	14.0	28.5	3.1	47.7
6.4	30.8	1.3	52.0	20.8	32.8	4.4	38.7
13.9	39.7	2.5	43.5	19.8	36.9	2.7	36.1
25.1	44.6	2.6	46.4	13.1	43.7	3.8	25.2
21.3	53.8	1.4	33.8	10.5	48.3	4.5	19.4
19.0	52.5	3.3	27.9	22.7	54.9	3.6	20.0
33.5	48.2	11.9	35.1	24.0	57.8	2.7	20.5
37.8	29.6	3.2	53.0	39.1	49.2	4.0	39.2
8.0	31.7	3.1	37.8	28.2	51.6	2.5	42.5
15.1	41.5	2.0	29.5	41.9	54.2	2.6	38.1
27.9	55.1	5.2	22.6	25.7	53.8	2.8	41.6
35.4	59.9	8.0	19.4	30.1	60.6	5.7	26.8
22.0	62.6	9.3	15.3	49.1	44.2	8.7	43.9
45.4	51.3	10.7	17.9	11.5	40.9	3.9	39.5
23.7	40.5	6.5	29.1	27.1	39.6	7.3	43.8

a. Let *y* represent PM, x_1 represent temperature, x_2 represent wind speed, and x_3 represent humidity. Construct the multiple regression equation $\hat{y} = b_0 + b_1x_1 + b_2x_2 + b_3x_3$. $\hat{y} = -42.354 + 0.65394x_1 + 3.662x_2 + 0.57857x_3$
b. Predict the particulate concentration on a day where the temperature is 40 degrees, the wind speed is 5 miles per hour, and the humidity is 30 percent. 19.47
c. Refer to part (b). Construct a 95% confidence interval for the particulate concentration. (13.44, 25.50)
d. Refer to part (b). Construct a 95% prediction interval for the particulate concentration. (−4.55, 43.50)
e. Are all the values in the prediction interval reasonable? Explain. No. The particulate concentration cannot be negative.
f. What percentage of the variation in particulate concentration is explained by the model? 53.8%
g. Is the model useful for prediction? Why or why not? Use the $\alpha = 0.05$ level. Yes
h. Test H_0: $\beta_1 = 0$ versus H_1: $\beta_1 \neq 0$ at the $\alpha = 0.05$ level. Can you reject H_0? Repeat for β_2 and β_3. Reject H_0 for β_1, β_2, and β_3.

12. Icy lakes: Following are data on maximum ice thickness in millimeters (y), average number of days per year of ice cover (x_1), average number of days the bottom temperature is lower than 8°C (x_2), and the average snow depth in millimeters (x_3) for 13 lakes in Minnesota.

y	x_1	x_2	x_3
730	152	198	91
760	173	201	81
850	166	202	69
840	161	202	72
720	152	198	91
730	153	205	91
840	166	204	70
730	157	204	90
650	136	172	47
850	142	218	59
740	151	207	88
720	145	209	60
710	147	190	63

a. Construct the multiple regression equation $\hat{y} = b_0 + b_1x_1 + b_2x_2 + b_3x_3$. *$\hat{y} = -372.98 + 3.5368x_1 + 3.7345x_2 - 2.1661x_3$*

b. Predict the ice thickness for a lake which is covered by ice an average of 140 days per year, the bottom temperature is less than 8°C an average of 190 days per year, and the average snow depth is 60 millimeters. *701.76*

c. Refer to part (b). Construct a 95% confidence interval for the ice thickness. *(656.34, 747.18)*

d. Refer to part (b). Construct a 95% prediction interval for the ice thickness. *(602.97, 800.55)*

e. What percentage of the variation in ice thickness is explained by the model? *72.9%*

f. Is the model useful for prediction? Why or why not? Use the $\alpha = 0.05$ level. *Yes*

g. Test H_0: $\beta_1 = 0$ versus H_1: $\beta_1 \neq 0$ at the $\alpha = 0.05$ level. Can you reject H_0? Repeat for β_2 and β_3. *Reject H_0 for β_1, β_2, and β_3.*

13. Percentage of variation: The following MINITAB output presents a multiple regression equation.

```
The regression equation is
Y = 5.3296 + 3.6204 X1 + 1.5721 X2 + 2.3328 X3

Predictor        Coef      SE Coef          T         P
Constant      -52.042       63.533    -0.8191     0.422
X1             1.4686       1.2076     1.2161     0.238
X2            -1.0333       1.1427    -0.9043     0.377
X3             2.0944       0.8783     2.3847     0.027

S = 44.953        R-sq = 40.0%        R-sq (adj) = 31.0%

Analysis of Variance

Source            DF         SS         MS       F        P
Regression         3     269.17     89.723    4.44    0.015
Residual Error    20     404.16     20.208
Total             23     673.33     29.275
```

a. What percentage of the variation in the response is explained by the multiple regression equation? *40.0%*

b. What percentage of the variation in the response is explained by the multiple regression equation, if we account for the number of explanatory variables in the model? *31.0%*

c. Is the multiple regression equation useful for prediction? Explain. Use the $\alpha = 0.05$ level. *Yes*

14. Confidence and prediction intervals: The following MINITAB output presents a confidence interval for a mean response and a prediction interval for an individual response.

```
New Obs       Fit      SE Fit         95.0% CI              95.0% PI
1          10.933      1.9789     (6.805, 15.061)     (0.6874, 21.178)

Values of Predictors for New Observations

New Obs        X1          X2           X3
1             2.5         4.1          3.2
```

a. Predict the value of y when $x_1 = 2.5$, $x_2 = 4.1$, and $x_3 = 3.2$. 10.933

b. We are 95% confident that an individual whose values are $x_1 = 2.5$; $x_2 = 4.1$, and $x_3 = 3.2$ will have a response between ________ and ________. 6.805, 15.061

c. We are 95% confident that the mean response when $x_1 = 2.5$, $x_2 = 4.1$, and $x_3 = 3.2$ is between ________ and ________. 0.6874, 21.178

15. Drop that variable: The following MINITAB output presents a multiple regression equation $\hat{y} = b_0 + b_1x_1 + b_2x_2 + b_3x_3 + b_4x_4$.

```
The regression equation is
Y = 0.742 + 3.58 X1 - 2.18 X2 - 0.818 X3 - 1.58 X4

Predictor        Coef     StDev        T        P
Constant       23.310    20.447   1.1400    0.263
X1             1.2736    0.3893   3.2713    0.003
X2             0.1954    0.3708   0.5270    0.602
X3            -0.6788    0.2830  -2.3981    0.023
X4            -4.5620    3.3893  -1.3460    0.188

S = 17.718       R-Sq = 35.0%      R-Sq (adj) = 26.6%

Analysis of Variance
Source            DF        SS        MS        F        P
Regression         4    523.37    130.84   4.1678    0.008
Residual Error    31    973.21    31.394
Total             35    1496.6    42.760
```

It is desired to drop one of the explanatory variables. Which of the following is the most appropriate action?

i. Drop x_1, then see whether R^2 increases.
ii. Drop x_2, then see whether adjusted R^2 increases.
iii. Drop x_4, then see whether R^2 increases.
iv. Drop x_1, then see whether adjusted R^2 increases.

Write About It

1. The quantity $\sum(x - \bar{x})^2$ measures the spread in the x-values. Explain why using x-values that are more spread out will result in a narrower confidence interval for β_1.

2. Suppose you are planning to buy a car that weighs 3500 pounds, and you have a linear model that predicts gas mileage (y) given the weight of the car (x). Which do you think would be more appropriate, a confidence interval for the mean gas mileage of cars that weigh 3500 pounds, or a prediction interval for the mileage of a particular car that weighs 3500 pounds? Explain.

In-Class Activities

1. Height and arm span: Your arm span is the distance between the middle fingertips on each hand when your arms are stretched out as far as you can reach. It is commonly thought that a person's arm span is approximately equal to his or her height. Investigate this by measuring each student's arm span and height and computing the least-squares regression line where x represents arm span and y represents height. Test the hypothesis $H_0: \beta_1 = 1$ versus $H_1: \beta_1 \neq 1$. What do you conclude?

2. Uncorrelated dice: Each student has two dice, one of which is thought of as x and the other as y. Each student rolls his or her dice 20 times, computes the correlation between x and y, and tests the hypothesis $H_0: \rho = 0$ versus $H_1: \rho \neq 0$. How many students reject H_0 at the $\alpha = 0.05$ level? Does this show that their dice are correlated?

Case Study: How Much Electric Power Is Generated By Wind Turbines?

In the chapter opener, we presented the following table of wind speeds and amounts of electric power generated by a windmill on 42 different days. It is important to be able to predict the amount of power that will be obtained from a given wind speed. To do this, we will compute an appropriate least-squares regression line to predict power from wind speed, then use the methods of this chapter to construct confidence intervals and prediction intervals for the amount of power produced.

Wind Speed	Power	Wind Speed	Power	Wind Speed	Power	Wind Speed	Power
4.2	10.3	4.2	5.3	2.6	9.1	4.6	5.0
1.4	1.4	3.7	1.1	7.7	46.1	8.0	65.1
6.6	39.4	5.9	29.2	6.1	26.9	7.7	54.6
4.7	12.3	6.0	24.4	5.5	19.7	1.6	0.6
2.6	9.0	5.3	13.9	4.7	5.4	5.1	17.9
5.8	20.6	5.1	9.5	4.0	12.4	6.6	32.5
1.8	3.8	4.9	2.9	2.3	5.0	2.3	9.4
5.8	21.1	8.3	63.7	8.6	67.4	5.8	10.8
7.3	41.3	7.1	37.1	5.6	15.2	6.9	29.3
7.1	45.2	9.2	87.5	4.2	11.0	6.2	21.4
6.4	28.4	4.4	0.5				

1. Compute the least-squares regression line for predicting power (y) from wind speed (x). $\hat{y} = -26.373 + 9.2915x$
2. Construct a residual plot, and explain how it shows that the assumptions of the linear model are violated.
3. We cannot use the least-squares regression line to predict y from x. In these situations, statisticians often try to manipulate one or both of the variables to try to find a linear fit. This is called **transforming the variables**. For each data point, compute the cube of the wind speed to obtain x^3, then compute the least-squares regression line for predicting power y from x^3. $\hat{y} = -0.01502 + 0.110265x^3$
4. Construct a residual plot to verify that the assumptions of the linear model are now satisfied.
5. Use the least-squares regression line computed in Exercise 3 to predict the power when the wind speed is 5.0 mph. (Note that the value of the explanatory variable is $5^3 = 125$.) 13.768
6. Use the least-squares regression line computed in Exercise 3 to compute a 95% confidence interval for the mean power on days when the wind speed is 5.0 mph. (12.023, 15.513)
7. Use the least-squares regression line computed in Exercise 3 to compute a 95% prediction interval for the power on a given day when the wind speed is 5.0 mph. (3.363, 24.173)

Analysis of Variance

chapter

14

crstrbrt/123RF

Introduction

Almost all electronic devices rely on integrated circuits, which are complex electrical circuits built onto wafers constructed of semiconductor material, typically silicon. The silicon wafers are coated with a very thin layer of a metal called tungsten. To determine whether the correct amount of tungsten has been applied, the wafers are weighed on an extremely accurate scale called a microbalance. In a study published in the collection *Statistical Case Studies for Industrial Process Improvement*, the performance of a microbalance was examined.

The policy in the laboratory where this study was conducted was that the balance was to be turned off when not in use and powered up only when it was about to be used. Each of three operators made two weighings of several silicon wafers. Results are presented in the following table for three of the wafers. All the wafers had weights very close to 54 grams, so the weights are reported in units of micrograms above 54 grams.

	Operator A		Operator B		Operator C	
Wafer 1	11	15	10	6	14	10
Wafer 2	210	208	205	201	208	207
Wafer 3	111	113	102	105	108	111

A new policy was then instituted to leave the balance powered up continuously. The three operators then made two weighings of three different wafers. The results are presented in the following table.

	Operator A		Operator B		Operator C	
Wafer 1	152	156	156	155	152	157
Wafer 2	443	440	442	439	435	439
Wafer 3	229	227	229	232	225	228

Which policy leads to more accurate results? In the case study at the end of this chapter, we will use the methods described in the chapter to find the answer.

Section 14.1 One-Way Analysis of Variance

Objectives

1. Describe the data and hypotheses for one-way ANOVA
2. Check the assumptions for one-way ANOVA
3. Perform a one-way ANOVA hypothesis test
4. Perform the Tukey–Kramer test of pairwise comparisons
5. Perform a one-way ANOVA test with technology

Objective 1 Describe the data and hypotheses for one-way ANOVA

Introduction to One-Way Analysis of Variance

In Section 11.1, we studied tests for a null hypothesis that two populations had equal means. In some situations, we have three or more populations to compare. In such situations, when appropriate assumptions are satisfied, we may test the hypothesis that all the population means are equal. This is done with a method called **analysis of variance** (ANOVA for short).

We illustrate the idea with an example. Welds, used to join metal parts, are made by heating a powdered material called flux. In a study conducted by G. Fredrickson at the Colorado School of Mines, four welding fluxes, with differing chemical compositions, were prepared. The purpose of the study was to determine whether the welds made from the fluxes would have different degrees of hardness. Five welds using each flux were made, and the hardness of each was measured in Brinell units; the higher the number, the harder the weld. The data are presented in Table 14.1.

NOTE TO INSTRUCTOR

When we have only two samples, we use the Student's *t*-test to test the hypothesis that the means are equal. One-way ANOVA can be thought of as an extension of this method to the situation where there are more than two samples.

Table 14.1 Hardness of Welds Using Four Different Fluxes

Flux	Sample Values					Sample Mean	Sample Standard Deviation
A	250	264	256	260	239	253.8	9.7570
B	263	254	267	265	267	263.2	5.4037
C	257	279	269	273	277	271.0	8.7178
D	253	258	262	264	273	262.0	7.4498

Source: G. Fredrickson, M.Sc. thesis, Colorado School of Mines

Each of the four samples in Table 14.1 can be thought of as coming from a different population. Figure 14.1 (page 663) presents dotplots for the sample hardnesses using the four fluxes. Each sample mean is marked with an "X." It is clear that the sample means differ. In particular, the welds made using flux C have the largest sample mean and those using flux A have the smallest. Of course, the sample means are random, and if the experiment were repeated, different fluxes might have the largest and the smallest means. The question we want to answer is this: Can we conclude that there are differences in the population means among the four flux types?

This experiment involves an outcome variable, hardness, and an explanatory variable, flux type. The explanatory variable is qualitative, and is called a **factor**. The outcome variable is called the **response variable**. The different values for the factor (in this example, the different fluxes) are called **treatments**. Because there is only one factor (flux type), this is called a **one-factor experiment**, Table 14.1 is called a **one-way table**, and the method of determining whether the population means differ is called **one-way ANOVA**.

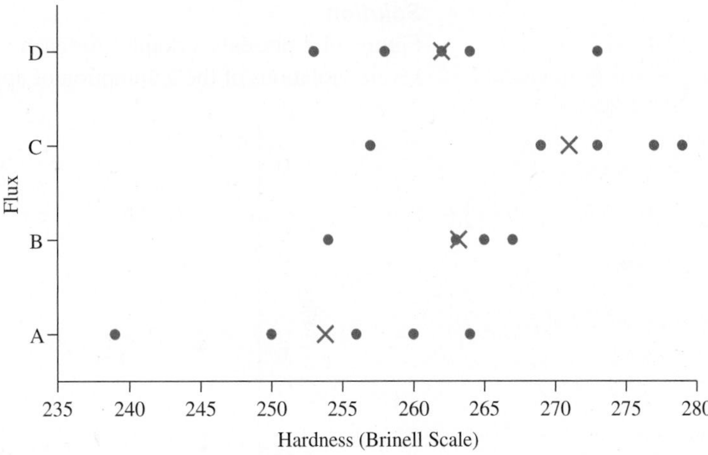

Figure 14.1 Dotplots for each sample in Table 14.1. Each sample mean is marked with an "X." The sample means differ somewhat, but the sample values overlap considerably.

Objective 2 Check the assumptions for one-way ANOVA

Assumptions for One-Way ANOVA

Data for a one-way analysis of variance consist of samples from several populations. The methods of one-way ANOVA require that the populations satisfy certain assumptions.

Assumptions for One-Way ANOVA

1. We have independent simple random samples from three or more populations.
2. Each of the populations must be approximately normal.
3. The populations must all have the same variance, which we will denote by σ^2.

EXPLAIN IT AGAIN

Balanced designs are better: Whenever possible, use a balanced design. If balance is not possible, an unbalanced design is acceptable in one-way ANOVA.

Balanced versus unbalanced designs

When the sample sizes are all the same, the data are said to have a **balanced design**. Although one-way analysis of variance can be used with both balanced and unbalanced designs, balanced designs offer a big advantage. A balanced design is much less sensitive to violations of the assumption of equality of variance than an unbalanced one. Because moderate departures from this assumption can be difficult to detect, it is best to use a balanced design whenever possible, so that undetected violations of the assumption will not seriously compromise the validity of the results. When a balanced design is impossible to achieve, a slightly unbalanced design is preferable to a severely unbalanced one.

SUMMARY

- With a balanced design, the effect of unequal variances is generally not great.
- With an unbalanced design, the effect of unequal variances can be substantial.
- The more unbalanced the design, the greater the effect of unequal variances.

Checking the assumptions

When sample sizes are small, which is often the case in ANOVA, there is really no very good way to check the assumptions of normality and equality of variance. Fortunately, the methods of ANOVA work well unless the assumptions are severely violated. We will check the assumption of normality with dotplots.

Example 14.1

Check the assumption of normality

Check the assumption of normality for the data in Table 14.1.

Solution

Figure 14.2 presents a dotplot for each sample in Table 14.1. There are no indications of severe violations of the assumption of approximate normality.

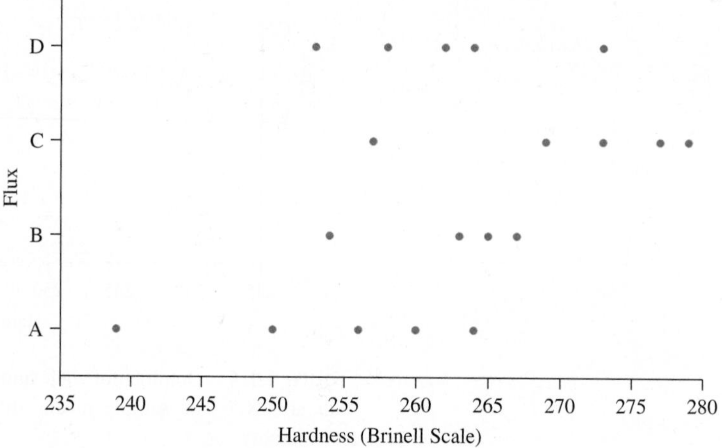

Figure 14.2 Dotplots for the samples in Table 14.1. There are no indications of severe violations of the assumption of approximate normality.

When the design is balanced, the method of ANOVA is not very sensitive to the assumption of equal population variances. When the design is unbalanced, we can check the assumption of equal variances by comparing the largest sample standard deviation to the smallest. If the largest sample standard deviation is no more than twice as large as the smallest, we conclude that the assumption of equal variances is satisfied.

When the design is unbalanced, we can check the assumption of equal population variances by computing the following quotient:

$$\frac{\text{Largest sample standard deviation}}{\text{Smallest sample standard deviation}}$$

If the result is less than 2, we conclude that the assumption of equal population variances is satisfied.

Objective 3 Perform a one-way ANOVA hypothesis test

The One-Way ANOVA Hypothesis Test

In practice, ANOVA tests are almost always performed with technology. To make clear the ideas behind the procedure, we will explain the hand calculations first. Then we will present examples in which technology is used.

Notation

We begin by describing our notation. We have I samples, each from a different population. The population means are denoted

$$\mu_1, \mu_2, \ldots, \mu_I$$

The sample means are denoted

$$\bar{x}_1, \bar{x}_2, \ldots, \bar{x}_I$$

The sample standard deviations are denoted

$$s_1, s_2, \ldots, s_I$$

The sample sizes are denoted

$$n_1, n_2, \ldots, n_I$$

The total number in all the samples combined is denoted by N:

$$N = n_1 + n_2 + \cdots + n_I$$

Finally, the average of all the items in all the samples taken together is called the **grand mean** and is denoted $\bar{\bar{x}}$.

Table 14.2 summarizes the notation.

Table 14.2 Notation for One-Way ANOVA

Sample	Population Mean	Sample Mean	Sample Standard Deviation	Sample Size
1	μ_1	$\bar{x}_1$	s_1	n_1
2	μ_2	$\bar{x}_2$	s_2	n_2
⋮	⋮	⋮	⋮	⋮
I	μ_I	$\bar{x}_I$	s_I	n_I

N is the total number of items in all samples combined.
$\bar{\bar{x}}$ is the grand mean, the average of all the items in all samples combined.

Example 14.2

Compute means and variances

For the data in Table 14.1, find $I, n_1, n_2, \ldots, n_I, \bar{x}_3, s_2, N, \bar{\bar{x}}$.

Solution

There are 4 samples, so $I = 4$. Each sample contains five observations, so $n_1 = n_2 = n_3 = n_4 = 5$. The quantity $\bar{x}_3$ is the sample mean of the third sample; this value is presented in Table 14.1, and $\bar{x}_3 = 271.0$. The total number of observations is $N = 20$. Finally, the grand mean $\bar{\bar{x}}$ is found by averaging all 20 observations; the grand mean is $\bar{\bar{x}} = 262.5$.

The hypotheses for one-way ANOVA

In one-way ANOVA, we test the null hypothesis that all the population means are equal versus the alternate hypothesis that two or more of the population means differ.

SUMMARY

The hypotheses in one-way ANOVA are

$$H_0: \mu_1 = \cdots = \mu_I \qquad H_1: \text{Two or more of the } \mu_i \text{ are different.}$$

Figure 14.3 presents the idea behind one-way ANOVA. The figure illustrates several hypothetical samples from different treatments, along with their sample means and the sample grand mean. The sample means are spread out around the sample grand mean. One-way ANOVA provides a way to measure this spread. If the sample means are highly spread out, then it is likely that the treatment means are different, and we will reject H_0.

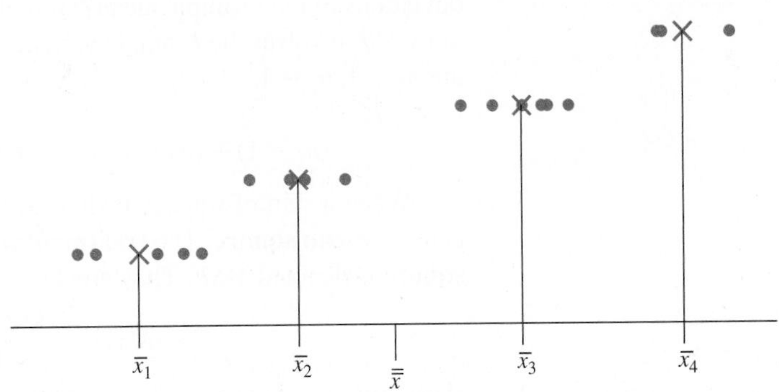

Figure 14.3 The spread within each sample is influenced only by the size of the population standard deviation σ. The spread of the sample means is influenced by both σ and the differences among the population means.

The spread of the sample means around the sample grand mean is measured by a quantity called the **treatment sum of squares** (*SSTr* for short), which is given by

$$SSTr = n_1(\bar{x}_1 - \bar{\bar{x}})^2 + n_2(\bar{x}_2 - \bar{\bar{x}})^2 + \cdots + n_I(\bar{x}_I - \bar{\bar{x}})^2$$

Each term in *SSTr* involves the distance from the sample mean to the sample grand mean. Note that each squared distance is multiplied by the sample size corresponding to its sample mean, so that the means for the larger samples count more. *SSTr* provides an indication of how different the treatment means are from each other. If *SSTr* is large, then the sample means are spread out widely, and it is reasonable to conclude that the treatment means differ and to reject H_0. If, on the other hand, *SSTr* is small, then the sample means are all close to each other, so it is plausible that the treatment means are equal.

EXPLAIN IT AGAIN

SSTr measures how spread out the treatment means are.

SSE measures how spread out the individual points are within the treatment groups.

To determine whether *SSTr* is large enough to reject H_0, we compare it to another sum of squares, called the **error sum of squares** (*SSE* for short). *SSE* measures the spread in the individual sample points around their respective sample means. This spread is measured by combining the sample variances. *SSE* is given by

$$SSE = (n_1 - 1)s_1^2 + (n_2 - 1)s_2^2 + \cdots + (n_I - 1)s_I^2$$

The size of *SSE* depends only on the standard deviations of the samples and is not affected by the location of treatment means relative to one another. *SSE* is influenced only by the population standard deviation σ.

Example 14.3

Compute *SSTr* and *SSE*

For the data in Table 14.1, compute *SSTr* and *SSE*.

Solution

The sample means are presented in Table 14.1. They are:

$$\bar{x}_1 = 253.8 \qquad \bar{x}_2 = 263.2 \qquad \bar{x}_3 = 271.0 \qquad \bar{x}_4 = 262.0$$

The sample grand mean was computed in Example 14.2 to be $\bar{\bar{x}} = 262.5$. The sample sizes are $n_1 = n_2 = n_3 = n_4 = 5$. We now compute *SSTr*:

$$\begin{aligned} SSTr &= 5(253.8 - 262.5)^2 + 5(263.2 - 262.5)^2 + 5(271.0 - 262.5)^2 + 5(262.0 - 262.5)^2 \\ &= 743.4 \end{aligned}$$

Now we compute *SSE*:

$$\begin{aligned} SSE &= (5 - 1)(9.7570)^2 + (5 - 1)(5.4037)^2 + (5 - 1)(8.7178)^2 + (5 - 1)(7.4498)^2 \\ &= 1023.6 \end{aligned}$$

The sums of squares *SSTr* and *SSE* have different degrees of freedom. *SSTr* involves the deviations of the I sample means around the grand mean, so *SSTr* has $I - 1$ degrees of freedom. *SSE* involves the I sample variances. The degrees of freedom for the sample variances are $n_1 - 1, n_2 - 1, \ldots, n_I - 1$. The degrees of freedom for *SSE* is the sum of these, which is $N - I$:

$$(n_1 - 1) + (n_2 - 1) + \cdots + (n_I - 1) = (n_1 + \cdots + n_I) - I = N - I$$

When a sum of squares is divided by its degrees of freedom, the quantity obtained is called a **mean square**. The **treatment mean square** is denoted *MSTr*, and the **error mean square** is denoted *MSE*. They are defined by

$$MSTr = \frac{SSTr}{I - 1} \qquad MSE = \frac{SSE}{N - I}$$

Now *MSE* measures how spread out the samples are around their own means. In particular, *MSE* is an estimate of σ^2, which is the variance of each of the populations. On the other hand, *MSTr* measures how spread out the sample means are around the grand mean. This spread is affected by σ^2, and also by how spread out the population means are. Now if H_0 is true, and all populations have the same mean, then *MSTr* is an estimate of σ^2, just as *MSE* is. When the population means differ, *MSTr* tends to be larger than σ^2, and thus larger than *MSE*.

We can use *MSTr* and *MSE* to construct a test statistic. The test statistic for testing H_0: $\mu_1 = \cdots = \mu_I$ is

$$F = \frac{MSTr}{MSE}$$

When H_0 is true, the numerator and denominator of F are, on average, the same size, so F tends to be near 1. In fact, when H_0 is true, this test statistic has an F distribution with $I - 1$ and $N - I$ degrees of freedom, denoted $F_{I-1,\,N-I}$. When H_0 is false, *MSTr* tends to be larger, but *MSE* does not, so F tends to be greater than 1.

We will now describe how to perform a hypothesis test using the critical value method and Table A.5. Later we will describe how to use the *P*-value method with technology.

The *F*-test for One-Way ANOVA

To test H_0: $\mu_1 = \mu_2 = \cdots = \mu_I$ versus H_1: Two or more of the μ_i are different:

Step 1: Check that the assumptions are satisfied.

Step 2: Choose a level of significance α.

Step 3: Compute *SSTr* and *SSE*.

Step 4: Compute *MSTr* and *MSE*.

Step 5: Compute the test statistic: $F = \dfrac{MSTr}{MSE}$.

Step 6: Find the critical value by using Table A.5 with $I - 1$ and $N - I$ degrees of freedom. If the value of the test statistic is greater than or equal to the critical value, reject H_0. Otherwise, do not reject H_0.

Step 7: State a conclusion.

We now apply the method of analysis of variance to the example with which we introduced this section.

Example 14.4

Perform a one-way ANOVA test

For the data in Table 14.1, test the null hypothesis that all the means are equal. Use the $\alpha = 0.05$ level of significance. What do you conclude?

Solution

Step 1: Check the assumptions. We checked the assumption of normality in Example 14.1. The design is balanced, so the procedure is not very sensitive to violations of the assumption of equal variances. In any event, the largest sample standard deviation is less than twice as large as the smallest. The assumptions are satisfied.

Step 2: Choose a level of significance. We will use $\alpha = 0.05$.

Step 3: Compute *SSTr* and *SSE*. We computed these quantities in Example 14.3: $SSTr = 743.4$ and $SSE = 1023.6$.

Step 4: Compute *MSTr* and *MSE*. We have $I = 4$ samples and $N = 20$ observations in all the samples taken together. Therefore,

$$MSTr = \frac{743.4}{4-1} = 247.8 \qquad MSE = \frac{1023.6}{20-4} = 63.975$$

Step 5: Compute the test statistic.

$$F = \frac{MSTr}{MSE} = \frac{247.8}{63.975} = 3.8734$$

Step 6: Find the critical value. The degrees of freedom are $I - 1 = 4 - 1 = 3$ and $N - I = 20 - 4 = 16$. We find the column corresponding to 3 degrees of freedom for the numerator, and the rows corresponding to 16 degrees of freedom for the denominator. We choose the row for which the tail area is 0.05. The critical value is 3.24. The value of the test statistic is $F = 3.8734$. Because $3.8734 > 3.24$, we reject H_0.

Denominator Degrees of Freedom	Area	Numerator Degrees of Freedom 1	2	3	4	5	6	7	8	9
15	0.100	3.07	2.70	2.49	2.36	2.27	2.21	2.16	2.12	2.09
15	0.050	4.54	3.68	3.29	3.06	2.90	2.79	2.71	2.64	2.59
15	0.025	6.20	4.77	4.15	3.80	3.58	3.41	3.29	3.20	3.12
15	0.010	8.68	6.36	5.42	4.89	4.56	4.32	4.14	4.00	3.89
15	0.001	16.59	11.34	9.34	8.25	7.57	7.09	6.74	6.47	6.26
16	0.100	3.05	2.67	2.46	2.33	2.24	2.18	2.13	2.09	2.06
16	0.050	4.49	3.63	3.24	3.01	2.85	2.74	2.66	2.59	2.54
16	0.025	6.12	4.69	4.08	3.73	3.50	3.34	3.22	3.12	3.05
16	0.010	8.53	6.23	5.29	4.77	4.44	4.20	4.03	3.89	3.78
16	0.001	16.12	10.97	9.01	7.94	7.27	6.80	6.46	6.19	5.98
17	0.100	3.03	2.64	2.44	2.31	2.22	2.15	2.10	2.06	2.03
17	0.050	4.45	3.59	3.20	2.96	2.81	2.70	2.61	2.55	2.49
17	0.025	6.04	4.62	4.01	3.66	3.44	3.28	3.16	3.06	2.98
17	0.010	8.40	6.11	5.18	4.67	4.34	4.10	3.93	3.79	3.68
17	0.001	15.72	10.66	8.73	7.68	7.02	6.56	6.22	5.96	5.75

EXPLAIN IT AGAIN

Interpreting the hypothesis test: If we reject the null hypothesis, we conclude that not all the population means are the same. However, we cannot tell which means are different.

Step 7: State a conclusion. We conclude that the mean hardness is not the same for all fluxes. There are at least two fluxes with different means.

Check Your Understanding

1. A certain experiment consisted of $I = 5$ treatments, with sample sizes $n_1 = n_2 = n_3 = n_4 = n_5 = 4$. The sums of squares were $SSTr = 43.7$ and $SSE = 79.8$.

a. How many degrees of freedom are there for $SSTr$? *4*

b. How many degrees of freedom are there for SSE? *15*

c. Find the treatment mean square $MSTr$. *10.925*

d. Find the error mean square MSE. *5.32*

e. Find the value of the test statistic for testing the hypothesis H_0: $\mu_1 = \mu_2 = \mu_3 = \mu_4 = \mu_5$. *2.05*

f. Find the level $\alpha = 0.05$ critical value. *3.06*

g. State your conclusion regarding H_0. Use the $\alpha = 0.05$ level of significance. *Do not reject H_0.*

2. A certain experiment consisted of $I = 3$ treatments, with sample sizes $n_1 = n_2 = n_3 = 5$. The sums of squares were $SSTr = 7.3$ and $SSE = 3.9$.

a. How many degrees of freedom are there for $SSTr$? *2*

b. How many degrees of freedom are there for SSE? *12*

c. Find the treatment mean square $MSTr$. *3.65*

d. Find the error mean square MSE. *0.325*

e. Find the value of the test statistic for testing the hypothesis H_0: $\mu_1 = \mu_2 = \mu_3$. *11.23*

f. Find the level $\alpha = 0.01$ critical value. *6.93*

g. State your conclusion regarding H_0. Use the $\alpha = 0.01$ level of significance. *Reject H_0.*

Answers are on page 677.

Objective 4 Perform the Tukey–Kramer test of pairwise comparisons

The Tukey–Kramer Test of Pairwise Comparisons

When we reject H_0 with the F-test, we conclude that the population means are not all equal. However, the F-test does not tell us which means are different from the rest. We can perform the **Tukey–Kramer test** on each pair of means to determine whether the two means in the pair are significantly different. Because this test is performed on pairs of means, it is called a test of **pairwise comparisons**.

NOTE TO INSTRUCTOR

You may wish to point out that we do not determine which pairs of means differ by performing separate hypothesis tests on each one. This would increase the frequency of Type I errors, due to the multiple testing problem. See Exercise 35.

The Tukey–Kramer Test Statistic

Let μ_i and μ_j represent two population means. The test statistic for the Tukey–Kramer test of $H_0\colon \mu_i = \mu_j$ versus $H_1\colon \mu_i \neq \mu_j$ is

$$q = \frac{|\bar{x}_i - \bar{x}_j|}{\sqrt{\dfrac{MSE}{2}\left(\dfrac{1}{n_i} + \dfrac{1}{n_j}\right)}}$$

We can perform the Tukey–Kramer test on every pair of population means. We conclude that the pairs of means for which H_0 is rejected are not equal.

When H_0 is true, the test statistic q has a distribution called the **Studentized range distribution**. Critical values for this distribution are given in Table A.6. The Studentized range distribution has two values for the degrees of freedom, I and $N - I$, where I is the number of samples and N is the total number of observations in all samples combined. Example 14.5 illustrates the method for finding a critical value for the Tukey–Kramer statistic q using the Studentized range table, Table A.6.

Example 14.5

Finding a critical value for the Tukey–Kramer statistic

Find the $\alpha = 0.05$ critical value for the Tukey–Kramer statistic for the data in Table 14.1.

Solution

We first find the degrees of freedom. The number of samples is $I = 4$, and the total number of observations in all samples combined is $N = 20$. Therefore, the degrees of freedom are $I = 4$ and $N - I = 20 - 4 = 16$. We find the column corresponding to $I = 4$ and the rows corresponding to $N - I = 16$. We choose the row for which $\alpha = 0.05$. The critical value is 4.05.

		I													
$N - I$	Area	2	3	4	5	6	7	8	9	10	11	12	13	14	15
15	0.10	2.48	3.14	3.54	3.83	4.05	4.23	4.39	4.52	4.64	4.75	4.84	4.93	5.01	5.08
15	0.05	3.01	3.67	4.08	4.37	4.59	4.78	4.94	5.08	5.20	5.31	5.40	5.49	5.57	5.65
15	0.01	4.17	4.84	5.25	5.56	5.80	5.99	6.16	6.31	6.44	6.55	6.66	6.76	6.84	6.93
16	0.10	2.47	3.12	3.52	3.80	4.03	4.21	4.36	4.49	4.61	4.71	4.81	4.89	4.97	5.04
16	0.05	3.00	3.65	4.05	4.33	4.56	4.74	4.90	5.03	5.15	5.26	5.35	5.44	5.52	5.59
16	0.01	4.13	4.79	5.19	5.49	5.72	5.92	6.08	6.22	6.35	6.46	6.56	6.66	6.74	6.82
17	0.10	2.46	3.11	3.50	3.78	4.00	4.18	4.33	4.46	4.58	4.68	4.77	4.86	4.93	5.01
17	0.05	2.98	3.63	4.02	4.30	4.52	4.70	4.86	4.99	5.11	5.21	5.31	5.39	5.47	5.54
17	0.01	4.10	4.74	5.14	5.43	5.66	5.85	6.01	6.15	6.27	6.38	6.48	6.57	6.66	6.73

If the value of I or $N - I$ is not in the table, use the closest smaller value.

Following are the steps for performing the Tukey–Kramer test.

Performing the Tukey–Kramer Test

To test H_0: $\mu_i = \mu_j$ versus H_1: $\mu_i \neq \mu_j$ for every pair of population means:

Step 1: Check that the assumptions are satisfied.

Step 2: Choose a level of significance α.

Step 3: Compute the sample means $\bar{x}_1, \ldots, \bar{x}_I$.

Step 4: Compute *MSE*.

Now perform Steps 5–7 for every pair of sample means $\bar{x}_i$ and $\bar{x}_j$.

Step 5: Compute the test statistic: $q = \dfrac{|\bar{x}_i - \bar{x}_j|}{\sqrt{\dfrac{MSE}{2}\left(\dfrac{1}{n_i} + \dfrac{1}{n_j}\right)}}$

Step 6: Find the critical value by using Table A.6 with I and $N - I$ degrees of freedom. If the value of I or $N - I$ is not in the table, use the closest smaller value. If the value of the test statistic is greater than or equal to the critical value, reject H_0. Otherwise, do not reject H_0.

Step 7: State a conclusion.

Example 14.6

Perform the Tukey–Kramer test

Perform the Tukey–Kramer test on the data in Table 14.1 to determine which pairs of population means, if any, can be concluded to differ. Use the $\alpha = 0.05$ level of significance.

Solution

We will illustrate the method in detail for $H_0: \mu_1 = \mu_2$, where μ_1 is the mean for flux A and μ_2 is the mean for flux B.

Step 1: Check the assumptions. We checked the assumption of normality in Example 14.1. Since the design is balanced, the procedure is not very sensitive to violations of the assumption of equal variances. In any event, the largest sample standard deviation is less than twice as large as the smallest. The assumptions are satisfied.

Step 2: Choose a level of significance. We will use $\alpha = 0.05$.

Step 3: Compute the sample means. The sample means appear in Table 14.1: $\bar{x}_1 = 253.8$, $\bar{x}_2 = 263.2$, $\bar{x}_3 = 271.0$, $\bar{x}_4 = 262.0$.

Step 4: Compute *MSE*. We computed *MSE* in Example 14.4: $MSE = 63.975$.

Step 5: Compute the test statistic. We will compare means μ_1 and μ_2. The sample sizes are $n_1 = 5$ and $n_2 = 5$. We compute the statistic

$$q = \frac{|\bar{x}_1 - \bar{x}_2|}{\sqrt{\dfrac{MSE}{2}\left(\dfrac{1}{n_1} + \dfrac{1}{n_2}\right)}} = \frac{|253.8 - 263.2|}{\sqrt{\dfrac{63.975}{2}\left(\dfrac{1}{5} + \dfrac{1}{5}\right)}} = 2.63$$

Step 6: Find the critical value. In Example 14.5, we found the critical value to be 4.05. The value of the test statistic is 2.63, which is not greater than 4.05. Therefore, we do not reject H_0.

We repeat Steps 5 and 6 for the other pairs of means. The following table presents the results for all pairs.

Means	Test Statistic	Critical Value	Decision
A and B	2.63	4.05	Do not reject H_0
A and C	4.81	4.05	Reject H_0
A and D	2.29	4.05	Do not reject H_0
B and C	2.18	4.05	Do not reject H_0
B and D	0.34	4.05	Do not reject H_0
C and D	2.52	4.05	Do not reject H_0

Step 7: State a conclusion. We conclude that the means for flux A and flux C differ. There is not enough evidence to conclude that any two other means differ.

Example 14.7

Use technology to perform the Tukey–Kramer test

Use technology to perform the Tukey–Kramer test on the data in Table 14.1 to determine which pairs of population means, if any, can be concluded to differ. Use the $\alpha = 0.05$ level of significance.

Solution

The following MINITAB output presents the results. MINITAB produces confidence intervals. For each pair of means, we reject H_0 if the confidence interval for that pair of means does not contain 0.

To interpret the output, the values labeled "Center" are the differences between pairs of treatment means. The quantities labeled "Lower" and "Upper" are the lower and upper bounds, respectively, of the confidence interval. The first group of confidence intervals is labeled "A subtracted from:". These intervals estimate the difference between the population mean for flux A and the population means for fluxes B, C, and D. The confidence interval "A subtracted from C" does not contain 0. Therefore, we conclude that the population means for fluxes A and C are different. The confidence intervals "A subtracted from B" and "A subtracted from D" do contain 0. Therefore, we cannot conclude that the population mean for flux A differs from the mean for B or D. In the other two groups of confidence intervals, all the intervals contain 0. Therefore, the only two fluxes for which we can conclude that the means are different are A and C.

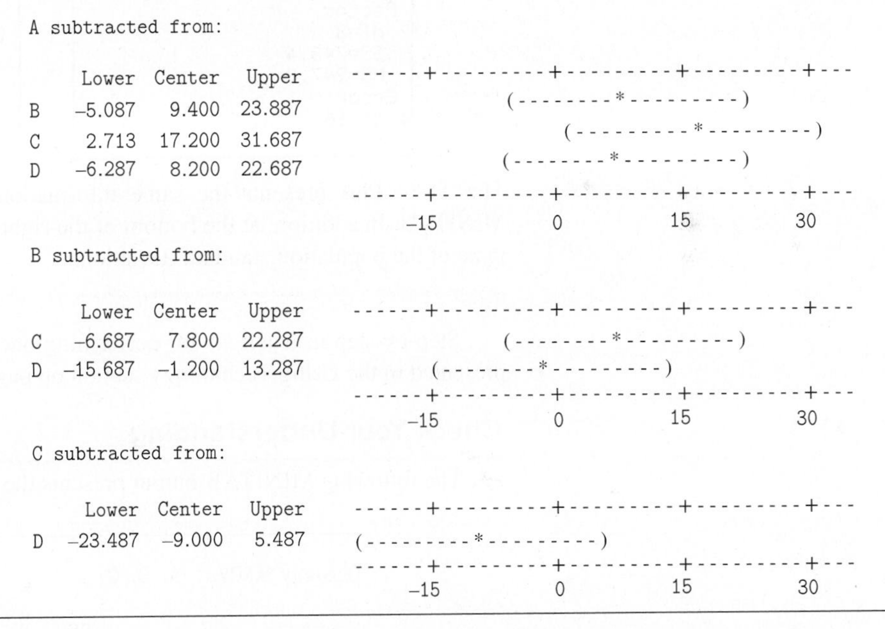

```
A subtracted from:

     Lower   Center   Upper   ------+---------+---------+---------+---
B   -5.087    9.400  23.887          (--------*---------)
C    2.713   17.200  31.687               (---------*---------)
D   -6.287    8.200  22.687         (--------*---------)
                              ------+---------+---------+---------+---
                                  -15         0        15        30

B subtracted from:

     Lower   Center   Upper   ------+---------+---------+---------+---
C   -6.687    7.800  22.287         (--------*---------)
D  -15.687   -1.200  13.287   (--------*---------)
                              ------+---------+---------+---------+---
                                  -15         0        15        30

C subtracted from:

     Lower   Center   Upper   ------+---------+---------+---------+---
D  -23.487   -9.000   5.487   (---------*---------)
                              ------+---------+---------+---------+---
                                  -15         0        15        30
```

Objective 5 Perform a one-way ANOVA test with technology

One-Way ANOVA with Technology: The ANOVA Table

In practice, ANOVA calculations are almost always performed with technology. The results can be summarized in an **analysis of variance (ANOVA) table**. The ANOVA table presents the degrees of freedom, the sums of squares $SSTr$ and SSE, the mean squares $MSTr$ and MSE, the value of the F statistic, and the P-value for testing the hypothesis that all the population means are equal.

Example 14.8

Construct an ANOVA table with technology

Use technology to construct an ANOVA table for the data in Table 14.1 and to test the null hypothesis that all the population means are equal.

Solution

We will present results from MINITAB and from the TI-84 Plus calculator. Following is the ANOVA table constructed by MINITAB:

One-way ANOVA: Hardness

Source	DF	SS	MS	F	P
Flux	3	743.40	247.800	3.87	0.029
Error	16	1023.60	63.975		
Total	19	1767.00			

In the row labeled "Flux," we find the degrees of freedom for the treatments, the treatment sum of squares *SSTr*, the treatment mean square *MSTr*, the value of the F statistic, and the P-value for the null hypothesis that all the population means are equal. The next row, labeled "Error," contains the degrees of freedom for error, the error sum of squares *SSE*, and the error mean square *MSE*. Finally, the row labeled "Total" contains the total number of degrees of freedom and the sum $SSTr + SSE$. This is called the **total sum of squares**. The P-value in the MINITAB output is 0.029, which indicates that we reject the hypothesis that all the population means are equal at the $\alpha = 0.05$ level. This agrees with the result calculated by hand in Example 14.4.

Following is the display from the TI-84 Plus calculator.

```
One-way ANOVA
F=3.873388042
p=.0294366508
Factor
 df=3
 SS=743.4
 MS=247.8
Error
↓ df=16
```

```
One-way ANOVA
↑ df=3
 SS=743.4
 MS=247.8
Error
 df=16
 SS=1023.6
 MS=63.975
Sxp=7.99843735
```

The TI-84 Plus presents the same information as the ANOVA table constructed by MINITAB. In addition, at the bottom of the right-hand screen, the quantity Sxp is an estimate of the population standard deviation.

Step-by-step instructions for performing one-way ANOVA tests with technology are presented in the Using Technology section on pages 673–674.

Check Your Understanding

3. The following MINITAB output presents the results of a one-way ANOVA.

One-way ANOVA: A, B, C

Source	DF	SS	MS	F	P
Factor	2	8.0150	4.0075	7.1057	0.007
Error	15	8.4598	0.5640		
Total	17	16.475			

a. State the null hypothesis. $H_0: \mu_1 = \mu_2 = \mu_3$

b. How many levels were there for the factor? 3

c. Assume the design was balanced. What was the sample size for each factor? 6

d. What are the values of *SSTr*, *SSE*, *MSTr*, and *MSE*? $SSTr = 8.0150$, $SSE = 8.4598$, $MSTr = 4.0075$, $MSE = 0.5640$

e. What is the value of the test statistic? 7.1057

f. What is the P-value? 0.007

g. State your conclusion regarding H_0. Use the $\alpha = 0.01$ significance level. Reject H_0.

4. The following TI-84 Plus display presents the results of a one-way ANOVA.

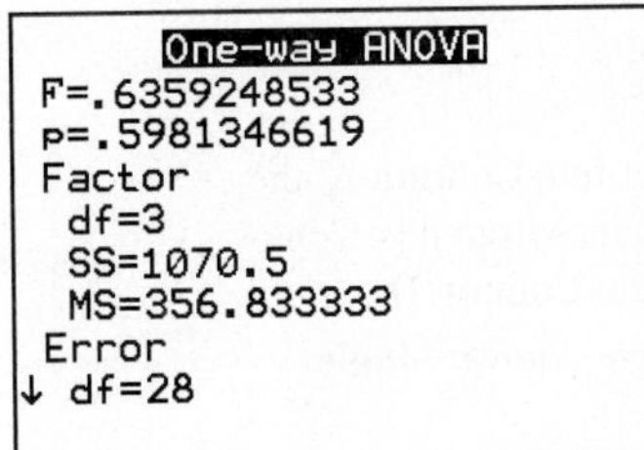

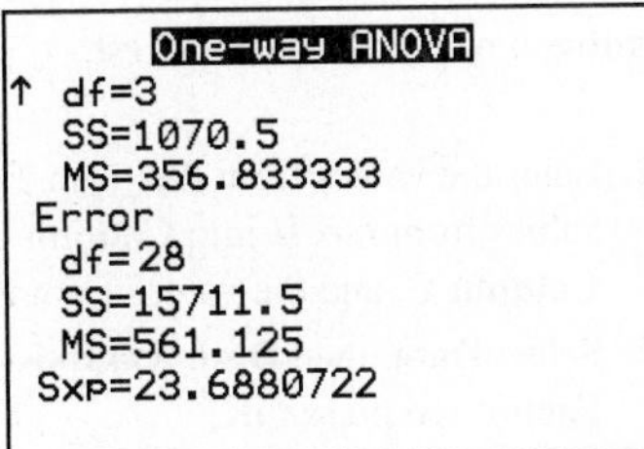

a. State the null hypothesis. *H_0: $\mu_1 = \mu_2 = \mu_3 = \mu_4$*

b. How many levels were there for the factor? *4*

c. Assume the design was balanced. What was the sample size for each factor? *8*

d. What are the values of *SSTr*, *SSE*, *MSTr*, and *MSE*? *SSTr = 1070.5, SSE = 15,711.5, MSTr = 356.833333, MSE = 561.125*

e. What is the value of the test statistic? *0.6359248533*

f. What is the *P*-value? *0.5981346619*

g. State your conclusion regarding H_0. Use the $\alpha = 0.05$ significance level. *Do not reject H_0.*

5. In a one-way ANOVA with three samples, the sample means were $\bar{x}_1 = 24.03$, $\bar{x}_2 = 14.88$, and $\bar{x}_3 = 12.76$. The sample sizes were $n_1 = n_2 = n_3 = 6$, and the error mean square was $MSE = 10.53$. Perform the Tukey–Kramer test on each pair of means. Which pairs can you conclude to be different? Use the $\alpha = 0.05$ level. *$\mu_1 \neq \mu_2$, $\mu_1 \neq \mu_3$*

6. In a one-way ANOVA with four samples, the sample means were $\bar{x}_1 = 86.8$, $\bar{x}_2 = 82.4$, $\bar{x}_3 = 85.8$, and $\bar{x}_4 = 89.1$. The sample sizes were $n_1 = n_2 = n_3 = n_4 = 4$, and the error mean square was $MSE = 1.3$. Perform the Tukey–Kramer test on each pair of means. Which pairs can you conclude to be different? Use the $\alpha = 0.01$ level. *$\mu_1 \neq \mu_2$, $\mu_2 \neq \mu_3$, $\mu_2 \neq \mu_4$, $\mu_3 \neq \mu_4$*

Answers are on page 677.

Using Technology

We use Table 14.1 and Example 14.4 to illustrate the technology steps.

TI-84 PLUS

Performing a one-way ANOVA test

Step 1. Enter the values from flux A in Table 14.1 into **L1**, the values from flux B into **L2**, the values from flux C into **L3**, and the values from flux D into **L4**.

Step 2. Press **STAT** and highlight the **TESTS** menu. Select **ANOVA** and press **ENTER**.

Step 3. Enter **L1, comma, L2, comma, L3, comma, L4** (Figure A), and then press **Enter** (Figures B and C).

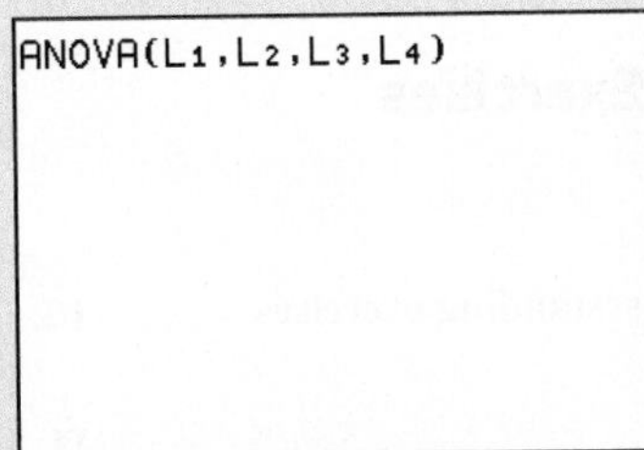

Figure A

```
One-way ANOVA
F=3.873388042
p=.0294366508
Factor
 df=3
 SS=743.4
 MS=247.8
Error
↓ df=16
```

Figure B

```
One-way ANOVA
↑ df=3
 SS=743.4
 MS=247.8
Error
 df=16
 SS=1023.6
 MS=63.975
Sxp=7.99843735
```

Figure C

EXCEL

Performing a one-way ANOVA test

Step 1. Enter the values from flux A in Table 14.1 into **Column A**, the values from flux B into **Column B**, the values from flux C into **Column C**, and the values from flux D into **Column D**.

Step 2. Select **Data**, then **Data Analysis**. Highlight **Anova: Single Factor** and press **OK**.

Step 3. Enter the range of cells containing the data in the **Input Range** field, and select the **Grouped by Columns** option (Figure D).

Step 4. Click **OK** (Figure E).

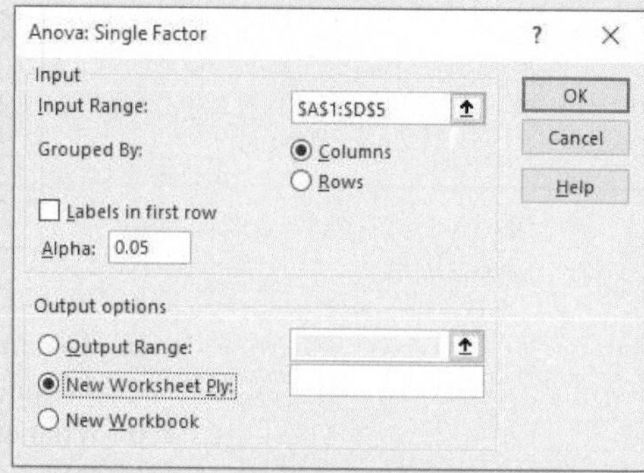

Figure D

ANOVA						
Source of Variation	*SS*	*df*	*MS*	*F*	*P-value*	*F crit*
Between Groups	743.4	3	247.8	3.873388	0.029437	3.238872
Within Groups	1023.6	16	63.975			
Total	1767	19				

Figure E

MINITAB

Performing a one-way ANOVA test

Step 1. Enter the values from flux A in Table 14.1 into **Column C1**, the values from flux B into **Column C2**, the values from flux C into **Column C3**, and the values from flux D into **Column C4**.

Step 2. Click on **Stat**, then **ANOVA**, and then **One-Way**. Select **Response data are in a separate column for each factor level.**

Step 3. In the **Responses** field, enter **C1, C2, C3, C4**.

Step 4. Click **OK** (Figure F).

Analysis of Variance

Source	DF	Adj SS	Adj MS	F-Value	P-Value
Factor	3	743.4	247.80	3.87	0.029
Error	16	1023.6	63.97		
Total	19	1767.0			

Figure F

Section 14.1 Exercises

Exercises 1–6 are the Check Your Understanding exercises located within the section.

Understanding the Concepts

In Exercises 7 and 8, fill in each blank with the appropriate word or phrase.

7. In one-way ANOVA, the null hypothesis states that all the population means are __________. *equal*

8. In one-way ANOVA, the effect of unequal variances can be substantial when the design is __________. *unbalanced*

In Exercises 9–12, determine whether the statement is true or false. If the statement is false, rewrite it as a true statement.

9. If the sample means are widely spread, the value of *SSTr* will tend to be large. *True*

10. In one-way ANOVA, all the samples must be the same size. *False*

11. In one-way ANOVA, we have samples from several populations. *True*

12. If the sample means are widely spread, the value of *SSE* will tend to be large. *False*

Practicing the Skills

13. In a one-way ANOVA, the following data were collected: $SSTr = 0.25$, $SSE = 2.11$, $N = 34$, $I = 4$.

a. How many samples are there? *4*

b. How many degrees of freedom are there for *SSTr* and *SSE*? *3, 30*

c. Compute the mean squares *MSTr* and *MSE*. *0.08333, 0.0703*

d. Compute the value of the test statistic F. 1.1848
e. Can you conclude that two or more of the population means are different? Use the $\alpha = 0.05$ level of significance. No

14. In a one-way ANOVA, the following data were collected: $SSTr = 145.34$, $SSE = 38.45$, $N = 9$, $I = 3$.
a. How many samples are there? 3
b. How many degrees of freedom are there for $SSTr$ and SSE? 2, 6
c. Compute the mean squares $MSTr$ and MSE. 72.67, 6.4083
d. Compute the value of the test statistic F. 11.3399
e. Can you conclude that two or more of the population means are different? Use the $\alpha = 0.01$ level of significance. Yes

15. Samples were drawn from three populations. The sample sizes were $n_1 = 8$, $n_2 = 6$, $n_3 = 9$. The sample means were $\bar{x}_1 = 1.04$, $\bar{x}_2 = 1.25$, $\bar{x}_3 = 1.87$. The sample standard deviations were $s_1 = 0.25$, $s_2 = 0.43$, $s_3 = 0.34$. The grand mean was $\bar{\bar{x}} = 1.42$.
a. Compute the sums of squares $SSTr$ and SSE. 3.1511, 2.2868
b. How many degrees of freedom are there for $SSTr$ and SSE? 2, 20
c. Compute the mean squares $MSTr$ and MSE. 1.5755, 0.11434
d. Compute the value of the test statistic F. 13.779
e. Can you conclude that two or more of the population means are different? Use the $\alpha = 0.01$ level of significance. Yes

16. Samples were drawn from four populations. The sample sizes were $n_1 = 9$, $n_2 = 7$, $n_3 = 10$, $n_4 = 8$. The sample means were $\bar{x}_1 = 73.5$, $\bar{x}_2 = 74.8$, $\bar{x}_3 = 75.1$, $\bar{x}_4 = 78.2$. The sample standard deviations were $s_1 = 2.5$, $s_2 = 4.5$, $s_3 = 3.8$, $s_4 = 3.2$. The grand mean was $\bar{\bar{x}} = 75.34$.
a. Compute the sums of squares $SSTr$ and SSE. 98.524, 373.14
b. How many degrees of freedom are there for $SSTr$ and SSE? 3, 30
c. Compute the mean squares $MSTr$ and MSE. 32.841, 12.438
d. Compute the value of the test statistic F. 2.6404
e. Can you conclude that two or more of the population means are different? Use the $\alpha = 0.05$ level of significance. No

Working with the Concepts

17. **Pesticide danger:** One of the factors that determines the degree of risk a pesticide poses to human health is the rate at which the pesticide is absorbed into skin after contact. An important question is whether the amount in the skin continues to increase with the length of the contact, or whether it increases for only a short time before leveling off. To investigate this, measured amounts of a certain pesticide were applied to 20 samples of rat skin. Four skins were analyzed at each of the time intervals 1, 2, 4, 10, and 24 hours. The amounts of the chemical (in micrograms) that were in the skin are given in the following table.

Duration	Amounts Absorbed			
1	1.7	1.5	1.2	1.5
2	1.7	1.6	1.9	1.9
4	1.9	1.7	2.1	2.0
10	2.2	1.9	1.7	1.6
24	2.1	2.2	2.5	2.3

a. Construct an ANOVA table.
b. Can you conclude that the amount in the skin varies with time? Use the $\alpha = 0.05$ level of significance. Yes

18. **Life-saving drug:** Penicillin is produced by the *Penicillium* fungus, which is grown in a broth whose sugar content must be carefully controlled. Several samples of broth were taken on three successive days, and the amount of dissolved sugars, in milligrams per milliliter, was measured on each sample. The results were as follows.

Day 1:	4.8	5.1	5.1	4.8	5.2	4.9	5.0
	4.9	5.0	4.8	4.8	5.1	5.0	
Day 2:	5.4	5.0	5.0	5.1	5.2	5.1	5.3
	5.2	5.2	5.1	5.4	5.2	5.4	
Day 3:	5.7	5.1	5.3	5.5	5.3	5.5	5.1
	5.6	5.3	5.2	5.5	5.3	5.4	

a. Construct an ANOVA table.
b. Can you conclude that the mean sugar concentration differs among the three days? Use the $\alpha = 0.05$ level of significance. Yes

19. **Pesticide danger:** Using the data in Exercise 17, perform the Tukey–Kramer test to determine which pairs of means, if any, differ. Use the $\alpha = 0.05$ level of significance. $\mu_1 \neq \mu_4$, $\mu_1 \neq \mu_{24}$, $\mu_2 \neq \mu_{24}$

20. **Life-saving drug:** Using the data in Exercise 18, perform the Tukey–Kramer test to determine which pairs of means, if any, differ. Use the $\alpha = 0.05$ level of significance. $\mu_1 \neq \mu_2$, $\mu_1 \neq \mu_3$, $\mu_2 \neq \mu_3$

21. **Maple syrup:** A sample of 18 maple trees in Vermont were treated with one of three amounts of fertilizer. The volume of sap (in mL) produced by each tree was recorded, with the following results.

Fertilizer	Sap Volume					
Low	76.2	80.4	74.2	79.4	87.9	86.9
Medium	87.0	95.1	93.0	98.2	94.7	96.2
High	84.2	87.5	83.1	90.3	89.9	93.2

a. Construct an ANOVA table.
b. Can you conclude that the mean amount of sap varies with the level of fertilizer? Use the $\alpha = 0.05$ level of significance. Reject H_0

22. **Solar power:** Four different types of solar energy collectors were tested. Each was tested at five randomly chosen times, and the power (in watts) was measured. The following table presents the results.

Collector	Power				
A	1.9	1.6	2.0	1.8	1.6
B	1.7	1.9	1.8	1.7	1.7
C	1.2	0.9	1.2	0.9	1.4
D	1.5	1.0	1.4	1.3	1.4

a. Construct an ANOVA table.
b. Can you conclude that the mean amount of power differs for different collectors? Use the $\alpha = 0.01$ level of significance. Reject H_0

23. **Maple syrup:** Using the data in Exercise 21, perform the Tukey-Kramer test to determine which pairs of means, if any, differ. Use the $\alpha = 0.05$ level. $\mu_1 \neq \mu_2$, $\mu_1 \neq \mu_3$

24. **Solar power:** Using the data in Exercise 22, perform the Tukey-Kramer test to determine which pairs of means, if any, differ. Use the $\alpha = 0.01$ level. $\mu_1 \neq \mu_3$, $\mu_1 \neq \mu_4$, $\mu_2 \neq \mu_3$, $\mu_2 \neq \mu_4$

25. **Artificial hips:** Artificial hip joints consist of a ball and socket. As the joint wears, the ball (head) becomes rough. Investigators performed wear tests on metal artificial hip joints. Joints with several different diameters were tested. The following table presents measurements of head roughness (in nanometers).

Diameter	Head Roughness
16	0.83 2.25 0.40 2.78 3.23
28	2.72 2.48 3.80
36	6.49 5.32 4.59

Source: Based on data in *Proceedings of the Institution of Mechanical Engineers* 215:483–493

a. Because the design is unbalanced, check that the assumption of equal variances is satisfied by showing that the largest sample standard deviation is less than twice as large as the smallest one.

b. Construct an ANOVA table.

c. Can you conclude that mean roughness varies with diameter? Use the $\alpha = 0.01$ level of significance. *Yes*

26. Floods: Rapid drainage of floodwater is crucial to prevent damage during heavy rains. Several designs for a drainage canal were considered for a certain city. Each design was tested five times, to determine how long it took to drain the water in a reservoir. The following table presents the drainage times, in minutes.

Channel Type	Drainage Time
1	41.4 43.4 50.0 41.2
2	38.7 48.3 51.1 38.3
3	32.6 33.7 34.8 22.5
4	26.3 30.9 33.3 23.8
5	44.9 47.2 48.5 37.1

a. Construct an ANOVA table.

b. Can you conclude that there is a difference in the mean drainage times for the different channel designs? Use the $\alpha = 0.01$ level of significance. *Yes*

Source: Don Becker/U.S. Geological Survey (USGS)

27. Artificial hips: Using the data in Exercise 25, perform the Tukey–Kramer test to determine which pairs of means, if any, differ. Use the $\alpha = 0.01$ level of significance. $\mu_{16} \neq \mu_{36}$

28. Floods: Using the data in Exercise 26, perform the Tukey–Kramer test to determine which pairs of means, if any, differ. Use the $\alpha = 0.01$ level of significance. $\mu_1 \neq \mu_4, \mu_2 \neq \mu_4, \mu_4 \neq \mu_5$

29. Polluting power plants: Power plants can emit levels of pollution that can lower air quality. For this reason, it is important to monitor the levels of pollution they produce. Investigators measured dust emissions, in milligrams per cubic meter, at four power plants. Thirty measurements were taken for each plant. The sample means and standard deviations are presented in the following table.

Plant	Mean	Standard Deviation	Sample Size
A	211.50	24.85	30
B	214.00	35.26	30
C	211.75	33.53	30
D	236.08	23.09	30

Source: *Journal of Air and Waste Management* 55:1042–1049

a. What are the sample sizes n_1, n_2, n_3, n_4? *30, 30, 30, 30*

b. What are the sample standard deviations s_1, s_2, s_3, s_4? *24.85, 35.26, 33.53, 23.09*

c. Compute *SSTr*. *12,712.7*

d. Compute *SSE*. *102,027.8*

e. How many degrees of freedom are there for *SSTr*? For *SSE*? *3, 116*

f. Compute *MSTr* and *MSE*. *4237.57, 879.55*

g. Compute the value of the test statistic for testing the null hypothesis that the population means are all equal. *4.818*

h. Can you conclude that there are differences among the mean emission levels? Use the $\alpha = 0.05$ level of significance. *Yes*

30. Breaking boards: The strength of wood products used in construction is measured to be sure that they are suitable for the purpose for which they are used. Measurements, in megapascals, of the strength needed to break green mixed oak boards were made. The sample means, standard deviations, and sample sizes for four different grades of lumber are presented in the following table.

Grade	Mean	Standard Deviation	Sample Size
Select	45.1	8.52	32
No. 1	42.0	5.50	11
No. 2	33.2	6.71	15
Below grade	38.1	8.04	42

Source: *Journal of Materials in Civil Engineering* 11:91–97

a. What are the sample sizes n_1, n_2, n_3, n_4? *32, 11, 15, 42*

b. What are the sample standard deviations s_1, s_2, s_3, s_4? *8.52, 5.50, 6.71, 8.04*

c. Compute *SSTr*. *1721.42*

d. Compute *SSE*. *5833.45*

e. How many degrees of freedom are there for *SSTr*? For *SSE*? *3, 96*

f. Compute *MSTr* and *MSE*. *573.808, 60.77*

g. Compute the value of the test statistic for testing the null hypothesis that the population means are all equal. *9.44*

h. Can you conclude that the strength differs for different grades of lumber? Use the $\alpha = 0.05$ level of significance. *Yes*

31. Small, medium, or large? Several large-sized sodas were ordered at each of several fast-food restaurants, and the volume of beverage in each was measured. The following TI-84 Plus display presents the results of a one-way ANOVA to determine whether the mean volume differs among the restaurants.

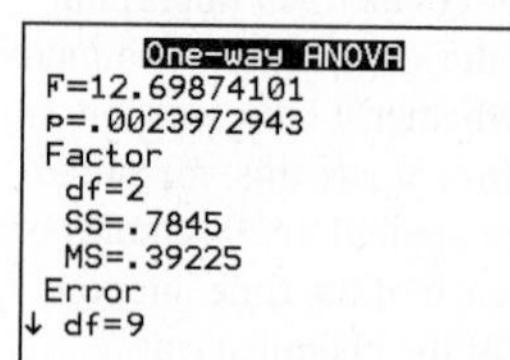


```
One-way ANOVA
↑ df=2
  SS=.7845
  MS=.39225
Error
  df=9
  SS=.278
  MS=.030888889
Sxp=.175752351
```

a. State the null hypothesis.

b. How many restaurants were involved in the study? *3*

c. Assume the design was balanced. How many sodas were measured in each restaurant? *4*

d. What are the values of *SSTr*, *SSE*, *MSTr*, and *MSE*? *0.7845, 0.278, 0.39225, 0.030888889*

e. What is the value of the test statistic? *12.69874101*

f. What is the *P*-value? *0.0023972943*

g. Can you conclude that the mean volume differs among restaurants? Use the $\alpha = 0.05$ level of significance. *Yes*

32. Spreadsheets: Several spreadsheet programs were tested by performing a certain task several times on each. The following TI-84 Plus display presents the results of a one-way ANOVA to determine whether the mean time required to perform a certain task differs among them.

```
One-way ANOVA
F=.4262152778
p=.7369008482
Factor
 df=3
 SS=.2455
 MS=.081833333
Error
↓ df=16
```

```
One-way ANOVA
↑ df=3
 SS=.2455
 MS=.081833333
Error
 df=16
 SS=3.072
 MS=.192
Sxp=.438178046
```

a. State the null hypothesis.

b. How many spreadsheets were involved in the study? 4

c. Assume the design was balanced. How many times was the task performed for each spreadsheet? 5

d. What are the values of *SSTr*, *SSE*, *MSTr*, and *MSE*? 0.2455, 3.072, 0.081833333, 0.192

e. What is the value of the test statistic? 0.4262152778

f. What is the *P*-value? 0.7369008482

g. Can you conclude that the mean time differs among spreadsheets? Use the $\alpha = 0.01$ level of significance. No

33. Watch your head: Automotive engineers compared three types of A-pillars in automobiles to determine which provided the greatest protection to occupants of automobiles during a collision. Following is a one-way ANOVA table. The response variable is the head injury criterion (HIC), which is a number that measures the amount of impact energy absorbed by the pillar.

One-way ANOVA: Pillar

Source	DF	SS	MS	F	P
Pillar	2	50946.6	25473.3	5.071	0.015
Error	24	120550.9	5023.0		
Total	26	171497.4			

a. What are the values of *SSTr* and *SSE*? 50,946.6, 120,550.9

b. What are the values of *MSTr* and *MSE*? 25473.3, 5023.0

c. What is the value of the test statistic? 5.071

d. Can you conclude that the mean energy absorbed varies with the type of pillar? Use the $\alpha = 0.01$ level of significance. No

Source: *Proceedings of the Institution of Mechanical Engineers* 215:1161–1169

34. Soil conservation: High levels of phosphorus in soil can lead to severe reductions in water quality and animal populations. Investigators treated soil specimens with six different treatments, and the acid phosphatase activity, which is related to the production of phosphorus, was recorded. Results are presented in the following one-way ANOVA table. (The units are micromoles of phosphatase per gram per hour.)

One-way ANOVA: Treatments A, B, C, D, E, F

Source	DF	SS	MS	F	P
Treatment	5	1.18547	0.23709	46.64	0.000
Error	6	0.03050	0.00508		
Total	11	1.21597			

a. What are the values of *SSTr* and *SSE*? 1.18547, 0.03050

b. What are the values of *MSTr* and *MSE*? 0.23709, 0.00508

c. What is the value of the test statistic? 46.64

d. Can you conclude that the mean level of acid phosphatase activity varies among the treatments? Use the $\alpha = 0.05$ level of significance. Yes

Source: *Journal of Agronomy and Crop Science* 184:137–142

Extending the Concepts

In Section 11.5, we studied the Bonferroni correction for multiple tests. Because the Bonferroni correction is valid for any collection of tests, it can be used as an alternative to the Tukey–Kramer test for pairwise comparisons. The Bonferroni test statistic is

$$\frac{|\bar{x}_i - \bar{x}_j|}{\sqrt{\text{MSE}\left(\frac{1}{n_i} + \frac{1}{n_j}\right)}}$$

The critical value for the Bonferroni test statistic comes from the Student's t distribution with $N - I$ degrees of freedom. If the desired level of significance is α, we reject H_0 if the P-value is less than $\alpha/(2K)$, where K is the number of differences $\mu_i - \mu_j$ to be tested.

35. More pairwise comparisons: Refer to Table 14.1, which presents data on the hardnesses of welds produced from four different fluxes.

a. There are six possible hypotheses to test regarding pairs of fluxes. They are: H_0: $\mu_1 - \mu_2 = 0$, H_0: $\mu_1 - \mu_3 = 0$, H_0: $\mu_1 - \mu_4 = 0$, H_0: $\mu_2 - \mu_3 = 0$, H_0: $\mu_2 - \mu_4 = 0$, and H_0: $\mu_3 - \mu_4 = 0$. Compute the value of the Bonferroni test statistic for each of these pairs.

b. Find the *P*-value for each of the tests, using the Student's *t* distribution with $N - I$ degrees of freedom. The alternatives are two-tailed.

c. Use the Bonferroni correction to determine which pairs of means, if any, differ at the $\alpha = 0.05$ level of significance. $\mu_A \neq \mu_C$

d. Compare the results of the Bonferroni correction with the results of the Tukey–Kramer test. Same

Answers to Check Your Understanding Exercises for Section 14.1

1. a. 4 **b.** 15 **c.** 10.925 **d.** 5.32 **e.** 2.05 **f.** 3.06 **g.** We do not reject H_0. It is possible that the means are all the same.

2. a. 2 **b.** 12 **c.** 3.65 **d.** 0.325 **e.** 11.23 **f.** 6.93 **g.** We reject H_0 and conclude that two or more of the means differ.

3. a. H_0: $\mu_1 = \mu_2 = \mu_3$ **b.** 3 **c.** 6 **d.** $SSTr = 8.0150$, $SSE = 8.4598$, $MSTr = 4.0075$, $MSE = 0.5640$ **e.** 7.1057 **f.** 0.007 **g.** We reject H_0 and conclude that two or more of the means differ.

4. a. H_0: $\mu_1 = \mu_2 = \mu_3 = \mu_4$ **b.** 4 **c.** 8 **d.** $SSTr = 1070.5$, $SSE = 15{,}711.5$, $MSTr = 356.833333$, $MSE = 561.125$ **e.** 0.6359248533 **f.** 0.5981346619 **g.** We do not reject H_0. There is not enough evidence to conclude that two or more of the means differ.

5. We can conclude that $\mu_1 \neq \mu_2$ and $\mu_1 \neq \mu_3$.

6. We can conclude that $\mu_1 \neq \mu_2$, $\mu_2 \neq \mu_3$, $\mu_2 \neq \mu_4$, and $\mu_3 \neq \mu_4$.

Section 14.2 Two-Way Analysis of Variance

Objectives

1. Interpret a two-way table
2. Describe main effects and interactions
3. Test hypotheses about interactions and main effects
4. Construct interaction plots

Objective 1 Interpret a two-way table

Two-Way Tables

Analysis of variance is concerned with data sets that consist of samples from several populations. In Section 14.1, we discussed one-factor experiments, in which each population corresponds to a different level of a single factor. In this section, we will discuss **two-factor experiments**, in which there are two factors that vary. Each combination of levels of the two factors corresponds to a population, and we have a sample from each population.

We will illustrate these ideas with an example. In a hypothetical study, a college instructor conducted an experiment to determine whether the instructional method and the class size affect the amount that students learn. Three instructional methods were used in the study: a traditional lecture method, an interactive method in which students worked with a computer program, and a third method that combined lectures with interactivity. Two classes were taught with each method. One class was large in size; the other was small. Four students were sampled from each class. Their final exam scores are shown in Table 14.3.

Andersen Ross/Getty Images

Table 14.3 Final Exam Scores

	Teaching Method		
	Lecture	**Interactive**	**Combination**
Large Class	62 74 68 69	87 75 69 84	66 66 81 75
Small Class	75 77 82 74	75 87 89 94	76 84 90 93

There are two factors, teaching method and class size, so this is a **two-way table**. The table contains two rows, corresponding to the two class sizes, and three columns, corresponding to the three teaching methods. Therefore, this is a 2×3 table. The two rows and three columns produce six **cells** in the table; each cell corresponds to a particular combination of class size and teaching method. The numbers in each cell are the exam scores of four students taught with that combination of teaching method and class size. For example, the exam scores for the students taught with a traditional lecture in a large class are 62, 74, 68, and 69. The exam scores for the students taught with the interactive method in a small class are 75, 87, 89, and 94. Because exam score is the quantity we are measuring, exam score is the response variable.

The purpose of the experiment that produced Table 14.3 would be to determine whether the factors, teaching method and class size, have an effect on the response variable, exam score. It turns out that there are two kinds of effects that a factor can have on the response: main effects and interaction effects (often referred to as interactions). We now describe these two kinds of effects.

Objective 2 Describe main effects and interactions

Main Effects and Interactions

We illustrate main effects and interactions with an example.

A chemist is running a chemical reaction in order to produce a chemical product that can be sold. She has two choices for the temperature at which to run the reaction: a high temperature or a low one. The reaction must be stirred, and the stirring rate can be set to either fast or slow. The chemist wants to determine whether the factors temperature and stirring rate have an effect on the yield, or amount of product, that is produced. She runs three reactions at each combination of temperature and stirring rate and measures the yield each time. The results are shown in Table 14.4 (page 679). The units of yield are grams.

Table 14.4 Yield of a Chemical Reaction

	Stirring Rate					
	Fast			Slow		
High Temperature	85	91	73	19	36	26
Low Temperature	34	25	20	52	45	42

To determine the effect of stirring rate on yield, we will compute the **main effect** of stirring rate. The main effect of stirring rate is the difference between the average yield at a fast rate and the average yield at the slow rate. There are six yields at the fast rate: 85, 91, 73, 34, 25, and 20. There are also six yields at the slow rate: 19, 36, 26, 52, 45, and 42. The main effect of stirring rate is therefore

NOTE TO INSTRUCTOR

Emphasize that main effects cannot be interpreted when large interactions are present.

$$\text{Main effect of stirring rate} = \frac{85+91+73+34+25+20}{6} - \frac{19+36+26+52+45+42}{6} = 18$$

The main effect of stirring rate is 18. This says that the mean yield with the fast rate of stirring is about 18 grams higher than the mean yield with the slow rate.

It turns out that this main effect is misleading. To see why, we will compute the effect of stirring rate when the temperature is high. When the temperature is high, there are three yields with the fast stirring rate: 85, 91, and 73; and three yields at the slow rate: 19, 36, and 26. The effect of stirring rate at high temperature is

$$\text{Effect of stirring rate at high temperature} = \frac{85+91+73}{3} - \frac{19+36+26}{3} = 56$$

This says that when the high temperature is used, the mean yield with the fast stirring rate is 56 grams higher than the mean yield with the slow stirring rate. This is quite a bit different from the main effect of 18. Now let's compute the effect of stirring rate when the low temperature is used.

When the low temperature is used, there are three yields at the fast stirring rate: 34, 25, and 20; and three yields at the slow stirring rate: 52, 45, and 42. The effect of stirring rate when the temperature is low is

$$\text{Effect of stirring rate at low temperature} = \frac{34+25+20}{3} - \frac{52+45+42}{3} = -20$$

The effect of stirring rate at the low temperature is negative. The effect of −20 says that when the low temperature is used, the mean yield with the fast stirring rate is 20 grams *less* than the mean yield with the slow stirring rate. This is also quite a bit different from the main effect, which says that the mean yield with fast stirring is 18 grams *higher*.

EXPLAIN IT AGAIN

Main effects and interactions: If a large interaction is present, the main effects may be misleading.

We see that the actual effect of stirring rate differs considerably, depending on whether we use high temperature or low temperature. Because of this, the main effect of stirring rate is misleading. The main effect says that the mean yield at the high stirring rate will be 18 grams higher than the mean yield at the low stirring rate. But if we use a high temperature, the mean yield will actually be about 56 grams higher, and if we use the low temperature, the mean strength will actually be 20 grams lower. So the main effect does not describe the actual effect of stirring rate for either choice of temperature.

When the effect of one factor depends on the level of another factor, we say that there is an **interaction** between the two factors. When a large interaction exists, we should not try to interpret the main effects, because they may be misleading.

Objective 3 Test hypotheses about interactions and main effects

Testing Hypotheses in Two-Way Analysis of Variance

There are three hypotheses to test in two-way analysis of variance. First, we perform a test about the interaction.

H_0: There is no interaction between the factors

H_1: There is an interaction between the factors

If we reject H_0, we conclude that there is an interaction, and we do not perform any tests on the main effects. We must choose a level of significance at which to test the hypothesis of no interaction. It is reasonable to choose a fairly large level of significance, for the following reason: If H_0 is false, and we do not reject H_0, then we will perform a test of the main effects whose results may be misleading. Therefore, we want to use a somewhat higher significance level than usual, so that this type of error will be less likely. We will use the significance level $\alpha = 0.10$ for the test of interaction.

If we do not reject the hypothesis of no interaction, we test the following hypotheses about the main effects.

H_0: There is no main effect of factor A on the response variable

H_1: There is a main effect of factor A on the response variable

and

H_0: There is no main effect of factor B on the response variable

H_1: There is a main effect of factor B on the response variable

Tests about main effects are usually conducted with a typical level of significance such as $\alpha = 0.05$ or $\alpha = 0.01$.

The tests for two-way ANOVA require certain assumptions.

Assumptions for Two-Way ANOVA

1. The populations must be approximately normal.
2. The sample sizes must all be the same.
3. The populations must all have the same variance (because the sample sizes must be the same, this assumption needs to be satisfied only very approximately).

The assumption of approximate normality can be checked with dotplots. The second assumption says that the design must be balanced. Although it is possible to analyze unbalanced designs, the analysis is much more difficult than for balanced designs. We present only the methods for balanced designs. The third assumption says that the populations have the same variance. As in one-way ANOVA, the methods are not very sensitive to violations of this assumption when the design is balanced.

Procedure for Performing Tests in Two-Way ANOVA

Step 1: Test the null hypothesis that no interaction is present. Use a fairly large level of significance such as $\alpha = 0.10$.

Step 2: If the hypothesis of no interaction is rejected, stop. Do not perform tests about the main effects, as the results could be misleading.

Step 3: If the hypothesis of no interaction is not rejected, then perform tests about the main effects, and interpret the results.

The calculations for two-way ANOVA tests are involved and are invariably performed with technology. We illustrate with some examples. Step-by-step instructions for performing two-way ANOVA tests with technology are presented in the Using Technology section on page 686.

Example 14.9

Test hypotheses in two-way ANOVA

The following table presents final exam scores for students taught by one of three teaching methods and one of two class sizes. These data were first presented in Table 14.3.

	Teaching Method		
	Lecture	**Interactive**	**Combination**
Large Class	62 74 68 69	87 75 69 84	66 66 81 75
Small Class	75 77 82 74	75 87 89 94	76 84 90 93

a. Test the null hypothesis of no interaction between teaching method and class size.

b. If the null hypothesis of no interaction is not rejected, test the hypothesis that there is no main effect of class size. Use the $\alpha = 0.05$ level of significance. Interpret the result.

c. If the null hypothesis of no interaction is not rejected, test the hypothesis that there is no main effect of teaching method. Use the $\alpha = 0.05$ level of significance. Interpret the result.

Solution

The MINITAB output is as follows.

```
Analysis of Variance for Exam Score

Source           DF        SS         MS          F        P
Class Size        1    600.00    600.000      12.84    0.002
Teaching Method   2    399.25    199.625       4.27    0.030
Interaction       2     43.75     21.875       0.47    0.634
Error            18    841.00     46.722
Total            23   1884.00
```

EXPLAIN IT AGAIN

Interpreting the hypothesis test: If we reject the null hypothesis of no main effects, we conclude that not all the levels of the factor have the same effect. However, we cannot tell which levels are different.

We use the output to perform the tests.

a. The P-value for interaction is 0.634, which is greater than 0.10. We therefore do not reject the null hypothesis of no interaction.

b. We continue by checking the P-values for the main effects. The P-value for the main effect of class size is 0.002. This value is less than 0.05, so we conclude that there are differences in mean exam scores between large and small classes.

c. The P-value for teaching method is $P = 0.030$. This value is less than 0.05, so we conclude that there are differences in mean scores among the three teaching methods.

Example 14.10

Test hypotheses in two-way ANOVA

A construction engineer is experimenting with three different concrete mixtures, A, B, and C. He forms nine blocks with each mixture. Three of the blocks are allowed to cure for two days, three others cure for four days, and the last three cure for six days. The engineer measures the strength of each block by measuring the pressure (in megapascals) needed to crush the block. The results are shown in the following table.

	Curing Time		
	2 Days	**4 Days**	**6 Days**
Mixture A	11 15 13	15 18 16	13 15 12
Mixture B	9 11 12	12 13 14	14 16 16
Mixture C	11 12 14	15 14 16	15 17 18

a. Test the null hypothesis of no interaction between curing time and mixture.

b. If the null hypothesis of no interaction is not rejected, test the hypothesis that there is no main effect of mixture. Use the $\alpha = 0.05$ level of significance. Interpret the result.

c. If the null hypothesis of no interaction is not rejected, test the hypothesis that there is no main effect of curing time. Use the $\alpha = 0.05$ level of significance. Interpret the result.

Solution

The MINITAB output is as follows.

Analysis of Variance for Strength

Source	DF	SS	MS	F	P
Mixture	2	13.407	6.7037	3.175	0.066
Curing Time	2	52.519	26.259	12.44	0.000
Interaction	4	29.037	7.2593	3.349	0.030
Error	18	38.000	2.1111		
Total	26	132.96			

We use the output to perform the tests.

a. The P-value for interaction is 0.030, which is less than 0.10. We therefore reject the null hypothesis of no interaction.

b. and **c.** Because the factors interact, a hypothesis test about the main effects could be misleading. Therefore, we do not perform hypothesis tests on the main effects.

Check Your Understanding

1. Dental researchers conducted an investigation of the manufacture of dental resins in which the response variable was the hardness of the resin. The factors were the distance between the curing tip and the surface of the resin during manufacture, and the depth of the resin at which the hardness was measured. The following MINITAB output presents the results.

Analysis of Variance for Hardness

Source	DF	SS	MS	F	P
Depth	5	3855.4	771.08	33.874	0.000
Distance	2	4111.8	2055.9	90.314	0.000
Interaction	10	47.486	4.7846	0.2102	0.994
Error	54	1229.2	22.764		
Total	71	9244.2			

a. Test the null hypothesis of no interaction between depth and distance. *Do not reject H_0.*

b. If the null hypothesis of no interaction is not rejected, test the hypothesis that there is no main effect of depth. Interpret the result. *Reject H_0.*

c. If the null hypothesis of no interaction is not rejected, test the hypothesis that there is no main effect of distance. Interpret the result. *Reject H_0.*

Source: *Brazilian Dental Journal* 11:11–17

2. Environmental scientists performed a two-factor experiment in which the response variable was the amount of an herbicide absorbed by a certain type of plant. The factors were the concentration of the herbicide and the ratio of the delivery rates of iron and water. The following MINITAB output presents the results.

Analysis of Variance for Sorption

Source	DF	SS	MS	F	P
Concentration	2	0.3793	0.1897	3.873	0.040
Delivery Ratio	2	7.3400	3.6700	74.95	0.000
Interaction	4	3.4447	0.8612	17.59	0.000
Error	18	0.8814	0.0490		
Total	26	12.045			

a. Test the null hypothesis of no interaction between concentration and delivery ratio. *Reject H_0.*

b. If the null hypothesis of no interaction is not rejected, test the hypothesis that there is no main effect of concentration. Interpret the result. *Do not perform a test.*

c. If the null hypothesis of no interaction is not rejected, test the hypothesis that there is no main effect of delivery ratio. Interpret the result. *Do not perform a test.*

Source: *Journal of Environmental Science and Health* 36:261–274

Answers are on page 691.

Objective 4 Construct interaction plots

Construct Interaction Plots

An **interaction plot** is a graph that allows us to visualize the magnitude of the interactions in a two-way ANOVA. Example 14.11 explains the steps in constructing an interaction plot.

Example 14.11

Construct an interaction plot

Construct an interaction plot for the data in Example 14.10.

Solution

We perform the following steps:

Step 1: For each cell, compute the mean of the values in that cell. This is called the **cell mean**. List the cell means in a two-way table. The following table presents the cell means for the data in Example 14.10.

	Curing Time		
	2 Days	**4 Days**	**6 Days**
Mixture A	13.00	16.33	13.33
Mixture B	10.67	13.00	15.33
Mixture C	12.33	15.00	16.67

Step 2: Construct a horizontal and a vertical axis. On the horizontal axis, place a tick mark for each level of the column factor (curing time in our example). The tick marks should be evenly spaced. Label the vertical axis with a scale that will allow the plotting of the cell means.

Figure 14.4 presents the axes for the data in Example 14.10. We have placed evenly spaced tick marks to represent the levels of curing time and have labeled the vertical axis with a range of values that will allow us to plot the cell means.

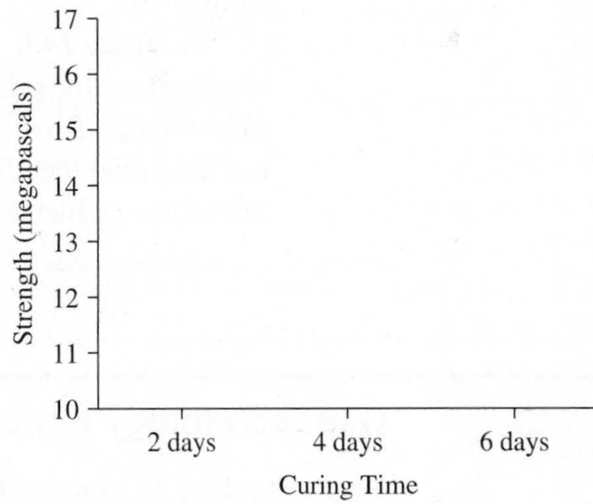

Figure 14.4

Step 3: For each mean in the first row of cell means, plot the value of the mean on the vertical axis above the tick mark for its column. Connect the points with straight lines. Figure 14.5 on page 684 presents the results for plotting the three means in the first row (corresponding to mixture A).

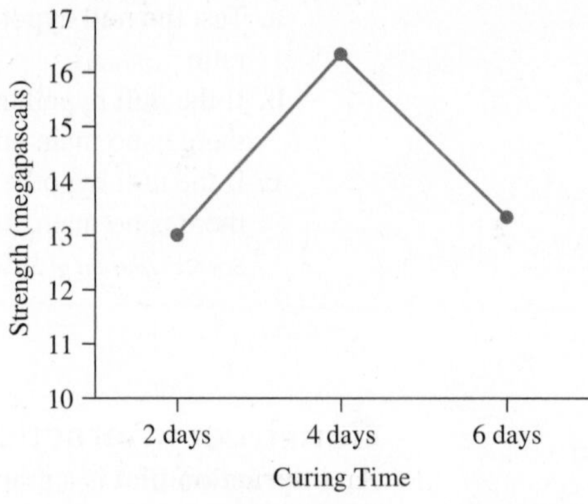

Figure 14.5

Step 4: Repeat for each row, adding the lines to the existing plot. The completed plot is shown in Figure 14.6.

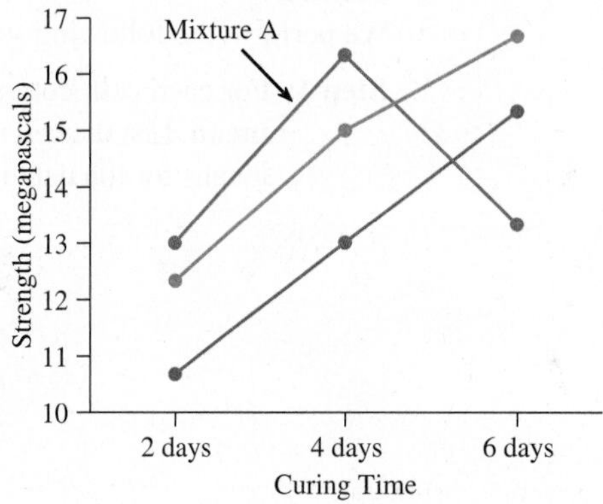

Figure 14.6

Step 5: Interpret the plot. The interaction plot contains a broken line for each row of the two-way table. The magnitude of the interactions is reflected in the directions of these lines. When the interactions are small, the lines are nearly parallel. When interactions are large, some of the lines will differ substantially in their direction from others.

In Figure 14.6, we can see that the line for mixture A has a very different direction from the other two lines. This indicates the presence of large interactions. In particular, it indicates that mixture A responds differently to the levels of curing time than the other mixtures. This is consistent with the results of the test for interaction, which showed that the interactions were statistically significant.

Example 14.12

Use technology to construct an interaction plot

Use technology to construct an interaction plot for the data in Example 14.9. Interpret the plot.

Solution

Following is the interaction plot from MINITAB. The lines are not very different in their directions, which suggests that the interactions are not large. This is consistent with the results of Example 14.9.

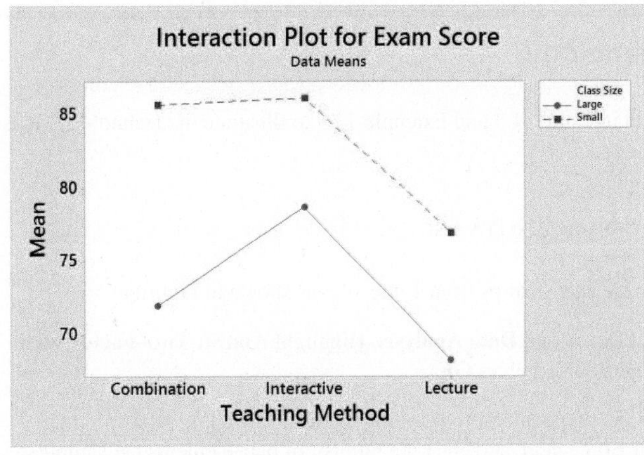

Step-by-step instructions for constructing interaction plots using MINITAB are presented on page 686.

Check Your Understanding

3. A psychiatrist studied the effects of three antidepressants on subjects in three different age groups. Each subject was rated on a scale of 0 to 100, with higher numbers indicating greater relief from depression. The following table presents the results.

	Age Group		
	18–30	**31–50**	**51–80**
Drug A	43 43 41 42	53 51 53 52	57 55 58 56
Drug B	52 51 51 50	62 61 60 61	65 66 66 67
Drug C	71 72 70 71	81 83 82 82	86 84 87 86

a. Compute the nine cell means.

b. Construct an interaction plot. Do there seem to be strong interactions?

c. Use technology to perform a test for interactions. Do you reject the hypothesis of no interactions? Are the results of the test consistent with the interaction plot? *No strong interactions*

4. Following are interaction plots for three data sets. Which data set has the largest interactions? Which has the smallest? *C has the largest; B has the smallest.*

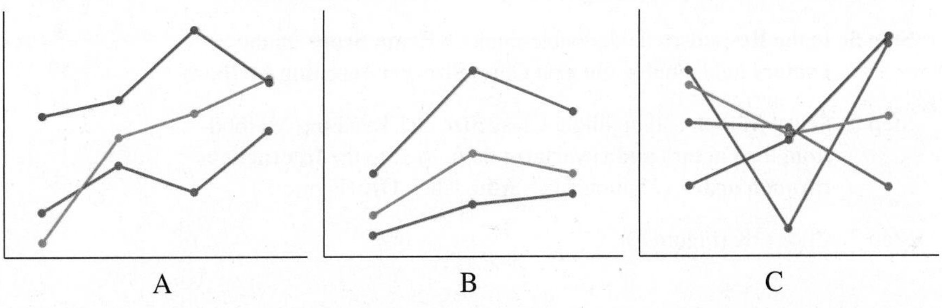

Answers are on page 691.

Using Technology

We use the data in Table 14.3 and Example 14.9 to illustrate the technology steps.

EXCEL

Performing a two-way ANOVA test

Step 1. Enter the exam scores from Table 14.3 as shown in Figure A.

Step 2. Select **Data**, then **Data Analysis**. Highlight **Anova: Two–Factor with Replication** and press **OK**.

Step 3. Enter the range of cells containing the data (including labels) in the **Input range** field, and enter the number of data elements per sample in the **Rows per sample** field. For Example 14.9, we use **4** (Figure B).

Step 4. Click **OK** (Figure C).

	A	B	C	D
1		Lecture	Interactive	Combination
2	Large Class	62	87	66
3		74	75	66
4		68	69	81
5		69	84	75
6	Small Class	75	75	76
7		77	87	84
8		82	89	90
9		74	94	93

Figure A

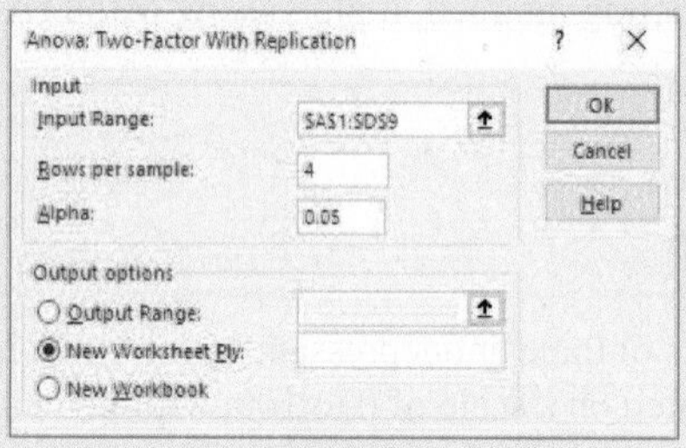

Anova: Two-Factor With Replication ? X
Input
Input Range: $A$1:$D$9
Rows per sample: 4
Alpha: 0.05
OK Cancel Help
Output options
Output Range:
New Worksheet Ply:
New Workbook

Figure B

ANOVA

Source of Variation	*SS*	*df*	*MS*	*F*	*P-value*
Sample	600	1	600	12.84185	0.002123
Columns	399.25	2	199.625	4.272592	0.03031
Interaction	43.75	2	21.875	0.468193	0.633547
Within	841	18	46.72222		
Total	1884	23			

Figure C

MINITAB

Performing a two-way ANOVA test

Step 1. Label Column C1 as **Exam Score**, Column C2 as **Class Size**, and Column C3 as **Teaching Method**.

Step 2. Enter all exam scores from Table 14.3 in the **Exam Score** Column.

Step 3. For each exam score, enter either **Large** or **Small** in the **Class Size Column** and either **Lecture**, **Interactive**, or **Combination** in the **Teaching Method Column**. The first 13 exam scores are shown in Figure D.

Step 4. Click on **Stat**, then **ANOVA**, then **General Linear Model**, and then **Fit General Linear Model**.

Step 5. In the **Responses** field, double-click on **Exam Score**. In the **Factors** field, double-click on **Class Size** and **Teaching Method**.

Step 6. Select **Model**, and highlight **Class Size** and **Teaching Method** from the **Factors and covariates** field. Next to the **Interactions through order...** option, click **Add**. Click **OK** (Figure E).

Step 7. Click **OK** (Figure F).

↓	C1	C2-T	C3-T
	Exam Score	Class Size	Teaching Method
1	62	Large	Lecture
2	74	Large	Lecture
3	68	Large	Lecture
4	69	Large	Lecture
5	87	Large	Interactive
6	75	Large	Interactive
7	69	Large	Interactive
8	84	Large	Interactive
9	66	Large	Combination
10	66	Large	Combination
11	81	Large	Combination
12	75	Large	Combination
13	75	Small	Lecture

Figure D

Note: To generate the interaction plot, click on **Stat**, then **ANOVA**, and then **Interaction Plot**. In the **Response field**, double-click on **Exam Score**. In the **Factors field**, double-click on **Class Size** and **Teaching Method**.

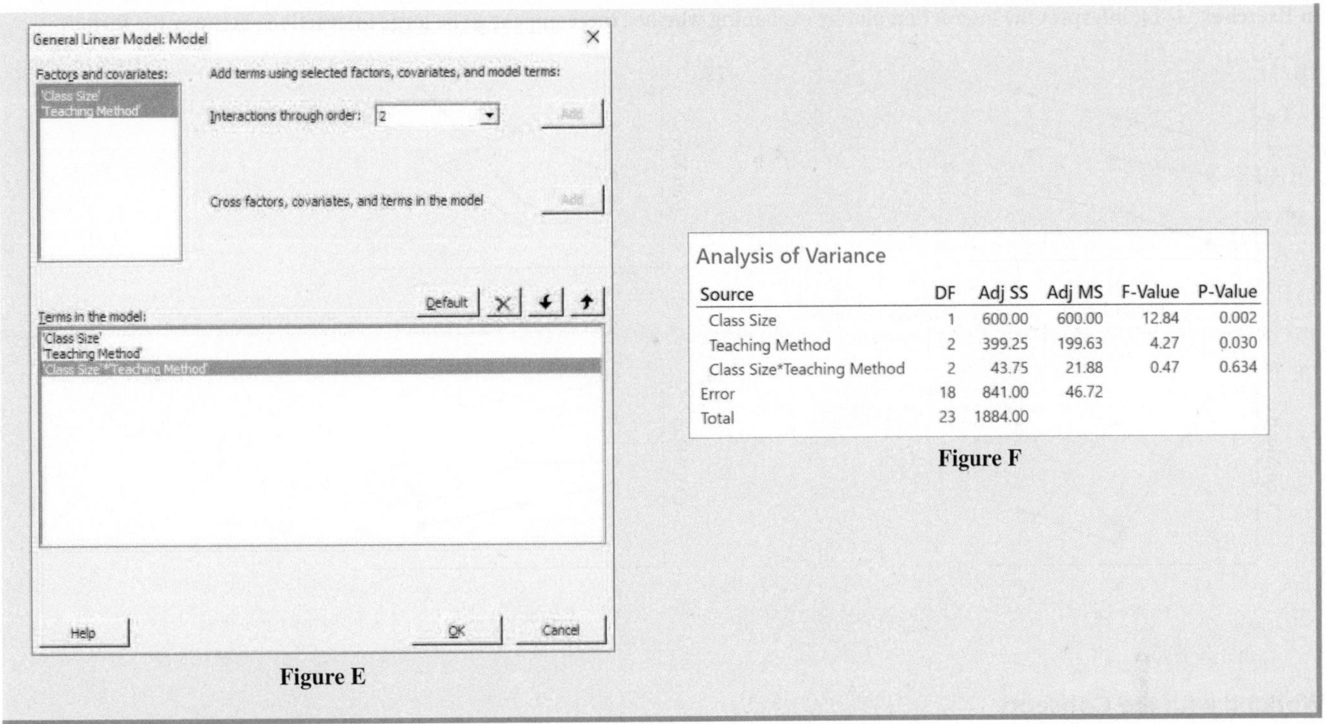

Figure E

Analysis of Variance

Source	DF	Adj SS	Adj MS	F-Value	P-Value
Class Size	1	600.00	600.00	12.84	0.002
Teaching Method	2	399.25	199.63	4.27	0.030
Class Size*Teaching Method	2	43.75	21.88	0.47	0.634
Error	18	841.00	46.72		
Total	23	1884.00			

Figure F

Section 14.2 Exercises

Exercises 1–4 are the Check Your Understanding exercises located within the section.

Understanding the Concepts

In Exercises 5 and 6, fill in each blank with the appropriate word or phrase.

5. When the effect of one factor depends on the level of another factor, an ________ is present. *interaction*

6. In two-way ANOVA, the factors can have two kinds of effects on the response: ________ and ________. *main, interaction*

In Exercises 7 and 8, determine whether the statement is true or false. If the statement is false, rewrite it as a true statement.

7. In two-way ANOVA, we should not interpret the main effects when there are no interactions. *False*

8. In two-way ANOVA, the results can be difficult to interpret if the design is unbalanced. *True*

Practicing the Skills

9. In a two-way ANOVA, the P-value for interactions is $P = 0.45$, and the P-values for the main effects are $P = 0.023$ for the row factor and $P = 0.34$ for the column factor.

a. Can you reject the hypothesis of no interactions? Explain. *No*

b. Is it appropriate to interpret the main effect of the row factor? If so, state whether there are differences in the mean response among different levels of the row factor, using the $\alpha = 0.05$ level of significance. If not, explain why not. *Yes; reject H_0.*

c. Is it appropriate to interpret the main effect of the column factor? If so, state whether there are differences in the mean response among different levels of the column factor, using the $\alpha = 0.05$ level of significance. If not, explain why not. *Yes; do not reject H_0.*

10. In a two-way ANOVA, the P-value for interactions is $P = 0.07$, and the P-values for the main effects are $P = 0.003$ for the row factor and $P = 0.014$ for the column factor.

a. Can you reject the hypothesis of no interactions? Explain. *Yes*

b. Is it appropriate to interpret the main effect of the row factor? If so, state whether there are differences in the mean response among different levels of the row factor, using the $\alpha = 0.05$ level of significance. If not, explain why not. *No*

c. Is it appropriate to interpret the main effect of the column factor? If so, state whether there are differences in the mean response among different levels of the column factor, using the $\alpha = 0.05$ level of significance. If not, explain why not. *No*

In Exercises 11–14, interpret the interaction plot by explaining whether there appear to be large interactions between the factors.

11.

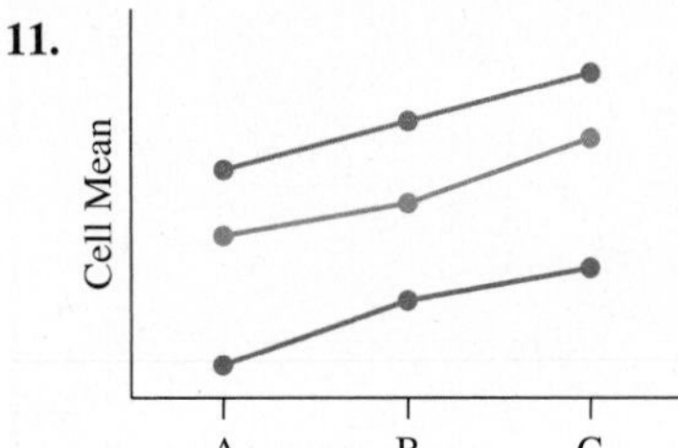

No large interactions

12.

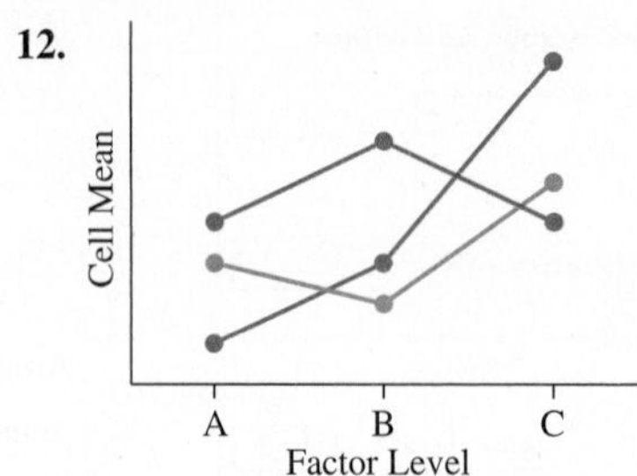

Large interactions

13.

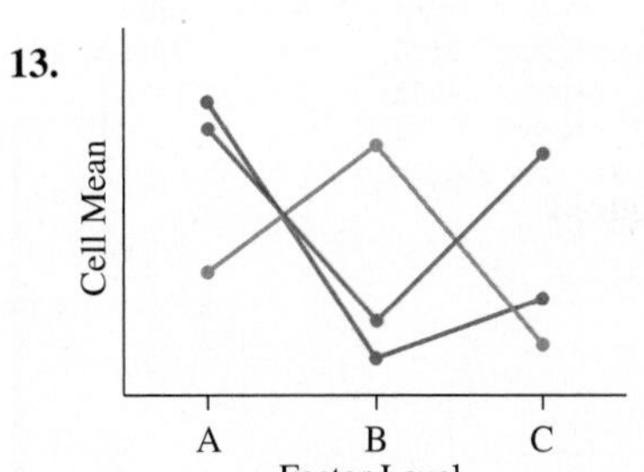

Large interactions

14.

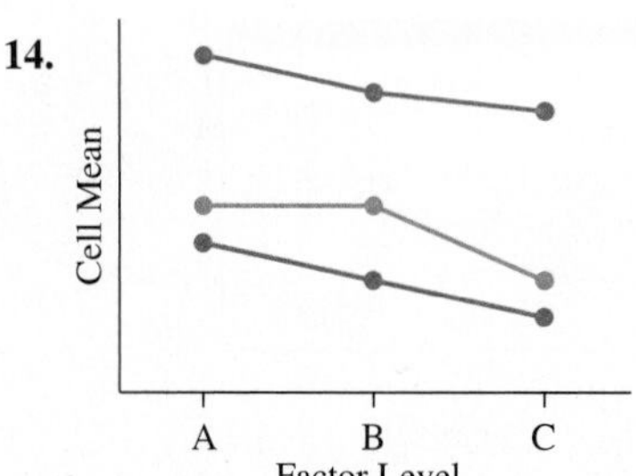

No large interactions

Working with the Concepts

15. What's your favorite subject? An education professor wants to find out which subject students find most difficult. She obtains final exam scores for three randomly selected freshmen and three randomly selected sophomores from three subjects, statistics, psychology, and history. The results are as follows.

	Statistics	Psychology	History
Freshmen	76 88 72	69 84 75	77 87 76
Sophomores	82 99 80	85 97 87	85 96 90

a. Can you reject the hypothesis of no interactions? Explain. *No*

b. Can the main effect of class (freshmen or sophomores) on final exam score be interpreted? If so, interpret the main effect, using the $\alpha = 0.05$ level of significance. If not, explain why not. *Yes; reject H_0.*

c. Can the main effect of subject on final exam score be interpreted? If so, interpret the main effect, using the $\alpha = 0.05$ level of significance. If not, explain why not. *Yes; do not reject H_0.*

16. The sap is running: Vermont maple sugar producers sponsored a testing program to determine the benefit of a new fertilizer. A random sample of 27 maple trees in Vermont were chosen and treated with one of three levels of fertilizer. In this experimental setup, nine trees (three in each of three climatic zones) were treated with each fertilizer level and the amount of sap produced (in milliliters) by the trees was measured. The results are as follows.

	Southern Zone	Central Zone	Northern Zone
Low Fertilizer	76.2 80.4 74.2	81.4 85.9 86.9	86.5 85.2 80.1
Medium Fertilizer	88.0 94.1 93.0	98.2 94.7 96.2	88.4 90.4 92.2
High Fertilizer	84.2 87.5 83.1	90.3 89.9 93.2	81.4 84.7 82.2

a. Can you reject the hypothesis of no interactions? Explain. *Yes*

b. Can the main effect of fertilizer on the amount of sap be interpreted? If so, interpret the main effect, using the $\alpha = 0.05$ level of significance. If not, explain why not. *No*

c. Can the main effect of climatic zone on the amount of sap be interpreted? If so, interpret the main effect, using the $\alpha = 0.05$ level of significance. If not, explain why not. *No*

17. What's your favorite subject? Construct and interpret an interaction plot for the data in Exercise 15.

18. The sap is running: Construct and interpret an interaction plot for the data in Exercise 16.

19. Lather up: A study was done to determine the effects of two factors on the latherability of soap. The two factors were type of water (tap or distilled) and glycerol (present or absent). The outcome measured was the amount of foam produced (in milliliters). The experiment was repeated three times for each combination of the factors. The results are as follows:

	No Glycerol	Glycerol
Distilled Water	165 181 168	170 197 190
Tap Water	155 142 139	139 160 160

a. Can you reject the hypothesis of no interactions? Explain. *No*

b. Can the main effect of water type on the amount of foam be interpreted? If so, interpret the main effect, using the $\alpha = 0.01$ level of significance. If not, explain why not. *Yes; reject H_0.*

c. Can the main effect of glycerol on the amount of foam be interpreted? If so, interpret the main effect, using the $\alpha = 0.01$ level of significance. If not, explain why not. *Yes; do not reject H_0.*

20. Cleaning solution: An experiment was performed to measure the effects of two factors on the ability of cleaning solutions to remove oil from a piece of cloth. The factors were the concentration of soap (in percent by weight) and the percentage of lauric acid in the solution. The experiment was repeated twice for each combination of factors. The outcome measured was the percentage of oil removed. The results are as follows.

	10% Acid	25% Acid
15% Soap	52.3 54.5	57.4 53.7
25% Soap	56.4 58.4	42.7 45.1

a. Can you reject the hypothesis of no interactions? Explain. *Yes*

b. Can the main effect of soap concentration on the amount of oil removed be interpreted? If so, interpret the main effect, using the $\alpha = 0.05$ level of significance. If not, explain why not. *No*

c. Can the main effect of lauric acid on the amount of oil removed be interpreted? If so, interpret the main effect, using the $\alpha = 0.05$ level of significance. If not, explain why not. *No*

21. Lather up: Construct and interpret an interaction plot for the data in Exercise 19.

22. Cleaning solution: Construct and interpret an interaction plot for the data in Exercise 20.

23. Let's stick together: Two parts will be fastened together with an adhesive. The adhesive will be applied to each part, and then the parts will be pressed together. Two types of adhesive are available. The amount of pressure applied can be either high or low. An experiment is run in which three sets of parts are glued together with each combination of adhesive and pressure. The strengths of the bond (in megapascals) are measured, with the following results.

	High Pressure	Low Pressure
Adhesive A	8 9 7	1 3 2
Adhesive B	3 3 2	5 4 4

a. Can you reject the hypothesis of no interactions? Explain. *Yes*

b. Can the main effect of adhesive on the strength be interpreted? If so, interpret the main effect, using the $\alpha = 0.05$ level of significance. If not, explain why not. *No*

c. Can the main effect of pressure on the strength be interpreted? If so, interpret the main effect, using the $\alpha = 0.05$ level of significance. If not, explain why not. *No*

24. Strong beams: The following table presents measurements of the tensile strength (in kilopascals) of asphalt-rubber concrete beams for three levels of binder content and three levels of rubber content.

	Low Rubber Content	Medium Rubber Content	High Rubber Content
Low Binder Content	212.5 138.5 246.0	200.9 144.9 236.1	280.9 97.8 152.2
Medium Binder Content	201.5 170.0 249.5	179.5 229.8 187.7	167.3 236.2 163.4
High Binder Content	184.1 132.6 211.3	114.0 180.3 209.7	77.5 161.9 183.5

Source: Based on data in *Journal of Materials in Civil Engineering* 11:287–294

a. Can you reject the hypothesis of no interactions? Explain. *No*

b. Can the main effect of the binder content be interpreted? If so, interpret the main effect. Use the $\alpha = 0.05$ level of significance. *Yes; do not reject H_0.*

c. Can the main effect of the rubber content be interpreted? If so, interpret the main effect. Use the $\alpha = 0.05$ level of significance. *Yes; do not reject H_0.*

25. Let's stick together: Construct and interpret an interaction plot for the data in Exercise 23.

26. Strong beams: Construct and interpret an interaction plot for the data in Exercise 24.

27. Fruit yields: An agricultural scientist performed a two-way ANOVA to determine the effects of three different pesticides on the yield of lemons, in pounds, from four different varieties of lemon tree. The following MINITAB output presents the results.

Analysis of Variance for Yield

Source	DF	SS	MS	F	P
Pesticide	2	1114.19	557.10	11.406	0.000
Variety	3	108.527	695.09	2.2219	0.103
Interaction	6	356.280	59.380	1.2157	0.321
Error	36	1758.37	48.844		
Total	47	5314.11			

a. Can you reject the hypothesis of no interactions? Explain. *No*

b. Can the main effect of pesticide on yield be interpreted? If so, interpret the main effect, using the $\alpha = 0.05$ level of significance. If not, explain why not. *Yes; reject H_0.*

c. Can the main effect of variety on yield be interpreted? If so, interpret the main effect, using the $\alpha = 0.05$ level of significance. If not, explain why not. *Yes; do not reject H_0.*

28. Fast shipping: An online retailer sends many packages to customers. To investigate the quality of shipping, the retailer employed three firms to ship packages. Each firm was given four packages to ship at each of three times of day, morning, afternoon, and evening. The outcome variable was the number of hours elapsed during shipment. The following MINITAB output presents the results.

Analysis of Variance for Hours Elapsed

Source	DF	SS	MS	F	P
Firm	2	47.494	23.747	4.6546	0.018
Time of day	2	62.779	31.390	6.1526	0.006
Interaction	4	65.170	16.293	4.1735	0.009
Error	27	137.75	5.1019		
Total	35	333.19			

a. Can you reject the hypothesis of no interactions? Explain. *Yes*

b. Can the main effect of firm on number of hours elapsed be interpreted? If so, interpret the main effect, using the $\alpha = 0.05$ level of significance. If not, explain why not. *No*

c. Can the main effect of time of day on number of hours elapsed be interpreted? If so, interpret the main effect, using the $\alpha = 0.05$ level of significance. If not, explain why not. *No*

Extending the Concepts

29. Watch those interactions: A component can be manufactured according to either of two designs, and with either of two types of material. Two components are manufactured with each combination of design and material, and the lifetimes of each are measured (in hours). The results are as follows.

	Material 1	Material 2
Design A	51 49	43 41
Design B	43 41	51 49

a. Explain why, if one wishes to maximize the lifetime, one should choose either design A and material 1, or design B and material 2.

b. Show that the main effect of design is equal to 0.

c. Show that the main effect of material is equal to 0.

d. Is it reasonable to conclude that neither design nor material has an effect on lifetime? Explain. *No*

Answers to Check Your Understanding Exercises for Section 14.2

1. **a.** $P = 0.994$. Do not reject the hypothesis of no interaction.
   b. $P = 0.000$. Reject the hypothesis of no main effect. We conclude that the mean hardness varies with depth.
   c. $P = 0.000$. Reject the hypothesis of no main effect. We conclude that the mean hardness varies with distance.
2. **a.** $P = 0.000$. Reject the hypothesis of no interaction.
   b. and **c.** Since the main effects interact, we do not perform hypothesis tests on the main effects.
3. **a.**

	18–30	**31–50**	**51–80**
Drug A	42.25	52.25	56.5
Drug B	51	61	66
Drug C	71	82	85.75

b.

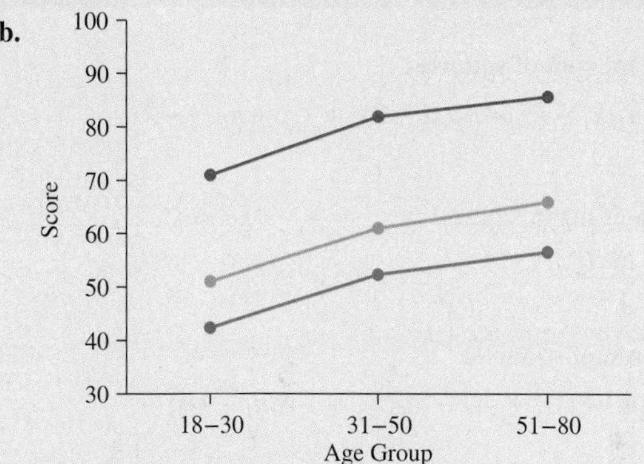

There do not appear to be strong interactions.

c. The MINITAB output is as follows.

```
Source       DF       SS        MS        F        P
Drug          2  5386.50   2693.25  2879.91   0.000
Age Group     2  1362.67   618.333  728.554   0.000
Interaction   4  2.33333   0.58333  0.62376   0.650
Error        27  25.2500   0.93519
Total        35  6776.75
```

The P-value for interactions is 0.650; we do not reject the hypothesis of no interaction. The results of the test are consistent with the interaction plot.

4. Plot C indicates the largest interactions. Plot B indicates the smallest.

Chapter 14 Summary

Section 14.1: Analysis of variance (ANOVA) is used to test the hypothesis that several population means are all equal. In one-way ANOVA, the populations correspond to different levels of a single factor. The levels of the factor are called treatments. We have a simple random sample from each population. Under the assumption that the populations are all approximately normal with the same variance, we construct a test statistic by comparing the spread of the sample means ($SSTr$) with the spread of the individual samples (SSE). When the quotient $\frac{MSTr}{MSE}$ is large, we reject H_0. Rejecting H_0 allows us to conclude that two or more of the population means are different, but it does not tell us which ones are different. We can perform the Tukey–Kramer test to determine which pairs of means differ from each other.

Section 14.2: In two-way ANOVA, there are two factors, and each population corresponds to a combination of levels of the two factors. The main effect of a factor is the difference in the mean response between two levels of that factor. When the effect of a factor depends on the level of the other factor, the factors are said to interact. When large interactions are present, the main effects can be misleading. We therefore begin by testing the hypothesis of no interaction. If we do not reject this hypothesis, we then test the hypothesis of no main effects. Interaction plots allow us to visualize the strength of the interactions.

Vocabulary and Notation

Important Formulas

Treatment sum of squares:

$$SSTr = n_1(\bar{x}_1 - \bar{\bar{x}})^2 + n_2(\bar{x}_2 - \bar{\bar{x}})^2 + \cdots + n_I(\bar{x}_I - \bar{\bar{x}})^2$$

Treatment mean square:

$$MSTr = \frac{SSTr}{I - 1}$$

Error sum of squares:

$$SSE = (n_1 - 1)s_1^2 + (n_2 - 1)s_2^2 + \cdots + (n_I - 1)s_I^2$$

Error mean square:

$$MSE = \frac{SSE}{N - I}$$

***F* statistic for one-way ANOVA:**

$$F = \frac{MSTr}{MSE}$$

Test statistic for Tukey–Kramer test:

$$q = \frac{|\bar{x}_i - \bar{x}_j|}{\sqrt{\frac{MSE}{2}\left(\frac{1}{n_i} + \frac{1}{n_j}\right)}}$$

Chapter Quiz

Exercises 1–4 refer to the following data:

At a certain college, random samples of students with various majors were taken, and their grade point averages (GPAs) were computed. The following table presents the sample means, standard deviations, and sample sizes.

Major	Mean	Standard Deviation	Sample Size
English	3.15	0.45	50
Statistics	3.05	0.56	48
Psychology	3.19	0.35	45
Business	3.01	0.49	55

1. How many degrees of freedom are there for *SSTr* and for *SSE*? *3, 194*
2. Compute the sums of squares *SSTr* and *SSE* and the mean squares *MSTr* and *MSE*. *SSTr = 1.0518, SSE = 43.017, MSTr = 0.35059, MSE = 0.22174*
3. Compute the value of the test statistic *F*. *1.5811*
4. Can you conclude that there are differences in mean GPA among the majors? Use the $\alpha = 0.05$ level of significance. *No*

Exercises 5–7 refer to the following data:

A machine shop has four machines used in precision grinding of bearings. Three machinists are employed to grind bearings on the machines. In an experiment to determine whether there are differences in output among the operators or the machines, each operator worked on each machine for three days. The response variable was the number of parts produced. The results are presented in the following table.

	Machine 1	Machine 2	Machine 3	Machine 4
Operator A	38 33 20	40 43 47	43 31 40	40 38 39
Operator B	33 34 34	39 37 47	35 43 31	36 46 46
Operator C	28 34 40	49 36 50	39 37 35	34 39 38

5. Can you reject the null hypothesis of no interactions? Explain. *No*
6. Can the main effect of operator on the number of parts produced be interpreted? If so, interpret the main effect, using the $\alpha = 0.05$ level of significance. If not, explain why not. *Yes; do not reject H_0.*
7. Can the main effect of machine on the number of parts produced be interpreted? If so, interpret the main effect, using the $\alpha = 0.05$ level of significance. If not, explain why not. *Yes; reject H_0.*

Exercises 8–10 refer to the following data:

As part of a study on weight loss, random samples of men and women were assigned to follow one of three diets. After several weeks, the amount of weight lost, in pounds, was recorded for each person. A two-way analysis was performed, with gender and diet as the factors, and weight loss as the response. The following MINITAB output presents the ANOVA table.

```
Analysis of Variance for Weight Loss

Source        DF      SS       MS         F        P
Gender         1   496.13   496.13    19.649    0.000
Diet           2     40.2     20.1   0.79604    0.463
Interaction    2   534.47   267.23    10.583    0.001
Error         24      606    25.25
Total         29   1676.8
```

8. Can you reject the null hypothesis of no interactions? Explain. *Yes*

9. Can the main effect of gender on weight loss be interpreted? If so, interpret the main effect, using the $\alpha = 0.01$ level of significance. If not, explain why not. *No*

10. Can the main effect of diet on weight loss be interpreted? If so, interpret the main effect, using the $\alpha = 0.01$ level of significance. If not, explain why not. *No*

Review Exercises

1. Soil quality: Soil scientists conducted an experiment in which the mineral gypsum was added in four different amounts to soil specimens to try to control the acid level of the soil. Three soil specimens received each amount added. The response variable is pH, which is a measurement of the acid level. The results are as follows. Construct an ANOVA table.

Gypsum Amount	pH
None	7.88 7.72 7.68
Low	7.81 7.64 7.85
Medium	7.84 7.63 7.87
High	7.80 7.73 8.00

Source: *Journal of The Soil Science Society of America* 66:92–98

2. Soil quality: Using the data in Exercise 1, can you conclude that the pH differs with the amount of gypsum added? Use the $\alpha = 0.05$ level of significance. *No*

3. Electrical conductor: The scientists referred to in Exercise 1 also considered the effect of gypsum on the electrical conductivity (in decisiemens per meter) of soil. Two types of soil were treated with three different amounts of gypsum. The results are as follows.

	Soil Type A	Soil Type B
No Gypsum	1.52 1.05	1.01 0.92
Low Gypsum	1.49 0.91	1.12 0.92
High Gypsum	0.99 0.92	0.88 0.92

Source: *Journal of The Soil Science Society of America* 66:92–98

a. Can you reject the hypothesis of no interactions? Explain. *No*

b. Can the main effect of the amount of gypsum be interpreted? If so, interpret the main effect. Use the $\alpha = 0.05$ level of significance. *Yes; do not reject H_0.*

c. Can the main effect of soil type be interpreted? If so, interpret the main effect. Use the $\alpha = 0.05$ level of significance. *Yes; do not reject H_0.*

4. Chemicals in the water: Earth scientists measured several chemical properties in water samples taken from wells at three locations. Following are sample means, standard deviations, and sample sizes for pH measurements.

Location	Mean	Standard Deviation	Sample Size
Upstream	6.8	0.6	16
Midstream	6.2	0.4	31
Downstream	6.4	0.6	30

Source: *Geosciences Journal* 6:35–46

a. What are the sample sizes n_1, n_2, n_3? *16, 31, 30*

b. What are the sample standard deviations s_1, s_2, s_3? *0.6, 0.4, 0.6*

c. Compute *SSTr*. *3.7995*

d. Compute *SSE*. *20.64*

5. **Chemicals in the water:** Refer to Exercise 4.
   a. How many degrees of freedom are there for *SSTr*? For *SSE*? *2, 74*
   b. Compute *MSTr* and *MSE*. *1.8997, 0.27892*
   c. Compute the value of the test statistic for testing the null hypothesis that the population means are all equal. *6.8111*
   d. Can you conclude that there are differences among the mean pH levels? Use the $\alpha = 0.05$ level of significance. *Yes*
6. **Chemicals in the water:** Using the data in Exercise 4, perform the Tukey–Kramer test to determine which pairs of means, if any, differ. Use the $\alpha = 0.05$ level of significance. $\mu_1 \neq \mu_2$, $\mu_1 \neq \mu_3$
7. **Electrical conductor:** The scientists referred to in Exercise 3 measured electrical conductivity (in microsiemens per centimeter) in several water samples. Following are sample means, standard deviations, and sample sizes.

Location	Mean	Standard Deviation	Sample Size
Upstream	463	208	49
Midstream	363	98	31
Downstream	647	878	30

Source: *Geosciences Journal* 6:35–46

Can a one-way ANOVA be used to determine whether conductivity varies with location? Or is one of the necessary assumptions violated? Explain. *The assumption of equal variances is violated.*

8. **Orange trees:** An agricultural scientist wants to determine how the type of fertilizer and the type of soil affect the yield of oranges in an orange grove. He has two types of fertilizer and three types of soil. For each of the six combinations of fertilizer and soil, the scientist plants four stands of trees and measures the yield of oranges (in tons per acre) from each stand. The data are shown in the following table.

	Soil Type 1	Soil Type 2	Soil Type 3
Fertilizer A	20 24 22 9	22 22 27 25	29 25 23 28
Fertilizer B	23 23 30 24	23 29 31 32	25 29 30 31

   a. Can you reject the hypothesis of no interactions? Explain. *No*
   b. Can the main effect of fertilizer on the yield of oranges be interpreted? If so, interpret the main effect, using the $\alpha = 0.05$ level of significance. If not, explain why not. *Yes; reject* H_0.
   c. Can the main effect of soil on the yield of oranges be interpreted? If so, interpret the main effect, using the $\alpha = 0.05$ level of significance. If not, explain why not. *Yes; reject* H_0.
9. **Orange trees:** Construct and interpret an interaction plot for the data in Exercise 8.
10. **Power plant pollution:** Sulfur dioxide (SO_2) is a chemical produced by many industrial processes. It is one of the ingredients in acid rain, so it is important to monitor the amount of sulfur dioxide that is released into the atmosphere. In an article published in the journal *Atmospheric Environment*, several measurements of the maximum hourly concentrations (in milligrams per cubic meter) of SO_2 were reported for each of four power plants. The results were as follows.

Plant	Concentration					
A	438	619	732	638		
B	857	1014	1153	883	1053	
C	925	786	1179	786		
D	893	891	917	695	675	595

   a. Because the design is unbalanced, check that the assumption of equal variances is satisfied by showing that the largest sample standard deviation is less than twice as large as the smallest one.
   b. Construct an ANOVA table.
11. **Power plant pollution:** Using the data in Exercise 10, perform the Tukey–Kramer test to determine which pairs of means, if any, differ. Use the $\alpha = 0.05$ level of significance. $\mu_A \neq \mu_B$, $\mu_A \neq \mu_C$
12. **Car tires:** Three brands of tires were tested on each of four makes of car. Three cars of each make were used. The number of miles traveled, in thousands, is given as follows for each car.

	Tire 1	Tire 2	Tire 3
Car Type A	56 58 59	59 61 62	54 56 56
Car Type B	60 61 62	61 64 65	58 59 59
Car Type C	57 59 59	60 62 62	56 57 58
Car Type D	58 61 59	58 59 59	55 57 56

   a. Can you reject the hypothesis of no interactions? Explain. *No*

b. Can the main effect of car type on number of miles traveled be interpreted? If so, interpret the main effect, using the $\alpha = 0.05$ level of significance. If not, explain why not. *Yes; reject H_0.*

c. Can the main effect of tire on number of miles traveled be interpreted? If so, interpret the main effect, using the $\alpha = 0.05$ level of significance. If not, explain why not. *Yes; reject H_0.*

13. Car tires: Construct and interpret an interaction plot for the data in Exercise 12.

14. Quality control: Four machines are used with three different materials to fabricate a certain part. Each combination of machine and material is run for three days, and the number of satisfactory parts made each day is counted. A two-way ANOVA is performed to determine whether the mean number of satisfactory parts differs among machines or materials. The following MINITAB output presents the ANOVA table.

```
Analysis of Variance for Parts

Source        DF      SS         MS        F       P
Material       2     39.056     19.528    5.618   0.010
Machine        3     120.11     40.037    11.52   0.000
Interaction    6     20.722     3.4537    0.994   0.452
Error         24     83.423     3.4760
Total         35     263.31
```

a. Can you reject the null hypothesis of no interactions? Explain. *No*

b. Can the main effect of material on number of satisfactory parts be interpreted? If so, interpret the main effect, using the $\alpha = 0.05$ level of significance. If not, explain why not. *Yes; reject H_0.*

c. Can the main effect of machine on number of satisfactory parts be interpreted? If so, interpret the main effect, using the $\alpha = 0.05$ level of significance. If not, explain why not. *Yes; reject H_0.*

15. Interactions: The following table presents the cell means for a two-way ANOVA. Construct and interpret an interaction plot.

		Column Factor		
		Low	**Medium**	**High**
	Low	6.4	26.7	62.6
Row Factor	**Medium**	17.1	42.3	71.0
	High	34.1	54.4	87.1

Write About It

1. Why does a large value of the F statistic provide evidence against the null hypothesis that all the population means are equal? Why doesn't a small value of F statistic provide evidence against H_0?
2. One-way analysis of variance requires the assumption that the populations are approximately normal. However, if the sample sizes are large ($n > 30$), this assumption is not so important. Why is this so?
3. Why are the main effects misleading when interactions are present?
4. When testing the null hypothesis of no interaction, describe the consequences of a Type I error and of a Type II error.
5. Explain why it is reasonable to use a larger level of significance for a test for interaction than for most other hypothesis tests.

In-Class Activities

1. **Lines and majors:** Each student draws a line by eye on a sheet of paper and tries to make the length as close to 3 inches as possible. Each student's line is measured. Record each student's major, and for each major, list the true lengths of the lines drawn by students with that major. Then construct an ANOVA table, and perform a hypothesis test to determine whether the mean length of lines varies among majors.
2. **Lines and semesters:** Same as Activity 1, but determine whether the mean length of the lines varies with the number of semesters or quarters the student has completed.

Case Study: Optimal Procedure For Using A Microbalance

In the chapter opener, we presented data collected in an experiment designed to evaluate the performance of a microbalance. The purpose of the experiment was to determine whether it was better to turn off the microbalance when not in use or to leave it powered up. The following table presents measurements of weights of three silicon wafers. The weight of each wafer was measured twice by each of three operators. These measurements were made under the policy that the balance was to be turned off when not in use and powered up only when it was about to be used.

	Operator A	Operator B	Operator C
Wafer 1	11 15	10 6	14 10
Wafer 2	210 208	205 201	208 207
Wafer 3	111 113	102 105	108 111

1. Construct an ANOVA table.

2. Can it be determined from the ANOVA table whether there are differences in the measured weights among the operators? If so, provide the value of the test statistic and the P-value. If not, explain why not. *Yes, $F = 13.52747$, $P = 0.00193977$*

It turned out that the measurements made by operator B were taken shortly after the balance had been powered up, while the others were taken after the balance had been powered up for some time. A new policy was then instituted to leave the balance powered up continuously. The three operators then made two weighings of three different wafers. The results are presented in the following table.

	Operator A	Operator B	Operator C
Wafer 1	152 156	156 155	152 157
Wafer 2	443 440	442 439	435 439
Wafer 3	229 227	229 232	225 228

3. Construct an ANOVA table.

4. Compare the ANOVA table in Exercise 3 with the ANOVA table in Exercise 1. Would you recommend leaving the balance powered up continuously? Explain your reasoning. *Yes*

Nonparametric Statistics

chapter

15

CUHRIG/Getty Images

Introduction

Modern water conservation efforts depend in large part on the treatment of sewage, or wastewater. Wastewater treatment is used to remove contaminants so that the water can be used for purposes such as irrigation. Microorganisms play an important role in wastewater treatment. In fact, wastewater treatment technology depends on beneficial microorganisms to remove organic material, transforming it into carbon dioxide and water.

An important feature in any wastewater treatment system is the average amount of time that solid organic matter resides in the treatment tank, which is called the solids retention time. A study was performed at the Colorado School of Mines to determine the relationship between the solids retention time and the diversity of the microorganism population. Because microorganisms feed on solid organic matter, it was thought that longer retention times could provide more opportunity for genetic mutations to occur, resulting in a greater diversity of microorganisms.

Following are counts of the numbers of species of microorganisms for solid retention times of 3 days, 12 days, and 30 days.

3-Day Cycle								
282	273	238	240	241	253	230	266	253

12-Day Cycle											
266	299	287	289	272	284	292	279	274	306	291	272

30-Day Cycle								
293	271	293	296	271	292	265	262	278
272	246	278	276	283	273	286	303	289

To determine whether there are differences in the numbers of species, we could perform t-tests, as described in Section 11.1. Of course, t-tests require the assumption that the populations are approximately normal. Tests that make assumptions about the distribution of a population are called **parametric** tests. As an alternative, we can use **nonparametric** tests. Nonparametric tests are tests that do not assume the data to have any specific distribution. In this chapter, we will discuss three nonparametric tests. In Section 15.1, we will discuss the **sign test**, which can be used to test hypotheses about the median of a population. The sign test is a nonparametric version of the t-test. Unlike the t-test, it does not require the assumption that the population is approximately normal. In Section 15.2 and Section 15.3 we discuss the **Wilcoxon rank-sum test** and **Wilcoxon signed-rank test**, respectively. These tests are nonparametric versions of the two-sample t tests for independent and for paired samples.

Differences Between Nonparametric Tests and Parametric Tests

- Nonparametric tests do not require the population to have any specific distribution.
- Parametric tests generally do require the population to have a specific distribution, such as the normal distribution.

Section 15.1 The Sign Test

Objective

1. Test hypotheses using the sign test

Objective 1 Test hypotheses using the sign test

The Sign Test

The sign test can be used to test a hypothesis about the median of a population. The corresponding parametric test is the t-test for a population mean discussed in Section 9.3. When samples are small, the t-test requires that the population be approximately normal. The primary advantage of the sign test, therefore, is that the only assumption required is that the population be continuous. The primary disadvantage is that when the population is approximately normal, the sign test is less likely to reject a false null hypothesis.

Primary Advantage and Disadvantage of the Sign Test

Advantage: The sign test is valid for any continuous population. It does not require the assumption of normality.

Disadvantage: When the population is approximately normal, the sign test is less likely to reject a false null hypothesis than the t-test.

Performing the Sign Test

Many of the hypothesis tests we have studied in previous chapters have been about the population mean. The sign test is used to test hypotheses about a population median m. The null hypothesis is of the form H_0: $m = m_0$. Sample values less than m_0 are assigned a minus sign, values above m_0 are assigned a plus sign, and values equal to m_0 are assigned a 0.

We then count the number of plus signs and the number of minus signs. If H_0 is true, and m_0 is the true population median, we would expect to see approximately equal numbers of plus and minus signs. If H_0 is false, we would expect to see more of one type of sign than the other.

To construct the test statistic, let x be the number of times the *less* frequent sign appears. For example, if there are 12 plus signs and 5 minus signs, then $x = 5$. The value of the test statistic is based on x, as follows:

Test Statistic for the Sign Test

Let x be the number of times the less frequent sign occurs. Let n be the total number of plus and minus signs.

- If $n \leq 25$, the test statistic is x.
- If $n > 25$, the test statistic is

$$z = \frac{x + 0.5 - n/2}{\sqrt{n}/2}$$

For $n \leq 25$, critical values of the test statistic are found in Table A.7.
For $n > 25$, critical values are found in Table A.2.

For sample sizes $n \leq 25$, the table is not detailed enough to allow the P-value to be computed. We therefore focus on the critical value method for the sign test. Following are the steps for performing the test:

Performing the Sign Test

Step 1: State the null and alternate hypotheses. The null hypothesis specifies a value for the population median m. We will call this value m_0. So the null hypothesis is of the form H_0: $m = m_0$.
The alternate hypothesis may be stated in one of three ways:

Left-tailed:	H_1: $m < m_0$
Right-tailed:	H_1: $m > m_0$
Two-tailed:	H_1: $m \neq m_0$

Step 2: Choose a significance level α.

Step 3: For each sample value greater than m_0, assign a plus sign. For each sample value less than m_0, assign a minus sign. For each sample value equal to m_0, assign a 0. Count the number of plus and minus signs.

Step 4: Let x be the number of times the less frequent sign appears. Let n be the total number of plus and minus signs (not counting any 0s). Compute the test statistic:
- If $n \leq 25$, the test statistic is x.
- If $n > 25$, the test statistic is $z = \dfrac{x + 0.5 - n/2}{\sqrt{n}/2}$

Step 5: Determine whether to reject H_0, as follows:
- If $n \leq 25$, find the critical value from Table A.7. If x is less than or equal to the critical value, reject H_0. Otherwise do not reject H_0.
- If $n > 25$, find the critical value from Table A.2. If z is less than or equal to the critical value, reject H_0. Otherwise do not reject H_0.

Step 6: State a conclusion.

EXPLAIN IT AGAIN

When to reject H_0: Because we use the frequency of the sign that occurs less often, we always reject when the test statistic is less than the critical value.

Example 15.1

Performing the sign test

Annual income, in thousands of dollars, was recorded for a random sample of 15 families in a certain town. The results were as follows.

75 10 92 49 58 25 96 40 90 87 10 49 20 200 34

An economist is interested in determining whether the median income in the town is greater than 40 thousand dollars. Perform a hypothesis test at the $\alpha = 0.05$ level of significance.

Solution

Step 1: State H_0 and H_1. The issue is whether the median is greater than 40. Therefore the null and alternate hypotheses are

$$H_0: m = 40 \qquad H_1: m > 40$$

Step 2: Choose a significance level. We choose a significance level of $\alpha = 0.05$.

Step 3: Assign plus and minus signs. The following table shows a plus sign for each value greater than 40 and a minus sign for each value less than 40.

75	10	92	49	58	25	96	40	90	87	10	49	20	200	34
+	−	+	+	+	−	+	0	+	+	−	+	−	+	−

Step 4: Compute the test statistic. There are 9 plus signs and 5 minus signs. The sample size is $n = 14$. Since $n \le 25$, the test statistic is the number of times the less frequent sign appears. The value of the test statistic is $x = 5$.

Step 5: Determine whether to reject H_0. Because $n \le 25$, we use Table A.7 to find the critical value. We have a one-tailed test, $n = 14$, and significance level $\alpha = 0.05$. The critical value is 3. Because $x > 3$, we do not reject H_0.

	One-tailed			
	$\alpha = 0.005$	$\alpha = 0.01$	$\alpha = 0.025$	$\alpha = 0.05$
	Two-tailed			
n	$\alpha = 0.01$	$\alpha = 0.02$	$\alpha = 0.05$	$\alpha = 0.10$
⋮	⋮	⋮	⋮	⋮
12	1	1	2	2
13	1	1	2	3
14	1	2	3	3
15	2	2	3	3
16	2	2	3	4
⋮	⋮	⋮	⋮	⋮

Step 6: State a conclusion. We cannot conclude that the median income is greater than 40 thousand dollars.

Example 15.2

Performing the sign test

The length of time, in minutes, needed to perform a certain type of surgery was recorded for 32 operations. The results were as follows.

149	144	218	153	134	152	148	144
178	107	199	135	171	110	160	119
86	127	106	153	169	153	153	173
156	145	205	132	169	174	130	175

A hospital administrator is interested in determining whether the median time for surgery is less than 170 minutes. Perform a hypothesis test at the $\alpha = 0.01$ level of significance.

Solution

Step 1: State H_0 and H_1. The issue is whether the median is less than 170. Therefore the null and alternate hypotheses are

$$H_0: m = 170 \qquad H_1: m < 170$$

Step 2: Choose a significance level. We choose a significance level of $\alpha = 0.01$.

Step 3: Assign plus and minus signs. The following table shows a plus sign for each value greater than 170 and a minus sign for each value less than 170.

149 −	144 −	218 +	153 −	134 −	152 −	148 −	144 −
178 +	107 −	199 +	135 −	171 +	110 −	160 −	119 −
86 −	127 −	106 −	153 −	169 −	153 −	153 −	173 +
156 −	145 −	205 +	132 −	169 −	174 +	130 −	175 +

Step 4: Compute the test statistic. There are eight plus signs and 24 minus signs. The number of times the less frequent sign appears is $x = 8$. The sample size is $n = 32$. Since $n > 25$, the test statistic is

$$z = \frac{x + 0.5 - n/2}{\sqrt{n}/2}$$

We substitute $x = 8$ and $n = 32$ to obtain

$$z = \frac{x + 0.5 - n/2}{\sqrt{n}/2} = \frac{8 + 0.5 - 32/2}{\sqrt{32}/2} = -2.65$$

Step 5: Determine whether to reject H_0. Because $n > 25$, we use Table A.2 to find the critical value. We have a one-tailed test with significance level $\alpha = 0.01$. The critical value is -2.326. Because $-2.65 < -2.326$, we reject H_0.

Step 6: State a conclusion. We conclude that the median time in surgery is less than 170 minutes.

Check Your Understanding

1. Prices for a gallon of regular gas were recorded for a sample of 12 gas stations, as shown below. Can you conclude that the median price is less than \$3.00? Use the $\alpha = 0.05$ level of significance.

2.93	2.95	2.76	2.89	2.57	3.06
2.61	2.66	2.98	2.79	2.96	2.74

a. State the null and alternate hypotheses. *$H_0: m = 3$, $H_1: m < 3$*

b. Compute the test statistic. *$x = 1$*

c. Find the critical value. *2*

d. State a conclusion. *Reject H_0.*

2. The registrar at a college sampled 30 students and recorded the GPA for each, as shown below. Can you conclude that the median GPA is less than 2.90? Use the $\alpha = 0.05$ level of significance.

2.55	2.07	3.35	2.86	2.90	3.22	3.12	2.63	3.55	3.39
2.25	2.26	2.18	2.02	2.85	3.31	3.45	3.06	2.22	3.26
2.25	2.27	2.20	2.28	2.34	3.39	2.63	2.63	2.44	2.50

a. State the null and alternate hypotheses. *$H_0: m = 2.90$, $H_1: m < 2.90$*

b. Compute the test statistic. *$z = -1.49$*

c. Find the critical value. *-1.645*

d. State a conclusion. *Do not reject H_0.*

Answers are on page 703.

Using Technology

We use Example 15.1 to illustrate the technology steps.

MINITAB

Performing the sign test

Note: MINITAB returns a *P*-value for the test.

Step 1. Enter the data in column **C1**.
Step 2. Select **Stat**, then **Nonparametrics**, and then **1-Sample Sign**.
Step 3. Enter **C1** in the **Variables** field. Select the **Test median** option, and enter the null hypothesis value for the median.
For Example 15.1, we enter 40.
Step 4. In the **Alternative field**, select the form of the alternate hypothesis. We use **greater than**.
Step 5. Click **OK** (Figure A).

Test

Null hypothesis H_0: η = 40
Alternative hypothesis H_1: η > 40

Sample	Number < 40	Number = 40	Number > 40	P-Value
C1	5	1	9	0.212

Figure A

Section 15.1 Exercises

Exercises 1 and 2 are the Check Your Understanding exercises located within the section.

Understanding the Concepts

In Exercises 3 and 4, fill in each blank with the appropriate word or phrase.

3. For the sign test, we reject H_0 when the value of the test statistic is ________ than the critical value. *less*

4. A disadvantage of the sign test is that when the population is approximately normal the sign test is ________ likely to reject a false null hypothesis than the *t*-test. *less*

In Exercises 5 and 6, determine whether the statement is true or false. If the statement is false, rewrite it as a true statement.

5. The sign test is valid for any continuous population. *True*

6. The test statistic for the sign test is the frequency of the most frequently appearing sign. *False*

Practicing the Skills

7. Compute the value of the test statistic for the following sample, for testing H_0: $m = 10$ versus H_1: $m \neq 10$. *$x = 4$*

14	11	9	3	4	2	14	11	9	5

8. Compute the value of the test statistic for the following sample, for testing H_0: $m = 50$ versus H_1: $m < 50$. *$x = 2$*

38	47	55	38	40	43	36	49
43	55	43	47	38	50	38	43

9. The sign test is performed to test H_0: $m = 35$ versus H_1: $m \neq 35$. There are 16 plus signs, 4 minus signs, and 3 zeros.
   a. What is the value of the test statistic? *$x = 4$*
   b. Is H_0 rejected at the $\alpha = 0.05$ level? *Yes*
   c. Is H_0 rejected at the $\alpha = 0.01$ level? *No*

10. The sign test is performed to test H_0: $m = 500$ versus H_1: $m < 500$. There are 9 plus signs, 26 minus signs, and 2 zeros.
   a. What is the value of the test statistic? *$z = -2.70$*
   b. Is H_0 rejected at the $\alpha = 0.05$ level? *Yes*
   c. Is H_0 rejected at the $\alpha = 0.01$ level? *Yes*

Working with the Concepts

11. **Weight loss:** A weight loss company claims that the median weight loss for people who follow their program is at least 15 pounds. Weight losses for a random sample of 20 people in the program were recorded, and the results were as follows. Can you conclude that the claim is false? Use the $\alpha = 0.05$ level of significance.

14	8	12	17	15	7	7	6	8	9
6	11	2	0	11	9	6	7	6	17

   a. State the appropriate null and alternate hypotheses. *H_0: $m = 15$, H_1: $m < 15$*
   b. Compute the value of the test statistic. *$x = 2$*
   c. Find the critical value. *5*
   d. State a conclusion. *Reject H_0.*

12. **At the movies:** A sample of 12 movies released in a given year had the following running times, in minutes.

118	125	113	156	168	121
115	136	194	149	105	161

Can you conclude that the median running time for movies released that year was greater than 120 minutes? Use the $\alpha = 0.01$ level of significance.
   a. State the appropriate null and alternate hypotheses. *H_0: $m = 120$, H_1: $m > 120$*
   b. Compute the value of the test statistic. *$x = 4$*
   c. Find the critical value. *1*
   d. State a conclusion. *Do not reject H_0.*

13. Tall trees: Heights, in feet, of a sample of 18 mature oak trees in a forest were measured. The results were as follows.

65	71	75	68	67	60
64	71	77	72	51	70
63	63	54	74	60	57

Can you conclude that the median height of oak trees in this forest is greater than 60 feet? Use the $\alpha = 0.05$ level of significance. Yes

14. Wheat yields: The highest non-irrigated wheat yields in the United States are in the Palouse region of Eastern Washington and Northern Idaho. Following are hypothetical yields, in bushels, for 15 randomly selected acres of farmland.

74	75	82	82	76	76	81	86
98	89	76	93	65	62	63	

Can you conclude that the median yield differs from 75 bushels per acre? Use the $\alpha = 0.01$ level of significance. No

15. Paying the rent: Monthly rents were recorded for a sample of 36 apartments in a certain city. The results were as follows.

660	980	1210	440	530	1110
610	350	810	670	700	680
1030	1100	910	920	1000	620
820	980	620	1300	360	1290
1090	1250	380	820	1050	1020
500	390	1150	1290	720	710

Can you conclude that the median rent is less than \$1000 per month? Use the $\alpha = 0.05$ level of significance. Yes

16. Morning commute: A random sample of 30 commuters were asked how long their typical morning commute is. The results, in minutes, were as follows.

8	10	5	33	20	43
69	63	63	26	14	14
35	33	60	46	31	36
2	39	19	66	2	4
4	35	42	43	61	20

Can you conclude that the median commute time is less than 60 minutes? Use the $\alpha = 0.05$ level of significance. Yes

17. Scam alert: The rates of fraud complaints (per 100,000 people) for the 20 largest metropolitan areas in the United States in a recent year are as follows.

404.4	525.6	409.2	385.9
461.7	486.4	525.3	390.7
451.6	426.9	370.7	485.8
532.0	432.5	506.3	398.3
453.5	393.7	607.8	402.4

Can you conclude that the median fraud complaint rate is less than 480? Use the $\alpha = 0.05$ level of significance. No

18. Stamp of approval: The Gallup poll approval ratings for the U.S. Congress each month during a three-year period are as follows.

12	17	18	17	11	17
9	16	18	15	13	24
11	15	21	17	13	17
19	13	13	12	15	18
14	15	10	10	13	23
15	14	16	13	18	20

Can you conclude that the median approval rating differs from 17? Use the $\alpha = 0.05$ level of significance. Yes

Extending the Concepts

Exercise 19 demonstrates that the sign test will generally not work when the population is discrete.

19. Discrete population: In a certain city, 45% of the households have no dog, 30% have one dog, 10% have two dogs, 10% have three dogs, and 5% have four dogs. The population median number of dogs is therefore 1. In a sample of 100 households, 45 have 0 dogs, 30 have one dog, 10 have two dogs, 10 have three dogs, and 5 have four dogs. Note that the sample reflects the population perfectly. Consider the hypotheses H_0: $m = 1$ versus H_1: $m < 1$. Compute the value of the test statistic for the sign test, and show that H_0 is rejected at the $\alpha = 0.05$ level. $z = -2.27$

Answers to Check Your Understanding Exercises for Section 15.1

1. a. H_0: $m = 3$, H_1: $m < 3$
b. $x = 1$ **c.** The critical value is 2.
d. Reject H_0. We conclude that the median price is less than \$4.00.

2. a. H_0: $m = 2.90$, H_1: $m < 2.90$
b. $z = -1.49$ **c.** $z = -1.645$
d. Do not reject H_0. We cannot conclude that the median GPA is less than 2.90.

Section 15.2 The Rank-Sum Test

Objectives

1. Assign ranks to a sample
2. Test hypotheses using the rank-sum test

Objective 1 Assign ranks to a sample

Ranks

The nonparametric tests we will discuss in this section are based on **ranks**. To rank the values in a sample, we order them from smallest to largest. The smallest value is assigned a rank of 1, the next smallest is assigned a rank of 2, and so forth. For values in the data set whose ranks are tied, each value is assigned the average of the corresponding ranks. Examples 15.3 and 15.4 illustrate the method.

Example 15.3

Assign ranks to a sample

The weights of six first graders, in pounds, are 35, 43, 38, 51, 45, 49. Assign ranks to these values.

Solution

We arrange the values in increasing order. The smallest value, 35, is assigned a rank of 1. The next smallest, 38, is assigned a rank of 2, and so forth. The results are as follows.

Weight	35	38	43	45	49	51
Rank	1	2	3	4	5	6

In some data sets, there are ties, that is, two or more values are the same. When two or more values are tied, each receives the average of the ranks. Example 15.4 illustrates the method.

Example 15.4

Assign ranks to a sample when ties are present

The heights of 10 fourth graders, in inches, are 48, 46, 45, 54, 46, 52, 50, 46, 52, 49. Assign ranks to these values.

Solution

We arrange the values in increasing order. We notice there are three values tied at 46, and two tied at 52. We assign the ranks 1 through 10 as preliminary ranks. The three values of 46 have preliminary ranks 2, 3, 4. We compute the average of these preliminary ranks: $(2 + 3 + 4)/3 = 9/3 = 3$. The rank assigned to each of these values is therefore 3. The two values of 52 have preliminary ranks 8 and 9. We assign each the rank $(8 + 9)/2 = 17/2 = 8.5$. The results are shown in the following table.

Height	45	46	46	46	48	49	50	52	52	54
Preliminary Ranks	1	2	3	4	5	6	7	8	9	10
Ranks	1	3	3	3	5	6	7	8.5	8.5	10

Objective 2 Test hypotheses using the rank-sum test

The Rank-Sum Test

Wilcoxon's rank-sum test, also known as the Mann-Whitney test, is a nonparametric test to determine whether two population medians are equal. It is a nonparametric alternative to the two-sample t-test, which is discussed in Section 11.1. The test requires that the populations have approximately the same shape. When two populations have the same shape and their medians are equal, they have the same distribution. For this reason the rank-sum test is sometimes described as a test to determine whether two populations differ.

The main advantage of the rank-sum test is that it does not require that the populations be approximately normal. The main disadvantage is that the null hypothesis can be falsely rejected when the population medians are equal, if the shapes of the two populations differ substantially.

Main Advantage and Disadvantage of the Rank-Sum Test

Advantage: The rank-sum test does not require the assumption of normality.

Disadvantage: The null hypothesis can be rejected when the two populations differ in shape, even if their medians are equal.

Performing the Rank-Sum Test

The rank-sum test is used to test the null hypothesis that two population medians are equal. Independent samples are drawn from each population. The combined samples are then ranked. The smallest value in the two samples combined is assigned a rank of 1, the second smallest is assigned a rank of 2, and so forth. If the two populations are the same, the ranks for the values assigned to one sample should not be substantially larger or smaller than the ranks assigned to the other. Thus the rank-sum test rejects the null hypothesis when the sum of the ranks for one of the samples is exceptionally large or small.

To construct the test statistic, let n_1 be the smaller of the two sample sizes and let n_2 be the larger. Let S be the sum of the ranks for the smaller sample. If both samples are the same size, then n_1 and n_2 are both equal to that size, and S may be the sum of the ranks of either sample. When H_0 is true, it can be shown that S has mean μ_S and standard deviation σ_S given by

$$\mu_S = \frac{n_1(n_1 + n_2 + 1)}{2} \qquad \sigma_S = \sqrt{\frac{n_1 n_2(n_1 + n_2 + 1)}{12}}$$

We construct the test statistic as

$$z = \frac{S - \mu_S}{\sigma_S}$$

When both samples sizes are 10 or more, the test statistic has an approximately standard normal distribution, so P-values and critical values can be found using the z-table (Table A.2). When one or both of the sample sizes is less than 10, special tables can be used. Since most samples in practice consist of at least 10 values, we do not consider this situation here; we will only consider the situation in which both sample sizes are at least 10.

Performing the Rank-Sum Test

Let m_1 be the population median corresponding to the sample of size n_1, and let m_2 be the population median corresponding to the sample of size n_2.

Step 1: State the null and alternate hypotheses.
The null hypothesis is $H_0\colon m_1 = m_2$.
The alternate hypothesis can be stated in one of three ways:

Left-tailed: $H_1\colon m_1 < m_2$
Right-tailed: $H_1\colon m_1 > m_2$
Two-tailed: $H_1\colon m_1 \neq m_2$

Step 2: Choose a significance level α.

Step 3: Combine the two samples, arrange the combined sample in increasing order, and assign ranks to the values.

Step 4: Let n_1 be the smaller sample size, and let n_2 be the larger. Compute S, the sum of the ranks of the smaller sample. If both samples are the same size, then n_1 and n_2 are both equal to that size, and S may be the sum of the ranks of either sample. Then compute μ_S and σ_S, as follows:

$$\mu_S = \frac{n_1(n_1 + n_2 + 1)}{2} \qquad \sigma_S = \sqrt{\frac{n_1 n_2(n_1 + n_2 + 1)}{12}}$$

Step 5: Compute the test statistic $z = \dfrac{S - \mu_S}{\sigma_S}$.

Step 6: Compute the P-value, as follows. $H_1\colon m_1 < m_2$: The P-value is the area to the left of z. $H_1\colon m_1 > m_2$: The P-value is the area to the right of z. $H_1\colon m_1 \neq m_2$: The P-value is the sum of the areas to the right of $|z|$ and to the left of $-|z|$.

Step 7: Interpret the P-value. If making a decision, reject H_0 if the P-value is less than or equal to the significance level α.

Step 8: State a conclusion.

Example 15.5

Performing the rank-sum test

A sample of 14 students took a statistics class online, and another sample of 12 students took an equivalent class in a traditional classroom. Both classes were given the same final exam at the end of the course. The scores were as follows.

Online	78	82	83	87	75	63	78	60	94	62	98	90	97	81
Traditional	73	72	92	100	74	90	64	84	77	89	70	64		

Can you conclude that there is a difference between the median scores from the two classes? Use the $\alpha = 0.05$ level of significance.

Solution

Step 1: State H_0 and H_1. The null hypothesis is that there is no difference in the scores between the two teaching methods. The alternate hypothesis is that there is a difference.

Step 2: Choose a significance level. We will use a significance level of $\alpha = 0.05$.

Step 3: Arrange the combined samples in increasing order, and assign ranks. Following are the results. It is important to label each value with the sample it came from.

Score	60	62	63	64	64	70	72	73	74	75	77	78	78
Sample	O	O	O	T	T	T	T	T	T	O	T	O	O
Rank	1	2	3	4.5	4.5	6	7	8	9	10	11	12.5	12.5

Score	81	82	83	84	87	89	90	90	92	94	97	98	100
Sample	O	O	O	T	O	T	T	O	T	O	O	O	T
Rank	14	15	16	17	18	19	20.5	20.5	22	23	24	25	26

Step 4: Compute S, μ_S, and σ_S. The sample sizes are $n_1 = 12$ and $n_2 = 14$. The smaller sample belongs to the traditional class, so S is the sum of the ranks corresponding to that sample. We compute $S = 154.5$, $\mu_S = 162$, and $\sigma_S = 19.442$.

Step 5: Compute the test statistic. The value of the test statistic is

$$z = \frac{S - \mu_S}{\sigma_S} = \frac{154.5 - 162}{19.442} = -0.39$$

Step 6: Compute the P-value. From Table A.2, we find that the P-value is 0.6965. (Technology gives 0.6997.)

Step 7: Interpret the P-value. Because $P > 0.05$, we do not reject H_0.

Step 8: State a conclusion. We cannot conclude that there are differences in the test scores between the two methods of instruction.

Check Your Understanding

1. Heart rates, in beats per minute, were measured for samples of 12 track athletes and 15 swimmers. The results are shown below. Can you conclude that the median heart rate is greater for swimmers than for track athletes? Use the $\alpha = 0.05$ level of significance.

Track	68	62	65	72	70	68	64	77	77	66	72	76			
Swim	82	81	71	69	79	65	66	70	80	78	75	82	75	63	79

a. State the null and alternate hypotheses. *H_0: $m_1 = m_2$, H_1: $m_1 < m_2$*

b. Compute the value of the test statistic. *−1.88*

c. Compute the P-value. *0.0301*

d. State a conclusion. *Reject H_0.*

2. Battery lifetimes, in hours, were measured for two types of batteries commonly used in laptop computers. Twelve batteries of one type and 13 of another type were tested. The results are shown below. Can you conclude that the lifetimes differ between the two types of batteries? Use the $\alpha = 0.01$ level of significance.

Type A							Type B							
	4.1	6.8	5.8	5.2	3.2	2.5		6.7	4.1	5.6	7.0	3.1	2.3	4.6
	3.9	2.1	4.9	5.2	2.1	4.0		3.8	4.4	4.9	7.1	4.7	4.0	

a. State the null and alternate hypotheses. *H_0: $m_1 = m_2$, H_1: $m_1 \neq m_2$*

b. Compute the value of the test statistic. *−0.90*

c. Compute the *P*-value. *0.3682 [Tech: 0.3695]*

d. State a conclusion. *Do not reject H_0.*

Answers are on page 709.

Using Technology

We use Example 15.5 to illustrate the technology steps.

MINITAB

Performing the rank-sum test

Note: MINITAB returns a *P*-value for the test.

Step 1. Enter the data for Sample 1 in column **C1** and the data for Sample 2 in column **C2**. For Example 15.5, we enter the online class data into **C1** and the traditional class data into **C2**.

Step 2. Click **Stat**, then **Nonparametrics**, then **Mann-Whitney**.

Step 3. Enter **C1** in the **First Sample** field and **C2** in the **Second Sample** field. In the **Alternative** field, select the form of the alternate hypothesis. We use **not equal**.

Step 4. Click **OK** (Figure A).

Test

Null hypothesis	H_0: $\eta_1 - \eta_2 = 0$	
Alternative hypothesis	H_1: $\eta_1 - \eta_2 \neq 0$	
Method	**W-Value**	**P-Value**
Not adjusted for ties	196.50	0.719
Adjusted for ties	196.50	0.719

Figure A

Section 15.2 Exercises

Exercises 1 and 2 are the Check Your Understanding exercises located within the section.

Understanding the Concepts

In Exercises 3 and 4, fill in each blank with the appropriate word or phrase.

3. If two or more values are the same when assigning ranks, we assign them the ________ of the preliminary ranks. *average*

4. To use the normal approximation to find the *P*-value for a rank-sum test, both sample sizes must be at least ________. *10*

In Exercises 5 and 6, determine whether the statement is true or false. If the statement is false, rewrite it as a true statement.

5. To compute the test statistic for the rank-sum test, we sum the ranks of the larger sample. *False*

6. The null hypothesis for the rank-sum test is that the two population medians are equal. *True*

Practicing the Skills

In Exercises 7–10, compute μ_S, σ_S, and the value of the test statistic z. Then find the *P*-value for the specified alternate hypothesis and values of n_1, n_2, and S.

7. $n_1 = 15, n_2 = 18, S = 244, H_1: \mu_1 < \mu_2.$ *$\mu_S = 255$, $\sigma_S = 27.659$, $z = -0.40$, $P = 0.3446$ [Tech: 0.3454]*

8. $n_1 = 20, n_2 = 25, S = 607, H_1: \mu_1 \neq \mu_2.$ *$\mu_S = 460$, $\sigma_S = 43.780$, $z = 3.36$, $P = 0.0008$*

9. $n_1 = 17, n_2 = 23, S = 272, H_1: \mu_1 \neq \mu_2.$ *$\mu_S = 348.5$, $\sigma_S = 36.550$, $z = -2.09$, $P = 0.0366$ [Tech: 0.0363]*

10. $n_1 = 14, n_2 = 27, S = 308, H_1: \mu_1 > \mu_2.$ *$\mu_S = 294$, $\sigma_S = 36.373$, $z = 0.38$, $P = 0.3520$ [Tech: 0.3502]*

Working with the Concepts

11. How's your gas mileage? Gas mileages, in miles per gallon, were measured for samples of compact cars and midsize cars. The results were as follows.

Compact				
34.5	33.7	26.1	28.5	27.4
30.6	31.1	28.0	33.0	33.0
32.8	28.5	25.5	32.1	34.9

Midsize				
21.6	21.1	29.1	24.8	28.5
28.1	21.9	22.5	20.5	26.1

Can you conclude that the median mileage is less for compact cars than for midsize cars? Use the $\alpha = 0.05$ level of significance.

a. State the null and alternate hypotheses. *H_0: $m_1 = m_2$, H_1: $m_1 < m_2$*

b. Compute the value of the test statistic. *$z = -3.24$*

c. Compute the *P*-value. *0.0006*

d. State a conclusion. *Reject H_0.*

12. Coffee prices: Prices of a 12-ounce latte (including tax) were determined at samples of coffee shops in two cities. The results were as follows.

City 1						
2.93	2.75	2.95	3.15	3.14	2.89	2.87
2.79	2.63	3.05	2.68	2.81	2.84	3.13

City 2						
2.41	2.91	2.57	2.55	2.69	2.84	2.52
2.53	2.99	2.76	2.58	2.98	2.94	2.51

Can you conclude that the median price differs between the two cities? Use the $\alpha = 0.01$ level of significance.

a. State the null and alternate hypotheses. *H_0: $m_1 = m_2$, H_1: $m_1 \neq m_2$*

b. Compute the value of the test statistic. *$z = 2.32$ (or -2.32)*

c. Compute the *P*-value. *0.0204 [Tech: 0.0203]*

d. State a conclusion. *Do not reject H_0*

13. Compressive strength: In an experiment to determine the effect of curing time on the compressive strength of concrete blocks, two samples of 15 blocks each were prepared identically except for curing time. The blocks in one sample were cured for 2 days, while the blocks in the other were cured for 6 days. The compressive strengths of the blocks, in tons per square inch, are presented below.

Cured 2 Days				
90	84	87	75	79
95	88	82	85	96
70	86	70	80	77

Cured 6 Days				
94	73	91	97	98
92	82	84	76	88
79	97	90	81	91

Can you conclude that the median strength differs between the two curing times? Use the $\alpha = 0.05$ level of significance. *No*

14. Recovery times: A new postsurgical treatment was compared to a standard treatment. Ten subjects received the new treatment, while 12 others (the controls) received the standard treatment. The recovery times, in days, were as follows.

Treatment				
12	20	15	22	20
21	27	13	19	21

Control					
30	23	33	30	32	35
40	18	32	24	29	28

Can you conclude that the median recovery time is less for treatment than for control? Use the $\alpha = 0.01$ level of significance. *Yes*

15. Let's watch a Blu-ray: Times, in seconds, for a Blu-ray to load were measured for samples of standard Blu-ray players and 3-D Blu-ray players. The results were as follows.

3-D Blu-ray Players			
27	21	22	26
21	24	28	24
28	19	19	22
25	32	22	24

Standard Blu-ray Players			
25	20	20	23
24	23	24	19
22	24	21	18

Can you conclude that the median load time for 3-D Blu-ray players is greater than for standard Blu-ray players? Use the $\alpha = 0.05$ level of significance. *No*

16. Laundry time: The costs, in dollars, per load of laundry were measured for samples of high-efficiency detergents like those used in many front-loading washing machines and standard detergents. The results were as follows.

Standard Laundry Detergents			
0.23	0.16	0.18	0.18
0.13	0.18	0.12	0.14
0.28	0.14	0.20	0.13
0.30	0.13	0.56	0.20
0.25	0.29		

High-Efficiency Laundry Detergents			
0.18	0.18	0.13	0.30
0.19	0.15	0.15	0.18
0.15	0.09	0.09	0.08
0.06	0.05	0.08	0.12
0.18	0.13		

Can you conclude that the median cost per load for high-efficiency detergents differs from the median cost per load for standard detergents? Use the $\alpha = 0.05$ level of significance. *Yes*

Extending the Concepts

The rank-sum test is valid under the assumption that the populations have similar shapes. Exercise 17 shows that when the shapes differ considerably, the null hypothesis may be rejected even when the sample medians are equal.

17. Populations of differing shapes: Consider the following two samples.

Sample 1				
11	12	13	14	15
16	17	30	50	60
70	80	90	100	110

Sample 2				
0	1	2	3	4
5	6	30	31	32
33	34	35	36	37

a. Compute the median of each sample to show that they are equal. *Both are 30*

b. Compute μ_S, σ_S, and the value of the test statistic z. $\mu_S = 232.5$, $\sigma_S = 24.109$, $z = -2.03$

c. Compute the P-value for testing H_0: $m_1 = m_2$ versus H_1: $m_1 \neq m_2$. Do you reject H_0? Use the $\alpha = 0.05$ level of significance. $P = 0.0424$ [Tech: 0.0421]; Reject H_0.

Answers to Check Your Understanding Exercises for Section 15.2

1. a. H_0: $m_1 = m_2$, H_1: $m_1 < m_2$ **b.** $S = 129.5$, $\mu_S = 168$, $\sigma_S = 20.49$, $z = -1.88$ **c.** $P = 0.0301$
d. Reject H_0. We conclude that the median heart rate for track athletes is less than the median for swimmers.

2. a. H_0: $m_1 = m_2$, H_1: $m_1 \neq m_2$ **b.** $S = 139.5$, $\mu_S = 156$, $\sigma_S = 18.38$, $z = -0.90$. **c.** $P = 0.3682$ [Tech: 0.3695]
d. Do not reject H_0. There is not enough evidence to conclude that the median lifetime differs between the two types of batteries.

Section 15.3 The Signed-Rank Test

Objective

1. Test hypotheses using the signed-rank test

Objective 1 Test hypotheses using the signed-rank test

The Signed-Rank Test

Wilcoxon's signed-rank test is a nonparametric test designed to determine whether population medians differ when the data are in the form of paired samples. It is therefore a nonparametric alternative to the t-test for matched pairs discussed in Section 11.3. As with other nonparametric tests, its main advantage is that it does not require that the population of differences have a specified distribution such as the normal distribution. However, it does require that the distribution be approximately symmetric.

NOTE TO INSTRUCTOR

The signed-rank test is more powerful than the sign test presented in Section 15.1, but it requires the population to be approximately symmetric.

The primary advantage and disadvantage of the signed-rank test are similar to those of the sign test. The advantage is that it does not require that the population of differences be approximately normal. The disadvantage is that when the population of differences is close to normal, the signed-rank test is less likely to reject a false null hypothesis than the t-test.

Primary Advantage and Disadvantage of the Signed-Rank Test

Advantage: The signed-rank test does not require the assumption of normality.

Disadvantage: The assumption of symmetry, while less restrictive than normality, is still required. When the population of differences is close to normal, the signed-rank test is less likely to reject a false null hypothesis than the t-test.

Performing the Signed-Rank Test

To perform the signed-rank test, we compute the differences between the values in each matched pair. It is likely that some of the differences will be positive and some will be negative. For each negative difference, we take the absolute value, so, for example, we would replace −3 with 3, −5 with 5, and so forth. If the true median of the differences is 0, the sum of the absolute values of the negative ranks should be close to the sum of the positive ranks.

Example 15.6 illustrates the method used to perform the signed rank test.

Example 15.6

Performing the signed-rank test

A group of eight individuals with high cholesterol levels participated in an exercise program that was designed to lower cholesterol levels. Cholesterol levels, in milligrams per deciliter, were measured before and after exercise for each individual, with the following results:

	Subject							
	1	2	3	4	5	6	7	8
Before	283	299	274	284	248	275	293	277
After	290	281	262	287	253	287	267	271

Can you conclude that the median cholesterol level after treatment is different from the median before treatment? Use the $\alpha = 0.05$ level of significance.

Solution

Step 1: State H_0 and H_1. The hypotheses are H_0: $m_d = 0$ and H_1: $m_d \neq 0$, where m_d is the population median of the differences.

Step 2: Choose a significance level. We choose a level of $\alpha = 0.05$.

Step 3: Compute the differences for matched pairs. The following table presents these differences.

	Subject							
	1	2	3	4	5	6	7	8
Before	283	299	274	284	248	275	293	277
After	290	281	262	287	253	287	267	271
Difference	−7	18	12	−3	−5	−12	26	6

Step 4: Compute the absolute values of the differences. We do this simply by making the negative differences positive. The following table presents the results.

	Subject							
	1	2	3	4	5	6	7	8
Before	283	299	274	284	248	275	293	277
After	290	281	262	287	253	287	267	271
Difference	−7	18	12	−3	−5	−12	26	6
Absolute Value	7	18	12	3	5	12	26	6

Step 5: Rank the absolute values, ignoring any differences of 0. The following table presents the results. There are no differences of 0 to ignore.

	Subject							
	1	2	3	4	5	6	7	8
Before	283	299	274	284	248	275	293	277
After	290	281	262	287	253	287	267	271
Difference	−7	18	12	−3	−5	−12	26	6
Absolute Value	7	18	12	3	5	12	26	6
Rank	4	7	5.5	1	2	5.5	8	3

Step 6: Give each rank a plus or minus sign, according to the sign of the difference it corresponds to. These are called the *signed ranks*. The following table presents the results.

	Subject							
	1	2	3	4	5	6	7	8
Before	283	299	274	284	248	275	293	277
After	290	281	262	287	253	287	267	271
Difference	−7	18	12	−3	−5	−12	26	6
Absolute Value	7	18	12	3	5	12	26	6
Rank	4	7	5.5	1	2	5.5	8	3
Signed Rank	−4	7	5.5	−1	−2	−5.5	8	3

Step 7: Compute the sum of the positive ranks and the sum of the negative ranks. The positive sum is $7 + 5.5 + 8 + 3 = 23.5$. The negative sum is $-4 - 1 - 2 - 5.5 = -12.5$.

Step 8: Compute the test statistic. The test statistic S is the sum that is smaller in absolute value. Because 12.5 is smaller than 23.5, the test statistic is $S = 12.5$.

Step 9: Find the critical value, and determine whether to reject H_0. Critical values are found in Table A.8. For a sample size $n = 8$ and a significance level of $\alpha = 0.05$, the critical value is 4. We reject H_0 if the value of the test statistic is less than or equal to the critical value. Because $12.5 > 4$, we do not reject H_0.

n	$\alpha = 0.10$	$\alpha = 0.05$	$\alpha = 0.02$	$\alpha = 0.01$
⋮	⋮	⋮	⋮	⋮
6	2	1	*	*
7	4	2	0	*
8	6	4	2	0
9	8	6	3	2
10	11	8	5	3
⋮	⋮	⋮	⋮	⋮

Step 10: State a conclusion. We cannot conclude that cholesterol levels before and after exercise are different.

In summary, the signed-rank test is performed using the following steps.

Performing the Signed-Rank Test

Step 1: State the null and alternate hypotheses. These are H_0: $m_d = 0$ and H_1: $m_d \neq 0$.

Step 2: Choose a significance level α.

Step 3: Compute the differences for the matched pairs.

Step 4: Compute the absolute values of the differences. To do this, simply make the negative differences positive.

Step 5: Rank the absolute values, ignoring any differences of 0.

Step 6: Compute the signed ranks. To do this, give each rank a plus or minus sign, according to the sign of the difference it corresponds to.

Step 7: Compute the sum of the positive ranks and the sum of the negative ranks.

Step 8: Compute the test statistic S. The test statistic S is the sum that is smaller in absolute value.

Step 9: Find the critical value, and determine whether to reject H_0. Critical values are found in Table A.8. Reject H_0 if the value of the test statistic is less than or equal to the critical value.

Step 10: State a conclusion.

Check Your Understanding

1. Following are the current prices and last year's prices of a gallon of regular gas at a sample of 14 gas stations. Can you conclude that the median price is different now from what it was a year ago? Use the $\alpha = 0.01$ level of significance.

	1	2	3	4	5	6	7	8	9	10	11	12	13	14
Current	3.85	3.88	3.81	3.58	3.48	3.55	3.59	3.80	4.11	3.51	3.86	3.93	3.64	3.54
Last year	3.90	3.79	3.73	3.55	3.42	3.54	3.62	3.78	3.99	3.52	3.85	3.93	3.68	3.50

a. Compute the value of the test statistic. *24*

b. Find the critical value. *13*

c. State a conclusion. *Do not reject H_0.*

2. Corn was grown on 12 plots of land. Half of each plot was treated with fertilizer A, and the other half was treated with fertilizer B. Yields in bushels per acre were as follows. Can you conclude that the median yield differs between the fertilizers? Use the $\alpha = 0.05$ level of significance.

	1	2	3	4	5	6	7	8	9	10	11	12
Fertilizer A	170	145	139	146	138	132	174	164	151	158	145	153
Fertilizer B	155	128	125	142	133	135	152	154	147	137	146	129

a. Compute the value of the test statistic. *3*

b. Find the critical value. *14*

c. State a conclusion. *Reject H_0.*

Answers are on page 714.

Using Technology

We use Example 15.6 to illustrate the technology steps.

MINITAB

Performing the signed-rank test

Note: MINITAB returns a *P*-value for the test.

Step 1. Enter the data for Sample 1 in column **C1** and the data for Sample 2 in column **C2**. For Example 15.6, we enter the before data into **C1** and the after data into **C2**.

Step 2. Calculate the differences by selecting **Calc**, then **Calculator**. Type **Difference** in the **Store result in variable** field and enter **C2 – C1** in the **Expression** field.

Step 3. Click on **Stat**, then **Nonparametrics**, then **1-Sample Wilcoxon**.

Step 4. Enter **Difference** in the **Variables** field, and select the **Test median** option.

Step 5. In the **Alternative** field, select the form of the alternate hypothesis. We use **not equal**.

Step 6. Click **OK** (Figure A).

Test

Null hypothesis H_0: η = 0
Alternative hypothesis H_1: η ≠ 0

Sample	N for Test	Wilcoxon Statistic	P-Value
Difference	8	12.50	0.484

Figure A

Section 15.3 Exercises

Exercises 1 and 2 are the Check Your Understanding exercises located within the section.

Understanding the Concepts

In Exercises 3 and 4, fill in each blank with the appropriate word or phrase.

3. For the signed-rank test, we reject H_0 when the value of the test statistic is ________ the critical value. *less than or equal to*
4. The signed-rank test does not require that the population be normal, but it does require that it be ________. *symmetric*

In Exercises 5 and 6, determine whether the statement is true or false. If the statement is false, rewrite it as a true statement.

5. When performing the signed-rank test, we rank the absolute values of the differences. *True*
6. When performing the signed-rank test, we reject H_0 if the test statistic is greater than or equal to the critical value. *False*

Practicing the Skills

In Exercises 7–10, compute the test statistic and the critical value, and determine whether to reject H_0 at the $\alpha = 0.05$ level.

7.

Sample 1	73	48	38	45	42	32	75	66	56	58	46	54
Sample 2	59	32	22	15	30	32	55	59	49	34	50	33

$S = 1$, Critical value is 14. Reject H_0.

8.

Sample 1	342	264	263	253	260	183	249	257	190
Sample 2	324	219	248	223	214	144	211	218	175

$S = 0$, Critical value is 6. Reject H_0.

9.

Sample 1	67	51	54	52	53	51	47	56	52	64	58
Sample 2	75	60	56	44	62	53	55	54	54	55	66

$S = 19$, Critical value is 11. Do not reject H_0.

10.

Sample 1	614	567	691	588	620	644	636	543
Sample 2	438	449	751	556	559	573	668	539

$S = 6.5$, Critical value is 4. Do not reject H_0.

Working with the Concepts

11. **Test scores:** A sample of 12 students participated in an online tutoring program in mathematics. Each student took a pretest before the program started and a posttest after completing the program. Following are the scores.

	Student											
	1	2	3	4	5	6	7	8	9	10	11	12
Pretest	73	76	45	77	65	44	51	53	78	70	46	79
Posttest	83	89	57	79	75	39	62	62	76	79	39	94

Can you conclude that the median scores differ between pretest and posttest? Use the $\alpha = 0.05$ level of significance. *Yes*

12. **The rent is due:** To determine whether rents are increasing in a certain town, an economist compared the current monthly rents for eight studio apartments with the rents for those same apartments one year ago. The results were as follows.

	Apartment							
	1	2	3	4	5	6	7	8
Now	975	710	810	640	1040	800	820	615
One Year Ago	950	690	800	630	1060	790	820	630

Can you conclude that the median rent now differs from that of a year ago? Use the $\alpha = 0.05$ level of significance. *No*

13. **Traffic:** To determine whether traffic levels differ between the morning and evening rush hours, a traffic engineer counted the number of cars passing through a certain intersection during five-minute periods in both the morning and evening for 10 days. The results were as follows.

	Day									
	1	2	3	4	5	6	7	8	9	10
Morning	65	68	44	64	63	38	36	55	78	47
Evening	70	51	59	49	63	50	59	83	81	25

Can you conclude that the median traffic level differs between morning and evening? Use the $\alpha = 0.01$ level of significance. *No*

14. **Exercise and weight:** A sample of nine men participated in a regular exercise program at a local gym. They were weighed both before and after the program. The results were as follows.

	Subject								
	1	2	3	4	5	6	7	8	9
Before	165	181	234	175	244	174	243	202	170
After	155	176	233	169	231	177	245	185	161

Can you conclude that the median weight differs before and after the exercise program? Use the $\alpha = 0.05$ level of significance. *Yes*

15. **Balance your checkbook:** A sample of 14 people attended a financial seminar. Each participant was given a 20-point financial literacy quiz both before and after the seminar with the following results.

	Subject													
	1	2	3	4	5	6	7	8	9	10	11	12	13	14
Before	3	15	2	17	20	16	4	9	13	4	2	15	3	16
After	6	15	6	18	18	15	11	7	19	9	1	18	5	10

Can you conclude that the median score on the quiz before the seminar differs from the median score after? Use the $\alpha = 0.05$ level of significance. *No*

16. **Hurricane Katrina:** Hurricane Katrina made landfall in August 2005 causing the evacuation of numerous people from the area. The total number of births was observed each month for a year following the hurricane in the counties directly affected by it. These were compared to the number of births in the same months for the previous year.

	Sep.	Oct.	Nov.	Dec.	Jan.	Feb.
Before Katrina	3113	3111	3000	3056	2926	2615
After Katrina	2288	2177	2134	2204	2424	2112

	Mar.	Apr.	May	Jun.	Jul.	Aug.
Before Katrina	2928	2602	2751	2848	2806	2446
After Katrina	2375	2105	2239	2334	2608	2629

Can you conclude that the median number of births differed after the hurricane from the median number of births before? Use the $\alpha = 0.05$ level of significance. *Yes*

Extending the Concepts

Exercise 17 demonstrates that the results of the signed-rank test may be misleading when the assumption of symmetry is violated.

17. Asymmetric differences: Consider the following paired samples and their differences.

Sample 1												
68	59	66	67	64	50	60	55	74	59	66	55	68
58	70	71	54	68	65	82	34	89	97	73	78	

Sample 2												
80	70	76	76	72	57	66	60	78	62	68	56	68
45	56	56	38	51	47	63	14	68	75	50	54	

Differences												
−12	−11	−10	−9	−8	−7	−6	−5	−4	−3	−2	−1	0
13	14	15	16	17	18	19	20	21	22	23	24	

a. Verify that the median of the sample differences is equal to 0.
b. Compute the test statistic for the signed-rank test. *78*
c. Do you reject the null hypothesis that the median is equal to 0? Use the $\alpha = 0.05$ level of significance. *Yes*

Answers to Check Your Understanding Exercises for Section 15.3

1. a. $S = 24$ **b.** The critical value is 13.
c. Do not reject H_0. We cannot conclude that the median price is different today from what it was a year ago.

2. a. $S = 3$ **b.** The critical value is 14.
c. Reject H_0. We conclude that the median yields differ between the two fertilizers.

Chapter 15 Summary

Section 15.1: In this section we presented the sign test, which is a nonparametric alternative to the t-test. The sign test is valid for any continuous population and does not require the assumption of normality. The major disadvantage of the sign test is that when the population is approximately normal, it is less likely to reject a false null hypothesis than the t-test.

Section 15.2: Wilcoxon's rank-sum test is a nonparametric test used to determine whether two population medians differ. The null hypothesis is that the medians are equal. The rank-sum test is a nonparametric alternative to the two-sample t-test. Its major advantage is that it does not require the assumption of normality. Its major disadvantage is that the null hypothesis can be rejected when the two populations differ in shape, even if their medians are equal.

Section 15.3: In this section we presented the signed-rank test, which is a nonparametric alternative to the t-test for paired differences. Although the signed-rank test does not require that the population of differences be normally distributed, it does require that the distribution be approximately symmetric. Like the sign test, the signed-rank test has the disadvantage that when the population is approximately normal, it is less likely to reject a false null hypothesis than the t-test.

Vocabulary and Notation

Important Formulas

Test statistic for the sign test:

$$z = \frac{x + 0.5 - n/2}{\sqrt{n}/2} \quad \text{if } n > 25$$

If $n \leq 25$, the test statistic is x, the number of times the less frequent sign occurs.

Mean of S, the sum of the ranks for the rank-sum test:

$$\mu_S = \frac{n_1(n_1 + n_2 + 1)}{2}$$

Standard deviation of S, the sum of the ranks for the rank-sum test:

$$\sigma_S = \sqrt{\frac{n_1 n_2(n_1 + n_2 + 1)}{12}}$$

Test statistic for the rank-sum test:

$$z = \frac{S - \mu_S}{\sigma_S}$$

Chapter Quiz

1. Following is a sample from a population with median m. Use the sign test to test H_0: $m = 30$ versus H_1: $m \neq 30$.

50	38	23	29	57	26	53	42	60	23

a. What is the value of the test statistic? *$x = 4$*
b. Is H_0 rejected at the $\alpha = 0.05$ level? *No*
c. Is H_0 rejected at the $\alpha = 0.01$ level? *No*

2. Following is a sample from a population with median m. Use the sign test to test H_0: $m = 20$ versus H_1: $m > 20$.

38	47	55	38	40	43	36	49	43	55	43	47	38	50	38	43

a. What is the value of the test statistic? *$x = 0$*
b. Is H_0 rejected at the $\alpha = 0.05$ level? *Yes*
c. Is H_0 rejected at the $\alpha = 0.01$ level? *Yes*

3. The sign test is performed to test H_0: $m = 42$ versus H_1: $m \neq 42$. There are 19 plus signs, 5 minus signs, and 4 zeros.
a. What is the value of the test statistic? *$x = 5$*
b. Is H_0 rejected at the $\alpha = 0.05$ level? *Yes*
c. Is H_0 rejected at the $\alpha = 0.01$ level? *Yes*

4. The sign test is performed to test H_0: $m = 250$ versus H_1: $m < 250$. There are 15 plus signs, 36 minus signs, and 6 zeros.
a. What is the value of the test statistic? *$z = -2.80$*
b. Is H_0 rejected at the $\alpha = 0.05$ level? *Yes*
c. Is H_0 rejected at the $\alpha = 0.01$ level? *Yes*

Exercises 5–7 present sample sizes and the sum of ranks for the rank-sum test. Compute μ_S, σ_S, and the value of the test statistic z. Then find the P-value.

5. $n_1 = 20, n_2 = 30, S = 400, H_1: m_1 < m_2$. *$\mu_S = 510, \sigma_S = 50.498, z = -2.18, P = 0.0146$ [Tech: 0.0147]*

6. $n_1 = 25, n_2 = 32, S = 850, H_1: m_1 \neq m_2$. *$\mu_S = 725, \sigma_S = 62.183, z = 2.01, P = 0.0444$*

7. $n_1 = 15, n_2 = 18, S = 280, H_1: m_1 > m_2$. *$\mu_S = 255, \sigma_S = 27.659, z = 0.90, P = 0.1841$ [Tech: 0.1830]*

In Exercises 8–10, use the signed-rank test to test the null hypothesis that the population medians are equal. Compute the test statistic, the critical value, and determine whether to reject H_0 at the $\alpha = 0.05$ level.

8.

Sample 1	61	37	36	28	59	30	16	52	31	23	32	15	17	61	62	41
Sample 2	56	35	48	18	73	19	27	57	47	37	53	36	15	74	55	27

$S = 37$, Critical value is 30. Do not reject H_0.

9.

Sample 1	159	193	213	295	160	156	181	273	272	276	239	250	209	279
Sample 2	187	217	237	317	180	184	206	287	301	287	254	270	227	304

$S = 0$, Critical value is 21. Reject H_0.

10.

Sample 1	73	74	64	75	61	56	81	80	64	70	69	49
Sample 2	72	66	55	72	61	54	78	75	62	65	63	48

$S = 0$, Critical value is 11. Reject H_0.

Review Exercises

1. **Freshman weights:** Do college students tend to gain or lose weight during their freshman year? Following are a sample of weights of 15 college freshmen at the beginning and at the end of their freshman year.

	Student														
	1	2	3	4	5	6	7	8	9	10	11	12	13	14	15
Beginning	174	150	148	190	161	162	186	181	158	118	124	189	103	149	117
End	193	161	153	194	153	172	177	173	164	111	139	204	115	143	127

Can you conclude that the median weight at the end of the year differs from the median at the beginning of the year? Use the $\alpha = 0.05$ level of significance. *No*

2. **Keep cool:** Following are the prices, in dollars, of a sample of 20 air conditioners.

377	559	430	490	352	343	500	203	223	239	326	503	491	194	330	381	337	433	421	287

Can you conclude that the median price differs from \$300? Use the $\alpha = 0.01$ level of significance. *No*

3. **Slow Mondays:** The owner of a restaurant believes that she serves the same number of customers on Mondays as on Tuesdays. For 12 weeks she counts the number of customers who come in on Monday and on Tuesday, with the following results.

	Week											
	1	**2**	**3**	**4**	**5**	**6**	**7**	**8**	**9**	**10**	**11**	**12**
Monday	163	122	218	137	131	157	140	169	154	215	212	125
Tuesday	173	113	215	139	149	154	159	161	162	222	214	133

Can you conclude that the median number of customers differs between Mondays and Tuesdays? Use the $\alpha = 0.01$ level of significance. *No*

4. **Exercise and blood pressure:** Systolic blood pressure, in millimeters of mercury, was measured for a sample of 15 subjects who reported exercising regularly and for a sample of 12 patients who reported having a sedentary lifestyle. The results were as follows.

Exercise														
127	107	105	140	107	101	122	135	127	108	115	118	139	106	134

Sedentary											
154	142	122	134	110	150	149	146	118	135	106	126

Can you conclude that the median systolic blood pressure is less for those who exercise than for those who are sedentary? Use the $\alpha = 0.05$ level of significance. *Yes*

5. **Scoring runs:** Following are the number of runs scored by both teams in a sample of 20 Major League Baseball games played in a recent season.

8	6	11	3	5	1	8	13	16	6	4	10	17	16	22	15	10	8	10	2

Can you conclude that the median is different from 8? Use the $\alpha = 0.05$ level of significance. *No*

6. **Recovery times:** The number of days spent in the hospital was determined for 12 patients who underwent coronary bypass surgery.

7	10	4	10	6	8	5	8	8	3	4	6

Can you conclude that the median is less than 5 days? Use the $\alpha = 0.01$ level of significance. *No*

7. **Boys and girls:** Following are weights, in pounds, for samples of 16 newborn boys and 14 newborn girls.

Boys															
8.3	7.8	8.2	8.0	8.8	8.4	8.5	7.9	7.5	7.0	6.6	6.4	5.9	6.7	7.4	7.6

Girls													
7.7	8.8	8.9	6.2	6.0	7.9	5.8	7.3	7.4	8.5	7.2	6.9	6.2	8.1

Can you conclude that the median weight differs between boys and girls? Use the $\alpha = 0.01$ level of significance. *No*

8. **Morning commute:** A woman who has moved into a new house is trying to determine which of two routes to work has the shorter driving time. Times in minutes for 12 trips on route A and 10 trips on route B are presented below.

Route A											
15.9	16.3	17.0	15.7	16.3	16.7	16.2	16.9	16.0	16.5	16.7	16.4

Route B									
17.0	16.6	16.9	19.8	17.4	20.0	17.2	16.1	17.6	16.7

Can you conclude that the median time differs between the two routes? Use the $\alpha = 0.05$ level of significance. *Yes*

9. **Tire tread:** Two gauges that measure tire tread depth were compared. Twelve different locations on a tire were measured once by each gauge. The results, in millimeters, were as follows.

	Location											
	1	2	3	4	5	6	7	8	9	10	11	12
Gauge 1	3.95	3.23	3.60	3.48	3.89	3.76	3.56	3.01	3.82	3.44	3.58	3.78
Gauge 2	3.80	3.30	3.59	3.61	3.88	3.73	3.45	3.02	3.77	3.49	3.42	3.66

Can you conclude that the median measurement differs between the gauges? Use the $\alpha = 0.05$ level of significance. *No*

10. **How much is in that can?** A machine that fills beverage cans is supposed to put 12 ounces of beverage in each can. Following are the amounts measured in a sample of 14 cans.

12.00	11.97	11.93	12.02	11.95	11.91	12.05	11.95	11.99	12.04	11.97	12.05	11.98	12.04

Can you conclude that the median amount differs from 12 ounces? Use the $\alpha = 0.01$ level of significance. *No*

Write About It

1. Would it be possible to use the number of times the more frequent sign appears as the test statistic for the sign test? Explain why you would reject the null hypothesis when the value of this test statistic is greater than a critical value.
2. Describe the assumptions necessary for the rank-sum test. How do they differ from the assumptions necessary for the two-sample t-test?
3. Describe a situation in which the signed-rank test would be preferable to the paired t-test. Describe a situation in which the paired t-test would be preferable to the signed-rank test.

In-Class Activities

1. **Rank-sum test and t-test:** The following data, taken from Exercise 12 in Section 11.1, are recovery times for seven patients who received a new postsurgical treatment and seven others (the controls) who received the standard treatment. Assume the populations are symmetric, so that their means and medians are equal.

Treatment:	12	13	15	19	20	21	24
Control:	18	23	24	30	32	35	39

Let m_1 denote the median recovery time for the patients who received the new treatment, and let m_2 represent the median recovery time for the controls. Use both the rank-sum test and the Student's t-test to test $H_0: m_1 = m_2$ versus $H_1: m_1 < m_2$. For the t-test, use the more accurate formula for the degrees of freedom presented in Section 10.1:

$$\text{Degrees of freedom} = \frac{\left[\dfrac{s_1^2}{n_1} + \dfrac{s_2^2}{n_2}\right]^2}{\dfrac{(s_1^2/n_1)^2}{n_1 - 1} + \dfrac{(s_2^2/n_2)^2}{n_2 - 1}}$$

Which test has the smaller P-value? Why would you expect this to be the case?

Case Study: Diversity of Microorganism Populations

Microorganisms play a crucial role in wastewater treatment by transforming solid organic matter into carbon dioxide and water. In a wastewater treatment system, the solids retention time is the average amount of time that solid organic matter resides in the treatment tank. A study was performed at the Colorado School of Mines to determine the relationship between the solids retention time and the diversity of the microorganism population. It was thought, because microorganisms feed on the solid matter, that longer retention times would provide more opportunity for genetic mutations to occur, resulting in a greater diversity of microorganisms.

Following are counts of the numbers of species of microorganism for retention times of 3 days, 12 days, and 30 days.

3 Days								
282	273	238	240	241	253	230	266	253

12 Days											
266	299	287	289	272	284	292	279	274	306	291	272

30 Days																	
293	271	293	296	271	292	265	262	278	272	246	278	276	283	273	286	303	289

1. Can you conclude at the $\alpha = 0.05$ level that the median number of species is greater for a retention time of 12 days than for a retention time of 3 days? Compute the appropriate test statistic, find the *P*-value, and state a conclusion. *$z = -3.23$, $P = 0.0006$. Reject H_0.*

2. Can you conclude at the $\alpha = 0.05$ level that the median number of species is greater for a retention time of 30 days than for a retention time of 3 days? Compute the appropriate test statistic, find the *P*-value, and state a conclusion. *$z = -3.06$, $P = 0.0011$. Reject H_0.*

3. Can you conclude at the $\alpha = 0.05$ level that the median number of species for a retention time of 30 days differs from the median for a retention time of 12 days? Compute the appropriate test statistic, find the *P*-value, and state a conclusion. *$z = 0.85$, $P = 0.3954$ [Tech: 0.3972]. Do not reject H_0.*

4. Based on the results of the tests performed in Exercises 1–3, which of the following is the best conclusion? *iii*

i. We can conclude that the median number of species increases as the solids retention time increases from 3 to 12 and from 12 to 30 days.

ii. We cannot conclude that the median number of species increases as the solids retention time increases.

iii. We can conclude that the median number of species increases as the solids retention time increases from 3 to 12 days, but it may not continue to increase as the retention time increases to 30 days.

iv. We can conclude that the median number of species increases as the solid retention time increases from 12 to 30 days, but it may not increase as the retention time increases from 3 to 12 days.

Appendix A

Tables

- **Table A.1** Binomial Probabilities
- **Table A.2** Cumulative Normal Distribution
- **Table A.3** Critical Values for the Student's *t* Distribution
- **Table A.4** Critical Values for the χ^2 Distribution
- **Table A.5** Critical Values for the *F* Distribution
- **Table A.6** Critical Values for the Studentized Range
- **Table A.7** Critical Values for the Sign Test
- **Table A.8** Critical Values for the Signed-Rank Test

Table A.1 Binomial Probabilities

$$P(x) = \frac{n!}{x!(n-x)!}p^x(1-p)^{(n-x)}$$

		p												
n	*x*	**0.05**	**0.10**	**0.20**	**0.25**	**0.30**	**0.40**	**0.50**	**0.60**	**0.70**	**0.75**	**0.80**	**0.90**	**0.95**
2	0	0.903	0.810	0.640	0.563	0.490	0.360	0.250	0.160	0.090	0.063	0.040	0.010	0.003
	1	0.095	0.180	0.320	0.375	0.420	0.480	0.500	0.480	0.420	0.375	0.320	0.180	0.095
	2	0.003	0.010	0.040	0.063	0.090	0.160	0.250	0.360	0.490	0.563	0.640	0.810	0.903
3	0	0.857	0.729	0.512	0.422	0.343	0.216	0.125	0.064	0.027	0.016	0.008	0.001	0.000+
	1	0.135	0.243	0.384	0.422	0.441	0.432	0.375	0.288	0.189	0.141	0.096	0.027	0.007
	2	0.007	0.027	0.096	0.141	0.189	0.288	0.375	0.432	0.441	0.422	0.384	0.243	0.135
	3	0.000+	0.001	0.008	0.016	0.027	0.064	0.125	0.216	0.343	0.422	0.512	0.729	0.857
4	0	0.815	0.656	0.410	0.316	0.240	0.130	0.063	0.026	0.008	0.004	0.002	0.000+	0.000+
	1	0.171	0.292	0.410	0.422	0.412	0.346	0.250	0.154	0.076	0.047	0.026	0.004	0.000+
	2	0.014	0.049	0.154	0.211	0.265	0.346	0.375	0.346	0.265	0.211	0.154	0.049	0.014
	3	0.000+	0.004	0.026	0.047	0.076	0.154	0.250	0.346	0.412	0.422	0.410	0.292	0.171
	4	0.000+	0.000+	0.002	0.004	0.008	0.026	0.063	0.130	0.240	0.316	0.410	0.656	0.815
5	0	0.774	0.590	0.328	0.237	0.168	0.078	0.031	0.010	0.002	0.001	0.000+	0.000+	0.000+
	1	0.204	0.328	0.410	0.396	0.360	0.259	0.156	0.077	0.028	0.015	0.006	0.000+	0.000+
	2	0.021	0.073	0.205	0.264	0.309	0.346	0.313	0.230	0.132	0.088	0.051	0.008	0.001
	3	0.001	0.008	0.051	0.088	0.132	0.230	0.313	0.346	0.309	0.264	0.205	0.073	0.021
	4	0.000+	0.000+	0.006	0.015	0.028	0.077	0.156	0.259	0.360	0.396	0.410	0.328	0.204
	5	0.000+	0.000+	0.000+	0.001	0.002	0.010	0.031	0.078	0.168	0.237	0.328	0.590	0.774
6	0	0.735	0.531	0.262	0.178	0.118	0.047	0.016	0.004	0.001	0.000+	0.000+	0.000+	0.000+
	1	0.232	0.354	0.393	0.356	0.303	0.187	0.094	0.037	0.010	0.004	0.002	0.000+	0.000+
	2	0.031	0.098	0.246	0.297	0.324	0.311	0.234	0.138	0.060	0.033	0.015	0.001	0.000+
	3	0.002	0.015	0.082	0.132	0.185	0.276	0.313	0.276	0.185	0.132	0.082	0.015	0.002
	4	0.000+	0.001	0.015	0.033	0.060	0.138	0.234	0.311	0.324	0.297	0.246	0.098	0.031
	5	0.000+	0.000+	0.002	0.004	0.010	0.037	0.094	0.187	0.303	0.356	0.393	0.354	0.232
	6	0.000+	0.000+	0.000+	0.000+	0.001	0.004	0.016	0.047	0.118	0.178	0.262	0.531	0.735
7	0	0.698	0.478	0.210	0.133	0.082	0.028	0.008	0.002	0.000+	0.000+	0.000+	0.000+	0.000+
	1	0.257	0.372	0.367	0.311	0.247	0.131	0.055	0.017	0.004	0.001	0.000+	0.000+	0.000+
	2	0.041	0.124	0.275	0.311	0.318	0.261	0.164	0.077	0.025	0.012	0.004	0.000+	0.000+
	3	0.004	0.023	0.115	0.173	0.227	0.290	0.273	0.194	0.097	0.058	0.029	0.003	0.000+
	4	0.000+	0.003	0.029	0.058	0.097	0.194	0.273	0.290	0.227	0.173	0.115	0.023	0.004
	5	0.000+	0.000+	0.004	0.012	0.025	0.077	0.164	0.261	0.318	0.311	0.275	0.124	0.041
	6	0.000+	0.000+	0.000+	0.001	0.004	0.017	0.055	0.131	0.247	0.311	0.367	0.372	0.257
	7	0.000+	0.000+	0.000+	0.000+	0.000+	0.002	0.008	0.028	0.082	0.133	0.210	0.478	0.698
8	0	0.663	0.430	0.168	0.100	0.058	0.017	0.004	0.001	0.000+	0.000+	0.000+	0.000+	0.000+
	1	0.279	0.383	0.336	0.267	0.198	0.090	0.031	0.008	0.001	0.000+	0.000+	0.000+	0.000+
	2	0.051	0.149	0.294	0.311	0.296	0.209	0.109	0.041	0.010	0.004	0.001	0.000+	0.000+
	3	0.005	0.033	0.147	0.208	0.254	0.279	0.219	0.124	0.047	0.023	0.009	0.000+	0.000+
	4	0.000+	0.005	0.046	0.087	0.136	0.232	0.273	0.232	0.136	0.087	0.046	0.005	0.000+
	5	0.000+	0.000+	0.009	0.023	0.047	0.124	0.219	0.279	0.254	0.208	0.147	0.033	0.005
	6	0.000+	0.000+	0.001	0.004	0.010	0.041	0.109	0.209	0.296	0.311	0.294	0.149	0.051
	7	0.000+	0.000+	0.000+	0.000+	0.001	0.008	0.031	0.090	0.198	0.267	0.336	0.383	0.279
	8	0.000+	0.000+	0.000+	0.000+	0.000+	0.001	0.004	0.017	0.058	0.100	0.168	0.430	0.663
9	0	0.630	0.387	0.134	0.075	0.040	0.010	0.002	0.000+	0.000+	0.000+	0.000+	0.000+	0.000+
	1	0.299	0.387	0.302	0.225	0.156	0.060	0.018	0.004	0.000+	0.000+	0.000+	0.000+	0.000+
	2	0.063	0.172	0.302	0.300	0.267	0.161	0.070	0.021	0.004	0.001	0.000+	0.000+	0.000+
	3	0.008	0.045	0.176	0.234	0.267	0.251	0.164	0.074	0.021	0.009	0.003	0.000+	0.000+
	4	0.001	0.007	0.066	0.117	0.172	0.251	0.246	0.167	0.074	0.039	0.017	0.001	0.000+
	5	0.000+	0.001	0.017	0.039	0.074	0.167	0.246	0.251	0.172	0.117	0.066	0.007	0.001
	6	0.000+	0.000+	0.003	0.009	0.021	0.074	0.164	0.251	0.267	0.234	0.176	0.045	0.008
	7	0.000+	0.000+	0.000+	0.001	0.004	0.021	0.070	0.161	0.267	0.300	0.302	0.172	0.063
	8	0.000+	0.000+	0.000+	0.000+	0.000+	0.004	0.018	0.060	0.156	0.225	0.302	0.387	0.299
	9	0.000+	0.000+	0.000+	0.000+	0.000+	0.000+	0.002	0.010	0.040	0.075	0.134	0.387	0.630

A value of 0.000+ indicates that the probability is 0.000 when rounded to three decimal places. The actual probability is slightly greater than 0.

Table A.1 Binomial Probabilities (continued)

		p												
n	*x*	**0.05**	**0.10**	**0.20**	**0.25**	**0.30**	**0.40**	**0.50**	**0.60**	**0.70**	**0.75**	**0.80**	**0.90**	**0.95**
10	0	0.599	0.349	0.107	0.056	0.028	0.006	0.001	0.000+	0.000+	0.000+	0.000+	0.000+	0.000+
	1	0.315	0.387	0.268	0.188	0.121	0.040	0.010	0.002	0.000+	0.000+	0.000+	0.000+	0.000+
	2	0.075	0.194	0.302	0.282	0.233	0.121	0.044	0.011	0.001	0.000+	0.000+	0.000+	0.000+
	3	0.010	0.057	0.201	0.250	0.267	0.215	0.117	0.042	0.009	0.003	0.001	0.000+	0.000+
	4	0.001	0.011	0.088	0.146	0.200	0.251	0.205	0.111	0.037	0.016	0.006	0.000+	0.000+
	5	0.000+	0.001	0.026	0.058	0.103	0.201	0.246	0.201	0.103	0.058	0.026	0.001	0.000+
	6	0.000+	0.000+	0.006	0.016	0.037	0.111	0.205	0.251	0.200	0.146	0.088	0.011	0.001
	7	0.000+	0.000+	0.001	0.003	0.009	0.042	0.117	0.215	0.267	0.250	0.201	0.057	0.010
	8	0.000+	0.000+	0.000+	0.000+	0.001	0.011	0.044	0.121	0.233	0.282	0.302	0.194	0.075
	9	0.000+	0.000+	0.000+	0.000+	0.000+	0.002	0.010	0.040	0.121	0.188	0.268	0.387	0.315
	10	0.000+	0.000+	0.000+	0.000+	0.000+	0.000+	0.001	0.006	0.028	0.056	0.107	0.349	0.599
11	0	0.569	0.314	0.086	0.042	0.020	0.004	0.000+	0.000+	0.000+	0.000+	0.000+	0.000+	0.000+
	1	0.329	0.384	0.236	0.155	0.093	0.027	0.005	0.001	0.000+	0.000+	0.000+	0.000+	0.000+
	2	0.087	0.213	0.295	0.258	0.200	0.089	0.027	0.005	0.001	0.000+	0.000+	0.000+	0.000+
	3	0.014	0.071	0.221	0.258	0.257	0.177	0.081	0.023	0.004	0.001	0.000+	0.000+	0.000+
	4	0.001	0.016	0.111	0.172	0.220	0.236	0.161	0.070	0.017	0.006	0.002	0.000+	0.000+
	5	0.000+	0.002	0.039	0.080	0.132	0.221	0.226	0.147	0.057	0.027	0.010	0.000+	0.000+
	6	0.000+	0.000+	0.010	0.027	0.057	0.147	0.226	0.221	0.132	0.080	0.039	0.002	0.000+
	7	0.000+	0.000+	0.002	0.006	0.017	0.070	0.161	0.236	0.220	0.172	0.111	0.016	0.001
	8	0.000+	0.000+	0.000+	0.001	0.004	0.023	0.081	0.177	0.257	0.258	0.221	0.071	0.014
	9	0.000+	0.000+	0.000+	0.000+	0.001	0.005	0.027	0.089	0.200	0.258	0.295	0.213	0.087
	10	0.000+	0.000+	0.000+	0.000+	0.000+	0.001	0.005	0.027	0.093	0.155	0.236	0.384	0.329
	11	0.000+	0.000+	0.000+	0.000+	0.000+	0.000+	0.000+	0.004	0.020	0.042	0.086	0.314	0.569
12	0	0.540	0.282	0.069	0.032	0.014	0.002	0.000+	0.000+	0.000+	0.000+	0.000+	0.000+	0.000+
	1	0.341	0.377	0.206	0.127	0.071	0.017	0.003	0.000+	0.000+	0.000+	0.000+	0.000+	0.000+
	2	0.099	0.230	0.283	0.232	0.168	0.064	0.016	0.002	0.000+	0.000+	0.000+	0.000+	0.000+
	3	0.017	0.085	0.236	0.258	0.240	0.142	0.054	0.012	0.001	0.000+	0.000+	0.000+	0.000+
	4	0.002	0.021	0.133	0.194	0.231	0.213	0.121	0.042	0.008	0.002	0.001	0.000+	0.000+
	5	0.000+	0.004	0.053	0.103	0.158	0.227	0.193	0.101	0.029	0.011	0.003	0.000+	0.000+
	6	0.000+	0.000+	0.016	0.040	0.079	0.177	0.226	0.177	0.079	0.040	0.016	0.000+	0.000+
	7	0.000+	0.000+	0.003	0.011	0.029	0.101	0.193	0.227	0.158	0.103	0.053	0.004	0.000+
	8	0.000+	0.000+	0.001	0.002	0.008	0.042	0.121	0.213	0.231	0.194	0.133	0.021	0.002
	9	0.000+	0.000+	0.000+	0.000+	0.001	0.012	0.054	0.142	0.240	0.258	0.236	0.085	0.017
	10	0.000+	0.000+	0.000+	0.000+	0.000+	0.002	0.016	0.064	0.168	0.232	0.283	0.230	0.099
	11	0.000+	0.000+	0.000+	0.000+	0.000+	0.000+	0.003	0.017	0.071	0.127	0.206	0.377	0.341
	12	0.000+	0.000+	0.000+	0.000+	0.000+	0.000+	0.000+	0.002	0.014	0.032	0.069	0.282	0.540
13	0	0.513	0.254	0.055	0.024	0.010	0.001	0.000+	0.000+	0.000+	0.000+	0.000+	0.000+	0.000+
	1	0.351	0.367	0.179	0.103	0.054	0.011	0.002	0.000+	0.000+	0.000+	0.000+	0.000+	0.000+
	2	0.111	0.245	0.268	0.206	0.139	0.045	0.010	0.001	0.000+	0.000+	0.000+	0.000+	0.000+
	3	0.021	0.100	0.246	0.252	0.218	0.111	0.035	0.006	0.001	0.000+	0.000+	0.000+	0.000+
	4	0.003	0.028	0.154	0.210	0.234	0.184	0.087	0.024	0.003	0.001	0.000+	0.000+	0.000+
	5	0.000+	0.006	0.069	0.126	0.180	0.221	0.157	0.066	0.014	0.005	0.001	0.000+	0.000+
	6	0.000+	0.001	0.023	0.056	0.103	0.197	0.209	0.131	0.044	0.019	0.006	0.000+	0.000+
	7	0.000+	0.000+	0.006	0.019	0.044	0.131	0.209	0.197	0.103	0.056	0.023	0.001	0.000+
	8	0.000+	0.000+	0.001	0.005	0.014	0.066	0.157	0.221	0.180	0.126	0.069	0.006	0.000+
	9	0.000+	0.000+	0.000+	0.001	0.003	0.024	0.087	0.184	0.234	0.210	0.154	0.028	0.003
	10	0.000+	0.000+	0.000+	0.000+	0.001	0.006	0.035	0.111	0.218	0.252	0.246	0.100	0.021
	11	0.000+	0.000+	0.000+	0.000+	0.000+	0.001	0.010	0.045	0.139	0.206	0.268	0.245	0.111
	12	0.000+	0.000+	0.000+	0.000+	0.000+	0.000+	0.002	0.011	0.054	0.103	0.179	0.367	0.351
	13	0.000+	0.000+	0.000+	0.000+	0.000+	0.000+	0.000+	0.001	0.010	0.024	0.055	0.254	0.513
14	0	0.488	0.229	0.044	0.018	0.007	0.001	0.000+	0.000+	0.000+	0.000+	0.000+	0.000+	0.000+
	1	0.359	0.356	0.154	0.083	0.041	0.007	0.001	0.000+	0.000+	0.000+	0.000+	0.000+	0.000+
	2	0.123	0.257	0.250	0.180	0.113	0.032	0.006	0.001	0.000+	0.000+	0.000+	0.000+	0.000+
	3	0.026	0.114	0.250	0.240	0.194	0.085	0.022	0.003	0.000+	0.000+	0.000+	0.000+	0.000+
	4	0.004	0.035	0.172	0.220	0.229	0.155	0.061	0.014	0.001	0.000+	0.000+	0.000+	0.000+
	5	0.000+	0.008	0.086	0.147	0.196	0.207	0.122	0.041	0.007	0.002	0.000+	0.000+	0.000+
	6	0.000+	0.001	0.032	0.073	0.126	0.207	0.183	0.092	0.023	0.008	0.002	0.000+	0.000+
	7	0.000+	0.000+	0.009	0.028	0.062	0.157	0.209	0.157	0.062	0.028	0.009	0.000+	0.000+
	8	0.000+	0.000+	0.002	0.008	0.023	0.092	0.183	0.207	0.126	0.073	0.032	0.001	0.000+
	9	0.000+	0.000+	0.000+	0.002	0.007	0.041	0.122	0.207	0.196	0.147	0.086	0.008	0.000+
	10	0.000+	0.000+	0.000+	0.000+	0.001	0.014	0.061	0.155	0.229	0.220	0.172	0.035	0.004
	11	0.000+	0.000+	0.000+	0.000+	0.000+	0.003	0.022	0.085	0.194	0.240	0.250	0.114	0.026
	12	0.000+	0.000+	0.000+	0.000+	0.000+	0.001	0.006	0.032	0.113	0.180	0.250	0.257	0.123
	13	0.000+	0.000+	0.000+	0.000+	0.000+	0.000+	0.001	0.007	0.041	0.083	0.154	0.356	0.359
	14	0.000+	0.000+	0.000+	0.000+	0.000+	0.000+	0.000+	0.001	0.007	0.018	0.044	0.229	0.488

A value of 0.000+ indicates that the probability is 0.000 when rounded to three decimal places. The actual probability is slightly greater than 0.

Table A.1 Binomial Probabilities (continued)

		p												
n	*x*	0.05	0.10	0.20	0.25	0.30	0.40	0.50	0.60	0.70	0.75	0.80	0.90	0.95
15	0	0.463	0.206	0.035	0.013	0.005	0.000+	0.000+	0.000+	0.000+	0.000+	0.000+	0.000+	0.000+
	1	0.366	0.343	0.132	0.067	0.031	0.005	0.000+	0.000+	0.000+	0.000+	0.000+	0.000+	0.000+
	2	0.135	0.267	0.231	0.156	0.092	0.022	0.003	0.000+	0.000+	0.000+	0.000+	0.000+	0.000+
	3	0.031	0.129	0.250	0.225	0.170	0.063	0.014	0.002	0.000+	0.000+	0.000+	0.000+	0.000+
	4	0.005	0.043	0.188	0.225	0.219	0.127	0.042	0.007	0.001	0.000+	0.000+	0.000+	0.000+
	5	0.001	0.010	0.103	0.165	0.206	0.186	0.092	0.024	0.003	0.001	0.000+	0.000+	0.000+
	6	0.000+	0.002	0.043	0.092	0.147	0.207	0.153	0.061	0.012	0.003	0.001	0.000+	0.000+
	7	0.000+	0.000+	0.014	0.039	0.081	0.177	0.196	0.118	0.035	0.013	0.003	0.000+	0.000+
	8	0.000+	0.000+	0.003	0.013	0.035	0.118	0.196	0.177	0.081	0.039	0.014	0.000+	0.000+
	9	0.000+	0.000+	0.001	0.003	0.012	0.061	0.153	0.207	0.147	0.092	0.043	0.002	0.000+
	10	0.000+	0.000+	0.000+	0.001	0.003	0.024	0.092	0.186	0.206	0.165	0.103	0.010	0.001
	11	0.000+	0.000+	0.000+	0.000+	0.001	0.007	0.042	0.127	0.219	0.225	0.188	0.043	0.005
	12	0.000+	0.000+	0.000+	0.000+	0.000+	0.002	0.014	0.063	0.170	0.225	0.250	0.129	0.031
	13	0.000+	0.000+	0.000+	0.000+	0.000+	0.000+	0.003	0.022	0.092	0.156	0.231	0.267	0.135
	14	0.000+	0.000+	0.000+	0.000+	0.000+	0.000+	0.000+	0.005	0.031	0.067	0.132	0.343	0.366
	15	0.000+	0.000+	0.000+	0.000+	0.000+	0.000+	0.000+	0.000+	0.005	0.013	0.035	0.206	0.463
16	0	0.440	0.185	0.028	0.010	0.003	0.000+	0.000+	0.000+	0.000+	0.000+	0.000+	0.000+	0.000+
	1	0.371	0.329	0.113	0.053	0.023	0.003	0.000+	0.000+	0.000+	0.000+	0.000+	0.000+	0.000+
	2	0.146	0.275	0.211	0.134	0.073	0.015	0.002	0.000+	0.000+	0.000+	0.000+	0.000+	0.000+
	3	0.036	0.142	0.246	0.208	0.146	0.047	0.009	0.001	0.000+	0.000+	0.000+	0.000+	0.000+
	4	0.006	0.051	0.200	0.225	0.204	0.101	0.028	0.004	0.000+	0.000+	0.000+	0.000+	0.000+
	5	0.001	0.014	0.120	0.180	0.210	0.162	0.067	0.014	0.001	0.000+	0.000+	0.000+	0.000+
	6	0.000+	0.003	0.055	0.110	0.165	0.198	0.122	0.039	0.006	0.001	0.000+	0.000+	0.000+
	7	0.000+	0.000+	0.020	0.052	0.101	0.189	0.175	0.084	0.019	0.006	0.001	0.000+	0.000+
	8	0.000+	0.000+	0.006	0.020	0.049	0.142	0.196	0.142	0.049	0.020	0.006	0.000+	0.000+
	9	0.000+	0.000+	0.001	0.006	0.019	0.084	0.175	0.189	0.101	0.052	0.020	0.000+	0.000+
	10	0.000+	0.000+	0.000+	0.001	0.006	0.039	0.122	0.198	0.165	0.110	0.055	0.003	0.000+
	11	0.000+	0.000+	0.000+	0.000+	0.001	0.014	0.067	0.162	0.210	0.180	0.120	0.014	0.001
	12	0.000+	0.000+	0.000+	0.000+	0.000+	0.004	0.028	0.101	0.204	0.225	0.200	0.051	0.006
	13	0.000+	0.000+	0.000+	0.000+	0.000+	0.001	0.009	0.047	0.146	0.208	0.246	0.142	0.036
	14	0.000+	0.000+	0.000+	0.000+	0.000+	0.000+	0.002	0.015	0.073	0.134	0.211	0.275	0.146
	15	0.000+	0.000+	0.000+	0.000+	0.000+	0.000+	0.000+	0.003	0.023	0.053	0.113	0.329	0.371
	16	0.000+	0.000+	0.000+	0.000+	0.000+	0.000+	0.000+	0.000+	0.003	0.010	0.028	0.185	0.440
17	0	0.418	0.167	0.023	0.008	0.002	0.000+	0.000+	0.000+	0.000+	0.000+	0.000+	0.000+	0.000+
	1	0.374	0.315	0.096	0.043	0.017	0.002	0.000+	0.000+	0.000+	0.000+	0.000+	0.000+	0.000+
	2	0.158	0.280	0.191	0.114	0.058	0.010	0.001	0.000+	0.000+	0.000+	0.000+	0.000+	0.000+
	3	0.041	0.156	0.239	0.189	0.125	0.034	0.005	0.000+	0.000+	0.000+	0.000+	0.000+	0.000+
	4	0.008	0.060	0.209	0.221	0.187	0.080	0.018	0.002	0.000+	0.000+	0.000+	0.000+	0.000+
	5	0.001	0.017	0.136	0.191	0.208	0.138	0.047	0.008	0.001	0.000+	0.000+	0.000+	0.000+
	6	0.000+	0.004	0.068	0.128	0.178	0.184	0.094	0.024	0.003	0.001	0.000+	0.000+	0.000+
	7	0.000+	0.001	0.027	0.067	0.120	0.193	0.148	0.057	0.009	0.002	0.000+	0.000+	0.000+
	8	0.000+	0.000+	0.008	0.028	0.064	0.161	0.185	0.107	0.028	0.009	0.002	0.000+	0.000+
	9	0.000+	0.000+	0.002	0.009	0.028	0.107	0.185	0.161	0.064	0.028	0.008	0.000+	0.000+
	10	0.000+	0.000+	0.000+	0.002	0.009	0.057	0.148	0.193	0.120	0.067	0.027	0.001	0.000+
	11	0.000+	0.000+	0.000+	0.001	0.003	0.024	0.094	0.184	0.178	0.128	0.068	0.004	0.000+
	12	0.000+	0.000+	0.000+	0.000+	0.001	0.008	0.047	0.138	0.208	0.191	0.136	0.017	0.001
	13	0.000+	0.000+	0.000+	0.000+	0.000+	0.002	0.018	0.080	0.187	0.221	0.209	0.060	0.008
	14	0.000+	0.000+	0.000+	0.000+	0.000+	0.000+	0.005	0.034	0.125	0.189	0.239	0.156	0.041
	15	0.000+	0.000+	0.000+	0.000+	0.000+	0.000+	0.001	0.010	0.058	0.114	0.191	0.280	0.158
	16	0.000+	0.000+	0.000+	0.000+	0.000+	0.000+	0.000+	0.002	0.017	0.043	0.096	0.315	0.374
	17	0.000+	0.000+	0.000+	0.000+	0.000+	0.000+	0.000+	0.000+	0.002	0.008	0.023	0.167	0.418

A value of 0.000+ indicates that the probability is 0.000 when rounded to three decimal places. The actual probability is slightly greater than 0.

Table A.1 Binomial Probabilities (continued)

		p												
n	*x*	**0.05**	**0.10**	**0.20**	**0.25**	**0.30**	**0.40**	**0.50**	**0.60**	**0.70**	**0.75**	**0.80**	**0.90**	**0.95**
18	0	0.397	0.150	0.018	0.006	0.002	0.000+	0.000+	0.000+	0.000+	0.000+	0.000+	0.000+	0.000+
	1	0.376	0.300	0.081	0.034	0.013	0.001	0.000+	0.000+	0.000+	0.000+	0.000+	0.000+	0.000+
	2	0.168	0.284	0.172	0.096	0.046	0.007	0.001	0.000+	0.000+	0.000+	0.000+	0.000+	0.000+
	3	0.047	0.168	0.230	0.170	0.105	0.025	0.003	0.000+	0.000+	0.000+	0.000+	0.000+	0.000+
	4	0.009	0.070	0.215	0.213	0.168	0.061	0.012	0.001	0.000+	0.000+	0.000+	0.000+	0.000+
	5	0.001	0.022	0.151	0.199	0.202	0.115	0.033	0.004	0.000+	0.000+	0.000+	0.000+	0.000+
	6	0.000+	0.005	0.082	0.144	0.187	0.166	0.071	0.015	0.001	0.000+	0.000+	0.000+	0.000+
	7	0.000+	0.001	0.035	0.082	0.138	0.189	0.121	0.037	0.005	0.001	0.000+	0.000+	0.000+
	8	0.000+	0.000+	0.012	0.038	0.081	0.173	0.167	0.077	0.015	0.004	0.001	0.000+	0.000+
	9	0.000+	0.000+	0.003	0.014	0.039	0.128	0.185	0.128	0.039	0.014	0.003	0.000+	0.000+
	10	0.000+	0.000+	0.001	0.004	0.015	0.077	0.167	0.173	0.081	0.038	0.012	0.000+	0.000+
	11	0.000+	0.000+	0.000+	0.001	0.005	0.037	0.121	0.189	0.138	0.082	0.035	0.001	0.000+
	12	0.000+	0.000+	0.000+	0.000+	0.001	0.015	0.071	0.166	0.187	0.144	0.082	0.005	0.000+
	13	0.000+	0.000+	0.000+	0.000+	0.000+	0.004	0.033	0.115	0.202	0.199	0.151	0.022	0.001
	14	0.000+	0.000+	0.000+	0.000+	0.000+	0.001	0.012	0.061	0.168	0.213	0.215	0.070	0.009
	15	0.000+	0.000+	0.000+	0.000+	0.000+	0.000+	0.003	0.025	0.105	0.170	0.230	0.168	0.047
	16	0.000+	0.000+	0.000+	0.000+	0.000+	0.000+	0.001	0.007	0.046	0.096	0.172	0.284	0.168
	17	0.000+	0.000+	0.000+	0.000+	0.000+	0.000+	0.000+	0.001	0.013	0.034	0.081	0.300	0.376
	18	0.000+	0.000+	0.000+	0.000+	0.000+	0.000+	0.000+	0.000+	0.002	0.006	0.018	0.150	0.397
19	0	0.377	0.135	0.014	0.004	0.001	0.000+	0.000+	0.000+	0.000+	0.000+	0.000+	0.000+	0.000+
	1	0.377	0.285	0.068	0.027	0.009	0.001	0.000+	0.000+	0.000+	0.000+	0.000+	0.000+	0.000+
	2	0.179	0.285	0.154	0.080	0.036	0.005	0.000+	0.000+	0.000+	0.000+	0.000+	0.000+	0.000+
	3	0.053	0.180	0.218	0.152	0.087	0.017	0.002	0.000+	0.000+	0.000+	0.000+	0.000+	0.000+
	4	0.011	0.080	0.218	0.202	0.149	0.047	0.007	0.001	0.000+	0.000+	0.000+	0.000+	0.000+
	5	0.002	0.027	0.164	0.202	0.192	0.093	0.022	0.002	0.000+	0.000+	0.000+	0.000+	0.000+
	6	0.000+	0.007	0.095	0.157	0.192	0.145	0.052	0.008	0.001	0.000+	0.000+	0.000+	0.000+
	7	0.000+	0.001	0.044	0.097	0.153	0.180	0.096	0.024	0.002	0.000+	0.000+	0.000+	0.000+
	8	0.000+	0.000+	0.017	0.049	0.098	0.180	0.144	0.053	0.008	0.002	0.000+	0.000+	0.000+
	9	0.000+	0.000+	0.005	0.020	0.051	0.146	0.176	0.098	0.022	0.007	0.001	0.000+	0.000+
	10	0.000+	0.000+	0.001	0.007	0.022	0.098	0.176	0.146	0.051	0.020	0.005	0.000+	0.000+
	11	0.000+	0.000+	0.000+	0.002	0.008	0.053	0.144	0.180	0.098	0.049	0.017	0.000+	0.000+
	12	0.000+	0.000+	0.000+	0.000+	0.002	0.024	0.096	0.180	0.153	0.097	0.044	0.001	0.000+
	13	0.000+	0.000+	0.000+	0.000+	0.001	0.008	0.052	0.145	0.192	0.157	0.095	0.007	0.000+
	14	0.000+	0.000+	0.000+	0.000+	0.000+	0.002	0.022	0.093	0.192	0.202	0.164	0.027	0.002
	15	0.000+	0.000+	0.000+	0.000+	0.000+	0.001	0.007	0.047	0.149	0.202	0.218	0.080	0.011
	16	0.000+	0.000+	0.000+	0.000+	0.000+	0.000+	0.002	0.017	0.087	0.152	0.218	0.180	0.053
	17	0.000+	0.000+	0.000+	0.000+	0.000+	0.000+	0.000+	0.005	0.036	0.080	0.154	0.285	0.179
	18	0.000+	0.000+	0.000+	0.000+	0.000+	0.000+	0.000+	0.001	0.009	0.027	0.068	0.285	0.377
	19	0.000+	0.000+	0.000+	0.000+	0.000+	0.000+	0.000+	0.000+	0.001	0.004	0.014	0.135	0.377
20	0	0.358	0.122	0.012	0.003	0.001	0.000++	0.000++	0.000++	0.000++	0.000++	0.000++	0.000++	0.000++
	1	0.377	0.270	0.058	0.021	0.007	0.000+	0.000+	0.000+	0.000+	0.000+	0.000+	0.000+	0.000+
	2	0.189	0.285	0.137	0.067	0.028	0.003	0.000+	0.000+	0.000+	0.000+	0.000+	0.000+	0.000+
	3	0.060	0.190	0.205	0.134	0.072	0.012	0.001	0.000+	0.000+	0.000+	0.000+	0.000+	0.000+
	4	0.013	0.090	0.218	0.190	0.130	0.035	0.005	0.000+	0.000+	0.000+	0.000+	0.000+	0.000+
	5	0.002	0.032	0.175	0.202	0.179	0.075	0.015	0.001	0.000+	0.000+	0.000+	0.000+	0.000+
	6	0.000+	0.009	0.109	0.169	0.192	0.124	0.037	0.005	0.000+	0.000+	0.000+	0.000+	0.000+
	7	0.000+	0.002	0.055	0.112	0.164	0.166	0.074	0.015	0.001	0.000+	0.000+	0.000+	0.000+
	8	0.000+	0.000+	0.022	0.061	0.114	0.180	0.120	0.035	0.004	0.001	0.000+	0.000+	0.000+
	9	0.000+	0.000+	0.007	0.027	0.065	0.160	0.160	0.071	0.012	0.003	0.000+	0.000+	0.000+
	10	0.000+	0.000+	0.002	0.010	0.031	0.117	0.176	0.117	0.031	0.010	0.002	0.000+	0.000+
	11	0.000+	0.000+	0.000+	0.003	0.012	0.071	0.160	0.160	0.065	0.027	0.007	0.000+	0.000+
	12	0.000+	0.000+	0.000+	0.001	0.004	0.035	0.120	0.180	0.114	0.061	0.022	0.000+	0.000+
	13	0.000+	0.000+	0.000+	0.000+	0.001	0.015	0.074	0.166	0.164	0.112	0.055	0.002	0.000+
	14	0.000+	0.000+	0.000+	0.000+	0.000+	0.005	0.037	0.124	0.192	0.169	0.109	0.009	0.000+
	15	0.000+	0.000+	0.000+	0.000+	0.000+	0.001	0.015	0.075	0.179	0.202	0.175	0.032	0.002
	16	0.000+	0.000+	0.000+	0.000+	0.000+	0.000+	0.005	0.035	0.130	0.190	0.218	0.090	0.013
	17	0.000+	0.000+	0.000+	0.000+	0.000+	0.000+	0.001	0.012	0.072	0.134	0.205	0.190	0.060
	18	0.000+	0.000+	0.000+	0.000+	0.000+	0.000+	0.000+	0.003	0.028	0.067	0.137	0.285	0.189
	19	0.000+	0.000+	0.000+	0.000+	0.000+	0.000+	0.000+	0.000+	0.007	0.021	0.058	0.270	0.377
	20	0.000+	0.000+	0.000+	0.000+	0.000+	0.000+	0.000+	0.000+	0.001	0.003	0.012	0.122	0.358

A value of 0.000+ indicates that the probability is 0.000 when rounded to three decimal places. The actual probability is slightly greater than 0.

Table A.2 Cumulative Normal Distribution

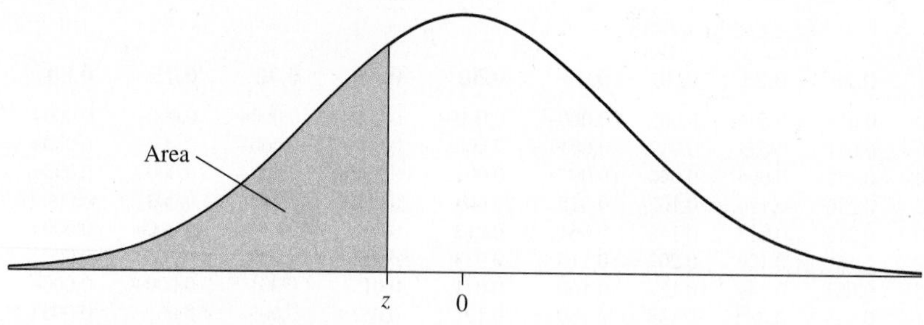

z	0.00	0.01	0.02	0.03	0.04	0.05	0.06	0.07	0.08	0.09
−3.7 or less	.0001									
−3.6	.0002	.0002	.0001	.0001	.0001	.0001	.0001	.0001	.0001	.0001
−3.5	.0002	.0002	.0002	.0002	.0002	.0002	.0002	.0002	.0002	.0002
−3.4	.0003	.0003	.0003	.0003	.0003	.0003	.0003	.0003	.0003	.0002
−3.3	.0005	.0005	.0005	.0004	.0004	.0004	.0004	.0004	.0004	.0003
−3.2	.0007	.0007	.0006	.0006	.0006	.0006	.0006	.0005	.0005	.0005
−3.1	.0010	.0009	.0009	.0009	.0008	.0008	.0008	.0008	.0007	.0007
−3.0	.0013	.0013	.0013	.0012	.0012	.0011	.0011	.0011	.0010	.0010
−2.9	.0019	.0018	.0018	.0017	.0016	.0016	.0015	.0015	.0014	.0014
−2.8	.0026	.0025	.0024	.0023	.0023	.0022	.0021	.0021	.0020	.0019
−2.7	.0035	.0034	.0033	.0032	.0031	.0030	.0029	.0028	.0027	.0026
−2.6	.0047	.0045	.0044	.0043	.0041	.0040	.0039	.0038	.0037	.0036
−2.5	.0062	.0060	.0059	.0057	.0055	.0054	.0052	.0051	.0049	.0048
−2.4	.0082	.0080	.0078	.0075	.0073	.0071	.0069	.0068	.0066	.0064
−2.3	.0107	.0104	.0102	.0099	.0096	.0094	.0091	.0089	.0087	.0084
−2.2	.0139	.0136	.0132	.0129	.0125	.0122	.0119	.0116	.0113	.0110
−2.1	.0179	.0174	.0170	.0166	.0162	.0158	.0154	.0150	.0146	.0143
−2.0	.0228	.0222	.0217	.0212	.0207	.0202	.0197	.0192	.0188	.0183
−1.9	.0287	.0281	.0274	.0268	.0262	.0256	.0250	.0244	.0239	.0233
−1.8	.0359	.0351	.0344	.0336	.0329	.0322	.0314	.0307	.0301	.0294
−1.7	.0446	.0436	.0427	.0418	.0409	.0401	.0392	.0384	.0375	.0367
−1.6	.0548	.0537	.0526	.0516	.0505	.0495	.0485	.0475	.0465	.0455
−1.5	.0668	.0655	.0643	.0630	.0618	.0606	.0594	.0582	.0571	.0559
−1.4	.0808	.0793	.0778	.0764	.0749	.0735	.0721	.0708	.0694	.0681
−1.3	.0968	.0951	.0934	.0918	.0901	.0885	.0869	.0853	.0838	.0823
−1.2	.1151	.1131	.1112	.1093	.1075	.1056	.1038	.1020	.1003	.0985
−1.1	.1357	.1335	.1314	.1292	.1271	.1251	.1230	.1210	.1190	.1170
−1.0	.1587	.1562	.1539	.1515	.1492	.1469	.1446	.1423	.1401	.1379
−0.9	.1841	.1814	.1788	.1762	.1736	.1711	.1685	.1660	.1635	.1611
−0.8	.2119	.2090	.2061	.2033	.2005	.1977	.1949	.1922	.1894	.1867
−0.7	.2420	.2389	.2358	.2327	.2296	.2266	.2236	.2206	.2177	.2148
−0.6	.2743	.2709	.2676	.2643	.2611	.2578	.2546	.2514	.2483	.2451
−0.5	.3085	.3050	.3015	.2981	.2946	.2912	.2877	.2843	.2810	.2776
−0.4	.3446	.3409	.3372	.3336	.3300	.3264	.3228	.3192	.3156	.3121
−0.3	.3821	.3783	.3745	.3707	.3669	.3632	.3594	.3557	.3520	.3483
−0.2	.4207	.4168	.4129	.4090	.4052	.4013	.3974	.3936	.3897	.3859
−0.1	.4602	.4562	.4522	.4483	.4443	.4404	.4364	.4325	.4286	.4247
−0.0	.5000	.4960	.4920	.4880	.4840	.4801	.4761	.4721	.4681	.4641

Table A.2 Cumulative Normal Distribution (continued)

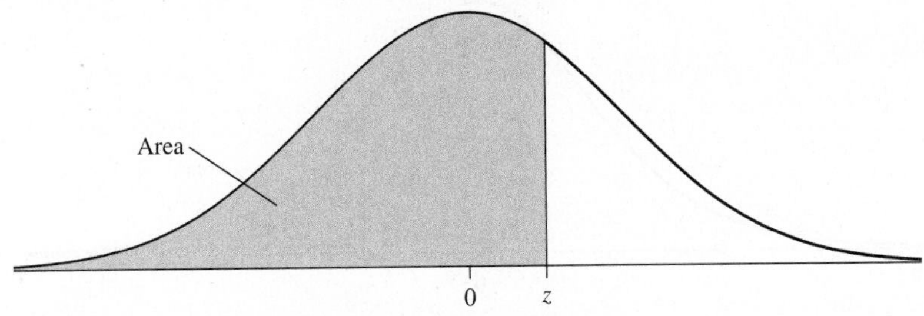

z	0.00	0.01	0.02	0.03	0.04	0.05	0.06	0.07	0.08	0.09
0.0	.5000	.5040	.5080	.5120	.5160	.5199	.5239	.5279	.5319	.5359
0.1	.5398	.5438	.5478	.5517	.5557	.5596	.5636	.5675	.5714	.5753
0.2	.5793	.5832	.5871	.5910	.5948	.5987	.6026	.6064	.6103	.6141
0.3	.6179	.6217	.6255	.6293	.6331	.6368	.6406	.6443	.6480	.6517
0.4	.6554	.6591	.6628	.6664	.6700	.6736	.6772	.6808	.6844	.6879
0.5	.6915	.6950	.6985	.7019	.7054	.7088	.7123	.7157	.7190	.7224
0.6	.7257	.7291	.7324	.7357	.7389	.7422	.7454	.7486	.7517	.7549
0.7	.7580	.7611	.7642	.7673	.7704	.7734	.7764	.7794	.7823	.7852
0.8	.7881	.7910	.7939	.7967	.7995	.8023	.8051	.8078	.8106	.8133
0.9	.8159	.8186	.8212	.8238	.8264	.8289	.8315	.8340	.8365	.8389
1.0	.8413	.8438	.8461	.8485	.8508	.8531	.8554	.8577	.8599	.8621
1.1	.8643	.8665	.8686	.8708	.8729	.8749	.8770	.8790	.8810	.8830
1.2	.8849	.8869	.8888	.8907	.8925	.8944	.8962	.8980	.8997	.9015
1.3	.9032	.9049	.9066	.9082	.9099	.9115	.9131	.9147	.9162	.9177
1.4	.9192	.9207	.9222	.9236	.9251	.9265	.9279	.9292	.9306	.9319
1.5	.9332	.9345	.9357	.9370	.9382	.9394	.9406	.9418	.9429	.9441
1.6	.9452	.9463	.9474	.9484	.9495	.9505	.9515	.9525	.9535	.9545
1.7	.9554	.9564	.9573	.9582	.9591	.9599	.9608	.9616	.9625	.9633
1.8	.9641	.9649	.9656	.9664	.9671	.9678	.9686	.9693	.9699	.9706
1.9	.9713	.9719	.9726	.9732	.9738	.9744	.9750	.9756	.9761	.9767
2.0	.9772	.9778	.9783	.9788	.9793	.9798	.9803	.9808	.9812	.9817
2.1	.9821	.9826	.9830	.9834	.9838	.9842	.9846	.9850	.9854	.9857
2.2	.9861	.9864	.9868	.9871	.9875	.9878	.9881	.9884	.9887	.9890
2.3	.9893	.9896	.9898	.9901	.9904	.9906	.9909	.9911	.9913	.9916
2.4	.9918	.9920	.9922	.9925	.9927	.9929	.9931	.9932	.9934	.9936
2.5	.9938	.9940	.9941	.9943	.9945	.9946	.9948	.9949	.9951	.9952
2.6	.9953	.9955	.9956	.9957	.9959	.9960	.9961	.9962	.9963	.9964
2.7	.9965	.9966	.9967	.9968	.9969	.9970	.9971	.9972	.9973	.9974
2.8	.9974	.9975	.9976	.9977	.9977	.9978	.9979	.9979	.9980	.9981
2.9	.9981	.9982	.9982	.9983	.9984	.9984	.9985	.9985	.9986	.9986
3.0	.9987	.9987	.9987	.9988	.9988	.9989	.9989	.9989	.9990	.9990
3.1	.9990	.9991	.9991	.9991	.9992	.9992	.9992	.9992	.9993	.9993
3.2	.9993	.9993	.9994	.9994	.9994	.9994	.9994	.9995	.9995	.9995
3.3	.9995	.9995	.9995	.9996	.9996	.9996	.9996	.9996	.9996	.9997
3.4	.9997	.9997	.9997	.9997	.9997	.9997	.9997	.9997	.9997	.9998
3.5	.9998	.9998	.9998	.9998	.9998	.9998	.9998	.9998	.9998	.9998
3.6	.9998	.9998	.9999	.9999	.9999	.9999	.9999	.9999	.9999	.9999
3.7 or more	.9999									

Table A.3 Critical Values for the Student's t Distribution

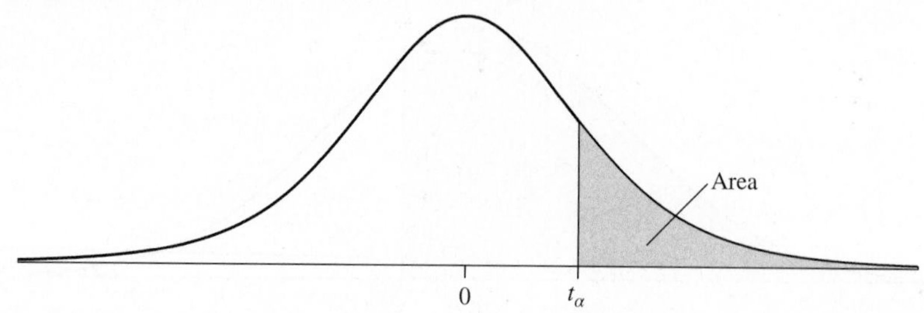

Degrees of Freedom	Area in Right Tail									
	0.40	**0.25**	**0.10**	**0.05**	**0.025**	**0.01**	**0.005**	**0.0025**	**0.001**	**0.0005**
1	0.325	1.000	3.078	6.314	12.706	31.821	63.657	127.321	318.309	636.619
2	0.289	0.816	1.886	2.920	4.303	6.965	9.925	14.089	22.327	31.599
3	0.277	0.765	1.638	2.353	3.182	4.541	5.841	7.453	10.215	12.924
4	0.271	0.741	1.533	2.132	2.776	3.747	4.604	5.598	7.173	8.610
5	0.267	0.727	1.476	2.015	2.571	3.365	4.032	4.773	5.893	6.869
6	0.265	0.718	1.440	1.943	2.447	3.143	3.707	4.317	5.208	5.959
7	0.263	0.711	1.415	1.895	2.365	2.998	3.499	4.029	4.785	5.408
8	0.262	0.706	1.397	1.860	2.306	2.896	3.355	3.833	4.501	5.041
9	0.261	0.703	1.383	1.833	2.262	2.821	3.250	3.690	4.297	4.781
10	0.260	0.700	1.372	1.812	2.228	2.764	3.169	3.581	4.144	4.587
11	0.260	0.697	1.363	1.796	2.201	2.718	3.106	3.497	4.025	4.437
12	0.259	0.695	1.356	1.782	2.179	2.681	3.055	3.428	3.930	4.318
13	0.259	0.694	1.350	1.771	2.160	2.650	3.012	3.372	3.852	4.221
14	0.258	0.692	1.345	1.761	2.145	2.624	2.977	3.326	3.787	4.140
15	0.258	0.691	1.341	1.753	2.131	2.602	2.947	3.286	3.733	4.073
16	0.258	0.690	1.337	1.746	2.120	2.583	2.921	3.252	3.686	4.015
17	0.257	0.689	1.333	1.740	2.110	2.567	2.898	3.222	3.646	3.965
18	0.257	0.688	1.330	1.734	2.101	2.552	2.878	3.197	3.610	3.922
19	0.257	0.688	1.328	1.729	2.093	2.539	2.861	3.174	3.579	3.883
20	0.257	0.687	1.325	1.725	2.086	2.528	2.845	3.153	3.552	3.850
21	0.257	0.686	1.323	1.721	2.080	2.518	2.831	3.135	3.527	3.819
22	0.256	0.686	1.321	1.717	2.074	2.508	2.819	3.119	3.505	3.792
23	0.256	0.685	1.319	1.714	2.069	2.500	2.807	3.104	3.485	3.768
24	0.256	0.685	1.318	1.711	2.064	2.492	2.797	3.091	3.467	3.745
25	0.256	0.684	1.316	1.708	2.060	2.485	2.787	3.078	3.450	3.725
26	0.256	0.684	1.315	1.706	2.056	2.479	2.779	3.067	3.435	3.707
27	0.256	0.684	1.314	1.703	2.052	2.473	2.771	3.057	3.421	3.690
28	0.256	0.683	1.313	1.701	2.048	2.467	2.763	3.047	3.408	3.674
29	0.256	0.683	1.311	1.699	2.045	2.462	2.756	3.038	3.396	3.659
30	0.256	0.683	1.310	1.697	2.042	2.457	2.750	3.030	3.385	3.646
31	0.256	0.682	1.309	1.696	2.040	2.453	2.744	3.022	3.375	3.633
32	0.255	0.682	1.309	1.694	2.037	2.449	2.738	3.015	3.365	3.622
33	0.255	0.682	1.308	1.692	2.035	2.445	2.733	3.008	3.356	3.611
34	0.255	0.682	1.307	1.691	2.032	2.441	2.728	3.002	3.348	3.601
35	0.255	0.682	1.306	1.690	2.030	2.438	2.724	2.996	3.340	3.591
36	0.255	0.681	1.306	1.688	2.028	2.434	2.719	2.990	3.333	3.582
37	0.255	0.681	1.305	1.687	2.026	2.431	2.715	2.985	3.326	3.574
38	0.255	0.681	1.304	1.686	2.024	2.429	2.712	2.980	3.319	3.566
39	0.255	0.681	1.304	1.685	2.023	2.426	2.708	2.976	3.313	3.558
40	0.255	0.681	1.303	1.684	2.021	2.423	2.704	2.971	3.307	3.551
50	0.255	0.679	1.299	1.676	2.009	2.403	2.678	2.937	3.261	3.496
60	0.254	0.679	1.296	1.671	2.000	2.390	2.660	2.915	3.232	3.460
80	0.254	0.678	1.292	1.664	1.990	2.374	2.639	2.887	3.195	3.416
100	0.254	0.677	1.290	1.660	1.984	2.364	2.626	2.871	3.174	3.390
200	0.254	0.676	1.286	1.653	1.972	2.345	2.601	2.839	3.131	3.340
z	0.253	0.674	1.282	1.645	1.960	2.326	2.576	2.807	3.090	3.291
	20%	50%	80%	90%	95%	98%	99%	99.5%	99.8%	99.9%
	Confidence Level									

Table A.4 Critical Values for the χ^2 Distribution

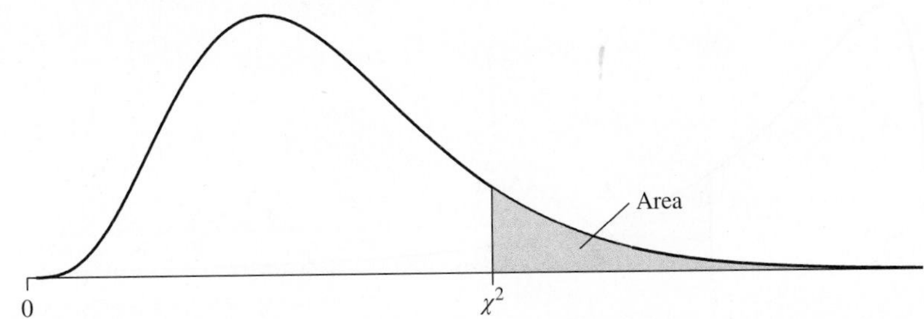

Degrees of	Area in Right Tail									
Freedom	**0.995**	**0.99**	**0.975**	**0.95**	**0.90**	**0.10**	**0.05**	**0.025**	**0.01**	**0.005**
1	0.000	0.000	0.001	0.004	0.016	2.706	3.841	5.024	6.635	7.879
2	0.010	0.020	0.051	0.103	0.211	4.605	5.991	7.378	9.210	10.597
3	0.072	0.115	0.216	0.352	0.584	6.251	7.815	9.348	11.345	12.838
4	0.207	0.297	0.484	0.711	1.064	7.779	9.488	11.143	13.277	14.860
5	0.412	0.554	0.831	1.145	1.610	9.236	11.070	12.833	15.086	16.750
6	0.676	0.872	1.237	1.635	2.204	10.645	12.592	14.449	16.812	18.548
7	0.989	1.239	1.690	2.167	2.833	12.017	14.067	16.013	18.475	20.278
8	1.344	1.646	2.180	2.733	3.490	13.362	15.507	17.535	20.090	21.955
9	1.735	2.088	2.700	3.325	4.168	14.684	16.919	19.023	21.666	23.589
10	2.156	2.558	3.247	3.940	4.865	15.987	18.307	20.483	23.209	25.188
11	2.603	3.053	3.816	4.575	5.578	17.275	19.675	21.920	24.725	26.757
12	3.074	3.571	4.404	5.226	6.304	18.549	21.026	23.337	26.217	28.300
13	3.565	4.107	5.009	5.892	7.042	19.812	22.362	24.736	27.688	29.819
14	4.075	4.660	5.629	6.571	7.790	21.064	23.685	26.119	29.141	31.319
15	4.601	5.229	6.262	7.261	8.547	22.307	24.996	27.488	30.578	32.801
16	5.142	5.812	6.908	7.962	9.312	23.542	26.296	28.845	32.000	34.267
17	5.697	6.408	7.564	8.672	10.085	24.769	27.587	30.191	33.409	35.718
18	6.265	7.015	8.231	9.390	10.865	25.989	28.869	31.526	34.805	37.156
19	6.844	7.633	8.907	10.117	11.651	27.204	30.144	32.852	36.191	38.582
20	7.434	8.260	9.591	10.851	12.443	28.412	31.410	34.170	37.566	39.997
21	8.034	8.897	10.283	11.591	13.240	29.615	32.671	35.479	38.932	41.401
22	8.643	9.542	10.982	12.338	14.041	30.813	33.924	36.781	40.289	42.796
23	9.260	10.196	11.689	13.091	14.848	32.007	35.172	38.076	41.638	44.181
24	9.886	10.856	12.401	13.848	15.659	33.196	36.415	39.364	42.980	45.559
25	10.520	11.524	13.120	14.611	16.473	34.382	37.652	40.646	44.314	46.928
26	11.160	12.198	13.844	15.379	17.292	35.563	38.885	41.923	45.642	48.290
27	11.808	12.879	14.573	16.151	18.114	36.741	40.113	43.195	46.963	49.645
28	12.461	13.565	15.308	16.928	18.939	37.916	41.337	44.461	48.278	50.993
29	13.121	14.256	16.047	17.708	19.768	39.087	42.557	45.722	49.588	52.336
30	13.787	14.953	16.791	18.493	20.599	40.256	43.773	46.979	50.892	53.672
31	14.458	15.655	17.539	19.281	21.434	41.422	44.985	48.232	52.191	55.003
32	15.134	16.362	18.291	20.072	22.271	42.585	46.194	49.480	53.486	56.328
33	15.815	17.074	19.047	20.867	23.110	43.745	47.400	50.725	54.776	57.648
34	16.501	17.789	19.806	21.664	23.952	44.903	48.602	51.966	56.061	58.964
35	17.192	18.509	20.569	22.465	24.797	46.059	49.802	53.203	57.342	60.275
36	17.887	19.233	21.336	23.269	25.643	47.212	50.998	54.437	58.619	61.581
37	18.586	19.96	22.106	24.075	26.492	48.363	52.192	55.668	59.893	62.883
38	19.289	20.691	22.878	24.884	27.343	49.513	53.384	56.896	61.162	64.181
39	19.996	21.426	23.654	25.695	28.196	50.660	54.572	58.120	62.428	65.476
40	20.707	22.164	24.433	26.509	29.051	51.805	55.758	59.342	63.691	66.766
41	21.421	22.906	25.215	27.326	29.907	52.949	56.942	60.561	64.950	68.053
42	22.138	23.650	25.999	28.144	30.765	54.090	58.124	61.777	66.206	69.336
43	22.859	24.398	26.785	28.965	31.625	55.230	59.304	62.990	67.459	70.616
44	23.584	25.148	27.575	29.787	32.487	56.369	60.481	64.201	68.710	71.893
45	24.311	25.901	28.366	30.612	33.350	57.505	61.656	65.410	69.957	73.166
50	27.991	29.707	32.357	34.764	37.689	63.167	67.505	71.420	76.154	79.490
60	35.534	37.485	40.482	43.188	46.459	74.397	79.082	83.298	88.379	91.952
70	43.275	45.442	48.758	51.739	55.329	85.527	90.531	95.023	100.425	104.215
80	51.172	53.540	57.153	60.391	64.278	96.578	101.879	106.629	112.329	116.321
90	59.196	61.754	65.647	69.126	73.291	107.565	113.145	118.136	124.116	128.299
100	67.328	70.065	74.222	77.929	82.358	118.498	124.342	129.561	135.807	140.169

Table A.5 Critical Values for the *F* Distribution

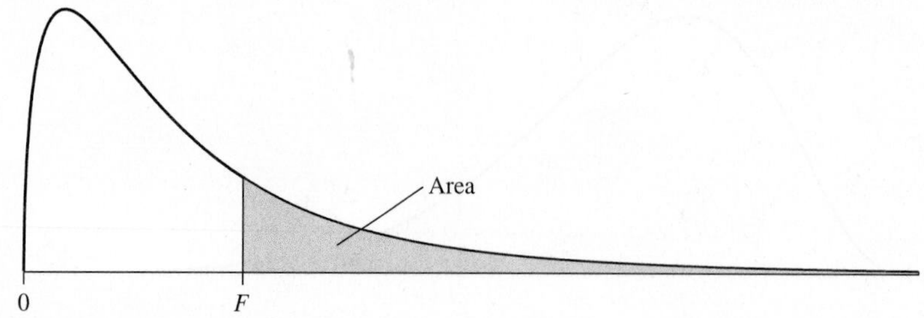

Denominator Degrees of Freedom	Area	Numerator Degrees of Freedom 1	2	3	4	5	6	7	8	9
1	0.100	39.86	49.50	53.59	55.83	57.24	58.20	58.91	59.44	59.86
1	0.050	161.45	199.50	215.71	224.58	230.16	233.99	236.77	238.88	240.54
1	0.025	647.79	799.50	864.16	899.58	921.85	937.11	948.22	956.66	963.28
1	0.010	4052.18	4999.50	5403.35	5624.58	5763.65	5858.99	5928.36	5981.07	6022.47
2	0.100	8.53	9.00	9.16	9.24	9.29	9.33	9.35	9.37	9.38
2	0.050	18.51	19.00	19.16	19.25	19.30	19.33	19.35	19.37	19.38
2	0.025	38.51	39.00	39.17	39.25	39.30	39.33	39.36	39.37	39.39
2	0.010	98.50	99.00	99.17	99.25	99.30	99.33	99.36	99.37	99.39
2	0.001	998.50	999.00	999.17	999.25	999.30	999.33	999.36	999.37	999.39
3	0.100	5.54	5.46	5.39	5.34	5.31	5.28	5.27	5.25	5.24
3	0.050	10.13	9.55	9.28	9.12	9.01	8.94	8.89	8.85	8.81
3	0.025	17.44	16.04	15.44	15.10	14.88	14.73	14.62	14.54	14.47
3	0.010	34.12	30.82	29.46	28.71	28.24	27.91	27.67	27.49	27.35
3	0.001	167.03	148.50	141.11	137.10	134.58	132.85	131.58	130.62	129.86
4	0.100	4.54	4.32	4.19	4.11	4.05	4.01	3.98	3.95	3.94
4	0.050	7.71	6.94	6.59	6.39	6.26	6.16	6.09	6.04	6.00
4	0.025	12.22	10.65	9.98	9.60	9.36	9.20	9.07	8.98	8.90
4	0.010	21.20	18.00	16.69	15.98	15.52	15.21	14.98	14.80	14.66
4	0.001	74.14	61.25	56.18	53.44	51.71	50.53	49.66	49.00	48.47
5	0.100	4.06	3.78	3.62	3.52	3.45	3.40	3.37	3.34	3.32
5	0.050	6.61	5.79	5.41	5.19	5.05	4.95	4.88	4.82	4.77
5	0.025	10.01	8.43	7.76	7.39	7.15	6.98	6.85	6.76	6.68
5	0.010	16.26	13.27	12.06	11.39	10.97	10.67	10.46	10.29	10.16
5	0.001	47.18	37.12	33.20	31.09	29.75	28.83	28.16	27.65	27.24
6	0.100	3.78	3.46	3.29	3.18	3.11	3.05	3.01	2.98	2.96
6	0.050	5.99	5.14	4.76	4.53	4.39	4.28	4.21	4.15	4.10
6	0.025	8.81	7.26	6.60	6.23	5.99	5.82	5.70	5.60	5.52
6	0.010	13.75	10.92	9.78	9.15	8.75	8.47	8.26	8.10	7.98
6	0.001	35.51	27.00	23.70	21.92	20.80	20.03	19.46	19.03	18.69
7	0.100	3.59	3.26	3.07	2.96	2.88	2.83	2.78	2.75	2.72
7	0.050	5.59	4.74	4.35	4.12	3.97	3.87	3.79	3.73	3.68
7	0.025	8.07	6.54	5.89	5.52	5.29	5.12	4.99	4.90	4.82
7	0.010	12.25	9.55	8.45	7.85	7.46	7.19	6.99	6.84	6.72
7	0.001	29.25	21.69	18.77	17.20	16.21	15.52	15.02	14.63	14.33
8	0.100	3.46	3.11	2.92	2.81	2.73	2.67	2.62	2.59	2.56
8	0.050	5.32	4.46	4.07	3.84	3.69	3.58	3.50	3.44	3.39
8	0.025	7.57	6.06	5.42	5.05	4.82	4.65	4.53	4.43	4.36
8	0.010	11.26	8.65	7.59	7.01	6.63	6.37	6.18	6.03	5.91
8	0.001	25.41	18.49	15.83	14.39	13.48	12.86	12.40	12.05	11.77
9	0.100	3.36	3.01	2.81	2.69	2.61	2.55	2.51	2.47	2.44
9	0.050	5.12	4.26	3.86	3.63	3.48	3.37	3.29	3.23	3.18
9	0.025	7.21	5.71	5.08	4.72	4.48	4.32	4.20	4.10	4.03
9	0.010	10.56	8.02	6.99	6.42	6.06	5.80	5.61	5.47	5.35
9	0.001	22.86	16.39	13.90	12.56	11.71	11.13	10.70	10.37	10.11

Table A.5 Critical Values for the *F* Distribution (continued)

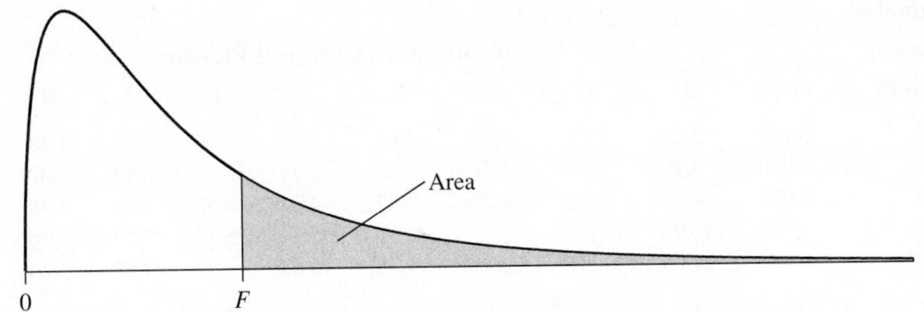

Denominator Degrees of Freedom	Area	Numerator Degrees of Freedom 10	12	15	20	25	30	40	50	60
1	0.100	60.19	60.71	61.22	61.74	62.05	62.26	62.53	62.69	62.79
1	0.050	241.88	243.91	245.95	248.01	249.26	250.10	251.14	251.77	252.20
1	0.025	968.63	976.71	984.87	993.10	998.08	1001.41	1005.60	1008.12	1009.80
1	0.010	6055.85	6106.32	6157.28	6208.73	6239.83	6260.65	6286.78	6302.52	6313.03
2	0.100	9.39	9.41	9.42	9.44	9.45	9.46	9.47	9.47	9.47
2	0.050	19.40	19.41	19.43	19.45	19.46	19.46	19.47	19.48	19.48
2	0.025	39.40	39.41	39.43	39.45	39.46	39.46	39.47	39.48	39.48
2	0.010	99.40	99.42	99.43	99.45	99.46	99.47	99.47	99.48	99.48
2	0.001	999.40	999.42	999.43	999.45	999.46	999.47	999.47	999.48	999.48
3	0.100	5.23	5.22	5.20	5.18	5.17	5.17	5.16	5.15	5.15
3	0.050	8.79	8.74	8.70	8.66	8.63	8.62	8.59	8.58	8.57
3	0.025	14.42	14.34	14.25	14.17	14.12	14.08	14.04	14.01	13.99
3	0.010	27.23	27.05	26.87	26.69	26.58	26.50	26.41	26.35	26.32
3	0.001	129.25	128.32	127.37	126.42	125.84	125.45	124.96	124.66	124.47
4	0.100	3.92	3.90	3.87	3.84	3.83	3.82	3.80	3.80	3.79
4	0.050	5.96	5.91	5.86	5.80	5.77	5.75	5.72	5.70	5.69
4	0.025	8.84	8.75	8.66	8.56	8.50	8.46	8.41	8.38	8.36
4	0.010	14.55	14.37	14.20	14.02	13.91	13.84	13.75	13.69	13.65
4	0.001	48.05	47.41	46.76	46.10	45.70	45.43	45.09	44.88	44.75
5	0.100	3.30	3.27	3.24	3.21	3.19	3.17	3.16	3.15	3.14
5	0.050	4.74	4.68	4.62	4.56	4.52	4.50	4.46	4.44	4.43
5	0.025	6.62	6.52	6.43	6.33	6.27	6.23	6.18	6.14	6.12
5	0.010	10.05	9.89	9.72	9.55	9.45	9.38	9.29	9.24	9.20
5	0.001	26.92	26.42	25.91	25.39	25.08	24.87	24.60	24.44	24.33
6	0.100	2.94	2.90	2.87	2.84	2.81	2.80	2.78	2.77	2.76
6	0.050	4.06	4.00	3.94	3.87	3.83	3.81	3.77	3.75	3.74
6	0.025	5.46	5.37	5.27	5.17	5.11	5.07	5.01	4.98	4.96
6	0.010	7.87	7.72	7.56	7.40	7.30	7.23	7.14	7.09	7.06
6	0.001	18.41	17.99	17.56	17.12	16.85	16.67	16.44	16.31	16.21
7	0.100	2.70	2.67	2.63	2.59	2.57	2.56	2.54	2.52	2.51
7	0.050	3.64	3.57	3.51	3.44	3.40	3.38	3.34	3.32	3.30
7	0.025	4.76	4.67	4.57	4.47	4.40	4.36	4.31	4.28	4.25
7	0.010	6.62	6.47	6.31	6.16	6.06	5.99	5.91	5.86	5.82
7	0.001	14.08	13.71	13.32	12.93	12.69	12.53	12.33	12.20	12.12
8	0.100	2.54	2.50	2.46	2.42	2.40	2.38	2.36	2.35	2.34
8	0.050	3.35	3.28	3.22	3.15	3.11	3.08	3.04	3.02	3.01
8	0.025	4.30	4.20	4.10	4.00	3.94	3.89	3.84	3.81	3.78
8	0.010	5.81	5.67	5.52	5.36	5.26	5.20	5.12	5.07	5.03
8	0.001	11.54	11.19	10.84	10.48	10.26	10.11	9.92	9.80	9.73
9	0.100	2.42	2.38	2.34	2.30	2.27	2.25	2.23	2.22	2.21
9	0.050	3.14	3.07	3.01	2.94	2.89	2.86	2.83	2.80	2.79
9	0.025	3.96	3.87	3.77	3.67	3.60	3.56	3.51	3.47	3.45
9	0.010	5.26	5.11	4.96	4.81	4.71	4.65	4.57	4.52	4.48
9	0.001	9.89	9.57	9.24	8.90	8.69	8.55	8.37	8.26	8.19

Table A.5 Critical Values for the *F* Distribution (continued)

Denominator Degrees of Freedom	Area	Numerator Degrees of Freedom								
		1	2	3	4	5	6	7	8	9
10	0.100	3.29	2.92	2.73	2.61	2.52	2.46	2.41	2.38	2.35
10	0.050	4.96	4.10	3.71	3.48	3.33	3.22	3.14	3.07	3.02
10	0.025	6.94	5.46	4.83	4.47	4.24	4.07	3.95	3.85	3.78
10	0.010	10.04	7.56	6.55	5.99	5.64	5.39	5.20	5.06	4.94
10	0.001	21.04	14.91	12.55	11.28	10.48	9.93	9.52	9.20	8.96
11	0.100	3.23	2.86	2.66	2.54	2.45	2.39	2.34	2.30	2.27
11	0.050	4.84	3.98	3.59	3.36	3.20	3.09	3.01	2.95	2.90
11	0.025	6.72	5.26	4.63	4.28	4.04	3.88	3.76	3.66	3.59
11	0.010	9.65	7.21	6.22	5.67	5.32	5.07	4.89	4.74	4.63
11	0.001	19.69	13.81	11.56	10.35	9.58	9.05	8.66	8.35	8.12
12	0.100	3.18	2.81	2.61	2.48	2.39	2.33	2.28	2.24	2.21
12	0.050	4.75	3.89	3.49	3.26	3.11	3.00	2.91	2.85	2.80
12	0.025	6.55	5.10	4.47	4.12	3.89	3.73	3.61	3.51	3.44
12	0.010	9.33	6.93	5.95	5.41	5.06	4.82	4.64	4.50	4.39
12	0.001	18.64	12.97	10.80	9.63	8.89	8.38	8.00	7.71	7.48
13	0.100	3.14	2.76	2.56	2.43	2.35	2.28	2.23	2.20	2.16
13	0.050	4.67	3.81	3.41	3.18	3.03	2.92	2.83	2.77	2.71
13	0.025	6.41	4.97	4.35	4.00	3.77	3.60	3.48	3.39	3.31
13	0.010	9.07	6.70	5.74	5.21	4.86	4.62	4.44	4.30	4.19
13	0.001	17.82	12.31	10.21	9.07	8.35	7.86	7.49	7.21	6.98
14	0.100	3.10	2.73	2.52	2.39	2.31	2.24	2.19	2.15	2.12
14	0.050	4.60	3.74	3.34	3.11	2.96	2.85	2.76	2.70	2.65
14	0.025	6.30	4.86	4.24	3.89	3.66	3.50	3.38	3.29	3.21
14	0.010	8.86	6.51	5.56	5.04	4.69	4.46	4.28	4.14	4.03
14	0.001	17.14	11.78	9.73	8.62	7.92	7.44	7.08	6.80	6.58
15	0.100	3.07	2.70	2.49	2.36	2.27	2.21	2.16	2.12	2.09
15	0.050	4.54	3.68	3.29	3.06	2.90	2.79	2.71	2.64	2.59
15	0.025	6.20	4.77	4.15	3.80	3.58	3.41	3.29	3.20	3.12
15	0.010	8.68	6.36	5.42	4.89	4.56	4.32	4.14	4.00	3.89
15	0.001	16.59	11.34	9.34	8.25	7.57	7.09	6.74	6.47	6.26
16	0.100	3.05	2.67	2.46	2.33	2.24	2.18	2.13	2.09	2.06
16	0.050	4.49	3.63	3.24	3.01	2.85	2.74	2.66	2.59	2.54
16	0.025	6.12	4.69	4.08	3.73	3.50	3.34	3.22	3.12	3.05
16	0.010	8.53	6.23	5.29	4.77	4.44	4.20	4.03	3.89	3.78
16	0.001	16.12	10.97	9.01	7.94	7.27	6.80	6.46	6.19	5.98
17	0.100	3.03	2.64	2.44	2.31	2.22	2.15	2.10	2.06	2.03
17	0.050	4.45	3.59	3.20	2.96	2.81	2.70	2.61	2.55	2.49
17	0.025	6.04	4.62	4.01	3.66	3.44	3.28	3.16	3.06	2.98
17	0.010	8.40	6.11	5.18	4.67	4.34	4.10	3.93	3.79	3.68
17	0.001	15.72	10.66	8.73	7.68	7.02	6.56	6.22	5.96	5.75
18	0.100	3.01	2.62	2.42	2.29	2.20	2.13	2.08	2.04	2.00
18	0.050	4.41	3.55	3.16	2.93	2.77	2.66	2.58	2.51	2.46
18	0.025	5.98	4.56	3.95	3.61	3.38	3.22	3.10	3.01	2.93
18	0.010	8.29	6.01	5.09	4.58	4.25	4.01	3.84	3.71	3.60
18	0.001	15.38	10.39	8.49	7.46	6.81	6.35	6.02	5.76	5.56
19	0.100	2.99	2.61	2.40	2.27	2.18	2.11	2.06	2.02	1.98
19	0.050	4.38	3.52	3.13	2.90	2.74	2.63	2.54	2.48	2.42
19	0.025	5.92	4.51	3.90	3.56	3.33	3.17	3.05	2.96	2.88
19	0.010	8.18	5.93	5.01	4.50	4.17	3.94	3.77	3.63	3.52
19	0.001	15.08	10.16	8.28	7.27	6.62	6.18	5.85	5.59	5.39

Table A.5 Critical Values for the *F* Distribution (continued)

Denominator Degrees of Freedom	Area	Numerator Degrees of Freedom 1	2	3	4	5	6	7	8	9
20	0.100	2.97	2.59	2.38	2.25	2.16	2.09	2.04	2.00	1.96
20	0.050	4.35	3.49	3.10	2.87	2.71	2.60	2.51	2.45	2.39
20	0.025	5.87	4.46	3.86	3.51	3.29	3.13	3.01	2.91	2.84
20	0.010	8.10	5.85	4.94	4.43	4.10	3.87	3.70	3.56	3.46
20	0.001	14.82	9.95	8.10	7.10	6.46	6.02	5.69	5.44	5.24
21	0.100	2.96	2.57	2.36	2.23	2.14	2.08	2.02	1.98	1.95
21	0.050	4.32	3.47	3.07	2.84	2.68	2.57	2.49	2.42	2.37
21	0.025	5.83	4.42	3.82	3.48	3.25	3.09	2.97	2.87	2.80
21	0.010	8.02	5.78	4.87	4.37	4.04	3.81	3.64	3.51	3.40
21	0.001	14.59	9.77	7.94	6.95	6.32	5.88	5.56	5.31	5.11
22	0.100	2.95	2.56	2.35	2.22	2.13	2.06	2.01	1.97	1.93
22	0.050	4.30	3.44	3.05	2.82	2.66	2.55	2.46	2.40	2.34
22	0.025	5.79	4.38	3.78	3.44	3.22	3.05	2.93	2.84	2.76
22	0.010	7.95	5.72	4.82	4.31	3.99	3.76	3.59	3.45	3.35
22	0.001	14.38	9.61	7.80	6.81	6.19	5.76	5.44	5.19	4.99
23	0.100	2.94	2.55	2.34	2.21	2.11	2.05	1.99	1.95	1.92
23	0.050	4.28	3.42	3.03	2.80	2.64	2.53	2.44	2.37	2.32
23	0.025	5.75	4.35	3.75	3.41	3.18	3.02	2.90	2.81	2.73
23	0.010	7.88	5.66	4.76	4.26	3.94	3.71	3.54	3.41	3.30
23	0.001	14.20	9.47	7.67	6.70	6.08	5.65	5.33	5.09	4.89
24	0.100	2.93	2.54	2.33	2.19	2.10	2.04	1.98	1.94	1.91
24	0.050	4.26	3.40	3.01	2.78	2.62	2.51	2.42	2.36	2.30
24	0.025	5.72	4.32	3.72	3.38	3.15	2.99	2.87	2.78	2.70
24	0.010	7.82	5.61	4.72	4.22	3.90	3.67	3.50	3.36	3.26
24	0.001	14.03	9.34	7.55	6.59	5.98	5.55	5.23	4.99	4.80
25	0.100	2.92	2.53	2.32	2.18	2.09	2.02	1.97	1.93	1.89
25	0.050	4.24	3.39	2.99	2.76	2.60	2.49	2.40	2.34	2.28
25	0.025	5.69	4.29	3.69	3.35	3.13	2.97	2.85	2.75	2.68
25	0.010	7.77	5.57	4.68	4.18	3.85	3.63	3.46	3.32	3.22
25	0.001	13.88	9.22	7.45	6.49	5.89	5.46	5.15	4.91	4.71
30	0.100	2.88	2.49	2.28	2.14	2.05	1.98	1.93	1.88	1.85
30	0.050	4.17	3.32	2.92	2.69	2.53	2.42	2.33	2.27	2.21
30	0.025	5.57	4.18	3.59	3.25	3.03	2.87	2.75	2.65	2.57
30	0.010	7.56	5.39	4.51	4.02	3.70	3.47	3.30	3.17	3.07
30	0.001	13.29	8.77	7.05	6.12	5.53	5.12	4.82	4.58	4.39
40	0.100	2.84	2.44	2.23	2.09	2.00	1.93	1.87	1.83	1.79
40	0.050	4.08	3.23	2.84	2.61	2.45	2.34	2.25	2.18	2.12
40	0.025	5.42	4.05	3.46	3.13	2.90	2.74	2.62	2.53	2.45
40	0.010	7.31	5.18	4.31	3.83	3.51	3.29	3.12	2.99	2.89
40	0.001	12.61	8.25	6.59	5.70	5.13	4.73	4.44	4.21	4.02
60	0.100	2.79	2.39	2.18	2.04	1.95	1.87	1.82	1.77	1.74
60	0.050	4.00	3.15	2.76	2.53	2.37	2.25	2.17	2.10	2.04
60	0.025	5.29	3.93	3.34	3.01	2.79	2.63	2.51	2.41	2.33
60	0.010	7.08	4.98	4.13	3.65	3.34	3.12	2.95	2.82	2.72
60	0.001	11.97	7.77	6.17	5.31	4.76	4.37	4.09	3.86	3.69
120	0.100	2.75	2.35	2.13	1.99	1.90	1.82	1.77	1.72	1.68
120	0.050	3.92	3.07	2.68	2.45	2.29	2.18	2.09	2.02	1.96
120	0.025	5.15	3.80	3.23	2.89	2.67	2.52	2.39	2.30	2.22
120	0.010	6.85	4.79	3.95	3.48	3.17	2.96	2.79	2.66	2.56
120	0.001	11.38	7.32	5.78	4.95	4.42	4.04	3.77	3.55	3.38

Table A.5 Critical Values for the *F* Distribution (continued)

Denominator Degrees of Freedom	Area	Numerator Degrees of Freedom								
		10	12	15	20	25	30	40	50	60
10	0.100	2.32	2.28	2.24	2.20	2.17	2.16	2.13	2.12	2.11
10	0.050	2.98	2.91	2.85	2.77	2.73	2.70	2.66	2.64	2.62
10	0.025	3.72	3.62	3.52	3.42	3.35	3.31	3.26	3.22	3.20
10	0.010	4.85	4.71	4.56	4.41	4.31	4.25	4.17	4.12	4.08
10	0.001	8.75	8.45	8.13	7.80	7.60	7.47	7.30	7.19	7.12
11	0.100	2.25	2.21	2.17	2.12	2.10	2.08	2.05	2.04	2.03
11	0.050	2.85	2.79	2.72	2.65	2.60	2.57	2.53	2.51	2.49
11	0.025	3.53	3.43	3.33	3.23	3.16	3.12	3.06	3.03	3.00
11	0.010	4.54	4.40	4.25	4.10	4.01	3.94	3.86	3.81	3.78
11	0.001	7.92	7.63	7.32	7.01	6.81	6.68	6.52	6.42	6.35
12	0.100	2.19	2.15	2.10	2.06	2.03	2.01	1.99	1.97	1.96
12	0.050	2.75	2.69	2.62	2.54	2.50	2.47	2.43	2.40	2.38
12	0.025	3.37	3.28	3.18	3.07	3.01	2.96	2.91	2.87	2.85
12	0.010	4.30	4.16	4.01	3.86	3.76	3.70	3.62	3.57	3.54
12	0.001	7.29	7.00	6.71	6.40	6.22	6.09	5.93	5.83	5.76
13	0.100	2.14	2.10	2.05	2.01	1.98	1.96	1.93	1.92	1.90
13	0.050	2.67	2.60	2.53	2.46	2.41	2.38	2.34	2.31	2.30
13	0.025	3.25	3.15	3.05	2.95	2.88	2.84	2.78	2.74	2.72
13	0.010	4.10	3.96	3.82	3.66	3.57	3.51	3.43	3.38	3.34
13	0.001	6.80	6.52	6.23	5.93	5.75	5.63	5.47	5.37	5.30
14	0.100	2.10	2.05	2.01	1.96	1.93	1.91	1.89	1.87	1.86
14	0.050	2.60	2.53	2.46	2.39	2.34	2.31	2.27	2.24	2.22
14	0.025	3.15	3.05	2.95	2.84	2.78	2.73	2.67	2.64	2.61
14	0.010	3.94	3.80	3.66	3.51	3.41	3.35	3.27	3.22	3.18
14	0.001	6.40	6.13	5.85	5.56	5.38	5.25	5.10	5.00	4.94
15	0.100	2.06	2.02	1.97	1.92	1.89	1.87	1.85	1.83	1.82
15	0.050	2.54	2.48	2.40	2.33	2.28	2.25	2.20	2.18	2.16
15	0.025	3.06	2.96	2.86	2.76	2.69	2.64	2.59	2.55	2.52
15	0.010	3.80	3.67	3.52	3.37	3.28	3.21	3.13	3.08	3.05
15	0.001	6.08	5.81	5.54	5.25	5.07	4.95	4.80	4.70	4.64
16	0.100	2.03	1.99	1.94	1.89	1.86	1.84	1.81	1.79	1.78
16	0.050	2.49	2.42	2.35	2.28	2.23	2.19	2.15	2.12	2.11
16	0.025	2.99	2.89	2.79	2.68	2.61	2.57	2.51	2.47	2.45
16	0.010	3.69	3.55	3.41	3.26	3.16	3.10	3.02	2.97	2.93
16	0.001	5.81	5.55	5.27	4.99	4.82	4.70	4.54	4.45	4.39
17	0.100	2.00	1.96	1.91	1.86	1.83	1.81	1.78	1.76	1.75
17	0.050	2.45	2.38	2.31	2.23	2.18	2.15	2.10	2.08	2.06
17	0.025	2.92	2.82	2.72	2.62	2.55	2.50	2.44	2.41	2.38
17	0.010	3.59	3.46	3.31	3.16	3.07	3.00	2.92	2.87	2.83
17	0.001	5.58	5.32	5.05	4.78	4.60	4.48	4.33	4.24	4.18
18	0.100	1.98	1.93	1.89	1.84	1.80	1.78	1.75	1.74	1.72
18	0.050	2.41	2.34	2.27	2.19	2.14	2.11	2.06	2.04	2.02
18	0.025	2.87	2.77	2.67	2.56	2.49	2.44	2.38	2.35	2.32
18	0.010	3.51	3.37	3.23	3.08	2.98	2.92	2.84	2.78	2.75
18	0.001	5.39	5.13	4.87	4.59	4.42	4.30	4.15	4.06	4.00
19	0.100	1.96	1.91	1.86	1.81	1.78	1.76	1.73	1.71	1.70
19	0.050	2.38	2.31	2.23	2.16	2.11	2.07	2.03	2.00	1.98
19	0.025	2.82	2.72	2.62	2.51	2.44	2.39	2.33	2.30	2.27
19	0.010	3.43	3.30	3.15	3.00	2.91	2.84	2.76	2.71	2.67
19	0.001	5.22	4.97	4.70	4.43	4.26	4.14	3.99	3.90	3.84

Table A.5 Critical Values for the *F* Distribution (continued)

Denominator Degrees of Freedom	Area	Numerator Degrees of Freedom 10	12	15	20	25	30	40	50	60
20	0.100	1.94	1.89	1.84	1.79	1.76	1.74	1.71	1.69	1.68
20	0.050	2.35	2.28	2.20	2.12	2.07	2.04	1.99	1.97	1.95
20	0.025	2.77	2.68	2.57	2.46	2.40	2.35	2.29	2.25	2.22
20	0.010	3.37	3.23	3.09	2.94	2.84	2.78	2.69	2.64	2.61
20	0.001	5.08	4.82	4.56	4.29	4.12	4.00	3.86	3.77	3.70
21	0.100	2.96	2.57	2.36	2.23	2.14	2.08	2.02	1.98	1.95
21	0.050	4.32	3.47	3.07	2.84	2.68	2.57	2.49	2.42	2.37
21	0.025	2.73	2.64	2.53	2.42	2.36	2.31	2.25	2.21	2.18
21	0.010	8.02	5.78	4.87	4.37	4.04	3.81	3.64	3.51	3.40
21	0.001	14.59	9.77	7.94	6.95	6.32	5.88	5.56	5.31	5.11
22	0.100	2.95	2.56	2.35	2.22	2.13	2.06	2.01	1.97	1.93
22	0.050	4.30	3.44	3.05	2.82	2.66	2.55	2.46	2.40	2.34
22	0.025	2.70	2.60	2.50	2.39	2.32	2.27	2.21	2.17	2.14
22	0.010	7.95	5.72	4.82	4.31	3.99	3.76	3.59	3.45	3.35
22	0.001	14.38	9.61	7.80	6.81	6.19	5.76	5.44	5.19	4.99
23	0.100	2.94	2.55	2.34	2.21	2.11	2.05	1.99	1.95	1.92
23	0.050	4.28	3.42	3.03	2.80	2.64	2.53	2.44	2.37	2.32
23	0.025	2.67	2.57	2.47	2.36	2.29	2.24	2.18	2.14	2.11
23	0.010	7.88	5.66	4.76	4.26	3.94	3.71	3.54	3.41	3.30
23	0.001	14.20	9.47	7.67	6.70	6.08	5.65	5.33	5.09	4.89
24	0.100	2.93	2.54	2.33	2.19	2.10	2.04	1.98	1.94	1.91
24	0.050	4.26	3.40	3.01	2.78	2.62	2.51	2.42	2.36	2.30
24	0.025	2.64	2.54	2.44	2.33	2.26	2.21	2.15	2.11	2.08
24	0.010	7.82	5.61	4.72	4.22	3.90	3.67	3.50	3.36	3.26
24	0.001	14.03	9.34	7.55	6.59	5.98	5.55	5.23	4.99	4.80
25	0.100	2.92	2.53	2.32	2.18	2.09	2.02	1.97	1.93	1.89
25	0.050	4.24	3.39	2.99	2.76	2.60	2.49	2.40	2.34	2.28
25	0.025	2.61	2.51	2.41	2.30	2.23	2.18	2.12	2.08	2.05
25	0.010	7.77	5.57	4.68	4.18	3.85	3.63	3.46	3.32	3.22
25	0.001	13.88	9.22	7.45	6.49	5.89	5.46	5.15	4.91	4.71
30	0.100	1.82	1.77	1.72	1.67	1.63	1.61	1.57	1.55	1.54
30	0.050	2.16	2.09	2.01	1.93	1.88	1.84	1.79	1.76	1.74
30	0.025	2.51	2.41	2.31	2.20	2.12	2.07	2.01	1.97	1.94
30	0.010	2.98	2.84	2.70	2.55	2.45	2.39	2.30	2.25	2.21
30	0.001	4.24	4.00	3.75	3.49	3.33	3.22	3.07	2.98	2.92
40	0.100	1.76	1.71	1.66	1.61	1.57	1.54	1.51	1.48	1.47
40	0.050	2.08	2.00	1.92	1.84	1.78	1.74	1.69	1.66	1.64
40	0.025	2.39	2.29	2.18	2.07	1.99	1.94	1.88	1.83	1.80
40	0.010	2.80	2.66	2.52	2.37	2.27	2.20	2.11	2.06	2.02
40	0.001	3.87	3.64	3.40	3.14	2.98	2.87	2.73	2.64	2.57
60	0.100	1.71	1.66	1.60	1.54	1.50	1.48	1.44	1.41	1.40
60	0.050	1.99	1.92	1.84	1.75	1.69	1.65	1.59	1.56	1.53
60	0.025	2.27	2.17	2.06	1.94	1.87	1.82	1.74	1.70	1.67
60	0.010	2.63	2.50	2.35	2.20	2.10	2.03	1.94	1.88	1.84
60	0.001	3.54	3.32	3.08	2.83	2.67	2.55	2.41	2.32	2.25
120	0.100	1.65	1.60	1.55	1.48	1.44	1.41	1.37	1.34	1.32
120	0.050	1.91	1.83	1.75	1.66	1.60	1.55	1.50	1.46	1.43
120	0.025	2.16	2.05	1.94	1.82	1.75	1.69	1.61	1.56	1.53
120	0.010	2.47	2.34	2.19	2.03	1.93	1.86	1.76	1.70	1.66
120	0.001	3.24	3.02	2.78	2.53	2.37	2.26	2.11	2.02	1.95

Table A.6 Critical Values for the Studentized Range

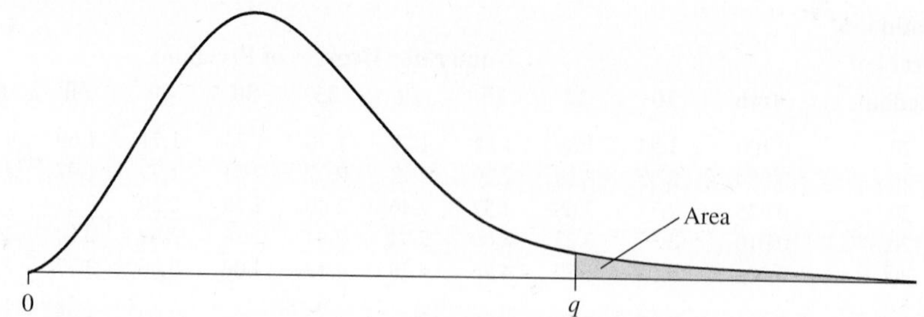

		I													
N − *I*	Area	2	3	4	5	6	7	8	9	10	11	12	13	14	15
1	0.10	8.93	13.44	16.36	18.49	20.15	21.51	22.64	23.62	24.48	25.24	25.92	26.54	27.10	27.62
1	0.05	17.97	26.98	32.82	37.08	40.41	43.12	45.40	47.36	49.07	50.59	51.96	53.20	54.33	55.36
1	0.01	90.03	135.0	164.3	185.6	202.2	215.8	227.2	237.0	245.6	253.2	260.0	266.2	271.8	277.0
2	0.10	4.13	5.73	6.77	7.54	8.14	8.63	9.05	9.41	9.72	10.01	10.26	10.49	10.70	10.89
2	0.05	6.08	8.33	9.80	10.88	11.74	12.44	13.03	13.54	13.99	14.39	14.75	15.08	15.38	15.65
2	0.01	14.04	19.02	22.29	24.72	26.63	28.20	29.53	30.68	31.69	32.59	33.40	34.13	34.81	35.43
3	0.10	3.33	4.47	5.20	5.74	6.16	6.51	6.81	7.06	7.29	7.49	7.67	7.83	7.98	8.12
3	0.05	4.50	5.91	6.82	7.50	8.04	8.48	8.85	9.18	9.46	9.72	9.95	10.15	10.35	10.52
3	0.01	8.26	10.62	12.17	13.33	14.24	15.00	15.64	16.20	16.69	17.13	17.53	17.89	18.22	18.52
4	0.10	3.01	3.98	4.59	5.03	5.39	5.68	5.93	6.14	6.33	6.49	6.65	6.78	6.91	7.02
4	0.05	3.93	5.04	5.76	6.29	6.71	7.05	7.35	7.60	7.83	8.03	8.21	8.37	8.52	8.66
4	0.01	6.51	8.12	9.17	9.96	10.58	11.10	11.55	11.93	12.27	12.57	12.84	13.09	13.32	13.53
5	0.10	2.85	3.72	4.26	4.66	4.98	5.24	5.46	5.65	5.82	5.97	6.10	6.22	6.34	6.44
5	0.05	3.64	4.60	5.22	5.67	6.03	6.33	6.58	6.80	6.99	7.17	7.32	7.47	7.60	7.72
5	0.01	5.70	6.98	7.80	8.42	8.91	9.32	9.67	9.97	10.24	10.48	10.70	10.89	11.08	11.24
6	0.10	2.75	3.56	4.07	4.44	4.73	4.97	5.17	5.34	5.50	5.64	5.76	5.87	5.98	6.07
6	0.05	3.46	4.34	4.90	5.30	5.63	5.90	6.12	6.32	6.49	6.65	6.79	6.92	7.03	7.14
6	0.01	5.24	6.33	7.03	7.56	7.97	8.32	8.61	8.87	9.10	9.30	9.48	9.65	9.81	9.95
7	0.10	2.68	3.45	3.93	4.28	4.55	4.78	4.97	5.14	5.28	5.41	5.53	5.64	5.74	5.83
7	0.05	3.34	4.16	4.68	5.06	5.36	5.61	5.82	6.00	6.16	6.30	6.43	6.55	6.66	6.76
7	0.01	4.95	5.92	6.54	7.01	7.37	7.68	7.94	8.17	8.37	8.55	8.71	8.86	9.00	9.12
8	0.10	2.63	3.37	3.83	4.17	4.43	4.65	4.83	4.99	5.13	5.25	5.36	5.46	5.56	5.64
8	0.05	3.26	4.04	4.53	4.89	5.17	5.40	5.60	5.77	5.92	6.05	6.18	6.29	6.39	6.48
8	0.01	4.75	5.64	6.20	6.62	6.96	7.24	7.47	7.68	7.86	8.03	8.18	8.31	8.44	8.55
9	0.10	2.59	3.32	3.76	4.08	4.34	4.54	4.72	4.87	5.01	5.13	5.23	5.33	5.42	5.51
9	0.05	3.20	3.95	4.41	4.76	5.02	5.24	5.43	5.59	5.74	5.87	5.98	6.09	6.19	6.28
9	0.01	4.60	5.43	5.96	6.35	6.66	6.91	7.13	7.33	7.49	7.65	7.78	7.91	8.03	8.13
10	0.10	2.56	3.27	3.70	4.02	4.26	4.47	4.64	4.78	4.91	5.03	5.13	5.23	5.32	5.40
10	0.05	3.15	3.88	4.33	4.65	4.91	5.12	5.30	5.46	5.60	5.72	5.83	5.93	6.03	6.11
10	0.01	4.48	5.27	5.77	6.14	6.43	6.67	6.87	7.05	7.21	7.36	7.49	7.60	7.71	7.81
11	0.10	2.54	3.23	3.66	3.96	4.20	4.40	4.57	4.71	4.84	4.95	5.05	5.15	5.23	5.31
11	0.05	3.11	3.82	4.26	4.57	4.82	5.03	5.20	5.35	5.49	5.61	5.71	5.81	5.90	5.98
11	0.01	4.39	5.15	5.62	5.97	6.25	6.48	6.67	6.84	6.99	7.13	7.25	7.36	7.46	7.56
12	0.10	2.52	3.20	3.62	3.92	4.16	4.35	4.51	4.65	4.78	4.89	4.99	5.08	5.16	5.24
12	0.05	3.08	3.77	4.20	4.51	4.75	4.95	5.12	5.27	5.39	5.51	5.61	5.71	5.80	5.88
12	0.01	4.32	5.05	5.50	5.84	6.10	6.32	6.51	6.67	6.81	6.94	7.06	7.17	7.26	7.36

Table A.6 Critical Values for the Studentized Range (continued)

		I													
$N-I$	**Area**	**2**	**3**	**4**	**5**	**6**	**7**	**8**	**9**	**10**	**11**	**12**	**13**	**14**	**15**
13	0.10	2.50	3.18	3.59	3.88	4.12	4.30	4.46	4.60	4.72	4.83	4.93	5.02	5.10	5.18
13	0.05	3.06	3.73	4.15	4.45	4.69	4.88	5.05	5.19	5.32	5.43	5.53	5.63	5.71	5.79
13	0.01	4.26	4.96	5.40	5.73	5.98	6.19	6.37	6.53	6.67	6.79	6.90	7.01	7.10	7.19
14	0.10	2.49	3.16	3.56	3.85	4.08	4.27	4.42	4.56	4.68	4.79	4.88	4.97	5.05	5.12
14	0.05	3.03	3.70	4.11	4.41	4.64	4.83	4.99	5.13	5.25	5.36	5.46	5.55	5.64	5.71
14	0.01	4.21	4.89	5.32	5.63	5.88	6.08	6.26	6.41	6.54	6.66	6.77	6.87	6.96	7.05
15	0.10	2.48	3.14	3.54	3.83	4.05	4.23	4.39	4.52	4.64	4.75	4.84	4.93	5.01	5.08
15	0.05	3.01	3.67	4.08	4.37	4.59	4.78	4.94	5.08	5.20	5.31	5.40	5.49	5.57	5.65
15	0.01	4.17	4.84	5.25	5.56	5.80	5.99	6.16	6.31	6.44	6.55	6.66	6.76	6.84	6.93
16	0.10	2.47	3.12	3.52	3.80	4.03	4.21	4.36	4.49	4.61	4.71	4.81	4.89	4.97	5.04
16	0.05	3.00	3.65	4.05	4.33	4.56	4.74	4.90	5.03	5.15	5.26	5.35	5.44	5.52	5.59
16	0.01	4.13	4.79	5.19	5.49	5.72	5.92	6.08	6.22	6.35	6.46	6.56	6.66	6.74	6.82
17	0.10	2.46	3.11	3.50	3.78	4.00	4.18	4.33	4.46	4.58	4.68	4.77	4.86	4.93	5.01
17	0.05	2.98	3.63	4.02	4.30	4.52	4.70	4.86	4.99	5.11	5.21	5.31	5.39	5.47	5.54
17	0.01	4.10	4.74	5.14	5.43	5.66	5.85	6.01	6.15	6.27	6.38	6.48	6.57	6.66	6.73
18	0.10	2.45	3.10	3.49	3.77	3.98	4.16	4.31	4.44	4.55	4.65	4.75	4.83	4.90	4.98
18	0.05	2.97	3.61	4.00	4.28	4.49	4.67	4.82	4.96	5.07	5.17	5.27	5.35	5.43	5.50
18	0.01	4.07	4.70	5.09	5.38	5.60	5.79	5.94	6.08	6.20	6.31	6.41	6.50	6.58	6.65
19	0.10	2.45	3.09	3.47	3.75	3.97	4.14	4.29	4.42	4.53	4.63	4.72	4.80	4.88	4.95
19	0.05	2.96	3.59	3.98	4.25	4.47	4.65	4.79	4.92	5.04	5.14	5.23	5.31	5.39	5.46
19	0.01	4.05	4.67	5.05	5.33	5.55	5.73	5.89	6.02	6.14	6.25	6.34	6.43	6.51	6.58
20	0.10	2.44	3.08	3.46	3.74	3.95	4.12	4.27	4.40	4.51	4.61	4.70	4.78	4.85	4.92
20	0.05	2.95	3.58	3.96	4.23	4.45	4.62	4.77	4.90	5.01	5.11	5.20	5.28	5.36	5.43
20	0.01	4.02	4.64	5.02	5.29	5.51	5.69	5.84	5.97	6.09	6.19	6.28	6.37	6.45	6.52
24	0.10	2.42	3.05	3.42	3.69	3.90	4.07	4.21	4.34	4.44	4.54	4.63	4.71	4.78	4.85
24	0.05	2.92	3.53	3.90	4.17	4.37	4.54	4.68	4.81	4.92	5.01	5.10	5.18	5.25	5.32
24	0.01	3.96	4.55	4.91	5.17	5.37	5.54	5.69	5.81	5.92	6.02	6.11	6.19	6.26	6.33
30	0.10	2.40	3.02	3.39	3.65	3.85	4.02	4.16	4.28	4.38	4.47	4.56	4.64	4.71	4.77
30	0.05	2.89	3.49	3.85	4.10	4.30	4.46	4.60	4.72	4.82	4.92	5.00	5.08	5.15	5.21
30	0.01	3.89	4.45	4.80	5.05	5.24	5.40	5.54	5.65	5.76	5.85	5.93	6.01	6.08	6.14
40	0.10	2.38	2.99	3.35	3.60	3.80	3.96	4.10	4.21	4.32	4.41	4.49	4.56	4.63	4.69
40	0.05	2.86	3.44	3.79	4.04	4.23	4.39	4.52	4.63	4.73	4.82	4.90	4.98	5.04	5.11
40	0.01	3.82	4.37	4.70	4.93	5.11	5.26	5.39	5.50	5.60	5.69	5.76	5.83	5.90	5.96
60	0.10	2.36	2.96	3.31	3.56	3.75	3.91	4.04	4.16	4.25	4.34	4.42	4.49	4.56	4.62
60	0.05	2.83	3.40	3.74	3.98	4.16	4.31	4.44	4.55	4.65	4.73	4.81	4.88	4.94	5.00
60	0.01	3.76	4.28	4.59	4.82	4.99	5.13	5.25	5.36	5.45	5.53	5.60	5.67	5.73	5.78
120	0.10	2.34	2.93	3.28	3.52	3.71	3.86	3.99	4.10	4.19	4.28	4.35	4.42	4.48	4.54
120	0.05	2.80	3.36	3.68	3.92	4.10	4.24	4.36	4.47	4.56	4.64	4.71	4.78	4.84	4.90
120	0.01	3.70	4.20	4.50	4.71	4.87	5.01	5.12	5.21	5.30	5.37	5.44	5.50	5.56	5.61
∞	0.10	2.33	2.90	3.24	3.48	3.66	3.81	3.93	4.04	4.13	4.21	4.28	4.35	4.41	4.47
∞	0.05	2.77	3.31	3.63	3.86	4.03	4.17	4.29	4.39	4.47	4.55	4.62	4.68	4.74	4.80
∞	0.01	3.64	4.12	4.40	4.60	4.76	4.88	4.99	5.08	5.16	5.23	5.29	5.35	5.40	5.45

Table A.7 Critical Values for the Sign Test

	One-tailed			
	$\alpha = 0.005$	$\alpha = 0.01$	$\alpha = 0.025$	$\alpha = 0.05$
	Two-tailed			
n	$\alpha = 0.01$	$\alpha = 0.02$	$\alpha = 0.05$	$\alpha = 0.10$
5	*	*	*	0
6	*	*	0	0
7	*	0	0	0
8	0	0	0	1
9	0	0	1	1
10	0	0	1	1
11	0	1	1	2
12	1	1	2	2
13	1	1	2	3
14	1	2	3	3
15	2	2	3	3
16	2	2	3	4
17	2	3	4	4
18	3	3	4	5
19	3	4	4	5
20	3	4	5	5
21	4	4	5	6
22	4	5	5	6
23	4	5	6	7
24	5	5	6	7
25	5	6	6	7

*It is impossible for the test statistic to be in the critical region.

Table A.8 Critical Values for the Signed-Rank Test

n	$\alpha = 0.10$	$\alpha = 0.05$	$\alpha = 0.02$	$\alpha = 0.01$
5	1	*	*	*
6	2	1	*	*
7	4	2	0	*
8	6	4	2	0
9	8	6	3	2
10	11	8	5	3
11	14	11	7	5
12	17	14	10	7
13	21	17	13	10
14	26	21	16	13
15	30	25	20	16
16	36	30	24	19
17	41	35	28	23
18	47	40	33	28
19	54	46	38	32
20	60	52	43	37
21	68	59	49	43
22	75	66	56	49
23	83	73	62	55
24	92	81	69	61
25	101	90	77	68
26	110	98	85	76
27	120	107	93	84
28	130	117	102	92
29	141	127	111	100
30	152	137	120	109

*It is impossible for the test statistic to be in the critical region.

Appendix B

TI-84 PLUS Stat Wizards

Stat Wizards is a feature on TI-84 PLUS calculators running operating system OS 2.55 MP or higher. This feature provides a wizard interface for selected commands and functions. The latest operating system may be downloaded from https://education.ti.com/en/downloads-and-activities.

The Stat Wizards may be turned on or off through an option on the **MODE** screen. The MATHPRINT/CLASSIC options affect the appearance of the calculator output. These options also appear on this screen. The screenshots in this text are generated using the **CLASSIC** setting.

```
MATHPRINT CLASSIC
NORMAL SCI ENG
FLOAT 0 1 2 3 4 5 6 7 8 9
RADIAN DEGREE
FUNCTION PARAMETRIC POLAR SEQ
THICK DOT-THICK THIN DOT-THIN
SEQUENTIAL SIMUL
REAL a+bi re^(θi)
FULL HORIZONTAL GRAPH-TABLE
FRACTION TYPE: n/d Un/d
ANSWERS: AUTO DEC FRAC-APPROX
GO TO 2ND FORMAT GRAPH: NO YES
STAT DIAGNOSTICS: OFF ON
STAT WIZARDS: ON OFF
SET CLOCK 03/01/17 8:18AM
```

Wizards are available for all commands in the [STAT]CALC menu and the DISTR menu.

```
EDIT CALC TESTS
1:1-Var Stats
2:2-Var Stats
3:Med-Med
4:LinReg(ax+b)
5:QuadReg
6:CubicReg
7:QuartReg
8:LinReg(a+bx)
9↓LnReg
```

```
DISTR DRAW
1:normalpdf(
2:normalcdf(
3:invNorm(
4:invT(
5:tpdf(
6:tcdf(
7:χ²pdf(
8:χ²cdf(
9↓Fpdf(
```

1-Var Stats

The 1-Var Stats wizard is accessed by selecting the 1-Var Stats option under the [STAT] CALC menu. L1 is the default setting for the List field. FreqList is an optional argument. FreqList accepts list names only. To run the command, select Calculate and press ENTER.

```
EDIT CALC TESTS
1:1-Var Stats
2:2-Var Stats
3:Med-Med
4:LinReg(ax+b)
5:QuadReg
6:CubicReg
7:QuartReg
8:LinReg(a+bx)
9↓LnReg
```

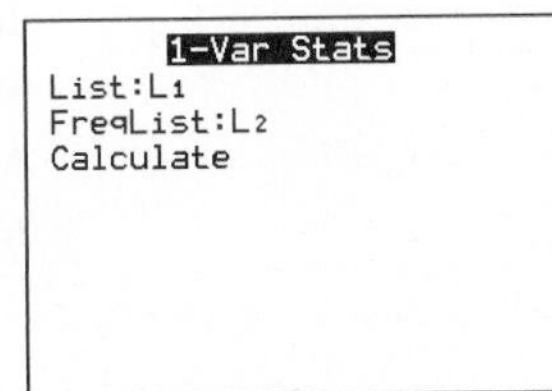

LinReg(a + bx)

The LinReg(a + bx) wizard is accessed by selecting the LinReg(a + bx) option under the [STAT] CALC menu. L1 and L2 are the default settings for the Xlist and Ylist fields. FreqList is an optional argument. A function name such as Y1 may be entered in the Store RegEQ field as a location to store the regression equation. To run the command, select Calculate and press ENTER.

```
EDIT CALC TESTS
1:1-Var Stats
2:2-Var Stats
3:Med-Med
4:LinReg(ax+b)
5:QuadReg
6:CubicReg
7:QuartReg
8:LinReg(a+bx)
9↓LnReg
```

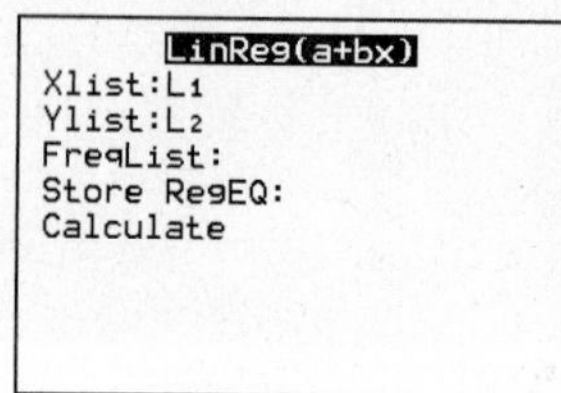

binompdf/binomcdf

The wizards for `binompdf` or `binomcdf` are accessed by selecting either the `binompdf` or `binomcdf` option under the `DISTR` menu. After entering the values for n, p, and x in the `trials`, `p`, and `x value` fields, select `Paste` and press ENTER to paste the command on the home screen.

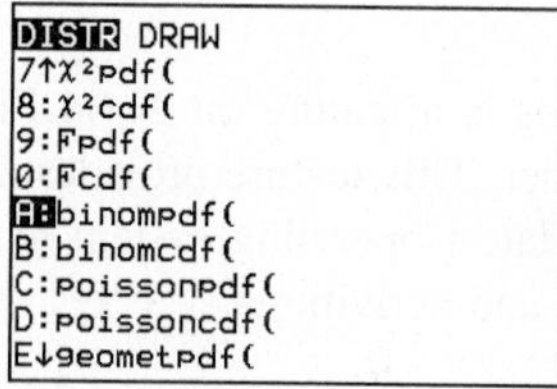

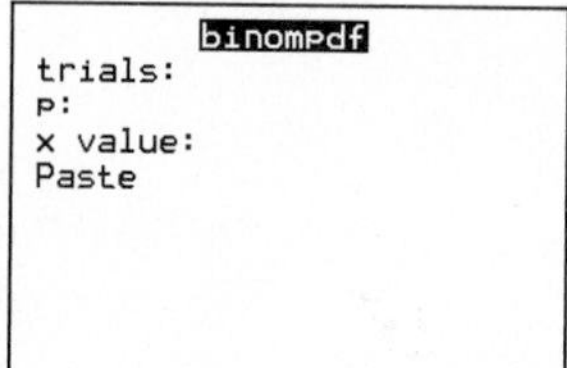

poissonpdf/poissoncdf

The wizards for `poissonpdf` or `poissoncdf` are accessed by selecting either the `poissonpdf` or `poissoncdf` option under the `DISTR` menu. After entering the values for λ and x in the λ and `x value` fields, select `Paste` and press ENTER to paste the command on the home screen.

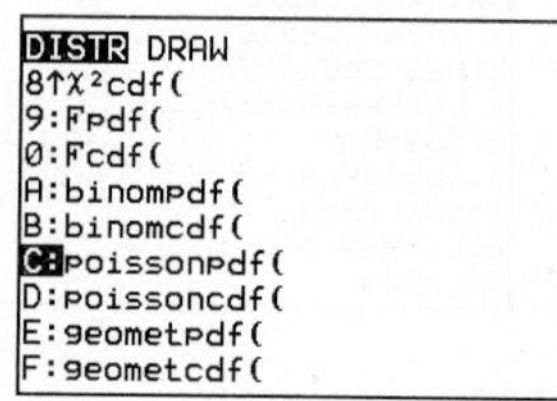

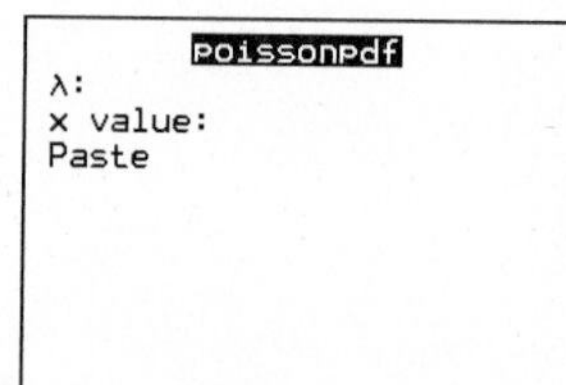

normalcdf

The `normalcdf` wizard is accessed by selecting the `normalcdf` option under the `DISTR` menu. The lower and upper endpoints are entered in the `lower` and `upper` fields. The default for the `lower` field is -1E99. The mean and standard deviation are entered in the μ and σ fields. The default for the μ field is 0, and the default for the σ field is 1. After entering all values, select `Paste` and press ENTER to paste the command on the home screen.

```
DISTR DRAW
1:normalpdf(
2:normalcdf(
3:invNorm(
4:invT(
5:tpdf(
6:tcdf(
7:χ²pdf(
8:χ²cdf(
9↓Fpdf(
```

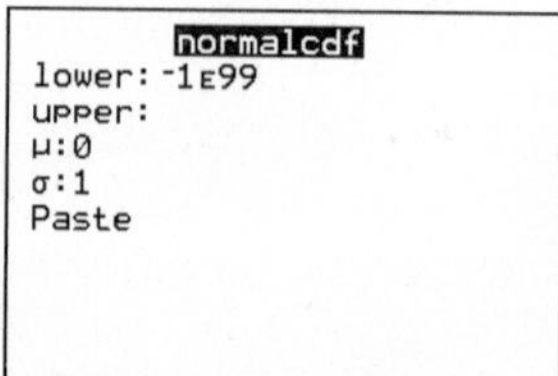

invNorm

The invNorm wizard is accessed by selecting the invNorm option under the DISTR menu. The area to the left of the desired value is entered in the area field. The mean and standard deviation are entered in the μ and σ fields. The default for the μ field is 0, and the default for the σ field is 1. After entering all values, select Paste and press ENTER to paste the command on the home screen.

```
DISTR DRAW
1:normalpdf(
2:normalcdf(
3:invNorm(
4:invT(
5:tpdf(
6:tcdf(
7:χ²pdf(
8:χ²cdf(
9↓Fpdf(
```

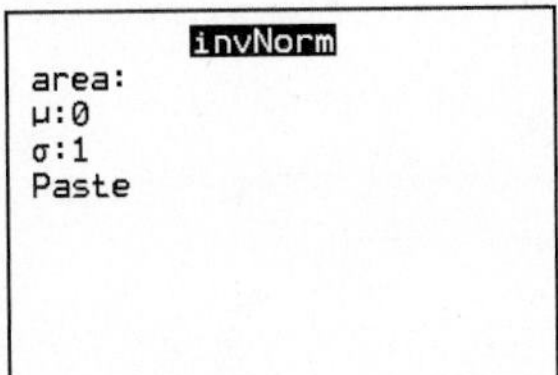

Answers to Selected Exercises

CHAPTER 1

Section 1.1

Exercises 1–6 are the Check Your Understanding exercises for this section. Answers to these exercises are on page 12.

7. population

8. sample

9. simple random sample

10. sample of convenience

11. cluster

12. stratified

13. False

14. False

15. True

16. True

17. Statistic

18. Parameter

19. Parameter

20. Statistic

21. Parameter

22. Statistic

23. Answers will vary.

24. Answers will vary. Stratified sample

25. Answers will vary. Cluster sample

26. Answers will vary. Systematic sample

27. Stratified sample

28. Sample of convenience

29. Cluster sample

30. Systematic sample

31. Voluntary response sample

32. Cluster sample

33. Sample of convenience

34. Simple random sample

35. Stratified sample

36. Cluster sample

37. Simple random sample

38. Stratified sample

39. Systematic sample

40. Voluntary response sample

41. Simple random sample

42. Voluntary response sample

43. It will be necessary to draw a sample of convenience. There is no list of all headache sufferers from which to draw a simple random sample.

44. It is feasible to draw a simple random sample of students from a list of all students enrolled.

45. Answers will vary. A simple random sample could be drawn from a list of all registered voters in the town.

46. Answers will vary. A systematic sample could be used in which every tenth item was sampled.

47. Answers will vary. A stratified sample, consisting of simple random samples of 100 men and 100 women, could be drawn.

48. Answers will vary. A simple random sample of tax forms could be drawn.

49. Answers will vary.

50. Let's say we will sample every kth item. This determines $k - 1$ clusters; the first cluster consists of items $1, k + 1, 2k + 1, \ldots$; the second cluster consists of items $2, k + 2, 2k + 2, \ldots$; and so on. Choosing a systematic sample is like choosing one of these clusters at random.

Section 1.2

Exercises 1–5 are the Check Your Understanding exercises for this section. Answers to these exercises are on page 19.

6. variables

7. data

8. qualitative

9. Quantitative

10. nominal; ordinal

11. discrete

12. Continuous

13. False

14. True

15. True

16. True

17. Qualitative

18. Quantitative

19. Quantitative

20. Qualitative

21. Quantitative

22. Quantitative

23. Qualitative

24. Qualitative

25. Qualitative

26. Quantitative

27. Ordinal

28. Nominal

29. Ordinal

30. Ordinal

31. Nominal

32. Nominal

33. Nominal

34. Ordinal

35. Continuous

36. Continuous

37. Discrete

38. Discrete

39. Continuous

40. Discrete

41. Discrete

42. Continuous

43. Interval

44. Ratio

45. Ratio

46. Interval

47. Interval

48. Ratio

49. Interval

50. Ratio

51. **a.** Ordinal
b. Qualitative

52. **a.** Discrete
b. Quantitative

53. Ordinal

54. Quantitative

55. Nominal

56. Ordinal

57. **a.** Game Title, Developer
b. Initial Release Year, Copies Sold
c. Developer **d.** Game Title

58. **a.** Movie Title, Studio
b. Release Year, Ticket Sales, Running Time
c. Studio **d.** Movie Title

59. **a.** Ordinal
b. Yes, it reflects a more favorable opinion of the construction of a new shopping mall.
c. No, we cannot say that Jason's opinion is twice as favorable.
d. Quantitative
e. Yes, Brenda's answer reflects the ownership of more cars, and specifically, the ownership of twice as many cars.
f. Nominal
g. No, Brenda's answer reflects neither more of something nor twice as much of something.

Section 1.3

Exercises 1–4 are the Check Your Understanding exercises for this section. Answers to these exercises are on page 27.

5. randomized

6. double-blind

7. observational

8. confounder

9. prospective

10. cohort

11. True

12. False

13. False

14. False

15. True

16. True

17. **a.** Randomized experiment
b. Yes, because the assignment to treatments is made at random, there is no systematic difference between the groups other than the drug taken that can explain the difference in pain relief.

18. **a.** Observational study
b. No, because this is an observational study, the results may be due to confounding. For example, those in the low-exercise group may be older and have higher blood pressure as a result of age.

19. **a.** Randomized experiment
b. Yes, because the assignment to treatments is made at random, there is no systematic difference between the groups other than the amount of exercise that can explain the difference in blood pressure.

20. **a.** The four types of fertilizer **b.** Randomized experiment
c. Yes, because the assignment to treatment is made at random.

21. An observational study will be necessary, because one can't assign people to live in areas with high pollution levels.

22. A randomized experiment is possible, with cold sufferers being assigned randomly to the old or new drug.

23. The result may be due to confounding. Areas with denser populations may have both more crime and more taxicabs.

24. The result could be due to confounding. For example, more older people may live in larger houses than younger people and also have less chance of recovering from a heart attack.

25. **a.** False **b.** True

26. Children with larger vocabularies tend to be older, and thus taller.

27. **a.** Heart rate **b.** Maternal smoking
c. Cohort **d.** Prospective
e. Yes. The level of prenatal care may differ between smoking and nonsmoking mothers.

28. **a.** Whether respiratory problems were experienced
b. The level of formaldehyde in the classroom
c. A cohort study **d.** Cross-sectional
e. It is unlikely. For confounding to be an issue, children with more sensitive respiratory systems would have to have been more likely to be assigned to classrooms with higher exposure.

29. **a.** Yes, because the subjects were randomly assigned to treatment.
b. If a doctor knew whether a child had received the vaccine, it might influence the diagnosis.
c. It could be due to confounding. The children who received the placebo were more likely to be middle- or upper-income than those who did not participate, and this may be the reason that the rate of polio was higher.

30. a. No b. No
    c. The treatment group consisted of children who had consent to participate; middle- and upper-income children were overrepresented in this group. Therefore, the children in the treatment group tended to be more susceptible to polio than those in the control group.
31. Smokers who quit are less healthy than smokers who don't.
32. Smokers drink more on the average than nonsmokers.

Section 1.4

Exercises 1 and 2 are the Check Your Understanding exercises for this section. Answers to these exercises are on page 31.

3. Voluntary response surveys
4. nonresponders
5. population
6. False
7. True
8. False
9. Nonresponse bias
10. Voluntary response bias
11. Self-interest bias
12. Self-interest bias
13. Voluntary response bias
14. Social acceptability bias
15. Nonresponse bias
16. Social acceptability bias
17. a. No
    b. No. Both questions are leading. The first question leads to a "yes" response, and the second leads to a "no" response.
18. Yes. People may be less likely to answer the phone when they don't recognize the number that is calling them.
19. Yes. People who do not have landline phones may tend to have different opinions on some issues than people who do have landline phones.
20. Yes. People may be more likely to pick the first choice.
21. a. The poll oversampled higher-income people.
    b. Nonresponse bias. The response rate was low—only 23%. This results in nonresponse bias.
    c. A sample that is not drawn by a valid method can produce misleading results, even when it is large.

CHAPTER 1 Quiz

1. Answers will vary.
2. Qualitative
3. True
4. Continuous
5. False
6. Stratified sample
7. acceptable
8. Sample of convenience
9. True
10. Observational study
11. Randomized experiment
12. differences in treatment
13. Seniors may be more likely to have better preparation for the class than sophomores.
14. True
15. Not reliable; this is a voluntary response survey, so the people who respond tend to hold stronger opinions than others.

CHAPTER 1 Review Exercises

1. Quantitative
2. Nominal
3. Continuous
4. a. True b. True c. False
5. Stratified sample
6. Voluntary response sample
7. Cluster sample
8. Simple random sample
9. a. Observational study
   b. Yes. People who live in areas with fluoridated water may have different dental habits than those who live in areas without fluoridated water.
10. a. Randomized experiment
    b. Because this is a randomized experiment, the results are unlikely to be due to confounding.
11. a. Observational study
    b. Yes. People who use the phone while driving may be more careless in general than those who do not.
12. a. Randomized experiment
    b. Because this is a randomized experiment, the results are unlikely to be due to confounding.
13. The sample is a voluntary response sample.
14. Nonresponse bias; people who are still alive are not included in the sample.
15. There is a considerable level of nonresponse bias.

CHAPTER 1 Case Study

1. 450
2. 41
3. 9.1%
4. 43
5. 2
6. 4.7%
7. Yes
8. The high-exposure people and the school-return people are the same people.
9. The low-exposure people and the mail-return people are the same people.
10. People who respond by mail will be responding during a period of lower PM.
11. People with symptoms may tend to respond earlier; therefore, people with symptoms are more likely to be school-return people.
12. There would be no tendency for people with symptoms to respond earlier.

CHAPTER 2

Section 2.1

Exercises 1–4 are the Check Your Understanding exercises for this section. Answers to these exercises are on pages 50–51.

5. frequency

6. relative frequency

7. Pareto chart

8. pie chart

9. False

10. True

11. True

12. False

13. a. Meat, poultry, fish, and eggs **b.** False **c.** True

14. a. Type O **b.** False **c.** True

15. a.

b.

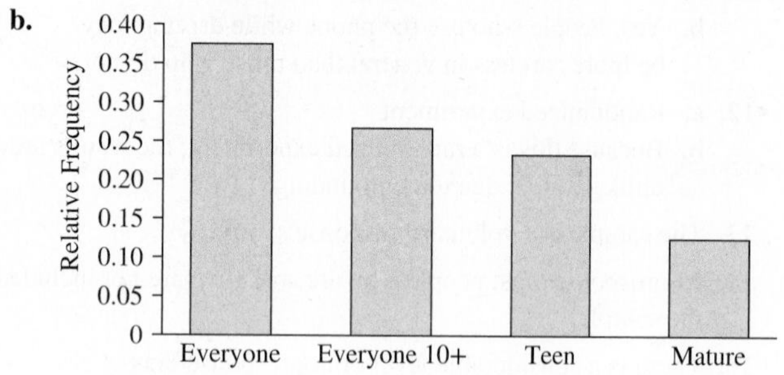

c. Everyone (E) **d.** False **e.** True

16. a.

b.

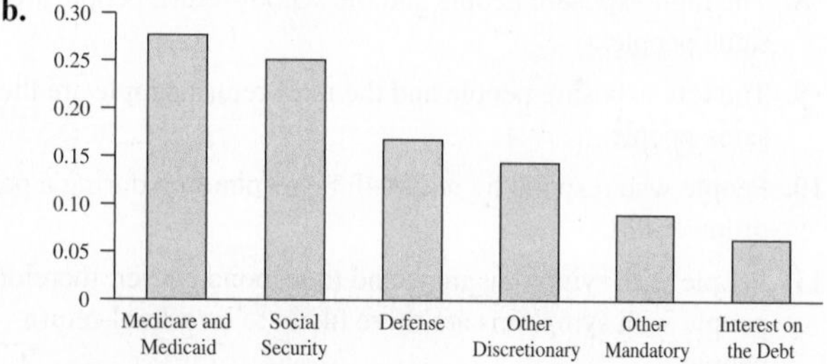

c. Medicare and Medicaid

d. 68.8%

17. a. West, South

b. Northeast, Midwest

c. True

d. False

18. a. The game

b. True

c. False

d. True

19. a.

b.

Quarter	Relative Frequency
Jan.–Mar. 2015	0.058
Apr.–Jun. 2015	0.058
Jul.–Sep. 2015	0.062
Oct.–Dec. 2015	0.069
Jan.–Mar. 2016	0.057
Apr.–Jun. 2016	0.060
Jul.–Sep. 2016	0.063
Oct.–Dec. 2016	0.074
Jan.–Mar. 2017	0.060
Apr.–Jun. 2017	0.060
Jul.–Sep. 2017	0.065
Oct.–Dec. 2017	0.068
Jan.–Mar. 2018	0.060
Apr.–Jun. 2018	0.059
Jul.–Sep. 2018	0.061
Oct.–Dec. 2018	0.065

c.

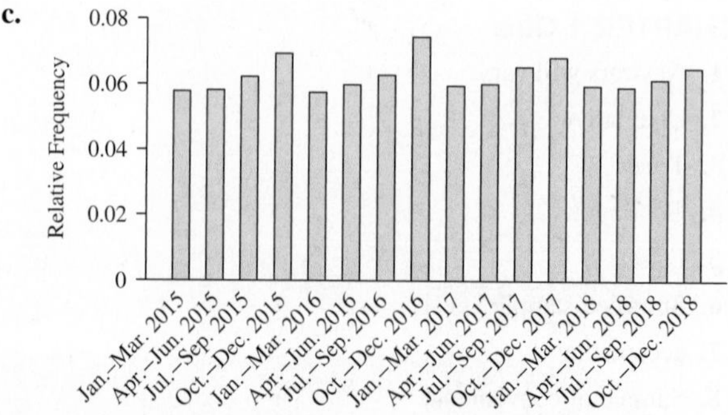

d. True

e. False

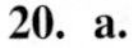

20. a.

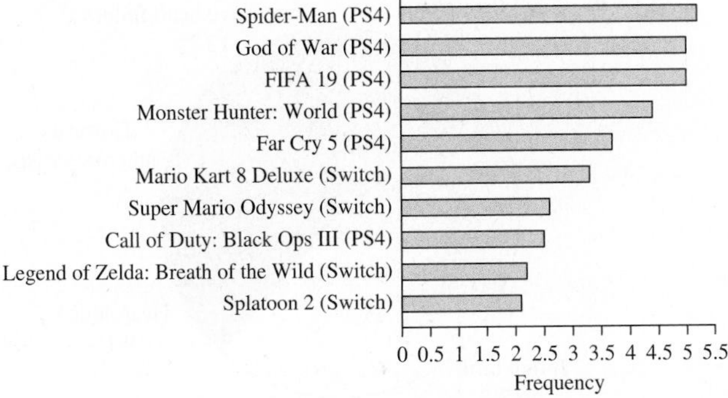

b.

Game	Relative Frequency
Spider-Man (PS4)	0.144
God of War (PS4)	0.139
FIFA 19 (PS4)	0.139
Monster Hunter: World (PS4)	0.122
Far Cry 5 (PS4)	0.103
Mario Kart 8 Deluxe (Switch)	0.092
Super Mario Odyssey (Switch)	0.072
Call of Duty: Black Ops III (PS4)	0.069
Legend of Zelda: Breath of the Wild (Switch)	0.061
Splatoon 2 (Switch)	0.058

c.

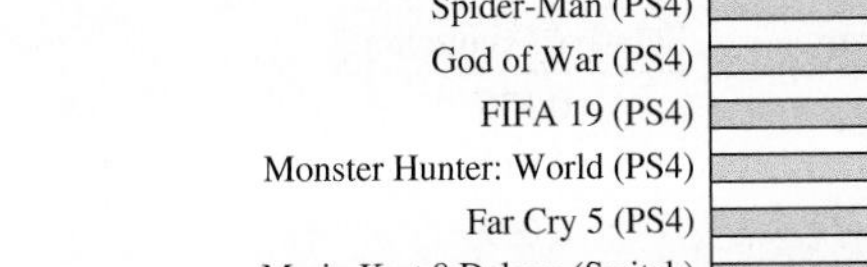

d. True

21. a.

Quarter	Frequency (millions)
Jan.–Mar.	998.1
Apr.–Jun.	1048.8
Jul.–Sep.	1144.1
Oct.–Dec.	1297.8

b.

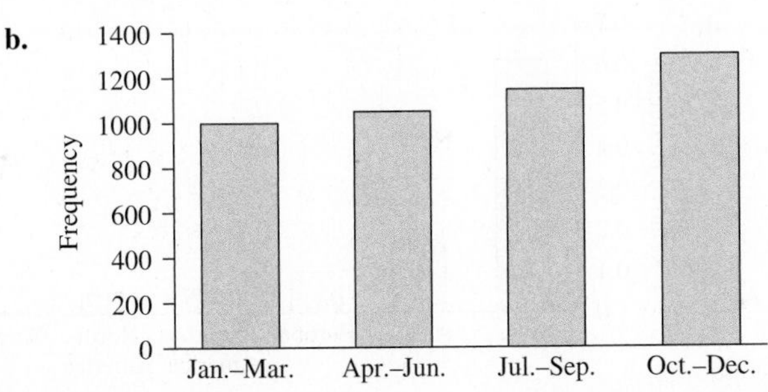

c.

Quarter	Relative Frequency
Jan.–Mar.	0.222
Apr.–Jun.	0.234
Jul.–Sep.	0.255
Oct.–Dec.	0.289

d.

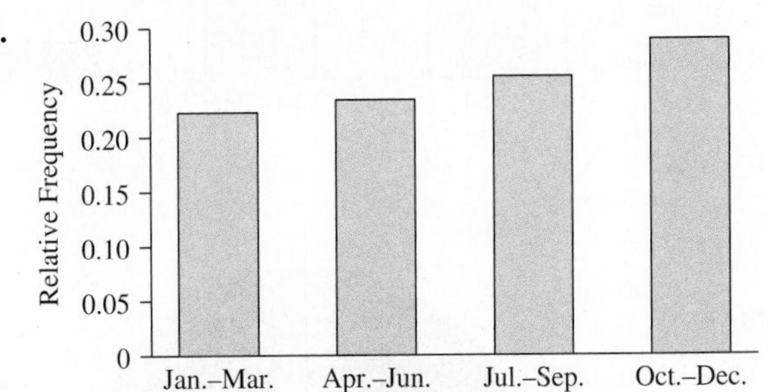

e.

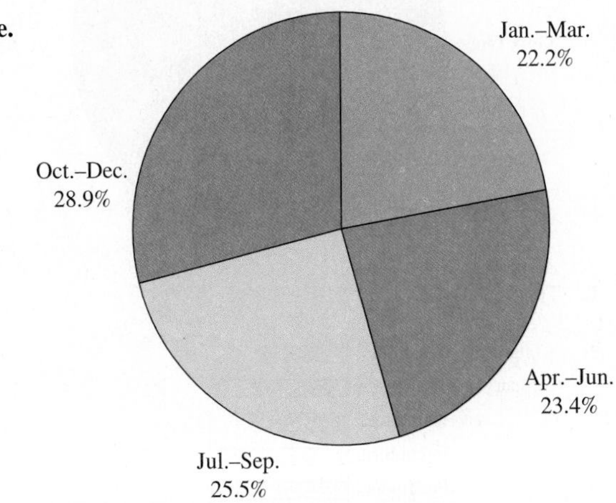

f. False

22. a.

Platform	Frequency (millions)
PS4	11.8
Xbox One	8.9
Wii U	1.5

b.

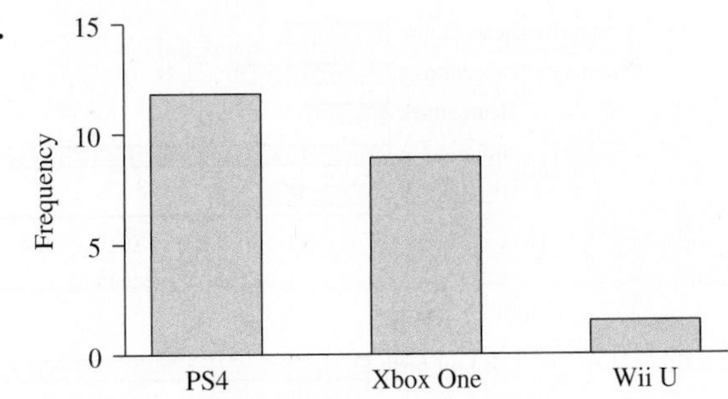

c.

Platform	Relative Frequency
PS4	0532
Xbox One	0.401
Wii U	0.068

d.

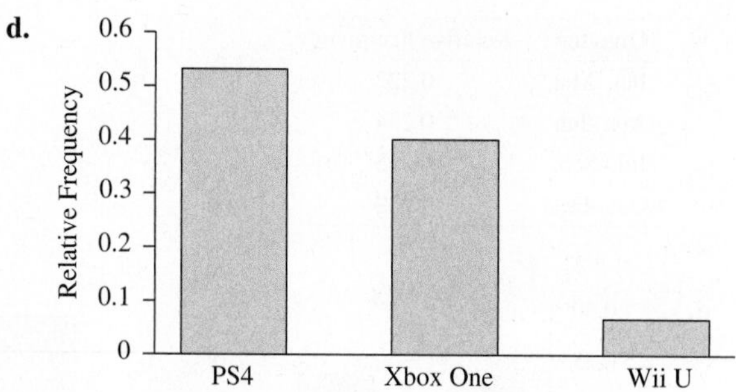

e.

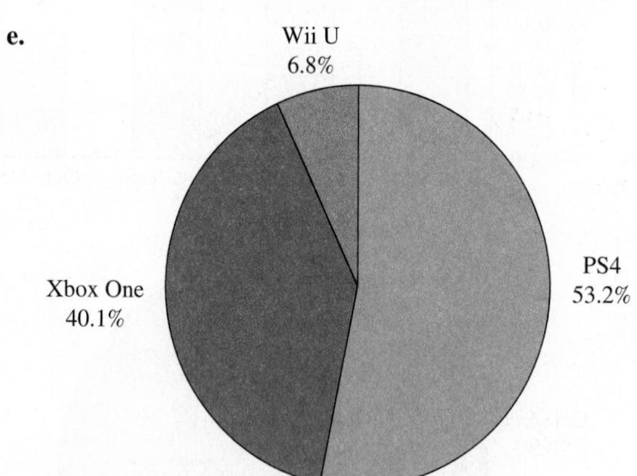

f. True

23. a.

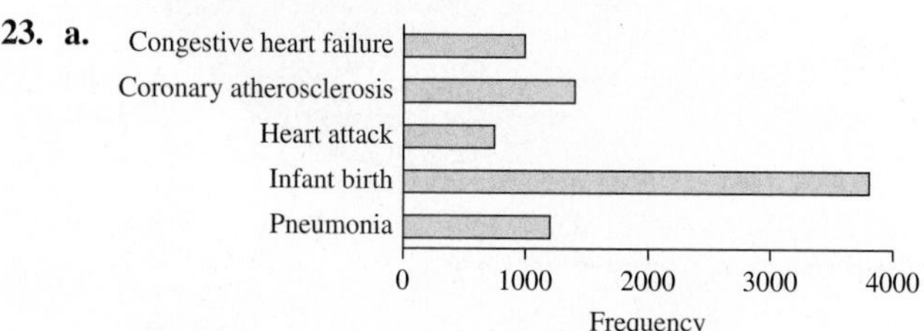

b.

Reason	Relative Frequency
Congestive heart failure	0.122
Coronary atherosclerosis	0.172
Heart attack	0.091
Infant birth	0.467
Pneumonia	0.148

c.

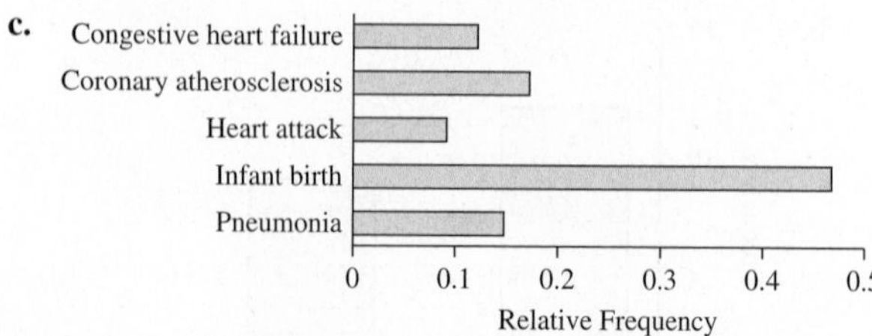

d.

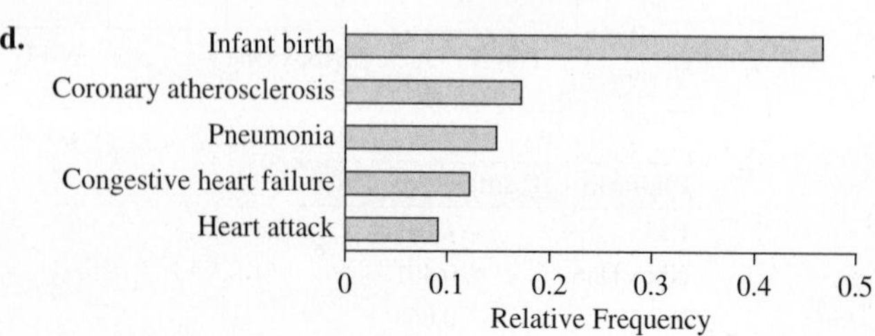

e.

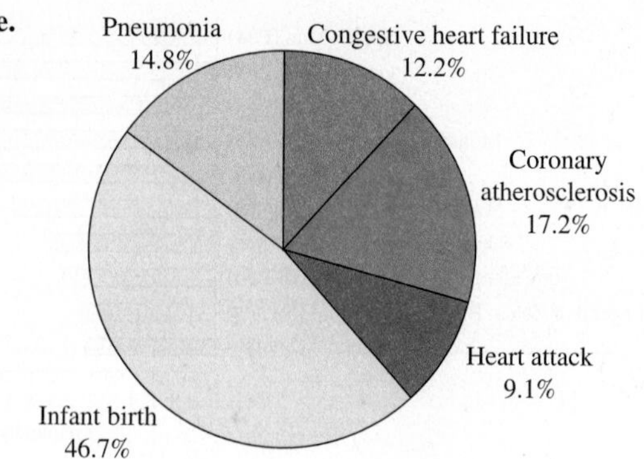

f. True

24. a.

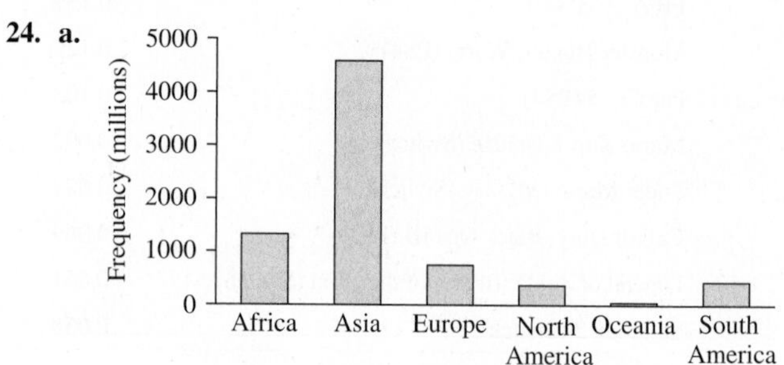

b.

Continent	Relative Frequency
Africa	0.176
Asia	0.612
Europe	0.099
North America	0.049
Oceania	0.006
South America	0.058

c.

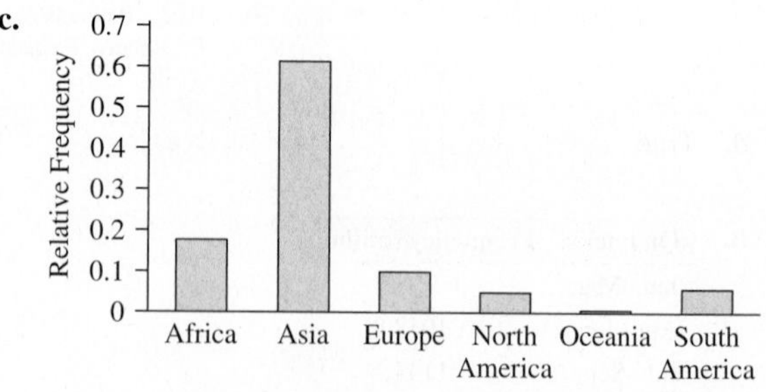

d.

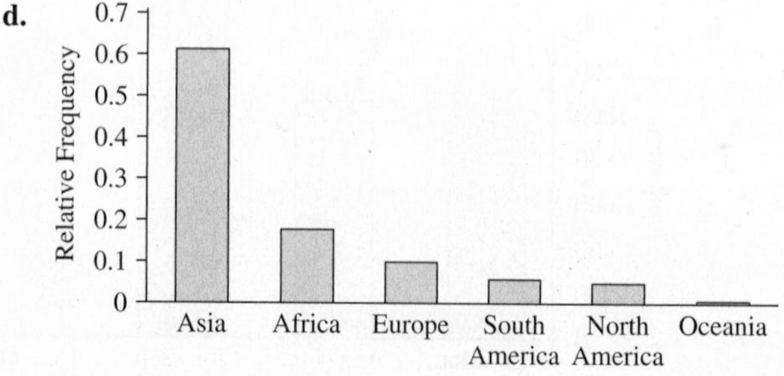

e.

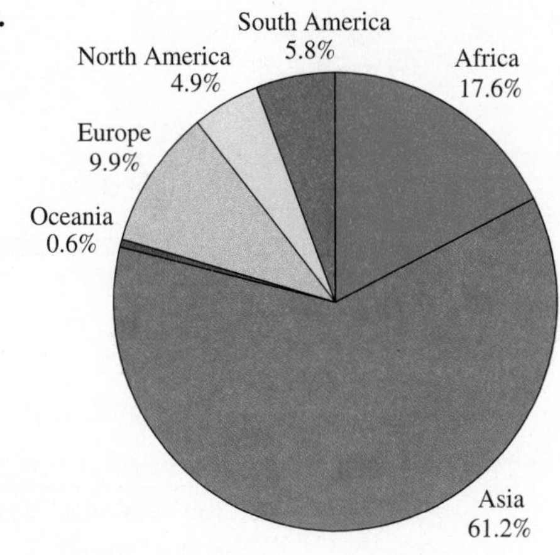

f. True

g. False

25. a.

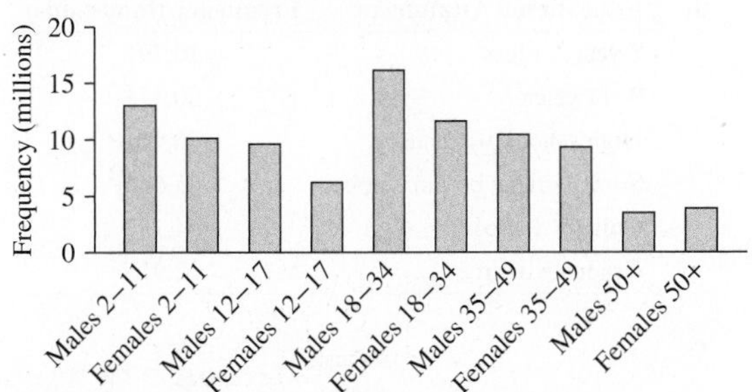

b.

Gender and Age Group	Relative Frequency
Males 2–11	0.139
Females 2–11	0.108
Males 12–17	0.102
Females 12–17	0.066
Males 18–34	0.172
Females 18–34	0.124
Males 35–49	0.111
Females 35–49	0.099
Males 50+	0.037
Females 50+	0.042

c.

Relative Frequency
0.20
0.15
0.10
0.05
0
Males 2–11
Females 2–11
Males 12–17
Females 12–17
Males 18–34
Females 18–34
Males 35–49
Females 35–49
Males 50+
Females 50+

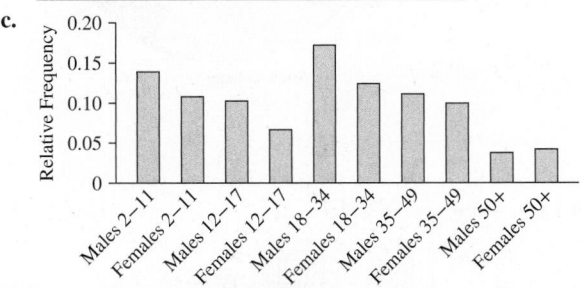

d.

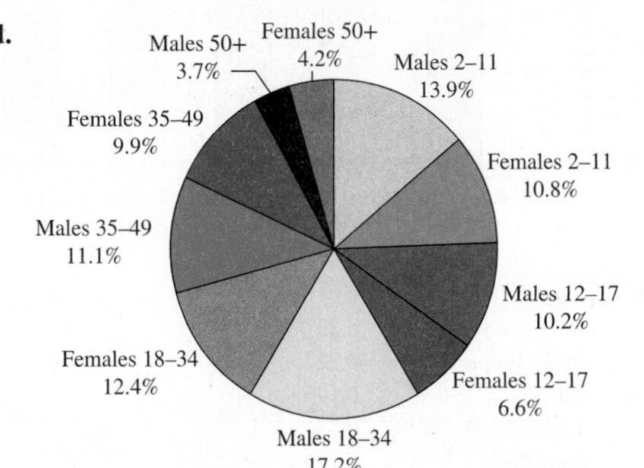

e. True **f.** True **g.** 0.289

26. a.

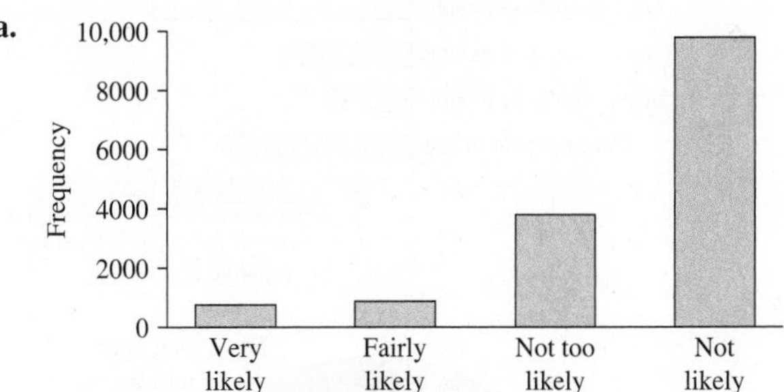

b.

Response	Relative Frequency
Very likely	0.049
Fairly likely	0.057
Not too likely	0.250
Not likely	0.645

c.

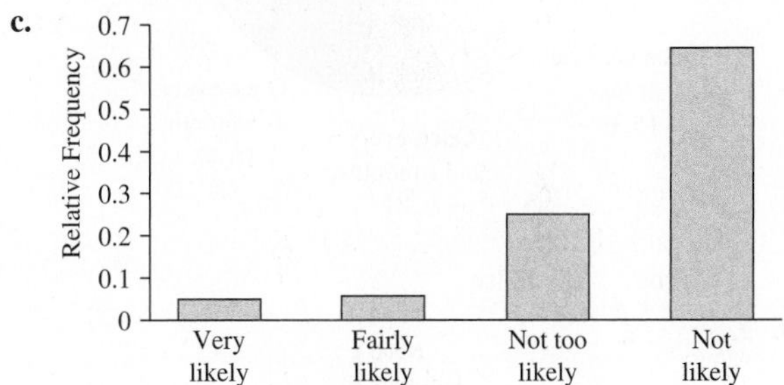

d.

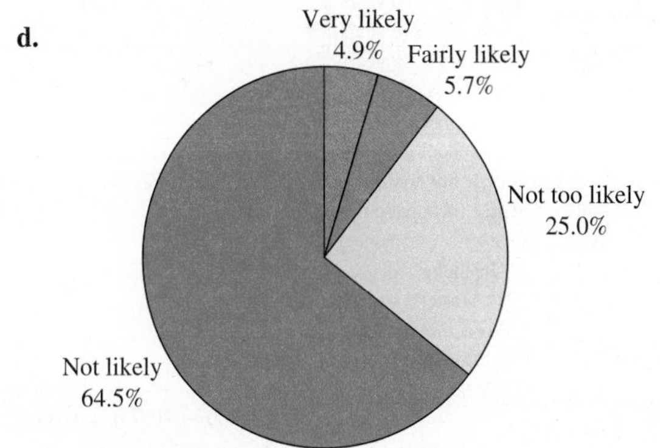

e. True **f.** 0.106

27. a.

More than once per month
Once every 1–3 months
Once every 4–6 months
Once every 7–11 months
Once per year or less
Never
0 100 200 300 400 500 600 700
Frequency

b.

Response	Relative Frequency
More than once per month	0.166
Once every 1–3 months	0.209
Once every 4–6 months	0.104
Once every 7–11 months	0.063
Once per year or less	0.153
Never	0.305

c.

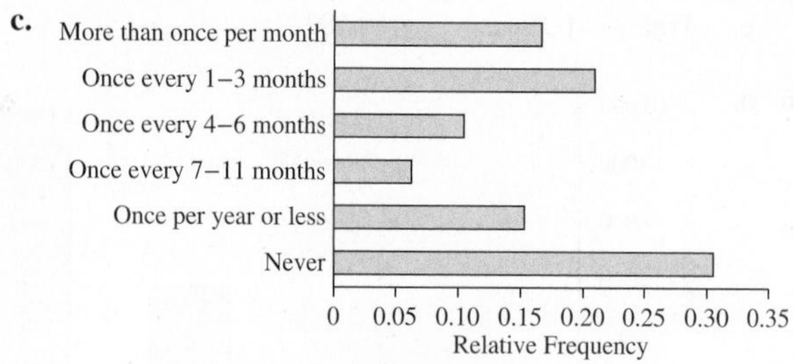

d.

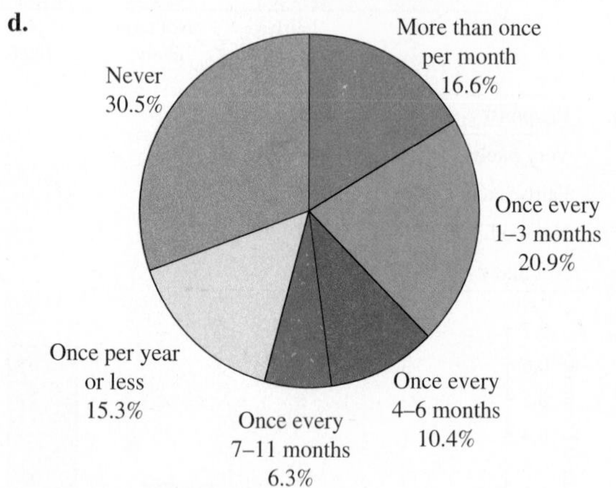

e. True **f.** False

28. a.

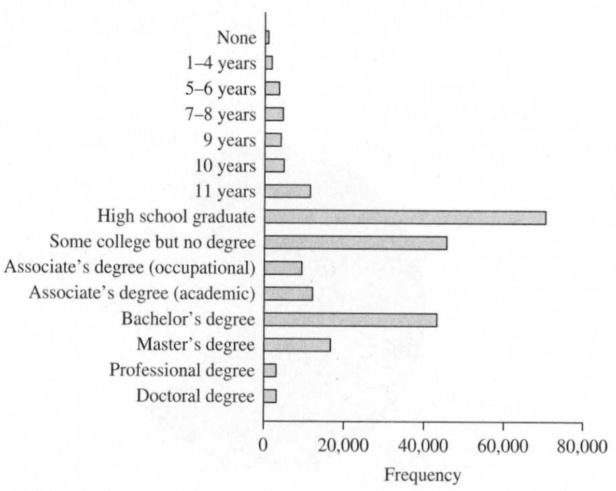

b.

Educational Attainment	Relative Frequency
None	0.004
1–4 years	0.008
5–6 years	0.015
7–8 years	0.019
9 years	0.017
10 years	0.020
11 years	0.049
High school graduate	0.300
Some college but no degree	0.194
Associate's degree (occupational)	0.040
Associate's degree (academic)	0.052
Bachelor's degree	0.184
Master's degree	0.071
Professional degree	0.013
Doctoral degree	0.014

c.

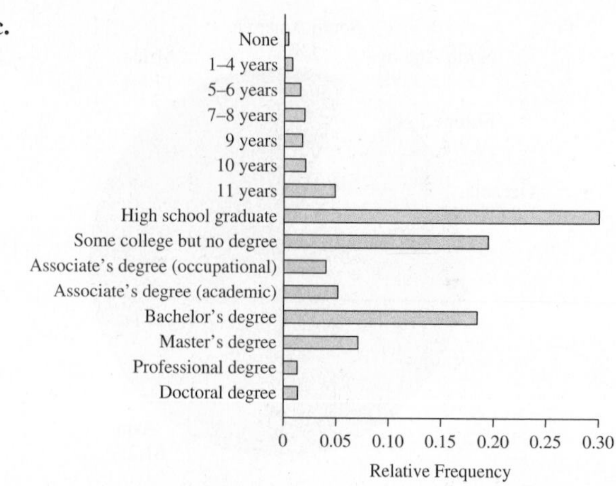

d.

Educational Attainment	Frequency (thousands)
8 years or less	10,791
9–11 years	20,311
High school graduate	70,441
Some college but no degree	45,645
College degree	64,757
Graduate degree	22,915

e.

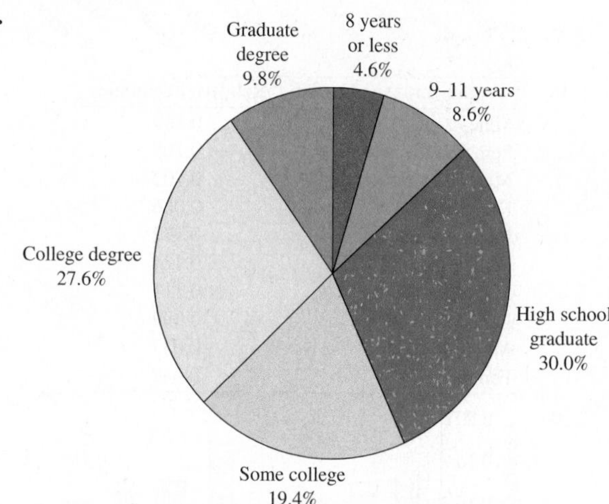

f. 0.132

29. a.

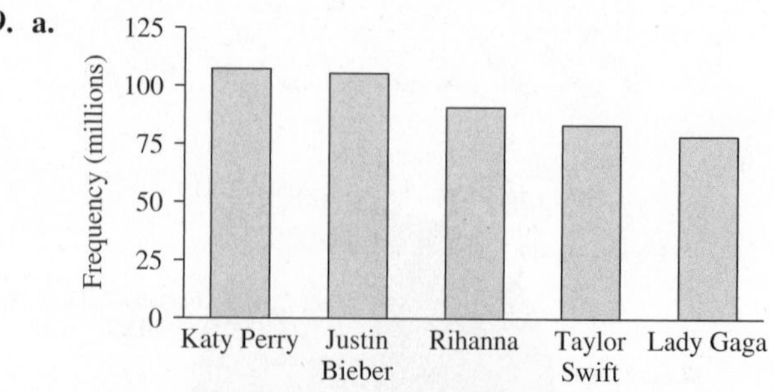

b.

Singer	Relative Frequency
Katy Perry	0.231
Justin Bieber	0.227
Rihanna	0.195
Taylor Swift	0.179
Lady Gaga	0.169

c.

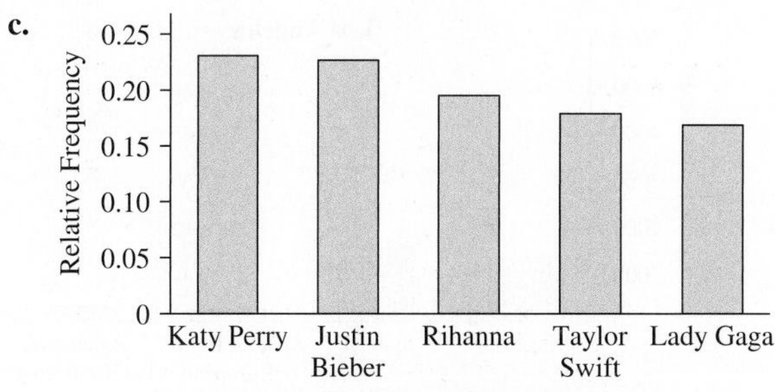

d.

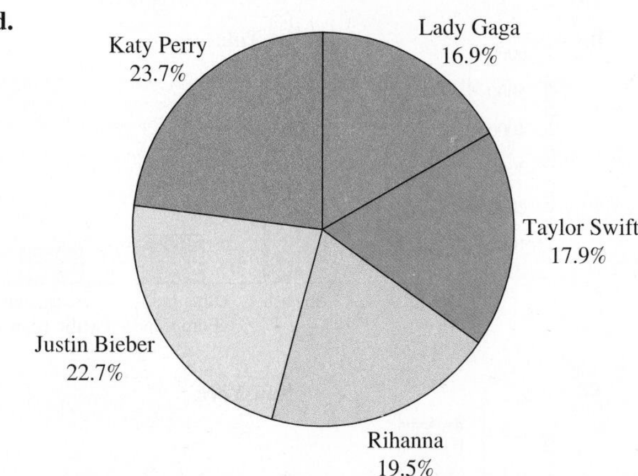

e. 0.179

30. a.

Type of Music	Relative Frequency
Digital subscription	0.482
Digital streaming	0.185
Digital downloads	0.157
Physical	0.176

b.

Type of Music	Relative Frequency
Digital subscription	0.565
Digital streaming	0.205
Digital downloads	0.109
Physical	0.121

c.

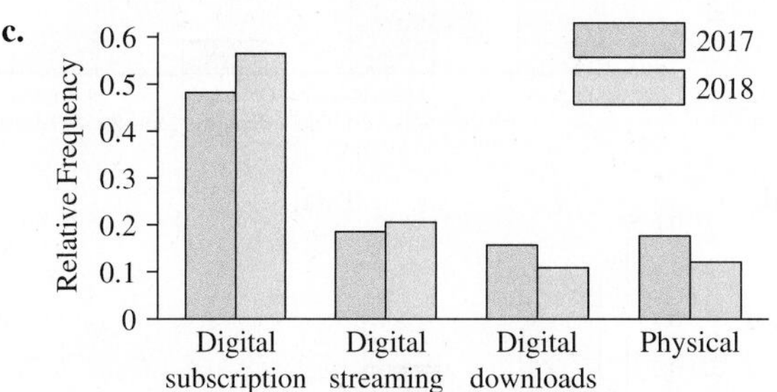

d. False

31. a.

Kardashian	Relative Frequency
Kim	0.363
Kendall	0.166
Kylie	0.163
Khloé	0.162
Kourtney	0.146

b.

Kardashian	Relative Frequency
Kim	0.245
Kendall	0.201
Kylie	0.245
Khloé	0.168
Kourtney	0.142

c.

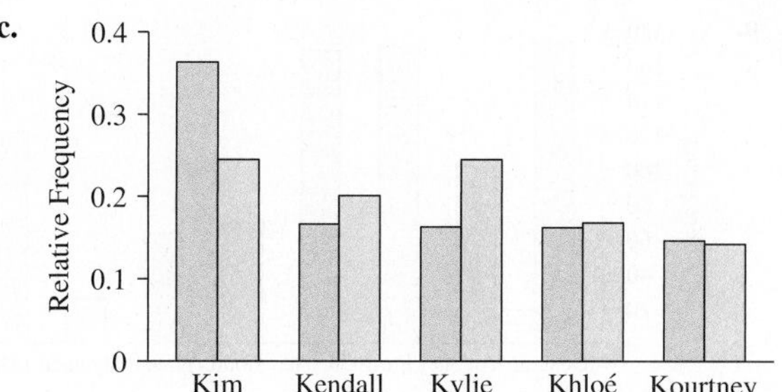

d. True

e. True

32. a.

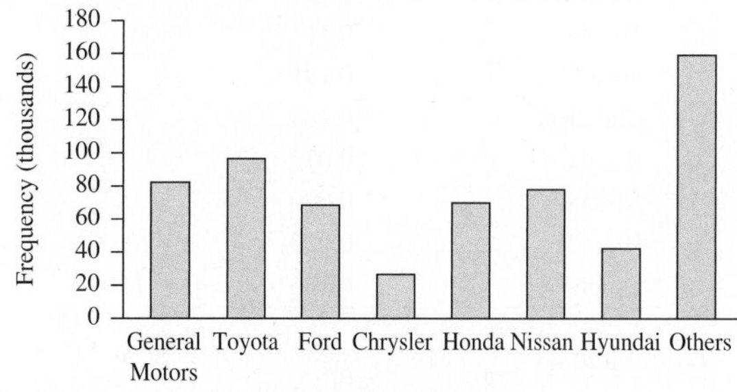

b.

Manufacturer	Relative Frequency
General Motors	0.132
Toyota	0.155
Ford	0.110
Chrysler	0.043
Honda	0.112
Nissan	0.125
Hyundai	0.067
Other	0.255

c.

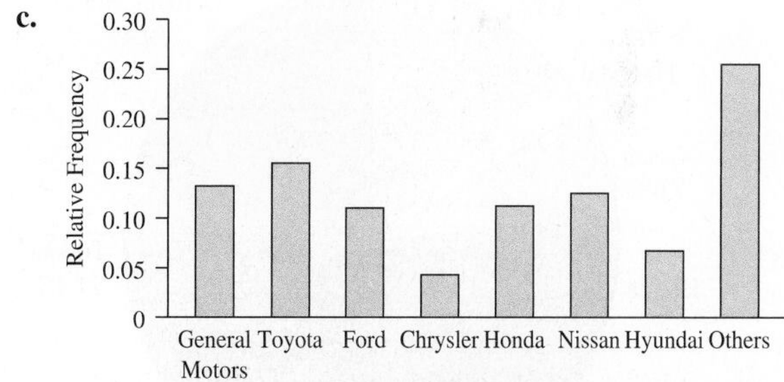

d.

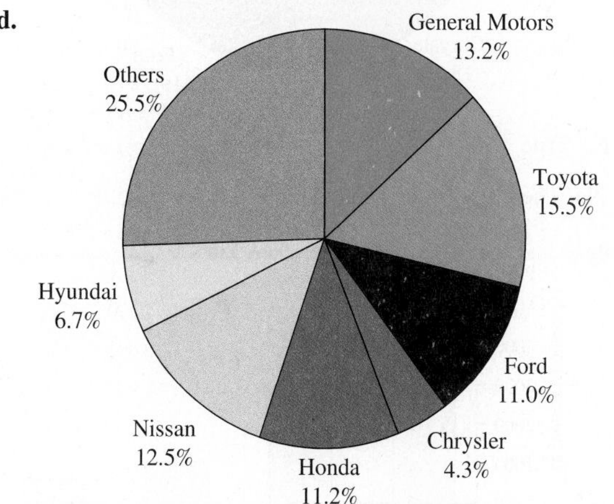

e. 0.132

33. a.

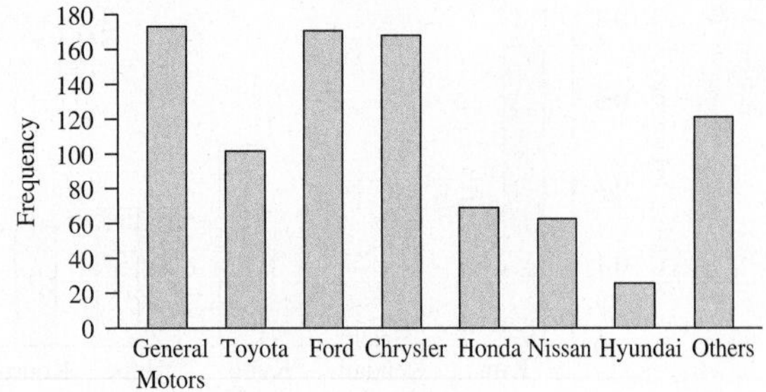

b.

Manufacturer	Relative Frequency
General Motors	0.194
Toyota	0.114
Ford	0.191
Chrysler	0.188
Honda	0.077
Nissan	0.070
Hyundai	0.029
Others	0.136

c.

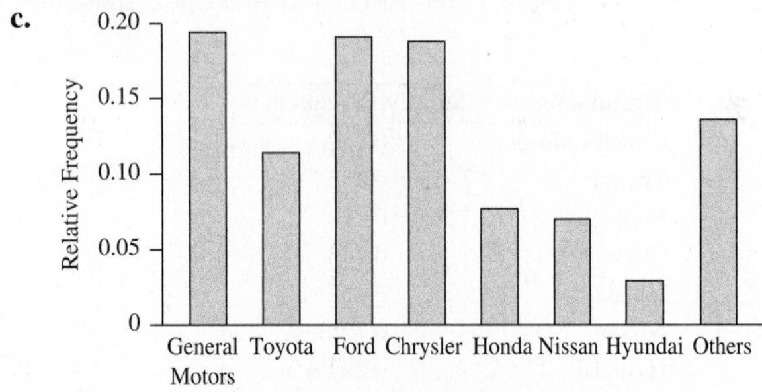

d.

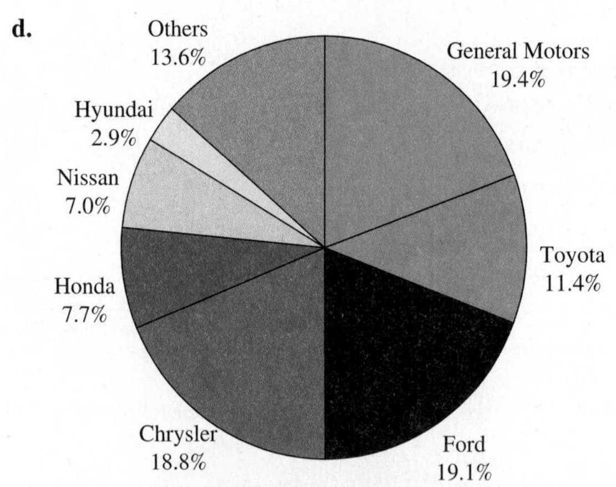

e. True

34. No. The proportions do not sum to 1.

35. a.

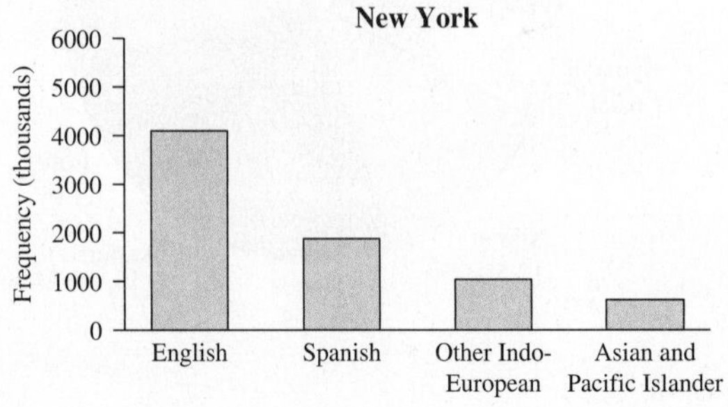

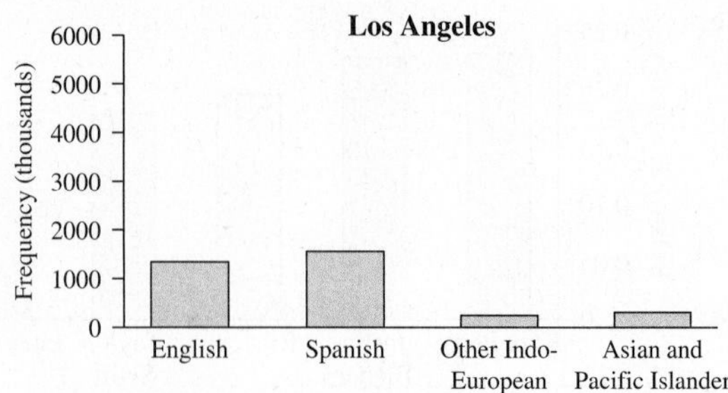

b.

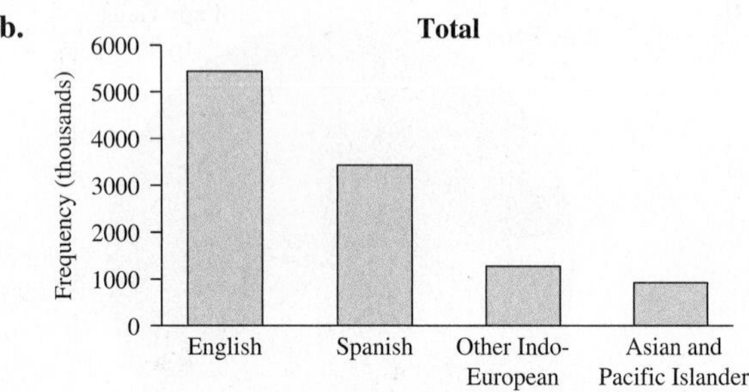

c.

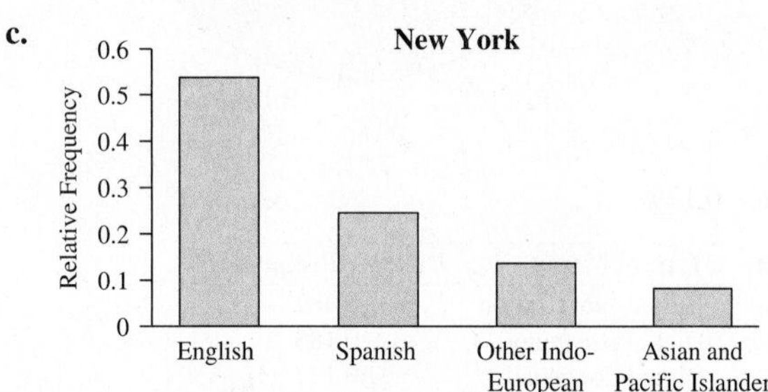

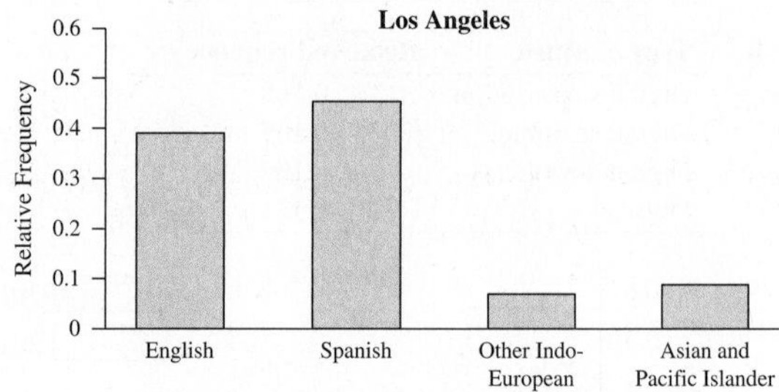

d.

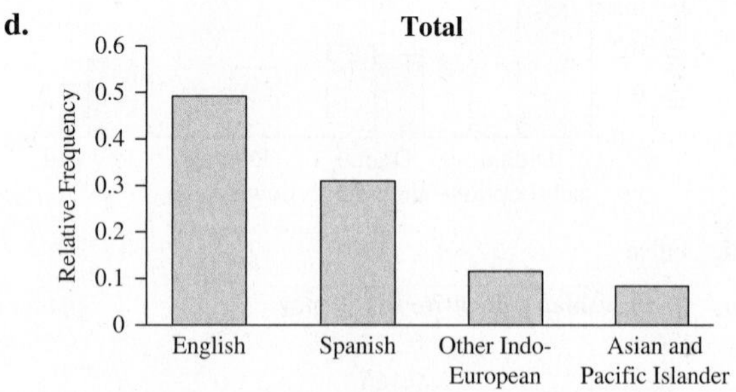

e. The total frequency is equal to the sum of the frequencies for the two cities.

f. The total relative frequency is the total frequency divided by the sum of all total frequencies. The relative frequency for each city is the frequency for that city divided by the sum of the frequencies for that city. Since the sum of the frequencies for each city is not the same as the sum of the total frequencies, the total relative frequency is not the sum of the relative frequencies for the two cities.

36. a. No. The proportions are not parts of a whole. In particular, they do not add up to 1.
b. No. This is not a frequency table.

Section 2.2

Exercises 1–4 are the Check Your Understanding exercises for this section. Answers to these exercises are on page 71.

5. symmetric
6. left, right
7. bimodal
8. cumulative frequency
9. False
10. False
11. True
12. True
13. Skewed to the left
14. Skewed to the right
15. Bell-shaped
16. Approximately uniform
17. Skewed right
18. Bell-shaped
19. Bimodal
20. Unimodal
21. a. 1 b. 20 c. 4 d. 55%
22. a. 4 b. 65 c. 20 d. 30.8%
23. a. 3–4 b. 500 c. 50 d. 20%
24. a. 2.5–3.0 b. 200 c. 80 d. 50%
25. a. 11 b. 1 c. 70–71
d. 9% e. Approximately symmetric
26. a. 3 b. 19 c. 3
d. Skewed to the right
27. a. 30%
b. 240–260
28. a. 50%
b. 130–135
29. a. False
b. True
c. 9
d. No
e. Skewed to the right
30. a. 50–55
b. 30%
c. 60–70
31. a. Right
b. Left
c. Left
32. a. Right
b. Left
c. Right
33. a. 9
b. 0.020
c. Lower limits: 0.180, 0.200, 0.220, 0.240, 0.260, 0.280, 0.300, 0.320, 0.340. Upper limits: 0.199, 0.219, 0.239, 0.259, 0.279, 0.299, 0.319, 0.339, 0.359.
d.

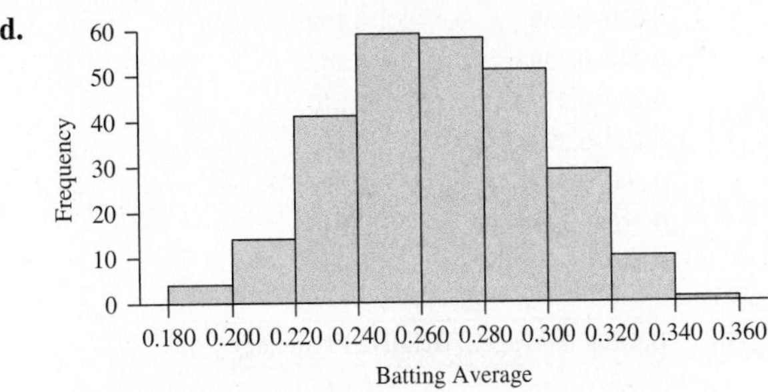

e.

Batting Average	Relative Frequency
0.180–0.199	0.015
0.200–0.219	0.052
0.220–0.239	0.154
0.240–0.259	0.221
0.260–0.279	0.217
0.280–0.299	0.191
0.300–0.319	0.109
0.320–0.339	0.037
0.340–0.359	0.004

f.

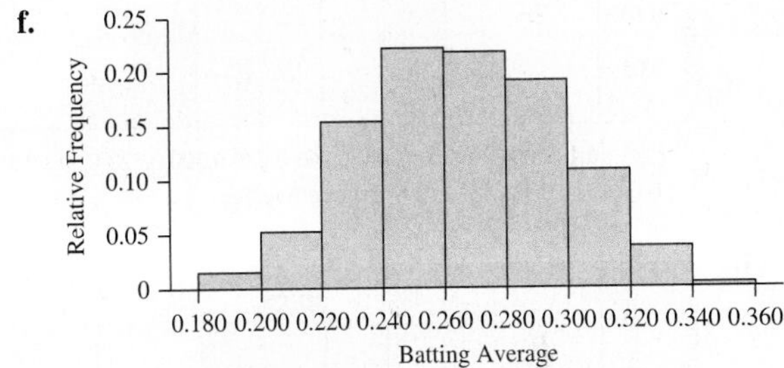

g. 15.0%
h. 6.7%

34. a.

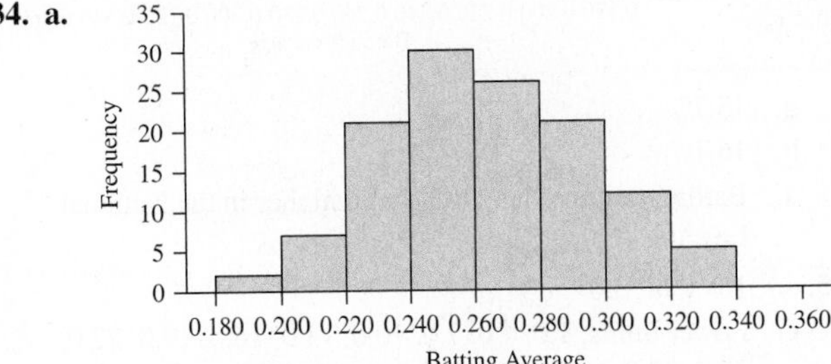

b.

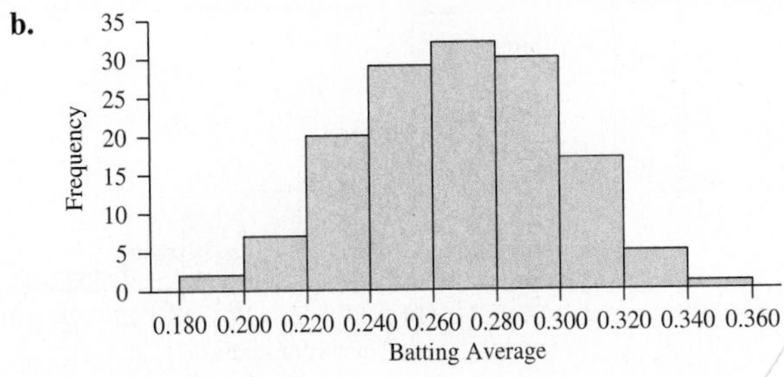

c.

Batting Average	Relative Frequency
0.180–0.199	0.016
0.200–0.219	0.056
0.220–0.239	0.169
0.240–0.259	0.242
0.260–0.279	0.210
0.280–0.299	0.169
0.300–0.319	0.097
0.320–0.339	0.040
0.340–0.359	0.000

d.

Batting Average	Relative Frequency
0.180–0.199	0.014
0.200–0.219	0.049
0.220–0.239	0.140
0.240–0.259	0.203
0.260–0.279	0.224
0.280–0.299	0.210
0.300–0.319	0.119
0.320–0.339	0.035
0.340–0.359	0.007

e.

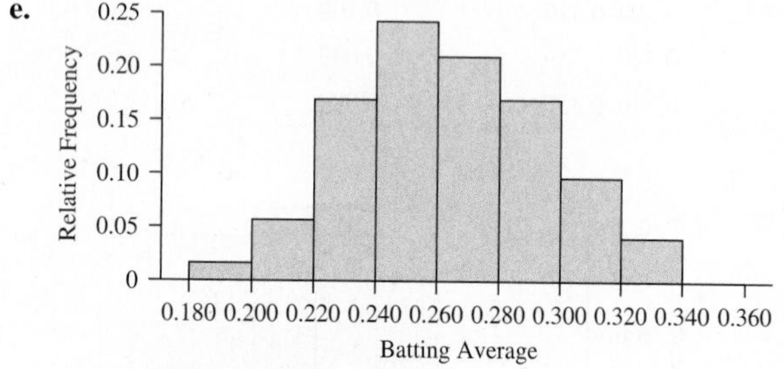

f.

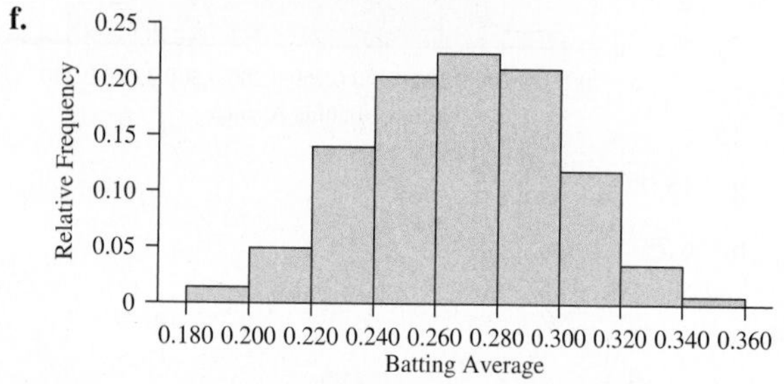

g. 13.7%

h. 16.1%

i. Batting averages tend to be a bit higher in the National League.

35. a. 10 b. 3.0

c. Lower limits: 1.0, 4.0, 7.0, 10.0, 13.0, 16.0, 19.0, 22.0, 25.0, 28.0. Upper limits: 3.9, 6.9, 9.9, 12.9, 15.9, 18.9, 21.9, 24.9, 27.9, 30.9.

d.

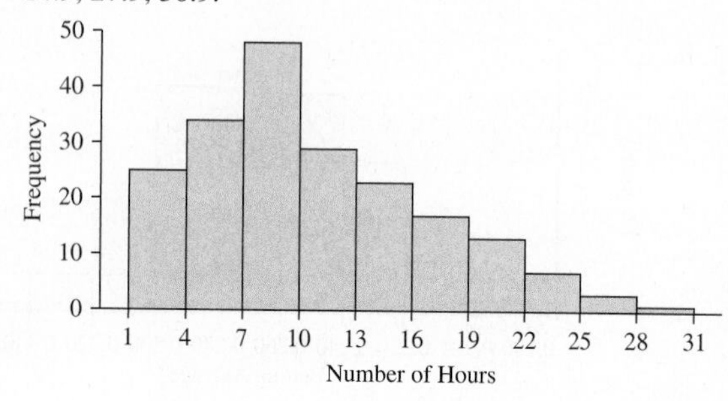

e.

Number of Hours	Relative Frequency
1.0–3.9	0.125
4.0–6.9	0.170
7.0–9.9	0.240
10.0–12.9	0.145
13.0–15.9	0.115
16.0–18.9	0.085
19.0–21.9	0.065
22.0–24.9	0.035
25.0–27.9	0.015
28.0–30.9	0.005

f.

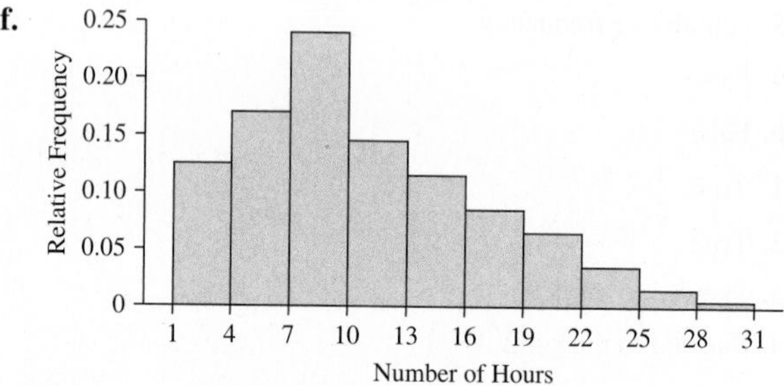

g. 53.5% h. 12.0%

36. a. 11

b. 5

c. Lower limits: 0.0, 5.0, 10.0, 15.0, 20.0, 25.0, 30.0, 35.0, 40.0, 45.0, 50.0. Upper limits: 4.9, 9.9, 14.9, 19.9, 24.9, 29.9, 34.9, 39.9, 44.9, 49.9, 54.9.

d.

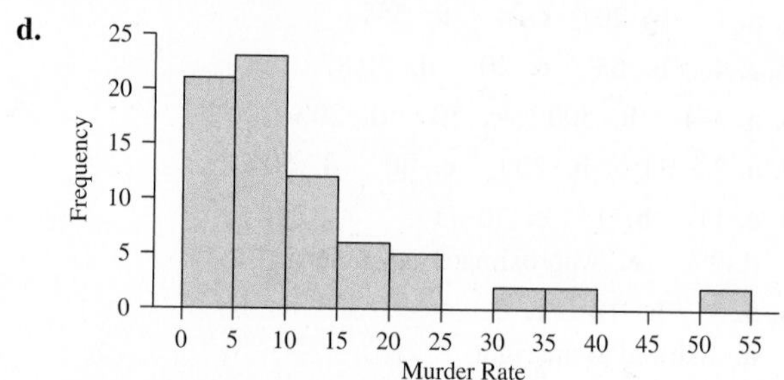

e.

Murder Rate	Relative Frequency
0.0–4.9	0.288
5.0–9.9	0.315
10.0–14.9	0.164
15.0–19.9	0.082
20.0–24.9	0.068
25.0–29.9	0.000
30.0–34.9	0.027
35.0–39.9	0.027
40.0–44.9	0.000
45.0–49.9	0.000
50.0–54.9	0.027

f.

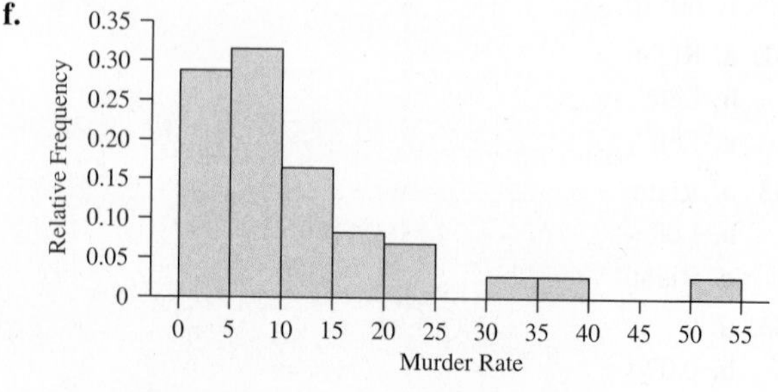

g. 60.3%

h. 8.2%

37. a.

Price ($1000s)	Frequency
30.0–39.9	2
40.0–49.9	8
50.0–59.9	12
60.0–69.9	4
70.0–79.9	3
80.0–89.9	3
90.0–99.9	1
100.0–109.9	2
110.0–119.9	0
120.0–129.9	0
130.0–139.9	0
140.0–149.9	0
150.0–159.9	1

b.

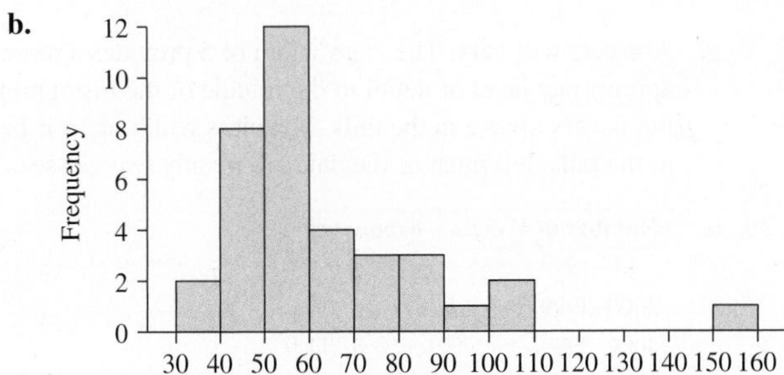

c.

Price ($1000s)	Relative Frequency
30.0–39.9	0.056
40.0–49.9	0.222
50.0–59.9	0.333
60.0–69.9	0.111
70.0–79.9	0.083
80.0–89.9	0.083
90.0–99.9	0.028
100.0–109.9	0.056
110.0–119.9	0.000
120.0–129.9	0.000
130.0–139.9	0.000
140.0–149.9	0.000
150.0–159.9	0.028

d.

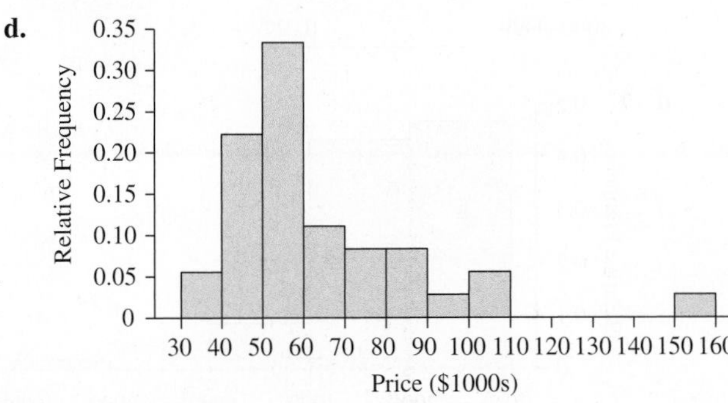

e. Unimodal

f.

Price ($1000s)	Frequency
30.0–49.9	10
50.0–69.9	16
70.0–89.9	6
90.0–109.9	3
110.0–119.9	0
130.0–149.9	0
150.0–169.9	1

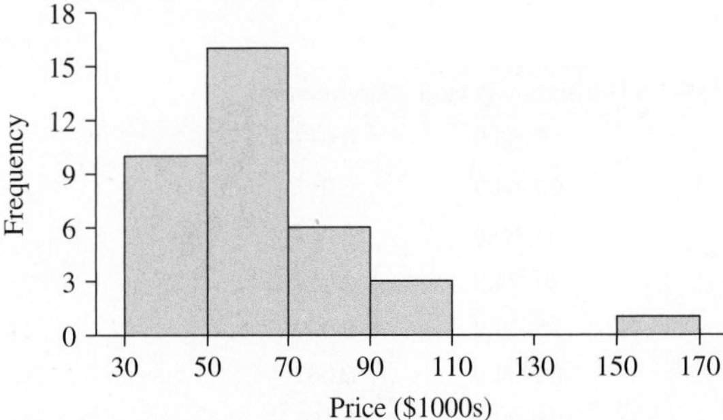

Price ($1000s)	Relative Frequency
30.0–49.9	0.278
50.0–69.9	0.444
70.0–89.9	0.167
90.0–109.9	0.083
110.0–119.9	0.000
130.0–149.9	0.000
150.0–169.9	0.028

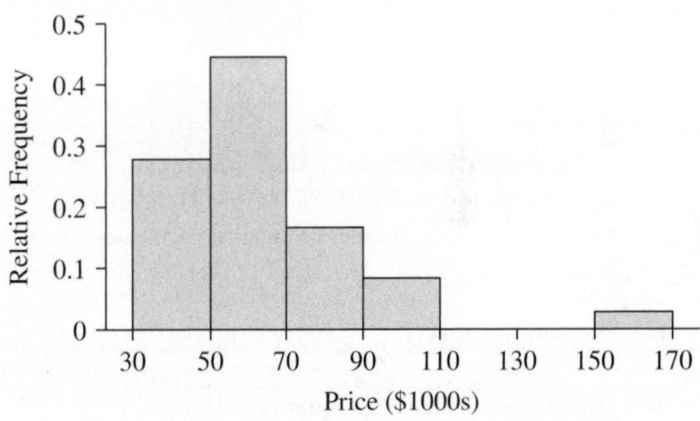

g. Answers will vary. Both choices seem reasonably good.

38. a.

Dormancy Period	Frequency
55–59.9	1
60–64.9	1
65–69.9	1
70–74.9	1
75–79.9	0
80–84.9	4
85–89.9	11
90–94.9	23
95–99.9	12
100–104.9	3
105–109.9	2
110–114.9	0
115–119.9	1

b.

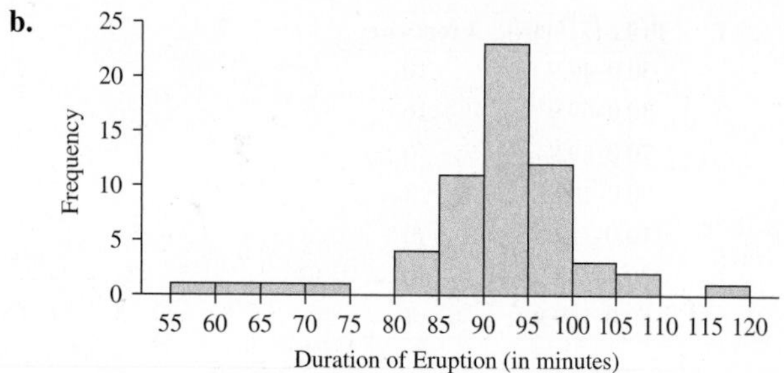

c.

Dormancy Period	Frequency
55–59.9	0.017
60–64.9	0.017
65–69.9	0.017
70–74.9	0.017
75–79.9	0.000
80–84.9	0.067
85–89.9	0.183
90–94.9	0.383
95–99.9	0.200
100–104.9	0.050
105–109.9	0.033
110–114.9	0.000
115–119.9	0.017

d.

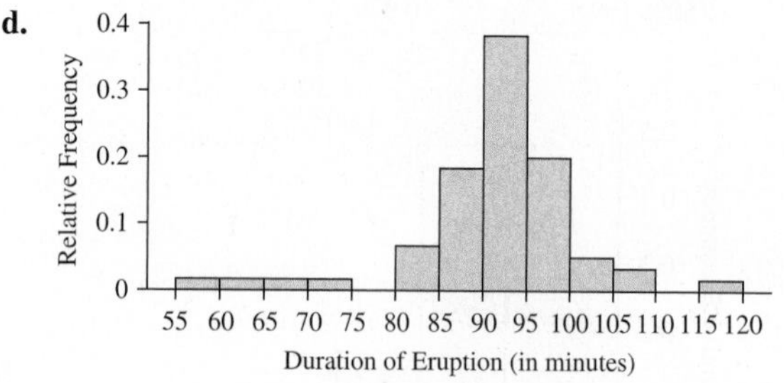

e. Skewed to the left

f.

Dormancy Period	Frequency
50–59.9	1
60–69.9	2
70–79.9	1
80–89.9	15
90–99.9	35
100–109.9	5
110–119.9	1

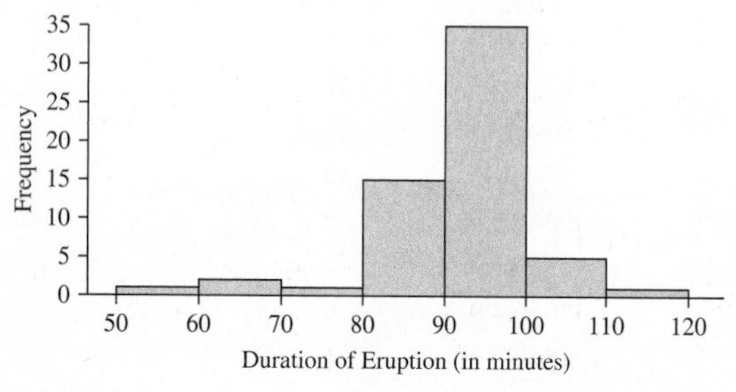

Dormancy Period	Frequency
50–59.9	0.017
60–69.9	0.033
70–79.9	0.017
80–89.9	0.250
90–99.9	0.583
100–109.9	0.083
110–119.9	0.017

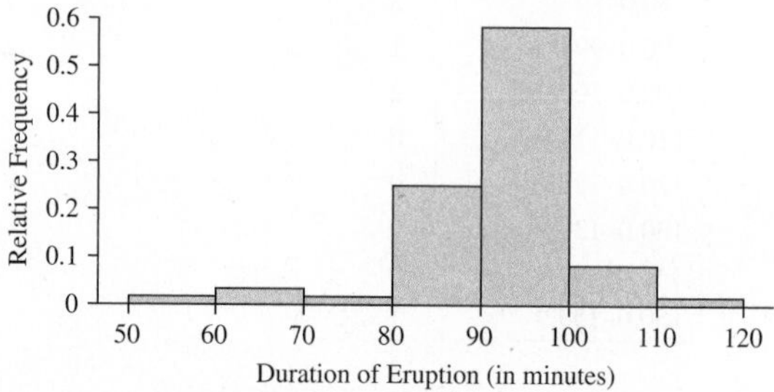

g. Answers will vary. The class width of 5 provides a more appropriate level of detail in the middle of the histogram but is very sparse in the tails. The class width of 10 is better in the tails, but most of the data are in only two classes.

39. a.

Number of Words	Frequency
0–1999	27
2000–3999	25
4000–5999	5
6000–7999	0
8000–9999	1

b.

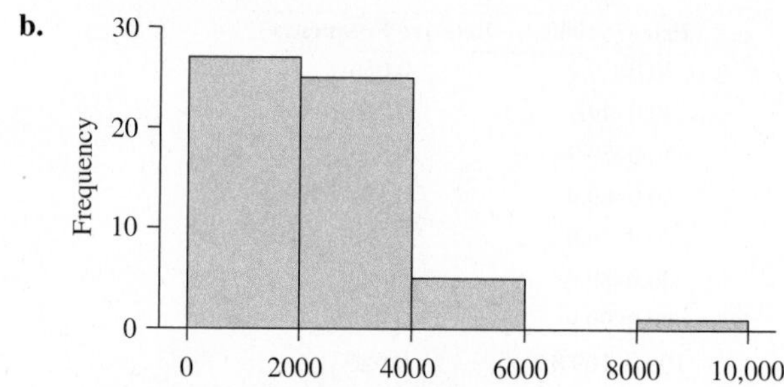

c.

Number of Words	Relative Frequency
0–1999	0.466
2000–3999	0.431
4000–5999	0.086
6000–7999	0.000
8000–9999	0.017

d.

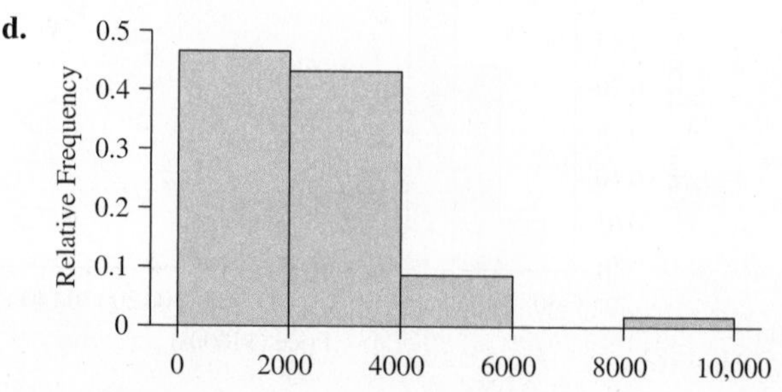

e. Skewed right

f.

Number of Words	Frequency
0–999	4
1000–1999	23
2000–2999	18
3090–3999	7
4000–4999	4
5000–5999	1
6000–6999	0
7000–7999	0
8000–8999	1

g.

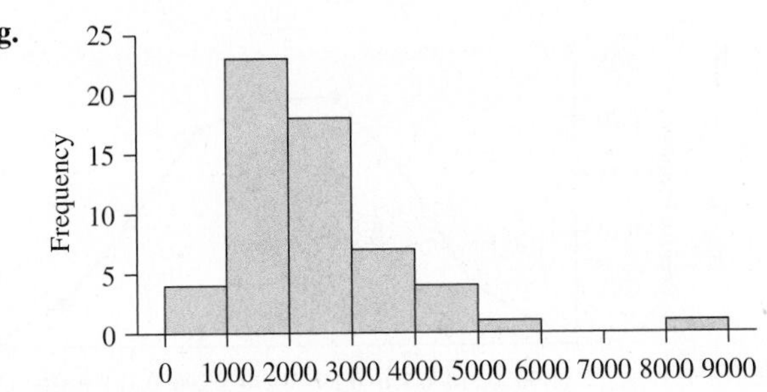

Number of Words	Relative Frequency
0–999	0.069
1000–1999	0.397
2000–2999	0.310
3090–3999	0.121
4000–4999	0.069
5000–5999	0.017
6000–6999	0.000
7000–7999	0.000
8000–8999	0.017

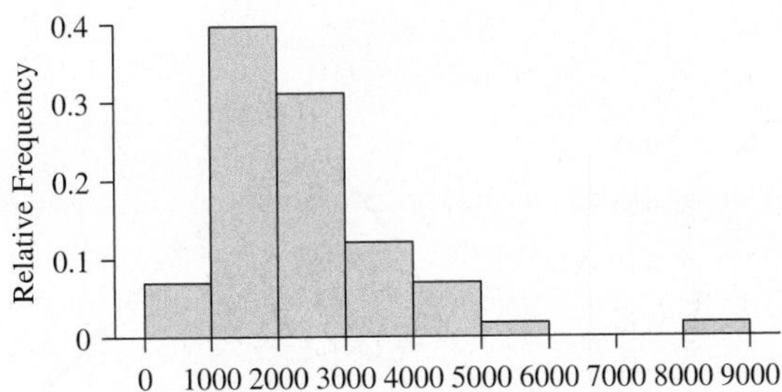

h. Answers will vary.

40. a.

Number of Hours	Frequency
0–4	5
5–9	6
10–14	8
15–19	9
20–24	6
25–29	2
30–34	0
35–39	3
40–44	0
45–49	0
50–54	1

b.

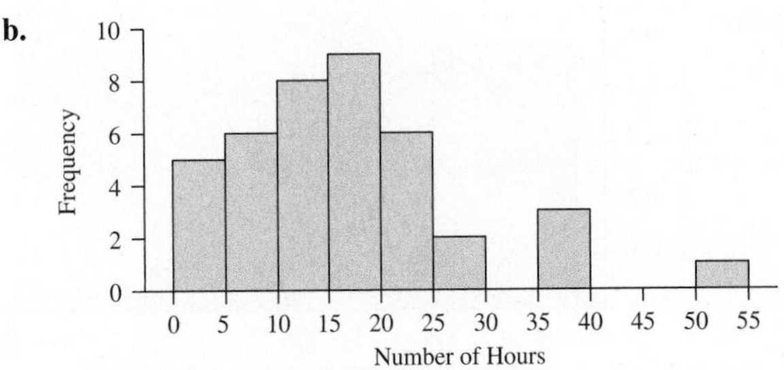

c.

Number of Hours	Relative Frequency
0–4	0.125
5–9	0.150
10–14	0.200
15–19	0.225
20–25	0.150
25–29	0.050
30–34	0.000
35–39	0.075
40–44	0.000
45–49	0.000
50–54	0.025

d.

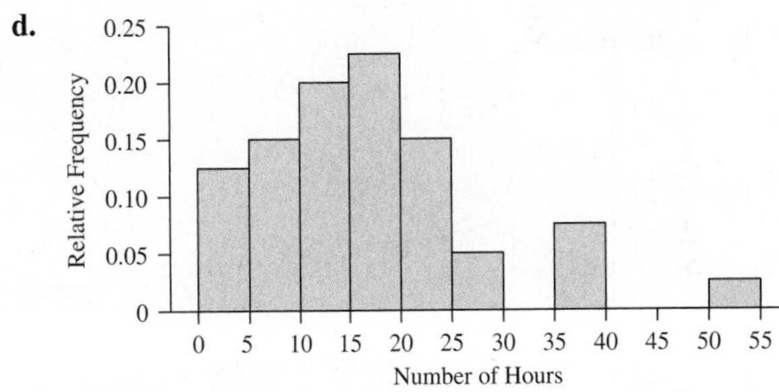

e. Skewed right

f.

Number of Hours	Frequency
0–14	19
15–29	17
30–45	3
46–60	1

g.

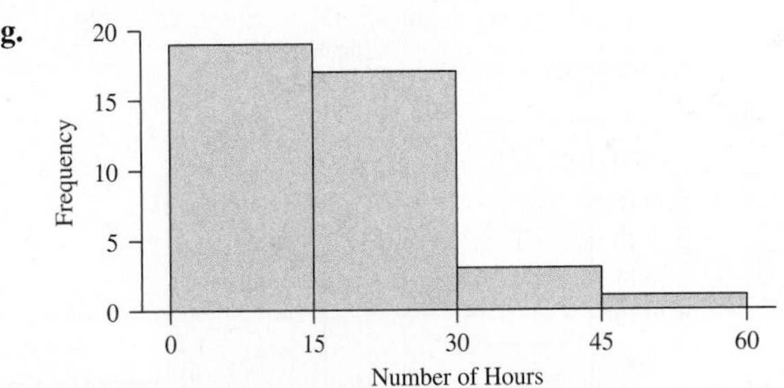

Number of Hours	Relative Frequency
0–14	0.475
15–29	0.425
30–45	0.075
46–60	0.025

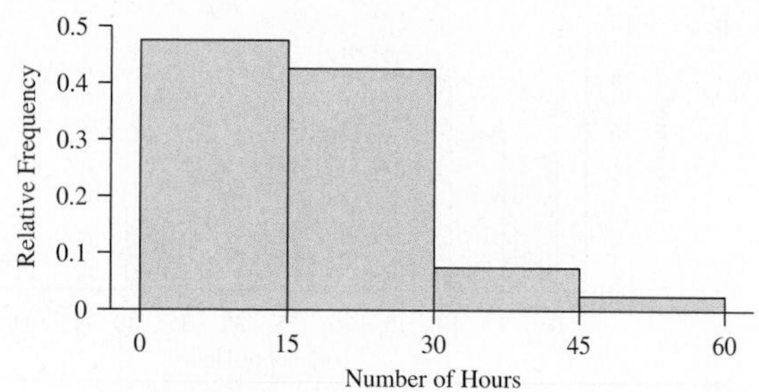

h. Answers will vary. Eleven classes provide a reasonable amount of detail, while four classes provide a more rough outline of the data.

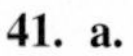

41. a.

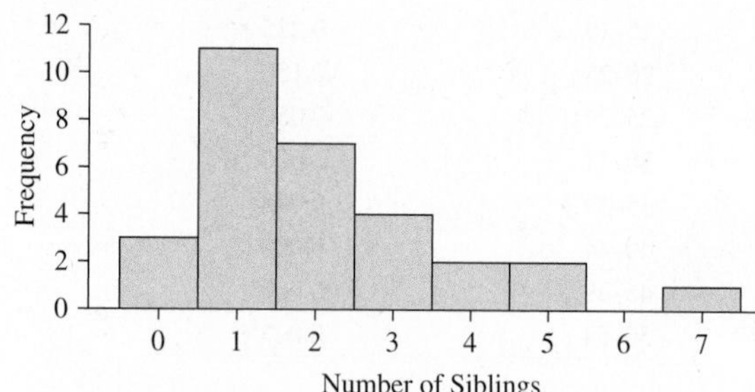

b.

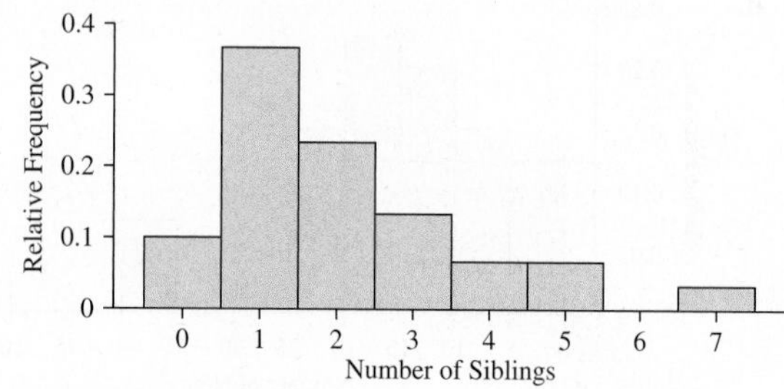

c. Skewed to the right

42. a.

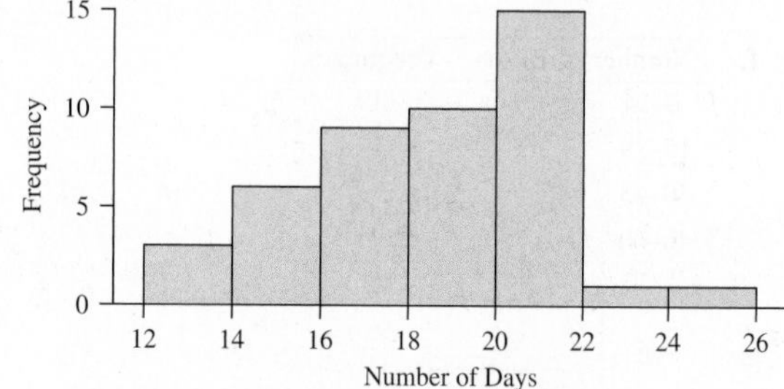

b.

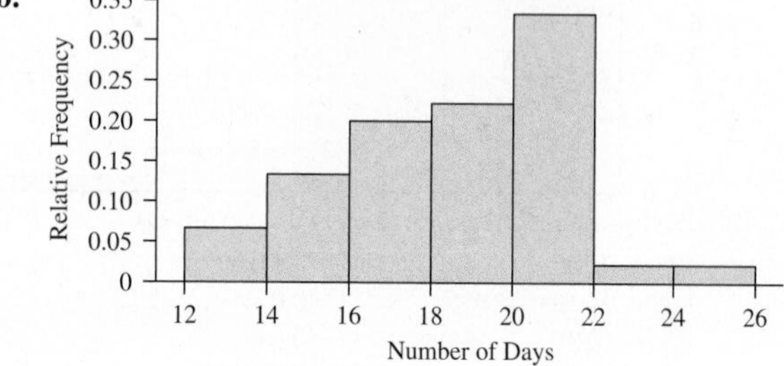

c. Skewed left

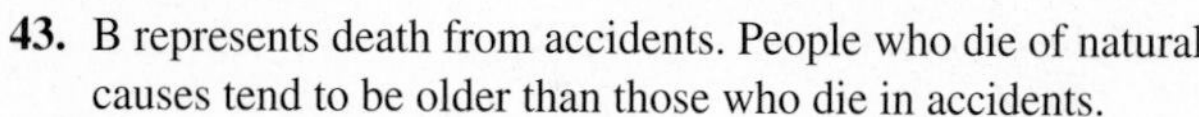

43. B represents death from accidents. People who die of natural causes tend to be older than those who die in accidents.

44. There is a gap between the top 15 scores and the others.

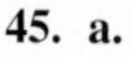

45. a.

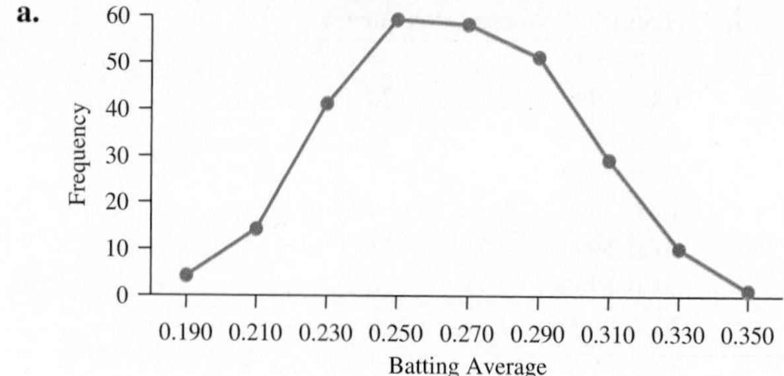

b.

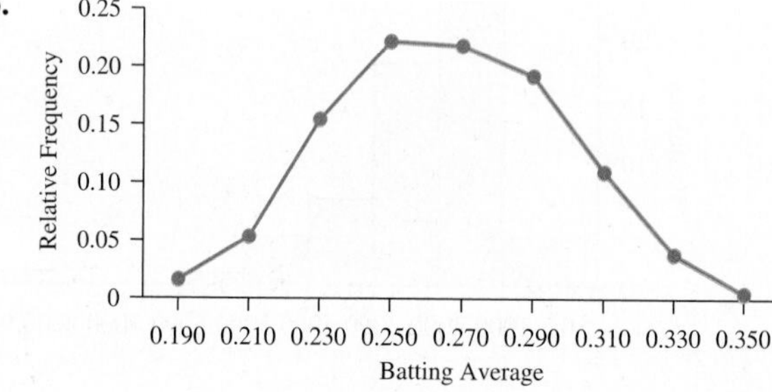

46. a.

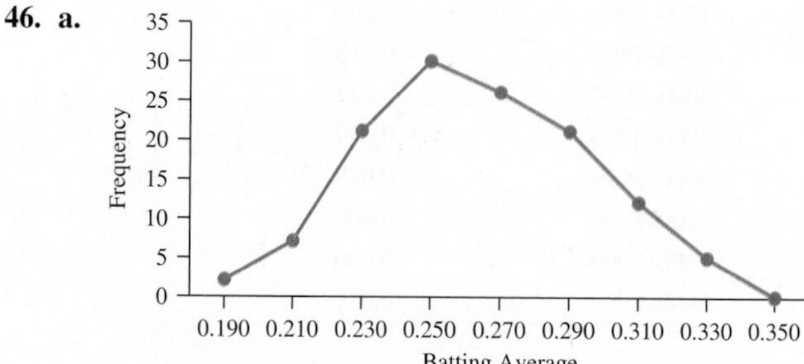

b.

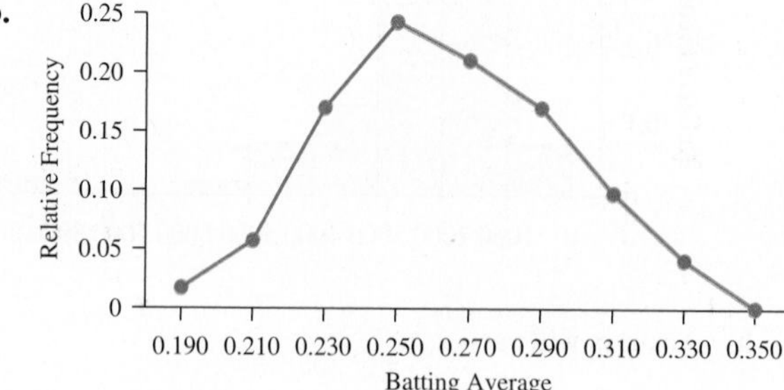

c.

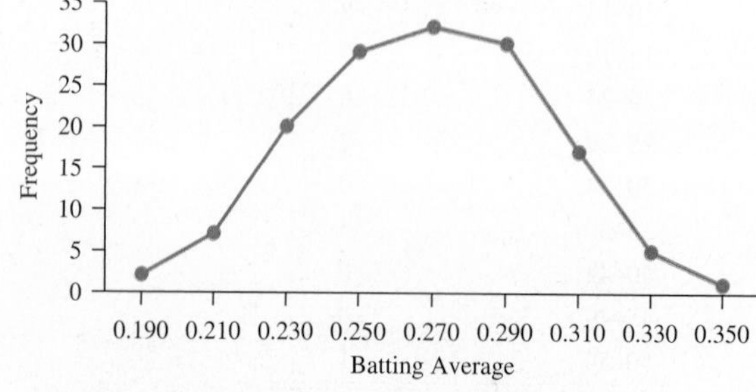

d.

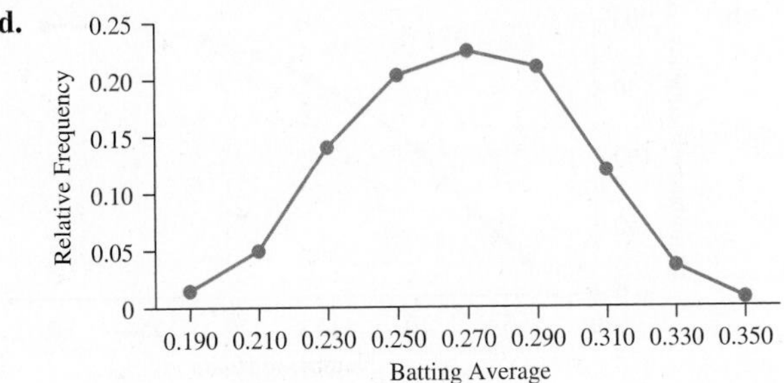

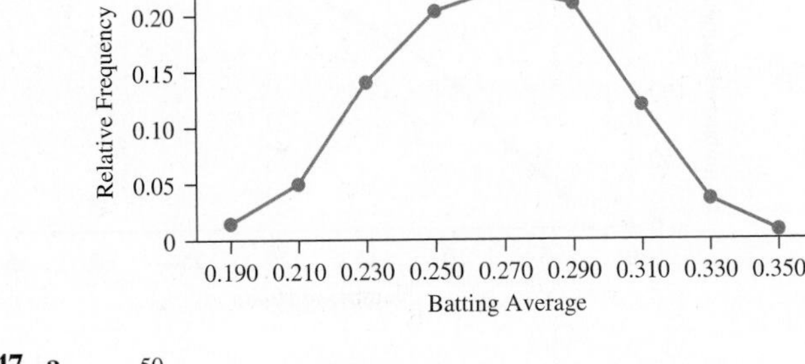

47. a.

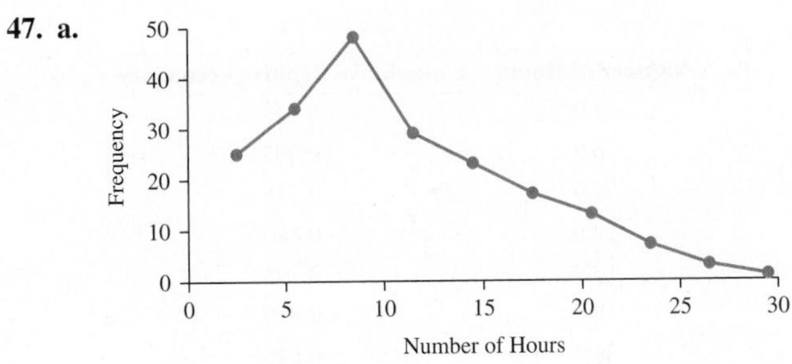

b.

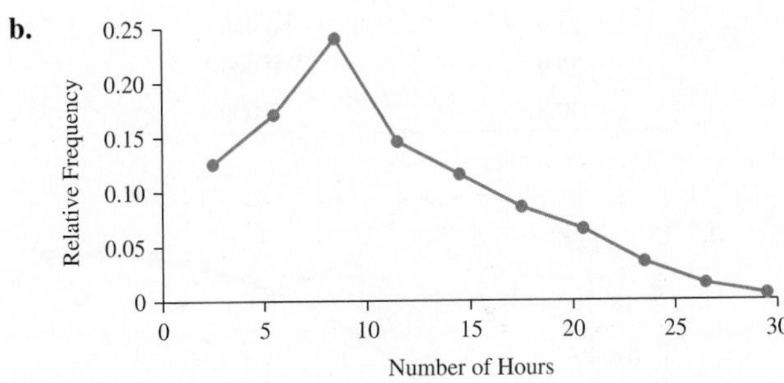

48. a.

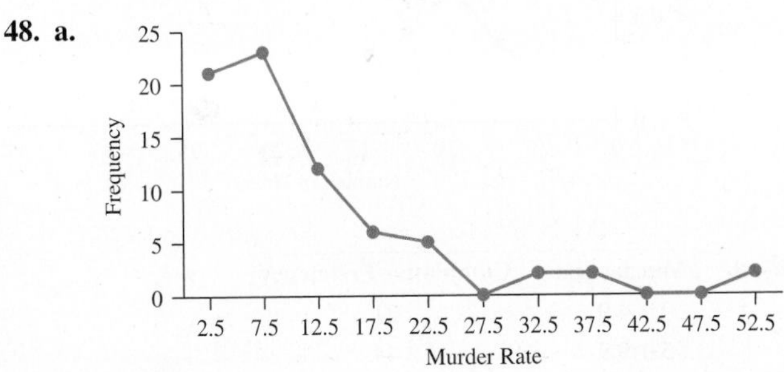

b.

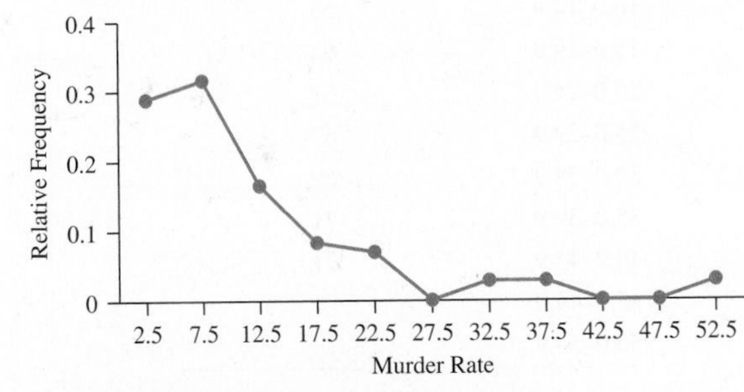

49. a.

Batting Average	Cumulative Frequency
0.180–0.199	4
0.200–0.219	18
0.220–0.239	59
0.240–0.259	118
0.260–0.279	176
0.280–0.299	227
0.300–0.319	256
0.320–0.339	266
0.340–0.359	267

b.

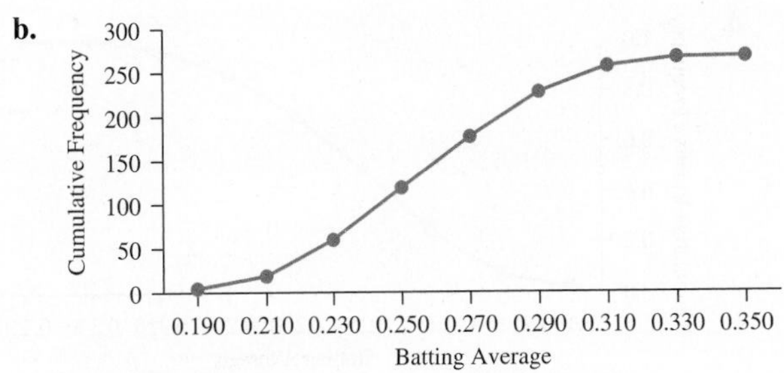

c.

Batting Average	Cumulative Relative Frequency
0.180–0.199	0.015
0.200–0.219	0.067
0.220–0.239	0.221
0.240–0.259	0.442
0.260–0.279	0.659
0.280–0.299	0.850
0.300–0.319	0.959
0.320–0.339	0.996
0.340–0.359	1.000

d.

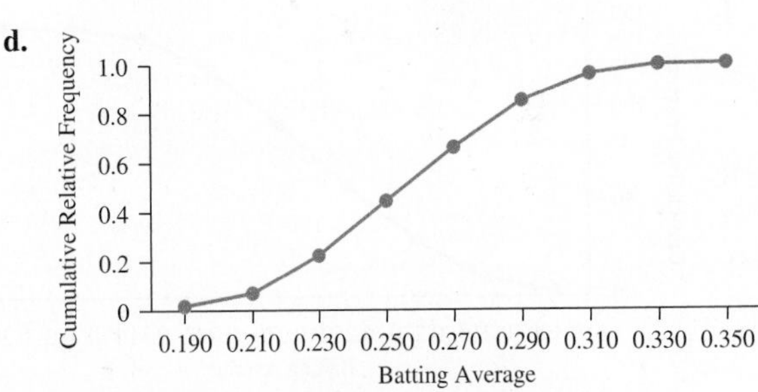

50. a.

Batting Average	Cumulative Frequency
0.180–0.199	2
0.200–0.219	9
0.220–0.239	30
0.240–0.259	60
0.260–0.279	86
0.280–0.299	107
0.300–0.319	119
0.320–0.339	124
0.340–0.359	124

b.

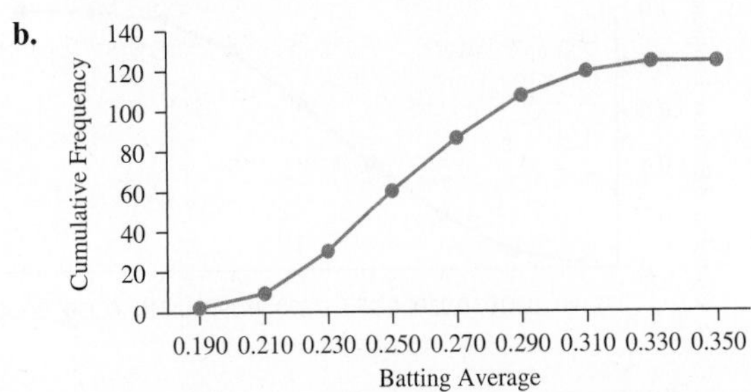

c.

Batting Average	Cumulative Relative Frequency
0.180–0.199	0.016
0.200–0.219	0.073
0.220–0.239	0.242
0.240–0.259	0.484
0.260–0.279	0.694
0.280–0.299	0.863
0.300–0.319	0.960
0.320–0.339	1.000
0.340–0.359	1.000

d.

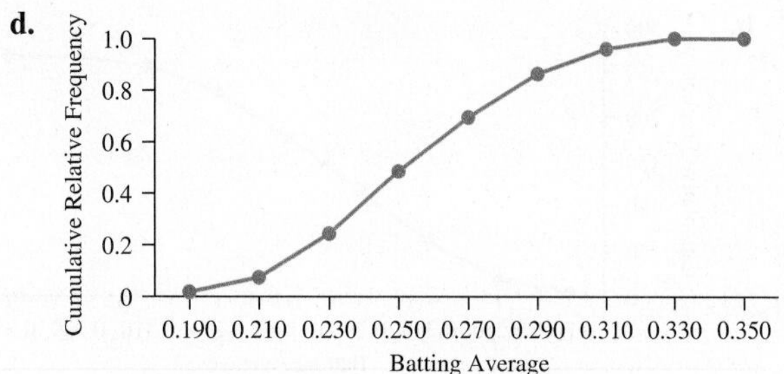

e.

Batting Average	Cumulative Frequency
0.180–0.199	2
0.200–0.219	9
0.220–0.239	29
0.240–0.259	58
0.260–0.279	90
0.280–0.299	120
0.300–0.319	137
0.320–0.339	142
0.340–0.359	143

f.

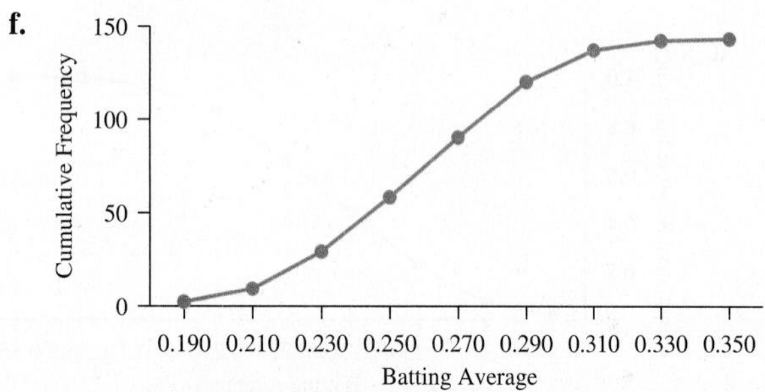

g.

Batting Average	Cumulative Relative Frequency
0.180–0.199	0.014
0.200–0.219	0.063
0.220–0.239	0.203
0.240–0.259	0.406
0.260–0.279	0.629
0.280–0.299	0.839
0.300–0.319	0.958
0.320–0.339	0.993
0.340–0.359	1.000

h.

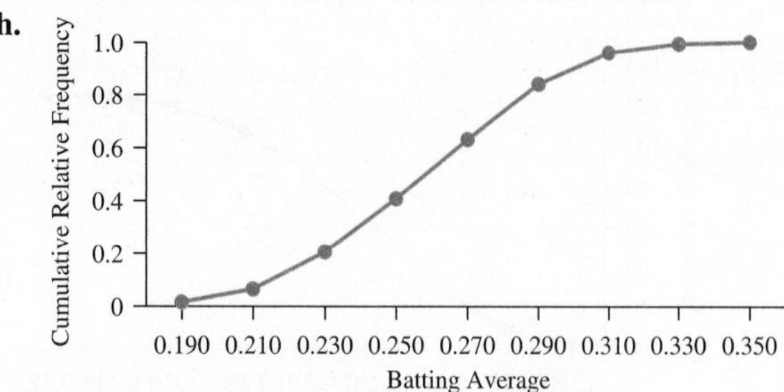

51. a.

Number of Hours	Cumulative Frequency
3.9	25
6.9	59
9.9	107
12.9	136
15.9	159
18.9	176
21.9	189
24.9	196
27.9	199
30.9	200

b.

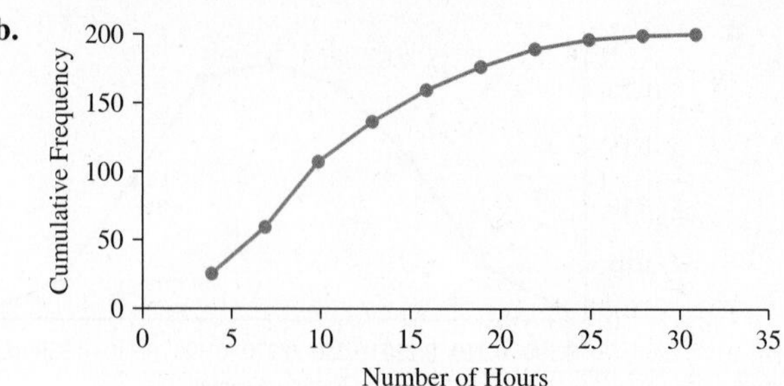

c.

Number of Hours	Cumulative Relative Frequency
3.9	0.125
6.9	0.295
9.9	0.535
12.9	0.680
15.9	0.795
18.9	0.880
21.9	0.945
24.9	0.980
27.9	0.995
30.9	1.000

d.

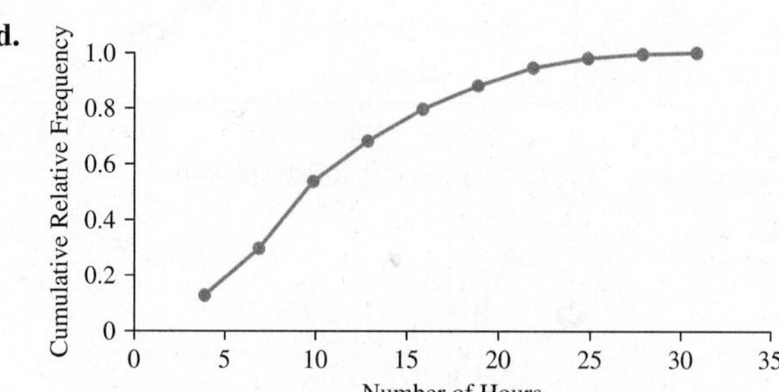

52. a.

Murder Rate	Cumulative Frequency
0.0–4.9	21
5.0–9.9	44
10.0–14.9	56
15.0–19.9	62
20.0–24.9	67
25.0–29.9	67
30.0–34.9	69
35.0–39.9	71
40.0–44.9	71
45.0–49.9	71
50.0–54.9	73

b.

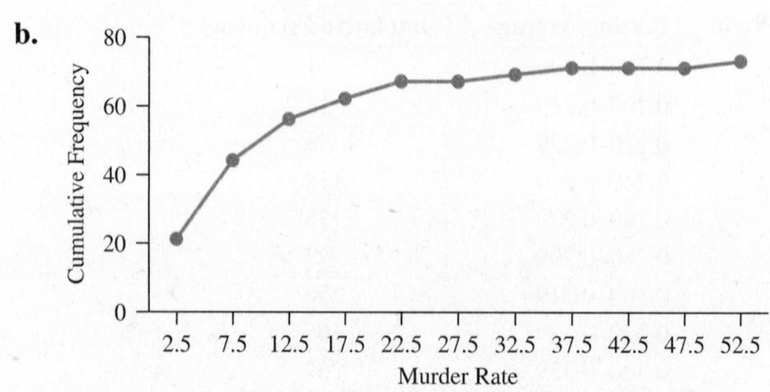

c.

Murder Rate	Cumulative Frequency
0.0–4.9	0.288
5.0–9.9	0.603
10.0–14.9	0.767
15.0–19.9	0.849
20.0–24.9	0.918
25.0–29.9	0.918
30.0–34.9	0.945
35.0–39.9	0.973
40.0–44.9	0.973
45.0–49.9	0.973
50.0–54.9	1.000

d.

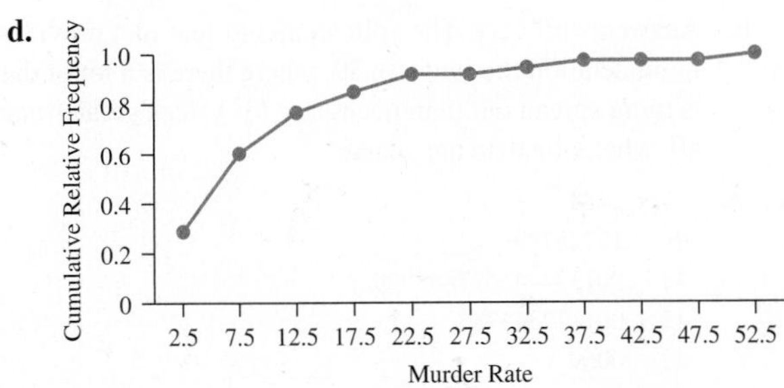

53. It is not possible to construct a histogram for this data set because the last class is open-ended.

54. Yes. The last class would become 30–34.9.

55. 0.15

56. a. The class widths are: 1, 0.5, 0.5, 1, 1, 3.

b.

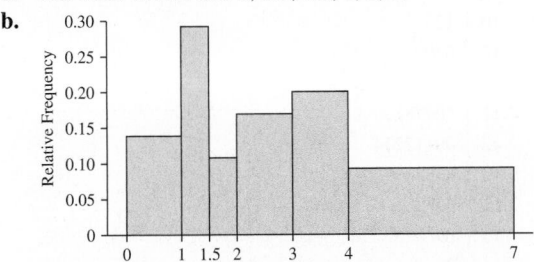

Relative Frequency
0.30
0.25
0.20
0.15
0.10
0.05
0
0 1 1.5 2 3 4 7

The histogram exaggerates the amount of data in the wider classes, particularly in the rightmost class.

c.

Class	Relative Frequency	Density
0.00–0.99	0.138	0.138
1.00–1.49	0.292	0.584
1.50–1.99	0.108	0.216
2.00–2.99	0.169	0.169
3.00–3.99	0.200	0.200
4.00–6.99	0.092	0.031

d.

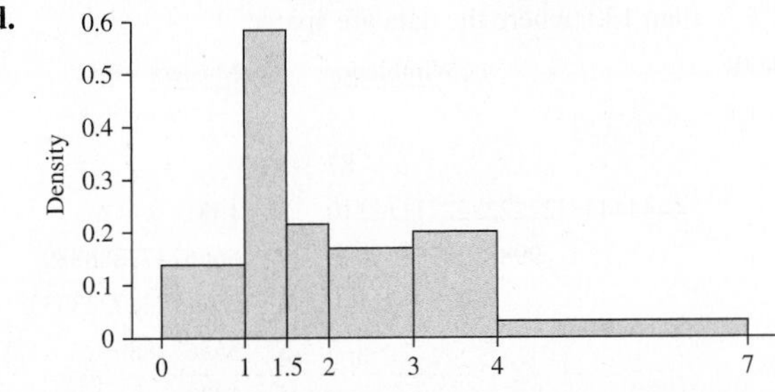

e. Dividing the relative frequency by the class width adjusts for the differing widths of the classes.

57. (i) is skewed; (ii) is approximately symmetric.

58. Skewed to the right

59. A and B are correct. C is not correct.

Section 2.3

Exercises 1 and 2 are the Check Your Understanding exercises for this section. Answers to these exercises are on page 83.

3. leaf

4. stems

5. time-series plot

6. time

7. True

8. False

9. True

10. False

11.

1	1225566
2	0012779
3	19
4	556
5	02578

12.

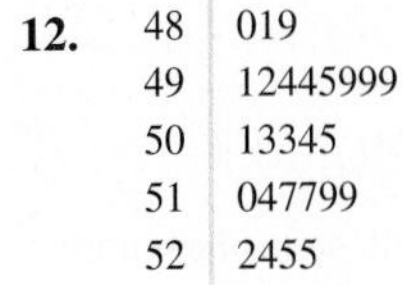

48 | 019
49 | 12445999
50 | 13345
51 | 047799
52 | 2455

13. The list is: 30, 30, 31, 32, 35, 36, 37, 37, 39, 42, 43, 44, 45, 46, 47, 47, 47, 47, 48, 48, 49, 50, 51, 51, 51, 52, 52, 52, 52, 54, 56, 57, 58, 58, 59, 61, 63

14. The list is: 14.4, 14.6, 14.8, 14.9, 15.1, 15.2, 15.2, 15.4, 15.5, 15.7, 15.7, 15.8, 16.0, 16.1, 16.1, 16.1, 16.2, 16.3, 16.7, 16.7, 16.9, 18.2, 18.3, 18.8

15.

10 20 30 40 50 60

16.

17. a.

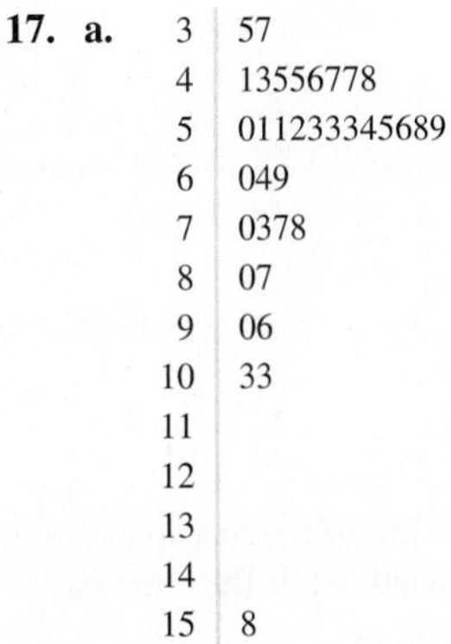

b.

3	
3	57
4	13
4	556778
5	01123334
5	5689
6	04
6	9
7	03
7	78
8	0
8	7
9	0
9	6
10	33
10	
11	
11	
12	
12	
13	
13	
14	
14	
15	
15	8

c. Answers will vary. The plot with the split stems has more detail than is needed for stems larger than 7.

18. a.

4	6688
5	000122244445555779
6	11111334444444666666688
7	00022222333355
8	12

b.

4	
4	6688
5	00012224444
5	5555779
6	1111133444444
6	66666688
7	000222223333
7	55
8	12
8	

c. Answers will vary. The plot with the split stems appears to have an appropriate amount of detail, while the other has too little.

19. a.

0	3
0	55669999
1	01111112222333344
1	555666889
2	11124
2	556777
3	0111334
3	555678
4	02
4	6
5	
5	9
6	
6	66

b. Answers will vary. The split stem-and-leaf plot provides more detail in the range 0–39, where there is a lot of data. It is more spread out than necessary for values greater than 40, where the data are sparse.

20. a.

9	2489
10	12346799
11	0013344456778889999
12	001122345788
13	0069
14	2
15	6
16	7
17	4

b.

9	24
9	89
10	1234
10	6799
11	00133444
11	5677888999
12	00112234
12	5788
13	00
13	69
14	2
14	
15	
15	6
16	
16	7
17	4
17	

c. Answers will vary. The split stem-and-leaf plot provides more detail in the range 90–130, where there is a lot of data. It is more spread out than necessary for values greater than 130, where the data are sparse.

21. a.

Wimbledon		Masters
	1	
87	1	
4444443332222222221111110	2	133
9988777766665555	2	56667777888889
21110	3	01111222222233333
5	3	5556778899
	4	1233
	4	6

b. The Wimbledon champions are generally younger than the Masters champions.

22. a.

PG or PG-13		R
	9	6
	10	2566
	11	7
54	12	26
653	13	24458
631	14	1
32	15	
420	16	
	17	0
2	18	
4	19	

b. R-rated movies tend to be shorter.

23.

Yes

24.

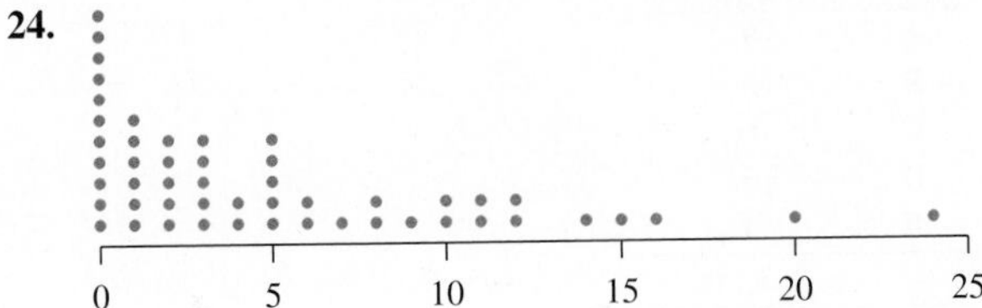

Skewed right

25. a.

b. Increasing: 2000–2003, 2007–2010.
Decreasing: 1996–2000, 2003–2007, 2010–2019.

26. a.

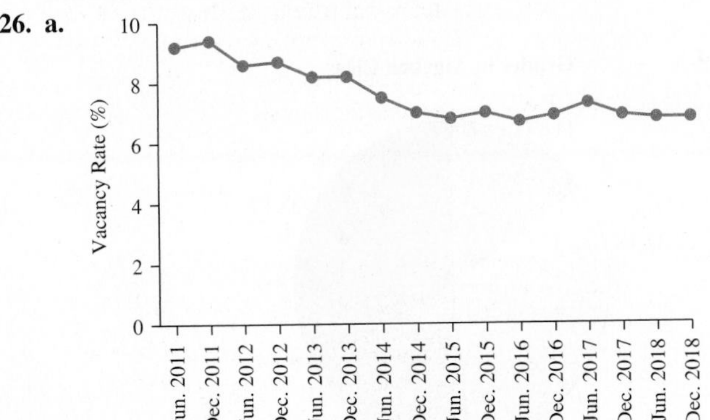

b. Decreasing

27. a.

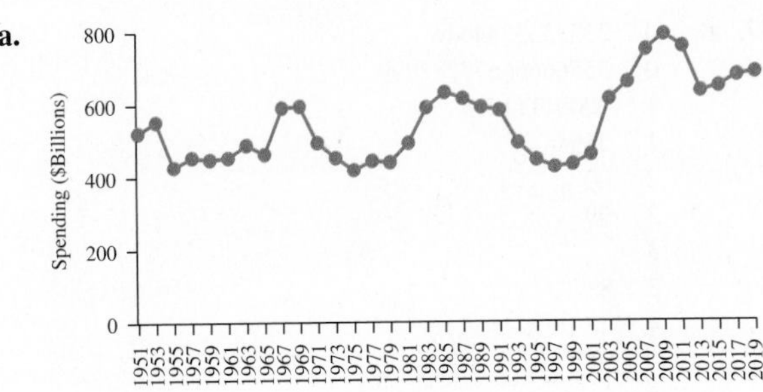

b. Increased: 1960s, 1980s, 2000s.
Decreased: 1950s, 1970s, 1990s, 2010s.

c. Decreased

d. Increased from 1965–1969, then decreased.

28. a.

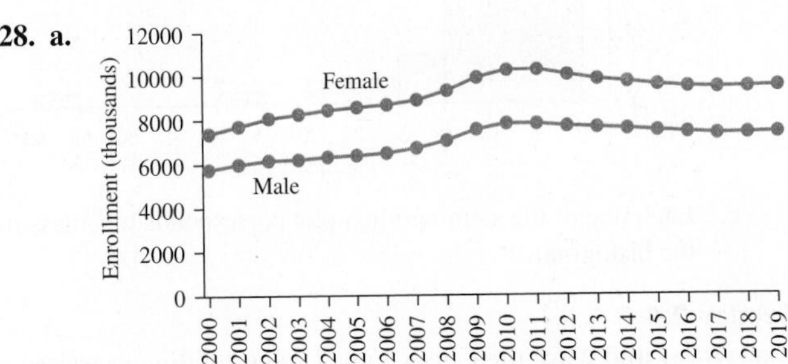

b. Female

29. a. $300,000

b. About the same

c. True

d. It decreased.

30. a. 1980

b. Approximately 85

c. About the same

31. a. 30 or 31

b. 1930

c. Less than

d. It increased.

e. True

32. The number of participating countries increased between 1952 and 1972.

33. a. 115

b. 1910

c. Less than

d. True

e. False

34. a. 2007

b. False

c. The number of shots increased, then decreased.

35. a. 4%

b. It increased.

c. It decreased.

d. It increased.

36. a. 2007

b. 2012

c. False

d. False

37. a.

Stem	Leaf
0	3333333344444
0	555666666677788999
1	00001111234
1	5668
2	00
2	99
3	
3	8
4	
4	
5	
5	5

b.

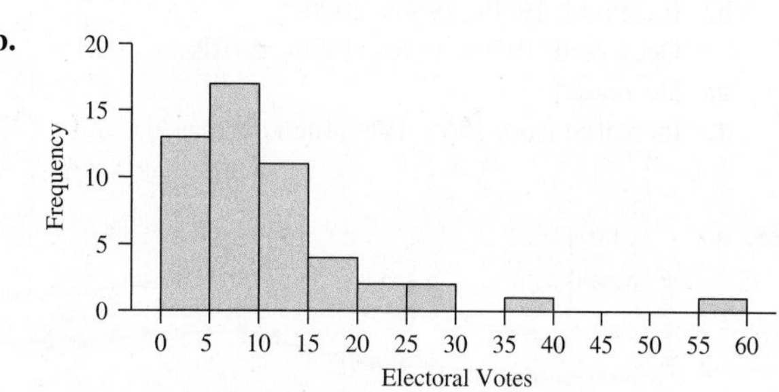

c. Each line of the stem-and-leaf plot corresponds to a class in the histogram.

Section 2.4

Exercises 1 and 2 are the Check Your Understanding exercises for this section. Answers to these exercises are on page 90.

3. zero

4. proportional

5. (i). Graph (A) presents an accurate picture, because the baseline is at zero. Graph (B) exaggerates the decline, because the baseline is above zero.

6. The bar graph presents the more accurate picture, because its baseline is at zero.

7. Graph (B) presents the more accurate picture. The baseline is at zero, and the bars are of equal width. The dollar bill graphic does not follow the area principle. The length and width of the smaller image are about 25% less than the length and width of the larger image, so the area of the smaller image is about 50% less than that of the larger image. This exaggerates the difference.

8. Graph (B) presents the more accurate picture, because it follows the area principle. In Graph (A), the area of the larger image is about six times that of the smaller image. This exaggerates the difference.

9. The graph is misleading, because the baseline is not at zero.

10. It presents an accurate picture, because the baseline is at zero.

11. It presents an accurate picture, because the baseline is at zero.

12. The graph is misleading, because the baseline is not at zero.

13. a. The bars appear shorter than they really are.

b.

Frequency (0, 20, 40, 60, 80); Male Senators, Female Senators

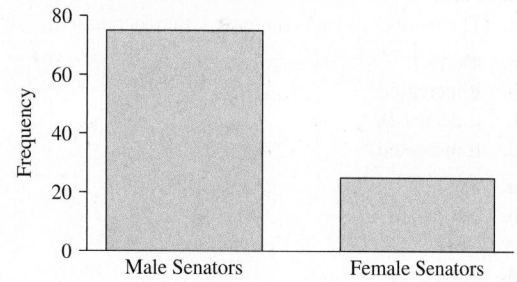

14. The increase is exaggerated because the baseline is not at zero.

15. (i) is more accurate. The plot on the left has its baseline at zero and presents an accurate picture. The plot on the right exaggerates the increase.

16. (ii) is more accurate. The plot on the left has its baseline at zero and presents an accurate picture. The plot on the right exaggerates the increase.

17. a.

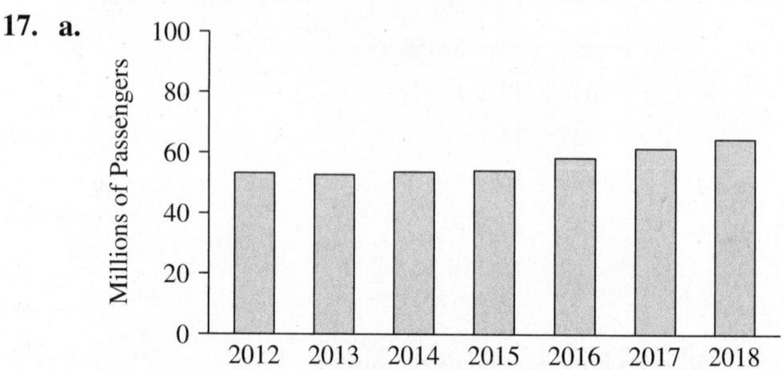

b. It makes the differences look smaller, because the scale on the *y*-axis extends much farther than the largest bar height.

c. Answers will vary. Figure 2.23 has the baseline at zero, and the scale on the *y*-axis is appropriate for the bar height.

CHAPTER 2 Quiz

1.

Grade	Frequency
A	9
B	5
C	6
D	3
F	4

2.

Grade	Relative Frequency
A	0.333
B	0.185
C	0.222
D	0.111
F	0.148

3.

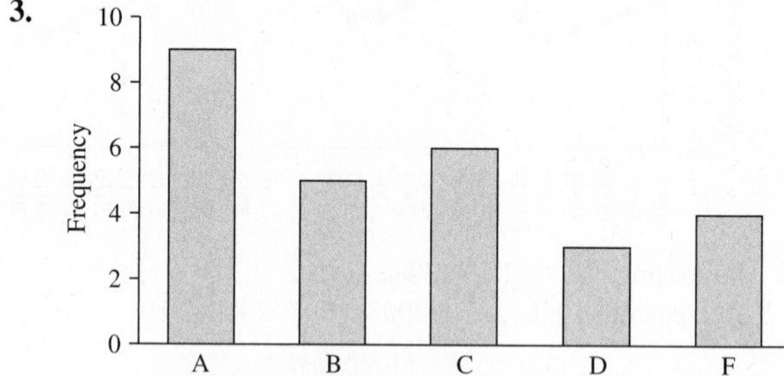

4.

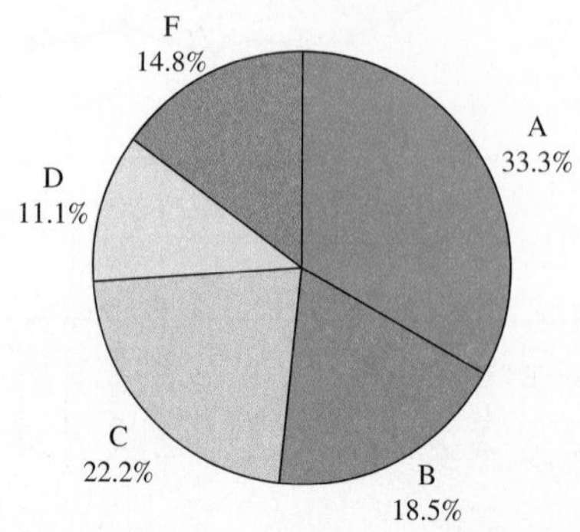

5. 5.0–7.9, 8.0–10.9, 11.0–13.9, 14.0–16.9, 17.0–19.9. The class width is 3.

6. True

7. a.

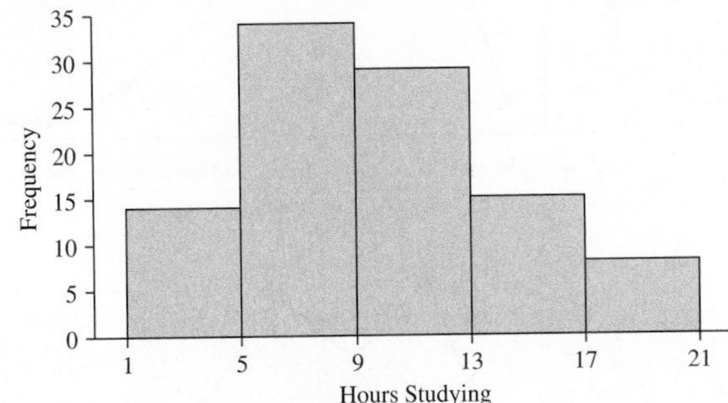

b.

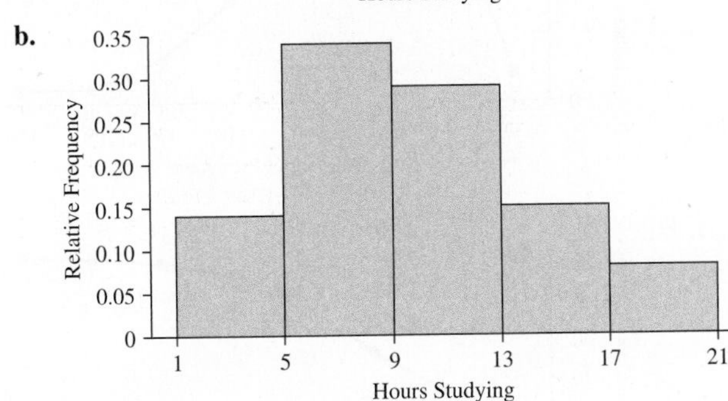

8.

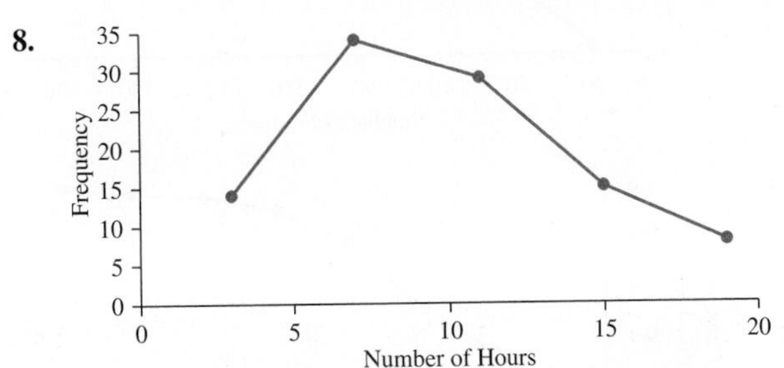

9.

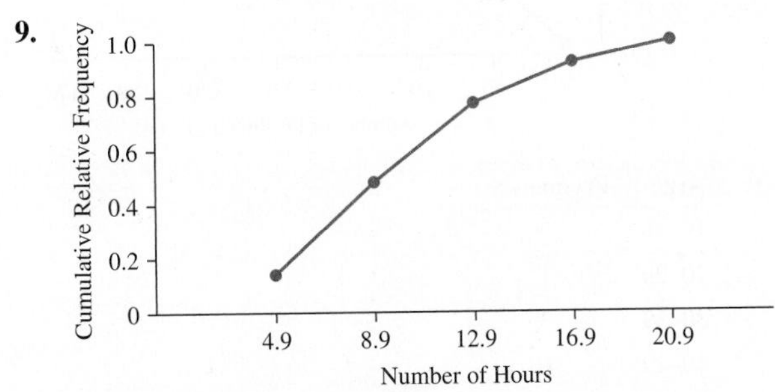

10. The list is: 11, 11, 15, 15, 19, 19, 19, 22, 22, 23, 25, 27, 28, 30, 30, 38, 44, 45, 47, 48, 50, 51, 53, 53, 55, 56, 58

11.

1	9
2	22889
3	579
4	1
5	
6	8

12.

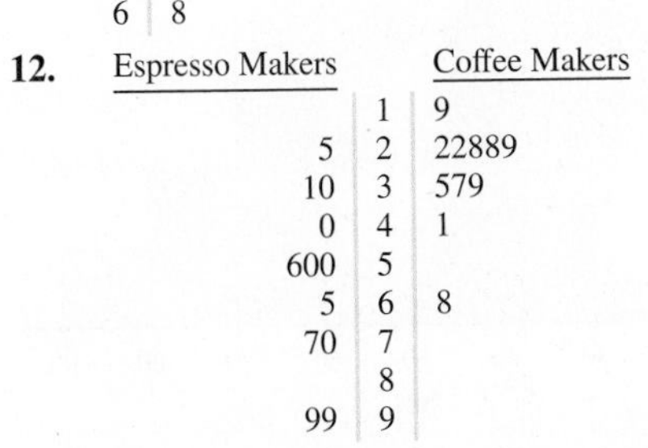

Espresso Makers		Coffee Makers
	1	9
5	2	22889
10	3	579
0	4	1
600	5	
5	6	8
70	7	
	8	
99	9	

13.

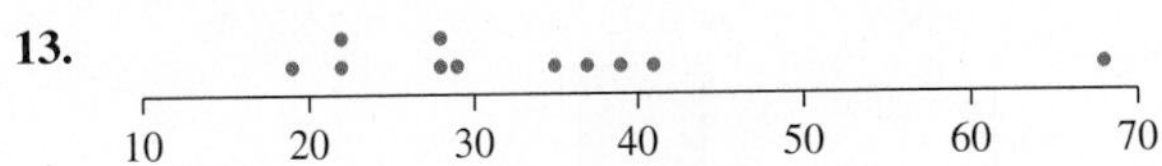

14.

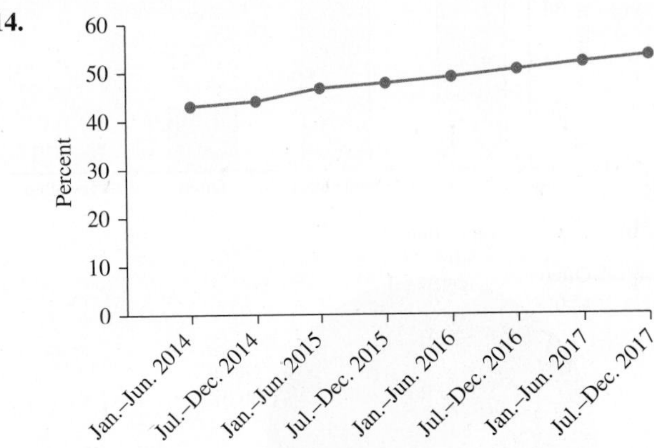

15. Twice

CHAPTER 2 Review Exercises

1. a. Somewhat **b.** True **c.** False **d.** True

2. a.

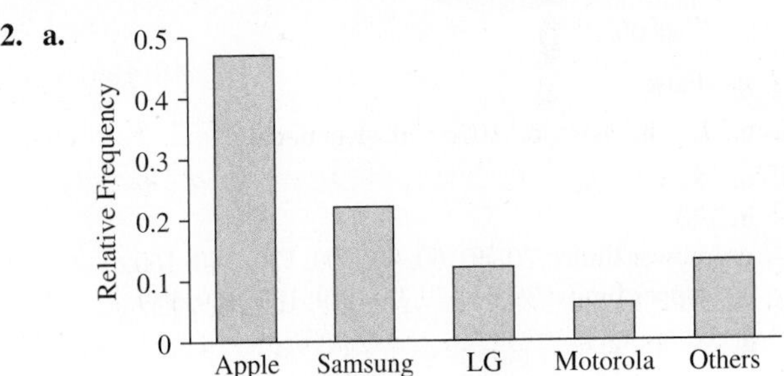

b.

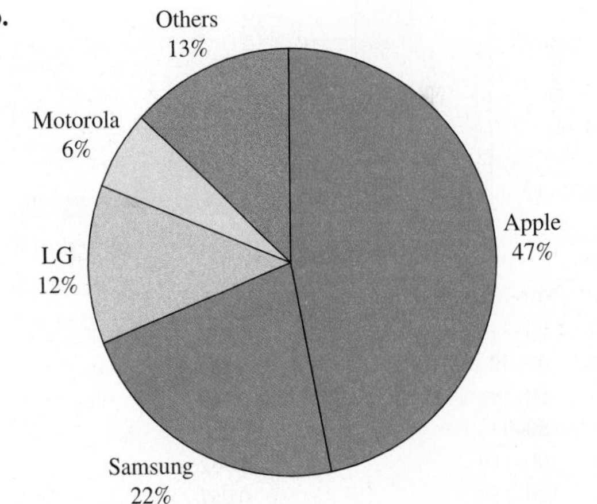

c. True

3. a.

Percent

Northeast Midwest South West

b. True

c. West

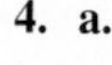

4. a.

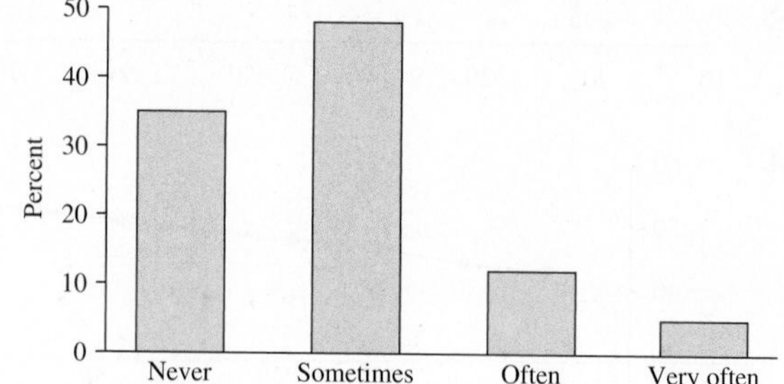

b.

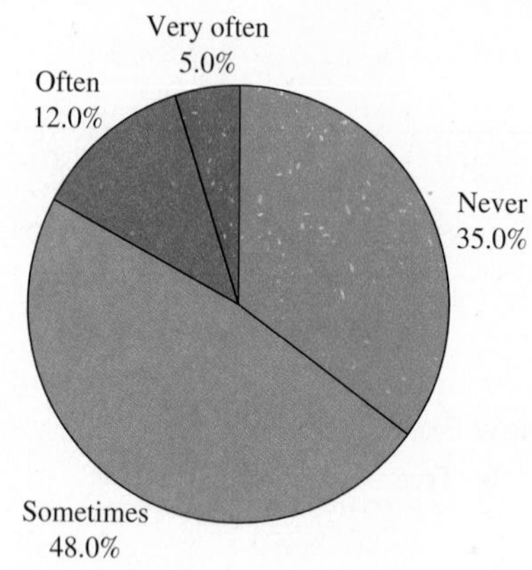

c. False

5. a. 7 **b.** 10 **c.** 10% **d.** Unimodal

6. a. 8

b. 20

c. Lower limits: 20, 40, 60, 80, 100, 120, 140, 160.
Upper limits: 39, 59, 79, 99, 119, 139, 159, 179.

d.

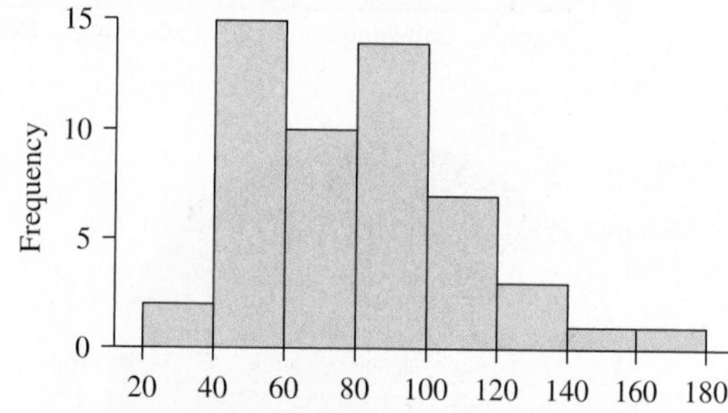

e.

Number of Freshmen	Relative Frequency
20–39	0.038
40–59	0.283
60–79	0.189
80–99	0.264
100–119	0.132
120–139	0.057
140–159	0.019
160–179	0.019

f.

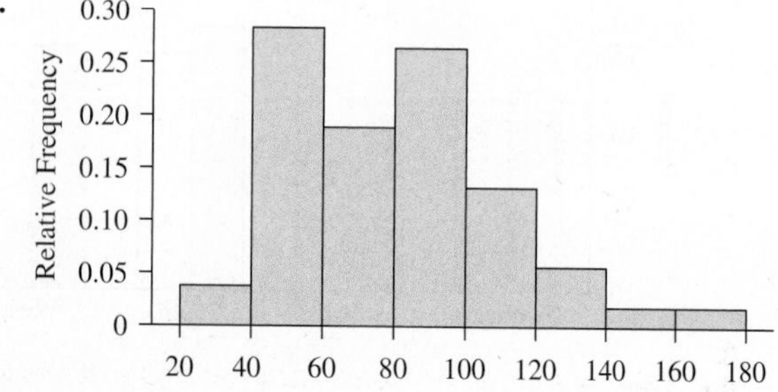

g. 22.6% **h.** 32.1%

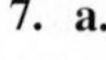

7. a.

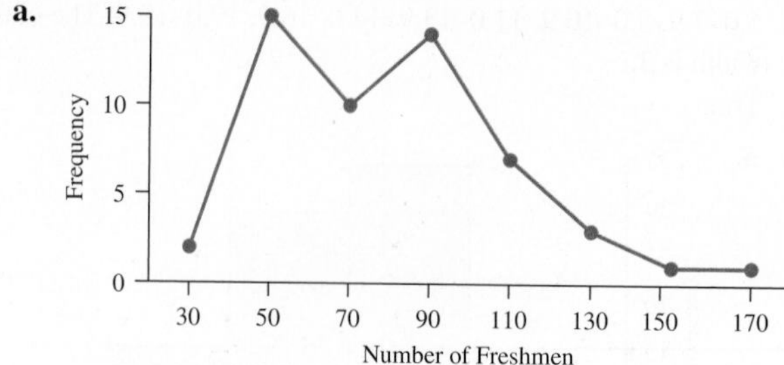

b.

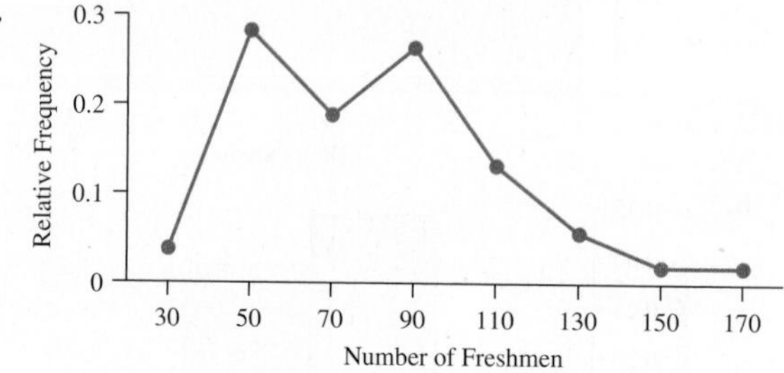

c.

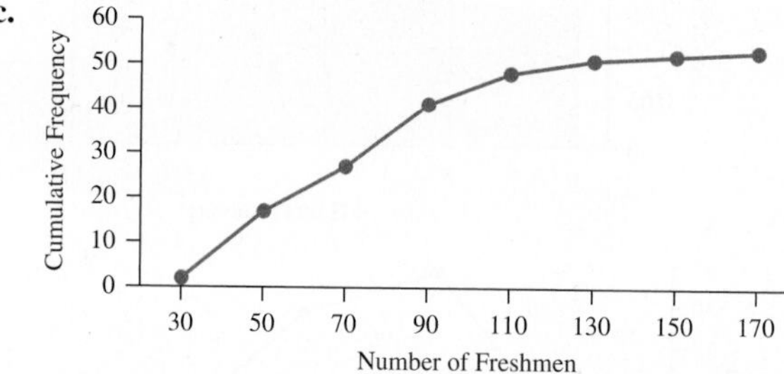

d.

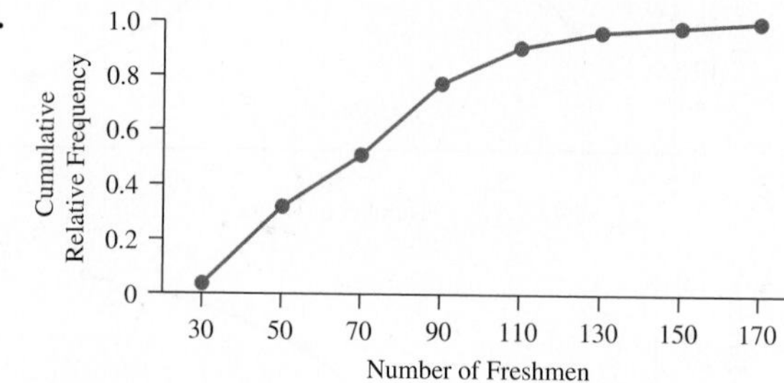

8. a.

Age	Frequency
10–19	2
20–29	1
30–39	3
40–49	10
50–59	9
60–69	9
70–79	4
80–89	2

b.

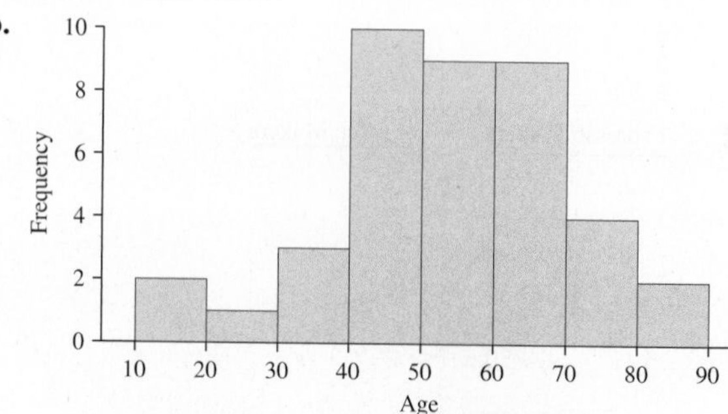

c.

Age	Relative Frequency
10–19	0.050
20–29	0.025
30–39	0.075
40–49	0.250
50–59	0.225
60–69	0.225
70–79	0.100
80–89	0.050

d.

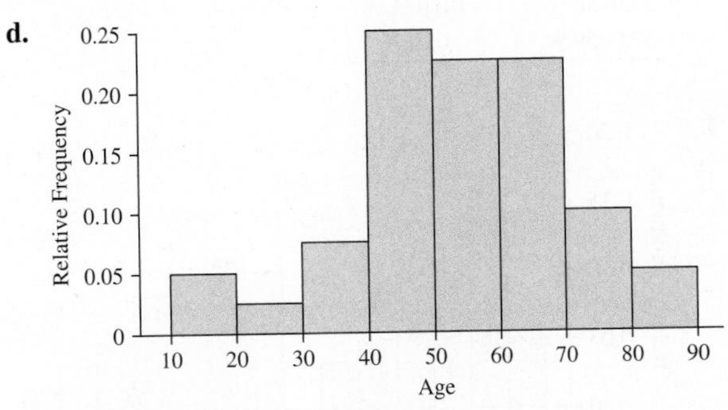

9.

1	25
2	8
3	235
4	0012368999
5	124566889
6	457777889
7	0167
8	11

10. a.

Age	Frequency
45–49	2
50–54	1
55–59	4
60–64	6
65–69	6
70–74	6
75–79	4
80–84	3
85–89	2
90–94	4

b.

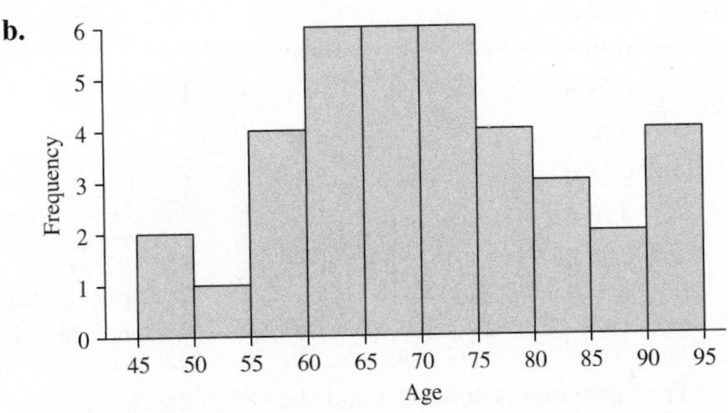

c.

Age	Relative Frequency
45–49	0.053
50–54	0.026
55–59	0.105
60–64	0.158
65–69	0.158
70–74	0.158
75–79	0.105
80–84	0.079
85–89	0.053
90–94	0.105

d.

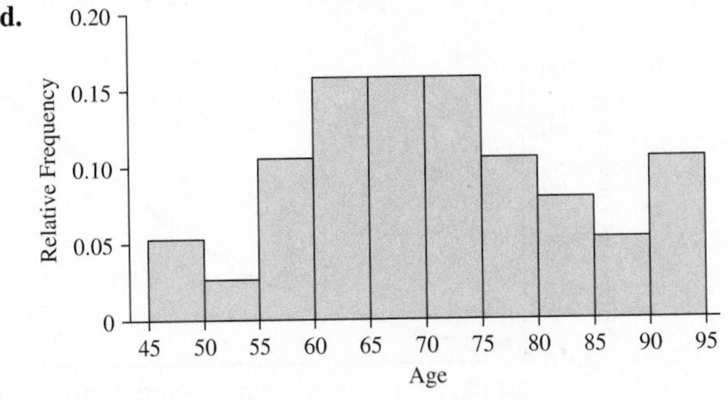

11. a.

Presidents		Monarchs
	1	25
	2	8
	3	235
96	4	0012368999
87763	5	124566889
877765443300	6	457777889
9887432110	7	0167
85310	8	11
3300	9	

b.

Presidents		Monarchs
	1	2
	1	5
	2	
	2	8
	3	23
	3	5
	4	00123
96	4	68999
3	5	124
8776	5	566889
443300	6	4
877765	6	57777889
432110	7	01
9887	7	67
310	8	11
85	8	
3300	9	
	9	

c. Answers will vary. The split stem-and-leaf plot provides a more appropriate level of detail.

12. a.

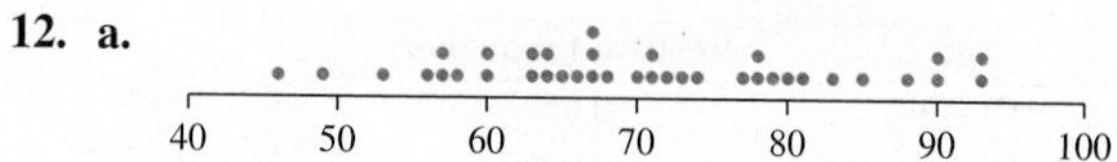

13. a.

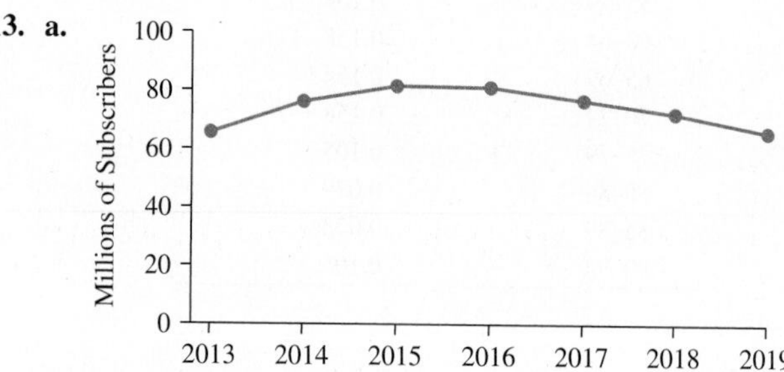

b.

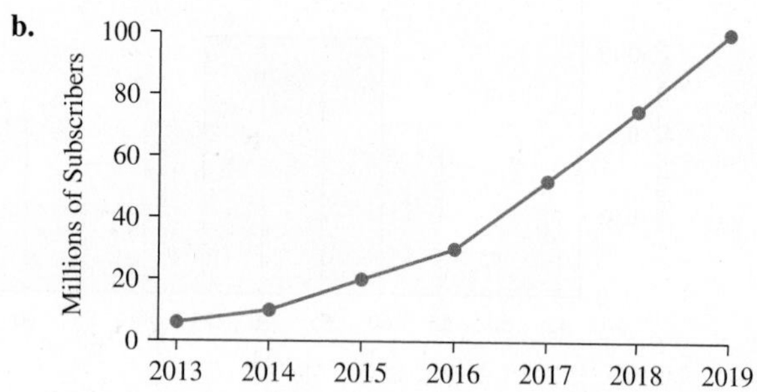

c. The number of Pandora subscribers increased from 2013 to 2015, then decreased. The number of Spotify subscribers increased steadily from 2013 to 2019.

14. a. Highest in 1990, lowest in 1930
b. Increasing
c. Life expectancy increased.

15. (i). The birth rate decreased somewhat between 2001 and 2017. The plot on the right is misleading, because the baseline is not at zero.

CHAPTER 2 Case Study

1.

Mileage	Frequency
18.0–19.9	3
20.0–21.9	3
22.0–23.9	7
24.0–25.9	4
26.0–27.9	4
28.0–29.9	6
30.0–31.9	3
32.0–33.9	1
34.0–35.9	1
36.0–37.9	0
38.0–39.9	1
40.0–41.9	5
42.0–43.9	5
44.0–45.9	2
46.0–47.9	3
48.0–49.9	3
50.0–51.9	2
52.0–53.9	3
54.0–55.9	1
56.0–57.9	1
58.0–59.9	1

2. There are 21 classes, which is too many for a data set with only 59 values.

3.

Mileage	Relative Frequency
18.0–20.9	0.051
21.0–23.9	0.169
24.0–26.9	0.119
27.0–29.9	0.119
30.0–32.9	0.051
33.0–35.9	0.034
36.0–38.9	0.000
39.0–41.9	0.102
42.0–44.9	0.119
45.0–47.9	0.051
48.0–50.9	0.085
51.0–53.9	0.051
54.0–56.9	0.034
57.0–59.9	0.017

4.

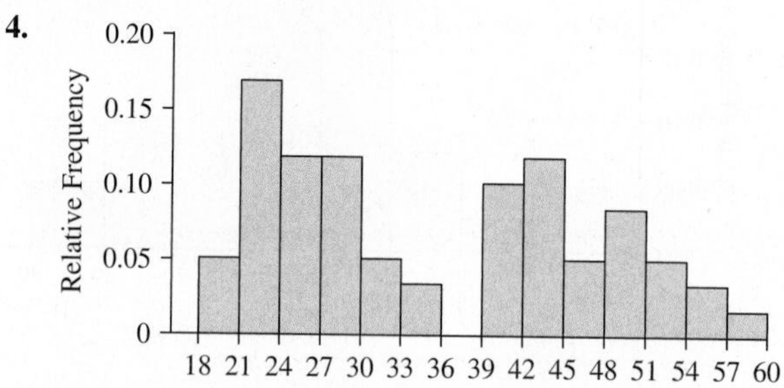

The histogram is bimodal, with little skewness.

5.

Mileage	Frequency
22.0–23.9	1
24.0–25.9	1
26.0–27.9	6
28.0–29.9	9
30.0–31.9	13
32.0–33.9	20
34.0–35.9	8
36.0–37.9	5
38.0–39.9	1

6.

Mileage	Relative Frequency
22.0–23.9	0.016
24.0–25.9	0.016
26.0–27.9	0.094
28.0–29.9	0.141
30.0–31.9	0.203
32.0–33.9	0.313
34.0–35.9	0.125
36.0–37.9	0.078
38.0–39.9	0.016

7.

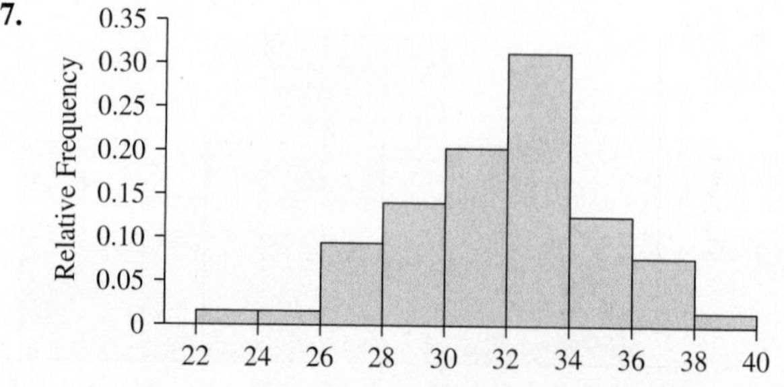

The histogram is unimodal and skewed to the right.

8. The mileages vary more for the hybrid cars.

9.

Hybrid		Non-hybrid
	1	
999	1	
443333222111	2	2
999888766655	2	5667777888999999
43100	3	000000011111222222222222333333334444
9	3	5555666679
443322211110	4	
988666	4	
22200	5	
86	5	

Answers may vary as to which illustrates the comparison more effectively. The back-to-back stem-and-leaf plot puts the data sets next to each other on the same axis. The histograms arguably illustrate the shapes of the distributions more clearly.

CHAPTER 3

Section 3.1

Exercises 1–8 are the Check Your Understanding exercises for this section. Answers to these exercises are on page 117.

9. mean

10. median

11. extreme values

12. mode

13. False

14. False

15. False

16. True

17. Mean: 23.4; median: 26; mode: 27

18. Mean: 3; median: 15; mode: −20

19. Mean: 5.5; median: 14; mode: 28

20. Mean: 81.5; median: 93.5; mode: 98

21. **a.** 22.7, underestimates
b. 27.3, overestimates
c. Answers will vary.

22. **a.** 321.67, overestimates
b. 206.33, underestimates
c. Answers will vary.

23. 24.6

24. 48.0

25. 145.0

26. 47.9

27. (ii)

28. (iv)

29. (i)

30. (iii)

31. Mean: 30.4; median: 29; mode: 27

32. Mean: 20.9; median: 24; mode: 25

33. Skewed to the right, because the mean is greater than the median.

34. Approximately symmetric, because the mean and median are nearly equal.

35. Skewed left

36. Skewed right

37. **a.** 290
b. 300
c. Median: 290; mean: 282.9

38. **a.** 200.2 **b.** 199.0

39. **a.** 9.9 **b.** 9.5 **c.** Mean: 9.2; median: 9

40. **a.** 1.272
b. 1.29
c. The mean would increase more, because it is not resistant.

41. **a.** 49.5 **b.** 24.9
c. Skewed to the right

42. **a.** 28.6
b. 30.1
c. Approximately symmetric

43. **a.** 786.4
b. 312.6
c. Mean: 1106.8; median: 483.6

44. **a.** 251.9
b. 229.55
c. The mean is less: 229.3 vs. 266.9.
The median is about the same: 229.6 vs. 228.1.

45. **a.** Mean: 15.08; median: 14.0
b. Mean: 13.93; median: 12.8
c. More people watched in 2013–2014.

46. **a.** 29.6
b. 56.3
c. 21
d. 30
e. The West has a lot more breweries per state than the East.
f. Mean: 31.4; median: 25.5
g. The means and medians for East and West are more similar after California is removed.

47. **a.** Mean: 5.267; median: 5.53
b. Mean: 5.884; median: 6.32
c. The median increased more than the mean.

48. **a.** Mean: 316.46; median: 254.6
b. Mean: 327.42; median: 258.3
c. The mean increased. **d.** The median increased.

49. **a.** Mean: 296; median: 302.5
b. Mean: 285.5; median: 285
c. Offensive linemen are somewhat heavier.

50. **a.** Mean: 872.51; median: 876.02
b. Mean 881.80; median: 878.88 **c.** The prices are about the same.

51. **a.** Mean: 2145; median: 999
b. Mean 528.5; median: 359 **c.** Yes, the data support this claim.

52. **a.** Mean: 355.07; median: 227.7
b. Mean 163.19; median: 73.7 **c.** No, the data do not support this claim.

53. **a.** 373.49 **b.** 335.65 **c.** The mean would increase to 1023.7; the median would be unchanged.

54. **a.** 20.117 **b.** 14.98 **c.** 14.98 **d.** 116.99
e. The mean would become 15.504. The median and mode would be unchanged.

55. **a.** 37.2 **b.** Too small, because the average age within the first class is greater than 16, but the midpoint is only 15.

56. a. 28.2
b. Too large. The approximate mean assumes that these people are 95 years old.

57. a. 81.1 **b.** Too large, because the mean would be 0, while the midpoint would be 10.

58. a. 45.9 **b.** Too small, because the mean would be 74, while the midpoint would be 72.5.

59. a. 37.5 **b.** 36 **c.** Skewed to the right
d. Answers will vary. Here is one possibility:

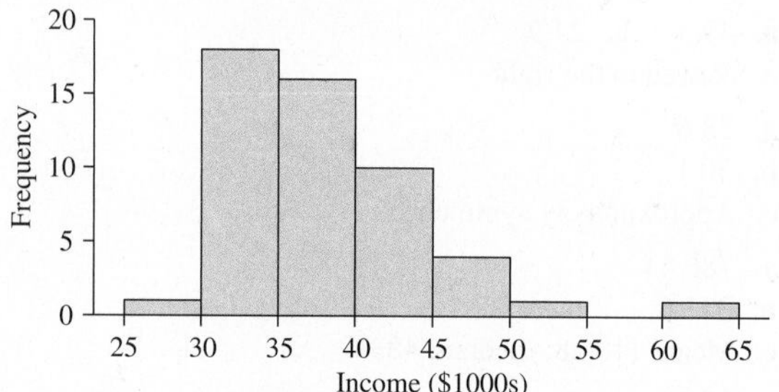

The results agree with the expectation.

60. a. Mean: 222.49; median: 229.95
b. Approximately symmetric, because the mean is close to the median.
c.

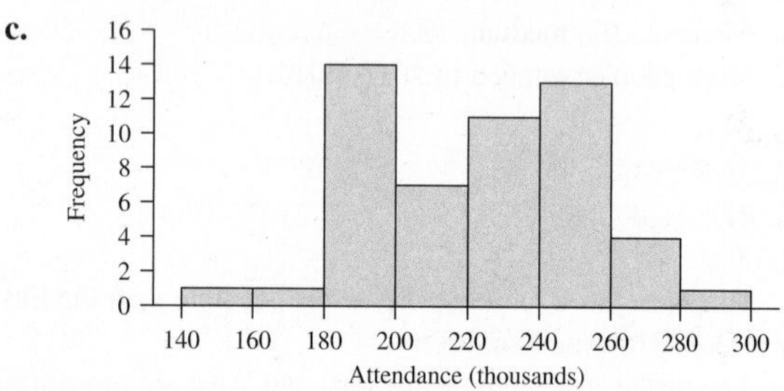

The results agree with the expectation.

61. Fiction

62. Television

63. 160 pounds. The mean would be greater than the median, because the histogram is skewed to the right.

64. $26,000. The median would be less than the mean, because the histogram is skewed to the right.

65. iii. Between 510 and 520

66. ii. Between 69 and 70

67. a. 161,000 **b.** 139,600 **c.** 66.8
d. There are more women than men.

68. a. 2050 **b.** 2340 **c.** 79.8
d. There are more students in section B than in section A.

69. 64.75 inches

70. 84

71. a. 13 **b.** 12

72. a. 17 **b.** 18

73. 208

74. 88

75. $12.14

76. 75

77. a. 220,600 **b.** 20,000
c. The mean is more appropriate, because it better reflects the amount of money the family now has.

78. a. 0.26
b. 0
c. The mean is more useful.

79. Answers will vary.

80. Answers will vary.

81. Answers will vary.

82. Answers will vary.

83. No. If the largest or smallest value is an outlier, the value of the midrange will be strongly affected.

84. Yes, all are equal to the average of the two numbers.

85. a. 68.4 **b.** 68 **c.** 5.417, 6.000, 5.667, 5.583, 5.833
d. 5.7; yes **e.** 5.667; yes

86. a. 45,000 **b.** 40,000 **c.** 46,000; yes **d.** 47,250; yes

87. a. They are both equal to 5.
b. The median. The mean increases to 6, and the median increases to 8.
c. The mean. The mean increases to 9.5, and the median increases to 8.
d. As the value becomes more extreme, the mean steadily increases, but the median stays the same. At some point, the mean becomes greater than the median.

88. a. Mean: 0.76; median: 1
b. Skewed to the left, because the mean is less than the median.
c.

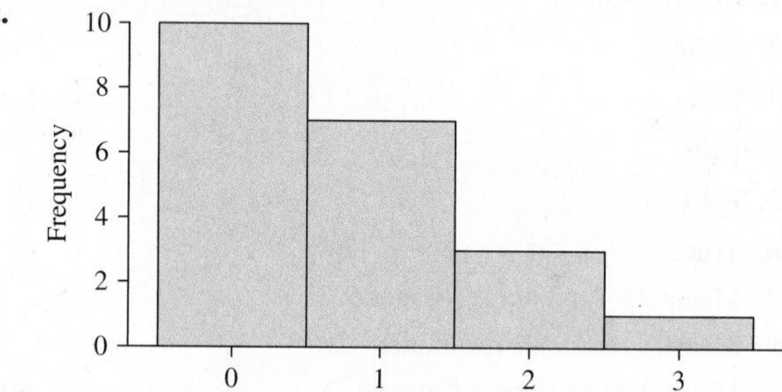

In fact, the histogram is skewed to the right.

Section 3.2

Exercises 1–10 are the Check Your Understanding exercises for this section. Answers to these exercises are on page 139.

11. zero

12. variance

13. 68%

14. $1 - 1/K^2$

15. False

16. True

17. False

18. True

19. Variance is 100; standard deviation is 10.

20. Variance is 225; standard deviation is 15.

21. Variance is 49; standard deviation is 7.

22. Variance is 64; standard deviation is 8.

23. Variance is 289; standard deviation is 17.

24. Variance is 36; standard deviation is 6.

25. 4

26. 1.5

27. 9

28. 12.25

29. 34

30. 9

31. 28 and 44

32. 39 and 75

33. Variance is approximately 228.41; standard deviation is approximately 15.11.

34. Variance is approximately 431.02; standard deviation is approximately 20.76.

35. Variance is approximately 5680; standard deviation is approximately 75.37.

36. Variance is approximately 1238.27; standard deviation is approximately 35.19.

37. **a.** Population
b. 942.4658037
c. 888,241.79

38. **a.** Sample
b. 1.663329993
c. 2.7666667

39. **a.** 68% **b.** 20 and 44 **c.** 75%

40. **a.** 95% **b.** 129 and 139 **c.** 88.9%

41. **a.** 95% **b.** 140 and 212 **c.** 88.9%

42. **a.** 68% **b.** 65 and 93 **c.** 75%

43. **a.** 64.98
b. 80.02
c. 2018

44. **a.** 2.1 **b.** 3.0 **c.** Sports cars

45. **a.** 9.22
b. 20.55
c. Older

46. **a.** 11.17
b. 16.09
c. Women

47. **a.** 26.8 **b.** 23.2 **c.** Offensive

48. **a.** 18.3 **b.** 87.4 **c.** East: 66; West: 300
d. West **e.** West
f. 30.5. The standard deviation decreased when the outlier was removed; this shows that it is not resistant.
g. 106; the range decreased when the outlier was removed, which shows that it is not resistant.

49. **a.** 2.91
b. 2.17
c. The range in 2013–2014 is 9.4. The range in 2018–2019 is 7.7.
d. The spread decreased.
e. The spread decreased.

50. **a.** 155.0
b. 159.2
c. The spread increased.

51. **a.** 18.528 **b.** 18.482 **c.** May

52. **a.** Stocks: 15.10; bills: 1.84; bonds: 8.35. Stocks are riskiest, bill are less risky.
b. Yes
c. Stocks: 5.96; bills: 1.84; bonds: 4.59. The results follow the theory.

53. **a.** 7.58 **b.** 2.75

54. **a.** 348.7 **b.** 18.7

55. **a.** Almost all **b.** 68% **c.** $38.98 and $50.22

56. **a.** 25 and 39 **b.** 95% **c.** Almost all

57. **a.** 422 **b.** 590

58. **a.** 570 **b.** 408

59. **a.** 16% **b.** 2.5% **c.** 34%

60. **a.** 16% **b.** 2.5% **c.** 47.5%

61. **a.** 521 or 522 **b.** 605 or 606 **c.** 295

62. **a.** 504 **b.** 585 **c.** 204

63. Not appropriate; histogram is skewed.

64. Not appropriate; histogram is skewed.

65. Approximately 68%

66. Approximately 95%

67. At least 75% of the days had temperatures between 56.2°F and 68.6°F.

68. 3. We expect that almost all of the data will be within 3 standard deviations of the mean. The largest and smallest values in the data set are between 9 and 10 inches from the mean. We therefore expect the standard deviation to be closest to 3.

69. 2.5. We expect that almost all of the data will be within 3 standard deviations of the mean. The largest and smallest values in the data set are between 7 and 8 inches from the mean. We therefore expect the standard deviation to be closest to 2.5.

70. At least 88.9% between 5.62 and 18.22.

71. **a.** Impossible **b.** Possible **c.** Impossible **d.** Possible

72. **a.** Impossible **b.** Possible **c.** Possible **d.** Impossible

73. Yes. Answers will vary.

74. Yes. Any list in which all the values are the same.

75. **a.** 0.045 **b.** 0.351 **c.** Weight

76. **a.** 0.13 **b.** 0.15 **c.** Calculus scores

77. **a.** 4.8

b.

x	$x - \bar{x}$	$(x - \bar{x})^2$	$\lvert x - \bar{x}\rvert$
1	−3.8	14.44	3.8
3	−1.8	3.24	1.8
4	−0.8	0.64	0.8
7	2.2	4.84	2.2
9	4.2	17.64	4.2

c. SD = 3.1937; MAD = 2.56
d. SD = 10.677; MAD = 7
e. The MAD is more resistant. Its value changed less when the outlier was added to the data set.

78. **a.** 17,795.13
b. 17,795.13; the standard deviation does not increase.
c. 18,684.89; the standard deviation does increase by 5%.

Section 3.3

Exercises 1–10 are the Check Your Understanding exercises for this section. Answers to these exercises are on page 158.

11. Quartiles

12. second

13. interquartile range

14. outlier

15. False

16. True

17. False

18. True

19. a. −1 **b.** 1.5 **c.** 11

20. a. −2.25 **b.** 1.5 **c.** 35

21. The outlier is 4.91. It seems certain to be an error.

22. The outlier is $1,200,000. It could conceivably be correct.

23. a. $Q_1 = 20; Q_3 = 44$ **b.** 24
c. Lower: −16; upper: 80 **d.** 82 is the only outlier.

24. a. $Q_1 = 3; Q_3 = 7$ **b.** 4
c. Lower: −3; upper: 13 **d.** 15, 25

25. a. 34 **b.** 14 **c.** 40 **d.** 8

26. a. 80 **b.** 44 **c.** 16 **d.** 61

27. Yes

28. Yes

29. a. 1.13 **b.** 1.16 **c.** SAT **d.** 25 **e.** 280

30. a. 1.33 **b.** 1 **c.** Anna's fish **d.** 147.6 **e.** 148

31. a. 19 **b.** Lower: 79.5; upper: 155.5 **c.** No

32. a. 15 **b.** Lower: 77.5; upper: 137.5 **c.** Yes

33. a. $Q_1 = 11; Q_3 = 32$ **b.** 15
c. Lower: −20.5; upper: 63.5 **d.** 67, 86, 97, 97, 116
e.

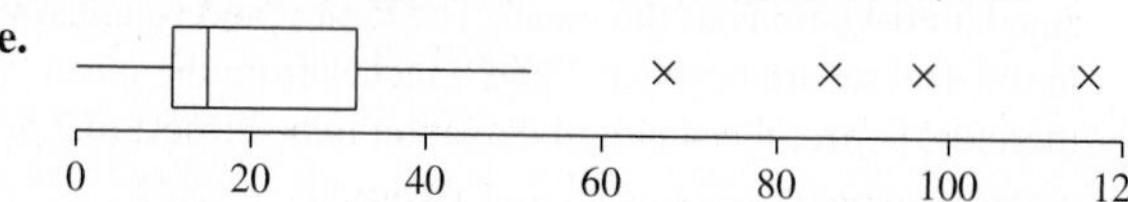

f. Skewed to the right **g.** 12 **h.** 49 **i.** 55th percentile

34. a. $Q_1 = 37; Q_3 = 65.5$ **b.** 56.5
c. Lower: −5.75; upper: 108.25 **d.** No outliers
e.

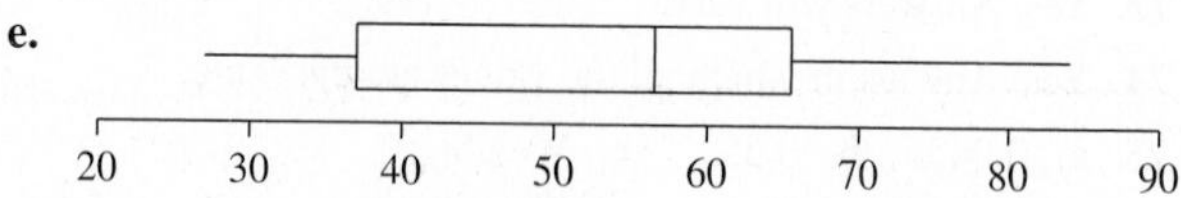

f. Approximately symmetric (slightly skewed to the left)
g. 37 **h.** 64 **i.** 54th percentile

35. a. $Q_1 = 19; Q_3 = 22$ **b.** 21
c. Lower: 14.5; upper: 26.5 **d.** 31, 36, 38, 39
e.

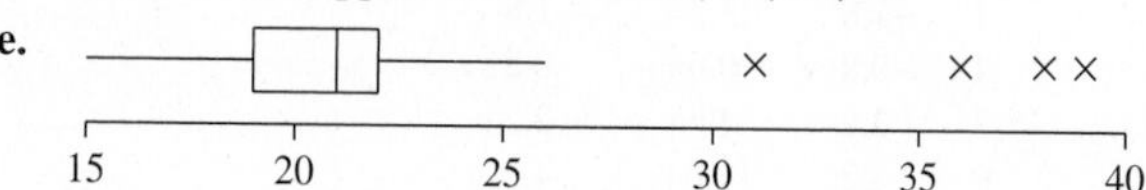

f. Skewed to the right **g.** 17 **h.** 26 **i.** 90th percentile

36. a. $Q_1 = 25.2; Q_3 = 38.1$ **b.** 30.8
c. Lower: 5.85; upper: 57.45 **d.** No outliers
e.

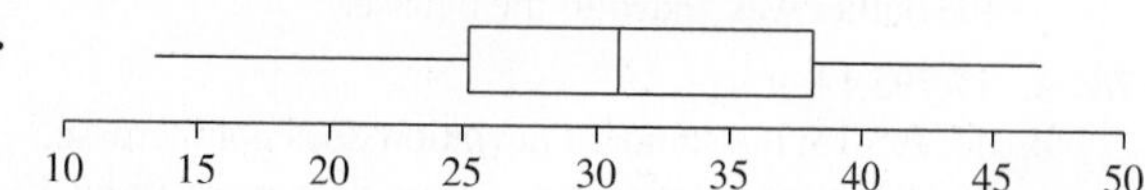

f. Approximately symmetric **g.** 27.7 **h.** 18.3
i. 47th percentile

37. a. $Q_1 = 14; Q_3 = 41$ **b.** 27
c. Lower: −26.5; upper: 81.5 **d.** No
e. No, neither is an outlier.
f.

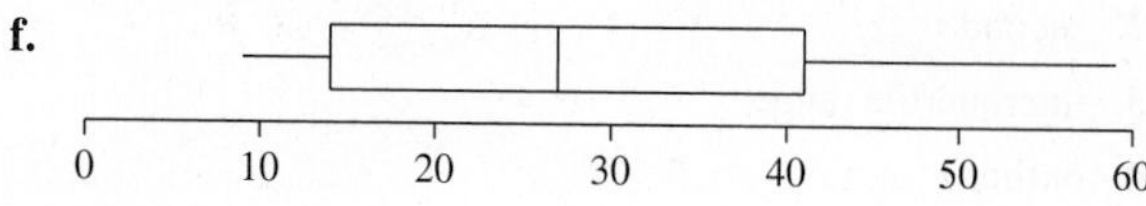

g. Skewed to the right **h.** 24 **i.** 39

j. 66th percentile

38. a. $Q_1 = 2; Q_3 = 18$ **b.** 5.5 **c.** Lower: −22; Upper: 42
d. France, Japan, United States
e.

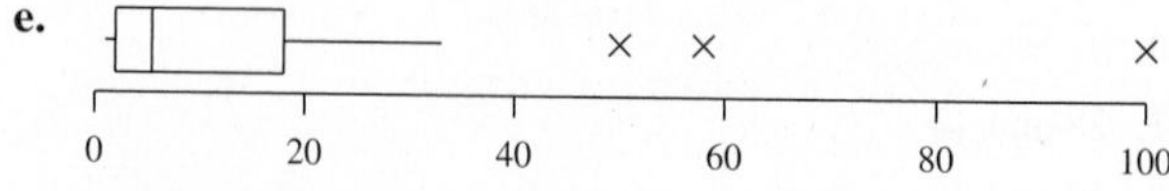

f. Skewed right **g.** 4 **h.** 33 **i.** 82nd

39. a. $Q_1 = 3; Q_3 = 21.5$ **b.** 7.5
c. Lower: −24.75; Upper: 49.25 **d.** 70, 114
e.

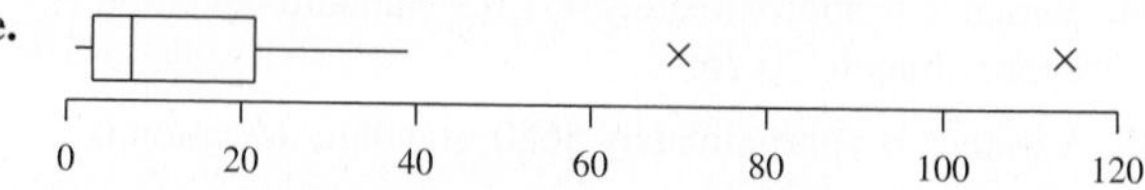

f. Skewed right **g.** 4 **h.** 31 **i.** 63rd

40. a. $Q_1 = 1431; Q_3 = 2915$ **b.** 2083.5
c. Lower: −795; Upper: 5141 **d.** No, neither is an outlier.
e. Yes, both are outliers.
f.

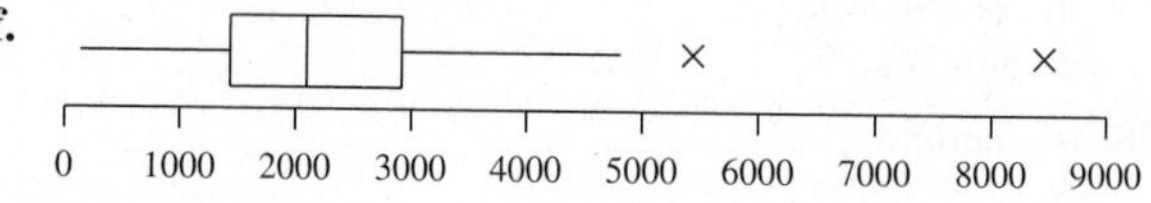

g. Skewed right **h.** 1177 **i.** 2395
j. 27th

41. a. 13th
b. 41st
c. 73rd

42. a. 79th
b. 11th
c. 85th

43. No, only 25% of the class scored lower than Ed.

44. Harry's was the highest; Tom's was the lowest.

45. a. Median: 1.60; $Q_1 = 0.56; Q_3 = 8.55$
b. Median: 0.71; $Q_1 = 0.57; Q_3 = 4.50$
c. Lower: −11.425; upper: 20.535
d. Lower: −5.325; upper: 10.395
e.

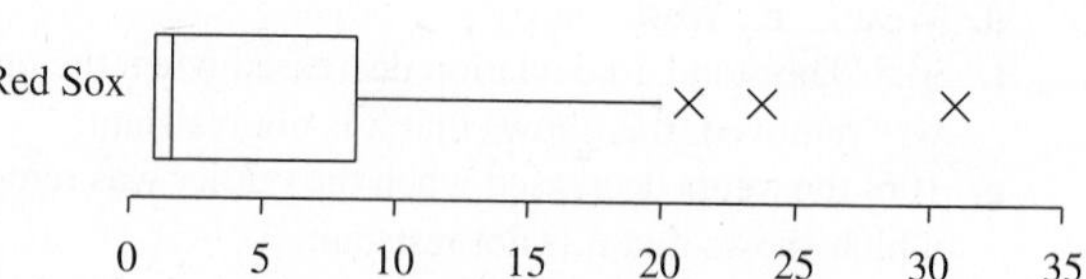

Both sets of salaries are skewed to the right. The Red Sox salaries are higher overall. The median salary for the Dodgers is only slightly larger than the minimum.

46. a. Median: 1.8; $Q_1 = 1.2; Q_3 = 3.1$
b. Median: 5.3; $Q_1 = 2.7; Q_3 = 6.5$
c. Lower: −1.65; upper: 5.95 **d.** Lower: −3; upper: 12.2
e.

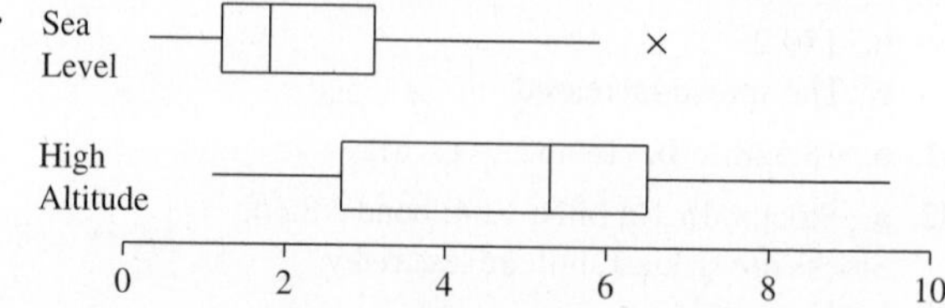

The high-altitude values are, on the whole, greater than the sea-level values. The sea-level values are somewhat skewed to the right and contain an outlier. The high-altitude values appear to be slightly skewed to the left.

47. a. Denver **b.** 25% **c.** Denver
d. Skewed right **e.** Skewed right

48. a. Over 60 **b.** 75% **c.** Over 60
d. Approximately symmetric
e. Approximately symmetric

49. a. Yes

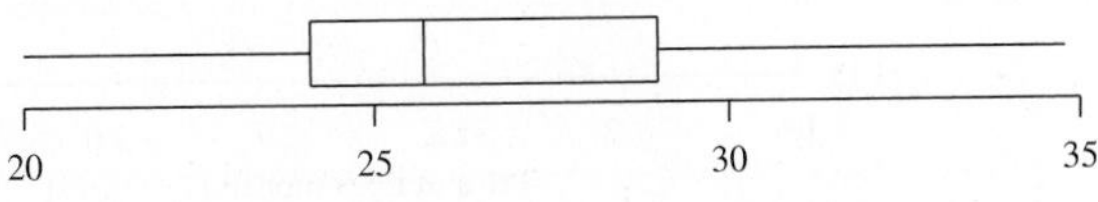

b. Skewed right

50. a. No. The maximum is greater than the upper outlier boundary. Therefore, it is impossible to determine the maximum data value that is less than the upper outlier boundary.
b. Skewed right

51. a. More than half the data values are equal to the minimum value of 0.

b.

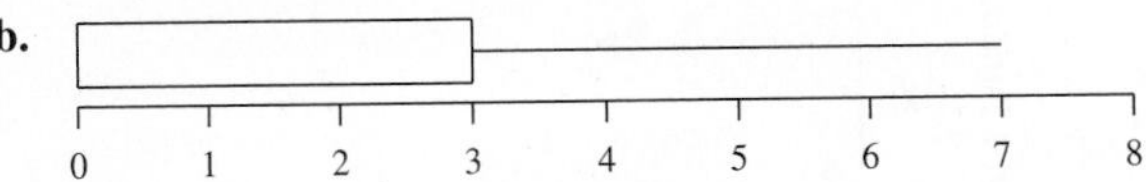

c. The first quartile is the same as the minimum value.

52. Answers will vary. Any data set with three or fewer distinct values will work.

53. Answers will vary.

54. Answers will vary.

55. a. $Q_1 = 20; Q_3 = 60;$ IQR $= 40$
b. Upper outlier boundary is $60 + 1.5 \cdot 40 = 120$. Since $150 > 120$, 150 is an outlier.
c. $Q_1 = 25; Q_3 = 105;$ IQR $= 80$ **d.** No
e. Both the third quartile and the IQR increased.

56. Follow the steps for finding a percentile, but compute the index as $L = \frac{i}{k} \cdot n$.

57. a. Computation
b. $z = \dfrac{0 - 3.660}{3.325} = -1.10$
c. $z = \dfrac{7.32 - 3.660}{3.325} = 1.10$
d. 2.2%
e. 13.3%
f. Less extreme
g. The right tail extends out farther than the left tail, so there are more data to the right of $z = 1.10$ than to the left of $z = -1.10$.

CHAPTER 3 Quiz

1. The mode

2. Mean: 520; median: 550; mode: 550

3. The mean

4. mean, median

5. a. 9.76 **b.** 4.51

6. 4

7. 3.96, 2.28

8. 95%

9. 75

10. 0.5

11. 0.037

12. False

13. 44

14. a. $Q_1 = 22; Q_3 = 39$
b. Outlier boundaries are −3.5 and 64.5.
c. 68 is the only outlier.

15.

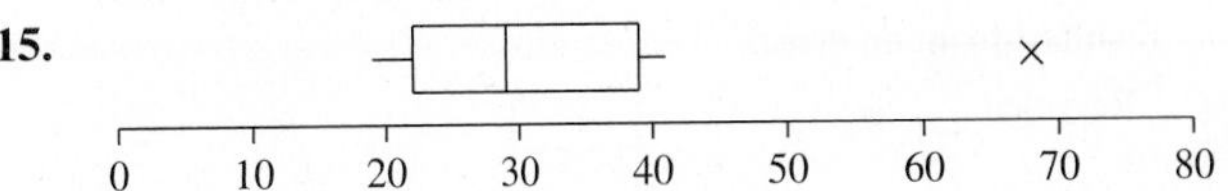

CHAPTER 3 Review Exercises

1. a. 150.43 **b.** 148.9

2. a. 10.08 **b.** 9.2 **c.** Approximately symmetric

3. a. Mean of process 1 is 92.87; mean of process 2 is 91.46.
b. Median of process 1 is 92.2; median of process 2 is 93.3.
c. They are about the same.

4. a. Variance of process 1 is 9.40; variance of process 2 is 53.36.
b. Standard deviation of process 1 is 3.07; standard deviation of process 2 is 7.30.
c. Process 1 produces a more uniform thickness.

5. a. Mean in May is 34.479; median in May is 34.67.
b. Mean in June is 34.052; median in June is 34.115.
c. They are about the same.

6. a. 0.791 **b.** 0.814 **c.** Greater in June

7. a. A: 2.87; B: 0.75
b. Method A. Estimating by eye is less precise than measuring with a ruler.
c. It is better to have a smaller standard deviation. With a small standard deviation, there is less need to remeasure, since all measurements will be reasonably close to the first one.

8. 153 and 172.6

9. 25%

10. 68%

11. At least 8/9 of the rents are between \$350 and \$1250.

12. a. 8.752 **b.** 7.365 **c.** 14.616 **d.** 3.82 **e.** 6.34
f. 8.98 **g.** 6.705 **h.** 8.66

13. a. (3) **b.** (1) **c.** (4) **d.** (2)

14. a. $Q_1 = 14.6; Q_3 = 17.4$ **b.** 16.2
c. Lower: 10.4; upper: 21.6 **d.** No outliers

e.

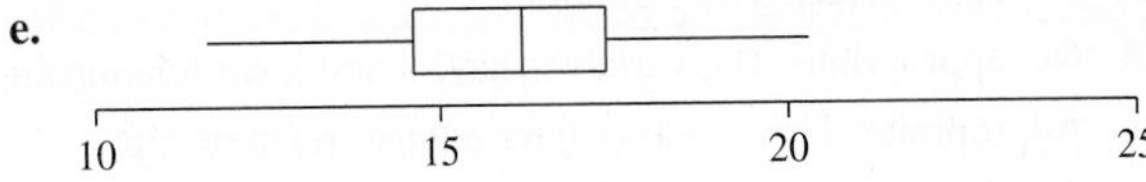

15. a. $Q_1 = 2.1; Q_3 = 15.7$ **b.** 9.2
c. Lower: −18.3; upper: 36.1 **d.** 41.1

e.

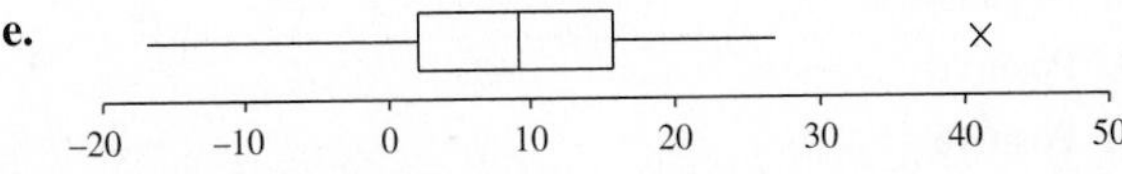

CHAPTER 3 Case Study

1.

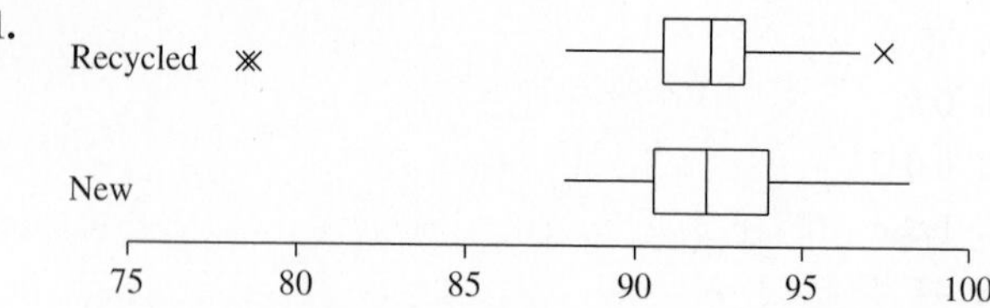

2. For the recycled wafers, 77.3, 77.5, and 97.4 are outliers. There are no outliers for the new wafers.

3. The outliers 77.3 and 77.5 should be deleted, because they resulted from an error.

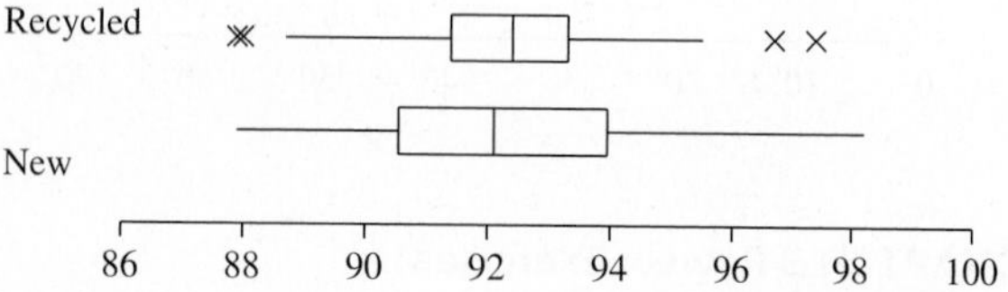

4. For the recycled wafers, 87.9, 88.0, 96.7, and 97.4 are outliers. There are no outliers for the new wafers. No outliers should be deleted, because they do not result from errors.

5. Approximately symmetric

6. New: 92.33; recycled: 92.31

7. New: 92.1; recycled: 92.4

8. New: 2.36; recycled: 2.21

9. Answers will vary. The interquartile range is more useful, because the standard deviation is not resistant to outliers.

CHAPTER 4

Section 4.1

Exercises 1–8 are the Check Your Understanding exercises for this section. Answers to these exercises are on page 178.

9. pairs
10. scatterplot
11. linear
12. linear
13. False
14. True
15. True
16. False
17. 0.824
18. −0.647
19. 0.515
20. 0.339
21. Appropriate. The variables have a weak linear relationship.
22. Not appropriate. There are outliers.
23. Not appropriate. The variables have a nonlinear relationship.
24. Appropriate. The variables have a linear relationship.
25. Positive
26. Negative
27. Positive
28. Negative
29. Positive
30. Positive

31. a.

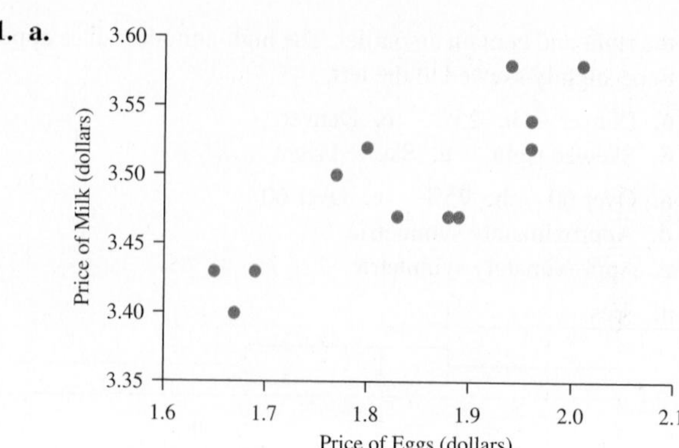

b. 0.845 **c.** Above average; r is positive. **d.** iii

32. a.

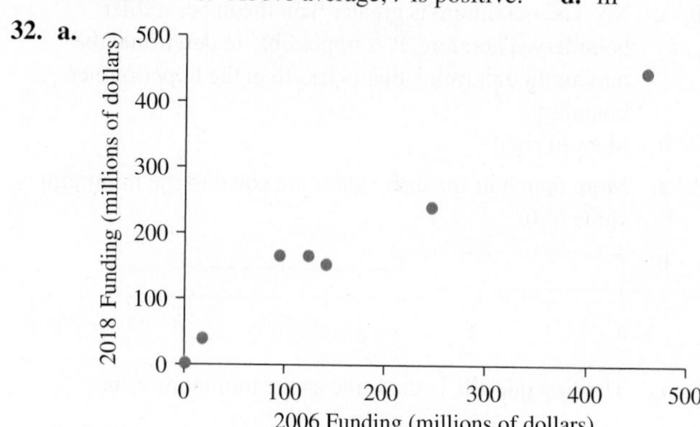

b. 0.984
c. Above average; r is positive.
d. iii

33. a.

Weight (pounds) 250 240 230 220 210 200 190
70 72 74 76 78
Height (inches)

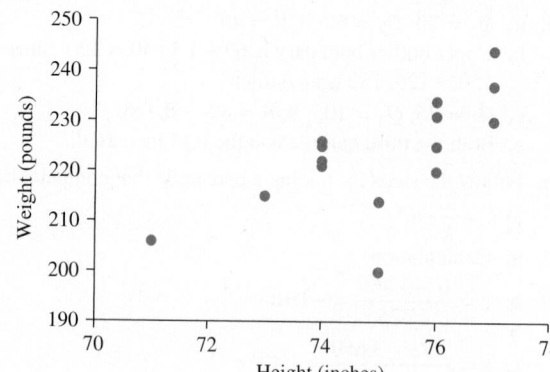

b. 0.674
c. Below average; r is positive.
d. ii

34. a.

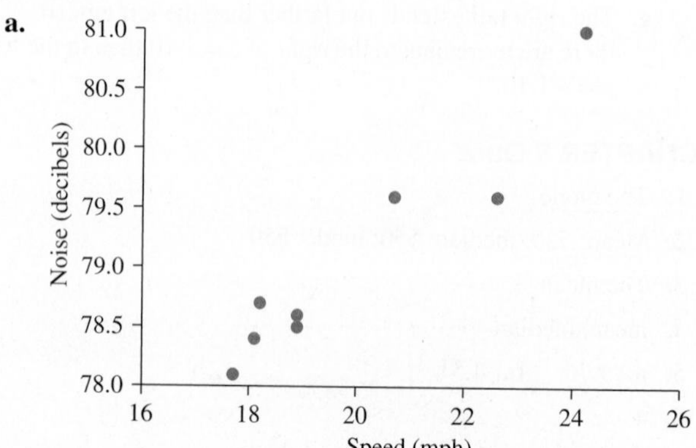

b. 0.958
c. Above average; r is positive.
d. iii

35. a.

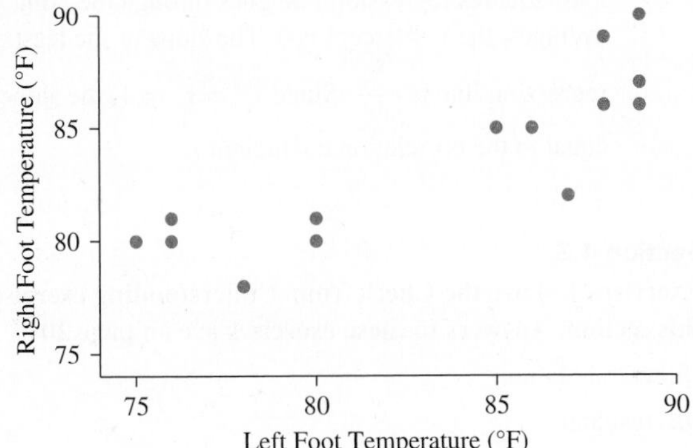

b. 0.812 **c.** Cooler; r is positive. **d.** i

36. a.

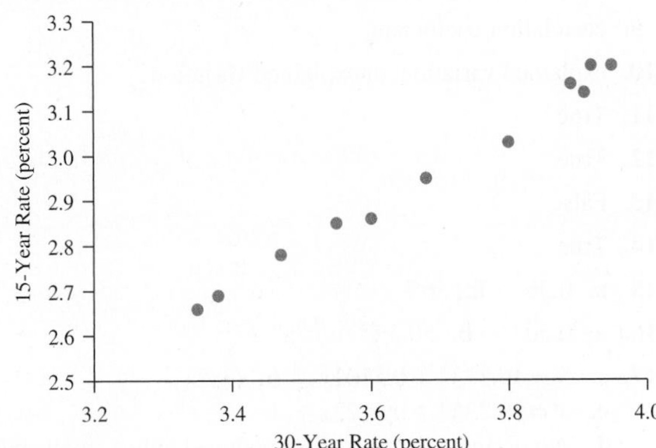

b. 0.996 **c.** Below average; r is positive. **d.** ii

37. a.

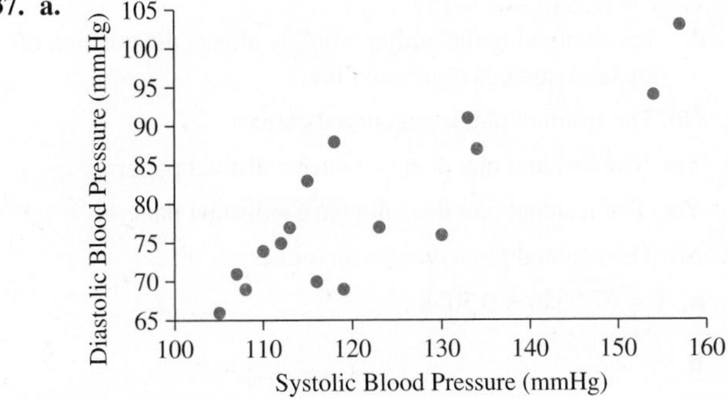

b. 0.857 **c.** Above average; r is positive.

38. a.

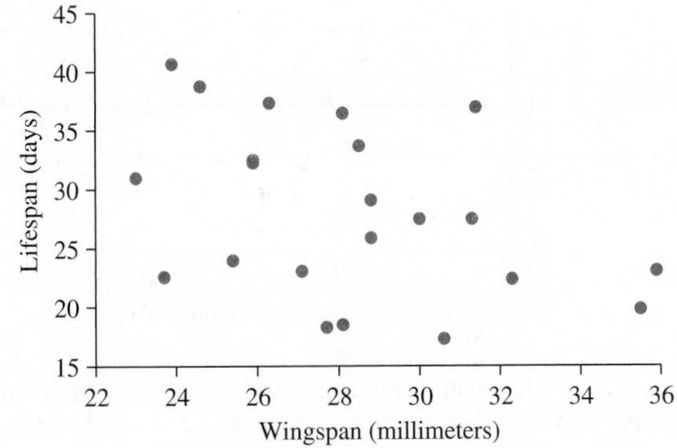

b. −0.412 **c.** Shorter; r is negative.

39. No. Larger cities have more police officers and also tend to have higher crime rates than smaller cities.

40. No. In the past, the proportion of people who graduated from high school or college was less than it has been recently. For this reason, the proportion of older people who are well educated is less than the corresponding proportion of younger people.

41. Close to −1. The difference between the ages of two people is in most cases very close to the difference between their graduation years, with the older person graduating in the earlier year. Therefore, the two variables have a nearly perfect negative linear relationship.

42. Close to 1. For most people, the year graduated from high school is close to the birth year plus 18. Therefore, the two variables have a nearly perfect positive linear relationship.

43. a. $\bar{x} = 1.75; \bar{y} = 10.828; s_x = 1.031; s_y = 1.981$
b. 0.906
c. $\bar{y} = 11.828; s_y = 1.981$
d. $\bar{y}$ increased by 1; s_y was unchanged.
e. 0.906. The quantities $y - \bar{y}$ are unchanged.
f. $\bar{x} = 21; s_x = 12.369$
g. Each was multiplied by 12.
h. 0.906. The quantities $(x - \bar{x})/s_x$ are unchanged.
i. unchanged
j. unchanged

Section 4.2

Exercises 1–4 are the Check Your Understanding exercises for this section. Answers to these exercises are on page 190.

5. explanatory, outcome
6. 0
7. 15
8. True
9. True
10. False
11. False
12. False
13. $\hat{y} = 5.2 + 0.6x$
14. $\hat{y} = 3.8441 + 1.4623x$
15. $\hat{y} = 0.8951 + 1.7674x$
16. $\hat{y} = 10.9386 - 1.5763x$
17. $\hat{y} = 1175 + 35x$
18. $\hat{y} = 41.301 - 1.3458x$
19. $\hat{y} = 86.5 + 0.0003x$
20. $\hat{y} = -9086.9565 + 3913.0435x$
21. a. $\hat{y} = 2.7592 + 0.3991x$ **b.** 0.10 **c.** 3.54
22. a. $\hat{y} = 32.672 + 0.90468x$
b. 9.05
c. 123.14
23. a. $\hat{y} = -121.67 + 4.6000x$
b. No. The x-values are all positive.
c. 9.2
d. 221.03
e. Less

24. a. $\hat{y} = -1917.1 + 8.8629x$
b. No. The x-values are all positive.
c. 88.629 **d.** 2159.9

25. a. $\hat{y} = 33.7754 + 0.5930x$
b.

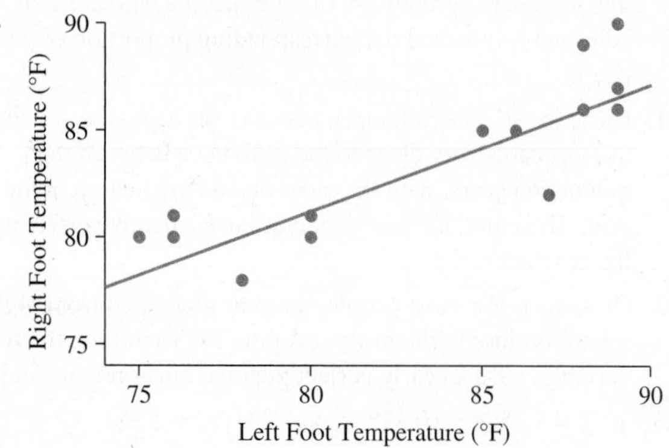

c. 1.186 **d.** 81.8

26. a. $\hat{y} = -0.2859 + 0.9421x$
b.

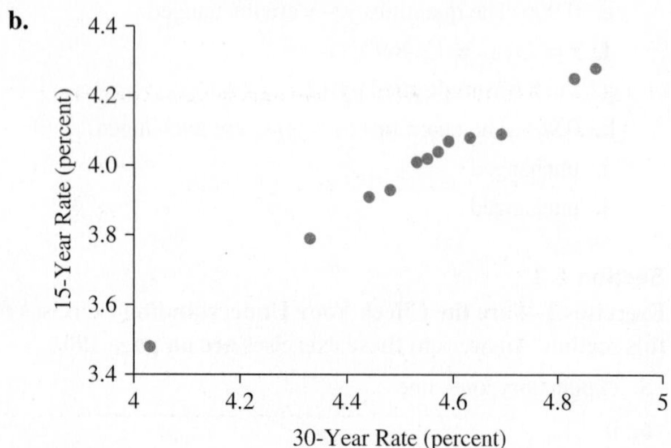

c. No
d. 0.2826 **e.** 3.0114

27. a. $\hat{y} = 9.1828 + 0.5748x$
b. No. The x-values are all positive.
c. 5.748 **d.** 81.0

28. a. $\hat{y} = 52.0434 - 0.8445x$
b. No. The x-values are all positive.
c. −1.689 **d.** 27.98

29. a. $\hat{y} = 49.7124 + 4.288759685x$
b. 0.9186464394
c. 92.6

30. a. $\hat{y} = 2.04528 + 0.4323287293x$
b. 0.6086336328
c. 4.2069

31. a. $\hat{y} = 33.8127 + 1.21015x$ **b.** 59.7

32. a. $\hat{y} = 23.7789 + 0.71384x$ **b.** 77.3

33. a. $\hat{y} = 55.91275257 + 2.58289361x$ **b.** 68.827

34. a. $\hat{y} = 3.79229595 + 0.08063922x$ **b.** 27.984

35. a. 0.750
b. $\bar{x} = 424$; $s_x = 123.26$
c. $\bar{y} = 499.33$; $s_y = 143.93$
d. $\hat{y} = 128.1812 + 0.8754x$
e. 0.0325, −0.3083, 1.8579, −0.8762, 0.1136, −0.8194
f. −0.8778, 0.4979, 1.297, −1.253, 0.7481, −0.4122
g. 0.750
h. $\hat{z}_y = 0.750z_x$. The means of z_x and z_y are both 0. Since the least-squares regression line goes through the point of averages, the y-intercept is 0. The slope of the least-squares regression line is $r\frac{s_{z_y}}{s_{z_x}}$. Since $s_{z_y} = s_{z_x} = 1$, the slope is equal to the correlation coefficient r.

Section 4.3

Exercises 1–4 are the Check Your Understanding exercises for this section. Answers to these exercises are on page 204.

5. extrapolation
6. residual
7. sum of squared residuals
8. an influential point
9. correlation coefficient
10. explained variation, unexplained variation
11. True
12. True
13. False
14. True
15. a. 0.36 **b.** 36%
16. a. 0.503 **b.** 50.3%
17. a. $\hat{y} = 19.3734 + 0.5101x$ **b.** (3, 21)
c. $\hat{y} = 19.2354 + 0.5123x$
d. No. Removing the outlier produced only a small change in the least-squares regression line.
18. a. $\hat{y} = 8.9888 - 0.9703x$ **b.** (1.8, 9.2)
c. $\hat{y} = 5.2915 - 0.1413x$
d. Yes. Removing the outlier strongly affects the position of the least-squares regression line.
19. No. The residual plot has a curved pattern.
20. Yes. The residual plot does not have a distinct pattern.
21. Yes. The residual plot does not have a distinct pattern.
22. No. The residual plot contains an outlier.
23. a. $\hat{y} = 67.0556 + 0.3058x$
b.

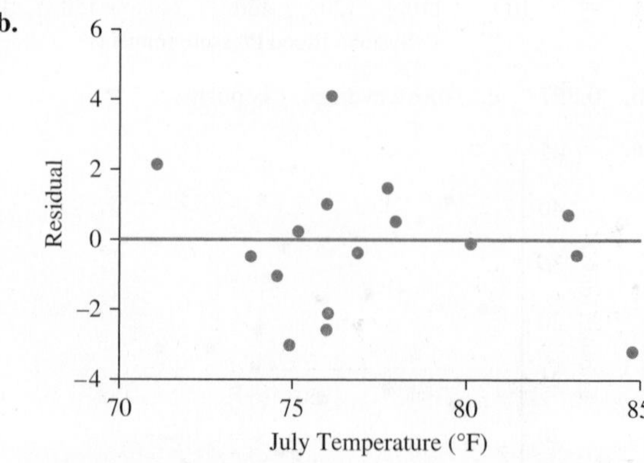

Yes. The relationship appears to be approximately linear.
c. 78.4
d. The prediction is 87.9. It is not appropriate because the value of 68.1 is outside the range of the data.

24. a. $\hat{y} = -1917.1 + 8.8629x$

b.

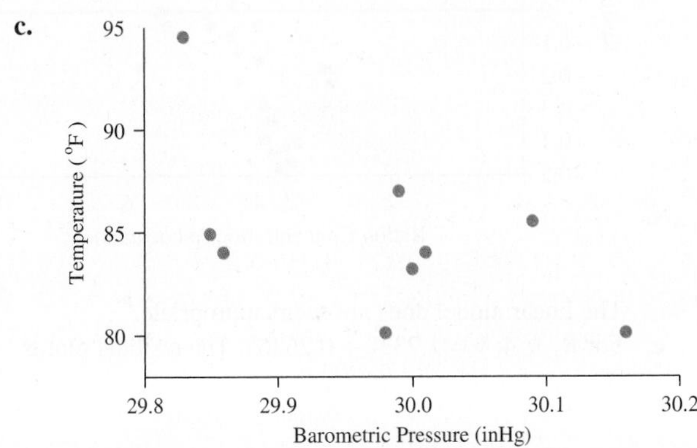

Yes. The relationship appears to be approximately linear.

c. 2071.2

d. The prediction is 554.8. It is not appropriate because the value of 3000 is outside the range of the data.

25. a. $\hat{y} = 740.3765 - 21.8708x$ **b.** 0.3225

c.

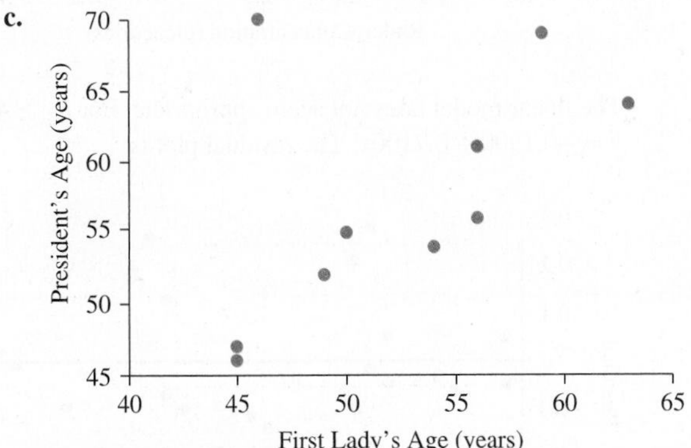

d. (29.83, 94.5) **e.** $\hat{y} = 320.54 - 7.8999x$

f. Yes. Removing the outlier strongly affects the position of the least-squares line.

g. 0.113. The proportion of variance explained is less without the outlier.

26. a. $\hat{y} = 19.701 + 0.7208x$

b. 0.287

c.

d. (46, 70)

e. $\hat{y} = -2.8506 + 1.1104x$

f. Yes. Removing the outlier strongly affects the equation of the least-squares regression line.

g. 0.826. The proportion of variance explained is greater without the outlier.

27. a. $\hat{y} = -76.2258 + 3.2499x$

b.

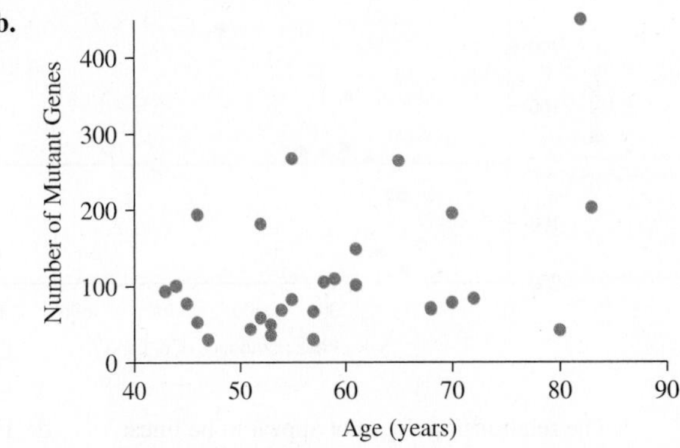

c. (82, 449) **d.** $\hat{y} = 31.8632 + 1.2282x$

e. Yes. Removing the outlier strongly affects the position of the least-squares regression line.

28. a. $\hat{y} = -19.2390 + 0.8868x$ **b.** 0.6425

c. $\hat{y} = 23.6335 + 0.6990x$. The result is noticeably different.

d. 0.4091 **e.** months 1 through 29

29. a. $\hat{y} = 1062.4048 + 0.3600x$ **b.** 0.9603

c.

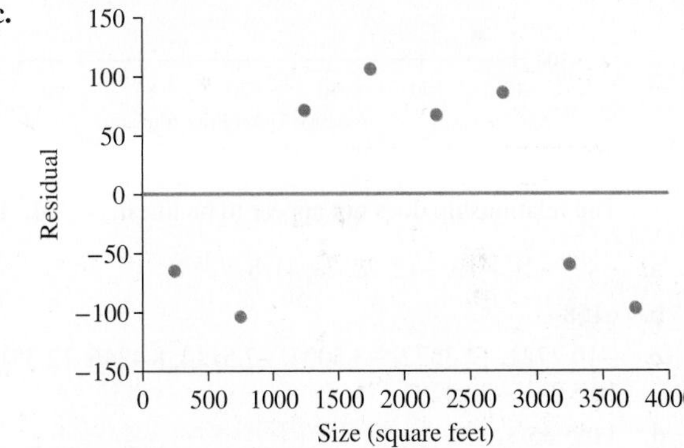

The relationship does not appear to be linear.

d. False

30. a. $\hat{y} = 30.7147 + 0.3728x$ **b.** 0.9479

c.

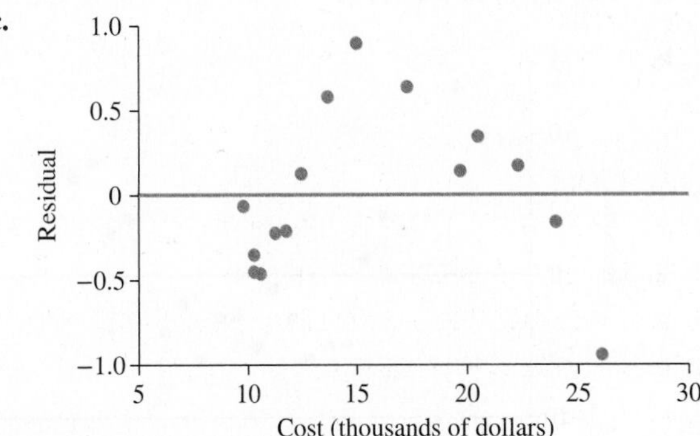

The relationship does not appear to be linear.

d. False

31. a. $\hat{y} = 757.2920 + 3.3009x$ **b.** 0.9043

c.

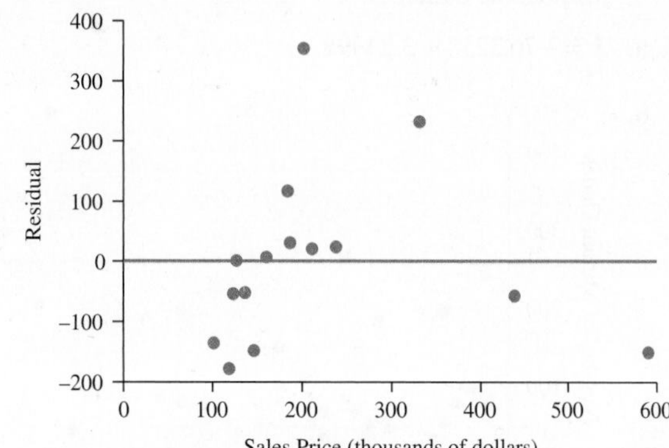

The relationship does not appear to be linear. **d.** False

32. a. $\hat{y} = -179.4279 + 14.2472x$ **b.** 0.9725

c.

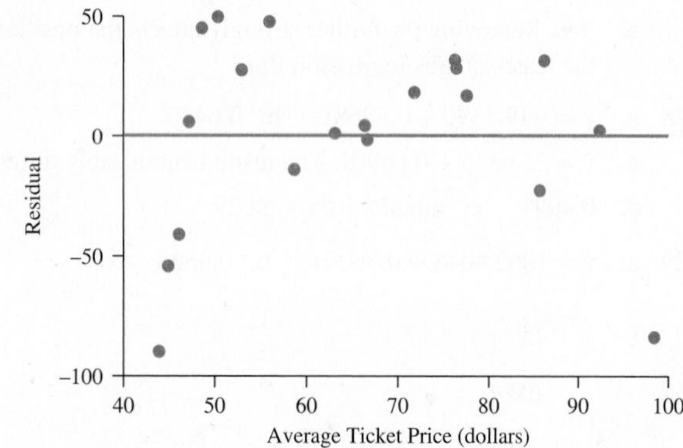

The relationship does not appear to be linear. **d.** False

33. a. −47, −21, −19, −12, 22, 28, 41, 8

b. 6168

c. −10.2721, 12.3552, −3.5004, −7.5114, 8.4346, 12.3514, 10.5709, −22.4291

d. 1165.35

e. 5002.65

f. 0.811

g. Compute the correlation coefficient to be $r = 0.9005918$. The coefficient of determination is $r^2 = 0.811$.

34. a. $\hat{y} = -17.1793 + 4.0119x$

b.

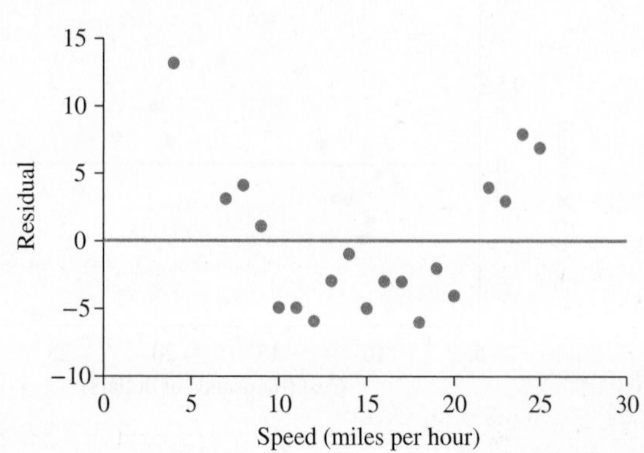

The plot has an obvious pattern. **c.** $\hat{y} = 8.3045 + 0.1332x^2$

d.

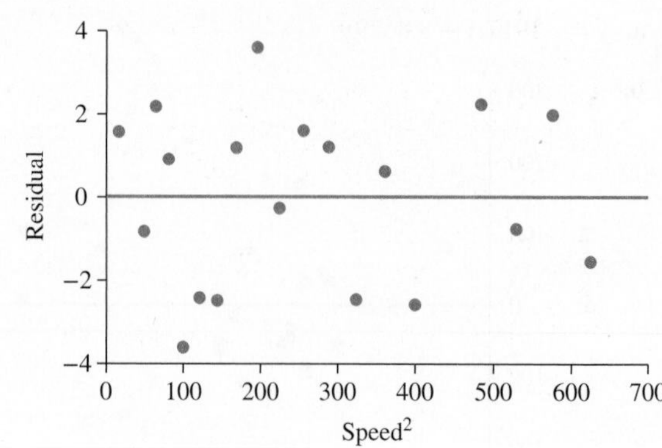

This line is an appropriate summary. **e.** 38.27 feet

35. a. $\hat{y} = 0.4636 + 0.5711x$

b.

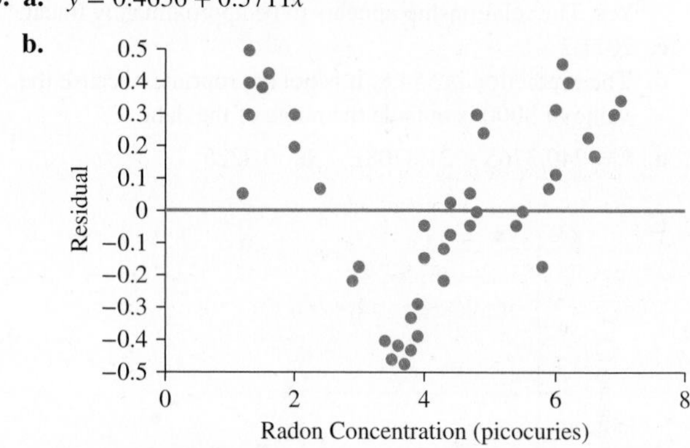

The linear model does not seem appropriate.

c. For $R_1 < 4$: $\hat{y} = 1.2330 + 0.2636x$. The residual plot is

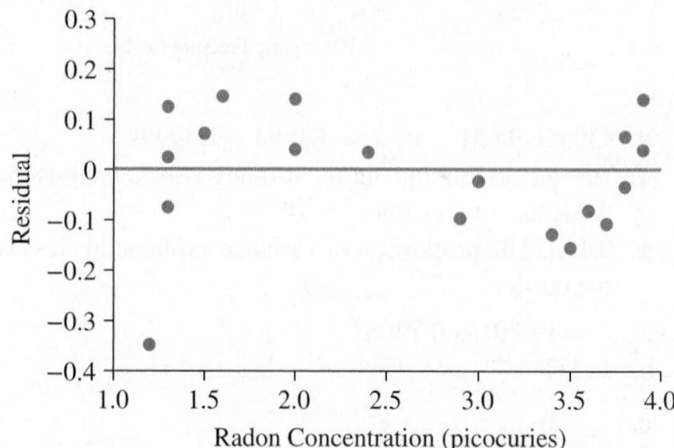

The linear model does not seem appropriate. For $R_1 \geq 4$: $\hat{y} = -0.1900 + 0.7100x$. The residual plot is

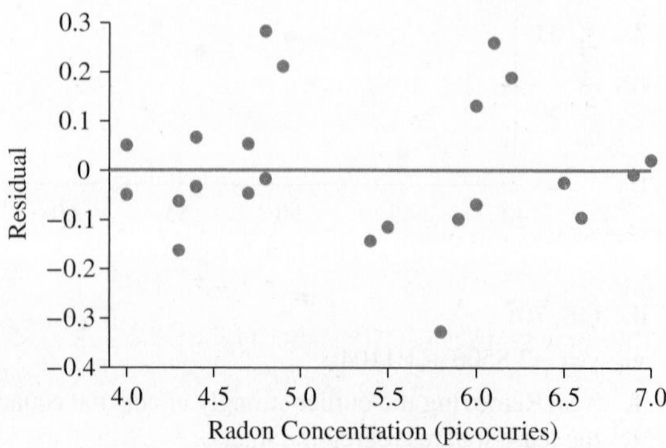

The line fits the data with $R_1 \geq 4$, but not the data with $R_1 < 4$.

d. A line describes the relationship well when $R_1 \geq 4$. A nonlinear model would be a better fit when $R_1 < 4$.

36. a. −0.8702 **b.** $\hat{y} = 814.8943 - 143.3279x$

c.

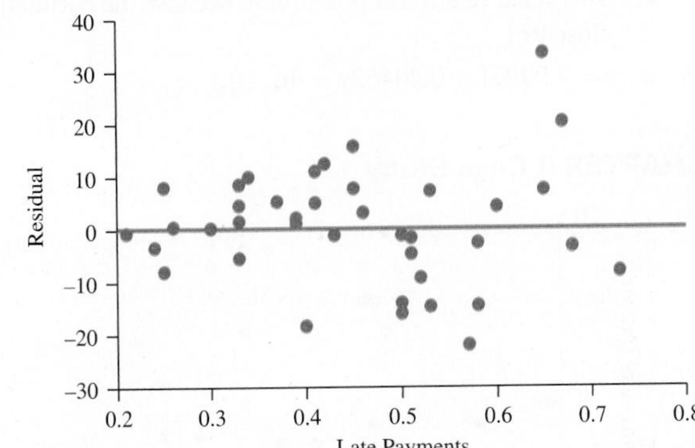

Yes, the residuals corresponding to $x \geq 0.5$ are more spread out than those corresponding to $x < 0.5$.

d. $\hat{y} = 805.8230 - 111.1299x$

e. $\hat{y} = 764.5173 - 60.5335x$

f. 0.5689 **g.** 0.1162 **h.** less than 0.5

CHAPTER 4 Quiz

1. −0.959

2.

3. −0.980; strong negative

4. The points lie along a line with negative slope.

5. No. Larger cities have more paved streets and more cars.

6. $\hat{y} = -6.195 + 1.2474x$

7. 66.78

8. 0.747

9. 2.495

10. 0

11. $\hat{y} = 6.0667 - 0.6x$

12. iii

13. No, 10 hours is outside the range of the data.

14. i

15. 36%

CHAPTER 4 Review Exercises

1. a. $\hat{y} = 44.534 + 2.5313x$

b. 2.5313

c. 68.581

2. a. $\hat{y} = -1.0295 + 0.0523x$

b. 0.4187

c. 1.2735

3. a. $\hat{y} = 8.5593 - 0.1551x$

b.

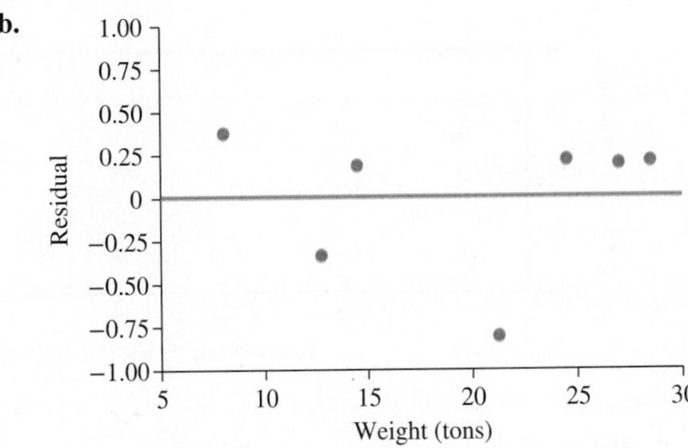

c. 0.7755 mile per gallon **d.** 6.2328

4. a. −0.945 **b.** 0.892 **c.** 89.2%

5. a. $\hat{y} = 2.8827 + 0.8895x$

b.

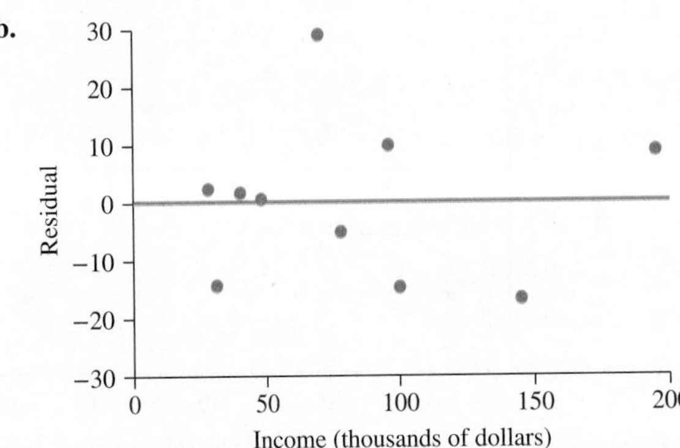

c. 10.674

d. 47.358

6. a. 0.960

b. 0.921

c. 92.1%

7. a. $\hat{y} = 7.5659 + 0.2015x$

b.

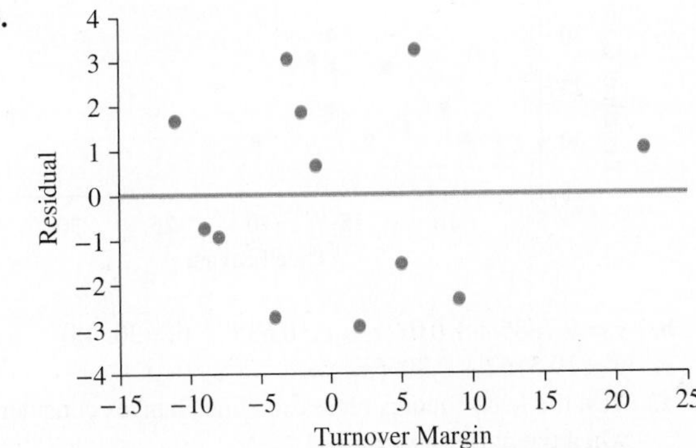

c. Florida, Alabama, Georgia, Ole Miss, LSU, and South Carolina

8. a. 0.643
b. 0.413
c. 41.3%

9. a. $\hat{y} = 417.2012 + 18.4756x$

b.

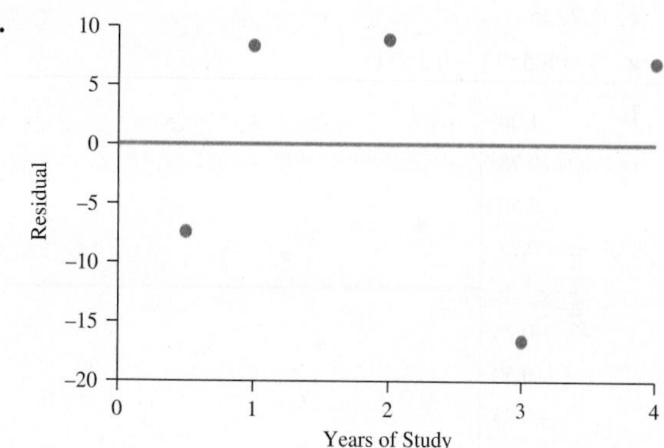

c. 463

10. a. 0.917
b. 0.8416
c. 84.16%

11. a. $\hat{y} = -3.9357 + 0.2857x$ **b.** 0.937

c.

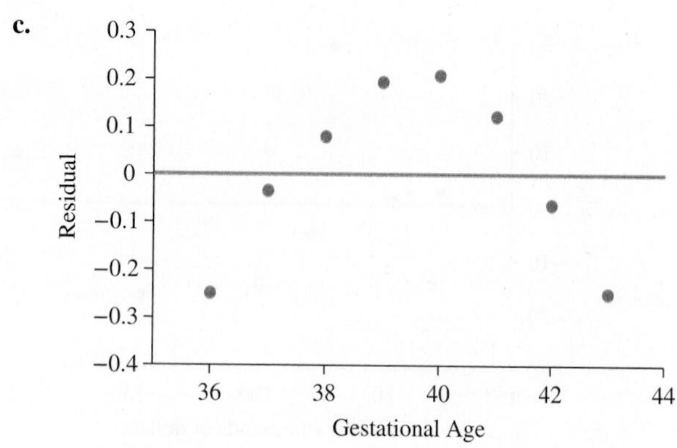

The relationship does not appear to be linear. **d.** False

12. a.

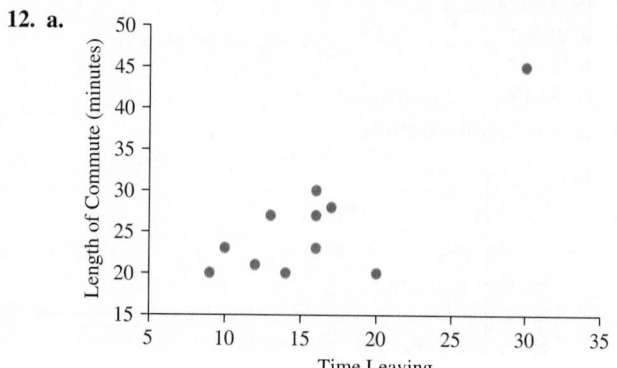

Length of Commute (minutes)
Time Leaving

b. $\hat{y} = 9.7885 + 1.0192x$ **c.** 0.635 **d.** (30, 45)
e. $\hat{y} = 19.5162 + 0.3066x$
f. Yes, the least-squares regression line changes considerably when the outlier is removed.
g. 0.0744; the relationship is weaker without the outlier.

13. a. $\hat{y} = 0.3946 + 3.643061791x$ **b.** 36.825 **c.** 0.064
d. The linear relationship is weak, because the correlation is close to 0.

14. a. $\hat{y} = -1.04349 + 3.045141227x$ **b.** 151.21 **c.** 0.994
d. The linear relationship is strong, because the correlation is close to 1.

15. a. $\hat{y} = 4.99971 + 0.20462x$ **b.** 10.1

CHAPTER 4 Case Study

1.

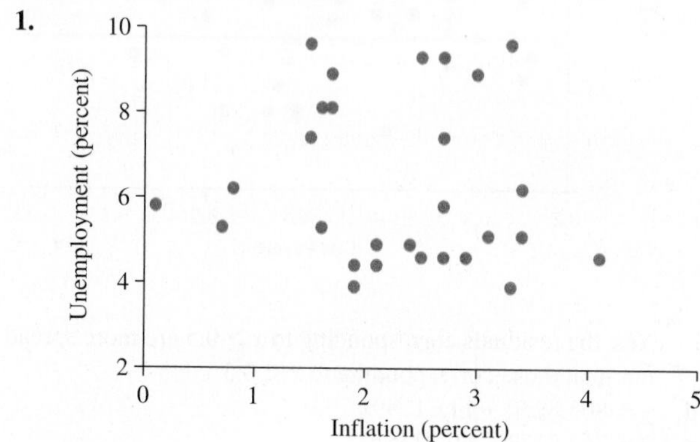

2. $\hat{y} = 6.1932 - 0.1160x$
3. 5.85
4. −0.0669
5. 0.447%
6. $\hat{y} = 6.5565 - 0.2857x$
7. 5.69
8. −0.166
9. 2.76%
10. $\hat{y} = 0.7864 + 0.8511x$
11. 4.19%
12. 0.829
13. 68.7%
14. Answers will vary. The model using unemployment in the previous year explains more of the variation than either of the other two models.
15. Underpredict

CHAPTER 5

Section 5.1

Exercises 1–4 are the Check Your Understanding exercises for this section. Answers to these exercises are on page 225.

5. 0
6. 1
7. sample space
8. event
9. True
10. False
11. False
12. True
13. 1/6

14. 1/2

15. 1/3

16. 2/3

17. 0

18. 1

19. **a.** No **b.** No **c.** Yes

20. **a.** No **b.** No **c.** Yes

21. No. The probabilities do not add up to 1.

22. Yes. The probabilities are all between 0 and 1, and they add up to 1.

23. Yes. The probabilities are all between 0 and 1, and they add up to 1.

24. No. A probability cannot be negative.

25. **a.** ii **b.** vi **c.** iv **d.** i
e. iii **f.** vii **g.** v **h.** vii

26. 88/100 = 0.88

27. **a.** 275/500 = 0.55
b. We estimate that 55% of all voters plan to vote to reelect the mayor.

28. **a.** 497/1769 = 0.2809
b. We estimate that 28.09% of people find their main satisfaction in their work.

29. **a.** {TTTT, TTTF, TTFT, TTFF, TFTT, TFTF, TFFT, TFFF, FTTT, FTTF, FTFT, FTFF, FFTT, FFTF, FFFT, FFFF}
b. 1/8 **c.** 1/4 **d.** 3/8

30. **a.** {HHH, HHT, HTH, HTT, THH, THT, TTH, TTT}
b. 1/8 **c.** 1/4 **d.** 3/8

31. 180/400 = 0.45

32. 85/600 = 0.1417

33. **a.** 1303/2515 = 0.5181 **b.** (385 + 815)/2515 = 0.4771

34. **a.** 1449/3727 = 0.3888 **b.** (1449 + 1168)/3727 = 0.7022

35. **a.** 7792/11,217 = 0.6947 **b.** 10,270/11,217 = 0.9156

36. **a.** 623/875 = 0.712 **b.** 0.288

37. **a.** 18/38 = 0.4737
b. The law of large numbers says that in the long run, the percentage of the time you win will approach 47.37%.

38. **a.** 2/38 = 0.0526
b. The law of large numbers says that in the long run, the percentage of the time the ball lands in a green pocket will approach 5.26%.

39. **a.** 0.8674 **b.** 0.3118 **c.** No

40. **a.** 0.2436 **b.** 0.6771 **c.** 0.0478 **d.** No

41. **a.** 0.1944 **b.** 0.2378 **c.** Yes

42. **a.** 0.2910 **b.** 0.4993

43. Answers will vary.

44. Answers will vary.

45.

(1, 1)	(1, 2)	(1, 3)	(1, 4)	(1, 5)	(1, 6)
(2, 1)	(2, 2)	(2, 3)	(2, 4)	(2, 5)	(2, 6)
(3, 1)	(3, 2)	(3, 3)	(3, 4)	(3, 5)	(3, 6)
(4, 1)	(4, 2)	(4, 3)	(4, 4)	(4, 5)	(4, 6)
(5, 1)	(5, 2)	(5, 3)	(5, 4)	(5, 5)	(5, 6)
(6, 1)	(6, 2)	(6, 3)	(6, 4)	(6, 5)	(6, 6)

46. 1/9

47. 1/6

48. 6

49. {(3, 1), (3, 2), (3, 3), (3, 4), (3, 5), (3, 6)}

50. 1/6; no

51. 1/6; yes

Section 5.2

Exercises 1–4 are the Check Your Understanding exercises for this section. Answers to these exercises are on page 235.

5. *P*(*A* and *B*)

6. 0

7. complement

8. 1 − *P*(*A*)

9. True

10. True

11. False

12. False

13. 0.9

14. 0.5

15. 0.7

16. 0.8

17. Yes

18. No

19. 0.65

20. 0.4

21. 0.73

22. 0.36

23. 1

24. 0.5

25. Not mutually exclusive

26. Mutually exclusive

27. Mutually exclusive

28. Not mutually exclusive

29. Not mutually exclusive

30. Mutually exclusive

31. **a.** 200 or fewer of them use Google as their primary search engine.
b. Fewer than 200 of them use Google as their primary search engine.
c. At least 200 of them use Google as their primary search engine.
d. The number that use Google as their primary search engine is not equal to 200.

32. **a.** The number of defective batteries was not equal to 24.
b. Fewer than 24 of the batteries were defective.
c. 24 or fewer of the batteries were defective.
d. At least 24 of the batteries were defective.

33. **a.** {RR, RY, RG, YR, YY, YG, GR, GY, GG}
b. {RR, YY, GG}
c. {RY, RG, YR, YG, GR, GY}
d. {RG, YG, GR, GY, GG}

e. Yes; they have no outcomes in common.
f. No; they both contain the event GG.

34. a.

(1, 1)	(1, 2)	(1, 3)	(1, 4)	(1, 5)	(1, 6)
(2, 1)	(2, 2)	(2, 3)	(2, 4)	(2, 5)	(2, 6)
(3, 1)	(3, 2)	(3, 3)	(3, 4)	(3, 5)	(3, 6)
(4, 1)	(4, 2)	(4, 3)	(4, 4)	(4, 5)	(4, 6)
(5, 1)	(5, 2)	(5, 3)	(5, 4)	(5, 5)	(5, 6)
(6, 1)	(6, 2)	(6, 3)	(6, 4)	(6, 5)	(6, 6)

b. {(1, 1), (2, 2), (3, 3), (4, 4), (5, 5), (6, 6)}
c. {(6, 1), (6, 2), (6, 3), (6, 4), (6, 5), (6, 6)}
d. {(6, 1), (1, 6)}
e. No; they both contain the outcome (6, 6).
f. Yes; they have no outcomes in common.

35. a. 0.11 **b.** 0.90

36. a. 0.17 **b.** 0.85

37. a. 336/800 = 0.42 **b.** 734/800 = 0.9175

38. a. 720/1200 = 0.6 **b.** 305/1200 = 0.2542

39. a. 0.226 **b.** 0.854

40. a. 17/30 = 0.5667 **b.** 5/30 = 0.1667 **c.** 0.6333 **d.** 0.8333

41. a. 0.6532 **b.** 0.4760 **c.** 0.5240 **d.** 0.7872

42. a. 0.075 **b.** 0.135 **c.** 0.925 **d.** 0.19

43. a. 18/25 = 0.72 **b.** 10/25 = 0.4 **c.** 0.88 **d.** 0.12

44. a. 55/500 = 0.11 **b.** 95/500 = 0.19 **c.** 0.26 **d.** 0.89 **e.** 0.74

45. a. 0.28 **b.** 0.70 **c.** 0.53 **d.** 0.47 **e.** 0.45 **f.** 0.02 **g.** 0.47

46. a. 0.0806 **b.** 0.7747 **c.** 0.2858 **d.** 0.7142 **e.** 0.0123 **f.** 0.2083

47. No. The events of having a fireplace and having a garage are not mutually exclusive.

48. Yes. The events of having one child and having two children are mutually exclusive.

49. Answers will vary.

50. A^c or B^c

Section 5.3

Exercises 1–6 are the Check Your Understanding exercises for this section. Answers to these exercises are on page 248.

7. conditional

8. $P(A)P(B \mid A)$ or $P(B)P(A \mid B)$

9. 5

10. independent

11. False

12. True

13. False

14. True

15. 0.12

16. 0.24

17. 0.18

18. 0.35

19. 0.16

20. 0.21

21. 0.28

22. 0.252

23. 1/16 = 0.0625

24. 1/16 = 0.0625

25. 1/216 = 0.0046

26. 1/216 = 0.0046

27. Mutually exclusive

28. Neither

29. Neither

30. Mutually exclusive

31. a. Yes; $P(A \text{ and } B) = P(A)P(B)$ **b.** 0.55 **c.** No; $P(A \text{ and } B) \neq 0$

32. a. No; $P(A \text{ and } B) \neq P(A)P(B)$ **b.** 1 **c.** No; $P(A \text{ and } B) \neq 0$

33. a. 0.3 **b.** No; $P(A \text{ or } B) \neq P(A) + P(B)$ **c.** No; $P(A \text{ and } B) \neq P(A)P(B)$

34. a. 0 **b.** Yes; $P(A \text{ and } B) = 0$ **c.** No; $P(A \text{ and } B) \neq P(A)P(B)$

35. 91/216 = 0.4213

36. 0.8704

37. 1/42 = 0.0238

38. 1/30 = 0.0333

39. a. 0.5660 **b.** 0.2333 **c.** 0.7391 **d.** 0.2537 **e.** 0.8113

40. a. 0.25 **b.** 0.53 **c.** 0.08 **d.** 0.32 **e.** 0.1509

41. a. 0.8 **b.** 0.7 **c.** 0.7 **d.** Yes

42. a. 0.1667 **b.** 0.8550 **c.** 0.1467 **d.** 0.88 **e.** 0.1715 **f.** No; $P(E_1 \text{ and } E_2) \neq P(E_1)P(E_2)$

43. a. 0.5123 **b.** 0.0555 **c.** 0.0316 **d.** 0.0617 **e.** 0.5691

44. a. 0.4637 **b.** 0.3652 **c.** 0.1855 **d.** 0.4000 **e.** 0.5079

45. 0.0015

46. 0.06

47. 0.0125

48. 0.0969

49. 0.05

50. 0.6336

51. a. 0.15 **b.** 0.05

52. a. 0.08 **b.** 0.06

53. a. 3/10 = 0.3 **b.** 2/9 = 0.2222 **c.** 1/15 = 0.0667 **d.** No; if the first component is defective, the second component is less likely to be defective.

54. a. 3/10 = 0.3 **b.** 299/999 = 0.2993 **c.** 0.0898 **d.** No. It is reasonable to treat *A* and *B* as though they were independent, because the sample size of 2 is less than 5% of the population of 1000.

55. No. If someone owns two vehicles, the types of vehicles are not independent.

56. Yes. The color of the light in the evening is independent of the color in the morning.

57. 0.7903

58. 0.8319

59. 0.743

60. 0.692

61. 0.98

62. 1

63. $1/3 = 0.3333$

64. The probability that a person has the disease given that the result is positive may be small.

65. Since $P(A) = P(B) = 0$, $P(A \text{ and } B) = 0$. Therefore, A and B are independent because $P(A \text{ and } B) = P(A)P(B)$. Also, A and B are mutually exclusive, because $P(A \text{ and } B) = 0$.

66. Since $P(A) = 0$, $P(A \text{ and } B) = 0$. Therefore, A and B are independent because $P(A \text{ and } B) = P(A)P(B)$. Also, A and B are mutually exclusive, because $P(A \text{ and } B) = 0$.

67. No. If A and B are mutually exclusive, then $P(A \text{ and } B) = 0$. But since $P(A) > 0$ and $P(B) > 0$, $P(A)P(B) > 0$. Therefore, $P(A \text{ and } B) \neq P(A)P(B)$, so A and B are not independent.

Section 5.4

Exercises 1–6 are the Check Your Understanding exercises for this section. Answers to these exercises are on page 256.

7. mn

8. $6! = 720$

9. False

10. True

11. 362,880

12. 120

13. 1

14. 479,001,600

15. 1

16. 6

17. 210

18. 8

19. 1190

20. 120

21. 1

22. 146,611,080

23. 126

24. 7

25. 2300

26. 10

27. 1

28. 1

29. 48

30. 144

31. **a.** 336 **b.** 56

32. **a.** 2730 **b.** 455

33. **a.** $10^4 \cdot 26^3 = 175{,}760{,}000$ **b.** $10^4 = 10{,}000$
c. $1/26^3 = 0.0000569$

34. **a.** 5040 **b.** 210 **c.** $1/5040 = 0.000198$
d. $1/210 = 0.00476$

35. **a.** ${}_{12}C_8 = 495$ **b.** 0.0303 **c.** 0.9697

36. **a.** $36^8 = 2.82 \times 10^{12}$ **b.** $26^8 = 2.09 \times 10^{11}$
c. $36^8 - 26^8 = 2.61 \times 10^{12}$
d. $(26/36)^8 = 0.0740$ **e.** 0.9260

37. **a.** $4^3 = 64$ **b.** ${}_4P_3 = 24$ **c.** 8
d. $24/64 = 0.375$ **e.** $1/8 = 0.125$

38. **a.** ${}_{18}C_3 = 816$ **b.** ${}_{10}C_3 = 120$ **c.** ${}_8C_3 = 56$
d. $120/816 = 0.1471$ **e.** 0.8529

39. **a.** 1326 **b.** ${}_4C_2 = 6$ **c.** $6/1326 = 0.00452$

40. **a.** 2652 **b.** 1326 **c.** 64 **d.** $64/1326 = 0.0483$

41. $\dfrac{1}{{}_{42}C_5} = 0.00000118$

42. $\dfrac{1}{{}_{42}C_6} = 0.000000191$

43. **a.** 8 **b.** $2/8 = 0.25$

CHAPTER 5 Quiz

1. (i)

2. **a.** The sample space is the population of one million voters.
b. 0.56

3. **a.** $P(A \text{ or } B) = P(A) + P(B) - P(A \text{ and } B)$
b. $P(A \text{ or } B) = P(A) + P(B)$
c. $P(A^c) = 1 - P(A)$
d. $P(A \text{ and } B) = P(A)P(B \mid A)$ or $P(A \text{ and } B) = P(B)P(A \mid B)$
e. $P(A \text{ and } B) = P(A)P(B)$

4. **a.** $132/400 = 0.33$ **b.** 0.865

5. (i)

6. $79/100 = 0.79$

7. $2/3 = 0.6667$

8. 0.48

9. $4/11 = 0.3636$

10. (i)

11. $0.38^3 = 0.0549$

12. 0.9084

13. **a.** $21^3 \cdot 5^3 \cdot 10 = 11{,}576{,}250$
b. $1/11{,}576{,}250 = 0.0000000864$

14. ${}_{24}C_5 = 42{,}504$

15. **a.** $15! = 1.308 \times 10^{12}$ **b.** ${}_{15}P_3 = 2730$

CHAPTER 5 Review Exercises

1. **a.** {R, W, W, B, B, B} **b.** $1/2$

2. $18/30 = 0.6$

3. **a.** 0.5 **b.** 0.6667

4. **a.** $570/1200 = 0.475$ **b.** $1010/1200 = 0.8417$

5. **a.** 0.03 **b.** 0.32

6. **a.** 0.9998 **b.** 0.0098

7. **a.** {DDD, DDG, DGD, DGG, GDD, GDG, GGD, GGG}
b. 0.6815
c. 0.9603
d. Yes

8. **a.** $5/10 = 0.5$ **b.** $4/9 = 0.4444$ **c.** 0.2222

9. a. 0.6 **b.** 0.75 **c.** 0.2727
d. No; P(Female and Business major) $\neq$ P(Female)P(Business major)
e. No; P(Female and Business major) $\neq 0$

10. a. 0.4035 **b.** 0.3647 **c.** 0.1066 **d.** 0.6616
e. 0.2922 **f.** 0.2641

11. a. That the two days are independent with regard to rain
b. If it rains on Saturday, it is more likely that it will rain on Sunday.
c. Too low. In the equation, P(Rain Sunday) should be P(Rain Sunday | Rain Saturday), which is greater than 0.1.

12. $_5C_3 = 10$

13. $1/10 = 0.1$

14. a. $_6P_3 = 120$ **b.** $_5P_3 = 60$

15. a. $1/120 = 0.00833$ **b.** $1/20 = 0.05$

CHAPTER 5 Case Study

1. 0.99613

2. 0.98739

3. 0.97801

4. 0.96533

5. 50: 0.93782; 60: 0.88164; 70: 0.76666; 80: 0.54385; 90: 0.22931; 100: 0.02809

6. 0.028444

7. 0.81749

8. The probability that a person aged 20 is still alive at age 50 is 0.94979. The probability that a person aged 50 is still alive at age 60 is 0.94010. It is more probable that a person aged 20 will still be alive at age 50.

9. 0.79419

CHAPTER 6

Section 6.1

Exercises 1–8 are the Check Your Understanding exercises for this section. Answers to these exercises are on page 278.

9. random variable

10. 1

11. Continuous

12. population

13. True

14. True

15. False

16. False

17. Discrete

18. Discrete

19. Continuous

20. Continuous

21. Discrete

22. Continuous

23. Discrete

24. Continuous

25. Continuous

26. Discrete

27. Continuous

28. Discrete

29. Represents a probability distribution

30. Represents a probability distribution

31. Does not represent a probability distribution; probabilities do not add up to 1.

32. Does not represent a probability distribution; a probability cannot be negative.

33. Does not represent a probability distribution; probabilities do not add up to 1.

34. Represents a probability distribution

35. Mean: 2.9; standard deviation: 2.042

36. Mean: 14.7; standard deviation: 3.494

37. Mean: 6.94; standard deviation: 2.087

38. Mean: 0.77; standard deviation: 1.548

39. Mean: 19.45; standard deviation: 3.232

40. Mean: 151.5; standard deviation: 23.511

41. 0.2

42. 0.40

43. a. 0.2 **b.** 0.3 **c.** 0.1 **d.** 0.1 **e.** 2
f. 1.095

44. a. 0.30 **b.** 0.35 **c.** 0.10 **d.** 0.35 **e.** 2.1
f. 1.3 **g.** 0.35

45. a. 0.1 **b.** 0.5 **c.** 0.2 **d.** 0.9 **e.** 0.8
f. 0.980 **g.** 0.8

46. a. 0.15 **b.** 0.05 **c.** 0.7 **d.** 0.95 **e.** 1.52
f. 0.933 **g.** No

47. a. 0.38 **b.** 0.96 **c.** 0.67 **d.** 0.10 **e.** 1.1
f. 1.054

48. a. 0.258 **b.** 0.888 **c.** 0.630 **d.** 0.112
e. 2.85 **f.** 1.107

49. a.

x	$P(x)$
0	0.0680
1	0.1110
2	0.2005
3	0.1498
4	0.1885
5	0.1378
6	0.0889
7	0.0197
8	0.0358

b. 0.282 **c.** 0.068 **d.** 3.36 **e.** 1.97

50. a.

x	$P(x)$
0	0.0326
1	0.5272
2	0.3180
3	0.0871
4	0.0230
5	0.0121

b. 0.087 **c.** 0.122 **d.** 1.58 **e.** 0.88

51. a.

x	$P(x)$
1	0.1278
2	0.1241
3	0.1236
4	0.1222
5	0.1227
6	0.1247
7	0.1266
8	0.1283

b. 0.122 **c.** 0.255 **d.** 4.51 **e.** 2.31

52. a.

x	$P(x)$
0	0.069
1	0.138
2	0.345
3	0.207
4	0.172
5	0.069

b. 0.2069
c. 0.7586
d. 2.4828
e. 1.3031

53. −\$0.50, an expected loss

54. −\$0.51, an expected loss

55. −1/6 = −\$0.17, an expected loss

56. −5/36 = −\$0.14, an expected loss

57. a. 0
b. Answers will vary. If you don't answer a question, your score is the same as the expected value of a random guess.

58. a. 0.0625
b. Yes, because your expected gain is greater than 0.

59. \$4500. It would be wise to make the investment.

60.

x	$P(x)$
1,000	0.9516
−19,000	0.0484

The expected profit is \$32.

61. a.

y	$P(y)$
0	0.5
1	0.25
2	0.125
3	0.125

b. 0.875 **c.** 1.0533

62. a.

x	$P(x)$
0	0.125
1	0.875

b. 0.875 **c.** 0.3307

63. a. 0, 1, 2, 3 **b.** 0.512 **c.** 0.128
d. $P(SFS) = P(SSF) = 0.128$ **e.** 0.384 **f.** 0.096
g. 0.008 **h.** 2.4 **i.** 0.6928

Section 6.2

Exercises 1–4 are the Check Your Understanding exercises for this section. Answers to these exercises are on page 290.

5. two

6. trials

7. $\sqrt{np(1-p)}$

8. True

9. False

10. True

11. Does not have a binomial distribution because the sample is more than 5% of the population.

12. Has a binomial distribution with 10 trials.

13. Has a binomial distribution with 7 trials.

14. Does not have a binomial distribution because it is not the number of successes in independent trials.

15. Does not have a binomial distribution because it is not the number of successes in independent trials.

16. Has a binomial distribution with 250 trials.

17. 0.3087, mean = 3.5, variance = 1.05, SD = 1.025

18. 0.2684, mean = 2, variance = 1.6, SD = 1.265

19. 0.0355, mean = 12, variance = 4.8, SD = 2.191

20. 0.0232, mean = 4.2, variance = 2.94, SD = 1.715

21. 0.2160, mean = 1.2, variance = 0.72, SD = 0.849

22. 0.2621, mean = 4.8, variance = 0.96, SD = 0.980

23. 0.7969, mean = 1.6, variance = 1.28, SD = 1.131

24. 0.5490, mean = 13.5, variance = 1.35, SD = 1.162

25. 0.8108, mean = 1.5, variance = 1.455, SD = 1.206

26. 0.4114, mean = 27, variance = 2.7, SD = 1.643

27. a. *ii* **b.** *iii* **c.** *i* **d.** *iv*

28. a. *iii* **b.** *ii* **c.** *iv* **d.** *i*

29. a. 0.2051 **b.** 0.0547
c. No, $P(7 \text{ or more}) = 0.1719$.

30. a. 0.2503 **b.** 0.4744
c. Yes, $P(7 \text{ or more}) = 0.0035$.

31. a. 0.0798 **b.** 0.2897 **c.** 0.5940
d. No, $P(11 \text{ or more}) = 0.3043$.

32. a. 0.0532 **b.** 0.9348 **c.** 0.0652
d. No, $P(0) = 0.3677$.

33. a. 0.175 **b.** 0.056 **c.** 0.248
d. Yes, $P(\text{Fewer than } 12) = 0.00059$.

34. a. 0.9521 **b.** 0.6570 **c.** 0.1900
d. Yes, $P(\text{More than } 12) = 0.0189$.

35. a. 0.4131 **b.** 0.9529 **c.** No, $P(0) = 0.2342$.
d. 1.4 **e.** 1.1411

36. a. 0.1780 **b.** 0.0920 **c.** 0.4065
d. Yes, $P(15) = 0.000073$. **e.** 7.95 **f.** 1.933

37. a. 0.1166 **b.** 0.9486 **c.** 0.0010 **d.** Yes, $P(\text{More than } 25) = 0.0101$. **e.** 19.8 **f.** 2.595

38. a. 0.1802 **b.** 0.0796 **c.** 0.1971
d. Yes, $P(\text{More than } 8) = 0.0075$. **e.** 4 **f.** 1.7321

39. a. 0.1472 **b.** 0.3231 **c.** 0.0332

d. No, $P(\text{More than } 10) = 0.0978$. **e.** 7.5 **f.** 2.2913

40. a. 0.1883 **b.** 0.8961 **c.** 0.0049

d. Yes, $P(\text{Fewer than } 4) = 0.00062$. **e.** 8.1 **f.** 1.2406

41. a. 0.000135 **b.** Yes

c. Yes, because if the shipment were good it would be unusual for 7 or more of 10 items to be defective.

d. 0.456 **e.** No

f. No, because if the shipment were good it would not be unusual for 2 of 10 items to be defective.

42. a. 0.00123 **b.** Yes

c. Yes, because if the claim were true it would be unusual for 2 or fewer of 8 people to have smoke detectors.

d. 0.497 **e.** No

f. No, because if the claim were true it would not be unusual for 6 of 8 people to have smoke detectors.

43. a. 0.16116 **b.**

x	$P(x)$
0	0.0134627
1	0.0724915
2	0.1756524
3	0.2522188
4	0.2376677
5	0.1535699
6	0.0689096
7	0.0212029
8	0.0042814
9	0.0005123
10	0.0000276

Section 6.3

Exercises 1–4 are the Check Your Understanding exercises for this section. Answers to these exercises are on page 296.

5. time, space

6. λt

7. False

8. True

9. 0.0378

10. 0.1804

11. 0.1839

12. 0.1353

13. 0.8647

14. 0.9093

15. 0.7851

16. 0.1665

17. 0.2834

18. 0.8088

19. a. 0.1339 **b.** 0.9826 **c.** 0.4589 **d.** 6 **e.** 2.4495

20. a. 0.1008 **b.** 0.0498 **c.** 0.1991 **d.** 0.8009

e. 3 **f.** 1.7321

21. a. 0.2231 **b.** 0.8647

22. a. 0.2019 **b.** 0.6454

23. a. 0.0916 **b.** 0.1048 **c.** 0.2381

24. a. 0.1033 **b.** 0.0724

25. a. 0.0844 **b.** 20 **c.** 4.4721

26. a. 0.0829 **b.** 15 **c.** 3.873

27. a. 0.1221 **b.** 0.0138 **c.** 0.1841 **d.** 8 **e.** 2.8284

28. a. 0.1714 **b.** 0.2017 **c.** 0.7983 **d.** 5.5 **e.** 2.3452

29. a. $P(X = 0) = e^{-\lambda}\frac{\lambda^0}{0!} = e^{-\lambda}$

b. $X = 0$ means that no events occurred in 1 second. Therefore, the amount of time that elapses until the next event will be greater than 1.

c. $P(T > 1) = P(X = 0) = e^{-\lambda}$

d. $P(X = 0) = e^{-2\lambda}\frac{(2\lambda)^0}{0!} = e^{-2\lambda}$

e. $X = 0$ means that no events occurred in 2 seconds. Therefore, the amount of time that elapses until the next event will be greater than 2.

f. $P(X = 0) = e^{-\lambda t}\frac{(\lambda t)^0}{0!} = e^{-\lambda t}$

g. $X = 0$ means that no events occurred in t seconds. Therefore, the amount of time that elapses until the next event will be greater than t.

h. $P(T > t) = P(X = 0) = e^{-\lambda t}$

CHAPTER 6 Quiz

1. The probabilities do not add up to 1.

2. 2

3. a. 9 **b.** 3

4. 0.19

5.

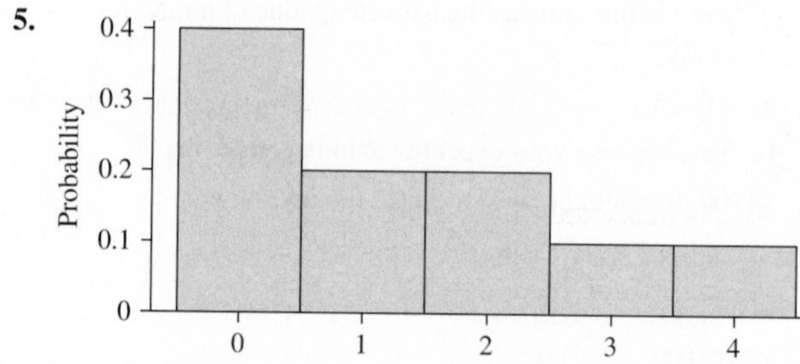

6. a. 0.4 **b.** 0.2 **c.** 0.9 **d.** 0.6

7. 1.3

8. 1.3454

9. a. 0.0769 **b.** 0.9231 **c.** 0.7604

10. Mean: 2.5; standard deviation: 1.5411

11.

x	$P(x)$
0	0.24
1	0.52
2	0.24

12. a. 0.1755 **b.** 0.0671 **c.** 0.9179

13. 80

14. Variance: 80; standard deviation: 8.9443

15. 0.1755

CHAPTER 6 Review Exercises

1. a. Yes **b.** No **c.** No **d.** Yes

2. a. 8.37 **b.** 3.3131 **c.** 1.8202

3. 0.41

4. a.

x	$P(x)$
1	0.5632
2	0.2500
3	0.1147
4	0.0473
5	0.0171
6	0.0053
7	0.0018
8	0.0006

b. 0.5632 **c.** 1.731 **d.** 1.049

5. a. 0.6630 **b.** 0.5760

6. a. 0.6943 **b.** 0.1353 **c.** Yes, $P(0) = 0.00564$.

7. −$0.11, an expected loss

8. a. 0.0480 **b.** 0.9854 **c.** Yes, $P(10) = 0.00511$.

9. a. 0.2756 **b.** 0.1844 **c.** Yes, $P(8) = 0.0360$.

10. No, the trials are not independent. If it rains on one day, it is more likely to rain the next day.

11. Yes; because the sample size is less than 5% of the population, X may be considered to have a binomial distribution.

12. a. 0.1606 **b.** 0.0174 **c.** 0.9826

13. 6

14. 2.4495

15. 0.0873

CHAPTER 6 Case Study

List (ii) is the fraud.

CHAPTER 7

Section 7.1

Exercises 1–10 are the Check Your Understanding exercises for this section. Answers to these exercises are on page 321.

11. density

12. 0

13. 1

14. equal

15. standard

16. negative

17. True

18. True

19. False

20. False

21. a. 0.25 **b.** 0.6 **c.** 0.4

22. a. 0.24 **b.** 0.09

23. a. 0.8 **b.** 0.4 **c.** 0.6

24. a. 0.5 **b.** 0.7 **c.** 0.3

25. a. 0.7995 **b.** 0.9015 **c.** 0.7010

26. a. 0.6498 **b.** 0.9381 **c.** 0.5879

27. a. ii **b.** iv **c.** i **d.** iii

28. a. iii **b.** ii **c.** iv **d.** i

29. a. 0.8944 **b.** 0.3594 **c.** 0.7085 **d.** 0.5633

30. a. 0.3520 **b.** 0.6844 **c.** 0.2744 **d.** 0.3853

31. a. 0.7704 **b.** 0.0154 **c.** 0.8461 **d.** 0.4404

32. a. 0.9948 **b.** 0.7019 **c.** 0.1736 **d.** 0.0968

33. a. 0.9394 **b.** 0.3745 **c.** 0.0007 **d.** 0.9916

34. a. 0.3192 **b.** 0.9982 **c.** 0.0207 **d.** 0.1379

35. a. 0.7288 **b.** 0.8492 **c.** 0.0744 **d.** 0.0936

36. a. 0.8128 **b.** 0.1311 **c.** 0.0374 **d.** 0.5136

37. a. 0.5059 **b.** 0.1153 **c.** 0.5078 **d.** 0.6402

38. a. 0.1342 **b.** 0.3044 **c.** 0.7173 **d.** 0.9209

39. 0.10

40. −1.13

41. 1.48

42. −0.67

43. −0.99

44. 1.08

45. 0.39

46. −1.41

47. −0.67 and 0.67

48. −1.04 and 1.04

49. −1.28 and 1.28

50. −2.33 and 2.33

51. 0.0401

52. 0.6950

53. 0.6956

54. 0.0590

55. 0.3

56. 0.58

Section 7.2

Exercises 1–10 are the Check Your Understanding exercises for this section. Answers to these exercises are on page 334.

11. standardization

12. −2

13. True

14. True

15. False

16. True

17. False

18. True

19. a. 0.3085 **b.** 0.1056

20. a. 0.9664 [Tech: 0.9666] **b.** 0.7486 [Tech: 0.7475]

21. a. 0.2061 [Tech: 0.2066] **b.** 0.0869 [Tech: 0.0863]

22. a. 0.0073 [Tech: 0.0074] **b.** 0.8830 [Tech: 0.8825]

23. a. 0.8314 [Tech: 0.8315] **b.** 0.1549 [Tech: 0.1548]

24. a. 0.2342 [Tech: 0.2356] **b.** 0.4714 [Tech: 0.4680]

25. 11.25 [Tech: 11.24]

26. 64.32 [Tech: 64.29]

27. 38.08 [Tech: 38.10]

28. 115.22 [Tech: 115.24]

29. **a.** 0.1446 [Tech: 0.1444] **b.** 0.5438 [Tech: 0.5421]
c. 0.1685 [Tech: 0.1686]
d. 0.0582 [Tech: 0.0587]; not unusual

30. **a.** 0.2877 [Tech: 0.2893] **b.** 0.9032 [Tech: 0.9026]
c. 0.5361 [Tech: 0.5335] **d.** 0.0207 [Tech: 0.0208]; unusual

31. **a.** 75.35 [Tech: 75.31] **b.** 84.86 **c.** 87.13 [Tech: 87.18]

32. **a.** 13.88 [Tech: 13.87] **b.** 8.04 **c.** 9.69 [Tech: 9.68]

33. **a.** 0.9192 **b.** 0.5834 [Tech: 0.5827] **c.** Yes

34. **a.** 0.7881 **b.** 0.1327 **c.** No

35. **a.** 524.15 **b.** 466.27 **c.** 502.23

36. **a.** 64.089 **b.** 118.16 **c.** 73.138

37. **a.** 0.2236 [Tech: 0.2248] **b.** 0.6533 [Tech: 0.6544]
c. Yes, *P*(Less than 400) = 0.0239 [Tech: 0.0241].

38. **a.** 0.0855 [Tech: 0.0854] **b.** 0.2420 [Tech: 0.2414]
c. No, *P*(More than 1550) = 0.1562 [Tech: 0.1557].

39. **a.** 489.20 [Tech: 489.28] **b.** 510.11 [Tech: 509.96]
c. 453.53 [Tech: 453.35] **d.** 438.36 [Tech: 438.51]

40. **a.** 1263.03 [Tech: 1262.68] **b.** 1625.28 [Tech: 1624.60]
c. 1279.13 [Tech: 1278.41] **d.** 1011.87 [Tech: 1012.46]

41. **a.** 0.3085 **b.** No **c.** 0.0062 **d.** Yes

42. **a.** 0.1611 [Tech: 0.1604] **b.** 0.8281 [Tech: 0.8279]
c. 0.6879 [Tech: 0.6890]

43. **a.** 4.022 [Tech: 4.023] **b.** 4.382 [Tech: 4.381] **c.** 4.1

44. **a.** 80.54 [Tech: 80.62] **b.** 116.34 [Tech: 116.25]
c. 110.14

45. **a.** 0.0548 **b.** 0.3811 [Tech: 0.3812] **c.** 0.8849

46. **a.** 0.2843 [Tech: 0.2839] **b.** 0.1949 [Tech: 0.1957]
c. 0.1367 [Tech: 0.1376]

47. **a.** 34.80 [Tech: 34.82] **b.** 42.35 [Tech: 42.34]
c. 36.65 [Tech: 36.63] **d.** 29.75 [Tech: 29.73]

48. **a.** 103.46 [Tech: 103.42] **b.** 126.56 [Tech: 126.51]
c. 121.38 [Tech: 121.44] **d.** 144.62 [Tech: 144.57]

49. **a.** 0.3085 **b.** 0.1525 [Tech: 0.1524]
c. Yes, *P*(Less than 12) = 0.0062.

50. **a.** 0.0162 [Tech: 0.0161] **b.** 0.6847 [Tech: 0.6859]
c. 0.7611 [Tech: 0.7625]

51. **a.** 12.055 **b.** 12.015 **c.** 12.011 and 12.089

52. **a.** 39.35 [Tech: 39.38] **b.** 21.71 [Tech: 21.73]
c. 16.04 and 33.96 [Tech: 16.03 and 33.97]

53. **a.** 0.1056 **b.** 0.0001 **c.** 0.8882 [Tech: 0.8881]

54. **a.** 0.0301 [Tech: 0.0304] **b.** 0.1056
c. 0.4714 [Tech: 0.4680]

55. **a.** 25.1200 [Tech: 25.1203] **b.** 25.0624 [Tech: 25.0626]
c. 24.9136 [Tech: 24.9138]
d. 25.0464 and 25.1536 [Tech: 25.0460 and 25.1540]

56. **a.** 73.00 [Tech: 72.97] **b.** 78.12 [Tech: 78.08]
c. 85.24 [Tech: 85.25]
d. 68.28 and 81.72 [Tech: 68.27 and 81.73]

57. 34%

58. 30%

59. They are not approximately normal. If they were normally distributed, approximately 26% of the wells would have negative concentrations, which is impossible.

60. 0.4045 [Tech: 0.4044]

61. Mean: 74.26; SD: 8.02 [Tech: 8.01]

62. 8:23 A.M.

Section 7.3

Exercises 1–4 are the Check Your Understanding exercises for this section. Answers to these exercises are on page 342.

5. sampling

6. Central Limit Theorem

7. True

8. False

9. **a.** 0.9232 [Tech: 0.9236] **b.** 8.56

10. **a.** 0.9998 **b.** 26.03

11. **a.** 0.7549 [Tech: 0.7540] **b.** 95.73

12. **a.** 0.0080 **b.** 45.28

13. **a.** 0.0465 [Tech: 0.0461] **b.** 31.81 [Tech: 31.80]

14. **a.** 0.9778 [Tech: 0.9779] **b.** 22.73

15. **a.** No, $n \leq 30$. **b.** Not appropriate
c. Not appropriate

16. **a.** Yes, $n > 30$. **b.** 0.1609 [Tech: 0.1608]
c. 51.82

17. **a.** Yes, the population is normal.
b. 0.9372 [Tech: 0.9376] **c.** 64.51 [Tech: 64.52]

18. **a.** No, $n \leq 30$. **b.** Not appropriate
c. Not appropriate

19. **a.** $\mu = 75.4$, $\sigma = 5.1225$
b. (69, 69), (69, 75), (69, 79), (69, 83), (69, 71), (75, 69), (75, 75), (75, 79), (75, 83), (75, 71), (79, 69), (79, 75), (79, 79), (79, 83), (79, 71), (83, 69), (83, 75), (83, 79), (83, 83), (83, 71), (71, 69), (71, 75), (71, 79), (71, 83), (71, 71)
c. $\mu_{\bar{x}} = 75.4$, $\sigma_{\bar{x}} = 3.62215$
d. $\mu_{\bar{x}} = 75.4 = \mu$; $\sigma_{\bar{x}} = 3.62215 = \sigma/\sqrt{2}$

20. **a.** $\mu = 22.4$, $\sigma = 0.96177$
b. (21, 21), (21, 22), (21, 22.5), (21, 23), (21, 23.5), (22, 21), (22, 22), (22, 22.5), (22, 23), (22, 23.5), (22.5, 21), (22.5, 22), (22.5, 22.5), (22.5, 23), (22.5, 23.5), (23, 21), (23, 22), (23, 22.5), (23, 23), (23, 23.5), (23.5, 21), (23.5, 22), (23.5, 22.5), (23.5, 23), (23.5, 23.5)
c. $\mu_{\bar{x}} = 22.4$, $\sigma_{\bar{x}} = 0.68007$
d. $\mu_{\bar{x}} = 22.4 = \mu$, $\sigma_{\bar{x}} = 0.68007 = \sigma/\sqrt{2}$

21. **a.** 0.9015 [Tech: 0.9016] **b.** 0.2966 [Tech: 0.2979]
c. Yes, *P*(Less than 26) = 0.0049.

22. **a.** 0.0202 [Tech: 0.0204] **b.** 0.3039 [Tech: 0.3034]
c. No, *P*(Less than 198) = 0.1539 [Tech: 0.1531].

23. **a.** 0.9671 [Tech: 0.9674] **b.** 0.0228 [Tech: 0.0229]
c. 2.17 **d.** Yes, *P*(Less than 2) = 0.0329 [Tech: 0.0326].
e. No, the Central Limit Theorem applies only to the sample mean, not to individual values.

24. **a.** 0.1685 [Tech: 0.1694] **b.** 0.6489 [Tech: 0.6490]
c. 526.3
d. Yes, *P*(Greater than 550) = 0.0069 [Tech: 0.0070].

e. No, the Central Limit Theorem applies only to the sample mean, not to individual values.

25. a. 0.4013 [Tech: 0.4001] b. 0.1840 [Tech: 0.1853]
c. 8000.5 [Tech: 7999.9]
d. Yes, P(Less than 7500) = 0.0003.
e. No, the Central Limit Theorem applies only to the sample mean, not to individual values.

26. a. 0.0838 b. 0.2632 c. 2643.5 [Tech: 2643.7].
d. Yes, P(Greater than 2800) = 0.0004.
e. No, the Central Limit Theorem applies only to the sample mean, not to individual values.

27. a. 200 pounds b. 0.2483 [Tech: 0.2472] c. 0.0001

28. a. 203 pounds b. 0.2005 [Tech: 0.2013] c. 0.0007

29. a. 0.1038 [Tech: 0.1030] b. 0.5646 [Tech: 0.5651]
c. 42.79 [Tech: 42.80]
d. Yes, P(Less than 35) = 0.0136 [Tech: 0.0134].
e. No, the Central Limit Theorem applies only to the sample mean, not to individual values.

30. a. 0.9994 b. 0.0150 [Tech: 0.0152] c. 14.82
d. Yes, P(Greater than 19) = 0.0006.
e. No, the Central Limit Theorem applies only to the sample mean, not to individual values.

31. No. We don't know the distribution of the page thicknesses. We only know that the mean of a large number of them will be approximately normally distributed.

32. No. We don't know the distribution of the weights of the bags. We only know that the mean of a large number of them will be approximately normally distributed.

33. a. 0.1922 [Tech: 0.1932] b. No
c. No, because this result would not be unusual if the claim were true.
d. 0.0047 e. Yes
f. Yes, because this result would be unusual if the claim were true.

34. a. 0.0013 b. Yes
c. No, because this result would be unusual if the claim were true.
d. 0.3446 e. No
f. Yes, because this result would not be unusual if the claim were true.

35. a. 2.01 b. smaller
c. The correction factor is 0, so the standard deviation is 0. If the whole population is sampled, the sample mean will always be equal to the population mean, and therefore constant.

Section 7.4

Exercises 1–4 are the Check Your Understanding exercises for this section. Answers to these exercises are on page 349.

5. proportion
6. sampling
7. True
8. False
9. 0.2358 [Tech: 0.2354]
10. 0.0823 [Tech: 0.0816]
11. 0.0918 [Tech: 0.0922]
12. 0.2341 [Tech: 0.2352]
13. Not appropriate; $np < 10$
14. 0.2219 [Tech: 0.2229]

15. a. 0.63 b. 0.0305 c. 0.0951
d. 0.8255 [Tech: 0.8261] e. 0.3707 [Tech: 0.3716]
f. Yes, P(Less than 0.57) = 0.0250 [Tech: 0.0247].

16. a. 0.73
b. 0.0362
c. 0.7967 [Tech: 0.7961]
d. 0.1765 [Tech: 0.1772]
e. 0.7088 [Tech: 0.7094]
f. No, P(Less than 68%) = 0.0838 [Tech: 0.0839].

17. a. 0.68
b. 0.0506
c. 0.0571 [Tech: 0.0569]
d. 0.7135 [Tech: 0.7145]
e. 0.0838 [Tech: 0.0832]
f. No, P(Less than 65%) = 0.2776 [Tech: 0.2766].

18. a. 0.82
b. 0.0331
c. 0.2743 [Tech: 0.2726]
d. 0.8016 [Tech: 0.8007]
e. 0.9830 [Tech: 0.9829]
f. Yes, P(Greater than 90%) = 0.0078.

19. a. No, $np < 10$. b. 0.1492 [Tech: 0.1503]
c. 0.3011 [Tech: 0.2997] d. 0.9901
e. Yes, P(Greater than 0.25) = 0.0099.

20. a. No, because $np = 3.75$, which is less than 10.
b. 0.9015 [Tech: 0.9011] c. 0.1949 [Tech: 0.1953]
d. 0.4999 [Tech: 0.5000]
e. No, P(Less than 4%) = 0.1949 [Tech: 0.1953].

21. a. Yes, $np \geq 10$ and $n(1 - p) \geq 10$. The probability is 0.3557 [Tech: 0.3559].
b. 0.3121 [Tech: 0.3118]
c. 0.2694 [Tech: 0.2689]
d. 0.9864
e. No, P(Less than 42%) = 0.1093 [Tech: 0.1099].

22. a. Yes, because $np = 11.4$, which is greater than 10. The probability is 0.0885 [Tech: 0.0893].
b. 0.0256 [Tech: 0.0255] c. 0.8109 [Tech: 0.8100]
d. 0.1635 [Tech: 0.1646]
e. Yes, P(More than 45%) = 0.0017.

23. a. 0.2709 [Tech: 0.2717] b. 0.2459 [Tech: 0.2456]
c. 0.2877 [Tech: 0.2874]
d. Yes, P(Less than 0.25) = 0.0375 [Tech: 0.0379].

24. a. 0.5 b. 0.1112 [Tech: 0.1103]
c. 0.9929 [Tech: 0.9928] d. 0.7776 [Tech: 0.7793]
e. No, P(Less than 45%) = 0.1112 [Tech: 0.1103].

25. 200

26. 112

27. a. Erin b. Baylee

28. a. School A b. School B

29. a. 0.0001 b. Yes
c. No, because a sample proportion of 0.075 would be unusual if the goal had been reached.
d. 0.3300 [Tech: 0.3317] e. No
f. Yes, because a sample proportion of 0.053 would not be unusual if the goal had been reached.

Section 7.5

Exercises 1–4 are the Check Your Understanding exercises for this section. Answers to these exercises are on page 355.

5. np, $\sqrt{np(1-p)}$
6. continuity
7. True
8. False
9. 0.0559 [Tech using normal approximation: 0.0557; tech using binomial: 0.0563]
10. 0.8340 [Tech using normal approximation: 0.8344; tech using binomial: 0.8342]
11. 0.6517 [Tech using normal approximation: 0.6524; tech using binomial: 0.6538]
12. 0.6406 [Tech using normal approximation: 0.6423; tech using binomial: 0.6418]
13. 0.3102 [Tech using normal approximation: 0.3097; tech using binomial: 0.3008]
14. 0.4792 [Tech using normal approximation: 0.4765; tech using binomial: 0.4714]
15. **a.** 0.9015 [Tech using normal approximation: 0.9011; tech using binomial: 0.8991]
    b. 0.4522 [Tech using normal approximation: 0.4534; tech using binomial: 0.4455]
    c. 0.7888 [Tech using normal approximation: 0.7900; tech using binomial: 0.7878]
16. **a.** 0.2843 [Tech using normal approximation: 0.2858; tech using binomial: 0.2822]
    b. 0.3520 [Tech using normal approximation: 0.3530; tech using binomial: 0.3582]
    c. 0.5546 [Tech using normal approximation: 0.5535; tech using binomial: 0.5607]
17. **a.** 0.7549 [Tech using normal approximation: 0.7558; tech using binomial: 0.7544]
    b. 0.0934 [Tech using normal approximation: 0.0929; tech using binomial: 0.0918]
    c. 0.2115 [Tech using normal approximation: 0.2103; tech using binomial: 0.2133]
18. **a.** 0.3936 [Tech using normal approximation: 0.3934; tech using binomial: 0.3869]
    b. 0.9808 [Tech using normal approximation: 0.9809; tech using binomial: 0.9783]
    c. 0.4474 [Tech using normal approximation: 0.4477; tech using binomial: 0.4381]
19. **a.** 0.0885 [Tech using binomial: 0.0887]
    b. 0.7389 [Tech using normal approximation: 0.7388; tech using binomial: 0.7392]
    c. 0.4598 [Tech using normal approximation: 0.4605; tech using binomial: 0.4593]
20. **a.** 0.7454 [Tech using normal approximation: 0.7450; tech using binomial: 0.7400]
    b. 0.0239 [Tech using normal approximation: 0.0241; tech using binomial: 0.0190]
    c. 0.6284 [Tech using normal approximation: 0.6272; tech using binomial: 0.6205]
21. **a.** 0.9192 [Tech using normal approximation: 0.9190; tech using binomial: 0.9163]
    b. 0.7389 [Tech using normal approximation: 0.7387; tech using binomial: 0.7341]
    c. 0.3232 [Tech using normal approximation: 0.3230; tech using binomial: 0.3340]
22. **a.** 0.1020 [Tech using binomial: 0.1038]
    b. 0.8980 [Tech using binomial: 0.9005]
    c. 0.1414 [Tech using normal approximation: 0.1417; tech using binomial: 0.1401]
23. 0.9744 [Tech using normal approximation: 0.9745; tech using binomial: 0.9833]
24. 0.7852 [Tech using normal approximation: 0.7854; tech using binomial: 0.7849]
25. 0.0796 [Tech using normal approximation: 0.0797]

Section 7.6

Exercises 1–6 are the Check Your Understanding exercises for this section. Answers to these exercises are on page 366.

7. outlier, skewness, mode
8. False
9. No. The sample is strongly skewed to the right.
10. Yes. There are no outliers, no strong skewness, and no evidence of multiple modes.
11. Yes. There are no outliers, no strong skewness, and no evidence of multiple modes.
12. No. The sample contains an outlier.
13. Yes. There are no outliers, no strong skewness, and no evidence of multiple modes.
14. No. The sample is strongly skewed to the right.
15. No. The sample is strongly skewed to the right.
16. Yes. There are no outliers, no strong skewness, and no evidence of multiple modes.
17. Yes. The points approximately follow a straight line.
18. No. The points do not approximately follow a straight line.
19. 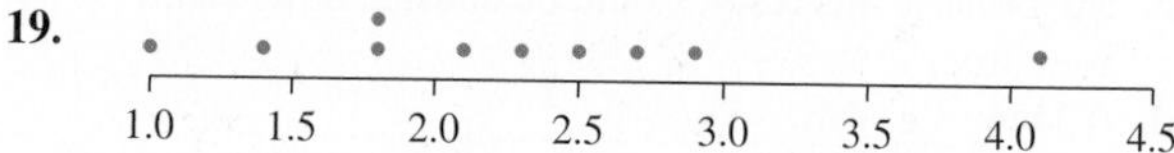

    No. The data contain an outlier.
20. 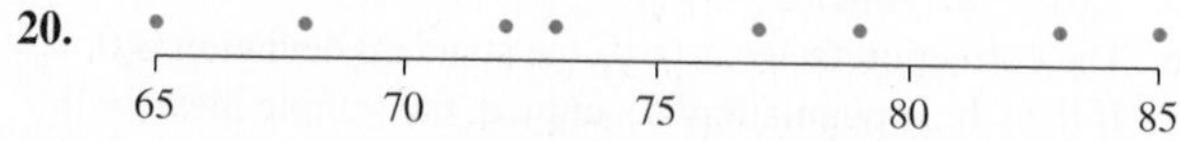

    Yes. There are no outliers, no strong skewness, and no evidence of multiple modes.
21. 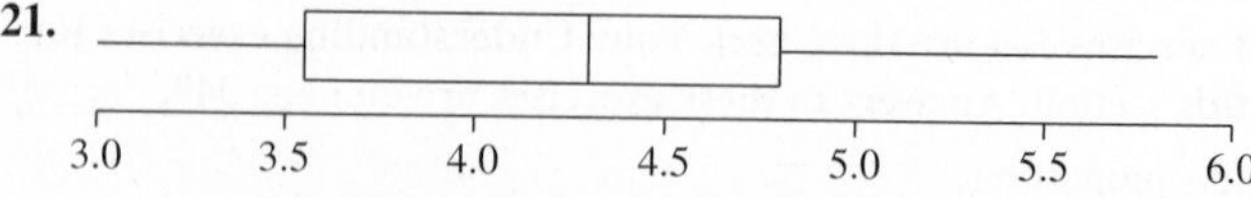

    Yes. There are no outliers, no strong skewness, and no evidence of multiple modes.
22. 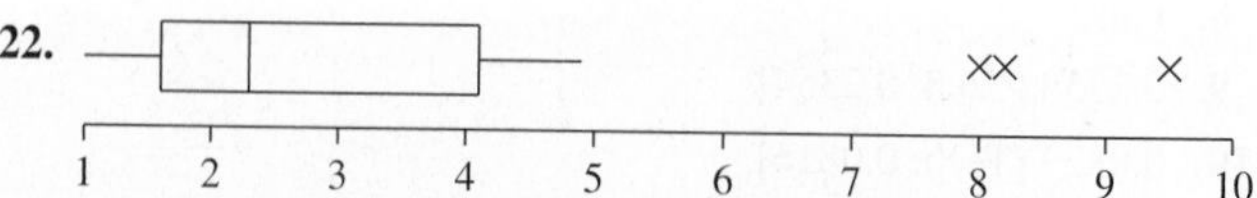

    No. The data contain outliers.

23. Answers will vary. Here is one possibility:

0	
0	99
1	0000111112234
1	69
2	1
2	79

The data are skewed to the right. It is not appropriate to treat this sample as coming from an approximately normal population.

24. Answers will vary. Here is one possibility:

7	
7	8
8	244
8	69
9	001
9	6999
10	1
10	58
11	34
11	

There are no outliers, no strong skewness, and no evidence of multiple modes. It is appropriate to treat this sample as coming from an approximately normal population.

25. Answers will vary. Here is one possibility:

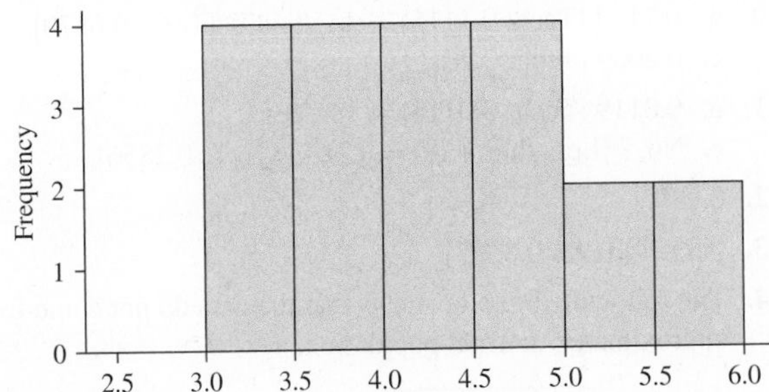

There are no outliers, no strong skewness, and no evidence of multiple modes. It is appropriate to treat this sample as coming from an approximately normal population.

26. Answers will vary. Here is one possibility:

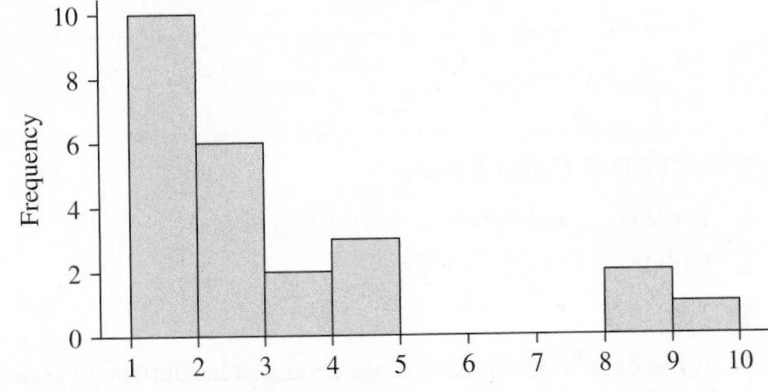

The data contain outliers. It is not appropriate to treat this sample as coming from an approximately normal population.

27.

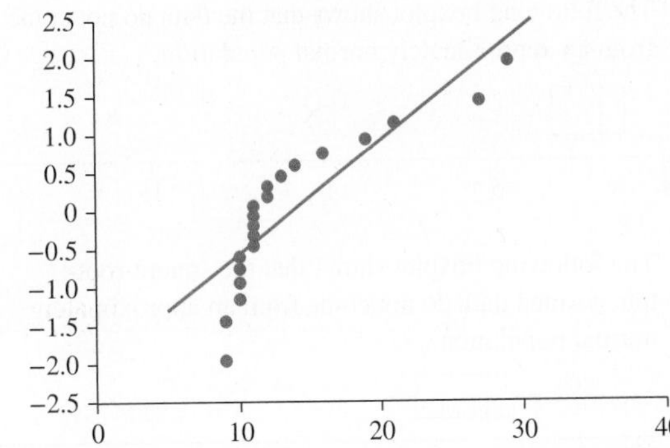

The points do not follow a straight line, so it is not appropriate to treat this sample as coming from an approximately normal population.

28.

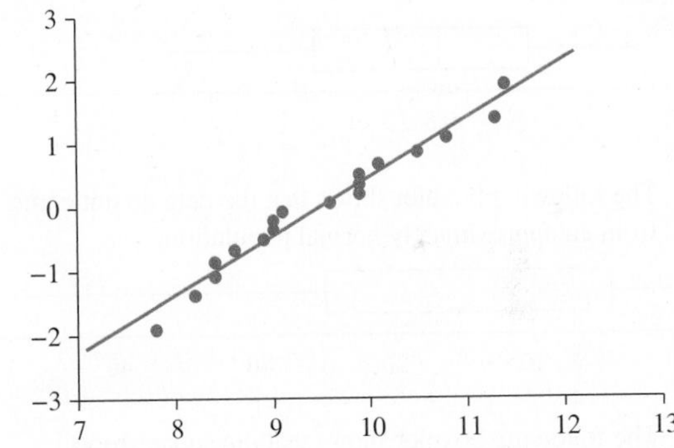

The points approximately follow a straight line, so it is appropriate to treat this sample as coming from an approximately normal population.

29. a. The following boxplot shows that the data do not come from an approximately normal population.

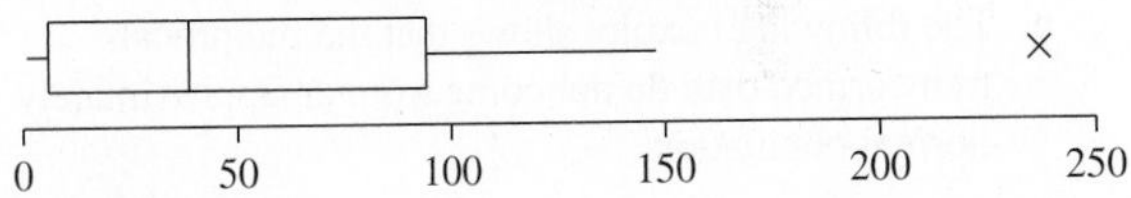

b. The following boxplot shows that the square-root-transformed data come from an approximately normal population.

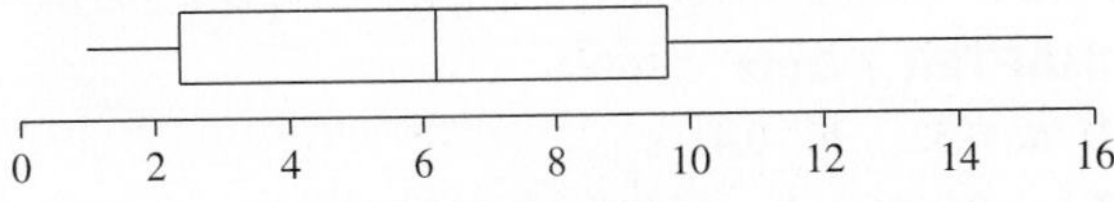

30. a. The following boxplot shows that the data do not come from an approximately normal population.

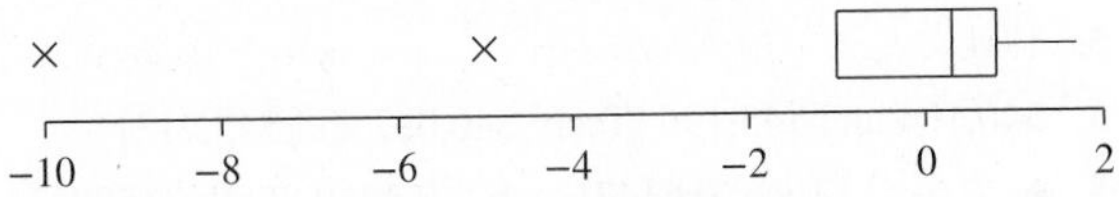

b. The following boxplot shows that the reciprocal-transformed data come from an approximately normal population.

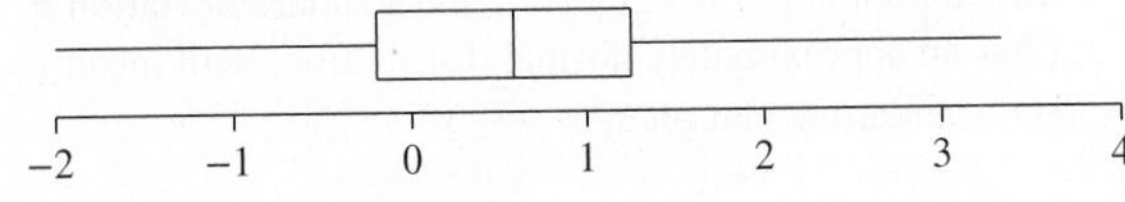

31. a. The following boxplot shows that the data do not come from an approximately normal population.

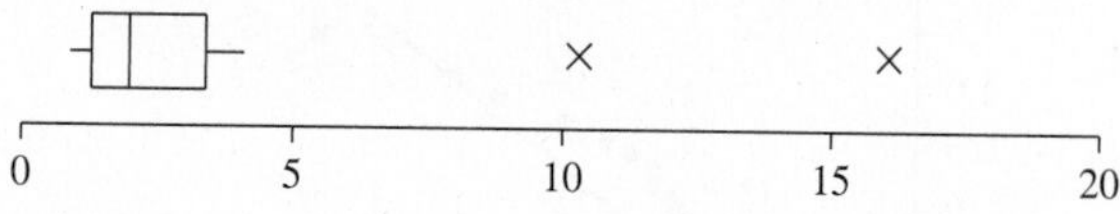

b. The following boxplot shows that the square-root-transformed data do not come from an approximately normal population.

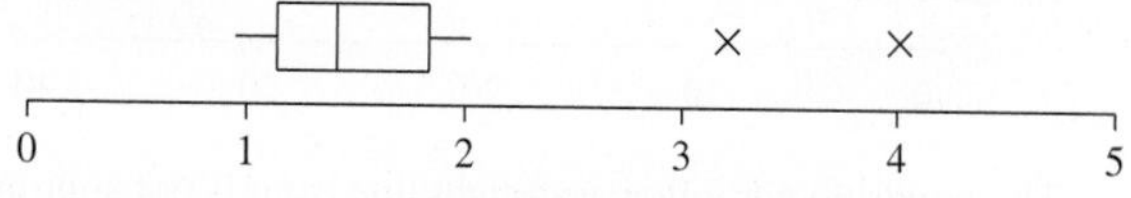

c. The following boxplot shows that the reciprocal-transformed data come from an approximately normal population.

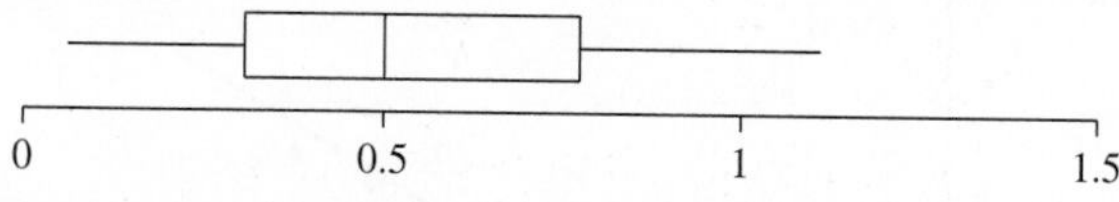

32. a. The following boxplot shows that the data do not come from an approximately normal population.

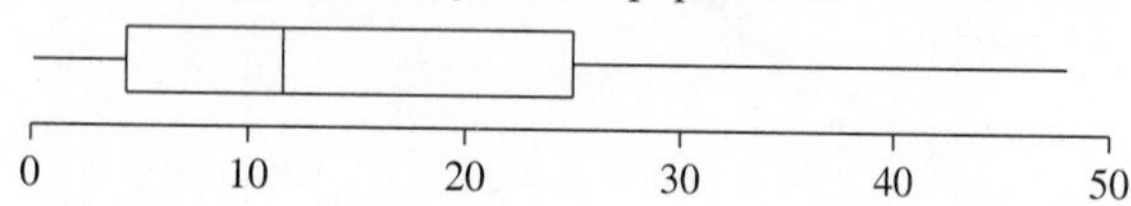

b. The following boxplot shows that the square-root-transformed data come from an approximately normal population.

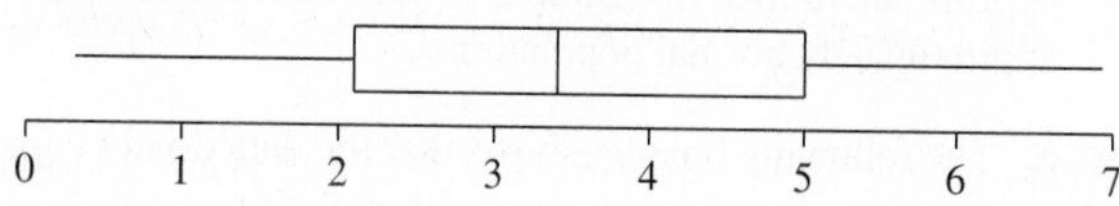

c. The following boxplot shows that the reciprocal-transformed data do not come from an approximately normal population.

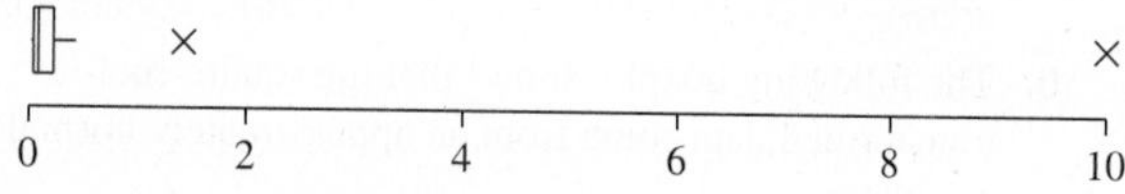

CHAPTER 7 Quiz

1. a. 0.32 **b.** 0.41

2. a. 0.9616 **b.** 0.3409 **c.** 0.8977

3. a. −0.44 **b.** −0.81

4. $z = -1.28$ and $z = 1.28$

5. 1.04

6. $40,104 and $45,196 [Tech: $40,087 and $45,213]

7. a. 0.6141 [Tech: 0.6142] **b.** 0.5910 [Tech: 0.5893]

8. 208.32 [Tech: 208.33]; because bowling scores are whole numbers, this can be rounded up to 209.

9. Let $\bar{x}$ be the mean of a large ($n > 30$) simple random sample from a population with mean μ and standard deviation σ. Then $\bar{x}$ has an approximately normal distribution, with mean $\mu_{\bar{x}} = \mu$ and standard deviation $\sigma_{\bar{x}} = \dfrac{\sigma}{\sqrt{n}}$.

10. $\mu_{\bar{x}} = 193$, $\sigma_{\bar{x}} = 5.25$

11. a. 0.5793 [Tech: 0.5803] **b.** 0.8997 [Tech: 0.9001]

12. a. $\mu_{\hat{p}} = 0.34$, $\sigma_{\hat{p}} = 0.063875$ **b.** 0.9793 [Tech: 0.9791] **c.** 0.8264 [Tech: 0.8262]

13. a. 0.2843 [Tech: 0.2842] **b.** Yes, P[More than 0.25] = 0.0228 [Tech: 0.0230].

14. a. 0.8461 [Tech: 0.8471] **b.** 0.7025 [Tech: 0.7031] **c.** 0.7157 [Tech: 0.7163]

15. The following boxplot shows that the data come from an approximately normal population.

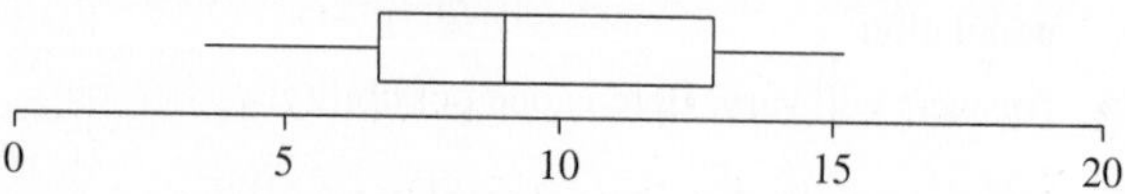

CHAPTER 7 Review Exercises

1. a. 0.6368 **b.** 0.9406 **c.** 0.3564

2. −1.23

3. 0.0122

4. a. 0.5621 [Tech: 0.5598] **b.** 13.280 [Tech: 13.286] **c.** 0.0009

5. a. 0.0793 [Tech: 0.0786] **b.** 0.9977 **c.** 0.7609 [Tech: 0.7600]

6. a. 0.0013 **b.** Yes

7. Yes, P(Greater than $2150) = 0.0037 [Tech: 0.0036].

8. a. 85 pounds **b.** 0.9429 [Tech: 0.9431]

9. a. 0.0985 [Tech: 0.0984] **b.** Yes, P[More than 70%] = 0.0019 [Tech: 0.0018].

10. a. 0.1112 [Tech: 0.1115] **b.** 0.6879 [Tech: 0.6870] **c.** 0.2009 [Tech: 0.2015]

11. a. 0.0119 [Tech: 0.0118] **b.** Yes **c.** No, P[Less than half] = 0.2843 [Tech: 0.2856].

12. 0.9495

13. 0.8554 [Tech: 0.8547]

14. The following boxplot shows that the data do not come from an approximately normal population.

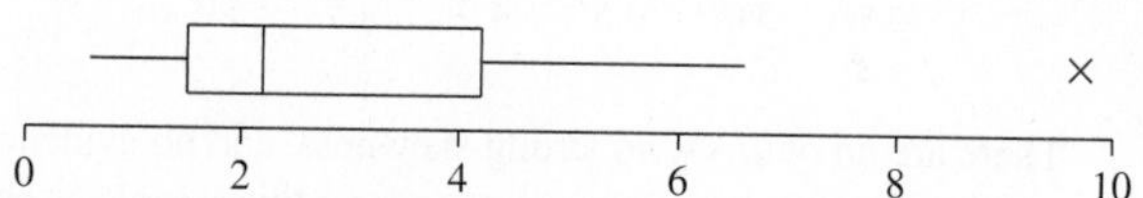

15. The following boxplot shows that the data come from an approximately normal population.

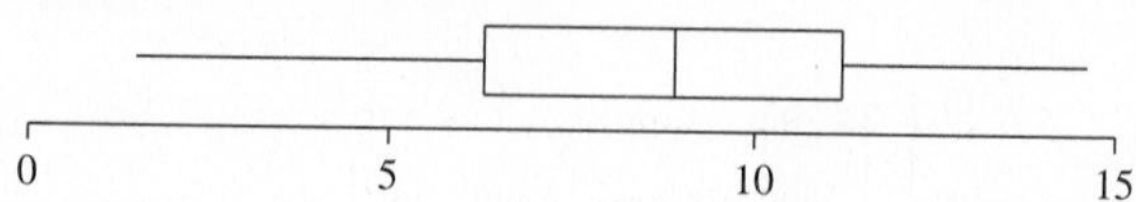

CHAPTER 7 Case Study

1. $\bar{x} = 99.6$, $s = 2.8363$

2. 0.0004

3. Yes

4. If the cans in each sample are arranged in increasing order of strength, each can in the second sample is stronger than the corresponding can in the first sample.

5. $\bar{x} = 101.9$, $s = 6.7404$

6. 0.0384 [Tech: 0.0387]

7. No

8. The following boxplot shows that the data do not come from an approximately normal population. For this reason, the method is not appropriate for the second shipment.

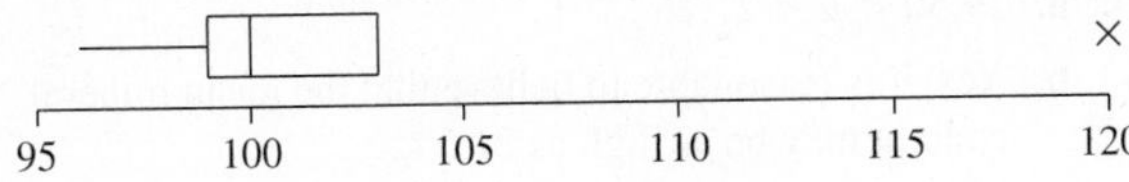

CHAPTER 8

Section 8.1

Exercises 1–10 are the Check Your Understanding exercises for this section. Answers to these exercises are on page 385.

11. point

12. critical value

13. margin of error

14. normal

15. True

16. False

17. No. The sample is skewed, and the sample size is less than 30.

18. Yes. The sample size is greater than 30.

19. Yes. The sample size is greater than 30.

20. No. The sample is only of seniors and does not represent all students.

21. 1.96

22. 1.44

23. 2.05

24. 2.97

25. 98%

26. 99%

27. 99.5%

28. 70%

29. 1.344

30. 6.049 [Tech: 6.048]

31. 4.192

32. 5.924

33. Yes. The sample is approximately normal.

34. Yes. The sample is approximately normal.

35. a. $6.37 < \mu < 9.47$

b. No, because the sample size is less than 30.

36. a. $39.16 < \mu < 41.66$

b. Yes, because the sample size is greater than 30.

37. a. $428 < \mu < 488$

b. Larger, because the sample size is smaller.

c. Smaller, because the confidence level is lower.

d. No, because the values in the confidence interval are all less than 500.

38. a. $1.25 < \mu < 1.39$

b. Larger, because the sample size is smaller.

c. Larger, because the confidence level is higher.

d. Yes, because all the values in the confidence interval are greater than 1.25.

39. a. $25.0 < \mu < 26.0$

b. No, because the sample consisted entirely of boys.

c. Yes, because all the values in the confidence interval are less than 28.

40. a. $0.92 < \mu < 0.96$

b. No, because the sample did not contain anyone over the age of 85.

c. No, because all the values in the confidence interval are greater than 0.90.

41. a. The sample size is small ($n \leq 30$).
b. Yes c. $202.6 < \mu < 303.1$

42. a. The sample size is small ($n \leq 30$).
b. No c. Not appropriate

43. a. The sample size is small ($n \leq 30$).
b. No c. Not appropriate

44. a. The sample size is small ($n \leq 30$).
b. Yes c. $480.93 < \mu < 575.22$

45. The students in the class are not a simple random sample of the students in the college.

46. There is a large nonresponse bias. The alumni who choose to respond are not a simple random sample of alumni.

47. a. True; this is the appropriate interpretation of a confidence interval.

b. False; the confidence interval is for the population mean, not the sample mean.

c. False; the confidence level is not the probability that the interval contains the true value.

d. False; the confidence level is about the population mean, not the proportion of the population contained in the interval.

48. a. False; the confidence interval is for the population mean, not the sample mean.

b. True; this is the appropriate interpretation of a confidence interval.

c. False; the confidence level is not the probability that the interval contains the true value.

d. False; the confidence level is about the population mean, not the proportion of the population contained in the interval.

49. a. 95%, 56.019, 60.881

b. Yes, because the sample size is large ($n > 30$).

50. a. 99%, 17.012, 20.048

b. No, because the sample size is small ($n \leq 30$).

51. a. 95%, 9.6956, 15.0084

b. $8.861 < \mu < 15.843$

52. a. 98%, 0.1233, 5.1007

b. $0.598 < \mu < 4.716$

Section 8.2

Exercises 1–12 are the Check Your Understanding exercises for this section. Answers to these exercises are on page 404.

13. 11

14. normal

15. decrease

16. increase

17. False

18. True

19. True

20. False

21. No. The sample size is less than 30, and there is an outlier.

22. Yes. The sample had no outliers or strong skewness.

23. No. The sample is small, and it contains outliers.

24. Yes. There are no outliers or strong skewness.

25. a. 2.074 **b.** 2.920 **c.** 2.567 **d.** 2.763

26. a. 2.015 **b.** 2.718 **c.** 2.040 **d.** 3.250

27. Smaller

28. Larger

29. a. $0.96 < \mu < 7.66$ **b.** No. The sample size is less than 30.

30. a. $11.2 < \mu < 14.2$ **b.** Yes. The sample size is greater than 30.

31. a. $1.2 < \mu < 3.0$
b. Narrower

32. a. $53.6 < \mu < 60.2$
b. Wider

33. a. $86.1 < \mu < 88.3$ **b.** Wider

34. a. $27.46 < \mu < 42.24$
b. Narrower

35. a. $4.7 < \mu < 16.3$ [Tech: $4.8 < \mu < 16.2$] **b.** 218
c. Smaller

36. a. $20.6 < \mu < 29.8$ **b.** 87 **c.** Larger

37. a. $17.7 < \mu < 18.9$ **b.** 79 **c.** Larger

38. a. $43.7 < \mu < 48.1$ **b.** 269 **c.** Larger

39. a. $5.42 < \mu < 5.64$
b. No, because 5.55 is contained in the confidence interval.

40. a. $13.21 < \mu < 13.53$
b. No, because 13.26 is contained in the confidence interval.

41. a. $11.7 < \mu < 16.5$
b. No, because all of the values in the confidence interval are greater than 10.

42. a. $118.19 < \mu < 132.21$
b. No, because 119.8 is contained in the confidence interval.

43. a. $132.9 < \mu < 140.9$
b. Wider, because the confidence level is higher.

44. a. $29.19 < \mu < 38.93$ [Tech: $29.21 < \mu < 38.91$]
b. Narrower, because the confidence level is lower.

45. a. $65.1 < \mu < 95.9$
b. Narrower, because the sample size is greater.

46. a. $14.32 < \mu < 18.70$ [Tech: $14.33 < \mu < 18.69$]
b. Wider, because the sample size is less.

47. a. $9.986 < \mu < 12.808$
b. Yes, because 11.5 is contained in the confidence interval.

48. a. $14.017 < \mu < 14.113$
b. No, because all the values in the confidence interval are greater than 14.

49. a. $19.50 < \mu < 21.26$
b. Yes, it is reasonable to believe that the mean mineral content may be as high as 21.3%.

50. a. $4.1190 < \mu < 4.5248$
b. Yes, because all the values in the confidence interval are greater than 4.050.

51. a. Yes; there are no outliers and no strong skewness.
b. $856.9 < \mu < 1168.9$

52. a. No; there are two outliers.
b. It is not appropriate to construct a confidence interval.

53. a. Yes; there are no outliers, and no strong skewness.
b. $2.91 < \mu < 4.26$
c. It does not contradict the claim, because the value 3.51 is contained in the confidence interval.

54. a. Yes; there are no outliers and no strong skewness.
b. $17.11 < \mu < 20.37$
c. It may be equal, because the value 19 is contained in the confidence interval.

55. a. Arya's confidence interval will have the larger margin of error because the confidence level is higher.
b. Arya's confidence interval is more likely to cover the population mean because the confidence level is higher.

56. a. Bertha's confidence interval will have the larger margin of error because the confidence level is higher.
b. Bertha's confidence interval is more likely to cover the population mean because the confidence level is higher.

57. a. Javier's **b.** Both are equally likely.

58. a. Emily's **b.** Both are equally likely.

59. a. Bob's confidence interval will have the larger margin of error because the standard error will be larger.
b. Both are equally likely to cover the population mean because they both have the same confidence level.

60. a. Martin's confidence interval will have the larger margin of error because the standard error will be larger.
b. Both are equally likely to cover the population mean because they both have the same confidence level.

61. The 90% confidence interval is $7.2 < \mu < 12.8$, the 95% confidence interval is $6.6 < \mu < 13.4$, and the 99% confidence interval is $5.6 < \mu < 14.4$.

62. The 95% confidence interval is $6.8 < \mu < 11.9$, the 98% confidence interval is $6.4 < \mu < 12.3$, and the 99.9% confidence interval is $5.1 < \mu < 13.6$.

63. a. $13.27 < \mu < 17.61$
b. $12.36 < \mu < 22.07$. The results are noticeably different. It is important to check for outliers in order to avoid misleading results.

64. a. $6.67 < \mu < 8.05$
b. $4.48 < \mu < 14.39$. The results are noticeably different. It is important to check for outliers in order to avoid misleading results.

65. 85.9%

66. 95.4%

67. We have data on the whole population of presidents, not just a sample. We know that the population mean is 70.8, so we don't need to construct a confidence interval.

68. We have data on the whole population of states, not just a sample. We know that the population mean is 6.3 million, so we don't need to construct a confidence interval.

69. **a.** 98%, 178.08, 181.58
b. Yes, because the sample is large ($n > 30$).

70. **a.** 95%, 0.85638, 2.2836
b. No, because the sample is small ($n \leq 30$).

71. **a.** 14
b. No, because the sample is small ($n \leq 30$).
c. 2.624
d. $4.5561 < \mu < 7.3185$ [Tech: $4.5558 < \mu < 7.3188$]

72. **a.** 70 **b.** Yes, because the sample is large ($n > 30$).
c. 2.000 [Tech: 1.994]
d. $22.9412 < \mu < 24.8108$ [Tech: $22.9438 < \mu < 24.8082$]

73. **a.** $69.3 < \mu < 100.7$
b. No, because we can't compute a sample standard deviation from a sample of size 1.

74. 75.3

75. $61.01

Section 8.3

Exercises 1–6 are the Check Your Understanding exercises for this section. Answers to these exercises are on page 417.

7. standard error

8. 0.5

9. True

10. False

11. No. The sample contains only six items in the damaged category.

12. No. This is not a simple random sample of days.

13. Yes. The assumptions for constructing a confidence interval are met.

14. Yes. The assumptions for constructing a confidence interval are met.

15. Point estimate: 0.1916; standard error: 0.01426; margin of error = 0.02794

16. Point estimate: 0.4742; standard error: 0.0507; margin of error = 0.1306

17. Point estimate: 0.4979; standard error: 0.02297; margin of error = 0.0378

18. Point estimate: 0.3625; standard error: 0.05375; margin of error = 0.0941

19. $0.325 < p < 0.550$

20. $0.618 < p < 0.846$

21. $0.341 < p < 0.448$

22. $0.302 < p < 0.369$

23. **a.** 0.429
b. $0.366 < p < 0.491$
c. No, because 0.45 is contained in the confidence interval.

24. **a.** 0.324
b. $0.275 < p < 0.373$
c. No, because 0.43 is not contained in the confidence interval.

25. **a.** 0.244
b. $0.190 < p < 0.297$
c. Yes, because all the values in the confidence interval are greater than 0.09.

26. **a.** 0.139
b. $0.114 < p < 0.163$
c. Yes, because all the values in the confidence interval are greater than 0.10.

27. **a.** 0.400 **b.** $0.373 < p < 0.426$
c. Yes, because all the values in the confidence interval are less than 0.50.

28. **a.** 0.350
b. $0.249 < p < 0.451$
c. No, because the value 0.39 is contained in the confidence interval.

29. **a.** $0.367 < p < 0.446$
b. $0.360 < p < 0.454$
c. Increasing the confidence level makes the interval wider.

30. **a.** $0.496 < p < 0.742$
b. $0.472 < p < 0.766$
c. Increasing the confidence level makes the interval wider.

31. **a.** $0.036 < p < 0.063$ **b.** 1825
c. Yes, because 0.10 is not contained in the confidence interval.

32. **a.** $0.470 < p < 0.505$ **b.** 7396
c. No, because 0.50 is contained in the confidence interval.

33. **a.** 2367 **b.** 2401
c. About the same. The necessary sample size does not depend on the population.

34. **a.** 455 **b.** 542
c. About the same. The necessary sample size does not depend on the size of the population.

35. **a.** 306 **b.** 1037

36. **a.** 958 **b.** 1068

37. **a.** $0.154 < p < 0.266$
b. Larger, because the sample size is smaller.

38. **a.** $0.337 < p < 0.483$
b. Larger, because the sample size is smaller.

39. $0.657 < p < 0.979$

40. $0.015 < p < 0.401$

41. **a.** 99%, 0.41911, 0.73714 **b.** $0.457 < p < 0.699$

42. **a.** 95%, 0.19525, 0.38253 **b.** $0.178 < p < 0.400$

43. **a.** 98%, 0.732082, 0.870128 **b.** $0.752 < p < 0.850$

44. **a.** 95%, 0.406111, 0.662854 **b.** $0.366 < p < 0.703$

45. This is not a sample; it is the whole population of senators.

46. This is a voluntary response sample, which is unreliable.

47. a. $0.357 < p < 0.802$ **b.** $0.357 < p < 0.801$

c. $0.352 < p < 0.848$

48. a. Small-sample method **b.** Yes

c. We divide by 15 rather than 19 when computing the standard error.

49. a. 90%: $0.393 < p < 0.777$; 95%: $0.357 < p < 0.802$; 99%: $0.296 < p < 0.842$

b. 90%: $0.393 < p < 0.765$; 95%: $0.357 < p < 0.801$; 99%: $0.287 < p < 0.871$

c. The 95% confidence interval, because $z_{\alpha/2} = 1.96$ for the 95% confidence interval, which is very close to 2.

Section 8.4

Exercises 1–4 are the Check Your Understanding exercises for this section. Answers to these exercises are on page 422.

5. 14

6. normal

7. True

8. False

9. 6.262, 27.488

10. 0.412, 16.750

11. $11.71 < \sigma < 20.87$

12. $4.41 < \sigma < 19.95$

13. a. $63.04 < \sigma < 137.26$

b. No, because the value 100 is contained in the confidence interval.

14. a. $6.25 < \sigma < 11.13$

b. Yes, because the value 15 is not contained in the confidence interval.

15. a. 1.798 **b.** $1.27 < \sigma < 3.05$

c. Yes, because the value 2.7 is contained in the confidence interval.

16. a. 0.0614 **b.** $0.04 < \sigma < 0.12$

c. No, because none of the values in the confidence interval are less than 0.03.

17. a. 0.838 **b.** $0.46 < \sigma < 2.92$

c. Yes, because some of the values in the confidence interval are less than 1.5.

18. a. 0.2263 **b.** $0.16 < \sigma < 0.38$

c. No, because none of the values in the confidence interval are greater than 0.5.

19. $\chi^2_{0.975} = 74.22$, $\chi^2_{0.025} = 129.56$; 95% confidence interval is $35.14 < \sigma < 46.43$.

20. $\chi^2_{0.975} = 72.28$, $\chi^2_{0.025} = 127.72$; 95% confidence interval is $35.39 < \sigma < 47.05$.

21. Lower bound differs by 0.25; upper bound differs by 0.62.

22. $\chi^2_{0.975} = 73.77$, $\chi^2_{0.025} = 129.07$; 95% confidence interval is $35.21 < \sigma < 46.57$.

23. Lower bound differs by 0.07; upper bound differs by 0.14.

Section 8.5

Exercises 1–4 are the Check Your Understanding exercises for this section. Answers to these exercises are on page 425.

5. Population mean, $63.47 < \mu < 79.17$

6. Population proportion, $0.486 < p < 0.614$

7. Population standard deviation, $0.99 < \sigma < 4.50$

8. Population mean, $2.45 < \mu < 4.29$

9. Population proportion, $0.686 < p < 0.794$

10. Population standard deviation, $4.01 < \sigma < 8.19$

11. Population mean, $7.32 < \mu < 9.58$

12. Population mean, $0.70 < \mu < 2.26$

13. $235.38 < \mu < 261.38$

14. $21.46 < \mu < 23.90$

15. $6.26 < \sigma < 16.62$

16. $0.531 < p < 0.589$

17. $37.91 < \mu < 52.11$

18. $0.66 < \sigma < 2.58$

19. $0.077 < p < 0.116$

20. $15{,}043 < \mu < 22{,}361$

CHAPTER 8 Quiz

1. a. A single number that is used to estimate the value of a parameter

b. An interval that is used to estimate the value of a parameter

c. The percentage of confidence intervals that will cover the true value of the parameter in the long run

2. 1.706

3. $21.7 < \mu < 24.7$

4. $7.60 < \sigma < 14.61$

5. $46.1 < \mu < 63.5$

6. $11.70 < \mu < 13.14$

7. $14.79 < \sigma < 45.97$ [Tech: $14.79 < \sigma < 45.96$]

8. 1.75 [Tech: 1.751]

9. 7.015, 34.805

10. a. 141 **b.** 2.53

11. a. 1.96 **b.** 4.958

c. $136 < \mu < 146$

12. 984

13. $0.288 < p < 0.373$

14. 230

15. 260

CHAPTER 8 Review Exercises

1. 271

2. a. $10.79 < \mu < 14.57$

b. No, because the value 13 is contained in the confidence interval.

c. 180

3. a. $0.235 < p < 0.410$ **b.** 1612 **c.** 1844

4. a. $56.3 < \mu < 98.4$

b. Yes, the value 27 is an outlier.

5. No, because the data do not come from a normal population.

6. a. $1.85 < \mu < 2.07$

b. No, because all the values in the confidence interval are greater than 0.5.

c. 737

7. a. $0.034 < p < 0.136$ **b.** 323 **c.** 1037

8. a.

15 20 25 30 35

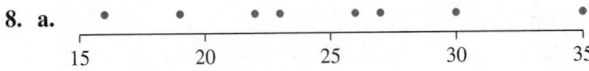

Yes. There are no outliers, no strong skewness, and no evidence of multiple modes.

b. $18.3 < \mu < 31.2$

9. $4.03 < \sigma < 12.39$

10. a. $0.453 < p < 0.618$

b. Yes, because all the values in the confidence interval are greater than 0.41.

c. 1529

11. a. $10.3 < \mu < 14.1$

b. Yes, because all the values in the confidence interval are less than 15.

12. a. $6.75 < \mu < 7.37$

b. No, because all the values in the confidence interval are less than 8.

c. 818

13. a. 0.168 **b.** $0.097 < p < 0.239$ **c.** 598

14. a. $36.3 < \mu < 37.5$

b. No, because the value 38.7 is not contained in the confidence interval.

15. The days are not independent trials. If it rains on one day, it is more likely to rain the next day.

CHAPTER 8 Case Study

1.

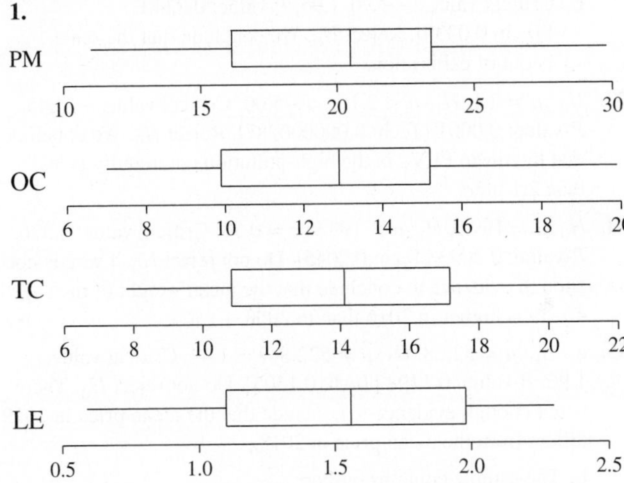

2. PM: $17.94 < \mu < 22.48$; OC: $11.08 < \mu < 14.19$;
TC: $12.428 < \mu < 15.847$; LE: $1.291 < \mu < 1.784$

3. It is reasonable to conclude that the mean levels were lower than the baseline for all the pollutants, because all the values in the confidence intervals are less than the corresponding baselines.

4.

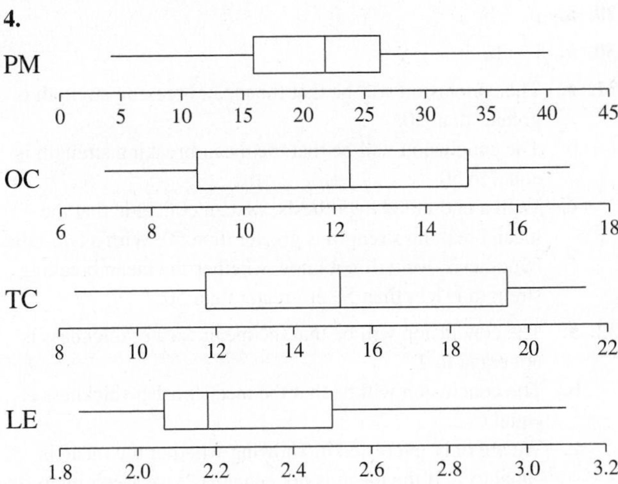

95% Confidence intervals:
PM: $17.33 < \mu < 25.00$; OC: $10.493 < \mu < 13.615$;
TC: $13.149 < \mu < 17.169$; LE: $2.144 < \mu < 2.496$

5. It is reasonable to conclude that the mean levels were lower than the baseline for all the pollutants, because all the values in the confidence intervals are less than the corresponding baselines.

CHAPTER 9

Section 9.1

Exercises 1–6 are the Check Your Understanding exercises for this section. Answers to these exercises are on page 438.

7. null, alternate

8. Type I, Type II

9. False

10. True

11. False

12. False

13. Left-tailed

14. Right-tailed

15. Two-tailed

16. Two-tailed

17. Correct decision

18. Type I error

19. Type II error

20. Correct decision

21. 0.9

22. 14

23. H_0: $\mu = 400$, H_1: $\mu > 400$

24. H_0: $\mu = 20$, H_1: $\mu \neq 20$

25. The mean amount spent by diners is greater than $30.

26. The mean caffeine content is not equal to 100 milligrams.

27. There is not enough evidence to conclude that the mean weight of adult German shepherd dogs differs from 75 pounds.

28. There is not enough evidence to conclude that the mean number days per year the sales representatives spend traveling is less than 50.

29. a. ii **b.** i

30. a. i **b.** ii

31. a. The conclusion will be that the mean breaking strength is greater than 50.
b. The conclusion will be that the mean breaking strength is equal to 50.
c. With a one-tailed hypothesis, we can conclude that the mean breaking strength is greater than 50. With a two-tailed hypothesis, we will not know whether the mean breaking strength is less than 50 or greater than 50.

32. a. The conclusion will be that the mean washer thickness is not equal to 2.
b. The conclusion will be that the mean washer thickness is equal to 2.
c. We are only interested in knowing whether the mean is equal to 2. If the mean is not equal to 2, we aren't interested in whether the mean is greater than or less than 2.

33. a. (ii) **b.** (iii) **c.** Type I error
d. No, a Type I error occurs if H_0 is rejected when it is true. Therefore, a Type I error cannot occur when H_0 is false.
e. Yes, if H_0 is not rejected, a Type II error will occur.

34. a. (i) **b.** (iii) **c.** Type II error
d. Yes, if H_0 is rejected, a Type I error will occur.
e. No, a Type II error occurs if H_0 is not rejected when it is false. Therefore, a Type I error cannot occur when H_0 is true.

35. a. 0.1 **b.** 0.6

Section 9.2

Exercises 1–18 are the Check Your Understanding exercises for this section. Answers to these exercises are on pages 457–458.

19. P-value

20. null

21. critical

22. significance

23. True

24. True

25. False

26. True

27. Yes. The sample size is greater than 30.

28. No. This is not a simple random sample.

29. Yes. The population is approximately normal.

30. No. The sample size is less than 30, and the sample is skewed.

31. No. This is not a simple random sample.

32. Yes. The sample size is greater than 30.

33. a. $z = 2.60$
b. Critical value: 1.645, P-value: 0.0047; H_0 is rejected.
c. Critical value: 2.326, P-value: 0.0047; H_0 is rejected.

34. a. $z = -2.31$
b. Critical values: −1.96, 1.96, P-value: 0.0208 [Tech: 0.0209]; H_0 is rejected.
c. Critical values: −2.576, 2.576; P-value: 0.0208 [Tech: 0.0209]; H_0 is not rejected.

35. a. $z = 0.99$
b. Critical values: −1.96, 1.96, P-value: 0.3222 [Tech: 0.3211]; H_0 is not rejected.
c. Critical values: −2.576, 2.576, P-value: 0.3222 [Tech: 0.3211]; H_0 is not rejected.

36. a. $z = -0.19$
b. Critical value: −1.645, P-value: 0.4247 [Tech: 0.4260]; H_0 is not rejected.
c. Critical value: −2.326, P-value: 0.4247 [Tech: 0.4260]; H_0 is not rejected.

37. 0.035

38. 0.28

39. a. True **b.** False **c.** True **d.** False

40. a. False **b.** True **c.** False **d.** True

41. a. Yes **b.** Type I error **c.** Correct decision

42. a. No **b.** Correct decision **c.** Type II error

43. a. No **b.** Type II error **c.** Correct decision

44. a. Yes **b.** Correct decision **c.** Type I error

45. iii

46. i

47. i

48. iii

49. a. H_0: $\mu = 19.3$, H_1: $\mu > 19.3$
b. $z = 2.75$ **c.** Critical value: 1.645, P-value: 0.0030. Reject H_0. We conclude that the mean time browsing Netflix has increased.

50. a. H_0: $\mu = 50$, H_1: $\mu > 50$ **b.** $z = 1.03$
c. Critical value: 2.326, P-value: 0.1515 [Tech: 0.1508]. Do not reject H_0. There is not enough evidence to conclude that the second graders have greater math skills than the nationwide average.

51. a. H_0: $\mu = 69.4$, H_1: $\mu < 69.4$ **b.** $z = -2.44$
c. Critical value: −2.326, P-value: 0.0073 [Tech: 0.0074]. Reject H_0. We conclude that the mean height of men aged 60–69 is less than the mean height of all U.S. men.

52. a. H_0: $\mu = 1000$, H_1: $\mu \neq 1000$ **b.** $z = 2.12$
c. Critical values: −1.96, 1.96, P-value: 0.0340 [Tech: 0.0339]. Reject H_0. We conclude that the scale is out of calibration.

53. H_0: $\mu = 2.1$, H_1: $\mu < 2.1$; $z = -5.00$; Critical value: −1.645, P-value: 0.0001 [Tech: 0.000000287]. Reject H_0. We conclude that the mean FEV_1 in the high-pollution community is less than 2.1 liters.

54. H_0: $\mu = 199.9$, H_1: $\mu > 199.9$; $z = 0.83$; Critical value: 2.326, P-value: 0.2033 [Tech: 0.2045]. Do not reject H_0. There is not enough evidence to conclude that the mean weight of men aged 40–59 is higher in 2016 than in 2008.

55. a. H_0: $\mu = 522.8$, H_1: $\mu \neq 522.8$; $z = 1.44$; Critical value: 1.96, P-value: 0.1498 [Tech: 0.1505]. Do not reject H_0. There is not enough evidence to conclude that the mean price in 2019 differs from the mean price in 2018.
b. The sample contains outliers.
c. The sample size is greater than 30.

56. a. Yes.
b. H_0: $\mu = 527$, H_1: $\mu \neq 527$; $z = 0.21$; Critical value: 1.96, P-value: 0.8336 [Tech: 0.8345]. Do not reject H_0. There is not enough evidence to conclude that the mean score in 2019 differs from the mean score in 2018.

57. a. The sample is small ($n \leq 30$).
b. Yes, there are no outliers and no evidence of strong skewness.
c. H_0: $\mu = 15$, H_1: $\mu < 15$; Test statistic: $z = -3.02$; Critical value: -2.326, P-value: 0.0013. Reject H_0. We conclude that the mean concentration meets the EPA standard.

58. a. Yes, there are no outliers and no evidence of strong skewness.
b. H_0: $\mu = 2.80$, H_1: $\mu \neq 2.80$; Test statistic: $z = 1.44$; Critical values: -1.96, 1.96, P-value: 0.1498 [Tech: 0.1499]. Do not reject H_0. There is not enough evidence to conclude that the mean GPA for business majors differs from the mean GPA at the whole university.

59. a. H_0: $\mu = 45$, H_1: $\mu \neq 45$
b. $z = 3.094063348$
c. 0.0019744896
d. Yes. We can conclude at the $\alpha = 0.05$ level that the mean age of books differs from 45 years.
e. Yes. We can conclude at the $\alpha = 0.01$ level that the mean age of books differs from 45 years.

60. a. H_0: $\mu = 125$, H_1: $\mu \neq 125$ b. $z = 1.73116377$
c. 0.0834224813
d. No. We cannot conclude at the $\alpha = 0.05$ level that the mean number of characters in a text message differs from 125.
e. No. We cannot conclude at the $\alpha = 0.01$ level that the mean number of characters in a text message differs from 125.

61. a. H_0: $\mu = 225$, H_1: $\mu \neq 225$
b. $z = 2.085$
c. 0.037
d. Yes. We can conclude at the $\alpha = 0.05$ level that the mean weight of white-fronted Amazon parrots differs from 225 grams.
e. No. We cannot conclude at the $\alpha = 0.01$ level that the mean weight of white-fronted Amazon parrots differs from 225 grams.
f. $z = 1.07$ g. 0.1423 [Tech: 0.1412]
h. No. We cannot conclude at the $\alpha = 0.05$ level that the mean weight of white-fronted Amazon parrots differs from 230 grams.

62. a. H_0: $\mu = 20$, H_1: $\mu > 20$
b. $z = 1.366$
c. 0.086
d. No. We cannot conclude at the $\alpha = 0.05$ level that the mean age of the students is greater than 20 years.
e. No. We cannot conclude at the $\alpha = 0.01$ level that the mean age of the students is greater than 20 years.
f. $z = -2.76$
g. 0.0058 [Tech: 0.0057]
h. Yes. We can conclude at the $\alpha = 0.05$ level that the mean age of the students differs from 24 years.

Section 9.3

Exercises 1–12 are the Check Your Understanding exercises for this section. Answers to these exercises are on page 478.

13. skewness, outliers
14. sample size
15. 0.05
16. decrease
17. increase
18. practical
19. True
20. False
21. False
22. False
23. True
24. True
25. No. The sample contains an outlier.
26. Yes. The sample has no outliers or strong skewness.
27. Yes. The sample size is greater than 30.
28. No. This is not a simple random sample.
29. Yes. The sample has no outliers or strong skewness.
30. Yes. The sample is skewed.

31. a. P-value is between 0.025 and 0.05 [Tech: 0.0399]
b. P-value is between 0.20 and 0.50 [Tech: 0.2086]
c. P-value is between 0.0025 and 0.005 [Tech: 0.0030]
d. P-value is between 0.10 and 0.25 [Tech: 0.1793]

32. a. P-value is between 0.10 and 0.20 [Tech: 0.1306]
b. P-value is between 0.005 and 0.01 [Tech: 0.0089]
c. P-value is between 0.10 and 0.20 [Tech: 0.1468]
d. P-value is between 0.01 and 0.025 [Tech: 0.0179]

33. a. -2.056, 2.056 b. 2.390 c. -1.753, 1.753
d. -1.812

34. a. 1.686 b. -2.445 c. -2.015, 2.015
d. -2.064, 2.064

35. a. 3.0594 b. 25
c. Critical value: 1.708. P-value is between 0.0025 and 0.005 [Tech: 0.002616]. Reject H_0.
d. Critical value: 2.485. P-value is between 0.0025 and 0.005 [Tech: 0.002616]. Reject H_0.

36. a. -2.0404 b. 50
c. Critical values: -2.009 and 2.009. P-value is between 0.02 and 0.05 [Tech: 0.0466]. Reject H_0.
d. Critical values: -2.678 and 2.678. P-value is between 0.02 and 0.05 [Tech: 0.00466]. Do not reject H_0.

37. a. 2.1651 b. 74
c. Critical values: -2.000 and 2.000. P-value is between 0.02 and 0.05 [Tech: 0.0336]. Reject H_0.
d. Critical values: -2.660 and 2.660. P-value is between 0.02 and 0.05 [Tech: 0.0336]. Do not reject H_0.

38. a. 0.4546 b. 161
c. Critical values: -1.984 and 1.984. P-value is between 0.50 and 0.80 [Tech: 0.6500]. Do not reject H_0.
d. Critical values: -2.626 and 2.626. P-value is between 0.02 and 0.05 [Tech: 0.6500]. Do not reject H_0.

39. 0.10
40. 0.02
41. a. True b. False c. True d. False
42. a. False b. True c. False d. True
43. a. No b. Correct decision c. Type II error
44. a. Yes b. Type I error c. Correct decision

45. **a.** No **b.** Type II error **c.** Correct decision

46. **a.** Yes **b.** Correct decision **c.** Type I error

47. ii

48. iii

49. **a.** H_0: $\mu = 178{,}258$, H_1: $\mu > 178{,}258$
b. $t = 2.464$; 54 degrees of freedom
c. Critical value: 1.676; P-value is between 0.005 and 0.01 [Tech: 0.00848]. Reject H_0. We conclude that the mean salary for family practitioners in Los Angeles is greater than the national average.

50. **a.** H_0: $\mu = 35{,}000$, H_1: $\mu \neq 35{,}000$
b. $t = 1.507$; 13 degrees of freedom
c. Critical values: -3.012, 3.012; P-value is between 0.10 and 0.20 [Tech: 0.1557]. Do not reject H_0. There is not enough evidence to conclude that the mean tuition and fees for private institutions in California differs from $35,000.

51. H_0: $\mu = 25$, H_1: $\mu > 25$, Test statistic: $t = 1.790$, Critical value: 2.326, P-value is between 0.025 and 0.05 [Tech: 0.0372]. Do not reject H_0. There is not enough evidence to conclude that the mean weight of one-year-old baby boys is greater than 25 pounds.

52. H_0: $\mu = 720$, H_1: $\mu < 720$, Test statistic: $t = -1.848$, Critical value: -1.664, P-value is between 0.025 and 0.05 [Tech: 0.0338]. Reject H_0. We conclude that the mean FICO score is less than 720.

53. H_0: $\mu = 25$, H_1: $\mu < 25$. Test statistic: $t = -1.687$, Critical value: -1.645; P-value is between 0.025 and 0.05 [Tech: 0.0459]. Reject H_0. We conclude that the mean commute time is less than 25 minutes.

54. H_0: $\mu = 3$, H_1: $\mu \neq 3$. Test statistic: $t = 1.1298$, Critical values: -2.576, 2.576; P-value is between 0.20 and 0.50 [Tech: 0.2588]. Do not reject H_0. There is not enough evidence to conclude that the mean amount of television watched differs from 3 hours per day.

55. **a.** Yes, there are no outliers and no evidence of strong skewness.
b. H_0: $\mu = 10$, H_1: $\mu > 10$, Test statistic: $t = 3.669$, Critical value: 1.771, P-value is between 0.001 and 0.0025 [Tech: 0.0014]. Reject H_0. We conclude that the mean weight loss is greater than 10 pounds.

56. **a.** Yes, there are no outliers and no evidence of strong skewness.
b. H_0: $\mu = 12$, H_1: $\mu \neq 12$, Test statistic: $t = 0.973$, Critical values: -2.365, 2.365, P-value is between 0.20 and 0.50 [Tech: 0.3632]. Do not reject H_0. There is not enough evidence to conclude that the mean volume differs from 12 ounces.

57. **a.** No, there is an outlier.
b. Not appropriate

58. **a.** No, the population is strongly skewed.
b. Not appropriate

59. **a.** Yes, there are no outliers and no evidence of strong skewness.
b. H_0: $\mu = 300$, H_1: $\mu > 300$, Test statistic: $t = 1.927$, Critical value: 2.821, P-value is between 0.025 and 0.05 [Tech: 0.0431]. Do not reject H_0. There is not enough evidence to conclude that the mean price is greater than $300.

60. **a.** Yes, there are no outliers and no evidence of strong skewness.
b. H_0: $\mu = 150$, H_1: $\mu > 150$, Test statistic: $t = 3.053$, Critical value: 2.718, P-value is between 0.005 and 0.01 [Tech: 0.0055]. Reject H_0. We conclude that the mean number of diners increased while the free dessert offer was in effect.

61. **a.**

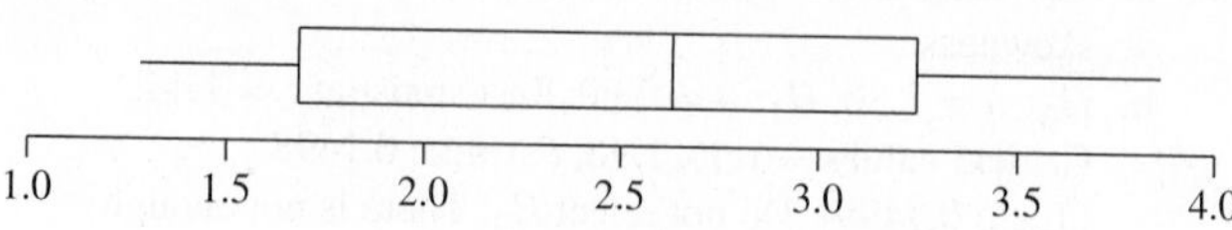

There are no outliers and no evidence of strong skewness. It is appropriate to perform a hypothesis test.
b. H_0: $\mu = 2$, H_1: $\mu > 2$, Test statistic: $t = 1.638$, Critical value: 1.895, P-value is between 0.05 and 0.10 [Tech: 0.0727]. Do not reject H_0. There is not enough evidence to conclude that the mean amount absorbed is greater than 2 micrograms.

62. **a.**

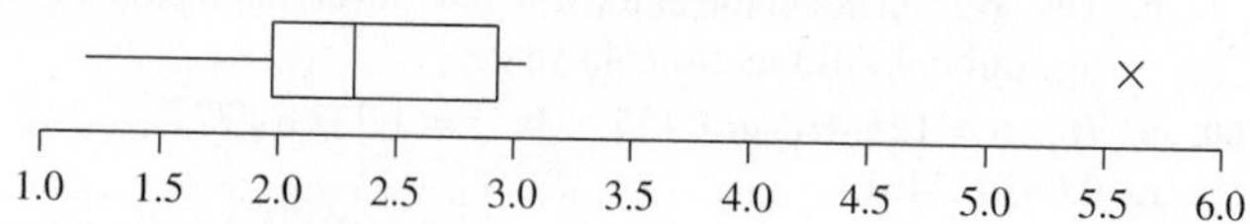

The data contain an outlier. It is not appropriate to perform a hypothesis test.
b. It is not appropriate to perform a hypothesis test.

63. **a.** H_0: $\mu = 2$, H_1: $\mu < 2$
b. 1.88
c. 0.21
d. 17
e. Yes. We can conclude at the $\alpha = 0.05$ level that the mean number of hours per day spent on social media is less than 2.
f. $t = 1.616$
g. The P-value is between 0.05 and 0.10 [Tech: 0.0622].
h. No. We cannot conclude at the $\alpha = 0.05$ level that the mean number of hours per day spent on social media is greater than 1.8.

64. **a.** H_0: $\mu = 20$, H_1: $\mu \neq 20$
b. 20.75
c. 3.93
d. 39
e. No. We cannot conclude at the $\alpha = 0.05$ level that the mean number of hours per day watching television differs from 20.
f. $t = -2.816$
g. The P-value is between 0.005 and 0.01 [Tech: 0.0076].
h. Yes. We can conclude at the $\alpha = 0.05$ level that the mean number of hours per day watching television differs from 22.5.

65. **a.** H_0: $\mu = 5.5$, H_1: $\mu > 5.5$
b. 5.92563
c. 0.15755
d. 4
e. Yes. We can conclude at the $\alpha = 0.05$ level that the mean weight of hamsters is greater than 5.5 grams.
f. $t = -8.152$
g. The P-value is between 0.0005 and 0.001 [Tech: 0.00062].

h. Yes. We can conclude at the $\alpha = 0.05$ level that the mean weight of hamsters is less than 6.5 grams.

66. a. H_0: $\mu = 16$, H_1: $\mu \neq 16$
b. 13.2874
c. 6.0989
d. 10
e. No. We cannot conclude at the $\alpha = 0.05$ level that the mean number of miles commuted to work differs from 16.
f. $t = 2.332$
g. The *P*-value is between 0.01 and 0.025 [Tech: 0.0210].
h. Yes. We can conclude at the $\alpha = 0.05$ level that the mean number of miles commuted to work is greater than 9.

67. a. $t = 1.25$
b. *P*-value is between 0.10 and 0.25 [Tech: 0.1071]. **c.** No

68. a. $t = 3.953$; larger
b. *P*-value is less than 0.0005 [Tech: 0.000041]. **c.** Yes
d. A weight loss of 0.5 pound is very small.

69. No, the people on a certain block may not be representative of the population of the city.

70. No, the people who have been treated in a certain hospital may not be representative of the population of all patients who have undergone the procedure.

71. The *P*-value will be less, because the sample mean is farther from the mean specified by the null hypothesis.

72. The *P*-value will be greater, because the sample mean is closer to the mean specified by the null hypothesis.

73. a. True **b.** False

74. a. False **b.** True

75. a. No, we can only conclude the mean is greater than 1500. The *P*-value does not tell us how much greater it is.
b. Yes, this is the correct interpretation of the *P*-value.

76. a. Yes, this is the correct interpretation of the *P*-value.
b. No, we can only conclude the mean is less than 135. The *P*-value does not tell us how much less it is.

77. a. 2.500
b. Critical value: 1.664. *P*-value is between 0.005 and 0.01. [Tech: 0.0070]. Reject H_0.
c. No. A difference of 2.5 points on the SAT is not of practical significance.

78. a. 2.2361
b. Critical value: 1.645. *P*-value is between 0.01 and 0.025. [Tech: 0.0129]. Reject H_0.
c. No. A weight loss of 0.5 pounds is not of practical significance.

79. Yes, the *P*-value should have been reported.

80. Yes, the value of the test statistic should have been reported.

81. If the value of 100 specified by the null hypothesis is not in the confidence interval, then the *P*-value must be less than 0.05.

82. a. Test statistic: $t = 1.000$, Critical value: 1.664, *P*-value is between 0.10 and 0.25 [Tech: 0.1599]. Do not reject H_0.
b. Test statistic: $t = 3.162$, Critical value: 1.645, *P*-value is between 0.0005 and 0.001 [Tech: 0.0008]. Do not reject H_0.
c. No.
d. The quantity $s/\sqrt{n}$ is smaller when the sample size is larger.

83. a. 1.645 **b.** 99 **c.** 0.0516

Section 9.4

Exercises 1–4 are the Check Your Understanding exercises for this section. Answers to these exercises are on page 488.

5. 10

6. 20

7. True

8. False

9. Yes. The assumptions for a hypothesis test for a proportion are satisfied.

10. Yes. The assumptions for a hypothesis test for a proportion are satisfied.

11. No. The sample is more than 5% of the population.

12. No. This is not a simple random sample.

13. a. 0.675
b. Yes, the number of individuals in each category is greater than 10.
c. $z = -2.80$
d. Yes. Critical value: −1.645, *P*-value: 0.0026.

14. a. 0.633
b. Yes, the number of individuals in each category is greater than 10.
c. $z = -1.13$
d. No. Critical values: −1.96, 1.96, *P*-value: 0.2584 [Tech: 0.2598].

15. a. 0.560
b. Yes, the number of individuals in each category is greater than 10.
c. $z = -0.71$
d. No. Critical values: −1.96, 1.96, *P*-value: 0.4778 [Tech: 0.4795].

16. a. 0.600
b. Yes, the number of individuals in each category is greater than 10.
c. $z = 2.45$
d. Yes. Critical value: 1.645, *P*-value: 0.0071 [Tech: 0.0072].

17. a. H_0: $p = 0.8$, H_1: $p \neq 0.8$
b. $z = -2.01$
c. Critical values: −1.96, 1.96, *P*-value: 0.0444 [Tech: 0.0442]. Reject H_0. We conclude that the percentage of emails that are spam differs from 80%.
d. Critical values: −2.576, 2.576, *P*-value: 0.0444 [Tech: 0.0442]. Do not reject H_0. There is not enough evidence to conclude that the percentage of emails that are spam differs from 80%.

18. a. H_0: $p = 0.15$, H_1: $p < 0.15$ **b.** $z = -3.83$
c. Critical value: −1.645, *P*-value: 0.0001 [Tech: 0.0000647]. Reject H_0. We conclude that less than 15% of U.S. adults have a great deal of confidence in banks.
d. Critical value: −2.326, *P*-value: 0.0001 [Tech: 0.0000647]. Reject H_0. We conclude that less than 15% of U.S. adults have a great deal of confidence in banks.

19. H_0: $p = 0.55$, H_1: $p > 0.55$, Test statistic: $z = 0.56$, Critical value: 2.326, *P*-value: 0.2877 [Tech: 0.2875]. Do not reject H_0. There is not enough evidence to conclude that more than 55% of children aged 8–12 have phones.

20. H_0: $p = 0.45$, H_1: $p < 0.45$, Test statistic: $z = -1.25$, Critical value: -1.96, P-value: 0.1056 [Tech: 0.1064]. Do not reject H_0. There is not enough evidence to conclude that less than 45% of U.S. adults favor an internet sales tax.

21. H_0: $p = 0.5$, H_1: $p < 0.5$, Test statistic: $z = -4.94$, Critical value: -1.645, P-value: 0.0001 [Tech: 0.00000040]. Reject H_0. We conclude that less than half of HIV-positive smokers have used a nicotine patch.

22. H_0: $p = 0.25$, H_1: $p > 0.25$, Test statistic: $z = 3.10$, Critical value: 2.326, P-value: 0.0010 [Tech: 0.000983]. Reject H_0. We conclude that more than 25% of video gamers prefer consoles.

23. H_0: $p = 0.25$, H_1: $p < 0.25$, Test statistic: $z = -4.17$, Critical value: -2.326, P-value: 0.0001 [Tech: 0.0000155]. Reject H_0. We conclude that less than 25% of adults have one or more tattoos.

24. H_0: $p = 0.6$, H_1: $p > 0.6$, Test statistic: $z = 2.21$, Critical value: 1.645, P-value: 0.0136. Reject H_0. We conclude that more than 60% of people undergoing the surgery recover from diabetes.

25. H_0: $p = 0.73$, H_1: $p > 0.73$, Test statistic: $z = 1.05$, Critical value: 1.96, P-value: 0.1469. Do not reject H_0. There is not enough evidence to conclude that more than 73% of technology companies have Twitter accounts.

26. H_0: $p = 0.5$, H_1: $p < 0.5$, Test statistic: $z = -3.39$, Critical value: -1.96, P-value: 0.0003. Reject H_0. We conclude that less than half of internet users have posted photos or videos online.

27. H_0: $p = 0.5$, H_1: $p < 0.5$, Test statistic: $z = -4.13$, Critical value: -2.326, P-value: 0.0001 [Tech: 0.000018]. Reject H_0. We conclude that less than half of patients prefer a doctor with high interpersonal skills to one with high technical skills.

28. H_0:$p = 0.5$, H_1:$p > 0.5$; $z = 1.50$; Critical value: 2.326, P-value: 0.0668. Do not reject H_0. There is not enough evidence to conclude that more than 50% of subscribers would continue subscribing.

29. H_0:$p = 0.35$, H_1:$p < 0.35$; $z = -1.22$; Critical value: -1.645, P-value: 0.1112 [Tech: 0.1111]. Do not reject H_0. There is not enough evidence to conclude that less than 35% of people with impaired fasting glucose have type 2 diabetes.

30. H_0:$p = 0.5$, H_1:$p > 0.5$; $z = 1.84$; Critical value: 1.645, P-value: 0.0329 [Tech: 0.0326]. Reject H_0. We can conclude that more than half of ACL injuries are to the left knee.

31. **a.** H_0: $p = 0.5$, H_1: $p < 0.5$
b. $\hat{p} = 0.3382352941$
c. Yes. We can conclude at the $\alpha = 0.05$ level that the proportion of individuals who support an increase in the sales tax is less than 0.5.
d. $z = 1.68$, P-value: 0.0930 [Tech: 0.0929]. Do not reject H_0. We cannot conclude at the $\alpha = 0.05$ level that the proportion of individuals who support an increase in the sales tax differs from 0.25.

32. **a.** H_0: $p = 0.4$, H_1: $p \neq 0.4$
b. $\hat{p} = 0.36$
c. No. We cannot conclude at the $\alpha = 0.05$ level that the proportion of mothers who had their first child before the age of 24 differs from 0.4.
d. $z = -2.42$, P-value: 0.0078 [Tech: 0.0077]. Reject H_0. We can conclude at the $\alpha = 0.05$ level that the proportion of mothers who had their first child before the age of 24 is less than 0.5.

33. **a.** H_0: $p = 0.6$, H_1: $p > 0.6$
b. $\hat{p} = 0.618829$
c. No. We cannot conclude at the $\alpha = 0.05$ level that the proportion of college students who have a job outside of school is greater than 0.6.
d. $z = -1.93$, P-value: 0.0268 [Tech: 0.0269]. Reject H_0. We can conclude at the $\alpha = 0.05$ level that the proportion of college students who have a job outside of school is less than 0.65.

34. **a.** H_0: $p = 0.7$, H_1: $p \neq 0.7$
b. $\hat{p} = 0.519231$
c. Yes. We can conclude at the $\alpha = 0.05$ level that the proportion of adults who are interested in economic news differs from 0.7.
d. $z = -1.19$, P-value: 0.2340 [Tech: 0.2345]. Do not reject H_0. We cannot conclude at the $\alpha = 0.05$ level that the proportion of adults who are interested in economic news differs from 0.6.

35. **a.** $z = 2.68$, $P = 0.004$ **b.** No **c.** Yes

36. **a.** $z = 2.48$, $P = 0.007$ **b.** Yes **c.** No

37. This is a voluntary response sample, not a simple random sample.

38. This is not a simple random sample of days.

39. **a.** $n = 10, X = 9$ **b.** $n = 10, p = 0.5$ **c.** $P = 0.0107$
d. Yes

Section 9.5

Exercises 1–4 are the Check Your Understanding exercises for this section. Answers to these exercises are on page 491.

5. 14

6. normal

7. False

8. True

9. Test statistic: $\chi^2 = 19.6$, Critical value: 18.307. Reject H_0.

10. Test statistic: $\chi^2 = 22.224$, Critical value: 13.565. Do not reject H_0.

11. Test statistic: $\chi^2 = 15.790$, Critical value: 10.196. Do not reject H_0.

12. Test statistic: $\chi^2 = 22.963$, Critical value: 21.026. Reject H_0.

13. Test statistic: $\chi^2 = 69.360$, Critical values: 12.401, 39.364. Reject H_0.

14. Test statistic: $\chi^2 = 15.750$, Critical values: 0.989, 20.278. Do not reject H_0.

15. H_0: $\sigma = 5$, H_1: $\sigma < 5$, Test statistic: $\chi^2 = 17.75$, Critical value: 13.848. Do not reject H_0. There is not enough evidence to conclude that the pediatrician's claim is true.

16. H_0: $\sigma = 2.87$, H_1: $\sigma < 2.87$, Test statistic: $\chi^2 = 18.939$, Critical value: 21.426. Reject H_0. We conclude that the standard deviation is less than 2.87.

17. H_0: $\sigma = 15$, H_1: $\sigma \neq 15$, Test statistic: $\chi^2 = 6.827$, Critical values: 12.401, 39.364. Reject H_0. We conclude that the standard deviation differs from 15.

18. H_0: $\sigma = 100$, H_1: $\sigma \neq 100$, Test statistic: $\chi^2 = 14.381$, Critical values: 8.907, 32.852. Do not reject H_0. There is not enough evidence to conclude that the standard deviation differs from 100.

19. H_0: $\sigma = 0.1$, H_1: $\sigma \neq 0.1$, Test statistic: $\chi^2 = 32.656$, Critical values: 2.700, 19.023. Reject H_0. We conclude that the standard deviation changed, and now differs from 0.1.

20. H_0: $\sigma = 1.43$, H_1: $\sigma > 1.43$, Test statistic: $\chi^2 = 26.407$, Critical value: 42.98. Do not reject H_0. There is not enough evidence to conclude that the standard deviation is greater than the value reported by the National Institutes of Health.

21. Test statistic: $\chi^2 = 177.778$, Critical values: 74.222, 129.561. Reject H_0.

22. Test statistic: $\chi^2 = 177.778$, Critical values: 72.282, 127.718. Reject H_0.

23. Test statistic: $\chi^2 = 177.778$, Critical values: 73.772, 129.069. Reject H_0.

Section 9.6

Exercises 1–4 are the Check Your Understanding exercises for this section. Answers to these exercises are on page 494.

5. The parameter is the population mean. H_0: $\mu = 9$, H_1: $\mu \neq 9$, Test statistic: $t = -2.5607$, Critical values: -3.055, 3.055, P-value is between 0.02 and 0.05 [Tech: 0.0250]. Do not reject H_0.

6. The parameter is the population mean. H_0: $\mu = 60$, H_1: $\mu < 60$, Test statistic: $z = -1.73$, Critical value: -1.645, P-value: 0.0418 [Tech: 0.0420]. Reject H_0.

7. The parameter is the population standard deviation. H_0: $\sigma = 5$, H_1: $\sigma > 5$, Test statistic: $\chi = 23.064$, Critical value: 24.996. Do not reject H_0.

8. The parameter is the population proportion. H_0: $p = 0.5$, H_1: $p < 0.5$, Test statistic: $z = -1.10$, Critical value: -2.326, P-value: 0.1357 [Tech: 0.1367]. Do not reject H_0. There is not enough evidence to conclude that less than half of law students are women.

9. The parameter is the population mean. H_0: $\mu = 40$, H_1: $\mu \neq 40$, Test statistic: $z = 2.32$, Critical values: -1.96, 1.96, P-value: 0.0204 [Tech: 0.0203]. Reject H_0.

10. The parameter is the population mean. H_0: $\mu = 4$, H_1: $\mu > 4$, Test statistic: $t = 1.540$, Critical value: 2.015, P-value is between 0.05 and 0.10 [Tech: 0.0921]. Do not reject H_0.

11. The parameter is the population proportion. H_0: $p = 0.6$, H_1: $p \neq 0.6$, Test statistic: $z = 2.72$, Critical values: -2.576, 2.576, P-value: 0.0066 [Tech: 0.0065]. Reject H_0. We conclude that the proportion of families with one or more pets differs from 0.6.

12. The parameter is the population standard deviation. H_0: $\sigma = 15$, H_1: $\sigma > 15$, Test statistic: $\chi = 34.766$, Critical value: 32.671. Reject H_0.

13. H_0: $p = 0.3$, H_1: $p > 0.3$, Test statistic: $z = 2.63$, Critical value: 1.645, P-value: 0.0043. Reject H_0. We conclude that more than 30% of U.S. adults with children have saved money for college.

14. H_0: $\mu = 2559$, H_1: $\mu > 2559$. Test statistic: $t = 1.511$, Critical value: 2.539; P-value is between 0.025 and 0.05 [Tech: 0.0358]. Do not reject H_0. There is not enough evidence to conclude that the mean area of homes built in 2020 is greater than that of homes built in 2018.

15. H_0: $\sigma = 0.1$, H_1: $\sigma < 0.1$, Test statistic: $\chi = 4.195$, Critical value: 1.239. Do not reject H_0. There is not enough evidence to conclude that the standard deviation of weights of cookie boxes is less than 0.1.

16. H_0: $p = 0.3$, H_1: $p > 0.3$, Test statistic: $z = 2.29$, Critical value: 1.645, P-value: 0.0110. Reject H_0. We conclude that more than 30% of people have considered a career in science.

17. H_0: $\mu = 50{,}000$, H_1: $\mu \neq 50{,}000$, Test statistic: $z = -1.50$, Critical values: -2.576, 2.576, P-value: 0.1336 [Tech: 0.1329]. Do not reject H_0. There is not enough evidence to conclude that the mean salary of public school teachers in Georgia differs from \$50,000.

18. H_0: $\sigma = 2$, H_1: $\sigma > 2$, Test statistic: $\chi = 15.556$, Critical value: 15.507. Reject H_0. We conclude that the standard deviation of highway speeds is greater than 2.

19. H_0: $\mu = 1$, H_1: $\mu > 1$, Test statistic: $t = 2.377$, Critical value: 1.943, P-value is between 0.025 and 0.05 [Tech: 0.0275]. Reject H_0. We conclude that the mean concentration is greater than 1 milligram per cubic meter.

20. H_0: $\mu = 8$, H_1: $\mu \neq 8$, Test statistic: $z = 2.00$, Critical values: -2.576, 2.576, P-value: 0.0456 [Tech: 0.0455]. Do not reject H_0. There is not enough evidence to conclude that the mean size of ladies' shoes differs from 8.

Section 9.7

Exercises 1 and 2 are the Check Your Understanding exercises for this section. Answers to these exercises are on page 499.

3. Type II

4. false

5. increase

6. False

7. False

8. True

9. **a.** True **b.** True **c.** False **d.** False

10. **a.** True **b.** False **c.** False **d.** False **e.** True **f.** False

11. **a.** 0.2578 [Tech: 0.2595] **b.** 0.6406 [Tech: 0.6388] **c.** 0.0918 [Tech: 0.0924] **d.** 0.3707 [Tech: 0.3721]

12. **a.** 0.4090 [Tech: 0.4088] **b.** 0.8810 [Tech: 0.8817] **c.** 0.1814 [Tech: 0.1808] **d.** 0.6915 [Tech: 0.6922]

13. **a.** 0.5319 [Tech: 0.5316] **b.** 0.9962 **c.** 0.2743 [Tech: 0.2735] **d.** 0.9761 [Tech: 0.9764]

14. **a.** 0.5517 [Tech: 0.5506] **b.** 0.9974 [Tech: 0.9973] **c.** 0.2912 [Tech: 0.2897] **d.** 0.9821 [Tech: 0.9823]

15. 0.5160

16. 0.8079 [Tech: 0.8074]

17. 0.9582 [Tech: 0.9586]

18. 0.2119 [Tech: 0.2107]

19. **a.** 1.645 **b.** 0.3342 **c.** 0.8186 [Tech: 0.8197]

CHAPTER 9 Quiz

1. 0.035

2. iv

3. **a.** True **b.** False **c.** True **d.** False
4. **a.** False **b.** True **c.** True **d.** True
5. **a.** No, because the population standard deviation is unknown.
   b. Yes, because the population is normal.
6. 1.711
7. True
8. **a.** $z = 2.50$ **b.** Yes
   c. We conclude that more than 85% of people pass their driver's test.
9. **a.** Yes, because the P-value is less than 0.05.
   b. No, because the P-value is greater than 0.01.
10. False
11. Either the P-value or the value of the test statistic should be reported.
12. Yes
13. True
14. Test statistic: $\chi^2 = 8.444$, Critical value: 10.117, reject H_0. We conclude that the population standard deviation is less than 15.
15. 0.9871

CHAPTER 9 Review Exercises

1. i.
2. **a.**

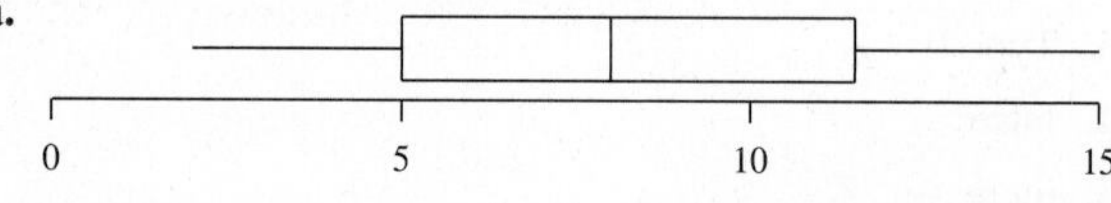

Yes

   b. H_0: $\mu = 8.87$, H_1: $\mu < 8.87$. Test statistic: $t = -0.6820$, Critical value: -1.714; P-value is between 0.25 and 0.40 [Tech: 0.2510]. Do not reject H_0. There is not enough evidence to conclude that the mean number of runs in 2019 is less than the mean number of runs in 2018.
3. **a.** H_0: $p = 0.6$, H_1: $p \neq 0.6$
   b. $z = 2.02$
   c. Critical values: -1.96, 1.96, P-value: 0.0434 [Tech: 0.0433]. Reject H_0.
   d. We conclude that the proportion of students who log in to Facebook daily differs from 0.6.
4. **a.** H_0: $\mu = 25$, H_1: $\mu \neq 25$
   b. t-test, because the population standard deviation is unknown.
   c. $t = 0.478$
   d. Critical values: -2.032, 2.032; P-value is between 0.50 and 0.80 [Tech: 0.6360]. Do not reject H_0.
   e. There is not enough evidence to conclude that the mean price differs from $25.00.
5. 0.2776 [Tech: 0.2762]
6. The power would be greater. When the significance level is increased, the probability of rejecting H_0 increases as well.
7. **a.** H_0: $\mu = 2.5$, H_1: $\mu < 2.5$
   b. We should perform a t-test, because the population standard deviation is unknown.
   c. $t = -1.633$
   d. Critical value: -2.364, P-value is between 0.05 and 0.10 [Tech: 0.0523]. Do not reject H_0.
   e. There is not enough evidence to conclude that the mean number of people per household is less than 2.5.
8. **a.** H_0: $p = 0.45$, H_1: $p > 0.45$
   b. Test statistic: $z = 3.14$
   c. Critical value: 2.326, P-value: 0.0008. Reject H_0.
   d. We conclude that more than 45% of employed people are completely or very satisfied with their jobs.
9. H_0: $\sigma = 2$, H_1: $\sigma > 2$, Test statistic: $\chi^2 = 15.75$, Critical value: 18.475. Do not reject H_0. There is not enough evidence to conclude that the population standard deviation is greater than 2.
10. **a.** H_0: $\mu = 4.7$, H_1: $\mu \neq 4.7$ **b.** $z = 2.074$
    c. P-value is 0.038. **d.** Yes **e.** No
11. **a.** Proportion **b.** H_0: $p = 0.2$, H_1: $p < 0.2$
    c. $z = -2.25$ **d.** P-value is 0.0122244334 **e.** Yes
    f. No
12. **a.** H_0: $\mu = 3$, H_1: $\mu > 3$
    b. We should perform a z-test, because the population standard deviation is known.
    c. $z = 0.52$
    d. Critical value: 2.326, P-value: 0.3015 [Tech: 0.3028]. Do not reject H_0.
    e. There is not enough evidence to conclude that the mean number of TV sets per household is greater than 3.
13. H_0: $\sigma = 10$, H_1: $\sigma < 10$, Test statistic: $\chi^2 = 6.00$, Critical value: 13.848. Reject H_0. We conclude that the population standard deviation is less than 10.
14. **a.** H_0: $\mu = 1000$, H_1: $\mu > 1000$
    b. We should perform a t-test, because the population standard deviation is unknown.
    c. $t = 2.108$
    d. Critical value: 1.685, P-value is between 0.01 and 0.025 [Tech: 0.0207]. Reject H_0.
    e. We conclude that the mean monthly rent is greater than $1000.
15. **a.** H_0: $p = 0.23$, H_1: $p > 0.23$
    b. Test statistic: $z = 3.36$
    c. Critical value: 2.326, P-value: 0.0004. Reject H_0.
    d. We conclude that more than 23% of students at the university watch cable news regularly.

CHAPTER 9 Case Study

1. Test statistic: $z = 4.16$, P-value: 0.000016. H_0 is rejected at any reasonable level, including $\alpha = 0.05$ and $\alpha = 0.01$. We conclude that the probability is greater than 0.5 that a record high occurred more recently than a record low.
2. Answers will vary. The days are not independent and thus do not constitute a simple random sample.
3. Test statistic: $t = 4.329$, P-value is less than 0.0005 [Tech: 0.000035]. H_0 is rejected at any reasonable level, including $\alpha = 0.05$ and $\alpha = 0.01$. We conclude that the probability is greater than 0.5 that the mean year of record high is greater than 1944.5.
4. Answers will vary. When two years are both record highs, only the later one counts.

5. Test statistic: $t = 3.664$, P-value is less than 0.0005 [Tech: 0.00035]. H_0 is rejected at any reasonable level, including $\alpha = 0.05$ and $\alpha = 0.01$. The conclusion does not change. We conclude that the probability is greater than 0.5 that the mean year of record high is greater than 1944.5.

6. Test statistic: $t = -6.941$, P-value is less than 0.0005 [Tech: 0.000000010]. H_0 is rejected at any reasonable level, including $\alpha = 0.05$ and $\alpha = 0.01$. We conclude that the probability is greater than 0.5 that the mean year of record low is less than 1944.5.

7. Answers will vary.

CHAPTER 10

Section 10.1

Exercises 1–4 are the Check Your Understanding exercises for this section. Answers to these exercises are on page 517.

5. independent

6. $n_1 - 1, n_2 - 1$

7. False

8. True

9. $8.5 < \mu_1 - \mu_2 < 14.9$ [Tech: $8.6 < \mu_1 - \mu_2 < 14.8$]

10. $128.10 < \mu_1 - \mu_2 < 184.54$
[TI-84 Plus: $129.10 < \mu_1 - \mu_2 < 183.54$]
[MINITAB: $129.07 < \mu_1 - \mu_2 < 183.57$]

11. $36.70 < \mu_1 - \mu_2 < 137.14$
[TI-84 Plus: $39.99 < \mu_1 - \mu_2 < 133.85$]
[MINITAB: $39.90 < \mu_1 - \mu_2 < 133.94$]

12. $-4.6 < \mu_1 - \mu_2 < 14.0$ [Tech: $-3.9 < \mu_1 - \mu_2 < 13.3$]

13. $-17.2 < \mu_1 - \mu_2 < 27.4$
[TI-84 Plus: $-16.5 < \mu_1 - \mu_2 < 26.7$]
[MINITAB: $-16.6 < \mu_1 - \mu_2 < 26.8$]

14. $7.8 < \mu_1 - \mu_2 < 29.9$ [Tech: $9.6 < \mu_1 - \mu_2 < 28.0$]

15. $8.0 < \mu_1 - \mu_2 < 14.0$ [Tech: $8.1 < \mu_1 - \mu_2 < 13.9$]

16. $0.5 < \mu_1 - \mu_2 < 8.7$ [Tech: $1.7 < \mu_1 - \mu_2 < 7.5$]

17. **a.** The sample sizes are small ($n \leq 30$).
b. Yes; there are no outliers and no evidence of strong skewness.
c. $31.4 < \mu_1 - \mu_2 < 94.9$
[TI-84 Plus: $34.1 < \mu_1 - \mu_2 < 92.2$]
[MINITAB: $34.0 < \mu_1 - \mu_2 < 92.3$]

18. **a.** The sample sizes are small ($n \leq 30$).
b. Yes; there are no outliers and no evidence of strong skewness.
c. $35.7 < \mu_1 - \mu_2 < 132.2$
[TI-84 Plus: $39.4 < \mu_1 - \mu_2 < 128.4$]
[MINITAB: $39.3 < \mu_1 - \mu_2 < 128.5$]

19. **a.** $-12.6 < \mu_1 - \mu_2 < 10.8$
[TI-84 Plus: $-11.9 < \mu_1 - \mu_2 < 10.1$]
[MINITAB: $-12.0 < \mu_1 - \mu_2 < 10.2$]
b. Because the confidence interval contains 0, it is reasonable to believe that the mean IQ may be the same for both firstborn and secondborn sons.

20. **a.** $-11.0 < \mu_1 - \mu_2 < 4.5$ [Tech: $-10.5 < \mu_1 - \mu_2 < 4.0$]
b. Because the confidence interval contains 0, it does not contradict the claim that both methods of instruction may be equally effective.

21. **a.** $0.3 < \mu_1 - \mu_2 < 1.5$
b. Because the confidence interval does not contain 0, it contradicts the claim that the mean weight is the same for both boys and girls.

22. **a.** $0.66 < \mu_1 - \mu_2 < 0.74$
b. Because the confidence interval does not contain 0, it contradicts the claim that the mean BMI is the same for both men and women.

23. **a.** $0.52 < \mu_1 - \mu_2 < 2.02$
b. Because the confidence interval does not contain 0, it is not reasonable to believe that the mean number of energy drinks consumed is the same for both men and women.

24. **a.** $-0.4 < \mu_1 - \mu_2 < 4.6$
b. Because the confidence interval contains 0, it does not contradict the claim that both diets are equally effective.

25. **a.** $0.799 < \mu_1 - \mu_2 < 12.401$ [Tech: $0.838 < \mu_1 - \mu_2 < 12.362$]
b. Because the confidence interval does not contain 0, it contradicts the claim that the mean systolic pressure is the same in both groups.

26. **a.** $-8.264 < \mu_1 - \mu_2 < 86.264$ [Tech: $-6.998 < \mu_1 - \mu_2 < 84.998$]
b. Because the confidence interval contains 0, it does not contradict the claim that the treadmill does not increase walking ability.

27. **a.** $-2.50 < \mu_1 - \mu_2 < 32.50$ [Tech: $-2.20 < \mu_1 - \mu_2 < 32.20$]
b. Because the confidence interval contains 0, it is reasonable to believe that the mean daily production is the same for both methods.

28. **a.** $8.960 < \mu_1 - \mu_2 < 11.040$ [Tech: $8.974 < \mu_1 - \mu_2 < 11.026$]
b. Because the confidence interval does not contain 0, it is not reasonable to believe that the mean number of calories is the same for both types of eggs.

29. **a.** $-8.0 < \mu_1 - \mu_2 < 2.2$ [Tech: $-7.9 < \mu < 2.1$]
b. Because the confidence interval contains 0, it does not contradict the claim that the mean scores are the same.

30. **a.** $-0.79 < \mu_1 - \mu_2 < 1.01$ [Tech: $-0.77 < \mu < 0.99$]
b. Because the confidence interval does not contain 1.5, it contradicts the claim that the mean for western cities is 1.5 years greater than the mean for eastern cities.

31. **a.** 47.519 **b.** 12.28537157 **c.** 20.904, 74.134

32. **a.** 3.81075
b. 86.51750655
c. 3.0101, 4.6114

33. **a.** 20.7429 **b.** 30 **c.** 98%, 12.9408, 28.5450

34. **a.** 23.455 **b.** 102 **c.** 95%, 20.019, 26.891

35. **a.** 9 degrees of freedom **b.** 2.262
c. $39.241 < \mu_1 - \mu_2 < 85.359$

Section 10.2

Exercises 1 and 2 are the Check Your Understanding exercises for this section. Answers to these exercises are on page 525.

3. paired

4. smaller

5. True

6. False

7. a. $-5.33 < \mu_d < 2.36$
b. Because the confidence interval contains 0, it does not contradict the claim that the mean speeds of the processors are the same.

8. a. $9.23 < \mu_d < 11.29$
b. No; it is reasonable to believe that the difference may be as large as 11,300.

9. a. $14.0 < \mu_d < 52.4$
b. Because the confidence interval does not contain 0, it contradicts the claim that the mean strengths are the same.

10. a. $-7.19 < \mu_d < 79.81$ [Tech: $-7.18 < \mu_d < 79.80$]
b. Because the confidence interval contains 0, it is reasonable to believe that the mean emissions may be the same.

11. a. $5.914 < \mu_d < 9.486$
b. Because the confidence interval contains 8, it does not contradict the claim that the difference in water loss is 8 g/m^2 per hour.

12. a. $16.49 < \mu_d < 22.31$
b. Because the confidence interval contains 20, it is reasonable to believe that the mean increase is 20 breaths per minute.

13. a. $19.166 < \mu_d < 43.634$
b. Because the confidence interval does not contain 50, it is not reasonable to believe that the mean reduction is 50 mmHg.

14. a. $0.3076 < \mu_d < 0.5109$
b. Because the confidence interval does not contain 0, it contradicts the claim that the absorption rates of the two ointments are the same.

15. a. $70.4 < \mu_d < 88.3$
b. No; it is reasonable to believe that the difference may be as large as 88.3.

16. a. $1.01 < \mu_d < 3.24$
b. Because the confidence interval does not contain 0, it is not reasonable to believe that the mean mileage may be the same.

17. a. $5.18 < \mu_1 - \mu_2 < 5.55$
b. Because the confidence interval contains 5.5, it does not contradict the claim that the mean increase is 5.5.

18. a. $8.0 < \mu_1 - \mu_2 < 17.8$
b. Because all of the values in the confidence interval are less than 20, the confidence interval contradicts the claim that the mean increase is greater than 20 points.

19. a. 8.7385 **b.** 13 **c.** 6.5788, 10.898

20. a. 2.56842 **b.** 6 **c.** −4.798, 9.9348

21. a. 2.5324 **b.** 5 **c.** 99%, −3.411394, 8.476194

22. a. 16.412 **b.** 11 **c.** 95%, 14.10815, 18.71585

23. a. $-0.48 < \mu_1 - \mu_2 < 4.72$
[TI-84 Plus: $-0.27 < \mu_1 - \mu_2 < 4.51$]
[MINITAB: $-0.28 < \mu_1 - \mu_2 < 4.53$]
b. The confidence interval using matched pairs is narrower, because the standard error is smaller.

24. a. Paired samples, because it is natural to pair each person with his or her twin.
b. Independent, because there is no natural way to determine pairs.
c. Independent, because there is no natural way to determine pairs.
d. Paired samples, because it is natural to pair the rent on an apartment this year with the rent on the same apartment last year.

Section 10.3

Exercises 1 and 2 are the Check Your Understanding exercises for this section. Answers to these exercises are on page 533.

3. independent

4. two, ten

5. True

6. False

7. $0.066 < p_1 - p_2 < 0.384$

8. $0.058 < p_1 - p_2 < 0.462$

9. $-0.489 < p_1 - p_2 < -0.193$

10. $-0.089 < p_1 - p_2 < 0.166$

11. $0.099 < p_1 - p_2 < 0.415$

12. $-0.227 < p_1 - p_2 < 0.303$

13. a. $0.038 < p_1 - p_2 < 0.079$
b. Because the confidence interval does not contain 0, it contradicts the claim that the proportion of rear-end accidents is the same at the two intersections.

14. a. $0.102 < p_1 - p_2 < 0.421$
b. Because the confidence interval does not contain 0, it contradicts the claim that the proportion of students who prefer the new method is the same for males and females.

15. a. $-0.027 < p_1 - p_2 < 0.272$
b. Because the confidence interval contains 0, it does not contradict the claim that the proportion of patients requiring pain medication is the same for both procedures.

16. a. $-0.497 < p_1 - p_2 < -0.151$
b. Because the confidence interval does not contain 0, it contradicts the claim that the proportion of customers purchasing pretzels is the same before and after the ad campaign.

17. a. $0.0171 < p_1 - p_2 < 0.0598$
b. Because the confidence interval contains 0.02, it does not contradict the claim that the difference between the proportions is 0.02.

18. a. $0.0357 < p_1 - p_2 < 0.0583$
b. Because the confidence interval does not contain 0, it is not reasonable to believe that the proportion of smokers is the same in both groups.

19. a. $-0.0679 < p_1 - p_2 < 0.2279$

b. Because the confidence interval contains 0, it is reasonable to believe that the proportion of seeds that germinate is the same for both concentrations of mushroom compost.

20. a. $0.073 < p_1 - p_2 < 0.295$

b. Because the confidence interval does not contain 0, it contradicts the claim that the proportions feeling relief are the same in both groups.

21. a. $0.120 < p_1 - p_2 < 0.178$

b. No, it does not contradict the claim. It is reasonable to believe that the proportion of defective parts may have decreased by as much as 17.8%.

22. a. $0.014 < p_1 - p_2 < 0.065$

b. Because the confidence interval contains 0.02, it does not contradict the claim that the proportion of people who are satisfied increased by more than 0.02.

23. a. $0.022 < p_1 - p_2 < 0.034$

b. Because the confidence interval does not contain 0, it contradicts the claim that the proportion of patients who have had colonoscopies is the same for those with and without colorectal cancer.

24. a. $-0.0124 < p_1 - p_2 < 0.0726$

b. Because all of the values in the confidence interval are less than 0.10, the confidence interval contradicts the claim that the proportion increased by more than 0.10.

25. a. 0.086666667 **b.** −0.0815, 0.25484

26. a. 0.19271 **b.** 0.04811, 0.33732

27. a. 0.177294 **b.** 99%, −0.051451, 0.406038

28. a. 0.010053 **b.** 95%, −0.006612, 0.026717

29. 301

CHAPTER 10 Quiz

1. Paired
2. Independent
3. 0.1183
4. 1.96
5. 0.0806
6. $0.038 < p_1 - p_2 < 0.199$
7. 8
8. 3.499
9. 8.0172 [Tech: 8.0183]
10. $-0.02 < \mu_d < 16.02$
11. 2543
12. 1.671 [Tech: 1.655]
13. 717.161 [TI-84 Plus: 710.262] [MINITAB: 710.269]
14. $1825.8 < \mu_1 - \mu_2 < 3260.2$ [Tech: $1832.7 < \mu_1 - \mu_2 < 3253.3$]
15. These are not independent samples.

CHAPTER 10 Review Exercises

1. $0.041 < p_1 - p_2 < 0.063$
2. The samples are not independent, because they are based on the same set of days.
3. $6.5 < \mu_d < 8.3$
4. $(-7153.4 < \mu_1 - \mu_2 < 5323.6)$ [TI-84 Plus: $(-6710.1 < \mu_1 - \mu_2 < 4880.3)$] [MINITAB: $(-6733.7 < \mu_1 - \mu_2 < 4903.9)$]
5. $0.7 < \mu_1 - \mu_2 < 21.3$ [TI-84 Plus: $(2.0 < \mu_1 - \mu_2 < 20.0)$] [MINITAB: $(1.8 < \mu_1 - \mu_2 < 20.2)$]
6. $-0.020 < p_1 - p_2 < 0.156$
7. $-2.44 < \mu_1 - \mu_2 < -2.16$
8. $-0.043 < \mu_d < 0.003$
9. $-0.241 < p_1 - p_2 < 0.018$

10. a. 11.402
b. 113.2701584
c. 9.8059, 12.998

11. a. 2.3515 **b.** 7 **c.** 95%, −2.02947, 6.73247

12. a. 0.31614 **b.** 0.12307, 0.5092

13. a. 0.249243 **b.** 95%, 0.135735, 0.362751

14. a. 8.533 **b.** 17 **c.** 7.7632, 9.3028

15. a. 9.8612
b. 14
c. 95%, 7.803957, 11.918443

CHAPTER 10 Case Study

1.

Characteristic	95% Confidence Interval
Age	$-2.03 < \mu_1 - \mu_2 < 0.03$
Systolic BP	$-4.65 < \mu_1 - \mu_2 < -1.35$
Diastolic BP	$-1.94 < \mu_1 - \mu_2 < -0.06$
Treatment for hypertension	$-0.050 < p_1 - p_2 < 0.040$
Atrial fibrillation	$-0.015 < p_1 - p_2 < 0.045$
Diabetes	$-0.039 < p_1 - p_2 < 0.047$
Cigarette smoking	$-0.017 < p_1 - p_2 < 0.045$
Coronary bypass surgery	$-0.048 < p_1 - p_2 < 0.036$

2. Systolic blood pressure and Diastolic blood pressure
3. No; the differences are too small to be of practical significance.

CHAPTER 11

Section 11.1

Exercises 1 and 2 are the Check Your Understanding exercises for this section. Answers to these exercises are on page 551.

3. skewness, outliers
4. $n_1 - 1, n_2 - 1$
5. False
6. True

7. a. 9 **b.** $t = -3.41$

c. Critical value: −1.833, *P*-value is between 0.0025 and 0.005 [Tech: 0.0039] [TI-84 Plus: 0.0010] [MINITAB: 0.0011]. Reject H_0.

8. a. 14 **b.** $t = 1.178$

c. Critical values: −2.145, 2.145, *P*-value is between 0.20 and 0.50 [Tech: 0.2584] [TI-84 Plus: 0.2457] [MINITAB: 0.2459]. Do not reject H_0.

9. a. H_0: $\mu_1 = \mu_2$, H_1: $\mu_1 < \mu_2$
 b. −3.637
 c. 1360
 d. Critical value: −1.645, $P < 0.0005$ [Tech: 0.00014] [TI-84 Plus: 0.00014] [MINITAB: 0.00014]. Reject H_0. We conclude that the mean number of hours spent on the internet increased between 2014 and 2018.

10. a. H_0: $\mu_1 = \mu_2$, H_1: $\mu_1 \neq \mu_2$ b. $t = 1.996$
 c. 76 [TI-84 Plus: 147.1] [MINITAB: 147]
 d. Critical values: −2.660, 2.660, P-value is between 0.05 and 0.10 [Tech: 0.0504] [TI-84 Plus: $P = 0.0477$] [MINITAB: $P = 0.0477$]. Do not reject H_0. There is not enough evidence to conclude that the mean weight loss differs between the two diets.

11. H_0: $\mu_1 = \mu_2$, H_1: $\mu_1 \neq \mu_2$, Test statistic: $t = -0.173$, Critical values: −3.250, 3.250, P-value is greater than 0.80 [Tech: 0.8662] [TI-84 Plus: $P = 0.8646$] [MINITAB: $P = 0.8647$]. Do not reject H_0. There is not enough evidence to conclude that there is a difference in mean IQ between firstborn and secondborn sons.

12. H_0: $\mu_1 = \mu_2$, H_1: $\mu_1 < \mu_2$, Test statistic: $t = -3.372$, Critical value: −1.943, P-value is between 0.005 and 0.01 [Tech: 0.0075] [TI-84 Plus: $P = 0.0036$] [MINITAB: $P = 0.0041$]. Reject H_0. We conclude that the mean recovery time for those receiving the new treatment is less than the mean for those receiving the standard treatment.

13. H_0: $\mu_1 = \mu_2$, H_1: $\mu_1 > \mu_2$, Test statistic: $t = 5.127$, Critical value: 2.132, P-value is between 0.0025 and 0.005 [Tech: 0.0034] [TI-84 Plus: $P = 0.00025$] [MINITAB: $P = 0.00031$]. Reject H_0. We conclude that the mean benzene concentration is less in treated water than in untreated water.

14. H_0: $\mu_1 = \mu_2$, H_1: $\mu_1 > \mu_2$, Test statistic: $t = 2.231$, Critical value: 1.684, P-value is between 0.01 and 0.025 [Tech: 0.0153] [TI-84 Plus: $P = 0.0141$] [MINITAB: $P = 0.0141$]. Reject H_0. We conclude that the mean distance walked for patients using a treadmill is greater than the mean for the controls.

15. H_0: $\mu_1 = \mu_2$, H_1: $\mu_1 < \mu_2$, Test statistic: $t = -1.043$, Critical value: −1.812, P-value is between 0.10 and 0.25 [Tech: 0.1607] [TI-84 Plus: $P = 0.1554$] [MINITAB: $P = 0.1557$]. Do not reject H_0. There is not enough evidence to conclude that the mean life span of those exposed to the mummy's curse is less than the mean of those not exposed.

16. H_0: $\mu_1 = \mu_2$, H_1: $\mu_1 \neq \mu_2$, Test statistic: $t = -1.156$, Critical values: −2.064, 2.064, P-value is between 0.20 and 0.40 [Tech: 0.2592] [TI-84 Plus: $P = 0.2535$] [MINITAB: $P = 0.2536$]. Do not reject H_0. There is not enough evidence to conclude that the mean weight differs between boys and girls.

17. H_0: $\mu_1 = \mu_2$, H_1: $\mu_1 \neq \mu_2$, Test statistic: $t = -2.559$, Critical values: −2.131, 2.131, P-value is between 0.02 and 0.05 [Tech: 0.0218] [TI-84 Plus: $P = 0.0164$] [MINITAB: $P = 0.0164$]. Reject H_0. We conclude that the mean score differs between men and women.

18. H_0: $\mu_1 = \mu_2$, H_1: $\mu_1 \neq \mu_2$, Test statistic: $t = -0.480$, Critical values: −1.96, 1.96, P-value is between 0.50 and 0.80 [Tech: 0.6314] [TI-84 Plus: $P = 0.6314$] [MINITAB: $P = 0.6314$]. Do not reject H_0. There is not enough evidence to conclude that the mean number of text messages differs between men and women.

19. H_0: $\mu_1 = \mu_2$, H_1: $\mu_1 \neq \mu_2$, Test statistic: $t = -1.166$, Critical values: −2.032, 2.032, P-value is between 0.20 and 0.50 [Tech: 0.2519] [TI-84 Plus: $P = 0.2479$] [MINITAB: $P = 0.2479$]. Do not reject H_0. There is not enough evidence to conclude that the mean score differs between paper and online tests.

20. H_0: $\mu_1 = \mu_2$, H_1: $\mu_1 < \mu_2$, Test statistic: $t = -0.248$, Critical value: −1.696, P-value is greater than 0.40 [Tech: 0.4029] [TI-84 Plus: $P = 0.4025$] [MINITAB: $P = 0.4025$]. Do not reject H_0. There is not enough evidence to conclude that the mean time is greater for western cities than for eastern cities.

21. a. Right-tailed
 b. 24.99965945
 c. 0.101223442
 d. No

22. a. Two-tailed
 b. 42.91558207
 c. 0.0209030051
 d. Yes

23. a. H_1: $\mu_1 \neq \mu_2$ b. Yes, the P-value is 0.003.
 c. 54 d. 35
 e. The P-value is between 0.002 and 0.005 [Tech: 0.0036].

24. a. H_1: $\mu_1 < \mu_2$ b. No, the P-value is 0.100.
 c. 38 d. 19
 e. The P-value is between 0.10 and 0.25 [Tech: 0.1046].

25. a. 1.833 b. 26.727 c. 0.039

Section 11.2

Exercises 1 and 2 are the Check Your Understanding exercises for this section. Answers to these exercises are on page 560.

3. skewness, outliers
4. mean
5. True
6. False

7. a. 1, −4, 6, 9, 7 b. $t = 1.614$
 c. Critical value: 2.132, P-value is between 0.05 and 0.10 [Tech: 0.0909]. Do not reject H_0.
 d. Critical value: 3.747, P-value is between 0.05 and 0.10 [Tech: 0.0909]. Do not reject H_0.

8. a. −6, −1, −9, −1, −5, −1, −4, −8, −2, −9
 b. $t = -4.399$
 c. Critical values: −2.262, 2.262, P-value is between 0.001 and 0.002 [Tech: 0.0017]. Reject H_0.
 d. Critical values: −3.250, 3.250, P-value is between 0.001 and 0.002 [Tech: 0.0017]. Reject H_0.

9. a. H_0: $\mu_d = 0$, H_1: $\mu_d > 0$ b. $t = 1.369$
 c. Critical value: 1.943, P-value is between 0.10 and 0.25 [Tech: 0.1100]. Do not reject H_0. There is not enough evidence to conclude that the mean pain level is less with drug B.

10. a. H_0: $\mu_d = 0$, H_1: $\mu_d \neq 0$ b. $t = -2.206$
 c. Critical values: −3.250, 3.250, P-value is between 0.05 and 0.10 [Tech: 0.0548]. Do not reject H_0. There is not enough evidence to conclude that the mean weight differs between the scales.

11. a. $H_0: \mu_d = 0, H_1: \mu_d \neq 0$ **b.** 6.082

c. Critical values: −2.571, 2.571, *P*-value is between 0.001 and 0.002 [Tech: 0.00174]. Reject H_0. We conclude that there is a difference in the mean concentration between the shoot and the root.

12. a. $H_0: \mu_d = 0, H_1: \mu_d > 0$

b. 2.072

c. Critical value: 3.143, *P*-value is between 0.025 and 0.05 [Tech: 0.0418]. Do not reject H_0. There is not enough evidence to conclude that the mean lifetime of Brand A is greater than the mean lifetime of Brand B.

13. a. $H_0: \mu_d = 0, H_1: \mu_d \neq 0$

b. 1.459

c. Critical values: −3.012, 3.012, *P*-value is between 0.10 and 0.20 [Tech: 0.1682]. Do not reject H_0. There is not enough evidence to conclude that the mean amount absorbed differs between the brand name and the generic drug.

14. a. $H_0: \mu_d = 65, H_1: \mu_d > 65$

b. 3.038

c. Critical values: −1.895, 1.895, *P*-value is between 0.005 and 0.01 [Tech: 0.00945]. Reject H_0. We conclude that the mean cholesterol level is reduced by more than 65 mg/dL after treatment.

15. a. $H_0: \mu_d = 0, H_1: \mu_d \neq 0$ **b.** $t = -4.790$

c. Critical values: −2.776, 2.776, *P*-value is between 0.005 and 0.01 [Tech: 0.0087]. Reject H_0. We conclude that the mean strength after three days differs from the mean strength after six days.

16. a. $H_0: \mu_d = 0, H_1: \mu_d > 0$ **b.** $t = 2.713$

c. Critical value: 1.833, *P*-value is between 0.01 and 0.025 [Tech: 0.0119]. Reject H_0. We conclude that the mean emissions are higher at 40°F.

17. a. $H_0: \mu_d = 5, H_1: \mu_d > 5$

b. $t = 4.009$

c. Critical value: 1.685, *P*-value is less than 0.0005 [Tech: 0.000133]. Reject H_0. We conclude that the mean increase is greater than 5 inches.

18. a. $H_0: \mu_d = 15, H_1: \mu_d < 15$

b. $t = -0.860$

c. Critical value: −1.696, *P*-value is between 0.10 and 0.25 [Tech: 0.1983]. Do not reject H_0. There is not enough evidence to conclude that the mean increase is less than 15 points.

19. a. Two-tailed **b.** 15 **c.** 0.0296591111

d. Yes, because the *P*-value is less than 0.05.

20. a. Right-tailed **b.** 17 **c.** 0.1745766435

d. No, because the *P*-value is greater than 0.05.

21. a. Right-tailed

b. Yes, because the *P*-value is less than 0.05.

c. No, because the *P*-value is greater than 0.01.

22. a. Two-tailed

b. No, because the *P*-value is greater than 0.05.

c. No, because the *P*-value is greater than 0.01.

23. a. *P*-value is between 0.05 and 0.10 [Tech: 0.0726] [TI-84 Plus: 0.0653] [MINITAB: 0.0658].

b. The *P*-value is greater because the standard error is larger.

Section 11.3

Exercises 1 and 2 are the Check Your Understanding exercises for this section. Answers to these exercises are on page 570.

3. 20

4. 10

5. False

6. True

7. a. $z = -1.00$ **b.** No **c.** No

8. a. $z = 3.29$ **b.** Yes **c.** Yes

9. a. $H_0: p_1 = p_2, H_1: p_1 \neq p_2$

b. $z = 0.62$

c. Critical values: −1.96, 1.96, *P*-value is 0.5352 [Tech: 0.5376]. Do not reject H_0. There is not enough evidence to conclude that the proportion of boys who are overweight differs from the proportion of girls who are overweight.

10. a. $H_0: p_1 = p_2, H_1: p_1 < p_2$ **b.** $z = -2.53$

c. Critical value: −2.326, *P*-value is 0.0057. Reject H_0. We conclude that the proportion of high-altitude vehicles exceeding the standard is greater than the proportion of low-altitude vehicles exceeding the standard.

11. $H_0: p_1 = p_2, H_1: p_1 > p_2$, Test statistic: $z = 3.11$, Critical value: 2.326, *P*-value is 0.0009. Reject H_0. We conclude that the proportion of patients suffering a heart attack or stroke is less for ticagrelor.

12. $H_0: p_1 = p_2, H_1: p_1 \neq p_2$, Test statistic: $z = 2.77$, Critical values: −1.96, 1.96, *P*-value is 0.0056 [Tech: 0.0055]. Reject H_0. We conclude that the proportion of people with elevated cholesterol levels differs between men and women.

13. $H_0: p_1 = p_2, H_1: p_1 > p_2$, Test statistic: $z = 3.26$, Critical value: 1.645, *P*-value is 0.0006. Reject H_0. We conclude that the proportion of patients who needed retreatment is less for those who received drug-coated stents.

14. $H_0: p_1 = p_2, H_1: p_1 > p_2$, Test statistic: $z = 2.33$, Critical value: 1.645, *P*-value is 0.0099 [Tech: 0.0098]. Reject H_0. We conclude that the proportion of residents with wheezing symptoms is greater among those who participated in the cleaning of flood-damaged homes.

15. $H_0: p_1 = p_2, H_1: p_1 < p_2$, Test statistic: $z = -2.71$, Critical value: −1.645, *P*-value is 0.0034 [Tech: 0.0033]. Reject H_0. We conclude that the method is more accurate for new age music than for pop music.

16. $H_0: p_1 = p_2, H_1: p_1 \neq p_2$, Test statistic: $z = 0.76$, Critical value: 2.326, *P*-value is 0.4472 [Tech: 0.4488]. Do not reject H_0. There is not enough evidence to conclude that the proportion of red light runners differs between the two intersections.

17. $H_0: p_1 = p_2, H_1: p_1 > p_2$, Test statistic: $z = 2.50$, Critical value: 2.326, *P*-value is 0.0062. Reject H_0. We conclude that the proportion of patch users is greater among European-Americans.

18. $H_0: p_1 = p_2, H_1: p_1 < p_2$, Test statistic: $z = -0.57$, Critical value: −1.96, *P*-value is 0.2843. Do not reject H_0. There is not enough evidence to conclude that the proportion of patients who fall is less among those who exercise.

19. These are not independent samples.

20. These are not random samples of children. Parents choose which sample their child will belong to.

21. **a.** Right-tailed **b.** 0.0037512809 **c.** Yes

22. **a.** Two-tailed **b.** 0.4155446361 **c.** No

23. **a.** Left-tailed **b.** 0.192

c. No, because the P-value is greater than 0.05.

24. **a.** Two-tailed **b.** 0.008

c. Yes, because the P-value is less than 0.05.

25. **a.** $H_0: p_1 - p_2 = 0.05, H_1: p_1 - p_2 > 0.05$
b. $z = 2.11$

c. Critical value: 1.645, P-value is 0.0174 [Tech: 0.0175]. Reject H_0.

d. The more expensive machine

Section 11.4

Exercises 1–4 are the Check Your Understanding exercises for this section. Answers to these exercises are on page 577.

5. normal

6. variances

7. False

8. True

9. 2.51

10. 13.27

11. **a.** The critical value is 4.00. Do not reject H_0.

b. The critical value is 7.72. Do not reject H_0.

12. **a.** The critical value is 3.97. Reject H_0.

b. The critical value is 7.46. Do not reject H_0.

13. **a.** $H_0: \sigma_1 = \sigma_2, H_1: \sigma_1 < \sigma_2$, sample standard deviation for day 1 is 0.13868, sample standard deviation for day 2 is 0.38481. $F = 7.7$. There are 12 and 12 degrees of freedom. Critical value is $f_{0.05} = 2.69$. Reject H_0. We conclude that the variability of the process is greater on the second day than on the first day.

b. $H_0: \sigma_2 = \sigma_3, H_1: \sigma_2 < \sigma_3$, sample standard deviation for day 2 is 0.38481, sample standard deviation for day 3 is 0.51826. $F = 1.8139$. There are 12 and 12 degrees of freedom. Critical value is $f_{0.01} = 4.16$. Do not reject H_0. There is not enough evidence to conclude that the variability of the process is greater on the third day than on the second day.

14. $H_0: \sigma_1 = \sigma_2, H_1: \sigma_1 \neq \sigma_2$, sample standard deviation for composite 1 is 14.238, sample standard deviation for composite 2 is 14.894. $F = 1.0942$. There are 12 and 8 degrees of freedom. Critical value is $f_{0.025} = 4.20$. Do not reject H_0. There is not enough evidence to conclude that the standard deviation of the breaking strengths differs between the two composites.

15. $H_0: \sigma_1 = \sigma_2, H_1: \sigma_1 \neq \sigma_2$, sample standard deviation for thread 1 is 6.0636, sample standard deviation for thread 2 is 5.2957. $F = 1.311$. There are 9 and 9 degrees of freedom. Critical value is $f_{0.025} = 4.03$. Do not reject H_0. There is not enough evidence to conclude that the standard deviation of the breaking strengths differs between the threads.

16. $H_0: \sigma_1 = \sigma_2, H_1: \sigma_1 < \sigma_2$, sample standard deviation for thread 1 is 2.6510, sample standard deviation for thread 2 is 6.4091. $F = 5.845$. There are 12 and 8 degrees of freedom. Critical value is $f_{0.01} = 4.50$. Reject H_0. We conclude that the sodium content is more variable for Brand B.

17. $H_0: \sigma_1 = \sigma_2, H_1: \sigma_1 < \sigma_2$, sample standard deviation for month 1 is 14.169, sample standard deviation for month 7 is 59.757. $F = 17.787$. There are 8 and 6 degrees of freedom. Critical value is $f_{0.01} = 8.10$. Reject H_0. We conclude that the time to freeze-up is more variable in the seventh month than in the first month.

18. $H_0: \sigma_1 = \sigma_2, H_1: \sigma_1 \neq \sigma_2$, sample standard deviation for firstborn is 13.6402, sample standard deviation for secondborn is 9.1433. $F = 2.2255$. There are 9 and 9 degrees of freedom. Critical value is $f_{0.025} = 4.03$. Do not reject H_0. There is not enough evidence to conclude that the standard deviation of IQ differs between firstborn and secondborn sons.

19. 0.177

Section 11.5

Exercises 1 and 2 are the Check Your Understanding exercises for this section. Answers to these exercises are on page 580.

3. less

4. tests

5. True

6. False

7. **a.** Hypothesis 4 can be rejected at the 0.05 level.

b. None of the hypotheses can be rejected at the 0.01 level.

8. **a.** Hypotheses 1, 3, and 4 can be rejected at the 0.05 level.

b. Hypothesis 4 can be rejected at the 0.01 level.

9. The P-value needs to be adjusted to $20(0.01) = 0.2$ because 20 tests were performed. The adjusted P-value is greater than 0.05, so the conclusion is not justified.

10. The P-value needs to be adjusted to $6(0.002) = 0.012$ because six tests were performed. The adjusted P-value is less than 0.05, so the conclusion is justified.

11. 0.0025

12. 0.008333

13. No. Several tests have been performed, so we cannot interpret the result in the way that we would if only one test had been performed.

14. The P-value needs to be adjusted to $6(0.03) = 0.18$. The adjusted P-value is greater than 0.05, so we cannot reject H_0 at the $\alpha = 0.05$ level.

15. The original P-value must be $0.05/20 = 0.0025$.

16. **a.** 0.424, 0.0092, 0.362, 0.0035, 0.104 **b.** i

17. **a.** 0.05

b. Binomial with 500 trials and success probability 0.05

c. 0.002701 (exact), 0.0014 using normal approximation

d. Yes, because the P-value for the hypothesis that all the null hypotheses are true is very small.

CHAPTER 11 Quiz

1. ii

2. iv

3. i

4. iii

5. Do not reject H_0. The difference $p_1 - p_2$ may be equal to 0.

6. No, the test statistic is $F = 3.199$ and the critical value is 3.21.

7. 14

8. Test statistic is 2.598. Critical value is 1.796, P-value is between 0.01 and 0.025 [Tech: 0.0124]. Reject H_0.

9. $H_0: p_1 = p_2, H_1: p_1 < p_2$

10. $z = -2.72$

11. Critical value: −2.326, P-value is 0.0033. Reject H_0.

12. $H_0: \mu_1 = \mu_2, H_1: \mu_1 \neq \mu_2$

13. $t = -1.971$

14. Critical values: −2.120, 2.120, P-value is between 0.05 and 0.10 [Tech: 0.0662] [TI-84 Plus: 0.0576] [MINITAB: 0.0577]. Do not reject H_0.

15. One

CHAPTER 11 Review Exercises

1. $H_0: \mu_1 = \mu_2, H_1: \mu_1 < \mu_2$. Test statistic: $t = -2.409$, Critical value: −2.374, P-value is between 0.005 and 0.01 [Tech: 0.0089] [TI-84 Plus: 0.0085] [MINITAB: 0.0085]. Reject H_0. We conclude that the mean number of days missed is less with flextime.

2. $H_0: p_1 = p_2, H_1: p_1 > p_2$. Test statistic: $z = 2.04$, Critical value: 1.645, P-value is 0.0207 [Tech: 0.0206]. Reject H_0. We conclude that the proportion of voters who favor the proposal is greater in county A than in county B.

3. $H_0: \sigma_1 = \sigma_2, H_1: \sigma_1 \neq \sigma_2$, Test statistic: $F = 1.4745$, Critical value: 9.20. Do not reject H_0. There is not enough evidence to conclude that the standard deviation of benzene concentration differs between treated water and untreated water.

4. $H_0: \mu_d = 0, H_1: \mu_d \neq 0$. Test statistic: $t = -2.945$, Critical values: −2.306, 2.306, P-value is between 0.01 and 0.02 [Tech: 0.0186]. Reject H_0. We conclude that the mean sales differ between the two programs.

5. $H_0: \sigma_1 = \sigma_2, H_1: \sigma_1 \neq \sigma_2$, Test statistic: $F = 1.3514$, Critical value: 5.12. Do not reject H_0. There is not enough evidence to conclude that the standard deviation of distances walked differs between treatment and control.

6. $H_0: \mu_1 = \mu_2, H_1: \mu_1 \neq \mu_2$, Test statistic: $t = -0.306$, Critical values: −2.000, 2.000, P-value is between 0.50 and 0.80 [Tech: 0.7606] [TI-84 Plus: $P = 0.7601$] [MINITAB: $P = 0.7601$]. Do not reject H_0. There is not enough evidence to conclude that the mean number of hours differs between men and women.

7. $H_0: \mu_d = 0, H_1: \mu_d \neq 0$. Test statistic: $t = -3.145$, Critical values: −2.201, 2.201, P-value is between 0.005 and 0.01 [Tech: 0.0093]. Reject H_0. We conclude that the mean time differs between the two applications.

8. **a.** $H_0: p_1 = p_2, H_1: p_1 < p_2$
   b. Test statistic: $z = -1.04$, Critical value: −1.645, P-value is 0.1492 [Tech: 0.1501]. There is not enough evidence to conclude that machine B produces a larger proportion of rods meeting specifications. Machine A should be used.

9. **a.** Right-tailed **b.** 21.18819537 **c.** 0.0012913382
   d. Yes, because the P-value is less than 0.05.

10. **a.** Left-tailed **b.** 99 **c.** 0.161
   d. No, because the P-value is greater than 0.05.

11. **a.** Two-tailed **b.** 0.2041558545
   c. No, because the P-value is greater than 0.05.

12. **a.** Right-tailed **b.** 0.003
   c. Yes, because the P-value is less than 0.05.

13. **a.** Right-tailed **b.** 14 **c.** 0.0021327513
   d. Yes, because the P-value is less than 0.05.

14. **a.** Two-tailed **b.** 11 **c.** 0.012
   d. Yes, because the P-value is less than 0.05.

15. iii

CHAPTER 11 Case Study

1. The P-values are 0.1164 [Tech: 0.1174], 0.2628 [Tech: 0.2640], 0.0052, 0.0802 [Tech: 0.0795], 0.0258.

2. Proportion with underground wiring and proportion with father Hispanic are significant at the $\alpha = 0.05$ level.

3. For proportion with underground wiring: 0.0260 [Tech: 0.0258]; for proportion with father Hispanic: 0.129.

4. Answers will vary. The evidence is strong only for underground wiring.

CHAPTER 12

Section 12.1

Exercises 1–8 are the Check Your Understanding exercises for this section. Answers to these exercises are on page 597.

9. expected

10. observed

11. False

12. True

13. 23.685

14. 15.086

15. 0.01

16. 0.10

17. **a.** 10.125 **b.** 4 **c.** Critical value is 9.488. Reject H_0.

18. **a.** 1.723 **b.** 5
   c. Critical value is 15.086. Do not reject H_0.

19. **a.**

Category	1	2	3	4
Expected	100	60	30	10

   b. 3.427 **c.** 3
   d. Critical value is 11.345. Do not reject H_0.

20. **a.**

Category	1	2	3	4	5
Expected	60	37.5	7.5	15	30

   b. 14.393 **c.** 4 **d.** Critical value is 9.488. Reject H_0.

21. **a.** Yes. The expected values are all greater than or equal to 5.
   b. Test statistic: 4.0496; 4 degrees of freedom; Critical value: 9.488. Do not reject H_0. There is not enough evidence to conclude that the distribution of the observed frequencies differs from that given by the expected frequencies.

22. **a.** No. One of the expected values is less than 5.
   b. The test is not valid.

23. **a.** No. One of the expected values is less than 5.
   b. The test is not valid.

24. **a.** Yes. The expected values are all greater than or equal to 5.
   b. Test statistic: 1.7333; 5 degrees of freedom; Critical value: 15.086. Do not reject H_0. There is not enough evidence to conclude that the distribution of the observed frequencies differs from that given by the expected frequencies.

25. a.

Category	1–3	4–6	7–9	10–12	13–15
Expected	68.8	68.8	68.8	68.8	68.8

b. 1.7267
c. 4
d. Critical value is 9.488. Do not reject H_0. There is not enough evidence to conclude that the lottery is unfair.

26. a. 29 got an A; 4 got an F.
b.

Grade	A	B	C	D	F
Expected	20	35	25	10	10

c. A and B were given out more often than expected; C, D, and F were given out less often.
d. 12.550 **e.** 4
f. Critical value is 9.488. Reject H_0. We conclude that the distribution of grades differs from what is claimed by the teacher.

27. Test statistic: 27.792; 11 degrees of freedom; Critical value: 24.725. Reject H_0. We conclude that fire alarms are more likely in some months than in others.

28. Test statistic: 6.027; 3 degrees of freedom; Critical value: 7.815. Do not reject H_0. There is not enough evidence to conclude that the proportions of crimes in the various categories in California differ from the United States as a whole.

29. Test statistic: 0.1533; 3 degrees of freedom; Critical value: 7.815. Do not reject H_0. There is not enough evidence to conclude that the proportions of people living in the various regions changed between 2010 and 2017.

30. Test statistic: 17.295; 3 degrees of freedom; Critical value: 11.345. Reject H_0. We conclude that the proportions of people giving the various responses changed between 2018 and 2019.

31. Test statistic: 18.747; 3 degrees of freedom; Critical value: 11.345. Reject H_0. We conclude that the proportions of people giving the various responses changed between 2018 and 2019.

32. Test statistic: 134.98; 4 degrees of freedom; Critical value: 9.488. Reject H_0. We conclude that students did not answer the questions by guessing.

33. a. Test statistic: 7.120; 5 degrees of freedom; Critical value: 11.070. Do not reject H_0. There is not enough evidence to conclude that the die is not fair.
b. 1: 0.1556 [Tech: 0.1544]; 2: 0.9124 [Tech: 0.9128]; 3: 0.5092 [Tech: 0.5110]; 4: 0.0376 [Tech: 0.0374]; 5: 0.3788 [Tech: 0.3808]; 6: 0.3222 [Tech: 0.3242].
c. The P-value is less than 0.05, so we reject H_0 at the $\alpha = 0.05$ level.
d. The P-value must be adjusted for the fact that multiple tests have been performed.
e. The adjusted P-value is $6(0.0376) = 0.2256$ [Tech: $6(0.0374) = 0.2244$]. The adjusted P-value is greater than 0.05, so we do not reject H_0 at the $\alpha = 0.05$ level.

Section 12.2

Exercises 1 and 2 are the Check Your Understanding exercises for this section. Answers to these exercises are on page 607.

3. grand
4. greater than or equal to
5. homogeneity
6. False
7. False
8. True

9. a. Row totals: 37, 25, 35; Column totals: 27, 35, 35; Grand total: 97
b.

	1	2	3
A	10.299	13.351	13.351
B	6.959	9.021	9.021
C	9.742	12.629	12.629

c. 6.481 **d.** 4
e. Critical value: 9.488. Do not reject H_0.

10. a. Row totals: 40, 10, 50; Column totals: 70, 10, 20; Grand total: 100
b.

	1	2	3
A	28	4	8
B	7	1	2
C	35	5	10

c. 14.432 **d.** 4
e. It is not appropriate, because some of the expected values are less than 5.

11. a.

	Shift		
	Morning	Evening	Night
Influenza	16.323	19.982	10.695
Headache	21.880	26.784	14.335
Weakness	11.114	13.605	7.281
Shortness of Breath	8.683	10.629	5.689

b. 17.572 **c.** 6
d. Critical value: 12.592. Reject H_0. We conclude that occurrence of symptoms and shift are not independent. The frequencies of the symptoms vary among the shifts.

12. a.

	Duration of Exposure		
	< 1	1 to < 5	≥ 5
Diseased	7.417	13.583	20.000
Sensitized	7.055	12.921	19.024
Normal	74.528	136.496	200.976

b. 10.829 **c.** 4
d. Critical value: 13.277. Do not reject H_0. There is not enough evidence to conclude that duration of exposure and disease status are not independent.

13. a.

	Household Size				
	1	2	3	4	5
Agree	66.443	105.581	52.335	44.144	35.497
No Opinion	32.784	52.096	25.823	21.781	17.515
Disagree	46.772	74.323	36.841	31.075	24.988

b. 6.377 **c.** 8
d. Critical value: 20.090. Do not reject H_0. There is not enough evidence to conclude that household size and opinion are not independent.

14. a.

	Number of Children					
Siblings	0	1	2	3	4	More Than 4
0	39.626	24.501	37.761	23.776	11.396	6.941
1	152.174	94.090	145.013	91.305	43.763	26.655
2	153.000	94.600	145.800	91.800	44.000	26.800
3	129.610	80.138	123.510	77.766	37.273	22.703
4	93.836	58.019	89.421	56.302	26.986	16.437
More Than 4	196.754	121.653	187.495	118.052	56.583	34.464

b. 126.444 **c.** 25
d. Critical value: 37.652. Reject H_0. We conclude that the number of siblings and the number of children are not independent.

15. Test statistic: $\chi^2 = 1.485$; Critical value: 7.815. Do not reject H_0. There is not enough evidence to conclude that people in some age groups are more likely to be promoted than those in other age groups.

16. Test statistic: 12.862; Critical value: 13.277. Do not reject H_0. There is not enough evidence to conclude that the number of visits to a science museum is related to interest in environmental issues.

17. a. 22 **b.** 39.86 **c.** C2 **d.** C3 **e.** 0.301
f. No, because the P-value is greater than 0.05.
g. No, we cannot conclude that the null hypothesis is true.

18. a. 136 **b.** 165.13 **c.** Frequently used and fair
d. Rarely used and fair **e.** 0.012
f. Yes, because the P-value is less than 0.05.

19.

	1	2	3	4	Row Total
A	10	52	29	7	98
B	25	38	10	19	92
C	30	27	44	46	147
Column Total	65	117	83	72	

20. a. 0.96455 **b.** 0.836 **c.** 0.821 **d.** 0.982115
e. $0.982115^2 = 0.96455$
f. P-values are 0.3260 for both tests.
g. When a contingency table has two rows and two columns, the chi-square test is equivalent to the test for the difference between proportions.

CHAPTER 12 Quiz

1. Yes. There are 12 degrees of freedom, and the critical value is 21.026.

2. False

3. True

4. H_0: The failure probabilities are the same for each line. H_1: The failure probabilities are not the same for all the lines.

5.

	Line			
	1	2	3	4
Pass	485.395	473.687	445.771	406.147
Fail	53.605	52.312	49.229	44.853

6. 4.676

7. 3

8. 7.815

9. Do not reject H_0. There is not enough evidence to conclude that the failure probabilities are not all the same.

10. H_0: The age distribution is the same at each site. H_1: The age distributions are not the same at all sites.

11.

	Ages of Skeletons		
Site	0–4 Years	5–19 Years	20 Years or More
Casa da Moura	32.349	60.219	121.433
Wandersleben	32.651	60.781	122.567

12. 2.123

13. 2

14. 9.210

15. Do not reject H_0. There is not enough evidence to conclude that the age distributions are not all the same.

CHAPTER 12 Review Exercises

1. H_0: $p_0 = 0.0625, p_1 = 0.25, p_2 = 0.375, p_3 = 0.25, p_4 = 0.0625$

2.

Number of Heads	0	1	2	3	4
Expected	12.5	50	75	50	12.5

3. Test statistic: $\chi^2 = 5.193$; Critical value: 9.488. Do not reject H_0. There is not enough evidence to conclude that the number of heads does not follow a binomial distribution.

4.

	Educational Level				
	No High School Diploma	High School Diploma	Associate's Degree	Bachelor's Degree	Graduate Degree
Men	165.749	653.433	101.999	230.864	127.954
Women	198.251	781.567	122.001	276.136	153.046

5. 17.419

6. Critical value is 13.277. Reject H_0. We conclude that education level and gender are not independent.

7.

	Number of Welds		
	High Quality	Moderate Quality	Low Quality
Day Shift	466.400	196.400	37.200
Evening Shift	433.086	182.371	34.543
Night Shift	266.514	112.229	21.257

8. 5.760

9. Critical value is 13.277. Do not reject H_0. There is not enough evidence to conclude that the quality varies among shifts.

10.

	Outcome		
Hospital	Substantial Improvement	Some Improvement	No Improvement
A	118	59	23
B	118	59	23
C	118	59	23
D	118	59	23

11. 17.524

12. Critical value is 12.592. Reject H_0. We conclude that the distribution of outcomes varies among the hospitals.

13.

	1	2	3	Total
A	8.333	3.333	13.333	25
B	3.333	1.333	5.333	10
C	13.333	5.333	21.333	40
D	25.000	10.000	40.000	75
Total	50	20	80	150

14. Test statistic: $\chi^2 = 72.544$; Critical value: 101.879. Do not reject H_0. There is not enough evidence to conclude that the numbers are not equally likely to come up.

15. Test statistic: $\chi^2 = 12.075$; Critical value: 9.488. Reject H_0. We conclude that absences are not equally likely on each day of the week.

CHAPTER 12 Case Study

1. Test statistic: 92.205; Critical value: 6.635. Reject H_0. We conclude that acceptance rates differ between men and women.

2. Test statistic: 778.907; Critical value: 15.086. Reject H_0. We conclude that acceptance rates differ among departments.

3. Test statistic: 1068.372; Critical value: 15.086. Reject H_0. We conclude that gender and department are not independent.

4. The critical value in each case is 6.635. A: Test statistic is 17.248. Reject H_0. B: Test statistic is 0.254. Do not reject H_0. C: Test statistic is 0.754. Do not reject H_0. D: Test statistic is 0.298. Do not reject H_0. E: Test statistic is 1.001. Do not reject H_0. F: Test statistic is 0.384. Do not reject H_0. In department A, 82.4% of the women were accepted, but only 62.1% of the men were accepted.

5. A: 65.4%; B: 63.2%; C: 35.1%; D: 34.0%; E: 25.2%; F: 6.4%

6. A: 88.4%; B: 95.7%; C: 35.4%; D: 52.7%; E: 32.7%; F: 52.2%

7. Yes

8. ii

CHAPTER 13

Section 13.1

Exercises 1–6 are the Check Your Understanding exercises for this section. Answers to these exercises are on page 629.

7. 18

8. straight line

9. False

10. True

11. **a.** 28 **b.** 2.048 **c.** 2.073 **d.** $-0.861 < \beta_1 < 3.285$

12. **a.** 18 **b.** 2.878 **c.** 1.231 **d.** $0.389 < \beta_1 < 2.851$

13. Test statistic: $t = 1.1975$, Critical values: −2.763, 2.763, P-value is between 0.20 and 0.50 [Tech: 0.2411]. Do not reject H_0.

14. Test statistic: $t = 3.7886$, Critical value: 1.734, P-value is between 0.0005 and 0.001 [Tech: 0.00067]. Reject H_0.

15. **a.** 3.9724 **b.** 1.1315 **c.** 277.33 **d.** 0.067946

e. 2.776 **f.** 0.1886 **g.** $3.7837 < \beta_1 < 4.161$

h. Test statistic: $t = 58.464$, Critical values: −2.776, 2.776, P-value is less than 0.001 [Tech: 0.00000051]. Reject H_0.

16. **a.** −0.15403 **b.** 0.71672 **c.** 159.88 **d.** 0.056684

e. 2.447 **f.** 0.1387 **g.** $-0.2927 < \beta_1 < -0.0153$

h. Test statistic: $t = -2.717$, Critical values: −2.447, 2.447, P-value is between 0.02 and 0.05 [Tech: 0.0347]. Reject H_0.

17. **a.** 3.1236 **b.** 0.19353 **c.** 71.2 **d.** 0.022935

e. 3.182 **f.** 0.0730 **g.** $3.0506 < \beta_1 < 3.1966$

h. Test statistic: $t = 136.19$, Critical values: −3.182, 3.182, P-value is less than 0.001 [Tech: 0.00000087]. Reject H_0.

18. **a.** −0.0625 **b.** 1.9843 **c.** 6.8571 **d.** 0.75777

e. 2.571 **f.** 1.9482 [Tech: 1.9479]

g. $-2.0107 < \beta_1 < 1.8857$ [Tech: $-2.0104 < \beta_1 < 1.8854$]

h. Test statistic: $t = -0.0825$, Critical values: −2.571, 2.571, P-value is greater than 0.80 [Tech: 0.9375]. Do not reject H_0.

19. **a.** $\hat{y} = 302.5212 - 0.5082x$

b. $-7.8848 < \beta_1 < 6.8684$ [Tech: $-7.8845 < \beta_1 < 6.8680$]

c. Test statistic: $t = -0.1461$, Critical values: −2.120, 2.120, P-value is greater than 0.80 [Tech: 0.8857]. Do not reject H_0. There is not enough evidence to conclude that the amount of protein is useful in predicting the number of calories.

20. **a.** $\hat{y} = 19.4093 + 0.7370x$ **b.** $0.3527 < \beta_1 < 1.1214$

c. Test statistic: $t = 4.1133$, Critical values: −2.145, 2.145, P-value is between 0.001 and 0.002 [Tech: 0.0011]. Reject H_0. We conclude that the father's height is useful in predicting the son's height.

21. **a.** $\hat{y} = 52.0434 - 0.8445x$

b. $-2.0322 < \beta_1 < 0.3432$ [Tech: $-2.0324 < \beta_1 < 0.3434$]

c. Test statistic: $t = -2.0229$, Critical value: −1.725, P-value is between 0.025 and 0.05 [Tech: 0.0283]. Reject H_0. We conclude that wingspan is useful in predicting lifespan.

d. Shorter, because we conclude that $\beta_1 < 0$.

22. **a.** $\hat{y} = 9.1828 + 0.5748x$

b. $0.2995 < \beta_1 < 0.8500$

c. Test statistic: $t = 6.2166$, Critical value: 2.624, P-value is less than 0.0005 [Tech: 0.000011]. Reject H_0. We conclude that systolic blood pressure is useful in predicting the diastolic blood pressure.

d. Higher, because we conclude that $\beta_1 > 0$.

23. **a.** $\hat{y} = 71.4363 + 0.2392x$

b. $0.1669 < \beta_1 < 0.3116$

c. Test statistic: $t = 8.0886$, Critical values: −3.707, 3.707, P-value is less than 0.001 [Tech: 0.00019]. Reject H_0. We conclude that speed is useful in predicting noise level.

24. **a.** $\hat{y} = 22.3600 + 0.8638x$

b. $0.3647 < \beta_1 < 1.3629$

c. Test statistic: $t = 3.991$, Critical values: −3.355, 3.355, P-value is between 0.002 and 0.005 [Tech: 0.0040]. Reject H_0. We conclude that visual response is useful in predicting auditory response.

25. **a.** $\hat{y} = 73.2662 - 0.4968x$

b. $-0.9521 < \beta_1 < -0.0415$ [Tech: $-0.9520 < \beta_1 < -0.0416$]

c. Test statistic: $t = -2.5806$, Critical values: −2.365, 2.365, P-value is between 0.02 and 0.05 [Tech: 0.0364]. Reject H_0. We conclude that horizontal expansion is useful in predicting vertical expansion.

26. **a.** $\hat{y} = 1.4314 + 0.1446x$

b. $-0.2163 < \beta_1 < 0.5056$ [Tech: $-0.2164 < \beta_1 < 0.5057$]

c. Test statistic: $t = 1.3443$, Critical values: −2.306, 2.306, P-value is between 0.20 and 0.50 [Tech: 0.2157]. Do not reject H_0. There is not enough evidence to conclude that temperature is useful in predicting evaporation.

27. **a.** $\beta_1 \neq 0$ **b.** 2.60388259 **c.** 6 **d.** 0.0404509768

e. Yes, because the P-value is less than 0.05.

28. **a.** $\beta_1 < 0$ **b.** −1.194401785 **c.** 6 **d.** 0.138695956

e. No, because the P-value is greater than 0.05.

29. **a.** Slope: −0.7524; Intercept: 88.761. **b.** Yes

30. **a.** Slope: 1.2499; Intercept: 162.15. **b.** Yes

31. $24.091 < \mu_{y|20} - \mu_{y|15} < 32.3025$

Section 13.2

Exercises 1 and 2 are the Check Your Understanding exercises for this section. Answers to these exercises are on page 635.

3. confidence
4. prediction
5. False
6. True
7. **a.** 6.35 < Mean response < 9.43 **b.** 0.43 < Individual response < 15.35
8. **a.** −4.59 < Mean response < 3.29 **b.** −17.82 < Individual response < 16.52
9. **a.** $b_0 = -12.2183, b_1 = 2.7919$ **b.** 21.284 **c.** 3.7885 **d.** 32.833 **e.** 2.776 **f.** 15.43 < Mean response < 27.14 **g.** 9.24 < Individual response < 33.32
10. **a.** $b_0 = 31.4955, b_1 = 0.6925$ **b.** 48.809 **c.** 17.753 **d.** 178.88 **e.** 2.447 **f.** 28.95 < Mean response < 68.67 **g.** 1.04 < Individual response < 96.57
11. **a.** 294.9 **b.** 245.69 < Mean response < 344.1 **c.** 294.9 **d.** 108.17 < Individual response < 481.63
12. **a.** 71.002 **b.** 69.855 < Mean response < 72.149 **c.** 71.002 **d.** 66.435 < Individual response < 75.57
13. **a.** 26.708 **b.** 23.367 < Mean response < 30.05 **c.** 26.708 **d.** 12.247 < Individual response < 41.17
14. **a.** 78.15 **b.** 75.099 < Mean response < 81.208 **c.** 78.15 **d.** 65.674 < Individual response < 90.634
15. **a.** 79.81 **b.** 79.29 < Mean response < 80.33 **c.** 79.81 **d.** 78.588 < Individual response < 81.032
16. **a.** 195.12 **b.** 178.97 < Mean response < 211.27 **c.** 195.12 **d.** 141.94 < Individual response < 248.3
17. **a.** 60.847 **b.** 53.103 < Mean response < 68.59 **c.** 60.847 **d.** 37.813 < Individual response < 83.88
18. **a.** 4.3244 **b.** 2.5604 < Mean response < 6.0884 **c.** 4.3244 **d.** −1.4665 < Individual response < 10.115
19. **a.** 43.62 **b.** 41.23 < Mean response < 46.00 **c.** 43.62 **d.** No, because the prediction interval is 20.86 < Individual response < 66.37.
20. **a.** 224.64 **b.** 211.40 < Mean response < 237.89 **c.** 224.64 **d.** No, because the prediction interval is 182.66 < Individual response < 266.42.
21. **a.** 9 **b.** 14 **c.** 10

Section 13.3

Exercises 1–6 are the Check Your Understanding exercises for this section. Answers to these exercises are on page 652.

7. residual
8. increases
9. True
10. False
11. **a.** $\hat{y} = 10.372 + 0.5215x_1 + 1.4426x_2 + 2.8834x_3$ **b.** 35.26 **c.** 94.1% **d.** Yes, $F = 31.98, P = 0.0004$. **e.** P-value for β_1 is 0.3531; do not reject H_0. P-value for β_2 is 0.0298; reject H_0. P-value for β_3 is 0.0004; reject H_0.
12. **a.** $\hat{y} = 7.502 + 0.3838x_1 + 0.2417x_2 + 0.5617x_3$ **b.** 28.15 **c.** 69.6% **d.** Yes, $F = 6.09, P = 0.0184$. **e.** P-value for β_1 is 0.0560; do not reject H_0. P-value for β_2 is 0.2459; do not reject H_0. P-value for β_3 is 0.0144; reject H_0.
13. **a.** $\hat{y} = -1.9446 + 1.1325x_1 + 3.0963x_2 + 0.4224x_3 + 1.2384x_4$ **b.** 39.263 **c.** 72.0% **d.** Yes, $F = 6.42, P = 0.0080$. **e.** P-value for β_1 is 0.0286; reject H_0. P-value for β_2 is 0.0007; reject H_0. P-value for β_3 is 0.4382; do not reject H_0. P-value for β_4 is 0.0162; reject H_0.
14. **a.** $\hat{y} = 63.14 + 0.532x_1 + 1.3975x_2 + 2.5277x_3 + 0.2046x_4$ **b.** 81.042 **c.** 33.7% **d.** No, $F = 1.27, P = 0.3438$. **e.** P-value for β_1 is 0.6313; do not reject H_0. P-value for β_2 is 0.2762; do not reject H_0. P-value for β_3 is 0.1059; do not reject H_0. P-value for β_4 is 0.8643; do not reject H_0.
15. **a.** 166.475 **b.** 113.46 **c.** Yes, $F = 15.469, P = 0.000$. **d.** 51.6% **e.** $P = 0.000$, reject H_0. **f.** $P = 0.013$, do not reject H_0.
16. **a.** 4.4242 **b.** 2.3504 **c.** Yes, $F = 51.689, P = 0.000$. **d.** 29.1% **e.** $P = 0.000$, reject H_0. **f.** $P = 0.797$, do not reject H_0.
17. ii
18. iii
19. **a.** $\hat{y} = 0.1210 + 0.5408x_1 + 0.002146x_2 + 0.0002816x_3$ **b.** 3.01 **c.** (2.8720, 3.1483) **d.** (2.4147, 3.6056) **e.** 72.5% **f.** Yes, $F = 14.04, P = 0.000095$. **g.** P-value for β_1 is 0.0075; reject H_0. P-value for β_2 is 0.1445; do not reject H_0. P-value for β_3 is 0.7475; do not reject H_0.
20. **a.** $\hat{y} = 29.292 + 0.031527x_1 - 0.20046x_2 - 0.084195x_3$ **b.** 15.183 years **c.** (13.343, 17.023) **d.** (11.609, 18.757) **e.** 77.1% **f.** Yes, $F = 7.85, P = 0.0124$. **g.** P-value for β_1 is 0.3308; do not reject H_0. P-value for β_2 is 0.0048; reject H_0. P-value for β_3 is 0.0209; reject H_0.
21. **a.** $\hat{y} = 8.8721 - 0.006795x_1 + 0.71632x_2 + 2.030x_3$ **b.** 42.202 **c.** (36.934, 47.470) **d.** (32.888, 51.517) **e.** 70.2% **f.** Yes, $F = 34.63, P = 0.00000000012$. **g.** P-value for β_1 is 0.9536; do not reject H_0. P-value for β_2 is 0.0078; reject H_0. P-value for β_3 is 0.000000026; reject H_0.
22. **a.** $\hat{y} = 120.15 + 0.89597x_1 - 0.74088x_2$ **b.** 70.979 **c.** (67.875, 74.084) **d.** (50.880, 91.079) **e.** 47.2% **f.** Yes, $F = 25.48, P = 0.00000012$. **g.** P-value for β_1 is 0.7067; do not reject H_0. P-value for β_2 is 0.00012; reject H_0.
23. **a.** $\hat{y} = 524.7948 + 0.0022154x_1 - 113.3986x_2 + 2.8074x_3 + 11.8733x_4$ **b.** 759.55 **c.** (750.368, 768.751) **d.** (741.385, 777.734) **e.** 90.0% **f.** Yes, $F = 45.16, P = 0.00000000097$. **g.** P-value for β_1 is 0.2202; do not reject H_0. P-value for β_2 is 0.00099; reject H_0. P-value for β_3 is 0.0056; reject H_0. P-value for β_4 is 0.3489; do not reject H_0.

CHAPTER 13 Quiz

1. 18
2. True
3. False

4. False
5. (65.3, 68.2) must be the confidence interval, because it is narrower.
6. $b_0 = 13.0508$, $b_1 = 1.0574$
7. $0.2247 < \beta_1 < 1.8902$
8. Test statistic: 3.107, Critical values: −3.707, 3.707, P-value is between 0.02 and 0.05 [Tech: 0.0209]. Do not reject H_0.
9. 29.195 < Mean response < 39.205
10. 19.327 < Individual response < 49.073
11. $\hat{y} = -13.542 + 0.204x_1 + 2.056x_2 + 0.181x_3$
12. 37.09
13. 78.2%
14. Yes, $F = 9.59$, $P = 0.0050$.
15. P-value for β_1 is 0.6397; do not reject H_0. P-value for β_2 is 0.00083; reject H_0. P-value for β_3 is 0.5681; do not reject H_0.

CHAPTER 13 Review Exercises

1. **a.** $\hat{y} = 8.5593 - 0.1551x$
   b. $-0.217 < \beta_1 < -0.093$
   c. Test statistic: $t = -6.437$, Critical value: −2.015, P-value is between 0.0005 and 0.001 [Tech: 0.00067]. Reject H_0. We conclude that weight is useful in predicting mileage.
2. **a.** 5.4567 **b.** 5.0064 < Mean response < 5.907
   c. 5.4567 **d.** 4.1857 < Individual response < 6.7278
3. **a.** $\hat{y} = -1.0123 + 0.0519x$ **b.** $0.0389 < \beta_1 < 0.0650$
   c. Test statistic: $t = 8.8546$, Critical value: 2.764, P-value is less than 0.0005 [Tech: 0.0000024]. Reject H_0. We conclude that diameter is useful in predicting volume.
4. **a.** 1.2733 **b.** 1.1224 < Mean response < 1.4241
   c. 1.2733 **d.** 0.8631 < Individual response < 1.6834
5. **a.** $\hat{y} = 12.1933 - 0.8333x$ **b.** $-1.078 < \beta_1 < -0.589$
   c. Test statistic: $t = -7.8523$, Critical value: −1.860, P-value is less than 0.0005 [Tech: 0.000025]. Reject H_0. We conclude that concentration is useful in predicting drying time.
6. **a.** 7.9433 **b.** 7.7945 < Mean response < 8.0922
   c. 7.9433 **d.** 7.4745 < Individual response < 8.4122
7. **a.** $\hat{y} = 2.9073 + 0.8882x$ **b.** $0.677 < \beta_1 < 1.100$
   c. Test statistic: $t = 9.6778$, Critical value: 2.896, P-value is less than 0.0005 [Tech: 0.0000054]. Reject H_0. We conclude that income is useful in predicting energy consumption.
8. **a.** 47.319 **b.** 34.48 < Mean response < 60.158
   c. 47.319 **d.** 10.954 < Individual response < 83.685
9. **a.** $\beta_1 \neq 0$ **b.** 3.461106178 **c.** 8
   d. 0.0085538598 **e.** Yes, $P < 0.01$.
10. **a.** Slope is 5.5582; Intercept is 1.9167.
    b. Yes, $P = 0.006$.
11. **a.** $\hat{y} = -42.354 + 0.65394x_1 + 3.662x_2 + 0.57857x_3$
    b. 19.47 **c.** (13.44, 25.50) **d.** (−4.55, 43.50)
    e. No, the particulate concentration cannot be negative.
    f. 53.8% **g.** Yes, $F = 13.20$, $P = 0.0000071$.
    h. P-value for β_1 is 0.0246; reject H_0. P-value for β_2 is 0.00000074; reject H_0. P-value for β_3 is 0.0322; reject H_0.
12. **a.** $\hat{y} = -372.98 + 3.5368x_1 + 3.7345x_2 - 2.1661x_3$
    b. 701.76 **c.** (656.34, 747.18) **d.** (602.97, 800.55)
    e. 72.9% **f.** Yes, $F = 8.06$, $P = 0.0064$.
    g. P-value for β_1 is 0.0160; reject H_0. P-value for β_2 is 0.0082; reject H_0. P-value for β_3 is 0.0316; reject H_0.
13. **a.** 40.0% **b.** 31.0% **c.** Yes, $F = 4.44$, $P = 0.015$
14. **a.** 10.933 **b.** 6.805, 15.061 **c.** 0.6874, 21.178
15. ii

CHAPTER 13 Case Study

1. $\hat{y} = -26.373 + 9.2915x$
2. 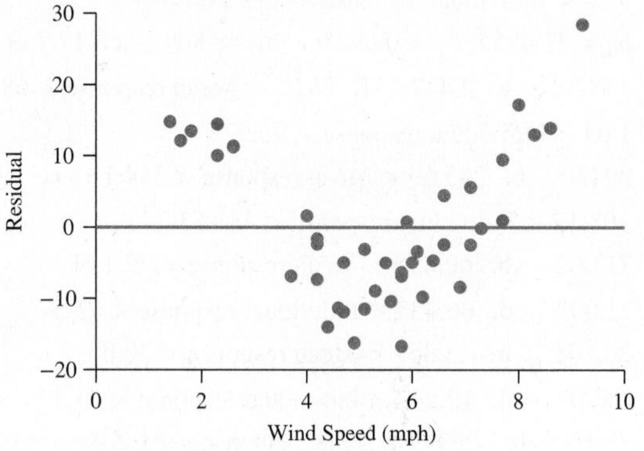 The plot exhibits a clear pattern, which indicates that the assumptions of the linear model are violated.
3. $\hat{y} = -0.01502 + 0.110265x^3$
4. 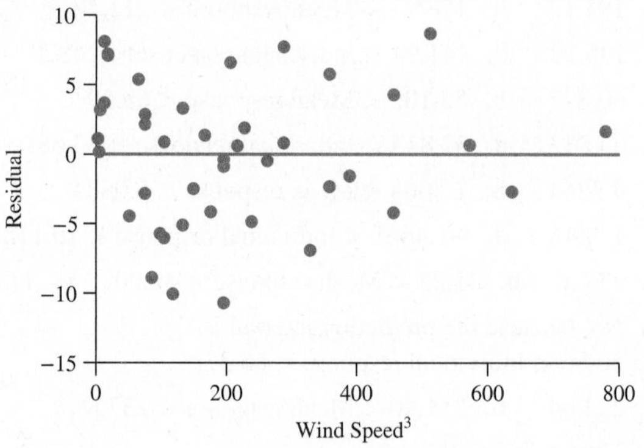 The assumptions of the linear model appear to be satisfied.
5. 13.768
6. 12.023 < Mean response < 15.513
7. 3.363 < Individual response < 24.173

CHAPTER 14

Section 14.1

Exercises 1–6 are the Check Your Understanding exercises for this section. Answers to these exercises are on page 677.

7. equal
8. unbalanced
9. True
10. False

11. True

12. False

13. a. 4 **b.** 3 for *SSTr*, 30 for *SSE*

c. $MSTr = 0.08333$, $MSE = 0.0703$ **d.** 1.1848

e. Critical value is 2.92, *P*-value is greater than 0.10 [Tech: 0.3321]. Do not reject H_0. There is not enough evidence to conclude that two or more of the population means are different.

14. a. 3 **b.** 2 for *SSTr*, 6 for *SSE*

c. $MSTr = 72.67$, $MSE = 6.4083$ **d.** 11.3399

e. Critical value is 10.92, *P*-value is between 0.001 and 0.01 [Tech: 0.0092]. Reject H_0. We conclude that two or more of the population means are different.

15. a. $SSTr = 3.1511$, $SSE = 2.2868$
b. 2 for *SSTr*, 20 for *SSE*

c. $MSTr = 1.5755$, $MSE = 0.11434$ **d.** 13.779

e. Critical value is 5.85, *P*-value is less than 0.001 [Tech: 0.00017]. Reject H_0. We conclude that two or more of the population means are different.

16. a. $SSTr = 98.524$, $SSE = 373.14$
b. 3 for *SSTr*, 30 for *SSE*

c. $MSTr = 32.841$, $MSE = 12.438$ **d.** 2.6404

e. Critical value is 2.92, *P*-value is between 0.05 and 0.10 [Tech: 0.0675]. Do not reject H_0. There is not enough evidence to conclude that two or more of the population means are different.

17. a.

Source	DF	SS	MS	*F*	*P*
Duration	4	1.328	0.332000	8.5862	0.00082
Error	15	0.580	0.038667		
Total	19	1.908			

b. Yes. $F = 8.5682$, $P = 0.00082$.

18. a.

Source	DF	SS	MS	*F*	*P*
Day	2	1.0908	0.54538	22.35	0.00000049
Error	36	0.8785	0.02440		
Total	38	1.9692			

b. Yes. $F = 22.35$, $P = 0.00000049$.

19. $\mu_1 \neq \mu_4$, $\mu_1 \neq \mu_{24}$, $\mu_2 \neq \mu_{24}$

20. $\mu_1 \neq \mu_2$, $\mu_1 \neq \mu_3$, $\mu_2 \neq \mu_3$

21. a.

Source	DF	SS	MS	*F*	*P*
Fertilizer	2	524.16	262.08	12.963	0.00054
Error	15	303.26	20.217		
Total	17	827.42			

b. Yes. $F = 12.963$, $P = 0.00054$.

22. a.

Source	DF	SS	MS	*F*	*P*
Collector	3	1.6135	0.53783	17.349	0.000028
Error	16	0.4960	0.03100		
Total	19	2.1095			

b. Yes. $F = 17.349$, $P = 0.000028$.

23. $\mu_1 \neq \mu_2$, $\mu_1 \neq \mu_3$

24. $\mu_1 \neq \mu_3$, $\mu_1 \neq \mu_4$, $\mu_2 \neq \mu_3$, $\mu_2 \neq \mu_4$

25. a. The three standard deviations are $s_{16} = 1.2309$, $s_{28} = 0.7031$, and $s_{36} = 0.9585$. The largest is less than twice as large as the smallest.

b.

Source	DF	SS	MS	*F*	*P*
Day	2	24.001	12.000	10.803	0.0053
Error	8	8.8867	1.1108		
Total	10	32.887			

c. Yes. $F = 10.803$, $P = 0.0053$.

26. a.

Source	DF	SS	MS	*F*	*P*
Channel	4	1011.7	252.93	9.2357	0.00057
Error	15	410.79	27.386		
Total	19	1422.5			

b. Yes. $F = 9.2357$, $P = 0.00057$.

27. $\mu_{16} \neq \mu_{36}$

28. $\mu_1 \neq \mu_4$, $\mu_2 \neq \mu_4$, $\mu_4 \neq \mu_5$

29. a. $n_1 = n_2 = n_3 = n_4 = 30$

b. $s_1 = 24.85$, $s_2 = 35.26$, $s_3 = 33.53$, $s_4 = 23.09$

c. 12,712.7 **d.** 102,027.8 **e.** 3 for *SSTr*, 116 for *SSE*

f. $MSTr = 4237.57$, $MSE = 879.55$ **g.** $F = 4.818$

h. Yes, $P = 0.003375$.

30. a. $n_1 = 32$, $n_2 = 11$, $n_3 = 15$, $n_4 = 42$

b. $s_1 = 8.52$, $s_2 = 5.50$, $s_3 = 6.71$, $s_4 = 8.04$ **c.** 1721.42

d. 5833.45 **e.** 3 for *SSTr*, 96 for *SSE*

f. $MSTr = 573.808$, $MSE = 60.77$ **g.** $F = 9.44$

h. Yes, $P = 0.000016$.

31. a. H_0: The mean volume is the same at all restaurants.

b. 3 **c.** 4

d. $SSTr = 0.7845$, $SSE = 0.278$, $MSTr = 0.39225$, $MSE = 0.030888889$

e. 12.69874101 **f.** 0.0023972943

g. Reject H_0. We conclude that the mean volume differs among restaurants.

32. a. H_0: The mean time is the same for all spreadsheets.

b. 4 **c.** 5

d. $SSTr = 0.2455$, $SSE = 3.072$, $MSTr = 0.081833333$, $MSE = 0.192$

e. 0.4262152778 **f.** 0.7369008482

g. Do not reject H_0. There is not enough evidence to conclude that the mean time differs among spreadsheets.

33. a. $SSTr = 50{,}946.6$, $SSE = 120{,}550.9$

b. $MSTr = 25{,}473.3$, $MSE = 5023.0$

c. 5.071

d. No, $P = 0.015$, so $P > 0.01$. Do not reject H_0. There is not enough evidence to conclude that the mean energy absorbed differs among pillars.

34. a. $SSTr = 1.18547$, $SSE = 0.03050$

b. $MSTr = 0.23709$, $MSE = 0.00508$ **c.** 46.64

d. Yes, $P = 0.000$, so $P < 0.05$. Reject H_0. We conclude that the mean level of acid phosphatase activity differs among treatments.

35.

	a.	b.
Pair of Means	Test Statistic	P-value
$\mu_A - \mu_B$	1.8582	0.0816
$\mu_A - \mu_C$	3.4001	0.0037
$\mu_A - \mu_D$	1.6210	0.1246
$\mu_B - \mu_C$	1.5419	0.1426
$\mu_B - \mu_D$	0.2372	0.8155
$\mu_C - \mu_D$	1.7791	0.0942

c. A pair of means differs if the *P*-value is less than $0.05/12 = 0.004167$. The only pair of means that we can conclude to be different is $\mu_A \neq \mu_C$.

d. In this example, the results of the Bonferroni correction are the same as the results of the Tukey–Kramer test.

Section 14.2

Exercises 1–4 are the Check Your Understanding exercises for this section. Answers to these exercises are on page 691.

5. interaction

6. main, interaction

7. False

8. True

9. a. No, the *P*-value for interactions is greater than 0.10.

b. Yes. We conclude that there are differences in the mean responses among levels of the row factor; $P = 0.023$.

c. Yes. There is not enough evidence to conclude that there are differences in the mean responses among levels of the column factor; $P = 0.34$.

10. a. Yes, the *P*-value for interactions is less than 0.10.

b. No. Because we reject the hypothesis of no interactions, it is not appropriate to interpret the main effect of the row factor.

c. No. Because we reject the hypothesis of no interactions, it is not appropriate to interpret the main effect of the column factor.

11. No large interactions

12. Some large interactions

13. Some large interactions

14. No large interactions

15. a. No, $P = 0.82946$.

b. Yes. We conclude that there are differences in the mean exam scores between freshmen and sophomores; $P = 0.010673$.

c. Yes. There is not enough evidence to conclude that there are differences in the mean exam scores among subjects; $P = 0.82946$.

16. a. Yes, $P = 0.033622$.

b. No. Because we reject the hypothesis of no interactions, it is not appropriate to interpret the main effect of fertilizer.

c. No. Because we reject the hypothesis of no interactions, it is not appropriate to interpret the main effect of climatic zone.

17.

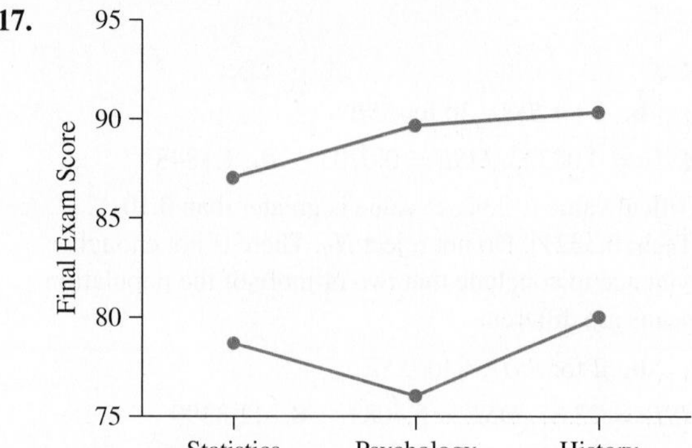

There are no strong interactions.

18.

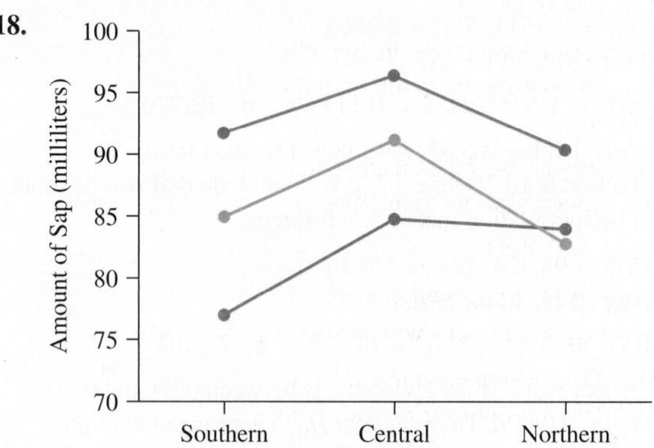

There is evidence of a strong interaction.

19. a. No, $P = 0.61534$.

b. Yes. We conclude that there are differences in the mean amounts of foam produced between distilled water and tap water; $P = 0.0017558$.

c. Yes. There is not enough evidence to conclude that the presence or absence of glycerol makes a difference in the mean amount of foam produced; $P = 0.12282$.

20. a. Yes, $P = 0.0041681$.

b. No. Because we reject the hypothesis of no interactions, it is not appropriate to interpret the main effect of soap concentration.

c. No. Because we reject the hypothesis of no interactions, it is not appropriate to interpret the main effect of the percentage of lauric acid.

21.

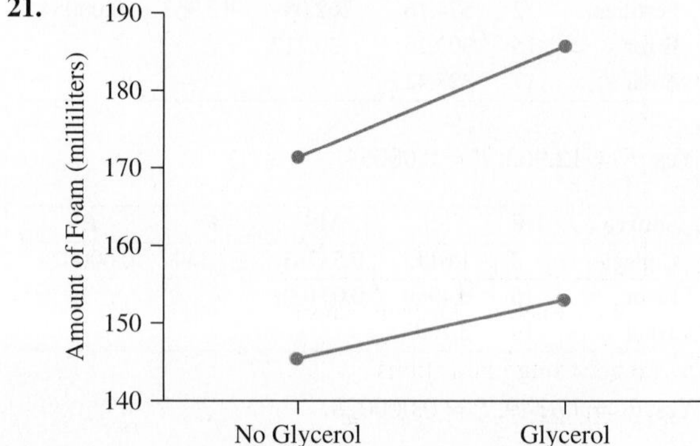

There are no strong interactions.

22.

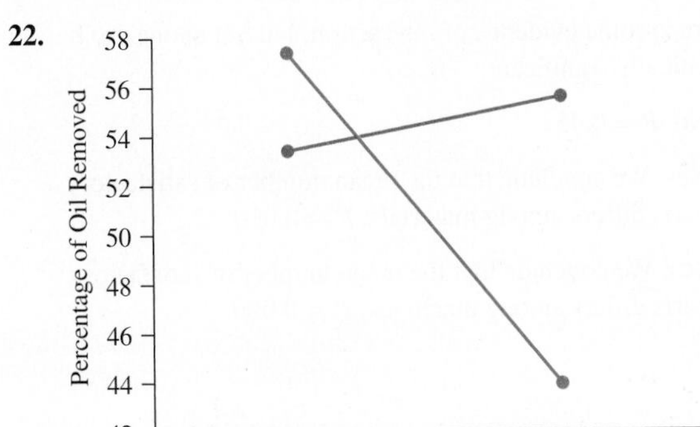

There is evidence of a strong interaction.

23. **a.** Yes, $P = 0.000039$.

b. No. Because we reject the hypothesis of no interactions, it is not appropriate to interpret the main effect of adhesive.

c. No. Because we reject the hypothesis of no interactions, it is not appropriate to interpret the main effect of pressure.

24. **a.** No, $P = 0.99836$.

b. Yes. There is not enough evidence to conclude that the mean strength varies with binder content; $P = 0.32687$.

c. Yes. There is not enough evidence to conclude that the mean strength varies with rubber content; $P = 0.59326$.

25.

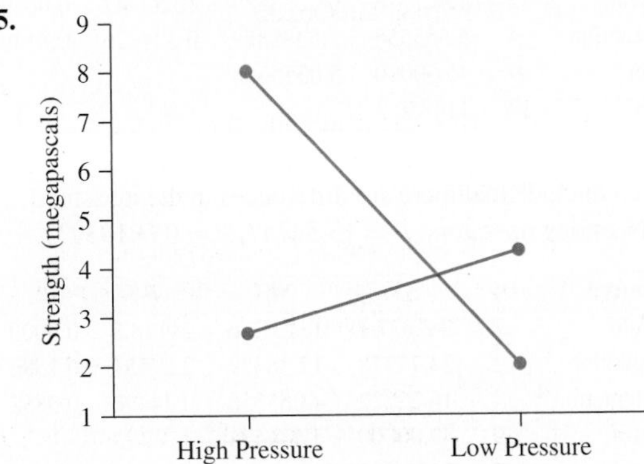

There is evidence of a strong interaction.

26.

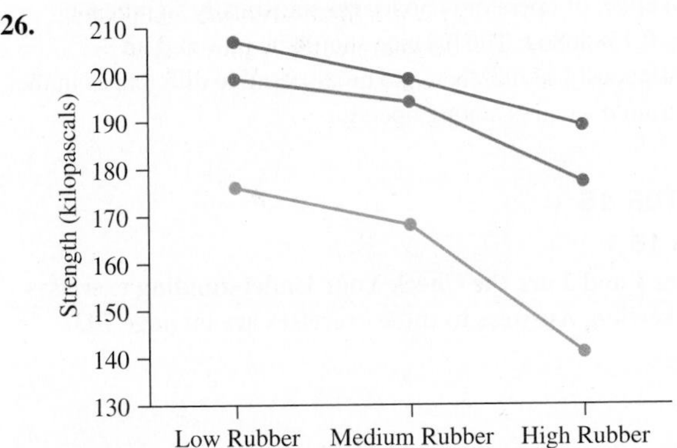

There are no strong interactions.

27. **a.** No, $P = 0.321$.

b. Yes. We conclude that the mean yield differs among pesticides; $P = 0.000$.

c. Yes. There is not enough evidence to conclude that the mean yield differs among varieties; $P = 0.103$.

28. **a.** Yes, $P = 0.009$.

b. No. Because we reject the hypothesis of no interactions, it is not appropriate to interpret the main effect of firm.

c. No. Because we reject the hypothesis of no interactions, it is not appropriate to interpret the main effect of time of day.

29. **a.** These have the largest cell means.

b. The main effect of design is $(51 + 49 + 43 + 41)/4 - (51 + 49 + 43 + 41)/4 = 0$.

c. The main effect of material is $(51 + 49 + 43 + 41)/4 - (51 + 49 + 43 + 41)/4 = 0$.

d. No. There is an interaction between design and material; $P = 0.00132$.

CHAPTER 14 Quiz

1. 3 for *SSTr*, 194 for *SSE*
2. $SSTr = 1.0518$, $SSE = 43.017$, $MSTr = 0.35059$, $MSE = 0.22174$
3. 1.5811
4. No, $P = 0.19527$.
5. No, $P = 0.74823$.
6. Yes. There is not enough evidence to conclude that the mean number of parts differs among operators; $P = 0.9376$.
7. Yes. We conclude that the mean number of parts differs among machines; $P = 0.0033212$.
8. Yes, $P = 0.001$.
9. No. Because we reject the hypothesis of no interactions, it is not appropriate to interpret the main effect of gender.
10. No. Because we reject the hypothesis of no interactions, it is not appropriate to interpret the main effect of diet.

CHAPTER 14 Review Exercises

1.

Source	DF	SS	MS	*F*	*P*
Gypsum	3	0.013092	0.0043639	0.28916	0.83215
Error	8	0.12073	0.015092		
Total	11	0.13383			

2. No, $P = 0.83215$.

3. **a.** No, $P = 0.72162$.

b. Yes. There is not enough evidence to conclude that the mean conductivity differs with the amount of gypsum; $P = 0.43865$.

c. Yes. There is not enough evidence to conclude that the mean conductivity differs between soil types; $P = 0.2057$.

4. **a.** $n_1 = 16, n_2 = 31, n_3 = 30$

b. $s_1 = 0.6, s_2 = 0.4, s_3 = 0.6$

c. 3.7995

d. 20.64

5. **a.** 2 for *SSTr*, 74 for *SSE*

b. $MSTr = 1.8997$, $MSE = 0.27892$

c. 6.8111

d. We conclude that there are differences among the mean pH levels, $P = 0.0019269$.

6. $\mu_1 \neq \mu_2$, $\mu_1 \neq \mu_3$

7. The largest standard deviation is more than twice as large as the smallest standard deviation. We cannot use ANOVA to determine whether conductivity varies with location.

8. a. No, $P = 0.6387$.

b. Yes. We conclude that the mean yield differs between fertilizers; $P = 0.011843$.

c. Yes. We conclude that the mean yield differs among soil types; $P = 0.024787$.

9.

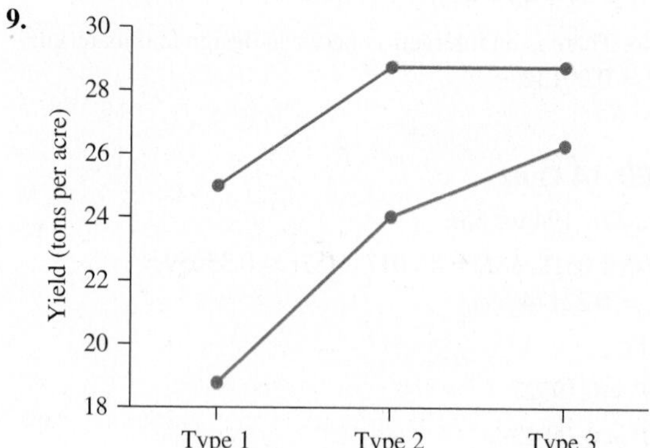

There are no strong interactions.

10. a. The standard deviations are $s_1 = 122.87$, $s_2 = 122.71$, $s_3 = 185.31$, $s_4 = 138.78$. The largest standard deviation is less than twice as large as the smallest standard deviation.

b.

Source	DF	SS	MS	F	P
Gypsum	3	378610.4	126203.5	6.210	0.0059
Error	15	304838.1	20322.5		
Total	18	683448.5			

11. $\mu_A \neq \mu_B$, $\mu_A \neq \mu_C$

12. a. No, $P = 0.19474$.

b. Yes. We conclude that the mean number of miles traveled differs among car types; $P = 0.00005992$.

c. Yes. We conclude that the mean number of miles traveled differs among tires; $P = 0.00000001$.

13.

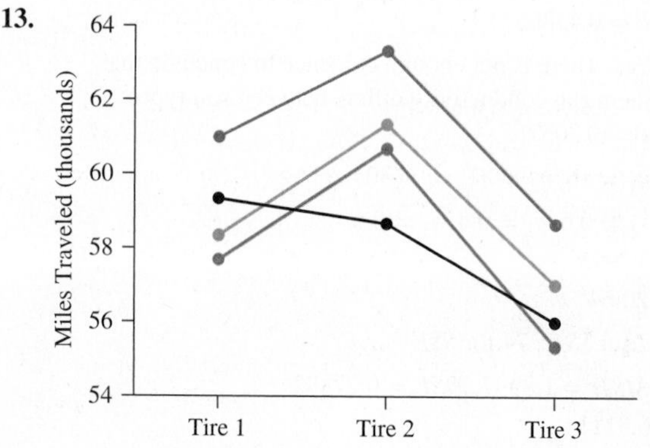

There is some evidence of interaction, but not enough to be statistically significant.

14. a. No, $P = 0.452$.

b. Yes. We conclude that the mean number of satisfactory parts differs among materials; $P = 0.010$.

c. Yes. We conclude that the mean number of satisfactory parts differs among machines; $P = 0.000$.

15.

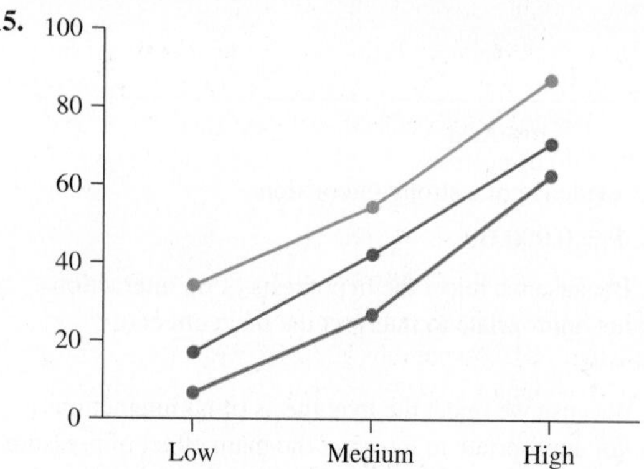

There are no strong interactions.

CHAPTER 14 Case Study

1.

Source	DF	SS	MS	F	P
Wafer	2	114661.4	57330.722	11340.14	0.0000
Operator	2	136.7778	68.388889	13.52747	0.0019
Interaction	4	6.555556	1.6388889	0.324176	0.8549
Error	9	45.50000	5.0555556		
Total	17	114850.3			

2. Yes, we conclude that there are differences in the measured weights among operators: $F = 13.52747$, $P = 0.00193977$.

3.

Source	DF	SS	MS	F	P
Wafer	2	262627.1	131313.6	24118.8	0.0000
Operator	2	24.77778	12.38889	2.27551	0.1586
Interaction	4	16.22222	4.055556	0.74490	0.5852
Error	9	49.00000	5.444444		
Total	17	262717.1			

4. The effect of operator is no longer statistically significant: $P = 0.15856864$. The balance should be powered up continuously so that there are no discernible differences in the measured weights among operators.

CHAPTER 15

Section 15.1

Exercises 1 and 2 are the Check Your Understanding exercises for this section. Answers to these exercises are on page 703.

3. less

4. less

5. True

6. False

7. $n = 10 \leq 25$, $x = 4$

8. $n = 15 \leq 25$, $x = 2$

9. a. $n = 20 \leq 25$, $x = 4$ **b.** Critical value is 5. Reject H_0.

c. Critical value is 3. Do not reject H_0.

10. a. $n = 35 > 25, x = 9, z = \dfrac{9 + 0.5 - 35/2}{\sqrt{35}/2} = -2.70$
b. Critical value is -1.645. Reject H_0. c. Critical value is -2.33. Reject H_0.

11. a. H_0: $m = 15$, H_1: $m < 15$ b. $n = 19 \le 25, x = 2$
c. Critical value is 5. d. Reject H_0. We conclude that the median weight loss is less than 15 pounds.

12. a. H_0: $m = 120$, H_1: $m > 120$ b. $n = 12 \le 25, x = 4$
c. Critical value is 1. d. Do not reject H_0. There is not enough evidence to conclude that the median running time is over 120 minutes.

13. H_0: $m = 60$, H_1: $m > 60$; $n = 16 \le 25$; Test statistic: $x = 3$. Critical value is 4. Reject H_0. We conclude that the median height of oak trees in this forest is greater than 60 feet.

14. H_0: $m = 75$, H_1: $m \ne 75$; $n = 14 \le 25$; Test statistic: $x = 4$. Critical value is 1. Do not reject H_0. There is not enough evidence to conclude that the median yield differs from 75 bushels per acre.

15. H_0: $m = 1000$, H_1: $m < 1000$; $n = 35 > 25$; Test statistic: $x = 12$, $z = -1.69$. Critical value is -1.645. Reject H_0. We conclude that the median rent is less than \$1000 per month.

16. H_0: $m = 60$, H_1: $m < 60$; $n = 29 > 25$; Test statistic: $x = 5$, $z = -3.34$. Critical value is -1.645. Reject H_0. We conclude that the median commute time is less than 60 minutes.

17. H_0: $m = 480$, H_1: $m < 480$; $n = 20 \le 25$; Test statistic: $x = 7$. Critical value is 5. Do not reject H_0. There is not enough evidence to conclude that the median fraud complaint rate is less than 480.

18. H_0: $m = 17$, H_1: $m \ne 17$; $n = 31 > 25$; Test statistic: $x = 9$, $z = -2.16$. Critical value is -1.96. Reject H_0. We conclude that the median approval rating differs from 17.

19. H_0: $m = 1$, H_1: $m < 1$; $n = 70 > 25$; Test statistic: $x = 25$, $z = -2.27$. Critical value is -1.645. We reject H_0 at the $\alpha = 0.05$ level. However, H_0 is true. The population median is 1, and the sample median is also 1. This shows that the sign test is unreliable for discrete populations.

Section 15.2

Exercises 1 and 2 are the Check Your Understanding exercises for this section. Answers to these exercises are on page 709.

3. average
4. 10
5. False
6. True
7. $\mu_S = 255$, $\sigma_S = 27.659$, $z = -0.40$, $P = 0.3446$ [Tech: 0.3454]
8. $\mu_S = 460$, $\sigma_S = 43.780$, $z = 3.36$, $P = 0.0008$
9. $\mu_S = 348.5$, $\sigma_S = 36.550$, $z = -2.09$, $P = 0.0366$ [Tech: 0.0363]
10. $\mu_S = 294$, $\sigma_S = 36.373$, $z = 0.38$, $P = 0.3520$ [Tech: 0.3502]

11. a. H_0: $m_1 = m_2$, H_1: $m_1 < m_2$
b. $S = 71.5$, $z = -3.24$
c. $P = 0.0006$
d. Reject H_0. We conclude that the mileage is less for compact cars than for midsize cars.

12. a. H_0: $m_1 = m_2$, H_1: $m_1 \ne m_2$
b. If City 1 is taken to be sample 1, then $S = 253.5$ and $z = 2.32$. If City 2 is taken to be sample 1, then $S = 152.5$ and $z = -2.32$.
c. $P = 0.0204$ [Tech: 0.0203]
d. Do not reject H_0. There is not enough evidence to conclude that there is a difference in coffee prices between the two cities.

13. H_0: $m_1 = m_2$, H_1: $m_1 \ne m_2$; $\mu_S = 232.5$, $\sigma_S = 24.109$. If Cured 2 days is taken to be sample 1, then $S = 194.5$ and $z = -1.58$. If Cured 6 days is taken to be sample 1, then $S = 270.5$ and $z = 1.58$. $P = 0.1142$ [Tech: 0.1150]. Do not reject H_0. There is not enough evidence to conclude that the strengths differ between the two curing times.

14. H_0: $m_1 = m_2$, H_1: $m_1 < m_2$; $\mu_S = 115$, $\sigma_S = 15.166$. $S = 64$, $z = -3.36$, $P = 0.0004$. Reject H_0. We conclude that the median recovery time is less for treatment than for control.

15. H_0: $m_1 = m_2$, H_1: $m_1 > m_2$; $\mu_S = 174$, $\sigma_S = 21.541$. $S = 141.5$, $z = -1.51$, $P = 0.0655$ [Tech: 0.0657]. Do not reject H_0. There is not enough evidence to conclude that the median load time for 3-D Blu-ray players is greater than for standard Blu-ray players.

16. H_0: $m_1 = m_2$, H_1: $m_1 \ne m_2$; $\mu_S = 333$, $\sigma_S = 31.607$. If Standard is taken to be sample 1, then $S = 411$ and $z = 2.47$. If High Efficiency is taken to be sample 1, then $S = 255$ and $z = -2.47$. $P = 0.0136$. Reject H_0. We conclude that the median cost per load differs between standard and high-efficiency laundry detergents.

17. a. The sample median is 30 for both groups.
b. $S = 183.5$ (or 281.5), $\mu_S = 232.5$, $\sigma_S = 24.109$, $z = -2.03$ (or 2.03)
c. $P = 0.0424$ [Tech: 0.0421]. Reject H_0.

Section 15.3

Exercises 1 and 2 are the Check Your Understanding exercises for this section. Answers to these exercises are on page 714.

3. less than or equal to
4. symmetric
5. True
6. False
7. Test statistic: $S = 1$, Critical value: 14. Reject H_0.
8. Test statistic: $S = 0$, Critical value: 6. Reject H_0.
9. Test statistic: $S = 19$, Critical value: 11. Do not reject H_0.
10. Test statistic: $S = 6.5$, Critical value: 4. Do not reject H_0.
11. Test statistic: $S = 8.5$, Critical value: 14. Reject H_0. We conclude that the median scores differ between pretest and posttest.
12. $n = 7$, ignoring one difference of 0. Test statistic: $S = 9.5$, Critical value: 2. Do not reject H_0. There is not enough evidence to conclude that the median rents now are different from a year ago.
13. Test statistic: $S = 17.5$, Critical value: 2. Do not reject H_0. There is not enough evidence to conclude that the median traffic times differ.
14. Test statistic: $S = 5$, Critical value: 6. Reject H_0. We conclude that the median weight after exercise differs from the median weight before.

15. Test statistic: $S = 25.5$, Critical value: 17. Do not reject H_0. There is not enough evidence to conclude that the median score before the seminar differs from the median score after.

16. Test statistic: $S = 1$, Critical value: 14. Reject H_0. We conclude that the median number of births before Hurricane Katrina differs from the median after.

17. **a.** The median is the middle value, which is 0.
b. $S = 78$
c. The critical value is 81. Reject H_0.

CHAPTER 15 Quiz

1. **a.** $n = 10 \le 25, x = 4$ **b.** Do not reject H_0 **c.** Do not reject H_0

2. **a.** $n = 16 \le 25, x = 0$ **b.** Reject H_0 **c.** Reject H_0

3. **a.** $n = 24 \le 25, x = 5$ **b.** Reject H_0 **c.** Reject H_0

4. **a.** $n = 51 > 25, x = 15, z = -2.80$ **b.** Critical value is -1.645. Reject H_0. **c.** Critical value is -2.33. Reject H_0.

5. $\mu_S = 510, \sigma_S = 50.498, z = -2.18, P = 0.0146$ [Tech: 0.0147]

6. $\mu_S = 725, \sigma_S = 62.183, z = 2.01, P = 0.0444$

7. $\mu_S = 255, \sigma_S = 27.659, z = 0.90, P = 0.1841$ [Tech: 0.1830]

8. $S = 37$, Critical value: 30. Do not reject H_0.

9. $S = 0$, Critical value: 21. Reject H_0.

10. $S = 0$, Critical value: 11. Reject H_0.

CHAPTER 15 Review Exercises

1. $S = 29.5$, Critical value: 25. Do not reject H_0. There is not enough evidence to conclude that the median weights differ.

2. $n = 20 \le 25, x = 5$, Critical value: 3. Do not reject H_0. There is not enough evidence to conclude that the median price differs from $300.

3. $S = 23$, Critical Value: 7. Do not reject H_0. There is not enough evidence to conclude that the median number of customers differs between Monday and Tuesday.

4. $S = 210.5, z = 2.07, P = 0.0192$ [Tech: 0.0190]. Reject H_0. We conclude that the median systolic blood pressure is less for people who exercise than for people who are sedentary.

5. $n = 17 \le 25, x = 7$, Critical value: 4. Do not reject H_0. There is not enough evidence to conclude the median is different from 8.

6. $n = 11 \le 25, x = 3$, Critical value: 1. Do not reject H_0. There is not enough evidence to conclude that the median is less than 5 days.

7. $S = 202, \mu_S = 217, \sigma_S = 24.055, z = -0.62, P = 0.5352$ [Tech: 0.5329]. Do not reject H_0. There is not enough evidence to conclude that the median weight differs between boys and girls.

8. $S = 157, z = 2.77, P = 0.0056$. Reject H_0. We conclude that the median times differ between the two routes.

9. $S = 24.5$, Critical value: 14. Do not reject H_0. There is not enough evidence to conclude the measurements differ between gauges.

10. $n = 13 \le 25, x = 5$, Critical value: 1. Do not reject H_0. There is not enough evidence to conclude that the median differs from 12 ounces.

CHAPTER 15 Case Study

1. $S = 53.5, \mu_S = 99, \sigma_S = 14.071, z = -3.23, P = 0.0006$. Reject H_0. We conclude that the median number of microorganisms is greater for a retention time of 12 days than for 3 days.

2. $S = 66.5, \mu_S = 126, \sigma_S = 19.442, z = -3.06, P = 0.0011$. Reject H_0. We conclude that the median number of microorganisms is greater for a retention time of 30 days than for 3 days.

3. $S = 206, \mu_S = 186, \sigma_S = 23.622, z = 0.85, P = 0.3954$ [Tech 0.3972]. Do not reject H_0. There is not enough evidence to conclude that the median number of microorganisms differs between a retention time of 30 days and a retention time of 12 days.

4. iii.

INDEX

Table A.2 Cumulative Normal Distribution

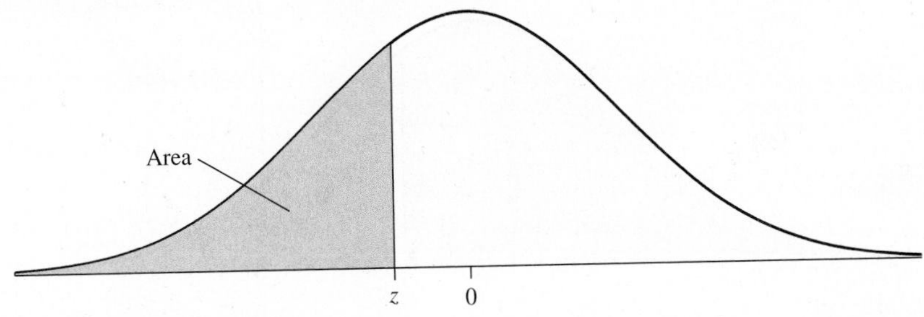

z	0.00	0.01	0.02	0.03	0.04	0.05	0.06	0.07	0.08	0.09
−3.7 or less	.0001									
−3.6	.0002	.0002	.0001	.0001	.0001	.0001	.0001	.0001	.0001	.0001
−3.5	.0002	.0002	.0002	.0002	.0002	.0002	.0002	.0002	.0002	.0002
−3.4	.0003	.0003	.0003	.0003	.0003	.0003	.0003	.0003	.0003	.0002
−3.3	.0005	.0005	.0005	.0004	.0004	.0004	.0004	.0004	.0004	.0003
−3.2	.0007	.0007	.0006	.0006	.0006	.0006	.0006	.0005	.0005	.0005
−3.1	.0010	.0009	.0009	.0009	.0008	.0008	.0008	.0008	.0007	.0007
−3.0	.0013	.0013	.0013	.0012	.0012	.0011	.0011	.0011	.0010	.0010
−2.9	.0019	.0018	.0018	.0017	.0016	.0016	.0015	.0015	.0014	.0014
−2.8	.0026	.0025	.0024	.0023	.0023	.0022	.0021	.0021	.0020	.0019
−2.7	.0035	.0034	.0033	.0032	.0031	.0030	.0029	.0028	.0027	.0026
−2.6	.0047	.0045	.0044	.0043	.0041	.0040	.0039	.0038	.0037	.0036
−2.5	.0062	.0060	.0059	.0057	.0055	.0054	.0052	.0051	.0049	.0048
−2.4	.0082	.0080	.0078	.0075	.0073	.0071	.0069	.0068	.0066	.0064
−2.3	.0107	.0104	.0102	.0099	.0096	.0094	.0091	.0089	.0087	.0084
−2.2	.0139	.0136	.0132	.0129	.0125	.0122	.0119	.0116	.0113	.0110
−2.1	.0179	.0174	.0170	.0166	.0162	.0158	.0154	.0150	.0146	.0143
−2.0	.0228	.0222	.0217	.0212	.0207	.0202	.0197	.0192	.0188	.0183
−1.9	.0287	.0281	.0274	.0268	.0262	.0256	.0250	.0244	.0239	.0233
−1.8	.0359	.0351	.0344	.0336	.0329	.0322	.0314	.0307	.0301	.0294
−1.7	.0446	.0436	.0427	.0418	.0409	.0401	.0392	.0384	.0375	.0367
−1.6	.0548	.0537	.0526	.0516	.0505	.0495	.0485	.0475	.0465	.0455
−1.5	.0668	.0655	.0643	.0630	.0618	.0606	.0594	.0582	.0571	.0559
−1.4	.0808	.0793	.0778	.0764	.0749	.0735	.0721	.0708	.0694	.0681
−1.3	.0968	.0951	.0934	.0918	.0901	.0885	.0869	.0853	.0838	.0823
−1.2	.1151	.1131	.1112	.1093	.1075	.1056	.1038	.1020	.1003	.0985
−1.1	.1357	.1335	.1314	.1292	.1271	.1251	.1230	.1210	.1190	.1170
−1.0	.1587	.1562	.1539	.1515	.1492	.1469	.1446	.1423	.1401	.1379
−0.9	.1841	.1814	.1788	.1762	.1736	.1711	.1685	.1660	.1635	.1611
−0.8	.2119	.2090	.2061	.2033	.2005	.1977	.1949	.1922	.1894	.1867
−0.7	.2420	.2389	.2358	.2327	.2296	.2266	.2236	.2206	.2177	.2148
−0.6	.2743	.2709	.2676	.2643	.2611	.2578	.2546	.2514	.2483	.2451
−0.5	.3085	.3050	.3015	.2981	.2946	.2912	.2877	.2843	.2810	.2776
−0.4	.3446	.3409	.3372	.3336	.3300	.3264	.3228	.3192	.3156	.3121
−0.3	.3821	.3783	.3745	.3707	.3669	.3632	.3594	.3557	.3520	.3483
−0.2	.4207	.4168	.4129	.4090	.4052	.4013	.3974	.3936	.3897	.3859
−0.1	.4602	.4562	.4522	.4483	.4443	.4404	.4364	.4325	.4286	.4247
−0.0	.5000	.4960	.4920	.4880	.4840	.4801	.4761	.4721	.4681	.4641

Table A.2 Cumulative Normal Distribution (continued)

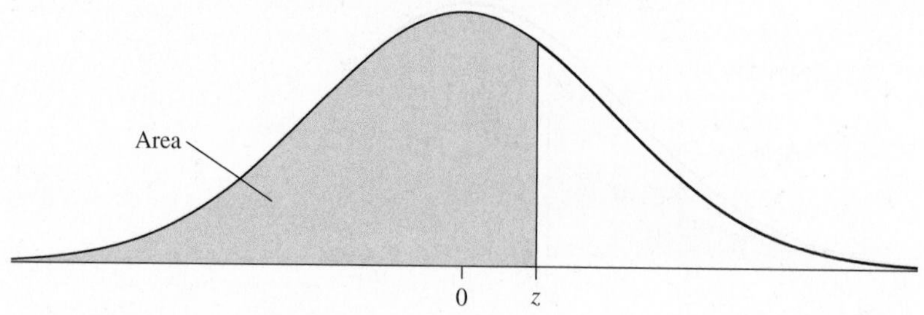

z	0.00	0.01	0.02	0.03	0.04	0.05	0.06	0.07	0.08	0.09
0.0	.5000	.5040	.5080	.5120	.5160	.5199	.5239	.5279	.5319	.5359
0.1	.5398	.5438	.5478	.5517	.5557	.5596	.5636	.5675	.5714	.5753
0.2	.5793	.5832	.5871	.5910	.5948	.5987	.6026	.6064	.6103	.6141
0.3	.6179	.6217	.6255	.6293	.6331	.6368	.6406	.6443	.6480	.6517
0.4	.6554	.6591	.6628	.6664	.6700	.6736	.6772	.6808	.6844	.6879
0.5	.6915	.6950	.6985	.7019	.7054	.7088	.7123	.7157	.7190	.7224
0.6	.7257	.7291	.7324	.7357	.7389	.7422	.7454	.7486	.7517	.7549
0.7	.7580	.7611	.7642	.7673	.7704	.7734	.7764	.7794	.7823	.7852
0.8	.7881	.7910	.7939	.7967	.7995	.8023	.8051	.8078	.8106	.8133
0.9	.8159	.8186	.8212	.8238	.8264	.8289	.8315	.8340	.8365	.8389
1.0	.8413	.8438	.8461	.8485	.8508	.8531	.8554	.8577	.8599	.8621
1.1	.8643	.8665	.8686	.8708	.8729	.8749	.8770	.8790	.8810	.8830
1.2	.8849	.8869	.8888	.8907	.8925	.8944	.8962	.8980	.8997	.9015
1.3	.9032	.9049	.9066	.9082	.9099	.9115	.9131	.9147	.9162	.9177
1.4	.9192	.9207	.9222	.9236	.9251	.9265	.9279	.9292	.9306	.9319
1.5	.9332	.9345	.9357	.9370	.9382	.9394	.9406	.9418	.9429	.9441
1.6	.9452	.9463	.9474	.9484	.9495	.9505	.9515	.9525	.9535	.9545
1.7	.9554	.9564	.9573	.9582	.9591	.9599	.9608	.9616	.9625	.9633
1.8	.9641	.9649	.9656	.9664	.9671	.9678	.9686	.9693	.9699	.9706
1.9	.9713	.9719	.9726	.9732	.9738	.9744	.9750	.9756	.9761	.9767
2.0	.9772	.9778	.9783	.9788	.9793	.9798	.9803	.9808	.9812	.9817
2.1	.9821	.9826	.9830	.9834	.9838	.9842	.9846	.9850	.9854	.9857
2.2	.9861	.9864	.9868	.9871	.9875	.9878	.9881	.9884	.9887	.9890
2.3	.9893	.9896	.9898	.9901	.9904	.9906	.9909	.9911	.9913	.9916
2.4	.9918	.9920	.9922	.9925	.9927	.9929	.9931	.9932	.9934	.9936
2.5	.9938	.9940	.9941	.9943	.9945	.9946	.9948	.9949	.9951	.9952
2.6	.9953	.9955	.9956	.9957	.9959	.9960	.9961	.9962	.9963	.9964
2.7	.9965	.9966	.9967	.9968	.9969	.9970	.9971	.9972	.9973	.9974
2.8	.9974	.9975	.9976	.9977	.9977	.9978	.9979	.9979	.9980	.9981
2.9	.9981	.9982	.9982	.9983	.9984	.9984	.9985	.9985	.9986	.9986
3.0	.9987	.9987	.9987	.9988	.9988	.9989	.9989	.9989	.9990	.9990
3.1	.9990	.9991	.9991	.9991	.9992	.9992	.9992	.9992	.9993	.9993
3.2	.9993	.9993	.9994	.9994	.9994	.9994	.9994	.9995	.9995	.9995
3.3	.9995	.9995	.9995	.9996	.9996	.9996	.9996	.9996	.9996	.9997
3.4	.9997	.9997	.9997	.9997	.9997	.9997	.9997	.9997	.9997	.9998
3.5	.9998	.9998	.9998	.9998	.9998	.9998	.9998	.9998	.9998	.9998
3.6	.9998	.9998	.9999	.9999	.9999	.9999	.9999	.9999	.9999	.9999
3.7 or more	.9999									

Table A.3 Critical Values for the Student's *t* Distribution

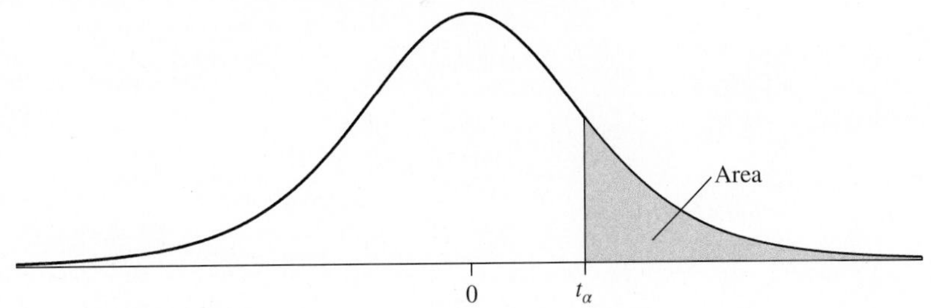

Degrees of Freedom	Area in Right Tail									
	0.40	**0.25**	**0.10**	**0.05**	**0.025**	**0.01**	**0.005**	**0.0025**	**0.001**	**0.0005**
1	0.325	1.000	3.078	6.314	12.706	31.821	63.657	127.321	318.309	636.619
2	0.289	0.816	1.886	2.920	4.303	6.965	9.925	14.089	22.327	31.599
3	0.277	0.765	1.638	2.353	3.182	4.541	5.841	7.453	10.215	12.924
4	0.271	0.741	1.533	2.132	2.776	3.747	4.604	5.598	7.173	8.610
5	0.267	0.727	1.476	2.015	2.571	3.365	4.032	4.773	5.893	6.869
6	0.265	0.718	1.440	1.943	2.447	3.143	3.707	4.317	5.208	5.959
7	0.263	0.711	1.415	1.895	2.365	2.998	3.499	4.029	4.785	5.408
8	0.262	0.706	1.397	1.860	2.306	2.896	3.355	3.833	4.501	5.041
9	0.261	0.703	1.383	1.833	2.262	2.821	3.250	3.690	4.297	4.781
10	0.260	0.700	1.372	1.812	2.228	2.764	3.169	3.581	4.144	4.587
11	0.260	0.697	1.363	1.796	2.201	2.718	3.106	3.497	4.025	4.437
12	0.259	0.695	1.356	1.782	2.179	2.681	3.055	3.428	3.930	4.318
13	0.259	0.694	1.350	1.771	2.160	2.650	3.012	3.372	3.852	4.221
14	0.258	0.692	1.345	1.761	2.145	2.624	2.977	3.326	3.787	4.140
15	0.258	0.691	1.341	1.753	2.131	2.602	2.947	3.286	3.733	4.073
16	0.258	0.690	1.337	1.746	2.120	2.583	2.921	3.252	3.686	4.015
17	0.257	0.689	1.333	1.740	2.110	2.567	2.898	3.222	3.646	3.965
18	0.257	0.688	1.330	1.734	2.101	2.552	2.878	3.197	3.610	3.922
19	0.257	0.688	1.328	1.729	2.093	2.539	2.861	3.174	3.579	3.883
20	0.257	0.687	1.325	1.725	2.086	2.528	2.845	3.153	3.552	3.850
21	0.257	0.686	1.323	1.721	2.080	2.518	2.831	3.135	3.527	3.819
22	0.256	0.686	1.321	1.717	2.074	2.508	2.819	3.119	3.505	3.792
23	0.256	0.685	1.319	1.714	2.069	2.500	2.807	3.104	3.485	3.768
24	0.256	0.685	1.318	1.711	2.064	2.492	2.797	3.091	3.467	3.745
25	0.256	0.684	1.316	1.708	2.060	2.485	2.787	3.078	3.450	3.725
26	0.256	0.684	1.315	1.706	2.056	2.479	2.779	3.067	3.435	3.707
27	0.256	0.684	1.314	1.703	2.052	2.473	2.771	3.057	3.421	3.690
28	0.256	0.683	1.313	1.701	2.048	2.467	2.763	3.047	3.408	3.674
29	0.256	0.683	1.311	1.699	2.045	2.462	2.756	3.038	3.396	3.659
30	0.256	0.683	1.310	1.697	2.042	2.457	2.750	3.030	3.385	3.646
31	0.256	0.682	1.309	1.696	2.040	2.453	2.744	3.022	3.375	3.633
32	0.255	0.682	1.309	1.694	2.037	2.449	2.738	3.015	3.365	3.622
33	0.255	0.682	1.308	1.692	2.035	2.445	2.733	3.008	3.356	3.611
34	0.255	0.682	1.307	1.691	2.032	2.441	2.728	3.002	3.348	3.601
35	0.255	0.682	1.306	1.690	2.030	2.438	2.724	2.996	3.340	3.591
36	0.255	0.681	1.306	1.688	2.028	2.434	2.719	2.990	3.333	3.582
37	0.255	0.681	1.305	1.687	2.026	2.431	2.715	2.985	3.326	3.574
38	0.255	0.681	1.304	1.686	2.024	2.429	2.712	2.980	3.319	3.566
39	0.255	0.681	1.304	1.685	2.023	2.426	2.708	2.976	3.313	3.558
40	0.255	0.681	1.303	1.684	2.021	2.423	2.704	2.971	3.307	3.551
50	0.255	0.679	1.299	1.676	2.009	2.403	2.678	2.937	3.261	3.496
60	0.254	0.679	1.296	1.671	2.000	2.390	2.660	2.915	3.232	3.460
80	0.254	0.678	1.292	1.664	1.990	2.374	2.639	2.887	3.195	3.416
100	0.254	0.677	1.290	1.660	1.984	2.364	2.626	2.871	3.174	3.390
200	0.254	0.676	1.286	1.653	1.972	2.345	2.601	2.839	3.131	3.340
z	0.253	0.674	1.282	1.645	1.960	2.326	2.576	2.807	3.090	3.291
	20%	50%	80%	90%	95%	98%	99%	99.5%	99.8%	99.9%
	Confidence Level									

IMPORTANT FORMULAS

Chapter 3: Numerical Summaries of Data

Sample mean:
$$\bar{x} = \frac{\sum x}{n}$$

Population mean:
$$\mu = \frac{\sum x}{N}$$

Weighted mean:
$$\bar{x}_w = \frac{w_1x_1 + \cdots + w_nx_n}{w_1 + \cdots + w_n} = \frac{\sum w_ix_i}{\sum w_i}$$

Range:
Range = largest value − smallest value

Population variance:
$$\sigma^2 = \frac{\sum(x-\mu)^2}{N}$$

Sample variance:
$$s^2 = \frac{\sum(x-\bar{x})^2}{n-1}$$

Coefficient of variation:
$$\text{CV} = \frac{\sigma}{\mu}$$

***z*-score:**
$$z = \frac{x-\mu}{\sigma}$$

Interquartile range:

$\text{IQR} = Q_3 - Q_1$ = third quartile − first quartile

Lower outlier boundary:
$Q_1 - 1.5\,\text{IQR}$

Upper outlier boundary:
$Q_3 + 1.5\,\text{IQR}$

Chapter 4: Summarizing Bivariate Data

Correlation coefficient:
$$r = \frac{1}{n-1}\sum\left(\frac{x-\bar{x}}{s_x}\right)\left(\frac{y-\bar{y}}{s_y}\right)$$

Slope of least-squares regression line:
$$b_1 = r\frac{s_y}{s_x}$$

***y*-intercept of least-squares regression line:**
$$b_0 = \bar{y} - b_1\bar{x}$$

Equation of least-squares regression line:
$$\hat{y} = b_0 + b_1x$$

Chapter 5: Probability

General Addition Rule:
$P(A \text{ or } B) = P(A) + P(B) - P(A \text{ and } B)$

Multiplication Rule for Independent Events:
$P(A \text{ and } B) = P(A)P(B)$

Addition Rule for Mutually Exclusive Events:
$P(A \text{ or } B) = P(A) + P(B)$

Rule of Complements:
$P(A^c) = 1 - P(A)$

General Method for Computing Conditional Probability:
$$P(B\,|\,A) = \frac{P(A \text{ and } B)}{P(A)}$$

General Multiplication Rule:
$P(A \text{ and } B) = P(A)P(B\,|\,A) = P(B)P(A\,|\,B)$

Permutation of *r* items chosen from *n*:
$${}_nP_r = \frac{n!}{(n-r)!}$$

Combination of *r* items chosen from *n*:
$${}_nC_r = \frac{n!}{r!(n-r)!}$$

Chapter 6: Discrete Probability Distributions

Mean of a discrete random variable:
$\mu_X = \sum[x \cdot P(x)]$

Variance of a discrete random variable:
$\sigma_X^2 = \sum[(x - \mu_X)^2 \cdot P(x)] = \sum[x^2 \cdot P(x)] - \mu_X^2$

Standard deviation of a discrete random variable:
$\sigma_X = \sqrt{\sigma_X^2}$

Mean of a binomial random variable:
$\mu_X = np$

Variance of a binomial random variable:
$\sigma_X^2 = np(1 - p)$

Standard deviation of a binomial random variable:
$\sigma_X = \sqrt{np(1 - p)}$

Mean of Poisson random variable:
$\mu_X = \lambda t$

Variance of Poisson random variable:
$\sigma_X^2 = \lambda t$

Standard deviation of Poisson random variable:
$\sigma_X = \sqrt{\lambda t}$

Chapter 7: The Normal Distribution

z-score:
$$z = \frac{x - \mu}{\sigma}$$

Convert z-score to raw score:
$x = \mu + z\sigma$

Standard deviation of the sample mean:
$$\sigma_{\bar{x}} = \frac{\sigma}{\sqrt{n}}$$

z-score for a sample mean:
$$z = \frac{\bar{x} - \mu}{\sigma_{\bar{x}}}$$

Standard deviation of the sample proportion:
$$\sigma_{\hat{p}} = \sqrt{\frac{p(1 - p)}{n}}$$

z-score for a sample proportion:
$$z = \frac{\hat{p} - p}{\sigma_{\hat{p}}}$$

Chapter 8: Confidence Intervals

Confidence interval for a mean, standard deviation known:
$$\bar{x} - z_{\alpha/2}\frac{\sigma}{\sqrt{n}} < \mu < \bar{x} + z_{\alpha/2}\frac{\sigma}{\sqrt{n}}$$

Sample size to construct an interval for μ with margin of error m, standard deviation known:
$$n = \left(\frac{z_{\alpha/2} \cdot \sigma}{m}\right)^2$$

Confidence interval for a mean, standard deviation unknown:
$$\bar{x} - t_{\alpha/2}\frac{s}{\sqrt{n}} < \mu < \bar{x} + t_{\alpha/2}\frac{s}{\sqrt{n}}$$

Sample size to construct an interval for μ with margin of error m, standard deviation unknown:
$$n = \left(\frac{z_{\alpha/2} \cdot s}{m}\right)^2$$

Confidence interval for a proportion:
$$\hat{p} - z_{\alpha/2}\sqrt{\frac{\hat{p}(1 - \hat{p})}{n}} < p < \hat{p} + z_{\alpha/2}\sqrt{\frac{\hat{p}(1 - \hat{p})}{n}}$$

Sample size to construct an interval for p with margin of error m:

$n = \hat{p}(1 - \hat{p})\left(\frac{z_{\alpha/2}}{m}\right)^2$ if a value for $\hat{p}$ is available

$n = 0.25\left(\frac{z_{\alpha/2}}{m}\right)^2$ if no value for $\hat{p}$ is available

Confidence interval for the variance of a normal distribution:
$$\frac{(n - 1)s^2}{\chi^2_{\alpha/2}} < \sigma^2 < \frac{(n - 1)s^2}{\chi^2_{1-\alpha/2}}$$

Confidence interval for the standard deviation of a normal distribution:
$$\sqrt{\frac{(n - 1)s^2}{\chi^2_{\alpha/2}}} < \sigma < \sqrt{\frac{(n - 1)s^2}{\chi^2_{1-\alpha/2}}}$$

Chapter 9: Hypothesis Testing

Test statistic for a mean, standard deviation known:

$$z = \frac{\bar{x} - \mu_0}{\sigma/\sqrt{n}}$$

Test statistic for a mean, standard deviation unknown:

$$t = \frac{\bar{x} - \mu_0}{s/\sqrt{n}}$$

Test statistic for a proportion:

$$z = \frac{\hat{p} - p_0}{\sqrt{\dfrac{p_0(1 - p_0)}{n}}}$$

Test statistic for a standard deviation:

$$\chi^2 = \frac{(n-1) \cdot s^2}{\sigma_0^2}$$

Chapter 10: Two-Sample Confidence Intervals

Confidence interval for the difference between two means, independent samples:

$$\bar{x}_1 - \bar{x}_2 - t_{\alpha/2}\sqrt{\frac{s_1^2}{n_1} + \frac{s_2^2}{n_2}} < \mu_1 - \mu_2 < \bar{x}_1 - \bar{x}_2 + t_{\alpha/2}\sqrt{\frac{s_1^2}{n_1} + \frac{s_2^2}{n_2}}$$

Confidence interval for the difference between two means, matched pairs:

$$\bar{d} - t_{\alpha/2}\frac{s_d}{\sqrt{n}} < \mu_d < \bar{d} + t_{\alpha/2}\frac{s_d}{\sqrt{n}}$$

Confidence interval for the difference between two proportions:

$$\hat{p}_1 - \hat{p}_2 - z_{\alpha/2}\sqrt{\frac{\hat{p}_1(1-\hat{p}_1)}{n_1} + \frac{\hat{p}_2(1-\hat{p}_2)}{n_2}}$$
$$< p_1 - p_2 < \hat{p}_1 - \hat{p}_2 + z_{\alpha/2}\sqrt{\frac{\hat{p}_1(1-\hat{p}_1)}{n_1} + \frac{\hat{p}_2(1-\hat{p}_2)}{n_2}}$$

Chapter 11: Two-Sample Hypothesis Tests

Test statistic for the difference between two means, independent samples:

$$t = \frac{(\bar{x}_1 - \bar{x}_2) - (\mu_1 - \mu_2)}{\sqrt{\dfrac{s_1^2}{n_1} + \dfrac{s_2^2}{n_2}}}$$

Test statistic for the difference between two means, matched pairs:

$$t = \frac{\bar{d} - \mu_0}{s_d/\sqrt{n}}$$

Test statistic for the difference between two proportions:

$$z = \frac{\hat{p}_1 - \hat{p}_2}{\sqrt{\hat{p}(1-\hat{p})\left(\dfrac{1}{n_1} + \dfrac{1}{n_2}\right)}}$$

where $\hat{p}$ is the pooled proportion $\hat{p} = \dfrac{x_1 + x_2}{n_1 + n_2}$

Test statistic for two standard deviations:

$$F = \frac{\text{Larger of } s_1^2 \text{ and } s_2^2}{\text{Smaller of } s_1^2 \text{ and } s_2^2}$$

Chapter 12: Tests with Qualitative Data

Chi-square statistic:

$$\chi^2 = \sum \frac{(O - E)^2}{E}$$

Expected frequency for goodness-of-fit:

$E = np$

Expected frequency for independence or homogeneity:

$$E = \frac{\text{Row total} \cdot \text{Column total}}{\text{Grand total}}$$

Chapter 13: Inference in Linear Models

Residual standard deviation:

$$s_e = \sqrt{\frac{\sum(y-\hat{y})^2}{n-2}}$$

Standard error for b_1:

$$s_b = \frac{s_e}{\sqrt{\sum(x-\bar{x})^2}}$$

Confidence interval for slope:

$$b_1 - t_{\alpha/2} \cdot s_b < \beta_1 < b_1 + t_{\alpha/2} \cdot s_b$$

Confidence interval for the mean response:

$$\hat{y} \pm t_{\alpha/2} \cdot s_e \sqrt{\frac{1}{n} + \frac{(x^* - \bar{x})^2}{\sum(x-\bar{x})^2}}$$

Test statistic for slope b_1:

$$t = \frac{b_1}{s_b}$$

Test statistic for correlation:

$$U = \frac{r\sqrt{n-2}}{\sqrt{1-r^2}}$$

Prediction interval for an individual response:

$$\hat{y} \pm t_{\alpha/2} \cdot s_e \sqrt{1 + \frac{1}{n} + \frac{(x^* - \bar{x})^2}{\sum(x-\bar{x})^2}}$$

Chapter 14: Analysis of Variance

Treatment sum of squares:

$$SSTr = n_1(\bar{x}_1 - \bar{\bar{x}})^2 + n_2(\bar{x}_2 - \bar{\bar{x}})^2 + \cdots + n_I(\bar{x}_I - \bar{\bar{x}})^2$$

Treatment mean square:

$$MSTr = \frac{SSTr}{I-1}$$

Error sum of squares:

$$SSE = (n_1 - 1)s_1^2 + (n_2 - 1)s_2^2 + \cdots + (n_I - 1)s_I^2$$

Error mean square:

$$MSE = \frac{SSE}{N-I}$$

***F* statistic for one-way ANOVA:**

$$F = \frac{MSTr}{MSE}$$

Test statistic for Tukey–Kramer test:

$$q = \frac{|\bar{x}_i - \bar{x}_j|}{\sqrt{\frac{MSE}{2}\left(\frac{1}{n_i} + \frac{1}{n_j}\right)}}$$

Chapter 15: Nonparametric Statistics

Test statistic for the sign test:

$$z = \frac{x + 0.5 - n/2}{\sqrt{n}/2} \quad \text{if } n > 25$$

If $n \leq 25$, the test statistic is x, the number of times the less frequent sign occurs.

Mean of *S*, the sum of the ranks for the rank-sum test:

$$\mu_S = \frac{n_1(n_1 + n_2 + 1)}{2}$$

Standard deviation of *S*, the sum of the ranks for the rank-sum test:

$$\sigma_S = \sqrt{\frac{n_1 n_2(n_1 + n_2 + 1)}{12}}$$

Test statistic for the rank-sum test:

$$z = \frac{S - \mu_S}{\sigma_S}$$